SUPERALLOYS 1996

SUPERALLOYS 1996

Proceedings of the Eighth International Symposium on Superalloys sponsored by the Seven Springs International Symposium Committee, in cooperation with TMS, the TMS High Temperature Alloys Committee and ASM International, held September 22-26, 1996, Seven Springs Mountain Resort, Champion, Pennsylvania.

Edited by
R. D. Kissinger • D. J. Deye • D. L. Anton
A. D. Cetel • M. V. Nathal • T. M. Pollock • D. A. Woodford

A Publication of
TMS
Minerals • Metals • Materials

A Publication of The Minerals, Metals & Materials Society
420 Commonwealth Drive
Warrendale, Pennsylvania 15086
(412) 776-9000

The Minerals, Metals & Materials Society is not responsible for statements or opinions and is absolved of liability due to misuse of information contained in this publication.

Printed in the United States of America
Library of Congress Catalog Number 95-77966
ISBN Number 0-87339-352-X

Authorization to photocopy items for internal or personal use, or the internal or personal use of specific clients, is granted by The Minerals, Metals & Materials Society for users registered with the Copyright Clearance Center (CCC) Transactional Reporting Service, provided that the base fee of $3.00 per copy is paid directly to Copyright Clearance Center, 27 Congress Street, Salem, Massachusetts 01970. For those organizations that have been granted a photocopy license by Copyright Clearance Center, a separate system of payment has been arranged.

© 1996

If you are interested in purchasing a copy of this book, or if you would like to receive the latest TMS publications catalog, please telephone 1-800-759-4867.

CONTENTS

Dedication xi

Memorial xii

Preface xiii

Committee Members xv

KEYNOTE PAPER

The Future Costs Less - High Temperature Materials
from an Aeroengine Perspective 3
    M. J. Goulette

ALLOY DEVELOPMENT

Best Paper Award

A New Type of Microstructural Instability in Superalloys - SRZ 9
    W. S. Walston, J. C. Schaeffer, and W. H. Murphy

René N4: A First Generation Single Crystal Turbine Airfoil Alloy with Improved Oxidation
Resistance, Low Angle Boundary Strength and Superior Long Time Rupture Strength 19
    E. W. Ross and K. S. O'Hara

René N6: Third Generation Single Crystal Superalloy 27
    W. S. Walston, K. S. O'Hara, E. W. Ross, T. M. Pollock, and W. H. Murphy

The Development and Application of CMSX®-10 35
    G. L. Erickson

The Development of the CMSX®-11B and CMSX®-11C Alloys for Industrial Gas Turbine Application 45
    G. L. Erickson

Development of a Hydrogen Resistant Superalloy For Single
Crystal Blade Application in Rocket Engine Turbopumps 53
    P. Caron, D. Cornu, T. Khan, and J. M. de Monicault

Hot Corrosion Resistant and High Strength Nickel-Based Single Crystal and
Directionally-Solidified Superalloys Developed by the d-Electrons Concept 61
    Y. Murata, S. Miyazaki, M. Morinaga, and R. Hashizume

The Control of Sulfur Content in Nickel-Base Single Crystal Superalloys and Its Effects on Cyclic Oxidation Resistance 71
    C. Sarioglu, C. Stinner, J. R. Blachere, N. Birks, F. S. Pettit, G. H. Meter, and J. L. Smialek

Dispersion Strengthened Sheet Alloys for Exhaust Component Applications 81
R. Darolia and J. R. Dobbs

Development of a Low Thermal Expansion, Crack Growth Resistant Superalloy 91
J. S. Smith and K. A. Heck

Phase Diagram Calculations For Ni-Based Superalloys 101
N. Saunders

Strength & Stability Considerations in Alloy Formulation 111
G. Durber

PHYSICAL METALLURGY

Mechanisms of Formation of Serrated Grain Boundaries in Nickel Base Superalloys 119
H. L. Danflou, M. Macia, T. H. Sanders, and T. Kahn

The Structure of René 88DT 129
S. T. Wlodek, M. Kelly, and D. A. Alden

Development of a Damage Tolerant Heat Treatment for Cast + HIP Incoloy 939 137
R. W. Hatala and J. J. Schirra

Precipitation Behavior in AEREX™ 350 145
C. M. Tomasello, F. S. Pettit, N. Birks, J. L. Maloney, and J. F. Radavich

Effects of Aluminum, Titanium and Niobium on the Time-
Temperature-Precipitation Behavior of Alloy 706 153
T. Shibata, Y. Shudo, and Y. Yoshino

Eta (η) and Platelet Phases in Investment Cast Superalloys 163
G. K. Bouse

Service Temperature Estimation For Heavy Duty Gas Turbine Buckets Based on Microstructure Change 173
Y. Yoshioka, N. Okabe, T. Okamura, D. Saito, K Fujiyama, and H. Kashiwaya

Directional Coarsening of Nickel Based Superalloys: Driving Force and Kinetics 181
M. Véron, Y. Bréchet, and F. Louchet

An Experimental Study of the Role of Plasticity in the Rafting Kinetics of a Single Crystal Ni-Base Superalloy 191
M. Fährmann, E. Fährmann, O. Paris, P. Fratzl, and T. M. Pollock

Investigation of γ/γ' Morphology and Internal Stresses in a Monocrystalline Turbine
Blade after Service: Determination of the Local Thermal and Mechanical Loads 201
H. Biermann, B. von Grossmann, T. Schneider, H. Feng, and H. Mughrabi

Role of Coherency Strain in Microstructural Development 211
J. K. Lee

Misfit and Lattice Parameters of Single Crystal AM1 Superalloy:
Effects of Temperature, Precipitate Morphology and γ-γ' Interfacial Stresses 221
A. Royer and P. Bastie

Elastic Properties and Determination of Elastic Constants of
Nickel-Base Superalloys by a Free-Free Beam Technique 229
W. Hermann, H. G. Sockel, J. Han, and A. Bertram

Local Order and Mechanical Properties of the γ Matrix of Nickel-Base Superalloys 239
N. Clément, A. Conjou, M. Jouiad, P. Caron, H. O. K. Kirchner, and T. Khan

Determination of Atomistic Structure of Ni-Base Single Crystal Superalloys
using Monte Carlo Simulations and Atom-Probe Microanalyses 249
    H. Murakami, Y. Saito, and H. Harada

Differential Thermal Analysis of Nickel-Base Superalloys 259
    D. L. Sponseller

CREEP & FATIGUE

Creep Anisotropy in Nickel Base γ,γ' and γ/γ' Superalloy Single Crystals 273
    D. M. Shah and A. Cetel

Creep Anisotropy in the Monocrystalline Nickel-Base Superalloy CMSX®-4 283
    V. Sass, U. Glatzel, and M. Feller-Kniepmeier

High Temperature Anisotropic Deformation of SRR99 Modeling and Microstructural Aspects 291
    M. B. Henderson, J.-Y. Buffiere, L.-M. Pan, B. A. Shollock, and M. McLean

Effect of Morphology of γ' Phase on Creep Resistance of a Single Crystal Nickel-Based Superalloy, CMSX®-4 297
    Y. Kondo, N. Kitazaki, J. Namekata, N. Ohi, and H. Hattori

Behavior of Single Crystal Superalloys under Cyclic Loading at High Temperatures 305
    H. Frenz, J. Kinder, H. Klingelhöffer, and P. D. Portella

Microstructural Behavior of a Superalloy under Repeated or Alternate LCF at High Temperature 313
    V. Brien, B. Décamps, and A. J. Morton

Interaction Between Creep and Thermo-Mechanical Fatigue of CM247LC-DS 319
    C. C. Engler-Pinto Jr., C. Noseda, M. Y. Nazmy, and F. Rézaï-Aria

Creep and Fatigue Properties of a Directionally Solidified Nickel Base Superalloy at Elevated Temperature 327
    M. Maldini, M. Marchionni, M. Nazmy, M. Staubli, and G. Osinkolu

Specific Aspects of Isothermal and Anisothermal Fatigue of
the Monocrystalline Nickel-Base Superalloy CMSX®-6 335
    H. Mughrabi, S. Kraft, and M. Ott

Prediction of Oxidation Assisted Crack Growth Behavior within Hot Section Gas Turbine Components 345
    G. Webb, T. Strangman, N. Frani, C. Daté, L. Wilson, R. Rana, and D. Fox

Creep Strength and Fracture Resistance of Directionally Solidified GTD111 353
    D. A. Woodford

The Influence of Inclusions on Low Cycle Fatigue Life in a P/M Nickel-Base Disk Superalloy 359
    E. S. Huron and P. G. Roth

Crack Initiation Studies of MA 760 During High Temperature Low Cycle Fatigue 369
    A. Hynnä, V.-T. Kuokkala, and P. Kettunen

Cyclic Deformation of Haynes 188 Superalloy under Isothermal and Thermomechanical Loadings 375
    M. G. Castelli and K. Bhanu Sankara Rao

Long Time Creep Rupture of Haynes™ Alloy 188 383
    R. L. Dreshfield

Effects of Grain Boundary Morphology and Dislocation Substructure on the Creep Behavior of Udimet 710 391
    Y. Zhang and F. D. S. Marquis

Dynamic Strain Aging Behavior of Inconel 600 Alloy 401
    S. H. Hong, H. Y. Kim, J. S. Jang, and I. H. Kuk

The Application of Neural Computing Methods to the Modeling of Fatigue in Ni-Base Superalloys 409
    J. M. Schooling and P. A. S. Reed

Neural Network Modeling of the Mechanical Properties of Ni-Base Superalloys 417
    J. Jones and D. J. C. MacKay

SOLIDIFICATION & CASTING TECHNOLOGY

Extending the Size Limits of Cast/Wrought Superalloy Ingots 427
    A. D. Helms, C. B. Adasczik, and L. A. Jackman

Coupled Macro-Micro Modeling of the Secondary Melting of Turbine Disc Superalloys 435
    P. D. Lee, R. Lothian, L. J. Hobbs, and M. McLean

Liquid Density Inversions During the Solidification of Superalloys and
Their Relationship to Freckle Formation in Castings 443
    P. Auburtin, S. L. Cockcroft, and A. Mitchell

The Effect of Phosphorus, Sulfur and Silicon on Segregation,
Solidification and Mechanical Properties in Cast Alloy 718 451
    J. T. Guo and L. Z. Zhou

Inclusion/Melt Compatibility in Pure Nickel and UDIMET 720 457
    M. Halali, D. R. F. West, and M. McLean

Primary Carbide Solution During the Melting of Superalloys 465
    F. Beneduce, A. Mitchell, S. L. Cockcroft, and A. J. Schmalz

Undercooling Related Casting Defects in Single Crystal Turbine Blades 471
    M. Meyer ter Vehn, D. Dedecke, U. Paul, and P. R. Sahm

The Columnar-to-Equiaxed Transition in Nickel-Based Superalloys 481
    J. W. Fernihough, S. L. Cockcroft, A. Mitchell, and A. J. Schmalz

Closed Loop Control Techniques for the Growth of Single Crystal Turbine Components 487
    M. E. Schlienger

Autonomous Directional Solidification (ADS), A Novel Casting Technique for Single Crystal Components 497
    I. A. Wagner and P. R. Sahm

The Engineering Applications of a Hf-Free Directionally Solidified Superalloy in the Aviation Industry of China 507
    S. Chuanqi, L. Qijiuan, W. Changxin, T. Shifan, and J. F. Radavich

On The Castability of Corrosion Resistant DS-Superalloys 515
    J. Rösler, M. Konter, and C. Tönnes

Advanced Airfoil Fabrication 523
    J. R. Dobbs, J. A. Graves, and S. Meshkov

Thermal Analyses from Thermally-Controlled Solidification (TCS) Trials on Large Investment Castings 531
    P. D. Ferro and S. B. Shendye

ALTERNATE MATERIALS

Gas Turbine Engine Implementation of Gamma Titanium Aluminide 539
    C. M. Austin and T. J. Kelly

Designing with Gamma Titanium - CAESAR Program Titanium Aluminide Component Applications 545
    D. E. Davidson

Microstructural Effects on Fatigue Crack Growth of Cast and Forged TiAl Alloys 555
  D. R. Clemens, C. I. Lobo, and D. L. Anton

NiAl Alloys for Turbine Airfoils 561
  R. Darolia, W. S. Walston, and M. V. Nathal

Ceramic Gas Turbine Technology Development 571
  M. L. Easley and J. R. Smyth

Titanium Metal Matrix Composites for Aerospace Applications 579
  S. A. Singerman and J. J. Jackson

POWDER METALLURGY & WROUGHT MATERIALS

Phosphorus-Boron Interaction in Nickel-Base Superalloys 589
  W. D. Cao and R. L. Kennedy

The Role of Phosphorus and Sulfur in Inconel 718 599
  X. Xie, X. Liu, Y. Hu, B. Tang, Z. Xu, J. Dong, K. Ni, Y. Zhu, S. Tien, L. Zhang, and W. Xie

The Effects of Ingot Composition and Conversion on the Mechanical
Properties and Microstructural Response of GTD-222 607
  T. Banik, T. C. Deragon, and F. A. Schweizer

Microstructure Modeling of Forged Waspaloy Discs 613
  G. Shen, J. Rollins, and D. Furrer

Effect of Cooling Rate From Solution Heat Treatment on Waspaloy Microstructure and Properties 621
  J. R. Groh

Effect of Stabilizing Treatment on Precipitation Behavior of Alloy 706 627
  T. Shibata, Y. Shudo, T. Takahashi, Y. Yoshino, and T. Ishiguro

Dual Alloy Disk Development 637
  D. P. Mourer, E. Raymond, S. Ganesh, and J. M. Hyzak

Development of HIP Consolidated P/M Superalloys for Conventional Forging to Gas Turbine Engine Components 645
  G. E. Maurer, W. Castledine, F. A. Schweizer, and S. Mancuso

Hot-Die Forging of P/M Ni-Base Superalloys 653
  C. P. Blankenship Jr., M. F. Henry, J. M. Hyzak, R. B. Rohling, and E. L. Hall

The Effect of High Temperature Deformation on Grain Growth in a PM Nickel Base Superalloy 663
  M. Soucail, M. Marty, and H. Octor

The Influence of Alloy Chemistry and Powder Production Methods on Porosity in a P/M Nickel-Base Superalloy 667
  E. S. Huron, R. L. Casey, M. F. Henry, and D. P. Mourer

Damage Tolerance of P/M Turbine Disc Materials 677
  M. Chang, A. K. Koul, and C. Cooper

Cooling Path Dependent Behavior of a Supersolvus Heat Treated Nickel Base Superalloy 687
  R. D. Kissinger

Development of Isothermally Forged P/M Udimet 720 for Turbine Disk Applications 697
  K. A. Green, J. A. Lemsky, and R. M. Gasior

Evaluation of P/M U720 for Gas Turbine Engine Disk Application 705
  H. Hattori, M. Takekawa, D. Furrer, and R. J. Noel

The Manufacture and Evaluation of a Large Turbine Disc in Cast and Wrought Alloy 720Li 713
    D. J. Bryant and G. McIntosh

Electroslag Refining as a Clean Liquid Metal Source for Atomization and Spray Forming of Superalloys 723
    M. G. Benz, W. T. Carter Jr., F. G. Müller, and R. M. Forbes Jones

Microstructure and Properties of Spray Atomized and Deposited Superalloys 729
    T. Shifan, Z. Xianguo, R. Liping, L. Zhikai, L. Zhou, M. Guofa, and J. F. Radavich

PrecadR, A Computer Assisted Design and Modeling Tool For Superalloy Powder Precision Molding 737
    D. Lasalmonie, L. Le Ber, C. Dellis, R. Baccino, F. Moret, J. C. Garcia, and J. P. Buhle

COATINGS, JOINING & REPAIR

Nondestructive Evaluation of High-Temperature Coatings for Industrial Gas Turbines 747
    G. L. Burkhardt, G. M. Light, H. L. Bernstein, J. S. Stolte, and M. Cybulsky

Effect of Postweld Heat Treatment on Ductility of Ni-Co-Cr Based Alloy Welds 753
    K.-K. Baek, C.-S. Lim, and J.-G. Youn

Low Cycle Fatigue Properties of LPM™ Wide-Gap Repairs in Inconel 738 763
    K. A. Ellison, J. Liburdi, and J. T. Stover

Subject Index 773

Alloy Index 779

Author Index 781

DEDICATION

The Symposium and these proceedings are dedicated to Dr. John F. Radavich in honor of his pioneering contributions to the superalloy and gas turbine engine industries.

IN MEMORY OF
CHESTER T. SIMS

Chester (Chet) T. Sims was a significant contributor to superalloys for many years. He began his career at Battelle in Columbus, Ohio in the 1950's where he investigated rhenium as an alloying element. (Back then, he was known as "Chet the jet, the best bet for rhenium yet!") Upon moving to the GE Gas Turbine Division in Schenectady, N.Y., he co-invented a still utilized cobalt base turbine vane alloy (FSX-414) and participated in refining the Phacomp method of determining the long-time stability of high temperature nickel base superalloys which is still incorporated into many current superalloy specifications.

Chet was a member of the committee that organized the first Seven Springs International Symposium on Superalloys in 1968. His contributions to the literature, history and understanding of superalloys were outstanding. He wrote (and illustrated) many articles and was the lead editor (and author or co-author of several chapters) of the two most utilized superalloy books in the industry, "The Superalloys" and "Superalloys II". Following his retirement from GE in 1986, he became an Adjunct Professor at RPI in Troy, N.Y., and a lecturer on superalloys in many parts of the world. He also made a video for this symposium committee, "Superalloys, A History".

Chet was an important contributor to the "community"; he had been an Eagle Scout, a wounded WWII infantryman in France, an expert skier and sailor, and he had a passion for the conservation of Lake George in New York state. Chet had lakefront property in Bolton Landing on Lake George and was very involved with the Nature Conservancy, and other groups, in protecting Lake George and its surrounding environment for future generations.

We will miss Chet Sims particularly for his knowledge of superalloys, for his spirit in discussions on any subject, and for his commitment to preserving the environment.

PREFACE

The purpose of the International Symposium on Superalloys, which is held once every four years, is to provide a forum for the researchers, producers and users to present their most recent technical information on the high temperature, high performance materials used in the gas turbine engine industry.

The initial symposium, twenty eight years ago, focused on the phase stability problems associated with superalloys. Since that time the breadth of subjects covered has greatly expanded to cover all aspects of superalloys research, development, production and applications. The sixth symposium, held in 1988, introduced a new area of interest, alternate materials for gas turbine applications. The Eighth International Symposium once again covers the most recent research and development in all aspects of superalloys and devotes a portion of the conference to those alternate materials that are being, or will soon be, used in development or production applications.

Starting with the second symposium in 1972, each symposium and corresponding published proceedings have been dedicated to an individual as a means of honoring that individual for his or her contributions to the superalloy industry. This Eighth International Symposium on Superalloys is dedicated to Dr. John Radavich for his contributions to superalloy development.

Dr. Radavich has had a distinguished career as an educator and researcher that spans over forty years. His principal contribution to high temperature materials technology has been his pioneering work in the transition of analytical procedures from optical to electron microscopy, and developing the preparation procedures that opened the window to all future metallographic studies of superalloys. The current state-of-the-art practices for phase extraction and identification procedures is dependent upon his fundamental understanding developed for a wide variety of superalloy compositions produced in wrought, cast and powder metallurgy forms.

He is the author or co-author of over thirty technical publications and scientific papers dealing primarily with phase identification and alloy element/microstructure interactions in high temperature superalloys. Dr. Radavich was briefly on the faculty of Purdue University from 1953-55. After completing his post doctoral studies at Cambridge (U.K.) in 1955-56 and at the Max Planck Institute in Dusseldorf, Germany in 1956, he rejoined the faculty and continued his association with Purdue University until his retirement last year. For the past 23 years at Purdue, he has pioneered and administered an interface program between the academic community and industry wherein students are guided through a comprehensive analytical program on subjects of interest to industry. This accomplished effective interfacing and exchange between students and industry and provided a focus for their research activities. A number of these students have gone on to positions in the superalloy industry and are contributors to these and previous proceedings.

Finally, this conference would not have been possible without the tireless work of the members of the International Symposium Committee and those who have laid the ground work at the previous seven symposia. The committee members listed on the following page were responsible for organizing and presenting this symposium.

July, 1996

R. D. Kissinger *M. V. Nathal*
D. J. Deye *T. M. Pollock*
D. L. Anton *D. A. Woodford*
A. D. Cetel

EIGHTH INTERNATIONAL SYMPOSIUM ON SUPERALLOYS
COMMITTEE MEMBERS

General Chairman	Doug Deye
Secretary	Dick Shiring
Treasurer	Gern Maurer
Program Chairman	Bob Kissinger
	Doug Deye
	Don Anton
	Al Cetel
	Mike Nathal
	Tresa Pollock
	David Woodford
Publication Chairman	David Woodford
Arrangements Chairman	Bob Gasior
	Jim Blair
	Sharon Miazga
International Publicity	Tasadduq Khan
U.S. Publicity	Ken Green
Awards Chairman	Bob Stusrud
	Chuck Kortovich
	Lou Lherbier
	Gern Maurer

The Committee would like to acknowledge Linda Lanning, GE Aircraft Engines, for her assistance in assembling these proceedings, and Vivienne Harwood Mattox, Management Plus, Inc., for providing the registration services for the symposium.

KEYNOTE PAPER

THE FUTURE COSTS LESS – HIGH TEMPERATURE MATERIALS FROM AN AEROENGINE PERSPECTIVE

M. J. Goulette
Project Director
Advanced Engineering
Aerospace Group – Rolls-Royce plc

Abstract

Aeroengines are sophisticated products resulting from over 50 years of research and development. During this time materials science has been a pacing technology, however the performances of current materials are reaching limits. The returns of research and development are diminishing, while the associated costs are continuously escalating. In todays highly competitive, cost driven marketplace this necessitates the definition of clear priorities for development work if future materials and engines are to satisfy the requirements of the market.

This paper addresses the changing market scenario for both civil and military aeroengines, addressing in particular the impact of economic circumstances upon the current and future technological drivers. In the light of this discussion, potential areas of future materials development will be discussed in terms of both performance and cost benefits.

Introduction

At the last Seven Springs conference in 1992, Jim Williams spoke about some of the rapid changes occurring in the Aerospace industry as a consequence of the collapse of Communism and the so called peace dividend. Since then these changes have continued and have been accelerated by the worst recession since the 1930s. The Aerospace market drivers have changed and consequently the perception of needs and priorities for materials have also changed. In this paper I will firstly review the business scene and then examine the consequences for aeroengine manufacturers, the high temperature materials industry and future materials requirements. My view will be from a European perspective, however, the conclusions should be universally relevant.

Military Aerospace

The global military market has seen a steady decline in turnover during the last fifteen years, partially as a result of the end of the Cold War, to a 1994 value only one third of that in 1980 (figure 1). In the aeroengine field there has been a dramatic reduction in demand for spares, as a result of reduced flying by the world's military and this trend has been exaggerated by the improved reliability of modern equipment. Another consequence has been delays and downsizing of new development programmes and significantly fewer new programme starts. Figure 1 also illustrates the difficulties in forecasting the military sector; predictions made in 1989 for the recovery of the market being seriously in error by 1995. In spite of this, military aeroengines remain a major market with a global annual turnover of around $5bn.

The downturn in military aircraft activity, together with national defence budget constraints, has led to an increasing emphasis on the cost of peacetime operation of military aircraft, thus putting pressure on both the lifetime and first cost. The historic demand for high

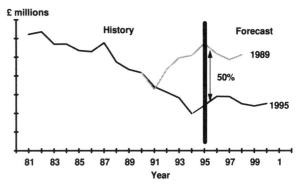

Figure 1: Military Aeroengine Business Turnover and Forecasts

technology levels and increased thrust to weight in military engines is still present, however this is now tempered by the requirement to achieve a reduction in total aircraft weight and hence cost. New product developments are rare and will continue to reduce in frequency. Consequently, technology demonstrator programmes such as JAST in the USA and AMET in Europe will be increasingly important in focusing the development and maintenance of technical capability.

Civil Aerospace

After a profitable period in the mid 1980s, the world civil engine business is now recovering from a difficult period of overall nonprofitability, during which time world air traffic fell for the only time since records began in 1929 and airline losses totalled over $15bn (figure 2). This inevitably had an impact on new aircraft orders; airlines being unwilling to commit themselves to large expenditure when confidence was low. The result was a reduction in new aircraft orders to their lowest level since 1983. In addition, due to bankruptcies and the difficulties experienced by airlines, orders were cancelled at a rate

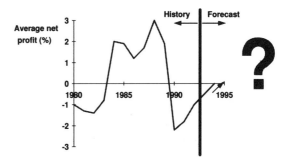

Figure 2: World Airlines – Profit Margins

comparable to that for new aircraft orders. In 1994 aircraft cancellations amounted to $10.6 billion, whilst new orders totalled only $18 billion. Nevertheless the industry is now emerging from recession and growth in traffic has resumed. Indeed the market for new engines is anticipated to grow by a factor of 2 over the next 20 years to meet the requirements of fleet replacement and new business. This is illustrated in figure 3 which shows the historic and predicted sales figures for the period 1988 to 2014.

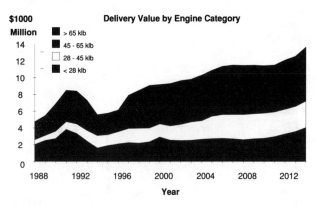

Figure 3: Market History and Forecast for Civil Aeroengines

Airlines do, however, remain under severe financial constraints, translating to an unrelenting pressure on aircraft purchase prices and operating costs. Although they achieved record profits in 1995, it is estimated that total airline profits for the period 1994 to 1996 will only equal about 85% of the losses in the previous three years and, as long as fuel prices remain relatively low, acquisition costs will be the major economic driver for the industry.

Figure 4 illustrates that during the period since 1972 the cost of new aircraft has risen at a rate generally twice that of inflation. This is due to the increased functionality of modern aircraft and includes a significant element attributable to more expensive materials and processes. While this period has also seen a continued increase in air traffic, the average airfare ticket price has remained broadly at the same level, representing a drop in real terms to one third of its 1972 value. This decreasing airline seat mile yield drives price pressure throughout the aerospace supply chain and consequently requires manufacturers to reduce their costs to maintain profitability. The net effect of this cost reduction requirement is a cascade of restructuring and cost cutting from the airlines down throughout the supply chain.

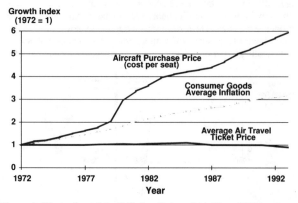

Figure 4: Illustration of the Inflation Rates of Air Travel Elements

The civil aeroengine market is becoming increasingly mature, with three major manufacturers producing similar products competing for market share. In such an environment there is a clear need to retain a competitive advantage. Historically this has centred around reduced fuel consumption with rapid gains being made in the early years largely through technology advances enabling higher operating temperatures and increased bypass ratios. This gave higher thermal and propulsive efficiencies. Today this improvement is slowing down as the industry and the technology become increasingly mature. For example the three engines offered for the Boeing 777 all give essentially the same fuel burn performance. In these circumstances the airline customer will discriminate on the basis of weight, reliability, quality of customer support and, most importantly, the financial package on offer. It is also possible that the customer will be technology averse, where it is perceived that new technology may cause unreliability or other problems. New technology must show an overall business benefit or it will not be applied.

Business Drivers

In the environment described above, the development of new technical capability has to be accurately focused on the needs of the business. Even the blue skies fundamental research needs to have a potential application driving it. Performance dominated military requirements no longer drive the development of new technology with civil engines benefiting from the spin off. Todays drivers are dominated by the civil engine market and the severe financial pressures experienced by the airline business. This shift to a civil dominated market demands dramatic reductions in both costs and product development times. While improved performance is still desirable, future competitive advantage is increasingly dependent upon unit cost and time to market.

This necessitates the adoption of effective business processes. Systems integration and the adoption of concurrent engineering principles are vital throughout the supply chain to enable the delivery of the right product at the right time and the right cost. It will also be necessary to ensure that the required technology is available prior to product validation as the simultaneous development of both technology and product carries too high a risk when timescales are reduced to a minimum.

The finance of research and technology is increasingly difficult, because of the costs to the business of restructuring and of concessions to secure the future market coupled with the reluctance of financial institutions to invest in long term programmes. A careful balance must therefore be achieved between cost and performance, technology push and product pull. Detailed cost benefit analysis is required together with a disciplined investment plan to select areas for research and technology investment before the initiation of work. This should result in fewer more promising development areas leading to reduced costs and development times. In this situation it will be essential to maintain some blue skies longer term research, particularly in universities.

Wherever possible technologies should be identified as dual use, that is applicable to both military and civil markets. The duplication of effort can thereby be minimised and development funding attracted from both military and civil sectors. Collaboration between competitors on projects of mutual benefit must also become more widespread enabling increasingly high development costs to be shared. This principle is well established at the engine project level (e.g. CFM International and IAE), in future we must find ways to make it work more effectively at the materials level. In most cases, materials are not product discriminators. We, therefore, need to find better mechanisms to share development, scaleup and database costs and to realise the benefits from industry standard materials.

The reduction in product development times will necessitate a parallel reduction in timescales for materials development. Together with the desire to reduce costs this will see the increasing importance of rapid prototyping, and modelling techniques both to guide research, cutting down on the amount of empirical testing required, and to better exploit materials in service.

Materials Supply Industry

The effects of these economic pressures have been felt throughout the materials supply industry. Here again the position today is one of upturn, with the future promising a period of growth as airlines return into profit, after a period of decline. Figure 5 illustrates these trends in superalloy billet volumes and selling prices.

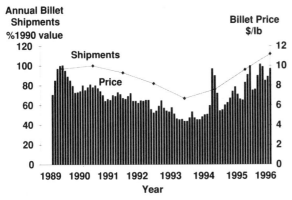

Figure 5: Tonnage and Price Trends for Superalloy Billet Supply

The industry responded to the recession with a period of consolidation, a number of mergers took place and manufacturing capacity was reduced, with some companies leaving the business. Companies also rationalised their operations and contracted down to their core businesses. The recession lead to a general loss of confidence in the aerospace market, as a result of which suppliers attempted to reduce their dependence upon aerospace through diversification into other product markets to enable them to survive through the lean periods. This process can be expected to continue.

The above issues have impacted upon the ability and willingness of the industry to respond to the recent upturn in orders. There is a reluctance to return lost capacity in what is still seen as an uncertain market, and significant difficulties exist in replacing lost personnel and expertise. With the rapid increase in the market there is also a problem with supporting the growing demand. The speed of the pick up has lead to problems of raw material supply up the food chain and an increase in lead times.

Economic pressure and the difficulty of competing in such an investment intensive, low volume industry, results in the requirement to cut costs. Joint ventures with risk and revenue sharing partners are increasingly required to expand limited resources. If investment in new processes is to be made a strong business case in terms of predicted volume usage must be made. In the light of these considerations there is still a future for conventional processes producing conventional materials with development focused on improved economics.

Materials Issues

Advanced materials have been a pacing technology throughout the development of the gas turbine. Higher temperatures and lower densities have been the twin driving forces for the materials community.

Current alloys have been developed over fifty years to the point where further advances are becoming increasingly difficult and expensive to achieve, at a time when resources are ever more limited. Nickel base superalloys are now operating at temperatures of up to 85% of their melting point (figure 6). It is clear that the melting point of Nickel imposes a natural ceiling to their potential and hence further improvement is limited. Continuing incremental development of these alloys is possible but this will require a deeper fundamental understanding of both the materials properties and processing routes, together with the increasing importance of computer modelling techniques. We will need better science to beat the laws of diminishing returns.

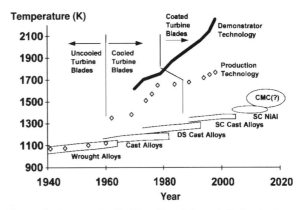

Figure 6: Progress in Turbine Materials and Technologies and Associated Increases in Turbine Entry Temperatures (K)

New materials are now becoming available which offer a step change in capability. However before these materials can displace conventional alloys they must demonstrate the ability to deliver cost effective performance benefits to the whole engine system.

Both of these approaches; the evolutionary development of existing alloys and the revolutionary development of new materials are discussed in greater detail below.

Evolutionary Development of Existing Materials.

Efforts to further exploit conventional superalloys continue although significant increases in temperature capability and strength beyond that of e.g. todays 3rd generation SX alloys will be difficult to achieve at acceptable cost and density.

It seems likely that further developments will concentrate largely on the adaptation of existing materials to niche applications and the production of cheaper alloys with performance parity. Such cost reductions are important in a situation where raw materials account for approximately 30% of the cost of turbine disc and 20% of a blade.

Further incremental advances in performance will require an improved understanding of material behaviour and the application of computer modelling techniques coupled with better processes and control. The growth of computing power has made the development of models of phase diagrams and processes such as melting, casting and forging a practical reality. Coupled with the prediction of properties from compositional and microstructural data this is enabling the prediction of the behaviour of a component from its chemical composition and processing route in a virtual design process. This in turn enables a move away from traditional empirical processes, reliant upon experience and experimentation, and hence dramatic reductions in cost and timescale, as a result of the need for less testing.

A detailed understanding of material properties and improved processing can also contribute through allowing more complete exploitation of existing properties and the safe employment of standard, cheaper, materials to higher stress levels. It may also be possible to control life cycle costs through the development of better life prediction methodologies.

Revolutionary Materials

There are two major classes of potential successor materials to the Nickel based superalloys, the intermetallics and the ceramic matrix composites.

Several barriers however exist to the implementation of any new material. It must be possible to demonstrate a cost effective performance benefit in a realistic time frame. There is a need for an existing manufacturing base, cost effective processing route and adequate design and lifing methodologies.

Intermetallics

Two main categories of intermetallics are available as superalloy successor materials. Titanium aluminides for use at lower temperatures and nickel aluminides for higher temperature components. Both offer the potential for cost competitive improvements in performance through significant density reductions, however the reluctance to accept the risk of designing with a material which is intrinsically brittle must be overcome.

The Gamma titanium aluminides are rapidly emerging as practical engineering materials which may be processed in a variety of conventional and, therefore, low cost processes. A combination of alloy modification and process refinement has increased ductility and defect tolerance to the point where none fatigue critical applications can be contemplated and are being demonstrated. Raw materials are inherently low cost for this system and there is no doubt that they will find significant applications in future engines.

The case for nickel intermetallics is much less mature. Again, there is low density and potentially low cost, however the ductility problem is more intractable and good strength and temperature capability are difficult to achieve without reinventing the superalloy.

Ceramic Matrix Composites

Ceramic matrix composites (CMCs) offer potentially significant temperature advantages over metals, together with a density typically one third that of Nickel. Currently they have a temperature capability of 1000 to 1200°C although this will need to be significantly increased if they are to see wide application. To date CMCs have been employed for high temperature components at low structural loads, for example the reheat systems in military engines and are now being considered as a problem solving material for niche applications such as turbine shroud seals. Here their excellent temperature capability can be exploited in order to avoid the need for cooling of metal components in difficult geometries. This can yield a cost benefit for the overall engine even where the use of CMCs is not justified on a component for component basis. CMCs may be employed for this kind of application within the next 25 years, enabling in service experience to be gained, facilitating ultimate wider application throughout the combustor and for static aerofoils. The use of CMCs in rotating components is, however, not currently envisaged due to their limited strength.

Despite the raw materials themselves being cheap a significant cost barrier does exist, largely due to processing difficulties, a component typically costing 1.5 to 2 times that of a metal part. Fibre manufacture, moulding and processing are all expensive requiring a large equipment investment. Machining is also problematic requiring a near net shape route and constraining production to simple shapes. These problems are compounded by the fact that the material has no scrap value.

There is a need to establish a large scale manufacturing base if a mature manufacturing process is to be developed with the consequent reduction in costs. The business case for investment in such a manufacturing base must however be justified by volume projections for the market and currently this is not the case. Indeed it is unlikely whether the aerospace industry itself would ever require the volumes necessary to constitute a sufficient market. In addition the generation of other applications and markets will be difficult before the manufacturing base is in place and the viability of the technology has been demonstrated.

Summary

The military engine scene has seen a steady decline in turnover since the early 1980s, with an increasing pressure to reduce first and operating costs in addition to the traditional requirement for increased thrust to weight ratio. As a result aeroengine technology is becoming increasingly driven by the cost based requirements of the civil market. The airlines, aided by the three cornered battle for market share between Rolls-Royce, General Electric and Pratt and Whitney, are putting ever increasing pressure on unit price and operating cost guarantees. Against this background, competitive advantage from further reductions in specific fuel consumption, via higher operating temperature and lower specific thrust is becoming increasingly difficult and costly to achieve, hence, increasingly, it is cost and timescale which will win contracts. The aeroengine has become a commodity and its supplier base must adapt to this reality. In the light of current economic constraints it is clear that the historic investment in materials research and development cannot be sustained. A balance must be struck between technological innovation and economic necessity. New technology must be able to demonstrate the potential for a clear competitive advantage before its adoption. In this context modelling techniques based upon a deeper physical understanding will become increasingly important to optimise material and manufacturing processes and reduce development costs and cycle times.

The result is that while new step change technologies are being developed the incremental development of existing alloy systems will remain vital. Past predictions of usage for new materials are now looking optimistic and superalloys will remain the high temperature material of choice for the foreseeable future. It is essential, however, that materials provide increased value to the OEM, this means both lower cost as well as increased functionality. Continued investment in materials science and engineering is vital if we are to achieve these goals and, increasingly, we will have to share this investment if we are going to be able to afford it.

ALLOY DEVELOPMENT

BEST PAPER AWARD

The following paper, "*A New Type of Microstructural Instability in Superalloys - SRZ,*" by W. S. Walston, J. C. Schaeffer and W. H. Murphy was selected by the Awards Subcommittee of the International Symposium on Superalloys as the Best Paper of the Eighth Symposium. The selection of the best paper was based on the following criteria: originality, technical content, pertinence to the superalloy and gas turbine industries, and clarity and style.

A NEW TYPE OF MICROSTRUCTURAL INSTABILITY IN SUPERALLOYS - SRZ

W.S. Walston, J.C. Schaeffer and W.H. Murphy
GE Aircraft Engines, Cincinnati, OH 45215

Abstract

A new type of instability in superalloys has been observed in advanced alloys containing high levels of refractory elements. One instability occurs under the diffusion zone of coatings and has been called secondary reaction zone or SRZ. Similar instabilities, in the form of cellular colonies, have been observed along grain boundaries and in dendrite cores. These microstructural instabilities are characterized and interpreted in terms of a nucleation and growth transformation. The similarities and differences between a similar phenomenon, cellular recrystallization, are outlined. The degradation of properties due to the SRZ and cellular colonies is described. Methods are shown that have successfully reduced or eliminated these instabilities. Finally, the implications of these new types of instabilities on superalloys in general are discussed.

Introduction

Recent advances in the creep rupture strength of single crystal superalloys have been accomplished by the addition of higher levels of refractory elements. These additions result in microstructural stability being even more important during alloy development. Precipitation of Topologically Close-Packed (TCP) phases in superalloys is well known and is a function of many variables, including temperature and alloy composition. Second and third generation single crystal superalloys all precipitate TCP phases under some conditions, however, in general, the quantity that precipitates does not significantly degrade properties. Thus, the occurrence of a moderate amount of TCP phases is not cause for general concern.

Most superalloy turbine airfoil components are put into service with an environmental coating. These coatings are typically either a diffusion aluminide or a MCrAlY. The interdiffusion of these coatings with advanced superalloy substrates causes phase instability at the surface. For many alloys, it is typical to observe TCP phases in the interdiffusion zone after high temperature exposures. Again, the occurrence of a moderate amount of TCP phases below the coating is not considered a problem.

A new type of instability in superalloys has been observed in alloys containing high levels of refractory elements. The new instability differs significantly from past TCP phases in morphology and effect on mechanical properties. This instability was first observed beneath the diffusion zone of an aluminide coating and was termed SRZ (secondary reaction zone). The occurrence of the SRZ-type of instability, however, is not limited to coating interdiffusion zones. In cases where the SRZ-type instability is observed along low angle grain boundaries and in dendrite cores away from the coating, it will be referred to as cellular colonies. This paper discusses the conditions under which the instability occurs, the effect on properties and methods of prevention.

Experimental Procedures

Many alloys have been evaluated for their propensity to form SRZ, however one particular alloy has been evaluated in-depth and is reported in this paper. This alloy is an experimental third generation single crystal superalloy with the composition given in Table 1.[1] In order to achieve the necessary creep rupture strength, this alloy contains a high amount of refractory elements compared to previous generation superalloys. Small additions of Hf, C and B were made to improve the strength of low angle grain boundaries, when present in the casting.[2]

Single crystal slabs measuring 1.3 x 5 x 10 cm were directionally solidified at commercial suppliers. The castings were solution heat treated at 1315°C for 2 hours followed by an aging heat treatment of 1120°C for 4 hours. Following heat treatment, the microstructure consisted of a γ matrix and 65 vol.% cuboidal γ' precipitates with an edge length of about 0.5 μm. A small number of MC carbides were also present in the interdendritic regions. Following specimen preparation, various diffusion aluminide coatings were applied. Final coating thicknesses were typically 50-75 μm thick. A diffusion heat treatment at 1080°C was performed, followed by the final alloy aging cycle at 870°C. Elevated temperature exposures were then conducted from 980-1150°C for times up to 400 hours to promote SRZ formation.

In an attempt to eliminate SRZ beneath coatings, specimens were coated with various elements prior to aluminization. Elements examined were Ni, Ta, Hf, B and C. The Ni and Ta were applied with a DC magnetron sputtering device using 6" diameter targets. The Hf, B and C were deposited in a small chemical vapor deposition (CVD) reactor with a 6" x 12" hot zone. After element application, pack aluminization was performed using the Codep process. SRZ exposures were performed to assess the effectiveness of each element.

Table 1. Major Elements in Alloy 5A, weight %.

Alloy	Ni	Co	Cr	Al	Ta	Re	W
Alloy 5A	Bal.	12.50	4.50	6.25	7.00	6.25	5.75

Results

SRZ Structure

The SRZ beneath coatings and the cellular colonies observed elsewhere in the microstructure have the same structural features and similar compositions. Figure 1 shows a schematic of SRZ beneath a coating with the individual phases labeled. The SRZ structure consists of a γ' matrix containing γ and P phase (TCP) needles. The γ and P phase needles tend to be aligned perpendicular to the growth interface. Figures 2 and 3 show the interface between the advancing SRZ and the γ/γ' microstructure. The matrix transforms from γ (superalloy) to γ' (SRZ) once the incoherent boundary passes. The P phase in the SRZ is continuous with the γ phase at the interface. Nucleation of the P phase on the γ phase is shown in Figure 4.

Table 2 shows the compositions of the phases ahead of the SRZ interface and within the SRZ. The phase compositions within the SRZ were determined by electron microprobe, and the phase compositions in the bulk alloy were determined by phase extraction.[3] The P phase is composed of nearly 50% Re with high levels of W, Cr and Co. P phase is similar in composition and structure to the sigma TCP phase with the major difference being a larger period on one axis. As a result of the high levels of Re and W in the P phase, the γ phase within the SRZ is depleted of these elements compared to the γ phase in the bulk alloy. The γ' phase within the SRZ is enriched with Al and Ta compared to the γ' phase in the bulk alloy, which may explain the stabilization of the γ' matrix. Other than these changes, the γ and γ' phases within the SRZ constituent have very similar compositions to the γ and γ' phases in the bulk alloy.

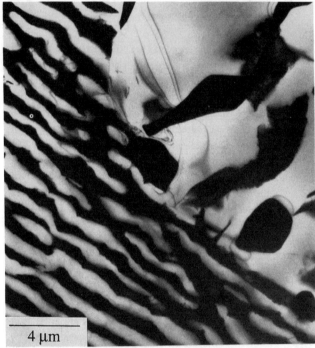

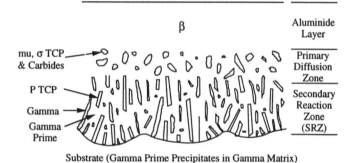

Figure 1. A schematic showing the secondary reaction zone under an aluminide coating.

Figure 2. TEM micrograph showing the SRZ/alloy interface. The SRZ has a γ' (gray) matrix, while the alloy has a γ (black) matrix.

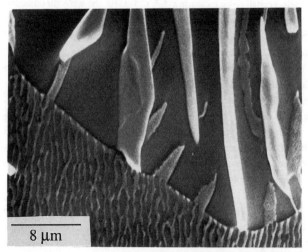

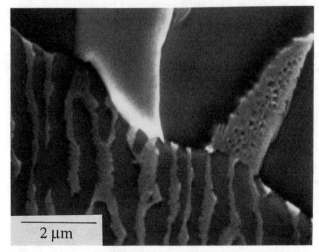

Figure 3. BSE SEM micrograph showing the SRZ/alloy interface. Note the relationship between the P phase (white) and the γ phase (gray).

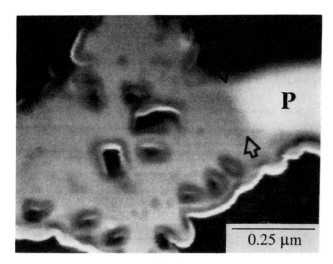

Figure 4. High magnification SEM micrograph showing the P phase nucleation from the γ phase (between the arrows).

Table 2. Composition of Phases Within and Adjacent to SRZ.

Phase	Ni	Co	Cr	Al	Ta	Re	W	Hf
P in SRZ	10.5	9.7	9.2	0.1	1.6	49.5	20.0	0.0
γ' in SRZ	64.1	11.2	2.7	8.7	8.2	1.1	4.4	0.1
γ in SRZ	50.6	18.8	9.7	4.5	2.6	8.3	5.3	0.0
γ' in Alloy	69.1	8.8	1.9	4.8	6.5	1.1	4.8	0.2
γ in Alloy	45.6	18.7	9.3	6.6	1.2	16.0	6.6	0.0

SRZ Under Coatings

Interdiffusion between the coating and the substrate alloy and mismatch strains in the alloy create an unstable situation in which γ and γ' are no longer the equilibrium phases beneath the coating. In many alloys, the diffusion zone consists of β' and TCP phases. However, in alloy 5A, SRZ occurs beneath the diffusion zone of simple aluminide, platinum aluminide and overlay coatings. In other, more stable alloys, SRZ may only occur under the diffusion aluminide coatings. Figure 5 shows a typical example of SRZ beneath a platinum aluminide coating in alloy 5A. Depending upon the coating characteristics and surface preparation, the SRZ can be continuous or occur in isolated cells. It is believed that the Al activity of the coating and the residual stress state of the surface play key roles in determining the propensity of a specimen to form SRZ.

Cellular Colonies at Grain Boundaries

A similar microstructure to the SRZ was observed along low and high angle grain boundaries in alloy 5A. Figure 6 shows that the morphology of the cellular colonies along grain boundaries is similar to the SRZ under coatings. Both constituents consist of a γ' matrix with needles of γ and P phase. Compositional analysis of the cellular colonies show phase compositions to be similar to those of the SRZ shown in Table 2. It was typical to observe the cellular colonies to form only on one side of a grain boundary. However, along the same grain boundary, the cellular colonies may form on either side of the boundary but never on both sides at the same time. It was also observed that the formation of the cellular colonies was more favorable on higher angle boundaries. In alloy 5A, grain boundaries with relative misorientations as low as 10° formed cellular colonies. However, in other more stable alloys, higher misorientations were required to form cellular colonies. In a thorough study on a similar alloy, Pollock and Nystrom found that the cellular colonies appeared to nucleate on P phase grain boundary precipitates.[4]

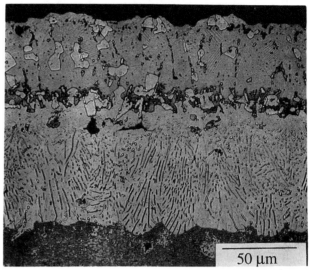

Figure 5. SRZ under a PtAl coating in alloy 5A following a 1093°F/400 hour exposure.

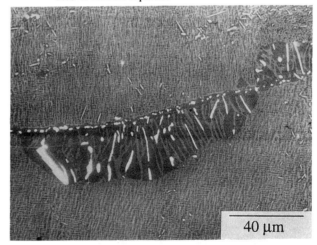

Figure 6. Cellular colonies along a grain boundary (~14° misorientation) in alloy 5A.

Cellular Colonies in Dendrites

Cellular colonies have been observed in dendrite cores with the same microstructure as those found along grain boundaries and the SRZ under coatings. The occurrence of the cellular colonies in the dendrite cores occurred to a much lesser extent than the other two reactions. The cellular colonies were primarily observed in either unstressed, as-cast specimens or in creep rupture specimens tested at temperatures near 1100°C. Figure 7 shows a longitudinal section of a failed creep rupture specimen showing cracking along one of the cellular colonies. Unlike the other two reactions, the cellular colonies in the dendrites were isolated occurrences without the presence of a boundary. A higher magnification view of one of these colonies is shown in Figure 8. It was common to observe a cracked interface along the cellular colonies in dendrite cores in creep rupture specimens. The effect of these cracked colonies on properties will be discussed later.

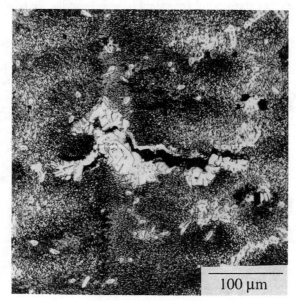

Figure 7. Cellular colony formed in a dendrite in a failed creep rupture specimen tested at 1093°C.

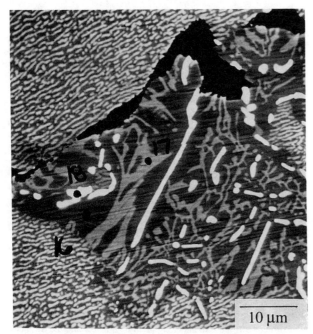

Figure 8. Crack running along the interface of a cellular colony in a failed creep rupture specimen tested at 1093°C.

Nucleation and Growth

The factors that affect the nucleation of the SRZ and cellular colonies have been studied in alloy 5A. These constituents have been observed after exposures at temperatures from 980 to 1150°C. Typical exposures were for 400 hours, however exposures as short as one hr at 1120°C have produced SRZ. One of the difficult aspects in studying the nucleation of these constituents is that there appears to be a large nucleation barrier to their formation. Thus, predicting nuclei formation as a function of time and temperature is difficult, and observing the earliest stages of nucleation is almost impossible.

The observation that isolated occurrences can occur under coatings, along boundaries or in dendrite cores is consistent with a high nucleation barrier. Otherwise, it would be more common to observe continuous cellular colonies along grain boundaries and cellular colonies in most dendrite cores. Contributions to the nucleation of the SRZ and cellular colonies can be described by the following equation for homogeneous nucleation:[5]

$$\Delta G = n(\Delta G_{\alpha/\alpha'} + \Delta G_\varepsilon) + \eta \gamma n^{2/3} \qquad (1)$$

where ΔG is the free energy of formation for an SRZ nuclei, $\Delta G_{\alpha/\alpha'}$ is the free energy difference between parent and product phases per unit volume (supersaturation), ΔG_ε is a strain energy term, η is a shape factor, γ is the surface free energy between the phases and n is a volume term. Nucleation is controlled by a number of factors, including supersaturation, surface energy, strain energy and the number of heterogeneous sites. Supersaturation can occur either by external (coating) or internal (segregation) chemistry imbalances. Strain energy can be introduced by surface preparation prior to coating or misfit strains along grain boundaries or between γ and γ'.

The growth of SRZ has been measured under a variety of coatings at 1093°C. Figure 9 shows the data for alloy 5A plotted as a function of the square root of time. The linear dependence shows that diffusion is controlling the rate of growth. The interdiffusion coefficient calculated from the SRZ layer thickness is 6.73×10^{-11} cm^2/sec. Janssen and Rieck have measured the diffusivity of Ni and Al in Ni-Al compounds and found the diffusivity at 1093°C of Ni to be 2.5×10^{-11} cm^2/sec in γ'; 4.0×10^{-11} cm^2/sec in γ; and 5.0×10^{-11} cm^2/sec in β.[6] The diffusion rate calculated from the SRZ growth is slightly higher than the volume diffusion rates for Ni and Al. This is likely due to enhanced diffusion in the SRZ along the growth interface.

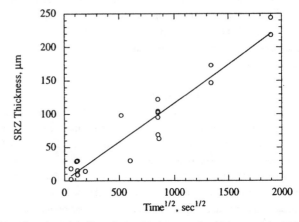

Figure 9. A straight line relationship between the thickness of the SRZ and the square root of time indicating a diffusion controlled process.

Effect of Composition

A large number of single crystal superalloys have been evaluated for their propensity to form SRZ and cellular colonies. Based on these evaluations, it is clear that composition plays a key role in the formation of these constituents. A systematic study of alloys similar to alloy 5A was conducted to determine the effect of various alloy additions on the formation of SRZ beneath a platinum aluminide (PtAl) coating. The same surface preparation and coating process was performed on each alloy since it was known that these factors could affect the amount of SRZ formation. Following PtAl coating the specimens were exposed at 1093°C for 400 hours. The total linear percent of SRZ around the periphery of the specimen was measured. A value of 100% meant that SRZ was continuous beneath the coating. The depth of the SRZ was not measured in this analysis. Statistical analysis of the results of these evaluations produced the following relationship for use in predicting the amount of SRZ which will form in an alloy:

$$[SRZ(\%)]^{1/2} = 13.88(\%Re) + 4.10(\%W) - 7.07(\%Cr) - 2.94(\%Mo) - 0.33(\%Co) + 12.13 \qquad (2)$$

The elements in this equation are in atomic percent, and this equation is valide for third generation single crystal superalloys. It is clear that Re is the most potent element for determining an alloy's propensity to form SRZ. Minor variations in the Al content of the alloy did not influence the formation of SRZ beneath the coating. However, significant Al enrichment occurs beneath the coating, and this plays a large role in the formation of SRZ.

While SRZ has been observed to some extent in many third generation single crystal superalloys, including René N6[9] and CMSX-10,[10] these alloys contain greater than 5 wt.% Re.[11,12] The overwhelming role of Re in SRZ formation in equation (1) suggests that it is not surprising that these alloys would form SRZ to some extent. However, even in alloys containing lower levels of refractory elements, including Re, SRZ has been observed. In rare cases, alloys with 3 wt.% Re have exhibited SRZ beneath aluminide coatings.[10] This surprising observation is most likely a result of extremes in alloy composition, surface preparation and coating parameters. Our experience suggests that alloys with less than 5 wt.% Re should rarely exhibit SRZ formation.

No quantitative expressions have been developed for the formation of cellular colonies along grain boundaries or in dendrite cores. However, it has been observed that it is easier to nucleate SRZ beneath the coating than it is to nucleate the cellular colonies along the grain boundaries or in dendrite cores. Thus, it is possible to screen alloys based on the above SRZ equation and obtain a qualitative indication of their propensity to form cellular colonies elsewhere.

Property Degradation

The effect of SRZ and cellular colonies has been evaluated in a wide range of mechanical property tests on bare and coated specimens. These constituents can form after exposures from about 980 to 1150°C, with the most favorable temperature around 1100°C. SRZ beneath a coating can affect test specimens and turbine airfoils by reducing the load bearing cross section or by crack initiation along the cell interface. In alloy 5A, slight losses in rupture strength were found at temperatures around 1100°C due to reduced cross section. Cracks emanating from SRZ were found in failed rupture specimens, however it was difficult to determine if these played a role in initiating premature failures. No decrease in fatigue properties has been attributed to SRZ, although it seems possible that SRZ could initiate cracks at early lives.

Cellular colonies along grain boundaries can reduce properties in a turbine airfoil. The magnitude of the reduction is a function of the grain boundary angle, the alloy's propensity to form the cellular colonies and the alloy's inherent grain boundary strength. In alloy 5A, stress rupture tests were conducted transverse to known low and high angle grain boundaries. It was found that above certain relative misorientations between grains, the presence of the cellular colonies reduced the rupture properties of the alloy.

The most detrimental form of these constituents are the cellular colonies in the dendrite cores. During rupture testing of alloy 5A at temperatures from 760-1150°C, a small number of tests at 1093°C had unusually low rupture lives. In many of the longer time tests at 1093°C, results were obtained that were as low as 30% of the expected rupture life, as shown in Figure 10. Cellular colonies formed in regions of high strain, cracked along the interface, and caused premature failure. Figure 11 compares a creep curve for a specimen with cellular colonies to a normal creep rupture curve. The unexpectedly low results only occurred in a small percentage of the rupture tests performed, however the effect of the cellular colonies was very dramatic when it did occur.

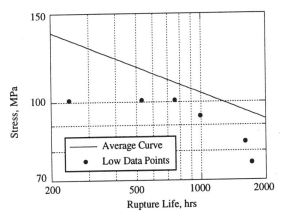

Figure 10. Rupture life of alloy 5A as a function of stress at 1093°C. Cellular colonies near the fracture surface were found in the specimens exhibiting very low rupture lives.

Prevention Methods

Clearly, the occurrence of SRZ and cellular colonies is undesirable, because of their effects on mechanical properties. The easiest method to reduce or eliminate the occurrence of these constituents is to change the alloy composition. Lowering the refractory content of the alloy, especially Re, will eventually eliminate the formation of these undesirable constituents. However, these alloys are designed for high creep rupture strength and reducing the refractory element content will have a direct negative effect on the strength of the alloy. A better understanding of the driving forces for SRZ and the effect of composition can lead to an alloy that balances strength and stability, as demonstrated by René N6.[9,11]

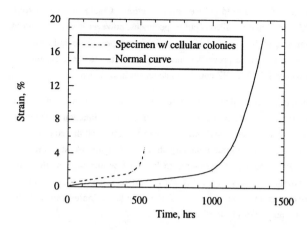

Figure 11. Comparison of a typical creep curve at 1093°C/103 MPa with a creep curve from a specimen containing cellular colonies.

Chemical supersaturation and surface residual stress are two important factors affecting the nucleation of SRZ beneath coatings. A set of experiments evaluated different surface preparations prior to coating ranging from electropolishing to shot peening. Following surface preparation, specimens were PtAl coated and exposed at 1093°C for 400 hours. The total linear percent of SRZ around the periphery of the specimen was measured. Figure 12 summarizes some of the data showing the effect of surface preparation on the amount of SRZ. It was found that electropolishing was effective at removing the surface stresses and subsequently eliminating the amount of SRZ beneath the coating after high temperature exposure. Low stress grinding, grit blasting or other moderate surface preparation techniques were sometimes effective at reducing the amount of SRZ compared to normal turbine airfoil production processing. However, these techniques produced significant scatter in the data, which is further evidence of the high nucleation barrier for SRZ. Shot peening and other aggressive surface preparation techniques resulted in complete coverage of SRZ due to a high contribution of strain energy to nucleation.

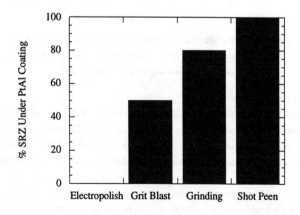

Figure 12. Effect of surface stress introduced by various methods on the occurrence of SRZ under a PtAl coating.

Another set of experiments were performed to modify the chemistry or microstructure of the surface of the superalloy prior to aluminide coating. Various elements were applied to the surface to stop the nucleation and growth of SRZ. Nickel was sputtered to reduce the concentration of refractories at the surface, while Ta was applied to decrease diffusivity. C and B were added by chemical vapor deposition (CVD) to form refractory boride and carbide precipitates that could reduce supersaturation and boundary mobility. Platinum was electroplated to determine if it exacerbated the formation of SRZ. Short anneals were given to the as-deposited specimens to determine if the addition of the new layer caused SRZ prior to aluminide coating. Additional specimens containing deposited surface elements were aluminized using the pack process. These specimens were evaluated for SRZ formation in the as-coated condition and following a 1120°C exposure for 50 hours, although little change in SRZ occurrence was observed with the 1120°C exposure.

Table 3 shows the qualitative results of this experiment. The Hf and Ta surface modifications resulted in SRZ formation under all conditions. These additions are γ stabilizers, which resulted in an unstable condition below the coating. The B treated specimens were extremely reactive to air and although the SRZ did not form, alloy 5A exhibited extensive boride formation and areas of local melting. The Pt plating by itself did not cause SRZ formation, however an abundant amount formed after aluminide coating. This is consistent with the observation that PtAl coatings promote SRZ formation more readily than simple aluminide coatings. It has been shown that Pt increases the amount of Al that assimilates into a coating.[13] The Ni surface modification showed some improvement compared to specimens with no surface modifications. The thin layer of Ni appears to have helped to reduce the supersaturation in the coating diffusion zone.

The most promising surface modification for preventing SRZ formation was the deposition of carbon prior to coating.[14] Table 3 shows that no SRZ formed even after the high temperature exposure following aluminizing. The sub-micron W- and Ta-rich carbides penetrated to a depth below the diffusion zone of the subsequent coating. These carbides accomplished two objectives. First, they tied up the refractory elements in stable compounds reducing the chemical driving force for SRZ nucleation. Second, they precipitated in sufficient amounts to preclude movement or growth of the SRZ colony. Both of these effects served to eliminate the formation of SRZ.

Table 3. Amount of SRZ Formation Following Substrate Surface Modifications.

Surface Modification	As Deposited + 1080°C/1 hr Anneal	Aluminized + 1080°C/1 hr Anneal
CVD Hf	Abundant	Abundant
Sputtered Ta	Abundant	Abundant
CVD B	None	Melted
Electroplated Pt	None	Abundant
Sputtered Ni	None	Little
CVD C	None	None

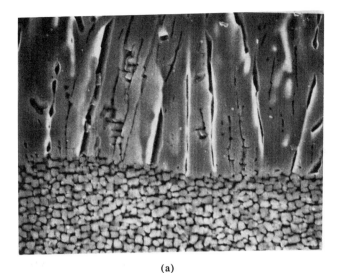 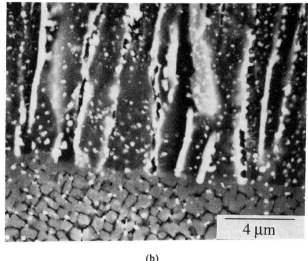

(a) (b)

Figure 13. An aluminide coated turbine airfoil showing (a) SRZ and (b) the absence of SRZ following a carburizing treatment. Small white particles in the primary diffusion zone are carbides.

Figure 13 shows the successful use of carburization to eliminate SRZ on an engine component of alloy 5A. This figure shows two turbine airfoils following a 1120°C/50 hour exposure. One turbine airfoil sample was carburized prior to PtAl application and the other airfoil was only PtAl coated. Figure 13a shows the SRZ/alloy interface, while Figure 13b shows the fine carbides present through the normal diffusion zone which prevented the formation of SRZ. The key condition for carburization to succeed was for the carbide precipitation depth to be greater than the depth of the coating's primary diffusion zone. Carburized specimens were tested in a cyclic oxidation/hot corrosion burner rig test with no detrimental effect from carburization.

Surface prevention methods can lead to reduced SRZ under the coating, but they do not affect the formation of cellular colonies along grain boundaries or in dendrite cores. The formation of these cellular colonies is a direct result of the supersaturation of the γ matrix with P or γ' forming elements. Short of changing alloy composition, heat treatment appears to be the only method to reduce the supersaturation. Solution heat treatment trials were performed on alloy 5A to reduce the segregation of Re and other refractory elements present in the dendrites. Rhenium is the most important element causing SRZ and the slowest diffusing element in superalloys. Thus, a parameter was developed to measure the segregation of Re in directionally solidified superalloys:

$$\text{Re } \Delta = \frac{\text{Wt.\% Re in dendrite core - Wt.\% Re in interdendritic region}}{\text{Wt.\% Re in dendrite core}}$$

Electron microprobe analysis of specimens was conducted following a series of heat treatments from 1310-1330°C for times from 2-25 hours. Figure 14 shows a summary of these data. As expected from diffusion theory, there is an initial rapid decrease in Re segregation followed by a more gradual decrease. In as-cast specimens of alloy 5A, Re levels as high as 9.5 wt.% were found in the dendrite core compared to the bulk level of 6.25 wt.%. This high level of Re, along with other refractory elements, leads to an unstable condition in the dendrite cores. Extended solution heat treatments can lower the level of Re in the dendrite core closer to the bulk alloy level so that the dendrite core is no longer unstable. For alloy 5A, it was found that a Re Δ of approximately 30% was necessary to eliminate the occurrence of cellular colonies in the dendrite cores. The analysis used in Figure 14 is valid for other third generation single crystal superalloys, however the appropriate Re Δ value to eliminate cellular colonies will vary for each alloy.

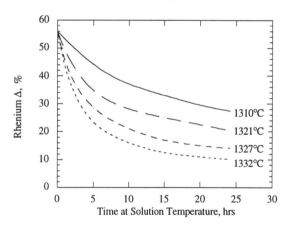

Figure 15. Effect of time at the maximum solution heat treatment temperature on the dendritic segregation of Rhenium.

Discussion

SRZ: Cellular Precipitation

The SRZ and cellular colony reactions that have been observed in alloy 5A and other superalloys containing high levels of Re are cellular precipitation reactions. This type of precipitation event has been observed in many alloys systems, including Pb-Sn,[15,16] Cu-In,[17] Ni-Al,[18] Cu-Ti[19] and Cu-Be.[20] There have been several reports of cellular precipitation in superalloys, mainly involving either carbides[21-23] or eta phase[24] at grain boundaries. The presence of grain boundary serrations in superalloys has also been attributed to cellular precipitation of γ' at the grain boundaries.[25] These serrations are reported to improve fatigue crack growth rate. It has also been

15

observed in turbine disk alloys that additions of Hf promoted a cellular precipitation reaction.[26] Pollock has previously reported on the occurrence of the cellular colonies along grain boundaries in alloy 5A,[4] but there have been no other reports of cellular precipitation in single crystal superalloys.

Cellular precipitation consists of the transformation of a supersaturated α' phase into a structurally identical α phase plus a lamellar β phase. The reaction visually resembles a eutectoid decomposition, such as pearlite in steels. The initial nucleation and growth theories originally proposed by Smith,[27] Turnbull[28] and Cahn[29] have only been modified slightly[30-32] since their inception in the 1950's. Nucleation of the cellular reaction occurs at grain boundaries, or more specifically, on favorably oriented precipitates along the grain boundaries. The driving force for nucleation is the supersaturation in the matrix adjacent to the nucleating particle. The presence of stress also aids the nucleation process.

Following the nucleation of a small grain boundary precipitate, the reaction grows in a cell morphology with lamellar precipitates. The growth of the cell boundary is driven by the difference in chemical potential between the supersaturated matrix ahead of the cell boundary and the matrix within the cell which contains the equilibrium structure. The lead interface is an incoherent boundary, while the lamallae boundaries within the cell are partially coherent. Thus, the dominant diffusion mechanism occurs along the advancing cell boundary. Volume diffusion in the supersaturated matrix is negligible. Growth of the cell can be slowed by precipitation in the matrix ahead of the cell, and growth will stop when the supersaturated condition driving the reaction have dissipated.

Comparison of the SRZ and the cellular colonies in superalloys with the observations in other systems leads to some interesting points. The observation that the lead interface is incoherent helps to explain the cracking in the cellular colonies in the dendrites that had such a detrimental effect on properties. Also, the fact that growth of the cellular precipitation reaction will stop when the driving force has been eliminated is observed in these superalloys. Under a coating, the growth of SRZ corresponds closely to the interdiffusion zone between the coating and substrate. The SRZ is rarely observed to extend deeply into the microstructure. The cellular colonies in the dendrites also are confined to the dendrite core, because there is little driving force outside of the supersaturated core.

The SRZ and cellular colonies shown in this paper differ in two important ways from the classical cellular precipitation discussed in the literature. First, in cellular precipitation the matrix ahead of the advancing cell is structurally identical to the matrix within the cell. In this paper, the matrix ahead of the cell is a γ matrix, while the matrix inside the cell is γ'. Theories on cellular precipitation state that the structure within the cell represents the equilibrium structure for the alloy. Thus, under certain conditions, the equilibrium microstructure in high γ' volume fraction superalloys with high refractory contents is a γ' matrix with γ and P phase precipitates. The other difference between the observations in this paper and classical cellular precipitation is the structure within the cells. Three phases co-exist in the SRZ and cellular colonies, while only two phases have previously been observed in cellular precipitation colonies. This is likely due to the complex phase relationships in these superalloys versus the simpler phase relationships in many of the alloys previously studied.

Cellular precipitation is described in the literature as occurring along grain boundaries. In the single crystal alloys described in this paper, the only grain boundaries present are those defects that form during solidification. It has been shown that cellular colonies form along these grain boundaries in the same manner as described in the literature on cellular precipitation. The SRZ beneath coatings also likely nucleates along a grain boundary in the coating or the coating primary diffusion zone. However, there are no apparent grain boundaries in the dendrite cores to serve as nucleation sites for the cellular colonies observed in this paper. Thus, while these cellular colonies structurally appear similar to the cellular colonies along grain boundaries of the SRZ under coatings, the nucleation mechanism may be different than classical cellular precipitation. Nucleation of the cellular colonies in the dendrite cores likely occurs from a heterogeneous site rather than undergoing homogeneous nucleation. The heterogeneous site could be a small TCP phase that has nucleated in this refractory-rich region. Other heterogeneous sites, such as carbides and γ/γ' eutectic, are confined to the interdendritic regions due to segregation of elements during casting. It appears that the nucleation barrier for the cellular colonies is very high because nucleation only occurs under special circumstances of high levels of supersaturation or high levels of strain energy. More work needs to be done to understand the nucleation of cellular colonies in the dendrite cores.

<u>Cellular Precipitation versus Cellular Recrystallization</u>
The cellular precipitation observed in alloy 5A closely resembles cellular recrystallization commonly observed in superalloys.[33] In fact, it is very difficult to determine the difference based on microstructural features. In high volume fraction γ' superalloys, both cellular precipitation and cellular recrystallization have γ' matrices with γ precipitates and sometimes a TCP phase precipitate. Both have a cellular structure with the precipitates within the cell aligned perpendicular to the growth front. The major difference is in the driving force for nucleation and growth. While cellular recrystallization is driven primarily by residual stress, cellular precipitation is driven by stress and composition (supersaturation). In the three cases of cellular precipitation presented in this paper, supersaturation plays a key role in their nucleation and growth. In the failed creep rupture specimens in which cellular colonies were observed very close to the fracture surface, the contribution of strain energy to the cell formation was very high. In this case, the distinction between cellular precipitation and cellular recrystallization becomes less clear.

<u>Implications of SRZ</u>
The presence of SRZ and cellular colonies is a serious issue for all advanced directionally solidified superalloys. The combination of a segregated solidification structure, high levels of refractory elements

and a high volume fraction of γ' make these alloys especially susceptible to cellular precipitation. This phenomenon has been observed in a large number of third generation single crystal alloys, including alloy 5A, René N6 and CMSX-10 to varying degrees. SRZ under coatings has even been observed in alloys with Re contents as low as 3 weight %.

The amount of SRZ under coatings can vary widely depending upon coating characteristics, surface preparation and exposure conditions. For these reasons, it is necessary to fully understand the effects of these variables and how they affect the processing window for each alloy. While PtAl coatings tend to promote the most SRZ, all aluminide coatings and MCrAlY coatings can cause SRZ. While drastic reductions in properties have not been observed for SRZ beneath coatings, there is a concern due to loss of load bearing cross section and the potential for crack initiation.

It has been found that cellular colonies along grain boundaries in directionally solidified or single crystal superalloys reduce rupture strength across the boundary. In single crystals, this effectively reduces the acceptable limit for low angle grain boundaries in castings. Traditional limits may not apply unless extensive testing across grain boundaries has been performed. Such testing was performed on alloy 5A, and it was found that the acceptable limit for grain boundary misorientation decreased by several degrees due to the presence of cellular colonies.

Cellular colonies in dendrite cores represent the most serious concern for advanced turbine airfoil alloys. It is difficult to screen for the presence of these colonies, and their dramatic impact on rupture strength may only be evident in long-time tests in certain temperature ranges. For alloy 5A, the loss in rupture strength was only observed in a small fraction of the total tests conducted. This experience and knowledge of the cellular precipitation reaction led to the successful development of René N6, which is free of cellular colonies in dendrites and has shown no property degradation in extensive testing.[9]

Prevention of SRZ

Some of the prevention methods for SRZ and cellular colonies discussed earlier are summarized in Table 4. For SRZ under coatings, there are several alternative prevention methods. Coating parameter changes, surface preparation and carburization all can be successfully employed. Prevention of cellular colonies along grain boundaries is difficult, except by changing alloy composition or screening castings based on grain misorientations. A balanced alloy composition is key to preventing the cellular colonies in the dendrite cores. Once a balanced alloy composition has been obtained, an extended solution heat treatment cycle to reduce the Re segregation is effective in ensuring the absence of this reaction.

Table 4. Summary of Prevention Methods for SRZ & Cellular Colonies.

Microstructural Instability	Method of Prevention
SRZ Under Coating	Change Alloy Composition Modify Coating Parameters Surface Carburization[14] Reduce Surface Stresses
Cellular Colonies Along Grain Boundaries	Change Alloy Composition Screen Castings for Grain Misorientation
Cellular Colonies in Dendrite Cores	Change Alloy Composition Extended Solution Heat Treatments[11]

Conclusions

1. A new type of instability in superalloys has been observed in alloys containing high levels of refractory elements. This instability can occur beneath coatings, along grain boundaries or in dendrite cores.

2. The instability has been termed secondary reaction zone (SRZ) for its occurrence under the primary diffusion zone of coatings. SRZ and the cellular colonies elsewhere in the microstructure are a form of cellular precipitation previously reported in a wide variety of alloy systems.

3. Cellular colonies in dendrite cores can reduce creep rupture properties over 50% at temperatures around 1100°C. SRZ under coatings can lower properties by reducing load-bearing area and serving as crack initiation sites.

4. Methods to reduce or eliminate SRZ under coatings have been developed, such as altering coating parameters or carburizing a thin layer of the substrate prior to coating.

5. Prevention of the cellular colonies in the dendrite cores is best accomplished by developing a balanced alloy composition, although an extended solution heat treatment cycle can reduce colony occurrence.

Acknowledgments

Many people at GE Aircraft Engines have contributed to this work over the past several years. Paul Fink, Dick McDaniel and Tresa Pollock performed many of the early studies on SRZ. Bob Field, Stan Wlodek and H.P Yan provided several of the fine micrographs and the compositional data in this paper. Tresa Pollock and Jeff Nystrom have developed a better understanding of cellular precipitation at Carnegie Mellon University and have been a great resource. Kevin O'Hara participated in many of the discussions and work on SRZ in alloy 5A and other advanced alloys. Joe Heaney, Steve Wilhelm, Ted Grossman, J. Moorhead and R. Knoerl helped with prevention techniques involving coatings.

References

1. C. M. Austin, R. Darolia, K. S. O'Hara and E. W. Ross, U.S. Patent 5,151,249, "Nickel-Based Single Crystal Superalloy and Method of Making" - Alloy 5A, 1992.
2. E. W. Ross, C. S. Wukusick and W. T. King, U.S. Patent 5,399,313, "Nickel-Based Superalloys for Producing Single Crystal Articles Having Improved Tolerance to Low Angle Grain Boundaries", 1995.
3. T. M. Pollock, (Unpublished Data, GE Aircraft Engines, 1991).
4. T. M. Pollock, "The growth and elevated temperature stability of high refractory nickel-base single crystals", Mat. Sci. & Eng. B, B32(1995), 255-266.
5. R. E. Reed-Hill, Physical Metallurgy Principles, (Van Nostrand Company, New York, 1973), 494.
6. M. M. P. Janssen and G. D. Rieck, "Reaction Diffusion and the Kirkendall Effect in the Ni-Al System", Metall. Trans., 239(1967), 1372.
7. R. Darolia, D. F. Lahrman and R. D. Field, "Formation of Topologically Closed Packed Phases in Nickel Base Single Crystal Superalloys", Superalloys 1988, ed. D. N. Duhl, et al., TMS, 1988, 255-264.
8. D. N. Duhl, "Alloy Phase Stability Requirements in Single Crystal Superalloys", Alloy Phase Stability and Design, ed. G. M. Stocks, D. P. Pope and A. F. Giamei, MRS, 1991, 389-399.
9. W. S. Walston, K. S. O'Hara and E. Ross, "René N6: Third Generation Single Crystal Superalloy", Superalloys 1996, ed. R. D. Kissinger, et al., TMS, 1996,
10. W. S. Walston and J. C. Schaeffer, (Unpublished data, GEAE, 1995).
11. W. S. Walston, E. W. Ross, K. S. O'Hara, T. M. Pollock and W. H. Murphy, U.S. Patent 5,455,120, "Nickel-Base Superalloy and Article with High Temperature Strength and Improved Stability" - René N6, 1995.
12. G. L. Erickson, U.S. Patent 5,366,695, "Single Crystal Nickel-Based Superalloy" - CMSX-10, 1994.
13. Jackson and Rairdon, (Unpublished Data, GE Aircraft Engines, 1990).
14. J. C. Schaeffer, U.S. Patent 5,334,263, "Substrate stabilization of diffusion aluminide coated nickel-based superalloys", 1994.
15. K. N. Tu and D. Turnbull, "Morphology of Cellular Precipitation of Tin from Lead-Tn Bicrystals", Acta Metall., 15(1967), 369-376.
16. K. N. Tu, "The Cellular Reaction in Pb-Sn Alloys", Metall. Trans., 3(1972), 2769-2776.
17. R. A. Fournelle and J. B. Clark, "The Genesis of the Cellular Precipitation Reaction", Metall. Trans., 3(1972), 2757-2767.
18. R. O. Williams, "Aging of Nickel Base Aluminum Alloys", Trans. TMS-AIME, 215(1959), 1026-1032.
19. R. W. Fonda and G. J. Shiflet, "Faceting of the Cellular Growth Front in Cu-3% Ti", Scripta Metall., 24(1990), 2259-2262.
20. H. Tsubakino, R. Nozato and A. Yamamoto, "Discontinuous and Continuous Coarsening of Lamellar Precipitates in Cu-Be", J. Mater. Sci., 26(11)(1991), 2851-2856.
21. M. K. Miller and M. G. Burke, "An APFIM/AEM Characterization of Alloy X750", Appl. Surf. Sci., 67(1993), 292-298.
22. T. M. Angeliu and G. S. Was, "Behavior of Grain Boundary Chemistry and Precipitates upon Thermal Treatment of Controlled Purity Alloy 690", Metall. Trans. A, 21A(1990), 2097-2107.
23. H. M. Tawancy, "High Temperature Creep Behavior of an Ni-Cr-W-B Alloy", J. Mat. Sci., 27(23)(1992), 6481-6489.
24. T. Takahashi, T. Ishiguro, K. Orita, J. Taira, T. Shibata and S. Nakato, "Effects of Grain Boundary Precipitation on Creep Rupture Properties of Alloys 706 and 718 Turbine Disk Forgings", Superalloys 718, 625, 706 and Various Derivatives, TMS, 1994, 557-565.
25. H. L. Danflou, M. Marty and A. Walder, "Formation of Serrated Grain Boundaries and Their Effect on the Mechanical Properties in a P/M Nickel Base Superalloy", Superalloys 1992, ed. S. D. Antolovich, et al, TMS, 1992, 63-72.
26. R. V. Miner, "Effects of C and Hf Concentration on Phase Relations and Microstructure of a Wrought Powder-Metallurgy Superalloy", Metall. Trans. A, 8A(1977), 259-263.
27. C. S. Smith, "Microstructure (1952 Edward deMille Campbell Memorial Lecture)", Trans. Am. Soc. Metals, 45(1953), 533-575.
28. D. Turnbull, "Theory of Cellular Precipitation", Acta Metall., 3(1955), 55-62.
29. J. W. Cahn, "The Kinetics of Cellular Segregation Reactions", Acta Metall., 7(1959), 18-27.
30. H. I. Aaronson and Y. C. Liu, Scripta Met., 2(1968), 1.
31. B. E. Sundquist, "Cellular Precipitation", Metall. Trans., 4(1973), 1919-1934.
32. M. Hillert, "An Improved Model for Discontinuous Precipitation", Acta Metall., 30(1982), 1689-1696.
33. J. M. Oblak and W. A. Owczarski, "Cellular Recrystallization in a Nickel-Base Superalloy", Trans. TMS-AIME, 242(1968), 1563-1568.

RENÉ N4: A FIRST GENERATION SINGLE CRYSTAL TURBINE AIRFOIL ALLOY WITH IMPROVED OXIDATION RESISTANCE, LOW ANGLE BOUNDARY STRENGTH AND SUPERIOR LONG TIME RUPTURE STRENGTH

Earl W. Ross and Kevin S. O'Hara
GE Aircraft Engines
Cincinnati, Ohio and Lynn, Mass.

Introduction

GE Aircraft Engine's first generation single crystal (SX) turbine airfoil alloy, René N,[1] was extensively tested and utilized in turbine blades of development engines. Factory engine testing then revealed that tips of the René N blades suffered from excessive oxidation. It was obvious that a more oxidation resistant single crystal alloy would be required. Two SX development programs were conducted; C. Wukusick and W. King to develop a more oxidation resistant SX alloy, and E. Ross to increase the low angle boundary (LAB) strength of the René N, which could, hopefully, be applied to the new more oxidation resistant SX alloy. This paper will describe the development of this improved first generation SX alloy, René N4,[2] which evolved from these two programs.

Improving the oxidation resistance of René N

René N, GE's first SX alloy was 100°F stronger (longitudinal rupture strength) than GE's workhorse conventionally cast turbine airfoil alloy, René 80.

The chemistry of René N is shown in Table I. It can be seen that René N has higher titanium (4.2%) than aluminum (3.7%). The higher level of titanium vs. aluminum imparted excellent hot corrosion resistance to the René N, but sacrificed oxidation resistance as revealed by the engine testing.

Wukusick and King evaluated the hot corrosion and oxidation resistance of many René N variations. Four elements (Mo, W, Cb, Co) always remained constant. The results of the other elements studied were:

1.   **Chromium:** The efficacy of Cr in environmental resistance is well known (as long as good stability is retained). Levels of chromium from the 9.25% of René N to 10.25% were evaluated in hot corrosion. As shown in Figure 1, when tested at 1600°F with 2 ppm salt, there were significant improvements in the hot corrosion resistance from the 9.25% to a level of 9.75% Cr with only slight changes at higher levels. Microstructural stability was retained in all alloys. The new replacement alloy would therefore be aimed at 9.75% Cr. Hot corrosion tests at 1700°F also showed significant improvement at 9.75% and 10.25% Cr.

Table I. Aim compositions of René N and René N4, wt.%

Element	René N	René N4
Cr	9.25	9.75
Mo	1.5	1.5
Ta	4.0	4.8
Ti	4.2	3.5
Al	3.7	4.2
W	6.0	6.0
Cb	0.5	0.5
Co	7.5	7.5
B	--	0.004
Hf	-	0.15
C	--	0.05

2.   **Carbon:** The grain boundary strengthening elements (B, Zr and Hf), as well as C, which were not present in René N, were studied to determine their effect on hot corrosion when added to René N. Of these elements only C significantly affected hot corrosion behavior. As shown in Figure 2, there was about an 8X improvement in 1700°F-5 ppm hot corrosion resistance when the C content was increased to 0.05%. The new alloy would then contain 0.05% C. It was also decided to reduce the titanium content to improve oxidation resistance. As a result, the Ta content was correspondingly increased to maintain the γ' volume fraction and account for the formation of Ta/Ti carbides.

The previously discussed increases in Cr, C and Ta will all increase the new alloy's hot corrosion resistance.

3.   **Al/Ti Ratio:** René N contains 4.2% Ti and 3.7% Al (approximately a 0.9 ratio). Increasing the Al content and lowering the Ti content improved the oxidation resistance of the alloy system, as shown in Figure 3. The hot corrosion resistance, however, was lowered as also shown in Figure 3, but the previously discussed addition of C

significantly reduced this degradation. The Al/Ti content of the new alloy was selected to be 4.2% Al and 3.5% Ti or a ratio of 1.2; this resulted in a 50% reduction in 100 hour oxidation metal loss in a 2150°F Mach 1 test.

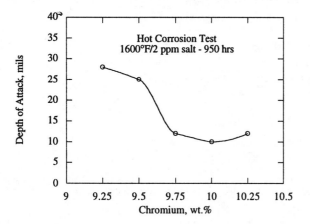

Figure 1. Effect of Increased Cr on Hot Corrosion of René N.

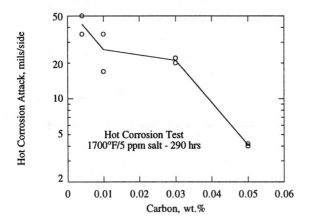

Figure 2. Effect of Carbon on Hot Corrosion of René N.

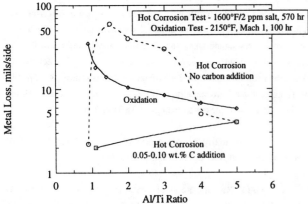

Figure 3. Effect of Al/Ti Ratio on Oxidation and Hot Corrosion. Also Showing Effect of Carbon on Hot Corrsion Resistance of René N Type Alloys.

4. **Hafnium:** Wukusick and King also found that the addition of a small amount of Hf, (0.10 to 0.20%) "caused a further reduction in the oxidation rate during the initial stages of oxidation. Hf tends to promote oxide adherence, providing an incubation period when virtually no metal loss occurs". The new alloy would therefore contain 0.15% Hf.

Summary: the New Alloy's Base Chemistry

Wukusick and King had developed a new alloy which not only maintained the strength of René N but also significantly improved the oxidation resistance, as shown in Figure 4. As shown in Table I, the new alloy, René N4, has 0.5% more Cr, 0.8% more Ta, 0.5% more Al, 0.7% less Ti plus 0.15% Hf and 0.05%C, than René N. This chemistry would be combined with the low angle boundary (LAB) work of Ross to realize the final René N4 composition.

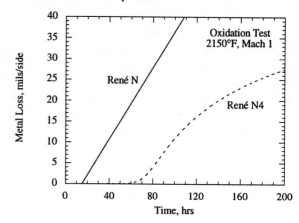

Figure 4. High Velocity Oxidation of René N vs René N4.

Low Angle Boundary (LAB) Strengthening

Conventionally cast/equiaxed turbine airfoil alloys such as René 80 contain approximately 150 ppm of B, 0.10% to 0.18% C, and a small amount of Zr. These elements, B particularly, are vital to high angle grain boundary (HAB) strengthening. Without these elements, the rupture strength of these alloys is dramatically reduced. Hafnium, approximately 1.5%, was added in the 1970's by Lund, et al to newer and stronger conventionally cast turbine airfoil alloys to improve castability and rupture strength, particularly lower temperature strength. Similar additions of B, Hf, C and Zr continued with directionally solidified (DS) alloys, such as MM200H and René 142, since these alloys were DS cast with HAB's parallel to the airfoil leading and trailing edges. These grain boundary strengtheners were required to provide transverse rupture strength and ductility.

When directionally solidified single crystal (SX) alloys were invented, B, Hf, C and Zr were not added, since these elements lowered the incipient melting temperature and the SX alloys were to be cast without grain boundaries. Therefore, GE's first SX alloy, René N, did not contain any purposeful addition of B, C, Zr or Hf (Table I).

In evaluating the first SX material cast with René N, it was found that often, the casting process resulted in material that had longitudinal grain boundaries. These boundaries were caused by perturbations in the casting process which occurs with all SX alloys.

The casting vendor detects these boundaries by carefully etching the parts after surface cleaning. The amount of misorientation of the boundary is determined by x-ray diffraction (small amounts of misorientation can be determined visually by the degree of reflectivity of the adjacent grains). Since the boundaries of René N contained no strengthening additives, it was necessary to determine what degree of the misorientation would be acceptable. GE, therefore, conducted a program on René N slabs which had formed boundaries with various misorientations. This program determined that the transverse rupture strength would not be reduced if the misorientation between the adjacent grains was 6° or less. (These 6° or less boundaries were called low angle boundaries or LAB's).

As a result of the 6° LAB misorientation requirement, a number of René N turbine airfoils were scrapped. A program was therefore initiated at GE to determine if small additions of grain boundary strengtheners could be added to René N (and later to René N4) which would increase the LAB grain boundary acceptance level to >6° without reducing the transverse rupture strength or heat treatability of the airfoils.

Strengthening LAB's

Round 1: Procuring slabs which contain LAB's was expensive, difficult, and had to be done at a casting vendor. Round 1 slabs were therefore cast at GE by DS casting (with high angle boundaries) 10 heats of René N to which varying amounts of C, B, Zr and Hf were added. 1800°F transverse rupture tests were then conducted across the HAB's. As shown in Table II, the best rupture lives were with 0.05%C, 0.01%B and 0.004% Zr additions (Heat 10). The 1800°F rupture life was 25 to 100X that of DS René N which had been previously tested with no grain boundary additions.

Round 2: In this round, René N slabs were again DS cast by GE with heat 10 as a base (now called heat 11) and with 1600°F rupture tests added. As can be seen in Table III, there were a few surprisingly low results in heats 12, 14 and 19, but heat 11 had good lives at both 1600°F and 1800°F. Heat 21 was conducted to note if B alone was as good as B C and Zr. It can also be seen in Table III that the B alone was very beneficial at 1600°F and 1800°F but not as efficacious as when the three elements were added. (It should be noted that in the first round C or C plus Hf, was not beneficial to strengthening).

Table II. Round 1 - Transverse Rupture Strength of DS René N with Elemental Additions.

Heat	Additions, wt.%				1800°F/30 ksi		1800°F/20 ksi	
	C	B	Zr	Hf	Life, hrs	R.A., %	Life, hrs	R.A., %
1	0.02	0.003	0.004	--	2.7	0.6	62.0	0.0
2	0.035	0.0035	0.007	--	1.8	0.0	30.4	1.2
3	0.05	0.005	0.01	--	4.6	0.6	63.9	0.6
4	0.035	0.0035	0.007	0.05	11.3	0.0	105.6	0.0
5	0.05	--	--	0.05	0.1	0.0	--	-
6	0.05	--	--	0.1	0.2	0.0	--	--
7	0.08	--	--	--	1.3	0.0	--	--
8	0.08	0.002	0.004	--	1.8	0.6	--	--
9	0.08	--	--	0.05	1.8	1.2	--	--
10	0.05	0.01	0.004	--	53.4	3.8	175.1	0.0
DS René N	--	--	--	--	< 0.5	--	~ 7	--

Table III. Round 2 - Transverse Rupture Strength of DS René N with Elemental Additions Made to a Production Heat.

Heat	Additions, wt.%				1800°F/30 ksi		1800°F/20 ksi		1600°F/50 ksi	
	C	B	Zr	Hf	Life, hrs	R.A.,%	Life, hrs	R.A.,%	Life, hrs	R.A.,%
11	0.05	0.01	0.004	--	42.6	1.2	223.3	2.2	498.6	2.0
12	0.05	0.01	--	0.05	23.3	1.2	211.7	2.2	F.O.L.	2.6
13	0.05	0.01	--	0.1	27.3	0.0	174.4	1.7	295.8	1.5
14	0.05	0.01	--	0.2	F.O.L.	0.6	164.9	1.4	52.8	1.2
15	0.05	0.015	0.004	--	18.3	0.6	332.4	1.5	601.4	3.4
16	0.07	0.015	0.004	--	22.4	1.2	328.3	2.0	392.7	3.2
17	0.07	0.015	--	0.05	22.3	1.3	162.3	1.0	213.7	2.0
18	0.07	0.015	--	0.1	34.8	2.0	300.5	1.8	8.5	2.5
19	0.07	0.015	--	0.2	F.O.L.	0.0	F.O.L.	1.5	--	--
20	0.07	0.02	--	--	41-46.6	0.0	74.6	1.7	F.O.L.	-
DS René N	--	--	--	--	< 0.5	--	~7	--	< 4	--
	Additions, wt.%				1800°F/23 ksi		1800°F/20 ksi		1600°F/55 ksi	
21	--	0.01	--	--	78.2	0.0	103.4	12	150.2	3.2
DS René N	--	--	--	--	< 3	--	~7	--	< 3	--

Round 3: A 300 lb. heat of the new (René N4) Wukusick/King chemistry which had the modified levels of Cr, Al, Ti, and Ta plus 0.05%C (but did not contain Hf) was procured, and then DS cast at GE with no additions of B, Zr, or Hf in one slab and with varying amounts of these additions in 5 heats. Transverse rupture tests were then conducted at 1600°F, 1800°F and 2000°F. As shown in Table IV, the base heat (with C the only addition) had 6 to 10X the transverse rupture life of DS René N, but the B and Hf additions increased the lives another 10 to 20X. The 0.0075% B plus 0.2% Hf (with the base 0.05%C) was particularly excellent at 1600°F and 1800°F. The results at 2000°F showed no significant improvement.

Round 4: Five 300 lb. heats of the modified chemistry (René N4) were then procured; all had 0.05%C and 0.15 or 0.2% Hf. These 5 heats had 0 (0 ppm), 0.002% (20 ppm), 0.003% (30 ppm), 0.004% (40 ppm) or 0.0075% (75 ppm) B. They were also <u>DS</u> cast into slabs at GE and transverse rupture tested at 1400°F, 1600°F, 1800°F and 2000°F. As shown in Table V, 0 ppm B and 20 ppm B were not beneficial to transverse rupture strength at 1400°F to 2000°F, but 30 ppm, 40 ppm and 75 ppm all had excellent transverse rupture strength at 1400°F to 1800°F, but again was not beneficial at 2000°F.

Table IV. Transverse Rupture Strength of DS René N4 with Elemental Additions Made to a Production Heat. All Contain 0.05 wt.% C.

Heat	Additions, wt.%			1600°F/50 ksi		1800°F/23 ksi		2000°F/14 ksi	
	B	Zr	Hf	Life, hrs	R.A., %	Life, hrs	R.A., %	Life, hrs	R.A., %
C1	--	--	--	43.6	1.0	18.5	1.1	3.1	1.2
C2	0.0075	--	--	320.7	4.7	325.2	0.6	11.4	0.8
C3	0.0125	--	--	397.0	18.2	73.4	0.0	--	--
						147.0	0.0		
C4	0.0125	0.006	--	379.0	8.7	343.1	1.2	--	--
C5	0.0075	--	0.2	520.3	11.3	536.7	1.0	4.0	0.0
C6	0.0125	--	0.2	455.4	7.9	409.0	0.6	--	--
DS René N4 (no C)	--	--	--	< 4	--	< 3	--	< 2 (coated)	--

Table V. Transverse Rupture Strength of DS René N4.*

B, ppm	Stress Rupture Properties				Compararison to René N
	Temp, °F	Stress, ksi	Life, hrs	R.A., %	Life, hrs
0	1400	90	4.0	0.0	N.A.
	1600	55	1.9	0.0	< 3
	1800	26	2.3	2.7	< 1
	2000	12	3.1	0.0	< 4
20	1400	90	3.3	0.0	N.A.
	1600	55	15.6	0.8	< 3
	1800	26	9.2	0.0	< 1
	2000	12	4.5	0.0	< 4
30	1400	90	184.4[1]	3.8	N.A.
	1600	55	69.2	0.0	< 3
	1800	26	65.6	0.0	< 1
	2000	12	9.1	1.3	< 4
40	1400	90	92.5[2]	6.2	N.A.
	1600	55	133.8	2.5	< 3
	1800	26	50.0	0.0	< 1
	2000	12	1.8	0.0	< 4
75	1400	90	92.4[3]	32.0	N.A.
	1600	55	54.1	0.0	< 3
	1800	26	98.1	0.6	< 1
	2000	12	4.1	0.6	< 4

* 300 lb. heats contain 0.05 wt.% C and 0.15-0.20 wt.% Hf
(1) Step loaded to 120 ksi (2) Step loaded to 150 ksi (3) Step loaded to 140 ksi

Table VI. 1500-1700°F Transverse Rupture Strength Across LAB's in René N4.*

	New Alloys					René N
B, ppm	LAB °	Temp, °F	Stress, ksi	Life, hrs	R.A., %	Life, hr; (LAB °)
40	14	1500	75	185.0[1]	2.5	N.A.
0	13	1600	58	24.6	0.0	2 (12°)
0	12	1600	58	10	1.2	2 (12°)
20	9	1600	58	146.0	0.0	15 (10°)
20	12	1600	58	77.7	0.0	2 (12°)
40	14	1600	58	304.0[2]	2.5	N.A.
40	15	1600	58	109.8	1.2	2 (12°)
30	12	1600	~55	175.1	1.8	2 (12°)
75	14	1600	58	347.9	1.8	2 (12°)
40	14	1700	45	92.2	0.7	N.A.

* Heats contain 0.05 wt.% C and 0.15-0.2 wt.% Hf
(1) Step loaded to 135 ksi to failure
(2) Step loaded to 78 ksi to failure

Table VII. 1800-1900°F Transverse Rupture Strength Across LAB's in René N4.*

	New Alloys					René N
B, ppm	LAB °	Temp, °F	Stress, ksi	Life, hrs	R.A., %	Life, hr; (LAB °)
0	12	1800	27.5	84.6	0.6	N.A.
0	13	1800	30	10.7	1.3	36 (11°)
0	12	1800	30	55.7	1.2	36 (11°)
20	12	1800	30	31.0	0.6	36 (11°)
20	9	1800	30	68.3	1.9	36 (11°)
20	12	1800	30	52.0	0.0	36 (11°)
30	12 Radius 4 Gauge	1800	~28 30	234.0[1] 234.0[1]	33.6	N.A. 134 (°4)
40	15	1800	24	124.7	0.6	N.A.
40	14	1800	30	108.7	1.3	N.A.
40	15	1800	30	33.3	0.0	N.A.
75	14	1800	30	73.0	0.6	N.A.
20	12	1900	14	140.8[2]	0.0	N.A.
40	15	1900	14	115.7	0.7	N.A.
75	14	1900	14	129.4	0.9	N.A.

* Heats contain 0.05 wt.% C and 0.15-0.2 wt.% Hf
(1) Step loaded to 50 ksi to failure
(2) Step loaded to 17 ksi to failure

Round 4A: The 300 lb. heats of René N4 from round 4 with varying B levels were <u>SX</u> cast into slabs at a casting vendor using a seeding method to produce slabs with a 9°-15° LAB down the center of the slab. The resultant LAB's were tested in transverse rupture at 1500°F to 2000°F. The 1500°F to 1800°F results (Table VI) show that 12° to 15° LAB's with 30-75 ppm of B had >50X improvement in transverse rupture over previously tested René N with 12° LAB's. The results at 1800°F to 1900°F (Table VII) were also excellent, while tests at 2000°F again showed no significant improvement.

Scale-Up Of René N4: It was determined from the previously discussed results that a B level of 30 ppm minimum with 0.05%C and 0.15%Hf would be necessary in René N4 to achieve a significant improvement in René N4's LAB capability from the 6° maximum of René N to possibly 12°. To keep the B level as low as possible to prevent incipient melting, a 3000 lb heat of René N4 was procured with 40 ppm of B (for an eventual range of 30-50 ppm), 0.05%C and 0.15%Hf.

Slabs were then <u>SX</u> cast at the casting vendor with 11° to 16° LAB's which were tested at 1500°F to 2000°F. As shown in Table VIII, the results (even at 2000°F) were excellent.

The 1600°F and 1800°F transverse rupture life across LAB's of SX René N and SX René N4 are plotted in Figures 5 and 6. It shows the LAB lives of René N (no grain boundary additions) which resulted in the 6° maximum René N LAB specification vs. the René N4 lives which resulted in the 12° LAB maximum René N4 specification. This LAB improvement was due primarily to the B and C added to René N4.

Table VIII. Transverse Rupture Strength Across LAB's in René N4.

3000 lb heat of René N4					DS René N
LAB °	Temp, °F	Stress, ksi	Life, hrs	R.A., %	Life, hrs
11	1600	58	380.1	24.9	< 3
	1800	30	118.8	0.6	< 0.4
	1900	14	175.8	0.0	5
	2000	12	89.3	0.0	4
14	1500	75	210.8(1)	7.4	10
	1600	58	171.4	2.5	< 3
	1800	30	51.0	2.5	< 0.4
	1800	24	296.1	0.0	2
	1900	14	148.7	0.0	5
	2000	12	47.9	1.9	4
16	1600	58	168.0	3.7	< 3
	1800	30	73.1	0.0	< 0.4
	1900	14	192.8	0.0	5
	2000	12	49.5	0.0	4

* (1) Step loaded to 135 ksi to failure

Conclusions (LAB Program)

- René N4 contains 0.004% (40 ppm) B + 0.05% C for LAB strengthening (plus 0.15% Hf)
- René N4 has 2X the maximum allowable LAB of René N (12° vs 6°)
- Second and third generation GE single crystal alloys have similar additions for LAB strengthening and 12° maximum allowable LAB's

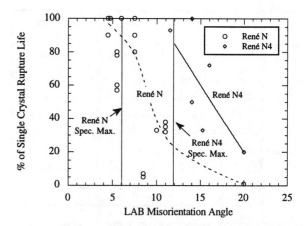

Figure 6. Effect of LAB Misorientation on 1800°F Rupture Life of René N vs. René N4.

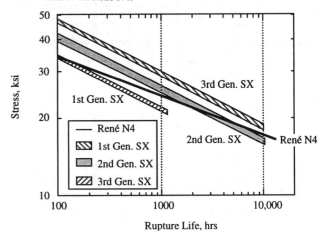

Figure 7. 1800°F Long Time Rupture Behavior of René N vs. Other Single Crystal Superalloys.

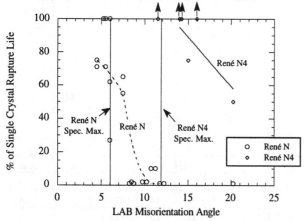

Figure 5. Effect of LAB Misorientation on 1600°F Rupture Life of René N vs. René N4.

René N4 Long Life Rupture Behavior at 1800°F

The "first generation" SX alloys are still preferred in several applications due to advantages of density, cost and surprisingly strength. In the 1800°F temperature range, alloys like René N4 are optimally balanced for prolonged rupture life; as Figure 7 shows, René N4 surpasses the "stronger" next generation alloys in the 7000 - 10000 hour life regime.

An interesting trend is noted in 1800°F rupture behavior by studying the rupture curves for several commercial and experimental alloys. Figure 8 shows alloy density plotted against the 1800°F isothermal rupture slope taken from regressing the logarithm of rupture life versus the logarithm of stress. Here a larger slope is a figure of merit for prolonged low stress, rupture life. Ironically as alloy density increases, the slope decreases. The alloy developer's traditional strengthening methodology has been to increase the refractory content largely in the matrix using rhenium additions. This approach unfortunately can compromise stable long term rupture behavior in the 1800°F temperature range.

An extensive internal research and development program[3] was undertaken at General Electric Aircraft Engines to study several superalloys with varying 1800°F slope behavior. Interrupted creep testing was performed over a range of temperatures and stresses from 10 to 1000 hours. Some specimens were continued to rupture. Optical and TEM metallographic studies were completed. The microstructural evolution of rafting was noted along with the attendant dislocation reactions, especially the formation of mismatch accommodating nets at the gamma-gamma prime interface. Particular attention was paid to the comparison between René N4 and a 3% Re alloy designated alloy 821, which displayed inferior life to René N4 at low stress owing to its steeply sloped isothermal rupture behavior when viewed with the traditional time axis as the horizontal one. Other second generation alloys were found to duplicate alloy 821's behavior in varying degrees.

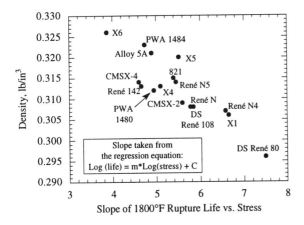

Figure 8. Slope of 1800°F Rupture Life vs. Stress as a Function of Density. Supplemental Data from Ref. 4 and 5.

The principal difference in 1800°F rupture behavior between René N4 and the second generation alloy 821 was associated with the rafting behavior. In René N4, well defined primary, secondary and tertiary stages of creep occurred, typical of high temperature creep behavior. Rafting occurred early, usually completed by the onset of the secondary regime. The gamma prime in secondary creep remained dislocation free. The second generation single crystal materials typically exhibited limited primary creep, a short secondary and prolonged tertiary creep response. Rafting mostly occurred during secondary creep; furthermore, as stress was decreased in alloy 821, a change in the stress exponent (n) occurred from ~ 10 to 3.5. This change in exponent with stress was attributed to the fact that rafting occurred earlier in life at lower stresses, so that the weaker rafted structure was present for a larger fraction of the total specimen life. The weakness of the gamma prime rafts was highlighted by pre-rafting experiments. Pre-rafting significantly degraded alloy 821 creep strength while being somewhat neutral to slightly degrading in René N4. This pre-rafting was performed at both 1800°F and 2000°F followed by 1800°F creep testing.

It was postulated that alloy 821 possessed a strength imbalance between the gamma and gamma prime phase. When the gamma matrix phase is strengthened without an accompanying strength increase in gamma prime, the gamma channel stresses can build up to higher levels (provided other factors such as gamma and gamma prime modulus difference, lattice mismatch and volume fraction are constant). This makes the gamma prime even more vulnerable to dislocation activity. To support this concept, it was shown that alloying the gamma prime in alloy 821 with Ta or Ti additions improved the 1800°F rupture life.[3] These alloys are shown in Figure 8 as X1-X6.

Apparently, at 2000°F a more overall favorable strength balance was maintained between the gamma and gamma prime in alloy 821. At this temperature, a more typical 3X rupture life advantage was noted between first and second generation materials (including alloy 821) without any tendencies for life crossover at low stress. Well defined primary, secondary and tertiary creep regimes were noted.

Interestingly, in third and fourth generation (with even more Re) single crystal alloys, there is a tendency for prolonged creep at 1800°F to induce a change to a gamma prime matrix during rafting.[6] Here the strength of the gamma prime is critical. Unique strengthening of the gamma prime phase can be induced along with exceptional alloy stability in a new class of superalloys recently discovered[7] called "Reverse Partitioning Ni-base Superalloys".

Acknowledgments

The authors particularly thank Carl Wukusick, now retired from GE Aircraft Engines, and Warren King who is now with GE Power Generation for the development of René N4's base chemistry. The authors gratefully acknowledge the contribution of Gary McCabe of GE Aircraft Engines in making the small heats of René N and René N4 and then the SX and DS (HAB) slabs. Other GE members of the alloy development and scale-up teams included Dick McDaniel, Peg Jones and Tom Berry. Of particular importance was Bob Allen who procured the funding and led the scale-up efforts for René N4. A key ingredient in the success of the program was the casting of the René N4 LAB slabs by Howmet, Whitehall, Michigan. The authors acknowledge the single crystal superalloy creep deformation studies which were performed at GE Aircraft Engines by Tresa Pollock, Wendy Murphy and Bob Field. We would like to also extend special thanks to Stephanie Boone of GE Aircraft Engines for being instrumental in the organization and execution, and Scott Walston for editing this paper.

References

1. Carl S. Wukusick and Leo Buchakjian, Jr., U.S. Patent 5,154,884, "Single Crystal Nickel-Base Superalloy Article And Method For Making", 1992.
2. E.W. Ross, C.S. Wukusick and W.T. King, U.S. Patent 5,399,313, "Nickel-Based Superalloys For Producing Single Crystal Articles Having Improved Tolerance To Low Angle Boundaries", 1995.
3. W. Murphy, T. Pollock and R. Field, Mechanism Studies in René N5 and Related Single Crystal Superalloys", (GE Aircraft Engines internal report, 1991).
4. A.D. Cetel and D.N. Duhl, "Second Generation Nickel-Base Single Crystal Superalloy", Superalloys 1988, ed. D.N. Duhl, et al, TMS, 1988, 235-244.
5. K. Harris, et al, "Development of Two Rhenium-Containing Superalloys for Single-Crystal Blade and Directionally Solidified Vane Applications in Advanced Turbine Engines", J. Mat. Eng. and Perf., (1993), 481-495.
6. W.S. Walston, K.S. O'Hara, E.W. Ross, T.M. Pollock and W.H. Murphy, "René N6: Third Generation Single Crystal Superalloy", Superalloys 1996, ed. R.D. Kissinger, et al, TMS, 1996.
7. K.S. O'Hara, W. S. Walston, E. W. Ross and R. Darolia, US Patent 5,482,789, "Nickel-Base Superalloy And Article", 1996.

RENÉ N6: THIRD GENERATION SINGLE CRYSTAL SUPERALLOY

W.S. Walston, K.S. O'Hara, E.W. Ross, T.M. Pollock[*] and W.H. Murphy

GE Aircraft Engines, Cincinnati, OH 45215

[*] Carnegie Mellon University, Pittsburgh, PA 15213

Abstract

A new third generation single crystal superalloy has been developed for aircraft engine turbine airfoil applications. This alloy, René N6, is microstructurally stable and is approximately 30°C stronger than the second generation single crystal, René N5. A new type of instability phenomena, SRZ, was encountered in the alloy development process. SRZ is a cellular precipitation reaction that primarily occurs beneath coatings, but can also be found along grain boundaries. René N6 was developed to avoid the detrimental SRZ, while maintaining excellent creep strength. The mechanical properties and environmental resistance of René N6 are covered and compared to similar superalloys. René N6 has undergone extensive engine testing and currently is being utilized for production engines.

Introduction

Since their development in the late 1980's, second generation, single crystal superalloys have attained success in both commercial and military aircraft engines. These alloys typically contain 3 wt.% Rhenium, which distinguishes them from first generation single crystal superalloys. The achievement of microstructural stability in these alloys involved control of topologically close packed (TCP) phases that could precipitate in moderate amounts after long times at temperatures above 900°C,[1,2] but were not detrimental to properties.

Development of third generation single crystal alloys proceeded with increases in refractory element content. The development of an early, experimental composition, alloy 5A,[3] showed that there were new stability issues associated with the additions of higher levels of refractory elements. Not only was the amount of TCP increased over prior alloys, but a new microstructural instability was discovered.[4] This detrimental instability occurred under coatings, along grain boundaries and within the microstructure. The instability under coatings was termed secondary reaction zone (SRZ) because it occurred beneath the primary diffusion zone between the coating and the alloy. A structurally similar instability occurred along grain boundaries and in dendrite cores. These features were called cellular colonies. The SRZ and cellular colonies consisted of a three-phase constituent with a γ' matrix containing γ and P phase (TCP) needles. The γ and P phase needles tended to be aligned perpendicular to the growth interface. The cellular colonies in the dendrite core were responsible for decreases in creep rupture life of over 50% in some cases.

The discovery of SRZ raised new issues in the further development of third generation single crystal superalloys. It was found that Re was the element most responsible for the formation of SRZ. Thus, the levels of Re and other refractory elements had to be carefully balanced to yield good microstructural stability and creep rupture strength. SRZ under coatings also caused coating compatibility to be a central focus of the alloy development effort. The goal in developing René N6 was to design an alloy that achieved a 30°C improvement in creep rupture strength over the second generation single crystal alloy, René N5, possessed both good environmental resistance and microstructural stability and did not form SRZ.

Alloy Development Approach

The development of René N6 was based upon work on alloy 5A and similar experimental alloys. Figure 1 shows a schematic of the alloy development practice. Initial single crystal castings of each alloy were produced at GE Aircraft Engines. Some alloy series were conducted using a "design of experiments" approach to evaluate individual and interactive elemental effects. Heat treatment trials were conducted on selected alloys to determine the proper solution heat treatment temperature for dissolution of the γ/γ' eutectic. Microstructural stability evaluations consisted of measuring the amount of TCP in 1093°C furnace exposures and in failed creep rupture specimens. Evaluations for SRZ consisted of evaluating different coatings and surface preparation techniques to determine an alloy's propensity to form SRZ. The susceptibility to formation of cellular colonies in dendrite cores was evaluated by using as-cast specimens to accentuate the amount of dendritic segregation.

Creep rupture specimens were tested at 982 and 1093°C in the uncoated and coated conditions. Environmental testing consisted of an 1177°C Mach 1 oxidation test, a 927°C/5 ppm salt hot corrosion test and a 816°C/899°C/2 ppm salt cyclic hot corrosion test. A limited number of alloys were evaluated further using 135 kg master heats. These alloys were evaluated using a wider range of test conditions, stability exposures and coating conditions. At this point, the castability and manufacturability evaluations were initiated.

Final alloy selection was based on the mechanical properties, TCP and SRZ stability evaluations, environmental resistance and castability of the candidate alloys. Following alloy selection, casting and manufacturing programs continued along with the initiation of efforts to obtain design data. Design data included a comprehensive compilation and analysis of physical and mechanical properties using uncoated and coated specimens. A wide variety of engine tests were conducted to assess the behavior of the alloy in various configurations. Finally, the production release of the alloy culminated the alloy development process.

René N6 Composition

The compositions of several second and third generation single crystal alloys are shown in Table I. The second generation alloys contain 3 wt.% Re, while the third generation alloys have significantly higher levels of Re. The Re content of René N6 was most important in determining the balance between creep rupture strength and microstructural stability with respect to both TCP phase formation and SRZ occurrence. Relatively high levels of Co and low levels of Cr were also selected to benefit stability. High levels of Ta were found to be beneficial to high temperature strength and castability.[5] While the elements Hf, C and B were initially not utilized in single crystal superalloys, it has been found by GE Aircraft Engines that they are beneficial to improving the tolerance for casting defects, such as low angle grain boundaries,[6] and for other properties, such as castability and environmental resistance. Yttrium was intentionally added to improve the adherence of the protective aluminum oxide layer formed in service.[7,8]

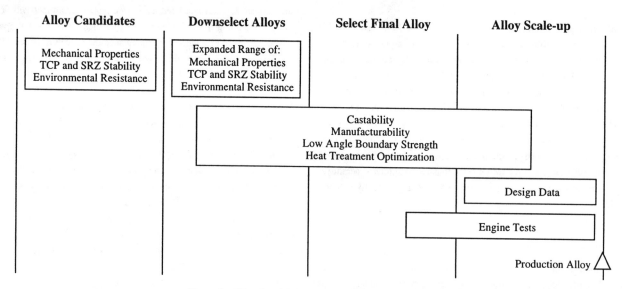

Figure 1. Alloy development approach for René N6.

Table I. Compositions of Second and Third Generation Single Crystal Superalloys (wt.%)

Alloy	Co	Cr	Mo	W	Re	Ta	Al	Ti	Nb	Hf	C	B	Y	ρ g/cc	Ref.
Second Generation Single Crystals															
René N5	7.50	7.00	1.50	5.00	3.00	6.50	6.20	0.00	0.00	0.15	0.05	0.004	0.01	8.63	[9]
CMSX-4	9.00	6.50	0.60	6.00	3.00	6.50	5.60	1.00	0.00	0.10	0.00	0.00	0.00	8.70	[10]
PWA 1484	10.00	5.00	2.00	6.00	3.00	8.70	5.60	0.00	0.00	0.10	0.00	0.00	0.00	8.95	[11]
SC180	10.00	5.30	1.70	5.00	3.00	8.50	5.20	1.00	0.00	0.10	0.00	0.00	0.00	8.84	[12]
Third Generation Single Crystals															
CMSX-10K	3.30	2.30	0.40	5.50	6.30	8.40	5.70	0.30	0.10	0.03	0.00	0.00	0.00	9.10	[13]
CMSX-10Ri	7.00	2.65	0.60	6.40	5.50	7.50	5.80	0.80	0.40	0.06	0.00	0.00	0.00	9.05	[13]
Alloy 5A	12.50	4.50	0.00	5.75	6.25	7.00	6.25	0.00	0.00	0.15	0.05	0.004	0.01	8.91	[3]
René N6	12.50	4.20	1.40	6.00	5.40	7.20	5.75	0.00	0.00	0.15	0.05	0.004	0.01	8.97	[14]

Microstructure and Heat Treatment

The solution heat treatment cycle for René N6 was carefully developed to provide acceptable levels of γ/γ' eutectic and incipient melting. Heat treatment trials to determine the levels of solutioning and incipient melting were conducted, resulting in over a 20°C window for acceptable solution heat treatment, as shown in Figure 2. It is well known that Re strongly segregates to the dendrites during solidification, and this can promote cellular colony formation.[4] It was found that longer solution heat treatment times can reduce the level of Re segregation and prevent cellular colony formation in the dendrite cores.[15] Based on this work, CMSX-10 also employs an extended solution heat treatment cycle in an attempt to improve stability.[16] The effect of the longer solution heat treatment cycles on the amount of TCP phase precipitation in René N6 was also studied. It was found that solution heat treatment time at maximum temperature had no effect on TCP formation.

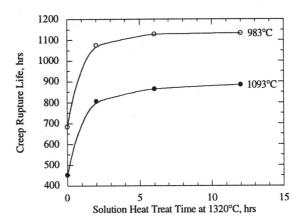

Figure 3. Rupture lives as a function of solution heat treatment time.

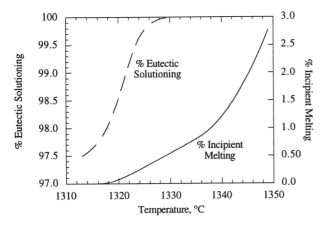

Figure 2. Effect of solution heat treat temperature on the amount of solutioning and incipient melting.

Creep rupture properties were also evaluated as a function of solution heat treatment time at 1320°C. Figure 3 shows that the rupture lives for as-cast specimens were about 60% of the life of solution heat treated specimens. Further, there were no detrimental effects of longer solution heat treatment times on rupture life. In fact, there was a slight increase in rupture life at both 982 and 1093°C. The need for longer heat treatment times must be balanced with the cost of performing the heat treatment. For René N6, the longer solution heat treatment served as insurance against the formation of the cellular colonies in the dendrite cores. Thus, extremely long times were not needed, as may be the case for less stable alloys. The optimum solution heat treatment cycle for René N6 was determined to be 1315-1335°C for approximately six hours. The effect of cooling rate from the solution temperature was evaluated, with the selected cooling rate yielding an initial γ' cube edge length of approximately 0.45 μm.

The effect of different primary age and simulated coating cycle temperatures was evaluated on alloy 5A to determine the effect on creep rupture properties within this class of alloys. Figure 4 shows that there was little change in the rupture life at 871°C as a function of the primary age temperature. This high tolerance to varying heat treatments is an advantage in developing manufacturing cycle parameters for turbine airfoils of René N6. The fully heat treated microstructure of René N6 consists of approximately 65 vol.% of cuboidal γ' precipitates in a γ matrix.

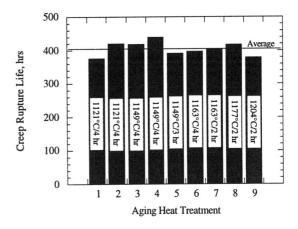

Figure 4. Effect of primary age heat treatment cycle on rupture life.

Microstructural Stability

The microstructural stability at temperatures above 1000°C was a key concern in the development of René N6. TCP phase formation is typically observed in many single crystal superalloys, especially those containing Re, although a small amount of TCP is not considered detrimental to creep rupture and other properties. Figure 5 shows the amount of TCP formed at 1093°C after 500 hours compared to CMSX-10Ri. The small amount of TCP phases that precipitated at elevated temperatures in René N6 did not detrimentally affect creep rupture properties, even at lives of 1000 hrs.

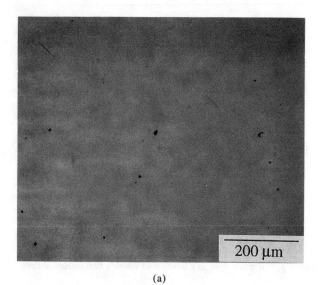 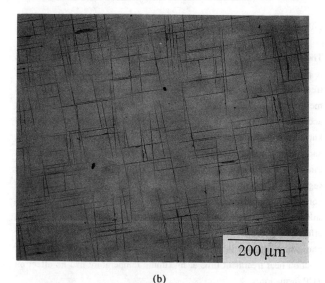

Figure 5. Amount of TCP phase precipitation after 500 hr at 1093°C in (a) René N6 and (b) CMSX-10Ri.

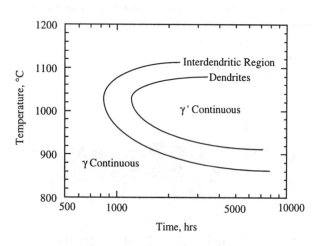

Figure 6. Schematic of times and temperatures at which the γ phase becomes continuous in creep rupture tests.

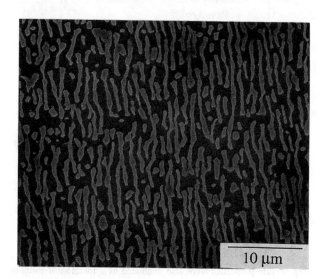

Figure 7. Example of the continuous γ phase following a 982°C creep rupture test.

The high volume fraction of γ' precipitates remained cuboidal in all rupture tests below 850°C, while precipitate rafting occurred at higher temperatures. Interestingly, there was an inversion such that the γ phase became the continuous phase in the temperature range of about 900-1100°C in many of the creep rupture specimens. Figure 6 summarizes the conditions under which the γ' phase became the continuous phase, and Figure 7 shows an example of the γ/γ' inversion at 982°C. Figure 6 also serves as a measure of the rafting kinetics of René N6. As long as the γ' phase remains cuboidal, the γ phase continues to be the matrix phase. When rafting initiates, the γ phase begins to become surrounded by the γ' phase. This inversion is a function of the dendritic structure because of the segregation present in the alloy. The interdendritic region contains a slightly higher volume fraction of γ', as well as a lower amount of refractory elements. Both of these factors cause the interdendritic region to initiate rafting and γ/γ' inversion at earlier times.

While the γ/γ' inversion is not a common phenomenon in superalloys, it has been observed in other single crystal superalloys.[17,18] Examination of the creep and rupture properties showed no detrimental effect of the γ/γ' inversion. Thus, the γ/γ' interface appears to still be the key barrier to creep deformation, as previously observed in single crystal superalloys.[19,20]

A stable microstructure with respect to SRZ and cellular colonies was a primary goal for René N6 since a prior experimental single crystal superalloy, Alloy 5A, was subject to the development of SRZ under coatings and occasional cellular colonies in the microstructure, which severely reduced creep rupture properties.[4] Developing screening tests for both the SRZ and cellular colonies required some knowledge of the mechanisms responsible for nucleation. Based on work on alloy 5A and similar alloys, it was known that Re played a key role in the formation of these instabilities. Surface preparation also seemed to affect the nucleation of the SRZ beneath the coating. The cellular colonies in the dendrite cores appeared to be caused by the large amount of Re in this region due to segregation during casting. The nucleation mechanism of the cellular colonies in the dendrites is unknown at this time. It is possible that a small TCP phase in the dendrite core could serve as a nucleation site.

The screening test developed for SRZ beneath a coating involved applying a coating to a cube which had each face prepared with different surface finishing processes. The specimen was then exposed at 982 and 1093°C for 400 hours to promote SRZ formation. The linear percent SRZ formed under the coating around the periphery of the specimen became the parameter used to rank alloys. A value of 100% indicated that SRZ was continuous beneath the coating around the entire periphery of the specimen. The depth of the SRZ was not measured in this analysis. Screening for cellular colonies in the dendrite cores was performed on as-cast specimens to accentuate the segregation. Thus, if an alloy did not form cellular colonies in this case, it would likely be free from cellular colonies in solution heat treated material. A 1093°C exposure for 400 hours was utilized on the as-cast specimens.

Each candidate alloy was screened using the above tests for SRZ and cellular colonies. Statistical analysis of these results produced the following empirical expression for the relationship between alloy chemistry and the linear % SRZ:

$$[SRZ(\%)]^{1/2} = 13.88(\%Re) + 4.10(\%W) - 7.07(\%Cr) \\ - 2.94(\%Mo) - 0.33(\%Co) + 12.13 \qquad (1)$$

The elements in this equation are in atomic percent. It is clear that Re is the most potent element for determining an alloy's propensity to form SRZ. Minor variations in the Al content of the alloy did not influence the formation of SRZ beneath the coating, however significant Al enrichment occurs beneath the coating, and it is believed that this plays a large role in the formation of SRZ. Interestingly, Cr and Mo had negative coefficients. Reasons for the beneficial effects of Cr and Mo are not well understood, however it has been observed that alloys that form typical sigma and mu TCP beneath coatings do not form SRZ as readily. This is likely because the TCP phases contain high levels of refractory elements, such as Re, W, Mo and Cr.[1,2] Precipitation of TCP phases in the diffusion zone of the coating, removes these elements from the matrix and reduces the chemical driving force for SRZ formation. Thus, Cr and Mo may assist in preferentially forming TCP phases in the diffusion zone instead of SRZ.

It became clear, however, that the amount of TCP in the microstructure away from the coating was not a good indicator of the amount of SRZ that may form under a coating, as shown in Figure 8. The SRZ measurement shown in this figure was made in the manner discussed above, while the TCP measurement was an empirical value of the amount of TCP in the microstructure based on a scale of 0-10. Thus, the stability of the alloy with respect to TCP formation did not correlate well with the alloy's tendency to form SRZ. This means that the driving forces for SRZ nucleation beneath a coating are different than the driving forces for TCP precipitation in the base alloy.

Equation 1 was used successfully to aid in developing alloys that had little or no SRZ under all types of environmental coatings. Control of the coating parameters was also found to be very important in reducing the amount of SRZ. By utilizing the correct coating process, René N6 does not readily form SRZ beneath aluminide, platinum aluminide or overlay coatings. Occasional SRZ has been observed under some conditions, but this has not been found to be detrimental. Using the screening tests described above, Figure 9 displays typical amounts of SRZ found beneath a PtAl coating on the third generation alloys, alloy 5A, CMSX-10Ri and René N6. CMSX-10K was not evaluated in this study because analysis of the composition suggested that CMSX-10Ri was the more stable alloy. The balance of the refractory elements in René N6 resulted in the most favorable coating compatibility compared to similar alloys.

The most important criteria for the successful development of René N6 was to avoid the formation of cellular colonies in the dendrite cores. Using the as-cast specimen described above, each candidate alloy was evaluated for the propensity to form cellular colonies. All failed creep rupture specimens were also evaluated for the presence of cellular colonies. By lowering the Re content and controlling the level of the other refractory elements, René N6 was absent of cellular colonies using the screening test. An extended solution heat treatment was applied to René N6 to further ensure the absence of the cellular colonies. Alloy 5A did show cellular colonies after the screening test, while CMSX-10 did not. It is believed that CMSX-10 did not precipitate cellular colonies in this test because of the relatively high amount of TCP phase precipitation that occurred in the as-cast and solution heat treated condition, as previously shown in Figure 5. A large amount of TCP precipitation reduces the driving force for cellular precipitation by removing the refractory elements from the supersaturated γ matrix.

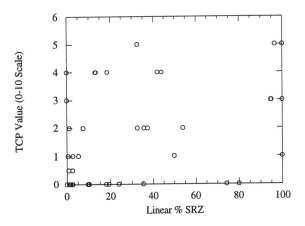

Figure 8. Measurements of SRZ beneath coatings and TCP in the microstructure showing no relationship between the two types of instability.

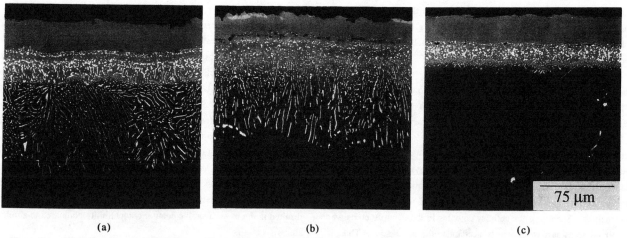

Figure 9. Typical amounts of SRZ observed beneath a PtAl coating after a 400 hour exposure at 982°C in (a) alloy 5A, (b) CMSX-10Ri and (c) René N6.

Properties

A comprehensive design data package containing a wide range of physical and mechanical properties has been generated. The physical properties of René N6 are very similar to other single crystal superalloys and will not be discussed further. Characterization of the mechanical properties of René N6 included over 450 creep rupture tests covering a wide variety of master heats, specimen configurations and coating variables. This included over 100 tests with rupture lives greater than 1000 hours and many over 5000 hours. The stress dependence of the minimum creep rates at several temperatures for René N6 are shown in Figure 10. Creep stress exponents (n) are in the range of 6.1 to 10.8, similar to other single crystal alloys.[21-23] René N6 retains good creep strength even at temperatures as high as 1200°C. Selected creep rupture properties at 982°C are shown in Figure 11. In this figure, the specimen configuration was a 30 mil sheet specimen with a platinum aluminide coating designed to better simulate the turbine airfoil compared to a larger, cylindrical specimen. René N6 displays about a 30°C benefit compared to René N5. The creep rupture properties are summarized in Figure 12 showing a Larson Miller Parameter comparison between René N5 and René N6. The 30°C advantage over René N5 is maintained even at the higher temperatures and longer times due to the good microstructural stability of René N6.

A wide variety of tensile, low cycle fatigue and high cycle fatigue tests have also been performed. The advantages shown in the creep rupture properties over René N5 generally held for the tensile and fatigue properties. For example, in LCF at 982°C and A=∞, René N6 had a 140 MPa advantage over René N5, and in HCF at 1093°C, A=1, René N6 had a 40 MPa advantage over René N5.

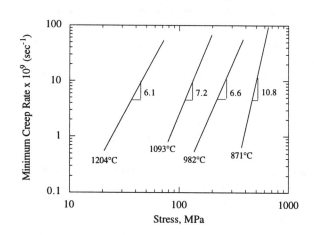

Figure 10. Stress dependence of the minimum creep rate for René N6 at several temperatures.

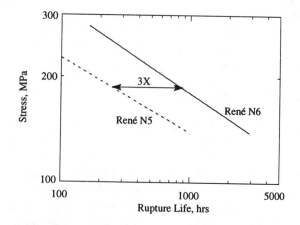

Figure 11. Creep rupture properties of thin wall, PtAl coated specimens tested at 982°C for René N5 and René N6.

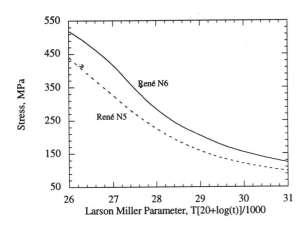

Figure 12. Comparison of the rupture strength of René N6 to René N5 using the Larson Miller Parameter.

The environmental resistance of René N6 is excellent relative to comparable single crystal and directionally solidified alloys. Despite the slightly lower Cr and Al content compared to previous alloys, René N6 develops a protective alumina scale that is enhanced by yttrium additions to the alloy. Unlike previous results on CMSX-4,[24] there were no detrimental effects of yttrium on mechanical properties. Figure 13 shows the results of an 1177°C Mach 1 burner rig oxidation test performed on uncoated specimens. It can be seen that René N6 compares favorably to the excellent oxidation resistance of René N5 and is superior to other single crystal superalloys, such as PWA 1484 and CMSX-10Ri. The hot corrosion resistance of René N6 has been measured in several different tests with favorable results. The hot corrosion resistance of René N6 is similar to the high strength, directionally solidified alloy, René 142, and slightly less than the excellent hot corrosion resistance of René N5. The application of environmental coatings further improves the oxidation and hot corrosion resistance. In addition, applying different techniques to reduce sulfur levels in René N6 has led to even better environmental resistance.[25,26]

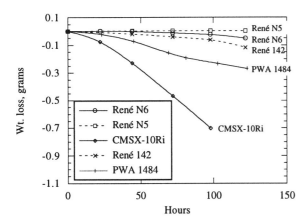

Figure 13. Comparison of several uncoated, single crystal superalloys in a 1177°C, Mach 1 oxidation test.

Production Experience

René N6 has undergone extensive engine testing in a variety of turbine airfoil applications. Low pressure and high pressure turbine blades and vanes have been tested in sizes ranging from relatively small (F414) to large (GE90) airfoils. The castability of René N6 has been excellent in each of these applications and at least equal to prior alloys. The castability of René N6 was enhanced by designed experiments which identified compositional effects[5] and processing changes to improve casting yield. The excellent manufacturability of René N6 has been demonstrated via a multitude of trials, such as brazing, hole drilling and machining.

The microstructural stability of the alloy allows the use of aluminide, platinum aluminide and overlay coatings. Thermal barrier coatings (TBC) have been successfully utilized on several turbine airfoil components. However, one of the advantages of René N6 is that the high strength of the alloy permits it to be used without TBC's with comparable performance to second generation single crystal superalloys with TBC. The high strength of René N6 has also been demonstrated through reduced creep and trailing edge bow of turbine airfoils following engine testing. For example, on one application of a large commercial engine turbine blade, there was no trailing edge bow on the René N6 blades, while René N5 blades had an average of 21 mils trailing edge bow. René N6 has completed extensive property and engine testing and currently is being used in production aircraft engines.

Summary

The development process for René N6 from the early stages of alloy development to production has been reviewed. The development of this alloy was faced with a unique challenge due to an instability phenomenon previously unknown to this class of alloys: SRZ. This is a cellular precipitation reaction that primarily occurs beneath coatings or along grain boundaries. A similar reaction also occurs in dendrite cores. Through better understanding of the mechanisms of formation of SRZ, alloy compositions were developed that avoided this instability. Thus, René N6 is compatible with all types of coatings and shows no property degradation from SRZ. In addition, René N6 does not form the very detrimental cellular colonies in the dendrite cores, which can significantly reduce creep rupture properties. Furthermore, good stability with respect to TCP phases was obtained, while improving the creep resistance of the material relative to second generation alloys. René N6 is approximately 30°C stronger than the second generation single crystal, René N5. Thus, René N6 represents the best combination of strength and stability in third generation single crystal superalloys. An extensive design data package has been generated, thorough engine testing with a variety of configurations has been completed and René N6 is currently being utilized for production engines.

Acknowledgments

The authors gratefully acknowledge the contribution of Gary McCabe of GE Aircraft Engines who produced the slabs for the numerous heats involved in the development and scale-up of René N6. Other members of the alloy development team included Paul Fink, Ron Rajala, Stan Wlodek and Don Kirch. Particular thanks go to Dick McDaniel and Bob Allen who sustained funding for the development of René N6 and led the scale-up and introduction of the alloy into production hardware. The authors also acknowledge the efforts of Howmet and PCC in successfully conducting the many castability trials and scale-up efforts involved with René N6.

References

1. D. N. Duhl, "Alloy Phase Stability Requirements in Single Crystal Superalloys", Alloy Phase Stability and Design, ed. G. M. Stocks, D. P. Pope and A. F. Giamei, MRS, 1991, 389-399.
2. R. Darolia, D. F. Lahrman and R. D. Field, "Formation of Topologically Closed Packed Phases in Nickel Base Single Crystal Superalloys", Superalloys 1988, ed. D. N. Duhl, et al., TMS, 1988, 255-264.
3. C. M. Austin, R. Darolia, K. S. O'Hara and E. W. Ross, U.S. Patent 5,151,249, "Nickel-Based Single Crystal Superalloy and Method of Making" - Alloy 5A, 1992.
4. W. S. Walston, J. C. Schaeffer and W. H. Murphy, "A New Type of Microstructural Instability in Superalloys - SRZ", Superalloys 1996, ed. R. D. Kissinger, et al., TMS, 1996,
5. T. M. Pollock, W. H. Murphy, E. H. Goldman, D. L. Uram and J. S. Tu, "Grain Defect Formation During Directional Solidification of Nickel Base Single Crystals", Superalloys 1992, ed. S. D. Antolovich, et al., TMS, 1992, 125-134.
6. E. W. Ross, C. S. Wukusick and W. T. King, U.S. Patent 5,399,313, "Nickel-Based Superalloys for Producing Single Crystal Articles Having Improved Tolerance to Low Angle Grain Boundaries", 1995.
7. E. W. Ross and K. S. O'Hara, "U.S. Patent 5,173,255, "Cast columnar grain hollow nickel base alloy articles and alloy and heat treatment for making" - René 142, 1992).
8. E. W. Ross and K. S. O'Hara, "René 142: A High Strength, Oxidation Resistant DS Turbine Airfoil Alloy", Superalloys 1992, ed. S. D. Antolovich, et al., TMS, 1992, 257-265.
9. C. S. Wukusick and L. Buchakjian, U.K. Patent Appl. GB2235697 "Improved Property Balanced Nickel-base Superalloys for Producing Single Crystal Articles" - René N5, 1991.
10. K. Harris and G. L. Erickson, U.S. Patent 4,643,782, "Single Crystal Alloy Technology" - CMSX-4, 1987.
11. D. N. Duhl and A. D. Cetel, U.S. Patent 4,719,080, "Advanced High Strength Single Crystal Superalloy Compositions" - PWA 1484, 1988.
12. X. Nguyen-Dinh, U.S. Patent 4,935,072, "Phase Stable Single Crystal Materials" - SC180, 1990.
13. G. L. Erickson, U.S. Patent 5,366,695, "Single Crystal Nickel-Based Superalloy" - CMSX-10, 1994.
14. W. S. Walston, E. W. Ross, K. S. O'Hara, T. M. Pollock and W. H. Murphy, U.S. Patent 5,455,120, "Nickel-Base Superalloy and Article with High Temperature Strength and Improved Stability" - René N6, 1995.
15. W. S. Walston, E. W. Ross, K. S. O'Hara and T. M. Pollock, U.S. Patent 5,270,123, "Nickel-Base Superalloy and Article with High Temperature Strength and Improved Stability", 1993.
16. G. L. Erickson, "A New, Third Generation, Single Crystal, Casting Superalloy", JOM, April 1995(1995), 36-39.
17. R. A. MacKay and L. J. Ebert, "The Development of γ/γ' Lamellar Structures in a Nickel-Base Superalloy During Elevated Temperature Mechanical Testing", Metall. Trans., 16A(1985), 1969-1982.
18. A. Fredholm and J. L. Strudel, "On the Creep Resistance of Some Nickel Base Single Crystals", Superalloys 1984, ed. M. Gell, et al., TMS, 1984, 211-220.
19. M. V. Nathal, J. O. Diaz and R. V. Miner, "High Temperature Creep Behavior of Single Crystal Gamma Prime and Gamma Alloys", High Temperature Ordered Intermetallics III, ed. C. T. Liu, et al., MRS, 1989, 269-274.
20. T. M. Pollock and A. S. Argon, "Creep Resistance of CMSX-3 Nickel Base Superalloy Single Crystals", Acta Metall. mater., 40(1)(1992), 1-30.
21. W. Murphy, T. Pollock and R. Field, "Creep Mechanism Studies in René N5 and Related Single Crystal Superalloys", (GE Aircraft Engines internal report, 1991).
22. M. V. Nathal and L. J. Ebert, "The Influence of Cobalt, Tantalum and Tungsten on the Elevated Temperature Mechanical Properties of Single Crystal Nickel-Base Superalloys", Metall. Trans., 16A(1985), 1863-1870.
23. M. V. Nathal and L. J. Ebert, "Elevated Temperature Creep-Rupture Behavior of the Single Crystal Nickel-Base Superalloy NASAIR 100", Metall. Trans., 16A(1985), 427-439.
24. H. Mueller-Largent and D. J. Frasier, "Advanced Single Crystal Superalloy Processing Development for Ultra High Performance Fabricated Transpiration Cooled Turbine Airfoils", (Allison Gas Turbine Division, U.S. Navy Final Report NAPC-PE-225-C, 1991).
25. J. C. Schaeffer and W. H. Murphy, Unpublished Data, GE Aircraft Engines, 1992-1996.
26. M. A. Smith, T. H. Mickle, W. E. Frazier and J. Waldman, "Development of a Hydrogen-based Annealing Process for the Desulfurization of Single Crystalline, Nickel-based Superalloy", (Naval Air Warfare Center, U.S. Navy Final Report NAWCADWAR-95001-4.3, 1994).

THE DEVELOPMENT AND APPLICATION OF CMSX®-10

G. L. Erickson
Cannon-Muskegon Corporation
(a subsidiary of SPS Technologies)
P.O. Box 506
Muskegon, MI 49443-0506 U.S.A.

Abstract

The CMSX®-10 alloy is a third generation single crystal (SX) casting material which is used in demanding turbine engine blading applications. The flight engine certified alloy is characterized by it's 6 wt. % rhenium content, high additive refractory element level, and relatively low chromium employment. Based on published data, the alloy is thought to exhibit the highest creep strength and resistance to fatigue of any production Ni-base, cast SX superalloy.

CMSX-10 alloy provides an approximate 30°C improved creep strength relative to second generation 3 wt. % containing SX alloys such as CMSX-4 and PWA 1484. Furthermore, it develops low cycle and high cycle fatigue (LCF and HCF) strengths as much as 2-3 times better than the best alternatives. Moreover, the alloy also develops an attractive blend of tensile and impact strengths, foundry performance, heat treatability and environmental properties characteristic. Most notably, the alloy provides surprisingly good hot corrosion resistance, despite its novel and relatively low chromium content (2-3 wt. %). Additionally, the alloy performs extremely well in both the aluminide and Pt - aluminide coated conditions.

Although the CMSX-10 alloy was developed to fulfill a perceived need in the aero-turbine industry, the alloy's long-term high strength, particularly at temperatures ranging from 850-950°C, has attracted significant industrial turbine interest. For this reason, longer term (currently to about 5000 hours) creep-rupture strength characterization is underway. Similarly, due to a continued need for materials with greater creep-strength, a higher strength CMSX-10 derivative, currently designated CMSX-10+, is under development.

This narrative characterizes the CMSX-10 alloy SX component castability, heat treatability, mechanical strength, environmental properties and coating characteristics. Active long-term creep-rupture programs are discussed, as well as preliminary results for a higher strength alloy, currently designated CMSX®-10+.

Introduction

The commercialization of the directional solidification casting process for turbine blade and vane manufacture resulted in the definition of many alloy designs seeking to maximize the benefit afforded with the directional structure. Foremost in this regard has been the definition of specific alloy formulations used in producing single crystal castings. Broadly, such materials began as simple modifications of polycrystalline and directionally solidified, columnar grained materials, with alloy complexity increasing with the ever-increasing results being sought.

The alloys initially defined were non-rhenium containing and generally afforded a 17-22 °C strength improvement relative to most directionally solidified, columnar grained materials. As the positive strengthening effects of Re alloying became more widely apparent, and the cost of Re metal became commercially viable, alloys containing about 3 wt. % Re ensued.

This category material containing 3 Wt. % Re (now referred to as second generation SX casting alloys), generally exhibited about 30-35°C improved strength in comparison to the so-called first generation SX alloys. These materials, such as CMSX-4, PWA 1484 and René N5, gained commercial significance in the latter part of the last decade, and continually realize new applications.

Although able to provide impressive strength, the second generation, 3 wt. % Re - containing superalloys were able to be improved upon through increased Re alloying. Partly stimulated by needs identified during new engine design for the recently introduced Boeing 777 wide body, twin engine aircraft, third generation SX alloys were defined.

Broadly, the new materials eventually found significance with Re level of about 6 wt. %. Successful alloys generally contain a high volume fraction of refractory elements while maintaining moderate Al + Ti content. Necessitated through concern for microstructural phasial stability, the alloys employ dramatically lower chromium content than second generation SX casting alloys.

CM 247 LC®, CM 186 LC®, CMSX-2®, CMSX-3®, CMSX-4®, CMSX-6®, CMSX®-10, CMSX®-11B AND CMSX®-11C are registered trademarks of the Cannon-Muskegon Corporation.

The resulting third generation class of materials generally achieve an approximate 30°C strength improvement relative to their second generation alloy counterparts. Strength, stability and hot corrosion resistance varies for the materials identified, however, all appear to exhibit acceptable hot corrosion resistance, despite their uniquely low chromium contents.

The availability of these higher strength materials has allowed engine designers to replace cooled turbine blades with un-cooled designs, thereby increasing turbine efficiency due to the attendant reduction in the cooling-air requirement. Demonstration of this occurrence is provided by the Rolls-Royce use of the CMSX-10 alloy (RR 3000) in a hollow, uncooled IP blade component of the TRENT 800 series engine (1,2), where rather than utilizing a cooled CMSX-4 blade, RR determined that they could eliminate the blade cooling requirement by utilizing the higher strength CMSX-10 alloy.

As the industrial TRENT is designed utilizing the turbine core technology developed for the aero-turbine TRENT 800 (2), CMSX-10 industrial turbine application is thereby realized. Along this line, CMSX-10 alloy application in other industrial turbines (marine and or land-based) is also envisioned due to the significant long-term creep-strength advantage the alloy provides in the lower temperature regime normally predominating in industrial turbine high pressure turbine sections, as illustrated in Figure 1.

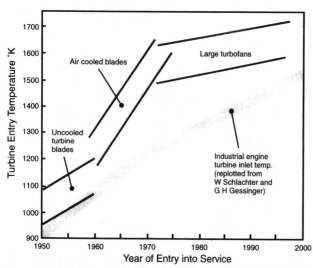

Figure 1: Comparison of civil aero engine take off temperatures and industrial engine turbine inlet temperatures (3).

ABB reports that relatively low turbine material temperatures (less than 900°C) prevail in their newly designed GT 24 and GT 26 engines (4). Similarly, other industrial turbine producing companies confirm typical hot section turbine operating temperatures ranging from 900 to 950°C (1652-1750°F), with relatively short-term exposure at higher temperatures (5).

As third generation alloys can exhibit phasial instability at elevated temperatures, the propensity for Topologically-Close-Packed (TCP) phase formation occurring during long-term industrial turbine operation has been an issue of concern. Most typically, third generation alloys tend to exhibit their greatest propensity for TCP phase formation with exposure at about 1090-1150°C (2000-2100°F). Moreover, as illustrated in Figure 2, TCP phase formation in the CMSX-10 alloy at 980°C and below is quite sluggish. Furthermore, microstructural review of specimens tested to about 5000 hours at 913°C shows a complete absence of TCP phase formation. Although longer-term test data continue to be developed, it appears at this point that third generation SX casting alloys such as CMSX-10 may, therefore, potentially offer great utility for usage in some industrial turbine designs.

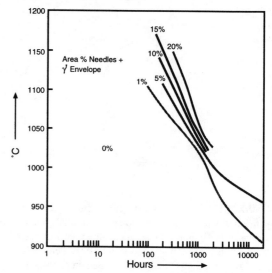

Figure 2: CMSX-10 alloy tendency for TCP phase formation.

To this end, the CMSX-10 alloy is under evaluation in the U.S. Department of Energy (DOE) funded Advanced Turbine Systems (ATS) Program. This effort is targeted toward developing and commercializing ultra-high efficiency, environmentally superior, cost-competitive gas turbine systems for base-load applications employing firing temperatures of 1427°C (2600°F) or greater. Moreover, the alloy system is also considered for evaluation by other individual industrial turbine engine producers throughout the world.

For aero-turbines (beyond the Rolls-Royce usage) the alloy is under consideration and evaluation by several turbine producers. As the strength requirements for the newest turbine designs continue to ascend, the need for a higher strength CMSX-10 alloy derivative has arisen, and is therefore under development. The higher strength alloy, currently designated CMSX-10+, provides an enhanced strength of about 8°C.

Alloy Design

Increasing alloy creep-strength beyond the levels exhibited by second generation SX casting superalloys requires an increase of Re alloying. Generally, it appears that moderate Re increases, ie., to levels of 4 or 5 wt. %, is ineffective toward promoting gains in creep capability sufficient to justify new alloy definition. Levels of 6 to 6.5 wt.. % appear the most feasible at this point, so the CMSX-10 alloy system is balanced around the inferred requirement. Along with the relatively high Re level employed, experience suggests that creep-strength optimization also requires increased levels of other refractory element hardeners such as W, Ta and Mo.

As W tends to be involved in TCP phase formation, its employment must be judiciously balanced with the predominating Re requirement. Moreover, since Cr also contributes to sigma phase formation, utilizing high Re + W content requires relatively low chromium alloying.

Since Cr is a significant contributor to alloy hot corrosion resistance, alloy Cr level selection must be carefully balanced to achieve the required alloy hot corrosion characteristic while being low enough to ensure adequate resistance to sigma phase formation.

Against these design constants, Mo and Ta alloying may be employed with increased latitude. However, as Mo is generally a negative factor for environmental properties, its employment is carefully scrutinized. Furthermore, since tantalum's atomic diameter is larger than Mo, it is a more efficient hardener. Significant Ta alloying is employed because of its beneficial effect in the SX casting process in reducing alloy freckle formation, plus it positively influences environmental property characteristics.

All told, the CMSX-10 alloy system employs the relatively high additive refractory element content of about 20.1 Wt.. %; this in comparison to CMSX-4 which is about 16.4% and CMSX-2 at 14.6%. Gamma prime formers such as Al and Ti are similar to second generation SX alloys and the CMSX-10 gamma prime chemistry is actually quite similar to that predominating for CMSX-4.

For this reason, creep-rupture properties of the CMSX-10 alloy at temperatures where gamma prime particle chemistry has a major influence to alloy creep resistance ie., relatively low temperatures (700-800°C), the alloy's properties are similar to CMSX-4. Additionally, the γ' particle chemistry characteristic implies that a significant level of Ta is distributed within the alloy matrix, thereby positively impacting solid solution strengthening.

This, in tandem with the high level of W + Re + Mo prevailing throughout the alloy matrix assists in the attainment of the extremely high creep-strength exhibited by the CMSX-10 alloy, particularly at elevated temperatures.

With TCP phase formation being a formidable issue in the design of third generation SX alloys, reduction in the tendency toward TCP phase formation is partially achieved through judicious selection of the cobalt aim composition. Prudent consideration of alloy Co + Cr content in a third generation alloy system can lead to attainment of acceptable alloy stability while maintaining high enough total refractory element content to achieve the desired alloy strength characteristics. Table I provides the CMSX-10 alloy chemistry in comparison to other first, second and third generation SX alloys.

Table I Nominal Compositions of Three Generations of Single-Crystal Superalloys (wt. %)

Alloy	Cr	Co	Mo	W	Ta	Re	V	Nb	Al	Ti	Hf	Ni	Density (kg/dm³)	Ref.
First Generation														
PWA 1480	10	5	–	4	12	–	–	–	5.0	1.5	–	Bal.	8.70	7
PWA 1483	12.8	9	1.9	3.8	4	–	–	–	3.6	4.0	–	Bal.	–	17
René N4	9	8	2	6	4	–	–	0.5	4.2	4.2	–	Bal.	8.56	8,9
SRR 99	8	5	–	10	3	–	–	–	5.5	2.2	–	Bal.	8.56	10,11
RR 2000	10	15	3	–	–	–	1	–	5.5	4.0	–	Bal.	7.87	10,11
AM1	8	6	2	6	9	–	–	–	5.2	1.2	–	Bal.	8.59	12
AM3	8	6	2	5	4	–	–	–	6.0	2.0	–	Bal.	8.25	13
CMSX-2®	8	5	0.6	8	6	–	–	–	5.6	1.0	–	Bal.	8.56	14
CMSX-3®	8	5	0.6	8	6	–	–	–	5.6	1.0	0.1	Bal.	8.56	14
CMSX-6®	10	5	3	–	2	–	–	–	4.8	4.7	0.1	Bal.	7.98	15
CMSX®-11B	12.5	7	0.5	5	5	–	–	0.1	3.6	4.2	0.04	Bal.	8.44	18
CMSX®-11C	14.9	3	0.4	4.5	5	–	–	0.1	3.4	4.2	0.04	Bal.	8.36	19
AF 56 (SX 792)	12	8	2	4	5	–	–	–	3.4	4.2	–	Bal.	8.25	16
SC 16	16	–	3	–	3.5	–	–	–	3.5	3.5	–	Bal.	8.21	20
Second Generation														
CMSX-4®	6.5	9	0.6	6	6.5	3	–	–	5.6	1.0	0.1	Bal.	8.70	21
PWA 1484	5	10	2	6	9	3	–	–	5.6	–	0.1	Bal.	8.95	22
SC 180	5	10	2	5	8.5	3	–	–	5.2	1.0	0.1	Bal.	8.84	23
MC2	8	5	2	8	6	–	–	–	5.0	1.5	–	Bal.	8.63	24
René N5	7	8	2	5	7	3	–	–	6.2	–	0.2	Bal.	NA	25
Third Generation														
CMSX®-10	2	3	.4	5	8	6	–	.1	5.7	.2	.03	Bal.	9.05	26
René N6	4.2	12.5	1.4	6	7.2	5.4	–	–	5.75	–	.15	Bal.	8.98	27

Alloy Manufacture

The CMSX-10 superalloy is manufactured in similar fashion to other single crystal casting alloys. The alloy's relatively high Re content does not present any new melting challenges other than raising the alloy liquidus temperature.

A standard CM VIM technique along the detail delineated in Ref 6 is utilized to produce the CMSX-10 superalloy. Greater than fifty (50) each 114-182 kg developmental VIM heats were produced and evaluated in the CMSX-10 alloy development program. The CMSX-10 alloy is a production status material with five production, 3.9 ton heats having been produced and sold (as of February 1996) as well as one 50% virgin/50% revert heat manufactured utilizing approximately 1800 kg of foundry process generated CMSX-10 alloy revert (runner systems and scrap blades).

As anticipated, the quality of the 50/50 product is nearly identical to that typical of 100% virgin. Table II offers a "window" toward defining alloy quality in that it presents the typical trace element levels achieved in the virgin and 50/50 production heat product. The only characteristics able to be differentiated among the alloy heats are in respective sulfur and heat oxygen contents. But of course, both levels are extremely good, and end-users realize significant alloy component cost improvements through the use of 50% virgin/50% revert alloy product.

Table II Typical Alloy Tramp Element Levels for 100% Virgin and 50/50 Revert Blends of CMSX-10 Alloy

Alloy	C	B	Zr	S	[N]	[O]
	wt. ppm					
CMSX®-10* (100% V)	20	<20	<10	1	1	1
CMSX®-10⁺ (100% V)	20	<20	<10	1	1	1
CMSX®-10⁺⁺ (50%/50% R)	22	<25	21	2	1	2

*Greater than 50 each developmental 114-182kg heats
⁺ 5 each 3.9 ton production heats
⁺⁺ 1 each 3.9 ton production, 50% virgin/50% revert heat

Earlier 140 kg heat work seeking to determine the suitability of CMSX-10 alloy 50% virgin/50% revert alloy product usage illustrated that identical alloy mechanical properties are achieved with the 50/50 product vs. the 100% virgin counterpart. Confirmation of this is anticipated with the production process produced 50/50 product.

Casting Experience And Heat Treatment

With the CMSX-10 alloy being flight engine certified in the TRENT 890 engine, considerable production foundry experience with the alloy system prevails. While some platform low angle boundary defect difficulties were occasionally encountered when producing TRENT 890 engine IP blade components with the CMSX-10 alloy, relatively simple SX grain extenders resolved the problem which, incidentally, had also occurred with the CMSX-4 alloy. Manufacture of other turbine blading components at several investment casting foundries around the world demonstrates that the alloy's castability is similar to the second generation, CMSX-4 superalloy.

Further confirmation of the alloy's castability characteristics was achieved with development casting trials undertaken with Allison Engine Co. T56 engine first stage HPT cooled blades plus solid, second stage blade components as illustrated in Figure 3. Excellent experience with first stage Solar Turbines MARS engine blades (Fig. 3) cast at Howmet Corporation in Whitehall, MI has been obtained, while castability experience developed in Japan has been equally favorable. Successful castability experience with demonstrator engine blades has also been achieved, thereby further confirming favorable CMSX-10 foundry characteristics.

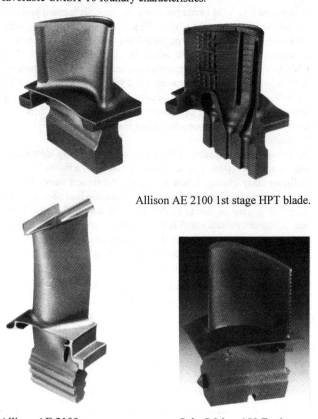

Figure 3: Selected turbine blade configurations used to demonstrate the CMSX-10 alloy castability characteristic.

Besides possessing good SX castability characteristics, CMSX-10 alloy also provides attractive solution heat treatment capability. The alloy provides a 21°C heat treatment window (numerical difference in °C between the alloys γ′ solvus and incipient melting point) and is thereby able to be fully solutioned. Solution heat treatment, undertaken with a final soak temperature of 1366°C, results in full coarse γ′ solutioning plus complete eutectic γ-γ′ dissolution. Re-precipitation of the dissolved γ′ into a more useful fine γ′ results, and primary aging of the fine γ′ then occurs at 1152°C.

The CMSX-10 alloy primary γ′ aging treatment helps develop an array of fine cubic γ′ with relatively regular alignment, exhibiting average edge dimensions of about 0.5 μm. (Fig. 4). The primary aging treatment is followed by two secondary aging treatments which promote the formation of finer γ′ within the alloy's matrix channels. These secondary aging treatments are preformed at 871°C for 24 hours and at 760°C for 30 hours.

The precipitates which develop through these aging conditions are relatively fine, and are likely dissolved within blade airfoil sections with normal service exposure at higher temperatures. However, as blade root sections do not normally experience exposures in the 760 to 870°C regime, the γ′ precipitates which are formed during the aforementioned aging processes prevail throughout turbine blade component life-cycles and are thought to contribute positively to component tensile and fatigue property characteristics. The small matrix channel precipitate occurring with the multiple step aging treatment is illustrated in Figure 5.

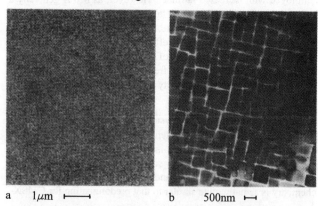

Figure 4: Two views of fully heat-treated CMSX-10 alloy.

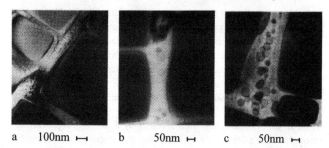

Figure 5: The CMSX-10 alloy aged at (a) 1,152°C, (b) 1,152°C + 871°C and (c) 1,152°C + 871°C + 760°C.

Further alloy developments seeking to identify a higher strength CMSX-10 alloy derivative led to the definition of the CMSX-10+ composition. Due to the design changes necessary to develop higher strength, the resulting alloy exhibits a narrower heat treatment window of approximately 8°C. However, as the producers and users of high temperature solution heat treatment furnaces have improved new furnace designs, extremely good process temperature control (± 4°C) within multiple layers or zones of vacuum heat treatment vessels are achieved. It thereby appears that, contrary to prior experience, alloys with relatively small heat treatment windows may be successfully heat treated on a production basis.

With successful solution heat treatment, relatively straight-forward aging cycles are applied to the newest, high-strength SX alloy system. Currently, the aging treatments applied to CMSX-10+ alloy articles are identical to those operative with CMSX-10, with nearly identical results.

Mechanical Properties

Increased superalloy Re alloying generally results in significant creep and fatigue strength improvements. However, as can be inferred from Figure 6, alloys with high solute element levels such as CMSX-10 do not necessarily exhibit improved alloy tensile properties. Similar to the CMSX-4 alloy, the CMSX-10 tensile strength peaks at about 760°C while maintaining relatively good ductility throughout the tested regime.

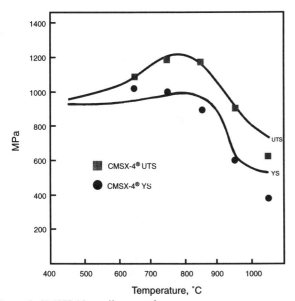

Figure 6: CMSX-10 tensile strength.

Similarly, superalloy impact properties do not appear to be dramatically influenced by increasing alloy Re content, as illustrated in Figure 7 with comparison of the CMSX-10 and CMSX-4 respective capabilities.

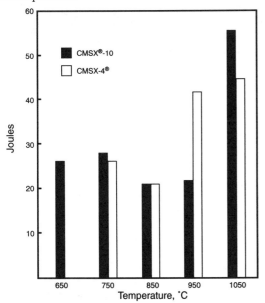

Figure 7: CMSX-10 impact strength. Cylindrical specimen, 7.2 mm diameter x 55 mm long.

Third generation SX casting alloys generally exhibit enhanced creep strength relative to their second generation predecessors. The relative improvement depends upon the alloy systems compared, but for CMSX-10, the advantage is about 30°C relative to CMSX-4. In terms of actual turbine engine operation, this advantage is shown quite dramatically through the difference in turbine blade growth occurring during an engine test for CMSX-10 vs CMSX-4 blades shown in Fig. 8.

In this particular test, identical CMSX-4 and CMSX-10 blades in a common blade ring were measured before and after the engine run. Analysis of the test data revealed average CMSX-4 blade growth

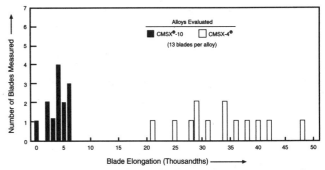

Figure 8: Post turbine engine test blade growth measurement results.

about eight times (8x) greater than with CMSX-10. Typically, the CMSX-10 alloy exhibits about a 3x to 5x advantage in creep strength, thereby suggesting the CMSX-4 alloy blades were in a tertiary creep regime while the CMSX-10 alloy blades were still in a primary creep mode. For CMSX-10, the alloy's 30°C strength advantage prevails to about 1100°C, where it's rupture strength begins to approximate CMSX-4, and with long term exposure, is actually lower. From the 1100°C to about 1160°C regime, the CMSX-10 alloy is not as strong as CMSX-4 on the basis of rupture. The longer the exposure at temperatures within this regime, the greater the alloy's debit, due to the propensity for TCP phase formation within the banded range of temperature. However, for creep-rupture tests above 1160°C, CMSX-10 alloy is again superior to CMSX-4. Moreover, metallographic review of samples tested to rupture at 1200°C reveals extremely good γ' particle stability after 400 hours of exposure.

The CMSX-10 alloy 1% creep-strength is compared to the respective CMSX-4, CM 186 LC and CM 247 LC capabilities in Figure 9. Unlike the alloy's rupture characteristics, CMSX-10 is shown to provide greater 1% creep strength than CMSX-4 throughout the temperature range typically of interest for turbine designers, in tests taken to about 1000 hours duration. This dynamic changes at certain temperatures, though, with longer term tests.

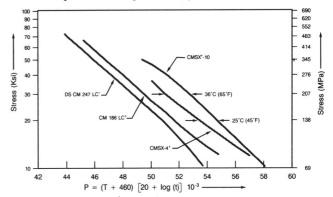

Figure 9: 1.0% Larson-Miller creep strength of several alloys.

The TTT diagram presented in Fig. 2 also illustrates that TCP phase forms relatively sluggishly in the CMSX-10 alloy system at 980°C and below. In this temperature region, the CMSX-10 alloy maintains it's 30°C advantage over CMSX-4 for extremely long duration. For this reason, the alloy is also of considerable interest to land-based industrial turbine producers whose predominating hot section blade component exposure temperatures are around 950°C and below. Like the turbine efficiency improvements realized through CMSX-10 aero-turbine application, some industrial turbine producers anticipate significant turbine efficiency gains with CMSX-10 alloy employment through the reduction of cooling air requirements and/or the increased component stressing which the alloy is able to endure. Such efficiency improvements are extremely desirable since it is reported that an increase of a single percentage point of efficiency can reduce operating costs by $15-20 million over the life of a typical gas-fired combined-cycle plant in the 400-500 megawatt range (28).

Industrial turbine producers require long-term creep-rupture data to accurately predict a material's performance within their turbines. To this end, significant numbers of creep-rupture tests have been completed with the CMSX-10 alloy to about 4000-4500 hours duration. Additional testing is underway, with many anticipated to reach at least 7000 hour lives, while others are planned which will achieve 15,000 hour lives.

For specimens which have ruptured, log-log plots of the "long-term" creep results are presented. While Fig. 10 presents the 982°C rupture strength of CMSX-10 in comparison to the CMSX-4, DS CM 247 LC and IN 738 LC alloys, Figure 11 presents the alloy's 1% creep strength at 982°C, and Figure 12 illustrates the CMSX-10 alloy strength for exposure temperatures ranging from 913 to 1010°C, and lives approaching 4500 hours. Generally, the CMSX-10 alloy develops significantly greater strength than CMSX-4 in these test conditions. However, at 982°C, the CMSX-10 benefit appears to narrow slightly with longer term exposures. This suggests that a point exists where the alloy doesn't provide any benefit relative to CMSX-4. As the narrowing improvement with longer times is a function of alloy TCP phase formation, it must also be noted that the CMSX-4 alloy also forms sigma at such temperatures. The crossover point in alloy capability may not be as early as initially thought, ie. the alloys may not reach parity at 982°C until 20,000 hour exposures occur, rather than 10,000 - 12,000 hours. Longer term creep-rupture tests will investigate the issue.

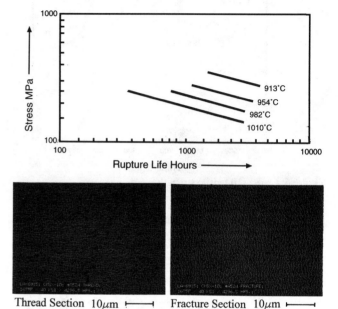

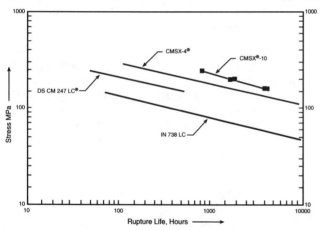

Figure 10: 982°C stress-rupture strength of several alloys.

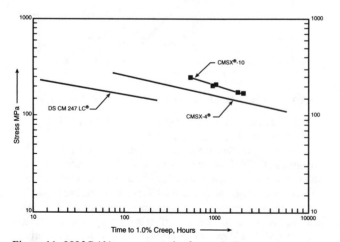

Figure 11: 982°C 1% creep strength of several alloys.

Thread Section 10μm ⊢——⊣   Fracture Section 10μm ⊢——⊣
913°C/275.8 MPa/4296.5Hr. Life Post-test specimen microstructure.
Figure 12: 913-1010°C stress-rupture strength of CMSX-10.

The same situation prevails with the alloy's 1% creep strength at 982°C. Although the TCP phase formation in CMSX-10 at 982°C is quite sluggish, it nonetheless does occur. Similarly, future tests will investigate the issue.

Tests performed at 954°C have run to about 4000 hours, while those performed at 913°C have run to about 4300 hours. Metallographic review of the failed rupture specimens reveal a slight amount of TCP phase forming in the 954°C tested specimen at 4000 hours, while no TCP is apparent in the 4300 hour, 913°C tested specimens.

Longer term creep-rupture tests are underway, all of which have realized at least 5800 hours exposure. The tests which are running at various stress conditions at temperatures ranging from 850°C to 1050°C, currently exhibit creep strains ranging from .03 to 3.0%, respectively suggesting likely rupture lives ranging from 7000 to 10000 hours. Additional tests slated for near-term commencement will explore the alloy's 15000 hour life characteristics at similar temperature ranges. The results of the current and future long-term creep tests are expected to reaffirm the alloy's significant strength advantage over the CMSX-4 alloy at temperatures of 982°C and below; a regime which is quite significant to the land-based power generating gas turbine engine community.

The alloy's high level of Re also positively influences fatigue properties. References 29 and 30 presented the results of LCF and HCF tests undertaken at various temperatures. Generally, the CMSX-10 alloy fatigue performance is at least as good as CMSX-4, and with some test conditions, 2-3 x better. For 50,000 cycle smooth

LCF, CMSX-10 performs similar to CMSX-4 at 700°C, however between 800-950°C, it exhibits a 30°C advantage, or 2.5 x improvement based on life or a 15% advantage based on strength. Notched LCF ($K_t=2$) tests at 950°C reveal similar results for the two alloys, while those undertaken at 750°C exhibit a 2.5 x advantage for CMSX-10. HCF comparative tests performed at 550°C and 950°C also reveal a 2.5 x improvement (See Fig. 13).

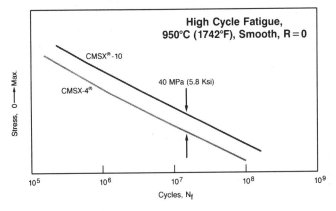

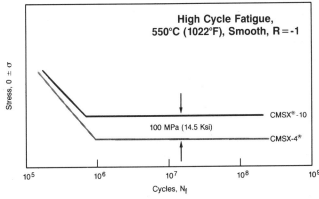

Figure 13: High cycle fatigue behavior of CMSX-10 and CMSX-4 at 550°C and 950°C.

Higher Strength CMSX-10 Derivatives

As the need for higher strength alloys continues, a higher strength derivative of the CMSX-10 alloy has been defined. The alloy currently designated CMSX-10+, exhibits about an 8°C improvement relative to CMSX-10. It continues to exhibit a tendency for TCP phase formation at high temperatures with long exposure, however, applications having service profiles able to accommodate such characteristics have been defined.

While creep-rupture properties are improved, alloy fatigue properties are similarly influenced through the chemistry modification. As greater refractory element levels are employed in the alloy design, alloy Cr and Co level adjustments required for stability jeopardized the derivative alloy's hot corrosion characteristics. However, similar tests to those previously defined for CMSX-10, show acceptable alloy hot corrosion resistance prevails.

The CMSX-10+ material is still in the process of characterization. Comparative creep-rupture test results show that the alloy generally develops about 1.3x - 1.7x greater creep and stress-rupture strengths than CMSX-10 in short-term tests, while longer term characterization (greater than 1000 hours) suggests improvements to 1.5x for tests performed at 1010°C and below. The derivative alloy does not offer significant improvements with long-term test conditions above 1010°C. Further alloy characterization is anticipated to continue through eventual gas turbine engine application.

Environmental Issues

Alloys designed for high temperature operation generally contain a fairly high Al + Ti content. Today, with it recognized that Ti alloying makes the attainment of adequate alloy solution heat treatment more difficult, most new alloys rely more on Al content for γ ' formation than Ti. Due to the higher level of Al thereby utilized, most of today's advanced materials exhibit relatively good oxidation resistance characteristics. Furthermore, industry investigations into the positive effect that rare earth elemental addition to superalloys provide, e.g., Y, La and Ce, is helping make the necessary superalloy oxidation characteristics easier to achieve.

For CMSX-10, Figs. 14a and 14b present the comparative results of salt enhanced oxidation testing performed at 1030°C and 1100°C under MACH 0.4 gas stream conditions. At the lower temperature, all alloys investigated provide similar results, however at 1100°C, only the CMSX-10 and CMSX-4 alloys perform similarly; with both significantly better than MM 002.

Since most high strength superalloy derivatives necessarily employ lower Cr content, most industry metallurgists intuitively expect third generation SX alloys to exhibit extremely poor hot corrosion resistance. But as shown in Fig. 14c, the CMSX-10 950°C, 2 ppm salt injected, MACH 0.4 hot corrosion resistance is similar to CMSX-4 and, as with the 1100°C comparative alloy oxidation test results, is much better than the MM 002 alloy corrosion resistance.

Until low sulfur, rare earth addition technology matures within the superalloy industry, all hot section blade and vane components will continue to require the application of protective coatings. For CMSX-10, the initial component application employed a plain aluminide coating. The relatively low temperature, high activity coating application provides acceptable coating lives (MM002, CMSX-4 level) without the occurrence of any TCP within the substrate/coating interface zone. Figure 15 illustrates the CMSX-10 alloy in the as-coated and test-soaked conditions. Such coatings are successfully applied to CMSX-10 through pack and chemical-vapor-deposition processes.

Pt aluminides are also applied to CMSX-10 successfully. Although not as straight-forward as aluminides, today's refined experimental process methods result in the attainment of Pt aluminide coatings which similarly do not cause the formation of TCP phase in the coating/substrate region.

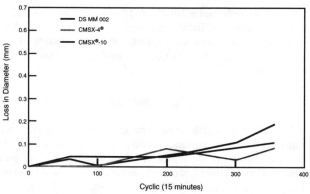

a 1030°C burner rig oxidation

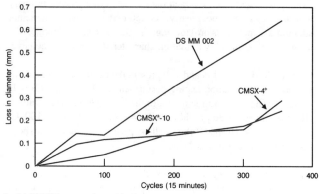

b 1100°C burner rig oxidation

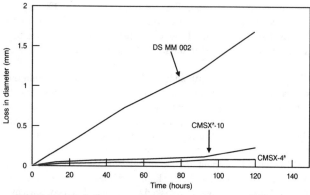

c 950°C burner rig hot corrosion

Figure 14: (a,b) Cyclic bare oxidation of three alloys at 1030°C and 1100°C, 0.25 ppm salt ingestion, and Mach 0.4. (c) Isothermal bare alloy burner rig corrosion of the alloys at 950°C, 2 ppm salt ingestion, and Mach 0.4 gas stream velocity.

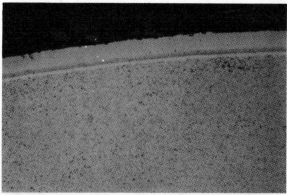

As coated. ⊢——⊣ .50μm

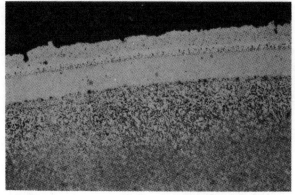

Soaked 150 hrs. at 1100°C. ⊢——⊣ .50μm
No needles.

Figure 15: CMSX-10 alloy aluminide coated specimens. No TCP phase present following a 150 hour soak in static air at 1100°C.

This has not always been the case, though. As Fig. 16 illustrates, the results experienced with early Pt aluminde trials were not acceptable since significant needle phase formed in the CMSX-10 substrate following exposure at 1080°C for 100 hours. Similar results prevailed with CoNiCrAlY overlay coatings, but with much greater needle zone formation (Fig. 16).

Subsequent experimental coating process modifications resulted in satisfactory Pt aluminide coating performance, as shown in Figure 17 where the early, standard Pt aluminide result is compared to the refined process result. Further CoNiCrAlY investigations have not occurred.

Summary

The CMSX-10 alloy is flight engine qualified and is a production status material exhibiting extremely good foundry functionality, heat treatment characteristics, high strength, environmental properties and coatability.

The alloy exhibits about 30°C greater creep and fatigue strengths than CMSX-4, particularly in 1000-2000 hour testing up to about 1100°C. For longer term exposures, such as those of interest to industrial turbine designers, the 30°C advantage is most prevalent at temperatures below 950-980°C. It's castability is similar to that associated with CMSX-4 while bare alloy hot corrosion and oxidation resistance is also similar to CMSX-4. Properly applied plain aluminide and Pt aluminide coatings perform well on the alloy system.

A higher strength CMSX-10 derivative is defined and continues under development.

The CMSX-10 alloy is utilized in at least two turbine engine applications while others are expected to be realized within the next two years.

Acknowledgement

The invaluable assistance of Dr. R.W. Broomfield of Rolls-Royce plc. is gratefully acknowledged.

References

1. S.C. Miller "Aero, Industrial and Marine Gas Turbines," Third International Charles Parsons Turbine Conference, (Newcastle upon Tyne, UK, April 1995) pps. 17-29.

2. Thomas Barker, "Rolls-Royce Industrial Trent Begins Rolling", *Turbomachinery International*, pps. 20-21, November/December 1995.

3. D.F. Betteridge, "Coatings for Aero and Industrial Gas Turbines," Prepared on behalf of Rolls-Royce for COST 501 - Materials for Power Engineering Components.

4. ABB, "First Firing ! - Progress Report on the GT24/GT26 Gas Turbine," *Sequence*, Report 1 from ABB, pps. 1-4, January 1996.

5. Private Communications With Various Turbine Producers, 1995.

6. G.L. Erickson, "Superalloy VIM and EBCHR Processes," *International Symposium on Liquid Metal Processing and Casting* (Santa Fe, New Mexico, 11-14 September 1994).

7. M. Gell, D.N. Duhl and A.F. Giamei, "The Development of Single Crystal Turbine Blades," *Superalloys 1980* (Warrendale, PA: TMS 1980) pps. 205-214.

8. C.S. Wukusick (Final Report NAVAIR/N62269-78-C-0315, 25 August 1980).

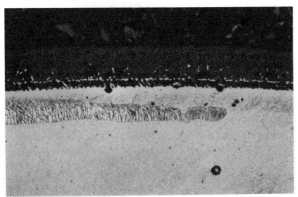

Standard Pt aluminide coating. ⊢⎯⎯⊣ .50μm
Soaked for 100 hrs. at 1080°C.
Needle phase present in 10% of specimen perimeter.

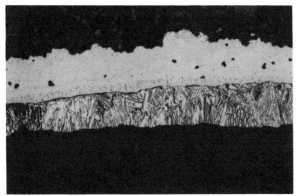

CoNiCrAlY overlay coating. ⊢⎯⎯⊣ .50μm
Soaked for 50 hrs. at 1100°C.
Continuous needle zone under coating.

Figure 16: Standard Pt Aluminide and CoNiCr AlY overlay coatings on CMSX-10.

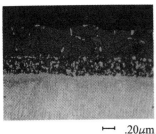 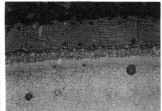

⊢⎯⊣ .20μm ⊢⎯⊣ .20μm

Standard Pt aluminide. | Experimental Pt aluminide.
Soaked 100 hrs. at 1100°C. Soaked 100 hrs. at 1100°C.
Extensive TCP phase region around entire specimen perimeter. | No significant TCP phase.

Figure 17: Standard and experimental Pt aluminide performance on the CMSX-10 alloy.

9. J.W. Holmes and K.S. O'Hara, *ASTM STP 942* (Philadelphia, PA: ASTM 1988), pps. 672-691.

10. M.J. Goulette, P.D. Spilling and R.P. Arthey, "Cost Effective Single Crystals" *Superalloys 1984* (Warrendale, PA: TMS, 1984) pps. 167-176.

11. D.A. Ford and R.P. Arthey, "Development of Single-Crystal Alloys for Specific Engine Application" *Superalloys 1984* (Warrendale, PA: TMS, 1984) pps. 115-124.

12. E. Bachelet and G. Lamanthe (Paper presented at the National Symposium on SX Superalloys, Viallard-de-Lans, France 26-28 February 1986).

13. T. Khan and M. Brun, *Symp. on SX Alloys* (Munich/Germany: MTU/SMCT, June 1989).

14. K. Harris and G.L. Erickson, Cannon-Muskegon Corporation, U.S. patent 4,582,548—CMSX-2 Alloy.

15. K. Harris and G.L. Erickson, Cannon-Muskegon Corporation, U.S. patent 4,721,540—CMSX-6 Alloy.

16. M. Doner and J.A. Heckler (Paper presented at the Aerospace Tech. Conf., Long Beach, CA, October 1985).

17. D.N. Duhl and M.L. Gell, United Technologies Corporation, G.B. patent 2112812A—PWA 1483 alloy.

18. G.L. Erickson, Cannon-Muskegon Corporation, U.S. patent 5,489,346—CMSX-11B Alloy.

19. G.L. Erickson, Cannon-Muskegon Corporation, U.S. patent pending—CMSX-11C Alloy.

20. T. Khan and P. Caron, "Development of a New Single Crystal Superalloy for Industrial Gas Turbine Blades," High Temperature Materials For Power Engineering 1990 (Kluwer Academic Publishers), pps. 1261-1270.

21. K. Harris and G.L. Erickson, Cannon-Muskegon Corporation, U.S. patent 4,643,782—CMSX-4 Alloy.

22. A.D. Cetel and D.N. Duhl, "Second Generation Nickel-Base Single Crystal Superalloy", *Superalloy 1988* (Warrendale, PA: TMS, 1988), pps. 235-244.

23. Garrett, U.S. patent 4,935,072—SC 180 Alloy.

24. P. Caron and T. Khan, "Development of a New Nickel Based Single Crystal Turbine Blade Alloy for Very High Temperatures," *Euromat '89* (1989), pps. 333-338.

25. C.S. Wukusick, L. Buchakjian, Jr., General Electric Company, U.K. patent application GB 2 235 697 A. Published March 13, 1991 "Improved Property-Balanced Nickel-Base Superalloy for Producing Single Crystal Articles."

26. G.L. Erickson, Cannon-Muskegon Corporation, U.S. patent 5,366,695—CMSX-10 Alloy.

27. W.S. Walston et al., General Electric Company, U.S. patent 5,455,120—René N6 Alloy.

28. Myra Pinkham, "Hot Competition propels thermal efficiency," American Metal Market, 23 January 1996, p. 10A.

29. G.L. Erickson, "The Development of CMSX-10, A Third Generation SX Casting Superalloy," Second Pacific Rim International Conference on Advanced Materials and Processing (PRICM-2), (Kyongju, Korea, 18-22 June 1995) pps. 2319-2328.

30. G.L. Erickson, "A New Third Generation, Single Crystal, Casting Superalloy," Journal of Metals, Vol. 47, No. 4, April 1995, pps. 36-39.

THE DEVELOPMENT OF THE CMSX®-11B AND CMSX®-11C ALLOYS FOR INDUSTRIAL GAS TURBINE APPLICATION

G. L. Erickson
Cannon-Muskegon Corporation
(a subsidiary of SPS Technologies)
P.O. Box 506
Muskegon, MI 49443-0506 U.S.A.

Abstract

Two different, non-Re containing single crystal (SX) superalloys are defined primarily for industrial turbine application. The alloys, CMSX®-11B and CMSX®-11C, contain respective chromium levels of about 12.5% and 14.5%. Both materials develop unique and extremely good blends of hot corrosion and oxidation resistance. They exhibit extremely good castability, employ relatively simple solution heat treatments and provide creep strength which is as good or better in comparison to other first generation SX materials such as CMSX-2/3®, PWA 1480 and René N4. Moreover, at certain engine-pertinent temperature/stress conditions, and particularly in long-term tests, the alloys appear to exhibit density corrected strengths which are similar or better than CMSX-4® and other second generation SX casting superalloys.

Introduction

The land-based, combustion turbine industry is experiencing tremendous growth, in part due to public utility commission rulings and environmental considerations. Prevailing rulings appear inconsistent with the long term, high capital expenditure associated with coal and nuclear power projects, thereby making the utility industry increasingly reliant on combustion turbine technology for its power generation requirements.

The General Electric Company forecasts that base-load electric power generation combustion turbines (CT) and combined cycle (CC) plants will account for about 45% of the new global orders and 66% of the new orders placed in the United States through the year 2001 (1). Advances in turbine design, materials, cooling technology and coatings have helped develop this market. Current advanced combustion turbines typically achieve more than 35% efficiency in the simple cycle mode and greater than 50% efficiency in the combined cycle mode. For example, the recent definition of its new gas turbine machine characteristics by Asea Brown Boveri (ABB) reveals that a simple cycle efficiency of 37.8% and combined cycle efficiency of 58.5% are anticipated for their GT 26 model; this being achieved through increasing power density and mass flow by doubling the pressure ratio (up from 15.0 in the GT 13E2 to 30.0) combined with sequential combustion at a relatively low maximum firing temperature (2).

And still another industry giant, Siemens, appears to rely on a more moderate increase to engine pressure ratio (eg., 16.0 for the Siemens V84.3 model versus 10.8 in its V84.2 machine) albeit in tandem with increased firing temperature, mass flow, compressor and turbine efficiencies. It's reported that the V84.3 firing temperature is expected to be 1310°C (2350°F) in contrast to the model V84.2 at 1120°C (2050°F). Furthermore, vane cooling increases from three to four stages while blade cooling is similarly increased from two to three stages. Further design improvements incorporated in the Siemens V84.3A engine are reported to allow an additional increase to firing temperature of about 22°C (40°F) while continuing to maintain acceptable NOx level. (4)

In order to achieve these increased firing temperatures which result in thermal efficiency improvement, large industrial turbine designers are beginning to utilize higher technology materials in their respective engine hot sections than previously applied. Where alloys such as IN 738 LC and IN 939 were previously employed, the search for greater engine efficiency has led the industrial turbine community to adopt technologies which have already been commercially applied within the aero-turbine and small industrial engine community, eg., directionally solidified columnar grain and single crystal components, serpentine and film cooling designs, as well as certain advanced coatings technology. While Siemens is apparently the first large frame turbine producer to announce the usage of SX components within their new turbines (5), it is also accepted that most other large turbine producers have designed SX components into their new products.

Complementary to the internal engine company R&D efforts toward improving industrial turbine efficiencies, industry collaborative programs targeted toward developing and commercializing ultra-high efficiency, environmentally superior, cost-competitive gas turbine systems for base-load applications [with a 1427°C (2600°F) or greater firing temperature] such as the U.S. Department of Energy (DOE) Advanced Turbine Systems (ATS) Program (6) are extremely active. While the ATS Program continues on target toward achieving the goals of greater than 60% net efficiency for utility scale combined cycles and a 15% jump in efficiency for small industrial machines (7), similarly aimed activities of the Collaborative Advanced Gas Turbine (CAGT) Program

---

CM 247 LC®, CM 186 LC®, CMSX-2®, CMSX-3®, CMSX-4®, CMSX-6®, CMSX®-10, CMSX®-11B AND CMSX®-11C are registered trademarks of the Cannon-Muskegon Corporation.

encompassing joint efforts between 17 parties in North America and Europe (8) further complement the the efforts of the COST program in Europe and the DOE sponsored ATS program.

To similar end, this narrative reports on the development of two single crystal casting alloys exhibiting characteristics which are hoped attractive to the industrial turbine community. Specifically, both alloys provide 738 LC - type hot corrosion resistance in tandem with CM 186 LC level oxidation resistance. The alloys' capability to exhibit good hot corrosion resistance concurrent to providing extremely good oxidation resistance is thought unique in the industry, as most alloys exhibit only one or the other. Combined with this positive characteristic, the alloys exhibit extremely good castability (in small or large components), are able to be solution heat treated in a relatively short period and provide extremely good creep-rupture strength. In certain engine-significant temperature/stress conditions, the alloys exhibit creep-strength, on a density-corrected basis, which is similar to second generation Re-containing alloys such as CMSX-4 and PWA 1484. Furthermore, the alloys appear to exhibit greater long-term rupture strength than CMSX-4 alloy. Moreover, since the alloys do not contain Re, they are at least 50% cheaper per pound of alloy purchased, in comparison to 3 wt.% Re-containing alloys.

Alloy Design

Industrial gas turbine engines have historically operated in temperature/pressure regime where Type II hot corrosion attack was the dominating environmental issue (Fig. 1). However, as the industry has sought to improve engine efficiency, engine firing temperatures have generally increased, thereby creating need for materials more able to endure exposures where a blending of Type I hot corrosion and oxidation predominates.

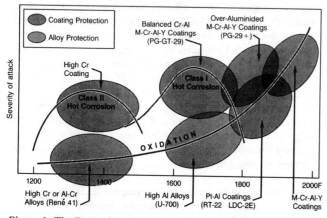

Figure 1: The Protection Scene (9).

To best achieve the perceived environmental need, moderate alloy chromium levels must be utilized in an alloy's design. As it appears that each turbine producer requires different blends of hot corrosion/oxidation/strength characteristic, two different chromium levels were used as starting points for the CMSX-11B (12.5%) and CMSX-11C (14.5%) alloy designs.

Along with relatively high chromium contributing to alloy environmental properties, other considerations applied to the designs include the selection of low molybdenum level, moderate Ta and elevated Ti: Al ratio. Relatively high Al+Ti levels are employed, which in tandem with moderate alloy Ta level, helps provide the high strength achieved.

Similarly, the alloy systems employ moderate tungsten levels for solid solution strengthening, however, the alloys also engage Ta: W ratios greater than unity to assist with SX component castability. Partly necessitated by the relatively high chromium levels employed, and the desired levels of W + Ta content, alloy cobalt levels are set relatively low to ensure adequate microstructural stability.

An overview of these alloy design considerations are provided in Figure 2, while more specific chemistry detail is provided in Table I, where the CMSX-11B and CMSX-11C alloy nominal compositions are compared to other first, second and third generation SX superalloys which have gained some commercial significance, albeit mostly in aero-turbine application. Note that the CMSX-11 derivatives do not rely on Re additions for strength attainment and that the alloys may, therefore, exhibit more desirable long-term lives and utility in certain components due to their inherently lower tendency for phasial instability in the temperature regime where Re containing alloys tend to form Topologically-Close-Packed (TCP) phase.

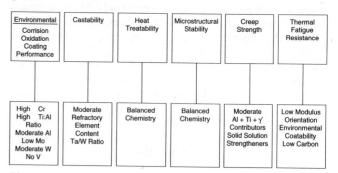

Figure 2: Alloy design criteria.

Table I Nominal Compositions of Three Generations of Single-Crystal Superalloys (Wt%)

Alloy	Cr	Co	Mo	W	Ta	Re	V	Nb	Al	Ti	Hf	Ni	Density (kg/dm³)	Ref.
First Generation														
PWA 1480	10	5	–	4	12	–	–	–	5.0	1.5	–	Bal.	8.70	10
PWA 1483	12.8	9	1.9	3.8	4	–	–	–	3.6	4.0	–	Bal.	–	31
René N4	9	8	2	6	4	–	–	0.5	3.7	4.2	–	Bal.	8.56	11,12
SRR 99	8	5	–	10	3	–	–	–	5.5	2.2	–	Bal.	8.56	13,14
RR 2000	10	15	3	–	–	–	1	–	5.5	4.0	–	Bal.	7.87	13,14
AM1	8	6	2	6	9	–	–	–	5.2	1.2	–	Bal.	8.59	15
AM3	8	6	2	5	4	–	–	–	6.0	2.0	–	Bal.	8.25	16
CMSX-2®	8	5	0.6	8	6	–	–	–	5.6	1.0	–	Bal.	8.56	17
CMSX-3®	8	5	0.6	8	6	–	–	–	5.6	1.0	0.1	Bal.	8.56	17
CMSX-6®	10	5	3	–	2	–	–	–	4.8	4.7	0.1	Bal.	7.98	18
AF 56 (SX 792)	12	8	2	4	5	–	–	–	3.4	4.2	–	Bal.	8.25	19
SC 16	16	–	3	–	3.5	–	–	–	3.5	3.5	–	Bal.	8.21	20
CMSX®-11B	12.5	7	0.5	5	5	–	–	0.1	3.6	4.2	0.04	Bal.	8.44	21
CMSX®-11C	14.9	3	0.4	4.5	5	–	–	0.1	3.4	4.2	0.04	Bal.	8.36	22
Second Generation														
CMSX-4®	6.5	9	0.6	6	6.5	3	–	–	5.6	1.0	0.1	Bal.	8.70	23
PWA 1484	5	10	2	6	9	3	–	–	5.6	–	0.1	Bal.	8.95	24
SC 180	5	10	2	5	8.5	3	–	–	5.2	1.0	0.1	Bal.	8.84	25
MC2	8	5	2	8	6	–	–	–	5.0	1.5	–	Bal.	8.63	26
René N5	7	8	2	5	7	3	–	–	6.2	–	0.2	Bal.	NA	27
Third Generation														
CMSX®-10	2	3	.4	5	8	6	–	.1	5.7	.2	.03	Bal.	9.05	28
René N6	4.2	12.5	1.4	6	7.2	5.4	–	–	5.75	–	.15	Bal.	8.98	29

The CMSX-11 alloy derivatives are moderate density materials due to their high chromium and Al+Ti levels. The density of CMSX-11B is 8.44 kg/dm³ while it is 8.36 kg/dm³ for CMSX-11C; this in comparison to CMSX-3 at 8.56 kg/dm³, CMSX-4 at 8.70 kg/dm³ and CMSX-10 at 9.05 kg/dm³.

Alloy Manufacture

The CMSX-11B and CMSX-11C alloys are VIM produced according to the process consideration detailed in Reference 30. Due to the relatively high alloy chromium contents employed, judicious Cr raw material selection is required to achieve the low levels of residual sulfur and phosphorus desired in the alloy product. Similarly, since low master alloy gas contents are always preferred, and high chromium and titanium containing alloys notoriously exhibit higher residual nitrogen levels, careful raw material selection in tandem with proper VIM procedure development are paramount to quality attainment.

Approximately twenty developmental 136 kg. heats have been produced through the development process. Consistency of major element heat chemistries are easily achieved, and typical tramp element levels prevailing in the CMSX-11 derivatives are illustrated in Table II. The levels achieved, except for gas content, are typical of those predominating in other SX alloys, such as CMSX-4 and CMSX-10. The abnormally high level of nitrogen in the CMSX-11 derivatives (4 ppm vs. 1 ppm) is a function of alloy chromium content, while the 4 ppm oxygen level (vs. 1 ppm in other CM product) is thought attributable to the alloys' relatively high Ti content. Although the developmental heat gas contents are higher than the other CM experience, they nonetheless have not adversely affected SX casting yields when measured in terms of defect formation tendency and/or non-metallic inclusion content. Furthermore, production alloy manufacture will likely provide improvement to each characteristic.

Table II Typical Tramp Element Levels in Development Heats of the CMSX-11B and CMSX-11C Alloys

Alloy	C ppm	B ppm	Zr ppm	S ppm	N ppm	O ppm	Si wt.%
CMSX®-11B	20	<20	<10	1	4	4	<.01
CMSX®-11C	20	<20	<10	1	4	4	<.01

Foundry and Heat Treatment Characteristics

Through February 1996, the CMSX-11B and CMSX-11C alloys have been cast successfully in six investment casting foundries located throughout the world. Items cast include test bars to 26 mm diameter, test slabs of varying size, plus both aero-turbine and industrial turbine blade components. Better than forty investment cast molds have been produced.

Through this experience, it is clear that both CMSX-11 alloy derivatives provide excellent SX castability. The alloys are not prone to formation of SX process defects such as freckles, slivers, high or low angle boundaries and/or stray grain formation. High production product yields are anticipated since every developmental mold produced has output nearly 100% satisfactory product. Moreover, no cleanliness problems, as measured through zyglo dye penetrant inspection and metallographic observation, have been experienced. An example of one of the test casting configurations utilized in the development is provided in Figure 3. The industrial turbine blade test configurations are considerably larger, with heavy platform/root-section blades to about 250mm length being successfully cast.

The test articles and components produced are solution heat treated and given a three step aging treatment. Slightly different peak solution heat treatment temperatures are employed for the CMSX-11B and CMSX-11C alloys. Both treatments, however, can be accomplished in 10 hours or less. Primary aging (pseudo coating treatment aging) is undertaken at 1121°C/5 Hr./AC condition for both alloys. Similarly, both alloys are further aged at 871°C/24 Hrs./AC + 760°C/30 Hrs./ AC condition.

Allison AE 2100 2nd Stage Blade.

Figure 3: One of the several turbine blade configurations used to characterize the CMSX-11B and CMSX-11C alloy castability.

The typical CMSX-11B and CMSX-11C fully heat treated microstructures are illustrated in Figures 4 and 5, as well as the specific respective solution heat treatments currently utilized. The respective solution heat treatments effect nearly 100% γ' and eutectic γ-γ' dissolution, while the primary aging treatments result in γ' growth and arrangement into fairly regularly aligned, cubic γ' precipitates of about 0.4 - 0.5 μm edge dimension. The secondary and tertiary aging treatments, as shown in Figure 6 for CMSX-11B, tend to promote the formation of relatively fine matrix channel γ' precipitates which likely enhance strength in blade root sections, since root section temperature exposures are generally lower than 760°C.

 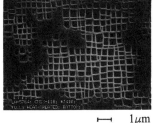

⊢ .25μm ⊢ 1μm

Solution: 1227°C/1 Hr. + 1249°C/1 Hr. + 1260°C/2 Hrs. + 1264°C/4 Hrs./AC.

Age: 1121°C/ 5 Hrs./AC.
871°C/24 Hrs./AC.
760°C/30 Hrs./AC.

Figure 4: Two views of fully heat treated CMSX-11B alloy.

Solution: 1204°C/1 Hr.. + 1227°C/1 Hr. + 1250°C/2 Hrs. + 1256°C/4 Hrs./AC.

Age: 1121°C/ 5 Hrs./AC.
     871°C/24 Hrs./AC.
     760°C/30 Hrs./AC.

Figure 5: Two views of fully heat treated CMSX-11C alloy.

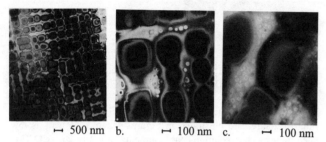

Figure 6: The CMSX-11B alloy aged at (a) 1121°C, (b) 1121°C + 871°C, and (c) 1121°C + 871°C + 760°C.

Mechanical Properties

Fairly extensive creep-rupture tests with temperature ranging 760-1038°C confirms the CMSX-11B and CMSX-11C alloys develop impressive respective strength levels. Alloy strengths are at least as good as commercialized SX casting alloys such as CMSX-2/3, PWA 1480 and René N4, plus as good or better, in certain tests, than creep-rupture strengths typically exhibited by the second generation, 3 wt. % Re containing CMSX-4, PWA 1484 and René N5 alloys.

Figure 7 illustrates the typical CMSX-11B/11C alloy strength in comparison to the DS René 80 and equiaxe IN 939 alloys. For a running stress of about 138 MPa, the CMSX-11 derivatives exhibit about 92°C greater strength than IN 939. Also indicated are the 871°C and 982°C capabilities of the PWA 1483 alloy (Ref. 31).

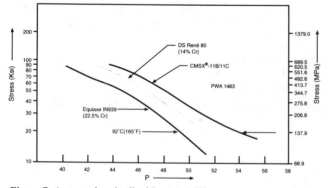

Figure 7: Average longitudinal Larson-Miller stress-rupture strength of several alloys.

While Figure 7 compared the CMSX-11B/11C materials to IN 939, DS René 80 and PWA 1483, Figure 8 illustrates the materials' advantage over the SC 16 alloy, an IN 738 SX alloy derivative developed by ONERA for the European Community Collaborative COST program (20). At about 207 MPa stress, the SC 16 alloy deficit is about 44°C and at 138 MPa, approximately 64°C. If compared on the basis of 1% creep strength, the SC 16 alloy exhibits an even greater deficit. Others have shown the SC 16 creep strength to be similar to equiaxed IN 738 (Ref. 32).

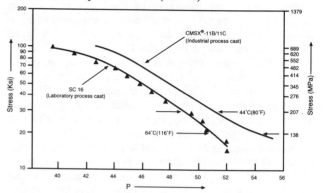

Figure 8: Average longitudinal Larson-Miller stress-rupture strength of three alloys.

As illustrated in Table I, the densities of the CMSX-11B and CMSX-11C alloys are moderately low in comparison to other commercially utilized DS and SX casting alloys. Figure 9 illustrates the specific or density corrected strengths of several commercial alloys in comparison to the CMSX-11 materials. Perhaps of most significance, the CMSX-11B alloy's strength is shown to equal or exceed that of CMSX-4, while the higher Cr containing CMSX-11C alloy appears superior to CMSX-4 only at higher temperatures such as 982-1038°C.

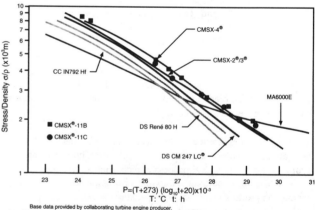

Figure 9: Density corrected stress-rupture strength of several alloys.

Along this line, non-density-corrected log stress vs. log time rupture strength comparison of the IN 738 LC, DS CM 247 LC, CMSX-4 and CMSX-11B alloys at 982°C is shown in Figure 10. Interestingly, the non-Re containing CMSX-11B alloy is shown to exceed the CMSX-4 alloy's rupture strength for tests run to between 1500-2000 hours life. Although not shown, the CMSX-11B alloy 1% average creep strengths demonstrated in these tests were not quite as good as the averaged CMSX-4 capability, however, they did not lag significantly. Similarly, Figure 11 compares the log stress vs. log rupture life of CMSX-4 and CMSX-11B for tests performed at 1038°C (also without density correction), and illustrates a significant advantage occurring with CMSX-11B for a test run to about 3000 hours. As comparative CMSX-4 alloy creep data is not available for the given test, future efforts will define the CMSX-4 alloy's creep-rupture characteristic, as well as expand the CMSX-11B data base in the 1038°C - 1100°C temperature regime.

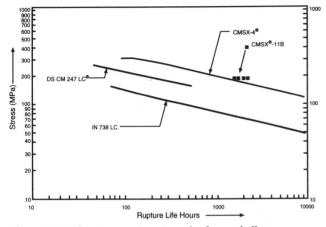

Figure 10: 982°C stress-rupture strength of several alloys.

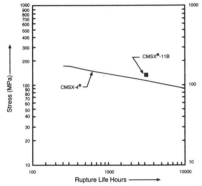

Figure 11: 1038°C stress-rupture strength comparison of CMSX-11B and CMSX-4.

Figure 12 illustrates the CMSX-11B γ' structure for a specimen tested at 1010°C/103 MPa condition and which ruptured at 17,278.8 hours. Fracture and thread section views are presented, with both illustrating an absence of TCP phase. This absence of TCP is significant since Re-containing alloys generally form TCP under similar condition and exposure time. Interestingly, this individual rupture result is near the extrapolated life prediction which could be made for the CMSX-10 alloy tested at identical condition.

Figure 12: Two post-test views of a CMSX-11B alloy specimen tested at 1010°C/103 MPa condition to rupture at 17,278.8 hrs.

A review of superalloy 10,000 hour rupture strength capability, on the basis of alloy chromium content, is presented in Figure 13. While traditional superalloy experience suggests that higher alloy chromium levels are accompanied by lower alloy strength, the CMSX-11B and CMSX-11C materials are shown to exhibit uniquely high relative strengths for their 12.5 and 14.5% respective Cr levels. Of particular significance is the positive strength comparison with the second generation, 3 wt. % Re containing superalloys such as PWA 1484 and CMSX-4. The data also illustrates that Re containing SX superalloys don't necessarily provide long-term rupture strength advantages since other non-Re containing SX alloys such as René N4, CMSX-2 and PWA 1480 apparently perform similarly for 10,000 hour rupture life at the 160 MPa stress level, as determined through 1000 to 3000 hour rupture data extrapolation undertaken by a collaborating gas turbine engine producer.

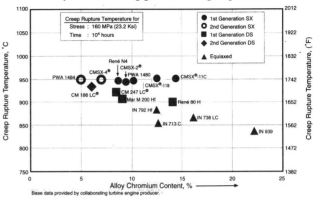

Figure 13: 10,000 hour rupture strength of several alloys.

Environmental Properties

Environmental tests performed on superalloy materials often give rise to varied results, depending on methods employed and sources employed. In this investigation, four different turbine producers performed both hot corrosion and oxidation tests on the CMSX-11 alloy derivatives. The tests performed were undertaken by both burner rig and crucible evaluation methods, with the significance of the results comparing favorably between three of the four investigative sources, thereby lending credibility to the results achieved.

Figure 14 illustrates the derivative alloys' 500 hour hot corrosion characteristics determined in comparative crucible tests undertaken at 750°C, 850°C and 900°C. Similarly, Figures 15 and 16 present results of hot corrosion tests performed at 732°C and 899°C, with comparison to the IN 738 LC alloy capability. The results appear to confirm that the CMSX-11 materials behave similar to the IN 738 LC material with exposure at around 750°C, and that with long term exposure at about 900°C, the higher Cr containing derivative, CMSX-11C, exhibits an advantage vs. the CMSX-11B alloy, while continuing to perform as well as IN 738 LC to at least 2400 hours.

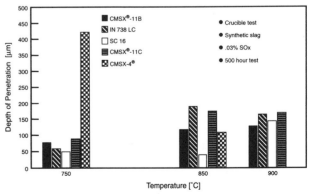

Figure 14: Hot corrosion of various alloys at 750, 850 and 900°C.

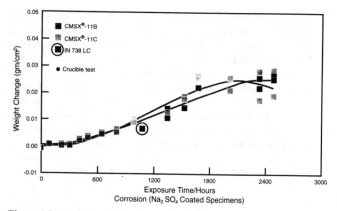

Figure 15: 732°C cyclic hot corrosion of several alloys.

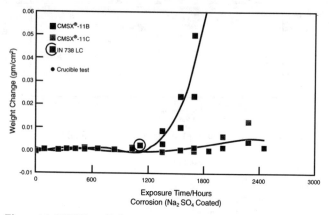

Figure 16: 899°C cyclic hot corrosion of several alloys.

The results of extremely aggressive burner rig hot corrosion tests performed to 500 hour duration at 1050°C are shown in Figure 17. Hot corrosion results are presented through comparison to other widely used gas turbine alloys such as FSX-414, DS René 80 H, DS IN 738 LC, DS IN 939 and DS CM 186 LC. The actual corrosion results are presented in terms of test specimen thickness loss, while the respective material's strength capabilities are expressed on the figure's y-axis as creep-rupture temperature capabilities for 1000 hour lives with a testing stress of 284.4 MPa. For this testing, the figure illustrates that the CMSX-11C alloy develops DS René 80 - type hot corrosion resistance with an attendant 25°C strength advantage. The CMSX-11B alloy doesn't provide quite as good hot corrosion capability in the test, but is significantly better than the DS CM 186 LC alloy, a material also considered for some industrial turbine applications.

Burner rig oxidation test results are shown in Figure 18. The alloys compared (along with the result presentation methods employed) are identical to those presented in Figure 17. For this run at 1200°C and exposure of 500 hours, the CM 186 LC alloy is shown to exhibit the best alloy oxidation resistance, with the CMSX-11B and CMSX-11C materials behaving quite similarly; a unique capability for materials exhibiting IN 738/IN 792/ René 80 type corrosion resistance. To that point, Figure 18 also exhibits the reduced oxidation resistance of the René 80 and IN 738 LC alloys in comparison to CMSX-11.

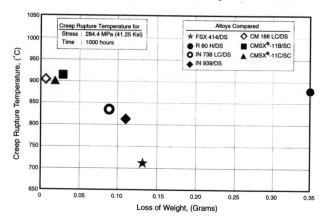

Figure 18: Alloy strength and 1200°C/500 hour burner rig oxidation comparison of several alloys.

Confirmation of the CMSX-11 oxidation characteristic is provided in Figure 19 where the results of a cyclic crucible test performed at 1000°C on the CMSX-11B, CMSX-11C and IN 738 LC alloys are presented. While the IN 738 LC alloy oxidation resistance is low, the two CMSX-11 alloy derivatives exhibit relatively good oxidation characteristic for the duration of the test, ie., 3000 hours. While not presented, the Figure 19 test data source has developed unpublished data at identical conditions for the PWA 1483 alloy which show it's capability appearing between the IN 738 LC and CMSX-11C test results. Similar oxidation testing undertaken at 1010°C at another turbine builder shows similar results for the CMSX-11 alloys, as illustrated in Figure 20.

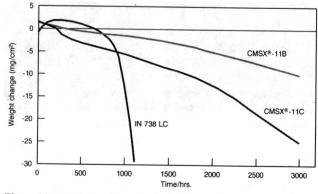

Figure 19: 1000°C cyclic oxidation of three alloys.

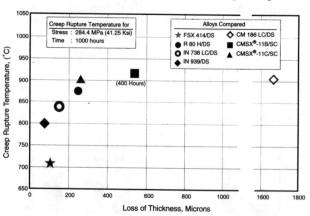

Figure 17: Alloy strength and 1050°C/500 hour burner rig hot corrosion comparison of several alloys.

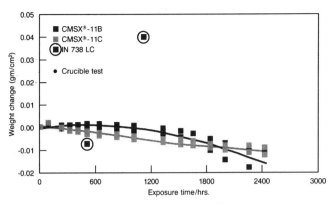

Figure 20: 1010°C cyclic oxidation of three alloys.

The alloys, therefore, provide surprisingly good oxidation characteristics, as confirmed through both burner rig and crucible tests performed by multiple gas turbine engine manufacturers. At the same time, similar multiple source testing confirms both alloys exhibit very good hot corrosion capabilities (IN 738 LC/René 80/IN 792 level). The ability of the CMSX-11B and CMSX-11C alloys to provide good hot corrosion and oxidation characteristics, in tandem, is thought unique among available turbine materials.

Summary

Two unique single crystal casting superalloys have been developed for gas turbine engine blade and/or vane application where demanding strength and environmental issues prevail. The non-rhenium containing alloys, CMSX-11B and CMSX-11C, exhibit short-term creep rupture strengths which are as good or better than other lower chromium-containing, first generation SX superalloys. Moreover, both alloys develop long-term creep-rupture strengths which are as good or better than Re-containing second generation SX alloys at conditions pertinent to gas turbine blade and vane component operation. Furthermore, the CMSX-11B and CMSX-11C alloys also provide uniquely attractive blends of bare hot corrosion and oxidation resistance; a characteristic thought unique among superalloys commercially available.

Acknowledegment

The invaluable assistance of several turbine engine companies, who wish to remain unidentified at this point, is gratefully acknowledged.

References

1. "G.E. Forecasts Worldwide Plant Orders of 925,000 MW Between now and 2001", Independent Power Report, p. 1, 3 July 1992.

2. "ABB's New Gas Turbine Decouple Efficiency and Temperature", Turbomachinery International, pps. 20-24, January/February 1994.

3. Robert Farmer, "First V84.3 to Met Ed for Joint 'B.O.T.' Demo Project", Gas Turbine World, pps. 24-33 July/August 1993.

4. Thomas Barker, "Siemens' New Generation", Turbomachinery International, pps. 20-22, January/February 1995.

5. Robert Farmer, "See 57% net efficiency combined cycles powered by 2400°F '3A' series turbines", Gas Turbine World, pps. 18-28, January/February 1995.

6. "Advanced Turbine Systems: Follow-up", Word Power Systems Intelligence, pps. 6-7, Issue No. 179, 22 September 1993.

7. Irwin Stambler, "Technology base in place for next phase of ATS development program", Gas Turbine World, pps. 10-14, September/October 1995.

8. "Future Gas Turbines for Power Generation", Turbomachinery International, pps. 30-31, January/February 1994.

9. Chester T. Sims, private communications during technical meeting at Cannon-Muskegon, 1989.

10. M. Gell, D.N. Duhl and A.F. Giamei, "The Development of Single Crystal Turbine Blades" *Superalloys 1980* (Warrendale, PA: TMS 1980) pps. 205-214

11. C.S. Wukusick (Final Report NAVAIR/N62269-78-C-0315, 25 August 1980).

12. J.W. Holmes and K.S. O'Hara, *ASTM STP 942* (Philadelphia, PA: ASTM 1988), pps. 672-691.

13. M.J. Goulette, P.D. Spilling and R.P. Arthey, "Cost Effective Single Crystals" *Superalloys 1984* (Warrendale, PA: TMS, 1984) pps. 167-176.

14. D.A. Ford and R.P. Arthey, "Development of Single-Crystal Alloys for Specific Engine Application" *Superalloys 1984* (Warrendale, PA: TMS, 1984) pps. 115-124.

15. E. Bachelet and G. Lamanthe (Paper presented at the National Symposium on SX Superalloys, Viallard-de-Lans, France 26-28 February 1986).

16. T. Khan and M. Brun, *Symp. on SX Alloys* (Munich/Germany: MTU/SMCT, June 1989).

17. K. Harris and G.L. Erickson, Cannon-Muskegon Corporation, U.S. patent 4,582,548—CMSX-2 Alloy.

18. K. Harris and G.L. Erickson, Cannon-Muskegon Corporation, U.S. patent 4,721,540—CMSX-6 Alloy.

19. M. Doner and J.A. Heckler (Paper presented at the Aerospace Tech. Conf., Long Beach, CA, October 1985).

20. T. Khan and P. Caron, "Development of a New Single Crystal Superalloy for Industrial Gas Turbine Blades," High Temperature Materials for Power Engineering 1990 (Kluwer Acedemic Publishers) pps. 1261-1270.

21. G.L. Erickson, Cannon-Muskegon Corporation, U.S. patent 5,489.346—CMSX-11B Alloy.

22. G.L. Erickson, Cannon-Muskegon Corporation, U.S. patent pending—CMSX-11C Alloy.

23. K. Harris and G.L. Erickson, Cannon-Muskegon Corporation, U.S. patent 4,643,782—CMSX-4 Alloy.

24. A.D. Cetel and D.N. Duhl, "Second Generation Nickel-Base Single Crystal Superalloy", *Superalloy 1988* (Warrendale, PA: TMS, 1988), pp. 235-244.

25. Garrett, U.S. patent 4,935,072—SC 180 Alloy.

26. P. Caron and T. Khan, "Development of a New Nickel Based Single Crystal Turbine Blade Alloy for Very High Temperatures," *Euromat '89* (1989), pps. 333-338.

27. C.S. Wukusick, L. Buchakjian, Jr., General Electric Company, U.K. patent application GB 2 235 697 A. Published March 13, 1991 "Improved Property-Balanced Nickel-Base Superalloy for Producing Single Crystal Articles."

28. G.L. Erickson, Cannon-Muskegon Corporation, U.S. patent 5,366,695—CMSX-10 Alloy.

29. W.S. Walston et al., General Electric Company, U.S. patent 5,455,120—René N6 Alloy.

30. G.L. Erickson, "Superalloy VIM and EBCHR Processes," *International Symposium on Liquid Metal Processing and Casting* (Santa Fe, New Mexico, 11-14 September 1994).

31. D.N. Duhl and M.L. Gell, United Technologies Corporation, G.B. patent 2112812A—PWA 1483 Alloy.

32. D. Goldschmidt, "Single Crystal Blades," Materials for Advanced Power Engineering 1994 (Kluwer Academic Publishers), pps. 661-674.

DEVELOPMENT OF A HYDROGEN RESISTANT SUPERALLOY FOR SINGLE CRYSTAL BLADE APPLICATION IN ROCKET ENGINE TURBOPUMPS

P. Caron*, D. Cornu**, T. Khan* and J.M. de Monicault**

* Office National d'Études et de Recherches Aérospatiales (ONERA)
BP 72 - 92322 Châtillon - France

** Société Européenne de Propulsion (SEP)
BP 802 - 27207 Vernon - France

Abstract

A new single crystal Ni-Fe based superalloy, designated THYMONEL 8, was developed in order to satisfy the specific requirements of rocket engine turbopump airfoils. The principal objective was to obtain an alloy with a high fatigue resistance together with a low susceptibility to hydrogen environment enbrittlement (HEE). The effects of the distribution of strengthening γ' precipitates on the tensile properties of the alloy were investigated. The best balance between tensile strength and ductility was obtained with a duplex dispersion of γ' particles. Smooth and notched tensile tests performed at 300 K showed a susceptibility to HEE significantly lower than that of single crystal superalloys designed for aircraft applications. THYMONEL 8 exhibited improved low cycle fatigue and stress-rupture properties compared to forged Superwaspaloy used by S.E.P. for turbopump rotors. In addition, THYMONEL 8 exhibited higher thermal conductivity and lower Young's modulus than the single crystal alloys for aircraft turbines, which is beneficial for the thermo-mechanical strength.

Introduction

High strength γ/γ' nickel base superalloys are used extensively in turbopump turbines for rocket engines. The operating conditions in hydrogen fueled rocket engines are, in some respects, more severe than for aircraft turbine engines (1). Indeed the turbopumps operate in a gaseous hydrogen environment at high pressure which requires the use of alloys with a low susceptibility to hydrogen environment embrittlement (HEE). Moreover the thermal transients experienced by the rocket engine turbines are more severe than those encountered in aircraft turbines, resulting in harmful low and high cycle fatigue conditions on the blades. On the other hand, the targeted life is only of the order of a few hours and the maximum use temperature is not higher than 1100K, a further increase of the turbine inlet temperature producing no significant power improvement.

Up to now, two separate approaches have generally been taken to improve the performance of the alloys constituting the turbopump blades : i) to develop specific chemistries of superalloys showing a low susceptibility to HEE (2), (ii) to select existing single crystal superalloys previously designed for aircraft turbine blade applications in order to take advantage of their unique thermo-mechanical strength, both due to the low value of the Young's modulus along the <001> growth direction and the absence of defects such as grain boundaries, carbides or coarse γ' particles (1, 3). However, this latter class of superalloys generally exhibit a substantial or strong susceptibility to HEE that makes them inadequate for applications in hydrogen environment (1, 4). Advanced rocket engine technology with improved efficiency and service life therefore relies on the development of specific chemistries for single crystal superalloys showing a high fatigue resistance together with a low susceptibility to HEE (1). This paper describes the approach adopted to define a new superalloy, designated THYMONEL 8, suitable for turbopump blade applications, as well as the results pertaining to the evaluation of its microstructural, physical and mechanical characteristics.

Alloy Development

The objective was to design an alloy with improved thermo-mechanical properties compared to the forged Superwaspaloy, used in turbine rotor of the turbopump designed by VOLVO from Sweden for the "Vulcain" engine developed by S.E.P., while maintaining a low susceptibility to HEE.

A series of experimental superalloys with compositions based on the Ni-Fe-Cr-Mo-Al-Ti-Nb system were first defined for screening tests. The choice of nickel-iron based alloys was justified by the good balance between strength and resistance to HEE generally exhibited by this class of alloys of which Incoloy 903 is the best example (5, 6). Chromium was added for improving the oxidation/corrosion resistance and molybdenum was introduced as a solid solution strengthener of the austenitic alloy matrix. Aluminium, titanium and niobium were added to enhance the precipitation of the strengthening, ordered $L1_2$-γ' phase (Ni, Fe)$_3$(Al, Ti, Nb). Additional objectives were to obtain an alloy characterized by a large solution heat treatment range and a good castability.

Small-scale laboratory heats of the experimental alloys were melted in a high vacuum induction furnace and then used for casting of <001> single crystal rods by the withdrawal process. In view of the results of microstructural and mechanical evaluations on these materials, the THYMONEL 8 alloy was selected as the one showing the best combination of properties. The nominal chemical composition (wt.%) of this alloy is reported in Table I.

Experimental Procedures

Single crystal castings used for extensive evaluation of the THYMONEL 8 superalloy were made from heats produced by vacuum induction melting by Aubert & Duval.

Table I Alloy compositions (wt.%)

Element	Ni	Fe	Co	Cr	Mo	Al	Ti	Nb	Ta
THYMONEL 8	Bal.	25	-	8	3	4	4	1.5	-
AM1	Bal.	-	6.5	7.8	2	5.2	1.1	-	7.9
AM3	Bal.	-	5.5	8	2.25	6	2	-	3.5

<001> oriented single crystal rods were directionally cast by the withdrawal process, either at O.N.E.R.A. using a seed or at industrial foundries using a selector, in low thermal gradient (G ≈ 40 K.cm^{-1}) directional solidification furnaces. Single crystal rods of the AM1 and AM3 <001> oriented superalloys designed for aircraft turbine blade applications (7) were cast using the same procedure in order to perform comparative tests. Their chemical compositions are reported in Table I.

Tensile tests performed for determining the effect of the γ/γ' microstructure were conducted in air at room temperature, and in argon at higher temperatures, on 4 mm diameter cylindrical specimens with an initial strain rate of $1.1 \times 10^{-4} s^{-1}$. Additional tensile tests were performed at room temperature both on smooth and notched ($K_T = 6.2$) <001> single crystal specimens in gaseous hydrogen at a pressure of 10 MPa and in air under atmospheric pressure.

Fully-reversed total strain controlled low cycle fatigue (L.C.F.) tests were performed in air at 1100 K with a frequency of 0.33 Hz on a MTS servo-hydraulic fatigue testing machine and using cylindrical specimens with a diameter of 4.65 mm. Constant load tensile stress-rupture tests were conducted in air at 950 and 1123 K on 3 mm diameter cylindrical specimens.

The thermal conductivity χ for single crystal alloy samples was calculated via the following relationship using experimentally determined values of thermal diffusivity λ, heat capacity C_p and density ρ:

$$\chi(T) = \lambda(T) C_p(T) \rho(T)$$

T denotes the temperature at which the relevant properties were determined. Thermal expansion measurements were performed using a ADAMEL DI 24 dilatometer with an argon atmosphere and a heating rate of 0.08 K.s^{-1}. Heat capacity measurements were made between R.T. and 873 K using a Perkin Elmer differential scanning calorimeter on small disks 4 mm diameter and 1 mm thick. Thermal diffusivity measurements were performed using the laser pulse method on small disks 10 mm diameter and 3 mm thick. The faces of the disks were normal to the <001> growing direction of the single crystals. Determination of density was performed at 300 K using simple volume measurement of a sample of known mass. The densities for temperatures up to 873 K were calculated from the room temperature values and the measured thermal expansion coefficients at the relevant temperature.

Microstructural observations of γ' precipitates and dislocation structures were performed by transmission electron microscopy (T.E.M.) using a Jeol 200 CX microscope operating at 200 KV. Thin foils were prepared by electrochemical polishing in a twin-jet Tenupol-3 polisher, using a solution of 45% acetic acid, 45% butyl cellosolve and 10% perchloric acid, cooled to 263 K and using a potential of 25 V.

Microstructure and Heat Treatments

The optical micrograph of Figure 1a illustrates the as-cast dendritic microstructure of a THYMONEL 8 <001> single crystal. A very low amount of $\gamma-\gamma'$ eutectic pools was formed in the interdendritic areas.

This alloy has a very large γ' solution heat treatment window (ΔT

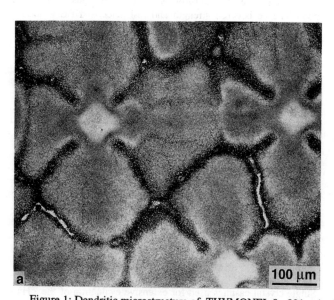
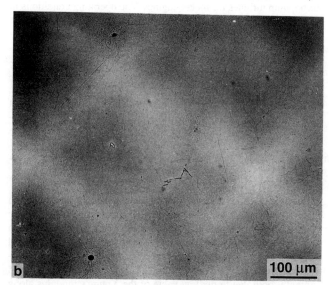

Figure 1: Dendritic microstructure of THYMONEL 8 <001> single crystal alloy : (a) as-cast, (b) solution heat treated for 3 hours.

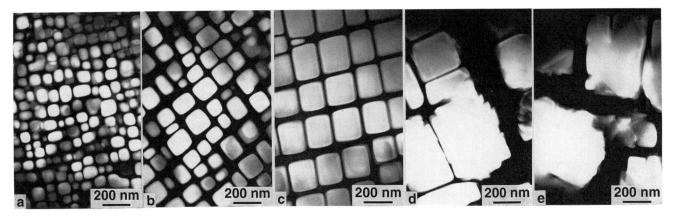

Figure 2: γ' precipitates in THYMONEL 8 single crystals after ageing procedures applied to achieve a monomodal distribution : (a) heat treatment MA , (b) heat treatment MB, (c) heat treatment MC, (d) heat treatment MD, (e) heat treatment ME (TEM dark field micrographs using γ' diffraction spots).

= 138 K) between the γ' solvus temperature and the solidus temperature. A one-step 3-hour solution heat treatment at a temperature within this window, followed by air cooling, is sufficient to homogenize the dendritic as-cast structure of the single crystals and to solution all the primary and secondary γ' particles (Figure 1b).

In order to achieve the best balance between tensile strength and ductility, various ageing procedures were applied on homogenized single crystals to produce either monomodal dispersions of γ' particles with different mean sizes, or duplex dispersions of precipitates with different ratios of small and large particles. Increasing the size of the γ' particles in single crystal superalloys has been demonstrated to reduce the yield strength but to increase the work hardening rate and the tensile ductility (8, 9). These effects result from a transition between the heterogeneous γ' cutting mechanism by pairs of a/2<110> matrix dislocations and the more homogeneous γ'-bypassing mechanism operating through a combination of slip and climb of γ matrix dislocations. On the other hand, a duplex ageing leads to a dispersion of large γ' particles, which renders slip more homogeneous, and a finer dispersion of γ' precipitates which confers a high yield strength to the alloy (10).

Monomodal γ' distributions were obtained in THYMONEL 8 using a primary ageing treatment below the γ' solvus temperature to achieve the desired size of γ' precipitates. The single crystals were then slowly cooled to a lower temperature and held during a few hours at this temperature, before cooling in air to ambient temperature. The typical γ/γ' microstructures obtained using this procedure are illustrated in Figure 2. The mean size of the γ' particles increased from 70 nm to 650 nm when the temperature of the primary ageing increased, while their shape changed continuously from spheroidal to cuboidal, and finally less regular cubes. With primary ageing at the highest temperatures (heat treatments MD and ME), hyperfine precipitates were however observed between the coarser γ' particles. This secondary precipitation occured, presumably during cooling, in the largest γ matrix regions separating coarse γ' precipitates formed during the high temperature ageing.

Two different duplex dispersions of γ' phase were produced in THYMONEL 8 using primary ageing treatments at a temperature below the γ' solvus temperature and secondary ageing treatments at a lower temperature, both followed by air cooling. The resulting γ/γ' microstructures are illustrated by the T.E.M. micrographs of Figure 3. The fraction of coarse γ' precipitates is higher with a lower primary ageing temperature. The respective mean size values for the coarse and fine precipitates are 315 and 15 nm for the heat treatment DA, while they are 430 and 60 nm for the heat treatment DB where the primary ageing temperature is higher. The heat treatment procedures are coded MX and DX, where M or D stand for monomodal or duplex dispersions of γ' precipitates.

Physical Properties

The density of THYMONEL 8 was measured to be 7.77 g.cm^{-3} that is significantly lower than the typical values for high-strength single crystal superalloys used today for aircraft applications which are close to 8.6-8.7 g.cm^{-3}.

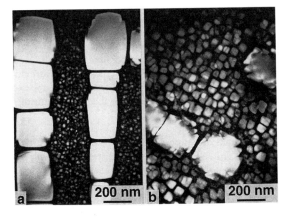

Figure 3: γ' precipitates in THYMONEL 8 single crystals after ageing procedures applied to achieve a duplex dispersion : (a) heat treatment DA, (b) heat treatment DB (T.E.M. dark field micrographs using γ' diffraction spots).

Figure 4 illustrates the variation with temperature of the thermal conductivity of THYMONEL 8 up to 873 K compared with the values determined by the same procedure for AM1, a typical high

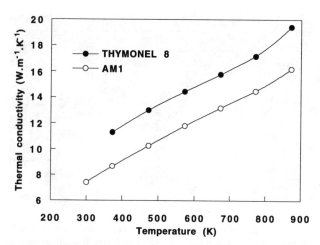

Figure 4: Thermal conductivity of single crystal superalloys THYMONEL 8 and AM1.

creep strength superalloy for single crystal aircraft turbine blades (7). The thermal conductivity of THYMONEL 8 is significantly higher than for AM1 in the whole temperature range.

Static Young's modulus measurements were performed between room temperature and 1123 K on <001>, <111> and <011> THYMONEL 8 single crystals using longitudinal extensometry on tensile specimens deformed in the elastic strain range. The values of the Young's modulus are compared in Figure 5 to those reported elsewhere for AM1 (11). The Young's modulus of these single crystal materials is highly anisotropic, the lowest values being measured along <001> directions, while the highest ones are obtained along <111> directions.

The value of the Young's modulus along the <001> growth direction of the blade is of prime importance because it directly influences the level of stress induced in it due to the thermo-mechanical strain. It is worth noting that the values of the Young's modulus are systematically lower for THYMONEL 8, which would result in an improvement of the thermo-mechanical strength of the single crystal blade.

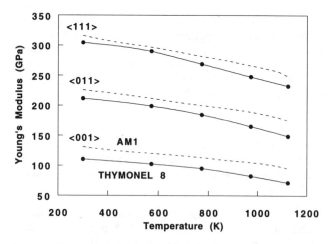

Figure 5: Young's modulus of single crystal superalloys THYMONEL 8 and AM1 along <001>, <011> and <111>.

Mechanical Properties

Effect of the γ/γ' Microstructure on the Tensile Properties

Tensile tests were performed between 300 and 1123 K on <001> THYMONEL 8 single crystal specimens with the different γ/γ' microstructures described previously. For the sake of clarity, only the most important and significant results are presented here. Figure 6 illustrates the wide range of values obtained for the 0.2% yield strength, the ultimate tensile strength and the elongation to rupture by applying the various ageing heat treatments.

The highest yield strength was obtained, in the whole temperature range, with the monomodal distribution of γ' precipitates exhibiting the finest mean size, i.e. 70 nm (heat treatment MA). The yield strength increased from 1250 to 1360 MPa between 300 and 873 K, then decreased to about 560 MPa when the temperature increased to 1123 K. However, in these conditions, the ductility was the lowest obtained in the whole temperature range, with a minimum observed at 873 K.

As expected, increasing the γ' size by increasing primary ageing temperatures strongly reduced the yield strength of THYMONEL 8, the lowest values being obtained with the largest γ' particles. However, the ductility of the alloy was improved significantly only by using the heat treatment ME. In this case, the ductility minimum shifted to 773 K with a value of 6.2%. Increasing the γ' precipitate size also led to an increase of the work hardening rate.

The best balance between strength and ductility was however obtained with the ageing procedures producing duplex dispersions of precipitates. Indeed, at 873 K, the yield strength of the alloy with the ageing heat treatment DB attained 1220 MPa with an elongation to rupture of 3.4% which represents a better compromise between strength and ductility than the alloy containing a fine dispersion of 70 nm precipitates. At 973 and 1123 K, the yield strengths obtained with the two types of microstructures are comparable.

As for the yield strength, the highest ultimate tensile strength between 300 and 873 K was obtained with the heat treatment MA producing 70 nm precipitates. In the case of monomodal distributions of precipitates, increasing the mean size of the γ' particles decreased the U.T.S., but to a less extent than for the yield strength which is indicative of a higher strain hardening rate. The single crystals with duplex γ' distributions exhibited comparable values for the U.T.S., irrespective of the temperature of the first ageing heat treatment. At 973 and 1123 K, all types of γ' distributions produced comparable values of ultimate tensile strength, except for the largest precipitates obtained with the heat treatment ME which gave lower values.

T.E.M. micrographs of Figure 7 illustrate the dislocation structures observed in THYMONEL 8 single crystal specimens, with a monomodal dispersion of 70 nm precipitates (heat treatment MA) and with a duplex dispersion of γ' particles (heat treatment DB), tensile strained up to 2% at room temperature. In the former case, the deformation is very heterogeneous with high densities of dislocations concentrated in narrow bands. The presence of paired 1/2<011> dislocations getting away from these dislocation bands is an indication of γ' precipitate shearing by <011> superdislocations. In the duplex γ' microstructure, the deformation is still heterogeneous, but the presence of the larger precipitates tend to homogenize the deformation in their vicinity. Straight <011> super-

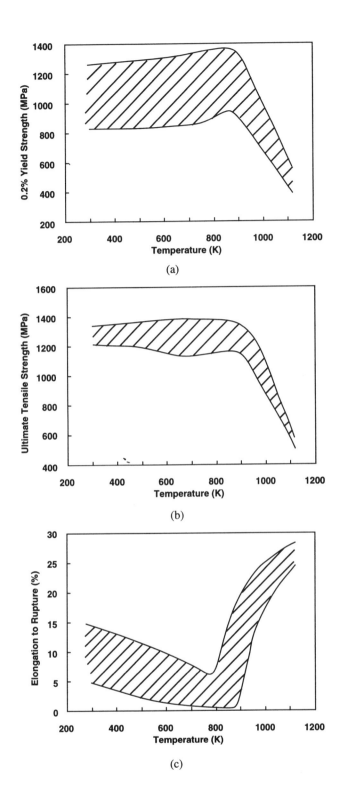

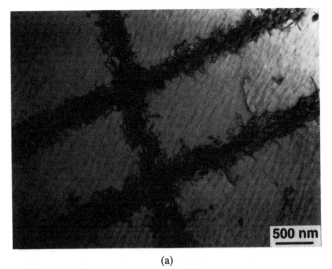

(a)

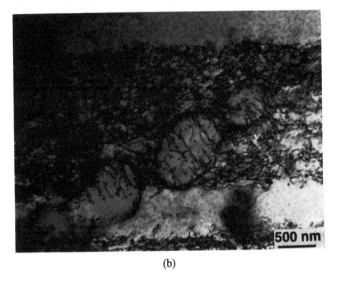

(b)

Figure 7 - Dislocation structures in THYMONEL 8 <001> single crystals tensile strained to 2% in air at 300 K: (a) monomodal dispersion of 70 nm γ' precipitates (MA), (b) duplex dispersion of γ' particles (DB).

dislocations were observed within the coarser precipitates indicating shearing of these γ' particles.

These results therefore confirm the benefits of a duplex dispersion of precipitates for achieving a balance of tensile properties of the THYMONEL 8 single crystal superalloy. Consequently, the selected heat treatment procedure DB was considered as optimum for the THYMONEL 8 alloy.

Smooth and Notched Tensile Properties in Hydrogen

Additional tensile tests in air and in hydrogen were performed at 300 K both on smooth and notched THYMONEL 8 <001> single crystal specimens with the monomodal dispersion of 70 nm precipitates (heat treatment MA) and with the optimized γ' duplex microstructure (heat treatment DB).

Figure 6 - Variation with temperature of tensile properties of <001> THYMONEL 8 single crystals with ageing heat treatment procedures producing either monomodal or duplex γ' precipitate dispersions: (a) yield strength, (b) ultimate tensile strength and (c) elongation to rupture.

Table II Smooth tensile properties at 300 K in air and in hydrogen of THYMONEL 8 <001> single crystals with monomodal dispersion of 70 nm γ' precipitates (MA) and duplex dispersion of γ' precipitates (DB).

Heat treatment	Air / atmospheric pressure			H_2 / P = 10 MPa		
	0.2% Y.S. (MPa)	U.T.S. (MPa)	Elongation to rupture (%)	0.2% Y.S. (MPa)	U.T.S. (MPa)	Elongation to rupture (%)
MA	1 250	1 320	5.4	1 182	1 249	5.4
"	1 211	1 271	5.1	1 239	1 293	4.7
DB	1 072	1 156	8.0	1 115	1 195	7.0
				1 090	1 135	8.0

There was no significant difference between the smooth tensile properties obtained in hydrogen and in air, for both ageing treatments (Table II). The presence of gaseous hydrogen generally does not affect the yield strength of superalloys, which was confirmed by our results. Hydrogen environment generally affects the ductility and, hence, the ultimate tensile strength. In the case of THYMONEL 8, the room temperature ductility was not affected by the presence of gaseous hydrogen independent of the microstructure, which indicated no or low susceptibility to HEE. It has been previously reported that smooth tensile tests performed in hydrogen on single crystals of various superalloys for aircraft turbine blades such as CMSX-2, CMSX-4, René N4 and PWA 1480 show a significant decrease of the UTS compared to tests conducted in air or helium, which was indicative of a loss a ductility (4).

A stronger indication of the susceptibility to HEE is generally given by the ratio of the notched strength in hydrogen to that in air, R_{H_2}/R_{Air}. Comparative notched tensile tests were also conducted on AM1 and AM3 (10) <001> single crystals. The respective mean values of the notched tensile strength in air, R_{Air}, and in hydrogen, R_{H_2}, together with the values of the ratio R_{H_2}/R_{Air} for the three alloys are compared in Table III. The ratios for THYMONEL 8 with monomodal and duplex dispersions of precipitates were 0.91 and 0.81 which are much higher than the values measured for AM1 and AM3, i.e. respectively 0.46 and 0.54. The notched tensile strength in air of THYMONEL 8 with a duplex γ' dispersion is very high and comparable to that of AM1 and AM3. The presence of hydrogen reduced the notched tensile strength to a level comparable to that measured in air for THYMONEL 8 with a monomodal γ' dispersion. It should be noted that the notched strength of THYMONEL 8 with optimized microstructure is higher than its ultimate smooth tensile strength regardless of the environment, whereas it is not the case for THYMONEL 8 with 70 nm γ' precipitates. This behaviour results from the higher strain hardening rate observed with the duplex γ' dispersion.

Low Cycle Fatigue Properties

Low cycle fatigue test results obtained at 1100 K on <001> single crystal smooth specimens of THYMONEL 8 containing 70 nm γ' precipitates (heat treatment MA) or with the optimized duplex microstructure (heat treatment DB) are compared in Figure 8 with strain controlled (R = -1) L.C.F. data obtained on forged Superwaspaloy tested at 1023 K with a frequency of 1 Hz (12). These results showed no significant effect of the γ/γ' microstructure on the L.C.F. strength of THYMONEL 8. Failures initiated primarily at the surface of the specimen from cracks formed within the external oxide scale (Figure 9a) and sometimes at pores located beneath the surface (Figure 9b). Shorter lives were generally observed when the failure initiated at pores.

Despite the higher testing temperature and the lower frequency, THYMONEL 8 exhibited a dramatic improvement in L.C.F. strength compared to Superwaspaloy. The single crystal superalloy showed a 50X life advantage over Superwaspaloy in the whole strain range.

Stress-Rupture Strength

Although the creep strength is not a critical property for turbopump applications, some stress-rupture tests were performed between 950 and 1123 K in air on <001> THYMONEL 8 single crystal specimens in order to establish a comparison with the Superwaspaloy. The alloy was tested both with the monomodal dispersion of precipitates providing the maximum strength and with the optimized γ' duplex microstructure. The times to rupture covered the range 1-1000 hours under stresses in the range 200-800 MPa.

The rupture lives are compared in a Larson-Miller plot (Figure 10) to data published on forged Superwaspaloy (13). The creep strength of THYMONEL 8 is much superior to that of the reference polycrystalline alloy Superwaspaloy, particularly at high temperatures. As an example, the rupture life at 1123 K and 400 MPa in-

Table III - Notched tensile strength (K_T = 6.2) at 300 K in air and in hydrogen of <001> single crystal superalloys.

Alloy	Notched strength (MPa)		Notched ratio
	R_{Air}	R_{H_2}	R_{H_2}/R_{Air}
THYMONEL 8 heat treatment MA	1 266	1 157	0.91
THYMONEL 8 heat treatment DB	1 580	1278	0.81
AM1	1 638	756	0.46
AM3	1580	850	0.54

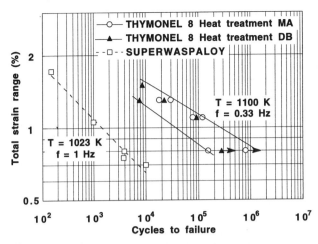

Figure 8 - Comparison of L.C.F. strengths of THYMONEL 8 <001> single crystals and forged Superwaspaloy (12).

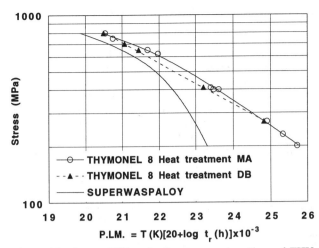

Figure 10 - Larson-Miller plot for stress-rupture lives of THYMONEL 8 and forged Superwaspaloy (13).

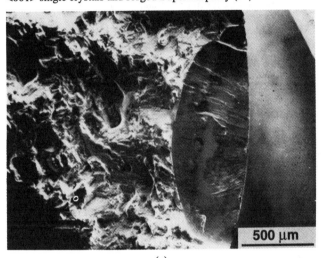

(a)

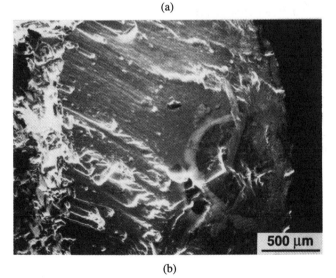

(b)

Figure 9 - SEM micrographs of L.C.F. fracture surfaces of THYMONEL 8 single crystals tested at 1100 K; (a) failure initiation at the specimen surface; (b) failure initiation at pores beneath the surface.

creased from 1 to 8 hours when comparing Superwaspaloy with THYMONEL 8.

Discussion of the Results

The smooth tensile behaviour of the THYMONEL 8 single crystal superalloy has been shown to be very sensitive to the distribution of the γ' precipitates which contribute to the strengthening of the alloy. The yield strength potential of the alloy is very high as demonstrated in the case of a monomodal dispersion of 70 nm sized precipitates. However, with this microstructure, the alloy suffered from a lack of ductility, especially in the temperature range 673-873 K. Attempts have therefore been made to improve the ductility of the THYMONEL 8 single crystal alloy, without significantly decreasing its tensile strength, by modifying the γ/γ' microstructure using various ageing procedures. In accordance with the results previously obtained on Incoloy 901 (10), the duplex dispersion of γ' precipitates in THYMONEL 8 confered to the alloy optimum properties in terms of strength and ductility, presumably because of a better homogeneity of the deformation process. The other approach for achieving a homogeneous deformation was to increase the size of the precipitates, while keeping a fairly monomodal dispersion of these particles. This procedure was successful in increasing the ductility of the single crystal, but too detrimental to the yield strength of the alloy, and it was therefore not retained for further evaluation. Moreover, the choice of a duplex γ' microstructure, instead of a monomodal dispersion of small γ' particles, led to a significant increase of the notched strength of THYMONEL 8 single crystals, both in air and in hydrogen, which is particularly advantageous while considering the residual life of airfoils containing cracks induced by thermo-mechanical stresses.

The susceptibility to HEE of the THYMONEL 8 alloy was investigated by means of both smooth and notched tensile tests at room temperature, and compared to that of competitive single crystal superalloys specifically designed for aircraft turbine blade applications. These tests revealed hardly any susceptibility to HEE for THYMONEL 8, irrespective of the choice of the ageing procedure. The reasons for this excellent behaviour are not really known, but the most important result is that the susceptibility of THYMONEL 8 to HEE is dramatically lower than that of the single crystal superalloys previously evaluated for use in turbopump airfoil applications.

The comparison between the L.C.F. strength of THYMONEL 8 single crystals and forged Superwaspaloy confirmed the expected advantage resulting from the low Young's modulus along the <001> growth orientation of the single crystals, together with the absence of defects such as massive primary γ', grain boundaries or carbides, likely to initiate failure under cyclic deformation. During L.C.F. testing of THYMONEL 8, failure was sometimes initiated on pores, which tends to reduce the life compared to the cases where the failure was initiated at the surface from cracks formed within the oxide scale. A further improvement of the LC.F. behaviour of THYMONEL 8 is possible by using hot isostatic pressing which closes the casting porosity, as previously demonstrated on other single crystal superalloys (3, 14). Moreover, some thermomechanical fatigue tests were also performed on hollow <001> single crystal specimens of THYMONEL 8 with duplex γ' precipitation (15). As expected, considering the excellent L.C.F. behaviour of this single crystal alloy, together with its rather low Young's modulus and high thermal conductivity, these tests have demonstrated a considerable superiority of THYMONEL 8 over the Superwaspaloy.

Conclusions

The combination of the various pertinent physical and mechanical properties (low density, high thermal conductivity, low susceptibility to HEE, high tensile strength, low Young's modulus and excellent low cycle fatigue resistance) of the single crystal superalloy THYMONEL 8 makes this material a very suitable candidate for turbompump blades of advanced rocket engines. An other interesting characteristic of THYMONEL 8 is its good creep strength compared to polycristalline superalloys, which makes this alloy a strong candidate for use in future reusable rocket engines. The future work on this alloy will be directed towards the scaling-up and industrial aspects as well as to the more complete understanding of its mechanical behaviour in air and in hydrogen.

Acknowledgments

A part of this work was supported by the C.N.E.S. The authors acknowledge Mr. J.L. Raffestin (O.N.E.R.A.) for casting of single crystal superalloys.

References

1. R.L. Dreshfield, and R.A. Parr, "Application of Single Crystal Superalloys for Earth-to-Orbit Propulsion Systems" (Paper presented at the 23rd Joint Propulsion Conference, San Diego, CA, June 29-July 2, 1987, Report NASA TM-89877, Lewis Research Center, Cleveland, OH).

2. W.B. McPherson, "A New High Strength Alloy for Hydrogen Fueled Propulsion Systems" (Paper presented at the 22nd Joint Propulsion Conference, Huntsville, Alabama, June 16-18, 1986, Report AIAA-86-1478, AIAA, New York, NY).

3. L.G. Fritzemeier, "The Influence of High Thermal Gradient Casting, Hot Isostatic Pressing and Alternate Heat Treatment on the Structure and Properties of a Single Crystal Nickel Base Superalloy," Superalloys 1988, ed. D.N. Duhl et al. (Warrendale, PA : The Metallurgical Society, Inc., 1988), 265-274.

4. R.A. Parr, W.S. Alter, M.H. Johnston, and J.P. Strizak, "High-Pressure Hydrogen Testing of Single Crystal Superalloys for Advanced Rocket Engine Turbopump Turbine Blades" (Report NASA CP 2372, June 1984), 150-163.

5. R.J. Schwinghamer, "Materials and Processes for Shuttle Engine, External Tank, and Solid Rocket Booster" (Report NASA TN D-8511, George C. Marshall Space Flight Center, Marshall Flight Center, AL, June 1977).

6. H.W. Carpenter, "Alloy 903 Helps Space Shuttle Fly," Metal Progress, August 1976, 25-29.

7. T. Khan and P. Caron, "Advanced Single Crystal Ni-base Superalloys," Advances in High Temperature Structural Materials and Protective Coatings, ed. A. K Koul et al. (Ottawa, Canada : National Research Council of Canada, 1994), 11-31.

8. A.A. Hopgood, and J.W. Martin, "The Effect of Aging on the Yield Stress of a Single-Crystal Superalloy," Mat. Sci. Eng., 91 (1987), 105-110.

9. D. Roux, "Rôle de la microstructure sur la sensibilité à la fragilisation par l'hydrogène de superalliages monocristallins à base de nickel" (Doctoral Thesis, Université Paris-Sud, France, 1993).

10. B. McGurran and J.W. Martin, "The Control of Slip Distribution by Duplex Dispersions of γ' Phase in a Nickel-based Superalloy," Z. Metallkde., 72 (1981), 538-542.

11. P. Mazot, and J. de Fouquet,"Elastic Constant Determination of Nickel Base Superalloy from Room Temperature to 1100°C by Dynamical Method,", Mém. et Ét. Sc. Rev. Met., March 1972, 165-170.

12. S.E.P., unpublished work.

13. D.J. Deyes, and W.H. Couts, "Super Waspaloy Microstructure and Properties," MiCon 78 : Optimization of Processing, Properties, and Service Performance Through Microstructural Control,ASTM STP 672, ed. G. N. Maniar et al. (American Society for Testing Material, 1979), 601-615.

14. T. Khan, and P. Caron, "Effect of Processing Conditions and Heat Treatments on Mechanical Properties of Single Crystal Superalloy CMSX-2," Mat. Sc. Technology, 2 (1986), 486-492.

15. S.E.P., unpublished work.

HOT CORROSION RESISTANT AND HIGH STRENGTH NICKEL-BASED SINGLE CRYSTAL AND DIRECTIONALLY-SOLIDIFIED SUPERALLOYS DEVELOPED BY THE d-ELECTRONS CONCEPT

Y.Murata[1], S. Miyazaki[2], M.Morinaga[1] and R.Hashizume[3]

1) Department of Materials Science and Engineering, School of Engineering, Nagoya University, Furo-cho, Chikusa-ku, Nagoya 464-01, JAPAN
2) Graduate School of Toyohashi University of Technology, (Present address: SMC Corp. Ltd., 4-2-2, Kinunodai, Yawara-mura, Tukuba-gun, Ibaraki 300-24, JAPAN)
3) Materials Research Section, Technical Research Center, The Kansai Electric Power Company, Inc., 11-20 Nakoji 3-chome, Amagasaki 661, JAPAN

Abstract

New nickel-based superalloys for industrial turbines with high Cr content were designed with the aid of the d-electrons concept. This concept has been developed on the basis of the molecular orbital calculations of the electronic structures of Ni alloys. Two electronic parameters are important in this concept. One is the bond order between an alloying element and nickel atoms, Bo, and the other is the d-orbital energy level of alloying element, Md. Employing these parameters, single-crystal(SC) superalloys were designed with the compositions : Ni-(9.5~12.0)Al-(16.5~18.0)Cr-(0.6~2.4)Ti-(0.3~1.8)Ta-(0~0.5)Nb-(0.9~1.6)W-(1.2~1.8)Mo-(0~0.25)Re [mol%]. A similar method was also employed for the design of directionally solidified(DS) alloys. Namely, alloy compositions without any grain-boundary strengthening elements were firstly designed and then a small amount of Hf was simply added into them to strengthen the grain-boundaries. The creep properties and hot corrosion resistance were evaluated experimentally for these SC and DS alloys. As a result, the SC alloys were found to be superior in the creep rupture life to a reference alloy, SC-16, while keeping hot corrosion resistance as high as SC-16 and IN738LC. The DS alloys also exhibited better creep properties than several alloys so far developed.

Introduction

There is a great demand for advanced nickel-based superalloys, mainly for the application to industrial gas-turbine blades. They should possess an excellent combination of hot corrosion resistance and high-temperature strength. Despite the recent innovation of coating technology, hot corrosion resistance is still important for industrial turbines which are for a long term service.

Superior nickel-based single-crystal superalloys (TUT alloys) for aircraft turbines [1, 2] have been developed successfully with the aid of the d-electrons concept. This concept has been devised on the basis of the molecular orbital calculations of electronic structures [3]. Two electronic parameters are utilized mainly for the alloy design following this concept. One is the bond order between an alloying element and nickel atoms (hereafter referred to as Bo), and the other is the d-orbital energy level of alloying element (referred to as Md). The values of Bo and Md are listed in Table 1 for each element in nickel-based superalloys. For simplicity, the units of Md, eV, are omitted in the following sections of this paper. For an alloy, the averaged values of Bo and Md are defined by taking the compositional average, and $\overline{Bo}$ and $\overline{Md}$ are denoted as follows:

$$\overline{Bo} = \Sigma X_i (Bo)_i ,$$
$$\overline{Md} = \Sigma X_i (Md)_i .$$

Here, X_i is the atomic fraction of component i in the alloy, $(Bo)_i$ and $(Md)_i$ are the Bo and Md values for component i, respectively.

The usefulness and the validity of the $\overline{Bo}$-$\overline{Md}$ diagram has been proved for the prediction of alloy properties. A target region for alloy design has been specified on the diagram in a concrete way, as explained later [4, 5].

Generally it is known that in order to increase hot-corrosion resistance, Cr content in the alloy should be

Table 1. List of Md and Bo values for various elements.

Element		Md(eV)	Bo	Element		Md(eV)	Bo
3d	Ti	2.271	1.098	4d	Zr	2.944	1.479
	V	1.543	1.141		Nb	2.117	1.594
	Cr	1.142	1.278		Mo	1.550	1.611
	Mn	0.957	1.001	5d	Hf	3.020	1.518
	Fe	0.858	0.857		Ta	2.224	1.670
	Co	0.777	0.697		W	1.655	1.730
	Ni	0.717	0.514		Re	1.267	1.692
	Cu	0.615	0.272	Others	Al	1.900	0.533
					Si	1.900	0.589

higher than 10mol%. The purpose of this study is to design single-crystal superalloys and directionally solidified superalloys both of which contain more than 10mol%Cr and possess an excellent combination of hot-corrosion resistance and creep-rupture life. For this design, the d-electrons concept has been employed to save time and cost for alloy development.

d-Electrons Concept

Target Region for Alloy Design

As shown in Fig.1 and Fig.2, it is interesting to note here that high performance nickel-based superalloys are located on a very small region in the $\overline{Bo}$-$\overline{Md}$ diagram. For example, as shown in Fig.1, the high temperature 0.2% yield stress of conventionally cast superalloys shows a maximum near the position of 0.67 for $\overline{Bo}$ and 0.98 for $\overline{Md}$. Also, as shown in Fig.2, the 100hrs creep-rupture strength at 1255K shows a maximum near this position. In addition, single crystal superalloys are located in the shadowed region near this position as shown in Fig.1 and Fig.2. Thus, there is definitely a target region for the design of high performance superalloys [4].

Alloying Vectors

Alloying vectors were also presented in Fig.1 and Fig.2. For an individual alloying element, the vector starts at the pure Ni position and ends at the position of a Ni-10mol%M alloy. Needless to say, the vector length increases with increasing composition of M in the alloy. The vector direction changes simply with the group number of the element in the periodic table. For example, all the 5A group elements, V, Nb and Ta, take the nearly same direction. This is also the case for the 4A group elements, Ti, Zr and Hf and for the 6A group elements, Cr, Mo and W. These periodic changes in the directions are simple due to the periodic changes in the respective parameters of Bo and Md with M, as shown in Fig.3. The knowledge of the periodic table of elements is indeed condensed in the two electronic parameters, Bo and Md.

It is also true that the γ' stabilizing elements such as Al, Ti, V, Nb, Ta take relatively lower θ angles than the γ stabilizing elements such as Cr, Mo, W, Re, as shown in the figure. So, the θ angle is an indication for the stability of the γ' phase, and it is related closely to the γ' volume fraction, as explained later.

A best combination of alloying elements and their amounts in the alloy can be selected readily using these alloying vectors so that the alloy composition falls on the target region in the $\overline{Bo}$-$\overline{Md}$ diagram. For a superalloy, a total vector is defined by the vector sum of each alloying vector, the length of which is proportional to the amount of the element in it. One example is shown in Fig.4 for the TUT92 alloy [4].

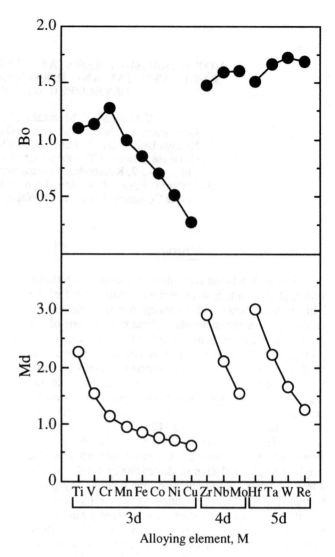

Fig.3. Changes in the Bo and Md values with alloying elements.

γ' Volume Fraction

In Fig.4, θ is defined as the angle between the total vector of TUT92 and the alloying vector of Al. For any superalloy, the value of θ can be calculated readily from the $\overline{Bo}$ and the $\overline{Md}$ values of the alloy. As shown in Fig.5 [5], they vary approximately linearly with the γ' volume fraction in the superalloys. Such a relationship holds even in a wide range of the γ' volume fraction of 20% to 70%. Using this relationship, the γ' volume-fraction in any multi-components superalloys can be estimated readily from the calculated θ angle of the total vector. The alloys locating in the target region in Fig.1 or Fig.2 have about 30° in θ, indicating that the γ' volume fractions of them are 55~65%.

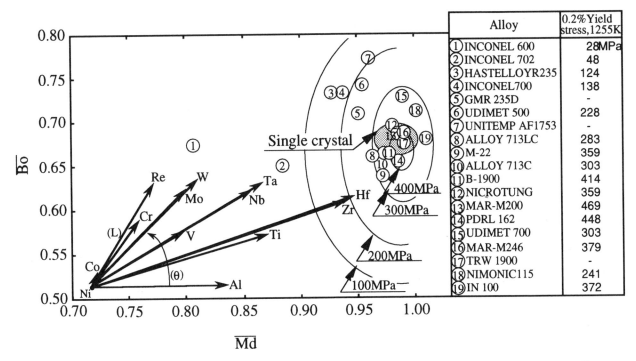

Fig.1 $\overline{Bo}$-$\overline{Md}$ diagram showing the locations of conventionally cast superalloys and the contour lines of 0.2% yield stress. The vectors represent the location of Ni-10mol% M alloys.

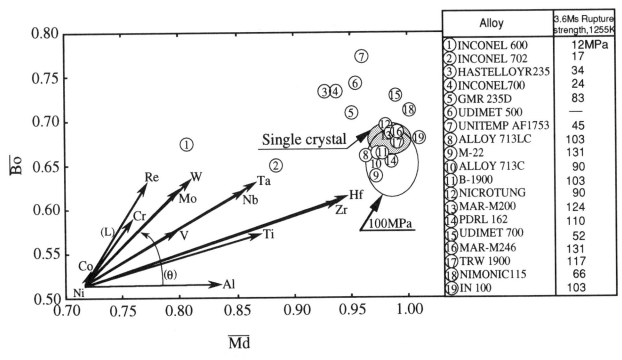

Fig.2. $\overline{Bo}$-$\overline{Md}$ diagram showing the locations of conventional cast superalloys and the contour lines of 3.6Ms rupture strength. The vectors represent the location of Ni-10mol% M alloys.

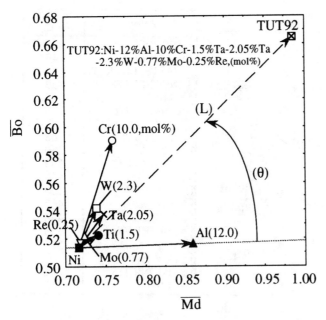

Fig.4. Representation of the alloying vectors for TUT92 alloy in the $\overline{Bo}$-$\overline{Md}$ diagram.

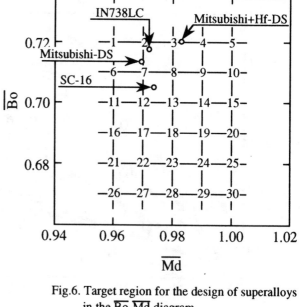

Fig.6. Target region for the design of superalloys in the $\overline{Bo}$-$\overline{Md}$ diagram.

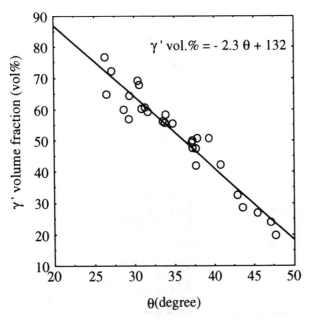

Fig.5. Correlation between the γ' volume fraction and the parameter, θ, of alloying vectors.

Criteria for Alloy Design

New superalloys were designed using the d-electrons concept together with the following six criteria; (1) containing more than 15mol%Cr in order to guarantee hot corrosion resistance superior or at least comparable to IN738LC which is a typical hot-corrosion resistant alloy used in corrosive environments, (2) increasing W+Mo+Re contents for strengthening the γ phase, (3) increasing the γ' volume fraction with the aid of the $\overline{Bo}$-$\overline{Md}$ diagram, (4) controlling the Ta+Ti(+Nb)/Al compositional ratio to be suitable for strengthening the γ' phase, (5) adjusting the W/Mo compositional ratio to a proper value in view of fabrication conditions, and (6) lowering alloy density calculated by the Hull's regression equation [6].

Calculation of Alloy Compositions

A suitable combination of alloying elements and their compositions in the alloy should be selected so that the alloy is located inside the target region on the $\overline{Bo}$-$\overline{Md}$ diagram shown in Figs.1 and 2, while holding all the six criteria. Following these fundamental objects in view, the target region was divided into 30 mesh points, as shown in Fig.6. The alloy compositions falling on each mesh point were then calculated following the steps shown in Fig.7. In this calculation, a clearance of the both $\overline{Md}$ and $\overline{Bo}$ values was limited to be within ±0.001 for each

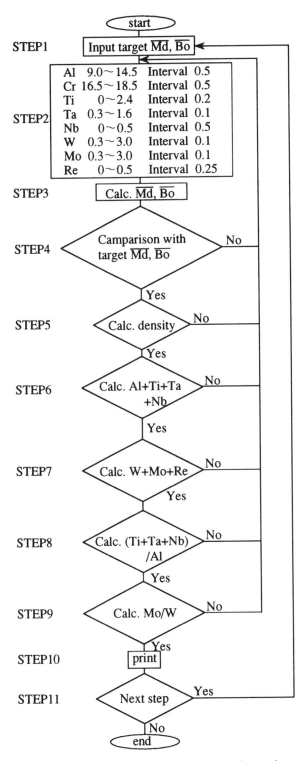

Fig.7. Flow chart for the design of hot-corrosion resistant and high strength nickel-based single crystal superalloys.

Fig.8. Typical SEM image of designed single-crystal superalloy, S-5, after the heat treatment.

value which was set primarily at the individual mesh point. The compositional range and the interval used for the calculation were also shown in Fig.7 for each element in the alloy.

As a result of the calculation, about 20 alloys were obtained as the first candidate alloys. With these alloys preliminary experiments were carried out on the phase stability at 1173K for 1.8Ms (500hrs) and on the Vickers hardness at elevated temperatures. Nine single-crystal (SC) superalloys were finally selected by counting on these results. Their compositions and the values of the parameters for alloy design are listed in Table 2, together with those values for reference alloys used in this study. The mesh numbers illustrated in Fig.6 and the calculated γ' volume fraction are also listed in the table.

The DS alloys were designed in the same way as the SC alloys, except for the calculation range of Cr content. That was set to be 15.0~18.5mol% for the DS alloys. At first, alloy compositions were obtained by the calculation without involving any grain-boundary strengthening elements, and then six directionally solidified(DS) alloys were designed simply by adding a small amount of Hf into them, but compensating a certain amount of alloying elements so that the alloy is still located well inside the target region.

A series of experiments was carried out in order to evaluate the properties of these designed alloys.

Table 2. Nominal compositions of the designed nickel-based alloys and reference alloys (mol%), and various parameters to characterize the alloys.

	Alloy	Cr	Al	Ti	Ta	Nb	W	Mo	Re	Ni	Co	C	B	Zr	Hf
SC-Alloy	S-1	17.00	9.50	0.60	1.50	0.50	1.50	1.80	-	Bal.	-	-	-	-	-
	S-2	17.50	10.00	1.20	1.20	0.50	1.20	1.80	-	Bal.	-	-	-	-	-
	S-3	17.00	11.00	2.00	0.60	0.50	1.20	1.40	-	Bal.	-	-	-	-	-
	S-4	17.00	11.50	2.40	0.60	0.50	1.20	1.20	-	Bal.	-	-	-	-	-
	S-5	17.50	11.00	2.40	0.30	0.50	1.60	1.60	-	Bal.	-	-	-	-	-
	S-6	17.00	10.50	1.60	0.60	0.50	1.30	1.60	-	Bal.	-	-	-	-	-
	SR-1	16.50	12.00	1.80	0.80	0.50	1.20	1.40	0.25	Bal.	-	-	-	-	-
	SR-2	17.50	11.00	1.10	1.40	0.50	1.20	1.30	0.25	Bal.	-	-	-	-	-
	SR-3	17.00	10.50	1.60	0.70	0.50	1.20	1.30	0.25	Bal.	-	-	-	-	-
DS-Alloy	D-1	16.50	10.77	1.96	1.57	-	1.60	1.60	-	Bal.	-	-	-	-	0.30
	D-2	17.50	10.77	2.35	0.78	-	1.60	1.60	-	Bal.	-	-	-	-	0.30
	D-3	15.00	11.76	2.35	0.78	-	1.20	2.00	-	Bal.	-	-	-	-	0.30
	DR-1	15.00	9.27	1.17	1.56	-	1.20	1.20	0.50	Bal.	-	-	-	-	0.30
	DR-2	15.00	13.72	-	0.78	-	0.80	1.60	0.25	Bal.	-	-	-	-	0.30
	DR-3	16.00	10.77	1.57	1.96	-	1.60	1.60	0.25	Bal.	-	-	-	-	0.30
Reference-Alloy	IN738LC	17.50	7.20	4.00	0.53	0.55	0.80	1.00	-	Bal.	8.20	0.52	0.05	0.03	-
	Mitsubishi-DS*	17.24	7.34	3.98	0.51	0.40	0.82	0.99	-	Bal.	8.07	0.47	0.058	0.02	-
	Mitsubishi+Hf-DS*	17.40	7.41	4.02	0.52	0.40	0.83	1.00	-	Bal.	8.14	0.48	0.058	0.03	0.45
	SC-16	17.46	7.36	4.15	1.10	-	-	1.77	-	Bal.	-	-	-	-	-

	Alloy	$\overline{Md}$	$\overline{Bo}$	Density (g/mm³)	Al+Ti+Ta+Nb (mol%)	W+Mo+Re (mol%)	(Ti+Ta+Nb)/Al	Mo/W	mesh No.	γ' (vol.%)
SC-Alloy	S-1	0.9696	0.7099	8.43	12.10	3.30	0.27	1.20	7	46
	S-2	0.9797	0.7102	8.29	12.90	3.00	0.29	1.50	8	49
	S-3	0.9894	0.6999	8.14	14.10	2.60	0.28	1.17	14	54
	S-4	0.9999	0.7002	8.11	15.00	2.40	0.30	1.00	15	56
	S-5	0.9987	0.7097	8.15	14.20	3.20	0.29	1.00	10	53
	S-6	0.9799	0.7009	8.19	13.20	2.90	0.26	1.23	13	52
	SR-1	1.0004	0.7004	8.16	15.10	2.85	0.26	1.17	15	56
	SR-2	0.9902	0.7096	8.30	14.00	2.75	0.27	1.08	9	51
	SR-3	0.9793	0.7005	8.23	13.30	2.75	0.27	1.08	13	52
DS-Alloy	D-1	1.0039	0.7117	8.39	14.30	3.20	0.33	1.00	10	53
	D-2	1.0023	0.7125	8.26	13.90	3.20	0.29	1.00	10	53
	D-3	1.0030	0.6931	8.18	14.89	3.20	0.27	1.70	20	60
	DR-1	0.9630	0.6919	8.54	12.00	2.90	0.29	1.00	16	50
	DR-2	0.9839	0.6735	8.08	14.50	2.62	0.06	2.00	28	62
	DR-3	1.0030	0.7131	8.49	14.30	3.45	0.33	1.00	10	53
Reference-Alloy	IN738LC	0.9719	0.7177	8.21	11.73	1.80	0.63	1.25	-	44
	Mitsubishi-DS*	0.9696	0.7136	8.21	11.83	1.13	0.61	0.38	-	46
	Mitsubishi+Hf-DS*	0.9828	0.7202	8.25	11.95	1.83	0.61	1.20	-	46
	SC-16	0.9741	0.7052	8.21	12.61	1.77	0.71	-	-	46

* after H.Kawai at al.[8]

Experimental Procedure

All the designed SC alloys had a two-phase structure consisting of the γ phase and the γ' phase, and any other phases did not exist. In fact, even the γ+γ' eutectic phase did not form in them. The typical morphology of the γ' phase is shown in Fig.8. Subsequently performed were two kinds of hot-corrosion tests and a creep-rupture test. Both the SC specimens and the DS specimens were prepared by the Bridgman method and supplied to the creep-rupture test. But polycrystalline specimens prepared by arc melting were employed for the hot-corrosion tests. In this study, SC-16 [7], IN738LC and Mitsubishi-DS alloys [8] were selected as the reference alloys for the purpose of property evaluation of the designed alloys.

All the alloys were prepared from raw materials of 99.95%Ni, 99.99%Al, 99.98%Cr, 99.8%Ti, 99.5%Ta, 99.9%Nb, 88.87%W, 99.96%Re and 99.5%Mo. They were heat treated as follows;

(i) for hot-corrosion tests:
  solution heat treatment : 1473K/86.4ks/A.C.
  aging : 1273K/57.6ks/A.C.+1123K/72ks/A.C.
(ii) for creep test:
  SC alloys:
  solution heat treatment : 1573K/14.4ks/A.C.
  aging : 1273K/57.6ks/A.C. + 1123K/72ks/A.C.
  DS alloys:
  solution heat treatment : 1498K/7.2ks/A.C.
  aging : 1373K/14.4ks/A.C. + 1143K/72ks/A.C.

Hot-Corrosion Test

Two kinds of methods were employed for the hot-corrosion test. One is the coating test, and the other is the immersion test. In both tests, hot corrosion resistance was evaluated by the amount of weight loss of the specimen after descaling the corrosion products on the surface.

Coating Method : The surface of the specimen with 2x5x10mm in size was coated with a solid solution of Na_2SO_4-25mol%NaCl salt by $0.2kg/m^2$, and then the specimen was inserted into a tube furnace for exposure to the static air at 1173K for 72ks (20hrs). After the exposure, corrosion products on the specimen surface were descaled with a brass-wire brush after boiling in a 18wt%NaOH-5wt%$KMnO_4$ aqueous solution for 1.8ks, followed by boiling in a 10wt%$(NH_4)_2HC_6H_5O_7$ aqueous solution for 1.2ks.

Immersion Method : A Na_2SO_4-25mass%NaCl solution of 45g was melted in a high purity alumina crucible, and then each specimen with 5x5x20mm in size was immersed into the solution at 1173K for 10.8ks. After the immersion, corrosion products on the specimen surface were descaled in the same way as mentioned above.

Creep-Rupture Test

For the designed SC alloys and a reference alloy, SC-16 [7], creep-rupture tests were carried out using the single-crystal specimens at 1223K under a constant load of 200MPa. On the other hand, for the designed DS alloys and two reference alloys, Mitubishi-DS and Mitubishi+Hf-DS [8], the tests were performed using directionally solidified columnar-crystal specimens at 1123K under a constant load of 264.6MPa. For both the SC and the DS specimens, gauge length and gauge diameter of the test piece were 20mm and 4mm, respectively.

Properties of Designed Alloys

SC Alloys

The results of creep-rupture tests are shown in Fig.9. In this figure, the present result on SC-16 was slightly different from that reported by Khan et al. [7], probably due to the difference in the specimen preparation between them.

All the designed SC alloys were superior in the rupture life to SC-16. In particular, SR-3 alloy showed about three times longer rupture-life than SC-16. On the other hand, there was a tendency for the elongation to become shorter for the designed alloys than for SC-16. But it is considered that the magnitude of the elongation for the designed alloys is still large enough for the practical use.

The results of hot-corrosion tests are shown in Fig.10. The weight changes for the designed alloys were comparable to those for two reference alloys, SC-16 and IN738LC. Therefore, it is concluded that the designed SC alloys have high hot-corrosion resistance as well as superior creep properties.

The γ' volume fraction is higher in the designed alloys than in the reference alloys, as shown in Table 2. It has been generally accepted that hot-corrosion resistance decreases with increasing γ' volume fraction, because of the lower hot-corrosion resistance of the γ' phase than the γ phase. However, it has been found recently that the hot corrosion resistance is nearly independent of the γ' volume fraction as long as alloy compositions vary along a γ-γ' tie line in the phase diagram [9]. This is interpreted as due to the formation of a new single phase region underneath the corrosion products on the surface in the course of corrosion time. This phase region is different in the composition from both the γ and γ' phases. The composition is nearly constant irrespective of the alloys as long as the alloy compositions lie on a γ-γ' tie line [9]. This means that hot-corrosion resistance is associated mainly with the composition of the single phase region formed underneath the corrosion products. However, it is also true that the composition of the single phase region varies when alloy compositions do not lie on the same tie line. In such a case, the hot-corrosion resistance changes largely with the alloy compositions.

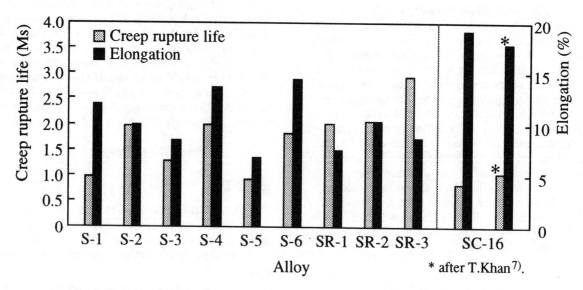

Fig.9. Results of the creep rupture life and the elongation for the designed single-crystal alloys and the reference alloy.

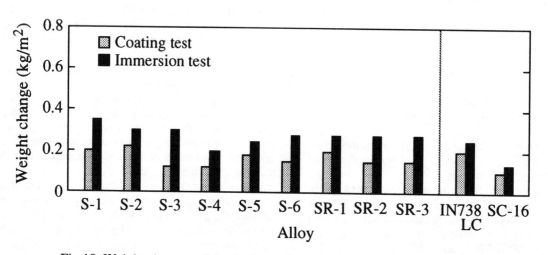

Fig.10. Weight changes of the designed single-crystal alloys and the reference alloys.

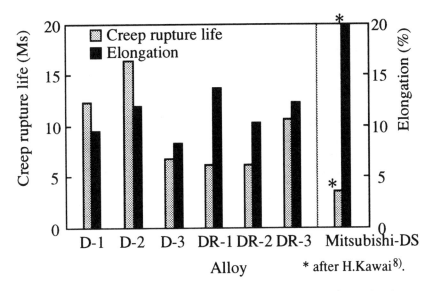

Fig.11. Results of the creep rupture life and the elongation for the designed DS alloys and the reference alloy.

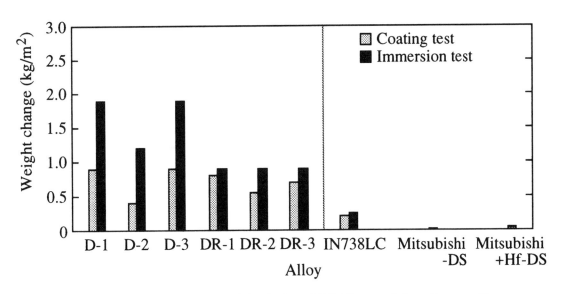

Fig.12. Weight changes of the designed DS alloys and the reference alloys.

DS Alloys

The creep-rupture properties of the designed DS alloys are shown in Fig.11, together with the result of a reference alloy, Mitsubishi-DS. The designed alloys showed more than 2 times longer rupture life than the reference alloy. In particular, D-2 alloy showed about 5 times longer rupture life. As is similar to the SC alloys, the designed DS alloys exhibited the shorter elongation than the reference alloy. But still it is large enough for the practical use.

Fig.12 shows the results of hot-corrosion resistance of the designed DS alloys and three reference alloys. The corrosion results depended on the method for the corrosion test. In case of the coating test, there were not significant differences in the corrosion resistance among the designed alloys and the reference alloys. However, the differences became remarkable in the immersion test. The corrosion resistance was poorer for the designed alloys than for the reference alloys. The reason for the poor resistance of the designed alloys is not unknown. But the addition of Hf, which is a strengthening element of the grain boundaries, may lead to the poor hot-corrosion resistance, since the Hf-free SC alloys showed excellent hot-corrosion resistance, as shown in Fig.10. The hot-corrosion resistance of the DS alloys will be improved by controlling the alloying of grain-boundary strengthening elements in a proper manner.

Conclusion

It is concluded that the present molecular orbital approach is very useful for the design and development of nickel-based superalloys for the applications to industrial turbines.

Acknowledgement

This research was supported in part by Grant-in Aid for Scientific Research from the Ministry of Education, Science and Culture of Japan.

References

1. N.Yukawa, M.Morinaga, Y.Murata, H.Ezaki and S.Inoue, " High Performance Single Crystal Superalloys Developed by the d-Electrons Concept ", Proc. of 6th Inter. Symp. on Superalloys, ed. by D.N.Duhl et al., The Metallurgical Society, Inc., (1988), 225-234.

2. K.Matsugi, R.Yokoyama, Y.Murata, M.Morinaga and N.Yukawa, " High Temperature Properties of Single Crystal Superalloys Optimized by an Electron Theory ", Proc. of 4th Int. Conf. on High Temperature Materials for Power Eng. 1990, ed. by E.Bachelet et al., Kluwer Academic Publishers, Dordrecht (Netherlands), (1990), 1251-1260.

3. M.Morinaga, N.Yukawa and H.Adachi, " Alloying Effect on the Electronic Structure of $Ni_3Al(\gamma')$ ", J. Phys. Soc. Japan, 53(2)(1984), 653-663.

4. K.Matsugi, Y.Murata, M.Morinaga and N.Yukawa, " Realistic Advancement for Nickel-based Single Crystal Superalloys by the d-Electrons Concept", Proc. of the 7th Inter. Symp. on Superalloys, Ed. by S.D. Antolovich et al., TMS, (1992), 307-316.

5. Y.Murata, S.Miyazaki and M.Morinaga, " Evaluation of the Partitioning Ratios of Alloying Elements in Nickel-Based Superalloys by the d-Electrons Parameters", Materials for Advanced Power Engineering 1994, ed. by D.Coutsouradis et al., Kluwer Academic Publishers, Dordrecht (Netherlands), (1994), 909-918.

6. F.C. Hull, " Estimating Alloy Density ", Metal Progress, (1996), Nov., 139-140.

7. T.Khan and P.Caron, " Development of a New Single Crystal Superalloy for Industrial Gas Turbine Blades ", Proc. of 4th Int. Conf. on High Temperature Materials for Power Eng. 1990, ed. by E.Bachelet et al., Kluwer Academic Publishers, Dordrecht (Netherlands), (1990), 1261-1270.

8. H.Kawai, I.Okada, I.Tsuji and K.Takahashi, " Development of Directionally Solidified Blade for Industrial Gas Turbine ", Report of the 123rd Committee on Heat-Resisting Metals and Alloys, Japan Soc. for the Promotion of Sci., 34(2)(1993), 223-228.

9. S.Miyazaki, Y.Kusunoki, Y.Murata and M.Morinaga, " A γ' Phase Volume-fraction at Effect on the Hot Corrosion Resistance of Ni-based Superalloys Varying the Compositions Along a γ-γ' Tie Line ", Tetsu-to-Hagane, 81,(12),(1995), 1168-1173.

THE CONTROL OF SULFUR CONTENT IN NICKEL-BASE, SINGLE CRYSTAL SUPERALLOYS
AND ITS EFFECTS ON CYCLIC OXIDATION RESISTANCE

C. Sarioglu, C. Stinner, J. R. Blachere, N. Birks, F. S. Pettit, and G. H. Meier
Department of Materials Science & Engineering
University of Pittsburgh

J. L. Smialek
NASA Lewis Research Center

Abstract

State-of-the-art superalloys are useful for high temperature applications, in large part, because they form protective alumina surface films by the selective oxidation of aluminum from the alloy. The adherence of the alumina to the alloy is crucial to maintaining oxidation resistance, particularly under thermal cycling conditions. It is now well established that small additions of reactive elements, such as yttrium, hafnium, and cerium, substantially improve the adherence of alumina films to alloy substrates. While the effects produced by the reactive elements are widely known the mechanisms whereby they improve adherence are not completely understood. Over the last fifty years a number of mechanisms have been proposed. However, it has recently become clear that a major effect of the reactive elements is to tie up sulfur in the alloy and prevent it from segregating to the alloy/oxide interface and weakening an otherwise strong bond.

This paper describes the results of a study of the control of sulfur content in alumina-forming nickel-base superalloys and NiAl by three methods:
1. Addition of Reactive Elements (Y and Hf).
2. Desulfurization in the solid state.
3. Desulfurization in the liquid state.

Additionally, calculations have been performed to determine how much sulfur is available to segregate to the scale/alloy interface and how this quantity is influenced by the type and amount of reactive element in the alloy and the level to which the alloy is desulfurized. Finally, the results from experiments to desulfurize the alloys are described and cyclic oxidation measurements are used to evaluate the calculations.

Introduction

State-of-the-art superalloys and aluminide coatings are useful for high temperature applications, in large part, because they form protective alumina surface films by the selective oxidation of aluminum from the alloy. The adherence of the alumina to the alloy is crucial to maintaining oxidation resistance, particularly under thermal cycling conditions. It is now well established that small additions of reactive elements, such as yttrium, hafnium, and cerium, substantially improve the adherence of alumina films to alloy substrates [1]. While the effects produced by the reactive elements are widely known the mechanisms whereby they improve adherence are not completely understood. Over the last fifty years a number of mechanisms have been proposed. These include:

1. Reactive elements act as vacancy sinks to suppress void formation at the alloy/oxide interface [2,3].
2. Reactive elements form oxide pegs at the alloy/oxide interface [4].
3. Reactive elements alter the growth mechanism of the oxide resulting in reduced growth stresses [5].
4. Reactive elements segregate to the alloy/oxide interface and form a graded seal [6] or otherwise strengthen the alloy/oxide bond [7].
5. Incorporation of the reactive element into the oxide increases its plasticity [8].
6. Reactive elements tie up sulfur in the alloy and prevent it from segregating to the alloy/oxide interface and weakening an otherwise strong bond [9,10].

The importance of the latter mechanism has been illustrated in experiments where hydrogen annealing of nickel-base single crystals decreased the sulfur contents to very low levels and resulted in dramatic improvements in the adherence of alumina films to the alloys [11,12].

This paper presents the results of studies of the effects of reactive element additions to alumina-forming single crystal Ni-base superalloys and the effect of lowering the sulfur content of these alloys and NiAl on the oxide adherence.

Experimental

The alloys studied include three alumina-forming Ni-base single crystal superalloys (PWA 1480, PWA 1484, and PWA 1487) and equiatomic NiAl. The single crystal alloys had an initial sulfur content of 8-10 ppm and were undoped (1480), doped with 0.1Hf (1484), and doped with 0.1Hf+0.1Y (1487). Additional polycrystalline alloys with compositions similar to PWA 1484 but containing 1, 3, and 45 ppm sulfur, provided by Howmet Corporation, were also studied. The NiAl contained 20 ppm sulfur. Most specimens had approximate dimensions 1cm X 1cm X 1mm, however, additional specimens of selected alloys with thicknesses ranging from 0.3 to 3.0mm were annealed in Zr-gettered hydrogen for 100 hours at 1200°C to remove sulfur. All specimens were polished through 600 grit SiC paper, washed in soapy water, and rinsed in ethanol prior to oxidation or hydrogen annealing. Hydrogen annealed specimens were oxidized with no further surface treatment. Cyclic oxidation was performed in air at 1100°C with one cycle consisting of 45 minutes in the hot zone and 15 minutes cooling above the furnace. Selected alloys were also oxidized isothermally in air at 1100°C. The oxidized specimens were examined normal to the surface and in cross-section using optical and scanning electron metallography.

The residual stress in the oxide scale was measured at room temperature via the classical $Sin^2\psi$ method[13] using a Philips X'pert diffractometer fitted with an open Eulerian cradle and parallel beam optics with a Ni-filter [14]. The x-ray source was a point source with a Cu target. The Tilting method combined with glancing incident angle ("Thin Film geometry") was employed to measure strain/stress in the α-alumina layer. The geometrical set up of this technique is schematically shown in Figure 1. This technique was chosen because this setup provides greater intensity of the

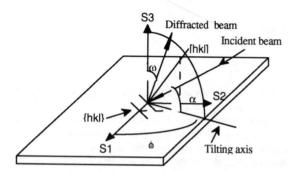

Figure 1: Schematic drawings of the set up of the Tilting technique

diffracted x-rays from the alumina layer as a result of increase in the interaction volume of the x-rays in the alumina layer. The specimen is tilted around an axis (parallel to the surface of the specimen and in the diffraction plane) which is the intersection of the diffraction plane with the surface of the specimen. The (113) reflection of the α-alumina was scanned with a fixed incidence of $\alpha = 1.75°$ for the tilt angles $\psi = 0°$, $28°$ and $33°$. $\psi = 0$ when the specimen normal is in the diffraction plane (Figure 1). The tilt angles were selected when (113) diffraction lines could be measured. With this type of geometry, ψ is not the angle of the tilt of the diffraction planes since these planes are inclined with respect to the surface of the specimen. The angle ψ_2 used in the $Sin^2\psi$ plots is measured between the normal to (113) planes and the normal to the specimen surface. It is given by equation (1) [15].

$$Cos\psi_2 = Cos\omega Cos\psi \quad (1)$$

in which $\omega = \theta_b - \alpha$. θ_b is the bragg angle for the (113) reflection. The absorption correction of the x-ray intensities is carried out by using the equation in reference 15. The line positions in the $2\theta_b$ scan were determined by fitting a parabola to the top of the peak [16].

Results And Discussion

Single Crystal Alloys

The results of cyclic oxidation experiments for the Ni-base single crystal alloys at 1100°C are presented in Figure 2. The oxidation resistance of the untreated alloys is in the order 1487 > 1484 > 1480 which suggests that doping with 0.1Hf improves adherence but that combined doping with 0.1Hf + 0.1Y is more effective. Unfortunately an alloy doped only with Y was not available.

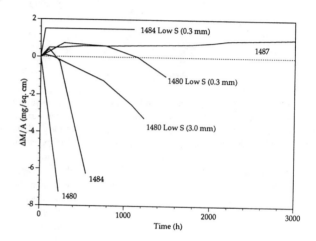

Figure 2: Cyclic oxidation kinetics in air at 1100 °C (1hour cycles) for 1480, 1484, 1487, desulfurized 1484, and desulfurized 1480.

Sulfur removal improved the cyclic oxidation resistance of both 1480 and 1484. These results support the observations that sulfur removal improves the alumina adherence to alloys which do not contain reactive elements (1480). However, the improvement for 1484, which contains Hf, indicates that the sulfur removal somehow makes the Hf more effective. Whether this effect will occur for an alloy containing Y is not yet clear.

Figure 2 also indicates a dependence on specimen thickness for the desulfurized PWA 1480. The hydrogen anneal improved the adherence for the 3.0 mm thick specimen but not to the extent for the 0.3 mm thick specimen. This implies a limitation on the section thickness which can be desulfurized in the solid state. This problem has been addressed by Smialek and coworkers [17,18]. For thin specimens the average sulfur content, C_S, may be calculated as a function of time during hydrogen annealing from

$$C_S/C_{S,i} = (8/\pi^2)*exp(-\pi^2 D_S t/L^2) \quad (2)$$

where: $C_{S,i}$ is the initial sulfur content.
D_S is the sulfur diffusivity in the alloy.
t is the annealing time.
L is the specimen thickness.

Equation (2) will be valid as long as the desulfurization process is diffusion controlled. Taking an initial sulfur content of 10 ppm and approximating D_S as that in pure Ni [19] one calculates from Eqn. (2)

that the 100 hrs., 1200°C anneal should lower the sulfur content to 3 ppm for a 3.0 mm thick specimen and to 10^{-42} ppm for a 0.3 mm thick specimen. While the latter number is clearly unrealistic in that diffusion control will not persist to such low concentrations, it indicates that the thinner specimen would be desulfurized to a very low level. On the other hand, the thicker specimen contains enough residual sulfur to segregate and degrade adherence of the alumina scale. Smialek et al [18] have considered the adsorption behavior of S to a Ni surface as a function of C_S and combined this with Eqn. (2) to express the relationhip between initial sulfur content and the hydrogen annealing conditions required to lower the sulfur content to a level that will result in some number of monolayers of sulfur, N_m, during saturation of the surface:

$$C_{S,i} = (8.27 \times 10^{-2} gm/cm^2)*(N_m A/W)*(\pi^2/8)*exp(\pi^2 D_{S,i} t/L^2) \quad (3)$$

where: $C_{S,i}$ is expressed in ppmw.
A is the specimen surface area in cm^2.
W is the specimen mass in gm.

Taking one monolayer ($N_m = 1$) as the amount of segregation necessary to degrade adherence allows calculation of the annealing time required at 1200°C to adequately desulfurize the specimens. These times are 2.6 h. for the 0.3 mm thick specimen and 492 h. for the 3.0 mm thick specimen. These times indicate that substantially longer hydrogen annealing would be required for the thicker specimen while the 0.3 mm specimen would be adequately desulfurized after 100 h. However, it is interesting to note that the 0.3 mm thick specimen of 1480 still did not have the cyclic oxidation resistance exhibited by the 0.3 mm thick specimen of hydrogen-annealed 1484 which was doped with Hf. A similar result has also been reported for CMSX-4 which is also doped with Hf [20]. This suggests that the removal of sulfur makes the Hf more effective. The unannealed 1487 which contained Hf+Y also showed better cyclic oxidation resistance than the annealed 1480. This result suggests Y is more effective than Hf as a "sulfur getter" or may provide some benefit to alumina adherence beyond sulfur-gettering and that optimum performance may be achieved by combining desulfurization and reactive-element doping.

Polycrystalline Alloys

The starting microstructures of the alloys with the varying sulfur levels were similar. They exhibited a dendritic γ+γ' microstructure with a grain size varying from 50 to 100μm. Small amounts of a Ta-rich phase, which was not present in single crystal 1484, were identified along the grain boundaries by EDS.

The isothermal oxidation rates of the single crystal and polycrystalline alloys were all similar at 1100°C yielding parabolic rate constants on the order of $5 \times 10^{-14} g^2/cm^4 s$. However, on cooling after 168 hours of oxidation the scale spalled completely from the 45 ppm S alloy and spalled partially from the 8 ppm S single crystal. The scale remained completely adherent to the 1 and 3 ppm S alloys. The surface and cross-section of the 3 ppm S alloy are shown in Figure 3. Transient oxidation resulted in thicker oxide on certain grains and dendrites (e.g. dark grains in the SEM surface micrograph) as opposed to others (light

A.

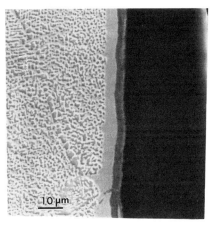

B.

Figure 3: SEM (SEI) surface (A) and cross-section (B) of a polycrystalline alloys (3ppm sulfur) after oxidation at 1100°C in air for 168 hours

grains). Bright spots in the SEM micrograph of the surface were Ta-rich oxides, as indicated by EDS. These apparently formed by the rapid oxidation of the Ta-containing phase on the grain boundaries. The cross-section indicates an adherent oxide in contact with an Al-depleted layer in the alloy along a wavy interface. The Al-depleted layer, which was observed for all the alloys, was indicated by XRD to contain only γ. X-ray diffraction and EDS indicated that the scale consisted of continuous α-Al_2O_3 covered with a layer containing NiO, Cr_2O_3, $Ni(Al,Cr)_2O_4$, and $NiTa_2O_6$ for all the alloys. The surface and cross-section of the 45 ppm S alloy are presented in Figure 4. The oxide has spalled from a large fraction of the specimen area during cooling and spalled oxide fragments are seen lying on the surface. The cross-section shows that the scale is separated from the alloy even in areas where it has not completely spalled.

Figure 5 shows the results of cyclic oxidation experiments on 1484 alloys with various sulfur contents at 1100°C. These data show that,

Figure 4: SEM (SEI) surface (A) and cross-section (B) of a polycrystalline alloy (45 ppm sulfur) after oxidation at 1100 °C in air

for this alloy, there is a critical sulfur content between 3 and 8 ppm below which adherent scales are formed. Comparison with the data for PWA 1480, in Figure 2, indicates that either the 0.1 wt% Hf in the polycrystalline specimens containing 1 and 3 ppm S lowers the content of sulfur which is free to diffuse to the alloy/oxide interface to below 0.1 ppm, or it provides a beneficial effect in addition to sulfur-gettering. Furthermore, the data for PWA 1487 (containing approximately 10 ppm S), in Figure 2, indicate that the combined presence of 0.1 wt% Hf and 0.1 wt% Y is able to mitigate the effects of higher sulfur contents. These results are significant in that they indicate that:

- the sulfur content must be reduced below 0.1 ppm if long-term scale adherence is to be achieved without the use of reactive elements and
- that, when reactive elements are used, the amount of sulfur which can be tolerated varies with the amount and type of reactive element used.

These two points will be discussed more quantitatively in the section on mechanisms.

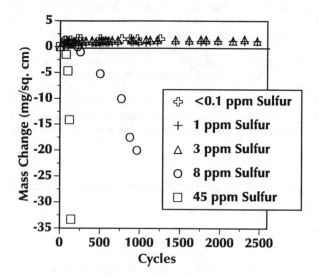

Figure 5: Cyclic oxidation kinetics in air at 1100 ° C (1 hour cycles) for polycrystalline alloys with different sulfur levels and desulfurized 1484 (< 0.1 ppm S).

Residual stress in the α-alumina layer underneath the transient oxide layer on a polycrystalline alloy containing 1ppm sulfur after 24 hours oxidation at 1100°C in air was measured by XRD at room temperature. Figure 6 shows the oxide scale consisting of an α-alumina layer (2 µm thick) between a transient oxide layer and an Al-depleted substrate (γ) layer. The oxide layer was adherent to the substrate after cooling to room temperature (there was no noticeable oxide spallation).

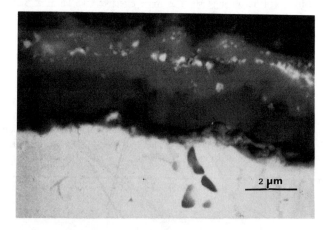

Figure 6: SEM (BEI) cross-section micrograph of a polycrystalline alloy (1 ppm sulfur) after isothermal oxidation at 1100 °C in air for 24 hours.

Assuming a biaxial (an isotropic) stress-state ($\sigma_{11}=\sigma_{22}=\sigma$, $\sigma_{33}=0$, $\sigma_{12}=\sigma_{23}=\sigma_{13}=0$) and a homogeneous elastic medium, the strain in a direction φ along the surface of the film for an inclination (tilt) ψ_2 of the selected (h,k,l) plane with respect to the surface of the film is:

$$(\varepsilon_3)_{\psi_2} = \frac{d_{\psi_2} - d_o}{d_o} = \frac{1+\nu}{E}\sigma \sin^2\psi_2 - \frac{2\nu}{E}(\sigma) \quad (4)$$

Here, $(\varepsilon_3)_{\psi_2}$ is the average strain in the direction normal to the diffracting planes (113) which have an angle ψ_2; d_{ψ_2} is the measured d-spacing of (113) planes and d_o is the unstrained value; the stress σ is the normal stress in the plane parallel to the surface of the film, and ν and E are the Poisson's ratio and Young's modulus of the film, respectively [13, 16, 21]. The data of Figure 7 were obtained from the value of φ=0 with stage tilt (ψ), 0°, 28° and 33°. Because of the strong texture of the alumina layer, a limited number of peaks is obtained as the specimen is tilted. Futher experiments

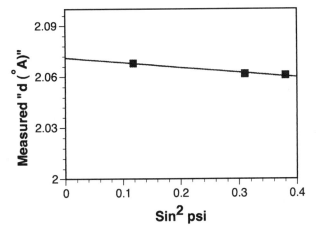

Figure 7: Plot of measured d-spacings for the (113) plane of α-alumina on polycrystalline alloy after isothermal oxidation at 1100 °C in air for 24 hours.

and more detailed analysis of the effect of texture on these data are underway. Nevertheless, the data showed a linear change indicating that the assumptions for equation (4) are satisfied and the stress gradient through the alumina layer is not significant (a stress gradient leads to curvature in the plot). The stress is obtained from the slope of the plot of $(d_{\psi_2} - d_0)/d_0$ or $d_{\psi 2}$ vs $\sin^2\psi_2$. The calculated value is -4051 MPa and the calculated strain in the alumina layer parallel to the surface of the specimen is -0.83%. The thermal stress in the alumina layer is calculated from the thermal expansion coefficients [22,23] and elastic moduli of the substrate and alumina [24], and the poisson ratio of alumina [24] as -3857 MPa. The range of uncertainty in the calculation of the thermal and measured stresses is approximately 15 %. Luthra and Briant [25] reported that the strain in the alumina layer formed on NiCrAlY and NiCoCrAlY at 1150-1225 °C in air for 72-96 hours was negligible (compressive strain) although scattering in the data was significant. C.Diot et. al. [23] reported that the residual stress in the alumina layer formed on a NiCoCrAlY coating at 1100 °C for 300 hours was calculated as -5700 MPa.

Residual stresses measured in the current study are almost equal to the calculated thermal stresses suggesting minor growth stresses or relaxation of the growth stresses. The alloy containing 1ppm sulfur withstood high residual stresses during cooling indicating that the alumina/ alloy interface is inherently strong (adherent) while the alloy containing 8ppm and 45 ppm sulfur showed extensive spallation during cooling indicating that sulfur weakened an otherwise strong interface. These results support the mechanism [9,10] by which the adherence of the alumina to the substrate is improved by addition of reactive elements such as Hf, and Y.

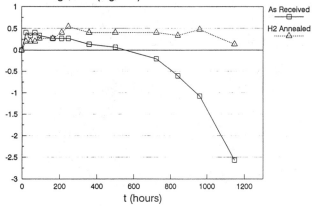

Figure 8: Cyclic oxidation kinetics in air at 1100 °C (1 hour cycles) for as- received and hydrogen annealed NiAl.

NiAl

Figure 8 shows the effect of desulfurization on the cyclic oxidation of NiAl. The as-received specimen contained 20 ppm S and exhibited extensive scale spallation. The sulfur content of the hydrogen-annealed specimen was estimated from Eqn. (2) to be on the order of 0.1 ppm and this alloy exhibited a substantial improvement in cyclic oxidation resistance.

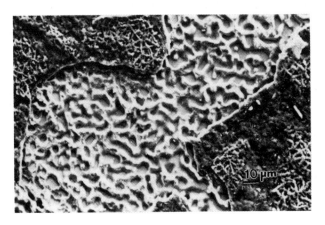

Figure 10: SEM (SEI) surface micrograph of as- received NiAl isothermally oxidized at 1100 °C in air for one week illustrating void formation.

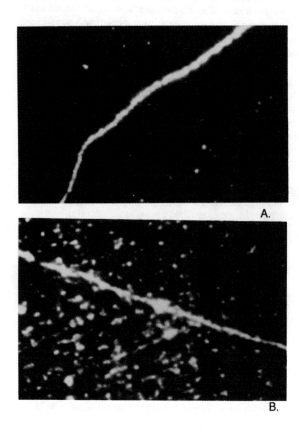

Figure 9: Polarized light optical surface micrographs of NiAl isothermally oxidized for 48 hours at 1100 °C. A. hydrogen annealed. B. as-received.

The hydrogen annealing also affected the morphology of the alloy/oxide interface. Figure 9 shows polarized light micrographs of the surfaces of as-received and hydrogen- annealed NiAl after isothermal oxidation at 1100°C for 48 hours. The bright areas are voids beneath the α-Al_2O_3 scale at the alloy/oxide interface. Voids have coalesced along grain boundaries in the NiAl in both specimens but there is profuse void formation over the grains in the as-received specimen whereas there are very few for the hydrogen annealed specimen. Figure 10 shows more detail of the voids in a spalled area of the as-received material after isothermal oxidation at 1100°C for one week. It is clear that most of the surface was separated from the scale by voids. Figure 11 shows one of the few spalled regions on the hydrogen-annealed specimen after the same oxidation treatment. The trench that formed on a grain boundary is evident, however, there are only a few voids within the adjacent grains. Most of the depressions on the surface of the grains are the imprints of oxide grains. The above results indicate that sulfur has detrimental effects on oxide adherence to NiAl, as it does to superalloys, and that contol of sulfur in diffusion aluminide coatings can be expected to increase coating life.

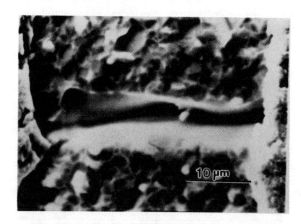

Figure 11: SEM (SEI) surface micrograph of hydrogen annealed NiAl isothermally oxidized at 1100 ° C in air for one week illustrating void formation at a grain boundary.

Mechanisms

The above results clearly indicate that removing sulfur improves the adherence of the alumina scales to both Ni-base superalloys and NiAl. This also supports the proposals in previous studies [9-12,17,18,20] that the improvements in scale adherence produced by reactive element additions, e.g. Figs. 2 and 5, are, at the very least, partly the result of sulfur being combined with the reactive element so that it cannot diffuse to the alloy oxide interface or enter the scale. However, the question remains as to how the sulfur actually degrades oxide adherence. Hou and Stringer [26] report Auger electron spectroscopy observations of sulfur segregation at in-tact FeCrAl/alumina interfaces and suggest that sulfur weakens the bond between the alloy and oxide. Grabke et al [27] argue that sulfur can only segregate to an alloy/gas interface and that, since sulfur lowers the surface free energy, it causes voids to grow along the alloy/oxide interface. No measurements of sulfur segregation were made in the present study but Figure 10 clearly shows that voids separated much of the scale from NiAl containing 20 ppm S whereas there are relatively few voids on desulfurized NiAl, Fig. 11, except at grain boundaries. These observations are more consistent with the void growth mechanism. However, the fact that desulfurization prevents voids, means that reactive elements acting as "vacancy sinks" [2,3] are not necessarily required to suppress interface voids. As long as sulfur is not free(or available) to diffuse to the alloy/oxide interface other sites in the alloy, such as grain boundaries and dislocations, become more effective sinks than the alloy/oxide interface.

The work of Smialek et. al. [17, 18] has pointed to a critical sulfur content which must be achieved by desulfurization in order to achieve good adherence. If the major effect of reactive element additions is to tie up sulfur the same concept should apply to doped alloys. The effect of sulfide stability on surface adsorption of sulfur has been analyzed by Luthra and Briant [25] for binary M-S alloys and they indicated how their treatment could be extended to reactive element-doped alloys. However, the activity coefficients and surface free energies required for the calculations are generally not available. The analysis used here will extrapolate existing solubility product data to indicate how reactive element additions affect the amount of sulfur available for transport to the alloy/oxide interface.

The equilibrium between a reactive element sulfide (MS) and the elements in solution in the alloy is shown schematically in Figure 12 and may be represented by

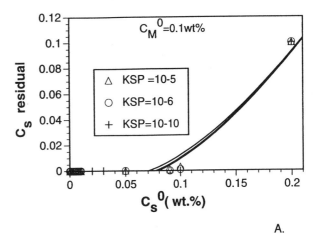

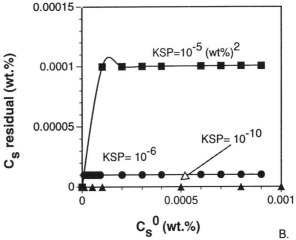

Figure 12: schematic illustration of the equlibrium between a reactive element sulfide (MS) and the elements in solution.

$$MS(s) = \underline{M} + \underline{S} \qquad (5)$$

If the activity of the sulfide is taken as unity the activity coefficients of the elements in solution (taken here as independent of concentration) can be combined with the equilibrium constant for Eqn. 5 to yield a solubility product for MS.

$$K_{SP} = C_M^{Resid} x C_S^{Resid} \qquad (6)$$

Here C_M^{Resid} refers to the concentration of reactive element (in at% or wt.%) left in solution and will be given by

$$C_M^{Resid} = C_M^o - C_M^{MS} \qquad (7)$$

where C_M^o is the total reactive element content of the alloy and C_M^{MS} is the amount of reactive element tied up in the sulfide. Similarly,

$$C_S^{Resid} = C_S^o - C_S^{MS} \qquad (8)$$

Figure 13: The effect of varying solubilty product of sulfide (MS) on residual sulfur content in solution. A: greater values of C_S^0. B. smaller values of C_S^0

where C_S^o is the total sulfur content of the alloy and C_S^{MS} is the amount of sulfur tied up in the sulfide. Note that $C_S^{MS} \approx C_M^{MS}$. Solubility products for sulfides in the alloys under study are not available. However, the use of representative values allows trends to be determined as indicated in Figure 13. (Note that sulfur concentrations have been converted to wt.%.) A K_{SP} value of $1.4 \times 10^{-6} (at\%)^2$ is reported for TiS at 1100°C in austenitic Fe [28] (which might approximate the solubility of TiS in a superalloy). The value of $10^{-10} (at\%)^2$ might approximate the dissolution of YS which is a much more stable sulfide than TiS. The other values were chosen to show the effect of varying the solubility product. The trends indicated in Figure 13 and 14 are:

    i. The residual sulfur content decreases with the amount of reactive element added.

    ii. The residual sulfur content decreases with increased stability of MS i.e. a decrease in K_{SP}

    iii. The residual sulfur content is essentially independent of the initial sulfur content between two critical values of C_S^o.

The two critical values of C_S^0 mentioned in item (iii) are a low value, below which the sulfur activity is too low to form a sulfide, and a high value, above which there is more than the stoichometric amount of sulfur to react with all the reactive element present.

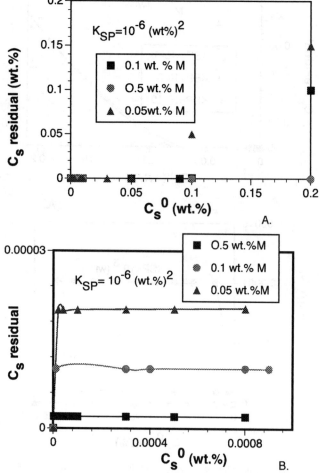

Figure 14: The effect of varying the concentration of reactive element added to an alloy on residual sulfur content in solution. A. For greater values and (B) smaller values of C_s^0

Some of the above results may be qualitatively explained using Figure 13 and 14. Firstly, at the levels of sulfur and reactive elements present in the alloys studied, the solubility products should be exceeded in all of the alloys, i.e. all of the alloys should contain sulfides unless the oxygen content is high enough to precipitate oxides. Point (i.) is qualitatively consistent with observations that additions of Y to PWA 1484 (with 2 ppm S) which resulted in a Y content of 2 ppm had no effect on adherence while those which resulted in Y contents of 19 ppm or greater produced dramatic improvements [29].

The results of Figure 13 and 14 are also consistent with studies in which alloys were intentionally doped with sulfur and reactive elements [25,30]. Smialek [30] found that a NiCrAl alloy doped with 1000 ppma Zr (0.15 wt%) and 1000 ppma S (0.054 wt%) showed extensive spalling which was essentially the same as that from undoped NiCrAl. If the solubility product of "ZrS" is close to 10^{-6}, Figure 14 indicates the doped alloy would contain approximately 1 ppmw (0.0001 wt%) of sulfur in solution, which is sufficient to cause spallings of the oxide. Similarly, alloys doped with 3000 ppma Zr (0.46wt%) and 1000 ppma S (0.054 wt%) or 1000 ppma Zr (0.15 wt%) and 100 ppma S (0.0054 wt%) showed similar amounts of spalling which were much less than that for the undoped alloy. Figure 14 indicates that both of these alloys would have nearly the same amount of sulfur left in solution ($0.2*10^{-5}$ wt% for 0.46 wt% Zr and $2*10^{-5}$ wt% for 0.15 wt% Zr) and, therefore, similar levels of scale adherence. Luthra and Briant [25] compared sulfur contents in NiCrAl of 500 (0.027 wt%) and 3000 ppma (0.16 wt%) along with 0.1 at% Y (0.15 wt%). The alumina was adherent to the lower sulfur alloy but spalled severely from the higher sulfur alloy. If the solubility product of YS may be estimated as 10^{-10}, Figure 13 indicates the alloy with good adherence had a negligible amount of residual sulfur while the higher sulfur alloy had a very high residual sulfur content. Therefore, for a given level of sulfur contamination anticipated, it should be possible to calculate the amount of a particular reactive element needed to reduce the residual sulfur content below some critical level which would prevent any sulfur-caused deterioration of alumina adherence. This has been done in Table 1. where, based on the

Table I Reactive Element Content Required for C_s^0=0.05 ppma:

K_{SP}=(at%M)* (at%S)=1*10^{-6}	
C_s^0 (ppma)	C_M^0 (at%)
0.1	0.28
1	0.28
10	0.28
50	0.28
100	0.29
500	0.33
1000	0.38

K_{SP}=(at%M)*(at%S)=1*10^{-10}	
C_s^0 (ppma)	C_M^0 (at%)
0.1	0.000005
1	0.00001
10	0.0001
50	0.005
100	0.01
500	0.05
1000	0.1

above described experimental results, a critical sulfur content of 0.05 ppma (0.027 wt%) was chosen. It will be noted that for the larger solubility product (thought to approximate additions of Ti, Zr, or Hf) the amount of reactive element required is essentially independent of bulk sulfur content for those usually encountered (1-100 ppm) and remains high even for low amounts of sulfur. For the smaller solubility product (thought to approximate Y additions) the amount of reactive element required is much less but varies with the amount of sulfur contamination. If accurate solubility products are obtained, such calculations should allow the choice of the optimum amount of reactive element doping without "overdoping" which can result in intermetallic phases which are detrimental to mechanical properties or large amounts of reactive element oxides in the scale.

The improvement in the cyclic oxidation resistance of PWA 1484 by lowering the initial sulfur content is qualitatively consistent with the above calculations. Similarly, the improvement in Hf-doped superalloys by hydrogen-annealing, PWA 1484 in Fig. 2 and CMSX-4 in Ref. 20, may be expected from an effective lowering of C_S^o by the anneal. However, Figure 14 indicates that the residual sulfur content, which would stand in equilibrium with "HfS", would not be lowered much by reducing C_S^o. The quantity which is lowered is C_S^{MS} i.e. the amount of sulfide in the alloy. Therefore, the improvements in PWA 1484 and CMSX-4 may result from decreasing the amount of sulfur which can be released from sulfide as the alloy/oxide interface encroaches and the Hf is oxidized. Also, as illustrated in Figure 15, the value of C_M^{Resid} increases continuously as C_S^o decreases. Therefore, lowering the initial sulfur content of PWA 1484 will leave more Hf in solution to tie up sulfur that is released during oxidation and to tie up other potentially detrimental impurities.

This approximate analysis is deficient in that accurate values for the

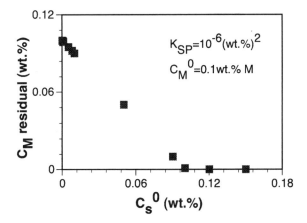

Figure 15: The concentration of reactive element (M) left in solution as a funtion of total sulfur content (C_S^0) in the alloy.

solubility products are not known. For example, the results in Figure 5 would tend to indicate that the steep portion in the top diagram of Figure 13 and 14 should occur at less than C_S^o =8 ppm i.e the solubility product of hafnium sulfide in PWA 1484 is actually more than an order of magnitude greater than the one used in Figure 14. However, it does indicate that evaluation of the sulfur-gettering capability of a given reactive element addition cannot be accomplished by a simple mass balance. The magnitude of K_{SP} and how the equilibrium may be affected by other reactive components, particularly oxygen and carbon, must be considered along with the amounts of sulfur present and reactive element added. This type of analysis has also been shown to be qualitatively explain the effects of reactive elements and sulfur on the cyclic oxidation behavior of alumina-forming FeCrAl alloys [31].

Liquid-Phase Desulfurization

The above results indicate the improvement in alumina adherence which can be achieved by removing sulfur and the need to reduce the sulfur content to the range of 1 ppm or lower to take full advantage of this approach. Although sulfur can be removed from solid alloys by annealing in purified hydrogen, this technique is both slow and inefficient. Recently an experimental program has been initiated to attempt to remove sulfur in nickel and nickel-based alloys to below 1 ppm in the liquid state. If successful, this will allow economical removal of sulfur from commercial sized heats.

The technique used is to melt 10 lb. of nickel in a vacuum induction furnace and allow the sulfur in the nickel to react with a reactive lining. The reactions will be of the type:

$$MO + \underline{S} = MS + \underline{O}$$
$$MO + \underline{S} + \underline{C} = MS + CO$$
$$M_2O_3 + \underline{S} = M_2O_2S + \underline{O}$$
$$M_2O_3 + \underline{S} + \underline{C} = M_2O_2S + CO$$

Under vacuum, the presence of carbon proceeding to form carbon monoxide will tend to push the reactions to the right. The presence of aluminum in the alloy would provide a similar impetus but would result in an oxide inclusion being formed. Thermodynamic calculations considering various active linings, such as CaO and Y_2O_3, indicate that it should be possible to remove the sulfur to well below the 1 ppm level.

Experiments carried out so far indicate that sulfur is maintained at 1 ppm and that the mechanism appears to proceed to produce calcium metal which bubbles through the liquid alloy, as indicated by the following reaction.

$$CaO + C = Ca(g) + CO$$

The reaction with sulfur in this case will involve

$$Ca(g) + \underline{S} = CaS$$

Experiments have not been carried out using Y_2O_3 linings and currently the vacuum furnace is being modified to allow metal samples for analysis to be taken during the run. This will allow mechanisms, kinetics, and limiting sulfur levels to be determined.

Summary And Conclusions

The results of this study have shown that desulfurization of alumina-forming Ni-base superalloys and NiAl by hydrogen annealing can result in improvements in cyclic oxidation comparable to that achieved by doping with reactive elements.

It is possible to estimate the amount of sulfur available to segregate to the alloy/oxide interface and how this is influenced by reactive element additions or hydrogen annealing. If these calculations can be made more quantitative it should be possible to engineer alumina-forming alloys for optimum resistance to cyclic oxidation. Preliminary experiments to desulfurize Ni-base alloys in the liquid phase have been discussed.

Acknowledgement

This work was supported by the University of Pittsburgh Materials Research Center, AFOSR Contract F49620-95-1-0167, C. Ward, monitor.

References

1. D. P. Whittle and J. Stringer, Phil. Trans. Roy. Soc. Lond., A295, (1980), 309.

2. J. Stringer, Met. Rev., 11 (1966), 113.

3. J. K. Tien and F. S. Pettit, Met. Trans., 3 (1972), 1587.

4. E. J. Felten, J. Electrochem. Soc., 108 (1961), 490.

5. F. A. Golightly, F. H. Stott, and G. C. Wood, Oxid. Met., 10 (1976), 163.

6. H. Pfeiffer, Werkst. Korros., 8 (1957), 574.

7. J. E. McDonald and J. G. Eberhardt, Trans. TMS-AIME, 233 (1965), 512.

8. J. E. Antill and K. A. Peakall, J. Iron Steel Inst., 205 (1967), 1136.

9. A. W. Funkenbusch, J. G. Smeggil, and N. S. Bornstein, Met. Trans., 16A (1985), 1164.

10. J. G. Smeggil, A. W. Funkenbusch, and N. S. Bornstein, Met. Trans., 17A (1986), 923.

11 B.K.Tubbs and J.K.Smialek, "Effect of Sulfur Removal on Scale Adhesion to PWA 1480" Symposium on Corrosion and Particle Erosion at High Temperatures, eds. V. Srinivasan and K. Vedula, TMS,(1989), 459.

12. "Oxidation of Low Sulfur Single Crystal Nickel-Base Superalloys", Superalloys 1992, eds.. S. D. Antolovich et sl.,. (TMS , 1992), 807.

13. J. C. Noyan and J. B. Cohen, Residual Stresses, (Springer Verlag 1987).

14.. C.S. Barrett et al., eds., Advances in X-Ray Analysis, vol. 135 (Plenum, 1992), 205-210.

15. C.S. Barrett et al., eds., Advances in X-ray Analysis, vol. 32 (1989), 285-292

16. B. D. Cullity, Elements of X-ray Diffraction, (2nd ed., Addison Wesley, Reading, MA, 1978).

17. J. L. Smialek, Met. Trans., 22A (1991), 739

18. J. L. Smialek et al., Thin Solid Films, 253 (1994), 285-292.

19. S. J. Wang and H. J. Grabke, Z. Metallk., 61 (1970), 597.

20. T. A. Kircher, A. Khan, and B. Pregger, cited in ref. 17.

21. Gilfrich et al., eds, Adances in X-Ray Analysis, vol. 37 (Plenum Press, New York, 1994), 189.

22. A.M.Huntz, J.L.Lubrun, and A.Boumaza, Oxid. Metals, 33 (1990), 321.

23. C.Diot, P.Choquet, and R.Mevrel, cited in ref.22.

24. M. Schutze , Ox. Metals. 44 (1995), 29.

25. K. L. Luthra and C. L. Briant, Oxid. Metals, 397 (1986), 26.

26. P. Y. Hou and J. Stringer, Oxid. Metals, 38 (1992), 323.

27. H. J. Schmutzler, H. Viefhaus, and H. J. Grabke, Surf. Interface Anal., 18 (1992), 581.

28. "Thermo-Kinetic Analysis of Precipitation Behaviour of Ti Stabilized Interstitial Free Steel", in Interstitial Free Steel Sheet: Processing, Fabrication and Properties, eds. L. E. Collins and D. L. Baragar (Canadian Inst. of Mining, Metallurgy, and Petroleum, Ottawa, 1991), 15.

29. P. R. Aimone and R. L. McCormick, "The Effects of Yttrium and Sulfur on the Oxidation Resistance of an Advanced Single Crystal Nickel Based Superalloy", Superalloys 1992, eds. S. D. Antolovich et al., (TMS , 1992), 817.

30. J. L. Smialek, "The Effect of Sulfur and Zirconium Co-doping on the Oxidation of NiCrAl", in High Temperature Materials Chemistry IV, ed. Z. A. Munir, D. Cubicciotti, and H. Tagawa, (The Electrochem. Soc., 1987),24.

31. G. H. Meier, F. S. Pettit, and J. L. Smialek, Materials and Corrosion, 46 (1995), 232.

DISPERSION STRENGTHENED SHEET ALLOYS FOR EXHAUST COMPONENT APPLICATIONS

R. Darolia and J. R. Dobbs*
GE Aircraft Engines, 1 Neumann Way, Cincinnati, Ohio 45215
*GE Corporate Research Center, Schenectady, New York 12301

Abstract

The development and properties of a new class of high-temperature sheet alloys are described in this paper. The alloys utilize rapid solidification processing and the dispersion strengthening concept in a Ni-base superalloy composition to achieve excellent high-temperature stress rupture properties. The processing, compositions, microstructures, physical and mechanical properties as well as component fabrication are discussed.

Introduction

Rapid solidification plasma deposition from alloy powders is an attractive processing approach for making thin section superalloy parts operating in the 1600 to 1800°F temperature range. In this process, parts are made by incrementally solidifying molten powder particles on a mandrel to form parts of essentially simple geometries. Several modifications of this approach have been evaluated for the past decade [1]. Rapid solidification plasma deposition (RSPD) and spray forming, commonly referred to as the Osprey process are two more commonly used processes. The process is particularly suitable when it is difficult to cast large thin section parts, or when subsequent processing is not possible or desirable. Composition homogeneity as well as higher strengthener concentration in the alloy are two other benefits derivable from rapid solidification. A fine grain microstructure (typical grain size 25 to 60 µm) is an inherent attribute of such a process. The fine grain structure provides several advantages as well as disadvantages. The main advantage of the fine grain size is that it can provide high tensile strengths based on the Hall Petch relationship. However, the major disadvantage is poor high-temperature stress rupture and creep properties which do not compare well with those of the cast superalloys of similar compositions. Due to the presence of a small amount of oxides in the deposits, it is difficult to grow grains to an appreciable size to improve high-temperature creep properties. Therefore, an alloy development program was carried out to improve creep and rupture properties by creating a dispersion of stable particles in the matrix as well as at the grain boundaries. Many types of dispersoids, such as Y_2O_3, Al_2O_3 and carbides were evaluated. The oxide dispersoids were found to be very inhomogeneously dispersed. In addition, alloys containing the oxide dispersoids had poor room temperature ductility and showed no improvement in high-temperature strength. Eutectic carbide dispersoids, principally TaC in the form of platelets and spheroids, were found to be effective strengtheners, resulting in alloys with rupture properties superior to the currently used sheet superalloys such as René 41, HS 188 and Hastelloy X.

Processing

The rapidly solidified plasma deposition (RSPD) process is generally carried out in a low vacuum, inert gas atmosphere chamber. Powder particles of a specified composition are fed into a high energy plasma flame, where they are melted to form droplets of liquid. The plasma flame directs the stream of droplets against a heated mandrel/substrate where they impact and rapidly solidify and adhere to the mandrel. The temperature of the mandrel is kept low enough to cause the droplets to freeze as soon they come in contact with the mandrel. The droplets typically flatten as they impact the mandrel. The desired deposit thickness is incrementally built up. The as-deposited densities are sufficiently high such that post deposition heat treatments are normally adequate to achieve near theoretical density. Near-net shapes can be formed by proper selection of the mandrel configuration and clever translation and rotation of the plasma gun and the mandrel. Deposition rates are high enough that the process is economically feasible. A schematic of the process is shown in Figure 1.

The deposit is removed from the mandrel which is generally made of a low alloy steel by chemically leaching, machining or mechanical separation. Because of the very rapid solidification rates (estimated to be about 10^7 °F/sec), a non-equilibrium microstructure is obtained. The deposits are heat treated to obtain an equilibrium structure.

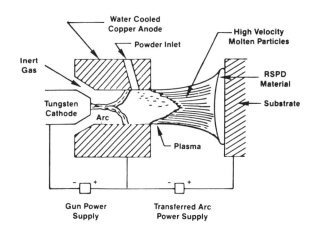

Figure 1. Schematic of a plasma spray process.

Another variation of this incremental solidification process is often called spray forming. The principles of spray forming differ from the RSPD processing only to the extent that the droplets of liquid are formed directly from the melt, rather than remelting powder particles. There is essentially no difference in the resulting microstructures from the two processes, and the mechanical properties are also identical. This paper describes the properties of alloys processed by the RSPD process. The properties of the spray formed material were also evaluated and found to be identical to those of the RSPD processed material.

There is a wide range of parameters under which acceptable alloy deposits can be made. Typical RSPD processing parameters are:

Plasma gun power	40-50 kW
Mandrel pre-heat temperature	1800 to 1950°F
Deposition chamber pressure	30 to 60 torr
Powder feed rate	10 lbs/hour
Gun to workpiece distance	10 to 12 inches
Primary gas - argon	50 to 120 liters/min
Secondary gas - helium	10 to 30 liters/min
Carrier gas	argon at 5-20 liters/min
Powder size	-400 mesh (< 38 μm), +10 μm

Powder particles of the specified alloy compositions are generally obtained by argon atomization. Based on poor mechanical properties of the plasma sprayed deposits containing oxides, particular emphasis was placed on obtaining powders with low oxygen levels. Oxygen level was specified at below 300 wppm.

Composition, Heat treatment and Microstructure

The initial work on the carbide dispersion strengthened alloy system identified an alloy designated RD-8A as the most promising alloy (2). The nominal composition of this alloy is shown in Table 1. This composition was based on earlier alloy development work on directionally solidified (DS) NiTaC alloys such as NiTaC-14B (3, 4). The alloy is essentially a gamma prime strengthened Ni-base superalloy which has been further strengthened by precipitation of TaC platelets and spheroids by an eutectoid reaction. The as-deposited microstructure does not contain these dispersoids. These dispersoids were found to significantly improve the rupture properties of the alloy.

Processing, microstructure, and properties of the RD-8A alloy were extensively evaluated. About forty composition variations of the RD-8A alloy which included C, Ta, Al, Mo, Re, Nb, Co, Ti, W and Hf were also evaluated. Selected composition variations are shown in Table 1. For each alloy, the composition variation is highlighted in bold in Table 1. The effect of these alloying variations on microstructure, especially carbide morphology and mechanical properties, will be discussed later in a separate section in this paper. The majority of the paper describes the microstructure and properties of the RD-8A alloy of the nominal composition.

As was mentioned earlier, the as-deposited microstructure does not contain dispersoids. A heat treatment schedule was developed to precipitate the dispersoids as well as to optimize their morphology. A typical heat treatment for the RSPD deposits consisted of 2100°F for 6 hours (for carbide precipitation), followed by 2400°F for 4 hours (for gamma prime solution), and followed by 1975°F for 16 hours and 1650°F for 16 hours (for gamma prime precipitation). The carbide platelets which were precipitated at 2100°F were stable at the 2400°F solution temperature, which was used to fully solution γ which subsequently precipitated during the 1975 and 1650°F aging treatments. An as-deposited microstructure of the RD-8A alloy is shown in Figure 2. The as-deposited microstructure does not show any dispersoids.

Figure 2. As-deposited microstructure of RD-8A alloy.

Table 1 Compositions of selected RD-8A alloys (in weight %). The alloys are arranged to show a systematic variation (typed in bold) in selected alloying elements.

Alloy	Al	Cr	Co	Mo	W	Re	Ta	C	B	Y	Hf	Ni
RD-8A	5.5	4.2	4.0	3.2	4.5	6.8	8.4	0.23	0.03	0.05		Bal
RD-8A-I	**5.2**	4.2	4.0	3.2	4.5	6.8	8.4	0.23	0.03	0.05		Bal
RD-8A-P	**6.0**	4.2	4.0	3.2	4.5	6.8	8.4	0.23	0.03	0.05		Bal
RD-8A-M	5.5	4.2	4.0	3.2	4.5	6.8	8.4	**0.12**	0.03	0.05		Bal
RD-8A-44	5.5	4.2	4.0	3.2	4.5	6.8	8.4	**0.19**	0.03	0.05		Bal
RD-8A-N	5.5	4.2	4.0	3.2	4.5	6.8	8.4	**0.30**	0.03	0.05		Bal
RD-8A-T	5.5	4.2	4.0	3.2	4.5	6.8	8.4	**0.30**	0.03	0.05	4.4 V	Bal
RD-8A-S	5.5	4.2	4.0	3.2	4.5	6.8	**9.5**	0.23	0.03	0.05		Bal
RD-8A-45	5.5	4.2	4.0	3.2	4.5	6.8	8.4	0.23	0.03	0.05	0.10	Bal
RD-8A-42	5.5	4.2	4.0	**4.0**	4.5	6.8	8.4	0.23	0.03	0.05		Bal
RD-8A-U	5.5	4.2	4.0	**4.0**	4.5	6.8	8.4	0.23	0.03	0.05	0.10	Bal
RD-8A-33	5.5	4.2	4.0	**1.0**	4.5	6.8	8.4	0.23	0.03	0.05	0.5 Nb	Bal
RD-8A-35	5.5	4.2	4.0	3.2	4.5	**5.0**	8.4	0.23	0.03	0.05	0.5 Nb	Bal
RD-8A-39	5.5	**5.0**	**7.5**	**0.0**	4.5	**6.0**	8.4	0.23	0.03	0.05	1.0 Ti	Bal
RD-8A-40	5.5	4.2	**7.5**	3.2	4.5	6.8	8.4	0.23	0.03	0.05		Bal
RD-8A-41	5.5	4.2	**12.5**	3.2	4.5	6.8	8.4	0.23	0.03	0.05		Bal
RD-8A-43	5.5	4.2	**12.5**	**4.0**	4.5	6.8	8.4	0.23	0.03	0.05		Bal
RD-8A-46	**5.2**	4.2	4.0	3.2	4.5	6.8	8.4	0.23	0.03	0.05	0.10	Bal
RD-8A-47	5.5	4.2	4.0	3.2	4.5	6.8	**6.0**	0.23	0.03	0.05	0.10	Bal
RD-8A-48	5.5	4.2	4.0	3.2	4.5	6.8	**7.5**	0.23	0.03	0.05	0.10	Bal
RD-8A-49	**5.2**	4.2	**12.5**	**4.0**	4.5	**5.5**	8.4	0.23	0.03	0.05	0.10	Bal

The microstructure after the deposit is fully heat treated is shown in Figure 3(a and b). In these micrographs, the platelet morphology of the dispersoids can be seen. The scanning electron micrographs in Figure 4 show the square morphology of the platelets. The micrographs in Figure 4 were taken after removing the $\gamma+\gamma'$ matrix by electrolytic etching. It was also observed that the platelets were dispersed predominantly at the grain boundaries. The electron microprobe analysis established these platelets to be TaC containing some amount of Mo, W and Re. For the purpose of this paper, these carbides will be referred to as TaC carbides. It is important to point out that a more traditional heat treatment schedule of solution and subsequent age typically utilized for Ni-base superalloys can also be used for these RSPD deposits for obtaining microstructures similar to the one shown in Figure 3, except with one exception. The carbide platelets following a traditional superalloy heat treatment cycle were larger in size (carbide length about 25 to 35 µm compared to about 17 µm for the 2100°F treatment). Heat treatments were carried out in flowing argon to avoid formation of a carbide denuded zone in the periphery of the specimens. The carbide morphology was also found to be a strong function of the alloy composition, as will be discussed later.

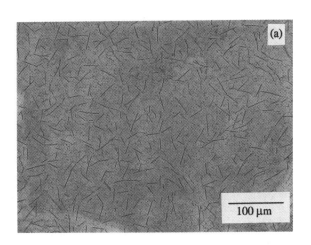

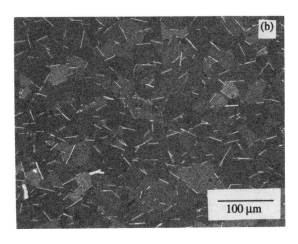

Figure 3. Microstructure of RD-8A alloy after heat treatment, (a) prior to etching, (b) after etching.

Figure 4. Scanning electron micrographs (three different magnifications) of RD-8A alloy after heat treatment.

Physical Properties

The incipient melting temperature of the RD-8A alloy was determined by differential thermal analysis (DTA) to be 2450°F which is about 50°F higher than the majority of Ni-base superalloys. The density was measured to be 0.328 lb/in^3. The mean coefficient of thermal expansion of the alloy RD-8A was measured and compared with several competitive alloys; René 41, HS-188, Hastelloy X, Haynes 214 and Haynes 230. As shown in Figure 5, RD-8A has a substantially lower coefficient of thermal expansion. The coefficient of thermal expansion is an important physical property influencing thermal fatigue behavior of Ni-base superalloys, especially in large thin structures for exhaust applications. A lower coefficient of thermal expansion is desirable to reduce thermal fatigue induced cracking at the 'hot spots'. The lower coefficient of thermal expansion of RD-8A appears to have contributed to the improved thermal fatigue lives demonstrated in several types of laboratory thermal fatigue tests discussed later.

Mechanical Properties

Tensile Properties

The 0.2% yield and ultimate tensile strengths of RD-8A are shown in Figure 6 in which the strengths are compared with three competitive alloys: René 41, René 80 and Hastelloy X. Whereas the RD-8A alloy is significantly superior to Hastelloy X, it has about 100 to 150°F advantage over René 41 and René 80 at temperatures above 1600°F. Tensile plastic elongation as a function of temperature is plotted and compared with René 41 in Figure 7. There is a ductility minima of about 1% at 1600°F. Alloying modifications and the carbide morphology (which was varied by heat treatments and by alloying modifications) were found to influence the tensile properties as discussed in a later section. Alloy modifications were especially directed towards improving the ductility of the alloy. Figure 8 shows plastic elongation as a function of temperature for several selected alloy modifications (compositions shown in Table 1). As can be seen in this figure, ductility was improved by alloy modifications. Highest ductilities were obtained in alloys with higher carbon levels. Several of the high ductility alloys had high-temperature strength properties comparable to RD-8A alloy of the nominal composition (see figure 14. The plastic elongation of these alloys, although adequate for most applications, is still an area of concern (compared to superalloys currently being used) which must be addressed by further alloy modifications and component and engine tests.

Stress Rupture Properties

The stress rupture and creep properties were measured at several temperatures of interest. The stress rupture behavior is plotted and compared with René 41 and HS-188 in Figure 9. Rupture properties of RD-8A obtained from two types of specimens are plotted in Figure 9: 20 mil as-deposited and heat treated sheet specimens and 160 mil gage diameter standard bar specimens machined from a thick deposit. The first few mils of the thin, 20 mil RSPD deposit had higher than normal porosity, which led to lower stress rupture lives shown in Figure 9. A subsequent surface finishing step may be required if these lower properties do not meet design requirements. The RD-8A type alloys exhibit about a 200°F rupture strength advantage over René 41 and HS 188, and though not shown in Figure 9, the RD-8A alloy was found to be equivalent to René 80 in rupture strength. The alloy was also found to be equivalent to René 80 in creep resistance.

Thermal Fatigue Resistance

Since the primary mode of failure for the exhaust components is thermal fatigue, thermal fatigue resistance was one of the critical properties evaluated for the RD-8A alloy. Since it is difficult to accurately simulate engine conditions in a laboratory test, five different types of tests with various conditions were run. LCF, LCF (Hold), thermal-mechanical low cycle fatigue (TMLCF, Hold) and TMLCF (Pulse) are standard fatigue tests. The test conditions are noted in Figure 10. In the burner rig test, the RD-8A tubes (1.5 inch in diameter and 40 mil thick wall) were cycled in front of a torch which generated an 1 inch diameter hot spot of about 2000°F in the middle of the tube. In this test, René 41 specimens failed at 1176 cycles, whereas the RD-8A specimens had an average life of over 4000 cycles. The results from these tests are summarized in Figure 10. Available data on René 41, cast René 80, MA 956 and Hastelloy X are also shown. In all these tests, except in the hold time LCF test, RD-8A was shown to be superior to these alloys. Two types of LCF tests were conducted at 1800°F and a strain range of 0.4%. Two cycles were used: 20 cpm triangular and 1/2 cpm with a compressive strain hold. At 20 cpm, the fatigue capability of RD-8A was found to be excellent; two specimens ran 18,819 and 18,148 cycles. This was much better than Hastelloy X (5,500 cycles) or cast René 80 (3,200) at this condition. However, when subjected to the 1/2 cpm compressive hold cycle, the life of RD-8A dropped drastically (555 and 668 cycles). At this condition, it was found to be inferior to Hastelloy X (1715 cycles), René 41 (1268 cycles) and cast René 80 (2600). Cracks were found to form in RD-

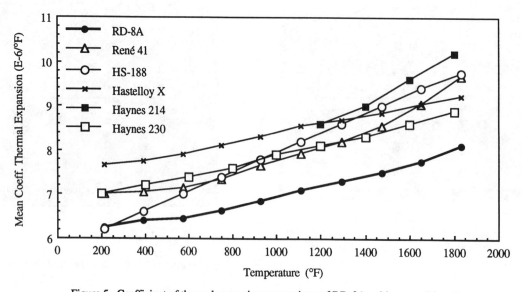

Figure 5. Coefficient of thermal expansion comparison of RD-8A with competitive alloys

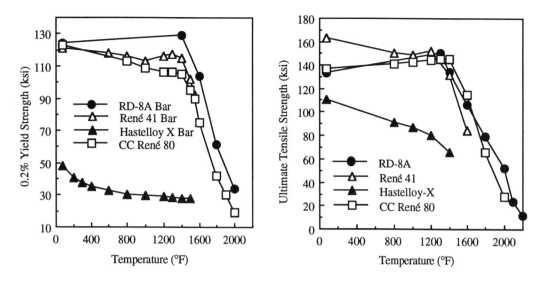

Figure 6. 0.2% yield strength and ultimate tensile strength of RD-8A as a function of temperature.

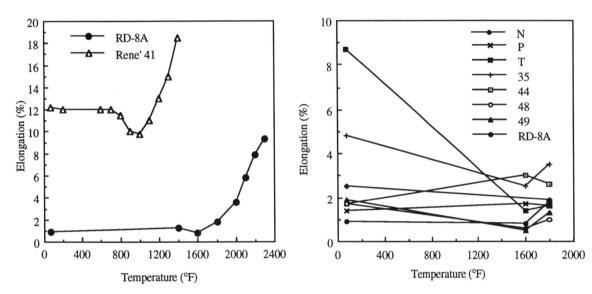

Figure 7. Tensile plastic elongation as function of temperature for RD-8A compared to René 41.

Figure 8. Tensile plastic elongation as a function of temperature for selected alloy modifications.

8A from small blisters which were probably formed by oxidation of the carbide platelets. It appears that, due to the limited ductility of RD-8A, cracks form exposing the TaC dispersoids to oxidation. However, in the TMLCF hold time testing, RD-8A was found to be superior to René 41 and Hastelloy X.

Oxidation Resistance

Oxidation pin specimens, 0.25 inch diameter by 3.5 inch long, of the RD-8-A alloy were tested for oxidation resistance in a Mach 1.0 gas velocity oxidation test rig at 2050 and 2150°F for up to 200 hours. Each test consisted of 200 cycles of one hour duration out of which the specimens were at the test temperature for 55 minutes. The results are compared with the oxidation behavior of René 41, René 80, René 125 and René N5 in Figure 11. The superior oxidation resistance of the RD-8A alloy compared to the conventional superalloys such as René 41, René 80 and René 125 is quite obvious. The oxidation resistance also approaches that of the single crystal alloy René N5 which was especially developed for good oxidation resistance. In the LCF (hold) test, however, oxidation susceptibility of the alloy was observed when a crack is formed due to the limited ductility of the alloy. Cracking of the alloy exposes TaC dispersoids which appear to have poor oxidation resistance.

Sliding Wear Behavior

In an application involving sliding wear, the RD-8A alloy was evaluated for wear resistance at 1200°F, and compared with several competing superalloys. In this test, a René 41 shoe was rubbed against a RD-8A block at 100 lbs load. The RD-8A alloy did not show any wear after 100,000 wear stroke cycles, whereas René 41 had an average wear of 3.1 mils. As shown in Figure 12, the alloy has demonstrated excellent wear resistance.

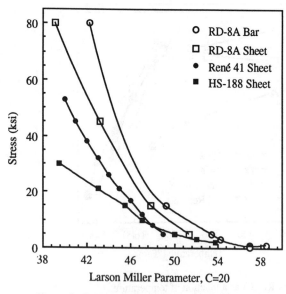

Figure 9. Stress rupture behavior of RD-8A compared with René 41 and HS-188.

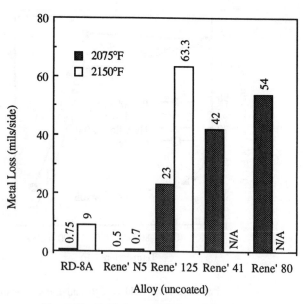

Figure 11. Oxidation resistance of RD-8A compared with several Ni-base superalloys.

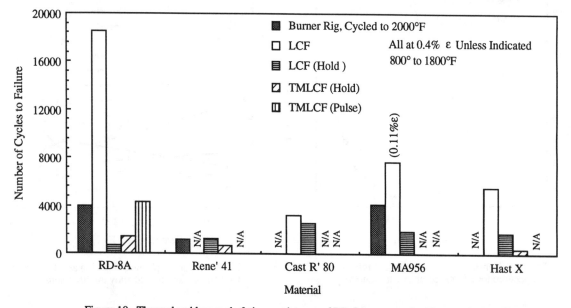

Figure 10. Thermal and low cycle fatigue resistance of RD-8A compared with several superalloys.

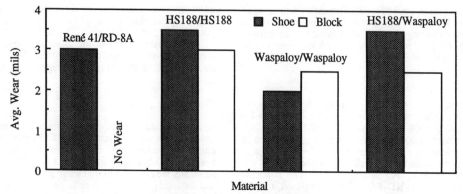

Figure 12. Sliding wear resistance of RD-8A compared with Ni-base superalloys.

Effect of Composition Variations on Microstructure and Mechanical Properties

About forty composition variations were evaluated to determine their effect on the carbide morphology and the mechanical properties. The main emphasis was to improve ductility of the alloy while maintaining high-temperature strength. Due to lack of space, it is not possible to describe in detail the effect of all the variations evaluated. Observations from a small number of selected compositions (shown in Table 1) will be described in this section. Figure 13 shows several representative microstructures with various carbide morphologies. The carbide morphology was found to vary from all platelets to all spheroids, as well as combinations of these two shapes. The spherical carbides formed a necklace structure at the grain boundaries, as clearly seen in Figure 13(a-d). Selected area diffraction in transmission electron microscopy was used to identify the spherical carbides as TaC carbides. The platelets were also identified as MC type (four MC carbides appear to be stacked together to form a tetragonal cell; crystal structure, however, was not clearly identified) carbides mainly containing Ta with varying amounts of Re, Mo, and W depending on the alloy composition. Occasionally, M_6C carbides were also seen. Depending on the alloy composition, TCP phases were also observed. Mo, Ta and C were seen to influence the carbide morphology. As expected, Al, Co, Re and Ta were found to influence alloy strength and stability.

Stress rupture properties of several representative alloys are shown in Figure 14. The alloy RD-8A-49 was shown to possess the best stress rupture properties. The microstructure of the alloy, as shown in Figure 13(d), consisted entirely of spherical carbides both at the grain boundaries and in the matrix. The tensile ductility of this alloy was slightly better than the baseline RD-8A alloy as shown in Figure 8.

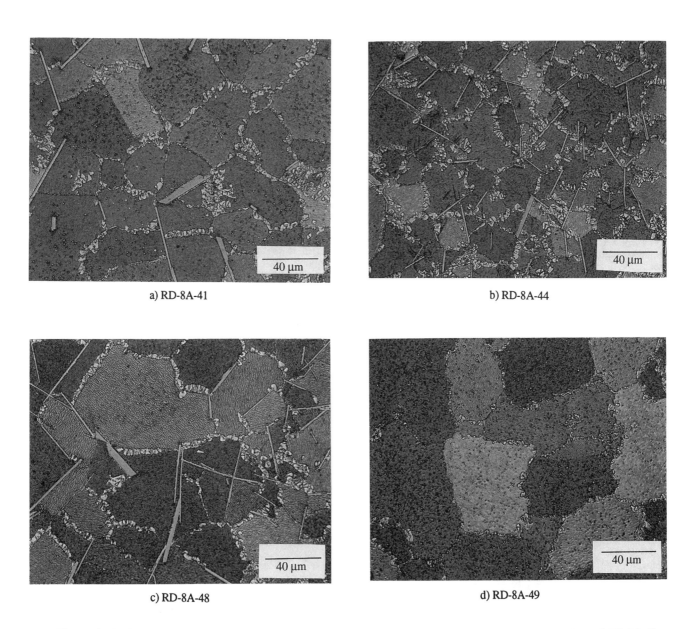

a) RD-8A-41

b) RD-8A-44

c) RD-8A-48

d) RD-8A-49

Figure 13. Optical micrographs showing various carbide morphologies, a) RD-8A-41, b) RD-8A-44, c) RD-8A-48 and d) RD-8A-49.

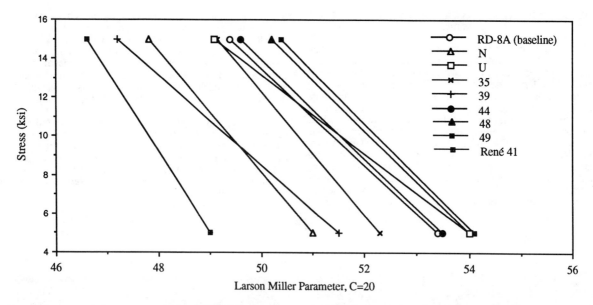

Figure 14. Stress rupture properties of selected alloy modifications.

Component fabrication

Exhaust parts of various configurations have been made out of the RD-8A alloy utilizing the rapid solidification plasma deposition technique. The deposition was carried out on mandrels of appropriate but simple configurations. After deposition, the mandrels were removed by using nitric acid, leaving free standing parts which were further shaped using hot dies. The hot die forming step to define the second or final shape was carried out at 2100°F for six hours. This step combined the forming and the carbide precipitation steps. During the first 15 minutes of this process, the final shape is easily formed because of the low strength of the alloy. After about 15 minutes, the alloy starts to harden due to the formation of the dispersoids and resists further deformation.

Several simple shapes made out of RD-8A are shown in Figure 15. An exhaust divergent seal face sheet made out of RD-8A is shown in Figure 16. This face sheet was fabricated by first depositing a 0.06 inch thick layer on an 8" diameter tube mandrel made out of a low alloy steel. The steel mandrel was removed by chemical leaching in nitric acid. The free standing tube was then cut open across the width, and straightened and simultaneously formed into the shape shown in Figure 16 by using a BN coated graphite die. All the subsequent heat treatments were done while the part was still in the die to prevent any warpage. Another exhaust component made entirely out of RD-8A by a combination of die forming and brazing various RD-8A parts is shown in Figure 17.

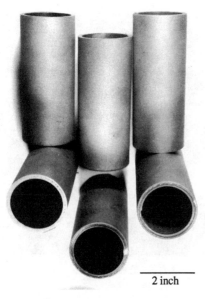

Figure 15(a). Simple tube shapes made out of RD-8A.

Figure 15(b). A transition duct made out of RD-8A.

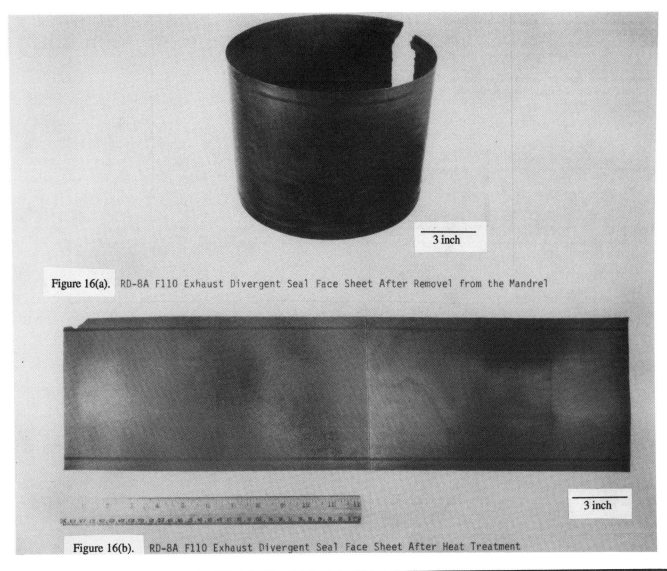

Figure 16(a). RD-8A F110 Exhaust Divergent Seal Face Sheet After Removal from the Mandrel

Figure 16(b). RD-8A F110 Exhaust Divergent Seal Face Sheet After Heat Treatment

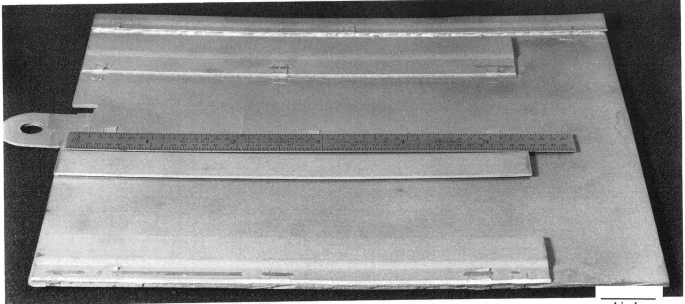

Figure 17. Another example of a RD-8A exhaust component.

In an attempt to obtain thin sheets from thick deposits, rolling experiments were also conducted. It was found that it was possible to roll 65 mil thick deposits down to 30 mil sheets at 2150°F. This was accomplished with the RD-8A alloy in the as-deposited condition. The microstructure evaluation revealed alignment of the carbides in the rolling direction as shown in Figure 18. This observation opens up possibilities to obtain directional properties, if required.

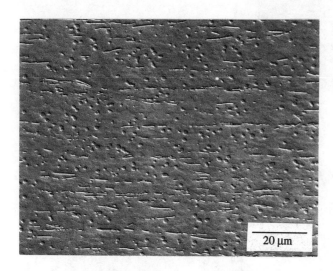

Figure 18. Microstructure of a rolled RD-8A showing alignment of carbides in the rolling direction.

Summary

A Ni-base superalloy has been developed based on the dispersion strengthening concept to take advantage of the attributes of the rapid solidification plasma deposition processing technique. The process and the alloy are suited for fabrication of large thin wall exhaust components. The high-temperature strength properties have been shown to be superior to currently used superalloys such as René 41 and HS-188. Process and alloy modifications have been evaluated to obtain a range of microstructures and properties.

References

1. Robert Mehrabian, ed., Rapid Solidification Processing, Principles and Technologies, III, National Bureau of Standards, Gaithersburg, Maryland, 1982.

2. U. S Patent 5,470,371; Dispersion Strengthened Alloy Containing In-Situ-Formed Dispersoids and Articles and Methods of Manufacture.

3. J. L. Walter et al., Conference on In-Situ Composites-III (Ginn Custom Publishing, Lexington, MA 1979).

4. M. F. Gigliotti and M. F. Henry, Conference on In-Situ Composites-II (Xerox Individualized Publishing, Lexington, MA 1976) 253.

DEVELOPMENT OF A LOW THERMAL EXPANSION, CRACK GROWTH RESISTANT SUPERALLOY

John S. Smith and Karl A. Heck
Inco Alloys International, Inc.
Huntington, WV 25705-1771

Abstract

Low thermal expansion superalloys have been used for a number of years in a variety of applications, including gas turbine engines. The low thermal expansion characteristics of the most widely used class of materials are derived from the ferromagnetic characteristics of Ni, Fe, and Co- based austenitic matrices containing little or no Cr. Over time, a progression of alloy developments ensued, aimed at improving the oxidation resistance and stress accelerated grain boundary oxygen (SAGBO) attack. While notch rupture tests have been used to screen for the SAGBO phenomenon, a more sensitive measure of this characteristic is the sustained load crack growth test performed in air.

This paper describes some final iterations in the development of a new class of low expansion superalloys utilizing high Al content and γ, γ', and β phases in the microstructure. Such alloys provide good general oxidation resistance, and rupture strength and ductility, and varying degrees of crack growth resistance. A number of designed factorial experiments were carried out to optimize 538°C crack growth resistance, yet maintain a balance of other important engineering properties. These experiments included examinations of Ni, Fe, Co, Cr, Nb, and Ti content combined with heat treatment studies. Al content remained essentially fixed on the basis of prior development work. Tests performed included thermal expansion, tensile, tensile and Charpy impact stability, stress rupture, creep, 538°C static crack growth, and microstructural analysis.

These studies showed, that for a given heat treatment cycle, a small amount of Cr combined with increased Co content in place of Ni provides a decrease in crack growth rate. Furthermore, the small Cr addition improves salt spray resistance, yet the addition is small enough as not to significantly affect thermal expansion performance. The crack growth rate was also reduced with increased Co content replacing Ni. The final alloy composition was designated INCONEL® alloy 783.

Crack growth rates were affected by heat treatment. Microstructural examinations showed heat treatment affected amounts of globular β phase present after hot working and annealing, and amounts of the phase re-precipitated within grain boundaries or intragranularly. Slower propagation rates correlated with increased volume percent of β phase with lower temperature anneals, or increased amounts of β phase precipitated in grain boundaries after high temperature anneals and "β-aging" at intermediate temperatures. A high temperature anneal was selected for compatibility with high temperature braze cycles without significantly coarsening grain structure. An appropriate β age was determined for good rupture and crack growth properties. Heat treatment studies further showed that higher yield strengths are achieved with treatments incorporating slow cooling within the γ' precipitation range. A final aging treatment compatible with other superalloys, such as alloy 718 was therefore selected for optimum tensile strength.

Alloy 783 has been successfully produced as VIM-VAR large diameter forging billet, and hot rolled small rounds and flats. Sustained load crack growth data at 538°C obtained from seamless rolled turbine engine rings are presented in this paper. Alloy 783 has been successfully welded and fabricated into gas turbine engine components that are under evaluation by gas turbine manufacturers.

® INCONEL and INCOLOY are trademarks of the INCO family of companies.

Introduction

For over two decades research has been directed at developing controlled low thermal expansion superalloys.[1,2] These efforts were initially successful in the commercial development of INCOLOY® alloys 903, 907, 908, and 909. This class of superalloys have found significant commercial usage in industries as diverse as gas turbine engine static parts and superconducting magnets for fusion reactors. However, it has not been possible to fully exploit the advantages of high strength, low expansion materials in elevated temperature designs due to the poor general surface oxidation resistance and their susceptibility to fast crack growth under sustained loading in air at intermediate temperatures.

Several attempts have been made in recent years to address the environmentally related property weaknesses. One alloy development effort ventured into a new class of γ-γ'-β superalloys.[3] This paper discusses the development of crack growth resistance under sustained load at intermediate temperatures in the Co-Ni-Fe-Al-Nb, γ-γ'-β alloy system. The studies described here are confined to the effects of varying Co, Ni, and Cr contents, and heat treatments on constant load crack growth at 538°C, with limited discussion of other relevant properties, that led to the development of INCONEL alloy 783. Other mechanical and physical properties are presented elsewhere.[4]

Procedure and Experimental Techniques

Development Procedure. The alloy development goal was to simultaneously optimize low thermal expansion, tensile strength and ductility at elevated temperatures, creep strength, stress rupture life and notch ductility, general oxidation and corrosion resistance, stability after long time exposure at intermediate service temperatures. Manufacturing and fabricating simplicity, and compatibility of heat treatments and joining parameters with other commonly used gas turbine superalloys were other factors that were considered advantageous and were evaluated throughout the development project.

In addition to the above superalloy characteristic goals, the primary goal was to achieve intermediate temperature, sustained load, crack growth resistance. Specifically, it was believed that a γ-γ'-β, controlled expansion superalloy should have da/dt at 538°C approaching that of thermomechanically processed, conventionally heat treated, INCONEL alloy 718. Past experience had shown that da/dt at 538°C was a critical alloy property to optimize if any new controlled expansion superalloy was to gain significant commercial usage. For static engine components, and other applications as well, it appeared feasible to achieve a commercial balance of physical, mechanical, thermal, manufacturing/fabricating, and crack growth properties within the γ-γ'-β alloy system.

Development of INCONEL alloy 783 was by necessity a concurrent development project. Compositions within the γ-γ'-β superalloy system were systematically explored via a series of interlocking factorial designs, and were screened for certain properties (such as da/dt). At appropriate points in the development process, certain compositional variations were also subjected to varying heat treatments using factorial designs to determine the compositional-heat treatment interactive and synergistic effects on selected properties. Response surface analyses were utilized to examine "the lay of the land", that is, the effect of two or more factors in combination on a given property. The significantly non-linear and interactive effects of Ni and Co on stress rupture life have been shown before, and demonstrate the need for this approach to alloy development.

Since optimization factors included manufacturing and fabricating simplicity, full scale commercial-sized melts were produced using γ-γ'-β compositions which were known to be non-optimal in properties. These full scale melts were subjected to various manufacturing processes and evaluated for manufacturing feasibility. Some portions of this approach have been described.[3]

This paper contains some examples of studies aimed at determining the effects of Co/Ni and Cr content, and the interaction of heat treatment on constant load crack growth, with discussion of other properties as relevant.

Compositions and Processing. The compositions evaluated in this paper are presented in Table I. Laboratory heats, designated by an HV prefix, were vacuum induction melted, 100 mm diameter, 22 kg ingots. These were homogenized, cut in half, and hot rolled on a 16 inch Birdsboro mill to 0.750 in diameter round bars and to 0.500 inch by 2 inch flats. Mechanical property test specimens were machined from the round bars and compact tension test specimens were machined from the flats. A few flats and round bars were subjected to mechanical property tests and microstructural evaluations to verify that structure and properties were similar within a given ingot.

Full scale melts, designated with a Y prefix, were vacuum induction melted and vacuum arc remelted. The process history for Y9342Y is described elsewhere.[3] Melt number Y9411Y was one of three vacuum arc remelted 457 mm diameter by 3810 mm long ingots. This ingot was homogenized and hot forged and rolled to 203, 254 and 305 mm diameter billets. Seamless rolled rings 50.8 mm thick by 101.6 mm high by 610 mm outside diameter were produced from a 203 mm diameter billet. These rings were subjected to property evaluations. Heat number Y9342Y is representative of non-commercially available Cr-free γ-γ'-β superalloys. Heat number Y9411Y represents the commercial, Cr-alloyed γ-γ'-β superalloy, INCONEL alloy 783.

Testing. Basic room and elevated temperature tensile, combination notch-smooth stress rupture, and thermal expansion testing were conducted in accordance with ASTM Standard Test Methods E8, E21, E139 and E228, respectively.

Static or constant load crack growth testing was conducted using standard compact tension specimens machined in conformance to ASTM E 647-91. Specifically, compact tension specimens were 7.6 mm thick by 25.4 mm wide. Overall outer dimensions were 30.5 mm by 31.8 mm. Specimens were fatigue pre-cracked in accordance with E647 to provide a starting nominal pre-crack plus notch length of 7.6 mm. Specimens satisfied the validity requirements of E 647-91 section 7.2.1. Static load crack growth testing was conducted at 538°C under induction heating in air at an initial stress intensity of nominally 27 MPa√m.

Crack length measurements were recorded using traveling optical microscopes, electrical potential, and compliance techniques. The potential and compliance techniques were calibrated using optical microscopy measurements. Crack growth testing was predominantly conducted at Martest, Inc., where both optical and compliance techniques were used. Some testing was conducted at Inco Alloys International, Inc., using both electrical potential and optical techniques. Crack growth results were found to be reproducible between the two laboratories and across the three testing techniques.

Transient behavior at the initiation of the crack propagation test was typically observed which was comparable to that often observed in superalloy crack growth testing. This portion of the crack growth raw data was ignored for this analysis.

Table I Chemical compositions, weight %.

HEAT	C	Fe	S	Si	Ni	Cr	Al	Ti	Nb	Co	B
HV7340	0.007	Bal	0.0025	0.10	29.9	<0.1	5.4	0.14	3.1	34.1	0.006
HV7341	0.006	Bal	0.0024	0.09	30.0	2.0	5.4	0.14	3.1	32.0	0.006
HV7342	0.007	Bal	0.0024	0.08	30.0	3.0	5.4	0.14	3.1	31.0	0.006
HV7346	0.017	Bal	0.0024	0.11	32.9	<0.1	5.4	0.13	3.1	31.0	0.007
HV7347	0.009	Bal	0.0022	0.10	32.9	2.0	5.5	0.14	3.1	28.9	0.006
HV7348	0.017	Bal	0.0022	0.08	33.0	3.0	5.4	0.14	3.1	27.9	0.006
HV7361	0.004	Bal	0.0025	0.12	27.4	2.9	5.4	0.23	3.0	34.9	0.008
HV7365	0.008	Bal	0.0025	0.10	27.0	4.0	5.4	0.10	3.1	33.1	0.008
HV7369	0.015	Bal	0.0023	0.11	30.0	4.0	5.3	0.10	3.1	30.1	0.009
HV7373	0.007	Bal	0.0020	0.10	33.0	4.0	5.4	0.09	3.1	27.0	0.008
HV7498	0.009	Bal	0.0027	0.09	24.1	3.0	5.4	0.09	2.9	38.7	0.007
HV7500	0.007	Bal	0.0037	0.10	26.9	3.0	5.4	0.10	3.0	35.8	0.007
HV7502	0.008	Bal	0.0042	0.10	29.9	3.0	5.4	0.10	3.0	32.7	0.007
HV7504	0.013	Bal	0.0029	0.10	33.1	3.1	5.3	0.09	2.9	29.6	0.008
Y9342Y	0.005	Bal	0.0014	0.02	32.9	<0.1	5.3	0.58	3.0	31.0	0.006
Y9411Y	0.004	Bal	0.001	0.03	28.6	3.0	5.4	0.01	3.0	33.9	0.004

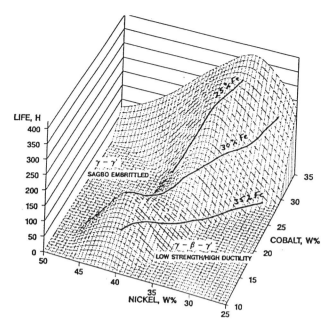

Figure 1 Effect of Ni and Co on the stress rupture life at 649°C and 510 MPa of Cr-free γ-γ'-β alloys.[3]

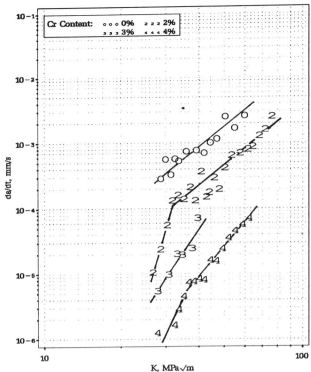

Figure 3 Effect of varying Cr content on da/dt at 538°C. Nominally 30% Ni, 27.5% Fe, 5.4% Al, 0.1% Ti, 3% Nb, balance Co. Cr added at expense of Co.

Results

Crack Growth in Cr-Free γ-γ'-β Superalloys. Depending on the selected Ni and Co content, γ-γ'-β superalloys exhibited good 649°C stress rupture life, as shown in Figure 1. Notch rupture lives in excess of the targeted goal were achieved. The full large scale melt Y9342Y also showed good 538°C notch (K_t= 2) rupture properties with lives exceeding 363 h under 827 MPa. These notch rupture properties were significant improvements over existing controlled thermal expansion superalloys and were thought to indicate probable resistance to sustained load crack growth.

However, sustained load crack growth testing at 538°C revealed a different situation. As shown in Figure 2, the range of da/dt at 538°C for Cr-free γ-γ'-β alloys was actually worse than INCOLOY alloy 909, despite these same alloys having superior notched stress rupture lives.

The crack growth rate was somewhat sensitive to heat treatment, but no significant improvements were found possible. This data illustrates the potential fallacy of using notched stress rupture data as indicators of sustained load crack growth resistance.

Effect of Cr on da/dt in γ-γ'-β Alloys. The effect of varying Cr content on da/dt at 538°C is shown in Figure 3. Cr was added at the expense of Co. Ni, Fe, Al, Ti and Nb contents were held constant at nominally 30%, 27.5%, 5.4%, 0.1% and 3%, respectively. Compact tension specimens were annealed at 1010°C for one hour and age hardened at 788°C for 16 hours, furnace cooled 55°C/h to 621°C, held for 8 h, and air cooled. Increasing Cr content from nil to 4% improved da/dt resistance by over 2 orders of magnitude.

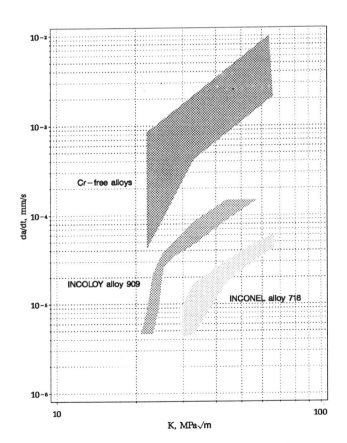

Figure 2 Comparison of da/dt at 538°C of Cr-free γ-γ'-β alloys, INCOLOY alloy 909, and INCONEL alloy 718.

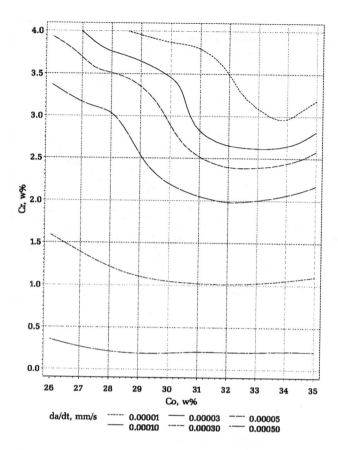

Figure 4 Effect of Cr and Co on da/dt at 538°C and 33 MPa√m.

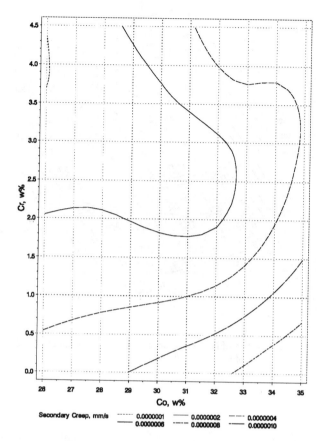

Figure 5 Effect of Cr and Co on secondary creep rate at 649°C under 380 MPa stress.

Response Surface Analysis: Cr-Co-da/dt. The preceding experiment led to expanded composition factorial studies to define the Cr-Co-da/dt response surface for alloy specimens annealed at 1010°C for one h, air cooled, and age hardened at 788°C for 16 h, furnace cooled 55°C/h to 621°C, held for 8 h and air cooled. The resulting da/dt isocontours versus Cr and Co, shown in Figure 4, reveal a Cr-Co interaction effect on da/dt in alloys containing greater than 2% Cr content. Da/dt decreases with Cr at all Co levels, but especially at Co contents greater than roughly 30%. The desired da/dt rates fell within a pocket of greater than 2.5% Cr and greater than about 30% Co content.

Response Surface Analysis: Cr-Co-ε. The effect of Cr and Co on the secondary creep rate of specimens tested at 649°C under 380 MPa is shown in Figure 5, as creep rate isocontours. Heat treatment was the same as noted above. Creep rate varied as a bowl relationship with Cr and Co, with minima occurring between 26 and 30% Co and 2 to 4% Cr. While increasing Co content increased the creep rate for any given Cr content, increasing Cr content up to 3.5% decreased the creep rate. It was therefore possible to offset losses in creep strength due to increasing Co content (for added da/dt resistance) by also increasing Cr content.

Effect of Co/Ni content and heat treatment on da/dt. Table II summarizes the effect of varying the Co/Ni content for alloys containing 3% Cr simultaneously with heat treatment on sustained load da/dt at 538°C and 33 MPa√m. Fe, Al, Ti and Nb were held constant at nominally 26%, 5.4%, 0.1% and 3%, respectively.

Annealing heat treatments were conducted at 982 and 1038°C for one hour and air cooled. Aging heat treatments were conducted at 732, 788, and 843°C for 16 h then furnace cooled 55°C/h to 621°C for 8 h and air cooled. Additionally, specimens were given a high temperature solution anneal at 1121°C for one hour and air cooled, followed by an intermediate heat treatment at 843°C for 2 h, air cooled, and age hardened at 732°C for 8 h furnace cooled to 621°C held for 8 h, and air cooled.

Sustained load crack growth was a strong function of the Co/Ni content and the heat treatment, as well as interactions between the composition and heat treatment. Regardless of the heat treatment, increasing the Co/Ni ratio reduced crack growth rates, though the amount of reduction in da/dt depended on both the anneal and aging temperatures. Specimens of 24% Ni content consistently failed to sustain crack growth, with cracking repeatedly stalling. When da/dt was measurable, rates were less than 0.9×10^{-6} mm/s. In most cases, the creep crack threshold was above 33 MPa√m. On the other hand, specimens of 33% Ni content consistently had the highest da/dt for any given combination of annealing and aging temperatures.

Both annealing and aging temperatures had significant effects on da/dt. Annealing at 982°C resulted in lowest crack growth rates for all aging temperatures. Crack growth rates increased significantly in specimens annealed at 1038°C, by nearly an order of magnitude in specimens containing 30% or more Ni content. Increasing the aging heat treatment temperature consistently decreased da/dt for all annealing temperatures and at all Ni contents, except for those specimens containing 24% Ni.

Table II Effect of Co/Ni content and heat treatment on da/dt (10^{-6} mm/s) at 538°C and 33 MPa$\sqrt{m}$.

Anneal:	982°C			1038°C			1121°C
Aging Temperatures:	732°C, FC 621°C, AC	788°C, FC 621°C, AC	843°C, FC 621°C, AC	732°C, FC 621°C, AC	788°C, FC 621°C, AC	843°C, FC 621°C, AC	843°C, AC 732°C, FC 621°C, AC
24% Ni (HV7498)	crack growth stalled	<0.2	crack growth stalled	crack growth stalled	crack growth stalled	not tested	crack growth stalled
27% Ni (HV7501)	0.7	precrack fractured	<1	5	1.8	0.5	2
30% Ni (HV7503)	2	1.3	<0.1	15	9	5	11
33% Ni (HV7505)	5.8	crack growth stalled	1.5	30	35	10	32

Notes:
1) Annealed at temperature shown for one hour, air cooled.
2) Two-step ages: Initial temperature held for 16 h, furnace cooled 55°C/h, to 621°C held for 8 h, air cooled.
3) Three-step age: 843°C held for 2 h, air cooled. **732°C** held for 8 h, furnace cooled 55°C/h to 621°C held for 8 h, air cooled.

The da/dt of the solutionized, intermediate and age hardened specimens also showed strong sensitivity to Ni content. This study demonstrated that da/dt performance is controlled by the judicious use of combined heat treatment temperatures and times. Exposing specimens to the 843°C heat treatment permitted the use of high temperature, solutionizing and grain coarsening anneal.

INCONEL alloy 718 da/dt. There is a considerable amount of da/dt data at 649°C published for alloy 718, but little da/dt data at 538°C. This was data determined by Sadananda and Shahinian from plate using non-optimal composition and thermomechanical processing.[5] To generate more relevant da/dt data, a fully heat treated turbine engine ring of 38.1 mm thick by 50.8 mm high was sacrificed for specimens.

The da/dt in the axial and radial orientations were determined at 538°C and are shown in Figure 6. Also plotted on this figure are the da/dt results obtained from fully heat treated plate. Compact tension specimens from both the ring and from the plate were heat treated at 954°C for 1 h, air cooled, 718°C for 8 h furnace cooled to 621°C, held for 8 h, and air cooled.

The da/dt of the engine ring was dependent on orientation to some degree, with the axial da/dt being lower than the radial da/dt. The ring had significantly better da/dt versus the plate. This was the result of the controlled thermomechanical ring-rolling process and composition intended to result in a controlled microstructure and good crack growth resistance. The da/dt of this ring matched the da/dt of a 50.8 mm diameter rod hot rolled and fully heat treated as produced at Inco Alloys International. This range of da/dt at 538°C was adopted as the goal.

Alloy 783 Engine Ring, Effect of Anneal Temperature. Based on the above examples and several other studies, a composition was selected for evaluating the large scale manufacturing feasibility. The aimpoint composition was 28.5% Ni, 34% Co, 5.4% Al, 3% Cr, 3% Nb, 0.006% B, with Ti and Si each less than 0.2%, and Fe being the balance. The actual composition of heat Y9411Y is given in Table I.

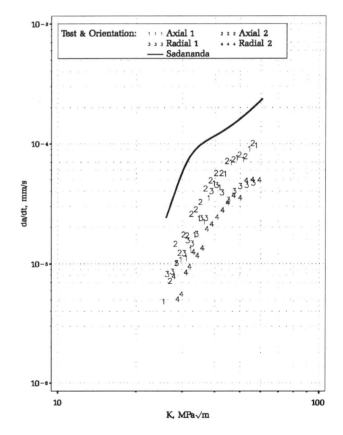

Figure 6 INCONEL alloy 718 da/dt at 538°C. Data determined from engine ring and from plate (Sadananda[5]).

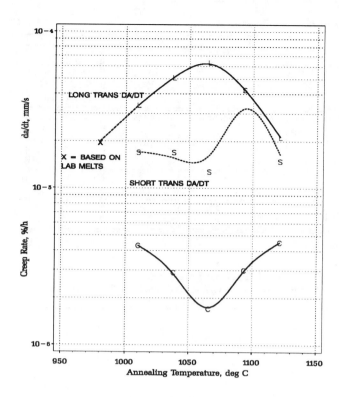

Figure 7 Effect of annealing temperature on da/dt at 538°C and 33 MPa√m, and secondary creep rate at 649°C and 380 MPa in annealed and two-step aged alloy 783 turbine ring.

For purposes of da/dt and other property evaluations, a 203 mm diameter billet was forged for seamless hot roll ring production. The billet was upset, pierced, and seamless ring rolled to 50.8 mm thick by 101.6 mm wide by 610 mm outer diameter, using a standard alloy 718-type thermomechanical ring-rolling practice. Both laboratory and full scale manufacturing experience had indicated that alloy 718 thermomechanical processes were also suited for Y9411Y. Sections from the ring were cut and subjected to da/dt testing at 538°C and creep testing at 649°C as a function of heat treatment. In most cases, both axial and radial orientations were subjected to crack growth testing to determine orientational effects.

The effect of annealing temperature with a predominantly γ' age hardening heat treatment (760°C for 12 h, furnace cooled 55°C/h to 621°C, held for 8 h, air cool) on da/dt at 538°C and creep at 649°C is shown in Figure 7. The ring da/dt was not tested with an 982°C anneal, but results from experimental heats are plotted. The long transverse da/dt varied log-parabolically with annealing temperature, reaching a maxima with a 1066°C anneal. The secondary creep rate varied inverse log-parabolically with annealing temperature, reaching a minima at 1066°C. The short transverse da/dt was essentially unchanged by varying annealing temperatures through 1066°C, increasing after the 1093°C anneal, then decreasing again after the 1121°C anneal.

Alloy 783 Engine Ring, Effect of β-Aging Heat Treatments. The effect of β-aging heat treatment temperature on da/dt at 538°C and 33 MPa√m as a function of annealing temperature is shown in Table III. Specimens were annealed at 1107, 1121, or 1135°C for one hour and air cooled. One specimen was directly age heat treated at 718°C for 8 h, furnace cooled 55°C/h to 621°C, held for 8 h, air cooled, after annealing at 1121°C. Other specimens were subjected to β precipitation (or "β-aging") heat treatment at either 829, 843, or 857°C for 2 h, air cooled, then heat treated at 718°C for 8 h, furnace cooled 55°C/h to 621°C, held for 8 h, air cooled.

Highest da/dt rates were obtained from specimens with either no intermediate β-aging heat treatment or a low temperature (829°C) β-aging heat treatment. Substantial reduction in da/dt rates occurred when intermediate β-aging heat treatments of 843°C and 857°C were used. The da/dt obtained after using the latter intermediate heat treatments were similar to those of INCONEL alloy 718. The da/dt at 538°C of axial and radial orientations after annealing at 1121°C for one h, air cooling, β-aging at 843°C for two hours, air cooling, and final γ' aging at 718°C for 8 h furnace cooling 55°C/h to 621°C, held for 8 h, then air cooling, is shown in Figure 8.

Table III Effect of annealing and β aging temperatures on da/dt (10^{-6} mm/s) at 538°C and 33 MPa√m in INCONEL alloy 783.

β Age: Anneal:	None	829°C	843°C	857°C
1107°C	-	30	-	11
1121°C	50	-	8, 15	-
1135°C	-	55	-	8
Alloy 718 da/dt: 11 to 120 Alloy 909 da/dt: 500 to 900				

Heat Treatment: Alloy 783 specimens heat treated as follows:
1) Annealed for one hour, air cooled.
2) β heat treated for 2 hours, air cooled.
3) Age hardened at 718°C for 8 h, furnace cooled 55°C/h to 621°C for 8 h, air cooled.

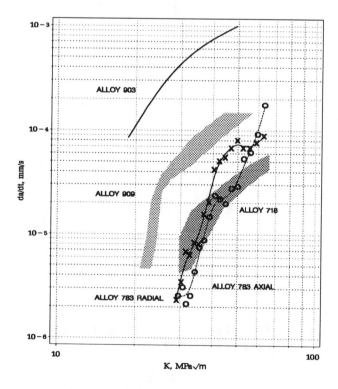

Figure 8 Comparison of fully heat treated INCONEL alloys 783, 718, and INCOLOY alloy 909 da/dt at 538°C and 33 MPa√m.

Discussion

General Metallurgy of INCONEL alloy 783. The general time-temperature-transformation diagram for INCONEL alloy 783 is shown in Figure 9. This diagram was determined from the processed seamless hot rolled engine ring produced from heat number Y9411Y. Specimens from the ring were solution heat treated at 1149°C for one hour, water quenched, exposed at the temperatures and times shown, and water quenched, and subjected to metallographic examination.

Some general fine grain boundary precipitates were observed at nearly all heat treatments. Gamma prime was observable after one hour at 788°C, and at temperatures as low as 700°C for exposure times as long as 100 h. The solvus is about 830°C. β precipitates (ordered BCC NiCoFeAl) were observed at temperatures above 750°C, with solutioning beginning above 1100°C.

Two types of β formation occur in these alloys. Primary β, which forms upon solidification or during prolonged exposure to hot working temperatures, occurs as coarse globular particles, more or less randomly scattered throughout the matrix. The primary β globules are soft at high temperatures and contribute to hot workability. These globules may not completely solution during high temperature solution anneals. Secondary β precipitates from solution during exposure to temperatures above 750°C. These precipitates tend to form intergranularly as fine discontinuous particulates and intragranularly as fine needles or platelets. About 10 to 15 volume % secondary β forms after solution annealing at 1120°C and exposure to 843°C for 2 h.[6] γ-γ'-β alloys attain their maximum resistance to sustained load cracking with the formation of intergranular secondary β precipitation.

Effect of Composition. The bulk alloy composition plays a major role in this precipitation. From the previous examples, it was seen that either reducing Ni content while increasing Co, or adding Cr in amounts exceeding 2% was necessary to reduce crack growth rates under sustained load. Although crack propagation was arrested in the alloy containing ~ 24% Ni, these compositions had little creep resistance or elevated temperature strength and long time exposure stability was suspect.

However, only 3% Cr was needed to significantly reduce da/dt levels to acceptable levels with Ni contents ranging from 27% to 30%. The relatively large effect of this small amount of Cr suggests that the effect is not merely one of providing oxidation resistance. For example, INCOLOY alloy 908 contains 4% Cr, but has comparatively poor cracking resistance.[7] Thus, it appears the Cr addition plays a synergistic role with the β phases, the Al and Ni content, and the environmental-intergranular microchemical reactions. Fortunately, this small Cr addition does not significantly increase thermal expansivity, allowing alloy 783 to have thermal expansion 20% below that of alloy 718, as shown in Figure 10.

Further improvement was achieved by increasing Co at the expense of Ni when Cr content is above 2%. This likely affects at least two factors. Reducing Ni content appears to help stabilize intergranular β phases. Secondly, increasing Co content decreased creep resistance at 649°C, for a given level of Cr. Increasing Co reduced strength compared to Cr-free γ-γ'-β alloys (yield strength at 538°C decreased from about 860 MPa to 740 MPa). Decreased strength and increased da/dt resistance (and vice versa) occurring simultaneously in superalloys is often observed. This observed fact has led some to conclude that sustained load crack growth in high strength superalloys is controlled by the competition between diffusional and deformational phenomenon.[5]

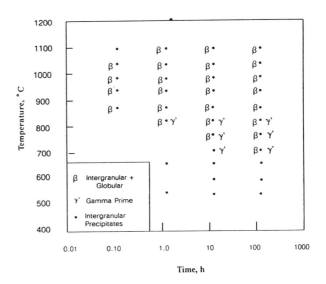

Figure 9 TTT diagram for INCONEL alloy 783.

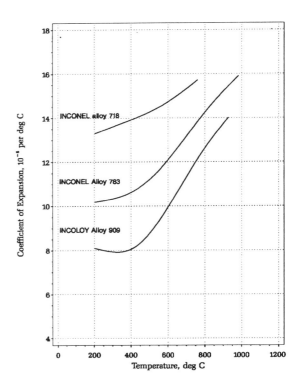

Figure 10 Mean coefficients of thermal expansion for alloys 783, 718, and 909. All alloys annealed and aged.

Effect of Heat Treatment. Sustained load crack growth performance of INCONEL alloy 783 is also dependent on the heat treatment. This dependency is illustrated by the annealing temperature-aging temperature-da/dt mean response surface shown in Figure 11. This response surface was constructed from the examples shown in this paper combined with data from other developmental studies, and is valid for compositions nominally containing 28% Ni, 34% Co, 3% Cr, 5.4% Al, 3% Nb, and balance Fe. The annealing time at temperature is one hour, and air cooled. The aging heat treatments are based on specimens aged at the temperatures shown for 8 to 16 h, furnace cooled 55°C/h to 621°C, held for 8 h, and air cooled. The da/dt shown is at 538°C and 33 MPa√m and represents mean values for the heat treatments.

The da/dt is maximized when annealed between about 1040 to 1080°C regardless of the aging temperature. However, the maximum da/dt is highest at aging temperatures around 750°C or lower, dropping significantly above 800°C. The minimum da/dt occurs after annealing below 1020°C or above 1100°C and aged at temperatures above 800°C.

A wide variety of heat treatment combinations may be used to achieve good da/dt resistance in alloy 783. Solution annealing at 1120°C followed by an intermediate β age at 845°C for 2 to 4 h, followed by a conventional γ' precipitation two-step heat treatment of 718°C for 8 h furnace cooling to 621°C, held for 8 h, and air cooling, has been found to offer a good combination of properties. This heat treatment provides for coarser grain size (ASTM #5 to #3) for creep resistance, reduces thermomechanically induced anisotropy, and provides for more uniform and controlled fine β precipitation throughout the grain boundaries and the microstructural matrix as a whole.

Microstructure. Low da/dt rates in low temperature (<1020°C) annealed alloy 783 is associated with fine grain and an abundance of globular and intergranular β in the microstructure. The microstructure for a specimen from a turbine ring heat treated at 1010°C for one hour, air cooled, 760°C for 12 h furnace cooled 55°C/h to 621°C, held for 8 h, air cooled, is illustrated in Figure 12. The globular β precipitates are abundant and uniformly distributed, and effectively pin grain boundaries. Intergranular β precipitation is also present. Overaged γ' is apparently darkening the grain interiors. Since the aging heat treatment temperatures for this specimen were below 800°C, the observed β is predominantly that which precipitated during the turbine ring thermomechanical processing.

The microstructure of an alloy 783 specimen given a high temperature anneal (1121°C for one hour, air cooled) and a γ' aging heat treatment (718°C for 8 h furnace cooled to 621°C, held for 8 h, air cooled), is shown in Figure 13. The microstructure of an alloy 783 specimen given the same heat treatment except with an intermediate β aging heat treatment, is shown in Figure 14. The microstructure of an alloy 718 compact tension specimen obtained from the engine ring is shown in Figure 15.

The annealed and γ' aged only alloy 783 specimen has essentially "clean" grain boundaries with some undissolved primary β particles scattered throughout the microstructure. In contrast, the annealed, β- and γ'-aged specimen contains extensive intergranular precipitates, as well as uniformly distributed intragranular lenticular β. The optical microstructure of the alloy 718 specimen is similar in appearance except the intergranular phase in this alloy is primarily δ (Ni_3Nb,Ti).

Fractographs. The fractographs of the latter three fractured compact tension specimens are also revealing. Creep resistant superalloys fracture intergranularly at 538°C when subjected to sustained loading in air. Likewise, fracture in all three specimens, alloy 783 in both heat treated conditions and alloy 718, followed an intergranular crack path, accompanied with frequent secondary intergranular branch cracks.

The fracture surface of the annealed and γ'-aged alloy 783 specimen, in Figure 16, has very clean grain facets devoid of any significant ductility. Nevertheless, this specimen had da/dt rates superior to alloy 909 (see Table III). The annealed, β- and γ'-aged alloy 783 specimen had significantly improved da/dt, yet the crack path remained the same, see Figure 17. However, the grain facets are rougher in appearance. The grain facet features are probably a combination of

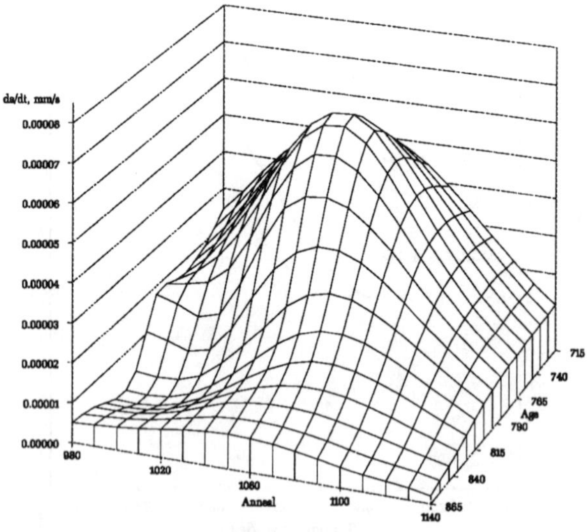

Figure 11 Effect of annealing and aging temperatures on the da/dt at 538°C and 33 MPa√m of INCONEL alloy 783.

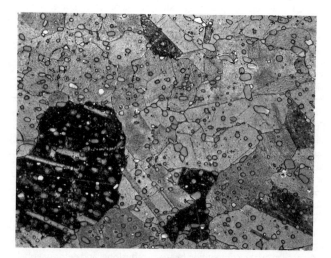

Figure 12 Alloy 783 heat treated 1010°C for 1 h, air cooled, 760°C for 12 h, cooled 55°/h to 621°C for 8 h, air cooled. Specimen from turbine ring, circumferential view, 500x.

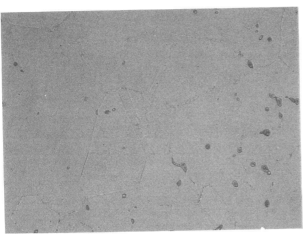

Figure 13 Alloy 783 annealed at 1121°C for 1 h, air cooled, aged 718°C for 8 h cooled 55°/h to 621°C for 8 h, air cooled. Specimen from turbine ring, circumferential view, 500x.

Figure 16 Alloy 783 da/dt at 538°C fracture surface. Heat treated at 1121°C for 1 h, air cooled, aged 718°C for 8 h, cooled 55°/h to 621°C for 8 h, air cooled. 500x.

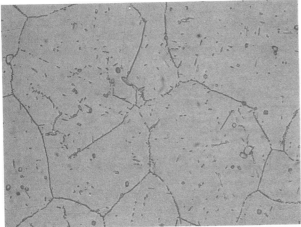

Figure 14 Alloy 783 heat treated 1121°C for 1 h, air cooled, 843°C for 2 h, air cooled, 718°C for 8 h cooled 55°C/h to 621°C for 8 h, air cooled. Specimen from turbine ring, circumferential view, 500x.

Figure 17 Alloy 783 da/dt at 538°C fracture surface. Heat treated 1121°C for 1 h, air cooled, 843°C for 2 h, air cooled, 718°C for 8 h cooled 55°C/h to 621°C for 8 h, air cooled. 500x.

Figure 15 Alloy 718 heat treated at 954°C for 1 h, air cooled, 718°C for 8 h cooled 55°/h to 621°C for 8 h, air cooled. Specimen from turbine ring, circumferential view, 500x.

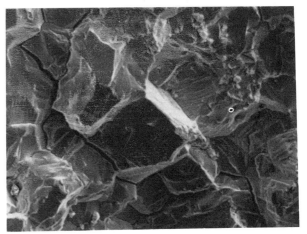

Figure 18 Alloy 718 da/dt at 538°C fracture surface. Heat treated at 954°C for 1 h, air cooled, 718°C for 8 h cooled 55°/h to 621°C for 8 h, air cooled. 500x.

the oxidation of the grain boundary phases and localized deformation. The alloy 718 specimen fracture surface, shown in Figure 18, is also rougher in comparison with the annealed and γ'-aged alloy 783 fracture surface, with perhaps some slight grain facet deformation observable.

Concluding Comments. It has been demonstrated that sustained load crack growth in air at temperatures between 450°C and 700°C is strongly driven by oxygen from the environment in most, if not all, superalloys. Although inert environmental da/dt performance was not been shown here, the same is true for INCONEL alloy 783. It is clear from these microstructures and fractographs that alloy 783 da/dt performance at 538°C is largely a result of grain boundary phase formation (or lack of it) and the interaction with the environment. Resistance to sustained load cracking in air is achieved by grain boundary phase engineering in these superalloys.

It is curious to note that the studies leading to the development of alloy 783 provide results congruous with the findings of Andrieu, et al, on alloy 718.[8] That work showed that Ni-rich oxides were "a prerequisite for a nickel-base alloy to be sensitive to the effect of environment." The studies on γ-γ'-β alloys showed that the reduction of bulk Ni content resulted in significant improvements in da/dt resistance, and thus grain boundary environmental resistance. While this certainly has to do with the stabilization of grain boundary β phases, alteration of γ' morphology, and reduction in creep resistance, it also may imply effects on environmental-grain boundary micro-oxidation interactions.

Secondly, it was also concluded that "high intergranular stresses resulting from strain incompatibilities due to either slip character or microstructural inhomogeneities, or both" were also a requirement for a strong environmental effect. The presence of intergranular precipitates as different in composition and structure as β (ordered BCC) in alloy 783 and δ (orthorhombic) in alloy 718 provide increased environmental da/dt resistance over that of precipitate-free grain boundaries in these alloys. It has been demonstrated in INCONEL alloy X-750 that even grain boundary $M_{23}C_6$ phases can aid in the reduction of environmental sensitivity to oxygen as measured by sustained load crack growth.[9] With intergranular precipitates so diverse in composition and morphologies, one is led to conjecture that some of the beneficial effect on da/dt at 538°C is due to altered (ie, reduced) intergranular strain incompatibilities. These observations offer future guidance (eg, the need for increased attention to practical grain boundary micro-engineering in elevated temperature superalloy development) and hope for new superalloys having improved environmental crack growth resistance.

Summary

INCONEL alloy 783 is a controlled low thermal expansion, oxidation resistant γ-γ'-β superalloy having sustained load crack growth resistance in air environments. Crack growth resistance was achieved by composition (optimized Co/Ni and Cr content) and heat treatment control, and is essentially equivalent to that of INCONEL alloy 718 in stress intensity ranges of 20 to 60 MPa√m. Crack growth resistance is attained by the controlled precipitation of fine β particulates in grain boundaries, either by thermomechanical processing with fine grain annealing (980°C), or preferably by solution annealing (1120°C) and using intermediate β re-precipitation heat treatments (845°C for 2 to 4 h, air cool). Primary strengthening is achieved by γ' aging heat treatment at 718°C for 8 h furnace cooled to 621°C, held for 8 h, and air cooled.

Acknowledgments

SEM fractographs were prepared by Frank J. Veltry. The authors also thank Bob Neugebauer and staff of Martest, Inc., for their efforts with crack growth testing. Also contributing to the development work were D. F. Smith, L. I. Stein, M. A. Moore, and M. A. Holderby.

References

1. D.F.Smith and J.S.Smith, "A History of Controlled, Low Thermal Expansion Superalloys," Physical Metallurgy of Controlled Expansion Invar-Type Alloys, ed. K.C.Russell and D.F.Smith, The Metallurgical Society, Warrendale, PA, 1990, pp 253-272.

2. E.A.Wanner, D.A.Antonio, D.F.Smith, J.S.Smith, "The Current Status of Controlled Thermal Expansion Superalloys," Journal of Metals, 43, 3, 1991, pp 38-43.

3. K.A.Heck, D.F.Smith, M.A.Holderby, J.S.Smith, "Three-Phase Controlled Expansion Superalloys with Oxidation Resistance," Superalloys 1992, ed. S.D.Antolovich, et.al., The Metallurgical Society, Warrendale, PA, 1992, pp 217-226.

4. K.A.Heck, J.S.Smith, R.Smith, "INCONEL alloy 783: An Oxidation Resistant Low Expansion Superalloy for Gas Turbine Applications," Turbo Expo '96, June, 1996, Birmingham, UK, to be published.

5. K. Sadananda and P. Shahinian, "Creep Crack Growth in Alloy 718," Met. Trans. A, Vol. 8A, pp. 439-449, March, 1977.

6. James Clawson Jr. and Jed Lyons, "The Effects of Heat Treatment on the Microstructure of INCONEL alloy 783," Scripta Materialia, in publication, 1996.

7. J. S. Smith, J. H. Weber, and H. W. Sizek, "Control of Stress-Accelerated Oxygen-Assisted Cracking of INCOLOY alloy 908 Sheath For Nb_3Sn Cable-In-Conduit," Advances in Cryogenic Engineering, in publication, 1996.

8. E. Andrieu, G. Hochstetter, R. Molins, and A. Pineau, "Oxidation and Intergranular Cracking Behaviour of Two High Strength Ni-Base Superalloys," Corrosion-Deformation Interactions, Fontainbleu, France, 5-7 Oct. 1992, Les Éditions de Physique, ed. T. Magnin and J. M. Gras, Avenue du Hoggar, Zone Industrielle de Courtaboeuf, B. P. 112, F-91944 Les Ulis Cedex A, France, pp. 477-491, 1993.

9. S. Floreen, "Effects of Environment on Intermediate Temperature Crack Growth in Superalloys," Micro and Macro Mechanics of Crack Growth, The Metallurgical Society of AIME, pp. 177-184, 1982.

PHASE DIAGRAM CALCULATIONS FOR NI-BASED SUPERALLOYS

N. Saunders

Thermotech Ltd, Surrey Technology Centre, The Surrey Research Park, Guildford GU2 5YG, U.K.
and
IRC in Materials for High Performance Applications, The University of Birmingham
Edgbaston, Birmingham B15 2TT, U.K

Abstract

At high temperatures, and when subjected to mid-range temperatures for long times, superalloys can reach states which approach equilibrium. Knowledge of stable phase structure at fabrication and working temperatures can, therefore, be very important and experimental determination of Ni-based binary and ternary systems has been reasonably extensive. However, when working with such highly alloyed multi-component materials as superalloys information based purely on experimental determination of lower order systems cannot always be directly applied to 'real' alloys. The field of computer aided thermodynamic phase diagram calculations holds substantial promise in this respect as it's possible to make predictions for phase behaviour of multi-component alloys based on models for the binary and ternary phase diagrams.

This paper will present a review of results which can now be obtained in alloys from the following multi-component system

Ni-Al-Co-Cr-Hf-Mo-Nb-Ta-Ti-W-Zr-B-C

Current work involving the extension of the database to include Fe and Re will also be presented. A further advantage of the CALPHAD route is that it is possible to predict properties other than those associated with equilibrium diagrams and the application to non-equilibrium solidification and the prediction of APB energies in γ' will be presented.

Introduction

The computer calculation of phase equilibria in multi-component alloys is becoming increasingly commonplace and it is now possible to make very accurate predictions for phase equilibria in a number of the more commonly used metallic and intermetallic alloys. These range from steels[1] to Ti-aluminides[2]. This paper will present results which can now be obtained in Ni-based superalloys giving a number of examples of where this methodology has been applied.

The CALPHAD method first requires that sound mathematical models exist for describing the thermodynamic properties of the various phases which can appear in an alloy. The coefficients used by the models are then held in databases which are accessed by software packages such as Thermo-Calc[3] which then perform a series of calculations, usually via Gibbs energy minimisation, to provide the user with detailed information on phase equilibria. These calculations can be augmented with kinetic modelling to provide answers for phase formation under conditions which can deviate substantially from equilibrium[4,5].

Early attempts at modelling of superalloys mainly concentrated on ternary sub-systems[6,7,8]. They provided some guidance in the search for high temperature eutectic reactions but the simple model types used in this early work inherently limited their more general usage. For example γ' was treated as a stoichiometric or line compound whereas it is substantially non-stoichiometric in practice. New models have been developed which now allow the full solubility range and thermodynamic properties of intermetallic compounds such as γ' to be modelled very accurately[9,10].

Some four years ago a development programme between Thermotech Ltd and Rolls-Royce plc was started to develop a database which could be used for CALPHAD and related calculations in Ni-based superalloys. Results[11,12] have demonstrated that the accuracy of calculated phase equilibria lies close to that obtained experimentally for commercial superalloys. Some examples of superalloys for which calculations have been made and validated are shown in Table 1. As can be seen they include all types of superalloys ranging from Nimonic types through to single crystal blade alloys. This paper will provide a brief background to the CALPHAD modelling and demonstrate the accuracy of results which are now obtained. Some applications of the database will then be shown including application to non-equilibrium solidification, σ phase formation and prediction of APB energies in γ'.

Table 1 Some alloys used in the validation of the Ni-based Superalloy database

Inconel 700	Nimonic 263	MAR-M247
Nimonic 115	EPK 55	PWA 1480
René 41	EPK 57	IN738LC
Udimet 500	Udimet 520	SRR 99
Udimet 700	CMSX-2	AF2 1DA
Waspaloy	IN939	AP1
Nimonic 80A	IN 100	APK6
Nimonic 81	Udimet 710	CH88-A
Nimonic 90	MXON	Udimet 720
Nimonic 105	B1900	MC2

Background to the Calculation Method

The roots of the CALPHAD approach lie in the mathematical description of the thermodynamic properties of the phases of interest. If they are stoichiometric compounds the composition is defined and a mathematical formula is then used to describe fundamental properties such as enthalpy and entropy. Where phases exist over a wide range of stoichiometries, which is the usual case for metallic materials, other mathematical models are used which account for the effect of composition changes on free energy. Details of modelling procedures can be found in the review of Ansara[13]. All types of models require input of coefficients which uniquely describe the properties of the various phases and these coefficients are held in databases which are either in the open literature or proprietary.

The main models used in the present work are the substitutional type model[13] and the multiple sublattice model[10]. Both of these models can broadly be represented by the general equation for a phase

$$\Delta G = \Delta G^\circ + \Delta G_{mix}^{ideal} + \Delta G_{mix}^{xs} \quad (1)$$

where ΔG° is the free energy of the phase in its pure form, ΔG_{mix}^{ideal} is the ideal mixing term and ΔG_{mix}^{xs} is the excess free energy of mixing of the components. It is not within the scope of the present paper to describe in detail these models, particularly the multiple-sublattice model, but it is useful to briefly discuss some of their aspects.

The free energy of the substitutional model (ΔG_m) for a many component system can be represented by the equation

$$\Delta G_m = \sum_i x_i \Delta G_i^\circ + RT \sum_i x_i \log_e x_i + \sum_i \sum_{j>i} x_i x_j \sum_v \Omega_v (x_i - x_j)^v \quad (2)$$

where x_i is the mole fraction of component i, ΔG_i° defines the free energy of the phase in the pure component i, T is the temperature and R is the gas constant. Ω_v is an interaction coefficient dependent on the value of v. When v=0, this corresponds to the regular solution model and when v=0 and 1 this corresponds to the sub-regular model. In practice the value for v does not usually rise above 2.

Eq.2 assumes higher order interactions are small in comparison to those which arise from the binary terms but this may not be always the case. Ternary interactions are often considered but there is little evidence of the need for interaction terms of a higher order than this. Various other polynomial expressions for the excess term have been considered, see for example the reviews by Ansara[13] and Hillert[14], however all are based on predicting the properties of the higher-order system from the lower-component systems.

The multi-sublattice model is substantially more complex and considers the phase to made up of multiple interlocking sublattices. There are then interaction terms to be considered (i) between the sublattices and (ii) on the sublattices themselves. For a 2-sublattice model as used to describe the γ' phase in Ni-Al the sublattice occupancy would be shown schematically as below:

$$(Ni,Al)_3(Ni,Al)$$

ΔG° would be written as

$$\Delta G^\circ = y_{Ni}^1 y_{Ni}^2 \Delta G_{Ni:Ni} + y_{Ni}^1 y_{Al}^2 \Delta G_{Ni:Al} + y_{Al}^1 y_{Ni}^2 \Delta G_{Al:Ni} + y_{Al}^1 y_{Al}^2 \Delta G_{Al:Al} \quad (3)$$

where

$$y_i^s = \frac{n_i^s}{\sum_i n_i^s}$$

and

$$\sum_i y_i^s = 1$$

y_i^s is the site fraction of component i, in this case Ni or Al, on sublattice s, n_i^s is the number of moles of constituent i on sublattice s. The ideal entropy of mixing is written as

$$\Delta G_{mix}^{ideal} = RT[3(y_{Ni}^1 \ln y_{Ni}^1 + y_{Al}^1 \ln y_{Al}^1) + (y_{Ni}^2 \ln y_{Ni}^2 + y_{Al}^2 \ln y_{Al}^2)] \quad (4)$$

The ΔG_{mix}^{xs} term considers the interactions between the components on the sublattice and can be quite complex, see for example Saunders[15]. As the level of complexity of the structure becomes more complex, more sublattices are needed to consider it Gibbs energy, for example the σ and μ phase may described using a three sublattice model.

Once the thermodynamics of the various phases are defined phase equilibria can be calculated using software packages such as Thermo-Calc[16] which is the programme used in this work. The main method of such programmes is usually a Gibbs free energy minimisation process and there are now a variety of such software packages which can perform complex multi-component calculations. For more information the recent review by Bale and Eriksson[17] provides a fairly comprehensive coverage of these.

Results

Binary and ternary systems

It is instructive to demonstrate how the current database has been built up by showing some of the lower order binary and ternary systems. This will also enable some aspects of γ/γ' elemental partitioning to be discussed. Fig.1 shows the calculated phase diagram for Ni-Al with data points for phase boundaries included. For references to the original studies the review of Nash et al[18] contains most these.

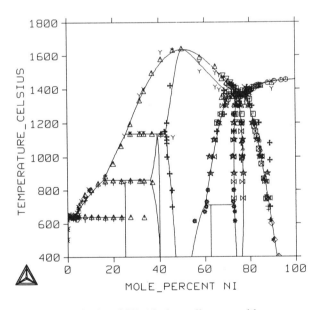

Fig.1. Calculated Ni-Al phase diagram with experimental phase boundaries superimposed

It is noted that the calculated diagram gives a eutectic reaction between γ and γ' rather than a peritectic as proposed in some recent work[19]. This is a consequence of matching all of the available data and it is noted that liquidus points from the Vorhoeven et al[19] are actually in very good agreement with the calculated boundaries. In practice whether the reaction is eutectic or peritectic is not important as the change in free energy to get either reaction is very small and its consequence on calculated equilibria in higher order systems is almost negligible.

The calculations for Ni-Al can then be combined with those of Ni-Ti and Al-Ti as part of the calculation for the Ni-Al-Ti ternary system. A calculated isothermal section at 1000°C is shown in Fig.2 below.

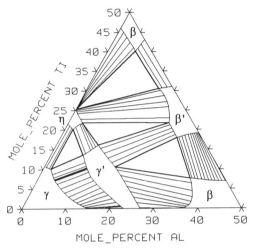

Fig.2. Calculated isothermal section for Ni-Al-Ti at 1000°C.

The diagram is in excellent agreement with that observed in practice. As is known in superalloys, Ti partitions preferentially to γ' but its value for $k^{\gamma/\gamma'}$ is about 0.5 with additions of 3-5at%Ti. This is substantially higher than observed in normal superalloys where the value is closer to 0.1[20].

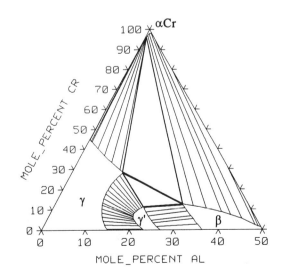

Fig.3. Calculated isothermal section for Ni-Al-Cr at 1000°C.

The behaviour of Cr in γ/γ' equilibria is interesting as the γ' intrusion into Ni-Al-Cr as well as pointing to Ni-Cr exhibits a tendency to go towards Cr as well (Fig.3). Cr partitions towards γ and $k^{\gamma/\gamma'}$ is slightly less than 2 in the range 10-20at%Cr. This is substantially lower than observed in superalloys[20] where the value lies closer to 7.

In both Ni-Al-Cr and Ni-Al-Ti the calculated tie-lines are in excellent agreement with those observed in practice and it is therefore clear that on alloying the respective Cr and Ti partition coefficients are substantially altered. This is something that cannot be predicted from the experimental determination of the ternary sections alone. The ability to predict the correct partitioning behaviour in both the ternary as well as higher order system is one of the strengths of the CALPHAD route and is something that automatically follows from a sound description of the underlying thermodynamics.

Before proceeding to look at superalloys themselves it is also instructive to look at the ternary system Ni-Cr-Mo. This with other elements such as Co, Mo and W forms the basis for σ and μ formation in superalloys. Due to its simple nature the use of PHACOMP does not allow for the true complexity of topologically close-packed (TCP) phase formation to be taken into account. This is a factor to consider even before questions concerning accuracy are discussed. There are three TCP phases which can form in Ni-Cr-Mo, σ, μ and P and Fig.4 shows a calculated isothermal section at 850°C which is in good agreement with the detailed study of Raghavan et al[21]. The P phase is not often seen in superalloys as Co tends to stabilise μ over P in the quaternary Ni-Co-Cr-Mo. The clear strength of the CALPHAD method is that as well as giving good levels of accuracy for TCP phase formation <u>and its temperature dependence</u> it can also consider the interplay between them.

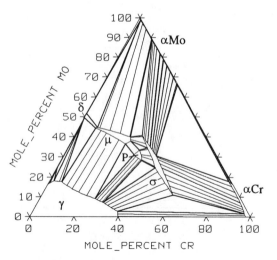

Fig.4. Calculated isothermal section for Ni-Cr-Mo at 850°C.

Multi-component alloys

As the database contains many of the elements seen in superalloys it is possible to validate calculated results against a substantial literature, particularly with respect to γ/γ' and liquid phase equilibria. Fig.5 shows a comparison for nearly 150 values of critical temperatures such as γ'_s, liquidus and solidus. The level of accuracy is excellent with the average difference ($\bar{d}$) between the calculations and experiment lying close to 10°C. The liquidus in particular is very well predicted with a $\bar{d}$ of 6-7°C. It is noted that the results come from all types of superalloys ranging from low γ' types such as Waspaloy through very highly alloyed types such as IN939 to single crystal alloys such as CMSX2 and SRR99.

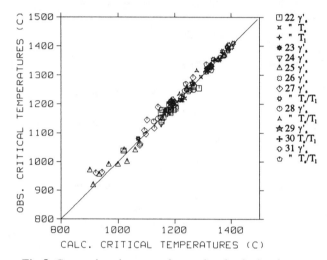

Fig.5. Comparison between observed and calculated critical temperatures for Ni-based superalloys.

There is a similarly large literature concerning amounts of γ' and compositions of both γ and γ'. Fig.6 shows a comparison between predicted and observed amounts of γ' in a variety of superalloys where the average difference between predicted values and those observed experimentally is of the order of 4%. In the comparison, results can be in either wt% or volume%. For the latter case, as lattice mismatches are so small, mole% values give almost identical values to volume%.

Figs.7-11 show some of the comparisons for the composition of γ and γ' where the high standards of results is maintained. Where experimental results have been quoted in wt% they have been converted to at% to allow for consistency of comparison. The average difference for elements such as Al,Co and Cr is close to 1at% while for Mo,Ta,Ti and W this value is close to 0.5at%. Too few experimental values for Hf and Nb were found to be statistically meaningful but where possible these were compared and results for average differences were found to be slightly better than obtained for Mo,Ta etc.

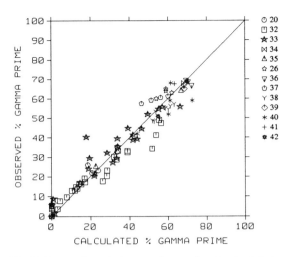

Fig.6 Comparison between observed and calculated amounts of γ' in Ni-based superalloys.

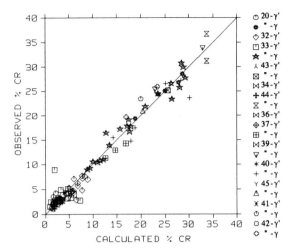

Fig.9 Comparison between calculated and observed Cr composition of γ and γ' in Ni-based superalloys.

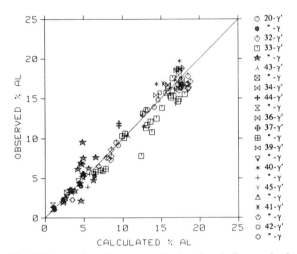

Fig.7 Comparison between calculated and observed Al composition of γ and γ' in Ni-based superalloys.

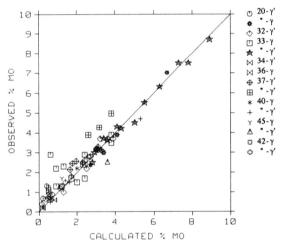

Fig.10 Comparison between calculated and observed Mo composition of γ and γ' in Ni-based superalloys.

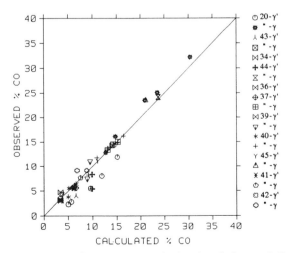

Fig.8 Comparison between calculated and observed Co composition of γ and γ' in Ni-based superalloys.

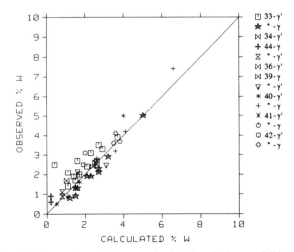

Fig.11 Comparison between calculated and observed W composition of γ and γ' in Ni-based superalloys.

One important thing to come from the results of calculated γ/γ' equilibria is that partition coefficients for elements such as Cr and Ti are well matched so the calculations have given the correct answer for both lower and higher order systems. The exaggeration of the lower order partitioning for Cr and Ti is due to the fact that they have repulsive thermodynamic interactions. Ti is very tightly bound to γ' which consequently causes Cr to be rejected from γ'. This increases levels of Cr in γ which in turn causes a rejection of Ti to γ'.

Apart from its success in predicting liquid phase and γ/γ' phase equilibrium the new database allows excellent predictions to be made for TCP phase formation. In a recent study at Rolls-Royce the database correctly predicted switches between $M_{23}C_6$ and M_6C formation and between σ and μ. The accuracy which can be obtained now allows for alloy design to be made with very specific targets. For example most superalloys are designed to be σ safe. Within a PHACOMP scenario this would mean the average electron hole number $\overline{N}_v$ of the alloy would be below a critical value. However, this critical value is not necessarily a fixed number and may often found by experience for each alloy.

Using the CALPHAD route a σ-solvus temperature can now be calculated below which σ will form and this value can be used to help define 'σ-safety'. A good example of this concept is in U720. This alloy was used in land based gas turbine engines and for long term use up to 900°C[46], but its excellent all round properties suggested that it could be used as a disc alloy. However, while long-term exposure at high temperatures produced only minor susceptibility to σ formation its use at 750°C led quickly to σ formation and in large amounts[46]. Clearly the alloy was either close to or above its σ-solvus at the higher temperature and it was found necessary to reduce Cr levels to destabilise σ at lower temperatures. This led to the development of U720LI with levels of Cr 2wt% less than for U720. Levels of C and B were also lowered to reduce the formation of borides and carbides which acted as nucleation sites for σ formation.

Fig.12 shows a calculated phase % vs temperature plot for U720 and it can be seen that its σ-solvus is indeed very close to 900°C and at 750°C the alloy would contain substantial levels of σ in excellent accord with experimental observation. Keefe et al[46] further determined TTT diagrams for both U720 and U720LI and these are shown in Fig.13.

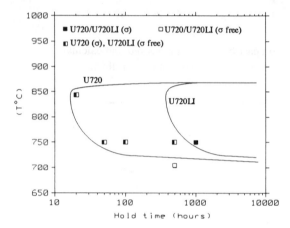

Fig.13   TTT diagrams for U720 and U720LI after Keefe et al[46].

As has been previously pointed out[12] decreasing the Cr levels must decrease the σ-solvus and, as the high temperature part of the TTT diagram asymptotes to the σ-solvus temperature, the two TTT diagrams should have distinct and separate curves. Taking the σ-solvus calculated for U720 and U720LI it was proposed[12] that the TTT diagrams should have the form as shown in Fig.14.

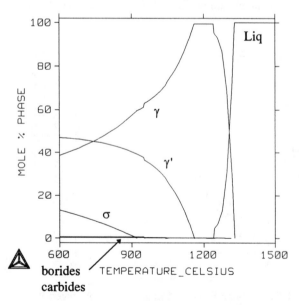

Fig.12 Phase % vs temperature plot for U720

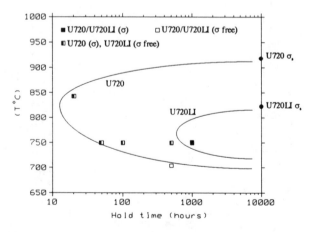

Fig.14   TTT diagrams for U720 and U720LI based on calculated σ-solvus temperatures

The corollary to such calculations is to calculate the how the σ-solvus varies as the composition of the different elements are altered. Fig.15 shows such variations in σ-solvus temperature for U720LI as each element is changed within its nominal composition specification.

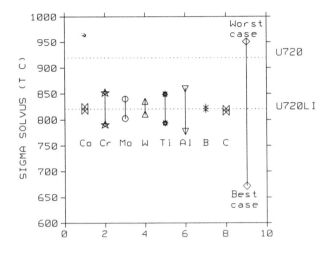

Fig.15 Variation in calculated σ-solvus temperature as elements in U720LI change between maximum and minimum specified limits

It is interesting to note that the greatest sensitivity is to Al, with Ti similar to Cr. This is because increases in Al and Ti increase the levels of γ' in the alloy and reject σ forming elements such as Cr, Mo and W into γ whose amount has decreased. This gives rise to very significant concentration increases of these elements in γ and leads to higher susceptibility to σ formation. Taking the information in Fig.15 it is now possible to define σ-sensitivity factors for each of the elements with a very simple mathematical formula. This can then be used to monitor σ-susceptibility of different heats during alloy production, replacing PHACOMP methods. Such sensitivity factors can be defined for all types of production features where phase equilibria can be important, for example γ' heat treatment windows, levels of γ' at heat treatment or forging temperatures, solidus and liquidus temperatures etc.

The results discussed so far have addressed issues mainly concerned with equilibrium behaviour, and in practice such issues are of importance to materials behaviour. However, one of the great strengths of a CALPHAD approach is that it can be combined with kinetic modelling to model features not usually associated with equilibrium calculations. For example growth rates[4], which can allow TTT diagrams to be produced, and modelling of non-equilibrium solidification[47]. Further the underlying thermodynamics controls some basic features which control mechanical properties such as stacking fault energies (SFEs)[48], APB energies in γ'[49] and even coefficients of thermal expansion[50]. The last part of this paper will discuss how such calculations can be tackled and give some examples of applications.

For some time it has been possible to use packages such as Thermo-Calc to model non-equilibrium solidification under 'Scheil' conditions, i.e. solidification assuming no back diffusion takes place in the solid phase(s) during growth of the solid phase(s)[3,51]. For Al-alloys in particular[52] it can be shown to give exceptionally good predictions for features such as fraction solid transformed/latent heat evolution as a function of temperature as well as the non-equilibrium phases which can appear. The technique has already been successful[12] in predicting the unexpected formation[22] of η in the interdendritic regions of a 'single crystal' U720 alloy in which the B and C had been removed.

As part of a progressive upgrading of the present database there is now sufficient data included for Fe to consider preliminary modelling of the solidification behaviour of IN718. Fig.16 shows plots of fraction solid transformed vs. temperature for an alloy with a standard composition specification[53] under both equilibrium and 'Scheil' conditions

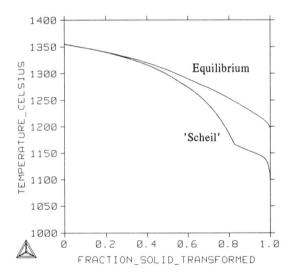

Fig.16 Fraction solid vs temp. plots for IN718 calculated for equilibrium and 'Scheil' conditions.

The final part of the 'Scheil' plot corresponds to the formation of the Laves phase as is commonly observed in this alloy, see for example Cao et al[54] and Murata et al[55]. The formation of the Nb-rich MC carbide is correctly predicted and some small amounts of Ni_3Nb and M_3B_2 (if B is present) may also form in the very last liquid to solidify. It is further noted that the Laves formed during solidification is metastable and will be dissolved on high temperature annealing as observed in practice. It is accepted that some back diffusion will occur and solidification will usually proceed somewhere between the two curves shown in Fig.16.

The extension of the database to model lower temperature solid state equilibria still requires some additional work. However, this should be completed very shortly and already very reasonable predictions for the solvus temperatures of δNi_3Nb, γ' and σ are obtained.

Apart from information concerning formation of phases it is also possible to calculate the heat evolution during solidification. This is of considerable importance in the modelling of heat flow during solidification and subsequent modelling of casting processes. By taking the latent heat evolved at each step of the calculation the latent Cp of solidification vs temperature plot of the 'Scheil calculation for IN718 was calculated and is shown in Fig.17. The C_P plot has been calculated by averaging the C_P over 2° steps and can be readily incorporated into casting simulation packages in this way.

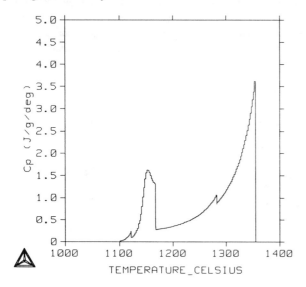

Fig.17 Plot of Latent Cp of solidification vs temp. for IN718 during 'Scheil' solidification.

Recent attempts at solidification modelling using data based on the early Kaufman work have been made for René N4[56] and IN718[57] and it is interesting to briefly discuss these results. The work on IN718 by Boettinger et al[57] is in quite reasonable agreement with the known behaviour of this alloy and the form of their 'Scheil' plot is very similar to the current work except that the liquidus is slightly lower and slightly less eutectic Laves is predicted to occur. The higher liquidus of the present calculations is more consistent with experimental observation[54] and the present work also considers C and B. But otherwise conclusions from both studies remain essentially similar.

The results for René N4, however, are quite different than obtained from the present work. The liquidus temperature from Chen et al[56] is some 45°C lower than would be calculated here which, considering the high level of accuracy for liquidus predictions from the present database, does raise some questions. Further their 'Scheil' simulation for this alloy produced a freezing range far greater than observed in practice. To account for this apparent discrepancy a new type of solidification mechanism was proposed and a modified 'Scheil' plot was produced[56]. However, it is noted that a straightforward 'Scheil' plot using the current database gives a very similar freezing range to the modified plot of Chen et al[56]. Clearly there is little need to resort to new and complex solidification mechanisms in this particular case.

From the above it is unclear within which composition range the Kaufman-base data can be used with confidence for input into solidification models. The composition of IN718 is based predominantly on the Ni-Cr-Fe ternary system with addition of other elements, notably Nb and Mo. The total of other elements is less than 2-3wt%. René N4, on the other hand is much more complex with high levels, greater than 4wt% each, of Al,Co,Cr,Ta,Ti and W. It is clear that a detailed validation exercise is necessary before this issue of confidence can be resolved.

It is sometimes forgotten that underlying thermodynamics of materials is inherently important in defining a number of basic materials properties. For example in a review of models concerned with fundamentals of strengthening in superalloys[58] two of the most important contributions were the stacking fault and APB energies. Calculations of SFEs in austenitic stainless steels have been performed very successfully[48] where the main input parameter has been the free energy difference between the austenitic FCC structure and the underlying CPH phase. Information can be extrapolated quite straightforwardly for the CPH phase in superalloys from the constituent lower order systems and therefore such techniques can now be applied to superalloys.

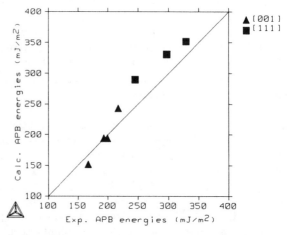

Fig.18 Comparison of calculated and experimentally measured [111] and [001] APB energies for some superalloys

Successful predictions for the APB energy in γ' from calculation of the γ→γ' ordering energy have been made[49]. The method directly utilises the database firstly to calculate the composition of γ' in the superalloy of interest. The formation energy of γ' is then calculated followed by that of γ at the same composition. This is then used to derive the ordering energy and by using a model which relates this term to the change in number and types of bonds at the boundary very good predictions for [111] and [001] APB energies can be made. Fig.18 shows the comparison between calculated and measured[59] values for a some model superalloys including U720.

Work is currently in progress to extend the database to include Re. This work entails substantial new work as Re phase diagrams have not been, to any at great degree, experimentally studied or thermodynamically assessed by CALPHAD methods. However, initial results look very interesting and it is worth examining a calculation of an isothermal section for Ni-Al-Re at 1000°C (Fig.19). Little or no experimental work is reported for this system and the calculations for this ternary are therefore preliminary. However, it is noted that the predicted solubility of Re in γ' at 1000°C is very limited ~1at% or 3-4wt%.

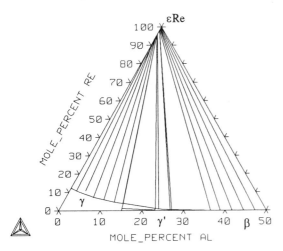

Fig.19 Calculated isothermal section for Ni-Al-Re at 1000°C

Although little experimental information exists for Ni-Al-Re there is one fairly well documented experimental feature of the ternary. There is very little solubility of Re in NiAl and a eutectic exists between NiAl and almost pure Re with a composition of ~1.5at%Re[60,61]. This feature is very well matched by the present calculation as is the binary system Ni-Re. The predicted results for the solubility of Re in γ/γ' alloys are therefore expected to be quite reasonable.

Interactions between Re and the other elements in an alloy such as CSMX4 will alter this solubility and perhaps enhance it, for example Ti and Re are quite tightly bound, it. However, it is clear that solubility levels will never be as high as for elements such W or Cr. Therefore, the initial conclusion is that with relatively small additions of Re one might expect Re particles to precipitate in superalloys with high volume fractions of γ'. This appears completely consistent with observations of clustering in Re containing superalloys[39]. It would be further interesting to observe if some segregation to stacking faults in γ occurred.

Summary

Some recent applications of CALPHAD and related techniques to Ni-based superalloys have been presented using a database recently developed by Thermotech Ltd and Rolls-Royce plc. The results of the calculations show that the CALPHAD route enables high quality predictions to be made for all aspects of phase equilibria in Ni-based superalloys from the following multi-component system

Ni-Al-Co-Cr-Hf-Mo-Nb-Ta-Ti-W-Zr-B-C

Results for calculated γ/γ' and liquid phase equilibria can be shown to approach the levels of accuracy usually associated with experimental measurement. Some applications particularly with respect to σ formation in U720 have been discussed and the extension to areas not usually associated with phase diagrams has been presented. These include modelling of non-equilibrium solidification and calculation of APB energies.

Examples of work in progress on the inclusion of Fe and Re to the current database have shown that valuable results can already be obtained for these elements. Based on calculations for the Ni-Al-Re ternary system a possible explanation for the experimentally observed clustering of Re atoms in single crystal blade alloys is suggested.

References

1. B.-J.Lee in "Applications of Thermodynamics in the Synthesis and Processing of Materials", eds.P.Nash and B.Sundman, (Warrendale, PA: TMS, 1995), 215
2. N.Saunders, to be published in "Titanium '95: Science and Technology", eds.P.Bleckinsop et al (London: Inst.Materials, 1996)
3. B.Jansson, M.Schalin, M.Selleby and B.Sundman, in "Computer Software in Chemical and Extractive Metallurgy", eds.C.W.Bale and G.A.Irins, (Quebec: Canadian Inst.Met., 1993), 57
4. J.Ågren, ISIJ International, 32, (1992), 291
5. N.Saunders and A.P.Miodownik, J.Mat.Research, 1, (1986), 38
6. L.Kaufman and H.Nesor, Met.Trans., 5, (1974), 1617
7. L.Kaufman and H.Nesor, Met.Trans., 5, (1974), 1623
8. L.Kaufman and H.Nesor, Met.Trans.A, 6A, (1975), 2115
9. M.Hillert and L.-I.Steffansson, Acta Chem.Scand., 24, (1970), 3618

10. B.Sundman and J.Ågren, J.Phys.Chem.Solids, **42**, (1981), 297
11. N.Saunders, in "Computer Aided Innovation of New Materials", eds.M.Doyama et al (Elsevier Publishers B.V., 1993), 731
12. N.Saunders, Phil.Trans.A, **351**, (1995), 543
13. I.Ansara: Int.Met.Reviews, **22**, (1979), 20
14. M.Hillert, CALPHAD, **4**, (1980), 1
15. N.Saunders, Z.Metallkde., **80**, (1989), 903
16. B.Sundman, "User Aspects of Phase Diagrams", ed.F.H.Hayes, (London: Institute of .Metals, 1991), 130
17. C.W.Bale and G.Eriksson: Canadian Metallurgical Quarterly, **29**, (1990), 105
18. P.Nash, M.F.Singleton and J.L.Murray, in "Phase Diagrams of Binary Nickel Alloys", ed.P.Nash (Metals Park, OH: ASM International, 1991), 4
19. J.D.Vorhoeven, J.H.Lee, F.C.Laabs and L.L.Jones, J.Phase Equilibria, **12**, (1991), 15
20. O.H.Kriege and J.M.Baris, Trans.ASM, **62**, (1969), 195
21. M.Raghavan et al, Met.Trans.A, **15A**, (1984), 783
22. C.Small, Rolls Royce plc, Derby DE24 8BJ, U.K, private communication 1993.
23. Y.Honnarat, J.Davidson and F.Duffaut, Mem.Sci.Rev., **68**, (1971), 105
24. E.H.van der Molen, J.M.Oblak and O.H.Kriege, Met.Trans., 2, (1971), 1627
25. W.Betteridge and J.Heslop, in "The NIMONIC Alloys and Other Ni-Based High Temperature Alloys: 2nd ed.", (Edward Arnold Ltd, 1974)
26. J.R.Brinegar, J.R.Mihalisin and J.Van der Sluis, in "Superalloys 1984", eds.M.Gell et al, (Warrendale, PA: Met.Soc.AIME., 1984),53
27. S.R.Dharwadkar et al, Z.Metallkde., 83, (1992), 744
28. S.K.Shaw, University of Birmingham, Edgbaston, Birmingham, UK, private communication, 1992
29. S.T.Wlodek, M.Kellu and D.Alden, in "Superalloys 1992", eds.S.D.Antolovich et al (Metals Park, OH: TMS, 1992), 165
30. J.Zou et al, in "Superalloys 1992", eds.S.D.Antolovich et al (Metal Park, OH: TMS, 1992), 165
31. J.S.Zhang et al, Met.Trans.A, **24A**, (1993), 2443
32. W.T.Loomis, J.W.Freeman and D.L.Sponseller, Met.Trans., **3**, (1972), 989
33. R.L.Dreshfield and J.F.Wallace, Met.Trans, **5**, (1974), 71
34. P.Caron and T.Khan, Mat.Sci.Eng., **61**, (1983), 173
35. M.Magrini, B.Badan and E.Ramous, Z.Metallkde., **74**, (1983), 314
36. T.Khan, P.Caron and C.Duret, in "Superalloys 1984", (Warrendale, PA: Met.Soc.AIME., 1984), 145
37. Z.-Y.Meng, G.-C.Sun and M.-L.Li, in "Superalloys 1984", (Warrendale, PA: Met.Soc.AIME., 1984), 563
38. M.V.Nathal and L.J.Ebert, in "Superalloys 1984", (Warrendale, PA: Met.Soc.AIME., 1984), 125
39. D.Blavette, P.Caron and T.Khan, in "Superalloys 1988", eds.S.Reichman et al, (Warrendale, PA: The Metallurgical Society, 1988), 305
40. H.Harada et al in "Superalloys 1988", eds.S.Reichman et al (Warrendale, PA: The Metallurgical Society, 1988), 733
41. R.Schmidt and M.Feller-Kniepmeier, Scripta Met.Mat., **26**, (1992), 1919
42. S.Duval, S.Chambreland, P.Caron and D.Blavette, Acta Met.Mat., **42**, (1994), 185
43. Y.Shimanuki, M.Masui and H.Doi, Scripta Met., **10**, (1976), 805
44. K.M.Delargy and G.D.W.Smith, Met.Trans.A, **14A**, (1983), 1771
45. K.Trinckhauf and E.Nembach, Acta Metall.Mater., **39**, (1991), 3057
46. P.W.Keefe, S.O.Mancuso and G.E.Maurer, in "Superalloys 1992", eds.S.D.Antolovich et al, (Warrendale PA: TMS, 1992), 487
47. W.Yamada and T.Matsumiya, Nippon Steel Tech.Rpt.No.52, January 1992, p31
48. A.P.Miodownik, CALPHAD, **2**, (1978), 207
49. A.P.Miodownik and N.Saunders, in "Applications of Thermodynamics in the Synthesis and Processing of Materials", eds.P.Nash and B.Sundman, (Warrendale, PA: TMS, 1995)
50. A.P.Miodownik, "Prediction of Thermal Expansion Coefficients from Thermodynamic Data", presented at CALPHAD XXIV, Kyoto, Japan, May 21-26, 1995
51. N.Saunders, "The prediction of microsegregation during solidification of a 7000 series Al-alloy", report to GKN Technology, Wolverhampton, UK, August 1988
52. N.Saunders, to be presented at the 5th International Conf. On Al-alloys, 1-5 July, Grenoble, France
53. C.T.Sims, N.S.Stoloff and W.C.Hagel, eds., "Superalloys II" (New York: J.Wiley&Sons, Inc., 1987)
54. W.D.Cao, R.L.Kennedy and M.P.Willis, in "Superalloys, 718, 625 and Derivatives", ed.E.A.Loria, (Warrendale PA: TMS, 1991), 147
55. Y.Murata et al, in "Superalloys, 718, 625 and Derivatives", ed.E.A.Loria, (Warrendale PA: TMS, 1994), 81
56. S.-L.Chen et al, Met.Mater.Trans.A, **25A**, (1994), 1525
57. W.J.Boettinger et al in "Modeling of Casting, Welding and Advanced Solidification Processes, VII", eds.M.Cross et al (Warrendale PA: TMS, 1995)
58. N.S.Stoloff, in "Superalloys II", eds. C.T.Sims, N.S.Stoloff and W.C.Hagel (New York: J.Wiley & Sons, Inc., 1987), 61
59. C.Small, Rolls-Royce plc, Derby DE24 8BJ, UK, private communication 1994
60. D.P.Mason, D.C.Van Aken and J.G.Webber, in "Intermetallic Matrix Composites", Mat.Res.Soc. Symp.Proc.Vol.194 (Pittsburgh, PA: Materials Research Society, 1990), 341
61. D.P.Mason and D.C.Van Aken, in "High Temperature Ordered Intermetallic Alloys", Mat.Res.Soc.Proc. Vol 213 (Pittsburgh, PA: Materials Research Society, 1991), 1033

STRENGTH & STABILITY CONSIDERATIONS IN ALLOY FORMULATION

Graham Durber [†]
Howmet Ltd., Exeter Alloy
England

Abstract

It is recognised that the development of new alloys requires the commitment of increasingly large amounts of money and time which few corporations can these days afford. Techniques for short-cutting the development path are needed, e.g. predictive models. Two such methods are presented here for pre-determining alloy stability and strength. The pursuit of this and similar approaches should help to minimise the expense, and time, incurred in manufacturing and testing many unsuccessful alloy formulations. New alloys are still needed with property combinations which have yet to be realised.

Introduction

New alloy design has traditionally required the assessment of multiple alloy formulations, which incurs considerable cost and time. However, it is reasoned that there exists sufficient non-proprietary data on a large number of superalloys for it to be feasible to estimate, at least to a first approximation, the likely strength of an alloy based on a consideration of its chemical composition and structure. The relationship between chemistry and other material properties such as corrosion resistance, castability and hot-tear resistance remains publically less well defined. Neither does alloy stability appear entirely predictable. Given the number of stability calculations that are routinely run to make such a determination, this statement may sound a little surprising. Particularly so when one considers that the development of the superalloys, from the simple precipitation hardened Ni-Cr system to those of today, where a high proportion of solid-solution strengtheners are used, has only been achieved by a careful balancing of strength versus microstructural stability.

A method that was developed in the 1960's and still used today for determining freedom from intermetallic phases such as σ, µ & χ phases is called "Phacomp" [1,2]. A critical electron vacancy or "N_v" value is determined for an alloy, below which it is judged stable over a set time and temperature. The problem with this system however is that a unique critical limit has to be established for each alloy that is assessed. This calculation is continually being refined with regard to the compositions of the various phases. More recently a modified approach involving the "d-electrons concept" [3] for SC alloy design has been extended to include also a measure of the bond strength of the alloying elements [4]. In this way, the relative stability of an alloy could be displayed on a map, but the boundaries remain undefined.

Electronic & Crystal Structures

Many of the physical properties of the transition metals (e.g. melting point, strength, high specific heat) can be explained by the ease with which electrons can be deflected from the outer shell orbitals into vacant d-states. This ability to absorb valency electrons enables alloys of the transition metals to exhibit a wide range of stability for a given crystal structure. It should be possible therefore to construct a map of crystal structure in a manner similar to the periodic table of elements, except that the two axes become the (s+p) and (d) orbital electron levels rather than elemental group and period. This has been done for the transition elements and its form is shown in Figure 1.

Within this figure a "ternary" boundary for Ni-Co-Cr can be seen, shown by the dotted lines, with Ni residing at the top right hand corner and Cr at the bottom. This ternary area is different however from a normal phase diagram in that allowance can be made for any element in the alloy. Alloy location within this field depends only on the summation of the (s+p) & (d) electrons in the residual matrix, having made due allowance for phase precipitation in the normal way as in any Phacomp calculation. The preferred crystal structure changes within this ternary area from fcc to fcc/hcp, depending on size factor considerations, to hcp as

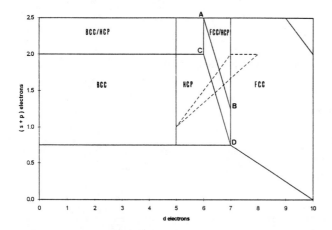

Fig.1: Crystal structure map in relation to (s+p) and (d) electrons. The dotted lines represent the ternary boundary for Ni-Cr-Co, as read clockwise from the top right hand corner.

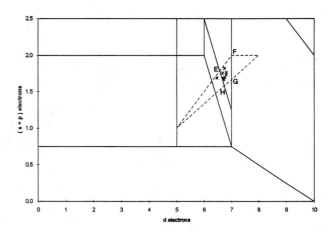

Fig.2: Positioning of various cast superalloys on the d-stable map

alloy location moves away from the nickel towards the chromium position. The lines AB & CD represent contours of constant electron hole numbers running across the ternary field.

The location of a number of cast superalloys (IN100, IN738LC, IN792mod5A, SRR99, Mar-M002, GTD111, CMSX4, R80) on this "d-stable" diagram is shown in Figure 2. With the exception of IN100, all of these plus many other superalloys lie within the quadrant EFGH.

The higher strength alloys amongst those displayed in Figure 2 tend towards the mid-span of the EF boundary. The lower strength alloys fall towards the lower quarter of the quadrant, towards the EH boundary. Many of the older alloys fall in this area, and tend to have their aim chemistries adjusted to the limit of their maximum permissible Phacomp value, to maximise stress-rupture strength, e.g. IN738LC. From a consideration of favoured crystal structure, alloy stability is expected to improve as location moves away from the EH boundary towards the FG boundary.

Stress-Rupture Strength Prediction

For the purpose of this evaluation, the creep strength of a superalloy is presumed to be controlled by the normal power law creep equation, plus a factor for alloy chemistry.

Table I Alloy Chemistries (Weight %)

	C	Al	B	Co	Cr	Hf	Mo	Nb	Re	Ta	Ti	V	W	Zr
IN713LC	.05	6.0	.01		12.0		4.2	2.0			0.7			.06
IN738LC	.10	3.5	.01	8.3	16.0		1.7	0.9		1.7	3.4		2.6	.05
IN100	.16	5.5	.014	15.0	10.0		3.0				4.7	1.0		.06
IN939	.15	1.9	.01	19.0	22.5			1.0		1.4	3.7		2.0	.09
IN792.5A	.08	3.4	.015	9.0	12.5		1.9			4.0	3.8		4.0	.02
MarM002	.14	5.7	.015	10.0	8.9	1.5				2.5	1.5		10.0	.04
SRR99	.02	5.5		5.0	8.6					2.8	2.2		9.5	
CMSX-4®		5.7		9.6	6.4	0.1	0.6		2.9	6.5	1.0		6.4	
R77	.05	4.3	.015	14.5	14.6		4.2				3.3			.04
Nim80	.06	1.4	.003		19.5						2.4			.06
Nim90	.07	1.5	.003	16.5	19.5						2.5			.06
MarM0011	.14	5.5	.015	10.0	8.3	1.5	0.7			3.0	1.0		10.0	.05
U500	.07	3.0	.008	18.0	19.0		4.0				3.0			.05

CMSX-4 is a Cannon-Muskegon alloy

Thus $\varepsilon = A.f(\text{chemistry}).(\sigma/E)^n.\exp(-Q/RT)$ (1)

And $t_r \approx B/\varepsilon$ (2)

where ε is the steady state creep rate at an applied stress σ, and t_r is the rupture life. E is the Young's modulus, Q is the apparent activation energy, R the gas constant and T the absolute temperature. A & B are constants. If $f(\text{chemistry})$ can be described in simple terms of alloy chemistry, then the strength of a new superalloy formulation can be largely predicted in advance of its manufacture.

By selecting an alloy data-base which contained a broad weight percentage range for the alloying elements and a broad spread in alloy strengths, the attempt was made to correlate alloy chemistry and strength. The alloy data-base that was chosen for this exercise is shown in Table I.

A Larson-Miller parameter *P(200MPa) actual* was determined for each alloy using stress-rupture data taken from various published sources. These values are shown in Table II.

Table II Larson-Miller Parameters, based on 1000h Stress-Rupture Data.

Alloy	Actual P(200MPa)	Calculated P(200MPa)
Nimocast 80	22.60	22.43
Nimocast 90	23.84	23.88
U500	25.50	25.44
IN713LC	26.25	26.19
IN939	26.26	26.07
IN738LC	26.33	26.27
R77	26.45	26.26
IN100	26.90	26.87
IN792mod5A	27.30	27.32
Mar-M-002	27.10	27.24
DS Mar-M-002	27.35	27.35
Mar-M-0011	27.40	27.36
SRR99	28.15	28.18
CMSX-4®	28.90	28.93

Table II also shows *P(200MPa) calculated* figures. These were obtained from multiple regression analysis of alloy chemistry (atomic percent) versus the *P(200MPa) actual* data.

The linear regression obtained between the calculated and real values is in excess of 0.99. The biggest error is below 1% and typically it is better than 0.5%. This correlation was achieved by using the multiplication factors for the alloying elements as shown in Table III, plus the following equation:

P(200MPa) calc $= 21.37 + 0.166\Sigma(M_n.A_n) + $ structure (3)

where M_n is the multiplication factor for the nth element and A_n is the atomic fraction of that same element. The incorporation of a structure factor to allow for the obvious differences between SC & DS versus equi-axed alloys assists the correlation.

The data presented in Table III is of course not definitive; the individual coefficients that are obtained will depend on the choice of alloys for the data-base. Different selections of alloys have been examined and it was found that variations were minor for most elements. As might be expected, fluctuations were greatest for carbon and boron. If only equi-axed alloys are considered, then the influences of these two elements greatly diminishes. Zirconium remains a very positive influence.

The weightings attributed in Table III to the 1st long period of elements (Ti, V, Cr, & Co) follow a similar dependency on (d + s) electrons as does the atomic volume (5) of these elements, i.e. atomic size differences could be contributing to the strengthening mechanisms. Continuing with this comparison, the 2nd & 3rd long period series of elements (Zr, Nb, Mo & Hf, Ta, W, Re) would be expected to show much higher weightings, as is indeed the case but for the exception of hafnium. The anomalous influence of this element on strength has been observed before (6).

The *P(200MPa) calculated* value of an alloy provides the first step in the determination of that particular alloy's $f(\text{chemistry})$ function. It was found that the actual Larson-Miller curves for the alloys investigated were best described by the dependency:

$t_r = f(\text{chemistry})/\{\sigma^{(A+BT/\sigma)}.\exp[-(Q+CT)/RT]\}$ (4)

where A, B, & C are constants

The stress exponent follows an inverse function of stress, Figure 3. The slight change in position of the curve from one alloy to another is a result of the temperature dependence of this factor. The apparent activation energy in

Table III Multiplication Factors used with Atomic Percentages of the Alloying Elements to produce *P(200MPa) Calculated*.

C	Al	B	Co	Cr	Hf	Mo	Nb
-15.5	1.4	15.5	0.67	-0.26	-0.6	1.9	8.5
Re	Ta	Ti	V	W	Zr	DS & SC	
1.0	4.4	3.0	1.0	3.9	70	0.2	

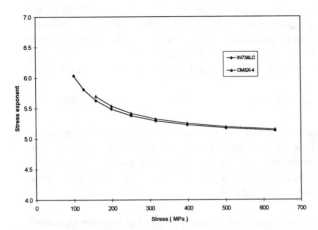

Fig.3 Variation of the stress exponent with stress. The slight shift in the curve from one alloy to another is a result of the temperature dependency of this factor.

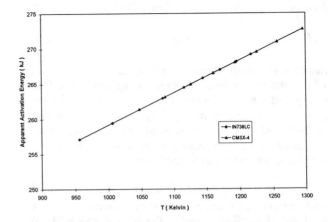

Fig.4 Dependence of apparent activation energy on temperature

equation 4 is linearly dependent on temperature, Figure 4. The final positions of the calculated curves were determined by inserting the calculated t_r values from equation 4 into equation 5 {with f(chemistry) assuming a value making P_f @ 200MPa = $P(200MPa)$ calculated}

$$P_f = T.\{20 + \log(t_r)\} \quad (5)$$

The results of this excercise can be seen in Figure 5, which compares actual v calculated Larson-Miller curves for the alloys.

The calculated curves tend to underestimate alloy stress-rupture life at combined high stress and high P values, by a factor of two. This level of discrepancy is of the same order as normal test scatter. It does indicate however that the stress exponent decreases slightly more quickly at high stresses with rising temperature than equation 4 allows.

Conclusions

A phase stability map has been constructed which describes the area within which superalloy chemistries should lie.

A single stress-rupture life v chemistry expression has been developed which predicts alloy performance over a working range of stress and temperature, for the alloys examined. Reliability will be improved as more alloys are added to the data-base.

Preliminary assessment of these two properties should enable a rapid selection of new chemistries for practical evaluation.

No attempt has been made in this report to establish alloy castability

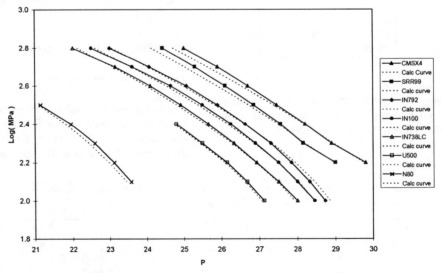

Fig.5 Comparison of calculated v actual Larson-Miller curves for various alloys.

References

1. W.J.Boesch & J.S.Slaney, Met. Prog. 86(1)(1964), 109

2. L.R.Woodyatt, C.T.Sims & H.J.Beattie, Trans AIME, 236(1966), 519

3. N.Yukawa et al, Superalloys 1988 (Warrendale, PA: The Metallurgical Society, 1988), 225

4. K.Matsugi et al, Superalloys 1992 (Warrendale, PA: The Minerals, Metals & Materials Society, 1992), 307

5. T.Lyman et al, eds., Metals Handbook, vol 1 (Metals Park, OH: American Society for Metals, 1985), 45

6. M.R.Winstone, "The Effect of Boron, Zirconium, Carbon & Hafnium on the Structure & Properties of Polycrystalline & Single Crystal IN792", (Report TM79018, National Gas Turbine Establishment, May 1979)

PHYSICAL METALLURGY

MECHANISMS OF FORMATION OF SERRATED GRAIN BOUNDARIES IN NICKEL BASE SUPERALLOYS

H. Loyer Danflou, M. Macia, T. H. Sanders, T. Khan*

Georgia Institute of Technology - School of Materials Science & Engineering
Atlanta, GA 30332-0245.
* Office National d'Etudes et de Recherches Aérospatiales (ONERA)
BP 72 - 92322 Chatillon cedex - France

Abstract

The growth kinetics of particles producing serrations on grain boundaries were studied on Astroloy and an experimental low misfit alloy. The serrations developed after applying an isothermal heat treatment a few degrees below the solvus temperature of the γ ' phase. In both materials, it was observed that arms which originated from dendritic particles pushed the grain boundaries on both sides and consequently produced serrations of a few microns in size (category A). In addition, in the alloy with low misfit, fans characteristic of discontinuous reaction which were approximatively a hundred microns in size were observed in regions of low particle density (category B). To identify the growth mechanisms of each category, the experimental kinetics were compared with those predicted by models of continuous and discontinuous precipitation. For the formation of large fans, it appeared that the model of discontinuous growth is in good agreement with the experimental results. However, for the second category of serrations, the continuous growth model provides a better description of the growth kinetics.

Introduction

In order to increase the resistance to crack propagation of disk alloys, it is necessary to reduce the grain boundary vulnerability at high temperatures. In this context, the development of microstructures containing grain boundary serrations shows promise for improving the mechanical properties of superalloys (1,2). The mechanisms of serration formation were therefore studied on Astroloy and on an alloy developed by Henry (3) made by P/M metallurgy and spray forming, respectively. The composition of the experimental alloy was chosen to reduce γ/γ ' misfit.

Serrated grain boundaries were developed first by slowly cooling through the γ ' solvus temperature and then were more recently formed by applying an isothermal heat treament just below the solvus temperature of the γ ' phase directly after solution heat treatment. The low undercooling results in a low density of particles hence favoring the growth of arms before supersaturation is lost and coarsening begins (2, 3). Because the arms of these precipitates push the boundary on both sides, serrations of several microns are produced. If the process of formation is presently well known, the growth mechanism itself is not clearly understood. The first satisfying attempt to describe the phenomena was proposed by Henry et al. (3). They identified the γ ' growth along the grain boundary with the dendritic growth. On the other hand, severals authors have shown that serration formation could be related to mechanisms of discontinuous reaction (1,2). Nevertheless, none of these hypotheses, taken separately, allows the contribution of each mechanism for the growth of the intergranular particles to be taken into account.

The purpose of this study is to analyse the growth kinetics of serrations by comparing them with kinetics predicted by models of discontinuous and continuous precipitation. To simplify the analysis of the phenomena, the serrations were produced by applying aging as proposed by Henry et al.(3). This analysis is also made easier by using a low misfit alloy where the effects of coherency strain are eliminated. Then, a model was developed to simulate the growth of particles controlled by both volume and grain boundary diffusion. This version allows the role of the grain boundary as a short circuit for diffusion for the solute to be taken into account. To help solve the problem, a modification to the solution of isolated spheres originally solved by Nolfi et al. was considered (4). On the other hand, the Hillert model was used to evaluate the kinetics of a discontinuous reaction (5). This model describes the migration rate of the interface as a function of microstructural characteristics : size and distribution of the precipitates, matrix and precipitate compositions, G.B. curvature and interfacial energy.

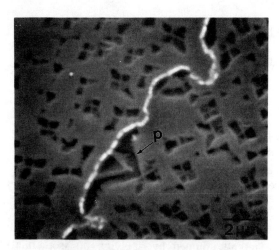

Figure 1 : Microstructure of the Astroloy heat treated at 1130°C/30' (+ thermal etching).
P : intergranular γ' particle

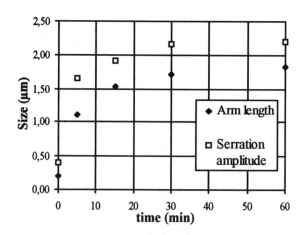

Figure 2 : Experimental growth kinetics of the intergranular γ' particle and serrations at 1130°C

Experimental Techniques and Procedures

The first alloy used in this study is Astroloy with the composition Ni-16.3Co-14.5Cr-4.35Al-3.65Ti-4.95Mo-0.018C-0.011B-0.047Zr (wt%). It was prepared by a prealloyed powder metallurgy process using the following steps : melting of the alloy in vacuum (VIM process), atomizing by a rotating electrode process (mean size of powder particle ~ 140μm) and consolidating the powder by extrusion using an 8:1 extrusion ratio. The second material is the experimental alloy developed by Henry whose composition is designed to minimize the coherency strain between the γ and γ' phases (3) : Ni-24Co-5Cr-4Al-4Ti-5Mo (wt %). It was prepared by spray forming and then was forged by General Electric. The consolidated material was homogenized by heat treatment above the γ' solvus temperature (1120°C) at 1150°C for 20h under argon.

Both superalloys were then heat treated at the temperature T_1 for 1.5h, around 20°C above the γ' solvus (Ts), before cooling down at 13°C/min until the temperature T_2 (below Ts). For generating the microstructure with serrations, the materials are maintained at T_2 during a period of time between 5 min. to a few hours before cold water quench. The temperatures T_1 used for this study were 1160°C for Astroloy and 1138°C for Henry's alloy. The temperature T_2 were selected between 1125°C and 1135°C for Astroloy and between 1090°C and 1110°C for the second. These microstructures were examined, both by optical and scanning electron microscopy on polished and etched samples. In order to made easier some observations of grain boundaries, a thermal etching (800°C/4h) is finally applied to Astroloy : this treatment allows the secondary carbides $M_{23}C_6$ to precipitate intergranuarly without modifying the microstructures of the coarse γ' particles.

Microstructural observations

To render easier the simulation of the γ' growth, the serrated microstructure was developed in two superalloys from an isothermal subsolvus heat treatement applied near Ts. In both materials, an optimum temperature for the treatment was selected in order to reduce the effects of chemical heterogeneities (T_2 = 1130°C in Astroloy and 1105°C in the alloy provided by Henry).

After a few hours of the heat treatment described above, the corresponding intragranular microstructures are characterized by a pattern of dentritic particles developed at high temperature and by a population of fine spherical particles (50nm) formed during quenching (figure 1). Depending on the heat treatment, the average lengh of the dendrite varies from 0.6 to 5 μm. In the case of the alloy provided by Henry, the dendrites do not have any preferential growth direction as observed in Fig.3. This observation confirms the small misfit existing between the γ and γ' phases.

As previously pointed out, the distribution of serrations in Astroloy is relatively homogeneous along grain boundaries, especially along random boundaries (see figure 1)(1). Because of the effects of lattice misfit strain, the γ' precipitates in Astroloy tend to grow along preferential directions and to split partially. Thus, the arms associated with serrations are unique and coarse in Astroloy, while they are several and thin in the experimental alloy. To study the growth kinetics of the γ' associated with serrations, the microstructures have been examined after 1,5,15,30 and 60 minutes of treatment at T_2 = 1130°C. This examination requires the evaluation of the global volume of the particle directly in contact with the grain

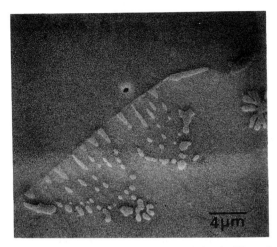

Figure 3 : γ' particles of category A in the Henry's alloy heat-treated at 1105°C/1h.

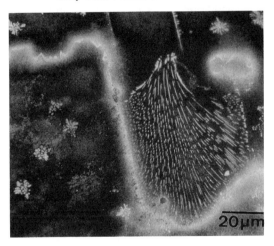

Figure 4 : exemple of fan (category B) in the Henry's alloy heat-treated at 1105°C/1h

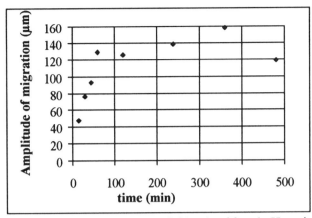

Figure 5 : Experimental growth kinetic of fans in Henry's alloy heat treated at 1105°C.

boundary. After sixty minutes, growth is over and coarsening begins. The results of this study are plotted in figure 2 which shows that the growth rate is drastically reduced after 30 minutes of treatment. Apparently, the growth rate changes with the level of supersaturation in the matrix.

As a consequence of compositional heterogeneities in Henry's alloy, the particle density varies from one region to another. Thus, two different categories of serrations can be distinguished on random grain boundaries as a function of the particle density. When the density around the grain boundaries is high enough, precipitates push the grain boundaries from both sides and consequently produce serrations of about ten μm in size, corresponding to category A (figure 3).

On the other hand, for the low density case (category B), the same rods produce fans on the order of a hundred μm in size (figure 4). It seems that the absence of obstacles allows the system to attain optimized growth rates ; the rod diameters hardly vary from one part to the other of a given fan (dia.= 0.4 μm).

For the category A (the equivalent of the intergranular population in Astroloy), the growth kinetics could not been determined because of the heterogeneity of the microstructure. Instead, a study of the growth kinetics was conducted for the B category at 1105°C. The resulting microstructures were observed by SEM during a period of treatment between 15 minutes and 8 hours. In order to generate reliable data, the measurements for the category B were made on well developed fans growing on the observation plane. The average amplitude of the fan versus the time t was plotted on figure 5.

Two domains can be distinguished : during the first hour, the growth rate, which is very fast, is stabilized around $v = 2.8 \times 10^{-6}$ cm/s, whereas in the second part, the growth slows down, due to the presence of other particles ahead of the growth front. In the first situation, the products of reaction and the observed growth rate seems compatible with a discontinuous reaction.

Theoretical Models

Continuous growth model

To describe the influence of the supersaturation on the growth of the intergranular γ' particles, a classical growth model for spherical particles, which takes into account soft impingement, has been adapted for the case of precipitates associated with serrations. The original Nolfi model (4) describes the solute profile around a growing or dissolving spherical precipitate as a function of the time assuming a stationary reaction interface. The growth of grain boundary serrations is more complex than the case originally treated by Nolfi et al. (4) since both volume diffusion to the grain boundary and grain boundary diffusion must be included.

Therefore, the use of Nolfi et al.'s equations does not determine the actual solute gradient at the reaction interface ; it only allows to evaluate an order of magnitude of the γ' growth rate as function of the remaining solute concentration. An effort was made in this model to simulate the effect of short circuit diffusion along the grain boundary through the introduction of the parameter δ.

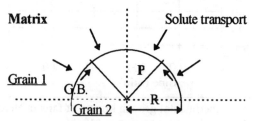

Figure 6 : Scheme of serration growth, R serration radius, P conic arm of the γ' particle, GB grain boundary

For the simulation, the serration surface S+ is represented by half of a sphere. This interface is pushed by a conic precipitate occupying a fraction δ of the serration volume Vs/2 (figure 6). The morphology of the precipitate is such that the γ' surface in contact with the grain boundary (S-) corresponds to a fraction δ of the total surface S+. It is worth noting that only the fraction of particle growing behind the grain boundary is described by the model (Vp= $2\pi R^3/3\delta$).

In order to express the growth rate of the particles, a flux balance across the reacting interface was performed. In this model, the solute diffusion is supposed to be carried out first by volume diffusion and then by grain boundary diffusion. In fact, the grain boundary represents a short circuit for the solute transport. Assuming S+ = $2\pi R^2$, the collecting surface of the solute and S- = S+/δ, the interface of reaction, the expression for the growth rate becomes :

$$v = \frac{dr}{dt} = \frac{\delta D}{(C_\beta - C_i)} \left(\frac{dC}{dr}\right)_R = \frac{\delta D}{(X_\beta - X_i)} \left(\frac{dX}{dr}\right)_R \quad (I)$$

with C_β the equilibrium solute concentration of the precipitate and C_i the solute concentration at the interface.

To evaluate the gradient of solute at the interface dX/dr as a function of the time and the size of the particle, the analytical expression of Nolfi for the solute concentration in the matrix was used (4). The matrix is assumed to be formed of equal cells delimited by a sphere of radius Rs and the particle, which is also spherical (radius R), grows from the center of the cell. The initial matrix is supersaturated with the composition C_0 and the diffusion coefficient D is assumed independant of time and composition. The only driving force for the movements of both solute and solvent elements is that due to concentration gradients. The solute concentration C(r,t) behaves according to Fick's second law.

For convenience, a new variable which represents the supersaturation is defined : C'(r,t)=C_{eq} - C(r,t). From the boundary conditions (equations VIII and IX in appendix A), Nolfi gives the analytical solution of the problem :

$$C'(r,t) = \sum_{n=0}^{\infty} \frac{A_n \exp(-t/\tau_n) C_n \sin(\lambda_n r - \delta_n)}{r} \quad (II)$$

where A_n, C_n are constants $\tau_n = (\lambda_n^2 D)^{-1}$ and λ_n, δ_n are the eigenvalues of the function (see expressions appendix A).

Then, to determine the concentration profile, a mass balance between the particle and the initial matrix has been carried out. The amount of solute in the initial matrix and the amount of solute in the particle are as follows (with R_0 the initial radius of particle) :

$$\Delta M_i = (C_0 - C_{eq}) 2\pi (R_s^3 - R_0^3)/3 \ ;$$
$$\Delta M_p = (C_\beta - C_0) 2\pi (R^3 - R_0^3)/3 \quad (III)$$

By calculating the remaining solute in the matrix (ΔM_i-ΔM_p), it becomes possible to determine the general profile of solute between R and Rs.

$$\Delta M_i - \Delta M_p = \int_R^{R_s} C'(r.t^*) 2\pi r^2 dr \quad (IV)$$

In this context, the parameter t* allows adjustment of the concentration profile as a function of the quantity of solute available.

It is assumed that the growth of the conic arms starts from a threshold value (R_0=200 nm in Astroloy). For each step of time Δt, the composition at the interface X_i is evaluated from the Gibbs Thompson expression (6). Then, the eigenvalues and constants of the function C'(r,t) are calculated (4). The concentration profile is determined by extracting the root t* of expression IV. Finally, the concentration gradient at the interface dC'(R,t*)/dr and the migration rate V can be deduced from expression XII in Appendix A).

For the numerical applications, the size of the cells R_s was determined from the experimental surface density of particle N_v in the material : $R_s = (S/\pi)^{1/2}$ and S = 1/N_v. In Astroloy during the first hour at 1130°C, R_s was approximated to 4 μm. The diffusion coefficient was approximated by using the expression of Swalin for aluminium in a nickel matrix (7): at 1130°C, the calculation gives Dv = 2.0×10^{-10} cm²/s.

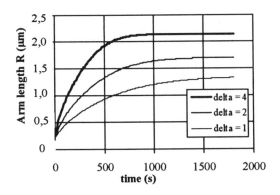

T =1130°C, Δt =10s , R_s = 4μm.

Figure 7 : Growth kinetics of serration, influence of δ

The parameter $\delta = S+/S-$ was choosen in this study as an adjustable coefficient which must reveal the effective area of grain boundary from which the particle draws solute. Through the theorical curves of kinetics (figure 7), it can be shown that the parameter δ is directly linked with the maximum limit of serration growth (Rmax). Thus, δ can be adjusted by using the experimental data Rmax and Rs.

Discontinuous growth model

To evaluate the migration rate of the interface under conditions of discontinuous precipitation, the Hillert model was choosen as a reference (5). Schematically, this model describes the reaction $\alpha_0 \longrightarrow \beta + \alpha_1$ where β and α_1 form as alternating lamellae growing edgewise into the α_0 grain. The purpose of this paragraph is to adapt this treatment describing lamellar growth to the case of rod-like growth. In our material, the β and α_1 phases could be replaced respectively by the γ' rods and the γ matrix.

The grain boundary is regarded as a thin film of boundary phase having its own molar free energy function for binary alloys (5). The free-energy losses during the movement of such an interface are evaluated from free energy diagrams. Two mechanisms of free energy dissipation are found, caused by diffusion ahead and inside the interface. For high migration rates of the interface, these retarding forces can almost disappear. The chemical deviation from local equilibrium at the interface β/α_0, results in a force pulling the grain boundary along with the precipitating phase allowing the formation of the β/α_1 interface.

In the simplified version of the model, the concentration profile of solute in α_1 and in β along the grain boundary is considered as constant for convenience. Consequently, Hillert has avoided Cahn's treatment of the solute profile along the α_1/α_0 interface to express the migration rate V of the grain boundary as a function of the thickness of the lamella (8). Instead, he has used the approximate treatment of the eutectoid reaction given by Zener (9) and applied it to the case of grain boundary diffusion control by Turnbull (flux balance method). Applied to the case of rod growth, this treatment gives an expression of the migration rate with the form (see appendix B equation XIII) : $V= f(r_\beta, r_{tot}, X_\alpha, X_{\alpha/\beta})$ with r_β the radius of the rods, r_{tot} the cell radius, X_α the composition of the $\alpha 1$ phase and $X_{\alpha/\beta}$ the composition at α/β interface along the grain boundary.

In order to express V as a function of r_β, Hillert has used the following assumptions (10) : the growth front is held back by the action of the surface tension σ in the α_1/β interface which acts perpendicular to the growth direction. The fraction L_α is carried by the edge of the α phase which must thus have a curvature which puts the α phase under an extra pressure ΔP_α. The balance of force requires that :

$$L_\alpha \sigma = \int_{r_\beta}^{r_\alpha + r_\beta} \Delta P_\alpha \, dr \quad \text{and} \quad L_\beta \sigma = \int_0^{r_\beta} \Delta P_\beta \, dr \quad (V)$$

The fraction L_β (= $1-L_\alpha$) is carried by the edge of the β rods.

The reaction gives chemical driving forces ΔGm_α and ΔGm_β which balance these extra pressures. But a part of this force can be lost by diffusion in the α_0 grain. Only a part of these contributions k_α and k_β can be used. We thus obtain for the case of rods (cylindrical symetry):

$$2L_\alpha \sigma V_m = k_\alpha \int_{r_\beta}^{r_\alpha + r_\beta} \frac{2r}{r_\beta} \Delta G_\alpha^m \, dr \quad \text{and}$$

$$2L_\beta \sigma V_m = \int_0^{r_\beta} \frac{2r}{r_\beta} \left(\Delta G_\beta^{m1} + k_\beta \Delta G_\beta^{m2} \right) dr$$

(VI)

From a molar Gibbs energy diagram, Hillert obtains approximate expressions of the chemical energies (5). By introducing them into the expressions (VI) and by using a mass balance and the expression of $\partial^2 G/\partial X^2$ (appendix B equations XIV and XV), it can be shown that the expression of the migration rate becomes :

$$\frac{v}{6K\delta \, D_b} = \left[\frac{r_\beta}{r_{tot}\left(r_{tot}^2 - r_\beta^2\right)} \right] \left[\left(\frac{k_\alpha \cdot \left(r_{tot}^2 - r_\beta^2\right)}{2L_\alpha r_\beta \cdot r_{rev}} \right)^{1/2} - 1 - \frac{k_\alpha \cdot L_\beta}{2L_\alpha} \right]$$

(VII)

According to Hillert's model, the values of L_α and L_β parameters have been approximated by expecting that grain boundary energies were equivalent for the α and β

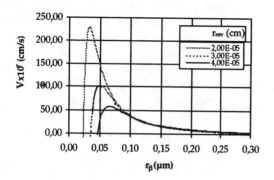

$T = 1105°C$; $L_\alpha = 0.5$; $k_\beta = 0$; $k_\alpha = 0.7$; $K = 2$; $D_b = 1.0 \times 10^{-7}$ cm^2/s

Figure 8 : Migration rate of the interface versus the radius r_β from the Hillert's model. Influence of r_{rev} (effect of the interface energy).

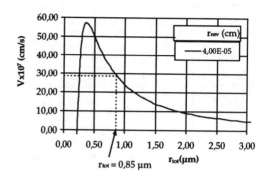

$T = 1105°C$; $L_\alpha = 0.5$; $k_\beta = 0$; $k_\alpha = 0.7$; $K = 2$; $D_b = 1.0 \times 10^{-7}$ cm^2/s

Figure 9 : Theorical kinetics predicted by the Hillert model in the case of rod growth

phases on a random grain boundary ($L_\alpha = L_\beta = 0,5$). On the other hand, assuming a high migration rate for the interface, the coefficient k_α was estimated around 1. In other words, it was assumed that the migration rate is fast enough to eliminate any volume diffusion ahead of the advancing reaction front.

The reversible radius r_{rev} has been evaluated from the Hillert formula : $r_{rev} = 2 \sigma V_m [X_s (1-X_s)] / ((RT(X_0 - X_{eq})^2)$ with $X_s = (X_0 - X_{eq})/2$; $V_m = 2.10^{-5}$ m^3 ; $X_0 = 0.061$; $X_{eq} = 0.0586$; $T = 1378$ °K ; $R = 8.31$ J.K^{-1}.mol^{-1}. Because the misfit is low in the Henry's alloy, σ is expected to be in the range $30 < \sigma < 90$ mJ/m^2. Consequently, r_{rev} has been estimated between $1.3 \times 10^{-5} < r_{rev} < 4.0 \times 10^{-5}$ cm. From the curves $V = f(r_\beta)$ (figure 8), it appeared that this parameter has a strong effect on the shape of the migration rate distribution. The maximum of the migration rate for the interface varies from 2.25×10^{-5} to 6.0×10^{-6} cm/s ($D_b = 1.0 \times 10^{-7}$ cm^2/s) into the interval $2.0 \times 10^{-5} < r_{rev} < 4.0 \times 10^{-5}$ cm.

The uncertainties in the grain boundary diffusion coefficient D_b has brought us to make the parameter adjustable with the experimental data. Knowing the level of the volume diffusion coefficient for aluminum in a nickel base matrix ($D_v = 2.0 \times 10^{-10}$ cm^2/s at 1135°C), the coefficient D_b was expected to be ten to one hundred times higher ($10^{-9} < D_b < 10^{-8}$ cm^2/s).

Discussions

Fan formation

As shown previously for the category B particles, the growth rate during the very fast regime was found to be around 2.8×10^{-6} cm/s. After one hour of treatment, the average diameter of the rods associated with the fans is about 0.17 µm to 0.21 µm. In fact, these observations are probably not representative of the real diameter during the reaction in contact with the grain boundary : after the discontinuous reaction, rods inside the fan continue to grow by consuming the remaining supersaturation. This is the reason the parameter $r_{tot} = r_\alpha + r_\beta$, which stays constant within the fans, was chosen as a reference for this study. In the second sample, the average r_{tot} was estimated to be 0.85 µm.

In order to determine a migration rate profile compatible with the experimental description, the level of r_{rev} was selected so that the peak of the curve $V = f(r_{tot})$ is close to the experimental r_{tot} (r_{rev} determines the width and the amplitude of the peak on the theorical curve). This is how r_{rev} was selected to be 4.0×10^{-5} cm (the maximum limit of the chosen interval) and the value for D_b as 1.0×10^{-7} which is five hundred time higher than Dv (with $k_\alpha = 0.7$). The corresponding calculated kinetics $V = f(r_{tot})$ are presented in figure 9 (to relate r_{tot} and r_β in the expression of V, the formula XVI in appendix B was used). When r_{tot} is equal to 0.85 µm, the calculated r_β corresponds to 0.11 µm with $X_\alpha = 0.060$ and the growth rate attains 3.0×10^{-6} cm/s. It is worth noting that the theoretical and experimental parameters (V, r_β) appear to be very close. But, the value of D_b, adjusted to fit the kinetics, seems to us a little bit overestimated. This simplified simulation only allows an estimation of the order of magnitude of the migration rate in steady state conditions. However, the results of this analysis tend to confirm that the experimental kinetics for the particles in category B is compatible with those of a discontinuous reaction. Despite that fans appear only in the low misfit alloy, the occurrence of the discontinuous reaction can not be connected with the absence of misfit : after similar heat treatment, such fans have already been observed in commercial alloys with a positive misfit (11).

Development of wavy grain boundaries

In Henry's alloy, when the particle density is high enough (particles of category A), the serrations formed are very similar to those in the Astroloy (several μm of amplitude). However, these similitudes do not concern the morphology of the γ' particles which depends on the misfit level of the alloy. And the size of the intergranular particles remains close to those of the intragranular γ' particles (figure 3).

Category	A	B	Bulk	Hillert
r_β (μm)	0.5	0.2	0.5-1.0	0.09
V (μm/s)	1.66×10^{-3}	2.6×10^{-2}	0.7×10^{-3}	2.6×10^{-2}

Table 1 : The different growth kinetics in Henry's alloy.

In table 1, a comparison of the average migration rate V and rod diameter have been made between the different categories of particles present in the material. Clearly, there is at least an order of magnitude (a factor 15) between the migration rates of the category A and those observed in the category B or in the simulation (Hillert data). The microstructural observations confirm that the characteristics of the category A particles are close to those of the intragranular ones (rods diameter = 0.5 μm). As in the grain, the growth of the A particles is affected by the presence of the neighbouring particles (overlapping of the diffusion fields). However, in this case, the phenomena is delayed by the grain boundary diffusion.

To identify the growth mechanism of this serration, the experimental results obtained in Astroloy at 1130°C, during the first hour of isothermal treatment, were exploited. The experimental data of growth kinetics were compared with two theorical conditions of growth, $\delta = 1$ and $\delta = 2$, calculated through the modified Nolfi's model (figure 10). The situation $\delta = 1$ represents a state where there is no short circuit of diffusion. This condition corresponds to a classical continuous precipitation. In fact, the comparison between the curve $\delta = 1$ and the experimental one (r exp) reveals that the contribution of the grain boundary is important. Despite the lack of experimental data, it appears clearly that the theoretical curve $\delta = 2$ fits very well with the experimental kinetics ; the growth limits Rmax, located around 1600 nm, are equivalent for both theorical and experimental descriptions. Indeed, these results, obtained for a homogeneous microstructure, suggest that neither a classical continuous model nor a discontinuous one can describe the serration formation. Consequently, the assumption of a growth rate, based on volume diffusion followed by grain boundary diffusion, appears realistic.

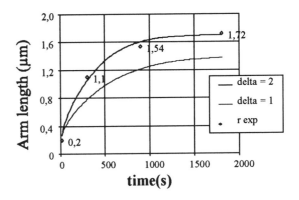

T = 1130°C ; $\Delta t = 10$ s ; Rs = 4 μm ; Dv = 2.0×10^{-10} cm²/s

Figure 10 : Comparison between the experimental results (r exp) and the theorical kinetics (delta =1,2) in Astroloy

In Henry's alloy, intermediate microstructures between A and B have also been observed. These microstructures can be the consequence of variations of the particle density in the material. Thus, an enlargement of the cell size in the matrix can explain an acceleration of the γ' particle growth. However, another assumption could explain this phenomena : as a function of the particle density, the discontinuous regime could represent a more or less important step in the growth process. In the present stage of our investigation, this migration scheme is not completely rejected .

Conclusion

Growth kinetics of particles producing serrations on the grain boundary have been studied on Astroloy and Henry's alloy. The development of serrations was achieved by applying an isothermal heat treatment, a few degrees below the solvus temperature of the γ' phase. In both materials, it was observed that arms orginating from dendritic particles push the grain boundary on both sides and consequently produce serrations of a few μm (category A). However, due to a certain chemical heterogeneity in the Henry's alloy, another category of serration (category B) develops simultan Arm lengh (nm) 1e former : " fans " of about 100μm size, characteristic of discontinuous reaction, are promoted in regions of low particle density.

The experimental growth rate of the category B were compared with those predicted by the Hillert's model. Thus, the mechanism of fan formation was clearly identified as a discontinuous growth process. On the other hand, a similar analysis operated on the category A did not allow a direct classification of the reaction ; neither the classical continuous reaction nor the discontinuous reaction describe the corresponding growth phenomena. To correlate the growth rate of these particles with the surrounding concentration profile, Nolfi's model was adapted for the case of a migrating reaction interface. It was assumed that the solute transport was controlled successively by volume and grain boundary diffusion. Indeed, the grain boundary is considered as a short circuit for the diffusion. Despite the simplicity of the present description, the results for the simulation in Astroloy are very encouraging. Overal, the comparison of the intergranular γ' particles between both materials has revealed that coherency strain influences the morphology of the particle but not the principal mechanism of serration growth itself.

The present investigation has an important implication for the control of the grain boundary geometry : it is now admitted that the presence of serrations homogeneously distributed along the grain boundary can improve the crack growth resistance in superalloys. On the other hand, the development of very large fans in the materials are considered as detrimental from the standpoint of mechanical properties. It is therefore important to understand the conditions which favor the formation of one or another type of grain boundary geometry. Indeed such reaction products have already been observed in some commercial superalloys.

Achnowledgments

This work was sponsored by the Direction des Recherches Etudes et Techniques (DRET). The authors wish to thank Dr. M.F. Henry from General Electric and Mr M. Marty from ONERA for supplying the material.

References

1. H. Loyer Danflou, M. Marty, A. Walder, " Formation of Serrated Grain Boundaries and their Effect on the Mechanical Properties in a P/M Nickel Base Superalloys ", Superalloys 92, (Seven Springs, PA ,1992).
2. H. Loyer Danflou, " Serrated grain boundaries and resulting properties at 750°C of a P/M nickel base superalloy", (Ph.D.thesis, Paris XI - Orsay, 1993).
3. M.F. Henry, Y.S. Yoo, D. Y. Yoon and J. Choi, Metall. Trans. A, 24 (1993), 1733-1743.
4. F.V. Nolfi, P.G. Shewmon, J.S. Forster, Trans. A.I.M.E., 245 (1969), 1427-1433.
5. M. Hillert, Acta. Metall., 30 (1982), 1689-1696.
6. D.A. Porter, K.E. Easterling, Phase Transformation in Metals Alloys, (Editor Van Nostrand Reinhold, London,1981), 46.
7. R.A. Swalin, A. Martin, Trans. A.I.M.E., 206 (1956), 567-572.
8. J.W. Cahn, Acta. Metall., 7 (1959), 18.
9. C. Zener, Trans. Am. Inst. Min. Engrs, 167 (1946), 550.
10. M. Hillert, in The Mechanism of Phase Transformations in Cristalline Solids , (Institute of Metals, London, 1969), 231.
11. D. Locq, H. Loyer Danflou, M. Marty, A. Walder, " Superalliages à base de nickel pour disques élaborés par métallurgie des poudres préalliées pour utilisation au delà de 700°C ",Rapport Technique ONERA RT46/1931 (1993)

Appendix A : Nolfi's model

The boundary conditions of the Nolfi's model (4) (with $C'(r,t)=C_{eq}-C(r,t)$, C_{eq} equilibrium concentration, $C(r,t)$ solute concentration in the matrix at the position r, R interface position, D coefficient of solute diffusion, K reaction rate constant) :

$$\left(\frac{\partial C'}{\partial r}\right)_{r=R_s} = 0 \quad \text{(VIII)} \; ; \; D\left(\frac{\partial C'}{\partial r}\right)_{r=R} - KC'(R) = 0 \quad \text{(IX)}$$

The eigenvalues λ_n, δ_n are deduced from the equations (X) obtained by applying the boundary conditions to the expression C'(r,t) (with $\beta = R/R_s$; $\alpha_n = \lambda_n.R_s$; $\beta.\alpha_n = \lambda_n.R$; σ a constant reflecting the rate controlling mechanism $\sigma = (1+KR/D)^{-1}$) :

$$\tan(\alpha_n(1-\beta) - n\pi) = \frac{\alpha_n(1-\sigma\beta)}{1+\sigma\beta\alpha_n^2} \quad n = 0,1,2,3$$

$$\delta_n = \alpha_n - \tan^{-1}(\alpha_n)$$

(X)

By using the orthogonal properties of the Fourier function and the definition of the Fourier coefficient, Nolfi gets the expression of the constant A_n, C_n :

$$A_n = \frac{4\pi R_0^2 \Delta C_0 C_n (1-\sigma)\cos(\beta\alpha_n - \delta_n)}{\beta\alpha_n}$$

(XI)

$$C_n = \left\{(2\pi R_0)\left[\frac{(1-\beta)\alpha_n^2 - \beta}{\beta(1+\alpha_n^2)} + \frac{\sigma}{1+(\sigma\beta\alpha_n)^2}\right]\right\}^{-1/2}$$

The expression of the gradient of solute at the interface from the Nolfi's model (with $\tau_n = (\lambda_n^2 D)^{-1}$) :

$$\left(\frac{dC'(r,t^*)}{dr}\right)_R =$$

$$\sum_{n=0}^{n\max} A_n C_n \exp(-\lambda_n^2 Dt^*)\left[\frac{\lambda_n}{r}\cos(\lambda_n r - \delta_n) - \frac{1}{r^2}\sin(\lambda_n r - \delta_n)\right]$$

(XII)

Appendix B : Hillert's model (5)

Expression of the migration rate of the interface for discontinuous precipitation of rods (with r_β the radius of the rods, r_{tot} the total radius, X_0 the initial matrix composition, X_α the composition of the $\alpha 1$ phase and $X_{\alpha/\beta}$ the composition at α/β interface along the grain boundary, K constant distribution coefficient, δ thickness of the grain boundary, D_b grain boundary diffusion) :

$$v = 4K\delta\, D_b \cdot \frac{r_\beta}{r_{tot} \cdot (r_{tot}^2 - r_\beta^2)} \cdot \frac{(X_\alpha - X_{\alpha/\beta})}{(X_0 - X_\alpha)} \quad \text{(XIII)}$$

The rods and matrix thicknesses are related by mass balance with the following lever rule (with X_β composition of the β phase) :

$$\frac{(X_0 - X_\alpha^{av})}{(X_\beta - X_0)} = \frac{r_\beta^2}{(r_{tot}^2 - r_\beta^2)} \quad \text{(XIV)}$$

Expression of the reversible radius with σ surface tension, V_m molar volume, X_{eq} equilibrium concentration, G^m_α free energy of the α phase :

$$r_{rev} = \frac{2\sigma.V_m}{(\partial^2 G_\alpha^m / \partial X^2).(X_0 - X_{eq})^2} \quad \text{(XV)}$$

In the situation of rod growth, r_{tot} can be deduced from r_β by the following expression (with k_α fraction of chemical energy contributing to the creation of the α/β interface, L_α fraction of pressure carried by the α phase, X_{eq} the equilibrium concentration and r_{rev} the reversible radius (5)):

$$\left(r_{tot}^2 - r_\beta^2\right) = \frac{k_\alpha r_\beta^3 (X_\beta - X_0)^2}{2L_\alpha (X_0 - X_{eq})^2 r_{rev}} \quad \text{(XVI)}$$

THE STRUCTURE OF RENE' 88 DT

S.T. Wlodek*, M. Kelly** and D. A. Alden***
This study was performed at the
Engineering Materials Technology Laboratory
GE Aircraft Engines
Cincinnati, OH 45215.

Abstract

As heat treated, Rene' 88 DT was found to contain some 42.5% γ', with both the cooling (24.5%) and aging (18.0%) forms exhibiting a very low positive (0.05%) mismatch to the matrix. Small amounts of grain boundary M_3B_2, TiNb[C], and traces of M_5B_3 were also present. The rate of cooling from the solutioning temperature could be related to the resultant cooling γ' diameter by a regression equation, as could the aging conditions. The rate of cooling, as well as the grain size, were found to effect the grain boundary serrations that were formed on cooling from the super - solvus annealing temperature. Both of these features, and thus the cooling rate, may effect mechanical properties. On prolonged high temperature exposure, the γ' phases grow, and more M_3B_2 precipitates above 650°C (1200°F), followed by $M_{23}C_6$ above 705 °C (1300°F), and an intragranular μ phase at 760 °C (1400°F). Simple regression equations were fitted to the kinetics of growth of both the cooling and aging γ' by the use of a Larsen-Miller parameter to normalize the time and temperature of exposure. This allowed the prediction of the growth of the γ' phases and the possible identification of the average service temperature that a part experienced. The growth of γ' during high temperature exposure, rather then any precipitation of deleterious phases, was found to limit the service range of this structurally stable alloy. Indeed, the formation of $M_{23}C_6$ and μ start only after the aging γ' had completely re - solutioned.

Rene' 88DT is an extremely stable alloy for prolonged 650°C (1200°F) service. Its superior properties reflect a high γ' content and a γ' whose low positive mismatch allows a highly coherent and finer γ' precipitation. These advantages are optimized by a super solvus annealing practice that eliminates large, sub-solvus γ' phases, which promote crack nucleation in fatigue, and interfere with the formation of serrated grain boundaries.

Introduction

Rene' 88 DT was developed[1] to be more damage tolerant than Rene' 95, hence the DT designation, while offering improved creep strength and fatigue crack growth resistance[2]. The nickel base chemistry can be given as: 13% Co, 16% Cr, 4% Mo, 4% W, 2.1% Al, 3.7 % Ti, 0.7% Nb, 0.03% C and 0.015% B. The production alloy is always processed through the powder metallurgy route. The standard heat treatment consists of: a super - solvus solution of 1.0 hr at 1150°C (2100°F), followed by a delayed oil quench, and aging for 8.0 hr at 760°C (1400°F). Its main structural characteristics are thus a fine grain size, achieved through PM consolidation, and a duplex distribution of γ', the coarser forming on cooling from the super solvus solution, the finer predominantly on aging. Rene' 88 DT is used in disk applications in advanced General Electric Co. engines.

Due to the super-solvus nature of the solution anneal, the microstructure of Rene' 88 DT contains no sub-solvus γ', which is usually termed primary[3]. Normal convention refers to the cooling γ' as secondary, and the γ' that forms predominantly on aging, as tertiary. That terminology will be used here.

This study completely characterized the structure of Rene' 88, including the effect of heat treatment and prolonged exposure on the structure of this alloy. Due to the well documented effect of γ' size[4], and the possible effect of grain boundary morphology[5] on properties, particular attention was paid to characterizing the effect of heat treatment on these features.

Experimental

Most of the structural studies reported here were performed on a full size production disc forging produced through the conventional PM route, with an average grain diameter of 60 μm. In order to ascertain the role of grain size on cooling rate effects, some limited work was also performed on a cast version of the alloy. This was produced by extruding a 43 cm. VAR ingot to a 15 cm diameter billet. The VAR material exhibited an average grain diameter of 130μm. The chemistries of both materials are given in Table I. Both chemistries were similar, and well within specification limits. Conventional metallographic, x-ray, and analytical electron microscopy procedures, as used in similar studies[3], were employed. All the image analyses and grain boundary morphology studies were performed on SEM images obtained by examining unmounted, electropolished, specimens in a field emission, scanning, electron microscope. As such morphological studies required very close control of specimen preparation [6], they will be described in greater detail.

Measurements of γ' were made on surfaces electropolished in cryogenic (-40 to -50 °C) 80% CH_3OH - 20% $HClO_4$ at 25 volts, for about 25 seconds, followed by immediate etching performed at 5 volts for about 3 - 5 seconds, in the same electrolyte and at the same temperature. This procedure produced almost a mirror level of flatness, with just enough relief to allow easy phase discrimination at the large magnifications that must be used to resolve the sub - micron γ'. Nevertheless, as the tertiary γ' is as small as 0.01 μm, a truly planar surface cannot be achieved on that scale. The observed γ' phases thus would not always lie in exactly the same plane. This is exacerbated by the excellent depth of field of a field emission SEM. The result is that, although the size and shape of the minute γ' phases are faithfully revealed, permitting the accurate measurement of the size of these features by image analysis, their volume is not. Attempts to determine the volume fraction of sub-micron precipitates by image analysis thus lead to an overstatement of the volume present. The amounts of phases were therefore determined by weighing electrolytic extractions, which is regarded as the preferred technique.

*Gamma Prime Consultants, San Diego, CA 92127-1272.
**Shepherd Color Co., Cincinnati, OH 45246.
***RITSS, Cincinnati, OH 45237-0383.

Due to the possibility that grain boundary morphology has some relation to mechanical properties, a simple procedure was also utilized to measure the straightness of grain boundaries. The grain boundaries were best revealed by electropolishing in 90% CH_3OH - 10% HCL, at -40 to -50 °C, for about 25 seconds at 25 volts, and etching for 3 to 5 seconds at 5 volts, in the same cycle, electrolyte and temperature. This etches out the γ', which can obscure the true grain boundary shape, but leaves the carbides, borides and oxides unaffected. It was then possible to record the grain boundary morphology at a triple point, at a magnification of 25,000. This allowed a characterization of the grain boundary curvature. That characterization consisted of measuring the true length of the three grain boundaries, meeting at the triple point, and dividing it by the straight line distance, measured for the same points of reference. Averaging six to ten such measurements yielded a grain boundary curvature ratio (GBCR) that served as a quantification of grain boundary morphology.

Analysis of Fully Heat Treated Structures

The microstructre of an as fully heat treated Rene' 88DT forging is shown in Figures 1 and 2, as revealed by both preparation procedures.

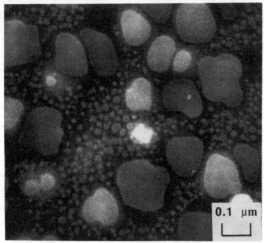

Figure 1. Morphology of γ' in as heat treated Rene' 88 DT, perchloric - metholic preparation. Note 0.14 μm cooling γ', and 0.017 μm, aging γ'.

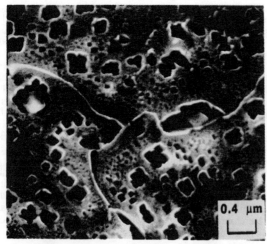

Figure 2. Triple point in fully heat treated Rene' 88 DT, hydrochloric - metholic preparation, revealing equiaxed borides and convoluted, or serrated, grain boundaries (GBCR = 1.16).

The γ' phases were extracted in the standard 1% ammonium persulfate + 1% citric acid solution. Using a combination of differential settling and centrifuging, the two size fractions of γ' were separated, to allow their chemical analysis and structural characterization by x-ray diffraction. The lattice parameter of the cooling, secondary γ' which, in the as heat treated, condition exhibited an average diameter of 0.14 μm, was 0.35917 nm. The 0.017 γ' aging, tertiary, γ' lattice parameter was found to be 0.35923 nm. Compared to the 0.3590 nm parameter of the γ matrix, both γ' phases reflected a positive mismatch of 0.05%. The chemical analysis of the cooling γ' is included in Table I, as is the analysis of the γ. The latter was calculated from the analysis of the liquor resulting from the phase extraction. The amount of the aging, tertiary, γ' that was recovered was insufficient to permit chemical analysis.

Weights of extracted residue indicated that the alloy contained a total of 42.5 wt. % γ', distributed amongst the cooling (60%), and aging (40 %) forms.

X-ray diffraction analyses, of the metholic-HCl+tartaric extraction residue, identified the presence of a tetragonal, high Cr plus Mo and W containing M_3B_2 ($D5_a$, a = b = 0.5787 nm, c = 0.31228 nm), and a high Ti, lower Nb, MC (B_1, a = 0.43514 nm). SAED procedures identified, as shown in Figure 3, a rare, tetragonal, high Mo and W plus Cr containing, M_5B_3 ($D8_l$, a = b = 0.572 nm, c = 1.08 nm). Some trace amounts of acicular eta (Ni_3Ti) were probably also present. All of these minor constituents amounted to about 0.2% of the total weight. All of these boride and carbide phases were similar in their microscopic appearance, and can be observed in Figure 2 as small, (0.1 to 0.5 μm) equiaxed, grain boundary phases. Despite their low concentration, their amount was sufficient to partially impair grain growth.

Table I Chemical Analyses

Weight %	PM Disk	VAR Bar	Secondary γ'	Matrix γ	Partitioned to Secondary γ'
Cr	15.70	16.08	2.1	22.0	0.095 : 1
Co	12.90	13.05	5.0	17.0	0.29 : 1
Mo	3.98	4.08	1.2	6.0	0.20 : 1
W	3.98	4.05	2.8	3.3	0.85 : 1
Ti	3.84	3.66	9.0	0.7	1 : 0.078
Al	2.21	2.33	4.0	0.7	1 : 0.175
Nb	0.72	0.69	1.6	0.3	1 : 0.19
Zr	0.04	0.05	*	*	
C	0.041	0.040	*	*	
B	0.015	0.015	*	*	
P	<0.01	0.01	*	*	
O	0.014	0.006	*	*	
N	0.029	0.006	*	*	
N_{v3}	2.26	2.30		2.17	

* Not analyzed. Partitioning ratio expresses concentration in γ' compared to γ. All γ' analyses refer only to secondary form.

The Rene' 88 that was consolidated by VAR, exhibited the same structural features and appreciable amounts of eta phase. Examination of the VAR ingot revealed center line segregation of eta and borides, that could not be homogenized. An ingot route could not thus be applied to this alloy.

Differential thermal analysis (DTA), backed up by microstructural examination, was used to determine the primary transitions in Rene' 88 DT. The more important were:

Liquidus, 1355°C (2471 °F),   Solidus, 1250 °C (2283 °F),
γ' solvus, 1130 °C (2065 °F),   Eta solvus, 1185 °C (2167 °F).

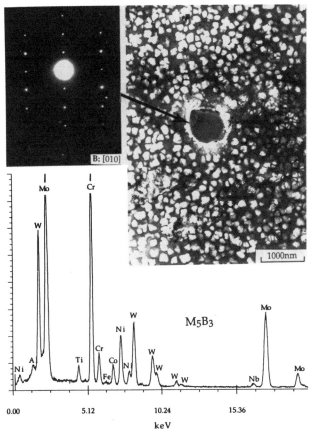

Figure 3. Identification of a spheroidal, intragranular, phase as the tetragonal M_5B_3 ($D5_a$). X-ray, energy dispersive, analysis indicated a high W and Mo chemistry with an appreciable Cr level.

Structural Response to Heat Treatment

Effective heat treatment of highly alloyed disk compositions required a controlled cooling rate from the solution annealing temperature, followed by an aging treatment.

During cooling from the super - solvus solution anneal of 1150°C, two structural changes occurred, the extent of both being a function of the cooling rate. The first is the precipitation of secondary γ', whose particle size varied inversely with the cooling rate. The second was the formation of convoluted grain boundaries, a process that partially depended on the as annealed grain diameter and became most pronounced within a specific range of cooling rates.

Both effects were studied by cooling 0.6 cm x 1.2 cm x 1.2 cm coupons, each spot welded to a thermocouple that recorded the cooling rate on a high speed recorder. In each case, cooling was initiated after a 1.0 hr. anneal at 1120°C and, for slow cooling rates, terminated at about 870°C by water quenching the specimen. The wide variation in cooling γ' morphology, that can occur, is illustrated by the microstructures that were observed on cooling at 30 °C /minute (Figure 4) and 2800 °C / minute (Figure 5). Note that copious γ' formed at even the fastest rates and, at very slow cooling rates, some tertiary " aging" γ' was observed .

The quantification of these factors is shown in Figure 6 for the effect of cooling rate on gamma prime size, and in Figure 7 for the effect of grain diameter, at two cooling rates, on grain boundary curvature. In Figure 6, each point reflects the average of some 100 γ' size measurements. It illustrates that at cooling rates faster then about 300 °C / min only one size of γ' formed. At slower cooling rates, two sizes were observed. The larger is believed to form

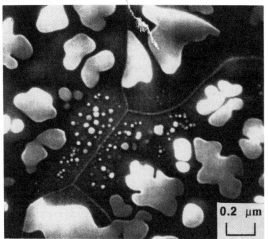

Figure 4. Cooled at 30°C / minute. Large 0.2 μm, secondary γ' formed on cooling approached almost a dendritic morphology, with trace amounts of tertiary (0.016 μm) γ' also present at curved grain boundaries.

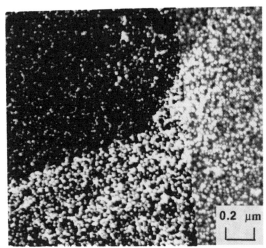

Figure 5. Cooled at 2800 °C / minute. Only a very fine 0.020 μm cooling γ' is present, and the grain boundaries are completely straight.

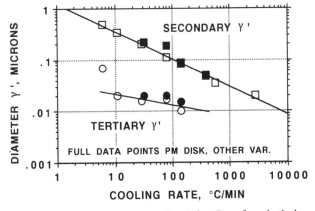

Figure 6. Effect of cooling rate on the γ' size. Data from both the production PM disk (full points), and VAR consolidated bar, show no effect of consolidation or grain diameter.

strictly on cooling and is termed secondary. It is believed that the very small γ' that forms at cooling rates slower then about 300° C / min, is precipitated in the latter part of the cooling cycle, through essentially an aging reaction. It is thus termed as tertiary, or aging γ'. The particle size of the tertiary γ' is not a strong function of cooling rate. It is always spherical. The secondary, cooling γ' is spherical at fast cooling rates, becoming cuboidal at about 100°C / min and tending to "dendritic" in the octahedral directions, almost cloverleaf in section, at even slower cooling rates. In all cases a curved, and probably coherent, interface is retained between the γ' precipitate and γ matrix.

The diameter D (μm) of the secondary, cooling γ', can be related to the cooling rate d°C / dt (°C per a minute) by the equation :

$$\log D\gamma'_2 = 0.178 - 0.551 (\log d°C / dt), \quad R^2 = 0.93,$$
(Eqn. 1)

The extent of grain boundary curving was measured on specimens electropolished and etched in metholic HCl, so as to etch out the γ', and clearly delineate the grain boundaries. The measurements were quantified by comparing the actual length of the three grain boundaries at a triple point, as measured at X25,000, to the straight line distance, to determine a quotient that was termed the Grain Boundary Curvature Ratio (GBCR). As shown in Figure 7, the degree of grain boundary curvature was found to be a function of both the cooling rate and the initial grain diameter.

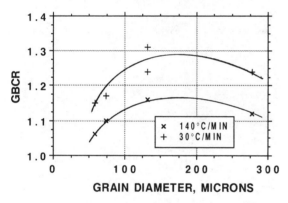

Figure 7. Variation of the ratio, actual grain boundary length divided by the straight line distance (GBCR), as a function of grain diameter, for two cooling rates. Slower cooling rates, and grain diameters in the 200 micron range, promote the formation of convoluted grain boundaries, indicated by higher GBCR ratios.

The effect of variations in aging conditions was studied by oil quenching coupons (0.6 cm x 1.2 cm x 1.2 cm) and aging them for 1 to 16 hr at 650° to 1040°C. Figure 8 shows the resultant relationship between the measured Rc hardness and the Larsen-Miller (L-M) parameter ($P = °K [25 + \text{Log } t] \times 10^{-3}$), where the aging temperature is expressed in °K and t is the aging time in hours. The variation in the diameter of the tertiary γ' that formed on aging, and the Larsen-Miller parameter P, is given in Figure 9. When the aging was performed at L-M parameters much above 29, the tertiary γ' was no longer observed. This suggested that a solvus for tertiary of γ' exists at this combination of time and temperature.

The data given in Figures 8 and 9 can be fitted to the following regression equations, which relate the Rc hardness and the diameter ($D\gamma'_3$) of the tertiary γ', in μm to the parameter of the aging exposure, $P = °K (25 + \log t) \times 10^{-3}$.

$$Rc = -253.92 + 21.371 P - 0.38722 P^2,$$
$$R^2 = 0.94 \quad \text{(Eqn. 2)}$$

$$D\gamma'_3 = 1.0286 - 0.085216 P + 0.0017691 P^2$$
$$R^2 = 0.99 \quad \text{(Eqn. 3)}$$

Since, at higher exposure conditions no tertiary γ' was observed, the validity of equation 3 is limited to P<29.

During aging, slight further growth occurred of the secondary, γ'. The amount of this growth increased inversely with the cooling rate. The main structural process that occurred during aging was, however, the precipitation of the tertiary γ' and it's growth. The precipitation was most concentrated along grain boundaries and continued until this form amounted to about 40% of the total γ'.

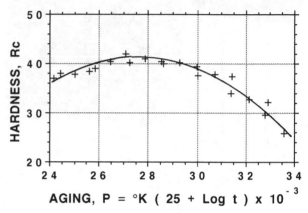

Figure 8. Relationship between hardness and the L - M parameter of the aging exposure. Note that at P>28 the hardness begins to decrease.

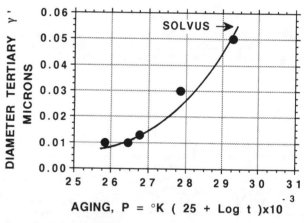

Figure 9. Relationship between tertiary γ' size and the L-M parameter of the aging exposure. The tertiary γ' was not observed in specimens subjected to aging conditions, whose time and temperature exposure > P = 29.

Effect of Long Time Exposure

The effect of long time exposure was studied for just the conventional PM version of Rene' 88 DT. As heat treated material was exposed for up to 1000 hours at 650° to 790°C, with shorter time exposures to 1040 °C. In order to evaluate the effect of stress, three unfailed creep specimens, that had been tested for up to 6400 hours in the temperature range of 680° to 760°C, and at stress levels of 276 to 689 MPa, were also examined.

The growth of γ' on long time exposure is portrayed in Figure 10. All the data, whether obtained in the presence or absence of stress, could be fitted to a Larsen-Miller relationship $P = °K (25 + \text{Log } t) \times 10^{-3}$, where the temperature of exposure is in °K, and t is the time, in hours.

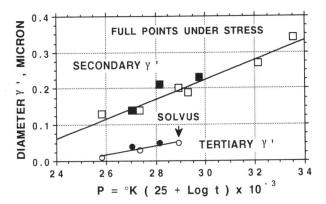

Figure 10. The change in diameter of both the secondary, cooling, γ' and the tertiary, aging, γ' can be correlated to the exposure conditions through a L - M parameter with C = 25. Note that the γ' that grew under stress, designated by full data points, falls in the same population as that which grew in the absence of stress.

This allowed the development of the following regression equations, that gave the diameter D, in μm, of both the secondary γ', that formed on cooling, and with less accuracy, the tertiary γ', that formed largely on aging, as a function of P.

For secondary γ', $D_{γ'2} = 0.02833\ P - 0.57557$, $R^2 = 0.93$ (Eqn. 4)

For tertiary γ', $D_{γ'3} = 0.01311\ P - 0.32435$, $R^2 = 0.82$ (Eqn. 5)

It should be noted that the growth of the aging, tertiary γ', stopped when this precipitate reached a diameter of some 0.05 μm. Exposure at L - M parameters of about 29, or larger, caused the tertiary form of γ' to completely dissolve. In addition to the above changes in γ' morphology, grain boundary films of γ' began to form at exposure conditions with a LM parameter of 25.8, or above. This is equivalent to exposures above 650°C and 1000 hr. Although the total amount of secondary and tertiary γ' does not change (< 5%) appreciably during high temperature exposure, changes do occur in the relative amounts of these two forms. At 650 °C the amount of tertiary γ' increases with exposure, reaching about 50% of the total γ' content after 1000 hr. At higher temperatures the growth of the tertiary γ' was accompanied by a decrease in the amount of this form. After 1000 hr at 705°C and 760°C, the tertiary γ' content was, respectively, 30% and 16% of the total. These are estimates, based on image analysis, and probably over-state the amount of tertiary γ'. Because of these complex changes, the variation of hardness with exposure is negligible, until a L-M parameter of 25 is exceeded. The scatter in hardness data was large, and a statistically significant equation could not be used to describe the change in hardness with high temperature exposure.

X-ray diffraction analysis of the metholic-HCl-tartaric electrolytic extraction residue, combined with analytical electron microscopy procedures, served to identify the phases that formed on exposure. Prolonged exposure (> 2,000 hr) at 650 °C, or for shorter times at higher temperatures, produced grain boundary precipitation of a high Cr and Mo + W, M_3B_2 ($D5_a$, a = 0.579 nm, b = c = 0.311 nm). For exposure conditions longer than 1000 hr. and at 705°C, or above, possibly associated with the solution of the tertiary γ', a high Cr, $M_{23}C_6$ ($D8_4$, a = 1.0706 nm) phase formed. Both the M_3B_2 and $M_{23}C_6$ precipitated largely at grain boundaries, often within the previously noted γ' films. In addition, an intragranular precipitation of small amount of MC (B1 a = 0.434 nm) which, being high in Zr, rather then Nb, differed in chemistry from that found in the as heat treated condition. It must be stressed that these precipitation reactions were observed above the expected temperature range for the long term utilization of Rene' 88 DT, and involved very small amounts of precipitation. For instance, the total amount of all borides and carbides recovered by extraction from a specimen that had been aged for 1000 hr at 760 °C, well above any expected service temperature, was only 0.7wt %.

In addition to these largely intergranular precipitation reactions, the formation of an intragranular, acicular, precipitate was observed in a specimen that had been aged for 6,300 hr at 760°C and 276 MPa. This phase, as shown in Figure 11, was identified using AEM procedures as a high Cr, Mo and W containing μ.

All of the results on alloy stability were obtained on production disk material with the chemistry given in Table I and $N_V3 = 2.26$.

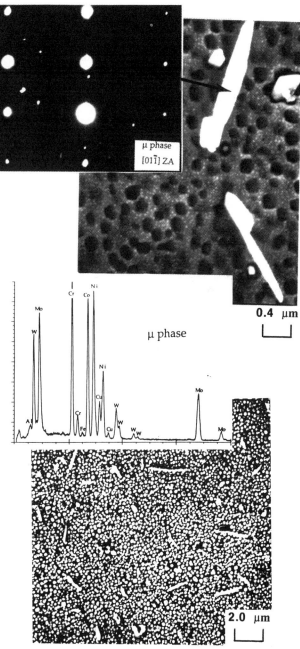

Figure 11. AEM identification of rhombohedral μ phase after 6300 hr. at 760°C and 276 MPa. Note high Cr and Mo+W chemistry.

The structural data for all the phases found in this study, for both the as heat treated condition and after prolonged elevated temperature exposure, are compiled in Table II.

Table II. Phases Found in Rene' 88 DT

Phase	Structure	Lattice Parameter nm
Gamma	A1	a = 0.35890
Secondary γ'	L1$_2$	a = 0.35917
Tertiary γ'	L1$_2$	a = 0.35923
M$_3$B$_2$ Cr, Mo, W	D5$_a$	a = b = 0.5787, c = 0.31228
M$_5$B$_3$ Mo, W, Cr	D8$_l$	a = b = 0.572, c = 1.08
Ti,Nb (C)	B1	a = 0.43514
Zr,Ti (C)*	B1	a = 0.434
M$_{23}$C$_6$ *	D8$_4$	a = 1.0706

* Found only after high temperature exposure. Eta in the as heat treated condition and, after prolonged exposure, a μ phase, were also identified but their exact lattice parameters were not determined.

Discussion

The processing of Rene' 88 DT through a powder metallurgy route allows greater alloying levels and a finer grain size. The use of a super-solvus anneal further benefits properties. This alloy exhibits the lowest γ' misfit of any commercial alloy [7]. Although the solute distribution between the γ' and γ phases is essentially the same as has been documented for other commercial alloys [8,9], the balance of elements is such, that effective alloying of both the γ and γ' phases has been achieved without introducing structural instabilities in it's normal operating range. In addition, the boron and carbon are low enough, so that the precipitation of grain boundary borides and carbides is not embrittling.

The most critical operation in the heat treatment of an advanced disk alloy is the achievement of an optimum cooling rate from the solution temperature. The faster the cooling rate, the smaller is the resultant secondary γ' size, and usually, the better the creep and tensile properties. The practical limit of this trend is that very fast cooling rates can cause high residual stresses, leading to quench cracking. Fortunately, Rene' 88 DT appears to be less sensitive to quench cracking then other disk alloys[2]. The relationship between the cooling rate dT/dt and the diameter of the secondary γ', can be calculated by equation 1. Previous studies[2], of Rene' 88 DT, have shown that preferred properties are associated with a cooling rate of some 140°C / min, a rate that should result in a secondary γ' size of some 0.1 μm. The availability of equation 1 permits the metallographic verification of the cooling rate that had been used in the production forging. Thus, the 0.13 μm γ' size of the disk which was used in this study, suggests that the cooling rate from the super - solvus anneal was about 85°C/min.

The effect of aging on tertiary γ' formation can be fully quantified, not only in terms of hardness (equation 2), but also in terms of the tertiary γ' size (equation 3). Both the size of the tertiary γ' and the Rc hardness can be related to the Larsen-Miller relationship that describes the time and temperature of aging. These equations may allow the verification of appropriate aging conditions in a production part through a metallographic examination. In addition, they can clarify some of the characteristics of the aging reactions in Rene' 88 DT. When solved, equations 2 and 3, predict a maximum hardness on aging at P = 27.4 or, for an 8 hr age, at 785°C, at which value the tertiary γ' diameter would be some 0.022 μm.

These measurements gave a good approximation of the specification aging conditions, the resultant hardness and the optimum tertiary γ' size, above which the hardness would drop. The absence of tertiary γ' after aging at L-M parameters >29, should be noted.

When solutioned and aged in the preferred processing range, Rene' 88 DT exhibits a structure containing some 42.5 % γ', distributed between the secondary (60 %) and tertiary (40 %) forms. Both types of γ' exhibit only a small (0.05 %), positive mismatch with the matrix. Because of this high level of coherency, the tertiary γ' is always spherical, and when the secondary γ' departs from spheroidicity at larger sizes, it tends to maintain a structure typified by a spheroidal morphology.

When compared to other disk alloys, such as Rene' 95[10] or N18[3], Rene' 88 DT forms a finer secondary γ', for an equivalent cooling rate. This characteristic benefits properties and results directly from the low mismatch of the γ', which not only produces a highly coherent precipitate, but also reduces the free energy change that must be overcome on precipitation, increasing the nucleation rate of the γ'.

The low γ'/γ mismatch, that is so beneficial to properties in the temperature range in which disk alloys are used, is a direct result of the judicious use of tungsten[13] which, unlike Mo [9], increases the lattice parameter of γ, without appreciably effecting γ'.

The beneficial effects of a coherent γ' may also reflect in the relatively low rate of growth of this phase during service. The growth of γ' during high temperature exposure can also be described by exploiting the L-M relationship in equations 4 and 5, to relate the time and temperature of exposure to both the secondary and tertiary γ' sizes. The measurement of the γ' size, after service, could thus allow the approximation of a parts average service temperature. Only a gross over temperature has any measurable effect on room temperature hardness, and usually no correlation of hardness and service temperature is possible for Rene' 88 DT. Measurements of γ' size, through a field metallographic procedure, and the use of equations 4 and 5, could prove particularly useful in any investigation where an estimate has to be made of the average service temperature.

It should also be noted that Rene' 88 DT achieves its property level at a much lower γ' content (42.5%) then Rene' 95 [10] (50%), or N18 [3] (58%).

The relationship between grain boundary curving, or formation of serrations, and propeties, is not well understood. Grain boundary curving is effected by both the cooling rate and grain diameter, but the relationships are not simple. The serrations are a result of grain boundary movement during sub- solvus cooling from the solution anneal. The serrations documented here are not due to the growth of γ' into a grain boundary. They are produced by the dynamic movement of a grain boundary into freshly nucleated, secondary, γ' precipitates. For a constant cooling rate, the force driving this movement appears to increase with the length of the unsupported grain boundary.

For an equivalent grain size, such grain boundary movement is much more pronounced in alloys that are given an anneal above, rather then below, the γ' solvus. Alloys such as Rene' 95 and N18, which are annealed below the γ' solvus, and thus contain large, primary, γ' during cooling from the annealing temperature, do not develop curved, or serrated, grain boundaries as readily as Rene' 88 DT. In addition, annealing above the γ' solvus removes the large primary γ' phases that occur in Rene' 95 and N 18, and serve as a preferred point of fatigue crack nucleation [3]. Annealing above the γ' solvus, however, introduces the possibility of grain growth. In the case of a sub-solvus annealed alloy, such grain growth is contained by the presence of large, primary, γ'.

During prolonged high temperature service the secondary and tertiary γ' grow and increase slightly in amount. Their growth, which, can be predicted by equations 4 and 5, does not vary greatly with the presence of stress during exposure. If the relationship between the mechanical properties and the γ' size could be established, and this has been achieved for other alloys [4, 11 and 12], these equations should prove useful in predicting the approximate level of properties that could be expected after prolonged service exposure.

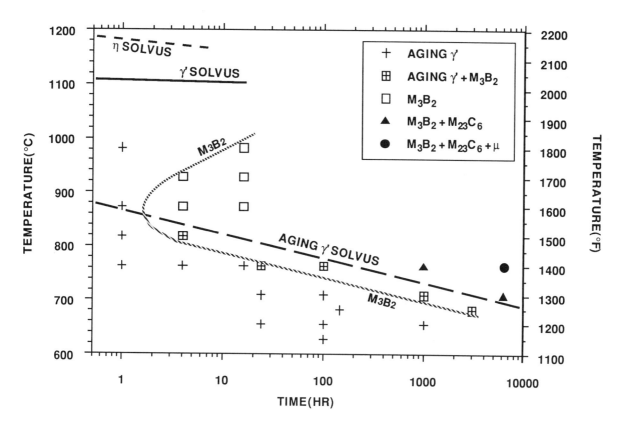

Figure 12. Schematic summary of phase reactions in Rene' 88 DT. Prolonged exposure produces the growth of both the secondary and the tertiary forms of γ' and the formation of grain boundary γ' films. Ultimately, at exposure conditions equivalent to LM = 29 the tertiary γ' that formed on aging dissolves. Continued exposure produces grain boundary M_3B_2 and $M_{23}C_6$ and, after prolonged exposure at 760°C, μ phase appears. Full points were stressed exposures.

In the meantime, it is possible to point out that when the L-M parameter exceeds 29, the tertiary γ' is completely dissolved and a very large reduction in load bearing properties should result. Some loss of properties should, of course, be also expected when the secondary and/or tertiary γ' diameter exceeds some critical size. If the relationship between any specific mechanical property and the sizes of the γ' phases were known, equations 4 and 5 could be used to calculate the L-M parameter associated with any minimum property level.

The tertiary γ' solvus, as well as the areas of stability for the various secondary phases found in Rene' 88 DT are indicated in Figure 12.

As heat treated, Rene' 88 DT contains only some 0.2 wt.% borides and carbides, predominantly M_3B_2, with some Nb rich MC, both largely at grain boundaries. Small amounts of eta phase and a M_5B_3 constituent were also observed, but in truly trace amounts. None of these phases are in any way unusual, but note should be taken of the fact that the major grain boundary constituents were borides and not carbides.

At the $N_v3 = 2.26$ chemistry used in these studies, the alloy is completely free of any deleterious precipitation reactions in its nominal service range. At 650 °C, a 2000 hr. exposure is required before even any additional M_3B_2 forms. Although, grain boundary films of $M_{23}C_6$ and intragranular μ appear at higher temperatures, this happens only after the tertiary γ' solvus is exceeded, at which point the alloy is well outside its long time service temperature capability. Rene' 95 and N - 18 are more prone to the formation of a topological close packed phase, than Rene' 88 DT. Both form μ, and N-18 is also subject to the precipitation of σ [3, 10].

Rene' 88 DT is, however, very highly alloyed and is an excellent example of the advantage of processing a highly alloyed composition through a powder metallurgy route. Due to the tendency to form segregation induced eta phase, in even small ingots, this alloy could never be processed through a conventional ingot practice.

Studies of this type allowed an understanding of an alloy's behavior, identification of the factors that must be controlled in it's processing and heat treatment, a method of identifying occurrences of inappropriate heat treatment, identification of service induced over-temperature. When combined with appropriate mechanical data, such studies may allow the prediction of behavior in service, particularly the degradation of mechanical properties. This could allow a more effective design philosophy, that need not be based completely on virgin, as heat treated, mechanical property levels.

In closing, one caution is required. The regression equations presented here may not reflect a level of great theoretical validity. Their use should not, therefore, be extrapolated past the range of the data on which they were based. Still, they are presented in the spirit of Lord Kelvin's admonition:

"When you can measure what you are speaking about, and explain it in numbers, you know something about it."

Acknowledgments

The authors are indebted to the management of the Engineering Materials Technology Laboratory, GE Aircraft Engines, for permission to publish this paper.

References

1. D.D. Krueger, R.D. Kissinger, R.D. Menzies and C. S. Wukusick, US Patent 4,957, 567.

2. D.D. Krueger, R.D. Kissinger and R.D. Menzies, "Development and Introduction of a Damage Tolerant High Temperature Nickel-Base Disk Alloy, Rene' 88 DT", Superalloys 1992, S.D.Antolovich et al Editors, TMS-AIME, Warrendale PA, (1992), 277- 286.

3. S.T. Wlodek, M.Kelly, and D. Alden, "The Structure of N 18", Superalloys 1992, S. D. Antolovich et al Editors, TMS-AIME, Warrendale PA, (1992), 467- 476.

4. P.R. Bhowal, E.F.Wright and E.L. Raymond, "Effects of Cooling Rate and Morphology on Creep and Stress Rupture of a Powder Metallurgy Superalloy", Met.Trans.A, 21, (1990), 1709-1717.

5. M. Zhiping, Y. Ruizeng, and G. Liang, " Effect of Zigzag Grain Boundary on Creep and Fracture Behavior of Wrought γ' Strengthened Superalloy", Materials Science and Technology, 4, (1988), 540 - 547.

6. S.T. Wlodek, "SEM Metallography of Superalloys", Microstructural Science, T.A. Place et al Editors, ASM, Metals Park OH, 18, (1990), 407- 429.

7. B.A. Parker and D.R.F. West, " The Influence of Particle/Matrix Interfacial Mismatch on the Size, Shape and Distribution of Coherent γ' Particles in Nickel Based Alloys", J. Australian Inst. of Metals", 14 , (1969), 102-110.

8. O.H. Kriege and J.M. Baris, " The Chemical Partitioning of Elements in Gamma Prime Seperated from Precipitation - Hardened, High - Temperature Nickel Base Alloys", Trans. ASM, 62, (1969), 195 - 200.

9. J.P. Collier, P.W. Keefe and J. K. Tien, "The Effects of Replacing the Refractory Elements W, Nb, and Ta with Mo in Nickel - Base Superalloys on Microstructural, Microchemistry, and Mechanical Properties", Met. Trans. A, 17A, (1986), 651 - 661.

10. S.T. Wlodek, Unpublished results.

11. J.J. Schirra and S.H.Goetschius, "Development of an Analytical Model Predicting Microstructure and Properties Resulting From the Thermal Processing of a Wrought Powder Nickel - Base Superalloy Component", Superalloys 1992, S.D. Antolovich et al editors, TMS-AIME, Warrendale PA, (1992), 437-446.

12. S.T.Wlodek and R.D. Field, "The Effects of Long Time Exposure on Alloy 718", Superalloys 718, 625, 706 and Various Derivatives", E.A.Loria editor, TMS-AIME, Warrendale PA, (1994),659-670.

13. A. Havalda, "Investigation of the Influence of Tungsten on the Morphology of the γ' Precipitation in Nickel - Base Alloys", Trans. ASM, 62, (1969), 477-480.

"DEVELOPMENT OF A DAMAGE TOLERANT HEAT TREATMENT FOR CAST + HIP INCOLOY 939"

Robert W. Hatala and John J. Schirra
Pratt & Whitney
United Technologies
East Hartford, CT

Summary

Current and advanced high thrust and improved efficiency turbine engines require alloys capable of operating at increased temperatures and improved durability. For cast diffuser case applications, additional requirements of weld reparability and castability are also imposed. The cast equiaxed turbine blade alloy Incoloy 939 demonstrated many of these requirements during initial component demonstrations, however preliminary mechanical property characterization showed that material processed with the standard heat treatment exhibited a significant acceleration (and associated intergranular fracture) in 649°C hold time crack growth capability. Metallographic evaluation showed the material to exhibit a continuous grain boundary $M_{23}C_6$ precipitate. A design of experiment approach (Taguchi L8) was utilized to develop a modified heat treatment. Key factors identified were the cooling rate from solution heat treatment, use of a secondary stabilization cycle and slight lowering of the stabilization cycle temperature. Application of the modified heat treatment resulted in suppression of the 649°C hold time crack growth acceleration and a transition to a transgranular fracture mode. Metallographic evaluation showed that the modified heat treatment produced a discrete grain boundary $M_{23}C_6$ precipitate. Additional mechanical property evaluations showed the modified heat treatment resulted in a slight (<10%) reduction in room and elevated temperature yield strength and improved ultimate strength and tensile elongation. There was no impact on stress rupture and fatigue (smooth and notch) capability.

Introduction

Increased thrust and performance demands of advanced turbine engines require materials capable of operating at higher temperatures and or stresses. For example, the current generation of very high thrust commercial engines required diffuser case materials with a 56°C to 111°C increase in operating capability over the alloy (cast + HIP Inconel 718) used for lower thrust models. Key characteristics of Inconel 718 that make it attractive for diffuser case applications are its relatively good castability and excellent weldability. Two shortcomings observed in the alloy are a propensity to segregate and a tendency to develop surface connected shrinkage porosity that does not heal during the HIP process. Inconel 718 tends to be limited to applications up to 555°C to 625°C.

Alloys considered for increased temperature applications include Mar-M-247, Inconel 713 and Incoloy 939. Wasploy is also a consideration but it does not offer the temperature capability of the previous three. It should be noted that each of the alloys considered for investment cast diffuser case (D/C) applications have all been extensively and successfully used as cast equiaxed turbine airfoils and/or buckets. Applying the weldability criteria of Shira (ref 1), Incoloy 939 appears to be have the best inherent weldablility of the alloys (Figure 1). Research has been conducted (ref 2 to 5) that

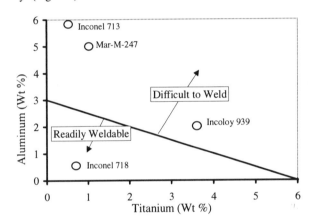

Figure 1
Weldability of Potential and Baseline Investment
Cast Structural Case Alloys Based on Shira [1]

suggests Inconel 713 could exhibit adequate weldability with Mar-M-247 expected to demonstrate the least weldability. An evaluation conducted by P&W verifed Incoloy 939 to be the most weldable of the alloys. A significant amount of development activity had also been undertaken by the structural casting industry in identifying the steps required to successfully produce and process investment cast Incoloy 939. Cast test plates were evaluated by P&W to assess the design capability of the material. The key promising result for the material was the lack of a dwell crack growth acceleration (relative to rapid cycle testing) at 649°C. Additional mechanical property testing showed adequate capability for diffuser applications. Based on this, additional characterization was conducted. Full scale evaluation of production PW4000 diffuser cases cast from Incoloy 939 showed the material to exhibit excellent castability and unlike Inconel 718, tended to develop centerline porosity resulting in an enhanced response to HIP. A typical PW4000 D/C cast from Incoloy 939 is presented in Figure 2. Consistent with preliminary

Figure 2
PW4000 Diffuser Case cast from Incoloy 939

results, weld trials showed the material to exhibit adequate weldability (ref 6,7). The heat treatment selected for the material was that applied to the cast test panels. The cycle consisted of a industry standard three step solution, stabilization & precipitation

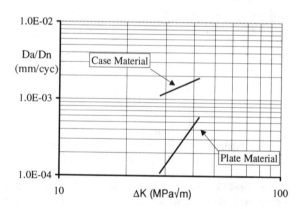

Figure 3
649°C Hold Time Fatigue Crack Growth Rate of Specimens Machined from the Diffuser Case Processed with the Standard Heat Treatment Exhibited a Significant Accelleration over Specimens Machined from the Plates.

process that was determined to produce the best balance of properties for blade application (ref 8). Evaluation work conducted by the casting industry (ref 9) also suggested that the three step heat treat process produced the best balance of monotonic properties. Hold time (2 minute dwell) crack growth testing at 649°C of specimens excised from the casting showed a significant acceleration over the initial cast test panel results (Figure 3). This behavior was confirmed by subsequent duplicate tests. Fractography of the tested specimens showed the case material (fast crack growth) to exhibit a predominantly intergranular/ interdendritic propagation mode with the test panel material showing predominantly transgranular fracture. (Figures 4 & 5). A comparison of the chemistries (Table I)

Figure 4
Case Material with Accelerated Crack Growth Rate Exhibited an Intergranular Propogation Mode Under 649°C, Dwell Conditions

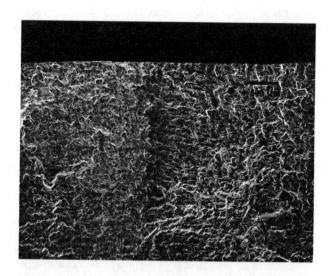

Figure 5
Plate Material with Desirable Crack Growth Rate Exhibited an Transgranular Propogation Mode Under 649°C, Dwell Conditions

showed no obvious reason for the difference in behavior. Standard metallographic characterization also failed to identify a cause for the accelerated crack growth with both the test panels and case material exhibiting similar microstructural characteristics (Figure 6). Additional characterization of the grain boundary structure

Table I
Chemistries (wt %) of Incoloy 939 Case and Plate Material

Element	Test Panels	Diffuser Case
Cr	22.7	22.7
Co	18.6	18.6
W	2	2
Al	2	2.1
Ti	4	3.6
Ta	1.4	1.3
Nb	1	1
C	.13	.13
Ni	balance	balance

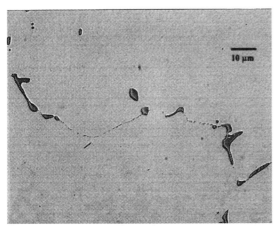

Figure 6-a
Typical Microstructure of Plate Material

Figure 6-b
Typical Microstructure of Case Material

was then initiated using a proprietary deep etching technique that has been found useful in investigating grain boundary precipitate networks. The technique consists of an elctropolish followed by an electroetch attacking the γ and γ' phases leaving intermetallic phases in relief. SEM analysis of the deep etch specimens showed the case material to exhibit a continuous grain boundary carbide decoration with the test panel material showing a discrete grain boundary carbide network (Figures 7 & 8). The hold time crack growth acceleration was attributed to the presence of the grain boundary carbide network.

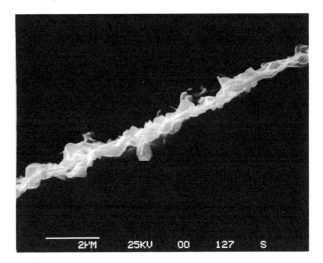

Figure 7
Continuous Grain Boundary Carbide Network Observed in Case Material with Accelerated Crack Growth Rate.

Figure 8
Discrete Grain Boundary Carbide Network Observed in Plate Material with Desirable Crack Growth Rate.

Heat Treatment Development

The probable cause for the different behavior observed for the case material with respect to the plate material in responding to the standard heat treatment is the effect of casting segregation on the local solvus of the $M_{23}C_6$ phase. The conjectured origins of the carbide network during the standard heat treatment for the case material includes suppressing the formation of discrete carbides by cooling rapidly from the solution and stabilization cycles (both above or at the $M_{23}C_6$ solvus) followed by a final age which is low enough to cause enhanced precipitation but too low for significant growth, thus a continuous network is produced. Previous experience with high Cr / high C alloys such as Waspaloy & IN100 suggested that the continuous grain boundary carbide network could be modified through heat treatment. Modifications that could be applied included using a controlled, slow cool from the solution heat treatment to produce carbide nuclei (and provide an additional benefit in serrating the grain boundary); applying an intermediate stabilization cycle after cooling from the solution heat treatment to grow the carbide nuclei into discrete particles; lower the stabilization temperature to ensure that the $M_{23}C_6$ solvus is not exceeded and increase the age temperature to decrease the propensity to form a carbide film. To establish the individual and interactive effects of these various heat treatment options on modifying the grain boundary carbide structure, a design of experiments approach was utilized. The matrix (based upon Taguchi methods) selected was a L8 (ref 10) to assess the effect of the four heat treatment modifications discussed above. The variables selected and the experimental levels used are listed in Table 2 together with the parameters for individual experimental runs. The material used for the experiment was sectioned from the fully heat treated case that exhibited the continuous grain boundary carbide. Upon completion of the heat treatment the samples were subjected to standard metallographic preparation procedures and evaluated. There was very little difference in the microstructure observed optically. However, when the samples were subjected to a deep etch / SEM evaluation it became evident that there were significant differences in the structure and distribution for the grain boundary carbide. Typical carbide structures varied from discrete particles to semi-continuous and continuous networks and they are presented in Figure 9. To provide a semiquantitative approach to analyzing the results, a grain boundary film factor was applied to each of the structures and used as the response variable. In addition, measurements were made using the photomicrographs to estimate interparticle spacing of the carbides and additional assessments of the grain boundary "filmines" solicited. A summary of the results is presented in Table III.

Table III
Heat treat development matrix results and rating of grain boundary carbide structure

Run #	Film Factor [1]	Film Factor [2]	Carbide Spacing(μm) [3]
1	10	10	2.42
2	10	10	4.02
3	10	4	5.26
4	10	8	3.00
5	5	3	0.06
6	3	1	1.87
7	2	2	1.50
8	10	6	2.81

1) Assessment by evaluator #1.
2) Assessment by evaluator #2.
10 = no film
0 = film
3) Distance between adjacent particles.

Table II
Heat treatment development matrix showing experimental variables and levels

Run #	Cooling Rate[1]	Intermediate Stabilization Cycle	Stabilization Cycle	Age Cycle
1	111 - 167°C/hr	none	954°C/6 hrs	802°C/4 hrs
2	111 - 167°C/hr	913°C/8 hrs	954°C/6 hrs	843°C/4 hrs
3	111 - 167°C/hr	913°C/8 hrs	999°C/6 hrs	802°C/4 hrs
4	111 - 167°C/hr	none	999°C/6 hrs	843°C/4 hrs
5	> 1000°C/hr	913°C/8 hrs	954°C/6 hrs	843°C/4 hrs
6	> 1000°C/hr	none	954°C/6 hrs	802°C/4 hrs
7	> 1000°C/hr	none	999°C/6 hrs	843°C/4 hrs
8	> 1000°C/hr	913°C/8 hrs	999°C/6 hrs	802°C/4 hrs

1) Cooling rate from 1163°C/4 hours solution heat treat cycle.
Heat treat procedure : Solution (& cool) + intermediate stabilization + stabilization + age.

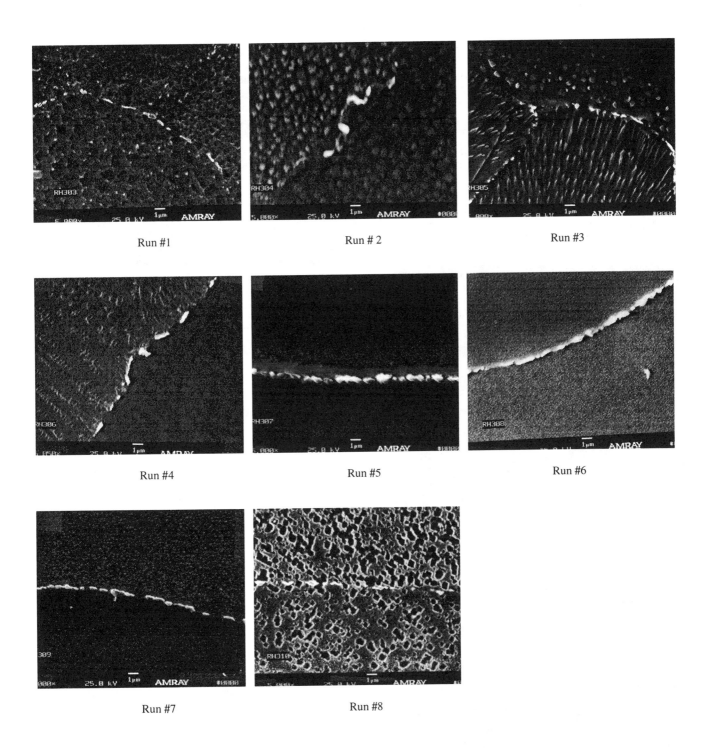

Figure 9
Grain Boundary Carbide Structures Observed in Incoloy 939 Processed Through the Experimental Heat Treatments Listed in Table II

Analysis showed that of the factors evaluated, slower cooling rate from solution heat treatment and use of an intermediate stabilization heat treatment resulted in the formation of a discrete grain boundary carbide structure. To verify the trends observed in the experiment, two confirmation heat treatments and a baseline heat treatment were conducted. The heat treatments are listed in Table IV. Confirmation heat treatment "Y" resulted in a discrete grain boundary carbide structure (Figure 10) with confirmation heat treatment "X" also producing an attractive grain boundary structure (Figure 11). The baseline heat treatment resulted in a

Table IV
Comparison of Standard and Modified Heat Treatments for Incoloy 939

Cycle	Heat Treatment		
	Standard	"X"	"Y"
Solution HT	1163°C/4 hrs	1163°C/4 hrs	1163°C/4 hrs
Cool from Solution	> 1000°C/hr	139°C/hr	139°C/hr
Intermediate Stabilization Cycle	none	none	912°C/6 hrs
Stabilization Cycle	999°C/4 hrs	968°C/4 hrs	982°C/4 hrs
Age Cycle	802°C/4 hrs	802°C/4 hrs	802°C/4 hrs

processed through the two confirmation heat treatments for specimen testing. Crack growth testing under hold time conditions at 649°C showed that both heat treatment "X" and heat treatment "Y" resulted in improved crack growth resistance compared to the baseline material. A comparison is presented in Figure 12. Post test specimen analysis showed that both modified heat treatments resulted in a transition to transgranular fracture from the intergranular / interdendritic fracture mode observed for the standard heat treatment. Because of the greater improvement in crack growth properties, heat treatment "Y" was selected for evaluation of additional specimens from multiple heats of material. All specimens tested continued to show improved hold time crack growth resistance. P&W's material specification was changed to reflect the modified heat treatment and patent protection (patent pending) solicited for the heat treatment. A comparison between the standard and modified heat treatments is presented in Table IV.

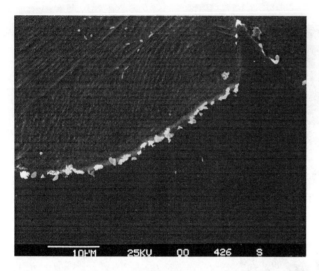

Figure 10
Modified Heat Treatment 'Y" Produced Discrete Grain Boundary Carbide Network

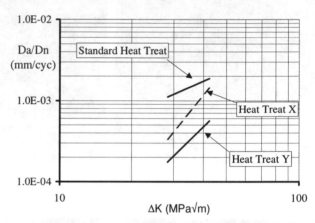

Figure 12
Modified Heat Treatments Improve the 649°C Hold Time Fatigue Crack Growth Rate of Incoloy 939

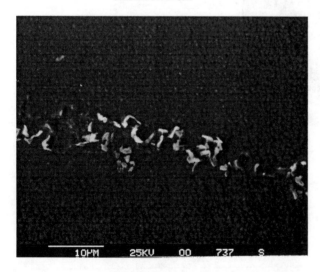

Figure 11
Modified Heat Treatment 'X" Produced Discrete Grain Boundary Carbide Network

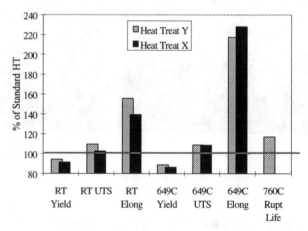

Figure 14
Modifed Heat Treatments Result in Improved Balance of Properties for Incoloy 939 Relative to Material Processed with the Standard Heat Treatment. Stress Rupture Testing of HT "X" was not Conducted.

continuous grain boundary carbide film. To determine if hold time crack growth properties were affected, additional case material was

The heat treatment modifications would be expected to, and did, alter the size and distribution of the γ' phase. A coarser secondary γ' developed due to the controlled cool from the solution heat treatment and there was a reduced volume fraction cooling γ' due to the lowering of the second stabilization cycle temperature. To address this, additional mechanical property characterization of the modified heat treatment has been conducted and the results indicate that an attractive balance of properties is attained. In general, the modified heat treatment results in a slight or no reduction in capability relative to the standard heat treatment. A comparison is presented in Figure 13. With respect to cast Inconel 718, Incoloy 939 with the modified heat treatment offers the potential for a 111°C improvement in operating temperature and a 56°C increase over cast Waspaloy.

Acknowledgments

Several people at Pratt & Whitney's Materials & Mechanics Engineering Organization contributed to the development of the modified heat treatment. Key contributors include Tony Cabral who conducted fractography of tested specimens, Tom Kenney who assisted in material processing and metallographic characterization and Klaus Gumz, Greg Levan and Al Nytch who provided deep etch / SEM analysis and documentation.

References

1. M. Prager and C. S. Shira, "Weldability of γ' Strengthened Alloys", Welding Research Council Bulletin #128, 1968

2. H. Ikawa, S. Shin, & Y. Nakao; "Weldability of Nickel-Base Heat Resisting Superalloy - Effect of Aluminum Content on the Weld Crackings in Inconel 713C - "; Osaka University Technology Reports; vol 18 (1968) ; 353-361

3. H. Ikawa, S. Shin, & Y. Nakao; "Weldability of Nickel-Base Heat Resisting Superalloy - Main Cause of Weld Crackings in Heat Affected Zone of Inconel 713C - "; Osaka University Technology Reports; vol 21 (1971) ; 101-120

4. H. Ikawa, S. Shin, & Y. Nakao; "Weldability of Nickel-Base Heat Resisting Superalloy - Improvement of Hot Cracking Sensitivity of Inconel 713C and Its Mechanical Properties "; Osaka University Technology Reports; vol 21 (1971) ; 461-474

5. A. Koren, et. al.; "The Effect of Weld Energy Input Parameters on the Crack Sensitivity of Alloy IN 713 C"; Welding Research Supplement ; November 1982; 137-149

6. J. Peng; "Weld Results of P&W Inco 939 B-Pads", Precision Castparts Corporation Internal Report, 11/29/89

7. J. Peng; "Weld Evaluation of Inco 939 Diffuser Case", Precision Castparts Corporation Internal Report, 5/21/91

8. S. Shaw, K. Delargy, G. Smith; "Effects of Heat Treatment on Mechanical Properties of High -Chromium Nickel-Base Superalloy IN 939"; Materials Science and Technology; October 1986; 1031-1036

9. J. Snow, O. Ballou; "Inco 939 Alloy Investigation - Final Report"; Precision Castparts Corporation Internal Report; October 1988

10. G. Taguchi; Introduction to Quality Engineering; American Supplier Institute - Dearborn, MI; 1986; pg 182

PRECIPITATION BEHAVIOR IN AEREX™ 350

C. M. Tomasello*, F. S. Pettit**, N. Birks**, J. L. Maloney*, and J. F. Radavich***

* Latrobe Steel Company, Latrobe, PA 15650
** University of Pittsburgh, Materials Science and Engineering Dept, Pittsburgh, PA 15261
*** Micro-met Laboratories, Inc. West Lafayette, IN 47906

Abstract

The precipitation behavior of a new nickel base superalloy, AEREX™* 350, is examined. An understanding of the microstructure as influenced by different heat treatments is necessary in order to realize the full potential of the alloy. This paper presents the effects of solution treating and aging treatments on the resulting microstructure. Samples are solution treated at 1093°C (2000°F) or 1052°C (1925°F) and then given various aging treatments. Optical and scanning electron microscopy are used to examine the phases in the microstructure. The precipitates are examined in the scanning electron microscope. In addition, the relationship between the microstructure and mechanical properties are elucidated.

Introduction

AEREX™ 350 is a new nickel-base superalloy with a nominal composition of 25 wt% cobalt, 17 wt% chromium, 3 wt% molybdenum, 2 wt% tungsten, 4 wt% tantalum, 2 wt% titanium, 1 wt% aluminum, 1.1 wt% niobium, 0.015 wt% carbon and 0.015 wt% boron. The alloy was designed to provide higher temperature strength than the currently available wrought superalloys and still maintain workability for ease of manufacturing. Initial property studies have shown that the alloy has superior properties for applications requiring operating temperatures of 704°C (1300°F) or higher.[1] It has been shown that, in the cold worked condition, the alloy can be used for fastener applications up to 732°C (1350°F).[2] Figure 1 illustrates the high temperature strength of AEREX™ 350 in a solution treated and aged condition as compared to other solution treated and aged superalloys.[3,4,5,6,7]

The composition of AEREX™ 350, predicts that its strengthening is derived from gamma prime (γ') and solid solution strengthening. The low "aluminum + titanium" level also predicts that the alloy is readily hot-workable. Forging and hot rolling trials done at Latrobe Steel Company have confirmed this. Preliminary studies performed at Micro-met Laboratories, Inc. (utilizing x-ray diffraction on extraction samples) have identified

---

* AEREX™ is a trademark of SPS Technologies, Inc.
** UDIMET is a registered trademark of Special Metals Corporation

the presence of the hcp phase, eta, Ni_3Ti (η).[8] This phase is common in nickel-iron base superalloys such as alloys 901 and 706 and iron-base alloys such as A-286.[9] Further studies including x-ray diffraction and transmission electron microscopy (TEM) have also verified the presence of both η and γ'.[10,11] It is the purpose of this paper to illustrate how heat treatment may effect the microstructure and how this, in turn, has an effect on the properties.

Procedure

Samples of hot rolled AEREX™ 350 were solution treated using either a 1093°C (2000°F) or 1052°C (1925°F) temperature for 1 hour in an air atmosphere furnace and water quenched. Samples from each solution treatment were aged for 4 hours at temperatures ranging from 732°C (1350°F) to 899°C (1650°F) and then air cooled to room temperature. Several samples were also heat treated using a double age (more commonly known as a stabilize and age heat treatment) to determine the effects on hardness and microstructure. These samples were then electropolished in a 20% sulfuric acid solution at 25 volts and electroetched in a solution consisting of 15g of chromium trioxide in 170ml of phosphoric acid and 10ml of sulfuric acid at 6 volts. The resulting microstructures were examined in the scanning electron microscope (SEM).

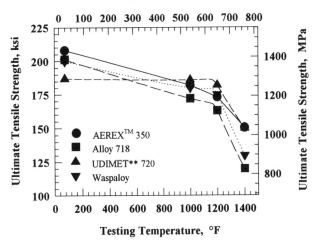

Figure 1. Tensile strength of selected superalloys.

At lower aging temperatures, γ' could not be resolved in the SEM. The observed increase in hardness, however, indicates that precipitation of a strengthening phase has occurred. Samples solution treated at 1093°C (2000°F) were aged at 732°C (1350°F) for longer times to reveal the precipitate. A sample aged for 64 hours was examined in the SEM. After reviewing the microstructure and related hardnesses of the heat treated samples, several heat treatments were chosen for further evaluation. Room and elevated temperature tensile tests were selected to study the relationship between mechanical properties and microstructure.

Results

Heat Treatment

Hardness measurements of solution treated and aged samples of AEREX™ 350 (composition illustrated in Table 1) are tabulated in Table II. Aging the solution treated samples provides a notable increase in hardness. The aging curves for both solution treatment temperatures are illustrated in Figure 2. A double aging treatment also contributes to the hardening response.

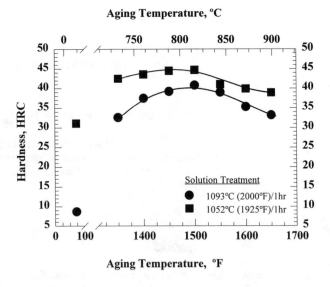

Figure 2. Aging curve for AEREX™ 350.

TABLE I. Composition of AEREX™ 350 - Heat G2389

	Ni	Co	Cr	Mo	W	Ti	Al	Ta	Nb	C	B	S	P
Ingot 1	44.69	25.15	17.03	3.05	2.17	2.07	0.84	4.02	1.13	0.019	0.021	0.001	0.004

TABLE II. Hardness Results (HRC) from Heat Treated Samples of AEREX™ 350

Aging Temperature (4hr/Air Cool)	Solution Treatment	
	1093°C (2000°F) (1hr/WQ)	1052°C (1925°F) (1hr/WQ)
As Solutioned	8.7	31.1
732°C (1350°F)	32.6	42.5
760°C (1400°F)	37.5	43.6
788°C (1450°F)	39.2	44.5
816°C (1500°F)	40.8	44.7
843°C (1550°F)	39.0	41.0
871°C (1600°F)	35.3	39.9
899°C (1650°F)	33.2	38.9
899°C (1650°F) + 732°C (1350°F)	40.0	42.7
899°C (1650°F) + 760°C (1400°F)	40.1	43.2
899°C (1650°F) + 788°C (1450°F)	39.9	42.4
899°C (1650°F) + 816°C (1500°F)	39.2	42.0
899°C (1650°F) + 843°C (1550°F)	36.8	40.1

Microstructure

Solution Treat. The microstructures obtained after solution treatment are shown in Figures 3 and 4. The grain size is fine (ASTM 8 or finer) in the samples solution treated at 1052°C (1925°F) and platelets of η appear at the grain boundaries (Fig. 3). Figure 4 shows the microstructure after solution treating at 1093°C (2000°F). The grains have grown (a grain size of ASTM 2, with an occasional 0) and the only apparent precipitates are primary carbides (MC, where M consists of combinations of tantalum, titanium and niobium).

Single Age. Aging at 732°C (1350°F) for 4 hours results in no apparent change in the microstructure when examined in the SEM (Figures 5 and 6). The hardness increase shown in Figure 2 illustrates, however, that some strengthening is occurring upon aging. Figure 7 shows a sample aged at 732°C (1350°F) after it has been aged for 64 hours; precipitates are now visible in the SEM.

SEM examination of samples heat treated at 788°C (1450°F) reveals a fine dispersion of γ′ particles beginning to appear in the grain interiors. Small platelets of η are starting to precipitate at the grain boundaries in the 1093°C (2000°F) solution treated material

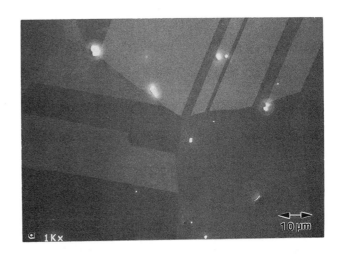

Figure 4. SEM micrograph of AEREX™ 350 solution treated for 1 hour at 1093°C (2000°F). (a) 1,000x and (b) 20,000x

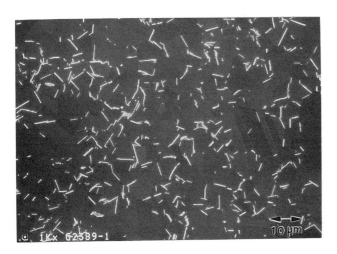

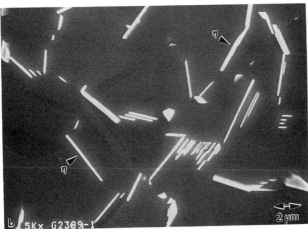

Figure 3. SEM micrograph of AEREX™ 350 solution treated for 1 hour at 1052°C (1925°F) illustrating the presence of eta (η) phase. (a) 1,000x and (b) 5,000x

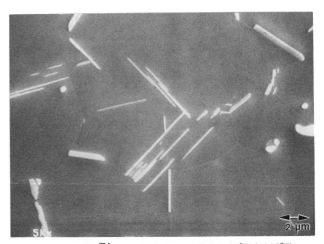

Figure 5. AEREX™ 350 solution treated at 1052°C (1925°F) and aged at 732°C (1350°F) for 4 hours (5000x).

as illustrated in Figure 8. Figure 9 shows the η phase is beginning to develop a Widmanstätten plate structure in the 1052°C (1925°F) solution treated sample aged at 816°C (1500°F). The γ' precipitates are now readily visible in the SEM.

Double Age. Figures 10 and 11 illustrate the effects of a double age. A sample solution treated at 1052°C (1925°F) and aged at 899°C (1650°F) for 4 hours plus 732°C (1350°F) for 4 hours is shown in Figure 10. The 1093°C (2000°F) solution treated sample aged at 871°C (1600°F) for 4 hours plus 760°C (1400°F) for 16 hours is shown in Figure 11. Samples with a double aging heat treatment reveal that the γ' precipitates have coarsened but remain evenly distributed. The η morphologies consist of Widmanstätten plate structures at the grain boundaries. Intragranular platelets of η are also apparent.

Mechanical Properties

The results from room temperature tensile tests are shown in Table III. The lower solution treatment temperature provides higher strength. Certainly, the smaller grain size has an influence on tensile strength. The Hall-Petch relationship, however, would predict a higher yield strength in the lower temperature solution treated material than is observed. This is determined by calculating a value for k in the Hall-Petch equation ($\sigma_{ys} = \sigma_o + kd^{-1/2}$). Given the yield strength in the high temperature solution treatment and an average grain size of 180 μm, the 22 μm grain size produced in the lower temperature solution treat would result in a very high yield strength. Although this is a very crude approximation which warrants further investigation, it illustrates that other mechanisms, such as precipitation, must be taken into consideration.

Table IV shows the elevated temperature tensile data. Once again, it is interesting to note the effect of precipitating phases on strength. Even of more interest is the ductility at elevated temperatures. It appears that the presence of η diminishes the ductility. The 1093°C (2000°F) solution treated material is more ductile. The effects of the solution treat temperature on ductility are illustrated in Figure 12. The ductility of AEREX™ 350 from the heat treatment providing the least ductility is compared to other superalloys in Figure 13.

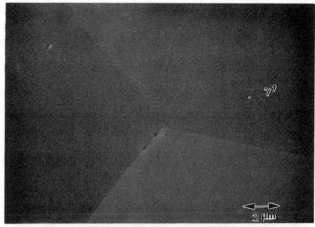

Figure 7. AEREX™ 350 solution treated at 1093°C (2000°F) and aged at 732°C (1350°F) for 64 hours (5000x).

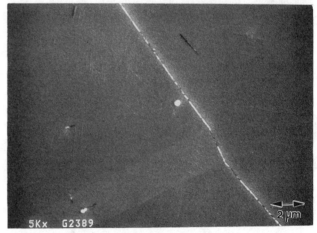

Figure 8. AEREX™ 350 solution treated at 1093°C (2000°F) for 1 hour and aged at 788°C (1450°F) for 4 hours (5000x).

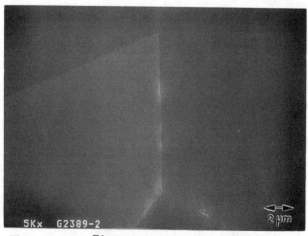

Figure 6. AEREX™ 350 solution treated at 1093°C (2000°F) and aged at 732°C (1350°F) for 4 hours (5000x).

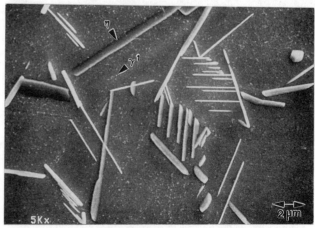

Figure 9. AEREX™ 350 solution treated at 1052°C (1925°F) for 1 hour and aged at 816°C (1500°F) for 4 hours (5000x).

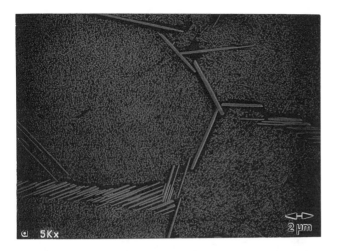
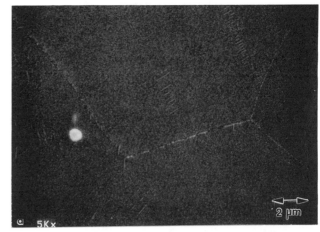
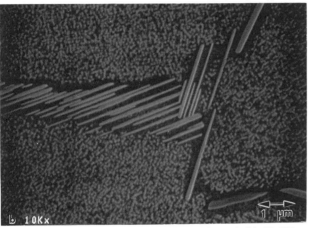
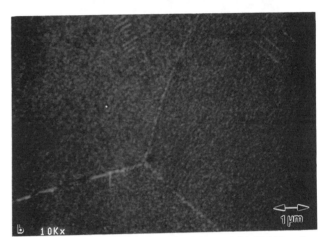

Figure 10. SEM micrograph at (a) 5,000x and (b) 10,000x illustrating AEREX™ 350 solution treated at 1052°C (1925°F) for 1 hour and aged at 899°C (1650°F) for 4 hours + 816°C (1500°F) for 4 hours.

Figure 11. SEM micrograph at (a) 5,000x and (b) 10,000x of AEREX™ 350 solution treated at 1093°C (2000°F) for 1 hour and aged at 871°C (1600°F) for 4 hours + 760°C (1400°F) for 16 hours.

Table III. Room temperature tensile results and precipitates present in the microstructure. (Note: Carbides are present in all heat treated conditions. gb=grain boundary, w=Widmanstätten structure, intra=intragranular)

Heat Treatment (ST/Age)	Figure No.	γ'	η	UTS (MPa)	YS (MPa)	Elong (%)	R.A. (%)
1093°C/1h+732°C/4h	Figure 6	fine	none	1091.4	751.5	52.3	54.9
1093°C/1h+816°C/4h		small	fine gb	1301.7	947.3	40.0	45.6
1093°C/1h+899°C/4h		large	gb (w)	1266.5	812.9	43.8	39.0
1093°C/1h+871°C/4h+760°C/16h	Figure 11	coarse	gb (w)	1456.2	1042.5	31.5	37.8
1093°C/1h+899°C/4h+816°C/4h		coarse	gb(w)/intra	1361.7	879.8	28.5	26.3
1052°C/1h+732°C/4h		fine	gb	1499.6	1088.0	30.0	34.7
1052°C/1h+816°C/4h	Figure 9	small	gb (w)	1515.5	1133.5	27.8	32.5
1052°C/1h+899°C/4h		large	gb(w)/intra	1441.0	992.8	27.3	29.9
1052°C/1h+899°C/4h+816°C/4h	Figure 10	coarse	lg. gb/intra	1436.2	855.6	25.3	27.2

Discussion of Results

AEREX™ 350 has been shown to precipitate both γ′ and η upon aging. The 1093°C (2000°F) solution treat temperature is above the eta solvus and the 1052°C (1925°F) is just below the eta solvus temperature. Samples from both solution treatments aged at 732°C (1350°F) are representative of an underaged condition. In the 1093°C (2000°F) solution treated sample, η is still not present and the ductility is at its greatest in both room and elevated temperature tensile tests. Material heat treated to the peak aging condition, which occurs at 816°C (1500°F) for both solution treatments, exhibits a maximum in both hardness and tensile strength. Overaging not only decreases strength, it decreases ductility as well.

Double aging increases the strength in the 1093°C (2000°F) solution treated material with a small sacrifice to ductility. The double age in the 1052°C (1925°F) solution treated material has very little effect on either strength or ductility. This behavior can be related back to the microstructure. The double age effectively coarsens the γ′ for strengthening in the 1093°C (2000°F) solution treated material with a good balance of the η phase. The η phase in the 1052°C (1925°F) solution treated material is becoming too coarse and continuous. This morphology would not be beneficial to the alloy.

Eta phase is an ordered geometrically close packed (GCP)-type precipitate with an HCP crystal structure. The η phase is common in iron- and nickel-iron-base superalloys but is generally not present in cobalt base superalloys. Although AEREX™ 350 is a nickel base alloy, it contains a significant amount of cobalt. GCP phases do not form easily in cobalt-base alloys, although efforts have been made to generate GCP phases in cobalt alloys for additional strengthening.[12]

Table IV. Elevated Temperature Tensile Results

Heat Treatment (ST/Age)	Figure No.	Test Temp (°C)	UTS (MPa)	YS (MPa)	Elong (%)	R.A. (%)
1093°C/1h+732°C/4h	Figure 6	704	901.8	678.5	43.0	42.4
1093°C/1h+732°C/4h	Figure 6	760	791.5	538.5	40.0	42.8
1093°C/1h+816°C/4h		704	1094.9	790.8	22.5	28.4
1093°C/1h+871°C/4h		704	946.0	704.0	12.5	19.5
1093°C/1h+899°C/4h		704	884.6	721.2	9.0	15.6
1093°C/1h+871°C/4h+760°C/16h	Figure 11	704	1140.4	823.3	10.0	12.4
1052°C/1h+732°C/4h		704	1148.0	939.8	13.5	14.9
1052°C/1h+816°C/4h	Figure 9	704	1228.0	966.6	12.5	11.9
1052°C/1h+816°C/4h	Figure 9	760	1037.0	845.3	6.5	9.2
1052°C/1h+899°C/4h		704	1176.9	861.8	14.0	11.5
1052°C/1h+899°C/4h+816°C/4h	Figure 10	704	1177.0	837.7	15.5	14.5

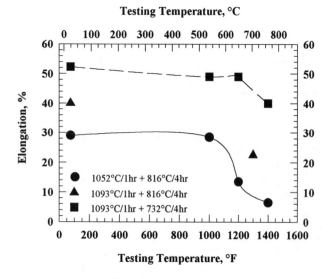

Figure 12. Tensile elongation values for AEREX™ 350.

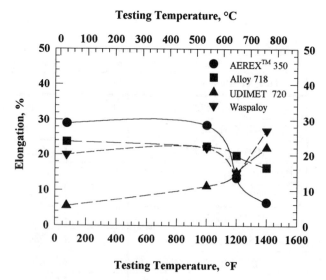

Figure 13. Tensile elongation values for selected superalloys.

Superalloys strengthened by γ′ (with sufficient titanium) are susceptible to η phase formation just as alloys with sufficient niobium and strengthened with γ″ may precipitate delta phase (as in alloy 718). Some studies have shown the presence of η phase to be deleterious to properties.[13,14] Degradation occurs when the η phase precipitates as a cellular structure or the volume fraction of η becomes too great in an overaged condition. Other studies have shown that η may enhance properties with the proper heat treatments to obtain an appropriate microstructure.[15,16] It has been demonstrated in this paper that the presence of η has an effect on both strength and ductility. The η in this alloy has been observed to have a Widmanstätten plate-like morphology. Therefore, the loss in ductility utilizing a subsolvus solution treatment temperature is not due to a cellular η precipitating at the grain boundaries.

Conclusions

The precipitating phases which occur in AEREX™ 350 include both γ′ and η. The presence of the η phase helps to provide high temperature strength to the alloy. It appears that the η phase precipitates as low as 788°C (1450°F) and the solvus temperature is between 1093°C (2000°F) and 1052°C (1925°F), whereas γ′ is present as low as 732°C (1350°F) and the solvus temperature is below 1038°C (1900°F).

The γ′ appears to be uniformly distributed in the matrix and coexists with η at temperatures above 788°C (1450°F). The η precipitates are generally plate shaped and appear predominately on the grain boundaries. Overaging temperatures, however, increase the apparent volume fraction of η, and η precipitates in the grain interiors on preferred crystallographic planes. The heat treatment is critical in determining the distribution and morphology of precipitating phases which, in turn, may be used for mechanical property optimization.

Work is in progress to further refine the heat treatment process parameters which will influence the final structure and properties of the alloy. It is anticipated that two heat treatments will be developed: one which will provide the best creep and stress rupture properties using a solution treatment above the η solvus temperature and one which will take advantage of the presence of η to provide elevated temperature strength. Long term, elevated temperature exposures will also be investigated to determine the effect on the microstructure.

References

[1] S. R. Buzolits, "New High Temperature Alloy Characterized by Superior Alloy Properties at Temperatures to 1350°F," Industrial Heating, 61 (12) (1994), 34-35.

[2] S. R. Buzolits and L. A. Kline, "Bolting Alloy Fills High Temperature Gap," Adv. Matls. Proc., 147 (2) (1995), 33-34.

[3] F. E. Scerzenie and G. E. Maurer, "Development of UDIMET 720 for High Strength Disk Applications," in Superalloys 1984, eds: M. Gell, C. S. Kortovich, R. H. Bricknell, W. B. Kent and J. F. Radavich (Warrendale, PA: TMS-AIME, 1984), 573-582.

[4] K. R. Bain, M. L. Gambone, J. M. Hyzak and M. C. Thomas, "Development of Damage Tolerant Microstructures in UDIMET 720," in Superalloys 1988, eds., S. Reichman, D. N. Duhl, G. Mauer, S. Antolovich and C. Lund (Warrendale, PA: TMS-AIME, 1988), 13-22.

[5] UDIMET 720, Special Metals Data Sheet, September 1978.

[6] Waspaloy, Alloy Digest, Ni-129, November 1967.

[7] UDIMET 718, Alloy Digest, Ni-258, November 1978.

[8] J. F. Radavich, Micromet Laboratories, Inc., unpublished (1994).

[9] D. R. Muzyka, in The Superalloys, ed., C. T. Sims and W. C. Hagel (New York, NY: John Wiley and Sons, Inc., 1972), 113.

[10] C. M. Tomasello, "Precipitation Behavior of a New Nickel-Base Superalloy, AEREX™ 350" (Masters Thesis, University of Pittsburgh, 1996).

[11] R. Doherty and S. Asgari, Drexel University, unpublished work (1995).

[12] C. T. Sims, in The Superalloys, ed., C. T. Sims and W. C. Hagel (New York, NY: John Wiley and Sons, Inc., 1972), 145.

[13] B. R. Clark and F. B. Pickering, "Precipitation Effects in Austenitic Stainless Steels containing Titanium and Aluminium Additions," JISI, 205 (1967), 70-84.

[14] J. A. Brooks and A. W. Thompson, "Microstructure and Hydrogen Effects on Fracture in the Alloy A-286," Met. Trans., 24A (1993), 1983-1991.

[15] J. H. Moll, G. N. Maniar, and D. R. Muzyka, "The Microstructure of 706, A New Fe-Ni-Base Superalloy," Met. Trans., 2 (1971), 2143-2151.

[16] J. H. Moll, G. N. Maniar, and D. R Muzyka, "Heat Treatment of 706 for Optimum 1200°F Stress-Rupture Properties," Met. Trans., 2 (1971), 2153-2160.

EFFECTS OF ALUMINUM, TITANIUM AND NIOBIUM ON THE TIME - TEMPERATURE - PRECIPITATION BEHAVIOR OF ALLOY 706

Takashi Shibata, Yukoh Shudo, and Yuichi Yoshino

Technology Research Center, The Japan Steel Works, Ltd.,
1-3 Takanodai, Yotsukaido, Chiba 284, Japan

Abstract

Ni-Fe-base superalloy 706 has been used for high temperature services. The time - temperature - precipitation (TTP) diagram is essential in the design of heat treatments for any precipitation strengthened superalloy. The TTP diagrams have been already presented for Alloy 706. However, effects of aluminum, titanium and niobium, important substitutional elements of γ' and γ'' precipitates, on the TTP behavior are not clear in the literature.

In this study, the TTP and the time - temperature - hardness (TTH) diagrams are presented for experimental alloys containing only one or two of Ti, Nb and Al, in a temperature range of 600 - 900℃. The observation by optical microscopy, scanning electron microscopy and transmission electron microscopy revealed that γ', γ'', γ'-γ'' co-precipitates and η form in alloys containing Ti.

Among the three elements, Ti plays the most important role in the precipitation strengthening behavior of Alloy 706. Furthermore, neither Al nor Nb can demonstrate their effects without Ti addition. Nb promotes γ'' formation and prevents η formation. Al enhances the formation of stable γ'-γ'' co-precipitates, more effectively in the co-existence of Ti and Nb.

Introduction

Ni-Fe-base superalloys are age-hardened by the precipitation of coherent γ' and/or γ'' in the austenitic matrix γ (1). Alloy 706 is a relatively new material and was developed from Alloy 718, a representative wrought superalloy. Compared with Alloy 718, it has a chemical composition of no molybdenum, reduced niobium, aluminum, chromium, nickel and carbon, and increased titanium and iron. This excellent balance of chemical composition results in superior characteristics to Alloy 718 in the segregation tendency, hot workability and machinability (2-4). Therefore, Alloy 706 is suitable for large forgings and has been used for high temperature services (5).

The time - temperature - precipitation (TTP) diagram is one of the essential tools for designing the heat treatment of precipitation strengthened superalloy. Especially for Alloy 706, complicated heat treatments are used to draw its full ability (6). In fact, its mechanical properties are greatly affected by the precipitation at the heat treatment (7-11). The TTP diagrams of Alloy 706 have already been presented (2,3), and updated recently (12).

Alloy 718 has been investigated thoroughly on its TTP behavior, which has led to many compositional modifications (13-22). However, effects of aluminum, titanium and niobium, which are all important substitutional elements in the precipitation of γ' and γ'', on the TTP behavior of Alloy 706 are not clear in the literature. In this study, the TTP diagrams are presented for six experimenyal 706 alloys, in order to clarify the role of Al, Ti and Nb in the TTP behavior.

Procedure

Material

Six heats of experimental alloys were melted in a 50 kg vacuum induction melting (VIM) furnace. The chemical composition of these six alloys is listed in Table I. Alloy 706 contains Al, Ti and Nb, but these experimental alloys contain only one or two of these elements. Nickel and chromium contents were nearly constant in all the alloys as shown in Table I, with iron being the balance. All the ingots were diffusion treated and subsequently forged to the billets with a cross section of $30^t \times 120^w$ mm^2. The billets were sectioned mechanically into samples of suitable sizes. For comparison, a commercial Alloy 706 forging, a large turbine disk, was also used as a sample.

Heat Treatment

The condition of solution treatment for each heat was determined by preliminary experiments so as to fully dissolve precipitates formed in the forging process and to obtain a mean grain size of ASTM #3-4. After the solution treatment, samples were isothermally heat treated in a temperature range of 600 - 900℃ for up to 100 h. In this study, the heating rate to the solution and aging temperatures was 50℃/h as shown in Figure 1, simulating a large forging.

Table I Chemical Composition of Experimental Alloys

Heat	Ni	Fe	Cr	Al	Ti	Nb	C	N	O	B	Si	Mn	P	S
No.1	43.36	39.23	16.43	0.34	0.01	0.05	< 0.003	0.0033	0.0020	< 0.005	0.035	0.010	< 0.003	0.0008
No.2	43.75	37.74	16.23	0.04	1.68	0.05	0.003	0.0012	0.0025	< 0.005	0.035	0.010	< 0.003	0.0010
No.3	43.96	36.89	15.95	0.05	0.01	2.56	< 0.003	0.0043	0.0145	< 0.005	0.040	0.010	< 0.003	0.0009
No.4	43.54	37.56	16.29	0.30	1.71	0.05	< 0.003	0.0014	0.0018	< 0.005	0.025	0.010	< 0.003	0.0006
No.5	43.62	35.05	16.13	0.05	1.75	2.82	< 0.003	0.0026	0.0025	< 0.005	0.045	0.010	< 0.003	0.0009
No.6	43.76	36.55	16.30	0.30	0.01	2.54	0.003	0.0025	0.0010	< 0.005	0.050	0.010	< 0.003	0.0008
706	43.64	34.81	16.27	0.25	1.78	2.87	< 0.003	0.0024	0.0019	< 0.005	0.010	0.010	< 0.003	0.0011

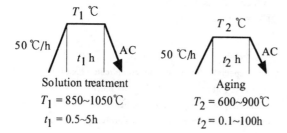

Figure 1 : Heat treatment program and conditions.

Evaluations of Precipitation Behavior

The heat treated samples were subjected to optical microscopy, scanning electron microscopy (SEM) and transmission electron microscopy (TEM) for their precipitation behavior. The sample preparation and observation conditions are described elsewhere. Hardness was measured by a Vickers hardness tester, in order to produce the time - temperature - hardness (TTH) diagram.

Result and Discussion

Solution Treatment Condition

The conditions of solution treatments were investigated in a range of 850 - 1050℃ and 0.5 - 5h. Figure 2 shows the relationship between Vickers hardness and the test temperature for the six experimental alloys aged for 2h. Hardness of all the alloys decreased rapidly over the temperature range from 900℃ to 950℃, and tailed off at about 120 Hv when temperature exeeds 950℃. It indicates that γ' and/or γ'' formed during the forging process are fully dissolved above 950℃. The dissolution temperature was practically the same for all the experimental alloys, suggesting the solvus temperature of these precipitates being unaffected by their chemical compositions. This is consistent with the early work (7).

In order to be sure if the grain boundary precipitate such as η and δ is dissolved, all of the solution treated samples were subjected to optical microscopy and SEM. The microscopic observations revealed no precipitate either in the grain or at the grain boundary in any of the alloys tested, as shown in Figure 3 as an example. Carbide and /or nitride have been reported to appear occasionally at the grain boundary (2-4, 7-10, 12). However they were not seen in this study due possibility to the relatively low carbon and nitrogen contents of these alloys.

When solution-treated at 980℃ for 2h, the grain size of all the alloys tested was within the range of ASTM #3 to 4. The grain grew rapidly to ASTM #1-2 above 1000℃, regardless of alloy composition. Therefore, the solution treatment was done at 980℃ for 2h for all the alloys.

TTH Behavior of Ti-Free Alloy

Hardness of alloys Nos.1, 3 and 6 changed little within the limit of this experiment. The strengthening element is Al for No.1, Nb and Al for No.3 and Nb for No.6, respectively. None of the alloys contain titanium. Neither 0.3% Al nor 2.5% Nb nor the combination of both is sufficient to produce noticiable age hardening, indicating that the Ti-free alloys are not hardenable. This is in part consistent with an early work on Nb containing Ni based alloy (23). Thus, Ti plays the most important role in the precipitation strengthening of Alloy 706. From this point of view, the following study was conducted with Ti-containing alloys, namely Nos.2, 4, 5 and commercial Alloy 706.

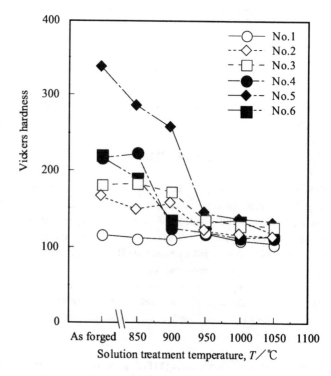

Figure 2 : Change in vickers hardness with the solution treatment of experimental alloys

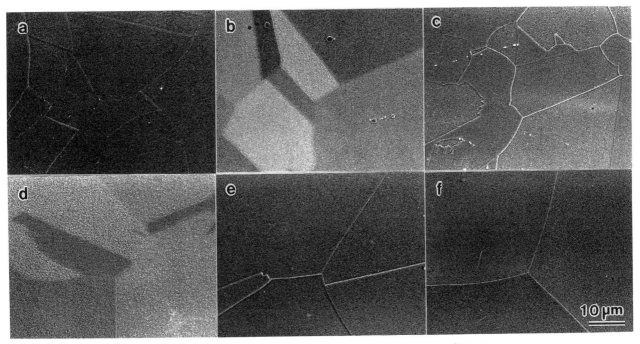

Figure 3 : SEM micrographs of experimental alloys solution -treated at 980 ℃ for 2h ;
(a) alloy No.1, (b) No.2, (c) alloy No.3, (d) No.4, (e) No.5 and (f) No.6.

TTH Diagram of Ti-Containing Alloy

The alloys containing Ti were all age-hardenable, especially at the temperatures between 700 - 800 ℃, indicating the formation of γ' and/or γ" phases. The TTH diagrams of three experimental alloys, Nos.2, 4 and 5, and Alloy 706 are shown in Figure 4. The highest hardness was about 400 Hv in No.5 and Alloy 706, but about 300 Hv in Nos.2 and 4. The higher hardness is attributed to the Nb content of those alloys, suggesting a synergestic effect between Nb and Ti.

The age hardening was fast at temperatures about 800 ℃, but the highest hardness was achieved below 700 ℃ in the Ti-containing alloys as seen in Figure 4. Fast over-aging prevents hardness from exceeding 300 Hv at 800 ℃. The over-aging is associated with the transformation of γ' to η as in A-286 or γ" to δ as in Alloy 718 (1). The softening occurs at 800 ℃ a little more extensively in alloys Nos.2 and 4 than the remainder. Likewise, No.5 appears slightly more sensitive to the over-aging than Alloy 706.

The TTH diagram of Alloy 706 is similar to the one in the previous report (12). However, the "nose" temperature of the diagram is higher in this study. This is thought to be due to the precipitation of γ' and/or γ" during the heating stage of the heat treatment that simulated slow heating of large ingots.

Identification of Precipitates

As an example, SEM micrographs of alloys Nos.2, 4, 5 and Alloy 706 aged at 730 ℃ for 10h and at 830 ℃ for 10h are shown in Figure 5. No precipitate was seen inside the grain despite the hardness increase in these samples. However, many cellular precipitates were observed at the grain boundary, except for No.5 and Alloy 706 aged at 730 ℃ for 10h.

TEM images and selected area diffraction patterns inside the grains of alloys Nos.2 and 4 aged at 730 ℃ for 100h are shown in Figure 6. The spherical precipitates were clearly observed in the grain interior. The micro-beam EDS revealed that the precipitates in No.2 consisted of Ni and Ti, and those in No.4 contained Ni, Ti and Al. The ratio of Ni to (Ti+Al) was nearly 3:1 for all the precipitates analysed. The intra-granular precipitates were identified γ' phase having FCC structure.

In the case of alloy No.5 and commercial Alloy 706, precipitates of different shape were observed. Figure 7 shows TEM micrograph, selected area diffraction pattern and micro beam diffraction patterns of alloy No. 5 aged at 730 ℃ for 10h. The disk shaped precipitates were observed inside the grain, appearing the same as γ" reported on Ally 706 (4) and on Alloy 718 (13-22). The diffraction patterns, prove that they are γ" phase. The precipitate designated C has a diffraction pattern of γ', but it may be a γ" disk that is viewed from its <001> direction. The precipitates in No.5 contained Ni, Nb and Ti, with the ratio of Ni to (Nb+Ti) being nearly 3:1.

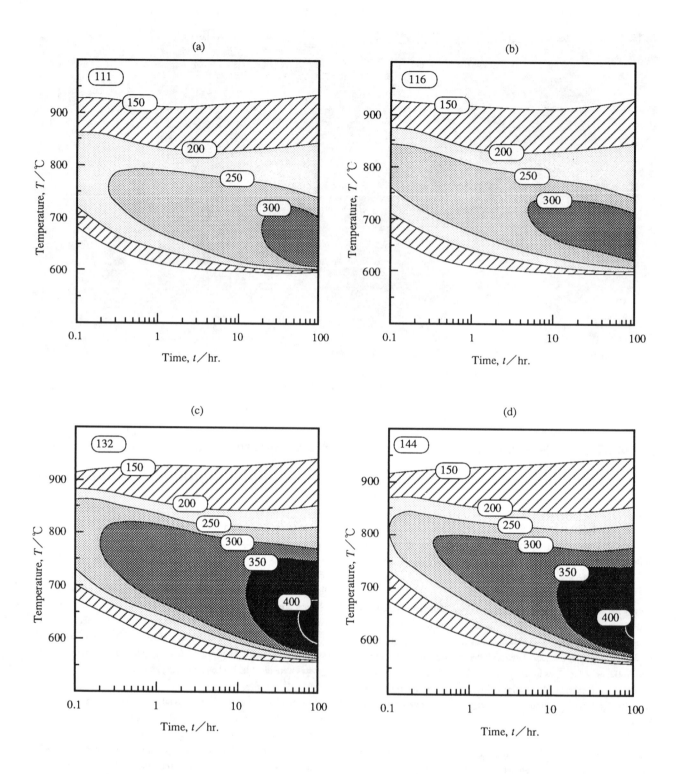

Figure 4: TTH diagrams of experimental alloys containing Ti : (a) alloy No.2, (b) No.4, (c) No.5 and (d) Alloy 706.

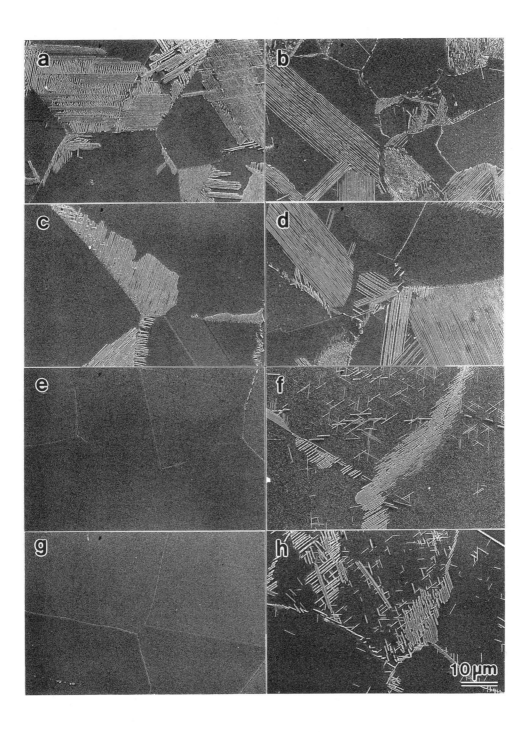

Figure 5 : SEM micrographs of experimental alloys after aging : (a) alloy No.2 aged at 730 ℃ for 10h, (b) No.2 aged at 830 ℃ for 10h, (c) No.4 aged at 730 ℃ for 10h, (d) No.4 aged at 830 ℃ for 10h, (e) No.5 aged at 730 ℃ for 10h, (f) No.5 aged at 830 ℃ for 10h, (g) Alloy 706 aged at 730 ℃ for 10h, and (h) Alloy 706 aged at 830 ℃ for 10h.

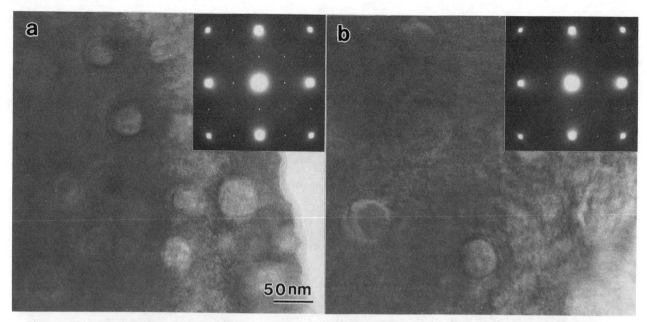

Figure 6 : TEM micrographs and selected area diffraction patterns of experimental alloys aged at 730℃ for 100h ; (a) alloy No.2 and (b) No.4

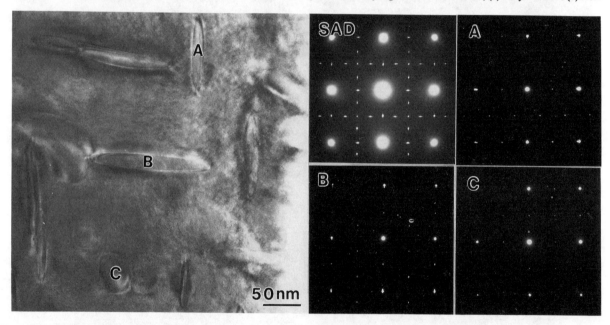

Figure 7 : TEM micrograph, selected area diffraction pattern and micro-beam diffraction patterns of experimental alloy No.5 aged at 730℃ for 10h.

A TEM micrograph, selected area diffraction pattern and micro beam diffraction pattern are shown in Figure 8 of Alloy 706 aged at 730℃ for 10h. Non-spherical precipitates were observed, but no-disk shaped γ ". From the diffraction patterns, the non-spherical precipitates are the co-precipitate having the core of γ ' phase being overlayed with γ " phase on its top and/or bottom, which is referred to as "non-compact morphology" (13-22). The co-precipitate is expressed here γ '-γ ". The results of micro-beam EDS revealed that the γ ' phase contained Ni, Ti, Nb and Al, and γ " phase Ni, Nb and Ti, and that the ratio of Ni to (Ti+Nb+Al) were nearly 3:1 for both γ ' and γ " phases. These co-precipitates were also found in alloy No.5.

The inter-granular precipitates as seen in Figure 5 were identified as η phase. Figure 9 shows TEM image and selected area diffraction pattern at the grain boundary of alloys No.4 aged at 730℃ for 10h. The η phase consisted of Ni and Ti in alloys Nos.2 and 4, while it consisted of Ni, Nb and Ti in alloy No.5 and Alloy 706. However, the ratio of Ni to (Ti+Nb) were maintained nearly 3:1 for all the alloys tested here. The η phase contained no aluminum because it has little solubility for Al (1). The selected area diffraction pattern indicates that the η phase has a specific orientation relationship with the γ matrix, as $[011]_\gamma$ // $[2\bar{1}\bar{1}0]_\eta$ and $\{11\bar{1}\}_\gamma$ // $\{0001\}_\eta$. This relationship is consistent with other work (9). The η phase appears parallel to each other as seen in Figure 5 in order to meet this orientation relationship.

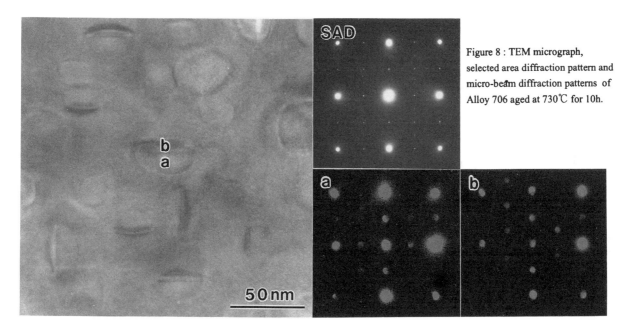

Figure 8 : TEM micrograph, selected area diffraction pattern and micro-beam diffraction patterns of Alloy 706 aged at 730℃ for 10h.

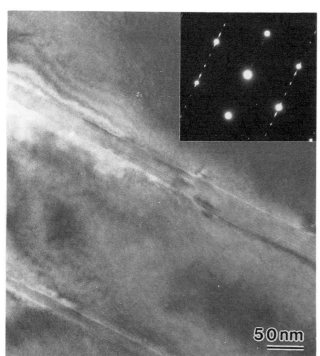

Figure 9 : TEM micrograph and selected area diffraction pattern near the grain boundary of experimental alloy No.2 aged at 730℃ for 10h.

TTP Diagram

As described above, four types of precipitates were identified in the alloys containing Ti. γ' and η were found in alloys Nos.2 and 4. In addition to them, γ'' and γ'-γ'' co-precipitate were found in alloy No.5 and Alloy 706. The TTP diagrams of the four alloys are shown in Figure 10. The regions of γ', γ'' and γ'-γ'' agreed well with the TTH diagram.

γ'' and γ'-γ'' are seen only in Nb-containing alloys, suggesting that the γ'' formation requires both Ti and Nb. These precipitates are the cause of the greater hardness of the Nb-bearing alloys, described ealier with respect to their TTH diagrams. That is, the γ'' phase reinforces the matrix more effectively than the γ' phase (1).

The region of η precipitation grows wider as aging time increases in all the alloys. In fact, the η phase was found to shoot out from the grain boundary as the aging time increased at about 800℃. The γ', γ'', γ'-γ'' all transform eventually to η when aged for long time at high temperatures in all the alloys tested. However, the region of η precipitation is much wider in the Nb-free alloys than in the Nb-containing alloys, suggesting that γ' transforms to η more readily than γ'' and γ'-γ''. That is, the Nb-containing alloys are thought to be more stable at high temperatures. This is supported by the aging response previously described of Figure 4. Thus, Nb not only reinforces the matrix by the γ'' precipitation, but also enhances the high temperature stability by delaying the transformation from the intra-granular precipitates to the grain boundary η phase.

The TTP behavior of alloy No.5 and Alloy 706 is more complicated than those of No.2 and No.4, especially at the temperatures between 700 - 800 ℃. Figure 11 demonstrates how the precipitate morphlogy develops with aging, when Alloy 706 is aged at 730℃. γ' phase appears faintly at 0.1h, which is characterized by the ordering spots in the diffraction pattern. As the exposure time increases, the γ'-γ'' begins to form replacing the γ' phase, and the γ'' phase becomes predominant. It should be noted that the size of the γ'-γ'' co-precipitate in Alloy 706 aged at 730℃ for 10h is much smaller than that of γ'' in alloy No.5 aged at the same condition. This suggests that the stability of the γ'-γ'' is greater than that of γ'' at high temperatures, as previously reported for Alloy 718 (13-22).

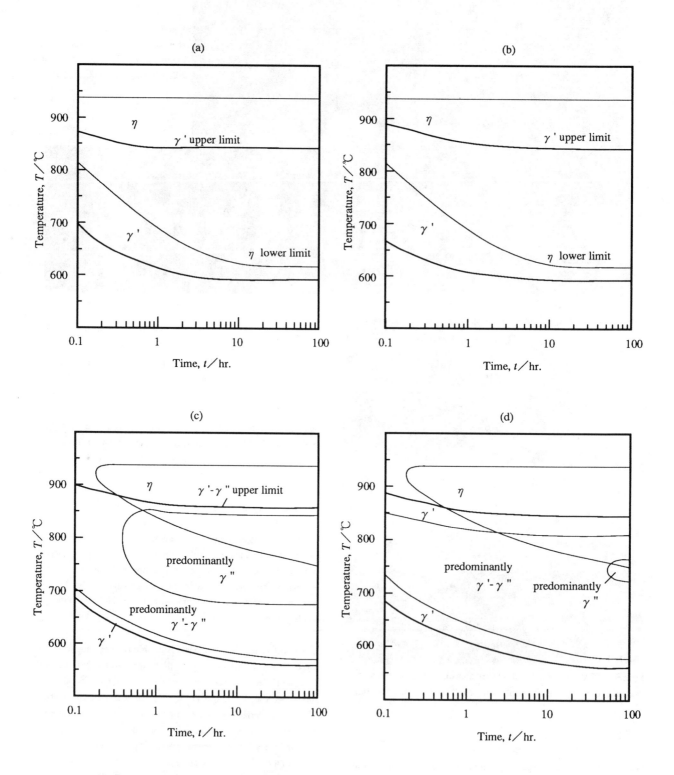

Figure 10 : TTP diagrams of experimental alloys containing Ti : (a) alloy No.2, (b) No.4, (c) No.5 and (d) Alloy 706.

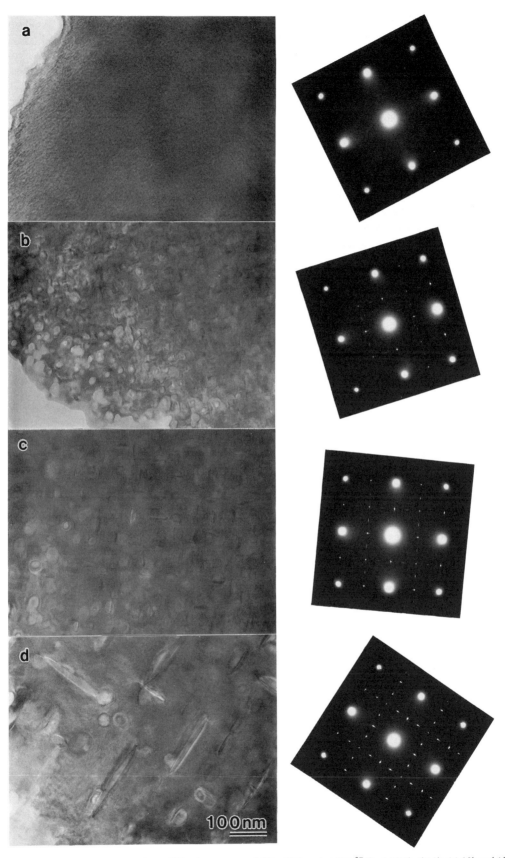

Figure 11 : TEM micrographs and selected area diffraction patterns of Alloy 706 aged at 730 ℃ for (a) 0.1h, (b) 1h, (c) 10h and (d) 100h.

The TTP diagrams of alloy No.5 and Alloy 706 are somewhat different, although those of alloys No.2 and No.4 are very similar. The difference between alloy No.5 and Alloy 706 is characterized especially by the region of γ'-γ''. The precipitation occurs in these alloys pass in the same sequence $\gamma' \rightarrow \gamma'$-$\gamma'' \rightarrow \gamma''$, but the region of γ'-γ'' is fairly wider in Alloy 706 than in alloy No.5. It indicates that the Alloy 706 has better thermal stability than alloy No.5. Such difference reflects a synergetic effect among Al, Nb and Ti. Therefore, the effect of Al addition is considered to form the stable γ'-γ''.

The solubility for Al in γ'' is extremely low whereas that for Nb in γ' is very high, hence a low Al content favors the γ'' formation whereas high Al content the γ' phase (4). This effect is seen in Figure 10. The dominant γ'-γ'' co-precipitation should be explained by the same effect of Al addition. Moreover, the same effect is expected in the transformation to η, since η has little solubility for Al (1). Further study is needed to shed more light on the stability of the precipitates as influenced by the chemical composition.

Conclusions

In order to help design the modification of Alloy 706, the isothermal TTP and TTH diagrams of experimental alloys and commercial Alloy 706 are presented. γ', γ'', γ'-γ'' co-precipitates and η were found in the alloys containing Ti.

Ti plays the most important role in the precipitation strengthening of Alloy 706. Al and Nb do not serve as a hardening agent without Ti. Nb is needed for the strengthening through the γ'' formation and the prevention of η formation. Al is useful for the formation of stable γ'-γ'' co-precipitates, and is effective when Ti and Nb are both present.

References

1. E.E.Brown and D.R.Muzyka, "Nickel-Iron Alloys", Superalloys II, ed., C.T.Sims, N.S.Stoloff, and W.C.Hagel (New York, John Willey & Sons, 1987), 165-188.
2. H.L.Eiselstein, "Properties of Inconel Alloy 706", ASM Technical Report, No.C 70-9.5 (1970), 1-21.
3. H.L.Eiselstein, "Properties of a Fabricable, High Strength Superalloy", Metals Engineering Quarterly, November(1971), 20-25.
4. E.L.Raymond and D.A.Wells, "Effects of Aluminum Content and Heat Treatment on Gamma Prime Structure and Yield Strength of Inconel Nickel-Chromium Alloy 706", Superalloys --Processing (Columbus, OH:Metals and Ceramics Information Center, 1972), N1-N21.
5. P.W.Schilke, J.J.Pepe, and R.C.Schwant, "Alloy 706 Metallurgy and Turbine Wheel Application", Superalloys 718,625,706 and Various Derivatives, ed., E.A.Loria (Pittsburgh, PA:TMS, 1994), 1-12.
6. Inconel 706 : Undated brochure obtained from The International Nickel Company, (1974).
7. J.H.Moll, G.N.Maniar, and D.R.Muzyka, "The Microstructure of 706, a New Fe-Ni-Base Superalloy", Metallurgical Transactions, 2(1971), 2143-2151.
8. J.H.Moll, G.N.Maniar, and D.R.Muzyka, "Heat Treatment of 706 Alloy for Optimum 1200 Stress-Rupture Properties", Metallurgical Transactions, 2(1971), 2153-2160.
9. L.Remy, J.Laniesse, and H.Aubert, "Precipitation Behavior and Creep Rupture of 706 Type Alloys", Materials Science and Engineering, 38(1979), 227-239.
10. G.W.Kuhlman et.al., "Microstructure - Mechanical Properties Relationships in Inconel 706 Superalloy", Superalloys 718, 625, 706 and Various Derivatives, ed., E.A.Loria (Pittsburgh, PA:TMS, 1994), 441-449.
11. T.Takahashi et.al., "Effects of Grain Boundary Precipitation on Creep Rupture Properties of Alloy 706 and 718 Turbine Disk Forgings", ibid., 557-565.
12. K.A.Heck, "The Time-Temperature-Transformation Behavior of Alloy 706", ibid., 393-404.
13. R.Cozar and A.Pineau, "Morphology of γ' and γ'' Precipitates and Thermal Stability of Inconel 718 Type Alloys", Metallurgical Transactions, 4(1973), 47-59.
14. J.P.Collier et.al., "The Effect of Varying Al, Ti, and Nb Content on the Phase Stability of Inconel 718", ibid., 19A(1988), 1657-1666.
15. J.P.Collier, A.O.Selius, and J.K.Tien, "On Developing a Microstructurally and Thermally Stable Iron - Nickel Base Superalloy", Superalloys 1988, ed., D.N.Duhl et.al. (Warrendale, PA: The Metallurgical Society, 1988), 43-52.
16. E.Andrieu, R.Cozar, and A.Pineau, "Effect of Environment and Microstructure on the High Temperature Behavior of Alloy 718", Superalloy 718 - Metallurgy and Applications, ed., E.A.Loria (Pittsburgh, PA:TMS, 1989), 241-256.
17. E.Gou, F.Xu, and E.A.Loria, "Effect of Heat Treatment and Compositional Modification on Strengthening and Thermal Stability of Alloy 718", Superalloys 718, 625 and Various Derivatives, ed., E.A.Loria (Pittsburgh, PA:TMS, 1991), 389-396.
18. E.Gou, F.Xu, and E.A.Loria, "Comparison of γ'/γ'' Precipitates and Mechanical Properties in Modified 718 Alloys", ibid., 397-408.
19. J.A.Manriquez et.al., "The High Temperature Stability of IN718 Derivative Alloys", Superalloys 1992, ed., S.D.Antolovich et.al. (Warrendale, PA: TMS, 1992), 507-516.
20. E.Andrieu et.al., "Influence of Compositional Modifications on Thermal Stability of Alloy 718", Superalloys 718, 625, 706 and Various Derivatives, ed., E.A.Loria (Pittsburgh, PA:TMS, 1994), 695-710.
21. X.Xie et.al., "Investigation on High Thermal Stability and Creep Resistant Modified Inconel 718 with Combined Precipitation of γ'' and γ'''", ibid., 711-720.
22. E.Gou, F.Xu, and E.A.Loria, "Further Studies on Thermal Stability of Modified 718 Alloys", ibid., 721-734.
23. I.Kirman, Precipitation in the Fe-Ni-Cr-Nb System", Journal of the Iron and Steel Institute, December(1969), 1612-1618.

ETA (η) AND PLATELET PHASES IN INVESTMENT CAST SUPERALLOYS

G. K. Bouse
Howmet Corporation Operhall Research Center
Whitehall, MI 49461

Abstract

The occurrence of platelet phases in equiax, directionally solidified (DS), and single crystal (SC) turbine blades, vanes, and integral wheels was investigated. Examples of platelet phases in turbine components made from low carbon PWA1480, C103 (or PWA1483, AF56, or SX792), IN792+Hf (or C101), IN939, IN6203, and GTD111 alloys were shown. For IN792+Hf, IN939 and GTD111 the phases formed internally, and for most of the alloys the phases also formed by reaction with the ceramic shell or facecoat, or from local segregation. The platelet phases were associated with casting scale or fluorescent penetrant indications. All of these alloys have 8-18 wt.% (15-28 at.%) Ti+Ta+Nb+Hf, compared to 2.3-5 wt.% (4-11 at.%) Al, which to some extent destabilizes the desired Ni_3Al (γ') phase and promotes Ni_3Ti (η) phase, or other platelet phases rich in Ta, Nb, and Hf.

Introduction

Since 1988, the occurrence of platelet phases in turbine hardware primarily for land-base engines has increased. A case study of several turbine parts that contained these phases will be discussed: equiax and directionally solidified (DS) IN792+Hf wheels and blades; IN939 vanes; DS GTD111 and IN6203 blades; single crystal (SC) C103 blades; and low carbon SC PWA1480 blades.

The compositions of these alloys are given in Table I [1-5]. Compared to older land-base turbine alloys such as U500, U700 and IN738, Ti, Nb, Ta, and noticeably Hf are included in increasing amounts for strength or corrosion considerations. During casting, these elements may segregate so that lower strength Ni_3Ti (η), or platelet phases form in preference to the traditional Ni_3Al (γ') phase.

These phases have usually been seen during casting process development. The phases were associated with the presence of scale (which can occur on either shell or core sides of the casting), or anomalies such as hot tears or other fluorescent penetrant (FPI) indications. These phases were usually found in the as-cast condition. In some cases, the platelet phases were found after a deviated process which required rejection of those castings. In most instances the normal grit blasting or belting operations removed the scale on external surfaces. However, this may not be true of internal surfaces used for weight reduction or air cooling.

To better understand these platelet phases, a study was conducted on six alloys in which the platelet phases had been reported by the casting operations.

Background

A review of the literature showed little documentation on the occurrence of platelet phases, other than topologically closed packed (TCP) phases such as σ or μ. The platelet phases that will be discussed here are different in several ways, notably they are usually visible at <100X and they are present in the as-cast condition. In addition, the platelet phases can usually be eliminated by process control, as opposed to control of alloying elements. Finally, the evidence suggests that small amounts of the platelet phases can be tolerated in today's turbine engines without degradation to mechanical properties.

Review of Ni_3Ti Eta (η) Phase

A review of platelet phases should start with eta (η) phase or Ni_3Ti which was identified in the late 1930's, according to The Superalloys and Superalloys II [6,7]. Eta has a hexagonal close-packed (HCP) structure that is non-coherent with the matrix and generally exists as large platelets which can extend across grains [8]. For awhile, the name "O'Hare" phase was used by some to describe η, because the η platelets resembled runways at Chicago's O'Hare airport.

In castings, η phase is seen in interdendritic regions or within γ' eutectic pools. In this form the phase confers "little benefit to the alloy and is therefore undesirable." In an environment of Ni and Cr (the "base" for most superalloys), when the Ti content exceeds the solubility limit of approximately 2.7 at.% [at 25 at.% Cr at 750C (1382F)], η phase of fixed composition is precipitated [8, p38]. Reference [9] indicates that η phase has "no solubility for other elements", however, more recent work indicates η phase does have solubility for Co, W, Al, and Cr [10]. Eta forms at high temperatures, but at lower temperatures there may be a solid state transformation to γ'.

Table I. Nominal Compositions (Wt. %) of Superalloys Studied

Element	IN792+Hf (C101)	IN6203	IN939	GTD111	C103*	Low C PWA1480
Process→	Eq or DS	DS	Equiax	DS	SC	SC
C	0.09	0.15	0.15	0.1	0.07	<0.01
B	0.015	0.01	0.01	0.01	-	-
Zr	0.06	0.1	0.1	-	-	-
Al	3.4	2.3	1.9	3	3.6	5
Ti	4	3.5	3.7	4.9	4.1	1.5
Ta	4.3	1.1	1.4	2.8	5	12
Hf	1	0.8	-	-	-	-
Nb	-	0.8	1	-	-	-
Ni	Bal.	Bal.	Bal.	Bal.	Bal.	Bal.
Co	9	19	19	9.5	8-9	5
Cr	12.6	22	22.5	14	12	10
W	4.3	2	2	3.8	3.8	4
Mo	1.9	-	-	1.5	1.9	-
Ref.	1	2	2	3	4, 5	4

* Also known as AF56, SX792, PWA1483

Table II. Summary of Observations For η and δ Phases

Morphology	Shape	Potential Problems
Precipitates	Tiny spheres usually only visible with SEM or TEM.	May not cause a problem.
Cells or cellular	Pearlitic	Can lower notched stress rupture strength.
Widmanstätten	Thin acicular needles intersecting at regular angles.	Reduces stress rupture strength but not ductility.
Platelet or globular, including η or δ phases	Some are parallel and others intersect at varying angles.	Platelets are associated with loss of RT tensile or stress rupture strength or ductility, and cracking. Delta phase is also associated with scale at core or shell surfaces, incipient melting, and recrystal-lized grains.

The difference in atomic stacking between the two phases, η and γ' is as follows:

- in γ', the close-packed planes are stacked upon each other in the ABCA... sequence, giving it the FCC structure.
- for η phase, the stacking sequence of the close-packed plane is ABAC... is obtained, which produces the HCP structure [8, p57].

Mihalisin and Decker [11] showed the γ' → η transformation is accelerated by deformation, and this is because the deformation facilitates the necessary shift of atomic planes described above. In the present work, the presence of cracks near platelet phases was not unusual.

Thus, in addition to a change in the lattice spacing, a shift of the atomic planes changes the stacking sequence determining γ' or η phase. This is similar to the γ" (body centered tetragonal) → δ Ni_3Nb (orthorhombic) transformation that occurs in IN718.

Beneficial Effects of η Phase. Eta has some beneficial effects in wrought alloys like A286 for structure control during forging. Because the transformation γ' → η occurs above 650C (1200F), the upper service temperature of A286 is defined by this transformation [7]. Eta phase is also found in high-Fe wrought superalloys like IN706 and IN901.

Deleterious Effects. Eta phase is known to be brittle with little strain tolerance [6,9]. Since η is non-coherent with the matrix, it may weaken the structure. However, when intergranular η precipitated in IN901 with long exposure times at 730C (1350F), neither strength nor ductility was affected. When η precipitated as cells (like pearlite) at grain boundaries, notch stress rupture strength and creep ductility was reduced. When η was precipitated intergranularly, in Widmanstätten fashion, stress-rupture strength but not ductility was reduced. Thus, when η phase precipitates from the matrix, the general experience has been some reduction of one or more mechanical properties. A summary of these observations are presented in Table II above.

For IN792+Hf castings discussed in [1], Fiedler mentioned that "η phase is a stable constituent and cannot be removed by solutioning at temperatures below the incipient melting point. Temperatures of 1190C (2175F) will only round-off the corners of an otherwise very angular appearance....η is most often seen in heavy sections, so that time for segregation to occur is an important factor in control of η." Freeman also notes that η can be found in the hubs of wheels or in alloys like IN792 that contains high levels of Ti and Ta [7, p432].

Elements Involved. It is well known that Ni and Al promote the Ni_3Al γ' phase. This γ' phase has also liberally been described as Ni_3(Al, Ti, Nb, Ta). But as the sum of Ti+Nb+Ta exceeds Al, the formation of η or platelet phases are favored. Some platelet phases are formed directly from the liquid, while others may have transformed from the γ' eutectic pools.

Fiedler [1] notes that Hf promotes η phase. However, Hf also readily forms Ni_5Hf, which in castings can be found with platelet and other eutectic phases. He also mentions that increased C reduces η. For example, η readily formed in IN792+Hf parts when the C was lowered from 0.15 wt.% to 0.08 wt.% (this is also discussed later). In [6 (p127), and 7 (p111)], Sims et. al. mention that the addition of trace amounts of B will retard the formation of η phase. In the same publication [6 (p127)] the addition of Si was stated to reduce η phase. However, by the time [7] was published, the effect of Si was not mentioned. In a paper by Decker [12], W is also said to retard the formation of η phase. Willemin [10] notes that "above 1250C (2282F), W becomes totally insoluble while the solubility of Cr is greatly reduced."

Identification of Other Platelet Phases

There has not been much discussion in recent literature about platelet phases. However, findings by Howmet researchers, which will be discussed later, have identified an orthorhombic platelet phase, tentatively identified as δ, or Ni_3Ta phase [13]. Besides containing Ta, this platelet phase has been found to contain measurable amounts of Al, Ti, Hf, and Nb, as well as Co and Cr. Work by [10] does not specifically describe platelet phases, but includes phase relationships of interest to the present work. Finally, Chinese researchers [14], show that P, Zr, and Si may be responsible for increasing the amount of eutectic or other solidification segregation occurring in castings.

Technical Approach

The morphology of the platelets was analyzed with optical metallography and with a Topcon scanning electron microscope (SEM). The chemistry was analyzed with a Cameca Camebax electron microprobe (EMP). This instrument utilized 20 kv and 10 nA beam current. Compositions were determined using pure element calibration standards. The crystal structure was analyzed with a Philips transmission electron microscope (TEM).

Results/Observations

IN792+Hf Wheels and Equiax/DS Blades

Platelet phases relevant to the current work were observed in the hubs of equiax cast integral wheels and roots of aero turbine blades in the early 1980's [15,16]. The wheel hubs and roots were susceptible to segregation and slower cooling rates, both of which promoted formation of globular and platelet phases. The phases became apparent after C in the alloy was reduced from ~0.15 wt.% to ~0.08 wt.%. Examples of these phases are shown in Figures 1 and 2.

Fiedler [1], who described the phase from the standpoint of a turbine engine producer, was comfortable with <5% η phase. It will be noted the amount of allowable η phase remains at this level today. However, the method of measuring the phase is not simple, requiring 1/2 hr or more of thermal tinting at 454C (850F). He mentioned there "was no apparent effect on conventional mechanical properties - of primary concern was the potential to cause small surface cracks by abrasive machining practices."

More recently, platelet phases were found in a DS land base turbine blade made of IN792+Hf as shown in Figure 3. The phase was within a crack-like void near the trailing edge (TE) and the platform. The "crack" was easily visible to the unaided eye, and affected several blades on the cluster.

The crack-like void formed parallel to the withdrawal direction, and platelets were seen to bridge the void. This led the author to conclude the platelets formed in the presence of a liquid. After the platelets formed, the liquid receded, leaving a gap. A brittle crack, which probably formed after the 1121C (2050F)/2 hr. solution heat treatment, is also seen to have severed some of the platelets. Despite this heat treatment, the platelet phases appeared to be stable. Adjacent to the platelets were recrystallized grains.

The compositions of the platelets in weight and atomic percentages, plus matrix, eutectic and other phases are shown in Table III. When compared to the average chemistry of the blade (determined by spectrographic methods), the platelet phases were shown to be enriched 10.5X in Hf and enriched 1.9X and 1.3X for Ti and Ta, respectively. In the tables, "γ' elements" Al, Ti, Ta, Hf, and Nb were summed *below* their listings, while "γ elements" Ni, Co, Cr, Mo, W, and Fe were summed *above* their listing. The γ' elements were also divided into the γ elements so the reader can readily see ratios of phases in both weight and atomic percentages.

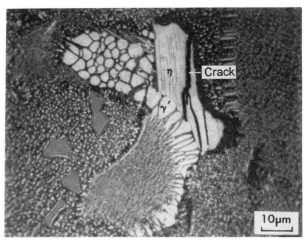

Figure 1. Platelet phases in a IN792+Hf blade. Fiedler [1] shows growth of η from a γ' eutectic pool. Note the phase is also cracked. 10% Nitric electrolytic etch.

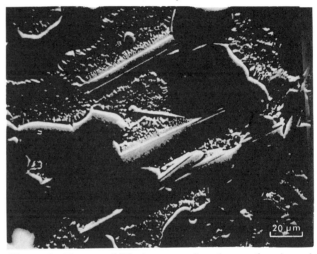

Figure 2. Emerson [15] shows an abundance of η phase in both platelet and globular shapes in the hub of an IN792+Hf wheel. Oxalic acid etch.

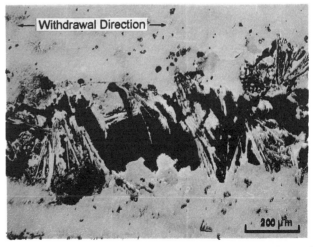

Figure 3. Platelet phases in the trailing edge of a DS IN792+Hf turbine blade are seen to span a crack-like void. Recrystallized grains are also visible. Kallings etch.

Table III. Average Compositions of Phases in IN792+Hf Alloy

	Phase	γ + γ'	Cr-Rich	Eutectic	Platelet	Platelet Alloy
	Element	Wt.%	Wt.%	Wt.%	Wt.%	Wt. Ratio
	Si	0.00	0.00	0.00	0.00	-
	S	0.00	0.00	0.00	0.00	-
	Zr	0.00	0.00	0.00	0.00	-
γ' elements	Al	3.80	1.07	5.60	2.91	0.9X
	Ti	6.00	1.93	8.60	7.58	1.9X
	Ta	4.20	0.77	5.80	5.38	1.3X
	Hf	1.10	0.43	3.20	10.43	10.5X
	Nb	0.00	0.00	0.00	0.00	-
	Total (γ')eq =	15.10	4.20	23.20	26.30	
	Ratio(γ/γ') =	5.91	22.41	3.53	2.89	
	Total (γ)eq =	89.30	94.13	81.80	76.05	
γ elements	Ni	64.10	20.97	71.10	66.21	1X
	Co	8.90	5.27	6.60	5.51	0.6X
	Cr	13.10	43.27	3.90	3.12	0.25X
	Mo	1.80	17.47	0.20	0.50	0.3X
	W	1.30	7.17	0.00	0.53	0.1X
	Fe	0.10	0.00	0.00	0.20	-
	Total	104.40	98.33	105.00	102.35	
	Element	At. %	At. %	At. %	At. %	At. Ratio
	Si	0.00	0.00	0.00	0.00	-
	S	0.00	0.00	0.00	0.00	-
	Zr	0.00	0.00	0.00	0.00	-
γ' elements	Al	7.75	2.19	11.30	6.51	0.9X
	Ti	6.89	2.43	9.77	9.55	2X
	Ta	1.28	0.27	1.74	1.79	1.4X
	Hf	0.34	0.16	0.98	3.59	15X
	Nb	0.00	0.00	0.00	0.00	-
	Total (γ')eq =	16.25	5.04	23.79	21.44	
	Ratio(γ/γ') =	5.15	18.83	3.20	3.66	
	Total (γ)eq =	83.75	94.96	76.21	78.56	
γ elements	Ni	60.06	20.09	65.92	68.58	1.1X
	Co	8.31	5.12	6.10	5.64	0.7X
	Cr	13.86	54.72	4.08	3.63	0.3X
	Mo	1.03	12.36	0.11	0.32	0.3X
	W	0.39	2.66	0.00	0.18	0.15X
	Fe	0.10	0.00	0.00	0.21	-
	Total	100.00	100.00	100.00	100.00	

Table IV. Average Compositions of Phases in IN6203 Alloy

	Phase	γ + γ'	White Phase	Eutectic	Platelet or Glob	Platelet Alloy
	Element	Wt. %	Wt. %	Wt. %	Wt. %	Wt. Ratio
	Si	0.03	0.27	0.15	0.00	-
	S	0.00	0.00	0.00	0.00	-
	Zr	0.00	4.07	2.04	0.47	~5X
γ' elements	Al	2.44	1.17	1.80	2.19	0.95X
	Ti	3.25	1.87	2.56	9.76	2.8X
	Ta	0.94	0.17	0.55	1.41	1.3X
	Hf	0.16	32.43	16.29	12.54	15.7X
	Nb	0.51	1.63	1.07	1.47	1.8X
	Total (γ')eq =	7.30	37.27	22.28	27.36	
	Ratio(γ/γ') =	12.70	1.58	3.40	2.65	
	Total (γ)eq =	92.64	59.07	75.85	72.61	
γ elements	Ni	48.32	44.10	46.21	59.76	1.3X
	Co	19.67	12.03	15.85	10.11	0.5X
	Cr	23.19	2.93	13.06	2.70	0.1X
	Mo	0.02	0.00	0.01	0.00	-
	W	1.37	0.00	0.68	0.00	0.01X
	Fe	0.09	0.00	0.04	0.04	-
	Total	99.97	100.67	100.32	100.44	
	Element	At. %	At. %	At. %	At. %	At. Ratio
	Si	0.06	0.71	0.38	0.00	-
	S	0.00	0.00	0.00	0.01	-
	Zr	0.00	3.33	1.67	0.32	~5X
γ' elements	Al	5.07	3.23	4.15	4.99	1X
	Ti	3.81	2.91	3.36	12.52	3X
	Ta	0.29	0.07	0.18	0.48	1.4X
	Hf	0.05	13.57	6.81	4.33	17.3X
	Nb	0.31	1.31	0.81	0.98	2X
	Total (γ')eq =	9.53	21.09	15.31	23.29	
	Ratio(γ/γ') =	9.49	3.58	5.42	3.28	
	Total (γ)eq =	90.46	75.58	83.02	76.38	
γ elements	Ni	46.18	56.11	51.15	62.60	1.4X
	Co	18.73	15.25	16.99	10.55	0.6X
	Cr	25.03	4.21	14.62	3.19	0.13X
	Mo	0.01	0.00	0.00	0.00	-
	W	0.42	0.00	0.21	0.00	0.01X
	Fe	0.09	0.00	0.00	0.05	-
	Total	100.06	100.71	100.38	100.00	

IN6203 DS Blades

A phenomenon similar to the IN792+Hf blade platelet phases was found *perpendicular* to the withdrawal direction in IN6203 DS turbine blades, as shown in Figure 4. But for these blades, which are 38 cm (15 in.) long, the platelet phases were associated with a crack-like void and recrystallized grains, both of which presented difficulties during production.

Problems in casting turbine blades from this high-Cr DS alloy have been mentioned in the literature [17,18], so considerable metallography was employed to find η-type phases during the casting process development [19]. It was found that platelet phases were prone to form in the airfoil/platform fillet. Here, a disturbance in the solidification front may have occurred due to the change in section size.

In addition to the platelet phases, eutectic and "white" phases were characterized. All of the phases were Hf-rich. The compositions are shown in Table IV and were similar to platelet phases in the IN792+Hf parts discussed above. The Hf content of the IN6203 platelets was elevated 15.7X over the concentration of Hf in the alloy. Titanium, Nb, and Ta were also elevated 2.8-1.3X over the alloy.

In addition to the presence of these platelet phases, eutectics, voids, and recrystallized grains were found after heat treatment. In [17], it was noted the platelet phases were taken into solution at >1600C (2912F). However, a temperature this high could not be used in production due to the incipient melting and grain growth that occurred. The platelets existed in both the as-cast and heat treated conditions. It was found that significant homogenization of these platelet phases did not occur during the solution cycle.

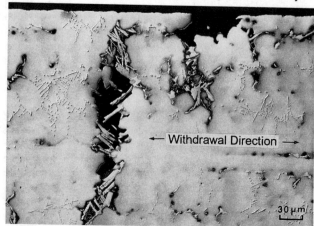

Figure 4. Photographs of platelet phases in an as-cast DS IN6203 blade. The platelet phase is seen to span a crack-like void that is perpendicular to the withdrawal direction.

IN939 Vane Segment

IN939 is a well-studied alloy having a composition close to IN6203 but without Hf. Several characterizations have been performed on cast bars and slabs, and some references to η phases with compositions similar to those discussed here have been mentioned [20-22]. Prevention of η phases was accomplished mainly through homogenization at temperatures between 1145-1160C (2090-2120F) for 4 hrs.

An example of a platelet phase from an as-cast equiaxed IN939 land base vane segment is shown in Figure 5. Platelet phases in DS blades appear similar. After etching the metallographic mount, a cluster of the platelet phases became visible to the unaided eye. Upon closer examination, some of the platelet phase was cracked, but the matrix *was not cracked*. This would imply the phase was brittle, as it was not able to withstand the stresses of normal processing, which in this case only included casting, cleaning, wet radiac cutting, and metallographic polishing. Remnants of the platelet phases have also been observed after the standard 1160C (2120F)/4 hr heat treatment, and may have been responsible for low tensile strength and ductility in test bars machined from another land-base nozzle segment that had no other apparent deficiencies [23].

The compositions of the platelet phases are given in Table V. The platelet phases were enriched by 3.7-1.8X for Ti, Nb, and Ta. A Cr-rich phase was also encountered near the platelet phases.

Another similar alloy used for vane applications is GTD222 [3]. This alloy, however, does not form the platelet phases in the as-cast condition. A comparison of the critical elements for GTD222 in Table VI shows that (Ti+Ta+Nb)/Al ratio is larger and the C content is less than for IN939, both of which might be expected to promote platelet phases. In this case the sum of the Ti+Ta+Nb is 2 wt.% less for GTD222, which may be the limiting factor here.

Table VI Comparison of C, Al, Ti, Ta and Nb in IN939 & GTD222

	Wt. % of Elements					
	C	Al	Ti	Ta	Nb	(Ti+Ta+Nb)/Al
IN939	0.15	1.9	3.7	1.4	1	6.1/1.9=3.21
GTD222	0.1	1.2	2.3	1	0.8	4.1/1.2=3.42

Figure 5. Cracked platelet phases in an as-cast equiax IN939 vane segment. Kallings etch.

Table V. Average Compositions of Phases in IN939 Alloy

Phase		γ + γ'	Cr-Rich	Platelet	Platelet/Alloy
	Element	Wt. %	Wt. %	Wt. %	Wt. Ratio
	Si	0.00	0.00	0.00	-
	S	0.00	0.00	0.00	-
	Zr	0.08	0.00	0.02	-
γ' elements	Al	2.30	0.70	1.97	1X
	Ti	5.50	1.08	13.53	3.7X
	Ta	1.33	0.08	2.53	1.8X
	Hf	0.00	0.00	0.00	-
	Nb	1.02	0.33	3.10	3.1X
	Total (γ')eq =	10.15	2.18	21.13	
	Ratio (γ/γ') =	8.76	45.33	3.73	
	Total (γ)eq =	88.94	98.60	78.80	
γ elements	Ni	49.52	10.78	60.13	1.2X
	Co	17.72	13.90	13.90	0.7X
	Cr	21.28	72.35	4.65	0.2X
	Mo	0.00	0.20	0.00	-
	W	0.36	1.33	0.11	0.06X
	Fe	0.06	0.05	0.00	-
	Total	99.17	100.78	99.95	
	Element	At. %	At. %	At. %	At. Ratio
	Si	0.00	0.00	0.00	-
	S	0.00	0.00	0.00	-
	Zr	0.05	0.00	0.01	-
γ' elements	Al	4.80	1.38	4.16	1X
	Ti	6.47	1.20	16.10	3.7X
	Ta	0.41	0.02	0.80	1.8X
	Hf	0.00	0.00	0.00	-
	Nb	0.62	0.19	1.90	3.2X
	Total (γ')eq =	12.30	2.79	22.96	
	Ratio(γ/γ') =	7.13	34.82	3.35	
	Total (γ)eq =	87.65	97.21	77.03	
γ elements	Ni	47.50	9.80	58.43	1.3X
	Co	16.93	12.59	13.45	0.7X
	Cr	23.05	74.28	5.11	0.2X
	Mo	0.00	0.11	0.00	-
	W	0.11	0.38	0.03	0.05X
	Fe	0.06	0.05	0.00	-
	Total	100.00	100.00	100.00	

GTD111 DS Turbine Blades

Both types of platelet phases have been found in GTD111 turbine blades [24,25]. Examples of these morphologies are shown in Figures 6 and 7. Both examples of the platelets were associated with cracks that were open to the external surfaces of the blades, which were subsequently detected by FPI. Figure 6 shows the coarse platelet phase in the as-cast condition, while Figure 7 shows the fine platelets after the 1121C (2050F)/2 hr. solution heat treatment.

In Figure 6, the coarse platelets were shown to span a crack-like void. In order for this to have happened, it is hypothesized the platelets formed from the liquid, which then receded to another location, leaving a void. Also, note the absence of eutectic. The compositions of the platelets are given in Table VII. The platelets were enriched 2.6X and 2.1X in Ti and Ta, respectively.

In the absence of cracking, the platelet phase most often seen in the GTD111 alloy is shown in Figure 7. These fine platelets, which in this case were associated with an interdendritic crack, may often be missed because the platelets are small and intimate with the γ' eutectic pools. To the untrained eye these fine platelets may resemble metallographic polishing artifacts. As such, it is difficult to measure quantities of fine platelet phases. The composition

Table VII. Average Compositions of Phases in GTD111 Alloy

	Phase	γ + γ'	Bright Phase	Eutectic	Platelet	Platelet/Alloy
	Element	Wt. %	Wt. %	Wt. %	Wt. %	Wt. Ratio
	Si	0.00	0.05	0.00	0.02	~2X
	S	0.00	0.00	0.00	0.00	-
	Zr	0.00	0.00	0.00	0.00	-
γ' elements	Al	3.41	2.05	4.12	3.24	1.1X
	Ti	4.25	4.30	7.67	11.30	2.3X
	Ta	1.50	1.65	1.80	4.28	1.5X
	Hf	0.00	0.05	0.00	0.00	-
	Nb	0.00	0.00	0.00	0.02	-
	Total (γ')eq =	9.16	8.05	13.59	18.83	
	Ratio(γ/γ') =	9.82	11.34	6.36	4.22	
	Total (γ)eq =	89.91	91.25	86.41	79.45	
γ elements	Ni	60.26	53.90	64.27	68.01	1.1X
	Co	10.05	10.30	8.86	6.67	0.7X
	Cr	16.18	21.80	12.16	4.11	0.3X
	Mo	1.40	3.05	1.02	0.54	0.4X
	W	2.00	2.20	0.11	0.10	0.02X
	Fe	0.03	0.00	0.00	0.00	-
	Total	99.07	99.35	100.00	98.30	
	Element	At. %	At. %	At. %	At. Pct.	At. Ratio
	Si	0.00	0.10	0.00	0.04	~2X
	S	0.00	0.00	0.00	0.00	-
	Zr	0.00	0.00	0.00	0.00	-
γ' elements	Al	7.19	4.38	8.42	6.84	1.15X
	Ti	5.05	5.18	8.83	13.62	2.4X
	Ta	0.48	0.53	0.55	1.38	1.7X
	Hf	0.00	0.02	0.00	0.00	-
	Nb	0.00	0.00	0.00	0.01	-
	Total (γ')eq =	12.73	10.11	17.80	21.86	
	Ratio (γ/γ') =	6.86	8.88	4.62	3.58	
	Total (γ)eq =	87.27	89.79	82.20	78.14	
γ elements	Ni	58.43	52.98	60.38	66.75	1.1X
	Co	9.70	10.08	8.29	6.52	0.75X
	Cr	17.67	24.20	12.91	4.52	0.3X
	Mo	0.83	1.84	0.59	0.33	0.4X
	W	0.63	0.69	0.03	0.03	0.03X
	Fe	0.03	0.00	0.00	0.00	-
	Total	100.00	100.00	100.00	100.04	

of these platelets could not be reasonably distinguished from the composition of the eutectic.

The mechanical properties for a wide range of land-base turbine blades made from GTD111 alloy were reviewed. These were reviewed because of potential concern for increasing amounts of platelet phases due to casting segregation or slower cooling rates due to increased size of the blades. It was found that there was no noticeable downward trend in properties. In a separate study performed at GE [26], where Ti and Ta were both at the high limits of the specification, there was no measurable effects of the platelet phases. These studies suggests that small amounts of platelets may not be deleterious to mechanical properties.

C103 SC Turbine Blade

This alloy is a derivative of IN792-type alloys used for single crystal turbine blade applications. Other alloys such as SX792, AF56, and PWA1483 are similar to C103, so the platelet phases described for this alloy might be found in the other alloys if processed similarly.

In this case, the platelets were found in the as-cast condition, and manifested themselves as scale on the airfoil near the platform. Usually this would not be a problem, because scale can be removed after heat treat. However, in this case, recrystallized grains were found with the scale after heat treatment. When metallographic sections were made through the airfoil, platelets like those shown in Figure 8 were found. These platelets were on the surface of the blade, and extended to a depth of about 0.1 mm (0.005 in.). The composition of the platelets is given in Table VIII.

The platelets were enriched by 2.6-2.1X in Ti and Ta, respectively, over the amounts found in the alloy. In addition

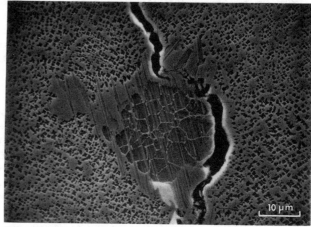

Figure 7. Fine platelet phases within a γ' eutectic pool, next to a crack in a DS GTD111 turbine blade that was given the solution heat treatment. Kallings etch.

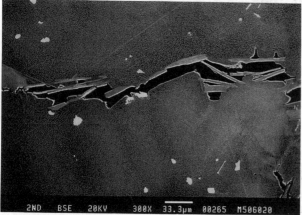

Figure 6. Coarse platelets spanning a crack-like void in an as-cast DS GTD111 turbine blade. Kallings etch.

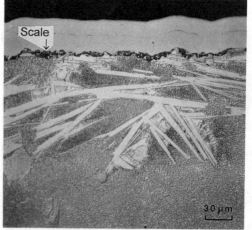

Figure 8. Platelet phases in an as-cast SC C103 turbine blade. From the casting surface, "scale" is seen to cover the platelets. Kallings etch.

Table VIII. Average Compositions of Phases in C103 Alloy.

		White (Si-rich)	Dark	Platelet	Platelet Alloy
	Element	Wt. %	Wt. %	Wt. %	Wt. Ratio
	Si	5.88	0.34	0.05	~5X
	S	0.00	0.00	0.00	-
	Zr	0.00	0.02	0.01	-
γ' elements	Al	0.26	4.98	3.39	0.95X
	Ti	4.88	11.00	10.84	2.6X
	Ta	10.65	4.61	10.41	2.1X
	Hf	0.00	0.00	0.00	-
	Nb	0.00	0.00	0.00	-
	Total (γ')eq =	15.79	20.59	24.64	
	Ratio (γ/γ')eq =	4.99	3.91	3.07	
	Total (γ)eq =	78.77	80.50	75.68	
γ elements	Ni	26.17	69.69	65.10	1.1X
	Co	11.60	7.04	6.86	0.8X
	Cr	16.83	3.51	3.17	0.3X
	Mo	17.87	0.26	0.46	0.24X
	W	6.30	0.00	0.09	0.02X
	Fe	0.00	0.00	0.00	-
	Total	100.44	101.45	100.38	
	Element	At. Pct.	At. Pct.	At. Pct.	At. Ratio
	Si	13.36	0.67	0.10	~5X
	S	0.00	0.00	0.00	-
	Zr	0.00	0.01	0.01	-
γ' elements	Al	0.62	10.16	7.39	0.93X
	Ti	6.50	12.64	13.30	2.8X
	Ta	3.76	1.40	3.38	2.1X
	Hf	0.00	0.00	0.00	-
	Nb	0.00	0.00	0.00	-
	Total (γ')eq =	10.88	24.20	24.07	
	Ratio (γ/γ')eq =	6.97	3.13	3.15	
	Total (γ)eq =	75.76	75.78	75.92	
γ elements	Ni	28.46	65.34	65.18	1.1X
	Co	12.57	6.58	6.84	0.75X
	Cr	20.66	3.72	3.58	0.26X
	Mo	11.89	0.15	0.28	0.26X
	W	2.19	0.00	0.03	0.03X
	Fe	0.00	0.00	0.00	-
	Total	100.00	100.67	100.10	

Table IX. Average Compositions of Phases in Low Carbon PWA1480 Alloy

		$\gamma + \gamma'$	Dark	White	Platelet (Core Side)	Platelet (Shell Side)	Platelet (Overall)	Platelet Alloy
	Element	Wt. %	Wt. %	Wt. Pct.	Wt. %	Wt. %	Wt. %	Wt. Ratio
	P	0.00	~20.0	0.00	0.00	0.00	0.00	-
	Si	0.00	0.00	0.60	0.08	0.03	0.06	~6X
	S	0.00	0.10	0.00	0.00	0.00	0.00	-
	Zr	0.00	0.00	0.00	0.03	0.00	0.01	~1X
γ' elements	Al	5.05	0.18	0.80	3.48	3.70	3.57	0.7X
	Ti	1.50	4.33	1.25	2.08	4.30	3.03	2X
	Ta	11.20	6.55	35.85	24.68	24.77	24.71	2.1X
	Hf	0.00	0.00	0.00	0.00	0.00	0.00	-
	Nb	0.00	0.00	0.00	0.00	0.00	0.00	-
	Total (γ')eq =	17.75	11.06	37.90	30.23	32.77	31.31	
	Ratio (γ/γ') =	4.59	6.22	1.61	2.31	2.02	2.18	
	Total (γ)eq =	81.45	68.82	61.20	69.78	66.03	68.17	
γ elements	Ni	62.15	36.25	26.10	60.33	57.60	59.16	0.9X
	Co	5.40	7.02	5.90	4.10	4.43	4.24	0.85X
	Cr	10.80	25.50	17.30	3.73	4.00	3.84	0.4X
	Mo	0.00	0.00	0.00	0.08	0.00	0.04	-
	W	3.10	0.00	11.90	1.50	0.00	0.86	0.2X
	Fe	0.00	0.05	0.00	0.05	0.00	0.03	-
	Total	99.20	99.98	99.70	100.10	98.83	99.56	
	Element	At. %	At. %	At. Pct.	At. %	At. %	At. %	At. Ratio
	P	0.00	32.12	0.00	0.00	0.00	0.00	-
	Si	0.00	0.00	1.78	0.18	0.08	0.14	~6X
	S	0.00	0.16	0.00	0.00	0.00	0.00	-
	Zr	0.00	0.00	0.00	0.02	0.00	0.01	~1X
γ' elements	Al	11.31	0.32	2.44	8.66	9.17	8.88	0.83X
	Ti	1.89	4.50	2.16	2.91	5.99	4.23	2.6X
	Ta	3.74	1.80	16.65	9.17	9.15	9.16	2.3X
	Hf	0.00	0.00	0.00	0.00	0.00	0.00	-
	Nb	0.00	0.00	0.00	0.00	0.00	0.00	-
	Total (γ')eq =	16.94	6.62	21.25	20.74	24.31	22.27	
	Ratio(γ/γ') =	4.90	9.23	3.71	3.82	3.11	3.49	
	Total (γ)eq =	83.06	61.09	78.75	79.24	75.69	77.72	
γ elements	Ni	63.96	30.72	36.92	69.08	65.51	67.55	1X
	Co	5.54	5.93	8.40	4.68	5.03	4.83	1X
	Cr	12.55	24.40	28.00	4.82	5.15	4.96	0.5X
	Mo	0.00	0.00	0.00	0.05	0.00	0.03	-
	W	1.02	0.00	5.44	0.55	0.00	0.31	0.25X
	Fe	0.00	0.04	0.00	0.06	0.00	0.03	-
	Total	100.00	99.99	101.78	100.18	100.08	100.14	

the Si was enriched several times, which undoubtedly is due to a contribution by the investment shell.

Two other phases contributed by the reactions with the shell were also found near the platelet phases. These phases, termed "white" and "dark," are listed in Table VIII, and contained up to ~6 wt.% Si in the "white" phase. This phase was also rich in Mo, Cr, Co, and Ta.

Low Carbon PWA1480 SC Turbine Blades

Low carbon versions of normal PWA1480 are used for non-commercial aerospace applications, so the presence of carbides could be eliminated as a source of fatigue cracks. Normal PWA1480 with about 0.04% carbon is not susceptible to the formation of platelet phases. However, when C is reduced to <0.01 wt.%, platelet phases may form on the surfaces adjacent to the investment shell or core.

The presence of the platelet phases was discovered below a layer of scale that was present on the turbine blades in the as-cast condition. The scale could not be easily removed. These platelet phases, shown in Figure 9, formed adjacent to the zircon facecoat of the investment shell. The compositions of the platelet phases are given in Table IX, and show ~2X increases in Ti and Ta over the base alloy. Silicon was also found to have increased several times over normal, undoubtedly due to reactions with the shell. In addi-

tion to the platelet phases, a white eutectic phase also formed which contained almost ~36 wt.% Ta.

Utilization of microdiffraction techniques by [13] revealed the platelets consisted of two phases. The continuous phase visible with metallography was an orthorhombic phase, identified as Ni_3Ta. Aligned precipitates within the Ni_3Ta phase had a hexagonal structure and were identified as Ni_3Ti or η phase, as shown in Figure 10.

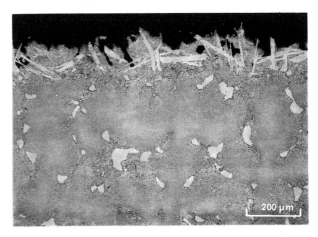

Figure 9. Platelet phases on the external surfaces of a low carbon PWA 1480 turbine blade [13]. Nitric electrolytic etch.

In other turbine blades, using an improved facecoat, platelets were found adjacent to some internal (core) surfaces [27]. An example of this platelet is shown in Figure 11. In this case, the silica-zircon core reacted with the alloy, and platelets were produced, differing only in Ti and W content from the platelets discussed above. In addition, a "dark" phase, as described in Table IX, was adjacent to the platelet phase. This phase contained about 20 wt.% P, which probably came from inadequate rinsing after being grain etched with H_3PO_4 acid, and then being given the full solution heat treatment. The presence of the H_3PO_4 acid probably contributed to the formation of the platelets.

Discussion

There may be some confusion about what the industry has traditionally termed η phase, and what recent Howmet researchers have presented. We know that η phase is HCP, and has solubility for several elements. Another phase that can take on the platelet appearance of η, identified as δ or Ni_3Ta, has an orthorhombic structure. This phase also has solubility for several elements. Both of these phases can coexist, with Ni_3Ti η precipitates being completely soluble within the Ni_3Ta orthorhombic phase. Based on Figure 10, it would appear the precipitates are aligned in the long direction of the platelet.

Effect of Major Alloying Elements on the Formation of Platelet Phases

When the compositions of the platelet are divided by the amount of element in the base alloy, the segregation for each element can be expressed in terms of a multiplication factor. These factors for several of the alloying elements that contribute precipitate strengthening are:

- hafnium, 10-15X (up to 15 times the amount in base alloy)
- titanium, 1.9-3.7X
- niobium, 1.8-3.1X
- tantalum, 1.3-2.1X

While it may not be possible to lower these elements without causing a loss of mechanical strength or corrosion resistance, there may be a trade-off between producibility and optimizing microstructures at all locations in large parts.

Aluminum and nickel are fairly neutral in partitioning to the platelet phases, with multiplication factors of 0.7-1.1X, and 0.9-1.3X, respectively. For those remaining elements that contribute solid solution strengthening, four were rejected from the platelets. Their multiplication factors were:

- cobalt, 0.5-0.85X
- molybdenum, 0.2-0.4X
- chromium, 0.1-0.4X
- tungsten, ~0.1X (almost totally rejected)

Effect of Carbon on the Formation of Platelet Phases

Carbon, which is tightly controlled during alloy-making, has a great influence on forming platelet phases. When C is reduced from 0.04 wt. % to <0.01 wt. % in PWA1480 in the presence of ceramic cores and shells from the investment casting process, platelet phases formed. When carbon was reduced from 0.15 wt.% to 0.08 wt.% in an alloy like IN792+Hf, the formation of platelet phases was promoted, especially for castings subject to segregation in thick sections or to slow solidification rates.

The traditional argument has been that carbon "ties-up" the platelet-forming elements. Thus, when carbon is removed from the alloy, more of platelet-forming elements are free to form the platelet phases. If this were the only mechanism, it becomes difficult to understand why so many platelet phases are created in PWA1480 after releasing only an additional 300 ppm carbon, when the Ta+Ti levels are already near 13.5 wt.%.

Figure 10. Brightfield image of an orthorhombic platelet phase as shown in Figure 9. Within the orthorhombic phase were aligned precipitates of hexagonal η phase [13].

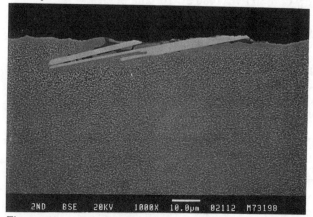

Figure 11. Platelets found next to the internal (core) surfaces on a solution heat treated SC PWA1480LC turbine blade. The dark phase contains P, probably from inadequate rinsing after the H_3PO_4 grain etch cycle. Kallings etch.

Effect of Minor/Trace Elements on the Formation of Platelet Phases

Even though the platelets were not found to contain high amounts of Si, Zr, or P, it is apparent they formed in the presence of enriched "white" or "dark" phases, which contained significant amounts of Si, Zr, and P. As suggested by [14], the amount of these phases could be reduced by lowering Si, Zr, and P. This, however, will not be easy, as investment shells and cores are full of SiO_2, Zr_2SiO_4 and ZrO_2, which eventually break down, especially for DS and SC castings [28]. This problem is intensified for the larger land-base castings, where liquid alloy may be in contact with the shells and cores for an hour or more. The P has the most likely chance of being reduced by proper rinsing of parts after H_3PO_4 acid etching or after use of FPI fluids which also contain P.

Summary

Based upon the results of this study, most of the platelet phases are believed to have formed due to:

- growth from γ' eutectic pools,
- casting segregation, and
- reactions with the investment shell and/or core.

Both coarse and fine platelet structures were observed. Coarse platelets were most often observed in the as-cast condition. After solution heat treatment, the shape of these platelets underwent little change. The presence of coarse platelets was usually associated with a crack-like void, scale, or recrystallized grains. Thus, these types of platelet phases were found by visual examination, grain etch, FPI, or x-ray inspection. The fine platelets were usually associated with γ' eutectic pools. These fine platelets were found in both the as-cast and fully heat treated condition. They were also associated with cracks that were found by FPI methods.

The concern for the presence of these platelet phases should be in the least inspectible areas of the parts, which have traditionally been on internal surfaces, where current cleaning methods may not adequately remove such phases. While small amounts of the platelet phases may not have caused noticeable losses of mechanical properties to date, it would be prudent to monitor the amount of the platelet phases as the parts become larger, or if the DS/SC casting process becomes longer.

Conclusions

1. Platelet phases were characterized in equiax IN939, equiax and DS IN792+Hf, DS IN6203 and GTD111, SC C103, and in low carbon versions of SC PWA1480.
2. The platelet phases are generally known to consist of Ni_3Ti (η) eta phase (which has an HCP structure), and-Ni_3Ta (δ) delta phase (which has an orthorhombic structure).
3. The platelet phases generally had 10-15X the concentration in the alloy for Hf, 2-4X the concentration in the alloy for Ti, 2-3X the concentration in the alloy for Nb, and 1.3-2X the concentration in the alloy for Ta.
4. Most of these platelet phases were associated with cracks or crack-like voids which would be detected by normal NDI techniques.
5. Most of these platelet phases were found in the as-cast condition. Once the coarse platelets formed, they were not readily dissolved during solution heat treatment. Fine platelet phases that existed within and near γ' eutectic pools can be partially solutioned, and were thought to be the least harmful form of this phase.
6. Several of the platelet phases formed due to a reaction with the investment shell or core, creating a visible scale. If these surfaces can be inspected by NDI techniques, the platelets can be found and removed. Note that most internal surfaces will not meet this criterion.
7. Platelet phases are known to slightly reduce tensile strength and ductility in wheels and large vane segments.

Recommendations

Reduction of the platelet phases is a desired goal. While there is only some evidence to suggest the platelets are undesirable, there is no evidence to suggest the platelets are beneficial. Since there are many constraints in the casting process (especially including post-cast cooling rates), the only practical solution to elimination of the platelet phases is to 1) reduce the alloying elements that promote the platelet phases, including Hf, Ti, Nb, and Ta, in that order, 2) keep C near specification maximums, 3) minimize Zr, Si, and P from the alloy, and 4) reduce or eliminate reactions between the alloy and the shell or core.

Acknowledgments

The author would like to acknowledge help from several personnel at the Howmet Technical Center: Metallography was performed by D. Bakos, SEM was performed by Bo-Ping Gu, and electron microprobe analyses were performed by P. Merewether. The expert work of these individuals is greatly appreciated. In addition, the author would also like to thank J. Mihalisin, P. Merewether, J. Grady, G. Cole, and F. Norris for several helpful discussions, and for reviewing the manuscript. Discussions with J. H. Wood of GE Power Generation and J. Radavich of Purdue Univ. (retired), were also helpful.

References

1. J. Fiedler, P. Follo, D. Wilson, unpublished data, AVCO Lycoming, 1980, used with permission.
2. K. Schneider (ABB Kraftwerke AG), "Advanced Blading", High Temperature Materials for Power Engineering 1990", Proceedings of a Conference at Liege, Belgium, September 1990, Kluwer Academic Publishers, p935.
3. P. W. Schilke, A. D. Foster, J. J. Pepe, and A. M. Beltran (GE-PG), "Advanced Materials Propel Progress in Land Base Gas Turbines", Adv. Mat. & Proc., April 1992, p22.

4. M. Doner, J. A. Heckler (Allison), "Effects of Section Thickness and Orientation on the Creep-Rupture Properties of Two Advanced Single Crystal Alloys", Aerospace Tech. Conf. Long Beach, CA, Oct. 1985.
5. United Technologies Alloy Reference List, Aug. 1992.
6. The Superalloys, Edited by C. T. Sims and W. C. Hagel, Published by John Wiley & Sons, 1972.
7. Superalloys II, Edited by C. T. Sims, N. S. Stoloff, and W. C. Hagel, John Wiley & Sons, 1987.
8. The Nimonic Alloys and Other Nickel-Base High Temperature Alloys, Edited by W. Betteridge and J. Heslop, Published by Edward Arnold, 1974.
9. Metals Handbook 9th Edition (1985), Metallography and Microstructures, Wrought and Heat Resistant Alloys, Published by ASM, p310-311.
10. P. Willemin, and M. Durand-Charre, "Phase Equilibria in Multicomponent Alloy Systems", Superalloys 1988, Pub. by TMS, p723.
11. J.R.Mihalisin & R.F.Decker, Trans AIME 218 (1960) p507.
12. R. F. Decker, "Strengthening Mechanisms in Nickel-Base Superalloys," presented at Steel Strengthening Mechanisms Symp., Zurich, Switzerland, May, 1969.
13. V. Biss and Bo-Ping Gu, Howmet Corp. (Tech. Center), unpublished data, 1988.
14. Y. Zhu, S. Zhang, et. al., "A New Way to Improve the Superalloys", Superalloys 1992, Pub. by TMS, p145.
15. M. Emerson, Howmet Corp. (Ceramic and Coating Div.), unpublished research, 1980.
16. S. A. Knight, Howmet Corp. (LaPorte, IN., Casting Div.), unpublished research, 1981.
17. S.W.K. Shaw and M. J. Fleetwood, "New Nickel Base Investment Casting Alloys IN6201 and IN6203", Materials Science and Technology, Sept. 1989, Vol. 5, page 925.
18. W. Esser (Siemens KWU), "Directional Solidification of Blades for Industrial Gas Turbines", Mat'ls for Advanced Power Engineering 1994, Edited by D. Coutsouradis et. al., Kluwer Academic Pub., 1994, p641.
19. G. K. Bouse, D. Bakos, and S. Renz, Howmet Corp. (Tech. Center), unpublished research, 1995.
20. S.W.K. Shaw, "Response of IN939 to Process Variations", Superalloys 1980, pub. by ASM, p275.
21. T.B.Gibbons and R.Stickler, "IN939: Metallurgy, Properties and Performance", High Temp. Alloys for Gas Turbines 1982, published by D. Reidel (1982), p369.
22. K. M. Delargy and G. D. W. Smith, "Phase Compositions and Phase Stability of Alloy IN939", ibid p705.
23. G. R. Cole II, Howmet Corp. (Tech. Center), unpublished research, 1995.
24. G. K. Bouse and D. Bakos, Howmet Corp., ibid, 1995.
25. S. A. Kriesel, Howmet Corp., ibid, 1992.
26. J. H. Wood, General Electric Co. Power Generation Division, private communication with the author, 1996.
27. G. K. Bouse and D. Bakos, Howmet Corp. unpublished research, 1995.
28. K. F. Lin, Y. L. Lin, and S. E. Hsu, "Interaction of Ceramic Shell Mold With Ni-Base Superalloys During Single Crystal Casting", Advanced Materials and Processing Techniques for Structural Applications, Ed. by T. Khan and A. Lasalmonie, Pub. by ONERA (1987), p285.

SERVICE TEMPERATURE ESTIMATION FOR HEAVY DUTY GAS TURBINE BUCKETS BASED ON MICROSTRUCTURE CHANGE

Yomei Yoshioka*, Nagatoshi Okabe*†, Takanari Okamura*, Daizo Saito*, Kazunari Fujiyama*, Hideo Kashiwaya*

*Heavy Apparatus Engineering Laboratory,
Toshiba Corporation, Yokohama, 236

†Present Address:Dept. of Mechanical Engineering
Ehime University, Matsuyama, 790

Abstract

This paper studies the microstructure change of Ni-base superalloy IN738LC in the temperature range of 750°-900°C, and develops methods for predicting the service temperature of a gas turbine bucket operated for around 20,000 hours. Specimens of IN738LC were aged in the temperature range up to 24,000 hours to obtain the data of microstructure change. Growth rate of γ' diameter in IN738LC was proportional to the 1/3 power of aging time and γ' density was inversely proportional to aging time. Accuracy of predicted temperature based on the γ' diameter and density increased with increasing the aging time and temperature. Metal temperature of the actual 20,000 hours serviced bucket estimated by the microstructure change agreed with the thermal analyses at the leading edge portion which is a most reliable point for the analysis. This shows the availability of the proposed methods.

Introduction

To keep the reliability of gas turbine hot-gas-path components, it is important to know accurate metal temperature. Especially the buckets are rotated and current ones have an air cooled system under high temperature and complex gas stream, which increase difficulty to analyze the metal temperature. In the case of high temperature components, microstructure changes are occurred during service due to overaging of the materials. By using this phenomena, it is thought to be possible to estimate the metal temperature.

This paper describes the development of temperature estimation methods by using the γ' coarsening law of gas turbine bucket alloy IN738LC and investigation results of accuracy of these methods. These methods are applied to the actual serviced bucket and the availability are demonstrated. The effect of stress on the γ' coarsening rate is also described.

Test Methods

The chemical composition of the cast Ni-base alloy IN738LC used for this investigation is shown in Table 1. The slab material of 200 mm × 80 mm × 20 mm was made by an investment casting process in vacuum and specified heat treatments of 1,120°C for 2 hours and 843°C for 24 hours followed by gas cooling were conducted. After the heat treatments, the material was aged at temperatures of 750°, 800°, 850°, and 900°C for 1000, 3000, 10,000, and 24,000 hours. After the agings, microstructural observations were performed. Metallographic specimens were prepared from the unaged and aged samples. After being sectioned, mounted, and polished, the specimens were etched using a marble's reagent consisting of 4g $CuSO_4$, 20ml HCl, and 80ml H_2O. Two stage replication technique was used for the observation of γ' phases by transmission electron microscopy (TEM). Quantitative image analysis was subsequently conducted to measure the mean diameter of γ' precipitates, number of precipitates per unit area (density), and area fraction by using a LUZEX III U NIRECO image analyzer.

Table I Chemical composition of IN738LC studied (mass %)

	C	Ni	Cr	Co	Al	Ti	W	Mo	Ta	Nb	B	Zr
Aging test material	.10	Bal.	15.82	8.60	3.58	3.40	2.90	1.68	1.83	.87	.011	.056
Creep test material	.09	Bal.	15.92	8.18	3.52	3.54	2.55	1.74	1.76	.89	.010	.034

Microstructural observation was conducted for new and around 20,000 hours serviced stage 1 buckets. Circumferential 12 points are observed on the mid-span cross-section of the airfoil of the buckets as shown in Figure 1. The surface observed is parallel to the centrifugal force.

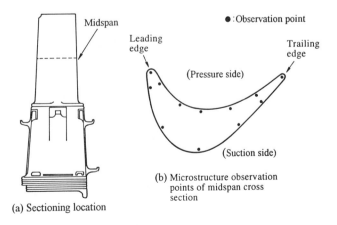

Figure 1: Sectioning location and microstructure observation points of buckets studied.

Test Results

Effect of Aging on Microstructural Change

Observation results of transgranular γ' precipitates were shown in Figure 2. The microstructure of as-specified heat-treated IN738LC shows around 0.4 μm diameter cubical γ' with around 0.02 μm fine spherical γ'. During the agings, fine γ' dissolved and cubical γ' is observed to have been coarsened and spheroidized.

Figure 3 shows the image analyses results of transgranular γ' which are mean diameter, density, and area fraction of γ'. During the agings, the mean diameter is increased and density decreased. This tendency accelerated with the temperature increased.

Effect of Stress on Microstructural Change

Observation results of transgranular γ' precipitates were shown in

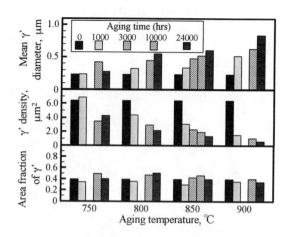

Figure 3: Image analysis results of γ' in aged IN738LC.

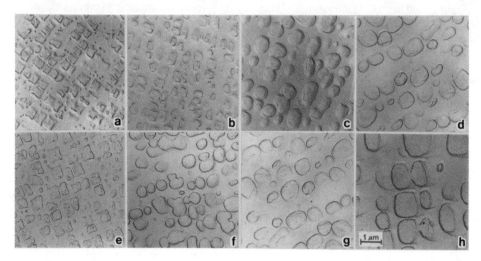

Figure 2: Microstructure of IN738LC aged at 850°C for (a)0 hrs, (b)1,000 hrs, (c)10,000 hrs, and for 24,000 hrs at (d)850°C, (e)750°C, (f)800°C, (g)850°C, (h)900°C.

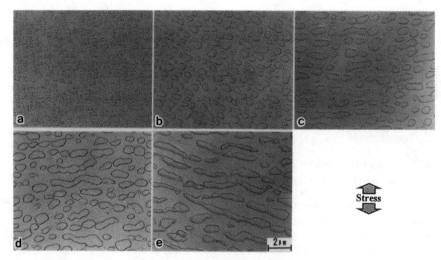

Figure 4: Microstructure of IN738LC creep-tested at 900°C under 98MPa interrupted at (a)0 hr, (b)2,500 hrs, (c)5,000 hrs, (d)7,500 hrs, (e)14,440.8 hrs (ruptured).

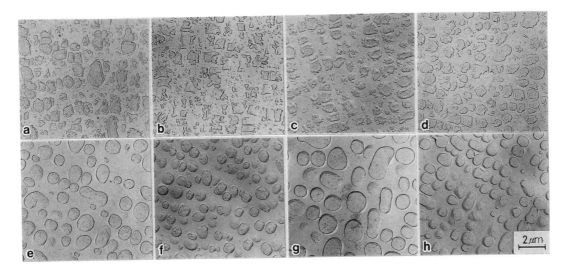

Figure 5: Microstructure of the new bucket at mid-span cross section of (a)leading edge portion, (b)mid-chord portion of suction side surface, (c)trailing edge portion, (d) mid-chord portion of pressure side surface, and of the 20,000 hrs serviced bucket at that of (e)leading edge portion, (f)mid-chord portion of suction side surface, (g)trailing edge portion, and (h)mid-chord portion of pressure side surface,

Figure 4. The microstructures of stress-free portion of the samples at the pre-tested and interrupted creep tested conditions was the same behaviors as aged samples. But, during creep testing, microstructures of stressed portions show a rafted structure which γ' grows at the direction perpendicular to the stress. This structure is growing with creep testing time increasing and most significant elongated one is observed in the ruptured sample.

Microstructural Observation Results of a Serviced Bucket

Figure 5 showed the microstructure at the airfoil surfaces of the new and around 20,000 hours serviced stage 1 buckets. Elimination of fine spherical γ' and coarsening of large cubical γ' are observed, but the degree of the coarsening in the serviced buckets are not so significant.

Discussions

γ' Coarsening Law of IN738LC

Coarsening of fine precipitates normally explained to be the volume diffusion-controlled coarsening theory formulated by Lifshitz, Slyozov (Ref.1), and Wagner (Ref.2). This model propose that the mean particle diameter increases according to the following equation.

$$\bar{d}^3 - \bar{d}_0^3 = K \cdot t \quad (1)$$

$$K = \frac{64\gamma_e D C_e V_m^2}{9kT} \quad (2)$$

where,

$\bar{d}$: average γ' mean diameter of pre-aged IN738LC
$\bar{d}_0$: average γ' mean diameter of aged IN738LC
γ_e : specific γ/γ' interfacial free energy
D : diffusion coefficient of γ' solutes in γ
C_e : equilibrium molar concentration of γ' solute in γ
k : Bolzmann constant
t : aging time
T : aging temperature
V_m : molar volume of γ'

The third power of the mean γ' diameter of the aging materials versus aging time are plotted as shown in Figure 6. The data of each aging temperature are scattered at the earlier stage of aging, but they are getting good linearity during aging, which indicates γ' coarsening of IN738LC follows the volume diffusion-controlled coarsening theory.

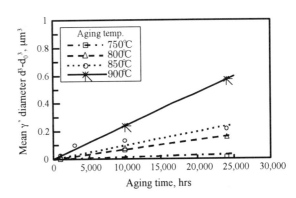

Figure 6: Change of γ' mean diameter in IN738LC during aging.

During aging, coarsening of γ' precipitates occurred with γ' density decreasing. In the equation (2), V_m is molar volume of γ', C_e is equilibrium molar concentration of γ' solute in γ and $\bar{d}^3$ is a mean volume of one precipitate, which induces the following equation.

$$N = \frac{V_m C_e}{d^3} \quad (3)$$

By using this equation, equation (1) can be modified to the following equation.

$$\overline{N}^{-1} - \overline{N}_0^{-1} = K' \cdot t \quad (4)$$

$$K' = \frac{64\gamma_e D V_m}{9kT} \quad (5)$$

Inverse of density versus aging time is also plotted in the Figure 7. Density versus time also shows good linearity at the longer aging time as well as mean γ' diameter does.

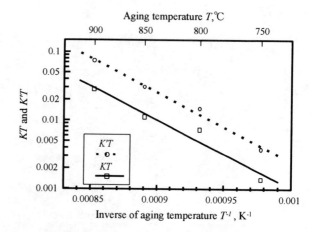

Figure 8: Temperature dependence of rate constant K and K' for γ' diameter and density in aged IN738LC.

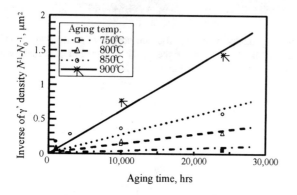

Figure 7: Change of γ' density in IN738LC during aging.

In both the equations of (2) and (5), diffusion coefficient D in the proportional constant K and K' is explained by the following equation.

$$D = D_0 \exp(-Q_d / kT) \quad (6)$$

where,

D_0 : frequency factor
Q_d : activation energy of diffusion

ln(KT) and ln(K'T) versus -1/T are plotted in Figure 8. Good linearity is also obtained and the activation energy is 192 kJ/mol, but the activation energy of Ti or Al in Ni is 257 to 270 kJ/mol (Ref.3) Those literal value is larger than the value we obtained.

By using Q_d=192kJ/mol, γ' diameter and density calculated from equation (1) and (4) are plotted in Figure 9 compared with measured value. Good correlation is obtained at the longer aging time and higher aging temperatures.

Effect of Stress on γ' Coarsening Rate

Figure 10 shows the image analyses results of transgranular γ' of the interrupted creep samples. Stressed (parallel portions in the test samples) and unstressed portions (attached portions) are separately evaluated to figure out the effect of stress on the microstructural change. No effects of stress on the area fraction, density, and mean diameter of γ' are observed, which concludes these microstructural methods can apply even to component materials which are under high stress and already has rafted microstructures.

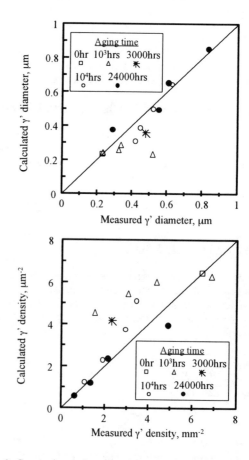

Figure 9: Comparison of measured γ' mean diameter and density with calculated ones.

Accuracy of Microstructural Estimation Methods

Changes of γ' diameter and density during aging are explained by the equation (1) and (4). By using these equations, it is thought to be possible to estimate the metal temperature of the components if γ' diameter and/or density are obtained. Accuracy of these estimation methods is investigated in this section by using the results of aging materials.

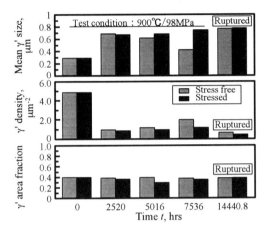

Figure 10: Change of γ' volume fraction, mean diameter, and density during creep testing under 98MPa at 900°C.

Aging temperature is estimated from the γ' diameter, density, and aging time, and then change of deviation between true aging temperature T_a and estimated one $T_{est.}$ during aging is investigated, where furnace temperature for the aging is defined as true aging temperature. To evaluate the deviation, variation coefficient C_v which is derived from the deviation divided by true temperature is introduced to eliminate the effect of absolute value of aging temperature. Average value of the coefficient for each aging time versus aging time are plotted in Figure 11. The coefficient which is derived by using both estimation methods is decreasing with the aging time increasing and shows the following equation.

$$C_v = a \cdot t^b \quad (7)$$

where

$$C_v = \frac{|T_a - T_{est.}|}{T_a}$$

a, b : constant

Those values of "a" and "b" are listed in Table II.

The solid line indicates a regression line made by the average coefficient derived by the estimation method from γ' diameter, and the broken line does a regression line made by the average coefficient derived from the method from γ' density. This figure shows that the estimation method by the γ' diameter is more accurate than the methods by γ' density up to 5,000 hours, but that by γ' density is more accurate over 5,000 hours. In the case of new and earlier aging time conditions, counting of fine γ' is thought to induce more observational error than measuring of γ' diameter. But in the case of longer aging time, γ' size increases and population of γ' observed decreases, which is thought to result in less accuracy to measure the diameter.

Table II Material constants of "a" and "b" in equation (Ref.7)

	a	b
Estimation method by γ' diameter	1.19	-0.371
Estimation method by γ' density	5.05	-0.556

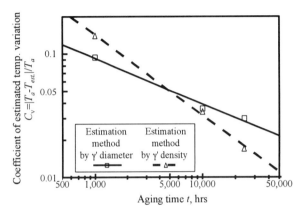

Figure 11: Effect of aging time on coefficient of estimated temperature variation.

Effect of aging temperature on accuracy of the temperature estimation methods is also investigated. By using equation (7), the value "a" is derived for each aging temperature and "a" versus aging temperature are plotted in Figure 12. No good relationship is observed at the 10,000 hours aging time, but at the 10,000 and 24,000 hours aging time, clear temperature dependency is observed and following equation is obtained. The value of "C" and "Q" are listed in Table III.

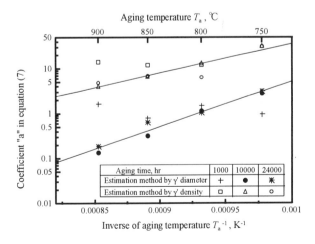

Figure 12: Temperature dependence of variation coefficient.

$$\ln a = C + Q/kT_a \quad (8)$$
$$C, Q : \text{constant}$$

Higher aging temperature shows less "a" value, which also means less variation coefficient. According to the temperature increasing, microstructure is getting homogeneous and observational error is thought to be decreasing.

The variation coefficient Cv is derived from aging time t and aging temperature T_a by using the equation (7), (8) as follows.

$$C_v = \exp(C + Q/kT_a) \cdot t^b \quad (9)$$

Table III Material constants of "C" and "Q" in equation (8).

	C	Q (J)
Estimation method by γ' diameter	-21.4	3.19×10^{-19}
Estimation method by γ' density	-11.4	2.07×10^{-19}

The scatter band of estimated temperature from these methods, therefore, is expressed as following equations.

$$T_{max.} = (1 + C_v) \cdot T_a \quad (10)$$
$$T_{min.} = (1 - C_v) \cdot T_a \quad (11)$$

In the case of the around 20,000 hours serviced bucket used in this investigation, both estimation methods indicate the same results, that is, $T_{max.}$=859 and $T_{min.}$=841℃ if true metal temperature of the bucket is supposed to be 850℃ and serviced time is 20,000 hours. Estimated temperature is thought to be within the scatter band of ±9℃, which means that quite accurate estimation can be obtained by these method at longer service time.

Applicability to the Actual Bucket

Microstructural temperature estimation results (estimated temperature) of the around 20,000 hours serviced bucket are shown in Figure 13 with temperature analyses results (analytical temperature) by using three different airfoil surface heat transfer rate calculations.

The transfer rate on the surface of buckets are calculated using both the integral method of boundary layer and Reynolds average Navier-Stokes equation (RANS). Calculation for an integral method were performed for two cases. One is based on the criterion of transition from laminar to turbulent in boundary layer (Analysis 1) and the other assumes the boundary layer as a fully turbulent boundary layer (Analysis 2). The analysis code of RANS (Analysis 3) is that the base code is STAN5 and two-equation turbulent model is incorporated (Ref.4, 5). These total three cases are calculated here.

The three analyses results show good coincidence with each other at the leading edge portion of the bucket. At the suction side, Analysis 2 shows higher temperature than the analysis 1 and 3 but almost the same values as estimated one. All the analysis results at the pressure side shows lower than the estimated one.

Temperatures estimated from the microstructure shows good coincidence at the leading edge portion, but higher values at the other portions. The applicability of the microstructural temperature estimation methods for the actual buckets is discussed hereafter.

The errors arisen from the microstructural estimations of the bucket are thought to be due to the differences of original microstructures and operating histories, and the microstructural measurement methods. To eliminate these error factors, the new bucket is also evaluated at the same locations as the serviced one. The accuracy of the bucket temperature estimation is, therefore, almost the same as that of the results from the aging material as we discussed in the previous section.

The accuracy of the analyses codes are also discussed here. The leading edge portion is regarded as a cylinder for calculation of the heat transfer rate. In this case, degree of acceleration due to the turbulence intensity is empirically known well, which means the accuracy of analyses is thought to be high. The good coincidence between analytical and estimated temperatures obtained at this point imply that this estimation method has quite high accuracy. The other portions shows analyses results are lower than the estimated ones. Turbine buckets are operated under the influences of the turbulence and the periodical fluctuations of wake which is generated from turbine nozzle vanes. This fluctuation of wake is also thought to accelerate the heat transfer rate. Recently, many researches on the influence of the wake to the heat transfer rate by using test turbines and the unsteady state analysis of nozzle vanes and buckets are performed. The enhancement effect for heat transfer rate on surface of a bucket have been recognized, but the accuracy of analysis is not so high. Large discrepancy between analytical and estimated values have been recognized at the surface regions from leading edge portion to the mid-chord of pressure and suction sides which are thought to be strongly influenced by the

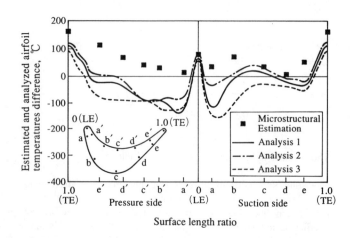

Figure 13: Temperature distribution of serviced bucket surface estimated by γ' mean diameter(mid-span cross section of bucket airfoil).

wake of nozzle vane. This discrepancy is, therefore, thought not to be due to the microstructural estimation methods, but to be due to less considerations on this wake. This analyses code does not take surface roughness into account, but it is reported that the heat transfer rate increases significantly after the characteristic value defined by surface roughness exceeds the critical value (Ref. 6, 7, 8). Roughness of the bucket surface is observed to increase during service and the pressure side surface is observed to be roughen significantly. The analyses should, therefore, be considered the roughness effect in this region.

Summary and Conclusions

Metallurgical metal temperature estimation methods were investigated and were verified to be applicable to actual buckets. The results are summarized and described as follows.

(1) Change of average γ' diameter in IN738LC is proportional to the one third power of aging time and density is inversely proportional to aging time.
(2) Accuracy of the estimation methods is explained by the function of aging time and temperature and increases with aging time and temperature increasing.
(3) Comparison between estimated and analyzed bucket surface metal temperature is conducted and good coincidence was observed at the most reliable analysis point, which verify these method are applicable to the actual bucket temperature estimation.
(4) Comparison between estimated and analyzed bucket surface temperature implies some possibility to analyze the actual phenomena of the buckets and also clarify the problem of analyses code

References

1. M.Lifshitz, V.V.Slyozov, "The Kinetics of Precipitation from Supersaturated Solid Solutions," J.Phys. Chem. Solids, 19(1961), 35-50.
2. C.Wagner, "Theorie der Alterung von Niederschlagen durch Umlosen," Z.Electrochem., 65(1061), 581-591.
3. The Japan Institute of Metals eds., Metals Data Book, (Maruzen, 1984), 27-28.
4. D.Biswas, Fukuyama, Araki, "Calculation of Transitional Boundary Layer with an Improved Low-Reynolds Version of k-ε Model of Turbulence Part 2," Journal of the Gas Turbine Society of Japan, Vol.20, No.77 (1992), 68-75.
5. D.Biswas,Fukuyama, Araki, "Calculation of Transitional Boundary Layer with an Improved Low-Reynolds Version of k-ε Model of Turbulence Part 2," Journal of the Gas Turbine Society of Japan, Vol.20 No.78 (1992), 25-31.
6. M.F.Blair, "An Experimental Study of Heat Transfer in a Large-Scale Turbine Rotor Passage," ASME Paper 92-GT-195(1992), 1-15.
7. M.H.Hosni, H.W.Coleman, R.P.Taylor, "Heat Transfer Measurements and Calculations in Transitional Rough Flow," ASME Paper 90-GT-53 (1990), 1-9.
8. R.P.Taylor, J.K.Taylor, M.H.Hosni, H.W.Coleman, "Heat Transfer in the Turbulent Boundary Layer with a Step Change in Surface Roughness," ASME Paper 91-GT-266(1991), 1-8.

DIRECTIONAL COARSENING OF NICKEL BASED SUPERALLOYS : DRIVING FORCE AND KINETICS

Muriel Véron, Yves Bréchet and François Louchet
Groupe Physique du Métal, LTPCM-CNRS (URA 29)
BP 74, 38402 St Martin d'Hères

Abstract :

The γ' morphology of a single crystal superalloy, that had first undergone a plastic prestrain, has been studied after having been annealed at high temperature. Observation of a directionaly coarsened microstructure has revealed the dominant role of plasticity, more precisely of dislocations at matrix/precipitates interfaces, in the driving force of directional coarsening. A model and a 3D computeur simulation for the phenomenon has been developed, based on a model that take into account plasticity induced during creep and misfit stresses. Results of the model are presented and compared to experimental observations on AM1 superalloy. Morphology maps describing the expected rafting geometry for other superalloys (other misfits) under a given applied stress are produced. Finally, considerations on the mechanical properties of the coarsened microstructure, in particular the ability of a dislocation to glide in the new morphology, are discussed.

Introduction :

Directional coarsening or "rafting" of γ' precipitates in single crystal Nickel based superalloys is usually observed to take place under stress at high temperatures (typically 1050°C). Though it has been extensively studied experimentally [1,2,3 for example] and theoretically [4,5,6], the exact nature of the driving force and the associated kinetics are still a matter of controversy. It is generally agreed that the rafting morphology and the kinetics are ruled by two key parameters, namely the misfit between the two phases, and the amplitude and sign of the uniaxial applied stress. Though most of the previous investigations were based on the elastic incompatibilities between the two phases, plasticity occurs in the stress-temperature domain where directional coarsening is observed [7,8]. Since plastic deformation is driven by the elastic energy stored in the material, it can participate significantly to its relaxation, and has to be taken into account in the modelling of directional coarsening.

The first part of the present contribution reports an experimental investigation of the possible role of plasticity in rafting mechanisms: the influence of a pre-strain on morphological evolution during annealing of a superalloy has been put forward. Then the second part is devoted to the physical basis and the results of a computer simulation, taking the local plastic strain as the origin of the main driving force for directional coarsening.
Finally, some properties of the coarsened structures after different creep-simulated times are discussed.

Driving force

Since dislocations have been observed at the matrix/precipitate interfaces in specimens tested under service conditions, we have investigated the influence of a pre-strain on the phenomenon. We have studied the morphological evolution during annealing of a pre-strained sample (in compression) of AM1 superalloys [9], provided by SNECMA.
The initial microstructure of this alloy is two-phased : γ' precipitates, of an average size of 0.45 μm and with a volume fraction about 70% at room temperature, are well aligned in a γ matrix. The average distance between two precipitates -i.e. the size of matrix channel- was determined using image analysis, and is about 0.09 μm. The misfit is defined as $\delta = 2(a_{\gamma'} - a_\gamma)/(a_{\gamma'} + a_\gamma)$, where a_γ and $a_{\gamma'}$ are respectively the lattice parameters of the γ and γ' phases. Its value was determined to be -3×10^{-3} at 1050° C [10].

Experimental procedure

Two compression samples were selected : both were introduced in a furnace and left 1 hour at 850° C (1123K). At this temperature the kinetics of coarsening are very low. After stabilisation, sample (b) was stressed in compression at 450 MPa up to a strain of about 1% (then the stress was removed), whereas sample (a) was left without stress. Then the temperature was increased, and both samples were left at 1323K during 20h.
The only difference between the two specimens is that sample (b) has undergone a strain of about 1% whereas

sample (a) was not strained. Then both samples were further heat treated, without any load, at 1323K during 20 hours and then air cooled down to room temperature.

Faces parallel to the various (100) directions (parallel or perpendicular to the stress axis) were mechanically polished, then etched to reveal the two phases. The microstructure was then observed in a scanning electronic microscope (SEM) (Jeol 6400). Thin foils have been also prepared for TEM observations of the interfaces.

Results and discussion.

TEM observations on sample (a) evidence a high density of dislocations, in the matrix corridors parallel to the stress axis. The corridors perpendicular to the stress axis are nearly dislocation free. Sample (b) shows no evidence of plastic deformation.

SEM observations have revealed differences in microstructures between the two samples. Though both of them were aged without applied stress at 1323K, sample (a) (which was not prestrained) has no sign of directional coarsening, the precipitates are still cuboidal with a size of about 0.45µm; whereas sample (b) which was prestrained shows pronounced directional coarsening : the precipitates are elongated into rafts of 3µm length in the (001) directions containing the compression axis (see fig. 1).

In absence of an externally applied stress, simply by prestraining the sample, directional coarsening was observed.

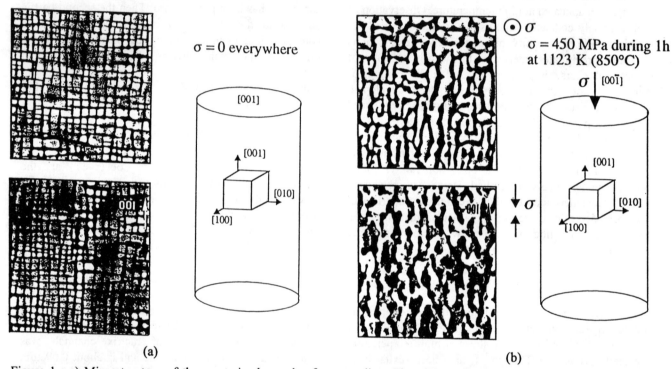

Figure 1 : a) Microstructure of the prestrained sample after annealing. The observations are made in two directions : perpendicular (first image) and parallel to the stress (at the bottom). b) Microstructure of the "non strain" sample, observed in the similar directions.

This results suggest that the dislocations, introduced during the prestrain and forming networks on specific interfaces as observed in TEM, have an effect on the matrix/precipitates interfaces, that is therefore probably to relax selectively the misfit stresses in the matrix corridors (the ones parallel to the applied stress in the example considered). This effect was already suggested for the dislocations networks observed during creep [7,11,12], but it had yet not been clearly separated from the effect of the applied stress.

As the net effect of prestrain is to relax the internal stresses only in some channels of the γ phase, coarsening is then more likely to be associated with inhomogeneous relaxation of internal stresses rather than to the coupling between the applied stress and the misfit stress. This experiment means clearly that the driving force due to plastic deformation is dominant as compared to other effects. A model for directional coarsening can be proposed : this anisotropic relaxation of coherency stresses induces a gradient in chemical potential (due to a gradient in elastic energy) which drives the diffusive flow through the channels, responsible for directional coarsening (fig. 2).

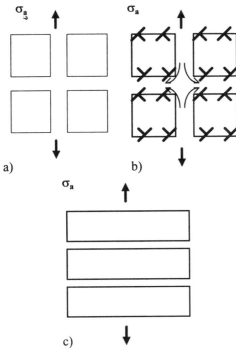

Figure 2 : a) In the initial condition, the misfit is isotropic on the γ/γ' interfaces. b) Under stress at high temperature, dislocations have developed and they relax selectively some γ/γ' interfaces. A diffusional flow occurs in the matrix (probably also in the precipitates), and c) the microstructure evolves.

A similar experiment has been conducted for a pre-strain in tension along a <001> direction, and the same observations were made : directional coarsening, with rafts perpendicular to the 'pre-strain axis', was observed [13].

Directional coarsening modelling

Physical basis

Due to the combination of misfit and applied stresses, the different types of γ phase channels are differently stressed. Under tension for instance, and in the case of a negative misfit, the "horizontal" channels (i.e. perpendicular to the stress axis) undergo a larger shear stress than the "vertical ones" (which contain the stress axis). As a consequence, both plastic flow and relaxation of misfit stresses are expected to take place first in horizontal channels, which is confirmed by transmission electron microscopy of dislocations at γ/γ' interfaces in these channels. A driving force therefore arises for diffusion between horizontal (relaxed) and vertical (non relaxed) channels. This driving force is obviously still present if the sample has been prestrained, and subsequently aged at high temperature without any applied stress, as reported in the first part.

The previous results allowed us to put forward a driving force for directional coarsening in the plastic regime. We have then modelled the phenomenon and developed a computer simulation taking into account the local plastic strain as the main driving force for coarsening. The simulation is at a mesoscopic scale, thus it provides the evolution of the entire precipitate population in the matrix, and the results can be directly compared to SEM images. Details of the simulation have been published elsewhere [14], in this contribution, we will present now only the main steps of the program, but some equations will be developed the Appendix.

First, the microstructure we have chosen to simulate consists of a network of matrix channels (not of precipitates), because the matrix undergoes much of the microstructure evolution and parameters such as the width of the channels is pertinent for directional coarsening. This network is a simple cubic lattice with a lattice constant of the average (γ channel plus γ' precipitate) size. The width of these channels is distributed according to an experimental distribution measured on SEM pictures of the AM1 alloy.

The principle of the simulation consists in calculating the flux of matter between the different matrix channels, due either to plastic flow, or to diffusional flow resulting from a chemical potential gradient. The different fluxes added together will lead too a redistribution of matter between the different channels (with a constant total volume fraction of precipitates) and to directional coarsening. In order to do so, we need to evaluate the local stresses inside each channel. As the value of the local stress in a channel depends on the width of this channel, the flux of matter is calculated for small time steps, then the width distribution is reset, and the new local stresses are re-estimated in all the corridors. Time steps were chosen to be small enough compare to the typical coarsening time.

Two stresses contribute to the local stress : on one hand the coherency stresses (misfit stresses) which may be relaxed on some interfaces due to the presence of dislocations; and on the other hand, the macroscopic applied stress, which is distributed between the channels and the precipitates according to a composite effect. The stresses in a given channel also change the local chemical potential of the elements in a channel and therefore a difference in local stresses in neighboring channels will result in a cross-diffusion flux contributing to the channel width evolution (this was observed in the prestrain sample experiment). This local stress in each channel also leads to plastic flow (and thus to mass transport from one channel to its neighbors).

In order to compute the local stresses induced by the applied stress in the channels, we have represented the microstructure as a complex composite structure. The equilibrium laws that govern this system are the partitioning of stresses between γ and γ' phases and the compatibility of strain rates of the matrix in the vertical and horizontal channels, assuming the γ' phase does not deform plastically.

From this two equations and from the relation between the strain rate and the stress in the matrix, the local stresses σ*v and σ*h in the vertical and horizontal channels respectively can be computed as functions of the applied stress and the coherency stresses, in each corridors, at each time step. The values of the coherency stresses can be tuned to account for the relaxation process. Thus in the "relaxed" channels (horizontal in our example), the value of the stresses is zero, and in the "non relaxed" ones (vertical), they remain at their initial value.

As a result of this partial relaxation, effective stresses in tensile creep are :

$$\sigma_v^* = \left(\frac{w_h}{l}\right)^{\frac{1}{n}} \frac{\sigma_a (l+w_v)^2 - E_v \delta_v (2l+w_v)}{\left(\frac{w_h}{l}\right)^{\frac{1}{n}} (2l+w_v) w_v + l^2} + E_v \delta_v \quad (1)$$

$$\sigma_h^* = \frac{\sigma_a (l+w_h)^2 - E_v \delta_v (2l+w_h)}{\left(\frac{w_h}{l}\right)^{\frac{1}{n}} (2l+w_h) w_h + l^2} \quad (2)$$

where l is the mean size of a precipitate, w_h and w_v are respectively the widths for the horizontal and vertical channels, σ_a is the applied stress intensity, E_v is the Young modulus of the composite structure and δ_v is the misfit at the vertical interfaces.

Once the stresses are determined in each channel, the next step of the program is to calculate the corresponding elastic energy densities.

In addition to this bulk term, another driving force stems from the capillary effect : the thinner channels tend to disappear and the thicker tend to grow, irrespectively of their orientation, simply to reduce the total surface in a process similar to Oswald's ripening.

Finally, in a given channel (i), the total energy density $E_{tot}(i)$ is :

$$E_{tot}(i) = E_{el} + E_i \quad (3)$$

where E_{el} is the elastic energy density stored in the channel due to the local stress, and E_i is the interface energy density.

The inhomogeneities in these energy densities will drive a diffusive flux between the channels, particularly between the vertical and horizontal channels for whom the elastic energy densities are different. The driving force carrying atoms from a given channel (i) to an other channel (j) is the chemical potential gradient which can be approximated by :

$$\nabla \mu(ij) = (E_{tot}(j) - E_{tot}(i)) / l \quad (4)$$

The characteristic distance l is of the order of precipitate size.

For a given element, the mobility M is related to the diffusion coefficient by the Einstein relation : M= D/kT, where D is the diffusion coefficient of a given element. As diffusion coefficients values are not accurate enough to make differences between diffusion coefficients of nickel, aluminium or chromium in γ phase which are the major elements of the alloy, D can be taken as an average diffusion coefficient. The response to the driving force gives the diffusional flux J_{ij}:

$$J_{ij} = -D/kT \, \nabla\mu(ij) = -D/kT \, \{[E_{tot}(j) - E_{tot}(i)] / l\} \times c \quad (5)$$

where c is the concentration of a given element [15].

After this step, we can calculate what is the algebraic change in channel width, per time step. For a given diffusional flux J_{ij}, the change in channel width depends on the surface S of the channel cross-section. Since a channel has 12 neighbors, the contributions of each flux have to be added and the width variation of channel (i) due to diffusion is :

$$W_d = \Omega / l^2 \sum_{12 \text{neighbors}} J_{ij} S_{ij} \quad (6)$$

where Ω is the atomic volume, and S_{ij} is the surface through which the diffusional flux goes. For S_{ij} we choose the cross section of the narrowest channel (i) or (j).

Direct plastic flow (transport of matter by dislocations) also contribute to the evolution of channel width. The plastic rate in both types of corridors is : $\varepsilon = K (\sigma^*)^n$, where σ^* is the local stress.

Then the channels width variation rate due to plastic flow is :

$$\dot{w}_{p,v} = K\left(\sigma_v^*\right)^n \frac{w_v}{l} \quad \text{for vertical channels} \quad (7)$$

$$\dot{w}_{h,v} = K\left(\sigma_v^*\right)^n w_h \quad \text{for horizontal channels} \quad (8)$$

The superposition of the contribution of plastic flow of this channel with the mass transfer with neighboring channels by diffusion.:

$$\dot{w} = \dot{w}_p + \dot{w}_d \quad (9)$$

The rate of evolution of a channel width is calculated for the microstructure at this level of evolution. After the microstructure has changed, local stresses have changed and new rates are calculated.

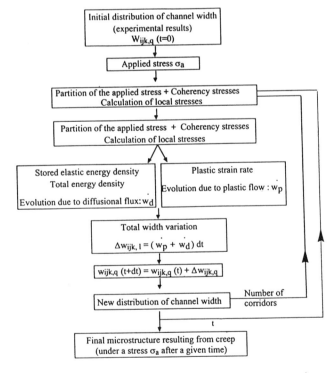

Figure 3 : Flow Chart of the simulation. The system size is typically 15x15x15 precipitates (10125 channels). The boundary conditions are periodic. Each channel is characterised by three indices (ijk) for its position, one index (q) for its type (vertical (v) or horizontal (h)) and its width $w_{ijk,q}$.

These local rules of evolution have been implemented in the computer simulation in order to describe the coupled evolution of a population of channels.

The flow chart of the program is presented in figure 3. At each time step, we calculate the variation in width of each channel according to equation (9). We then compute the new width distribution and the associated stress states which will allow to compute at the next step the new rate of evolution of channels. The iteration of this procedure allows us to compute the morphological evolution of the system with a computer time which is unambiguously related to the physical time since our evolution equations are fully deterministic.

The value of the constants (see Table 1) used were found in literature [10,11,16], provided by SNECMA (Norton's coefficient for AM1 in tensile creep) or are own results (Norton's coefficient for AM1 in compressive creep, width distribution of channels).

Volume fraction [10] :	f_v = 50 % at 1323K
Elastic constants:	$E\gamma$ = 70 Gpa;
	$E\gamma'$ = 90 Gpa
Creep parameters :	n_t = 5 (tension),
(Norton's law coefficients)	n_c = 2.8 (comp.)
Interface energy [16]:	Γ = 20 mJ.m^{-2}
Diffusion coefficient in γ [11] :	D= 10^{-14} m^2.s^{-1}
Mean channel width :	0.095 µm
Mean channel length	0.45 µm

Table 1 : Data used in the simulation

Results
Morphology maps :
Simulation has been run to study the behaviour of AM1 superalloy under different stress conditions. The results, both on the predicted morphologies and kinetics of the evolution, are in good agreement with experimental results (fig. 4). For tensile and compressive stresses, we predict coarsening in opposite directions as observed experimentally.
Moreover the first coarsening event is predicted to occur after about 5h as observed experimentally in this range of stress. Thus it has been developed to predict the behaviour of other superalloys, i.e. superalloys with other misfit values (other parameters, such as elastic constants, or creep exponents, have been kept constant in order to simplify the calculation and also because they were not available for all cases). As a result, the predictions give a qualitative idea of the evolution of microstructures under stress at high temperature. These results are presented in a morphology map (fig. 5), that shows -as a function of

the applied stress σ and the misfit value δ– what will be the microstructure of a given alloy after 20h of creep. The map is devised in regions, in which the plastic and diffusive fluxes can either contribute to the same type of evolution or to opposite ones.

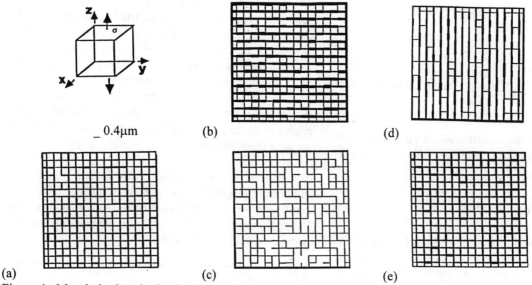

Figure 4 : Morphologies obtained with the simulation : (a) initial state (b) & (c) after 10h of tensile creep, (d) & (e) after 20h of compressive creep

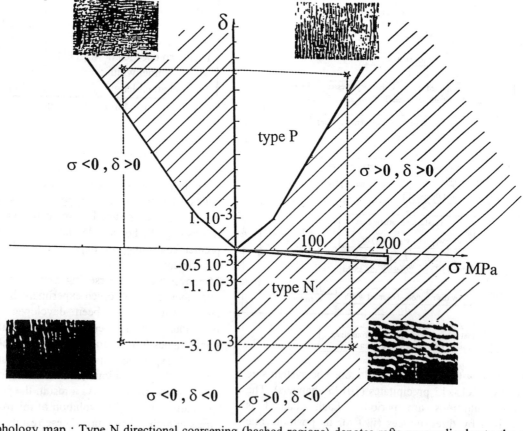

Figure 5 : Morphology map : Type N directional coarsening (hashed regions) denotes rafts perpendicular to the stress axis, type P directional coarsening (blank regions) corresponds to rafts parallel to the stress axis.

When those two contributions where competing, we have determine when one of them became dominant, and we assume that the microstructure will evolve in the direction of this contribution. In particular, for alloys with positive misfit, the fluxes were opposite : for example in tension, the plastic flow would have tend to lead to a coarsening perpendicular to the stress axis, but the diffusive one to rafts parallel to the axis. We have then determine for what value of stress the plastic contribution would overcome the diffusive one : this line has been drawn to be the limit between two different modes of coarsening. As a result, two behaviours are predicted for alloys with positive misfit, in relation with the intensity of the applied stress. Some regions of the map have been validated with experiments found in literature, but some other characterisations should be done, especially for alloys with positive misfit : when the competition between the driving forces exists, the results suggests the existence of a region of isotropic coarsening.

Mechanical properties of the coarsened structures :

In superalloys, the yield stress is due to three mechanism: hardening by solid solution (mainly in the γ phase), precipitation hardening (high volume fraction of γ' precipitates), and internal stresses (namely coherency stresses). The contribution to the critical resolved shear stress of both solid solution hardening and γ' precipitation have been investigated in alloys with a low volume fraction of precipitates ($f_v < 50\%$) and no strong anisotropic coherency effects [17]. In this study, we will only consider the contribution of the precipitation of a coherent, harder phase in high volume fraction ($f_v > 50\%$). Moreover, we will concentrate in the regime where precipitates are circumvented. This regime corresponds to the stationnary creep observed at high temperature in commercial superalloys [18], when the Orowan process operates. Since the simulation provides a 3D matrix channel structure, it is a good tool to estimate the evolution of the matrix channel width in the <111> planes, and therefore to make some prediction on the evolution of the yield stress when the microstructure becomes rafted.

When loaded in the [001] direction (the main direction of stress during service), the four <111> slip planes are similarly activated in the single crystal. To take into account the microstructure effect on some mechanical properties of the alloy, we have investigated what is the stress for a dislocation to go through a matrix corridors, in one of this four planes. Therefore we have taken cuts of the structure corresponding to different coarsening time, in the <111> directions (fig. 6).

In the cuts corresponding to the initial state, the geometry of precipitates depends strongly on the chosen level of the cut : if the level of cut corresponds to the middle of the precipitates, the precipitates are mostly hexagonal with acute angles (fig. 6.a); but if it corresponds to 3/4 of the precipitates, then in the slip planes, they have triangular shapes, with obtuse angles (fig. 6.b). Thus the behaviour of dislocations can be expected to be different, especially concerning the shearing of precipitates [19], and this is to take into account for plastic localisation [20]. On the contrary, after directional coarsening is completed, no such differences are observed in cuts.

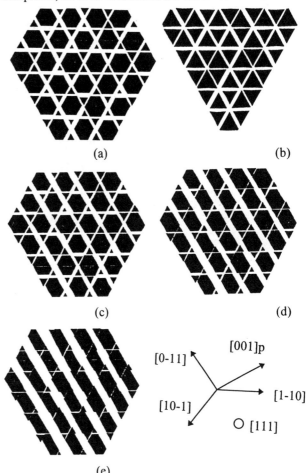

Figure 6 : Cuts of the simulated γ/γ' network in the {111} plane : (a) initial state, cut realized in the middle of the precipitates, (b) initial state, cut realized near an edge of the precipitates, (c) after 5h under stress at 1050°C, (d) after 10h and (e) 20 h in the same conditions.

As far as the Orowan process is concerned, the behaviour of dislocations is dependant on the obstacles volume fraction and on the distance between obstacles. The stress required for a dislocation to circumvent obstacles, is given by :

$$\tau_{or} = \mu b/L$$

where τ_{or} is the Orowan stress, μ is the shear modulus, b, the Burgers vector of the dislocation, and L, the distance between the obstacles.

In superalloys, the distance between precipitates is then of crucial importance : for a given applied stress, dislocations will pass through matrix channels only if their width is large enough.

We have run the simulation to calculate the channel width evolution with time, in the gliding plane for a stress along [001]. The results are presented in figure 7.a, which shows the size of the matrix channels as a function of time, distinguishing the ones who are thinning (type 1 and 3, the vertical ones) and the ones who are thickening (type 2, the horizontal ones).

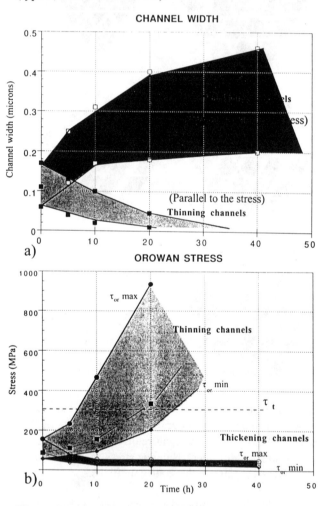

Figure 7 : (a) Channel width evolution versus time (b) Orowan stress evolution versus time. τ_t is an estimation (FEM calculation) of the effective stress in the channels, taking into account coherency stresses, for an applied stress as high as 300 MPa.

On this graph, the extrema and average of the distribution width of the channels are presented. First, the microstructure seems to evolve fast, and then, the kinetics become slower. We calculated that after 20h under a tensile stress of 120 MPa, the structure is almost completely rafted - the "vertical" channels have disappeared, and no further strong evolution is observed.

We have then estimated at a temperature of 1050°C what was the evolution of the Orowan stress when the microstructure is evolving (fig. 7.b).

For the channels which are thickening, the Orowan stress is decreasing and it reaches a plateau when the width is stabilising. On the contrary and as expected, it increases for the thinning channels, and very fast, it reaches prohibitive values. For example, after 10 hours, some channels are smaller than 20 nm, and the theoritical Orowan stress is then 470 MPa

For an applied stress of the range of 120 MPa, even with the contribution of coherence stresses, this level of stress will not been achieved in this channel, and no dislocation will be able to overpass it. In figure 7, we have indicated the value of the effective stress in the vertical channels for an applied stress of 300 MPa, it is about 100 MPa. As a result, after a couple of hours, the fraction of penetrable channels is strongly reduced.

After typically 20 hours, dislocations are gliding mainly in the horizontal channels between two γ' rafts, each dislocation building and contributing to a back-stress for the new one.

The bypassing mechanism of the precipitates at 1050° C combines glide and climb process. However, the strong aspect ratio of the rafted precipitates will tend to prevent the climb of dislocations. Then when the rafts are well formed, glide is the main deformation mecanism. Therefore to propagate a dislocation into the rafted microstructure, dilsocations have now to shear the precipitates. As some shearing events have been reported in litterature in a similar alloy after 20 h of creep [18], the stress in the horizontal channels must increase and reach the shear stress of the precipitates.

Since the shearing process of the ordered intermetallic phase involves pairs of matrix a/2<110> dislocations coupled with an anti-phase boundary (APB), a better knowledge of the APB's energy in the system would be useful to study this phenomenon. It would then be possible to simulate the occurrence of the shearing process, and to study a coupling of the shearing of rafts with the change of connectivity of the phases.

Conclusion

With a simple experiment, we were able to put forward the crucial role of dislocations in the directional coarsening process. Dislocations are at the origin of an anisotropy in elastic energy in different types of channels with respect to the applied stress. This anisotropy drives a diffusion flux, responsible for the microstructural evolution. The directional coarsening process has been simulated, and the results are in a good agreement with experimental observations. Moreover, it has been possible to predict the rafting behaviour of others superalloys under different conditions. particularly, a possible isotropic coarsening behaviour for alloys with a positive misfit has been predicted, due to the competition of the difference flows driving the morphological evolution. With simple microstructural considerations, based on the γ/γ' patterns provided by the simulation, some preliminary investigations have been conducted to study the deformation mechanism. With a better knowledge of the shear stress in the commercial superalloys, it will be possible to calculate as a function of time, temperature and applied stress, when the shearing of the rafts occurs. The 3-D simulation we have developed appears as a very promising tool to study high temperature mechanical properties of superalloys, with respect to the complex evolution of the microstructure.

Acknowledgements :

This work was supported by the French CPR "SSM" of CNRS. The author would like to thank SNECMA for providing the material, and for supporting this work through a research grant (Muriel Véron). We also would like to thank the french GDR "Simulation numérique" for support.

References :

[1] J.K. Tien and S. M. Copley, "The Effect of Uniaxial Stress on the Periodic Morphology of Coherent Gamma Prime Precipitates in Nickel-base Superalloys Crystals", Met. Trans., 2 (1971), 215-219

[2] J. K. Tien and R. P. Gamble, "The Influence of applied Stress and Stress Sense on Grain Boundary Precipitate Morphology in a Nickel-base Superalloy During Creep", Met. Trans., 2 (1971), 1663-1667

[3] J. K. Tien and R. P. Gamble, " Effects of Stress Coarsening on Coherent Particle Strengthening", Met. Trans., 3 (1972), 2157-2162

[4] T. Miyazaki, K. Nakamura, and H. Mori, "Experimental and Theoretical investigations on morphological changes of γ' Precipitates in Ni-Al single Crystals During Uniaxial Stress-Annealing", J. Mat. Sci., 14 (1979),1827-1837

[5] A. Pineau, "Influence of Uniaxial Stress on the morphology of Coherent Precipitates During Coarsening", Acta Metall., 24 (1976), 559-564

[6] S. Socrate and D. M. Parks, "Numerical Determination of the Elastic Driving Force for Directional Coarsening in Ni-Superalloys", Acta Metall. Mater, 41 (1993), 2185-2209

[7] C. Carry and J. L. Strudel, "Apparent and Effective Creep parameters in Single Crystals of a Nickel Base Superalloy- II Secondary Creep", Acta Metall., 26 (1978), 859-870

[8] T. M. Pollock and A. S. Argon, "Directional Coarsening in Nickel-Based Single Crystals with High Volume Fractions of Coherent Precipitates", Acta Metall. Mater., 42 (1994), 1859-1874

[9] M. Véron, Y. Bréchet, F. Louchet , "Directional Coarsening of γ' Precipitates in Nickel Based Superalloys : Driving Force and Influence of a Prestrain", Phil. Mag. Letters, (1996), (in press)

[10] A. Royer et al., "Temperature Dependence of the Lattice Mismatch of the AM1 Superalloy. Influence of the γ' Precipitates Morphology", Phil. Mag. A, 72 (1995), 669-689

[11] J. Y. Buffière and M. Ignat, "Dislocation Based Criterion for the Raft Formation in Nickel-Based Superalloys Single Crystals", Acta Metall. Mater., 43 (1995),1791-1797

[12] A. Fredholm, "Monocristaux d'Alliages Base Nickel. Relation entre composition, Microstructure et Comportement en Fluage à haute température", (PhD thesis, Ecole Nationale des Mines de Paris, 1987).

[13] M. Véron and P. Bastie,"Strain Induced Directional Coarsening in Nickel Based Superalloy. Investigation on Kinetics Using SANS Technique", Scripta Metall. Mater., submitted

[14] M. Véron, Y. Bréchet, F. Louchet , "Directional Coarsening of Nickel Based Superalloys : Computer Simulation at Mesoscopic Scale", Acta Metall. Mater. , (1996), in press

[15] J. Philibert, "Diffusion et transport de matiere dans les solides", (Les Ulis Cedex, France; Editions de Physique, 1985), 13.

[16] A. Ardell, " The Effect of Volume fraction on Particle Coarsening: Theoretical Considerations", Acta Metall., 20 (1972), 61-71

[17] E. Nembach and G. Neite, "Precipitation Hardening of Superalloys by Ordered γ'-Particles", Prog. Mater. sci., 29 (1985),177-319

[18] P. Caron and T. Khan, "Improvement of Creep Strength in a Nickel base Single-Crystal Superalloy by Heat Treatment", Mat. Sci. Eng., 61 (1983) 173-184

[19] J. Courbon and F. Louchet, "On the Mechanism of Formation of Superlattice Stacking Faults in L12 Precipitates Embedded in a F.C.C Matrix", Phys. Stat. Sol. (a), 137 (1993), 417-428

[20] L.P. Kubin, B. Lisiecki and P. Caron, "Octahedral slip instabilities in γ/γ' superalloy single crystals CMSX-2 and AM3", Phil. Mag. A, 71 (1995),991-1009

Appendix :

Compatibility of strain rate assuming no deformation of γ', leads to :

$$l (\sigma_v)^n = w_h (\sigma_h)^n \qquad (A.1)$$

and the partitioning of the applied stress gives :

$$\sigma_v[(l+w_v)^2-l^2] + \sigma_h l^2 = \sigma_a (l+w_v)^2 \qquad (A.2)$$

where l is the mean size of the precipitates, σ and w are the stress and the width of a channel; the 'v' indices denotes a vertical one, whereas 'h' denotes a horizontal one; n is the Norton's law coefficient of the matrix.

The relation between σ and ε in the γ channels has been assumed to follow Norton's law :

$$\varepsilon = K \sigma^n \qquad (A.3)$$

The capillary energy density driving this process is given by :

$$E_i = 2 \Gamma / w \qquad (A.4)$$

where Γ is the γ/γ' interfacial energy, and w is the width of the channel (horizontal or vertical).

AN EXPERIMENTAL STUDY OF THE ROLE OF PLASTICITY IN THE RAFTING KINETICS OF A SINGLE CRYSTAL NI-BASE SUPERALLOY

M. Fährmann*, E. Fährmann*, O. Paris**, P. Fratzl**, and T. M. Pollock*

* Department of Materials Science and Engineering, Carnegie Mellon University,
Pittsburgh, PA 15213-3890, U.S.A.

** Institut für Festkörperphysik, Universität Wien,
Strudlhofgasse 4, Wien A-1090, Austria

Abstract

Directional coarsening in a single crystal Ni-Al-Mo alloy has been studied with varying pre-strain paths that alter the initial state of the $\gamma - \gamma'$ interfaces. These microstructures were generated by pre-straining experiments at intermediate temperatures, maintaining the equiaxed morphology of the precipitates. Their subsequent evolution at high temperatures during ordinary aging as well as stress-annealing was examined by electron microscopy (SEM, TEM) and small-angle X-ray scattering (SAXS). These results are discussed in terms of the connection between the nature of the $\gamma - \gamma'$ interfaces, the local stress state and the resulting driving force for rafting.

Introduction

Rafting, i.e. the directional coarsening of the γ'-precipitates at sufficiently high temperatures under an applied stress, is a well-known phenomenon in single crystal Ni-base superalloys observed under both service and laboratory conditions. Compelling experimental evidence (recently summarized in [1]) shows that the direction and extent of rafting for the given loading condition is controlled by the sign and the magnitude of the $\gamma - \gamma'$ lattice mismatch. This has led to the idea that the superposition of the applied stress with the coherency stresses establishes stress gradients in the microstructure, causing mass flow resulting in rafting. Hence, rafting is expected to occur under conditions where both the γ- and γ'-phases respond elastically to the applied load as modeled in [1-5].

In contrast, in most experiments conducted on alloys of technological significance, rafting has been found to be associated with a finite amount of creep strain [6-14]. Under these conditions, the matrix has been subject to plastic deformation and the $\gamma - \gamma'$ interfaces are no longer fully coherent in the rafted microstructures. Based on these observations, various models accounting for the presence of dislocations have been proposed [7,12,15,16].

Thus, it is still a matter of controversy to what extent the interfacial dislocations affect the driving force and the kinetics of rafting, compared to the behavior when purely elastic conditions prevail. The present work aims to address this problem experimentally by employing various initial microstructures in terms of the state of the $\gamma - \gamma'$ interfaces (fully coherent, partially coherent, isotropically relaxed), and by examining and comparing their subsequent evolution during aging and stress-annealing at high temperatures. This approach permits us to draw important conclusions concerning the microstructural stability of single crystal Ni-base superalloys under creep conditions.

Experimental

Material

The Ni-13.3 at%Al-8.8 at%Mo model alloy employed in this study was provided by PCC Airfoils, Inc. in the form of single crystal slabs approximately 6 mm thick, 100 mm wide, and 150 mm long. The [001] dendritic growth direction was in all instances within 10° of the long axis of the slab. Samples measuring about 25 mm in width were cut from these slabs, and homogenized for 48 hours at 1300 °C in a protective atmosphere.

On overaged and subsequently quenched samples of the homogenized material the γ'-volume fraction was measured to be 0.60 ± 0.06 at 980 °C. The $\gamma - \gamma'$ lattice mismatch in the same overaged material, where the coherency strains were largely relieved, was measured by X-ray diffraction to be − (0.45 ± 0.05) % in the temperature range of the experiments conducted. Details of the experimental hot stage X-ray diffraction technique will appear elsewhere [17]. Qualitative phase analysis by powder X-ray diffraction did not reveal any detectable amounts of undesirable third phases. However, electron microscopy did show occasional evidence of a needle-like third phase precipitate (presumably δ-NiMo [9]).

The blanks were ground to about 3 mm in thickness, resolutionized in air, quenched, and subsequently aged for 20 minutes at 980 °C. The resulting microstructure consisted of fine scale cuboidal γ'-precipitates of approximately 100 nm mean edge length, embedded coherently in the γ-matrix. Tensile creep specimens with 25 mm gauge length, and compression specimens of approximately 6 mm height, were sectioned from these thin plates by electric discharge machining, and machined to final shape by low-stress grinding.

Pre-straining Experiments

The pre-straining experiments at intermediate temperatures of 850 °C as well as the stress-annealing experiments at the higher temperature of 980 °C were conducted in lever arm creep machines in air. Creep strains in case of the tensile tests were measured by extensometry attached to the shoulders of the specimen. For the compression tests the accumulated creep strain was derived from the reduction in specimen height. The temperature was controlled by two thermocouples in direct contact with the gauge. All samples were cooled under load.

Four different initial microstructures were sought in terms of the state of the γ/γ' interfaces :
(a) fully coherent,
(b) partially coherent, where predominantly the misfit strains in the horizontal (001) matrix channels (with respect to the future [001] loading axis) are largely relieved,
(c) partially coherent with relaxed vertical (100) and (010) matrix channels,
(d) isotropically relaxed, where all {100} type interfaces contain dislocation networks in equilibrated configurations.

State (a) could readily achieved by short-term aging, adjusting the precipitate size to the desired level of ≈ 150 nm suitable for characterization. States (b) and (c) may in principle be accomplished by uniaxial loading in tension or compression along [001]. Based on finite element calculations [12,18], the resolved shear stress for <110>{111} slip in a [001] oriented specimen is expected to be highly dependent on the sense of the load (Fig. 1). An applied tensile stress increases the resolved shear stress for slip in the horizontal matrix channels while reducing the stresses in the vertical matrix channels, for a negative misfit alloy. The situation is reversed under compression. Thus, it was estimated that for an applied stress of 500 MPa along [001], the resolved shear stress would exceed the Orowan stress [7,19] needed to bow slip dislocations into only one set (horizontal or vertical) of the narrow 30 nm matrix channels. The time under load was adjusted such as to approach the secondary creep stage. As will be shown later, under these loading conditions the resolved shear stresses in the ordered γ'-precipitates were still sufficiently low to avoid shearing of the precipitates.

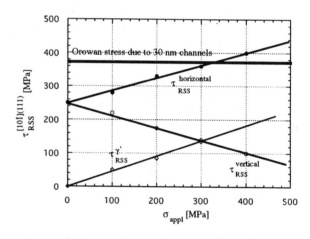

Fig. 1 : Plot of the resolved shear stresses for <110>{111} slip in the horizontal and vertical matrix channels, and the γ'-precipitates, as a function of the applied stress along [001]. Calculations are based on a finite element code [12].

State (d) with its isotropically relaxed interfaces was approximately realized by a two-step thermomechanical treatment : pre-straining in tension, followed by additional pre-straining in compression along the same [001] loading axis. Here the compression specimens were machined from the gauge length of the tensile-tested specimen. Table 1 summarizes the pre-straining conditions employed and the corresponding creep strains accumulated.

Table 1 : Summary of the pre-straining conditions employed to establish different initial microstructures

State	Conditions	Creep Strain
(a)	4 h @ 850 °C	
(b)	4 h @ 850 °C @ 500 MPa^T	≈ 0.03
(c)	4 h @ 850 °C @ 500 MPa^C	≈ 0.05
(d)	4 h @ 850 °C @ 500 MPa^T + 7 h @ 850 °C @ 500 MPa^C	≈ 0.04

T = tension   C = compression

Characterization

For all samples tested, sections parallel and transverse to the actual or prior (corresponding to the pre-straining experiment) loading axis were prepared metallographically (electrolytic etching in a solution of 1 wt% ammonium sulfate and 1 wt% citric acid in water), and surveyed in a SEM. Subsequently, TEM specimens were prepared from the same slices by electropolishing employing a solution of 4 % sulfuric acid in methanol at -24 °C. The TEM foils of (110) orientation permitted location of the [001] loading axis in the corresponding micrograph. The state of the interfaces was assessed by employing various 2-beam diffraction conditions.

In addition to the microscopic examinations, small-angle X-ray scattering was utilized to obtain statistical information about precipitate morphologies in the various micostructures. The SAXS measurements were performed using Cu-K$_\alpha$ radiation and a point collimation system with a sample to detector distance of about 1 m. The intensity scattered in an angular range below 2°, the domain of this technique, was collected with a two-dimensional position sensitive detector with either the (100) or (010) crystallographic plane of the approximately 30 μm thin specimen perpendicular to the incoming X-ray beam. Details of the experimental set-up and the procedure of data evaluation will be published in a forthcoming paper [20]. It is important to note that the SAXS spectra contain information about a much larger number (6 to 7 orders in magnitude) of precipitates than typically sampled by electron microscopy.

Results

Microstructural Evolution during Aging

Fig. 2 displays the characteristic pattern of microstructural evolution during ordinary aging, starting from the various initial microstructures. The inserts are Fourier transforms, calculated by Interactive Data Language software (IDL version IV) from traced and subsequently scanned SEM images, reflecting the crystallographic orientation of the interfaces and the amount of interfacial area perpendicular to a given direction. These Fourier transforms are the 2D analogue to the SAXS spectra from the actually 3D scattering objects, i.e. the precipitates.

The initially fully coherent microstructure (a) coarsens in the capillarity-driven fashion increasing the length scale of the system, with retention of the cuboidal shape of the precipitates. This morphological feature, i.e. the {100} type habit planes, is reflected in the corresponding Fourier transform by two rather sharp streaks into both <100> type directions. Only occasional irregular morphologies are discernible, indicating the beginning of the loss of coherency. Furthermore, the four-fold symmetry of the transform implies that the *average* precipitate shape is square-like in the (100) plane of the image.

The semicoherent initial microstructure (d) displays in the aged condition rather irregular precipitate morphologies. Although the precipitates are not equiaxed, no bias in the direction of coarsening is observed. This is consistent with the symmetry of the corresponding Fourier transform.

In contrast to interfacial conditions (a) and (d), samples with initially partially coherent interfaces exhibit directional coarsening during aging. This is most clearly visible for the sample pre-strained in tension (b) : qualitatively the microstructure evolves *as if* a tensile load was present (for comparison the typical microstructure of samples stress-annealed in tension is shown in Fig. 2e). Due to the emerging predominantly plate-like precipitate morphology oriented perpendicular to the tensile axis, the Fourier transform is now quite asymmetric. It consists mainly of a single streak perpendicular to the broad faces of the plates (the plate-like nature was confirmed by examining all three {100} sections).

In the case of the sample pre-strained in compression (c) the situation is less obvious : precipitates extended along the former [001] compression axis as well as precipitates extended in both <100> directions in the (100) image plane are observed. This complies with the findings of stress-annealing tests in compression [2,12,24], that is, rod-like as well as *two* orientational variants of plate-like precipitates *parallel* to the compression axis, their normals being of <100> type, are found. As a consequence, the asymmetry of the Fourier transforms is less pronounced as compared to the tensile case. Finally we note that in both cases (pre-straining in tension and compression) the resulting aged microstructures appeared to be less perfect compared to their stress-annealed counterparts.

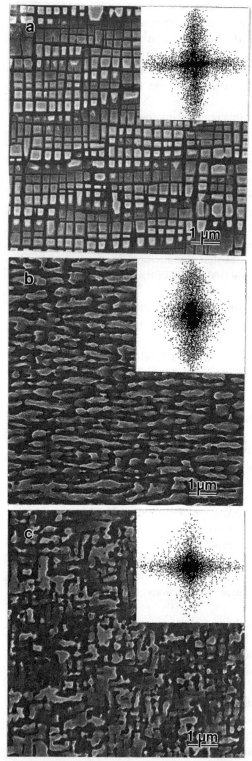
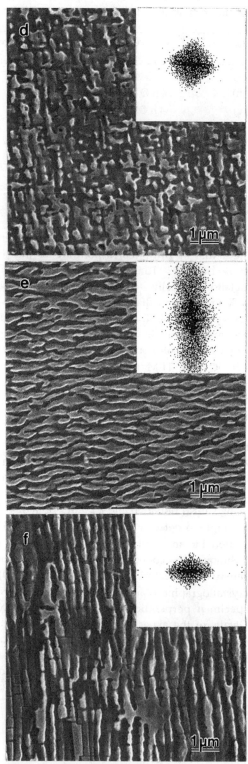

Fig. 2 : SEM micrographs of (100) sections of samples aged 24 hours at 980 °C, starting with (a) fully coherent $\gamma - \gamma'$ interfaces, (b) partially coherent interfaces due to pre-straining in tension, (c) partially coherent interfaces due to pre-straining in compression, (d) semicoherent interfaces. In (e) and (f) typical microstructures of this alloy obtained after stress-annealing in tension and compression, respectively, are shown for comparison. The magnification is the same in all the micrographs.

Microstructural Evolution during Stress-Annealing

In Fig. 3, the evolution of an initially fully coherent microstructure (left column), and of a microstructure pre-strained in tension at 850 °C (middle column), during subsequent stress-annealing at 980 °C under *compression* is shown. The compression axis is vertical in all micrographs. In addition, for comparison purposes, in the third right column the microstructural evolution of the same pre-strained sample during ordinary aging is presented. The corresponding SAXS spectra are arranged in the same order below the micrographs. The components of the scattering vector k_x perpendicular to, and k_y parallel to the [001] compression axis, are indicated. The scattered intensity is plotted in a logarithmic gray-coded scale.

Starting with a fully coherent microstructure, stress-annealing under compression for one hour under the given loading conditions did not result in any significant morhological changes of the γ'-cuboids. Accordingly, the SAXS spectrum appears four-fold symmetric, and displays streaks into the <100> type directions. After 8 hours under load, however, the cuboids started to transform into plate/rod-like particles parallel to the compression axis. This leads to an asymmetry in the corresponding SAXS spectrum, with the [100] streak now being more pronounced than the [001] streak [25]. For a sample stress-annealed in tension the situation would be reversed [20], with the asymmetry, however, being much more obvious for comparable loading conditions.

The microstructure of the sample pre-strained in tension evolves during the first hour under compression apparently such that still a slight directional coarsening *perpendicular* to the [001] compression axis occurs, as also suggested by the SAXS spectrum. Later on, the microstructure appears disintegrated, showing no signs of directional coarsening. Consistent with this microscopic observation, the corresponding SAXS spectrum is then nearly isotropic. The microstructure of the same pre-strained sample during ordinary aging follows the pattern described in the previous section, i.e. evolving into plate-like particles perpendicular to the *former* tensile axis. This is corroborated by the SAXS spectrum showing a pronounced streak into the [001] direction.

Interestingly, all SAXS spectra obtained with the pre-strained samples are much broader in terms of streak widths as compared to the spectra obtained with the initially fully coherent samples. This indicates [20] that the $\gamma - \gamma'$ interfaces in the pre-strained and subsequently annealed / stress-annealed samples are crystallographically less planar. TEM observations showed that the interfaces in the formerly coherent microstructure were still predominantly of {100} type even after 8 hours under compression (Fig. 4a), whereas starting with a pre-strained sample led to rounded irregular interfaces decorated with interfacial dislocations (Fig. 4b).

Despite the less defined nature of the $\gamma - \gamma'$ interfaces in the pre-strain series, it was still possible to extract useful information from the SAXS spectra in case of the 1 hour samples. The data evaluation is based on the concept of the total interfacial areas $S^{\{100\}}$ parallel or nearly parallel to the {100} planes. The ratio $S^{(001)} / S^{(100)} \approx S^{(001)} / S^{(010)}$ can then be identified with an *average* aspect ratio of the precipitates referred to as ρ_{SAXS} [20]. In addition, we performed semi-automatic image analyses (also described in greater detail in [20]) of TEM DF images, taken from the same microstructures, to obtain a similar parameter ρ_{TEM}. Thus, for plates evolving normal to the [001] applied stress, above definitions of ρ result in values larger than 1.

In terms of the evolution of precipitate aspect ratios (Table 2), the two most notable findings are : (1) the *initially greatly enhanced* rate of rafting in the pre-strained sample even under no applied stress, compared to the sample exhibiting initially coherent interfaces, (2) the, albeit small, but resolvable *initial increase* in ρ for the pre-tensioned sample subject to subsequent compression. This trend is *opposite* to what is usually expected in a stress-annealing test in compression with this alloy. We note that the pre-strained sample was heated in the creep frame under full compressive load before conducting the 1 hour isothermal stress-annealing treatment, thus prohibiting rafting due to thermal exposure as described in the previous section.

Table 2 : Summary of the quantitative evaluation of the SAXS spectra and TEM micrographs regarding the average precipitate aspect ratio after 1 hour stress-annealing. The first value refers to SAXS, the second (in parenthesis) to TEM.

	Fully coherent Compression / Tension	Pre-strained in tension Compression / Aged
ρ_{start}	1.2 ± 0.2 (1.3 ± 0.1)	1.6 ± 0.2 (1.4 ± 0.1)
ρ_{end}	1.2 ± 0.2 / <1.8	2.1 ± 0.2 / 2.7 ± 0.2 (1.8 ± 0.2) / (2.3 ± 0.2)

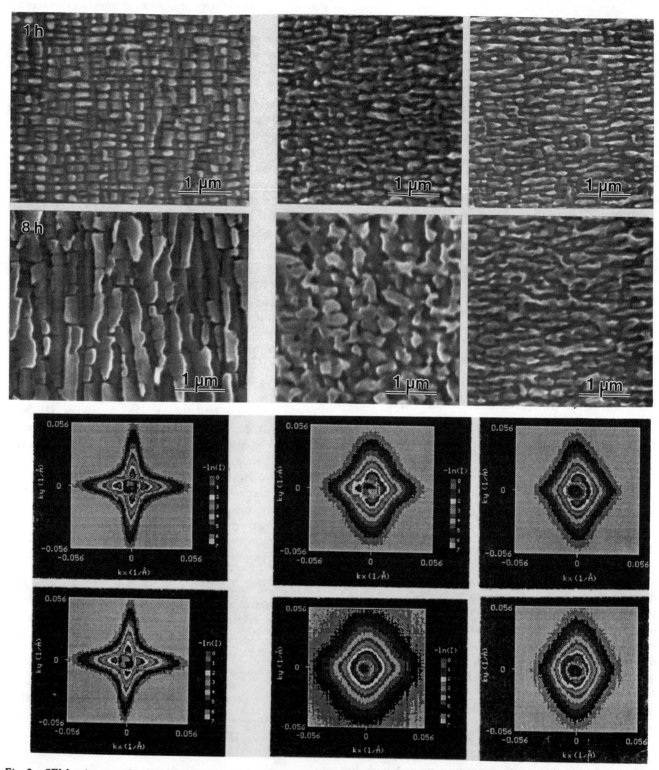

Fig 3 : SEM micrographs ((110) sections) and corresponding SAXS spectra of samples stress-annealed in compression. Left column : initially fully coherent interfaces. Middle column : initially partially coherent interfaces due to pre-straining in tension. The right column shows the microstructure and SAXS spectra obtained after ordinary aging of the same pre-strained sample. The magnification is the same in all the micrographs. The loading direction, if applicable, is perpendicular.

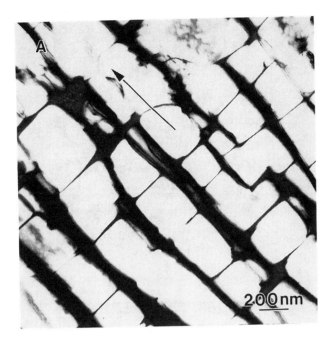

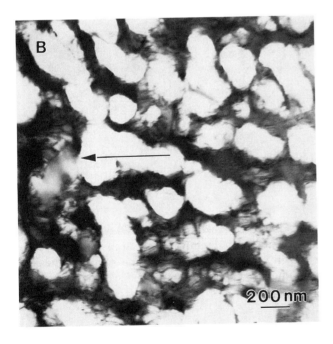

Fig. 4 : Dark-field (DF) images (g=110) of samples stress-annealed for 8 hours at an applied compressive stress of 130 MPa, starting with (a) fully coherent interfaces, (b) partially coherent interfaces due to pre-straining in tension. The direction of the compression axis is indicated by arrows in the micrographs.

Discussion

The experimental results presented provide strong evidence that the kinetics *and* the driving force of rafting are both greatly affected by the state of the $\gamma - \gamma'$ interfaces which in turn is related to the creep deformation. This conclusion is primarily based the observations that

(a) the microstructure in the pre-strained samples rafted even under *no* applied stress in a direction corresponding to the *former* loading axis.
This observation corroborates recent findings by Veron et al. [21] with another single crystal superalloy, and extends their pre-strain compression experiments to all four conceivable initial situations in terms of the state of the $\gamma - \gamma'$ interfaces.

(b) the microstructure in a sample pre-strained in tension rafted under an applied compressive stress *initially* in a direction *opposite* to what is generally observed in compression in this alloy.

Thus, any discussion of the effect of matrix plasticity in rafting in single crystal superalloys *only* in terms of an enhanced kinetics due to the supply of fast diffusion paths is *incomplete*, and misses the important fact that matrix plasticity can profoundly alter the magnitude and direction of the driving force for rafting. Moreover, it is not the presence of dislocations per se, but their arrangement in the various matrix channels which determines the magnitude of the effect. This was realized first by Carry and Strudel [7], and has been subsequently refined in [12,15,16]. The main idea is that slip dislocations deposited on the $\gamma - \gamma'$ interfaces during creep and their subsequent reaction and rearrangement by climb leads, at least partially, to a reduction of the misfit strains in certain sets of matrix channels, while aggravating them in other sets.

In this context the experimental findings regarding the microstructural evolution of the various pre-strained samples during stress-free aging (Fig. 2) can be rationalized : the total strain energy of the single crystal is substantially reduced by eliminating the highly stressed matrix channels. This is accomplished by directional flow of "γ' - material" into those matrix channels, leading to the observed rafting as shown schematically in Fig. 5. In the case of the fully coherent (Fig. 2a) and ideally fully relaxed interfaces (Fig. 2d) no such strain energy gradient exist. Hence, no directional coarsening is observed during stress-free aging.

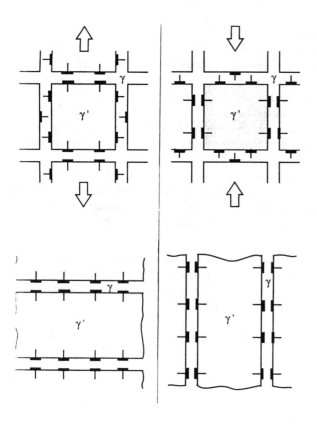

Fig. 5 : Schematic sketch of the dislocation arrangement in the pre-strained samples, and the subsequent microstructural evolution during aging.

This picture (a similar version has been proposed by Veron et al. [21] assuming, however, one set of channels to be dislocation-free) is quite simplistic. We note that the schematic of the initial partially coherent microstructures (Fig. 5) assumes that the misfit strains in certain matrix channels are and remain ideally relieved by virtue of misfit dislocation networks residing at the interfaces. Under the choosen conditions (see Table 1), a high density of interfacial dislocations on e.g. the horizontal (001) interfaces of the pre-tensioned sample is observed as expected (Fig. 6a). However, the reaction and rearrangement of the slip dislocations by climb processes has rarely led to the characteristic misfit strain relieving networks as described in [22,23]. After sufficient thermal activation, e.g. aging for 20 hours at 980 °C (Fig. 6b), significant rearrangement by climb has occurred. Then, however, the precipitates were no longer of cuboidal shape. Apparently only a fairly narrow experimental window in time and strain exists for a given alloy to study such pre-strain effects.

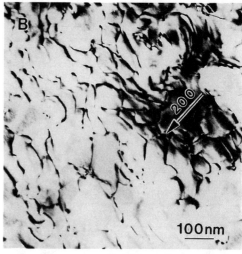

Fig. 6 : TEM bright-field images (g=200) of the dislocation arrangement on (001) interfaces in the sample pre-strained in tension (a), and the same sample annealed additionally for 20 hours at 980 °C.

The interpretation of the second crucial experimental result, i.e. the initially "wrong" direction of rafting in the sample pre-strained in tension under subsequent compression, follows the same line of thinking. Due to the pre-straining in tension, many dislocations deposited onto the (001) interfaces and their reaction products will to some extent relieve the misfit strain locally. Thus, they are arranged in an energetically favorable fashion, and are not expected to leave the interface upon reversal of the applied stress. The interfacial dislocations on the (100) and

the applied stress, misfit strain diminishing arrangements of interfacial dislocations will then gradually build up also at the vertical interfaces.

If this is true, a certain initial time span under compressive applied stress can be expected during which the relaxation of the misfit strains in the various channels is still dominated by the former tensile pre-strain experiment. This is the time span where principally rafting in the "wrong" direction could take place. During prolonged stress-annealing in compression this initial gradient will be substantially reduced, if not reversed. Apparently, after 8 hours of stress-annealing (Fig. 4b), an intermediate stage has been reached in this alloy, the interfaces being uniformly wrapped by dislocations.

In spite of the simplifying assumptions which were made, we believe that this model reflects essential elements of rafting in single crystal Ni-base superalloys. A thorough thermodynamic treatment would have to couple the chemical potential of the various alloying elements with the local stress state in the material, and to account for multicomponent diffusion in the mass flow problem. This was beyond the scope of the present work. Nevertheless, several technologically relevant conclusions can been drawn regarding the microstructural stability of single crystal superalloys under creep conditions :
(a) it is not the high magnitude of the coherency stresses built into the microstructure, but their anisotropic relaxation during creep deformation which ultimately generates the gradients driving rafting;
(b) alloying additions which slow down Ostwald ripening in the alloy, would at the same time be potent candidates to retard rafting since in both cases crossing fluxes of "γ - formers" (such as Cr, Mo, Re ...) and "γ' - formers (such as Al, Ti, Ta ...) are required to accomplish the redistribution of the phases in the microstructure.

Summary

Directional coarsening in a single crystal Ni- 13.3at%Al- 8.8at%Mo alloy has been studied with varying pre-strain paths that alter the initial state of the $\gamma - \gamma'$ interfaces. These microstructures were generated by pre-straining experiments in uniaxial tension / compression along [001] at intermediate temperatures of 850 °C, maintaining the equiaxed morphology of the precipitates. Electron microscopy and small-angle X-ray scattering were employed to examine the microstructural evolution during subsequent ordinary aging and stress-annealing at the higher temperature of 980 °C.
The two most important observations are :
(a) the microstructure in samples pre-strained in tension or compression rafted during subsequent aging in a direction as if the former load was still present, whereas a sample with isotropically relaxed interfaces did not show directional coarsening;
(b) the microstructure in a sample pre-strained in tension rafted under an applied compressive stress initially in a direction opposite to what is generally observed in compression for this alloy.

These experimental results provide strong evidence that the driving force for rafting can be greatly affected by matrix plasticity. The anisotropic relaxation of misfit strains in the initially highly stressed matrix channels due to deposition of slip dislocations onto the interfaces and their subsequent rearrangement, appears to be the key for an understanding of rafting during creep of single crystal superalloys.

Acknowledgments

Financial support by the "Österreichischer Fonds zur Förderung wissenschaftlicher Forschung" FWF S5601, the Deutsche Forschungsgemeinschaft, Grant FA 290/1-1, General Electric Company, and the National Science Foundation through Grant DMR 9258297, is gratefully acknowledged. We wish to thank J. Wolf and A. Boegli for conducting X-ray diffraction work and measuring the γ'- volume fraction, and S. Mahajan for helpful discussions.

References

[1] J. C. Chang and S. M. Allen, *J. Mater. Res.* **6**, 1843 (1991).
[2] J. K. Tien and S. M. Copley, *Metall. Trans.* **2**, 219 (1971).
[3] A. Pineau, *Acta metall.* **24**, 599 (1976).
[4] W. C. Johnson, *Metall. Trans.* **A 18**, 233 (1987).
[5] T. Miyazaki, K. Nakamura, and H. Mori, *J. Mater. Sci.* **14**, 1827 (1979).
[6] C. Carry and J. L. Strudel, *Acta metall.* **25**, 767 (1977).
[7] C. Carry and J. L. Strudel, *Acta metall.* **26**, 859 (1978).
[8] D. D. Pearson, B. H. Kear, and F. D. Lemkey, in : *Creep and Fracture in Engineering Materials Structures*, eds. B. Wilshire and D. R. J. Owens, Pineridge Press Ltd., Swansea, UK, 1981, p. 213.
[9] R. A. MacKay and L. J. Ebert, *Scripta metall.* **17**, 1217 (1983).
[10] M. V. Nathal, *Metall. Trans.* **A 18**, 1961 (1987).

[11] U. Glatzel and M. Feller-Kniepmeier, *Scripta metall.* **23**, 1839 (1989).
[12] T. M. Pollock and A. S. Argon, *Acta metall. mater.* **42**, 1859 (1994).
[13] H.-A. Kuhn, H. Biermann, T. Ungar, and H. Mughrabi, *Acta metall. mater.* **39**, 2783 (1991).
[14] Z. Peng, U. Glatzel, T. Link, and M. Feller-Kniepmeier, *Scripta Materiala* **34**, 221 (1996).
[15] S. Socrate and D. M. Parks, *Acta metall. mater.* **41**, 2185 (1993).
[16] J. Y. Buffiere and M. Ignat, *Acta metall. mater.* **43**, 1791 (1995).
[17] M. Fährmann, J. Wolf, and T. M. Pollock, *Mat. Sci. Eng.* A **210**, 8 (1996).
[18] L. Müller, U. Glatzel, and M. Feller-Kniepmeier, *Acta metall. mater.* **41**, 3401 (1993).
[19] T. M. Pollock and A. S. Argon, *Acta metall. mater.* **40**, 1 (1992).
[20] O. Paris, M. Fährmann, E. Fährmann, T. M. Pollock, and P. Fratzl, submitted to Acta Materiala, (1996).
[21] M. Veron, Y. Brechet, and F. Louchet, *Scripta Materiala*, (1996), in press.
[22] R. D. Field, T. M. Pollock, and W. H. Murphy, in : Superalloys 1992, eds. S. D. Antolovich et al., TMS Society, Warrendale, PA (1992), p. 557.
[23] R. R. Keller, H. J. Maier, and H. Mughrabi, *Scripta metall. mater.* **28**, 23 (1993).
[24] M. V. Nathal and L. J. Ebert, *Scripta metall.* **17**, 1151 (1983).
[25] M. Fährmann, *unpublished results*.

INVESTIGATION OF THE γ/γ' MORPHOLOGY AND INTERNAL STRESSES IN A MONOCRYSTALLINE TURBINE BLADE AFTER SERVICE: DETERMINATION OF THE LOCAL THERMAL AND MECHANICAL LOADS

Horst Biermann, Berthold von Grossmann, Thomas Schneider, Hua Feng and Haël Mughrabi

Institut für Werkstoffwissenschaften, Lehrstuhl I, Universität Erlangen-Nürnberg,

Martensstr. 5, D-91058 Erlangen, Fed. Rep. Germany

Abstract

In the present work, a procedure is presented which allows the determination of the locally acting loads during service of a turbine blade by detailed investigations of its microstructure. The microstructural characterization of a monocrystalline turbine blade of the nickel-base superalloy CMSX-6 subjected to service-like conditions was performed with different, complementary techniques. The morphology of the turbine blade was investigated with conventional scanning and transmission electron microscopy. Lattice parameter changes and the lattice distortions were measured by a special high-resolution X-ray diffraction method and by convergent beam electron diffraction.

The investigation of the microstructure of the turbine blade showed in some regions of the blade the build-up of a marked raft structure, especially near the leading and trailing edges. The investigations of local lattice parameter changes yielded lattice distortions in both phases γ and γ' which are attributed to residual deformation-induced and thermally induced long-range internal stresses. Using finite element calculations, lattice distortions were calculated for characteristic positions within the blade on the basis of the actual microstructure. The calculated lattice parameter distributions were compared with X-ray line profiles measured locally at the same points. The investigations showed that the microstructure and the lattice parameter distributions are determined mostly by the centrifugal stresses and the high material temperatures. A result of the convergent beam electron diffraction investigations obtained on the turbine blade was that the evaluation of the higher order Laue zone lines yielded local lattice distortions of the two phases.

Introduction and Objectives

Two-phase nickel-base superalloy turbine blades are subjected locally to complex triaxial stresses during service. These stresses result from the superposition of centrifugal stresses which are acting in the [001] direction, the internal stresses on a microstructural scale and thermomechanical fatigue. Due to the difference of the lattice parameters a^γ and $a^{\gamma'}$ and/or the different thermal expansion coefficients of the phases γ and γ', coherency stresses and thermal stresses exist already in the initially undeformed state, e.g. [1-3]. During service of nickel-base superalloy turbine blades, changes of the γ/γ' morphology occur. In those regions of the turbine blades exposed to high temperatures, especially at the leading and trailing edges, directional coarsening of the γ' precipitates occurs, and a raft-like γ/γ' plate structure can develop [4-7].

Thermal stresses, coherency stresses and deformation-induced long-range internal stresses were evaluated from room temperature and high-temperature X-ray and neutron line profile measurements in undeformed and creep-deformed samples [2, 3, 8-13]. Investigations of the line profiles measured on samples creep deformed to different strains showed that the change of the lattice parameters due to internal stresses is completed after the creep rate has reached its minimum, i.e. after strains which are comparable to those experienced by the blade [5]. The investigations [5-7] showed that internal stresses can be measured by X-ray diffraction in turbine blades exposed to service. These internal stresses were attributed mostly to creep deformation experienced by the blade due to the centrifugal stresses during service. Further work proved the existence of internal stresses in nickel-base superalloys on a microscopical scale by the analysis of interfacial dislocation networks [14] and by convergent beam electron diffraction (CBED) [15, 16]. Finally, long-range internal stresses in the phases γ and γ' in creep-deformed samples were calculated by the finite element method (FEM) [17, 18].

This background provides the motivation for the present work on turbine blades that had been exposed to service-like conditions. The approach consists of a detailed quantitative microstructural characterization whose results are compared with available data on well-characterized laboratory specimens (with defined mechanical history) and are then used as input data for a FEM evaluation of the local triaxial stress and strain states in critical parts of the blade. The ultimate aim is to quantify the locally acting load states (i.e. the local temperatures and stresses) and to relate them to the locally experienced thermal and mechanical loading history.

In this paper, results of 3-dimensional FEM calculations of local lattice parameters and lattice mismatch values (the lattice mismatch δ is defined as the relative difference of the lattice parameters of the phases γ' and γ, $\delta = 2(a^{\gamma'}-a^{\gamma})/(a^{\gamma'}+a^{\gamma})$) in the principal ⟨100⟩ directions in a CMSX-6 turbine blade which has been exposed to service are reported. For this purpose, the γ/γ' microstructure of the turbine blade is evaluated quantitatively by scanning electron microscopy. The results of the modelling are compared with X-ray line profiles, which were measured at different positions within the turbine blade. The comparison of the X-ray line profiles and the calculated lattice parameter distributions of the turbine blades with data obtained on samples creep-deformed in tension provides evidence on the type of deformation experienced by the investigated parts of the turbine blades. Thus, the local loads under service of the turbine blade can be estimated from an adaptation of the input of the FEM model with the aim to obtain a good correlation between the calculated lattice parameters and the measured X-ray profiles. The existence of local strains in the two phases is investigated with high lateral resolution by convergent beam electron diffraction experiments by the analysis of the higher order Laue zone lines (HOLZ lines).

Internal stresses

Measurements of local strains indicate a complex interaction of different contributions. These are i) coherency stresses which are present even in the initial state, ii) thermal stresses arising during cooling and iii) residual deformation-induced internal stresses. In the following, these different contributions are discussed in more detail.

Coherency and thermal stresses

In the undeformed, standard heat-treated state, the ordered γ' phase is coherently precipitated in the γ matrix. Since most nickel-base superalloys have a misfit at room temperature in the order of 10^{-3}, there are marked internal coherency stresses. At the high deformation temperatures, the misfit decreases due to the smaller thermal expansion coefficient of the γ' phase and is negative in the alloys used in practice. In the case of a negative lattice misfit, the γ phase is under the action of compressive stresses and the γ' phase under the action of tensile stresses. After long-term annealing at high temperatures, the coherency stresses are reduced by interfacial dislocations. The spacings of these dislocations which are arranged in regular dislocation networks can be evaluated to obtain the lattice misfit at the annealing temperature based on a relation stated by Brooks [19]. With the Brooks formula the lattice misfit can be determined from the Burgers vector b and the dislocation spacings d: $\delta = |\underline{b}|/d$.

Anisotropic thermal stresses arise after raft formation due to the different thermal expansion coefficients of the two phases γ and γ' [2, 3]. The thermal stresses after deformation and cooling to room temperature are of compressive (tensile) nature in the γ' phase (γ phase), since the γ' phase has the smaller thermal expansion coefficient. Theses thermal stresses act in the plane of the γ and γ' plates, i.e. in the directions [100] and [010]. Therefore, the thermal stresses cause tetragonal lattice distortions of the two phases at room temperature in the raft structure.

Deformation-induced internal stresses

Due to the different strengths of the two phases γ and γ' (see e.g. [20]), deformation-induced long-range internal stresses arise during deformation which prevail after unloading (cf. [8-12]). They are built up by interface dislocations [8, 16, 21] which form dislocation networks during deformation from strains of about 1 % onwards.

After tensile creep deformation in the [001] direction, the interface dislocations at (001) interfaces have the same nature as dislocations which relieve the coherency stresses in the case of a negative lattice misfit, see Figure 1(a). The faint lines in the figure represent the traces of the glide planes of the active slip systems. The lines are dashed in the ordered γ' phase, since dislocation activity concentrates mostly in the softer γ phase. The actual dislocation networks at the (001) interfaces reduce the coherency stresses. These dislocation networks also build up additional deformation-induced stresses which are of opposite sign compared to the coherency stresses (the coherency stresses are of compressive nature in the directions [100] and [010], the deformation-induced internal stresses, e.g., of tensile nature in γ, cf. Figure 1(b)). The interface dislocations at (100)/(010) interfaces do not reduce the coherency stresses, but they

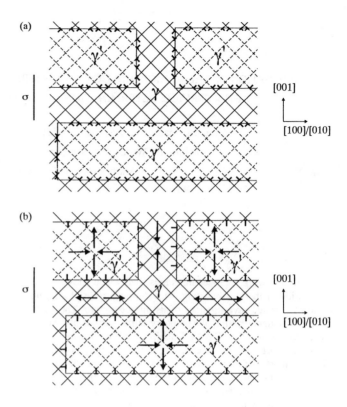

Figure 1: Idealized γ/γ' microstructure with resultant interface dislocations. σ shows the axis of the centrifugal stress (i.e. the [001] direction). (a) Accumulation of glide dislocations at the γ/γ' interfaces. (b) Resultant interface dislocations which put the two phases under internal stresses as indicated schematically by arrows. After [8].

cause additional deformation-induced internal stresses. Figure 1(b) shows the effect of the interfacial dislocation networks on the lattice parameters of the two phases. The addition of the Burgers vectors of adjacent pairs of the edge dislocations from symmetrical slip systems (Figure 1(a)) gives the resultant dislocations of Figure 1(b) with Burgers vectors parallel to the interfaces. These resultant edge dislocations are of different nature: i) Dislocations at the interfaces perpendicular to the stress axis have their extra half-plane in the γ' phase, ii) dislocations at the interfaces parallel to the stress axis in the γ phase. The γ' phase (γ phase) is under tensile strain (compressive strain) in the direction [001] and under compressive strain (tensile strain) in the directions [100] and [010], as indicated schematically by arrows. It has to be noted that in Figure 1 a microstructure is shown, where γ' plates are embedded in a γ matrix phase. This is in contrast to the drawing for creep-deformed samples in Reference [8], where the situation is reversed and γ plates are embedded in the γ' phase.

Experimental

Turbine blade

A monocrystalline turbine blade of the "light" nickel-base superalloy CMSX-6 with an orientation near [001] from a developmental turbine was investigated. The blade had been exposed to service for several hundred hours in two test turbines. The turbines have been operated in so-called accelerated mission tests. Local stresses acting during service were calculated analytically. The calculated local centrifugal stresses in the [001] direction experienced by the blade during service at the investigated (001) section II (see Figure 2) were between 40 MPa and 230 MPa. In Figure 3, the local stress data are given for the investigated sites of section II. The turbine blade was sectioned and polished carefully mechanically parallel to the (001) lattice planes (the investigated cross sections are indicated in Figure 2) and to the (100) and (010) lattice planes.

Electron microscopy

Scanning electron microscopy. The sections mentioned in the last paragraph were etched with a solution of H_2O, HCl, MoO_3 and HNO_3 which etches preferentially the γ' phase. Subsequently, the sections were sputtered with gold. The investigations were performed with a scanning electron microscope at an accelerating voltage of 20 kV using back scattered electrons and evaluated for the particle size distributions of the phases γ' and γ using a digitizing table.

Transmission electron microscopy and convergent beam electron diffraction (CBED). From the sections investigated by SEM, thin slices with the orientations (001) and (100) were prepared for the TEM investigations. They were polished mechanically to a thickness of about 100 μm and thinned electrolytically by a mixture of methanol, butanol and perchloric acid at a temperature of -40 °C. Since the CBED technique is very sensitive on thickness variations, some of the TEM samples were subsequently thinned by an ion thinning device. The TEM investigations were performed with an electron microscope of the type Philips CM 200. The TEM micrographs were taken at an accelerating voltage of 200 kV.

The CBED investigations were performed in the microscope at an accelerating voltage of 120 kV using a double tilt cooling stage at a temperature of -170 °C. The use of the cooling stage improves the contrast of the HOLZ line and reduces contamination of the samples. Cooling has no noteworthy effect on the strains in the TEM foil, since the two phases γ and γ' have equal thermal expansion coefficients below about 500 to 600 °C. The electron beam was focussed to a diameter of about 20 nm. In the investigations presented in this paper, the [100] zone axis was used for different reasons (see also the work of Keller at al. [15]). i) The ⟨100⟩ directions are distinguished directions of the raft structure. ii) The principal strains and stresses act in these directions (the centrifugal stress acts in [001]). iii) Relaxation of the internal stresses which could lead to angular distortions of the angles α, β and γ of the unit cells are minimal for foil preparation in the {100} planes. Due to these advantages, the [100] zone axis was investigated. In a first approximation, the observed patterns were simulated using a kinematical computer simulation [22]. Preliminary dynamical simulations, however, showed pronounced effects of dynamical scattering on the line positions in the ⟨100⟩ zone axes. Therefore, in the present work, the measured patterns are discussed qualitatively, and no quantitative results are drawn.

X-ray diffraction

X-ray line profile measurements of the {200} type Bragg reflections were performed on a special high-resolution double-crystal diffractometer with high angular resolution and negligible instrumental line broadening, using the $CuK_{\alpha 1}$ radiation. The cross section of the X-ray beam was limited in order to obtain high lateral resolution. Therefore, the investigated surface of the blade was about 0,5 mm × 0,5 mm. The measured X-ray line profiles were separated into the subprofiles of the phases γ and γ' by a simple mirror technique. The setup of the diffractometer and the evaluation of the line profiles are described in detail in Reference [8].

Finite Element Calculations

The FEM calculations were performed on workstations using the FEM program MARC K6.1/MENTAT 2.0. One eighth of a γ' particle with the surrounding matrix was divided in a 3-dimensional model by 10 × 10 × 10 elements consisting of 8 integration points. In the present case of the turbine blade, the γ phase still surrounds the γ' particles, in contrast to the creep-deformed samples investigated in Reference [18]. A further difference to the calculations of Ref. [18] is that in the present case plastic back flow of the γ phase during cooling is not considered, since the blades cool down to lower temperatures under centrifugal stresses.

Relevant input parameters of the FEM model are the dimensions of the finite element mesh, the thermal expansion coefficients of the phases γ and γ', the anisotropic elastic constants of the two phases and the lattice misfit obtained from the spacings of dislocations in the interfacial dislocation networks on the γ/γ' interfaces. These parameters are explained in the next paragraph in detail.

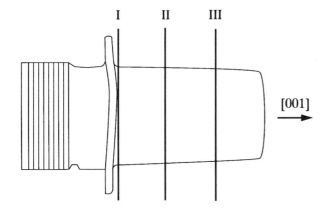

Figure 2: Schematic representation of the turbine blade. The marked sections I, II and III have been investigated in detail.

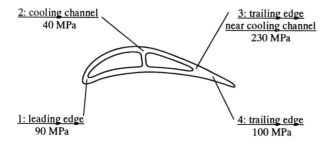

Figure 3: Local load data at the sites 1 (leading edge) to 4 (trailing edge) calculated analytically for section II, see Figure 2 (schematic drawing of the cross section).

Input parameters. The dimensions of the FEM models for the different modelled points were obtained from a quantitative evaluation of the microstructure at these positions. Available thermal expansion coefficients of the phases γ and γ' of the alloy SRR 99 were obtained from X-ray measurements of bulk SRR 99 samples. The use of bulk superalloy specimens has the advantage that the dissolution of the γ' phase at high temperatures is taken into consideration. This is not possible in the case of separate γ and γ' samples. The difference of the expansion coefficients of the phases γ and γ' between the two alloys SRR 99 and CMSX-6 is assumed to be small, because the expansion coefficients of the bulk alloys differ only slightly [23]. The temperature-dependent anisotropic elastic constants of the phases γ and γ' were taken from References [24] and [25], respectively.

The interfacial dislocation networks in the turbine blades, which are the physical origins of the deformation-induced internal stresses [8], were not analyzed experimentally. Rather, they were determined indirectly from the values obtained by the analysis of spacings of the dislocations in the interfacial dislocation networks of samples of the alloy SRR 99 which had been tensile creep deformed at 1050 °C at different stresses [14]. The dislocation networks at (001) interfaces reduce the coherency stresses and build up the deformation-induced internal stresses which are of opposite sign compared to the coherency stresses (see Figure 1(b)). Therefore, the measured dislocation spacings of these interfaces had to be corrected for the spacings of dislocation networks which just relieve the coherency stresses. The temperature dependence of the used input value of the lattice misfit was not determined directly from the analysis of dislocation networks, but from the well-determined correlation of the change of the lattice parameters after tensile creep deformation obtained by X-ray diffraction on samples deformed at different temperatures [2, 3, 5, 8, 9]. The local stress data were taken from the analytically calculated values, see Figure 3.

The local temperatures were used as fit parameters for the calculations and were adjusted by systematical variations in order to obtain a good correlation of the calculated lattice parameters in three dimensions with the X-ray line profiles. From this, the local temperatures in the turbine blades can be inferred. A second possibility to determine the local temperatures is the comparison of the measured lattice misfit values with the dependence of the lattice misfit on local temperatures and stresses determined by a systematical variation of the input of the FEM calculations.

Procedure of the FEM calculations. In the FEM model, the internal stresses are considered by a modification of the actual thermal expansion coefficients of the two phases γ and γ' in a similar way as in the case of creep-deformed samples by Feng et al. [18]. This modification serves to introduce the resultant strains caused by the extra half-planes of the deformation-induced interfacial dislocations. These dislocations are of different character at the interfaces parallel and perpendicular to the stress axis (see Figure 1 and the experimental results of Keller et al. [14]) and serve to introduce the coherency stresses prevailing after deformation (only at the (100)/(010) interfaces, because the coherency stresses are relaxed at the (001) interfaces).

In order to obtain the complete stress state for room temperature, a cooling procedure from the service temperature to room temperature was used in the FEM calculations with the modified expansion coefficients. From the strain distributions at room temperature, the lattice-parameter distributions and the average lattice-mismatch values are determined in the ⟨100⟩ directions.

Results

Initial state

The microstructure of an original turbine blade prior to service was characterized by SEM and TEM. The cuboidal γ' precipitates are clearly visible in the SEM micrograph of Figure 4(a) and in the dark field TEM micrograph of Figure 4(b). An X-ray line profile of the "virgin" turbine blade will be discussed later together with line profiles of the turbine blade subjected to service-like conditions (Figure 6).

Turbine blade subjected to service-like conditions

Microstructure. In Figure 5, a schematical drawing of the turbine blade is shown with some micrographs of local γ/γ' microstructures. Near the cooling channels, the originally cuboidal shape of the γ' particles is preserved. The size, however, is larger than in the initial state, and the precipitates exhibit a rounded shape. Near the leading and the trailing edges, directional coarsening can be observed. The γ phase surrounds at all investigated positions the γ' phase.

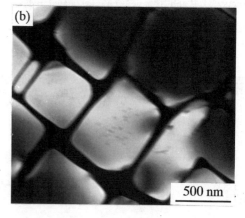

Figure 4: γ/γ' microstructure of the original turbine blade (before service). (a) SEM, γ': dark; (b) TEM dark field, γ': bright.

The mean edge lengths of the two phases in the ⟨100⟩ directions were determined with a digitizing table and used as a basis for the dimensions of the FEM mesh. In the case of the modelling near the trailing edge (site 4 in Figure 5), the mean size of the γ' phase was determined as 1370 nm in the direction [100] (and in the equivalent direction [010]) and 570 nm in the direction [001]. The FEM mesh, which models only one eighth of the γ' particle together with the corresponding half of the surrounding γ phase, has a rectangular shape with the dimensions 773 × 773 × 373 nm³ (i.e. the whole γ channel width for site 4 is assumed to be 176 nm).

The investigations on section III show microstructural changes which are qualitatively similar to those obtained on section II. The degree of coarsening, however, is less pronounced. The situation is more complicated at section I, since the middle of the section in the region near the cooling channels is directly above the platform, while the leading and trailing edges have a distance of some mm from the platform. This situation arises from the curved shape of the platform, see Figure 2. Therefore, the observation of a γ/γ' raft structure near the edges of the blade in section I is plausible. The region near the cooling channels exhibits γ' particles which are coarsened compared to the γ' precipitates of the "virgin" blade.

X-ray line profiles. (002) and (200)/(020) X-ray line profiles measured on cross section II and on different longitudinal sections are presented in Figure 6. The locations, where the individual measurements were made, correlate with the sites corresponding to the micrographs given in Figure 5. The intensities of the profiles are normalized to their maximum and plotted versus the glancing angle θ.

In addition, a typical line profile of a turbine blade in the initial state is given.

The profile of the initial state is symmetrical, indicating a vanishing lattice mismatch ($\delta \approx 0$). The profiles of the loaded and the unloaded blade are different in two ways: i) The profiles of the blade subjected to service are broadened compared to the profile of the blade prior to service and ii) the profiles of the blade subjected to service are asymmetrical. The reasons are, first of all, crystal distortions (e.g. dislocations) which lead to line broadening, and, in addition, long-range internal stresses which lead to the shift of the subprofiles of the two phases. The (002) profiles show in all cases a shoulder on the right side of the maximum. The shoulder can be attributed to the γ phase and the maximum to the γ' phase. The shape of the profiles is equivalent to the case of tensile creep-deformed samples [2, 3, 8, 9]. The shoulder (or second peak), however, is more pronounced the larger the distance from the cooling channels is. The values of the lattice misfit which are evaluated from the separation of the two subprofiles are positive for all (002) profiles (see Table I). This situation is similar to the investigations on other turbine blades reported earlier [5-7].

The shapes of the (200) and (020) profiles are more complex. The (020) profile of site 1 has two maxima, with the higher maximum at the right side. The type of asymmetry is again the same as for tensile creep-deformed samples. The same holds for the (200) profile of site 2, where the second (γ) profile only builds up a shoulder on the left side. The profile of site 3, however, is symmetrical, and the profile of site 4 shows an inverse asymmetry compared to creep-de-

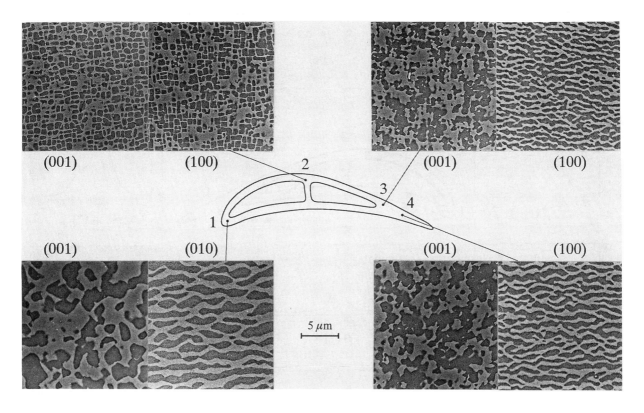

Figure 5: Microstructure of a turbine blade subjected to increased temperature tests for several hundred hours, nickel-base superalloy CMSX-6. The SEM micrographs show the γ/γ' microstructure at the positions indicated in the schematical drawing, section II. The (100)/(010) micrographs belong to adjacent longitudinal sections. γ': dark. The stress axis is vertical in the case of the longitudinal sections (100) and (010).

Table I: Lattice misfit obtained by the analysis of the X-ray line profiles of Figure 6. The lattice misfit of the undeformed state of the alloy CMSX-6 is $\delta \approx 0$.

site no.	$\delta_{[100]/[010]}$	$\delta_{[001]}$
1	-1.7×10^{-3}	$+2.6 \times 10^{-3}$
2	-1.5×10^{-3}	$+1.8 \times 10^{-3}$
3	-1.5×10^{-3}	$+2.5 \times 10^{-3}$
4	-1.7×10^{-3}	$+2.8 \times 10^{-3}$

formed samples. At present, the shapes of these profiles can only been understood by considering also the (300) super-lattice reflection of the γ' phase. This super-lattice reflection is asymmetrical, with a considerably more slowly decreasing tail on the (left) lower angle side. Therefore, the measured (200) profiles consist of the asymmetrical γ' peak (on the right) and, on the left, of the γ peak. The superposition of the asymmetrical γ' peak and the γ peak can give rise to the shape of the (200) profiles of sites 3 and 4. The values of the lattice misfit are also given in Table I. The results obtained on the sections I and III are qualitatively equivalent to those of section II. The magnitude of the changes of the lattice misfit, however, is smaller at sections I and III.

Complementary investigations of the X-ray line profiles on other turbine blades [5-7] yielded similar results as those reported in the present work. One blade, however, was used under so-called emergency conditions. A (002) X-ray line profile measured near the leading edge of that blade showed a reversed asymmetry which is not understood in detail. The microstructure of that part of the blade had a bimodal size distribution of the γ' particles, thus indicating that the temperature during service exceeded normal operation conditions. In the case of the (200) and (020) line profiles, the investigations reported in [5-7] showed an almost symmetrical shape. This is in contrast to the results of the present work. One possible reason is that the blade investigated in the present work was in service for 400 hours, whereas the blades of Refs. [5-7] for only about 70 hours.

Transmission electron microscopy. Figure 7(a) shows a TEM micrograph of a (100) oriented sample at site 4 near section II with dislocation networks at the γ/γ' interfaces (i.e. at the interfaces which are more or less parallel to the (100) lattice planes). Due to the strong curvature of the (100) and (010) γ/γ' interfaces, the interfacial dislocation networks cannot be analyzed in order to obtain the dislocation spacings in the principal ⟨100⟩ directions which would be necessary as input for the FEM calculations, if the real microstructure of the blade should be used. Complementary investigations on TEM samples with (001) orientation yielded also dense interfacial dislocation networks, see Figure 7(b).

Convergent beam electron diffraction (CBED). The CBED patterns of the [100] zone axis observed in the phases γ and γ' of a TEM slice prepared near site 4 (see Figure 7) are shown in Figures 8(a) and (b), respectively. The HOLZ lines clearly indicate that the fourfold symmetry of the undistorted crystal lattice is disturbed, and only a twofold symmetry remains. A systematic variation of the input lattice parameters for a kinematical simulation of the HOLZ lines of the two phases was performed. Fitting the patterns with calculated HOLZ lines indicates that the crystal lattice of the γ phase is strained in tension in the directions [100] and [010] and in com-

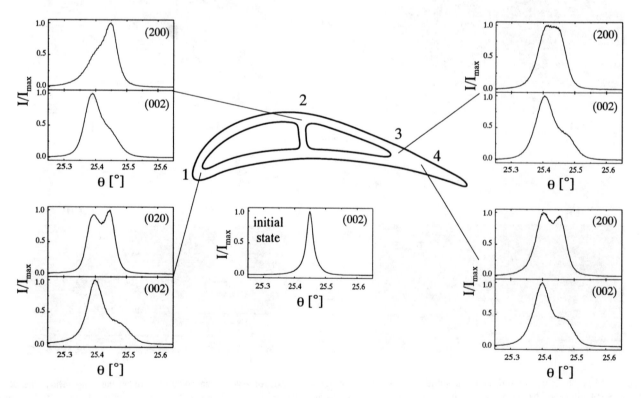

Figure 6: (002) and (200)/(020) X-ray line profiles of the turbine blade subjected to increased temperature tests, measured at the sites given in Figure 5. The line profile of the undeformed state of the alloy CMSX-6 is plotted in addition. θ: glancing angle.

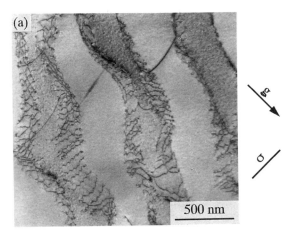
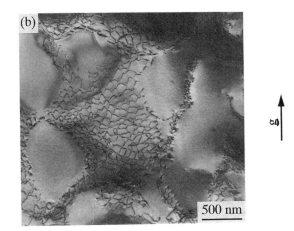

Figure 7: TEM micrographs of the γ/γ' structure with interfacial dislocation networks, turbine blade subjected to service-like conditions, site 4. (a) (100) section, g = (020) and (b) (001) section, g = (020).

pression in the [001]-direction. In the γ' phase, the inverse is true: The lattice parameter in the [001] direction is strained in tension, the lattice parameters in the two perpendicular directions are strained in compression. Hence, the investigations prove the tetragonal lattice distortion of the two phases.

FEM Calculations. Systematical variations of the volume fraction of the γ' phase and of the dimensions of the γ/γ' plate structure proved that these two input parameters have only a minor effect on the lattice-parameter distributions and lattice-mismatch values at room temperature. The volume fraction of the γ' phase used for the modelling was 55 Vol.%, based on a quantitative evaluation of the SEM micrographs. The strains in the two phases are essentially determined by the temperatures, from which the samples are cooled to room temperature in the calculations, and the local stresses, which determined the dislocation networks at the interfaces. This finding is not surprising, since the starting temperatures determine the amount of thermal stresses at room temperature, and the local stresses determine the deformation-induced internal stresses.

In Figure 9, calculated lattice parameters (solid lines in Figure 9) of the phases γ and γ' in the directions [001] (Figure 9(a)) and [100] (Figure 9(b)) are compared with the (002) and (200) X-ray line profiles (dashed lines), respectively, measured near the trailing edge of the turbine blade at site 4. Our computations do not take into account line broadening due to dislocations, etc.. A good correlation of the positions of the peaks of the calculated lattice parameter distributions with the measured profiles was obtained under the assumption of a local stress state equivalent to that after creep deformation at a temperature of 1050 °C under a stress of 100 MPa. The determined local temperatures (for the given stress values, see Figure 3) at sites 1, 3 and 4 are given in Table II. In the table, the local

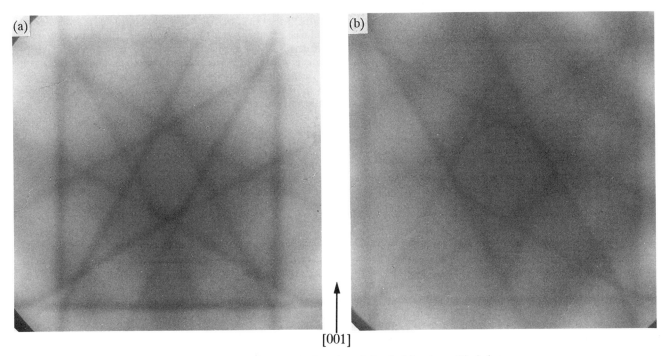

Figure 8: HOLZ line patterns of the [100] zone axis (longitudinal section of site 4). (a) γ phase, (b) γ' phase.

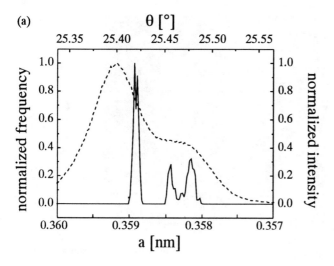

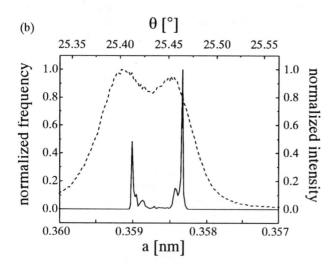

Figure 9: Comparison of calculated lattice parameters (solid lines) in the directions [001] (a) and [100] (b) with (002) and (200) X-ray line profiles (dashed lines) from a region near the trailing edge of a CMSX-6 turbine blade, which has been exposed to service; site 4, as indicated in Figure 5. θ: glancing angle.

temperatures determined from the systematical variation of the temperature to obtain a good correlation of the shape of the lattice parameter distributions with the line profiles are called "fit of the shape".

The temperatures determined by the use of the measured misfit values and their comparison with the dependencies calculated from systematical variations of the FEM input parameters are also given in Table II and called "comparison of δ". In the case of the use of δ, the temperatures are up to 100 °C higher that the values determined from the comparison of the shape of the lattice parameter distributions. At site 2, where the γ' particles have cuboidal shape, no valid temperature could be estimated by the FEM calculations. The reason is that the present input parameters are based on the dislocation spacings of dislocation networks of the γ/γ' raft structure.

Table II: Comparison of temperatures obtained by the fit of the shape of the lattice parameter distribution and from the comparison of δ.

site no.	σ	T (fit of the shape)	T (comparison of δ)
1	90 MPa	1050 °C	1169 °C
2	40 MPa	—	—
3	230 MPa	940 °C	977 °C
4	100 MPa	1050 °C	1146 °C

Discussion

γ/γ' morphology

The changes of the γ/γ' morphology are understandable with regard to the loads acting in the blade during service. The region near the cooling channels shows coarsened γ' precipitates with cuboidal shape. In these regions, the combination of the relatively low material temperatures and the local stresses did not lead to rafting. In the regions with higher material temperatures near the leading and trailing edges, however, the material temperature was higher and, in combination with the external centrifugal stress, rafting occurred. This observation of rafting in a turbine blade after engine operation is in agreement with the results of Draper et al. [4]. These authors reported, for example, that rafting occurred in up to 60 % of the span length in blades of the nickel-base superalloy NASAIR 100. The degree of rafting depended on the test conditions, i.e. the operation time, the temperatures and the externally applied (centrifugal) stresses. Earlier investigations of other turbine blades in the authors' group [5-7] showed similar γ/γ' morphologies in turbine blades already after about 70 hours of service.

In the turbine blade investigated in the present work, the γ phase remains the surrounding matrix phase with included γ' plates. The investigations of γ/γ' morphologies of creep-deformed samples of the alloys SRR 99 and CMSX-4, however, showed that the connectivity had changed after about 1 % plastic strain. The reason may lie in the details of the thermal and mechanical exposure and, perhaps more important, in the fact that the volume fraction of the γ' phase is smaller in the alloy CMSX-6 compared to the two other alloys (SRR 99 and CMSX-4). A final explanation of the reason for these different behaviours cannot be given yet.

Lattice distortions

The lattice distortions were measured with two complementary techniques, i.e. high-resolution X-ray diffraction and CBED. These methods have different lateral resolutions. The X-ray method integrates the lattice parameter distributions over a volume of 0.5 mm × 0.5 mm × 5 - 10 μm and hence provides a mean value of the lattice parameter changes with high accuracy. In the way applied here, the CBED technique yields no absolute but only relative lattice parameter values, but has an excellent lateral resolution with a cross section of the electron beam of about 20 nm. Therefore, the two methods provide complementary information. It is satisfying to note, that both methods yield qualitatively the same tetragonal lattice distortions, namely tensile and compressive distortions in the γ' and the γ phase in the direction [001], respectively. Therefore, both methods give experimental evidence of lattice distortions compatible with those shown schematically in Figure 1.

For the CBED investigations performed so far (see e.g. Figure 8), the [100] zone axis was used. Since low indexed zone axes (as the ⟨100⟩ type zone axes) are highly distorted by dynamical scattering effects, only limited quantitative information can be obtained from the kinematical simulation of the HOLZ line patterns of such zone axes. In the present case, the change of the line pattern from a four-fold symmetry in the initially undeformed state to a pattern with two-fold symmetry in the blade exposed to service confirms the tetragonal lattice distortion of both phases γ and γ'. The HOLZ line patterns (Figure 8) are qualitatively the same as the patterns observed by Keller et al. [15], who investigated a sample of the nickel-base superalloy SRR 99 creep-deformed in tension at a temperature of 1050 °C at a stress of 305 MPa. In the present case, the degree of the tetragonal distortion is somewhat smaller than that observed in the sample of Keller et al.. This indicates that the lattice distortions are smaller in the present case. A comparison of preliminary kinematical HOLZ line simulations of the patterns of Figure 8 with dynamical simulations showed a difference of the positions of the calculated HOLZ lines. Future work will be performed which will simulate the ⟨100⟩ zone axes with dynamical computer simulations of the HOLZ lines in combination with experimental investigations on the ⟨114⟩ zone axes.

Estimation of the local loads

The comparison of X-ray investigations with the results of FEM calculations allows the estimation of either the stresses or the temperatures present locally during service, if the other of the two parameters is known. In the present case, values of the stresses are given which had been calculated analytically (see Figure 3). These values should be more reliable than the computed temperatures, since during service of the turbine blade, several factors can lead to temperature fluctuations. Therefore, in our studies, the determination of the local temperatures seem to be more appropriate.

The local temperatures determined in the present work by the fit of the shape of the lattice parameter distribution are in good correlation with the specific test conditions of the engine. This fit yields lower temperatures than the use of the averaged misfit values (see Table II). The temperatures determined by the latter method at sites 1 and 4 are too high. This shows that the comparison of more global values such as the misfit gives unsatisfactory results, especially in the case, where the X-ray line profiles have shapes which cannot be explained in a straightforward manner. In the case of creep-deformed samples [18], the values of the lattice misfit could be used for the comparison of measured profiles with the calculated lattice parameter distributions, since the shape of the profiles agreed fairly well with the calculated results. This may be explained by the different morphologies of the two cases: In the turbine blade, the γ phase surrounds γ', in the creep samples, the γ' phase surrounds γ.

Figure 9 shows some differences between the measured X-ray profiles and the calculated lattice parameter distributions: i) The measured profiles are much more broadened than the calculated distributions, because different line broadening contributions, i.e. dislocations in the two phases and at the interfaces, residual coherency strains, the particle size effect of the small coherently scattering particles and the difference between the idealized model and the real microstructure (Figure 5) were not taken into account. ii) The intensity relation between the left (γ) and the right (γ') peak of the lattice parameter distribution in the [100] direction corresponds not to the (200) line profile (Figure 9(b)). The reason for this discrepancy may be connected with the asymmetry of the (300) line profile mentioned above. The shape of this profile, however, is not understood yet. Therefore, only the separation of the peaks of the calculated distribution can be compared with the splitting-up of the measured profile into two peaks. iii) The FEM calculations of this work consider the dislocation structure which was investigated in samples after pure tensile creep deformation. The blade, however, is subjected to the superposition of complex loads such as centrifugal stresses, bending or thermomechanical fatigue. These factors cause complex triaxial stress states which were not taken into consideration.

The calculations are based on an assumption which was not discussed until now. The dislocation network data used for the modelling were obtained on samples creep deformed until fracture. The dislocation structure in the turbine blade, where the total allowed strain is 1 or 2 % and the maximal experienced local strain is in the order of up to 4 or 5 %, may be different compared to samples deformed to fracture with strains of about 20 %. Since earlier investigations showed that the change of the line profiles of creep-deformed samples is completed after strains of about 1 % [5], the local strains experienced by the blade seem to be sufficient for the full build up of the internal stress state. Another problem arises from the possible differences between the dislocation structure in the blade and in a creep sample which would experience equal stresses. The quite good agreement between the measured line profiles and the calculated lattice parameters shows that the method is successful. In further work, the real dislocation structure of the turbine blade will be investigated in detail.

Conclusions

The main conclusions of the investigations on a turbine blade subjected to service-like conditions are:
1. In the regions exposed to high metal temperatures and (centrifugal) stresses, directional coarsening and rafting occur.
2. X-ray line profiles indicate elastic, tetragonal lattice distortions.
3. CBED investigations confirm the predicted tetragonal lattice distortions of the two phases γ and γ'.
4. From the comparison of calculated lattice parameters with experimentally measured X-ray line profiles, the locally acting temperatures are determined for given stress values. These temperatures were determined to be up to about 1050 °C near the leading and trailing edges.
5. It can be concluded from the microstructural characterization that the leading and trailing edges of the turbine blade are under stress states which are similar to those of samples creep-deformed in tension.
6. In summary, it has been demonstrated that a sufficiently detailed microstructural characterization of blades exposed to service conditions permits an estimate of the local stress states and temperatures experienced during service.

Acknowledgment

The authors thank Dr.-Ing. D. Goldschmidt (formerly MTU, Munich), now Siemens AG/KWU, Mülheim, and Mr. W. Buchmann, MTU, for support of this work, and the company MTU for providing the investigated turbine blades. The authors are grateful to Mr. S. Mechsner for performing some of the X-ray measurements.

Literature

1. T. M. Pollock and A. S. Argon, "Creep Resistance of CMSX-3 Nickel Base Superalloy Single Crystals", Acta metall. mater., 40 (1992), 1-30.

2. H. Biermann, M. Strehler and H. Mughrabi, "High-Temperature X-Ray Measurements of the Lattice Mismatch of Creep-Deformed Monocrystals of the Nickel-Base Superalloy SRR 99", Scripta metall. mater., 32 (1995), 1405-1410.

3. H. Biermann, M. Strehler and H. Mughrabi, "High-Temperature Measurements of Lattice Parameters and Internal Stresses of a Creep-Deformed Monocrystalline Nickel-Base Superalloy", Met. Trans. A, 27A (1996), 1003-1014.

4. S. Draper, D. Hull and R. Dreshfield, "Observations of Directional Gamma Prime Coarsening during Engine Operation", Metall. Trans. A, 20A (1989), 683-688.

5. S. Spangel, "Röntgenographische Untersuchung lokaler Gitterparameter in Labor- und Turbinenschaufelproben aus einkristallinen Nickelbasis-Superlegierungen", (diploma thesis, Universität Erlangen-Nürnberg, Germany, 1994).

6. H. Biermann, S. Spangel and H. Mughrabi, "Investigation of Local Lattice Parameter Changes in Monocrystalline Nickel-Based Turbine Blades after Service", in: Proc. EUROMAT 95, Padua/Venice, Italy, Vol. 2, (ed. Associatione Italiana di Metallurgica), Milano, Italy (1995), 255-260.

7. H. Biermann, S. Spangel and H. Mughrabi, "Local Lattice Parameter Changes in Monocrystalline Turbine Blades Subjected to Service-Like Conditions", Z. Metallkde., 87 (1996), 403-410.

8. H.-A. Kuhn, H. Biermann, T. Ungár and H. Mughrabi, "An X-Ray Study of Creep-Deformation Induced Changes of the Lattice Mismatch in the γ'-Hardened Monocrystalline Nickel-Base Superalloy SRR 99", Acta metall. mater., 39 (1991), 2783-2794.

9. H. Mughrabi, H. Biermann and T. Ungár, "X-Ray Analysis of Creep-Induced Local Lattice Parameter Changes in a Monocrystalline Nickel-Base Superalloy", in "Superalloys 1992", Proc. 7th Int. Symp. on Superalloys (eds. S. D. Antolovich, R. W. Stusrud, R. A. MacKay, D. L. Anton, T. Khan, R. D. Kissinger and D. L. Klarstrom), TMS, Warrendale, Pennsylvania (1992), 599-608.

10. U. Glatzel and A. Müller, "Neutron Scattering Experiments with a Nickel Base Superalloy, Part I: Material and Experiment", Scripta metall. mater., 31 (1994), 285-290 and U. Glatzel, "Neutron Scattering Experiments with a Nickel Base Superalloy, Part II: Analysis of Intensity Profiles", Scripta metall. mater., 31 (1994), 291-296.

11. U. Glatzel and A. Müller, "Calculated and Measured Internal Stresses of Creep-Deformed Single Crystal Nickel-Based Superalloys", in: "Numerical Predictions of Deformation Processes and the Behaviour of Real Materials", Proc. 15th Risø Int. Symp. on Materials Science, (eds. S. I. Andersen, J. B. Bilde-Sørensen, T. Lorentzen, O. B. Pedersen and N. J. Sørensen), Risø National Laboratory, Roskilde, Denmark (1994), 319-324.

12. T. Gnäupel-Herold and W. Reimers, "Stress States in the Creep Deformed Single Crystal Nickelbase Superalloy SC16", Scripta metall. mater., 33 (1995), 615-621.

13. A. Royer, P. Bastie, D. Bellet and J. L. Strudel, "Temperature Dependence of the Lattice Mismatch of the AM 1 Superalloy. Influence of the γ' Precipitates' Morphology", Phil. Mag. A, 72 (1995), 669-690.

14. R. R. Keller, H. J. Maier and H. Mughrabi, "Characterization of Interfacial Dislocation Networks in the Creep-Deformed Nickel-Base Superalloy SRR 99", Scripta metall. mater., 28 (1993), 23-28.

15. R. R. Keller, H. J. Maier, H. Renner and H. Mughrabi, "Local Lattice Parameter Measurements in a Creep-Deformed Nickel-Base Superalloy by Convergent Beam Electron Diffraction", Scripta metall. mater., 27 (1992), 1167-1172, and Scripta metall. mater., 28 (1993), 661.

16. J. Li and R. P. Wahi, "Investigation of γ/γ' Lattice Mismatch in the Polycrystalline Nickel-Base Superalloy IN 738 LC: Influence of Heat Treatment and Creep Deformation", Acta metall. mater., 43 (1995), 507-517.

17. L. Müller, U. Glatzel and M. Feller-Kniepmeier, "Calculation of the Internal Stresses and Strains in the Microstructure of a Single Crystal Nickel-Base Superalloy During Creep", Acta metall. mater., 41 (1994), 3401-3411.

18. H. Feng, H. Biermann and H. Mughrabi, "3D Finite Element Modelling of Lattice Misfit and Long-Range Internal Stresses in Creep-Deformed Nickel-Base Superalloy Single Crystals", Mater. Sci. Eng., in press.

19. H. Brooks, "Metal Interfaces", American Society of Metals (1952), 20-64.

20. M. V. Nathal, J. O. Diaz and R. V. Miner, "High Temperature Creep Behaviour of Single Crystal Gamma Prime and Gamma Alloys", in Proc. MRS Symp., Vol. 133, (eds. C. T. Liu, A. I. Taub, N. S. Stoloff, C. C. Koch), MRS, Pittsburgh, Pennsylvania (1989), 269-274.

21. C. Carry, S. Dermarkar, J. L. Strudel and B. C. Wonsiewicz, "Internal Stresses due to Dislocation Walls Around Second Phase Particles", Metall. Trans. A, 10A (1979), 855-860.

22. P. A. Stadelmann, "EMS — A Software Package for Electron Diffraction Analysis and HREM Image Simulation in Materials Science", Ultramicroscopy, 21 (1987), 131-146.

23. W. Hermann, "Elastizität und Anelastizität technisch wichtiger Hochtemperaturlegierungen", (doctorate thesis, Universität Erlangen-Nürnberg, Germany, 1995).

24. H.-A. Kuhn, "Anwendung von Grenzwertkonzepten und Phasenmischungsregeln auf die elastischen Eigenschaften von Superlegierungen zwischen Raumtemperatur und 1200 °C", (doctorate thesis, Universität Erlangen-Nürnberg, Germany, 1987).

25. R. W. Dickson, J. B. Wachtmann and S. M. Copley, "Elastic Constants of Single-Crystals Ni_3Al from 10 °C to 850 °C", J. Appl. Phys., 40 (1969), 2276-2279.

ROLE OF COHERENCY STRAIN IN MICROSTRUCTURAL DEVELOPMENT

Jong K. Lee
Department of Metallurgical and Materials Engineering
Michigan Technological University
Houghton, MI 49931, USA

Abstract

With a purely dilatational misfit strain, an elastically-soft, coherent precipitate imbedded in an infinite matrix tends to have a plate-like equilibrium morphology, whereas a hard particle tends to take on a round shape of high symmetry. Shape evolution proceeds through dynamic activities of coherency-induced interfacial waves. These interfacial waves seem to be responsible for the protrusions often observed along elastically hard directions in coherent particles of nickel-based superalloys. Soft particles with a positive misfit strain become plates perpendicular to an applied tensile stress, while hard particles elongate along the stress direction. If the elastic interaction between the applied stress and the coherency strain is strong enough, soft precipitates often split into smaller particles and then follow coarsening. If the applied stress increases further, coherent particles tend to dissolve into the matrix — in agreement with the theory of coherent phase equilibria. As expected, a coherent particle with a positive misfit strain migrates to the tension region of an edge dislocation, whereas a particle with a negative strain diffuses to the region of compression. Morphological change is, however, caused by the dislocation as the particle tries to capitalize on the dislocation stress field. The results are analyzed by means of a discrete atom method (DAM), which is predicated upon Hookean atomic interactions and Monte Carlo diffusion under the condition of a plane strain, a purely dilatational misfit strain, and no dislocation climbing.

Introduction

The microstructural development of nickel-based superalloys or other age-hardenable alloys has been a subject of great interest, as it is closely tied to the performance during application. For unstressed two-phase alloy systems, the equilibrium shape of a precipitate is established solely by the interfacial free energy and its dependence on crystallographic orientation. On the other hand, the equilibrium morphology of second-phase, coherent precipitates is dictated by both the interfacial free energy and the elastic strain energy. Consequently, there has been a need for a computational technique, through which one can analyze the elastic state associated with *arbitrarily-shaped* precipitates whose elastic constants are *different* from those of the matrix phase.

Eshelby (1) was the pioneer in the field of coherency strain who devised the seminal equivalency method and thus brought much understanding to the coherency strain problem; however, the method is limited to a single ellipsoidal precipitate (2-4). Since his work, several numerical techniques have been developed, but most involve either computations of an *elastically homogeneous* state, or approximate solutions for integro-differential equations when faced with an *inhomogeneous* system (5-13). In an effort to develop a tool to study general shape evolution of a coherent precipitate, a Discrete Atom method (DAM) was recently developed on the basis of a statistical approach (14,15). The method has been applied to a number of coherency strain problems (16-18), demonstrating that an elastically inhomogeneous, multi-particle system can be readily analyzed. In this work, the DAM is reviewed to examine morphological evolution of second-phase, coherent particles with a purely dilatational transformation strain.

Discrete Atom Method

A two-dimensional triangular lattice with Hookean nearest neighbor interaction is elastically isotropic (19). If the atomic bond energy is expressed in the form of $k(r - a)^2/2$, both Lamé elastic constants, $\lambda (= C_{12})$ and $\mu (= C_{44})$, are equal to $0.433k$, where k is the spring constant, r is the interatomic distance, and a is the lattice parameter of the lattice. A coherent precipitate is designated with a different spring constant and lattice parameter, k^* and $a^* = a(1 + \varepsilon)$, where ε is the dilatational misfit strain. Directional, instead of uniform, spring constants make the crystal elastically anisotropic. Cubic systems with Zener's anisotropy ratio, $A = 2C_{44}/(C_{11} - C_{12}) = 2.33$ are studied to mimic nickel-based superalloys. Since the major concern is coherency effects, an isotropic interface energy is assumed. An interface atom is defined as the precipitate atom having unlike nearest neighbor bonds, and has a specific interfacial energy depending on the number of unlike bonds. Morphological evolution is then examined through a Monte Carlo process, which, by exchanging precipitate and matrix atoms, generates a Boltzmann-weighted chain of the configurations of a given system. The details of the DAM was described elsewhere (14,15).

All the computations are performed under the conditions of a constant number of precipitate atoms, a pure dilatational misfit, and a plane strain. The 'diffusion' temperature is kept low enough so that surface roughening is negligible and yet atoms have sufficient mobility for shape change. Here diffusion means a process of atomic site exchange, which may imply vacancy mechanism, but no real vacancies are introduced into the system. Thus the work is intended to mimic low temperature phenomena in which the presence of point defects is insignificant to relax the coherency strain itself. With $k = 1.38 \times 10^{-18} J/a^2$ and isotropic interfacial energy, $\gamma_o = 2.5 \times 10^{-21} J/a$, a typical temperature of 50K is used for a system in which the melting point of the matrix phase is about 1000K.

The results in the first four parts deal with a dislocation-free system. In the last part, effects of edge dislocations are examined. Dislocation gliding is permitted, but no dislocation climbing is allowed as it requires a substantial amount of point defects which may relax coherency strain. The introduction of edge dislocations is accomplished with a truncated Hookean potential (19). The potential is made of three interaction zones: $k[(r-a)^2/2 - w^2]$ if $r \leq a + w$, $-k(r - a - 2w)^2/2$ if $a + w < r \leq a + 2w$, and 0 for $r > a + 2w$, where w/a is equal to 0.15 for the interaction with first nearest neighbors and 0.10 for second nearest neighbors. In the text, if elastic stiffness is described in terms of shear modulus, μ, it indicates an elastically isotropic system, whereas if the stiffness is measured in terms of C_{ijkl}, an anisotropic system is meant. Both isotropic and anisotropic elasticity results are discussed to understand intriguing coherency strain effects.

Results and Discussion

Shape Bifurcation

If one carefully examines the shape evolution of typical Ni_3Al precipitates in nickel-based superalloys (20), it is easy to recognize a series of shape bifurcations beginning from a spherical shape (radial symmetry) to a cuboidal shape (four-fold symmetry), then to a plate-like shape (two-fold symmetry), as their size increases. This is a straight manifestation for the competition between the γ–γ' interfacial energy and the elastic strain energy, as was shown elegantly by Johnson and Cahn (21). Let us examine a shape bifurcation behavior as displayed through the DAM. In Fig. 1, soft, isotropic precipitates with $\mu^* = \mu/2$, $\varepsilon = 0.05$, and $\gamma = \gamma_o$ undergo shape evolution at T = 50K. Size, R, along the horizontal axis indicates the particle radius in units of a, while β on the vertical axis marks the aspect ratio with $\beta = 1$ indicating a circle. J-C denotes the bifurcation point predicted from the Johnson-Cahn theory. For each particle size, a circle is the initial shape, and snap-shot morphologies during the evolution are exhibited downward. With the DAM, the bifurcation from a circle to an ellipse-like shape occurs at $R \cong 9a$, which is close to the Johnson-Cahn analytical value, $7a$. Note also that the particles with $R > 14a$ experience transient state(s) from a circular to a polygon-like shape, before reaching a plate-like equilibrium shape.

In Fig. 2, fractional changes in energy versus Monte Carlo steps (MCS) are plotted for the precipitate with $R = 14a$ of Fig. 1. The broken curve at the top represents change in the interfacial energy, the solid curve at the middle is for the sum of the interfacial and elastic strain energy, and the bottom, dotted curve is for the elastic stain energy. Two distinctive plateaus are clearly displayed on the top, interfacial energy curve: the first plateau between MCS = 5M (M for a million steps) and 20M stands for a triangle-like shape of a three-fold symmetry, while the second one from MCS = 28M and thereafter indicates a plate-like shape of a two-fold symmetry. At the final equilibrium, the interfacial energy shows an increase of 31%, the strain energy registers a decrease of 12%, and the total energy is reduced by 3%.

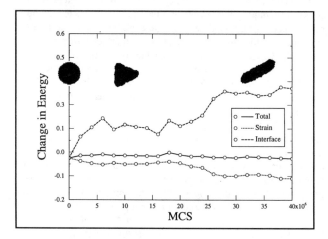

Figure 2: Fractional changes in energy versus Monte Carlo steps for a soft, isotropic precipitate with $R = 14a$ and $\varepsilon = 0.05$. The top, broken curve represents change in the interfacial energy, the middle, solid curve is for the sum of the interfacial and strain energy, and the bottom, dotted curve is for the elastic stain energy. Note two distinctive plateaus on the interfacial energy curve, indicating shape transitions from a radial to a three-fold, then to a two-fold symmetric shape.

The fact that all the soft particles with a pure dilatational misfit attain a plate-like equilibrium shape confirms the well-known Eshelby's inclusion theory that the stress field within a thin plate is one-dimensional along its major axis, and consequently all the strain energy tends to be contained in the soft plate itself (2,3). Similarly, a separate DAM work upheld Crum's theorem that for the homogeneous case ($\mu^* = \mu$), the strain energy of a particle is independent of its shape (1). If the particle is stiffer than the matrix as in a case with $\mu^* = 2\mu$, however, the strain energy is proven to drive the particle morphology toward a circle. Shape transition is found to begin with interfacial waves induced by the coherency strain, whose wavelengths depend on elastic constants, particle geometry, anisotropy, misfit strain, and diffusion temperature. The wave activities are more pronounced with large precipitates, an example of which is demonstrated in Fig. 3. Conditions similar to those of Fig. 1 are placed to a soft particle with $R = 35a$. A circular,

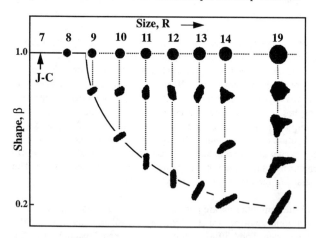

Figure 1: A shape bifurcation diagram for soft, isotropic particles with $\mu^* = \mu/2$, $\varepsilon = 0.05$, and $\gamma = \gamma_o$. Size, R, along the horizontal axis indicates the particle radius in units of a, while Shape, β, on the vertical axis marks the aspect ratio with $\beta = 1$ indicating a circle. J-C denotes the bifurcation point predicted from the Johnson-Cahn theory. For each particle size, a circle is the initial shape, and snapshot morphologies during the evolution are exhibited downward.

soft particle initially develops a spectrum of wavelengths, but waves with a maximum growth rate soon dominate the process, creating a number of distinctive, small lobes. These lobes then coarsen into a lower density of larger lobes. All but two larger lobes eventually recede as the plate-like equilibrium shape is approached.

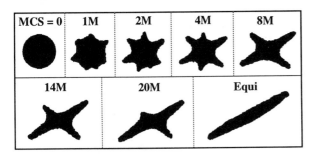

Figure 3: Morphological evolution of a soft particle with $R = 35a$, $\mu^* = \mu/2$, and $\varepsilon = 0.05$ in an isotropic system.

Obviously, morphological evolution depends on the diffusion temperature. For example, the series of shape bifurcations from a circular to the final, plate-like shape observed for the isotropic particle with $R = 19a$ in Fig. 1 or $R = 35a$ in Fig. 3 are products at a temperature as low as 50K. When the temperature is raised to 300K or 500K, as shown in Fig. 4, large thermal fluctuations allow a direct transformation from a circle to a plate. Though there occurs significant surface roughening at these high temperatures, plate-like shapes are clearly recognized when quenched to 30K. Therefore, in all the simulations, the diffusion temperature is kept low enough so that the evolution is hoped to follow a surface of lowest energy and yet atoms have sufficient mobility for shape change.

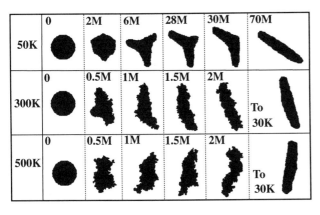

Figure 4: Temperature effects on the shape evolution of a soft particle with $R = 19a$, $\mu^* = \mu/2$, and $\varepsilon = 0.05$ in an isotropic system.

Directional, instead of uniform, spring constants make the triangular lattice elastically anisotropic. For the precipitates in Fig. 5, the anisotropy ratio A is equal to 2.33, a value similar to that of nickel. Other ancillary data are: $R = 40a$, $\varepsilon = 0.05$, and $C_{12} = C_{44} = 0.606$ k. Soft <100> directions are marked with arrows, indicating diagonal directions of <110> are elastically hard. The left column shows the early stage of shape evolution for a soft precipitate ($C^*_{ijkl} = C_{ijkl}/2$), the middle column for the homogeneous case ($C^*_{ijkl} = C_{ijkl}$), and the right column for a hard case ($C^*_{ijkl} = 2C_{ijkl}$). All three cases display a symmetry-breaking transition from a radial to four-fold symmetry, consistent with the cubic anisotropy of nickel-based superalloys. Further evolution, however, reveals that the soft particle transforms into a two-fold shape, attaining an ellipse-like shape stretched along a soft <100> direction. The four-fold form of the homogeneous case is also unstable with respect to a two-fold shape. The particle stretches, at a very sluggish rate, again along a soft <100> direction, and is eventually transformed into a plate shape with blunt edges (15). On the other hand, the four-fold shape of the hard particle at MCS = 4M is an equilibrium morphology, as it is reproduced by the evolution process of an initially rectangular particle. Notice the rounded corners along the hard <110> direction, whose curvature deceases with increase in the precipitate stiffness. Though two-dimensional, the overall evolution sequence observed in this simulation is consistent with many experimental observations in typical Ni-based superalloys (22-25).

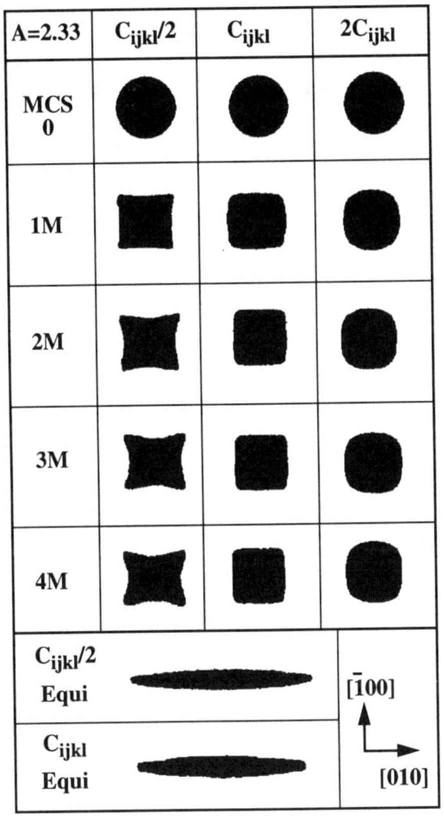

Figure 5: Morphological evolution of a soft (left column), an elastically homogeneous (middle column), and a hard particle (right column) in an anisotropic matrix with A = 2.33.

Elastic effects on the coarsening behavior of a two-particle system with A = 2.33 are examined in Fig. 6. Two precipitates, initially circular, are separated along the [110] direction. The first row, a case of zero misfit strain ($\varepsilon = 0$), shows a classic, Ostwald ripening process driven by capillarity alone. Because of the low temperature employed, each particle maintains a circular shape during the process. In the second row, the two soft particles ($C^*_{ijkl} = C_{ijkl}/2$) with $R = 20a$ and $\varepsilon = 0.05$ fuse together along the elastically soft <100> directions. On the other hand, if the precipitates are hard as in the third row ($C^*_{ijkl} = 2C_{ijkl}$), the repulsive elastic interaction along the [110] direction is seen to stabilize the precipitates against coarsening. When the precipitate elastic constants are intermediate as in the fourth and fifth row ($C^*_{ijkl} = C_{ijkl}$), the elastic interaction also appears to stabilize the particles. But it is only transient: the particles migrate slowly along the [010] to lower the energy, and eventually coalesce to yield a plate-like equilibrium shape. Even with two-particle systems shown here, one can easily recognize that there occurs a significant shape change and realignment for soft particles, and to a lesser degree, for the homogeneous particles. Continuous fusion and migration processes among the particles are clearly seen to promote linear stringers along elastically favorable directions. On the other hand, hard particles display strong coherency-stabilizing effects.

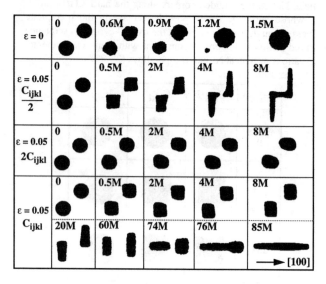

Figure 6: Coarsening behavior of two precipitates in an anisotropic matrix with A = 2.33. The first row with zero misfit strain shows a classical, capillary-driven ripening process, whereas the second row displays fusing of two soft particles along elastically-soft directions. The third row shows a coherency-stabilized state for the hard particles. The homogeneous case of the fourth and fifth row portrays a stabilized state in the early stage, but an ensuing coalescence forming a plate-like equilibrium shape.

Effects of Applied Stress

To control microstructural development, external stresses are often applied to alloy systems. For example, Tien and Copley (26) showed that when a tensile stress was applied along the [001] direction, γ' plates in the Udimet 700 precipitated on the (001) plane. When the applied stress direction was reversed, the precipitates became rods parallel to the [001] direction. Miyazaki et al., however, refuted these results by showing that Ni_3Al rods and plates aligned along the tensile [001] axis in Ni-15 at.% Al alloys (27). Equally contradicting morphological evolution is also reported for other alloy systems such as Al-Cu alloys (28,29). Though unsettling and confusing, these experimental results are certainly calling for further understanding of the applied stress effects. Several theoretical investigations which treat the change in strain energy with particle shape were reported (1,3, 30-33). In this section, we examine the shapes predicted from the DAM under applied stresses.

The effect of a pure dilatational misfit strain should be similar to that of a hydrostatic pressure on an inhomogeneous particle with ε = 0 imbedded in an infinite matrix. This is demonstrated in Fig. 7, where both soft ($\mu^* = \mu/2$) and hard ($\mu^* = 2\mu$) particles with R = 25a are placed under a hydrostatic pressure with a magnitude equal to 5% strain. The shape transition for the circular, soft particle starts with dynamics of interfacial waves induced by the inhomogeneity. The waves of a maximum growth rate dominate the early evolution process, creating several large lobes. A coarsening process follows, leading the morphology to an elliptical equilibrium shape. Compared to the misfit strain case of Fig. 3, the lobe tips are quite sharp, indicating a high stress concentration. For the hard, rectangular particle, there are two driving forces to drive the shape toward a circle: the isotropic interfacial energy and the strain energy due to the inhomogeneity effect.

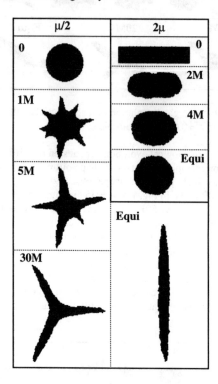

Figure 7: Morphological evolution of both soft ($\mu^* = \mu/2$) and hard ($\mu^* = 2\mu$) isotropic particles with ε = 0 under a hydrostatic pressure. Similar to the dilatational misfit strain case, the soft particle becomes a plate, while a hard particle attains a circular shape at equilibrium.

In Fig. 8, an isotropic, soft particle with R = 17a, ε = 0.05, and $\mu^* = \mu/2$ is placed under a *uniaxial compressive* stress with a magnitude equal to 4% strain along the vertical direction. With a positive misfit strain, this soft particle has to align its major axis parallel, instead of perpendicular, to the applied stress axis. Being misplaced, the particle must reorient itself. Instead of rotating on a high energy path, however, the precipitate undergoes splitting at first, then coalescence before reaching its equilibrium state. At the birth, the small platelets position themselves properly with the applied stress. During the follow-up coarsening, some smaller platelets sustain diffusional migration before their demise. If a precipitate begins a shape transition from a highly non-equilibrium state, such an evolution of splitting-coalescence is found to be a common feature (15).

To illuminate the effects of an applied stress, a schematic shape stability diagram is constructed in Fig. 9, where the alignment of both soft ($\mu^* = \mu/2$, gray) and hard ($\mu^* = 2\mu$, black) particles is shown under a uniaxial applied stress with a magnitude equal to

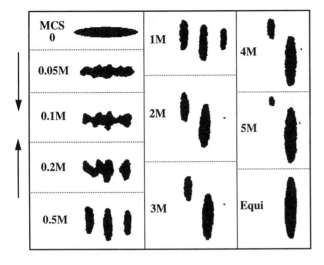

Figure 8: Morphological evolution of an isotropic, soft particle with R = 17a, ε = 0.05, and μ* = μ/2 under a compressive applied stress along the vertical direction.

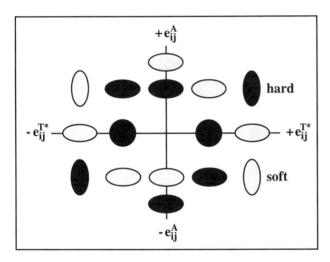

Figure 9: Shape stability diagram of coherent precipitates under a uniaxially-applied tensile or compressive stress. The horizontal axis indicates the sense of the misfit strain with + e_{ij}^{T*} for a positive misfit strain, while the vertical axis represents the sense of the applied stress with +e_{ij}^A for a tensile stress directed along the vertical axis. The applied stress magnitude is equivalent to 4% strain, and the stiffness is at the level of μ* = μ/2 for a soft (gray), and μ* = 2μ for a hard (black) particle.

4% strain. The positive x-axis, + e_{ij}^{T*}, indicates a positive misfit strain (5%), while a negative misfit (5%) is marked with -e_{ij}^{T*}. The positive ordinate, marked with +e_{ij}^A, stands for a tensile stress, whereas a compressive stress is indicated by -e_{ij}^A. Thus a hard particle with a positive misfit strain aligns its major axis parallel to the tensile stress axis as indicated in the 1st quadrant. If the hard particle has a negative misfit strain, however, its major axis becomes perpendicular to the stress axis (2nd quadrant). In the absence of applied stress, a hard particle commands a circular shape regardless of the misfit strain sign, as shown along the abscissa.

What makes such a drastic difference in the alignment? Following Eshelby (1), we may write the energy of the system containing an *elliptical* precipitate, Φ, as:

$$\Phi = \frac{1}{2}C^*_{ijkl}(e^{T^*}_{kl} - e^{c1}_{kl})e^{T^*}_{ij}V - \sigma^A_{ij}e^{T1}_{ij}V - \frac{1}{2}\sigma^A_{ij}e^{T2}_{ij}V + \gamma S \quad (1)$$

Here V and S are the area and perimeter of the elliptical precipitate, respectively, and γ is the interfacial energy per unit length. The constrained strain, e^{c1}_{ij}, is the product of Eshelby's tensor and an equivalent stress-free transformation strain, $S_{ijkl}e^{T1}_{kl}$, and is related to the misfit strain, $e^{T^*}_{ij}$, in the form of:

$$C^*_{ijkl}(e^{c1}_{kl} - e^{T^*}_{kl}) = C_{ijkl}(e^{c1}_{kl} - e^{T1}_{kl})$$

The second equivalent stress-free transformation strain, e^{T2}_{kl}, arises due to the applied stress, $\sigma^A_{ij} = C_{ijkl}e^A_{kl}$, and is obtained from the following relationship:

$$C^*_{ijkl}(e^A_{kl} + e^{c2}_{kl}) = C_{ijkl}(e^A_{kl} + e^{c2}_{kl} - e^{T2}_{kl})$$

The first term in Eq. (1) is the self strain energy due to the misfit strain, $e^{T^*}_{ij}$, and the second term is the interaction energy between the applied stress and the misfit strain. The third term indicates the inhomogeneity effect due to the applied stress, and the last represents the capillary effect. As pictured in the first quadrant in the stability diagram of Fig. 9, let us consider the shape change of a *hard* precipitate with ε = +0.05 and μ* = 2μ from a circle to an ellipse-like under a tensile stress. Let the aspect ratio of an ellipse be β = b/a, where a and b are the semi-axes along the horizontal and vertical direction, respectively. For a hard particle with a pure dilatation (1,2), the self strain energy prefers a circular shape. The inhomogeneity term increases slightly with β, but the interaction energy decreases more rapidly with β, and thus the sum of these elastic energy terms sets its minimum at β > 1. If the applied stress is large enough, the decrease in the interaction term can offset more than the increase in the summed energy of self strain energy, inhomogeneity term, and interfacial energy, thus leading the particle shape to an ellipse-like with β > 1. If the particle is *soft* (μ* = μ/2), on the other hand, the interaction energy decreases with decrease in β. Thus for a soft particle, all the elastic energy terms favor a shape with β → 0. Obviously, if the decrease in the elastic energy terms makes up more than the increase in the interfacial energy, the particle achieves an ellipse-like shape (β < 1) perpendicular to the applied stress axis.

The interaction behavior between an applied stress and a coherency strain is intriguing and certainly complicated as it depends not only on the sense of the various strain terms but also on their magnitude. Put in a simplified picture, it could be argued that a hard, elliptical particle accommodates more of its misfit strain along the direction of the major axis, whereas a soft particle settles more strain along its minor axis. Thus, in order to benefit a maximum interaction, a hard particle aligns its major axis parallel to the tensile axis, while a soft particle lines up its minor axis. More detailed descriptions will be reported elsewhere.

In Fig. 10, a soft particle with A = 2.33, $C^*_{ijkl} = C_{ijkl}/2$, R = 40a, and ε = 0.05 is placed under a tensile stress of 4% strain directed along the [010] direction. The elastic interaction between the applied stress and the coherency strain is strong enough to split the particle into two platelets at MCS = 7M. The two platelets, aligned along the soft [100] direction and perpendicular to the tensile stress axis, are quite stable, although some coarsening can be induced by further aging.

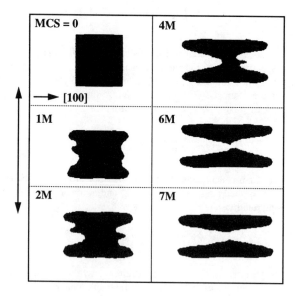

Figure 10: Splitting of a soft precipitate with its effective $R = 40a$ and $\varepsilon = 0.05$ under a tensile stress along [010] in an anisotropic system with $A = 2.33$.

In Fig. 11, two hard particles with $A = 2.33$, $C^*_{ijkl} = 2C_{ijkl}$, $R = 40a$, and $\varepsilon = 0.05$ are tested under an applied stress of 4% strain directed along the [010] direction: the left particle under a tension and the right one under a compression. As in the isotropic case, the interaction between the coherency strain and the applied stress polarizes the particle orientation in the opposite direction. Further, the particle under a compression is shown to be on the verge of splitting.

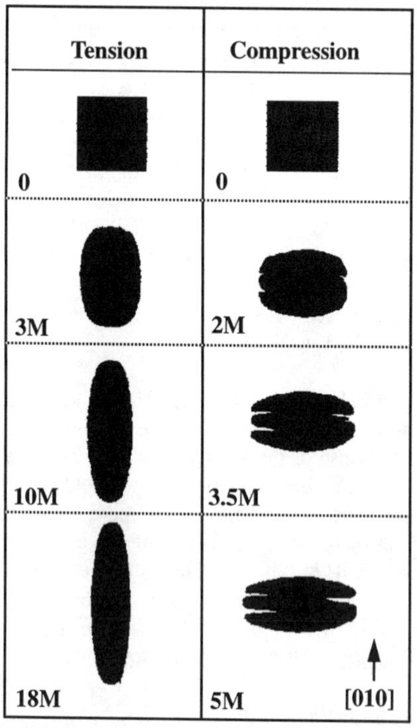

Let us now consider the effect of an applied stress on a multi-particle system. In Fig. 12, four, soft precipitates with $R = 20a$, $\varepsilon = 0.05$, and $C^*_{ijkl} = C_{ijkl}/2$ undergo coarsening in anisotropic media with $A = 2.33$. In the upper row, the particles experience a usual aging process stretching along the elastically soft <100> directions. In the bottom row, however, the same particles are placed under a tensile stress with a magnitude equal to 4% strain along the [010] direction. The elastic interaction between the applied stress and the coherency strain now polarizes the particle orientation only in the [100] direction, which is perpendicular to the applied stress axis. Such a raft structure would undoubtedly impede dislocation motions along the direction of the tensile stress, consequently promoting its creep properties.

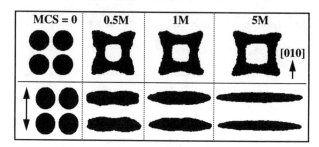

Figure 12: Shape evolution of four soft particles with $C^*_{ijkl} = C_{ijkl}/2$, $\varepsilon = 0.05$, and $A = 2.33$. In the upper row, the particles undergo coarsening without the influence of an applied stress. In the bottom row, they are placed under a tensile stress with a magnitude equal to 4% strain along the [010] direction.

The elastic stress fields of the four-particle system of Fig. 12 are examined in Fig. 13. In (a) and (b), the three stress components, σ_{xx}, σ_{yy}, and σ_{xy} (in units of $\mu\varepsilon$) are plotted as a function of distance (in units of a) along the [100] and [010] direction, respectively. Both directions pass through the center of the four particles at MCS = 0 and without an applied stress, $\sigma^A_{ij} = 0$. Both σ_{xx} in (a) and σ_{yy} field in (b) show tensile as well as compressive nature in the opening area between the particles. (c) and (d) are the counterparts of (a) and (b), respectively, for the case with a tensile applied stress along the [010] direction at MCS = 0: this is evidenced by the large σ_{yy} component. Significant changes are noticed in the σ_{xx} field. Finally, (e) and (f) are the stress profiles at MCS = 5M under the applied stress influence. Obviously, the particle reshaping has drastically changed the stress field. Note also that the plate-like particles still maintain a strong σ_{xx} component.

Interaction with a Free Surface

Thus far we have considered precipitates imbedded in an infinite matrix. A coherent particle is known to interact with a free surface (1,3). As a dislocation is attracted by an image force toward a free surface, so is a coherent precipitate. This is demonstrated in Fig. 14, where both soft ($\mu^* = \mu/2$) and hard ($\mu^* = 2\mu$) isotropic particles with $R = 25a$ and $\varepsilon = 0.05$ are initially surrounded by matrix phases of a finite size. The free surface is located at $R = 100a$ for the soft case and at $R = 50a$ for the hard case. In the simulations, the atomic layer of the free surface is kept to maintain matrix atoms, and consequently a reduction of the precipitate-matrix

Figure 11: Effect of applied stress along the [010] direction on hard particles with $C^*_{ijkl} = 2C_{ijkl}$ and $\varepsilon = 0.05$ in an anisotropic system with $A = 2.33$. The left column shows the influence of a tensile tress, while the right column displays that of a compressive stress.

interfacial energy is eliminated as a driving force for the migration. Consequently, at the end of the diffusional drift, the soft case shows that the interfacial energy increases 9% but the strain energy decreases 68%, resulting in 58% reduction in the total energy. Si-

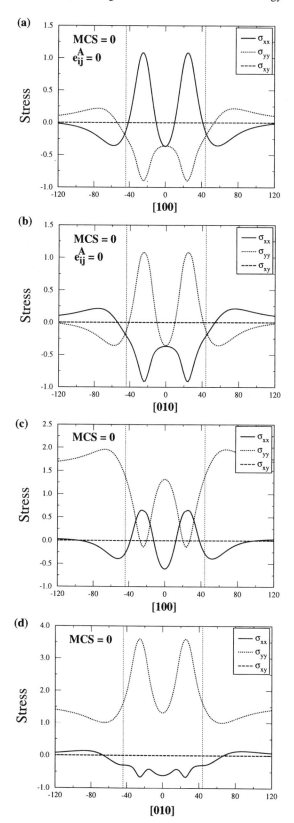

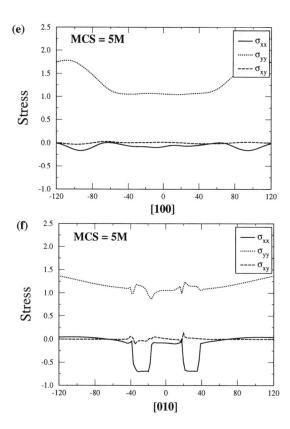

Figure 13: Stress field of the four-particle system passing through its center. Both (a) and (b) are for the case of zero applied stress at MCS = 0: (a) is along the [100] and (b) along the [010] direction. (c) and (d) are the counterparts of (a) and (b), respectively, for the case with a tensile applied stress along the [010] direction at MCS = 0, while (e) and (f) are the stress profiles at MCS = 5M under the applied stress influence.

milarly, the hard case records a 10% increase in the interfacial energy, 68% decease in the strain energy, and 60% reduction in the total energy. Before reaching the free surface, the circular soft particle sustains significant shape change. Though the effect may be weaker, small coherent particles should migrate toward high-angle grain boundaries.

Clustering

To mimic an aging process under the influence of coherency strain, clustering behavior is studied in Fig. 15. Solute atoms of 7 atom% are randomly distributed initially (MCS = 0), and then allowed to cluster through diffusion at 300 K. Elastic constants of the matrix phase are $C_{11} = 167$ k, $C_{12} = 108$ k, $C_{44} = 85$ k, and A = 2.85, while those of the precipitate phase are $C_{11} = 81$ k, $C_{12} = 54$ k, $C_{44} = 45$ k, and A = 3.33. A misfit strain of $\varepsilon = 0.01$ is given for a system of $200a \times 200a$ under periodic boundary conditions. As $C_{12} \neq C_{44}$, small volume-dependent terms are necessary to construct these lattices (34). In the left column, the interfacial energy of $\gamma = 4\gamma_0$ is relatively small, resulting in clusters of low aspect ratio. In the absence of misfit strain, the system undergoes a classic Ostwald ripening, yielding a circular shape at equilibrium, as shown in the inset with $\varepsilon = 0$. As the γ value increases from $4\gamma_0$ to $10\gamma_0$ (middle column) and then to $30\gamma_0$ (right column), the average cluster aspect ratio increases. In each sequence, shape bifurcation phenomena from circular to rectangular shapes are demonstrated.

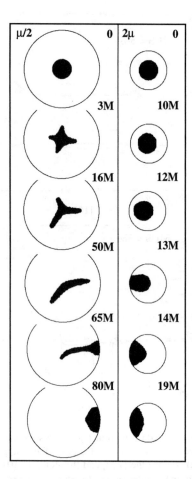

Figure 14: Migration of coherent particles due to an image force at a free surface. In the left column, a soft isotropic particle with $\mu^* = \mu/2$, $R = 25a$, and $\varepsilon = 0.05$ migrates to a free surface located at $R = 100a$, while a hard particle ($\mu^* = 2\mu$) drifts toward a free surface at $R = 50a$ in the right column.

Interaction with an Edge Dislocation

The core configuration of an [100] edge dislocation in a Hookean solid is first examined. In Fig. 16, the left figure with $A = 1$ portrays the atoms located within $5a$ from the central core atom marked with ⊥ in an isotropic solid, while the right one shows those in an anisotropic case with $A = 2.33$. Clearly visible is that the core size with $A = 2.33$ is smaller than that of the isotropic case, as the [100] direction is elastically soft in the anisotropic case. As one might have expected with a harmonic force model, dislocation core regions are relatively small. Note a bending of the crystal due to the dislocation introduction.

Let us examine how a coherent precipitate interact with an edge dislocation. In Fig. 17, a soft, circular particle with $R = 25a$, $\mu^* = \mu/2$ and $\varepsilon = 0.05$ contains initially an edge dislocation (marked with ⊥) at its center. The particle is then allowed, as before, for Monte Carlo diffusional relaxation. As the number of Monte Carlo steps increases, the precipitate migrates toward the region of tension, and reaches an equilibrium at MCS = 5.5M. Note that no dislocation climbing is allowed in this model. As expected, the elastic interaction between the positive misfit strain and the tensile stress field of the edge dislocation promotes such a migration. The particle morphology at MCS = 5.5M appears to follow an iso-stress

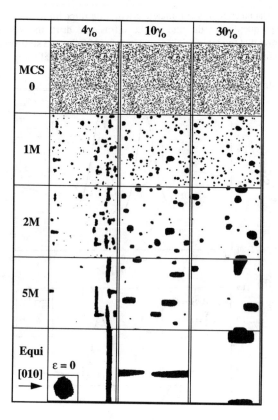

Figure 15: Clustering behavior in an anisotropic system with 7 solute atom% and $\varepsilon = 0.01$ under periodic boundary conditions.

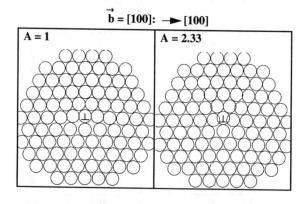

Figure 16: Atomic configuration around an edge dislocation in an isotropic ($A = 1$) and an anisotropic ($A = 2.33$) Hookean solid.

contour for the dislocation's σ_{xx} component (35), but it is also due to the softness of the particle under the influence of the dislocation tensile stress (see Fig. 3). Final configurations for the homogeneous case ($\mu^* = \mu$) and a hard particle ($\mu^* = 2\mu$) are given in Fig. 18. Also pictured is the soft particle case ($\mu^* = \mu/2$) with a negative misfit strain, which exhibits a mirror image to the MCS = 5.5M configuration of Fig. 17.

As compared to the isotropic case of Figs. 17 and 18, the particle shape in an anisotropic system maintains features of the cubic, anisotropic symmetry. In Fig. 19, particles with $A = 2.33$, $R = 25a$, and $\varepsilon = 0.05$ show their equilibrium morphology under the influence of an edge dislocation. Each particle follows a morphology as dictated by its stiffness, anisotropy, and interaction with the dislocation stress field.

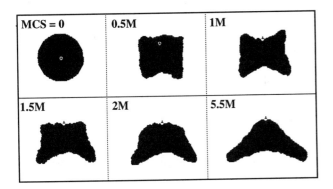

Figure 17: Shape evolution of a soft precipitate with $R = 25a$, $\mu^* = \mu/2$ and $\varepsilon = 0.05$ under the influence of an edge dislocation. The location of the edge dislocation is marked with $\perp$.

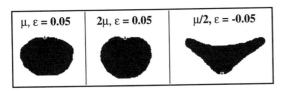

Figure 18: Equilibrium configuration of an isotropic particle under the influence of an edge dislocation. On the left, a homogeneous particle with a positive misfit strain, in the middle a hard particle, and on the right, a soft particle with a negative misfit strain are shown.

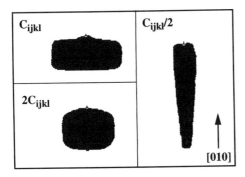

Figure 19: Equilibrium shape of an anisotropic particle with $A = 2.33$ under the influence of an edge dislocation. On the left, homogeneous ($C^*_{ijkl} = C_{ijkl}$) and hard ($C^*_{ijkl} = 2C_{ijkl}$) particles are displayed, and a soft particle ($C^*_{ijkl} = C_{ijkl}/2$) is shown on the right.

Even if they are perfectly matching with the matrix phase, i.e., $\varepsilon = 0$, precipitates interact with dislocations through their inhomogeneity effect. Soft particles would be attracted to the dislocation core region in order to reduce the dislocation energy, whereas hard particles would be rejected by dislocations. The role of the dislocation image force is demonstrated in Fig. 20, where isotropic particles with $R = 25a$ but $\varepsilon = 0$ contain edge dislocations initially. For the soft case with $\mu^* = \mu/2$, the dislocation is seen to remain at the particle center. For the hard case with $\mu^* = 2\mu$, however, the particle is shown to run away from the dislocation. This is the same phenomenon in which a dislocation slips away from a hard phase such as an oxide film.

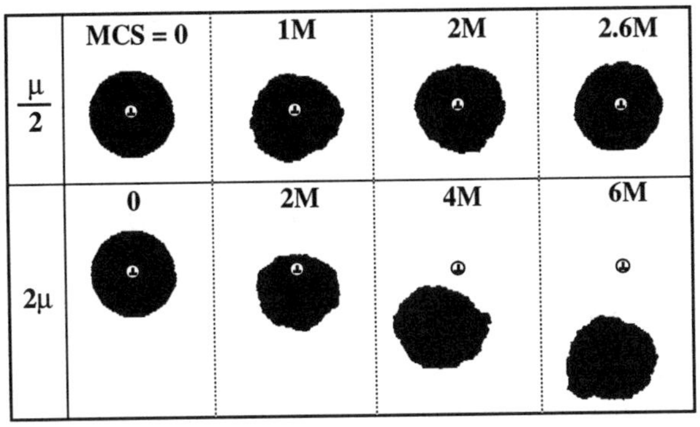

Figure 20: Role of a dislocation image force. In the top row, an edge dislocation remains at the center of a soft particle with $\mu^* = \mu/2$ and $\varepsilon = 0$, while in the bottom row, the hard particle with $\mu^* = 2\mu$ and $\varepsilon = 0$ wanders away from the dislocation.

Summary

In general, a soft precipitate imbedded in an infinite matrix tends to have an equilibrium morphology of low symmetry such as a plate shape, whereas a hard particle tends to take on a shape of high symmetry such as a circle. In an anisotropic system with comparable elastic constants between precipitate and matrix phase, however, the equilibrium morphology depends sensitively on the degree of anisotropy, size, misfit strain, and interfacial energy. Shape evolution undergoes through dynamic actions of coherency-induced interfacial waves and these waves seem to be responsible for the protrusions often observed along elastically hard directions in γ' particles of nickel-based superalloys. Coherent precipitates of finite misfit strain, regardless of their elastic stiffness, are attracted toward a free surface through an image force.

Under an applied tensile stress, soft particles with a positive misfit strain tend to become plates perpendicular to the applied stress axis, while hard particles elongate along the stress direction. If the elastic interaction between the applied stress and the coherency strain is strong enough, soft precipitates often split into smaller particles and then follow coarsening. If the applied stress increases further, coherent particles tend to dissolve into the matrix — in agreement with the theory of coherent phase equilibria (36-39). The applied stress lowers the relative chemical potential of solute atoms, which in turn increases their solubility. Elastic interaction of a misfitting particle with an edge dislocation shows no surprise: a particle with a positive misfit strain migrates to the tension region of the dislocation, whereas a particle with a negative strain diffuses to the region of compression. Morphological change is, however, caused by the dislocation as the particle tries to capitalize on the dislocation stress field.

The discrete atom method is still under development. Some aspects of transition from a coherent to a semi-coherent state were reported elsewhere (16), and the role of non-dilatational misfit strains such as Bain strain is just completed. Future work is planned including nucleation and growth of γ' precipitates and an extension to 3-dimensional elasticity problems.

Acknowledgment

The research was supported by the U.S. Dept. of Energy under Grant DE-FG02-87ER45315 and by a Research Excellence Fund from the State of Michigan, for which much appreciation is expressed.

References

1. J. D. Eshelby: *Prog. Solid Mech.*, 1961, vol. 2, p. 89.

2. J. K. Lee, D. M. Barnett, and H. I. Aaronson: *Metall. Trans. A*, 1977, vol. 8A, p. 963.

3. T. Mura: *Micromechanics of Defects in Solids*, 2nd ed., Martinus Nijhoff, Dordrecht, 1987, p. 177.

4. A. G. Khachaturyan: *Theory of Structural Transformations in Solids*, Wiley & Sons, New York, 1983, p. 213.

5. Y. Wang and A. G. Khachaturyan: *Acta Metall.*, 1995, vol. 43, p. 1837.

6. Y. Wang, L. Q. Chen and A. G. Khachaturyan: in *Solid → Solid Phase Transformations*, W. C. Johnson et al., eds., TMS, Warrendale, 1994, p. 245.

7. P. Fratzl and O. Penrose: *Acta Metall.*, 1995, vol. 43, p. 2921.

8. M. E. Thompson, C. S. Su and P. W. Voorhees: *Acta Metall.*, 1994, vol. 42, p. 2107.

9. Z. A. Moschovidis and T. Mura: *J. Appl. Mech.*, 1975, vol. 42, p. 847.

10. S. Satoh and W. C. Johnson: *Metall. Trans. A*, 1992, vol. 23A, p. 2761.

11. P. H. Leo and R. F. Sekerka: *Acta Metall.*, 1989, vol. 37, p. 3119.

12. J. Gayda and D. J. Srolovitz: *Acta Metall.*, 1989, vol. 37, p. 641.

13. J. K. Lee: *Metall. Trans. A*, 1991, vol. 22A, p. 1197.

14. J. K. Lee: *Scripta Metall.*, 1995, vol. 32, p. 559.

15. J. K. Lee: *Metall. Trans. A*, in press.

16. J. K. Lee: in *Micromechanics of Advanced Materials*, S. N. G. Chu et al., eds., TMS, Warrendale, 1995, p. 41.

17. J. K. Lee: in *Phase Transformations during the Thermal/Mechanical Processing of Steel*, E. B. Hawbolt and S. Yue, eds., The Metallurgical Society of CIM, Montreal, 1995, p. 49.

18. J. K. Lee: *Mat. Res. Soc. Symp. Proc.*, 1995, vol. 356, p. 63.

19. W. G. Hoover, W. T. Ashurst, and R. J. Olness: *J. Chem. Phys.*, 1974, vol. 60, p. 4043.

20. Y. S. Yoo, D. N. Yoon and M. F. Henry: *Metals and Mat.*, 1995, vol. 1, p. 47.

21. W. C. Johnson and J. W. Cahn: *Acta Metall.*, 1984, vol. 32, p. 1925.

22. S. J. Yeom, D. Y. Yoon, and M. F. Henry: *Metall. Trans. A*, 1993, vol. 24A, p. 1975.

23. A. Maheshwai and A. J. Ardell: *Scripta Metall.*, 1992, vol. 26, p. 347.

24. T. Miyazaki and M. Doi: *Mater. Sci. Eng.*, 1989, vol. A110, p. 175.

25. M. Meshkinpour and A. J. Ardell: *Mater. Sci. Eng.*, 1994, vol. A185, p. 153.

26. J. K. Tien and S. M. Copley: *Metall. Trans.*, 1971, vol. 2, p. 215.

27. T. Miyazaki, K. Nakamura and H. Mori: *J. Mater. Sci.*, 1979, vol. 14, p. 1827.

28. W. F. Hosford and S. P. Agrawal: *Metall. Trans. A*, 1975, vol. 6A, p. 487.

29. B. Skrotzki, E. A. Starke, Jr., and G. J. Shiflet: in *Microstructures and Mechanical Properties of Aging Materials*, P. K. Liaw et al., eds., TMS, Warrendale, in press.

30. A. Pineau: *Acta Metall.*, 1976, vol. 24, p. 559.

31. J. K. Lee and W. C. Johnson: in *Solid → Solid Phase Transformations*, H. I. Aaronson et al., eds., TMS, Warrendale, 1982, p. 127.

32. W. C. Johnson, M. B. Berkenpas and D. E. Laughlin: *Acta Metall.*, 1988, vol. 36, p. 3149.

33. S. Socrate and D. M. Parks: *Acta Metall.*, 1993, vol. 41, p. 2185.

34. M. I. Baskes and C. F. Melius: *Phys. Rev.*, 1979, vol. 20B, p. 3197.

35. J. P. Hirth and J. Lothe: *Theory of Dislocations*, McGraw-Hill, New York, 1968, p. 74.

36. R. O. Williams: *Metall. Trans.*, 1980, vol. 11A, p. 247.

37. J. W. Cahn and F. C. Larche: *Acta Metall.*, 1984, vol. 32, p. 1915.

38. W. C. Johnson and P. W. Voorhees: *Metall. Trans.*, 1987, vol. 18A, p. 1987.

39. J. K. Lee and W. Tao: *Acta Metall.*, 1994, vol. 42, p. 569.

MISFIT AND LATTICE PARAMETERS OF SINGLE CRYSTAL AM1 SUPERALLOY: EFFECTS OF TEMPERATURE, PRECIPITATE MORPHOLOGY AND γ-γ' INTERFACIAL STRESSES

A. Royer and P. Bastie

Laboratoire de spectrométrie physique
BP 87, 38402 St Martin d'Hères cedex, France

Abstract :

The lattice mismatch between the γ and γ' phases and the tetragonal distortion of the lattice cells of single crystal specimens of the nickel-based superalloy AM1 have been measured respectively by high resolution neutron diffraction and γ-ray diffraction techniques. Both temperature evolution on heating and time evolution during annealing of the diffraction profiles were analysed. The misfit depends on the precipitates morphology and is not an intrinsic property of the material. Tetragonal distortions of the γ phase have been evidenced and are mainly related to internal stresses at the γ-γ' interfaces. A transition from an elastic accommodation to a plastic relaxation was experimentally observed at the γ-γ' interfaces. To interpret the temperature evolution of the misfit and of the distortion of the cubic cells, a model is proposed for the evolution of the lattice parameters of the two phases. Similar description explains the temperature behaviour of the lattice parameters of the CMSX-2 superalloy showing that γ-γ' lattice mismatch is negative at room temperature.

Introduction :

The value and the sign of the lattice parameter mismatch δ between the γ and γ' phases of single crystal superalloys are often considered as comparison criteria between superalloy species. Their influence on the directional coarsening under stress at high temperature is very important as the shape and the size of the precipitates are strongly dependent on the misfit [1, 2]. However the determination of the lattice parameter mismatch usually defined as $\delta = 2(a_{\gamma'} - a_\gamma)/(a_\gamma + a_{\gamma'})$, has been the subject of numerous controversies and its measurement at high temperature (the using temperature of these alloys) is always an open and difficult question [3, 4, 5, 6, 7]. This is due to the facts that :
- δ is small ($\delta \approx 10^{-3}$)
- samples present segregation and chemical inhomogeneities even after standard heat treatments
- single crystals are not perfect (typical mosaicity of about 30 min of arc or more) [8, 9]
- internal stresses at the interfaces between matrix and precipitates modify the lattice parameters and distort the lattice cell. It is important to notice that difference must be done between lattice parameters of isolated phases and of the real biphasic materials [10]. Furthermore tetragonal distortion has been also reported for some superalloys [11, 12, 13].

In order to determine the origin of these controversies, systematic study of temperature dependence of the misfit and of the tetragonal distortion was undertaken. The influence of the thermomechanical history of the sample was also analysed. In situ experiments were performed in bulk samples using neutron and γ-ray radiations in order to avoid perturbation related to quenching, to oxidation or surface relaxation and also in order to average over dendritic and interdendritic regions.

Experimental procedure:

High resolution neutron diffraction experiments were carried out for determination of the lattice parameter distribution using a two nearly parallel crystal arrangement. This set up allows to minimize the instrumental contribution to the broadening of the diffraction peaks [14] and it becomes possible to operate at large Bragg angle ($\approx 80°$) in order to increase the sensitivity. Experiments were carried out on the Institut Laue Langevin (ILL, France) facility S21. Special cares and detailed procedure used in analysing the superalloy data are given in [15]. More details concerning the experimental conditions of the measurements reported in this paper are given in [7].

γ-ray diffractometry is usually used to measure the mosaicity of the sample [16]. In the particular case of single crystal superalloys a special feature occurs related to the cuboidal morphology of the precipitates. The full width at half maximum (FWHM) of the reflections (200) and (220) should have the same value if the lattice cells are cubic. The measured values are different and reflect a tetragonal distortion of the cells, the tetragonal axis being equivalently distributed along the three <h00> axes. The principle of these non standard measurements is fully described in [13]. The value of the tetragonal distortion is related to the difference 2ε = FWHM(220) - FWHM(200). In the present study measurements were performed on the γ-ray diffractometer of ILL with a radioactive gold source (λ = 0.03 Å).

Materials:

This paper focuses on AM1 single crystal superalloy. Similar studies have been done on CMSX-2 and are in agreement with results reported below. The composition of these two superalloys are given table I.

Table I: Chemical composition (in weight %) of AM1 and CMSX-2 superalloys.

element	Ni	Co	Cr	Mo	W	Al	Ti	Ta
AM1	balance	6.5	7.5	2.0	5.5	5.3	1.2	8.0
CMSX-2	balance	4.6	8.0	0.6	7.9	5.6	0.9	5.8

Two different types of samples were studied:
- crept sample (140 MPa, 1050°C, ε=0.58%) with rafted precipitates
- reference samples cut in an "as cast" bar. Samples selected (Ø=3mm, h =10 mm for γ-ray diffraction experiment and 10*15*3 mm^3 for neutron experiments) were chosen for their low mosaicity (typically 10 min of arc).

Measurements were performed
- on the crept sample for increasing temperature from room temperature up to 1300°C (temperature of complete solutionizing of the γ' phase), and then for decreasing temperature. Two different directions were analysed, parallel and perpendicular to the rafting plane.
- on reference samples with different initial thermal history
    - as cast
    - homogenized (1300°C/30min)
    - heat treated (as homogenized + 1050°C/16 h)

for increasing and decreasing temperature and during annealing at chosen temperatures between 1000°C and 1250°C after a complete solutionizing of the γ' phase (1300°C/30min).

γ-ray and neutron rocking curves were recorded in less than 20 minutes. This duration is short enough compared with the microstructure evolution of the superalloy.

Results:

Measurements at room temperature on the crept sample reveal a first surprise. The value of the misfit depends on the crystallographic direction. Its value is +3.3 10^{-3} along the <002> axis perpendicular to the rafts and is -1.4 10^{-3} along the <200> and <020> axis parallel to the rafts. For the reference sample the misfit value is the same for the three <h00> directions and is negative, close to zero in the case of the homogenized sample. These facts are sufficient to explain the controversies about the misfit values given in the literature. The value and even the sign of the misfit depend on the thermomechanical history of the sample. The same measurements were performed in temperature up to complete solutionizing of the γ' phase. Figure 1 reports the results for the crept sample (a) and the homogenized reference one (b). A detailed analysis of these results is given in [7].

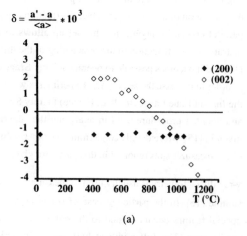
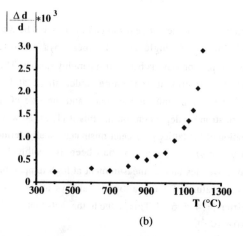

Figure 1: Temperature dependence of (a) lattice mismatch for the crept sample parallel (200) and perpendicular (002) to the rafts and (b) difference between the largest and the smallest lattice parameters for the homogenized reference sample

The most striking observation concerns the temperature behaviour of the misfit. It is hugely dependent on the morphology of γ' precipitates and on the crystallographic direction analysed. However we have shown that a crystallographic quantity almost independent of the history of the sample can be found in considering the value of the misfit averaged over the three cube directions:

$$\langle\delta\rangle = 1/3\,(\delta_{200} + \delta_{020} + \delta_{002}) \approx 1/3\,\Delta V/V$$

where $\Delta V/V$ is the relative change in unit cell volume between the two phases.

After solutionizing of the γ' phase at high temperature and precipitation, the anisotropy of the diffraction pattern for the crept sample is lost; it "has forgotten" its mechanical and thermal history, from the lattice parameter point of view.

γ-ray diffraction measurements were performed in the same temperature range. The FWHM of the (200) and (220) reflections are reported in figure 2 for the homogenized reference sample.

Below 900°C a constant difference 2ε of about 1 minute of arc is observed. For higher temperature this difference increases, reaches a maximum close to 3 minutes of arc at 1150°C and then decreases and disappears with the complete solutionizing of the γ' phase.

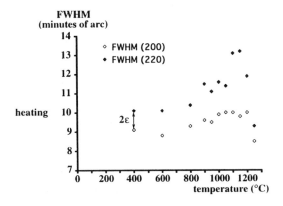

Figure 2: Behaviour versus temperature of the FWHM of the γ-ray rocking curves for the (200) and (220) reflections in the case of the homogenized reference sample

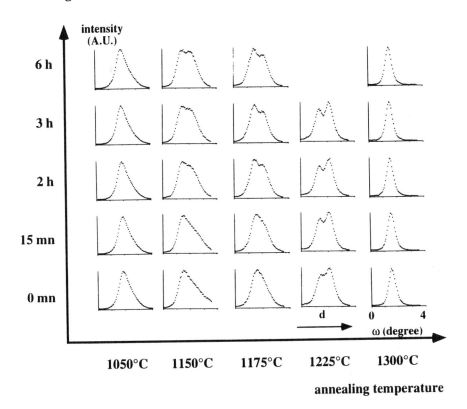

Figure 3: Time evolution of the neutron profile for annealing at different temperature

The behaviours of |Δd/d| and 2ε are similar up to 1150°C. These two quantities are strongly correlated through the coherent nature of the γ-γ' interface [4]. The different behaviours observed above 1150°C suggest a lost of the interface coherency. In order to confirm this hypothesis, evolutions of the neutron and γ-ray rocking curves were analysed during annealing at constant temperature. The sample was solutionized at 1300°C during 30 minutes and then quenched to the studied temperature. Figure 3 shows the time evolution of the neutron profile during 6 hours at different temperatures.

At 1050°C no significant evolution is observed during the experiment time. In opposite at 1150°C the diffraction profiles evolve strongly: at the beginning one peak is observed with a tail spreading over the larger lattice parameters. With annealing time this tail tends to disappear and the lattice parameter distribution focuses around two different values. This fact is reflected by the occurrence of two well defined peaks in the diffraction profiles.

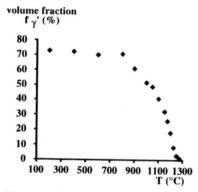

Figure 4: Temperature evolution of the γ' phase volume fraction of the AM1 superalloy

Above 1150°C two well split peaks are rapidly observed while at 1300°C only one peak corresponding to the γ phase diffraction is present. At 1175°C the higher peak corresponds to the lower lattice parameter value while it corresponds to the larger one at 1225°C. This confirms the negative sign of the misfit in accordance with the measured evolution of the γ' phase volume fraction for the AM1 superalloy as a function of temperature shown figure 4 [17] [1]. However it is important to notice that the evolution of the shape of the diffraction profiles is mainly related to a modification of the distribution of the lattice parameters around two main values rather than to a change of these two values. But because of the observation of a tetragonal distortion of the lattice cells by γ-ray diffractometry, cares have to be taken before to associate a diffraction peak to only one phase.

The FWHM of the (200) and (220) γ-ray reflections measured during annealing at 1050°C and 1150°C are reported figure 5. The difference 2ε shows the existence of a tetragonal distortion. Above this temperature this distortion is not evidenced in agreement with the result given by the figure 2. At 1150°C, 2ε which is about 2 minutes of arc at the beginning of the annealing tends to disappear after 10 hours. This duration is comparable with the annealing time after which the neutron profiles show two well defined peaks. Let us notice that this 2ε value is smaller than that obtained during the heating experiments on the homogenized sample. This difference is probably due to a different morphology of the γ' precipitates.

---

(1) The γ' phase volume fraction was measured on a bulk sample "in situ" from room temperature to 1300°C using neutron diffraction technique. It was shown that the temperature evolution of the γ' phase volume fraction is not very sensitive to the thermomechanical history of the sample.

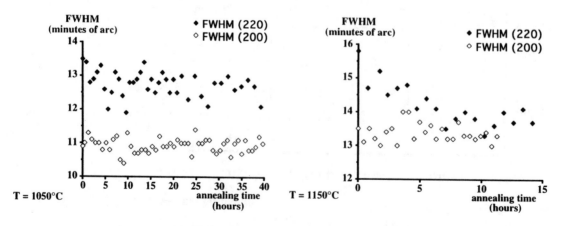

Figure 5: Behaviour of the FWHM of the γ-ray rocking curves for the (200) and (220) reflections during annealing at 1050°C and 1150°C

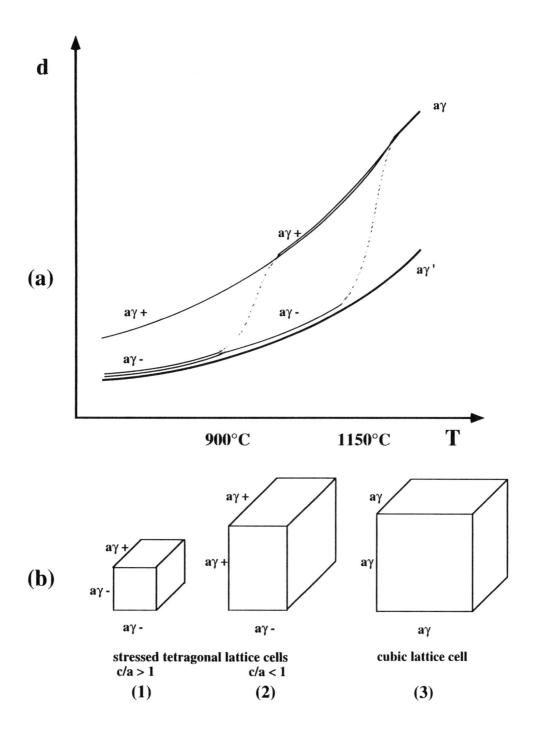

Figure 6: (a) Schematic view of the temperature evolution of lattice parameters of the γ and γ' phase
(b) Schematic view of the distortion of the lattice cells

The decreasing of the 2ε value also occurs at 1050°C but much slower as seen in figure 5. At this temperature the characteristic time would be of the order of few hundred hours. These observations show that the lattice parameter spreading and the tetragonal distortion are related to internal stresses due to the γ-γ' phase interfaces. During annealing a relaxation of these internal stresses occurs. Each phase tends to recover its cubic structure; the coherency of the interfaces disappears progressively.

Discussion and interpretation:

Due to the internal stresses at the γ-γ' interface, induced by the negative lattice mismatch, the cuboidal precipitates are triaxially expanded while the γ corridors are rather biaxially compressed due to their platelet shape. So in a first approximation, one can assume that the unit cell of the γ' phase remains cubic and that of the γ phase becomes tetragonal explaining the distortion observed by γ-ray diffractometry at room temperature. On heating, the volume of the unit cell of the γ phase increases faster than that of the γ' phase [7]; so the internal stresses increase too. The lattice coherency is maintained at the interface until a threshold temperature, between 900°C and 1000°C and no significant change is observed both on the diffraction profile and for the 2ε values. Above this threshold, the elastic accommodation is progressively replaced by a plastic relaxation provided by incoming dislocations in an attempt to reduce interfacial stresses. As shown by the annealing measurements the relaxation rate depends strongly on the temperature. So in fact when the rate is slow, this plastic relaxation does not occur simultaneously on the six faces of each precipitate but sequentially and at random. This explains the rapid increase of both |Δd/d| and the tetragonal distortion observed during the analysis of increasing temperature up to 1150°C, because some interfaces are relaxed and some others not. This relaxation becomes faster and faster when the annealing temperature becomes higher. This explains why above 1150°C the diffraction profiles show two well split peaks since the beginning of the annealing and why the cell distortion disappears as relaxation becomes rapidly complete in all directions. The two phases recover their cubic structure.

From these observations it becomes possible to explain the temperature behaviour of the misfit parameter on reference sample during heating. Figure 6 gives a schematic description of this qualitative model. It is based on a transition from purely elastic accommodation to a progressively plastic relaxation provided by the development of regular dislocation networks at the matrix-precipitate interfaces. For the clarity of the presentation, we assume that γ' phase kept always its cubic structure. This is not exactly true, in particular when only some interfaces are relaxed, but the major tetragonal distortion occurs in the γ phase. Furthermore it remains possible to add the γ' phase distortion for a more quantitative analysis of the neutron and γ-ray diffraction profiles.

On figure 6 (a) the temperature evolution of the lattice parameters of the γ and γ' phase is schematized. At low temperature (below 900°C) the interfaces are coherent so the γ phase is biaxially compressed by the bounding to the γ' phase. The two lattice parameters of the γ phase parallel to the interface are close to those of the γ' phase ($a\gamma-$). The third one expands because of the Poisson coefficient ($a\gamma+$). The cell of the γ phase is tetragonal with a c/a ratio larger than one (figure 6 (b1)). Between 900°C and 1150°C some of the interfaces become partially incoherent by plastic relaxation along one of the two stressed directions. The unit cell of the γ phase remains tetragonal but now only one direction is compressed and the c/a ratio becomes smaller than one (figure 6 (b2)). Both the Δd/d and the tetragonal distortion increase. Above 1150°C the plastic relaxation is complete, the unit cells of the γ and γ' phases recover their cubic structure. The tetragonal distortion disappears while Δd/d continue to increase (figure 6 (b3)). This interpretation is conforted by the observations made during the annealings. The progressive vanishing of the tetragonal distortion and the better resolution of the two peaks in the neutron diffraction profiles are in accordance with the plastic relaxation of interfacial stresses. Below 1150°C the relaxation time is large and only a partial relaxation is observed. This fact is due to the slowness of the process. For a given interface only one direction is relaxed (figure 6 (b)) leading to a large tetragonal distortion which decreases with annealing time. This temperature evolution of the lattice parameters explains why it is not possible to attribute a priori one diffraction peak to one phase.

Several other observations are consistent with this description. Indeed, the existence of partial relaxation at intermediate temperatures is compatible with the observation, after long time annealing, of periodic dislocation networks on some of the interfaces [18]. This may also explain the occurrence of the "bamboo" texture reported in the CMSX-2 after a heat treatment of 1000h at 1000°C [12], the oriented coalescence of the precipitates being favoured by the presence, locally, of a particular set of dislocations [19, 20].

Conclusion:

From these high resolution neutron diffraction and γ-ray diffraction measurements on reference and crept samples, the following remarks and conclusions can be drawn:

- The thermomechanical history and then the morphology of the γ' precipitates is an important parameter to be taken into account for measurements of lattice parameter mismatch. Mismatch measured only along one crystallographic direction does not provide intrinsic characteristic for a particular family of superalloy. But the "average" misfit for the three <h00> directions is almost independent of the morphology of precipitates and then can be chosen as a relevant parameter for a structural characterisation. For alloy AM1, it is close to zero at room temperature (slightly negative) and becomes negative on heating due to the larger thermal expansion of the γ phase compared to that of the γ' phase.

- A qualitative model explaining the temperature behaviour of the lattice parameters during heating and annealing is given. The transition from purely elastic accommodation to a progressively plastic relaxation provided by the development of regular dislocation networks at the matrix-precipitate interfaces has been demonstrated. The previously observed distortion of the cubic unit cells and the shape of neutron diffraction profiles are correctly interpreted by this model which is coherent with previous electron microscopy observations.

Annex:

Similar experiments made on the CMSX-2 single crystal superalloy have shown that the lattice parameters have the same behaviours as those of the AM1 superalloy. The same mechanism is involved for the stress relaxation. The average misfit measured is negative from room temperature to the temperature of the complete γ' phase solutionizing. Its value is $-1.4 \cdot 10^{-3}$ at room temperature and decreases to $-3.0 \cdot 10^{-3}$ at 1200°C. In opposition to the usually admitted misfit sign at room temperature, the measured value is negative and then there is no inversion of the misfit sign around 800°C. Properties of the material in this temperature range cannot be related to a null value of the misfit. This negative value at room temperature suggested in [7] has been confirmed by recent X-ray measurements performed on a Philips MRD diffractometer [21].

References:

1. T.M. Pollock and A.S. Argon, "Directional Coarsening in Nickel-Based Single Crystals with High Volume Fractions of Coherent Precipitates", Acta Metall. Mater., 42 (1994), 1859-1874

2. M. Veron, Y. Brechet and F. Louchet, "Directional Coarsening of Nickel Based Superalloys: Computer Simulation at Mesoscopic Scale", Acta Metall. Mater. (1996), in press

3. D.F. Lahrman et al., "Investigation of Techniques for Measuring Lattice Mismatch in a Rhenium Containing Nickel Base Superalloy", Acta Metall., 36 (1988), 1309-1320

4. H.-A. Kuhn et al., "An X-Ray Study of Creep-Deformation Induced Changes of the Lattice Mismatch in the γ' - Hardened Monocrystalline Nickel-Base Superalloy SRR 99", Acta Metall., 39 (1991), 2783-2794

5. U. Glatzel and A. Muller, "Neutron Scattering Experiments with a Nickel Base Superalloy. Part I: Material and Experiment", Scripta Metall. Mater., 31 (1994), 285-290.

6. U. Glatzel, "Neutron Scattering Experiments with a Nickel Base Superalloy. Part II: Analysis of Intensity Profiles", Scripta Metall. Mater., 31 (1994), 291-296

7. A. Royer et al., "Temperature Dependence of the Lattice Mismatch of the AM1 Superalloy. Influence of the γ' Precipitates Morphology", Phil. Mag. A, 72 (1995), 669-689,

8. D. Bellet, P. Bastie and J. Baruchel, "White Beam Synchrotron Topography and γ-ray Diffractometry Characterisation of the Crystalline Quality of Single Grain Superalloys: Influence of the Solidification Conditions", J. Phys. D: Appl. Phys., 26 (1993), A50-A52

9. N. Siredey et al., "Dendritic Growth and Crystalline Quality of Nickel-Base Single Grains", J. Crystal Growth, 130 (1993), 132-146

10. D. Grose and G. Ansell, "The Influence of Coherency Strain on the Elevated Temperature Tensile Behaviour of Ni-15Cr-Al-Ti-Mo Alloys", Met. Trans. A, 12 (1981), 1631-1645

11. A.J. Porter et al., "The application of Convergent-Beam Electron Diffraction to the Detection of Small Symmetry Changes Accompanying Phase Transformations, part II: Recrystallization of Superalloy", Phil. Mag. A, 44 (1981), 1135-1148

12. R. Bonnet and A. Ati, "Mise en Evidence par MET d'une Phase Ordonnée Légèrement Quadratique dans le Superalliage CMSX-2 Recuit", J. Microsc. Spectrosc. Electron., 14 (1989), 169-180, .

13. D. Bellet and P. Bastie, "Temperature Dependence of the Lattice Parameter of the γ and γ' Phases in the Nickel-Based Superalloy CMSX-2, Part I : Observation of a Tetragonal Distortion of the γ' Phase at High Temperature by γ-ray Diffractometry", Phil. Mag. B, 64 (1991), 135-141

14. A.H. Compton and S.K. Allison, "X-rays in Theory and Experiment", Van Nostrand Compagny Inc, Princeton, New Jersey, Toronto and London, p. 718, 1967.

15. D. Bellet and P. Bastie, "Temperature Dependence of the Lattice Parameter of the γ and γ' Phases in the Nickel-Based Superalloy CMSX-2, Part II : Neutron Diffraction Study of the Lattice Parameter Mismatch", Phil. Mag. B, 64 (1991), 143-152

16. J.R. Schneider, "A γ-ray Diffractometer: A Tool for Investigating Mosaic Structure", J. Appl. Cryst., 7 (1974), 541-546, ibid 7 (1974) 547-554

17. A. Royer et al., "Mesure par Diffraction Neutronique de la Fraction de Phase γ' dans le Superalliage Monocristallin AM1 entre 20 et 1300°C", Revue de Métallurgie, Science et Génie des

Matériaux, (1996), in press

18. A. Fredholm, "Monocristaux d'Alliages Base Nickel. Relation entre Composition, Microstructure et Comportement en Fluage à Haute Température", (PhD thesis, Ecole Nationale des Mines de Paris, 1987).

19. M. Veron, Y. Brechet and F. Louchet," Directional Coarsening of γ' Precipitates in Nickel Based Superalloys: Driving Force and Influence of a Prestrain", Phil. Mag. Letters, (1996), in press

20. M. Veron and P. Bastie," Strain Induced Directional Coarsening in Nickel Based Superalloy. Investigation on Kinetics Using SANS Technique", Scripta Metall. Mater. , submitted

21. O. Straudo and P. Bastie, private communication with authors, July 1994.

Acknowledgements:

This work was supported by the French CPR "SSSM" of CNRS. Samples were provided by SNECMA which is gratefully acknowledged for its interest to this study. Experiments were performed at the Institut Laue Langevin at Grenoble (France). We thank its staff and particularly P. Andant and P. Martin from the High Temperature Laboratory and R. Chagnon and P. Ledebt for their help during experiments. We are grateful to D. Bellet from LSP (Grenoble, France), C. Zeyen from ILL and J.L. Strudel from ENSMP (Evry, France) for fruitful discussions.

ELASTIC PROPERTIES AND DETERMINATION OF ELASTIC CONSTANTS OF NICKEL-BASE SUPERALLOYS BY A FREE-FREE BEAM TECHNIQUE

W. Hermann[*], H.G. Sockel[*], J. Han[**], and A. Bertram[***]

[*] Institut für Werkstoffwissenschaften, Lehrstuhl I, Universität Erlangen-Nürnberg, Martensstr. 5, D-91058 Erlangen, F.R. Germany,

[**] Bundesanstalt für Materialforschung und -prüfung, Unter den Eichen 87, D-12205 Berlin, F.R. Germany,

[***] Institut für Mechanik, Otto-von-Guericke-Universität, Universitätsplatz 2, D-39106 Magdeburg, F.R. Germany,

Abstract

The elastic properties of several monocrystalline and textured Nickel-base alloys were determined as a function of orientation in single crystals and of direction in textured materials by a dynamic resonance technique between 20°C and 1200°C. For the quantitative description of the elastic behavior of anisotropic solids the elastic single crystal constants, the texture, and the orientation are needed. The texture of polycrystalline materials is described by the orientation distribution function (ODF). In single crystals the orientation is given by the Eulerian angles. The determination of the elastic constants from single crystal measurements by a regression and a Finite Element (FE) method is shown in this paper to be very accurate. A new regression method allows the determination of single crystal elastic constants from strongly textured materials. This method is based on the measured elastic moduli in different directions with regard to the direction of rolling, growth or recrystallization. The evaluation of the elastic constants by regression and FE methods and the results for several Nickel-base superalloys are presented and discussed.

Introduction

Advanced industrial gas turbines must be operated at increasing high temperatures to improve the efficiency and to enhance power output. In order to meet the requirements for this application, turbine blades are manufactured from monocrystalline alloys or alloys with special textures, such as directionally solidified (DS) alloys having fibre textures, or directionally recrystallized (DR) alloys exhibiting sheet textures. With a decreasing number of transverse grain boundaries, improved thermal fatigue resistance and creep strength of these materials are achieved, allowing higher stresses and temperatures in service. Nickel-base superalloys with γ'-(Ni$_3$(Al,Ti)) precipitates, such as CMSX-4, CMSX-6, IN 738 LC or SRR 99 are of great interest for these applications.

For the design of statically and dynamically loaded components, the knowledge of the elastic moduli and their variations with temperature and orientation or direction, respectively, is a fundamental requirement, especially for the analyses of vibrations, strength, fracture, or stability of structures. The elastic moduli (Young's modulus E and shear modulus G) of anisotropic solids show a strong variation with orientation in single crystals and with direction in textured materials (Fig. 1 and 2). Therefore, it is necessary to describe the elastic behaviour of each anisotropic material by the determination of the orientation- or direction-independent elastic constants. Furtheron mathematical descriptions of the elastic properties as functions of the texture and the direction are needed. The direction-independent elastic constants (elastic single crystal constants or elastic material constants) are the elastic compliances S_{ij} and the elastic stiffnesses C_{ij} in Hooke's law. These constants can be determined with high accuracy by resonant frequency measurements with regression methods or with a newly developed FE analysis.

Materials

Out of the great number of γ'-precipitation hardened Ni-base alloys investigated by the authors, measurements on SRR 99, CMSX-4, CMSX-6 and IN 738 LC, and also on near-γ' Ni$_3$Al and on a precipitation-free near-γ matrix material are presented. The elemental compositions of these alloys are given in Table I.

The γ'-precipitation hardened Nickel-base alloys IN 738 LC, SRR 99, CMSX-4 and CMSX-6 were obtained as monocrystalline sheets. Several cylindrical specimens of about 4 - 5 mm in diameter and 40 - 50 mm in length were machined. The orientations of these single crystal specimens are distributed statistically in the standard stereographic triangle. In addition the alloy IN 738 LC was available in form of a directionally solidified turbine blade with a mean grain diameter of 4 mm, and a grain length larger than 100 mm. This DS material exhibits a strong <100>-fibre texture. Several specimens were machined in such a way that the angle between the specimen axis and the fibre axis was varied between 0° and 90° in steps of 15°.

Table I: Elemental composition of the investigated Nickel-base alloys in weight-%.

	Co	Ta	Cr	W	Al	Re	Ti	Mo	Hf	Fe	Nb	Ni
SRR 99	19.3	2.9	9.0	9.5	5.5	-	1.8	-	-	1.0	0.7	bal.
CMSX-4	9.5	6.5	6.4	6.3	5.7	2.9	1.0	0.6	0.1	0.1	0.1	bal.
CMSX-6	5.0	2.0	10.0	-	4.9	-	4.8	3.0	-	-	-	bal.
IN 738 LC	8.6	1.9	16.0	2.7	3.4	-	3.4	1.8	-	0.1	0.9	bal.
Matrix alloy	16.7	-	15.8	8.0	-	5.0	-	1.8	-	-	-	bal.

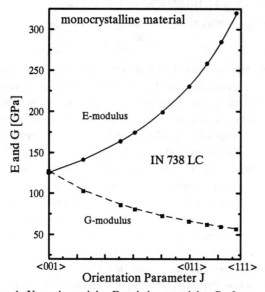

Figure 1: Young's modulus E and shear modulus G of monocrystalline IN 738 LC as a function of the orientation parameter J (equ. 8), describing the orientation.

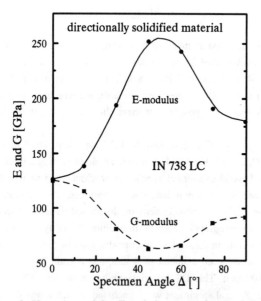

Figure 2: Young's modulus E and shear modulus G of directionally solidified IN 738 LC as a function of the angle Δ between the specimen axis and the direction of solidification.

Experimental

For the determination of the elastic properties from resonant frequency measurements a modified Förster resonance technique is used. Long thin rods with approximately 50 mm in length and 5 mm in diameter are excited to vibrations by piezo-electrical transducers and the resonant frequencies are measured (Fig. 3). The transducers are coupled to the specimen by suspension of the rod into carbon wire loops. By the Gain-Phase Analyzer the amplitude and the phase angle are measured as a function of the applied frequency. In the case of resonance, a maximum in amplitude and a characteristic phase shift are measured. The measurements were carried out in vacuum in the 4 to 150 kHz frequency range between 20°C and 1250°C. The temperature was controlled by a Pt/PtRh-thermocouple located 1 mm away from the middle of the specimen.

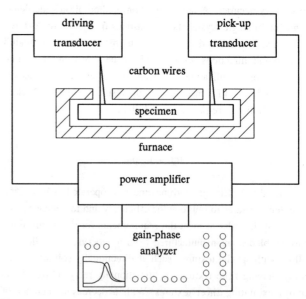

Figure 3: Schematic representation of the experimental equipment for measuring the resonant frequencies.

Three vibrational modes, the flexural, longitudinal and torsional, are excited in the fundamental vibrations and in several overtones. Flexural and longitudinal vibrations supply information about the Young's modulus E(T), and the torsional vibrations about the shear modulus G(T). From the measured resonant frequencies the Young's modulus and the shear modulus are calculated via the following theoretical relationships /1,2,3/:

$$E(T) = \frac{4 \cdot \pi^2 \cdot \rho(T) \cdot l^4(T) \cdot f_{n,flex}^2(T)}{\chi^2(T) \cdot (m_n)^4} \cdot K_b, \quad (1)$$

$$G(T) = 4 \cdot \rho(T) \cdot l^2(T) \cdot \left(f_{n,tors}^2(T)/n^2\right) \cdot K_t, \quad (2)$$

where $\rho(T)$ is the density, $l(T)$ the length, $\chi(T)$ the radius of inertia, n the order of vibration, and $f_n(T)$ the resonant frequency of the n-th mode of vibration of flexure or torsion. K_b, K_t, and m_n are vibration mode dependent correction factors, tabulated in /1/.

For measurements at temperatures above 20°C the specimen dimension and the density had to be corrected by the coefficient of thermal expansion.

Elastic Behaviour of Anisotropic Solids with Cubic Crystal Structure

Single Crystals

The proportionality of stress and strain is described by the two forms of the generalized Hooke's law. It relates the stress vector σ to the strain vector ε by the equations /4/

$$\sigma_i = \sum C_{ij} \cdot \varepsilon_j, \quad (3)$$

$$\varepsilon_i = \sum S_{ij} \cdot \sigma_j, \quad (4)$$

where C_{ij} is the stiffness matrix and S_{ij} is the compliance matrix. The matrices C_{ij} and S_{ij} have in general twenty-one independent components with respect to an arbitrarily selected coordinate system, but this number can be reduced drastically in the presence of crystal symmetries. For instance, single crystals with cubic, hexagonal and orthorhombic symmetries have three, five and nine independent elastic stiffnesses C_{ij} (compliances S_{ij}), respectively. In the case of cubic crystal symmetry the three independent elastic single crystal constants are S_{11}, S_{12} and S_{44} or C_{11}, C_{12} and C_{44}. The stiffnesses C_{ij} can be calculated from the compliances S_{ij} and vice versa /5/.

If the single crystal specimens are long and thin rods, then orientation-dependent elastic moduli are connected to the compliances by the following equations /4,6/:

$$E_{mes}(\theta,\varphi) = [S_{11} - 2SJ]^{-1}, \quad (5)$$

$$G_{mes}(\theta,\varphi) = [(S_{44} - 4SJ) \cdot (1-\delta)]^{-1}, \quad (6)$$

where $E_{mes}(\theta,\varphi)$ and $G_{mes}(\theta,\varphi)$ are the measured Young's and shear modulus for a given orientation, θ and φ are the Eulerian angles according to Fig. 4. The factor δ corrects coupling effects between torsion and bending /6,7/. S and the orientation parameter J are given by the following equations:

$$S = S_{11} - S_{12} - S_{44}/2, \quad (7)$$

$$J = \sin^2\theta \cdot \cos^2\theta + \frac{1}{8}\sin^4\theta \cdot (1-\cos(4\varphi)). \quad (8)$$

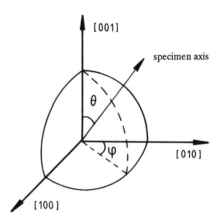

Figure 4: Definition of the Eulerian angles θ and φ.

Textured Materials

A description of the elastic behavior of a textured material is based on a consideration of the symmetry of the whole polycrystalline aggregate. Fibre textured materials, e.g. directionally solidified Nickel-base alloys, exhibit hexagonal symmetry, while rolled sheets can be described by assuming an orthorhombic symmetry. Hence, the elastic behavior of these textured materials is determined by five or nine elastic polycrystal constants S_{ij}^{pc} and C_{ij}^{pc} in the cases of hexagonal and orthorhombic symmetry, respectively. Then Hooke's law can be expressed by equations (9) and (10), where the polycrystal constants of the textured material correlate the mean stresses $\overline{\sigma}$ with the mean strains $\overline{\varepsilon}$. S_{ij}^V and C_{ij}^V are the polycrystal constants according to Voigt /4/ and Reuss /8/, respectively.

For the condition of constant strain in all grains an upper bound of the elastic properties according to Voigt is obtained, while for the condition of constant stress in all grains one gets a lower bound of the elastic properties according to Reuss:

$$\overline{\sigma}_i = \sum C_{ij}^V \cdot \varepsilon_j, \quad (9)$$

$$\overline{\varepsilon}_i = \sum S_{ij}^R \cdot \sigma_j. \quad (10)$$

The mean values of these upper and lower bounds given by

$$S_{ij}^H = \tfrac{1}{2}\left[S_{ij}^R + S_{ij}^V\right] \quad (11)$$

according to Hill /9/ are very close to the real elastic properties of the textured material. S_{ij}^V and C_{ij}^V are connected via $S_{ij}^V = [C_{ij}^V]^{-1}$.

These polycrystal constants are related to the single crystal constants by the orientation distribution function (ODF) of the texture by equations given in the appendix (A1 and A2) for hexagonal texture symmetry /10,11/. The relations between the direction-dependent elastic moduli E and G and the elastic polycrystal constants S_{ij}^{pc} are given by equations (12) and (13) for hexagonal symmetries as in the fibre textures:

$$E(\theta) = \left[S_{11}^{pc}\sin^4\theta + S_{33}^{pc}\cos^4\theta + \left(2S_{13}^{pc} + S_{44}^{pc}\right)\sin^2\theta \cdot \cos^2\theta\right]^{-1}, \quad (12)$$

$$G(\theta) = \left[S_{44}^{pc} + \left(S_{11}^{pc} - S_{12}^{pc} - (S_{44}^{pc}/2)\right)\sin^2\theta + 2\left(S_{11}^{pc} + S_{33}^{pc} - 2S_{13}^{pc} - S_{44}^{pc}\right)\cos^2\theta \sin^2\theta\right]^{-1} \cdot [1-\kappa]^{-1}, \quad (13)$$

where Δ denotes the angle between the specimen axis and the fibre axis.

Using the equations (12) and (13) with the polycrystal constants according to Voigt, Reuss or Hill we obtain upper and lower bounds of the elastic moduli or the Hill mean values. Fig. 5 shows the dependence of Young's and shear modulus versus the angle Δ between the specimen axis and the fibre axis of directionally solidified IN 738 LC. The Voigt-Reuss-Hill values calculated according to eqns. (12), (13), (A1), (A2) are compared with the measured ones. The Hill mean values, based on the elastic constants determined from single crystal measurements and the ODF show a good agreement with the measurement data.

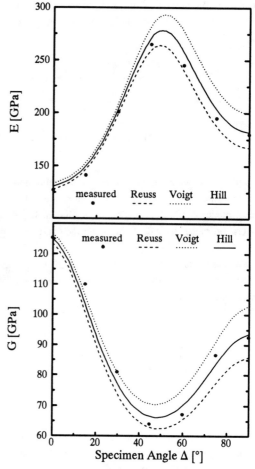

Figure 5: Measured Young's and shear modulus and calculated Voigt-Reuss-Hill values for IN 738 LC DS, as a function of the angle Δ between the fibre axis and the specimen axis.

Determination of Elastic Constants

As mentioned above the knowledge of the elastic single crystal constants is very important to describe the elastic behavior of anisotropic solids. For this reason several methods have been developed to determine the elastic constants from measured resonant frequencies. In the following the determination of the elastic constants from resonance measurements of single crystals and of textured materials will be presented. Two different methods are used, based on regression and on Finite Elements (FE), respectively.

Single Crystals

The regression method for cubic single crystals, based on the equations (5) to (8), is the most simple way to determine the elastic compliances. According to them, the reciprocal values of the measured moduli, $(E_{mes})^{-1}$ and $(G_{mes}\cdot(1-\delta))^{-1}$, of several specimens, each with different orientation, are plotted versus the orientation parameter J, given in equation (8), (Fig. 6). A linear regression with respect to J=0 leads to the elastic constants S_{11} and S_{44}, while S_{12} can be calculated by equation (7) from the slope of the regression lines. The slope of the regression lines from measured reciprocal elastic moduli versus J is equal to -2S and 4S, respectively.

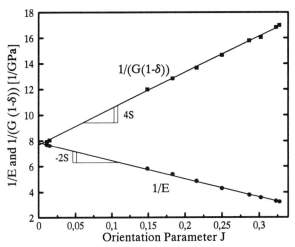

Figure 6: Determination of S_{11}, S_{44} and S_{12} by linear regression from single crystal measurements for the alloy CMSX-4.

Using the Finite Element Method (FEM), the resonant frequencies of a cylindrical specimen are calculated by an FE-model for given elastic constants. The specimen is modelled by 16 isoparametric three-dimensional 20-node-elements /12/. The desired resonant frequencies are obtained from solutions of the following corresponding eigenvalue equations:

$$(-\omega_i^2 M + K) \cdot u_i = 0; \quad i = 1,2,... \quad , \quad (14)$$

where ω_i and u_i are the i-th natural frequency and mode, M and K are the mass and stiffness matrices of the FE-model, respectively.

The strategy for the determination of the elastic constants and the orientation is the adjustment of these constants in an optimization procedure, where the differences between the measured and the calculated frequencies of several specimens are minimized. The eigenvalue equations (14) are solved by the FEM code ADINA /13/. The optimizer is from MINIPACK /14/. It uses a modified Levenberg-Marquard algorithm. In every iteration step of the optimization procedure, the calculated frequencies must be assigned to the measured ones. This is carried out by the use of a frequency assignment algorithm which is based on the orthogonality of the natural modes. In the optimization procedure also the orientation of the specimens can be determined beside the elastic constants. The details about the frequency assignment algorithm and the numerical simulation of the dependence of the resonant frequencies on the orientation and on the elastic constants are given in reference /15,16/.

Textured material

Problems can arise in obtaining the elastic constants of the single crystal in the cases where the latter is difficult to grow, expensive or cannot be prepared at all, as for instance in the case of ODS-alloys. In all these cases the elastic constants of the single crystal have to be determined from measurements on textured polycrystalline material. This determination is only possible if the polycrystalline material exhibits elastic properties which depend on the direction. These properties have to be measured in several directions with regard to the direction of rolling, growth or recrystallization. From such experimental results the elastic constants of the single crystal can be calculated by connecting the properties of the textured polycrystalline and the monocrystalline material as shown above. Using the concept of Hill, a new empirical evaluation method allows the determination of the elastic single crystal constants from measurements on textured material.

This method is formally based on the relations between the elastic properties and the orientation parameter J for a single crystal of cubic crystal structure /17/. In the following this method is presented for fibre textured material. From the input data, which are the coefficient of the ODF and approximately assumed elastic constants S_{11}, S_{12} and S_{44}, the five elastic constants S_{ij}^H for hexagonal symmetry are calculated by equation (12), (13), (A1) and (A2). Then these are used for the determination of theoretical values $E(\Delta)$ and $G(\Delta)$, where Δ denotes the specimen angle as introduced above. In the next step these values are introduced into the equations (15) and (16) for the determination of theoretical orientation parameters $J_e(\Delta)$ and $J_g(\Delta)$:

$$1/E(\Delta) = S_{11} - 2SJ_e(\Delta) , \quad (15)$$

$$1/[G(\Delta)(1-\delta)] = S_{44} + 4SJ_g(\Delta) . \quad (16)$$

After the calculation of $J_e(\Delta)$ and $J_g(\Delta)$ the measured moduli are introduced in the equations (15) and (16). A linear regression with respect to $J_e(\Delta)$ and $J_g(\Delta)$ leads to the compliances of the single crystal, analogous to the regression in the case of cubic single crystals. Using these single crystal constants as new input data for the next iteration step, improved values of $J_e(\Delta)$ and $J_g(\Delta)$ can be determined as described above. It is necessary to repeat this procedure until there is no further significant change in the calculated constants from the (n-1)-th and n-th cyclus. In practice not more than five iteration steps are necessary.

The Finite Element Method can also be applied to sheet or fibre textured material. According to the regression method it is assumed that textured material macroscopically exhibits a hexagonal or orthorhombic symmetry. In these cases five or nine elastic constants must be adjusted /18,19/. In the FE-model of the specimen the exact macroscopic constitutive law for the elastic anisotropic material is used. This method can be regarded as a suitable averaging method which does not use any restrictions on the distributions of the stresses or the strains and any prior knowl-

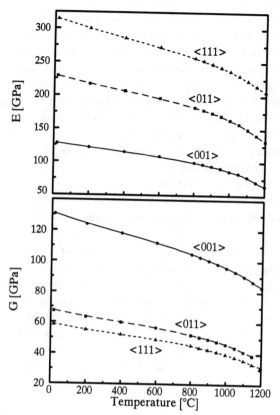

Figure 7: Young's and shear modulus for three different monocrystalline specimens of the alloy CMSX-4.

edge about the orientation distribution of a textured alloy. Unfortunately, the successful determination of the elastic single crystal constants can be carried out only in the case of single crystals or strongly fibre textured materials. In the other cases only the elastic polycrystal constants depending on the texture can be determined.

Results and Discussion

Elastic Moduli

The elastic moduli of several monocrystalline specimens with different orientations of the alloys SRR 99, CMSX-6, CMSX-4 and IN 738 LC were determined between 20°C and 1200°C. As an example Figure 7 shows the measured Young's and shear moduli in <001>-, <011>- and <111>-direction as a function of the temperature for the alloy CMSX-4. Similar orientation dependencies of the elastic moduli were observed for all other monocrystalline Ni-base alloys.

Young's and shear modulus in Figure 7 show a linear decrease with temperature up to about 900°C and a very strong decrease between 900°C and 1100°C. Extensive investigations of the damping behaviour of the alloy CMSX-4 reveals two different damping maxima in the temperature range in which the strong decrease of the elastic moduli occurs /20/ (Fig. 8).

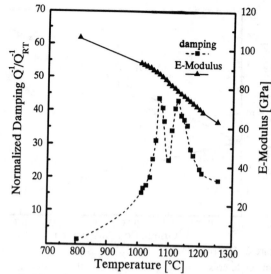

Figure 8: Young's modulus and normalized damping for the monocrystalline alloy CMSX-4.

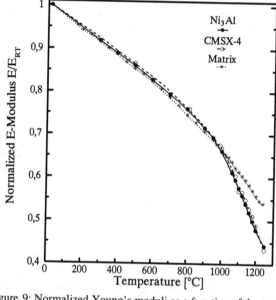

Figure 9: Normalized Young's moduli as a function of the temperature for CMSX-4, Ni$_3$Al and the matrix material.

Further investigations performed on precipitation-free material, which is near to the composition of the γ-matrix of CMSX-4 /21/, and on ordered Ni$_3$Al, which is near to the composition of the γ'-precipitates, leads to the assumption, that these damping maxima are caused by diffusion processes in the γ'-phase /20/.

From Figure 9 it is clearly visible, that the temperature dependence of the normalized Young's moduli of CMSX-4 and Ni$_3$Al agree very well, while for the γ'-free matrix material no stronger modulus decrease at higher temperatures was found. This indicates that the stronger modulus decrease is caused by the γ'-phase. Further details about the damping behaviour of Ni-base alloys are published in /20/.

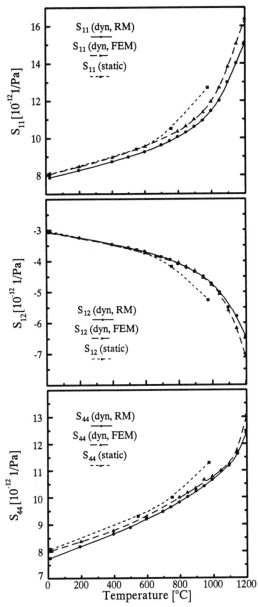

Figure 10: Elastic compliances S_{11}, S_{12} and S_{44} for the Ni-base alloy SRR 99.

<u>Elastic Constants</u>

The elastic compliances S_{ij} of the alloys CMSX-4, CMSX-6 and SRR 99 were determined from the measured resonant frequencies of several monocrystalline specimens by the regression- and the FE-method. In order to test the accuracy of these methods the evaluated constants S_{11}, S_{12} and S_{44} of the alloy SRR 99 are compared in Figure 10. In addition the data obtained from tensile tests up to 980°C are also included in Figure 10. The comparison of the constants S_{ij} of different independent origins shows good agreement between them. The deviations lie in the range of 5% percent. Only at higher temperatures the constants from static tests show deviations up to 10%, which is probably due to measurement effects.

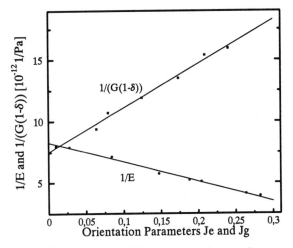

Figure 11: Determination of the elastic single crystal constants from measurements for fibre textured material of IN 738 LC.

In order to determine the elastic single crystal constants from textured material measurements according to the regression method described above, <100>-fibre textured IN 738 LC was investigated. In Figure 11 the experimentally determined $[E(\Delta)]^{-1}$ and $[G(\Delta)(1-\delta)]^{-1}$ of the textured material at room temperature are plotted versus the calculated $J_e(\Delta)$ and $J_g(\Delta)$, respectively. The plotted properties exhibit a good linear dependence on $J_e(\Delta)$ or $J_g(\Delta)$, respectively.

Figure 12 shows a comparison of elastic compliances S_{11}, S_{12} and S_{44} of IN 738 LC determined from <100>-fibre texture and from single crystal measurements. The comparison of the constants S_{ij} of different independent origins shows good agreement between them. The deviations lie in the range of 5% percent. This result confirms the high accuracy of the S_{ij} values obtained from the investigation of textured materials by the new regression method.

In Figure 13 the elastic compliances of the alloys CMSX-4, CMSX-6, SRR 99 and IN 738 are shown as a function of the temperature. The constants of these four alloys are very similar, with deviations lieing in the range of 10%. Hence, the elastic constants of most of the Ni-base alloys can be regarded as nearly identical.

<u>Conclusions</u>

The elastic moduli (Young's modulus E and shear modulus G) of anisotropic materials exhibit a strong variation with orientation, direction and temperature. They were measured by a modified Förster resonance method between 25°C and 1200°C. At temperatures between 900°C and 1100°C the elastic moduli of all investigated Ni-Base alloys show a strong decrease, which is most probably caused by diffusion processes in the γ'-precipitation phase.

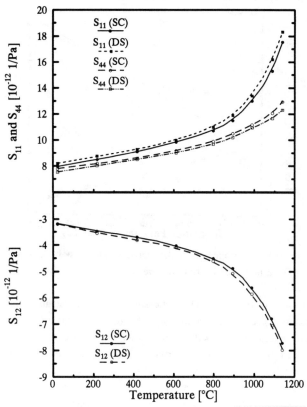

Figure 12: Elastic constants S_{11}, S_{44} and S_{12} for IN 738 LC, determinated from single crystal and DS-material measurements.

In order to describe the elastic behavior of an anisotropic material the direction-independent elastic single crystal constants (compliances S_{ij} or stiffnesses C_{ij}) must be known. In the case of textured material the elastic polycrystal constants (S_{ij}^{pc} and C_{ij}^{pc}) are needed. They can be calculated from the single crystal constants and the orientation distribution function (ODF).

The determination of the elastic constants from resonance measurements of single crystals and textured materials by regression and by Finite Element methods were presented. All methods require measurements of the elastic properties on specimens prepared with different orientations of single crystals or with their axes at different angles to the direction of solidification, recrystallization or rolling in textured materials. These methods were applied to monocrystalline and textured Nickel-base alloys (SRR 99, CMSX-6, CMSX-4, IN 738 LC). The good agreement of the results from different methods and materials with different anisotropic behaviour demonstrate convincingly the efficiency and the usefulness of the regression and FE methods for the determination of the elastic single crystal constants.

Experiences with these methods for different materials show that a strong direction dependence of the elastic properties in the textured materials is a fundamental requirement for a successful application of the presented methods.

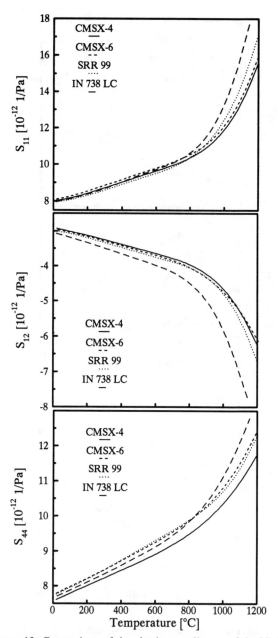

Figure 13: Comparison of the elastic compliances of the Nickel-base alloys CMSX-4, CMSX-6, SRR 99 and IN 738 LC.

Acknowledgements

The authors are grateful to the Forschungsgemeinschaft Verbrennungskraftmaschinen (FVV), the Arbeitsgemeinschaft Industrieller Forschungsvereinigungen (AIF) and the Deutsche Forschungsgemeinschaft (DFG) for the financial support of this work.

References

1. E. Goens, "Über die Bestimmung des Elastizitätsmoduls von Stäben mit Hilfe von Biegeschwingungen," Annalen der Physik 11 (1931), 6, 649-678.

2. J. Spinner and W.E. Tefft, "A method for determinating mechanical resonance frequencies and for calculating elastic moduli from these frequencies," Proc. ASTM 61 (1961), 1229-1238.

3. G. Pickett, "Equations for computing elastic constants from flexural and torsional resonant frequencies of vibration of prisms and cylinders," Proc. ASTM 45 (1945), 846-865.

4. W. Voigt, Lehrbuch der Kristallphysik (Leipzig: Teubner, 1928).

5. J.F. Nye, Physical Properties of Crystals, Oxford University Press, New York, 1960.

6. R.F.S. Hearmon, "The elastic constants of anisotropic materials," Rev. Modern Physics, 18 (1946), 409-440.

7. E. Goens, "Über die Biegungs- und Drillungsschwingungen eines dünnen kreiszylindrischen Kristallstabes von beliebiger kristallographischer Orientierung," Annalen der Physik, 15 (1932), 455-484.

8. A. Reuss, "Berechnung der Fließgrenze von Mischkristallen auf Grund der Plastizitätsbedingungen für Einkristale", Zeitschrift für angewandte Mathematik und Mechanik, 9 (1929), 49-58.

9. R. Hill, "The elastic behavior of a crystalline aggregate", Proc. Phys. Soc., A65 (1952), 349-354.

10. H.-J. Bunge, Mathematische Methoden der Texturanalyse (Berlin: Akademie Verlag, 1969).

11. R.A. Adamesku et al., "Invarianten der Anisotropie elastischer Eigenschaften von texturierten kubischen Metallen," Zeitschrift für Metallkunde, 76 (1985), 11, 747-749.

12. A. Bertram et al., "Bestimmung der elastischen Konstanten anisotroper Festkörper mittels FE-Simulation der Eigenschwingungen," Zeitschrift für angewandte Mathematik und Mechanik, 70 (1990), 322-323.

13. ADINA Users Manual, ADINA R&D, Inc., 71 Elton Avenue Watertown, MA 02172 USA, 1990.

14. J.J. Moré, B.S. Garbow and K.E. Hillstrom, "User Guide for MINPACK-1," (Report ANL-80-74, Argonne National Laboratory, Argonne, IL, 1980).

15. A. Bertram et al., "Identification of Elastic Constants and Orientation of Single Crystals by Resonance Measurements and FE-Analysis," Special Issue of Int. Journal of Computer Applications in Technology, 7 (1994), 3/4, 285-292.

16. J. Han et al., "Identification of Crystal orientation by Resonance Measurements, Zeitschrift für angewandte Mathematik und Mechanik, 74 (1994), 4, 138-143.

17. U. Bayerlein and H.-G. Sockel, "Determination of single crystal elastic constants from DS- and DR-Ni-based superalloys by a new regression method between 20°C and 1200°C," In: Superalloys 1992, Proc. of the Seventh Internat. Symposium on Superalloys, S.D. Antolovich et al. (eds.), TMS, Warrendale, PA., USA, 1992, 599-608.

18. J. Han et al., "Identification of Elastic Constants of Directionally Solidified Superalloys based on Resonance Measurement and FE-Analysis," In: Procceed. of the Academia Sinica Conference on Scientific and Engineering Computing, Beijing, China, Aug. 16-21, 1992 (1993).

19. J. Han et al., "Identification of Elastic Constants of Alloys with Sheet and Fibre Textures Based on Resonance Measurements and FE-Analysis," Mat. Sci. Eng.A, 191 (1995), 105-111.

20. W. Hermann and H.-G. Sockel, "Investigation of the High-Temperature Damping of the Nickel-Base Superalloy CMSX-4 in the kHz-Range," (Paper presented at the International Symposium on M^{3}D III: Mechanics and Mechanisms of Material Damping, Norfolk, Virginia, USA, 15.-17.11.1995).

21. V. Sass, W. Schneider and H. Mughrabi, "On the Orientation Dependence of the Intermediate-Temperature Creep Behaviour of a Monocrystalline Nickel-Base Superalloy," Scripta Metallurgica et Materialia, 31, (1994), 7, 885-890.

Appendix

Calculation of the elastic polycrystal constants after Voigt and Reuss in the case of fiber-textured material exhibiting cubic lattice structure:

$$\begin{aligned} S_{11}^R &= S_{11} - 0{,}4S + 3aC_4^1 S \\ S_{33}^R &= S_{11} - 0{,}4S + 8aC_4^1 S \\ S_{12}^R &= S_{12} + 0{,}2S + aC_4^1 S \\ S_{13}^R &= S_{12} + 0{,}2S - 4aC_4^1 S \\ S_{44}^R &= S_{44} + 0{,}8S - 16aC_4^1 S \end{aligned} \qquad (A1)$$

$$C_{11}^V = C_{11} - 0{,}4C + 3aC_4^1 C$$
$$C_{33}^V = C_{11} - 0{,}4C + 8aC_4^1 C$$
$$C_{12}^V = C_{12} + 0{,}2C + aC_4^1 C$$
$$C_{13}^V = C_{12} + 0{,}2C - 4aC_4^1 C$$
$$C_{44}^V = C_{44} + 0{,}2C - 4aC_4^1 C \tag{A2}$$

where:
a = 0,006155
$S = S_{11} - S_{12} - S_{44}/2$
$C = C_{11} - C_{12} - 2C_{44}$
C_4^1: coefficient of the ODF.

Calculation of the elastic stiffnesses C_{11}, C_{12} and C_{44} from the compliances S_{11}, S_{12} and S_{44} for cubic lattice structure:

$$C_{11} = \frac{S_{11} + S_{12}}{(S_{11} - S_{12}) \cdot (S_{11} + 2S_{12})}, \tag{A3}$$

$$C_{12} = \frac{-S_{12}}{(S_{11} - S_{12}) \cdot (S_{11} + 2S_{12})}, \tag{A4}$$

$$C_{44} = \frac{1}{S_{44}}. \tag{A5}$$

LOCAL ORDER AND MECHANICAL PROPERTIES OF THE γ MATRIX OF NICKEL-BASE SUPERALLOYS

N. Clément[1], A. Coujou[1], M. Jouiad[1],

P. Caron[2], H.O.K. Kirchner[3], and T. Khan[2]

[1] CEMES C.N.R.S., 29 rue Jeanne Marvig, F-31055 Toulouse, France
[2] ONERA, 29 avenue de la Division Leclerc, B.P. 72, F-92322 Châtillon, France
[3] INSTITUT DE SCIENCES DES MATERIAUX, Université Paris-Sud, F-91405 Orsay, France

Abstract

For the first time γ single crystals, closely matching with the chemical composition of the matrices of the industrial single crystal superalloys AM3 and MC2 have been made. Macroscopic tensile tests as well as "in situ" deformation tests show that in both alloys a strong localization is the main characteristic of the deformation at different scales of observation. Collective movements of dense planar arrays of dislocations are observed related to the presence of order which appears to be different depending on the alloy. These ordering phenomena introduce high friction forces larger than the applied stress ($\tau_f \approx 2.4\ \tau_{el}$) opposing dislocation movement.

The MC2 matrix has a tensile strength 15 % higher than that of the AM3 matrix. This improvement in strength is partially due to the higher content in W. From deformation and neutron diffraction experiments, it appears that the hardening mechanism is not only a solid solution one, but also associated with a local order hardening which depends on the alloy. The superior mechanical properties of the MC2 γ phase are attributed presumably to the presence of a DO_{22} short range order due to W.

Introduction

In nickel base superalloys, most of the studies reported in the literature have, up to now, been devoted to the mechanical properties of the two phase γ/γ' material or the strengthening γ' phase. Even though the creation and propagation of dislocations in those materials, especially at high temperature, occurs in the γ matrix, the hardening contribution of the γ matrix has been neglected or underestimated. With the exception of the paper by Beardmore et al. [1] very little information is available on the intrinsic characteristics of this f.c.c. phase, often considered to be disordered. Two main reasons can explain this :
- Direct microscopic observations within the very narrow γ channels are not easy,
- The precise composition of the γ phase is difficult to determine and due to solidification difficulties no matrix single crystals were available until now.

In order to improve the mechanical properties of superalloys, it is necessary to understand the hardening mechanisms operating in the γ phase itself, which, of course, are related to its structure and composition.

For the first time, we succeeded in the fabrication of γ single crystals, almost exactly of the chemical composition of the matrices of AM3 and MC2 industrial nickel-base superalloys [2]. These alloys were recently developed at ONERA for single crystal turbine blade applications. AM3 is a low density (d = 8.25 g.cm^{-3}) superalloy which exhibits a mechanical strength comparable to that of the first generation single crystal superalloys with densities close to 8.6 g cm^{-3}. MC2 (d = 8.62 g.cm^{-3}) is a very high strength single crystal superalloy showing a 50°C operating temperature advantage over AM3. With a chemical composition quite similar to those of many conventional superalloys, its creep behaviour is comparable to that of some recent superalloys which contain rhenium.

The aim of this study is to elucidate the reasons for such a difference considering the microstructural aspects and the microscopic deformation processes occurring in the γ phase . Here we have to deal with the question of order vs. disorder complicated by the fact that in the matrix of complex industrial alloys, not only short range order, but also long range order (for example, DO_{22} as well as $L1_2$ type) may be present. For Cr (20 - 30 at %) rich Ni-Cr alloys, neutron scattering, deformation experiments and in situ electron microscopy indicate the presence of short range order, of the type Ni_2Cr [4, 6] or Ni_3Cr [3, 5, 7, 8, 9]. Short range order of the Ni-Cr pairs has also been found by X-ray scattering in Ni_2CoCr, and the same was confirmed to exist in the MC2 matrix single crystals used in the present study [3, 10].

Experimental procedure

The microchemistries of the γ MC2 and γ AM3 phases in the two phase materials were determined by atom probe nanoanalysis [11]. It must, however, be noted that the compositions chosen and prepared correspond to the equilibrium composition at 850°C (the temperature of the final ageing treatment of the alloys). The composition of the corresponding γ single crystals were checked to be very similar to those of the matrices of the two-phase materials. They are given in Table I. The W content is significantly higher in the MC2 matrix, while the concentrations of Al, Ta and Ti are higher in the AM3 matrix. As Al, Ti and Ta are considered to favour precipitation of the γ' phase and in order to check their influence on ordering, MC2 type matrix crystals without these γ' formers were also grown. All these specimens were homogenized 3h at 1300°C and air cooled.

Table I : Composition (at.%) of the three different matrices investigated, as checked by chemical analysis.

	Ni	Cr	Co	Al	Ta	W	Ti	Mo
γ AM3 matrix	52.19	27.1	11.81	3.69	0.42	1.53	0.36	2.9
γ MC2 matrix	54.93	25.58	9.21	2.84	0.29	4.23	0.25	2.67
γ MC2 matrix without γ' formers	57.42	26.5	9.16	-----	-----	4.19	----	2.72

In order to determine unambiguously the controlling mechanisms in such complex industrial alloys, experiments were conducted at different scales :
- Macroscopic tensile tests,
- S.E.M. observations of the surfaces,
- "In situ" deformation tests in a T.E.M.,
- "Post mortem" T.E.M. observations of macroscopically deformed samples,
- Neutron diffuse scattering.

Experimental Results
Tensile tests

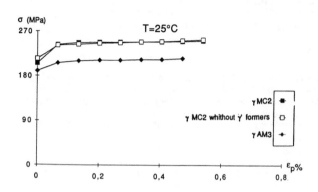

Figure 1 : True stress-strain curves of three different matrices. [001] tensile axis, room temperature.

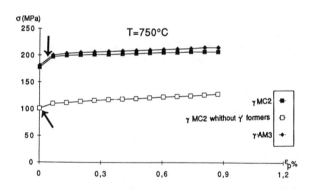

Figure 2 : True stress-strain curves of three different matrices. [001] tensile axis, T = 750°C.

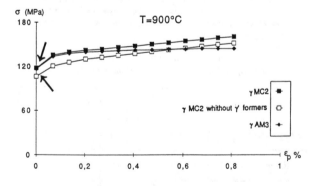

Figure 3 : True stress-strain curves of three different matrices. Tension [001] tensile axis, T = 900°C.

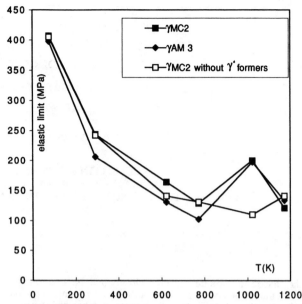

Figure 4 : Yield stress as a function of temperature for γ MC2, γ AM3 and γ MC2 without γ' formers. [001] tensile axes.

Tensile tests were conducted on single crystals with [001] orientation. The test temperature was varied from -196°C to 900°C, and the crosshead speed corresponded to a strain rate of 10^{-4} s^{-1}. The deformation was interrupted at around 2.5 % maximum plastic strain. Samples were then cut from the tensile specimens for observation in a JEOL 200CX transmission electron microscope.

The "true" curves corresponding to instantaneous values of section and length during the test are shown in Figures 1, 2 and 3. Plastic instabilities corresponding to load drops exist on these curves. The onset of the serrations indicated by arrows in Figures 1, 2 and 3, is seen to appear at decreasing strains, while their amplitude increases. The temperature dependence of the yield stress is shown in Figure 4.

From these stress-strain curves it appears that :

- The strength of the γ phase is higher than expected [12], i.e. 243 MPa at 20°C and 121 MPa at 900°C (Figures 1 and 3).

- The strength of γ MC2 is about 15% higher than that of γ AM3, from room temperature to 500°C (Figures 1 and 4).

- A stress peak exists at 760°C in γ alloys containing γ' elements while it disappears in the alloy without these γ' formers (Figure 4) where a continuous decrease of the flow stress is observed as expected for thermally activated phenomena.

The comparison of the data obtained on the two MC2 matrices with and without γ' forming elements shows that Al, Ti and Ta do not have a significant influence on the yield strength below the stress peak.

Scanning Electron Microscopy

Strong localization is the main characteristic of the deformation observed at different scales :

At a mesoscopic scale (Cambridge S.E.M., 20 kV), Figure 5 shows that, independent of the temperature, for a low deformation of about 2 %, slip bands of bundles of {111} slip lines are created at the sample surface. They are very heterogeneous: slip zones 2.5 to 10 μm large separated by 50 to 100 μm wide zones, free of any traces, are observed.

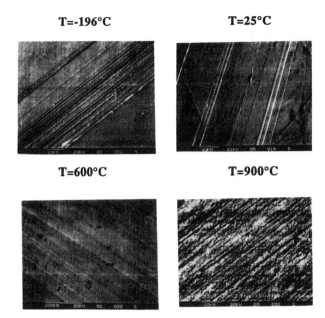

Figure 5 : Observation of slip traces at the surface of deformed samples (S.E.M.) : a) T = -196°C ; b) T = 25°C ; c) T = 600° ; d) T = 900°C.

"In situ "Deformation Tests in a T.E.M.

"In situ" deformation tests were performed at various temperatures in a JEOL 200 CX T.E.M., at 200 kV. The tensile axis of the microsamples was <001> (Schmid factor ≈ 0.43 for {111} planes).

As in the macroscopic observations at low and medium temperatures, the deformation appears to be very localized. The microstructure developed during these deformation tests consists of dense planar arrays of mobile dislocations. In Figure 6, such a moving group is observed with three pairs clearly visible at the head of the group. Here the average velocity is 85 nm.s^{-1} and the movement is observed to be correlated for dislocations belonging to the same pair. This microstructure is very different from a F.C.C. disordered phase microstructure where individual dislocation movements create an homogeneously distributed deformation. Several physical mechanisms have been suggested to explain this behaviour observed in concentrated solid solutions [13, 14, 15]. Here the presence of tiny ordered coherent precipitates, or the presence of short range order is emphasized and corroborated by the neutron diffraction experiments.

"Post mortem" T.E.M. Observations of Macroscopically Deformed Samples

Samples cut parallel to {111} planes in deformed tensile specimens allow large dislocation arrays to be imaged in their glide planes. Up to 900°C, as in the "in situ" experiments, the deformation is very heterogeneous; above this temperature, on the contrary, it is homogeneous (Figure 7).

In Figure 12a such a planar group in γ AM3 is shown : a source located at S emitted 16 dislocations on the (111) plane, the sense of the Burgers vectors b = 1/2[-1 0 1], being different on each side of the source.

Figures 8, 11a, 12a, 13a and 14a correspond to static and relaxed configurations : unlike a standard pile-up, the spacing between dislocations inside the first group does not follow the expected scheme. Instead of a regular increase of the spacing from the front to the back of the pile-up, pairs exist at the head of the pile-up. Measurements of the distance between two successive dislocations, $d_{ij} = f(i)$, allow their identification when a minimum appears on the distribution (in γ MC2, Figures 11c, 13b and 14c, in γ AM3 Figure 14c).

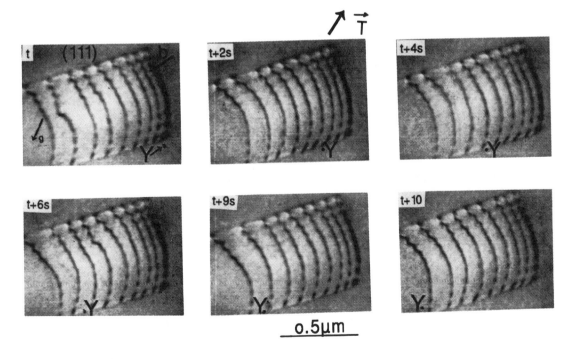

Figure 6 : T = 350°C. "In situ" dynamic sequence in γ MC2. Y is a fixed point.

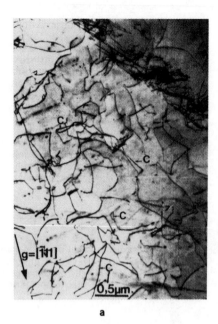
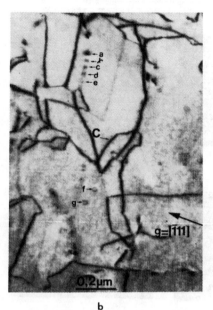

Figure 7 : T = 900°C ; "Post mortem" observations in γ MC2 macroscopically deformed samples. Homogeneous distribution of dislocations. Presence of loops.

When the γ MC2 and γ AM3 samples are deformed at increasing temperatures, the number of pairs increases (Figure 8) : 2 pairs at 25°C, 3 pairs at 350°C, 5-6 pairs at 600°C. On the contrary, in γ MC2 samples devoid of γ' formers, only one pair is observed at the head of the moving pile-ups independent of the temperature between 25°C and 600°C. (Figure 14a, T = 500°C).

From these observations one can deduce that some tiny ordered particles exist in the γ MC2 and γ AM3 phases even at room temperature. With this hypothesis the reason for pairing is clear : if during the deformation a dislocation glides through a long range ordered particle, its order is destroyed. Dislocations therefore move in pairs, the leading one creates an antiphase boundary and the trailing one restores order. The complete shearing of a microprecipitate, whose radius is r, is obtained after the crossing of r/b dislocations, (b = 0.25 nm is the Burgers vector of the moving dislocations). Measuring the pair number (from 2 to 6 depending on the temperature) allows the magnitude of the ordered zone to be determined (here d varies from 0.5 to 3.0 nm). This value is in good agreement with the dimensions of ordered zones estimated by atom-probe techniques which were found to be around 4 to 5 x $2d_{002}$ = 1.5 to 2 nm [3, 11]. Since they are so tiny, they do not give any superstructure spots in the electron diagrams at room temperature.

Neutron Diffuse Scattering

As these pairings indicate the presence of order, different types of ordering seem to exist in different matrices. In order to evaluate and compare them, all these experiments were completed with neutron diffraction on γ AM3 and γ MC2 single crystals. They prove that, contrary to the widely accepted idea, these solid solutions are not homogeneously disordered [5].

In agreement with T.E.M. experiments, $L1_2$ long range order appears to exist in both samples : the special points for this type of ordering are (1 0 0) and (1 1 0). In both the MC2 (Figure 9a) and AM3 specimens (Figure 9c) well defined (100) peaks are present. This is in good agreement with the microscopic observations (Figure 10) where beyond 700°C superstructure <100> and <110> diffraction spots are clearly visible on electron diffraction patterns (Figures 10b and 10d) of both samples annealed in the microscope. When they are selected they image small precipitates appearing in dark field contrast conditions. The existence of γ' $L1_2$ microprecipitates growing when the temperature is raised is therefore corroborated, their dimension being d ≈ 15 nm at 760°C (Figure 10c).

Short range order seems to exist too : taken in [4 2 0] direction, (1 1/2 0) diffuse maxima are observed in both samples (Figures 9b and 9d). They can only come from DO_{22} ordering.

As in the γ MC2 alloy (Figures 9a and 9b), the three characteristic reflections (1 0 0), (1 1/2 0) and (2 1 0) are present, while (1 1/2 0) and (2 1 0) only are visible in γ AM3, (Figures 9c and 9d), it seems that γ MC2 has a greater propensity for DO_{22} ordering than γ AM3.

If one considers that Ni-Cr pairs only are responsible of such an order, it must be recalled that in binary Ni 75 at.% - Cr 25 at.% the order parameters at 530°C agree with a Ni_3Cr [16] type order with a critical transition temperature of less than 150°C. Due to the chemical composition of the alloy it is also possible in terms of off-stoichiometric compounds based on Ni-Cr, Ni-W (Tc = 820°C) [17] or Ni-Mo (Tc = 700°C) pairs participating in DO_{22} short range ordering type [8].

In any case, these ordering phenomena introduce friction forces opposing dislocation movement and thus increase the flow stress. Close inspection of the arrangement of moving dislocations in the pile-ups allows to evaluate them.

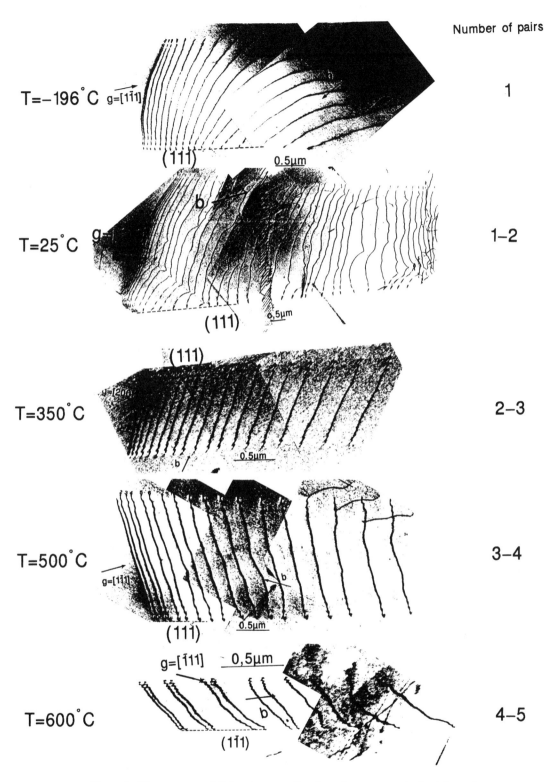

Figure 8 : Planar groups of dislocations as a function of temperature, in the deformed γ matrix of MC2 superalloy. "Post mortem" observations. An increase of the number of pairs at the head of the pile-up is observed : a) one at T = -196°C ; b) one at T = 25°C ; c) three at T = 350°C ; d) four at T = 500°C ; e) five at T = 600°C.

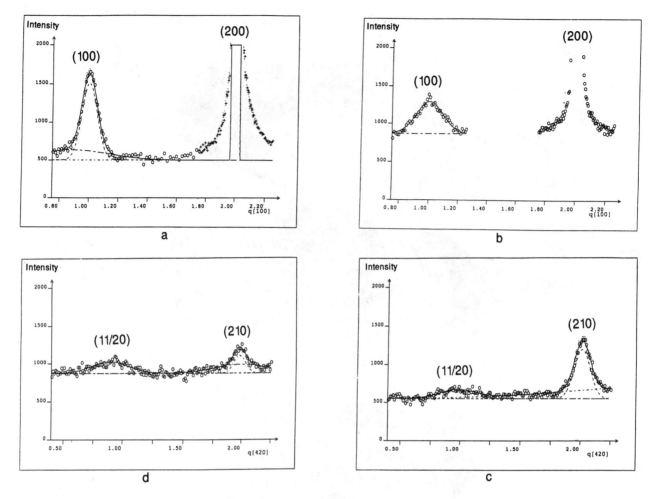

Figure 9 : Diffuse neutron scattering
In [1 0 0] direction : (a) γ MC2 ; (b) γ AM3.
In [4 2 0] direction : (c) γ MC2 ; (d) γ AM3.

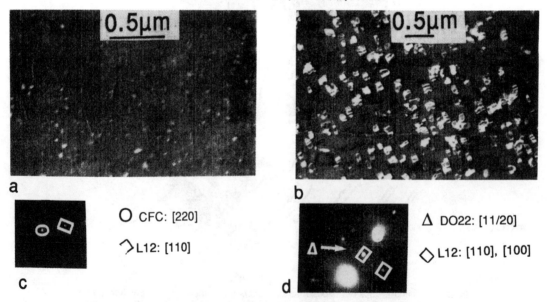

Figure 10 : Annealing of the MC2 matrix in the T.E.M. :
T = 700°C : a) diffraction pattern ; b) corresponding dark field image with a (110) superstructure spot.
T = 760°C : c) diffraction pattern ; d) dark field image with a (110) superstructure spot.

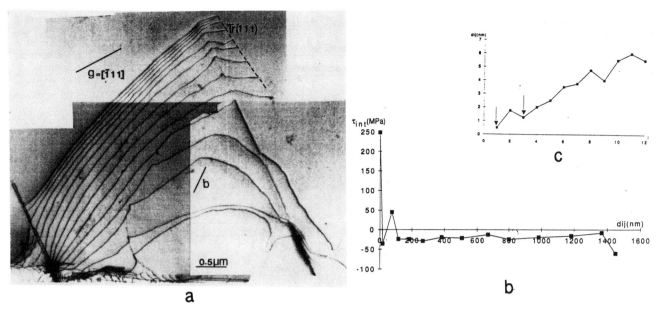

Figure 11 : γ MC2 ; T = 25°C
a) Experimental pile-up.
b) Determination of the friction stress, from the experimental distribution of dislocations within the pile-up.
c) Determination of the number of pairs
$\tau_{f/i}$ max = 249 MPa τ_{ss} = 27 MPa $\tau_{f/i} \approx 2.4\ \tau_{el}$

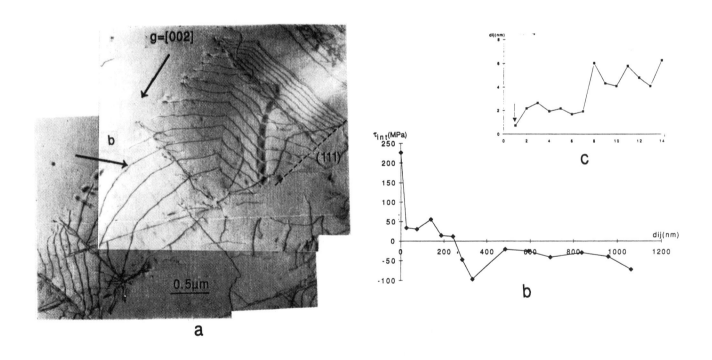

Figure 12 : γ AM3 ; T = 25°C
a) Experimental pile-up
b) Determination of the friction stress, from the experimental distribution of dislocations within the pile-up.
c) Determination of the number of pairs
$\tau_{f/i}$ max = 221 MPa τ_{ss} = 15 MPa $\tau_{f/i} \approx 2.4\ \tau_{el}$

Evaluation of Friction Stresses

The friction stress τ_f opposing the moving dislocations can be deduced by measuring the equilibrium position of each dislocation in the pile-ups [18]. Here we use an improved method compared with the estimation made in the case of Ni-Cr binary alloys [4].

As the observed pile-ups are relaxed, the applied stress has disappeared, $\tau_a = 0$. Since they are observed in a thin foil, the long range stress due to other dislocations inside the samples can be neglected. The position of each dislocation i in a pile-up is the result of two opposing forces which are in equilibrium:

$$b\tau_{f/i} = - b\tau_{int/i}$$

- $b\tau_{int/i}$ is the force due to the elastic interaction of all the dislocations of the pile-up on dislocation i with:

$$\tau_{int/i} = \sum_{j \neq i} \frac{A}{x_i - x_j}$$

and

$$A = \frac{\mu b}{2\pi}\left[\cos^2\phi + \frac{\sin^2\phi}{1 - \nu}\right]$$

μ : Shear modulus ν : Poisson's ratio

- $b\tau_{f/i}$ is the friction force opposing the movement.

From the experimental positions x_i and x_j, measured in the observed pile-ups, $\tau_{int/i}$ can be calculated (Figures 11b, 12b and 13c) and the corresponding friction stress $\tau_{f/i}$ deduced. The results are given in Table II.

Table II : Friction stresses τ_f and solid solution stresses τ_{ss} compared with the elastic limit τ_{el}. T = 25°C

	γ AM3	γ MC2	γ MC2 without γ ' formers
τ_f (MPa)	220	249	240
τ_{ss} (MPa)	10 to 20	20 to 30	10 to 20
τ_{el} (MPa)	91	104	54

T = 25°C

On the average the friction stresses were found to be larger at the head of the pile-up than the yield stress τ_{el}, with $\tau_f \approx 2.4\ \tau_{el}$ for the three matrices at 25°C, while they are smaller, $\tau_f \approx 0.2\ \tau_{el}$ at the tail. This latter value corresponds to τ_{ss}, the solid solution stress, which is constant all along the pile-up.

This is in agreement with the observed mechanical behaviour. At the elastic limit, a single dislocation cannot move under the applied stress alone and pile-ups are necessary to enhance the stress and allow the deformation to proceed. The planar glide observed in the groups is associated with a softening mechanism. The first dislocations destroy the ordered arrangement existing in the glide plane, the following ones face a lower resistance to their movement. As a consequence they remain in the same glide plane.
At room temperature both short range order and long range order coexist in γ AM3 and γ MC2 samples as shown by neutron diffraction, while only S.R.O. can exist in γ MC2 without γ ' formers. But it is not easy to separate their contribution to the measured friction stresses since only one pair of dislocations exists independent of the state of order.

T = 500°C

At this temperature a difference appears depending on the nature of the matrix. Several pairs of dislocations are observed in γ MC2 and γ AM3 pile-ups, serrations corresponding to these pairs exist in $\tau_{int/i}$ curves (Figure 13c) which are a consequence of the shearing of L.R.O. small particles. On the contrary, in γ MC2 without γ ' formers only one pair is visible at this temperature (Figure 14).

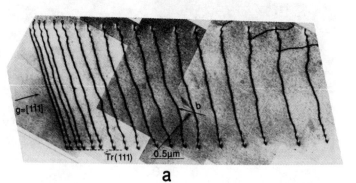

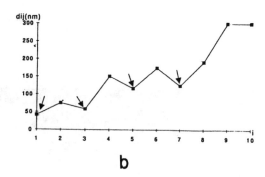

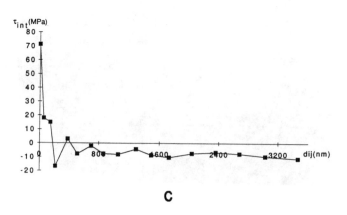

Figure 13 : γ MC2 ; T = 500°C : a) Experimental pile-up; b) Determination of the number of pairs; c) Determination of the friction stress from the experimental distribution of dislocations within the pile-up.

$\tau_{f/i}$ max = 72 MPa τ_{ss} = 10 MPa $\tau_{f/i} \approx 1.3\ \tau_{el}$

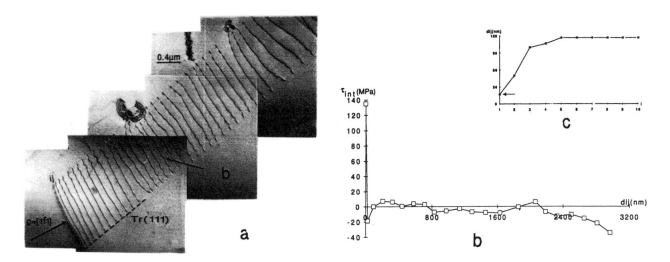

Figure 14 : T = 500°C : a) "Post mortem" observation in γ MC2 without γ' formers.
Macroscopically deformed samples.
One pair is observed at the head of the pile-up.
b) Determination of the friction stress, from the experimental distribution of dislocations within the pile-up.
c) Determination of the number of pairs
$\tau_{f/i}$ max = 135 MPa τ_{ss} = 10 MPa $\tau_{f/i} \approx 2.5\ \tau_{el}$

As proposed by Schwander et al. [8] for binary Ni-Cr alloys, we use the concept of diffuse antiphase boundary energy, γ_{DAPB}, left between two successive dislocations moving in a S.R.O. structure. We denote by Γ, the total A.P.B. energy existing in a sample when both L.R.O. and S.R.O. coexist :

$$\Gamma = \gamma_{S.R.O.} + f\ \gamma_{L.R.O.}$$

f is the surface fraction of L.R.O. precipitates encountered by the dislocation during shearing. Two successive dislocations within a pair are then linked by a force $b\ \gamma_{D.A.P.B.}$ or $b\ \Gamma$ and these different antiphase energies can be separated [18] assuming that S.R.O. only acts on the first pair while L.R.O. exerts its influence on the four following ones. In this case :

$$b\ \tau_f = b\ \tau_{int} = b\ (\tau_{ss} + \Gamma / b)$$

Figures 13c and 14b which compare γ MC2 and γ MC2 without γ' formers are analyzed with this hypothesis. The results of this analysis are given in Table III.

Table III : Friction stresses and solid solution stresses compared with the elastic limits at 500°C for the MC2 matrices.

	γ MC2	γ MC2 without γ' formers
τ_f (MPa)	i=1 72 i=2 20 i=3 18 i ≥ 4 20 $\tau_{f\ total}$ = 130	i=1 135 i ≥ 2 0 $\tau_{f\ total}$ = 135
τ_{el} (MPa)	55	55
τ_{ss} (MPa)	≥ 10	≥ 10

Finally, through the variation of τ_{int} (i) calculated from the observed relaxed pile-ups ($\tau_a = 0$), we are able to identify and evaluate three different contributions to the friction stress in the matrices investigated :

$$\tau_f = \tau_{S.R.O.} + \tau_{L.R.O.} + \tau_{ss}$$

At 500°C comparing γ MC2 and γ MC2 without γ' formers it appears that :

1 - the contribution of the solid solution is the same ($\tau_{ss} \geq 10$ MPa).

2 - $\tau_{S.R.O.}$ in γ MC2 without γ' formers is ≥ 135 MPa, it acts only on the first pair.

3 - summarizing the resulting stresses due to L.R.O. which exist on the fourth first pairs in γ MC2 a similar value is obtained ($\tau_{L.R.O.} \geq 130$ MPa).

Points 2 and 3 are in good agreement with the observed mechanical properties of these two alloys, where similar elastic limits are observed in both samples at this temperature (Figure 4 : $\tau_{el} \approx 55$ MPa). S.R.O. effect in γ MC2 without γ' formers appears to balance the effect of L.R.O. found in γ MC2.

Conclusion

According to our knowledge, for the first time, the strengthening mechanism of different γ matrices of industrial superalloys has been identified and compared on several levels, by various means of investigations (chemical analysis, tensile tests, T.E.M. "in situ" deformation tests at various temperatures, "post-mortem" observations of macroscopically deformed samples, neutron diffraction).

Regardless of the scale of observation and temperature, in the range 25°C - 600°C, the deformation in these single crystals appears to be very localized. The deformation proceeds through the collective movements of planar groups of dislocations. At the head of these groups dislocations clearly move in pairs which indicates the presence of local order (S.R.O. and L.R.O.). By a close inspection of these planar arrays the friction forces τ_f involved in each material have been calculated. They appear to be very high at the head of the moving groups, $\tau_f \approx 2.4\ \tau_{el}$ and much lower at the tail, $\tau_f \approx 0.2\ \tau_{el}$, in agreement with a softening mechanism postulated if the local order is destroyed by the moving groups.

In γ MC2 and γ AM3 alloys an hyperfine precipitation can occur when the sample is annealed. This is due to a supersaturation of the γ' forming elements (Al, Ti, Ta) at room temperature allowing a demixing to appear when the temperature is raised. Such a phenomenon has probably to be also considered when a γ/γ' superalloy is cooled from high temperature.

From tensile tests on γ single crystals, it appears that the γ phase of the MC2 and AM3 superalloys have a higher strength than generally expected. Below 500°C, in γ MC2, the presence or absence of γ' formers does not contribute significantly to the yield stress. Its high strength therefore appears to be due to the presence of elements such as Cr, Co, W and Mo. The strengthening of these matrices is not only due to solid solution hardening, but also to a local order hardening which depends on the alloy. The better mechanical properties of the MC2 γ phase compared to AM3 γ phase can be partially attributed to the occurence of a DO_{22} short range type order due to the presence of W.

Acknowledgements

This research was carried out in the framework of the CPR "Stabilité Structurale des Superalliages Monocristallins", sponsored by D.R.E.T. (French Ministry of Defense). The authors thank J.L. Raffestin (O.N.E.R.A.) for the growth of the γ matrix single crystals.

References

1. P. Beardmore, R. G. Davies, and T. L. Johnston, "On the Temperature Dependence of the Flow Stress of Nickel-Base Alloys," Trans. Aime, 245 (1969), 1537-1545.

2. T. Khan, and P. Caron, "Advanced Single Crystal Ni-base Superalloys," Advances in High Temperature Structural Materials and Protective Coatings, ed. A.K. Koul et al. (Ottawa, Canada: National Research Council of Canada, 1994), 11-31.

3. N. Clément et al., "Local Order and Associated Deformation Mechanisms of the γ Phase of Nickel Base Superalloys," to be published in Microscopy, Microanalysis, Microstructures (1995).

4. N. Clément, D. Caillard, and J.L. Martin, "Heterogeneous Deformation of Concentrated Ni-Cr F.C.L. Alloys : Macroscopic and Microscopic Behaviour," Acta Met., 32 (1984), 961-975.

5. R. Glas et al."Order in the γ matrix of superalloys," to be published in Acta metall. mater.(1996).

6. V. V. Rtishchev, "Structure Transformations and Property Changes of Ni-base Superalloys on Ageing," Materials for Advanced Power Engineering 1994, Part I, ed. D. Coutsouradis et al. (Dordrecht, The Netherlands : Kluwer Academic Publishers, 1994), 889-898.

7. B. Schönfeld, L. Reinhard, and G. Kostorz, "Short-Range Order and Atomic Displacements in Ni-20at% Cr Single Crystals," Phys. Stat. Sol. b, 147 (1988), 457-470.

8. P. Schwander, B. Schönfeld, and G. Kostorz, "Configurational energy change caused by slip in short-range order Ni-Mo," Phys. Stat. Sol. b, 172 (1992), 73-85.

9. K. Wolf, H.A. Calderon, and G. Kostorz, "Fatigue and Dislocation Structure in Short-Range Ordered Alloys," Strength of Materials, ICSMA 10, ed. Okawa et al. (Tokyo, Japan, The Japan Inst. of Metals, 1994), 485-488.

10. F. Guillet, "Etude de l'Ordre Chimique dans les Alliages Ternaires Base Nickel," (Doctoral thesis, Université Paris VI, France, 1993).

11. S. Duval, S. Chambreland, P. Caron, and D. Blavette, "Phase Composition and Chemical Order in the Single Crystal Nickel Base Superalloy MC2," Acta metall. mater., 42, n°1 (1994), 185-194.

12. T. M. Pollock, and A. S. Argon, "Creep Resistance of CMSX3 Nickel Base Superalloys Single Crystals," Acta Metall. Mat., 40 n°1, (1992), 1-30.

13. J. F. Cohen, and M. E. Fine, "Some Aspects of Short-Range Order," J. Phys. Rad., 23 (1962), 749-769.

14. N. Clément, "Influence de l'ordre à courte distance sur les mécanismes de déformation des solutions solides," paper presented at Ordre et Désordre dans les Matériaux, Ecole d'Hiver d'Aussois, France (1984), 167-182.

15. V. Gerold, and H. P. Karnthaler, "On the Origin of Planar Slip in F.C.C. Alloys," Acta Met., 37 (1988), 2177-2183.

16. R. Caudron et al., "In Situ Diffuse Scattering of Neutrons in Alloys and Application to Phase Diagram Determination," J. Phys. I. France, 2 (1992), 1145-1171.

17. N. S. Mishra, and S. Ranganathan, "Electron Microscopy and Diffraction of Ordering in an off-stoichiometric Ni-W Alloy," Scripta Met., 27, n°10, (1992), 1337-1342.

18. M. Jouiad, N. Clément, and A. Coujou "Origin and Determination of Friction Stresses in the γ Phase of a Superalloy," to be published.

DETERMINATION OF ATOMISTIC STRUCTURE OF Ni-BASE SINGLE CRYSTAL SUPERALLOYS USING MONTE CARLO SIMULATIONS AND ATOM-PROBE MICROANALYSES

H. Murakami, Y. Saito* and H. Harada

National Research Institute for Metals, Computational Materials Science Division,
1-2-1, Sengen, Tsukuba Science City, 305, Japan
*Department of Materials Science and Engineering, School of Science and Engineering,
Waseda University, 3-4-1, Okubo, Shinjuku-ku, Tokyo 169, Japan.

Abstract

The atomic locations of alloying elements in some Ni-base single crystal superalloys have been investigated using Monte Carlo Simulations (MCS), and the predictions have been compared with the experimental results obtained from atom-probe field ion microscopy (APFIM). The γ and γ' phase compositions and the site occupancy of alloying elements in the γ' phase at equilibrium conditions were predicted using MCS. The predictions were then compared with estimates obtained by the Cluster Variation Method (CVM) and experimental results obtained by APFIM, so as to verify the applicability of MCS. It was found that the MCS estimations were generally in good agreement with both CVM predictions and APFIM results. For multicomponent systems, alloying elements may be classified into three general groups: (1) those which preferentially partition to the γ' phase and substitute for the Al site in the γ' phase, such as Ti and Ta; (2) those which preferentially partition to the γ phase and substitute for the Ni site in the γ' phase, such as Co; (3) those which generally partition to the γ phase and substitute for the Al site in the γ' phase, such as Re, Mo and W. In this paper, virtual experiments were also performed to investigate the chemistry of γ/γ' interfaces.

Introduction

Although nickel-base single crystal superalloys have already been developed extensively for turbine blade applications, there are continuing efforts to improve the mechanical properties of these alloys for next-generation aeroengines. The superalloys have $L1_2$ ordered γ' precipitates in γ matrices. Their mechanical properties are thus expected to be controlled by the microstructural parameters of the two phases, such as their inherent mechanical properties, and the interaction between the two phases. It is therefore of the utmost importance to investigate the atomic configurations of alloying elements which accordingly determine the characteristics of the two phases and the mechanical properties.

The Cluster Variation Method (CVM) has been applied to predict the atomic configurations of alloying elements in nickel-base superalloys[1,2]. Using the CVM, the atomic location of solute atoms in both phases at equilibrium is estimated as a function of alloy composition and temperature.

The interaction between the γ and γ' phases also affects the mechanical properties of superalloys. It is thus important to understand the chemistry of γ/γ' interfaces, where atomic configurations are not expected to reach the equilibrium state. Monte Carlo Simulations (MCS)[3] have been employed to obtain such interface information. This method enables predictions of temporal and spatial evolution of the atomic arrangement, and the kinetics of ordering of nickel-base superalloys. An additional advantage of MCS is that, since the position of every single atom in the system is determined, three dimensional representation of the atomic configurations is possible. Therefore, the distinction between ordered and disordered regions, and the chemistry at interfaces may be visualised.

However, the theoretical models will not be of practical use until the agreement between predictions and experimental results is demonstrated. In order to verify theoretical estimates obtained from MCS and CVM, experimental analysis using Atom-Probe Field Ion Microscopy (APFIM) has been carried out. APFIM, which has sub-nanometer spatial resolution, is a powerful tool for experimentally obtaining atomic information of materials. The main objective of this study is to understand the atomic configurations of some multicomponent nickel-base superalloys by comparing the predictions of MCS with experimentally obtained APFIM results, so that a novel alloy design program, which will predict not just the atomic structure but also estimate the mechanical properties of these superalloys, may be established in the near future.

In this paper, the atomic configurations at equilibrium conditions, such as the γ and γ' phase compositions and the site occupancy behaviour of alloying elements in the γ' phase, determined by MCS, CVM and APFIM in five multicomponent nickel-base single crystal superalloys under equilibrium conditions, are compared so as to verify the applicability of MCS and CVM. Additionally, the chemistry at the γ/γ' interface simulated by MCS is compared with that obtained by APFIM.

TABLE 1 COMPOSITION OF ALLOYS (in at. %)

ALLOY	Ni	Al	Ti	Cr	Co	Mo	Hf	Ta	W	Re
CMSX2	67.4	12.5	1.2	9	5	0.4	-	2	2.5	-
CMSX4	63.0	12.6	1.3	7.6	9.8	0.4	0.03	2.2	2.1	1.0
MC2	65.9	11.2	2.5	9.3	5.1	1.3	-	2.0	2.6	-
TMS-63	72.0	12.8	-	7.8	-	4.6	-	2.8	-	-
TMS-71	66.6	12.7	-	6.9	6.2	4.0	-	2.8	-	0.8

Experimental

The materials investigated in this study are listed in Table 1. Among these alloys, first generation CMSX2[4] and second generation CMSX4[5], provided as a 20 mm diameter bar provided by Rolls Royce plc, are in commercial use. TMS-63[6,7] and TMS-71[8] were developed by the National Research Institute for Metals (NRIM) employing the regression analysis based Alloy Design Program (ADP)[6,7,9]. Details of thermal treatment, sample preparation and the experimental results for alloys CMSX2, CMSX4, MC2, TMS-63 and TMS-71 have been described in previous papers[8,10-13]. Note that for all the alloys, the final ageing treatment was conducted at about 1313 K. The temperature of the system used for both CVM and MCS was thus set to 1313 K.

Procedure of Numerical Simulations

Energy calculation

Both the CVM and MCS require appropriate modelling of atomic interactions in order to calculate total energy change in the systems. Lennard-Jones pair potentials were employed in order to simplify the numerical simulations. The Lennard-Jones pair potentials $e_{ij}(r)$ are described by:

$$e_{ij}(r) = e_{ij}^0 \left[(r_{ij}/r)^{m_{ij}} - (m_{ij}/n_{ij})/(r_{ij}/r)^{n_{ij}} \right]$$

where r_{ij} is the interatomic distance between atomic species i and j for which $e_{ij}(r)$ reaches is minimum value i.e.. $e_{ij}(r) = -e_{ij}^0$.

It has been shown by Sanchez et al.[14] that the best values for the exponents n_{ij}, m_{ij} of the attractive and repulsive potentials of metals are 4 and 8, respectively. Details of the method for determining pair potential parameters and the procedure for the CVM are described in previous papers[1,2].

Algorithm for MCS data acquisition

For MCS, the initial structure may be generated by randomly assigning atoms to lattice sites. The kinetics of ordering may be controlled by the direct exchange of a randomly selected atom with one of its neighbouring atoms (Kawasaki dynamics[15]). The probability W that an exchange trial is accepted is given by:

$$W = \exp(-\Delta H/kT)/[1 + \exp(-\Delta H/kT)]$$

where ΔH is the change in energy associated with the exchange of the atoms, k is the Boltzmann's constant and T is the absolute temperature. ΔH is calculated by taking account of contributions from only the first nearest neighbour atoms.

The simulations were performed on a 16x16x16 unit cell mesh (16,384 atoms) mainly for visualisation, and on a 32x32x32 unit cell mesh (131,072 atoms) mainly for determination of the γ and γ' phase compositions and for the investigation of the site occupancy behaviour of alloying elements in the γ' phase. Periodic boundary conditions were employed. It should be noted that all the lattice positions in the system were fixed: i.e. no relaxation effects were considered in the calculation. A Hewlett Packard Model 712 / 80 workstation system was used for the calculations. The simulations were carried out until the atomic configurations were regarded as equilibrium (typically 20,000 Monte Carlo steps): the total enthalpy change of the system proved to be negligibly small after 20,000 Monte Carlo steps.

Fig. 1 shows a 3-dimensional representation of the atomic arrangement of solute atoms in TMS-71, obtained by MCS after 20,000 Monte Carlo steps. Here, the 22nd atomic layer of the system is highlighted. The distribution of atoms in the system may be thus described. Fig. 2 shows the 2-dimensional arrangement of atoms on the 22nd to the 25th layers, normal to the <100> direction. The atom species in TMS-71 may be identified by their size and brightness. Since pure Ni layers and Ni-Al mixed layers must appear in turn along <100> for pure fully-ordered Ni₃Al, the atomic locations on the (100) layers in the γ' phase for multicomponent Ni-base superalloys will be represented as shown in Fig. 3. From MCS, it is thus possible to identify the ordered γ' phase, as indicated by arrows in Fig. 2, the atomic configurations of which are almost identical to the expected representation (Fig. 3). From Fig. 2, the site occupancy of alloying elements may also be understood.

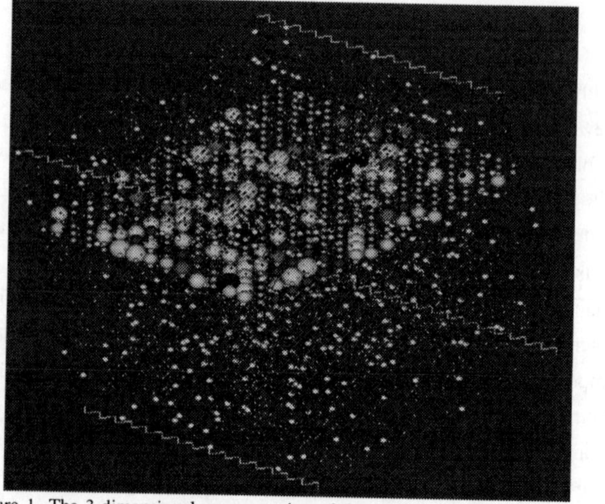

Figure 1. The 3-dimensional representation of the atomic arrangement of solute atoms in TMS-71, obtained by MCS after 20,000 Monte Carlo steps

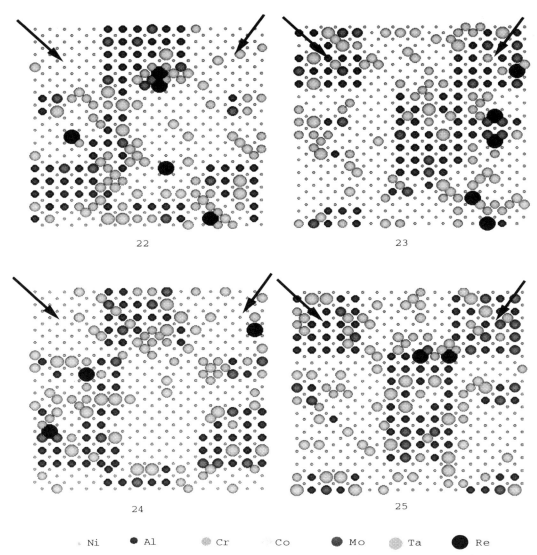

Figure 2. The 2-dimensional arrangement of atoms on the 22nd to the 25th layers, normal to the <100> direction.

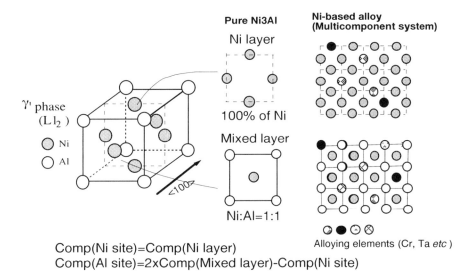

Comp(Ni site)=Comp(Ni layer)
Comp(Al site)=2×Comp(Mixed layer)-Comp(Ni site)

Figure 3. The expected atomic configurations of (100) layers in the γ' phase.

Determination of γ and γ' phase compositions and site occupancy behaviour in the γ' phase.

Although the γ and γ' phases may be visualised from MCS, the determination of the γ and γ' phase compositions requires analysis based on the identification of neighbouring atoms. Consider a selected atom: if its 1st neighbouring atoms are mainly Ni, and the second neighbouring atoms are mainly Al, the selected atom is deemed to occupy the Al site in the γ' lattice. Fig. 4 shows the flow chart for the first step of analysis. Atoms which remain unidentified after step 1 are further analysed. Finally, γ' regions are determined by utilizing a cluster analysis method[16]. Since all the atoms are positioned in either the Al site or Ni site when determining the γ' regions, site substitution behaviour is also determined. Note that in this study, regions which are not assigned as γ' regions are regarded as the γ phase.

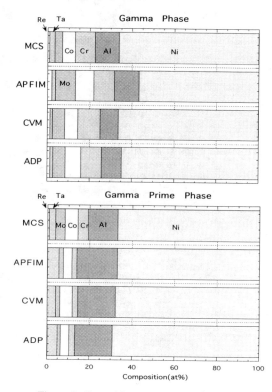

Figure 5. Compsition analysis for TMS-71.

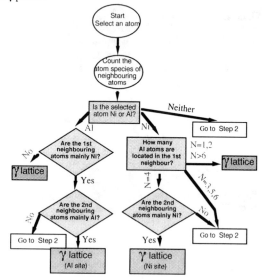

Figure 4. The flow chart for the first step of analysis.

Results and Discussion

γ and γ' phase compositions

The experimentally-determined γ and γ' phase compositions obtained by APFIM were compared with data estimated by CVM and MCS. Fig. 5 shows a typical composition analysis for TMS-71; compositions estimated by the ADP are also shown. Excellent agreement is obtained, particularly for the γ phase.

In order to investigate the partitioning behaviour of an alloying element 'i' into the γ or γ' phase, a partitioning parameter k_i is defined as:

$$k_i = c_{i\gamma} / c_{i\gamma'}$$

where $c_{i\gamma}$ and $c_{i\gamma'}$ are the concentrations of an alloying element 'i' in the γ and γ' phases, respectively. Thus, for example, when the element 'i' has a tendency to partition into the γ phase. Fig. 6 summarises the partitioning behaviour of alloying elements for

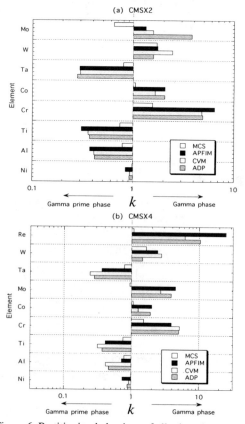

Figure 6. Partitioning behaviour of alloying elements.

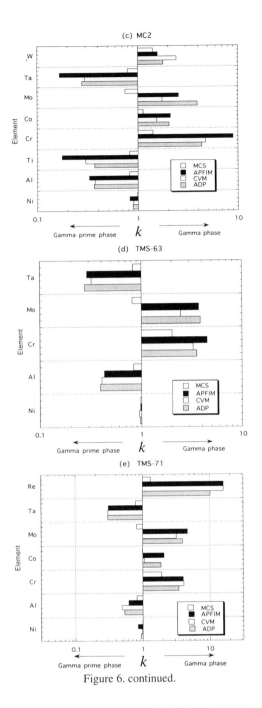

Figure 6. continued.

(a) CMSX2; (b) CMSX4; (c) MC2; (d) TMS-63 and (e) TMS-71 as a function of k_i. Hf in CMSX4 is omitted in this comparison because of its low solute content. Good agreement is shown except for the partitioning behaviour of Mo, obtained by MCS. It is also found that the partitioning tendencies predicted by MCS are not as pronounced as those obtained by other methods.

Further composition analysis revealed that Ti and Ta preferentially partition into the γ' phase while Co, Cr, W and Re have preference to partition into the γ phase. CVM, ADP and APFIM analysis showed that Mo tends to partition into the γ phase.

Site occupancy of alloying elements in the γ' phase

It is important to investigate the site occupancy of alloying elements in the γ' phase since whether the alloying element prefers the Ni sites or the Al sites dramatically alters the γ' volume fraction. Fig. 7 illustrates a simplified ternary phase diagram[8] for a Ni-Al-M system (M = Ti, Cr, Co, Mo, Ta etc.). It is well known that alloying elements which prefer the Al sites have the γ' / γ+γ' phase boundary along the line AB. On the other hand, alloying elements which prefer the Ni sites have the AC phase boundary. Therefore, from this figure, it may be seen that alloy X whose composition is pointed as ● may be in the γ' single phase when the alloying element substitutes for the Al site, whereas it may be almost 50%γ + 50%γ' two-phase structure when the alloying element substitutes for the Ni site. Hence, alloying elements which substitute for the Al site increase the γ' volume fraction. Understanding the substitution behaviour of solute atoms in the γ' phase is thus essential in order to estimate the γ' volume fraction, which is one of the main factors in determining the mechanical properties of Ni-base superalloys. In the case of a multicomponent system, however, the site preference of an alloying element might differ from that in a ternary system, due to the 'site competition' of other solute elements. Accordingly, it is difficult to determine the site occupancy in a multicomponent system by analysis methods such as Mössbauer spectroscopy[17]. Layer-by-layer analysis employed in APFIM enables experimental determination of the site occupancies of alloying elements in multicomponent alloys. The detailed experimental procedure for layer-by layer analysis of multicomponent alloys has been reported previously[11,12].

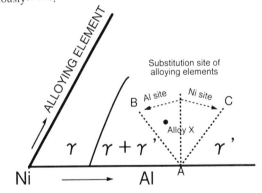

Figure 7. A simplified ternary phase diagram[8].

Fig. 8 shows the occupancy probabilities of alloying elements on (a) Al sites and (b) Ni sites for TMS-63, estimated by MCS and CVM, and experimentally determined by APFIM. Good agreement is obtained especially for the Ni site.

In order to determine the site occupation behaviour, a parameter s_i may be defined:

$$s_i = p_{iAl} / c_{i\gamma'}$$

where p_{iAl} and $c_{i\gamma'}$ are the occupancy probabilities for the Al site and concentration in the γ' phase of an alloying element 'i', respectively.

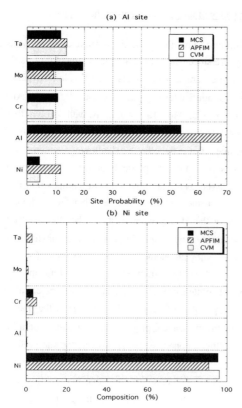

Figure 8. the occupancy probabilities of alloying elements on (a) Al sites and (b) Ni sites for TMS-63.

Since the number of Ni sites is three times that of Al sites for Ni$_3$Al, s_i =1, 2 and 4 correspond to the occupancy behaviour of an alloying element 'i' for the cases in which atoms are randomly distributed in the two sites, an equal number of atoms enter the two lattices and all atoms substitute for the Al sites, respectively. Fig. 9 summarises the site occupancy behaviour of alloying elements in (a) CMSX2; (b) CMSX4; (c) MC2; (d) TMS-63 and (e) TMS-71. Here, Mo in CMSX2 and Hf in CMSX4 are omitted because of the experimental uncertainties[11,12]. In addition, APFIM analysis of TMS-71 has not yet been conducted. Generally, good agreement is observed except for some discrepancies between estimated and experimental data in the cases of Co and Ta, and estimated site substitution behaviour of W obtained by MCS and CVM.

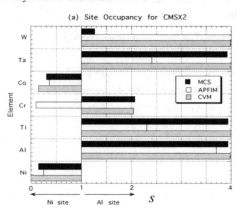

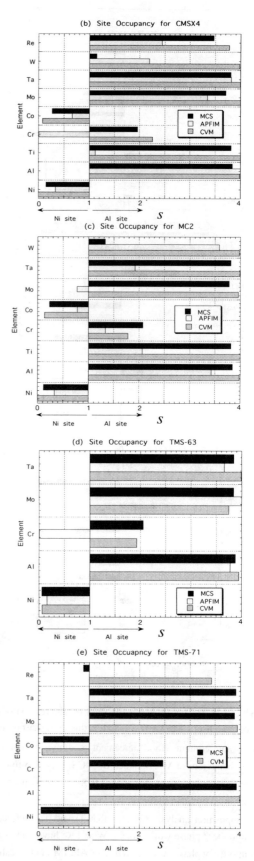

Figure 9. The site occupancy behaviour of alloying elements in 5 Ni-base single crystal superalloys.

Site occupancy determination for other alloys revealed that generally Mo, Ta, W, Re and Ti tend to occupy the Al sites whilst Co has a preference for the Ni sites. The site occupancy of Cr is composition dependent.

From the investigations of phase compositions and site occupancy of alloying elements in the γ' phase, two types of discrepancies have been found: (1) the discrepancy between experimental analysis and numerical simulations, such as the site occupancy of Cr in the γ' phase, and (2) the discrepancy between CVM and MCS, such as partitioning behaviour of alloying elements (typically Mo) and the site occupancy of Re in the γ' phase for TMS-71 and the site occupancy of W in the γ' phase for all the other alloys. Discrepancy (1) may be attributed to the determination of potential parameters for Cr. Further investigation of the experimental conditions should also be conducted.

Discrepancy (2) is caused by the difficulty in distinguishing between the γ and γ' phases from MCS. As discussed above, the MCS were carried out using fixed lattice positions and identical pair potentials in both the γ and γ' phases. In addition, the number of atoms used for the calculation may not be large enough to describe the atomic configurations of the two phases for multicomponent alloys. Fig. 10 shows an example of the two dimensional atomic configurations on a (100) layer for TMS-71 obtained by MCS for the 32x32x32 unit cell system. Here, some Mo-Re clusters with 3-8 atom numbers are indicated by the arrows. Since the pair potential between Mo and Re is very binding, Mo-Re atom pairs easily form in TMS-71 during MCS. W atoms behave similarly to Re for all the other alloys: *e.g.* Mo-W clusters are formed during calculations. These clusters frequently appear in the vicinity of γ/γ' interfaces or even embedded in γ' precipitates.

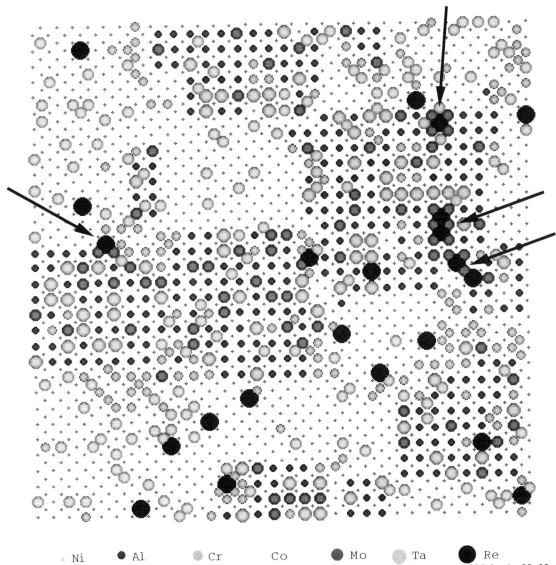

Figure 10. An example of the two dimensional atomic configurations on a (100) layer for TMS-71 obtained by MCS for the 32x32x32 unit cell system in which some Mo-Re clusters with 3-8 atom numbers are indicated by the arrows.

Superalloys containing large amount of solute elements tend to form a topologically close-packed (TCP) phase such as the μ phase, which has very high concentration of W and Mo. For example, it has been reported for MC2[18] that the μ phase precipitates from the γ matrix, preferentially in the region where the W level is maximum and that the μ phase particles are embedded in the γ' phase, and that the depletion of the matrix of elements Cr, Co and Mo leads to the transformation $r \rightarrow r' + \mu$. Our MCS predictions describe these results well if we consider Mo-W clusters to be the μ phase. Alternatively, the clusters might show the composition fluctuation in the γ phase at which precipitation of the μ phase might occur. Due to the small number of atoms which forms a cluster, the analysis algorithm regards these clusters (Re, Mo) as parts of the γ' phase, resulting in the fact that additional Mo atoms appear to be positioned in the Al sites in the γ' phase, and the neighbouring atoms regarded to be positioned in the Ni sites in the γ' phase, These results may cause the discrepancies between CVM and MCS. Employing a super computer facility, which accepts the increased system size for calculation than a conventional workstation system does, may be of help to overcome some of these limitations. The authors are also examining a modified MCS algorithm employing a local relaxation parameter[19]. It should be noted, however, that semi-quantitative agreement for compositional analysis has been obtained by MCS.

In this study, it has been shown that elements such as Mo, W and Re have a tendency to partition to the γ phase and occupy the Al site in the γ' phase. These alloying elements therefore cause the lattice misfit between the γ and γ' phases to tend towards a negative direction[6,9] and thus, enhance the creep resistance[5,7]. However, it should be noted that the addition of these elements will increase the γ' volume fraction since they tend to substitute for the Al site. It is thus important to determine the optimum solute content in order to design alloys having superior high-temperature capabilities.

Chemistry change at γ/γ' interfaces

The composition changes at γ/γ' interfaces may be determined by APFIM analysis, since the analysis is of atoms successively field-evaporated from the specimen surface and chemically identified. When the cumulative number of each solute atom is plotted against the total number of detected atoms in a 'ladder diagram', the horizontal axis includes depth information and the gradient corresponds to the local composition of that alloying element. Fig. 11 shows a typical ladder diagram of the composition change at a γ/γ' interface. Note the enrichment of Re in the γ phase in the vicinity of the interface.

This ladder diagram may be simulated using MCS by analysing a cylinder along the <100> direction (containing, for example, 40 atoms per atomic layer). Fig. 12 shows such a ladder diagram across a γ/γ' interface, similar to Fig. 11 except for that the number of atoms 'detected' is smaller by a factor of 10.

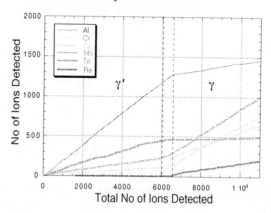

Figure 11. A typical ladder diagram of the composition change at a γ/γ' interface.

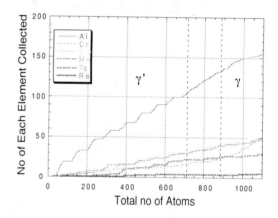

Figure 12. A 'virtual' ladder diagram obtained by the MCS.

When Al atoms are plotted in a ladder diagram, pure Ni-layers appear as horizontal lines whereas mixed Ni / Al layers appear as inclined lines in the ordered γ' region. It is thus clear that the left side in Fig. 12 represents the ordered γ' phase and the right represents the γ phase. The composition change between the γ and γ' phases may also be observed for Co, Cr and Ta. As discussed in the previous section, however, the composition change between the γ and γ' phases in Fig. 12 is not as clear as that experimentally obtained by APFIM. It is again suggested that further investigation such as layer by layer analysis using APFIM, MCS with a much larger system with local relaxation factor considered *etc*. will be of help in obtaining more detailed comparisons between the numerical simulations and the experimental results.

Summary

The γ and γ' phase compositions and the site occupancy of alloying elements in the γ' phase were investigated by MCS. It was found that the estimates obtained by MCS are in good agreement with experimental results obtained by APFIM and CVM predictions. For the multicomponent system, alloying elements may be classified into three groups: (1) elements which preferentially partition into the γ' phase and substitute for the Al sites in the γ' phase, such as Ti and Ta; (2) elements which preferentially partition into the γ phase and substitute for the Ni sites in the γ' phase and (3) elements which generally partition into the γ phase and substitute for the Al sites in the γ' phase, such as Re, Mo and W, dependent upon alloy composition. The simulation of APFIM analysis has been demonstrated by MCS, by use of a 'virtual' ladder diagram. It is found that ordered and disordered regions in the system are clearly identified by this representation.

The employment of a supercomputer which has much higher performance than workstations will allow an increase in the size of the calculated system and the introduction of a local relaxation factor will allow the investigation of some of the discrepancies between MCS and APFIM, or MCS and CVM, such as the partitioning behaviour of Mo and the substitution behaviour of W and Re.

Acknowledgment

Part of this work was carried out under the auspices of the U.K.-Japan Cooperative Science Program, sponsored by The Royal Society and The Japan Society of the Promotion of Science. The authours are also very grateful to Prof. M.Enomoto at Ibaraki University and Dr. H.G.Read at NRIM for many useful suggestions.

References

1. M.Enomoto and H.Harada, "Analysis of γ'/γ Equilibrium in Ni-Al-X Alloys by the Cluster Variation Method with the Lennard-Jones Potential," Metall. Trans., 20A(1989), 649-664.
2. M.Enomoto, H.Harada and M.Yamazaki, "Calculation of γ'/γ Equilibrium Phase Compositions in Nickel-base Superalloys by Cluster Variation Method," CALPHAD, 15(1991) 143-158.
3. Y.Saito, submitted to Mater. Sci. Eng. A.
4. P.Caron and T.Khan, "Improvement of Creep Strength in a Nickel-base Single-crystal Superalloy by Heat Treatment," Mat. Sci. Eng. 61(1983), 173-184.
5. D.J.Frasier et al., "Process and Alloy Optimization for CMSX-4 Superalloy Single Crystal Airfoils," Proceedings of the conference "High Temperature Materials for Power Engineering 1990," Liège, Belgium (1990), 1281-1300.
6. H. Harada et al., "Design of High Specific-strength Nickel-base Single Crystal Superalloys," Proceedings of the conference "High Temperature Materials for Power Engineering 1990," Liège, Belgium (1990), 1319-1328.
7 H. Harada et al, "Computer Analysis on Microstructure and Property of Nickel-base Single Crystal Superalloys," Proceedings of the 5th International Conference on Creep and Fracture of Engineering Materials and Structures, Swansea, U.K,.(1993) 255-264.
8 H.Murakami, P.J.Warren and H.Harada, "Atom-probe Microanalyses of Some Ni-base Single Crystal Superalloys," Proceedings of the 3rd International Charles Parsons Turbine Conference "Materials Engineering in Turbines and Compressors", New Castle, U.K., (1995) 343-350.
9 T.Yokokawa et al., "Towards an Intelligent Computer Program for the Design of Ni-base Superalloys," Proceedings of the 5th International Conference on Creep and Fracture of Engineering Materials and Structures, Swansea, U.K,.(1993) 245-254.
10 H.Harada et al., "Atom-probe Microanalysis of a Nickel-base Single Crystal Superalloy," Appl. Surf. Sci., 67(1993) 299-304.
11   H.Murakami, H.Harada and H.K.D.H.Bhadeshia , "The Location of Atoms in Re- and V-containing Multicomponent Nickel-base Single-crystal superalloys," Appl. Surf. Sci., 76/77(1994)177-183.
12 D.Blavette and A.Bostel, "Phase Composition and Long Range Order in γ' phase of a Nickel Base single Crystal Superalloy CMSX2: An Atom Probe Study," Acta Metall., 32(1984) 811-817.
13 S.Duval et al., "Phase Composition and Chemical Order in the Single Crystal Nickel Base Superalloy MC2," Acta Metall. Mater., 42(1994) 185-194.
14 J.M.Sanchez et al., "Modeling of γ/γ' Phase Equilibrium in the Nickel-aluminum System," Acta Metall., 33(1984) 1519-25.
15 K.Kawasaki, "Diffusion Constants near the Critical Point for Time-Dependent Ising Models. I," Physical review, 145(1966) 224-230.
16 S.Sakamoto and F.Yonezewa, "The Cluster Analysis Method using Computers and its Applications," Kotaibutsuri, 24(1989), 219-226 (in Japanese).
17 J.R.Nicholls and Rees D.Rawlings, "A Mössbauer Effect Study of Ni3Al with Iron Additions," Acta Metallurgica, 25(1976), 187-194.
18 M.Pessah, P.Caron and T.Khan, "Effect of μ Phase on the Mechanical Properties of a Nickel-base Single Crystal Superalloy," Proceedings of Superalloys 1992, Seven Springs, U.S.A., (1992), 567-576.
19 Y.Saito H.Murakami and H.Harada, submitted to the proceedings for the International Workshop on Computer Modelling and Simulation for Materials Design, Tsukuba Science City, Japan, (1996).

DIFFERENTIAL THERMAL ANALYSIS OF NICKEL-BASE SUPERALLOYS

D.L. Sponseller

ERIM Transportation & Energy Materials Laboratory
4683 Freedom Drive
Ann Arbor, Michigan 48108

Abstract

Differential thermal analysis was performed on over 200 nickel-base superalloys that included model, experimental, developmental, and commercial alloys. About one-half of these alloys were studied during melting and solidification, and about three-fourths were studied using solid specimens. Large samples were employed to permit the use of internal thermocouples and to minimize both the effects of segregation and the changes in chemical composition resulting from the loss of strong oxide-forming components. The features of thermograms for solid specimens were identified by two calibration techniques: (1) Inserting thin bar specimens of four cast alloys into a gradient furnace so that all parts of the test specimen were heated at the same rate as in a DTA test to the temperature range of interest; water quenching preserved the entire range of microstructures for metallographic evaluation. (2) Heating (and cooling, when applicable) 35 disc specimens of various alloys to points of interest on the heating thermogram and especially on the cooling thermogram, at which point each specimen was water quenched and evaluated metallographically. The effects of alloying elements on liquidus temperature, carbide-formation temperature, and the extent of carbide dispersion in the cast structure, γ' solvus temperatures (for DTA heating/cooling rates and for practical heat treatments and soaking treatments), incipient-melting temperature, and the width of the temperature "window" for solution-heat-treating of single-crystal alloys are presented. Some examples of problem solving by DTA related to cracking during casting caused by very-low-melting liquid and to microporosity related to the γ + γ' eutectic constituent are presented.

Introduction

Phase reactions play a crucial role in many aspects of the processing and service of high-strength nickel-base superalloys. Before 1968, however, the determination of the most important phase-reaction temperature, that at which the γ' strengthening precipitate dissolves, required the laborious heat-treating and metallography of many specimens, and begged for a simpler technique. The ability of differential thermal analysis (DTA) to simply and accurately measure the temperatures of phase reactions in materials was first discovered by LeChateliér in 1887, and perfected for metals by Roberts-Austen in 1899 (1,2). Considering its great utility in the study of phase reactions, DTA has generally been underutilized, however, by the metallurgical community. Because the author had successfully used DTA to help identify elevated-temperature ductile-iron compositions that could be used to 815 C (1500 F) without re-austenitization, DTA was investigated at the Climax Molybdenum (AMAX) laboratory in 1968 as a means of directly measuring the γ' solvus temperature in nickel-base alloys (3,4). It was found that DTA gives a clear and accurate indication of the γ' solvus temperature during both heating and cooling. Following this discovery, DTA was used in many other superalloy studies at the Climax laboratory. Also, DTA was performed on as many as five sets of samples received from each of the following companies seeking DTA-based data and interpretation in their R&D and problem-solving.

Allied Signal (Garrett Engine)	Pratt & Whitney
Allison Gas Turbine	Precision Castparts
Cannon-Muskegon	Rolls Royce
Carpenter Technology	Ross & Catherall
Cytemp	Sorcery Metals
General Electric	Special Metals
Haynes International	Westinghouse
Howmet	

Results obtained on more than 200 model, experimental, developmental, and commercial alloys to date have demonstrated the power of DTA to provide invaluable guidance in the alloy and process development, production, and characterization of superalloys (5,6).

The purpose of this paper is to document the main findings from the above studies. It is hoped that the sharing of this information with others in the superalloy community may help to further advance the state of the art in the metallurgy of gas-turbine alloys for the transportation, energy, industrial, and military sectors.

Solidification Studies

Experimental Procedure

Test specimens for studies of melting and solidification weighed 9 grams, about 100 times heavier than the 50 to 150-mg samples used in most commercial DTA units. This avoided the variation of results associated with segregation, especially for samples taken from ingots and investment castings, and minimized changes in chemical composition associated with oxidation of strong oxide formers such as aluminum, titanium, and hafnium that would show up especially on resolidification. An additional advantage is that the large specimen, shown in Figure 1, permitted the use of the pictured small immersion thermocouple (W-W/Re type) to avoid errors in sensing the specimen temperature. The test specimen, and two similarly shaped pure-molybdenum specimens (one containing the reference thermocouple and the other containing the furnace-control

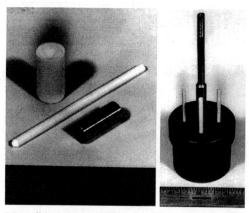

Figure 1. DTA cell for melting/solidification studies, showing: (left) 9-g machined specimen with drilled axial thermocouple hole, high-purity recrystallized alumina thermocouple tube and crucible and (right) assembly ready for thermocouple insertion and test, showing pure molybdenum block with thick lid that keeps the thermocouple tube centered in the sample when molten, and threaded hanger rod.

thermocouple) all were similarly situated in high-purity recrystallized alumina crucibles contained within a pure molybdenum block, with a thick, close-fitting lid, that was suspended from the top of a resistance-heated vacuum furnace. A 4-wire thermocouple was used in the specimens for all but some of the early runs, so that this thermocouple could both oppose the reference thermocouple and drive the X axis, eliminating the need for any temperature corrections in "reading" the thermograms. In each DTA experiment, the assembly was first heated and then cooled through the melting range at 10 C/min, followed by a similar run at 20 C/min. Plots of ΔT vs specimen temperature were obtained with an X-Y recorder. The system was straight-forward to use, and gave consistent and reliable results. Tested specimens were normally examined metallographically. Approximately 100 different superalloys were so tested; not all of these alloys, however, are represented in this paper.

Alloying Effects

Model Alloys. A series of eight model alloys was melted, to delineate the effects of key alloying elements on the melting/solidification behavior of nickel-base alloys. To a base composition of Ni-15Cr-10Co-5Mo-3Al were added 0.16 and 0.32% C, an additional 3% Al, 3 and 6% Ti, and 6% Ta, in various combinations. Thermograms for the base alloy and three modifications, along with final microstructures, are shown in Figure 2a. The base alloy shows the expected strong endothermic effect during melting and exothermic effect during solidification at 10 C/min, with no unusual features except for some (up to 23 C [42 F]) undercooling and recalescence at the start of solidification. All three thermograms for the model alloys in Figure 2a show inflections associated with the melting/solidification of the carbide eutectic. Of special interest is the position of this eutectic reaction with respect to the type of carbide involved. The reaction comes very late in the solidification process for the alloy containing no titanium, but much earlier for the titanium-containing alloys. Tantalum was found to have a similar effect to that of titanium. This simply reflects the fact that these two alloying elements are much stronger carbide formers than is chromium. Correspondingly, the chromium carbide is more clearly segregated at dendritic interstices than are the titanium or tantalum carbides. (Though not presented in this paper, a thermogram for IN-738 containing

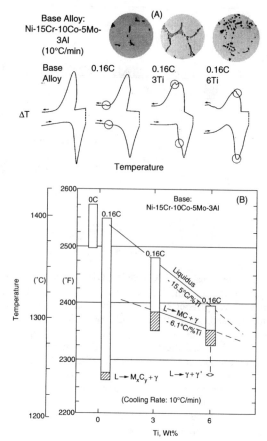

Figure 2. Results of melting/solidification studies of four model alloys of indicated compositions. (a) Thermograms obtained during melting and solidification, showing undercooling and recalescence of the two alloys without titanium and showing the carbide eutectic reactions (encircled) in the carbon-containing alloys. Microstructures of the three alloys with carbon, after solidification during the second run (20 C/min), are shown. (b) Solidification ranges obtained from the thermograms of (a), showing the effect of the carbon addition and then of titanium additions on the liquidus temperature. Note the very low temperature of the carbide eutectic reaction in the absence of titanium, and the reaction's progressive approach toward the liquidus temperature with increasing titanium content.

0.17% C and 0.10% Zr, in which the carbide eutectic inflection occurs at a much earlier point during solidification--midway up the rising part of the exotherm in the solidification thermogram--has carbides that are distributed much more widely throughout the structure than in the model alloy containing no titanium, the carbide eutectic in IN-738 occupying roughly one-half of the specimen volume. This shows that the thermogram can serve as an indicator of whether carbides are highly segregated or well dispersed in the cast structure.) The model alloy containing 0.16% C, 3% Al, 3% Ti, and 6% Ta formed some carbides 75 C (135 F) above the start of metallic dendrite formation.

The effect of the stability of the carbide on the temperature of its eutectic reaction may be seen more quantitatively in Figure 2b. The solidification range for the alloy containing carbon but no strong carbide formers is approximately 155 C (275 F), but the ranges for the titanium-containing alloys are less than one-half this amount. Figure 2b also shows that the liquidus temperatures are depressed, first mildly by the carbon addition, then more strongly by the additions of 3 and 6% Ti. The relationship between chemical composition and liquidus temperature is seen more clearly in Figure 3, a plot of liquidus temperature vs hardener content. The slopes of the line segments connecting the data points representing the individual alloys are labeled with the specific melting-point-depressing effect (ΔT per 1 wt pct) of the one element that varies between the two alloys. Titanium exhibits the strongest specific effect, with an average value of 14.2 C (25.5 F)/wt pct. That of aluminum is somewhat weaker, 12.2 C (22 F)/1 wt pct, while that of tantalum is the weakest of the three hardeners, with an average value of 4.8 C (8.5 F)/1 wt pct.

Commercial Alloys. Thermograms taken during melting and solidification of the commercial alloy IN-100 and its experimental niobium-free modification are presented in Figure 4a. The main thermal effects during melting and solidification are shifted to the right by removal of niobium, with the exact changes being denoted more clearly in the bar graph of Figure 4b. Removal of niobium also brings out a carbide eutectic reaction at the end of solidification. At the left ends of the thermograms are endotherms revealed during heating and exotherms revealed during cooling caused by the dissolution and precipitation, respectively, of γ'. Removing niobium from IN-100 causes a wide flat spot in the thermogram reflecting the fact that the loss of niobium moves the melting/solidification and γ' dissolution/precipitation reactions in opposite directions.

A bar graph showing the melting ranges of six well-known commercial superalloys during heating at 10 C/min is presented in Figure 5. The alloys are arranged in order of increasing Al + Ti content, since these elements are the dominant metallic melting-point depressants, as shown above. The highest-melting of these alloys is the iron-base superalloy A-286, its high liquidus temperature (1410 C [2570 F]) resulting not just from its low (2.2%) Al + Ti content but from its high (54%) iron content. Liquidus temperatures drop more-or-less regularly in going to the progressively higher Al + Ti contents of Waspaloy, Alloy 713LC, IN-738, and René 80, which has a liquidus temperature of just 1323 C (2413 F). Owing to its high cobalt content, the melting range for IN-100 lies above the trend line for the other alloys. Liquidus temperatures agree rather closely with those for the model alloys (dashed line) of Figure 3.

Table I presents the liquidus temperatures of two heats of IN-738, one having the original levels of carbon and zirconium, and the other having reduced levels of these elements. Whether measured during tests at 10 C/min or 20 C/min, and during heating or cooling, all increases lie between 7 and 9 C (13 and 16 F), the mean increase being 8 C (15 F). This consistent assessment of the change in liquidus temperature resulting from a minor change in chemical

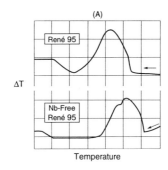

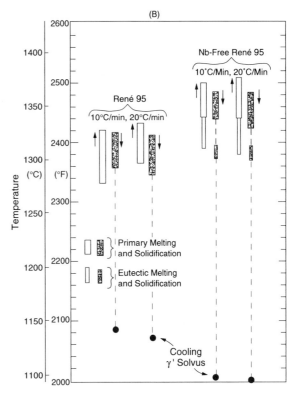

Figure 4. Effect of removal of niobium from René 95 on melting and solidification. (a) Comparison of solidification thermograms obtained at 10 C/min. (b) Melting/solidification ranges, showing increase in liquidus temperature and increase in temperature interval between the solidification range and the start of γ' precipitation during cooling, resulting from removal of niobium.

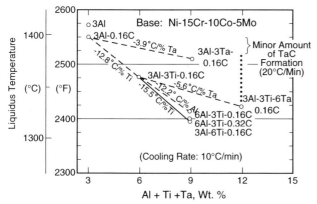

Figure 3. Graphical summary of the effects of carbon, aluminum, titanium, and tantalum on liquidus temperature of model alloys. Note indication of carbide formation well above the liquidus in the 6% Ta alloy when cooled at 20 C/min.

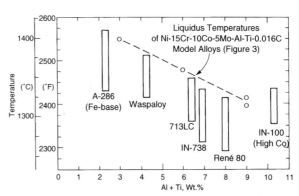

Figure 5. Variation of melting ranges of some commercial alloys with Al+Ti content during heating at 10 C/min. Liquidus temperatures agree closely with those for the Ni-0.16C-15Cr-10Co-5Mo-Al-Ti model alloys of Figure 3.

Table I. Effects of Small Changes in Composition on Liquidus Temperature of IN-738

Rate, C/Min	Mode	Liquidus Temperature, C (F)			
		Standard IN-738 (0.17C-0.10Zr)	Low C, Zr IN-738 (0.11C-0.04Zr)	Difference, C (F)	
10	Heating	1325 (2417)	1334 (2433)	+9 (+16)	
	Cooling	1324 (2415)	1331 (2428)	+7 (+13)	Mean Difference, +8 C (+15 F)
20	Heating	1330 (2425)	1339 (2442)	+9 (+17)	
	Cooling	1322 (2412)	1330 (2427)	+8 (+15)	

Table II. Critical Temperatures for Six Commercial Heats of René 80 During Solidification at 10 C/Min

	Temperature, C							Mean Deviation, C (F)
	Heat A	Heat B	Heat C	Heat D	Heat E	Heat F	Mean	
Liquidus	1317	1317	1318	1323	1321	1323	1321	3 (5)
Start of MC Formation	1297	1299	1299	1299	1299	1301	1299	1 (2)
Solidus	1273	1277	1277	1281	1280	1283	1278	3 (5)
Start of γ + γ' Eutectic Formation	--	1207	1206	1207	1206	--	1206	2 (3)
γ' Solvus	1149	1157	1153	1160	1160	1154	1156	3 (6)

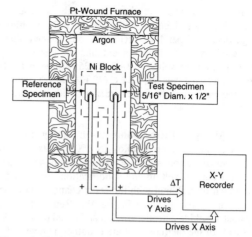

Figure 6. Physical arrangement for DTA tests of solid specimens.

composition, so as to provide guidance in setting pouring temperatures and possibly controlling other operations, illustrates the value of DTA in process control of superalloy production.

A more complete characterization of the solidification process is presented in Table II for six heats of René 80 remelt ingot. Reported are (1) the liquidus temperature, (2) the start of the MC + γ eutectic reaction, (3) the solidus temperature, (4) the start of the γ + γ' eutectic reaction, and (5) the γ' solvus temperature during cooling, marking the start of γ' precipitation, all measured during cooling at 10 C/min. The mean deviations in these temperatures are 1 to 3 C (2 to 6 F). Since these differences reflect the heat-to-heat differences in chemical composition of the René 80, as well as any errors in the DTA measurements, it is obvious from the small mean variations in the critical temperatures of the five reported phenomena that the DTA measurements are quite precise.

Studies of Solid Specimens

Experimental Procedure

Measurements of γ' solvus temperature, incipient-melting point, and several other effects were made with a different type of specimen and test instrument than in the preceding section, Figure 6. Cylindrical specimens weighing 4 g and with an axial thermocouple hole drilled to midlength were tested to a maximum temperature just above the incipient-melting point. This afforded greater precision than with the small specimens of commercial systems described above because of (1) the ability to place the thermocouple bead within and in direct contact with the specimen and (2) the higher magnification of the differential temperature that is made possible by avoiding the need to accommodate the large thermal effects encountered during full melting and solidification. The DTA specimen containing a 4-wire thermocouple, and a high-purity alumina reference specimen, both positioned in a cylindrical nickel block parallel to its axis and 180 degrees apart, were heated in a platinum-wound furnace at a constant rate (usually either 2 or 10 C/min) to a certain maximum temperature. The peak temperature was higher than the γ' solvus temperature, of course, and generally was about 40 C (70 F) above the incipient-melting temperature for the 10 C/min heating rate, and was somewhat lower for the 2 C/min rate. No sagging of the specimen nor bleeding of eutectic liquids occurred unless the specimen had been heated a greater amount beyond the incipient-melting temperature. The specimen was normally cooled at a natural rate (with power off) of approximately 25 C/min through the γ' solvus temperature range, so as to achieve a very "clean" cooling curve, free of any slight thermal disturbances that can occur with the low power inputs of the cooling mode. In all, more than 150 different model, experimental, developmental, and commercial superalloys, often with multiple runs on each, have been so-tested to date.

Interpretation of Thermograms

The features of thermograms were identified in two ways: (1) by gradient-furnace experiments and (2) by quenching experiments, as illustrated in Figure 7. In the former method, a thin bar specimen, 64 mm (2.5 inches) in length and containing thermocouples embedded near each end, was manually advanced along the horizontal hearth in the slot-like chamber of a SATEC gradient furnace at a rate that achieved a heating rate of 10 C/min. As soon as the temperature range within the specimen had encompassed all the critical temperatures of interest, the specimen was quenched vigorously in water. This procedure achieved the same microstructure at any point along the length of the specimen as if that point were contained in a normal DTA specimen being heated at 10 C/min, and "froze" that microstructure for subsequent metallographic identification. Four cast alloys (IN-100, B-1900, B-1900 + Hf [MM-007], and Mar-M246 + Hf [MM-006]) were studied by this method. Obviously, a gradient specimen could not readily be physically arranged to approach the temperature span of critical temperatures from above the liquidus temperature, and thus the cooling portion of thermograms was "calibrated" by a series of quenching experiments. (A smaller number of points on the heating thermograms also were so-checked.) In each test, a thin disc was removed from the top of a normal specimen and, with a thin wire "handle" spot-welded to it, was placed in its original position on top of the shortened and thermocoupled DTA specimen. Covered by an insulating lid atop the nickel block, the test disc was taken through the DTA run to the point of interest. The furnace was quickly opened (a possibility facilitated by the design of the Harrop furnace), and the specimen was immediately (transfer time approximately 1/2 second) quenched in water. Such quenching was done from the points denoted by the heavy dots on the thermogram of alloy Mar-M246 + Hf (MM-006), as shown in Figure 7, and on some specimens of other alloys, for a total of 35 specimens. The quenched discs were subsequently studied by conventional, and in some cases SEM, metallography.

Based on the results of the above "calibration" studies on solid specimens, (annotated) thermograms for Mar-M246 + Hf, with the various critical temperatures and other features labeled, are presented in Figure 8. Most striking, of course, are the large endotherm during heating and exotherm during cooling caused by the dissolution and precipitation, respectively, of γ'. In this paper, the peaks in these thermal perturbations are called the "maximum thermal effect" and correspond to the temperature at which there is the maximum time-rate-of-change in the volume of γ' precipitate. The location of this peak reflects the combined effects of (1) the shape of the γ' solvus curve in the quasi phase diagram of temperature versus combined hardener content, (2) the degree of chemical segregation or "coring," especially in cast structures, and (3) heating or cooling rates and diffusion effects. In cast structures, the maximum thermal effect corresponds rather closely to the γ' solvus temperature at the centers of dendrite arms, these locations being leanest in hardener content. (Actually, the centers of dendrite arms reach their "solvus" during heating somewhat before the maximum thermal effect occurs, as indicated in Figure 8a.) During heating, the "front" marking total dissolution of matrix γ' precipitate advances in a wave that eventually has consumed each entire dendrite arm by the time the specimen reaches the γ' solvus temperature. During cooling, Figure 8b, the reverse process occurs. The γ' solvus temperatures during heating and cooling, then, are represented by the corners at the right ends of these large thermal effects, where the curve joins the baseline. The γ' solvus temperature during cooling is always lower than that during heating, due to diffusion effects--the equilibrium solvus temperature value lying somewhere between the heating and cooling values.

Heating thermograms for specimens containing large amounts of the $\gamma + \gamma'$ eutectic constituent have a small inflection to the right of the main endotherm, caused by the dissolution of the coarser γ' particles that are characteristic of this constituent. The right end of this perturbation in the curve denotes the eutectic γ' solvus, Figure 8a. Micrographs of such zones in disc samples quenched from temperatures just below and just above this small inflection during heating are shown in Figure 9, upper. Following the specimen's temperature excursion above the incipient-melting point, during which time the zones containing the former $\gamma + \gamma'$ eutectic had remelted, disc specimens were quenched from just above and just below the corresponding inflection on the cooling curve. The hotter specimen exhibits pockets of quenched liquid (fine structure), while the cooler specimen contains the familiar $\gamma + \gamma'$ eutectic nodules (separated by a small amount of very low-melting liquid), thus showing that the inflection represents solidification of this eutectic, having an appearance much like in the original casting.

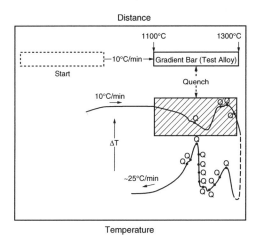

Figure 7. Method of obtaining metallographic samples for calibration of thermograms by means of gradient bars and selected quenched samples for heating thermograms (top and middle) and quenched samples for cooling thermograms (bottom).

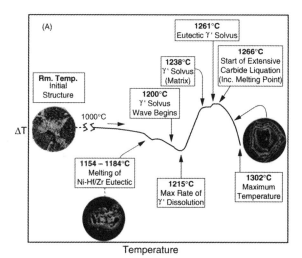

Figure 8A. Heating thermogram for Mar-M246 + Hf (MM-006), with critical temperatures and other key features labeled. Micrographs (SEM) at bottom and right show quenched liquid at Hf/Zr-enriched zone and at partially liquated carbide particle, respectively. Heating rate: 10 C/min.

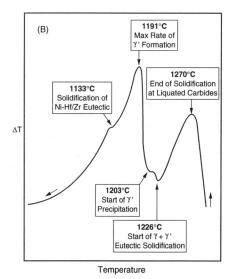

Figure 8B. Same as Figure 8A, but during natural cooling after DTA specimen had been heated about 40 C (70 F) above the incipient-melting point.

One of the most useful features of the present type of DTA test performed on solid specimens is its very sensitive detection of the incipient-melting point, or very beginning of the main melting process (ignoring the melting of the trace quantities of very low melting patches). The incipient-melting point is denoted by the corner at the right end of the heating thermogram where the curve breaks sharply downward. Because the thermocouple makes direct contact with the specimen, and because the differential temperature is highly expanded, as described previously, the incipient-melting point can be measured precisely, thus providing reliable guidance in heat-treating and other considerations. Physically, the incipient-melting point for typical superalloys containing carbon is the temperature at which carbides begin to liquate, i.e., to undergo the carbide eutectic solidification reaction in reverse. Such partial liquation of a carbide particle is seen in the SEM micrograph at the right-hand side of Figure 8a. Though "calibration" experiments have not been done on carbon-free superalloys, the incipient-melting point in such alloys presumably corresponds to the melting of any undigested $\gamma + \gamma'$ eutectic, or to the melting of the segregated zones of lowest melting point at dendrite interstices.

The final features of interest sometimes seen in thermograms of solid specimens are small inflections representing the melting and solidification at very low temperatures ($\approx$1150 C [2100 F]) of the tiny pockets of eutectic described above, containing such elements as zirconium, hafnium, boron, and phosphorus, combined with nickel. Zhu et al. have observed small pockets of such liquid even at 1100 C (2010 F) in slowly solidified nickel-base superalloys (7). These are located at interdendritic interstices and grain boundaries, between or adjacent to the $\gamma + \gamma'$ eutectic nodules, when present, as seen in Figure 9, lower.

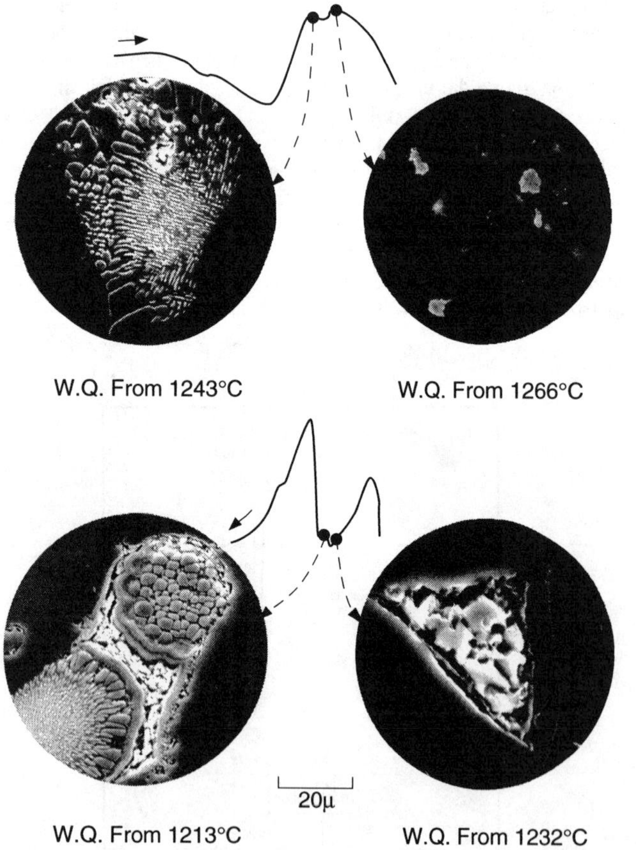

Knowledge of γ' solvus temperatures is often important in the control of solution-heat-treatment temperatures prior to precipitation hardening, and in the control of soaking temperatures for hot working. Thus it is important to know the relationship between γ' solvus temperature measured conveniently by DTA and the practical γ' solvus temperature for a heat treatment of some fixed duration at a constant temperature. Research at the AMAX laboratory and that of Maurer et al. has shown that the γ' solvus temperature measured by DTA drops moderately as heating rate decreases (8). In the present study, DTA results obtained at two heating rates for cast AF2-1DA and two of its modifications were compared with results for specimens heat treated two hours at six temperatures spaced 5.5 C (10 F) apart, and examined metallographically to determine the "practical" solvus temperature. The results, presented in Figure 10, show that the γ' solvus temperatures measured at 10 C/min lie above the "practical" solvus temperatures determined metallographically, the average temperature difference being about 15 C (27 F). Results for DTA specimens heated at 2 C/min, however, agree closely with the metallographic study. This indicates that DTA tests should be made at lower heating rates than the commonly used 10 C/min rate when seeking guidance for practical heat treatments. Alternatively, subtracting a correction factor of about 15 C (27 F) for as-cast alloys (and presumably somewhat less for specimens with fine γ' particles that dissolve more readily) from the γ' solvus temperature determined by DTA at 10 C/min would most likely provide sufficient accuracy for control of heat-treating or soaking temperatures.

γ' Solvus Temperatures

Model alloys. The ability of DTA to accurately detect the γ' solvus temperature during heating and cooling was discovered during tests on a series of 16 model alloys (four molybdenum levels at each of four hardener levels) at the Climax Molybdenum (AMAX) laboratory in April 1968 by Loomis, Freeman, and Sponseller (4). The series of wrought alloys with controlled additions of molybdenum to a Ni-Cr base (Ni/Cr ratio = 4.6 on a weight basis) at four aluminum levels (one with titanium, not presented here) show progressive raising of the γ' solvus temperature by molybdenum as it dissolves in, and increases the quantity of, γ', Figure 11. The molybdenum

Figure 9. Scanning electron micrographs of Mar-M246 + Hf (MM-006) samples quenched from temperatures bracketing the thermogram features that denote the solvus temperature for eutectic γ' during heating (top) and the temperature for solidification of the $\gamma + \gamma'$ eutectic during cooling (bottom). Note that only carbides remain above the eutectic γ' solvus.

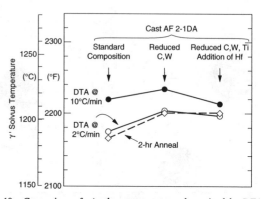

Figure 10. Comparison of γ' solvus temperatures determined by DTA during heating at 2 C/min and 10 C/min with those obtained metallographically from specimens held two hours at temperatures spaced 5.5 C (10 F) apart, for cast AF 2-1DA and two of its modifications. The slower heating rate is seen to provide "practical" solvus temperatures.

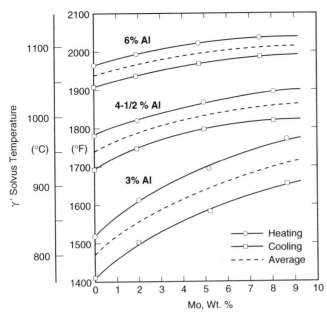

Figure 11. Effect of molybdenum on γ′ solvus temperature, as determined by DTA during heating and cooling at 10 C/min of model Ni-Cr-Al-Mo alloys having three levels of aluminum content (Ref. 4).

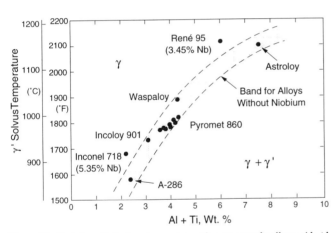

Figure 12. Variation of γ′ solvus temperatures of seven wrought alloys with Al + Ti content. Niobium raises the data points for Inconel 718 and René 95 above the band for the niobium-free alloys. Values were obtained by DTA during heating at 10 C/min.

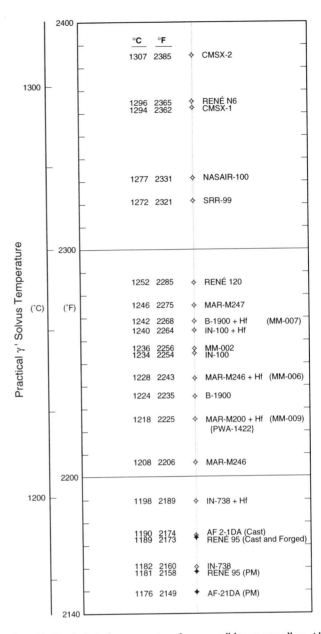

Figure 13. Practical γ′ solvus temperatures for some well-known superalloys. All but three data points represent cast alloys, the wrought alloys being represented by solid symbols. The five alloys with the highest solvus temperatures are single-crystal alloys. Note the effect of hafnium in raising the solvus temperatures of four alloys. Data were determined by DTA of solid-type specimens at a heating rate of 2 C/min.

effect decreases as solvus temperature increases, suggesting a kind of exhaustion or compression effect, as the melting range is approached. The equilibrium γ′ solvus temperature is taken as the average of those measured during heating and cooling at 10 C/min; the decreasing interval between the two temperatures as hardener level increases is due to faster diffusion at higher temperatures.

Commercial Alloys. The γ′ solvus temperature of seven wrought alloys, including commercial disc alloys and wrought Pyromet 860, measured during heating at 10 C/min, are presented in Figure 12. This plot of solvus temperature versus Al + Ti content shows the progressive increase in solvus temperatures over the years, starting with the early-generation disc Alloy A-286, and proceeding to the higher-hardener alloys such as Inconel 718, Waspaloy, and René 95. An increase in solvus temperature of over 500 F is evident. (The reason Inconel 718 and René 95 lie above the band is that their solvus temperatures are raised by high niobium content.)

The range of practical γ′ solvus temperatures among 21 commercial or developmental alloys is shown in Figure 13. As described above, "practical" solvus temperatures are those obtained at a very slow heating rate, 2 C/min, and closely represent the solvus temperature for a 2-hr heat treatment or soak prior to hot working. The alloys represented are primarily of the cast type, but the wrought alloys AF2-1DA (PM), René 95 (PM), and René 95 (cast and forged) are included. Because γ′ is the primary high-temperature strengthener in nickel-base superalloys, and because the quantity of γ′ tends to increase with increasing solvus temperature, the height of an alloy within Figure 13 serves as a general indicator of its high-temperature strength. Of special interest are the significant boosts in γ' solvus temperature of four of the alloys--IN-738, Mar-M246, B-1900, and IN-100--caused by the addition of 1.5-2% Hf. Alloys with γ' solvus temperatures above 1260 C (2300 F) are single-crystal alloys; mainly first-generation alloys are shown, but the state-of-the-art single-crystal alloy René N6 is included.

Numerous superalloys containing hafnium have found their place in modern gas-turbine engines. The curves of Figure 14 illustrate a "fingerprint" of hafnium that shows up in the cooling thermogram of solid specimens after they have been heated to about 40 C (70 F) above the incipient-melting point. Shown are the cooling thermograms for IN-738 and 0.5, 1.0, and 1.5% Hf modifications thereof. At 1.0% Hf, and especially at 1.5% Hf, distinct sidepeaks are noted on the higher-temperature side of the γ' precipitation exotherm. These are caused by solidification of the γ + γ' eutectic constituent and by γ' precipitation that begins in the so-called "swirly" γ' regions before general precipitation of γ' begins in the rest of the matrix. This is attributed to enrichment of the matrix in the swirly zones by hafnium, a γ' former. Kotval et al. have shown that the boundaries of the "swirly" zones are enriched in boron (9). (The prominence of the hafnium sidepeak diminishes, owing to homogenization, as repeated DTA tests are made on the same specimen, not shown here). Because this phenomenon extends the γ' precipitation process over a wider temperature range, the peak height of the main γ' precipitation exotherm is progressively reduced by the hafnium additions.

The processing method can have a marked effect on the dissolution and precipitation of γ' in a given alloy. This is illustrated in Figure 15a, which compares heating and cooling thermograms for IN-738 in the cast and powder-metallurgy-consolidated conditions. During heating at 10 C/min, the curve for the PM specimen rises much more sharply during the final stages of γ' dissolution than does that for the cast specimen. This reflects the greater uniformity of composition in the PM material because of the vastly finer dendritic structure in atomized powder particles than in an investment casting. Following consolidation, the PM-processed IN-738 is very homogeneous in chemical composition, while the casting is quite segregated. In a sense, the casting can be thought of as a composite of many alloys of closely spaced compositions. The thermogram for the metal of each composition would have a sharp rise to the γ' solvus point, similar to that for a PM sample. Because the curve for each of the aforementioned alloys would be displaced somewhat from the curves for the others, the combined effect of these displaced curves would be to spread out the rising portion of the thermogram near the γ' solvus. Thus, this portion is more than twice as wide for the cast specimen as for the PM specimen, 58 C (105 F) vs 28 C (50 F). (That this portion of the γ' dissolution endotherm for the PM specimen has a finite width should not be taken as an indication that the specimen retains significant segregation. Even a perfectly homogeneous specimen would have definite width at this stage, because of the curvature in the γ' solvus surface, as described previously.) Figure 15a reveals a similar effect in the γ' precipitation exotherm of the cooling thermogram. The difference between the two specimens is lower, however, because of some homogenization that occurs in the cast specimen during its excursion above the incipient-melting point. The effect of homogenization may be seen more clearly in Figure 15b. This shows the progressive march to the right of the peak in the γ' precipitation exotherm as one cast specimen is tested during three successive runs. The flow of hardener into the lean areas at the skeleton of each dendrite raises its

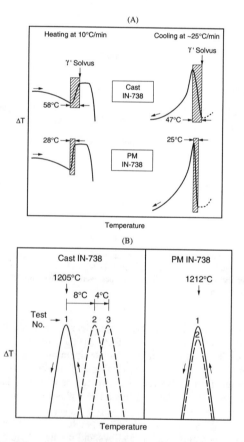

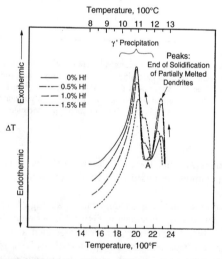

Figure 14. Effect of hafnium additions to IN-738 on the thermograms obtained during cooling after the solid-type specimens had been heated approximately 40 C (70 F) above the incipient-melting point. Hafnium additions of 1% and 1.5% cause progressively larger sidepeaks at A on the high-temperature side of the main γ' precipitation exotherm, and hafnium additions progressively lower the height of the main exotherm, for reasons discussed in the text. Cooling rate: ~25 C/min.

Figure 15. Effect of processing method on the thermograms of IN-738. (a) The temperature interval (crosshatched band) between the maximum thermal effect and the γ' solvus point is markedly lower, during both heating and cooling, for the PM specimen than for the cast specimen. (b) The peak of the γ' precipitation exotherm during cooling of the cast specimen shifts progressively to higher temperatures during successive DTA tests of the same specimen, because of incremental homogenization. No shift occurs for the PM specimen, because of its high homogeneity.

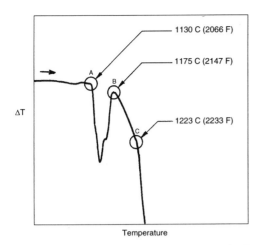

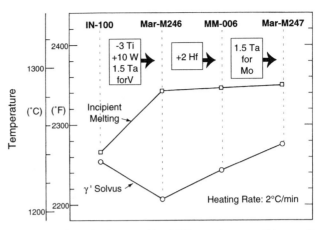

Figure 16. Distinct low-temperature melting reaction in Inconel-718, occurring at 1130 C (2066 F) during heating at 10 C/min. The more-general melting reaction begins at a moderate rate at 1175 C (2147 F), and at a stronger rate at 1223 C (2233 F).

Figure 17. Effects of main compositional differences between well-known alloys on practical γ' solvus temperature and on incipient-melting temperature during heating at 2 C/min, and thus on the width of the solution-heat-treating "window." The window is changed most sharply by titanium.

γ' solvus and, hence, shifts the exotherm for the leanest region to higher temperatures. Two successive runs for the PM specimen, however, reveal no shift in the peak, indicating that the starting structure was quite homogeneous.

<u>Incipient Melting</u>

Knowledge of the temperature marking the very onset of melting is crucial to the control of superalloy processing and utilization. It has long been known that small amounts of liquid play havoc with the workability of superalloys; hot shortness thus causes severe cracking. This severe sensitivity to the presence of a small amount of liquid is undoubtedly aggravated by the "stiffness" and poor thermal conductivity associated with the high alloy content of superalloys. Lack of space does not permit a full exposition here of the limits posed on the workability of superalloys by incipient-melting. The most dramatic example of this effect, however, is presented by Inconel 718, for which forging temperatures must be strictly limited to a maximum of 1120 C (2050 F). This is because the 5% Nb content of this alloy causes some eutectic liquid to persist to very low temperatures, at which point Laves phase is formed. During heating at 10 C/min, DTA reveals a distinct melting reaction in this region at 1130 C (2066 F), Figure 16. The more general melting process begins at a moderate rate at 1175 C (2147 F) and at a stronger rate at 1223 C (2233 F).

The DTA method has been especially valuable in coping with the problem of unwanted partial melting during the solution treating of alloys possessing high γ' solvus temperatures. The difference in width of the so-called heat-treating "window" for solution heat-treating among some familiar alloys is shown in Figure 17, for which the critical temperatures were determined during heating at a very slow rate, 2 C/min, to obtain a realistic assessment of the γ' solvus and incipient-melting temperatures. Alloy IN-100, with its very low incipient-melting point and high γ' solvus temperature, both caused by high hardener--especially titanium--content, has virtually no window, the width being just 7 C (12 F). Removing 3% Ti from the IN-100 composition, while adding 10% W and replacing 1% V with 1.5% Ta, gives essentially the composition of Mar-M246. Inasmuch as tungsten raises both the γ' solvus temperature and incipient-melting point, the dominant effect is that of removing 3% Ti, which strongly lowers the solvus temperature

and strongly raises the incipient-melting point. This movement of the two critical temperatures in opposite directions thereby leads to a very wide window--76 C (136 F). Adding about 2% Hf (1.9% in the present study) narrows the window appreciably [to 57 C (103 F)] for the resulting MM-006 alloy, mainly because hafnium raises the solvus temperature significantly. The further substitution of 1.5% Hf for about an equal amount of molybdenum (and slight reduction in hafnium content) leads to the Mar-M247 composition. Because the tantalum addition raises the solvus temperature considerably more than it does the incipient-melting temperature, a significantly narrower window results--43 C (74 F).

Control of the heat-treating window is most crucial for single-crystal superalloys, a class of alloys for which the maximum amount of hardener has been "packed in" to maximize high-temperature strength. Strong melting-point depressants have been replaced, in part, by strengtheners with a milder effect, or that even tend to raise the incipient-melting point, thus permitting a raising of the solvus temperature and/or a widening of the window.

Figure 18 presents DTA results for the state-of-the-art production turbine-blade Alloy René N6 (Ni-12.5Co-4.2Cr-1.4Mo-6W-5.9Re-7.2Ta-5.75Al-0.15Hf-0.01Y) (10) and three other (experimental) single-crystal alloys, all fully heat treated, supplied by GE Aircraft Engines. Though their compositions are proprietary, all experimental alloys are identical in composition, except for having different levels of cobalt, namely, 5, 12.5, and 20%. The critical temperatures obtained by DTA at a heating rate of 4 C/min are shown in Figure 18. Besides presenting the γ' solvus temperature and incipient-melting temperature, the graph shows the temperature at which the maximum thermal effect (minimum in the main endotherm during heating) occurs. This represents the temperature at which the maximum rate of γ' dissolution occurs for each alloy. Also shown, connected to the heating data points by dashed lines, are the temperatures of the corresponding γ' reactions during natural cooling at about 25 C/min, i.e., the start of γ' precipitation, and the maximum thermal effect, representing the maximum rate of γ' formation (peak of the main exotherm). The following observations are noteworthy:

(1) René N6 has rather high γ' solvus and incipient-melting temperatures, 1299 C (2370 F) and 1340 C (2444 F), respectively, the difference giving it an adequately wide heat-treating window, 41 C (74 F). Alloy X-1 has exceptionally high γ' solvus and

incipient-melting temperatures, 1340 C (2444 F) and 1372 C (2502 F), respectively, and a 32 C (58 F) window.

(2) Among the three experimental alloys, increasing the cobalt content lowers the γ' solvus temperature sharply and the incipient-melting temperature mildly, thereby widening the window to 46 C (82 F) at 12.5 % Co and to 68 C (122 F) at 20 % Co, intervals that are greater than necessary for ease of solution heat treating. This raises the possibility of increasing the hardener content and strength of these two alloys.

(3) The width of the interval between the temperature of the maximum thermal effect and the γ' solvus temperature during heating provides a useful qualitative indicator of the degree of homogenization in the casting. This is seen in the following comparison of the temperature intervals during heating:

René N6	X-1	X-2	X-3	X-1,2,3 Avg
17 C (30 F)	23 C (42 F)	22 C (39 F)	20 C (36 F)	22 C (39 F)

The interval for René N6 is considered to be close to the minimum possible for a fully homogenized alloy, the minimum reflecting the curvature in the γ' solvus surface, as discussed previously. The interval for each of the experimental alloys is significantly wider, the average difference being 5 C (9 F). Similar differences are indicated by the cooling data points of Figure 19. This suggests more segregation in the experimental alloys than in René N6. Such a difference was clearly confirmed metallographically, as illustrated in Figure 19, and reflects the longer solution heat treatment that had been used for René N6 than for the experimental alloys. This effect is similar to that described earlier for the difference, between cast (segregated) and powder metallurgy (nearly homogeneous) IN-738, with regard to the width of this temperature interval, Figure 15. The foregoing suggests that careful measurement of this interval can be used to monitor the degree or progress of homogenization in nickel-base superalloys.

The solution-heat-treating windows of 24 single-crystal alloys evaluated by the author are presented in Figure 20. Represented are five commercial alloys, including the René N6 described above, and 19 proprietary, experimental, or developmental alloys from different sources, including the experimental alloys X-1, X-2, and X-3 described above. The bottom of each bar represents the alloy's practical γ' solvus temperature, and the top represents the incipient-melting temperature. The bars of Figure 20 reveal quite a large range of γ' solvus temperatures, especially because of the low solvus temperature of MM-002. The widths of the windows vary from about zero to 71 C (127 F). Thus some alloys would be impossible to solution heat treat without incurring incipient melting, while other alloys have excessively wide windows--larger than needed for ease of heat treating--thus indicating the possibility of raising hardener content for the sake of improving high-temperature strength.

Problem-Solving

In the following are given a few examples of how DTA of solid specimens has been used to solve practical problems in the casting of superalloys. The first one has to do with a serious and longstanding hot-tearing problem in the casting of medium-sized Alloy 713C integral turbocharger wheels. A DTA specimen taken from a cracked wheel containing 0.15% Zr revealed the presence of a very-low-melting (about 1100 C [2000 F]) liquid in both the heating and cooling thermograms, as encircled in Figure 21a. Because pockets of high-zirconium constituent were detected near

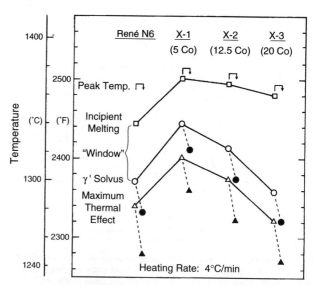

Figure 18. Critical temperatures of the single-crystal Alloy René N6 and three experimental alloys, as determined by DTA during heating at 4 C/min. The window is widened markedly by cobalt. Critical temperatures obtained during natural cooling (~25 C/min) are denoted by solid symbols and are connected to their corresponding temperatures obtained during heating by dashed lines.

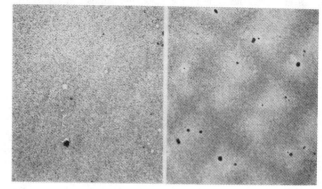

Figure 19. Microstructures of René N6 (left) and Experimental Alloy X-3 (right) in the heat-treated condition. The former alloy exhibits negligible segregation, while the latter is distinctly cored. The existence of this difference was indicated by DTA. 75X.

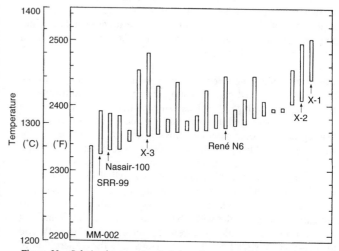

Figure 20. Solution-heat-treating "windows," represented by open bars, for 24 experimental and first-generation single-crystal alloys. State-of-the-art alloy René N6 also is shown. Alloys are arranged in order of increasing practical γ' solvus temperatures.

the crack, test castings made from alloys doped with various levels of zirconium were examined for hot tears. The results show a striking correlation between the incidence of hot tears and zirconium content:

0.09% Zr	0% hot tears	(0 of 12)
0.15% Zr	50% hot tears	(6 of 12)
0.19% Zr	100% hot tears	(12 of 12)

The powerful effect of zirconium in promoting the very-low-melting liquid is seen in the thermograms of Figure 21b, representing unusually high (0.34%) zirconium content in hafnium-modified B-1900; very prominent low-melting reactions (encircled) are evident during heating and cooling. Subsequent control of the zirconium content in the Alloy 713C remelt ingot eliminated the costly hot tearing of the wheels.

Microporosity in castings of IN-100 led to a comparison of good and bad castings by DTA, Figure 22. Perturbations (encircled in Figure 22b) in the thermograms were noted for the most porosity-prone castings. These inflections indicate the presence of $\gamma + \gamma'$ eutectic. The development of this constituent, and the porosity itself, could have been influenced by a higher-than-normal concentration of nitrogen detected in the porous castings.

A somewhat similar problem of microporosity was observed in castings of MM-002. Both the heating and cooling portions of the thermograms for porous castings, obtained during heating at 2 C/min followed by natural cooling, possessed strong indications of the inflections associated with the $\gamma + \gamma'$ eutectic constituent. Such inflections are marked B and C in Figure 23. It was noted that the reaction was so strong that it even continued above the incipient-melting point for some castings (not illustrated here). As an indication of the intensity of the reaction, the ratio of the height of the eutectic γ' inflections (B and C) to the height of the main endotherm/exotherm for the dissolution/precipitation, respectively, of matrix γ' marked A and D, respectively, are as follows:

Ratio	Sound Casting	Porous Casting
B/A, Heating	0.49	3.27
C/D, Cooling C/D	0.15	0.24

Ratios are smaller during cooling, partly because of partial absorption of the eutectic γ' by the matrix during the hottest part of the DTA test. Nevertheless, it is seen that higher ratios are associated with, and apparently indicative of, the tendency toward excessive microporosity.

Summary

Differential thermal analysis (DTA) performed on specimens of more than 200 model, experimental, developmental, and commercial nickel-base superalloys, using large specimens with internally positioned thermocouples, has revealed much useful information about the critical phase changes and microstructures governing the processing and performance of this important class of alloys.

Studies of eight Ni-15Cr-10Co-5Mo-C-Al-Ti-Ta model alloys during melting and solidification show that the carbide eutectic reaction occurs at very low temperatures in the absence of a strong carbide former, and that carbide formation occurs progressively earlier in the solidification process as the quantities of strong carbide-forming reactants increase, an alloy with 0.16C-3Al-3Ti-6Ta forming carbides well before (75 C [135 F] above) the start of metallic dendrite formation. The specific effects (per 1 wt pct) of aluminum, titanium, and tantalum on liquidus temperature during solidification are

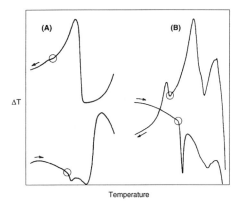

Figure 21. Thermograms illustrating the use of DTA to identify the cause of hot tearing in Alloy 713C turbocharger wheel castings. (a) Small inflections A and B in the heating and cooling thermograms, respectively, for only the cracked castings suggested that low-melting liquid caused by zirconium was responsible. (b) The powerful effect of zirconium in promoting such low-melting liquid is seen in these thermograms of modified B-1900 containing 1.35% Hf and 0.34% Zr, which have very prominent indications of low-melting liquid (encircled).

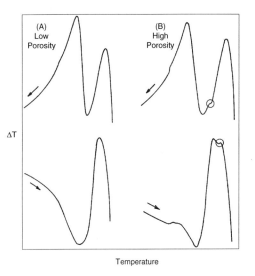

Figure 22. Thermograms for IN-100 castings having low and high amounts of porosity. The encircled inflections related to the $\gamma + \gamma'$ eutectic constituent were found only in castings of high porosity.

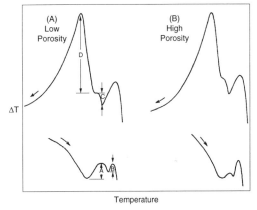

Figure 23. Thermograms for MM-002 castings having low and high amounts of porosity. Both castings had strong DTA indications of the $\gamma + \gamma'$ eutectic reaction, but porosity was worse in those castings with higher B/A and C/D ratios, denoting more of the $\gamma + \gamma'$ constituent.

presented. Solidification ranges, and alloying effects thereon, of some well-known commercial alloys are given.

Measurements of γ' solvus temperature (for matrix precipitate and for eutectic γ') and of incipient-melting temperature are best performed on large solid specimens with internal thermocouple contact, giving direct and precise indications of these critical temperatures and providing other useful information as well. Features of thermograms were identified by calibration studies of thin bars of four cast alloys heated at 10 C/min in a thermal gradient furnace and quenched, and of 35 small disc specimens of various cast alloys quenched from the DTA furnace at temperatures corresponding to various points on heating and especially on cooling thermograms. Indicated γ' solvus temperatures decrease moderately as heating rate decreases; values determined at 2 C/min agree closely with those for practical 2-hour annealing and soaking treatments. Guided by the above results, the effects of alloying elements and of processing on γ' solvus and/or incipient-melting temperatures of alloys are presented. Of special interest is the unique ability of DTA to directly and precisely reveal the width of the temperature interval, or "window," for the solution heat treating of single-crystal alloys; windows ranging from essentially zero to 71 C (127 F) in width are reported.

Examples are presented of using DTA to solve costly problems of hot tearing and excessive microporosity in castings.

In the 28 years since γ' solvus temperatures were first measured by DTA, this technique has established itself widely in the superalloy industry, owing to its great utility in monitoring the critical phase changes that are important in the metallurgical development and process-control of superalloys. The value of DTA in the metallurgy of superalloys already is considered by the author to be at least as great as that of dilatometry in the metallurgy of transformable and hardenable steels. Hopefully, the results described herein may further advance the usefulness of DTA in the metallurgy of superalloys.

Acknowledgements

The author is deeply indebted to: W.T. Loomis, whose doctoral research created the opportunity for discovering that γ' solvus temperatures could be measured directly by DTA; G.A. Timmons, D.V. Doane, and the late M. Semchyshen, for maintaining financial support for DTA studies; W.P. Danesi, R.G. Dunn, and C.C. Clark, for providing many interesting sample sets of superalloys for DTA analysis; W.S. Walston, for supplying single-crystal alloys; W.C. Hagel, W.J. Boesch, D.R. Muzyka, C.R. Whitney, J.A. Petrusha, J.H. Wood. L.S. Taylor, P.D. Spilling, G.L. Erickson, J. Domingue, R. Egbert, and the late G.E. Marshall, for enlightening discussions; R.C. Binns, D.E. Schumacher, E.K. Ohriner, M.C. LaPine, R.R. Lannin, and R.W. McConnell, for assistance with DTA experiments; and D.J. Kruzich, K.S. Smalinskas, D.J. Pipe, and M.S. Snell for assistance in manuscript preparation. The author is most indebted to, and dedicates this paper to his wife Mary, a "DTA widow" on too many nights and weekends, for her considerable patience and steadfast support throughout the very long research stage and the preparation of the manuscript, adding immeasurably to the success of this endeavor.

References

1. H. LeChateliér, "The Action of Heat on Clays," Compt. rend. 104 (1887), pp. 1443-46.

2. W.C. Roberts-Austen, "Fifth Report to the Alloys Research Committee," Proc. Inst. Mech. Engrs. (London) 1 (1899), pp. 35-102.

3. D.L. Sponseller, W.G. Scholz, and D.F. Rundle, "Development of Low-Alloy Ductile Irons for Service at 1200-1500 F," AFS Transactions 76 (1968), pp. 353-68.

4. W.T. Loomis, J.W. Freeman, and D.L. Sponseller, "The Influence of Molybdenum on the γ' Phase in Experimental Nickel-Base Superalloys," Metallurgical Transactions 3 (1972), pp. 989-1000.

5. D.L. Sponseller, "Differential Thermal Analysis of Superalloys," presented at the session Physical Metallurgy of Superalloys," ASM/AIME Materials Science Symposium, Oct 21-24, 1974, Detroit, MI.

6. D.L. Sponseller, "Use of Differential Thermal Analysis in the Melting, Casting, and Heat Treating of Superalloys," presented at the 1975 Vacuum Metallurgy Conference, June 23-25, 1975, Columbus, OH.

7. Y. Zhu, S. Zhang, T. Zhang, J. Zhang, Z. Hu, X. Xie, and C. Shi, "A New Way to Improve the Superalloys," Superalloys 1992, pp. 145-154.

8. G.E. Maurer, J. Domingue, and W.J. Boesch, "Superalloy Design with Differential Thermal Analysis," Proceedings of the 27th Annual Meeting of the Investment Casting Institute (1979), Rosemont, IL, pp. 16:01-16:23.

9. P.S. Kotval, J.D. Venables, and R.W. Calder, "The Role of Hafnium in Modifying the Microstructure of Cast Nickel-Base Superalloys," Metallurgical Transactions 3 (1972), pp. 453-458.

10. W.S. Walston, E.W. Ross, K.S. O'Hara, T.M. Pollock, and W.H. Murphy, U.S. Patent 5,455,120, "Nickel-Base Superalloy and Article with High-Temperature Strength and Improved Stability," René N6, 1995.

CREEP & FATIGUE

CREEP ANISOTROPY IN NICKEL BASE γ, γ' AND γ/γ' SUPERALLOY SINGLE CRYSTALS

Dilip M. Shah and Alan Cetel

Pratt & Whitney
M/S 114-45, 400 Main Street, E. Hartford, CT 06108
Tel: (860)565-8499 / 3485
FAX: (860)565-5635

Abstract

A comprehensive study of creep deformation as a function of orientation for a broad range of single crystal nickel base alloys, is presented. The alloys include a wide range of precipitation hardenable γ/γ' superalloys, as well as single phase γ and γ'. To provide a complete overview, results of other critical studies are also cited. In general it is observed that superalloys, predominantly strengthened with elements such as Mo, partitioning strongly to the γ matrix, behave very anisotropically, as do single phase γ alloy. In contrast most superalloys strengthened with refractory elements such as Ta, which partition to the γ' phase, tend to display isotropic creep behavior, as do most single phase γ' alloys. The seemingly wide variation in creep anisotropy with composition, as well as with temperature, and microstructure, is in most cases quantitatively rationalized based on a model requiring only a single parameter. The parameter is defined as the ratio of creep resistance of cube slip to octahedral slip for a given alloy, at the temperature and microstructure of interest. The model shows that increased participation of cube slip is responsible for the isotropic behavior of single phase γ' and several superalloys. An eight-fold enhancement in creep resistance for <111> orientation over <100> orientation is predicted and observed, for a γ alloy, which predominantly deforms by octahedral slip. A concept of threshold shear stress is proposed, to account for occasional observation of unusually high creep anisotropy at low stresses.

Introduction

It is well recognized, that elastic behavior, as well as plastic behavior of single crystal superalloys, is a strong function of orientation. This aspect is of some practical interest from a design standpoint, for limiting the spread in mechanical properties with variation in primary orientation of cast single crystal components. Of all the mechanical properties studied, the orientation dependence of creep behavior or creep anisotropy, has been most intriguing. A wide variation in creep rupture life with tensile axis orientation - especially along <111> compared to <100> - has been observed. These variations exist over a wide range of temperature, stress, precipitate size, and alloy composition, even at concentrations of some elements below 1000 ppm. A number of studies of creep anisotropy[1-5], attest to this perspective. For clarity of discussion, we shall define "creep anisotropy" as a ratio of creep resistance in a high modulus direction such as <111>, to the lowest modulus orientation <100>. The creep resistance may be measured variously as rupture life, time to a fixed strain, or the inverse of minimum creep rate.

For many superalloys, the potential of achieving significantly higher creep resistance in off-<100> orientations, makes the study of creep anisotropy even more attractive. With the evolving application of single crystal superalloys to supersonic transport and land based turbines, very long time creep durability is becoming a critical engineering requirement. In this context the ability to enhance creep resistance by factor of 5-10X in off-<100> orientations may have practical implications. From an application standpoint, as well as from a desire to gain a better insight into the creep behavior of superalloys, a comprehensive grasp of the nature of creep anisotropy in nickel base alloys is desirable. Especially significant, is determining the range of alloy compositions and microstructures that lead to high creep anisotropy. This paper is a first such approach, providing a framework for understanding the creep anisotropy of all nickel base alloys.

Results of our study covering a broad range of nickel base single crystal alloys are presented and analyzed. The alloys include single phase γ-nickel base solid solution, γ' - Ni_3Al base intermetallics, as well as a wide range of complex superalloys. For the sake of completeness, relevant data from other sources are also cited. It is shown that a model originally developed for single phase γ', is able to capture much of the nature of creep anisotropy of both simple γ alloys, as well as complex superalloys. The model is conceptually based on the orientation dependence of creep strain distribution among octahedral and cube slip systems.

The effects of alloying, temperature, and microstructure on creep anisotropy, are linked through the manner in which relative strengthening of the two slip systems is affected. It is demonstrated that, as predicted by the model, γ alloys such as Hastelloy-X, with presumed absence of cube slip, show higher creep anisotropy than γ'- Ni_3Al and many γ/γ' superalloys with dual slip modes. Consistent with the model, experimental observations show that γ' strengthening elements such as Ta and W reduce creep anisotropy; while alloying elements such as Mo which largely partition to γ, increase the creep anisotropy. Variations in creep anisotropy with temperature and γ' size are also explained by the model in terms of changes in the participation of cube slip. Very high creep anisotropy in some alloys, is rationalized by introducing a modified concept of threshold shear stress for each family of slip systems in the model.

Alloy Selection

This work originated as a scientific study of a binary γ'-Ni_3Al single crystal alloy, for which a model was proposed to normalize the observed creep anisotropy[2]. To verify the general validity of the model, a ternary γ'-$Ni_3(Al, Ta)$ alloy was evaluated. The proposed model helped rationalize the reduced creep anisotropy in the ternary alloy, though a subtraction of a back-stress was necessary to fully account for the high stress exponents. Around the same time, some high Mo containing alloys, were observed to display very high creep anisotropy. This behavior differed from that of single crystal alloys such as PWA1480, which were primarily strengthened with Ta additions. It was evident that alloy composition was playing a significant role in controlling the nature of creep anisotropy. To gain a better insight into compositional effects, a series of superalloys were selected, with wide variations in refractory metal content. Alloy selection was not constrained by usual practical considerations of stability, castability, heat treatability, or oxidation and corrosion resistance. Surprisingly, in spite of the complex precipitation hardening contribution in these alloys, most of the observed behavior could be rationalized based on the model developed for single phase γ' alloys. The successful application of the model, indicated that single phase γ nickel base solid solution hardened alloys should display relatively high creep anisotropy. To confirm this prediction, and for the sake of scientific completeness, Hastelloy-X was selected as a model γ alloy.

All of the selected γ, γ' and complex superalloy compositions evaluated in this study are listed in Table I. For purposes of comparison and discussion, compositions of several alloys of interest from other sources are also listed.

Evaluation Procedure

The experimental alloys in this study were directionally solidified as nominally <100> oriented single crystals by a standard investment casting technique. Hastelloy-X, as a representative γ-nickel base solid solution alloy, as well as binary Ni_3Al and ternary $Ni_3(Al, Ta)$-γ' alloys, were cast as 15 mm (5/8") diameter bars. For each alloy, compressive creep specimens of dimensions 5 mm x 5 mm x 13 mm (0.2" x 0.2" x 0.5"), were machined in various crystallographic orientations, by properly aligning the bars using a goniometer stage in conjunction with back reflection X-ray Laue analysis. All alloys were evaluated in compressive creep, to determine the minimum creep rate as a function of orientation at several temperatures of interest. At a given temperature, typically a single specimen was used per orientation. Once the steady state creep was established, the specimen was uploaded to determine minimum creep rate at another stress level.

The selected superalloy compositions were cast into 90 mm (3.5") diameter single crystal ingots. Tensile creep specimens with desired crystallographic orientations were machined using 6 mm dia. x 50 mm (0.25" dia. x 2") cylindrical slugs. The slugs were extracted from the ingots, using electric discharge machining(EDM), at pre-designed angles. The required machining angles were determined using an average crystallographic orientation of the ingot, derived from Laue analysis at several locations. Subsequently, the exact orientation for each specimen was individually verified. Deviations up to 10° from the target orientation were considered insignificant for this study. The required heat treatment for each alloy was carried out prior to final machining. Most alloys were solution heat treated in the range of 1260-1315°C (2300-2400°F), avoiding any incipient melting, and were subsequently aged for 4 hours at 1079°C (1975°F) followed by 32 hours at 871°C (1600°F). No special effort was made to optimize the heat treatment. The exact heat treatment is listed in Table I along with nominal compositions of the alloys. Specimens were creep tested to failure at stress levels producing rupture within a few hundred hours. For most alloys, at least two specimens were tested per creep condition. For some alloys, yield strength was measured in compression to provide confirmation of the proposed model.

Recognizing that γ' morphology and γ/γ' lattice mismatch could be critical parameters controlling creep anisotropy, replica TEM and precision X-ray diffraction techniques were used, respectively, to quantify these aspects. The observed average γ' size and morphology in the dendritic core, as well as lattice misfit values measured at room temperature are also listed in Table I.

Results

The compressive creep results for a γ alloy, three γ' alloys, and a superalloy are presented in Table II. All the tensile creep results for multicomponent superalloys are presented in Table III. Results for alloys from other sources are also listed in the tables for comparison, and the sequence of alloys listed in Table I is preserved. For the compression test results in Table II, the creep resistance is measured in terms of steady state minimum creep rate. Since higher creep rate implies lower creep resistance, the relative creep anisotropy is calculated as the ratio of the inverse minimum creep rates in <hkl> to <100> orientation.

In Table III, both time to 1% creep and rupture life are listed for superalloys. In the last column the relative creep anisotropy is calculated as a ratio of rupture life in <hkl> to the rupture life in <100> orientation. Note that the relative creep anisotropy could have been calculated using the time to 1% creep as well, but the ratio of rupture life was preferred as a better approximation of the steady state creep rate. In many alloys, high primary creep results in unrealistically short times to 1% creep, which consequently does not reflect any significant time spent in steady state creep. In contrast, rupture life is a better, albeit inverse approximation of the steady state creep rate because (a) in most cases the total elapsed time in primary and tertiary creep is a small fraction of the rupture life; and (b) for most single crystal alloys, the strain to failure is approximately the same.

While creep anisotropy was not measured at varying stresses at constant temperature in all alloys, the available results for several alloys suggest no strong effect of stress on the extent of isothermal creep anisotropy. See for example, data for Ni_3Al in Table II and for Alpha-4, Alpha-6, Alloy-490, and SC7-14-6. in Table III. This is not surprising if the power law relationship, $\dot{\varepsilon} \propto \sigma^n$, between the stress($\sigma$) and the minimum creep rate($\dot{\varepsilon}$) is expected to be valid. However, this may also be a consequence of the stress regime in which creep tests are generally carried out to achieve reasonable rupture failures between 100 and 1000 hours. As we shall discuss later, in tests at low stresses with very long lives, this assumption may be incorrect. For further discussion, it is sufficient to consider an average isothermal relative creep anisotropy in <110> and <111> orientations as summarized in the last columns of Table II and Table III.

Table I Alloy Compositions, Heat Treatment, and Microstructural Characterization of Several Nickel Base γ, γ′, and γ/γ′ Superalloys.

Alloy	Ni	Co	R	Cr	Mo	W	Ta	Ti	Al	Other	Hf	Zr	C	B	Heat Treatment	Microstructure γ′ size/morphology	γ/γ′ misfit (aγ′-aγ)/aγ %
γ′ Alloys																	
Ni₃Al	Bal.	-	-	-	-	-	-	-	12.4		-	-	-	-	As Cast	Single Phase	
Ni₃(Al,Ta)	Bal.	-	-	-	-	-	15.4	-	9.4		-	-	-	-	2200F/24 hrs	Single Phase	
GPB-3LA	Bal.	2	0.5	2	1.5	1.5	4.5	2.5	13.8		0.1	-	-	-	2150F/24 hrs	10-15% Interdend. second phase	
γ Alloys																	
Hastelloy-X	Bal.	1.5	-	22	9.0	0.6	-	-	-	18.5Fe	-	-	-	-	2150F/4 hrs	Single phase	
Superalloys																	
Mar M-200X	Bal.	10	-	9	-	12.5	-	2	5.0	1Nb	-	-	-	-	2300F/4 hr +P		Ref. [3]
Mar M-200	Bal.	10	-	9	-	12.5	-	2	5.0	1Nb	2.0	0.05	0.14	.015	2300F/4 hr +P		Ref. [3]
Mar M-247	Bal.	10.3	-	8.4	0.75	9.9	3.1	1	5.5		1.5	0.1	0.2	0	2250F/2 hr +P		Ref. [1]
CMSX-2	Bal.	5	-	8	0.6	8.0	6	1	5.5		-	-	-	-	2372F/4 hr+.....	0.23, 0.3 and 0.45 μm	Ref [5]
MXON	Bal.	5	-	8	2.0	8.0	6	-	6.1		-	-	-	-	2372F/4 hr+.....	0.2, and 0.38 μm	Ref. [5]
SC 7-14-6	Bal.	-	-	-	13.5	6.1	-	-	6.8		-	-	-	-	2350F/4 hr +P	1.5 μm + α-Mo	Ref. [4]
PWA1480	Bal.	5	-	10	-	4.0	12	1.5	5.0		-	-	-	-	2350F/4 hr +P	0.3 μm cuboidal	0.00
Alloy-606	Bal.	-	-	3	11.4	6.2	1.5	-	7.1		-	-	-	-	2350F/4 hr +P	0.45 μm semi cont.	-1.01
Alpha-3	Bal.	-	-	10	-	4.0	12	1.5	5.0		-	-	-	-	2380F/4 hr +P	0.45 μm cuboidal	-
Alpha-4	Bal.	5	-	10	-	16.0	-	1	5.0		-	-	-	-	2350F/4 hr +P	0.27 μm cuboidal	-
Alpha-4A	Bal.	5	-	10	-	16.0	-	1	5.0		-	.05	-	.015	2300F/4 hr +P	0.27 μm cuboidal	-
Alpha-5	Bal.	5	-	10	-	-	16	1	5.0		-	-	-	-	2350F/4 hr +P	0.5 μm non-cuboid.	-
Alpha-6	Bal.	5	-	10	4.0	4.0	4	1.5	5.0		-	-	-	-	2350F/4 hr +P	0.45 μm cuboidal	-
Alloy-610	Bal.	10	4	10	4.0	-	4	1.5	5.0		-	-	-	-	2325F/4 hr +P	0.3 μm cuboidal	0.25
Alloy-620	Bal.	10	2	10	4.0	2.0	4	1.5	5.0		-	-	-	-	2300F/4 hr +P	0.3 μm cuboidal	0.21
Alloy-630	Bal.	10	4	10	2.0	2.0	4	1.5	5.0		-	-	-	-	2325F/4 hr +P	0.27 μm cuboidal	-
Alpha-631	Bal.	10	4	7.5	2.0	2.0	4	1.5	5.0		-	-	-	-	2325F/4 hr +P	0.36 μm semi cub.	0.26
Alpha-632	Bal.	10	4	5	2.0	2.0	4	1.5	5.0		-	-	-	-	2325F/4 hr +P	0.5 μm non-cub.	0.28
Alpha-7	Bal.	5	-	-	-	4.0	12	1.5	5.0		-	-	-	-	2440F/4 hr +P	0.85 μm semi cont.	0.89
Alpha-8	Bal.	-	-	3	10.0	6.0	1.5	-	6.8		-	-	-	-	2337F/4 hr +P	0.45 μm semi cont.	-
Alloy-488	Bal.	10	3	7.5	2.0	4.0	8	-	5.25		-	-	-	-	2350F/4 hr +P	0.36 μm cuboidal	0.26
Alloy-489	Bal.	10	3	7.5	2.0	-	12	-	5.25		-	-	-	-	2325F/4 hr +P	0.45 μm non-cub.	0.51
Alloy-490	Bal.	10	3	7.5	4.0	-	12	-	5.25		-	-	-	-	2400F/4 hr +P	0.36 μm cuboidal	-
Alloy-491	Bal.	10	3	7.5	4.0	-	10	-	5.5		0.1	-	-	-	2350F/4 hr +P	0.32 μm non-cub.	
Alloy-492	Bal.	10	3	10	4.0	-	10	-	5.5		0.1	-	-	-	2300F/4 hr +P	0.3 μm non-cub.	
Alloy-493	Bal.	10	3.5	10	4.0	-	6	-	6.0		0.1	-	-	-	2300F/4 hr +P	0.3 μm non-cub.	
Alloy-494	Bal.	10	4	10	5.0	-	6	-	6.0		0.1	-	-	-	2350F/4 hr +P		
Alloy-495	Bal.	10	3	7.5	2.0	6.0	9	-	5.5		0.1	-	-	-	2360F/4 hr +P	0.25 μm non-cub.	
Alloy-497	Bal.	10	3.5	7.5	2.0	6.0	6	-	6.0		0.1	-	-	-	2350F/4 hr +P		
Alloy-498	Bal.	10	4	10	2.0	6.0	6	-	6.0		0.1	-	-	-	2350F/4 hr +P		
PWA1484	Bal.	10	3	5	1.9	5.9	8.7	-	5.65		0.1	-	-	-	2385F/4 hr +P		

P=1975F/4 hrs+1600F/32 hrs

DISCUSSION OF RESULTS

Since the objective of this paper is to understand the nature of creep anisotropy and not the absolute comparison of creep strength per se; the listing of relative isothermal creep anisotropy in the last columns of Tables II and III, provides a clear summary of all the pertinent information.

High Temperature Creep Anisotropy

As seen in Tables II and III, the majority of the test data are around 982°C(1800°F), and we shall first focus on high temperature creep anisotropy. For this discussion we shall consider the anisotropy in <111> orientation, listed in the last column of Tables II and III.
A comprehensive analysis of the data in Tables II and III, in conjunction with alloy compositions and microstructural characterization presented in Table I, is summarized graphically in Fig. 1. Besides a bar chart representing relative creep anisotropy in <100>, <110>, and <111> orientations, γ′ microstructure, γ/γ′ misfit, and a table of alloy compositions are also included in Fig. 1. A more detailed analysis of the results reveals that in general, high creep anisotropy -that is <111> being 6-16X more creep resistant than <100> - occurs in superalloys with high Mo concentrations and low levels of Co. Such alloys (Alloy-606, Alpha-8 and SC7-14-6), also typically show semi-continuous γ′ morphology and tend to have a large negative γ/γ′ mismatch. Interestingly enough, the single phase γ alloy (Hastelloy-X), with comparable Mo and Co concentration also displays high creep anisotropy. Note that neither of the γ′ alloys, nor any of the superalloys with high concentration of W (Alpha-4, CMSX-2) or Ta(Alpha-5), or W+Ta (PWA1480), fall in this category. As per the study by Caron et al.[5] alloys(CMSX-2) with typically discrete γ′ morphology also

Table II Effect of Orientation on Compressive Creep Results for Several Single Crystal Nickel Base γ, γ' and Superalloys.

Alloy	Creep Temp.		Stress		Compressive Minimum Creep Rate 10^{-5} hr^{-1}					Relative Creep Anisotropy				
	°F	°C	ksi	MPa	<100>	<110>	<123>	<223>	<111>	<100>	<110>	<123>	<223>	<111>
γ' Alloys														
Ni$_3$Al	1400	760	30	207	1.8	-	-	-	-	1.00	-	-	-	-
			40	276	11.0	6.0	15.0	-	8.0	1.00	1.83	0.73	-	1.38
			45	310	42.0	-	24.0	-	-	1.00	-	0.57	-	-
			50	345	-	-	-	-	20.0	1.00	-	-	-	-
			55	379	-	49.0	-	-	48.0	1.00	-	-	-	-
										1.00	**1.83**	**0.65**	**-**	**1.38**
	1600	871	10	69	11.0	6.8	5.7	-	4.7	1.00	1.62	1.93	-	2.34
			15	103	59.0	25.0	28.0	-	13.0	1.00	2.36	2.11	-	4.54
			20	138	114.0	67.0	80.0	-	48.0	1.00	1.70	1.43	-	2.38
										1.00	**1.89**	**1.82**	**-**	**3.08**
	1800	982	5	34	39.0	-	-	-	-	1.00	-	-	-	-
			7	48	170.0	47.0	50.0	-	-	1.00	3.62	3.40	-	-
			10	69	400.0	175.0	180.0	122.0	111.0	1.00	2.29	2.22	3.28	3.54
			15	103	-	-	-	580.0	470.0	1.00	-	-	-	-
										1.00	**2.95**	**2.81**	**3.28**	**3.54**
Ni$_3$(Al,Ta)	1400	760	60	414	0.9	1.3	0.9	-	3.0	1.00	0.69	1.00	-	0.30
			62	427	-	-	-	-	10.4	1.00	-	-	-	-
			70	483	1.0	-	-	-	-	1.00	-	-	-	-
										1.00	**0.69**	**1.00**	**-**	**0.30**
	1600	871	25	172	4.3	4.7	3.8	-	9.5	1.00	0.91	1.13	-	0.45
			35	241	34.5	25.0	15.0	-	50.0	1.00	1.38	2.30	-	0.69
			40	276	-	-	-	-	103.0	1.00	-	-	-	-
										1.00	**1.15**	**1.72**	**-**	**0.57**
	1800	982	8	55	3.5	5.5	4.8	-	-	1.00	0.64	0.73	-	-
			10	69	13.6	8.4	10.0	-	9.2	1.00	1.62	1.36	-	1.48
			12	83	-	-	-	-	14.8	1.00	-	-	-	-
			15	103	58.8	-	43.1	-	33.0	1.00	-	1.36	-	1.78
			25	172	-	-	-	-	1120.0	1.00	-	-	-	-
										1.00	**1.13**	**1.36**	**-**	**1.78**
GPB-3LA	1800	982	32	221	26.0	-	-	-	10.0	**1.00**	**-**	**-**	**-**	**2.60**
γ alloys														
Hastelloy-X	1400	760	17	117	22.5	-	-	-	1.8	**1.00**	**-**	**-**	**-**	**12.50**
	1800	982	5	34	244.0	118.0	-	-	36.4	**1.00**	**2.07**	**-**	**-**	**6.70**
Superalloy														
Mar M200X[3]	1400	760	110	758	500	10	-	-	180	**1.00**	**50.00**	**-**	**-**	**2.78**

show no significant influence of γ' size at high temperatures. Alloys which display moderately high creep anisotropy (in the range of 4-6), typically contain Re, with a combination of other refractory elements, but little W. See for example Alpha-632, Alloy-489 and Alloy-494.

The overall trends described in the preceding, is well illustrated by the composite in Fig. 1. It is clear from the behavior of a single phase γ' alloy with low creep anisotropy (extreme left on the bar chart), to that of a single phase γ alloy with high creep anisotropy (extreme right on the bar chart), that the nature of creep anisotropy transcends the γ' microstructure, and the γ/γ' misfit in superalloys. As highlighted in the companion table of compositions in Fig. 1, the alloys with large concentrations of Mo, exhibiting high creep anisotropy, behave as the γ alloy, and are therefore lumped on the right side of the bar chart. In contrast, the alloys primarily strengthened with W and Ta, show isotropic behavior similar to that of γ', and are lumped to the left. Indeed, as we shall show with the proposed model, the creep anisotropy is fully rationalized in terms of the relative strength of the octahedral and cube slip modes, and hence the composition.

In simple terms, isotropic creep behavior in γ' alloys is a result of increased multiplicity of slip systems with the participation of six(6) {100}<110> cube slip systems in addition to twelve(12) {111}<110> octahedral slip systems; whereas high creep anisotropy in the γ alloy is a geometric manifestation of slip mode restricted to octahedral slip alone. Qualitatively, superalloys containing stronger γ', enriched with Ta and W, behave isotropically, whereas alloys with the γ matrix strengthened by Mo or Re, display high creep anisotropy.

Table III Tensile Creep Results for a Wide Range of Nickel Base Multi Component, Single Crystal Superalloys in Three Principal Orientations.

Alloy	Creep Condition Temperature		Stress		<100> Time to 1% Creep hrs	Rupt. Life hrs	<110> Time to 1% hrs	Rupture Life hrs	<111> Time to 1% hrs	Rupture Life hrs	Relative Creep Anisotropy		
	°F	°C	ksi	MPa							<100>	<110>	<111>
Superalloys													
Mar M200X[3]	1400	760	110	758	-	986.3	-	6.4	-	48.4	1.00	0.01	**0.05**
MAR M-200[3]	1400	760	110	758	-	373.0	-	3.3	-	240.2	1.00	0.01	**0.64**
Mar M-247[1]	1425	774	105	724	-	65.8	-	1.0	-	1242.0	1.00	0.02	**18.89**
CMSX-2[5]	1400	760	109	750	(0.23 μm γ') 321.0		-	-	-	1107.0	1.00	-	**3.45**
					(0.3 μm γ') 569.0		-	3.3	-	480.0	1.00	0.01	**0.84**
					(0.45 μm γ') 1138.0		-	44.0	-	36.0	1.00	0.04	**0.03**
	1922	1050	17.4	120	(0.3 μm γ') 705.0		-	536.0	-	578.0	1.00	0.76	**0.82**
					(0.45 μm γ') 1055.0		-	381.0	-	522.0	1.00	0.36	**0.49**
MXON[5]	1562	850	72.5	500	(0.2 μm γ') 305.0		-	1150.0	-	-	1.00	-	**3.77**
			72.5	500	(0.38 μm γ') 402.0		-	93.0	-	-	1.00	-	**0.23**
SC 7-14-6[4]	1400	760	100	690	4.0	47.0	2.0	5.0	5.0	183.0	1.00	0.11	3.89
			110	758	0.3	11.0	1.0	2.0	0.2	42.0	1.00	0.18	3.81
											1.00	0.14	**3.85**
	1800	982	36	248	6.0	58.0	23.0	62.0	249.0	523.0	1.00	1.07	**9.02**
	2000	1093	20	138	4.0	28.0	86.0	382.0	402.0	450.0	1.00	13.64	**16.07**
PWA1480	1800	982	32	221	-	175.0	-	171.2	-	301.5	1.00	0.98	**1.72**
	1900	1038	25	172	-	96.1	-	61.3	-	257.2	1.00	0.64	**2.68**
Alloy-606	1800	982	32	221	4.4	40.0	24.8	77.1	152.3	211.9	1.00	1.93	5.30
			36	248	2.5	25.9	9.8	38.1	56.5	130.6	1.00	1.47	5.05
											1.00	1.70	**5.17**
	1900	1038	25	172	-	25.0	-	67.5	-	188.4	1.00	2.70	**7.54**
Alpha-3	1800	982	32	221	59.0	129.3	102.7	173.9	44.4	236.8	1.00	1.34	**1.83**
Alpha-4	1800	982	32	221	15.1	121.3	55.9	76.2	39.7	118.2	1.00	0.63	0.97
			36	248	10.4	75.8	24.6	36.9	15.6	59.5	1.00	0.49	0.78
											1.00	0.56	**0.88**
Alpha-4A	1800	982	32	221	-	143.4	-	142.9	-	211.0	1.00	1.00	1.47
			36	248	8.9	108.0	32.9	81.3	23.0	104.8	1.00	0.75	0.97
											1.00	0.87	**1.22**
Alpha-5	1800	982	32	221	7.2	66.3	33.9	128.9	23.9	233.4	1.00	1.94	**3.52**
Alpha-6	1800	982	32	221	33.0	317.0	128.8	150.0	25.6	212.9	1.00	0.47	0.67
			36	248	13.4	104.1	45.0	58.5	24.1	87.9	1.00	0.56	0.84
											1.00	0.52	**0.76**
Alloy-610	1800	982	36	248	22.1	107.0	42.5	46.0	82.8	286.6	1.00	0.43	**2.68**
Alloy-620	1800	982	36	248	18.9	116.9	51.7	57.4	53.5	193.7	1.00	0.49	**1.66**
Alloy-630	1800	982	36	248	28.1	127.4	74.0	81.2	213.6	335.3	1.00	0.64	**2.63**
Alpha-631	1800	982	36	248	2.7	39.4	30.8	45.9	8.3	116.1	1.00	1.16	**2.95**
Alpha-632	1800	982	36	248	0.5	10.7	11.0	29.8	2.3	48.7	1.00	2.79	**4.57**
Alpha-7	1800	982	32	221	4.2	29.0	9.8	50.3	10.0	92.3	1.00	1.74	**3.19**
Alpha-8	1800	982	32	221	54.9	119.6	342.9	383.4	>600	>600.0	1.00	3.21	**>5.0?**
	1800	982	36	248	21.3	63.7	53.9	131.2	740	883.7	1.00	2.06	**13.6**
Alloy-488	1800	982	36	248	31.5	144.5	106.0	115.4	82.0	366.5	1.00	0.80	**2.54**
Alloy-489	1800	982	36	248	4.0	80.3	95.9	124.9	73.0	439.7	1.00	1.56	**5.48**
Alloy-490	1800	982	36	248	75.4	193.3	134.1	174.3	213.7	487.8	1.00	0.90	2.52
			38	262	-	128.4	-	-	-	313.4	1.00	-	2.44
											1.00	0.90	**2.48**
Alloy-491	1800	982	38	262	41.5	126.1	-	-	97.1	210.9	1.00	-	**1.67**
Alloy-492	1800	982	38	262	-	119.9	-	-	-	221.5	1.00	-	**1.85**
Alloy-493	1800	982	38	262	30.4	99.7	-	-	63.6	140.9	1.00	-	**1.41**
Alloy-494	1800	982	38	262	3.5	18.0	-	-	20.5	83.9	1.00	-	**4.66**
Alloy-495	1800	982	38	262	42.4	125.6	-	-	84.5	205.6	1.00	-	**1.64**
Alloy-497	1800	982	38	262	21.6	67.6	-	-	54.7	131.6	1.00	-	**1.95**
Alloy-498	1800	982	38	262	5.0	22.6	-	-	17.7	61.3	1.00	-	**2.71**
PWA1484	1850	1010	36	248	45	120.0	-	-	122	234.9	1.00	-	**1.96**

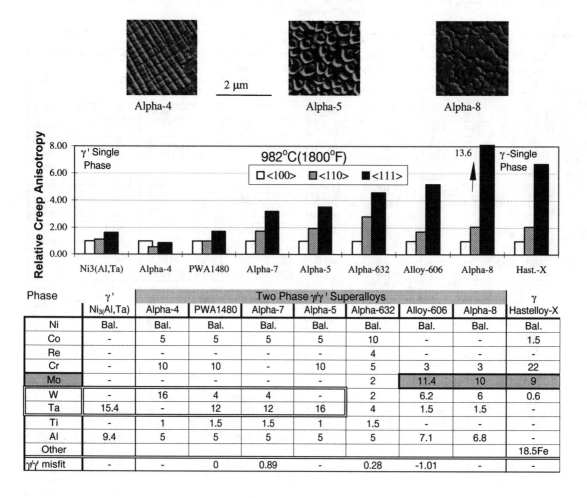

Figure 1: High temperature creep anisotropy behavior for nickel base single phase γ and γ', and two phase γ/γ' superalloy single crystals.

Low Temperature Creep Anisotropy

The nature of creep anisotropy in superalloys at low temperature around 760°C (1400°F) is even more interesting and varied. A representation of available data from Table IV, presented in bar chart form in Fig. 2, depicts this vividly. To show the extreme sensitivity of creep behavior to various parameters, a table of compositions and other relevant information is included in the figure. First, note that, as at high temperature, the single phase γ alloy, Hastelloy-X is also highly anisotropic at 760°C, with <111> being a factor of 13X stronger than <100>. Similar behavior also persists with the high Mo containing alloy SC7-14-6. In contrast, the creep anisotropy of single phase γ'-Ni$_3$(Al,Ta) is reversed with <111> becoming a factor of 3X weaker than <100>. The nature of creep anisotropy in Mar M200 and CMSX-2 (coarse γ'), with significant concentrations of W and W+Ta, respectively, is similar to this. However, the behavior of Mar M247 with a comparable composition does not follow suit. As the results from Caron et al. for CMSX-2 in Fig. 2 reflect, the creep anisotropy at low temperature is extremely sensitive to the γ' size. Finer γ' tends to increase the creep resistance of <111> compared to <100>, while the coarser γ' renders <111> weaker than <100>. This may be rationalized, recognizing that superalloys with coarse γ' tend to behave as single phase γ' at low temperatures[6]. Alternatively, the finer γ' may be viewed as promoting Orowan looping, thereby restricting deformation to the γ matrix and enhancing octahedral slip. Of course this qualitative reasoning cannot be extended too far, but the fact is that fine γ' enhances the low temperature yield strength of <111> much more than it does for <100> [6]. Since 760°C creep tests are typically carried out at stresses closer to the yield strength of the material, with little contribution of diffusional controlled climb, the parallel between creep and yield strength anisotropy is expected.

With 760°C(1400°F) being a relatively low temperature in superalloys, complications due to a superdislocation constriction stress enters the picture. This has been amply shown elsewhere[6,3] and has dual implications. First the critical resolved shear stress(CRSS) for octahedral slip is no longer a constant, but becomes a function of orientation. Secondly, the orientation dependence reverses with the sign of the resolved shear

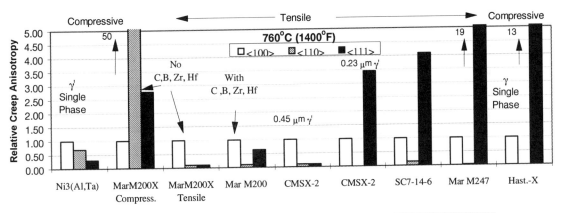

Figure 2: Low temperature creep anisotropy behavior for nickel base single phase γ and γ', and two phase γ/γ' superalloy single crystals.

stress, or in practice going from a tensile test to a compressive test. Thus <100> and <110> orientations with identical Schmid factors for octahedral slip become unequal in yield strength, with the <100> stronger than <110> in tension, and <110> stronger than <100> in compression[6]. This is further exaggerated in a creep test at 760°C, at a stress close to the yield strength. It is remarkable that all available data for superalloys show the <110> to have a very short rupture life in tension around 760°C(1400°F). However, in compression the situation reverses as shown in Fig. 2, by the behavior of Mar M200X, a single crystal version of Mar M200. For a more complete discussion see reference [3]. Finally, at 760°C (1400°F), minor elements are also known to affect the yield strength anisotropy, and in a magnified fashion the creep anisotropy. A comparison of results for Mar M200 and Mar M200X, in Fig. 2, bear out this fact. A more detailed description of this aspect has been presented previously [3].

These seemingly complex qualitative explanations of creep anisotropy are more easily comprehended by the proposed model which attempts to quantitatively normalize the creep behavior as a function of orientation. Without such a model, it is difficult to bring forth the geometrical relationships between the principal stress direction and the operative slip systems in a rigorous manner.

Creep Anisotropy Model

Following the arguments presented in reference [2], it is proposed that in a fundamental sense a power law creep relationship must exist between the minimum shear strain rate $\dot{\gamma}$, and the resolved shear stress τ, as

$$\dot{\gamma} = A \tau^n \tag{1}$$

where A and n are constants. Further, if we assume that no strong interaction occurs between different operating slip systems up to a few percent strain, then we can write the total axial strain rate $\dot{\varepsilon}_{ii}$ as a linear superposition of shear strain rates, $\dot{\gamma}_{jk}$, for all operative slip systems as

$$\dot{\varepsilon}_{ii} = \sum a_{ij} a_{jk} \dot{\gamma}_{jk} \tag{2}$$

where $(a_{ij} a_{jk})$ is a product of direction cosines of the slip plane normal and slip directions with reference to the stress axis - or more commonly known as the Schmid factor. Similarly (1) may be expressed in terms of applied stress σ_{ii} as

$$\dot{\gamma}_{jk} = A_{jk} (a_{ij} a_{jk} \sigma_{ii})^n \tag{3}$$

which when combined with (2) yields

$$\dot{\varepsilon}_{ii} = A \sum (a_{ij} a_{jk})^{n+1} \sigma_{ii}^n \tag{4}$$

Within a family of slip systems such as {111}<110> (octahedral slip), there is no fundamental reason for the constant A to be different, and hence the subscripts of A are dropped.

In general, for any nickel base single crystal alloy deforming by both octahedral {111}<110> and cube slip {100}<110> systems, the steady state strain rate in any direction <hkl> may be expressed simply as

$$\dot{\varepsilon}_{<hkl>} = (A_o G_o + A_c G_c)\sigma_{<hkl>}^n \quad (5)$$

where A_o and A_c are characteristic creep parameters for the octahedral and cube slip systems, respectively, and are material properties specific to a given alloy.

However, G_o and G_c in equation (5) are purely geometric factors and can be calculated by appropriate summation of Schmid factors as a function of stress exponent, n, for 12 octahedral and 6 cube slip systems, respectively, as

$$G_o = \sum_1^{12} (a_{ij} a_{jk})^{n+1} \quad \text{and} \quad G_c = \sum_1^{6} (a_{ij} a_{jk})^{n+1} \quad (6)$$

For a symmetric orientation such as <100>, with an identical Schmid factor of 0.41 for 8 operative octahedral slip systems, the value of G_o can be easily estimated as $8 \times (0.41)^{4.5} = 0.14$, assuming a reasonable stress exponent of 3.5. For other less symmetric orientations, the calculations are tedious but straightforward with the aide of a computer. For the purpose of further discussion, values of both geometric factors for some common orientations are listed in Table IV at n =3.5.

Table IV Some Typical Values of Geometric Factors G_o and G_c Assuming a Stress Exponent of 3.5.

Orientation	<100>	<110>	<123>	<223>	<111>
			n=3.5		
G_o	0.1420	0.0710	0.0580	0.0330	0.0172
G_c	0	0.0372	0.0480	0.0850	0.1017

With the aide of Table IV, creep anisotropy at a constant stress can be estimated using equation (5). Consistent with our definition of creep anisotropy used throughout this paper, the relative creep resistance R of any <hkl> orientations with reference to <100>, is a ratio of the inverse of respective steady state creep rates at a constant stress. Accordingly, and using the values of G_o and G_c for <100> orientation from Table IV, it can be shown that

$$R_{<hkl>/<100>} = \frac{\dot{\varepsilon}_{<100>}}{\dot{\varepsilon}_{<hkl>}} = \frac{0.1420}{\{G_o + (A_c/A_o) G_c\}_{<hkl>}} \quad (7)$$

The only unknown in equation (7) is the ratio (A_c/A_o), which can be determined using only one additional creep test in any orientation other than <100>. If the model is correct, this should allow prediction of creep behavior of any and all orientations. The model is very successful in normalizing creep behavior of the single phase γ' alloys Ni_3Al and $Ni_3(Al, Ta)$. For a γ alloy such as Hastelloy-X, where cube slip participation is not expected, A_c can be set to zero and the creep anisotropy can be predicted a priori from equation (7). An eight fold increase in creep resistance for <111> compared to <100> orientation was predicted by the model and was experimentally confirmed to be seven fold at 982°C(1800°F).

Instead of demonstrating the validity of the model for a specific alloy system, it will be more meaningful to show the generic applicability of the model. The ratio (A_c/A_o), may be construed as a measure of relative creep rate of cube slip in comparison to octahedral slip at a constant stress. We shall consider three cases: (a) no participation of cube slip, (b) moderate participation of cube slip, and (c) extensive participation of cube slip. As listed in Table V, these cases are somewhat arbitrarily assigned the values of 0, 0.2, and 2, respectively, for the ratio (A_c/A_o). In each case the relative creep anisotropy is calculated using equation (7) for <110> and <111> orientations. To provide a comprehensive view of the decreasing creep anisotropy with increasing participation of cube slip, plots of contours of creep anisotropy within a standard stereographic triangle are presented in Fig. 3.

For comparison, some typical alloys with their experimentally observed values of creep anisotropy at 982°C(1800°F) are also listed in Table V. Note the parallel trend between the theoretical and experimental values. However, if the correlation is valid, it also implies that alloys showing isotropic creep behavior should somehow manifest increased participation of cube slip. To verify this in a most direct manner, yield strengths in <111> and <100> orientations were measured at 982°C(1800°F), for several alloys as listed in Table V. Recall that using a Schmid factor of 0.41 for octahedral slip with <100> orientation, and 0.47 for cube slip with <111> orientation, CRSS for both slip systems can be estimated. The ratio of CRSS for cube to octahedral slip systems, listed in the last column of Table V, clearly shows cube slip to be stronger than octahedral slip in alloys such as Alpha-8, which also displays high creep anisotropy. As the CRSS for cube slip becomes comparable or lower than that for octahedral slip, the alloys behave more isotropically in creep. Interestingly, testing of alloy SC7-14-6 at 1093, 982, and 760°C, cited from [4], shows a decrease in creep anisotropy, as the CRSS for cube slip comes closer to that for octahedral slip, with decreasing temperature.

In spite of a strong correlation between creep anisotropy and yield strength anisotropy, there is no simple way to predict the extent of creep anisotropy based on yield strengths measured at high strain rates. However, it is an inescapable fact that a small relative variation in CRSS for the two slip modes is vastly magnified in creep. Also note that the model correctly predicts the behavior of <110> orientation in most cases.

Concept of Threshold Shear Stress

While the proposed model correctly predicts a high level of anisotropy in alloys with little participation of cube slip, it sets an upper limit of 8X for <111> creep rupture life, relative to <100>. However, in some instances, anomalously high anisotropy has been observed. For example in the case of SC7-14-6, not only is rupture life for <111> increased by a factor of 16 versus <100>, but time to 1% creep is increased by a factor of 100. It is proposed that such behavior may be rationalized by introducing a concept of threshold shear stress. To account for a very high stress exponent, especially in ODS alloys, the concept of back stress or threshold stress has been suggested. The underlying reasoning is that in some alloys, there is a stress threshold below which virtually no creep occurs, and it is appropriate to subtract this component from the applied stress. If such a stress were hydrostatic in nature and were

Table V Comparison of Theoretically Predicted and Experimentally Determined Values of Relative Creep Anisotropy for Several Alloys, in Relation to Available Yield Strength Data.

Theoretical					Experimental								
Cube slip Participation	A_c/A_o	Relative Creep Anisot.			Alloy	Temp.	Relative Creep Anisotropy			0.2 % Yield Str.		CRSS Octahed. Slip	CRSS Cube Slip	CDRSS Ratio cube/octa.
		<100>	<110>	<111>			<100>	<110>	<111>	<100>	<111>			
						°C				MPa	MPa	MPa	MPa	
None	0.0	1.00	2.00	8.26	Hastelloy-X	982	1	2.07	6.7	-	-	-	-	-
					Alpha-8	982	1	2.06	13.6	560	543	230	255	1.11
					SC7-14-6	1093	1	13.64	16.07	409	452	168	212	1.27
						982	1	1.07	9.02	477	495	196	233	1.19
						760	1	0.14	4.08	968	817	397	384	0.97
Moderate	0.2	1.00	1.81	3.78	Alpha-5	982	1	1.94	3.52	431	429	177	202	1.14
	0.5	1.00	1.58	2.07	PWA1480	982	1	1	1.72	496	386	203	181	0.89
Strong	2.0	1.00	0.98	0.64	Alpha-4	982	1	0.56	0.88	487	365	200	172	0.86
					$Ni_3(Al,Ta)$	871	1	1.15	0.57	-	-	-	-	-

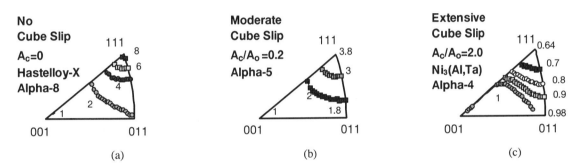

Figure 3: Contours of relative creep anisotropy within a standard stereographic triangle showing the effect of increasing participation of cube slip.

subtracted from the applied tensile stress σ_{ij} in equation (5), it will have no effect on the nature of creep anisotropy. However, fundamentally if such a stress were a shear stress - a creep equivalent of CRSS for a high strain rate test - it can lead to interesting consequences. Consider a hypothetical situation illustrated in Fig. 4, where a tensile stress of $3\tau_o$ is applied to two specimens with [001] and [$\bar{1}$11] axial orientations. The specimens are assumed to be made from a material with a threshold shear stress of τ_o for octahedral slip. If the deformation in the material were limited to octahedral slip alone, the resolved shear stress must exceed the threshold stress for at least some octahedral slip modes, for any finite creep to take place. It follows from Fig. 4, that while that occurs in the [001] oriented specimen, owing to unfavorable geometry, resolved stress has not exceeded the threshold stress in the [$\bar{1}$11] oriented specimen. Thus while significant creep deformation is expected for the [001], ideally no creep should occur in the [$\bar{1}$11]. In practice, however, because of crystallographic imperfections due to dendritic structure or localized notch effects, creep deformation will occur for the [$\bar{1}$11] specimen, manifesting a very high creep anisotropy.

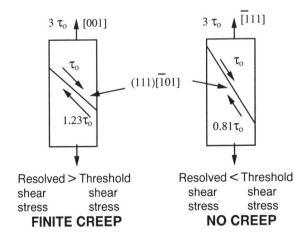

Figure 4: Potential effect of threshold shear stress on creep anisotropy.

This concept can be easily introduced in the proposed model as follows, where τ_o and τ_c are threshold stresses for octahedral and cube slip systems respectively.

$$\dot{\varepsilon}_{<hkl>} = A_o \sum_{1}^{12} a_{ij} a_{jk} (\sigma_{ii} a_{ij} a_{jk} - \tau_o)^n + A_c \sum_{1}^{6} a_{ij} a_{jk} (\sigma_{ii} a_{ij} a_{jk} - \tau_c)^n \quad (8)$$

Obviously with the proposed modification, the creep anisotropy cannot be expressed in the simple form as in equation (7), but can be easily calculated using a computer. Fig. 5 presents creep anisotropy contours for a specific case of a hypothetical material with an applied tensile stress $\sigma = 5\tau_o$, with no cube slip participation. A factor of 80 enhancement of creep resistance in <111> in comparison to <100> is consistent with the behavior of SC 7-14-6 at 1093°C up to 1% strain. However, the model fails to normalize the behavior of <110>.

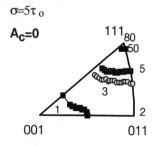

Figure 5: Creep anisotropy contours for a hypothetical superalloy with high threshold shear stress on octahedral slip systems and no participation of cube slip.

Of course it is premature to take the implication of the model too far, but the concept illustrated in Fig. 4 is worth considering for creep tests at low stresses. We recognize that the creep behavior of superalloys is a complex phenomenon and we have not addressed many other issues such as effect of rafting and pre-rafting of γ' [7].

SUMMARY

1. Creep anisotropy - as measured by the ratio of creep resistance in <111> to <100> orientations- in all single crystal nickel base single phase γ, and γ', and precipitation hardened γ/γ' superalloys, is primarily controlled by the relative creep strength of cube to octahedral slip systems, at the temperature of interest.

2. Single phase γ alloys, deforming exclusively by octahedral slip, behave more anisotropically than single phase γ' alloys, where multiplicity of slip systems is increased with additional participation of cube slip. The increase in number of operative slip systems, leads to a more isotropic creep behavior.

3. Precipitation hardened γ/γ' superalloys, primarily strengthened by elements such as Mo, partitioning principally to the γ matrix, behave more anisotropically, similar to γ alloys. If the strengthening is predominantly attained through strengthening of γ', with the addition of elements such as Ta and W, isotropic creep behavior develops, akin to that in γ' alloys. Complex superalloys strengthened with a balanced combination of refractory elements, display a moderate level of creep anisotropy.

4. Creep anisotropy can be modeled at a given temperature in terms of a single parameter reflecting the relative creep resistance of cube to octahedral slip. Axial strain rate may be approximated as a linear superposition of shear strain rates on all operative slip systems, assuming a stress exponent of 3.5 for a power law relationship between the shear strain rate and resolved shear stress for individual slip systems.

5. The proposed model, correctly predicts creep resistance in the <111> orientation to be a factor of 8 better than that in <100> orientation, for alloys similar to γ alloys, where little participation of cube slip is expected. Allowing for increased participation of cube slip, the model is able to project the observed trend of decreasing creep anisotropy, for a series of complex alloys, or with temperature for a given alloy. The presumed association of increasing participation of cube slip, with the decrease in creep anisotropy, either with alloy composition or temperature, is empirically confirmed by a decreasing ratio of CRSS for cube to octahedral slip.

6. Based on the model, it is projected, that the existence of a threshold shear stress, can lead to unusually high creep anisotropy at low stresses. This is offered as a plausible explanation for unusually high creep anisotropy, occasionally observed at low stresses.

REFERENCES

1. R. A. MacKay, R. L. Dreshfield, and R. D. Maier, "Anisotropy of Nickel-Base Superalloy Single Crystals" (SUPERALLOYS 1980 ed. by J. K. Tien et al. TMS-AIME, PA., 1980) 385.

2. D. M. Shah, "Orientation Dependence of Creep Behavior of Single Crystal γ' (Ni_3Al)," Scripta Metall., 17(1983), 997-10002.

3. D. M. Shah and D. N. Duhl, "Effect of Minor Elements on the Deformation Behavior of Nickel-Base Superalloys" (SUPERALLOYS 1988 ed. by D. N. Duhl et al. TMS, PA., 1988) 693.

4. R. P. Dalal, C. R. Thomas, and L. E. Dardi, "The Effect of Crystallographic Orientation on the Physical and Mechanical Properties of an Investment Cast Single Crystal Nickel-Base Superalloy" (SUPERALLOYS 1984 ed. by Maurice Gell et al. TMS-AIME, PA., 1984) 185.

5. P. Caron et al., "Creep Deformation Anisotropy in Single Crystal Superalloys" (SUPERALLOYS 1988 ed. by D. N. Duhl et al. TMS, PA., 1988) 215.

6. D. M. Shah and D. N. Duhl, "The Effect of Orientation, Temperature and Gamma Prime Size on the Yield Strength of a Single Crystal Nickel-Base Superalloy" (SUPERALLOYS 1984 ed. by Maurice Gell et al. TMS-AIME, PA., 1984) 105.

7. D. D. Pearson, F. D. Lemkey and B. H. Kear, "Stress Coarsening of γ' and Its Influence on Creep Properties of a Single Crystal Superalloy" (SUPERALLOYS 1984 ed. by J. K. Tien et al. TMS-AIME, PA., 1980) 513.

CREEP ANISOTROPY IN THE MONOCRYSTALLINE NICKEL-BASE SUPERALLOY CMSX-4

V. Sass, U. Glatzel and M. Feller-Kniepmeier

Technische Universitaet Berlin, Institut fuer Metallforschung,
Sekr. BH 18, Str. des 17. Juni 135,
D10623 Berlin, Germany

Abstract

The orientation dependence of creep strength of the single crystal superalloy CMSX-4 was studied in the temperature range 1123 - 1253K. At 1123K creep behavior and lifetimes for small strains are highly anisotropic with the highest variations occurring in the primary creep regime. Secondary creep strength decreases in the order [001] - [011] - [$\bar{1}$11].
Under the present conditions plastic deformation in the primary and at the beginning of secondary creep stage occurs mainly in the matrix phase. The motion of matrix dislocations is strongly influenced by coherency stresses, which restrict dislocations from entering certain types of matrix channels and control the number of active slip systems. Extensive cutting of the γ'-phase is observed only in [011] oriented samples and in [001] orientation under the highest applied stress and results in pronounced primary creep. The poor creep strength of [$\bar{1}$11] oriented crystals is related to poor work hardening and a cross slip mechanism of matrix screw dislocations.
At the higher temperature of 1253K creep behavior is affected by the instability of the γ'-morphology and the extent of anisotropy is highly reduced. The creep strength of the [$\bar{1}$11] orientation, however, remains poor.

Introduction

In order to take full advantage of the high temperature capabilities of single crystal superalloys, the strong anisotropy of various properties including high temperature creep strength must be considered. Although turbine blades are mainly stressed in the [001] direction, which is their natural solidification direction, their complex shapes and high temperature gradients produce locally multiaxial stress states and high thermal stresses acting in various directions.
A comprehensive study of the influence of orientation on the creep behavior of the single crystal alloys MAR-M200 and MAR-M247 at 1047 K was presented by MacKay and Maier [1]. They reported long lifetimes for orientations near [$\bar{1}$11] or [001] and short lives for samples oriented close to [011]. The creep strength was found to depend mainly on the amount of lattice rotation required to produce intersecting slip on {111}<112> type slip systems, which is necessary for the transition from primary to secondary creep.
Caron et al. [2] showed for some advanced superalloys (CMSX-2, Alloy 454, MXON, CMSX-4) the strong influence of the γ'-particle size on creep strength and rupture lives at intermediate temperatures (1033 K–1123 K). For a mean particle size of about 0.45µm optimum creep strength is obtained from orientations close to [001], whereas rupture lives for [$\bar{1}$11] are drastically reduced. For more elevated temperatures they reported a rapid decrease of the creep anisotropy due to a change in the prevailing deformation mechanism from heterogeneous cooperative shearing of the γ/γ'-structure by {111}<112> slip to a more homogenous {111}<110> type slip in the matrix phase and creep induced morphology changes of the γ'-phase. The authors also suggested, that the rhenium content of second generation superalloys may reduce the extent of anisotropic creep, however, no microscopic reasons for this behavior were given.
Generally most studies of creep anisotropy were mainly performed under testing conditions, which promoted extensive shearing of the γ'-phase by {111}<112> slip [1-4]. This mode of deformation results in a high degree of anisotropy especially in the primary creep stage and accordingly in lifetimes for small strains. The transition to more homogenous matrix deformation with increasing temperature and decreasing stress tends to reduce orientation dependence [2]. At temperatures above 1173K, however, the instability of the γ'-morphology makes it difficult to distinguish between the effects of different deformation mechanisms and raft formation. Unfortunately there is a lack of creep data and knowledge of deformation mechanisms in the temperature range between 1073K and 1173K, where extensive shearing of the γ'-phase disappears and the γ'-morphology remains stable during creep tests.
Therefore the main target of this paper is the investigation of creep anisotropy in the advanced Re-containing alloy CMSX-4 at a temperature of 1123K. Analysis of creep data was focused on small strains up to 5% in order to study the creep behavior in primary and secondary creep stage, which are most relevant for technical applications. Additional tests at 1253K were carried out to explore the influence of

Table I: Nominal Chemical Composition of CMSX-4 (wt.-%) [5].

Cr	Al	Ti	Mo	W	Ta	Co	Re	Hf	Ni
6.5	5.6	1.0	0.6	6.0	6.0	10.0	3.0	0.1	bal.

morphological changes of the γ´-phase on creep behavior. Extensive transmission electron microscopy (TEM) investigations were performed in order to identify the microscopic deformation mechanisms as function of crystal orientation and to provide an explanation for the macroscopic behavior.

Material and Experimental Procedures

The chemical composition of the superalloy CMSX-4 is given in Table I. The addition of approximately 3 wt.-% rhenium improves the creep strength of the material significantly [5,6].

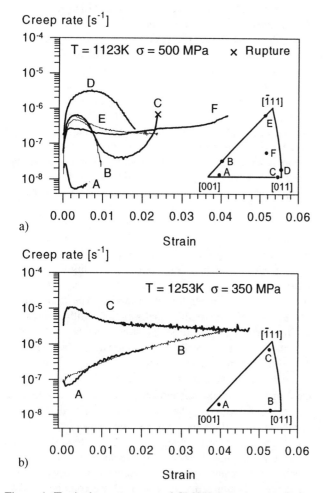

Single crystal rods were grown using the withdrawal process with a selector technique for [001] orientations by Thyssen (Bochum) and a seeding technique for other orientations by VIAM (Moscow). All specimens received a standard heat treatment comprising a three-step solution treatment (1h/1553K, 2h/1563K, 6h/1573K) and a two-step aging treatment (6h/1413K, 16h/1143K). The heat treatment produced cuboidal γ´-particles with a mean edge length of about 0.45μm taking up a volume fraction of about 70% [7]. A careful examination of the fully heat treated microstructures by scanning electron microscopy (SEM) and TEM showed no significant differences between crystals produced by the selector and the seeding technique.

Creep specimens with a gauge length of 55 mm and a diameter of 9 mm for [001] oriented crystals and a gauge length of 62 mm and a diameter of 6 mm for other orientations were prepared by mechanical machining and grinding. The creep tests were performed in air under constant loads ranging from 350 MPa to 650 MPa at 1123 K and 1253 K. The strain was measured continuously with extensometers. The creep tests were usually interrupted in different creep stages before rupture. The specimens were then cooled under load in order to preserve the typical microstructure for subsequent TEM-investigations. TEM analysis was carried out using a Philips CM 30 microscope operated at 300kV and a Jeol 200C microscope operated at 200kV. The initial orientation of each single crystal was determined by the Laue back-reflection X-ray technique.

Results

Tensile creep test were performed in order to investigate the creep strength of CMSX-4 single crystals as a function of orientation, stress and temperature. At an intermediate temperature of 1123K lifetimes for smaller strains were found to be governed by the primary creep behavior, which is highly sensitive to the crystal orientation. Typical creep curves for the main orientations under a load of 500 MPa plotted as strain rate versus strain are shown in Fig. 1a. Although the shape of the creep curves is similar for all orientations, the strain rates and the amount of primary creep strain vary dramatically.

Pronounced primary creep is observed for all crystals except those close to the [001] orientation with a misorientation towards [011] (e.g. sample A). No clear transition from primary to secondary creep is visible for crystals oriented near [$\bar{1}$11] and in the center of the stereographic triangle. This is reflected in the strain-lifetimes for 0.5% and 2.0% plastic strain given in Fig. 2a. Creep strength decreases strongly in the order [001] - [011] - [$\bar{1}$11].

The highest strain-lifetimes are obtained by near [001] oriented samples due to the small amplitude of primary

Figure 1: Typical creep curves of CMSX-4 single crystals for the main orientations plotted as strain rate $\dot{\varepsilon}$ vs. plastic strain ε. The exact orientations are given in the stereographic standard triangles. a) T = 1123K, σ = 500 MPa. b) T = 1253K, σ = 350 MPa.

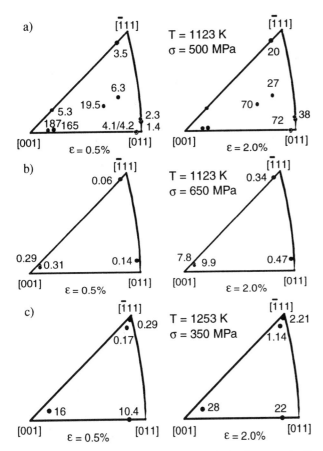

Figure 2: Strain lifetimes of CMSX-4 single crystals for 0.5 and 2.0% strain in hours plotted as function of orientation. a) T = 1123K and σ = 500 MPa, b) T = 1123K and σ = 650 MPa, b) T = 1253K and σ = 350 MPa.

creep and the lowest secondary creep rates. All other samples needed much larger strains to reach the secondary creep stage which explains the extremely short lifetimes for 0.5% strain.

At higher strains the secondary creep rates become more important and accordingly the lifetimes for the [$\bar{1}$11] orientation are the poorest. The lifetimes of crystals oriented near [001] and [011] are highly sensitive to misorientations away from the [001]-[011] boundary, which is a single slip orientation for {111}<112> slip. Increasing the load to 650 MPa leads to very similar results shown in Fig. 2b.

The effect of temperature on creep anisotropy is much more pronounced than the influence of stress. At a temperature of 1253K and a load of 350 MPa the creep curves for [001] and [011] orientations (Fig. 1b) have an altered shape with an overall minimum of the creep rates at small strains followed by a continuos softening with increasing strain. The same behavior has previously been reported to be typical for [001] oriented CMSX-4 samples at high temperatures [8]. The very similar lifetimes in Fig. 2c confirm, that the anisotropy between [001] and [011] virtually disappears at high temperatures. For [$\bar{1}$11] the characteristic shape of the creep curve at lower temperatures with a pronounced maximum of the creep rate is preserved leading to very poor lifetimes for small strains.

Discussion

The creep tests of CMSX-4 single crystals showed a high degree of anisotropy, which is retained up to high temperatures. The most prominent features of creep in CMSX-4 are the concentration of plastic deformation in the matrix phase at the smaller strains investigated here and the poor creep resistance of [$\bar{1}$11] oriented crystals at both temperatures of 1123K and 1253K. This has not been reported for other alloys [2]. The comparison of our result with other findings, however, is not straightforward, since we chose to evaluate creep strength by strain-lifetimes instead of total rupture lives. In the latter case the results depend strongly on rupture elongation and tertiary creep behavior, which may possess a different orientation dependence than the creep behavior in the primary and secondary creep stage we were interested in. Strain-lifetimes for small strains depend strongly on the primary creep behavior, which proved to be far more anisotropic at 1123K than secondary creep behavior. It is therefore convenient to discuss both regimes separately.

Primary Creep at 1123K

The strong sensitivity of the primary creep behavior of crystals in [001] and [011] orientation to small misorientations is generally explained by the multiplicity of active {111}<112> slip systems [1,3,4]. In this mode of deformation the γ'-precipitates are sheared by a/3<112> partial dislocations under formation of a superlattice stacking faults (SSF). Single slip orientations promote high primary creep rates and strains due to the absence of strain hardening. Under the present conditions extensive {111}<112> slip was observed only for [001] at 650 MPa and particularly for [011] at 500 MPa as shown in Fig. 4b,c. In the latter case misorientations towards [$\bar{1}$11] promote coplanar slip on the (111) slip plane as shown in Fig. 4b and result in exceedingly high primary creep rates (see sample D in Fig. 1a). For crystals oriented on the [001]-[011] boundary (e.g. sample C in Fig. 1a), which is a duplex slip orientation for {111}<112> slip, intersecting slip on two {111} slip planes prevails from the start of plastic deformation on as can be seen from Fig. 4c. Here primary creep rates and strains are strongly reduced.

The predominant occurrence of γ'-cutting in [011] orientation indicates, that the shear stresses required for the generation of superlattice stacking faults are higher for [001] than for [011], since the Schmid factors for {111}<112> slip are equal in both cases. TEM investigations revealed, that in [011] cutting of the γ'-phase occurs by a specific mechanism shown in Fig. 3.

Two dipoles of {111}<110> matrix dislocations marked as 1 and 2 expand in a (001) matrix channel on the ($\bar{1}$11) plane.

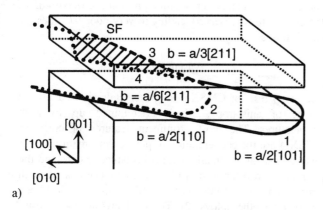
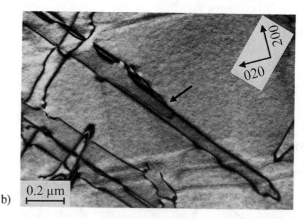

Figure 3: Generation of stacking faults inside γ′-particles in [011] oriented crystals by the reaction of two matrix dislocation loops. a) Schematic sketch. b) TEM micrograph taken after 0.14% strain at T = 1123K and σ = 500 MPa. The arrow marks the location of the dislocation reaction.

Behind the tip of the trailing dislocation 2 a reaction with the leading dislocation 1 takes place:

$$\frac{a}{2}[110] + \frac{a}{2}[101] \rightarrow \frac{a}{3}[211] + SSF + \frac{a}{6}[211] \quad (1)$$

The partial dislocation a/3[211] (3) cuts into the γ′-particle and generates a superlattice stacking fault (SSF), while the second partial dislocation a/6[211] (4) remains in the γ/γ′-interface. As can be seen from Fig. 3b, the cutting into the γ′-particles takes places periodically along the dislocation line where the matrix dislocations encounter a γ′-cube. This mechanism is therefore very effective to generate stacking faults. A similar mechanism has been observed by Feller-Kniepmeier and Kuttner in [011] samples of SRR 99 tested at 1033 K and 680 MPa [9]. The characteristic feature of this mechanism is the reaction of two coplanar matrix dislocations with different Burgers vectors. This condition would require the operation of two dislocation sources on the same {111} plane, which seems quite unlikely at least in the early stages of deformation. In [011] orientation this obstacle can be overcome by cross slip of interfacial dislocations with screw character on {001} planes toward the immobile second interfacial dislocation with 60° character [9], whereas in [001] orientations all interfacial dislocations in the predominantly deformed (001) horizontal channel have 60° character [10]. Therefore the easier generation of stacking faults in [011] orientation appears to be a major cause for the pronounced primary creep and the poor strain-lifetimes compared to [001].

The deformation of the matrix-phase occurs independent of orientation by the well know mechanism of the expansion of a/2<110>{111} dislocation loops on {111} slip planes in the narrow matrix channels [5,12,13] (see Fig. 4). The propagation of dislocation loops starting from grown-in sources is directed by the effective shear stresses in the matrix channels, which vary with the orientation of the matrix channels with respect to the load axis. The effective stress level in a matrix channel is determined by the superposition of external load and internal coherency stresses. In [001] orientation this effect leads to a stress concentration in horizontal matrix channels and a reduced stress level in vertical channels [5,10,14]. In contrast to that in [011] orientation two types of matrix channels, the so called "roof channels", which are inclined at an angle of 45° to the stress axis, are highly stressed and only the (100) "gable channels" aligned parallel to the load axis experience lower stresses [9]. For the highly symmetric [$\bar{1}$11] orientation all three types of matrix channels are equally stressed.

In the absence of intense γ′-particle cutting primary creep in [001] and [$\bar{1}$11] orientation at 500 MPa is controlled by matrix deformation. In [001] orientation the plastic deformation of the matrix phase is confined to the highly stressed horizontal matrix channels. The activation of multiple <110>{111} slip systems results in a rapid build-up of dense dislocation networks in the γ/γ′ interfaces, which inhibit further deformation. Accordingly the primary creep stage is completed after small strains in crystals oriented near [001]. The pronounced primary creep in sample B with a large misorientation of 13° towards [$\bar{1}$11] was caused by coplanar slip on the ($\bar{1}$11) plane.

In [$\bar{1}$11] the multiplicity of <110>{111} slip systems and the low value of the Schmid factor for octahedral slip of 0.27 should result in a homogenous matrix deformation and superior creep strength similar to [001]. The actual primary creep rates, however, are more similar to those of [011] oriented crystals, where primary creep deformation occurs in a much more heterogeneous way. Poor creep strength of [$\bar{1}$11] oriented crystals was reported by Caron et al. [2] for several alloys containing large γ′-precipitates, which promoted homogenous matrix deformation instead of the heterogeneous shearing of the γ′-phase by {111}<112> slip observed for small γ′-particles. The reasons for the negative effect of homogenous matrix deformation lie in the influence of coherency stresses on dislocation movements. TEM

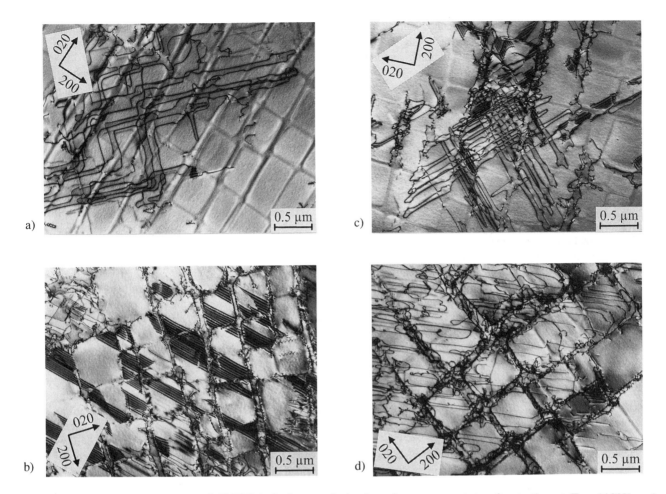

Figure 4: Typical microstructures of CMSX-4 single crystals in the primary creep stage after testing at T = 1123K and σ = 500 MPa: a) [001] orientation: Expansion of dislocation loops in a horizontal (001) matrix channel (0.14% strain). b) [011] orientation: Sample oriented for coplanar slip on the (111) plane. The γ'-phase is cut by extended bands of stacking faults (1.8% strain). c) [011] orientation: Intersecting slip in a sample oriented for slip on two {111} planes (0.5% strain). d) [111] orientation: Interfacial dislocation segments with predominantly parallel line vectors (0.14% strain).

analysis of the dislocation structures in deformed [$\bar{1}$11] samples revealed a selection mechanism for matrix dislocations caused by coherency stresses. In each channel type four {111}<110> slip systems on two slip planes are active, producing dislocations with either 60° or predominantly screw character [15]. Due to the crystallography in [$\bar{1}$11] orientation the line vectors of the dislocations in a given channel type are parallel as can be seen from Fig. 4c. Accordingly the probability of dislocation interaction is low and strain hardening is poor despite the operation of multiple slip systems.

Repeated cross slip of interfacial screw dislocations [15], which were produced by the usual Orowan by-passing mechanism, on two {111} planes was also observed exclusively in [$\bar{1}$11] orientation as schematically shown in Fig. 5. An expanding dislocation loop in a matrix channel deposits long segments with screw character in the γ/γ'-interfaces. The long screw segments are glissile by cross slip to a second {111}-type slip plane. This mechanism is possible due to equal shear stresses on the two possible {111} slip planes of the screw dislocation segments and explains the high secondary creep rates typical for [$\bar{1}$11]. The macroscopic deformation associated with this mechanism is identical to that of {100}<110> cube slip.

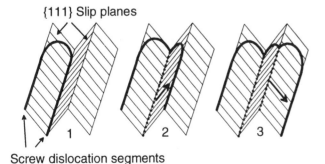

Figure 5: Schematic sketch of repeated cross slip of interfacial screw dislocation in [111] orientation.

This may explain, why Pan et al. [16] found changes in the shape of cross sections of creep deformed SRR 99 single crystals in [$\bar{1}$11] orientation, which are only compatible with deformation on cubic slip systems.

Secondary Creep at 1123K

The secondary creep rates, which mainly control the total rupture lives, are much less sensitive to small misorientations than the primary creep behavior. Regardless of orientation and stress deformation occurs mainly in the matrix phase, even in samples with prominent cutting of the γ´-phase by stacking faults during primary creep. TEM investigations on [001] oriented samples revealed, that the number of stacking faults remains virtually constant after the maximum of the creep rate up to strains of about 6%. The depletion of the stacking fault mechanism with increasing strain can be explained by the loss of coherency between the two phases caused by interfacial dislocation networks [17]. Coherency stresses were shown to facilitate the dissociation of matrix dislocations required for the generation of stacking faults. With increasing strain cutting of the γ´-phase by individual pairs of a/2<110>{111} dislocation pairs is observed. Pollock and Argon [11] related this behavior to the build-up of stresses in the γ´-particles caused by the plastic flow of the matrix phase between the undeformable γ´-cubes.

The rate-controlling mechanism, however, is the deformation of the matrix phase. TEM observation showed for samples of all orientations a quasi-stationary dislocation structure, which is retained in the strain range investigated in this study. This dislocation structure still reflects the influence of coherency stresses discussed above.

In [001] specimens the dislocation density in horizontal matrix channels is still considerably higher than in vertical channels. The concentration of matrix deformation on horizontal matrix channels contributes significantly to the superior creep resistance of the [001] orientation. Other beneficial features are the multitude of active {111}<110> type slip systems and the high stresses required for the dislocation loop process in the narrow matrix channels. The outcome of these processes is a quick increase in dislocation density after small strains and the build-up of dense dislocation networks in the γ/γ´-interfaces.

In [011] orientation the creep resistance of the matrix phase is clearly weaker than for [001]. Here two types of matrix channels, the so called "roof channels" are highly stressed and therefore subjected to strong deformation. The number of active slip systems is only four instead of eight for [001] and from Schmid factor considerations no cross slip on {111} planes is possible. Therefore the probability of dislocation interactions with a hardening effect is reduced. The lower resistance to matrix deformation together with the pronounced γ´-particle shearing result in higher secondary creep rates as compared to the [001] orientation.

In [$\bar{1}$11] orientation the creep curves (see Fig. 1a) show no clear transition from primary to secondary creep stage. Since deformation in the matrix phase occurs by {111}<110> multiple slip already during primary creep, the amount of strain hardening following the maximum of the creep rate in specimen F is rather limited (see Fig. 1a), resulting in the highest secondary creep rates of all orientations. TEM observations showed, that the basic deformation mechanisms are the same during primary and secondary creep [17]. Therefore the poor creep strength of [$\bar{1}$11] oriented crystals can be understood as a combined effect of the uniform matrix deformation in all three types of matrix channels, poor strain hardening due to a selection mechanism for active slip systems in each channel type and the high mobility of interfacial screw dislocations by repeated cross slip on {111} planes.

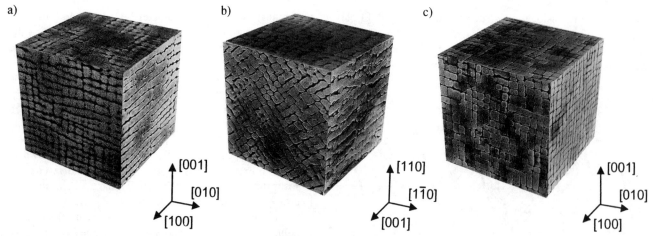

Figure 6: Morphological changes of the γ´-phase (light color) in creep deformed CMSX-4 crystals tested at T = 1253K and σ = 350 MPa. SEM micrographs taken from three cuts perpendicular to each other were arranged in order to provide a 3-dimensional view of the γ´-structure. a) After creep testing in [001] orientation for 28h. b) After creep testing in [011] direction for 28h. c) After creep testing in [111] orientation for 4h.

Creep at 1253K

The main features of creep deformation of superalloys at high temperatures are the instability of the γ´-morphology and a homogenous character of plastic deformation due to the absence of extended stacking fault bands and a higher mobility of dislocations. A comprehensive study on the creep behavior of several alloys by Caron et al. [2] suggests, that these effects reduce the orientation dependence of creep strength. This tendency is also reflected in the creep curves of [001] and [011] oriented CMSX-4 single crystals (see Fig. 1b). The primary creep stage, which showed the highest degree of anisotropy at lower temperatures, is here reduced to a short transient. In the following distinct minimum of the creep rate a quasi-steady state is reached. It should be noted, that in the representation of creep curves as creep rate vs. strain plots sections with low strain rates appear compressed. Actually the specimens spend a great fraction of their total lifetime in this stage. The minimum creep rates appear to be quite independent of orientation, although plastic deformation is concentrated in highly stressed matrix channels in the same way as described above for lower temperatures. The effect of coherency stresses is here aggravated by the gradual process of directional coarsening of the γ´-phase, which results in the disappearance of lower stressed matrix channels.

The evolution of the γ´-morphology results in characteristic spatial arrangements of the γ´-phase shown in Fig. 6. The initially cuboidal γ´-precipitates form rafts perpendicular to the load axis in [001] orientation and bars aligned parallel to the [100] direction for [011] oriented samples. This process is almost completed in the samples deformed to about 4% strain. The evolution of the γ´-morphology corresponds with a drastic reduction of creep strength. The reasons for this are the broadening of the remaining matrix channels and a reduction of the threshold stress for the cutting of the γ´-phase by a/2<110>{111} dislocation pairs [18].

In [$\bar{1}$11] orientation the typical shape of the creep curves with a pronounced maximum of the creep rate is retained up to high temperatures and results in very short lifetimes for small strains. The microstructure in the primary creep stage, where the initial cuboidal γ´-morphology is still intact, shows the same features as at lower temperatures, i.e. homogenous deformation in all types of matrix channels, cross slip of screw dislocations and a reduced number of active slip systems. In the following secondary creep stage the γ´-particles coarsen irregularly parallel to the cubic directions. Due to the high creep rates in [$\bar{1}$11] the extent of particle coarsening after equal strains cannot be easily compared to other orientations due to the large differences in creep test duration. Careful inspection of the creep curves of [$\bar{1}$11] oriented crystals (see Fig. 1b), however, revealed, that the creep rates are still decreasing at strains as large as 5%. This indicates, that the undirected coarsening of the γ´-phase in [$\bar{1}$11] orientation tends to promote a reduction of creep rates rather than the intense softening associated with the morphological changes typical for [001] and [011] orientations. The reasons for this unusual behavior are not clear and require further investigations.

Conclusions

Tensile creep tests of CMSX-4 single crystals at temperatures of 1123K and 1253K showed a high degree of anisotropy, which is most pronounced in the primary creep stage.

At 1123K creep lifetimes for small strains of 0.5 and 2.0% decrease drastically in the order [001] - [011] - [$\bar{1}$11]. For most orientations plastic deformation occurs predominantly in the matrix phase and is strongly influenced by internal coherency stresses, which control the fraction of deformable matrix volume and lead to deviations from Schmid´s law for the active slip systems. Extensive cutting of the γ´-phase by {111}<112> slip is observed only during primary creep of [011] oriented crystals and for [001] orientation under the highest level of applied stress. In these cases a pronounced primary creep stage and a high sensitivity of primary creep behavior to small misorientations are observed.

At 1253K the creep anisotropy is generally reduced and creep of [001] and [011] oriented crystals is controlled by the instability of the γ´-morphology. The creep strength of the [$\bar{1}$11] orientation remains poor.

References

1. R.A. MacKay and R.D. Maier, "The Influence of Orientation on the Stress Rupture Properties of Nickel-Base Superalloy Single Crystals", Metall. Trans., 13A (1982), 1747-1754.

2. P. Caron, Y. Ohta, Y.G. Nakagawa and T. Khan, "Creep Deformation Anisotropy in Single Crystal Superalloys", in: Superalloys 1988, eds. S. Reichmann et al. (Warrendale, PA: The Metall. Society of AIME, 1988), 215-224.

3. A.A. Hopgood and J.W. Martin, "The Creep Behaviour of a Nickel-Based Single Crystal Superalloy", Mater. Sci. Engng., 82 (1986), 27-36.

4. V. Sass, W. Schneider and H. Mughrabi, "On the Orientation Dependence of the Intermediate Temperature Creep Behaviour of a Monocrystalline Nickel-Base Superalloy", Scripta metall. mater., 31 (1994), 885-890.

5. D.J. Frazier, J.R. Whetstone, K. Harris, G.L. Erickson and R.E. Schwer, "Process and Alloy Optimization for CMSX-4 Superalloy Single Crystal Airfoils", Cost Conf. Liège, Sept. 24.-27. 1990, Proc. Part II, 1281-1300.

6. D. Blavette, P. Caron and T. Khan, "An Atom-Probe Study of Some Fine-Scale Microstructural Features in Ni-Based Single Crystal Superalloys", in: Superalloys 1988,

eds. S. Reichmann et al. (Warrendale, PA: The Metall. Society of AIME, 1988), 305-314.

7. U. Glatzel, "Microstructure and Internal Strains of Undeformed and Creep Deformed Samples of a Nickel-Base Superalloy" (Habilitation thesis, Technical University Berlin, 1994).

8. H. Mughrabi, W. Schneider, V. Sass and C. Lang, "The Effect of Raft Formation on the High Temperature Creep Deformation Behaviour of the Monocrystalline Nickel-Base Superalloy CMSX-4", in: Strength of Materials, Proc. ICSMA-10, eds. H. Oikawa et al., The Japan Institute of Metals, (1994), 705-708.

9. M. Feller-Kniepmeier and T. Kuttner, "[011] Creep in a Single Crystal Nickel Base Superalloy at 1033K", Acta metall. mater., 42 (1994), 3167-3174.

10. M. Feller-Kniepmeier, U. Hemmersmeier, T. Kuttner and T. Link, "Analysis of Interfacial Dislocations in a Single Crystal Nickel-Base Superalloy after [001] Creep at 1033K. Evolution of Internal Stresses", Scripta metall. mater., 30 (1994), 1275-1280.

11. T.M. Pollock and A.S. Argon, "Creep Resistance of CMSX-3 Nickel Base Superalloy Single Crystals", Acta metall. mater., 40 (1992), 1-30.

12. A. Fredholm and J.L. Strudel, "On the Creep Resistance of some Nickel Base Single Crystals", in: Superalloys 1984, Proc. 5th Int. Symp. on Superalloys, eds. M. Gell et al, (Warrendale, PA: The Metall. Society of AIME, 1984), 211-220.

13. T. Link and M. Feller-Kniepmeier, "Elektronenmikroskopische Untersuchungen von γ/γ´-Phasengrenzen in der einkristallinen Nickelbasislegierung SRR 99 nach Hochtemperaturkriechen", Z. Metallkunde, 79 (1988), 381-387.

14. L. Mueller, U. Glatzel and M. Feller-Kniepmeier, "Calculation of the Internal Stresses and Strains in the Microstructure of a Single Crystal Nickel-Base Superalloy During Creep", Acta metall. mater., 41 (1993), 3401-3411.

15. R. Voelkl, U. Glatzel and M. Feller-Kniepmeier, "Analysis of Matrix and Interfacial Dislocations in the Nickel-Base Superalloy CMSX-4 after Creep in [111] Direction", Scripta metall. mater., 31 (1994), 1481-1486.

16. L.-M. Pan, L. Scheibli, M.B. Henderson, B.A. Shollock and M. McLean, "Asymmetric Creep Deformation of a Single Crystal Superalloy", Acta metall. mater., 43 (1995), 1375-1384.

17. T. Link and M. Feller-Kniepmeier, "Shear mechanisms of the γ´-Phase in Single Crystal Superalloys and their Relation to Creep", Met. Trans., 23A (1992), 99-105.

18. P. Caron, P.J. Henderson, T. Khan and M. McLean, "On the Effects of Heat Treatment on the Creep Behaviour of a Single Crystal Superalloy", Scripta metall. mater., 20 (1986), 875-880.

HIGH TEMPERATURE ANISOTROPIC DEFORMATION OF SRR99 MODELLING AND MICROSTRUCTURAL ASPECTS

M.B. Henderson[1], J.-Y. Buffiere[2], L.-M. Pan[3], B.A. Shollock and M. McLean

Department of Materials, Imperial College of Science, Technology and Medicine,
Prince Consort Road, London SW7 2BP U.K.
Now at: [1]Structure and Materials Department of Defence Research Agency,
Pyestock, Hants. GU14 0LS U.K.
[2]INSA Lyon, 20 av. Albert Einstein, F69621 Villeurbanne France
[3]European Gas Turbines SA, 3 av. des Trois Chenes, 90018 Belfort France

Abstract

A predictive model for the anisotropic deformation of a single crystal superalloy has been applied to the alloy SRR99. Model parameters are derived from a database of creep curves. The model predictions are compared with both characteristic shape and crystal rotations that occur during high temperature deformation and with the results of mechanical tests. Model and experiment are shown to agree and exceptions are discussed with respect to macroscopic microstructural features and the operation of cube and octahedral slip systems.

Introduction

Single crystal nickel-base superalloys comprise an important class of engineering alloys for high temperature applications. Their primary use in turbine blades relies on their resistance to elevated temperature deformation. In order to exploit fully the creep capabilities of these alloys, both microstructural and engineering approaches must be developed. Even for isotropic materials, generation of a complete database of all types of mechanical behaviour required for design is impractical and crystallographic anisotropy introduces additional complexity. To address the design issues, a unified approach combining engineering modelling backed by an underlying knowledge of the micromechanisms controlling the material behaviour is required if these alloys are to be used to their full potential. Both approaches must be employed to ensure effective extrapolation to conditions not covered by available data; an understanding of the deformation mechanisms and the influence of microstructure must be presented in a form that can be used in the design process, and from the engineering aspect, predictive models for creep performance must be developed to allow full implementation of these alloys in their desired application.

The aim of the present work has been to continue the development, validation and use of a constitutive description of anisotropic creep deformation that can be quantified by the analysis of a limited database of creep tests but is capable of extrapolation to other conditions. The model has built upon previous work based on continuum damage mechanics (1) to represent the deformation rate at any given time in terms of state variables that are related to the mechanisms and microstructural evolution that controls the deformation rate (2, 3). This paper will assess the representation from the anisotropic model of a database of creep for the single crystal alloy SRR99 and compare the predictions of high temperature behaviour with experimental data. Validation of the model by measurement of changes in crystal orientation using electron back scattering patterns and changes in specimen shape will be summarised. In addition, microstructural variations including those associated with casting porosity and γ' morphology relative to the model predictions will be described.

Anisotropic Model - Data Analysis

The model for anisotropic creep used in this study was originally described by Ghosh, Curtis and McLean (4) and its subsequent development has been described in detail elsewhere (5, 6) and are summarised in the Appendix. The approach taken in the present assessment of the model has included prediction of mechanical behaviour as well as of crystallographic and macroscopic geometrical changes in specimen characteristics during deformation. The anisotropic model accounts for the deformation observed in single crystals by restricting deformation to specific slip systems and to occur at a given rate (4). This rate depends on the resolved shear stress, temperature and state variables incorporating damage parameters. Individual creep curves can be used to derive these model parameters as described by Dyson and McLean (3). Data analysis was performed on a data set consisting of uniaxial creep tests for two orientations <001> and <111> of the single crystal superalloy SRR99 conducted by the Defence Research Agency, Farnborough. Tests were at constant stress ranging from 150 to 950 MPa and temperatures were between 750 to 1050°C. Figures 1 and 2 present experimental creep curves for these two orientations. Although all curves show a tertiary creep region, differences in the shape of these curves are apparent such as the limited range of primary creep at 1050°C for both orientations. At lower temperatures, marked differences between the responses of the two orientations can be seen.

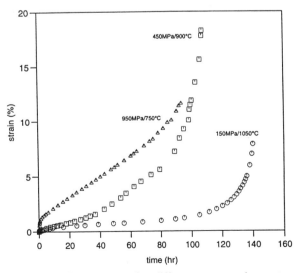

Figure 1: Creep curves for different stress and temperature conditions for <001> oriented specimens of SRR99.

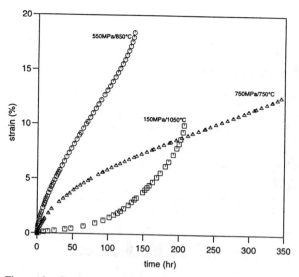

Figure 2: Creep curves for different stress and temperature conditions for <111> oriented specimens of SRR99.

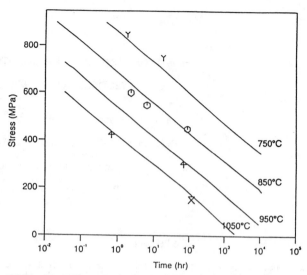

Figure 3b: Comparison of the measured and predicted times to reach 2% creep strain for <111> specimens of SRR99.

These differences pose a challenge to the data analysis but the model can cope with these variations by using the approach for the derivation of model parameters first proposed by Dyson and McLean (3).

The approach has been to analyse each individual creep curve to determine the optimum set of model parameters. Simple representations of these parameters as a function of stress and temperature were then derived. Figure 3 presents a comparison of the model predictions based on this global parameter set with the entire creep database for SRR99. At temperatures of 750°C, 850°C and generally 900°C, the model predictions show consistency with the experimental data. However, as the temperature increases to 950°C and 1050°C, this excellent agreement between experiment and prediction decreases and may result from the microstructural changes to be described and from alterations in the active slip systems.

Anisotropic Model - Predictions and Experimental Validation

Scanning electron micrographs generated using secondary electrons from electroetched specimens showing the typical γ' morphology evolved at 850°C and 950°C are presented in Figure 4. At the lower temperature although the γ' has lost its initial cuboidal shape, it typically exists as discrete particles. As the test temperature increases to 950°C, extensive rafting is observed. This change in precipitate morphology has been proposed to influence the deformation mechanism in superalloys and may account for the model deviations described in the previous section and shown in Figure 3.

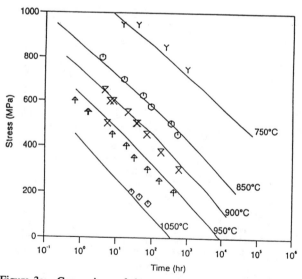

Figure 3a: Comparison of the measured and predicted times to reach 5% creep strain for <001> specimens of SRR99.

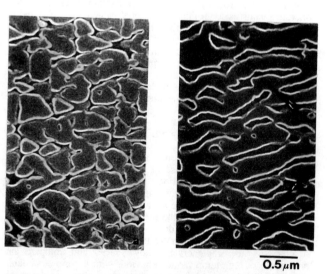

Figure 4: Secondary electron micrographs showing the difference in γ' morphology for a <210> oriented specimen after creep testing at:
(a) 850°C, 550MPa, time to failure 149.5h.
(b) 950°C, 300MPa, time to failure 112.5h.

Validation from Macroscopic Measurements of Creep Specimens

As a result of the high levels of creep strain before fracture, the tensile creep specimens are highly necked on failure. Consequently, there is a considerable gradient in local strain as measured by the reduction in area. Macroscopic measurements of deformation such as crystal rotation and changes in specimen shape from the original circular cross-section characteristic of the gauge length prior to deformation provide a challenging test of the model validity. In the latter case, after careful determination of the initial orientation using Laue techniques, diameters of the gauge at various distances along the specimen length were measured as a function of rotation and related to the local reduction in area and to a reference transverse direction on creep fractured specimens that were part of the database. Using the model predictions shown by the uninterrupted line, good agreement with the experimental data is obtained as can be seen in Figure 5. In some cases, such as shown in Figure 6, experiment and model agree only when cube shear is allowed to dominate. This provides strong justification for the initial premise of the model that both cube and octahedral shear can occur. A summary of the observations is presented in Figure 7 indicating that cube shear is restricted to orientations in the vicinity of <111> (7). It must be noted, however, that the summary covers a range of temperatures and does not describe the slip behaviour at a given temperature.

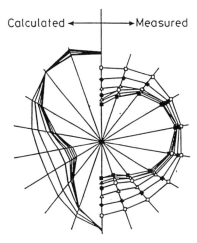

Figure 6b: Comparison of the measured change in specimen shape of a failed creep specimen of SRR99 with a <112> orientation with model predictions assuming only octahedral slip.

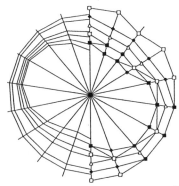

Figure 5: Comparison between the measured and predicted creep specimen cross-section at various levels along the gauge length of a nominal <110> oriented specimen of SRR99. The experimental data are shown on the right, the prediction is shown by the solid lines on the left.

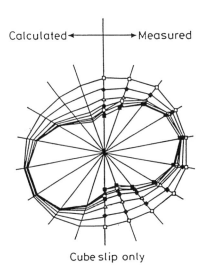

Figure 6a: Comparison of the measured change in specimen shape of a failed creep specimen of SRR99 with a <112> orientation with model predictions assuming only cube slip.

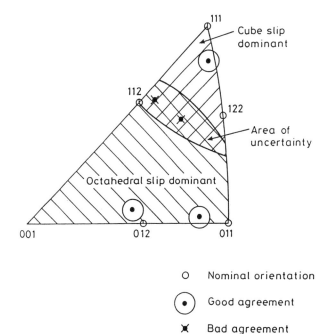

Figure 7: Summary of experimental and predicted changes in cross-sectional area for SRR99 creep specimens.

The same specimens were sectioned parallel to the gauge length and the resulting surface was mechanically polished followed by electropolishing in 10% orthophosphoric acid in water. The local orientations were determined along intervals of the length using electron back scattering patterns (EBSP) and in some cases, a matrix of orientations was obtained from a region of the specimen. Figure 8 presents the results of an EBSP analysis obtained along the gauge length of a <110> oriented specimen that was tested to failure at 850°C, 550MPa with a total strain to failure of 5.8%. It is important to note that the EBSP technique has a high spatial resolution (approximately $1\mu m^2$) and is able to determine the orientation close to the fracture surface of the specimen at high levels of strain where X-ray measurements are not possible. The model predictions for rotation are summarised in Figure 9. On this representation, initial orientations near <001> and <111> rotate to <001> and <111> respectively. In contrast, orientations near <011> are not stable and in general rotate away from this orientation, as shown by the example in Figure 8. The predicted rotations in Figure 9 are also consistent with cube slip dominating for orientations close to <111> and octahedral slip dominating for other orientations.

Other factors must be considered when analysing the results of EBSP. For example, Figure 10 presents the results of an analysis for a specimen with an initial orientation close to <112>. Scatter in the measurements increases with distance along the gauge length approaching the deformation neck and fracture surface and these large rotations were associated with interdendritic porosity, such that presented in Figure 11. Orientations measured in the vicinity of the pores showed large deviations from the average.

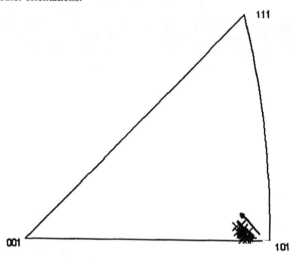

Figure 8: Stereographic triangle showing EBSP measurements for a <110> specimen of SRR99 creep tested at 850°C, 550MPa. The direction of rotation is indicated by the arrowed line.

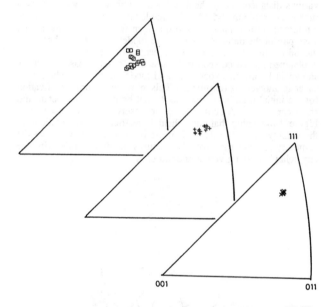

Figure 10: Stereographic triangle showing EBSP measurements for a <112> specimen of SRR99 taken from the threaded end, middle and fractured end (right to left) showing the increasing degree of scatter as the fracture surface is approached.

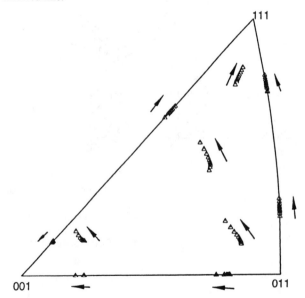

Figure 9: Model predictions for crystallographic rotation of the tensile axis.

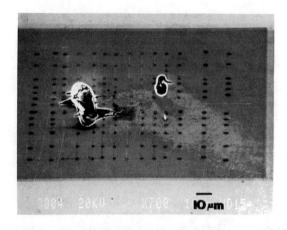

Figure 11: Secondary electron image showing the matrix of points used to obtain the EBSP data.

Validation from High Temperature Mechanical Testing

If the model is to be used for other design purposes, it must be capable of representing generalised deformation under a wide range of loading conditions. Various types of high temperature mechanical behaviour have been examined and compared to model predictions. Here we compare the results of slow strain rate controlled tensile tests on specimens of various orientations with simulations from the model with parameters determined from creep tests.

Figure 12a shows the experimentally determined tensile curves for <011> oriented specimens tested at 950°C at various strain rates; the model simulations are shown in Figure 12b. Clearly the general shapes of the curves are correctly modelled; after yield the ultimate tensile strength is achieved at 1 to 2% strain and thereafter there is a steady decline in stress. This softening is a direct parallel to the extensive tertiary creep behaviour that is observed. The qualitative agreement is good at the lowest strain rate ($10^{-7}s^{-1}$), but the model over estimates the tensile strength at higher strain rates. Since the database from which the model parameters were derived had creep rates of less than or equal to $10^{-7}s^{-1}$, the result indicates difficulties in extrapolating to different strain rates.

The effect of orientation on the tensile behaviour is shown in Figure 13 for tests at a strain rate of $10^{-7}s^{-1}$ and 950°C. In fact, very little anisotropy is predicted, and the results for <001> and <110> specimens are accurately represented by the model. The stress for the <112> falls off much more rapidly than predicted.

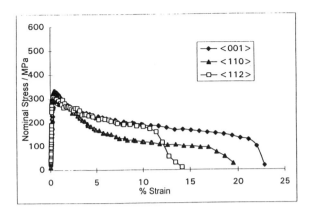

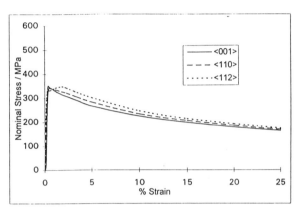

Figure 13: Experimental and modelled stress-strain curves for three different orientations of specimen tested at $10^{-7}s^{-1}$ at 950°C.

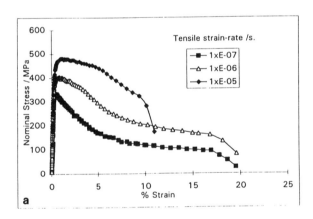

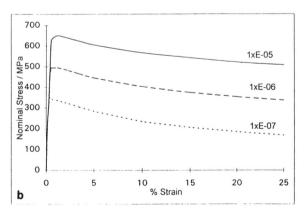

Figure 12: (a) Experimental stress-strain curves for <110> oriented specimens tested at 950°C at different strain rates.
(b) Model simulations for the conditions of the experimental curves presented in Figure 12a.

Conclusions

A model for anisotropic creep deformation has been applied to the single crystal superalloy SRR99. The model has been validated using both macroscopic measurements of specimen deformation such as changes in cross-sectional shape and crystallographic rotation and high temperature strain-rate controlled mechanical tests. Agreement between the model predictions and experimental data is good; however, microstructural features such as evolution of γ' morphology and interdendritic porosity may dominate the deformation of the materials and the model is currently being evaluated with respect to these features.

Acknowledgements

The authors wish to thank the Engineering and Science Research Council (grant GR/J 02667), The Royal Society and the Defence Research Agency for financial support and Dr. M. R. Winstone for the provision of specimens on which measurements were performed.

References

1. L.M. Kachanov, Izv Akad. Nauk. S.S.R. 26(8) (1958).
2. J. Ion, A. Barbosa, M. Ashby, B. F. Dyson,. and M. McLean, "The Modelling of Creep for Engineering Design - I," NPL Report DMA A115, The National Physical Laboratory, Teddington, U.K. (1986).
3. B. F. Dyson and M. McLean, "Creep Deformation of Engineering Alloys: Development from Physical Modelling," ISIJ International 30(10) (1990) 802-811.
4. R. N. Ghosh, R. V. Curtis and M. McLean, "Creep Deformation of Single Crystal Superalloys - Modelling the Crystallographic Anisotropy," Acta Metall. Mater. 38(10) (1980) 1977-1992.
5. M. McLean, R. N. Ghosh, R. V. Curtis, U. Basu-Conlin and M. R. Winstone, "Anisotropy of High Temperature Deformation in Single Crystal Superalloys - Constitutive Laws, Modelling and Validation," Superalloys 1992, ed. S. D. Antolovich et al. The Minerals, Metals and Materials Society (1992) 609-618.
6. L.-M. Pan, R, N. Ghosh and M. McLean, "Load and Strain Controlled Deformation of a Single Crystal Superalloy - Modelling and Validation," Proceedings 10th International Conference on the Strength of Materials, Sendai Japan, eds. H. Oikawa, K. Maruyama, S. Takeuchi and M. Yamaguchi, Japan Institute of Metals, (1994) 583-586.
7. L.-M. Pan, I. Scheibli, M. B. Henderson, B. A. Shollock and M. McLean, "Asymmetric Creep Deformation of a Single Crystal Superalloy," 43(4) (1995) 1375-1384.

Appendix: Constitutive equations for anisotropic creep in single crystal superalloys

The constitutive equations used in the present study are required to represent two important aspects of the creep deformation of single crystal superalloys; the anisotropic deformation characteristics associated with crystallographic orientation and the highly non-linear variation of creep rate with either strain or time.

A crystallographic model has been adopted in which deformation is considered to occur by shear on a restricted number of slip systems, $(n_1 n_2 n_3)(b_1 b_2 b_3)$. In particular, for single crystal superalloys slip is only considered here to be allowed on the families of systems of the type $(111)\langle 1\bar{1}0 \rangle$ and $(001)\langle 110 \rangle$. The rate of shear deformation $\dot{\gamma}^k$ on any of the k allowed slip systems will be a function of the resolved shear stress in that direction and the total strain of the body can be computed by summing all possible shear displacements:

$$\varepsilon_{ij} = \sum_{k=1}^{N} \gamma^k b_i^k n_j^k$$

This leads to both a change in orientation of an arbitrary direction from $[x_1 x_2 x_3]$ to $[X_1 X_2 X_3]$:

$$\begin{bmatrix} X_1 \\ X_2 \\ X_3 \end{bmatrix} = \begin{bmatrix} 1+\varepsilon_{11} & \varepsilon_{12} & \varepsilon_{13} \\ \varepsilon_{21} & 1+\varepsilon_{22} & \varepsilon_{23} \\ \varepsilon_{31} & \varepsilon_{32} & 1+\varepsilon_{33} \end{bmatrix} \begin{bmatrix} x_1 \\ x_2 \\ x_3 \end{bmatrix}$$

and a linear strain in that direction of $\frac{X-x}{x}$. The shear strain rate $\dot{\gamma}^k$ on each of the two families of slip systems is considered to depend on two state variables; an internal stress S^k that dominates at low strains leading to primary creep deformation and a damage parameter ω^k that is most important at higher strains leading to tertiary creep deformation. The specific equation set used in the present study is:

$$\boxed{\begin{aligned} \dot{\gamma}^k &= \dot{\gamma}_i^k (1-S^k)(1+\omega^k) \\ \dot{S}^k &= H^k \dot{\gamma}_i^k \left(1 - \frac{S^k}{S_{ss}^k}\right) \\ \dot{\omega}^k &= \beta^k \dot{\gamma}^k \end{aligned}}$$

The shear deformation on the particular slip system for a fixed temperature and shear stress is determined by the four constants $(\dot{\gamma}_i^k, H^k, S_{ss}^k, \beta^k)$. This parameter set was determined for the octahedral shear system by analysis of some 35 constant stress tensile creep tests at a range of temperatures and stresses for <001> orientations of SRR99. Cube shear parameters were determined from a smaller database creep curves for <111> specimens.

Each of the model parameters, $P=(\dot{\gamma}_i^k, H^k, S_{ss}^k, \beta^k)$ are represented by exponential functions of shear stress τ^k and temperature T:

$$P = a \exp\left(b\tau^k - \frac{Q}{RT}\right)$$

The database examined is well represented by the values of the parameter listed in the Table.

Table
Parameters representing the shear creep behaviour on octahedral and cube slip systems in SRR99

Parameter P	a	b (MPa^{-1})	Q (J mol^{-1})
Octahedral shear			
$\dot{\gamma}_i^k$	1.39×10^8(s^{-1})	4.49×10^{-2}	4.325×10^5
H^k	327.0	0	0
S_{ss}	0.60	0	0
β^k	3.42×10^2	-1.20×10^{-2}	0
cube shear			
$\dot{\gamma}_i^k$	1.94×10^3(s^{-1})	3.73×10^{-2}	2.98×10^5
H^k	374.0	-6.90×10^{-3}	0
S_{ss}	0.93	0	0
β^k	525.0	-8.88×10^{-3}	0

EFFECT OF MORPHOLOGY OF γ' PHASE ON CREEP RESISTANCE OF A SINGLE CRYSTAL NICKEL-BASED SUPERALLOY, CMSX-4

Yoshihiro KONDO, Naoya KITAZAKI, Jirou NAMEKATA, Narihito OHI* and Hiroshi HATTORI*

The National Defense Academy, Yokosuka 239, Japan

*Ishikawajima-Harima Heavy Industries Co., Ltd. Tokyo 188, Japan

Abstract

The influence of the formation of a lamellar γ-γ' structure on creep resistance was investigated using the single crystal nickel-base superalloy, CMSX-4. The evolution of lamellae, or rafts was controlled by the duration of the prior-creep test conducted at 1273K-160MPa up to 3.24×10^6s. Under these creep condition, the cuboidal γ' phase in the as-heat treated specimens turned to the rafted one through a transient creep stage. The stress enhanced creep tests were conducted at 1273K-250MPa to evaluate creep resistance of the prior-creep tested specimens, and the minimum creep rates decided by the stress enhanced creep test were compared with that of the as-heat treated specimen. The minimum creep rates of the prior-creep tested specimens were always larger than that of the as-heat treated one. The dislocation substructure was not formed in the cuboidal γ' and rafted γ', and was observed at the γ/γ' interface and γ channel. The thickness of the γ channel increased with increasing the prior-creep testing time. The correlation between the thickness of the γ channel and the minimum creep rate in the stress enhanced creep test shows linear and is independent of the shape of γ'. The TEM observation of specimens interrupted the stress enhanced creep test at the minimum creep rate showed that the radius of dislocation curvature was directly proportional to the thickness of the γ channel. Consequently, the loss of creep resistance through rafting of γ' is caused by the increase in the radius of dislocation curvature, not by the shape and the size of γ'.

Introduction

Excellent creep resistance at high temperature of advanced single crystal nickel-based superalloys is derived from the large volume fraction more than 70% of intermetallic $Ni_3(Al,Ti)$ precipitate, known as γ', and from the regular array of the cuboidal γ' precipitate. By submitting to the creep deformation under the conditions of higher temperatures and relatively lower stresses, the cuboidal γ' precipitates in the [001] oriented single crystal turn their shape into platelets, which are called as the rafted structure and are oriented normal to the tensile stress axis[1]-[5]. This directional coarsening has been believed to enhance the creep resistance[6][7]. Pearson et al. presented that the creep rupture life of a Ni-13Al-9Mo-2Ta(at%) single crystal with rafted γ' was four times longer than that of the as-heat treated one[6]. In contrast to this, Nathal et al. indicated that a Ni-9.5Cr-5.5Al-1.2Ti-3.2Ta-10.0W-1.0Mo(wt%) single crystal submitted to the creep deformation at 1273K-148MPa for 1.5×10^5s showed two times larger creep rate than the as-heat treated one[8].

To evaluate creep resistance of the specimens subjected to the creep deformation, a stress enhanced creep test is applied to the prior creep tested specimen. To avoid the microstructural change, the stress enhanced creep test is preferable to the temperature enhanced creep test. Unfortunately, in both studies done by Pearson and by Nathal, temperatures in the stress enhanced creep test were reduced, thereby the enhanced stress seems to be high enough to allow the dislocations to cut the cuboidal γ' precipitate. The effect of γ' morphology on creep resistance should be discussed under the conditions where dislocation climb in the γ channel is predominant, because the prior-creep tests done to make the rafted specimens must be undertaken at the condition where the dislocation climb is predominant. Therefore the stress levels in the stress enhanced creep tests done by Pearson and Nathal are both unsuitable. Systematic research programs interrupting the creep tests at various times ranging from a transient creep stage to an accelerating creep stage have been desired to elucidate the correlation between the formation of the rafted structure and creep resistance.

In this study, the creep tests for the single crystals oriented in the [001] orientation were interrupted in the wide region ranging from a transient creep stage to an accelerating creep stage. The stress enhanced creep tests were performed only by increasing the applied stress without changing the testing temperature. The change in creep resistance of the specimens submitted to the creep was discussed by correlating with the change in γ' precipitate morphology to identify that the rafted structure of γ' precipitates acts as a creep strengthener or as a creep weakener.

Experimental procedure

Single crystals of CMSX-4 (analyzed composition in weight percent ; 6.4Cr, 9.3Co, 5.5Al, 0.9Ti, 6.3Mo, 6.2Ta, 6.2W, 2.8Re, 0.1Hf, balance Ni) were prepared in the form of bars 13mm in diameter by directional solidificated casting. The exact orientations were determined by the Laue back-reflection technique; longitudinal axes of single crystals selected for this study were within 5deg of the [001] orientation. After employing a eight steps solid solution treatment*1 and a two steps aging heat treatment*2, the prior-creep tests were carried out at 1273K and 160MPa(creep

---

*1 Solid solution treatment: 1550Kx7.2ks → 1561Kx7.2ks → 1569Kx10.8ks → 1577Kx10.8ks → 1586Kx7.2ks → 1589Kx7.2ks → 1591Kx7.2ks → 1594Kx7.2ks → GFC

*2 Aging heat treatment: 1413Kx21.6ks → 1144Kx72ks → AC

Fig. 1. Scanning electron micrograph of a single crystal nickel-based superalloy, CMSX-4.

rupture life is 3.60×10^6s), using the specimens with a gauge diameter of 8.0mm and a gauge length of 50mm, and were interrupted at various times ranging from the transient creep stage to the accelerating creep stage. All the creep interrupted specimens were followed by cooling under load. The prior-creep tested specimens were re-machined into the specimens for the stress enhanced creep tests with a gauge diameter of 6.0mm and a gauge length of 30mm to remove the surface cracks. The stress enhanced creep tests were performed at 1273K and 250MPa. Creep strain was measured automatically through linear variable differential transformers (LVDT's) attached to extensometers. Microstructural examinations by SEM and TEM were carried out on specimens cut parallel to (100) planes. Specimens for the SEM observation were prepared metallographically and electroetched with a supersaturated phosphoric acid-chromic acid solution. TEM foils were electropolished with a 5% solution of perchloric acid in alcohol. The volume fraction of the γ' phase was measured by use of image processing facilities. The ratio of the average width to the length in the γ' phase was defined as the aspect ratio and measured as a function of the prior creep testing time.

Results

Microstructure of the as-heat treated specimen

The microstructure of the as-heat treated specimen of CMSX-4 is shown in Fig. 1. No eutectic γ' phase was observed and the cuboidal γ' phase was regularly present in the γ matrix. The average edge length of the cuboidal γ' particles was controlled to be about 0.5μm and the width of the γ channel was estimated approximately 0.09μm.

Prior-creep test

The creep rate-time curve in the prior-creep test at 1273K-160MPa of the as-heat treated specimen is shown in Fig. 2. The open circles correspond to time where the prior-creep tests were interrupted to prepare the specimens for the stress enhanced creep test. The minimum creep rate was attained at about 10^6s, and the prior-creep testing time extended over a wide range from the latter half of the transient creep stage to the late accelerating creep stage.

Microstructure of the prior-creep tested specimens

The scanning electron micrographs of the specimens prior-creep tested for 1.08×10^6 and 3.24×10^6s are shown in Fig. 3. The γ' phase coalesced with each other and the lamellar γ-γ' structures, that is rafted γ' structures, formed perpendicular

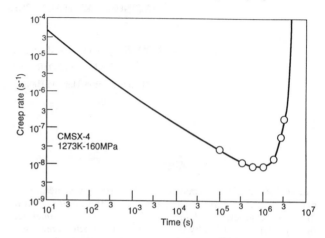

Fig. 2. Creep rate-time curve of the specimen crept at 1273K-160MPa. The open circles correspond to time where tests were interrupted for the microstructural examination and stress enhanced creep test.

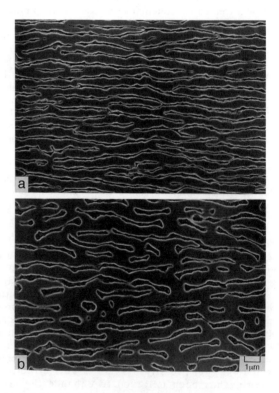

Fig. 3. Scanning electron micrographs of the specimens prior-creep tested at 1273K-160MPa for (a) 1.08×10^6 and (b) 3.24×10^6s. Stress axis is vertical on these photos.

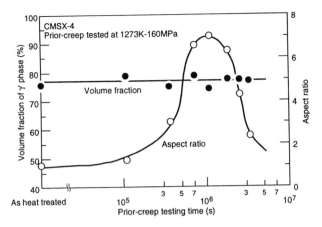

Fig. 4. Changes in the volume fraction and the aspect ratio of γ' of the prior-creep tested specimen with the prior-creep testing time at 1273K.

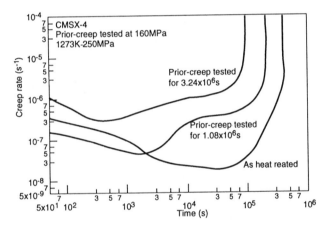

Fig. 5. Creep rate-time curves at 1273K-250MPa of the specimens as-heat treated and prior-creep tested for 1.08×10^6 and 3.24×10^6s.

to the stress axis was observed in the specimen prior-creep tested for 1.08×10^6s. However, the further coarsening of the γ' phase was detected in the specimen prior-creep tested for 3.24×10^6s. And the lamellar γ-γ' structure was out of shape. The changes in the volume fraction and the aspect ratio of the γ' phase with the prior-creep testing time are shown in Fig. 4. The volume fraction of the γ' phase shown in the solid symbol was approximately 76%, independent of the prior-creep testing time. Precipitation of the γ' phase seems to be finished through the pre-heat treatment. The aspect ratio of the γ' phase increased after subjecting the prior-creep tests beyond 1.08×10^5s, and reached to the maximum value of 7 at 1.08×10^6s. However, this maximum value of the aspect ratio did not remain, and soon the aspect ratio of the γ' phase decreased with increasing the prior-creep testing time and attained to about 2 at 3.24×10^6s. The lamellar γ-γ' structures showing the maximum aspect ratio were formed at the stage where creep rate reached the minimum. Therefore, the lamellar structure was defined as the unstable structure, and the change in the shape of γ' phase was proceeding with the prior-creep testing time through the accelerating creep stage.

Stress enhanced creep test

The creep rate-time curves at 1273K-250MPa of the specimens as-heat treated and prior-creep tested for 1.08×10^6 and 3.24×10^6s are shown in Fig. 5. The minimum creep rate of the specimen prior-creep tested for 1.08×10^6s was about three times that of the as-heat treated one. And the minimum creep rate of the specimen prior-creep tested for 3.24×10^6s showed twenty times larger value than that of the as-heat treated one.

The minimum creep rates obtained by the stress enhanced creep test were plotted as a function of the prior-creep testing time in Fig. 6. The minimum creep rate of the prior-creep tested specimens increased gradually up to 3×10^6s with increasing the prior-creep testing time. Namely, the minimum creep rate of the specimen prior-creep tested for 7×10^5 to 2×10^6s with the lamellar γ-γ' structures was three to five times that of the as-heat treated one with the cuboidal γ' phase. And the minimum creep rate of the specimen prior-creep tested for 3.24×10^6s when the lamellar γ-γ' structures was out of shape, showed twenty times that of the as-heat treated one. Therefore, creep resistance of CMSX-4 decreased remarkably with increasing the prior-creep testing time.

Discussion

In general, the loss of strength of the polycrystalline heat-resistant alloys due to high temperature creep is thought to result from mechanical damage such as initiation and propagation of cracks at grain boundary[9][10]. CMSX-4 used in this study, however, is the single crystal alloy and the surface layer of 1mm in thickness of the prior-creep tested specimens was eliminated by re-machining after the prior-creep tests. The rupture elongation of the prior-creep tested specimens was constant, independent of the prior-creep testing time. Therefore, it was difficult to interpret that the loss of creep resistance due to the prior-creep test caused by mechanical damage. On the other hand, it was reported that the loss of strength of the polycrystalline heat-resistant alloys due to high temperature creep was attributed to deterioration such as degradation of deformation resistance caused by coalescence and coarsening of carbides and precipitation of

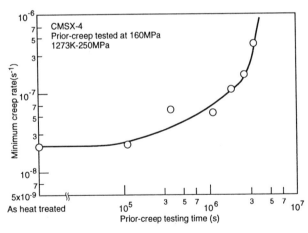

Fig. 6. Change in the minimum creep rate at 1273K-250MPa of the as-heat treated and the prior-creep tested specimens with the prior-creep testing time.

Fig. 7. Transmission electron micrograph of the as-heat treated specimen crept up to the minimum creep rate at 1273K-250MPa. Stress axis is vertical on this photo.

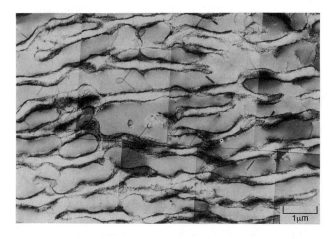

Fig. 8. Transmission electron micrograph of the specimen prior-creep tested for 1.08×10^6s and crept up to the minimum creep rate at 1273K-250MPa. Stress axis is vertical on this photo.

the TCP phase such as the sigma phase[11)-20)]. The loss of creep resistance of the prior-creep tested specimens in this study was thought to result from deterioration based on the structural changes which take place uniformly over the whole structure. The loss of creep resistance due to the prior-creep tests appears to be attributed to the formation of the lamellar γ-γ' structures. Therefore, substructural examinations by transmission electron microscopy were carried out on the as-heat treated and the prior-creep tested specimens interrupted the stress enhanced creep test at the time showing the minimum creep rate.

The transmission electron microstructure of the as-heat treated specimen interrupted the stress enhanced creep test at the time showing the minimum creep rate is shown Fig. 7, where the electron beam direction, B, was close to [100].

Few dislocations were observed in the γ matrix and a small number of dislocations were present on the γ/γ' interface.

The transmission electron microstructure of the specimen prior-creep tested for 1.08×10^6s and interrupted the stress enhanced creep test at the time showing the minimum creep rate, which was about five times that of the as-heat treated one, is shown in Fig. 8, where B=[100]. The dark regions were locally formed in the γ matrix and was derived from the dislocation density. A number of dislocations were observed on the γ/γ' interface.

The transmission electron microstructure of the specimen prior-creep tested for 3.24×10^6s and interrupted the stress enhanced creep test at the time showing the minimum creep rate, which was about twenty times that of the as-heat treated one, is shown in Fig. 9, where B=[100]. In the γ matrix, the dark regions were also observed. By comparing the area of the dark region with that of the specimen prior-creep tested for 1.08×10^6s, the dark region increased with increasing the prior creep testing time. A number of dislocations were also

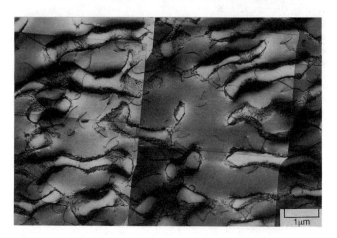

Fig. 9. Transmission electron micrograph of the specimen prior-creep tested for 3.24×10^6s and crept up to the minimum creep rate at 1273K-250MPa. Stress axis is vertical on this photo.

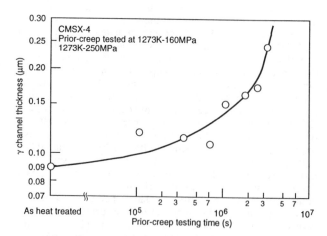

Fig. 10 Change in the γ channel thickness of the specimens as-heat treated and the prior-creep tested with the prior-creep testing time.

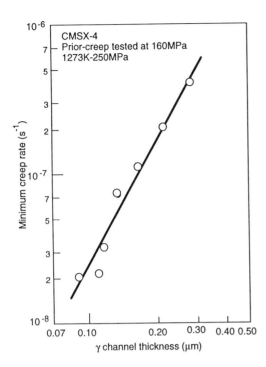

Fig. 11. Relation between the minimum creep rate and the γ channel thickness of the as-heat treated and the prior-creep tested specimens.

present on the γ/γ' interface. From the results that few dislocations existed within rafted γ', it was supposed that the creep deformation within the γ channel was predominant, compared with the rafted γ'. So, the movement of the dislocation within the thin γ channel appeared to be strongly restricted. From the SEM observation of the specimens as-heat treated and prior-creep tested for 1.08×10^6 and 3.24×10^6s as shown in Figs. 1 and 3, the γ channel thickness seems to increase with increasing the prior-creep testing time. Releasing the dislocations from the restriction of the thin γ channel through increasing the γ channel thickness seems the substantial mechanism to increase the creep rate. The γ channel thickness was measured on the as heat treated and the prior-creep tested specimens.

The γ channel thickness of the as heat treated and the prior-creep tested specimens interrupted at the time showing the minimum creep rate was plotted as a function of the prior-creep testing time as shown in Fig. 10. The specimens subjected the longer prior-creep tests provide the thicker γ channel. The γ channel thickness of the specimens prior-creep tested for 2×10^6 and 3×10^6s were approximately twice and three times that of the as-heat treated one with the cuboidal γ' phase, respectively.

The minimum creep rate of the prior-creep tested specimens was plotted as a function of the γ channel thickness, as indicated in Fig. 11. The correlation between the minimum creep rate and the γ channel thickness showed linear, independent of the shape of γ'. Consequently, creep resistance of single crystal nickel-based superalloys depends on the γ channel thickness. The reason was discussed why creep resistance decreased with an increase in the γ channel thickness.

The high magnification transmission electron microstructures of the specimens prior-creep tested for 1.08×10^5, 1.08×10^6 and 3.24×10^6s are shown in Fig. 12, where B=[100]. The bent dislocation with the small radius of curvature was observed within the γ channel of the specimen prior-creep tested for 1.08×10^5s. The radius of dislocation curvature of the specimen prior-creep tested for 1.08×10^6s was larger than that of the specimen prior-creep tested for 1.08×10^5s and smaller than that of the specimen prior-tested for 3.24×10^5s. The radius of dislocation curvature seems to increase with an increase in the prior-creep testing time. It is well known that

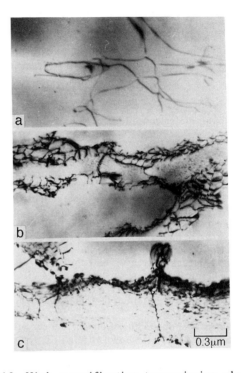

Fig. 12. High magnification transmission electron micrographs of the specimens prior-creep tested for (a) 1.08×10^5, (b) 1.08×10^6 and (c) 3.24×10^6s and crept up to the minimum creep rate at 1273K-250MPa.

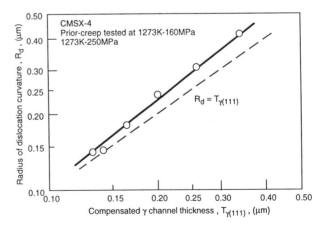

Fig. 13. Relation between the radius of dislocation curvature and the compensated γ channel thickness of the as-heat treated and the prior-creep tested specimens.

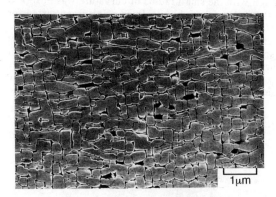

Fig. 14. Scanning electron micrograph of a single crystal nickel-based superalloy, CMSX-4 with the fine γ' phase.

an applied shear stress, τ, is inversely proportional to the radius of dislocation curvature, R_d, as shown in the following equation : $\tau = A/R_d$, if the dislocation is bent in the slip plane by an applied shear stress[21)-23)]. The smaller radius of dislocation curvature needs the greater applied shear stress. It can be speculated that as the radius of dislocation curvature increases with an increase in the prior-creep testing time, the applied shear stress decreases and then the creep rate increases. The measurements of the radius of dislocation curvature were made on the stress enhanced creep tested specimens sectioned parallel to (111) planes which are the slip planes of fcc metals.

The relation between the radius of dislocation curvature and the γ channel thickness is shown in Fig. 13 where the γ channel thickness was converted to the spacing between two neighboring γ' precipitates on (111) planes. The radius of dislocation curvature was directly proportional to the γ channel thickness. From these results, it was concluded that creep resistance of single crystal nickel-based superalloys depends on the radius of dislocation curvature which is based on the γ channel thickness.

To confirm the above results, a similar investigation is made on the single crystal nickel-base superalloy, CMSX-4, with the fine γ' phase which was obtained by a two steps aging heat treatment for γ' refining (1323K×21.6ks → 1144K×72ks → Water quench) after a standard eight steps solid solution treatment and the results of the fine-γ' specimen are compared with those of the standard-γ' one.

The microstructure of the fine-γ' specimen in the as-heat treated condition is shown in Fig. 14. The cuboidal γ' phase was regularly present in the γ matrix. The average edge length of the cuboidal γ' particles was shorter and the γ channel thickness was thinner than those of the standard-γ' CMSX-4 which was shown in Fig. 1.

The γ channel thickness of the fine-γ' specimen was plotted as a function of the prior-creep testing time compared with that of standard-γ' specimen, as shown in Fig. 15. The γ channel thickness of the as-heat treated specimens with fine γ' was approximately 2/3 of that of the as-heat treated one with standard γ'. The both specimens subjected the longer prior-creep tests provided the larger γ channel thickness. The magnitude of the increase in the γ channel thickness of the fine-γ' specimens with an increase in the prior-creep testing time was larger than that of standard-γ' specimen, and by the prior-creep test for 2.52×10^6s, the γ channel thickness of the fined-γ' specimen was approximately equal to that of the standard γ' specimen.

The minimum creep rate of the fine- and the standard-γ' specimens obtained by the stress enhanced creep test at 1273K-250MPa was plotted as a function of the prior-creep testing time as shown in Fig. 16. The minimum creep rate of the as-heat treated specimens with fine γ' was approximately 1/10 of that of the as-heat treated one with standard γ'. The minimum creep rate of the fine- and the standard-γ' specimens increased with increasing the prior-creep testing time. The increasing ratio of the minimum creep rate with increasing the prior-creep testing time in the fine-γ' specimen was larger than that of standard-γ' specimen, and the minimum creep rate of the fine-γ' specimen prior-creep tested for 2.52×10^6s was approximately equal to that of the standard-γ' specimen. The minimum creep rate in the fine- and the standard-γ' specimens, was plotted as a function of the γ channel

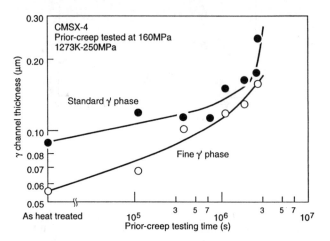

Fig. 15. Change in the γ channel thickness of the fine-γ' specimen as-heat treated and prior-creep tested with the prior-creep testing time compared with that of the standard-γ' specimen.

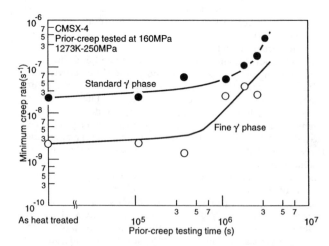

Fig. 16. Change in the minimum creep rate of the fine-γ' specimen as-heat treated and prior-creep tested with the prior-creep testing time compared with that of the standard-γ' specimen.

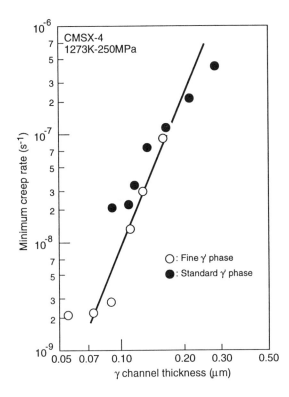

Fig. 17. Relation between the minimum creep rate and the γ channel thickness of the as-heat treated and prior-creep tested specimens with the fine and the standard γ' phase.

thickness, as indicated in Fig. 17. The correlation between the minimum creep rate and the γ channel thickness showed linear, independent of the shape and the size of γ'. Consequently, creep resistance of single crystal nickel-based superalloys directly depends on the γ channel thickness.

Conclusions

High temperature creep resistance of the prior-creep tested single crystal nickel based superalloy, CMSX-4, was investigated in connection with the morphology of the γ' phase. The results can be summarized in the following:
1) The γ-γ' lammelar structures were formed perpendicular to the stress axis during the transient creep stage, but regularity of the rafted γ' diminished during the long time prior-creep test.
2) The volume fraction of the γ' phase was constant independent of the prior-creep testing time. The aspect ratio of the γ' phase of the prior-creep tested specimens increased with an increase in the prior-creep testing time and achieved the maximum value 7, and then it decreased to about 2.
3) The minimum creep rate obtained by the stress enhanced creep test using the prior-creep tested specimens increased with an increase in the prior-creep testing time.
4) The TEM observation of specimens interrupted the stress enhanced creep test showed that the radius of dislocation curvature was directly proportional to the γ channel thickness.
5) The γ channel thickness increased with an increase in the prior-creep testing time and the correlation between the minimum creep rate and the γ channel thickness showed linear, independent of the shape and the size of γ'.
6) It was concluded that the creep resistance of single crystal nickel-based superalloys depends on the radius of dislocation curvature which was based on the γ channel thickness, independent of the shape and the size of γ'.

Reference

1) J. K. Tien and R. P. Gamble, "Effects of Stress Coarsening on Coherent Particle Strengthening," Metall. Trans A, 3A(1972), 2157-2162
2) M. V. Nathal and L. J. Ebert, "Elevated Temperature Creep-Rupture Behavior of the Single Crystal Nickel-base Superalloy NASAIR 100," Metall, Trans A, 16A(1985), 427-439
3) N. Kitazaki, Y. Kondo, J. Namekata, N. Ohi, and H. Hattori, "Creep Resistance of Single crystal Ni-based Superalloy, CMSX-4, at 1273K," 123rd Committee on Heat Resisting Metals and Alloys Rep., 35(1994), 353-360
4) K. Ishibashi, Y. Kondo, J. Namekata, N. Ohi, and H. Hattori, "Effect of Tensile Orientation Rafting Structure of Gamma Prime Precipitates in Single Crystal Ni-base Superalloy CMSX-2," 123rd Committee on Heat Resisting Metals and Alloys Rep., 34(1993), 165-172
5) K. Ishibashi, Y. Kondo, J. Namekata, N. Ohi, and H. Hattori, "Long-term Creep Rupture Properties of Single Crystal Ni-based Superalloy, CMSX-4," 123rd Committee on Heat Resisting Metals and Alloys Rep., 37(1996), 1-10
6) D. D. Pearson, F. D. Lemkey and B. H. Kear, "Stress Coarsening of γ' and its Influence on Creep Properties of a Single Crystal Superalloy," Proc. of the 4th Int'l Conf. Superalloys 1980, (1980), 513-520
7) R. A. MacKay and L. J. Ebert, "The Development of γ-γ' Lamellar Structures in a Nickel-Base Superalloy during Elevated Temperature Mechanical Testing," Metall. Trans A, 16A(1985), 1969-1982
8) M. V. Nathal, R. A. MacKay and R. V. Miner, "Influence of Precipitate Morphology on Intermediate Temperature Creep Properties of a Nickel-Base Superalloy Single Crystal," Metall. Trans A, 20A(1989), 133-141
9) D. A. Woodford, "Creep Damage and the Remaining Life Concept," J. Eng. Mater. Technol., 101(1979), 311-316
10) N. Shin-ya and S. R. Keown, "Correlation between Rupture Ductility and Cavitation in Cr-Mo-V Steels," Met. Sci., 13(1979), 89-93
11) K. R. Williams and B. Wilshire, "Effects of Microstructural Instability on the Creep and Fracture Behaviour of Ferritic Steels," Mater. Sci. Eng., 28(1977), 289-296
12) K. R. Williams and B. Wilshire : Mater. Sci. Eng., 38(1979), 199-210
13) C. J. Bolton, B. F. Dyson and K. R. Williams, "Metallographic methods of Determining residual Creep Life," Mater. Sci. Eng., 46(1980), 231-239
14) L.P. Stoter, Thermal Ageing Effects in AISI Type 316 Stainless Steel," J. Mater. Sci., 16(1981), 1039-1051
15) J. M. Leitnaker and J. Bentley, " Precipitate Phases in Type321 Stainless Steel After Again 17 years at ~600℃," Metall. Trans A, 8A(1977), 1605-1613
16) J. H. Hoke and F. Eberle, "Experimental Superheater for steam at 2000Psi and 1250F-Report after 14,281 Hours of Operation," Trans. ASME, 79(1957), 307-317
17) Y. Kondo, T. Matsumura, J. Namekata, Y. Yamaguchi, M. Tanaka, and F. Hangai, "Degradation of SUS 304 by High Temperature Exposure and Recovering Heat-treatment

18) M. Tanaka, F. Hangai, Y. Kondo, and J. Namekata, "Study on Microstructure and Mechanical Properties of SUS304 Used for a Long Time at Elevated Temperature," *123rd Committee on Heat Resisting Metals and Alloys Rep.*, 24(1983), 373-384

19) Y. Kondo, T. Matsumura, J. Namekata, M. Tanaka, and F. Hangai, "Effect of Recovery Treatment on the Mechanical Properties of SUS304 Serviced for Prolong Time at Elevated Temperature," *123rd Committee on Heat Resisting Metals and Alloys Rep.*, 26(1985), 133-139

20) Y. Yamaguchi, M. Tanaka, F. Hangai, Y. Kondo, and J. Namekata, "Microstructures and Mechanical Properties of SUS304 Serviced for Prolonged Time at 520〜610℃," *Trans. ISIJ.*, 24(1984), B-61

21) A. Kelly, and R. B. Nicholson, Strengthening Methods in Crystals(Amsterdam: Elsevier, 1971), 9

22) F. R. N. Nabarro, Theory of Crystal Dislocations(London: Oxford, 1967), 53

23) J. Weertman, and J. R. Weertman, ElementaryDislocation Theory(New York, NY: Macmillan, 1964)

BEHAVIOUR OF SINGLE CRYSTAL SUPERALLOYS
UNDER CYCLIC LOADING AT HIGH TEMPERATURES

Holger Frenz, Joachim Kinder, Hellmuth Klingelhöffer and Pedro D. Portella

Federal Institute for Materials Research and Testing (BAM)

Unter den Eichen 87, D-12205 Berlin, Germany

Abstract

The LCF-behaviour of two single crystal superalloys developed as blade materials for the use in aircraft engines, viz. SRR99 and CMSX-6, was investigated at 980°C. Special attention was given to the influence of hold periods at either the maximum (tension phase) or the minimum (compression phase) strain level. The introduction of a hold period in the compression phase led for both alloys to a pronounced reduction of the cycle number for crack initiation, N_A, when compared with the values obtained at the same total strain range, $\Delta\epsilon_t$, without hold periods. This reduction was accentuated by reducing $\Delta\epsilon_t$. On the other hand, the influence of a hold period in the tension phase on N_A was much less pronounced. The evolution of the γ/γ'-microstructure and the fracture mechanisms characteristic for the different loading conditions were investigated using mainly SEM. First results concerning the LCF-behaviour at 950°C of SC 16, a single crystal superalloy developed for the use in land-based industrial gas turbines, are also presented.

Introduction

Under normal service conditions, gas turbine blades are subjected to very complex stress-strain-temperature loading cycles in a non-inert environment. The identification of the predominant damage mechanisms requires careful testing under controlled conditions. In this sense, the introduction of more realistic thermal-mechanical fatigue testing contributes to a better understanding of the material behaviour in real blades [1]. However, the high costs of such complex tests and the difficulties in interpretating their results keep a living interest in isothermal low-cycle fatigue testing [1,2].

In this work we present some results concerning the LCF-behaviour of single crystal nickel-base superalloys with and without hold periods. Due to their relatively simple microstructure, single crystal superalloys permit a more detailed investigation of the microstructural modifications as well as of the fracture mechanisms and how they depend on cyclic loading.

Experimental details

The specimens were cast by Thyssen Guß AG in Bochum, Germany, using a high gradient DS/SX-furnace [3]. Six specimens were produced in each run by using helix selectors. The chemical composition of each master charge as determined by X-ray spectrometry or atomic absorption spectrometry is given in Table I. The heat treatment consisted of a multi-step solution treatment followed by a two-step ageing treatment according to the standard specification of each alloy.

The resulting microstructure, Figure 1, consisted of a monomodal distribution of cuboidal γ'-particles for SRR99 and CMSX-6 and a bimodal distribution of fine spherical and larger cuboidal γ'-particles for SC 16. Table II summarizes the microstructural parameters of each batch as determined either by point analysis with a light microscope or by area analysis of SEM pictures using a computer assisted system. Metallographic specimens were conventionally prepared paying attention to the orientation of the specimen surface and by using etchant solutions which preferentially etched the γ'-phase. Both SE- and BSE-modes of SEM were used to observe the γ/γ'-microstructure. The problems associated with the determination of the γ' volume fraction based on the measurement of the area fraction in etched metallographic

Table I Chemical composition of the master charges (Mass fractions in %)

Alloy	Ni	Cr	Al	Ti	Mo	W	Co	Ta	others
SRR99	66.05	8.53	5.57	2.22	0.03	9.53	5.03	2.85	0.18
CMSX-6	70.48	9.76	4.78	4.68	3.00	0.03	5.02	2.08	0.17
SC 16	70.93	16.37	3.46	3.50	2.84	0.09	0.05	3.60	0.16

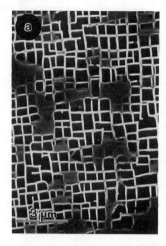 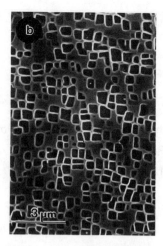 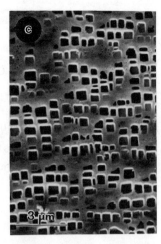 

Figure 1: Initial microstructure of: a) SRR99 (SEM); b) CMSX-6 (SEM); c) SC 16 (SEM); d) SC 16 (TEM)

Table II Microstructural parameters of the specimens after the standard heat treatment

Alloy	γ'-phase			porosity		γ/γ'-eutetics
	V_V in %	edge length [1] in µm	diameter [2] in µm	V_V in %	maximal feret in µm	V_V in %
SRR99	60	0.52	-	< 0.5	15	< 0.1
CMSX-6	62	0.57	-	< 0.5	10	2
SC 16	36	0.45	0.08	< 1.0	30	nearly 0

[1] cuboidal particles
[2] spheroidal particles

specimens at the SEM were discussed elsewhere (4,5). The orientation of the specimens was determined by Thyssen Guß using a SCAMP system (6). The angle between the specimen symmetry axis and the [001] direction, χ, varied between 2° and 8° with a median of 5°. The specimens used in the mechanical tests had a total length of 115 mm and a gauge length of 26 mm with 9 mm diameter. Their final shape was given by circumferential grinding.

Mechanical testing was carried out using closed loop, computer assisted servohydraulic or electromechanical systems with a load capacity of 100 kN. The specimen grips were specially developed and allow a precise axial loading in both tension and compression. To achieve low levels of specimen bending during a test series, constant accurate manufacturing and centric clamping of specimens have to be ensured. For the determination of superimposed bending, a cylindrical reference specimen with three sets of four strain gauges placed symmetrically along the specimen gauge length was loaded to a maximum of 0.1% axial strain in tension and compression (7). According to the french standard NF A 03-403 1990, a maximum of 10% bending strain at this point can be accepted. In round robin experiments we further observe the formation of surface cracks on the whole surface of the gauge length after LCF-testing at high temperatures for a large variety of materials, even for those with a high crack sensitivity (8). This is consistent with the low levels of specimen bending measured at room temperature. All testing systems are equipped with three-zone furnaces in order to achieve temperature differences less than 3 K along the gauge length up to 1150°C during testing. The effective gauge length of the extensometers was 21 mm.

The tests were carried out with uncoated specimens in air under total strain control with a triangular waveform ($R_\epsilon = -1$). The absolute value of the strain rate was in most experiments 10^{-3} s^{-1}, but some results for 10^{-4} s^{-1} and 10^{-5} s^{-1} are also available. Hold periods of 300 s were introduced at either the maximum or the minimum strain level and are represented as t_t and t_c respectively. For each hysteresis loop the mean stress σ_m was defined as $\sigma_m = \frac{1}{2}(\sigma_{max} + abs(\sigma_{min}))$ and the total stress range as $\Delta\sigma = 2\,\sigma_m$. The specimen lifetime, N_A, was determined through a macroscopic crack initiation criterion from the diagram total stress range versus cycle number, N (7,8).

Mechanical behaviour

Figure 2 shows the dependence of the cycle number for crack initiation, N_A, on the total strain range, $\Delta\epsilon_t$, for SRR99 at 980°C. This log-log-diagramm shows the results of LCF tests without hold periods (▲), with hold periods of $t_t = 300$ s at the maximum tensile strain (□) or with hold periods of $t_c = 300$ s at the maximum compressive strain (⊠). For the sake of simplicity, the results for these three loading types were connected by straight lines, which do not involve any theoretical consideration. The relatively narrow scatter in N_A observed for tests without hold periods can be exemplified by the results of seven tests with $\Delta\epsilon_t = 0.7\%$. The angle χ for these specimens varied between 2.5° and 5.5°. Three of them were ground in longitudinal direction, the other four as usual in circumferential direction.

The introduction of a hold period in the compressive phase led in comparison to the tests without hold periods to a drastic reduction in N_A up to one order of magnitude in the range investigated. On the other hand, the introduction of a hold period in the tensile phase led to a moderate reduction in N_A for large values of $\Delta\epsilon_t$, whereas for lower values of the total strain range an increase in lifetime was observed.

The mean stress during tests without hold periods was less than 1% of the total stress range. The introduction of hold periods in the compressive phase led to a shift of the hysteresis loop in the tensile stress region, so that the mean stress reached values of about 15% of the total stress range after few cycles and kept approximately constant up to N_A. The inverse tendency was observed for tests with a hold period in the tensile phase. The total stress ranges for tests with the same value of $\Delta\epsilon_t$ was nearly independent of the specific cycle waveform.

The LCF-behaviour of CMSX-6 at 980°C is shown in Figure 3. There is an evident similarity to the response of SRR99 under the same conditions, which includes all other observations made above.

Figure 4 shows the dependence of N_A on $\Delta\epsilon_t$ for SC 16 at 950°C. The LCF tests were carried out in air with strain rates of 10^{-3} s^{-1}, 10^{-4} s^{-1} and 10^{-5} s^{-1}. There is no clear evidence of a strain rate influence on the lifetime of the specimens. The effect of hold periods on the LCF-behaviour of SC 16 is currently being investigated and will be presented elsewhere.

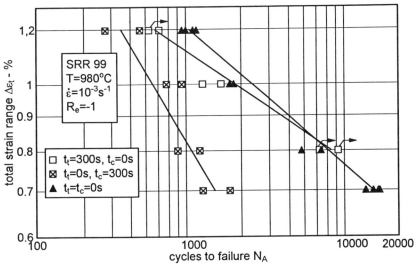

Figure 2: SRR99 LCF-tests at 980°C

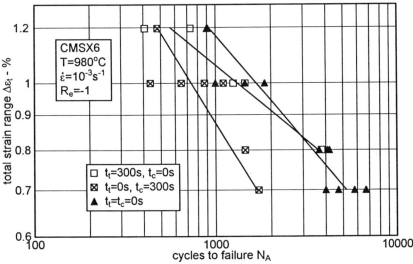

Figure 3: CMSX-6 LCF-tests at 980°C

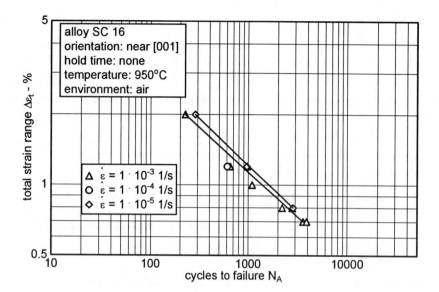

Figure 4: SC 16
LCF-tests at 950°C

Microstructural evolution

Figure 5 shows the microstructural evolution for SRR99 under LCF loading without hold periods. There is a slight tendency of the γ'-particles to interconnect and form plates parallel to the specimen axis. These changes are accompanied by an almost linear reduction in $\Delta\sigma$ with increasing N. The sharp softening observed at the beginning of the tests, which was completed after few cycles, can not be associated with any change in the γ/γ'-microstructure. On the other hand, the large reduction in $\Delta\sigma$ after N_A is due to the growth of the main crack and will be discussed below. Since the presence of a large crack leads to the unloading of large parts of the specimen, we do not observe any signifcant changes in the γ/γ'-microstructure between crack initiation and final fracture, as can be seen in Figure 5 by comparing the two SEM micrographs on the right side.

The changes in the γ/γ'-microstructure of SRR99 under LCF loading with $\Delta\epsilon_t = 1.2\%$ are presented in Figure 6. The upper curve shows for a strain rate of 10^{-3} s^{-1} the same evolution as reported above for $\Delta\epsilon_t = 0.7\%$. On the other hand, the much longer tests with a strain rate of 10^{-5} s^{-1} show a much more pronounced softening with increasing N and drastic microstructural changes. After few cycles with the lower value of the strain rate, we observed (left SEM micrograph) nearly the same γ/γ'-microstructure as after the test with 10^{-3} s^{-1}. The duration of these both tests was nearly the same. The pronounced softening observed for larger N values is associated with a notable coarsening of the γ'-particles, which assume the form of folded plates, the segments of which lie on (111) planes. This unusual configuration was confirmed by preparing two longitudinal sections perpendicular to each other and a transversal section for SEM and by preparing thin foils from a tranversal section for TEM, Figure 7.

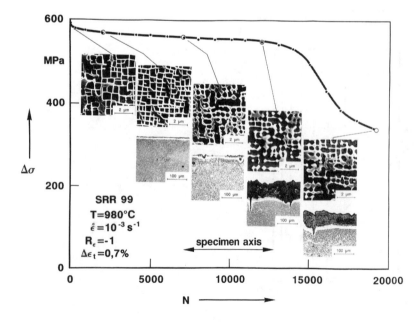

Figure 5: SRR99
Microstructural evolution during a LCF-test

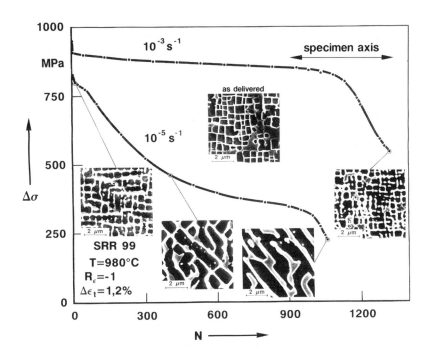

Figure 6: SRR99 Microstructural evolution during LCF-tests with different strain rates

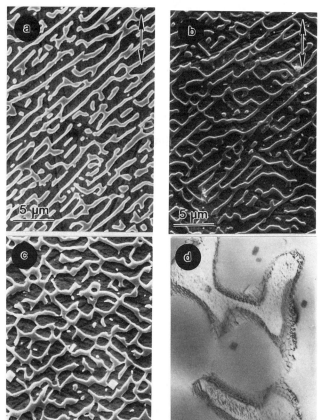

Figure 7: SRR99
Microstructure after a LCF-test at 980°C with $\Delta\varepsilon_t = 1.2\%$, $R_\varepsilon = -1$ and $\dot{\varepsilon} = 10^{-5}$ s^{-1}
a,b) longitudinal section (SEM); c) tranversal section (SEM); d) tranversal section (TEM)

The introduction of hold periods in the tensile phase leads to the formation of γ-plates in a γ'-matrix oriented perpendicular to the specimen axis, Figure 8a, whereas hold periods in the compressive phase lead to γ-plates parallel to the specimen axis, Figure 8b.

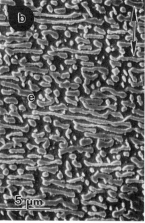

Figure 8: SRR99
Microstructure after LCF-tests at 980°C with $\Delta\varepsilon_t = 1.0\%$, $R_\varepsilon = -1$, $\dot{\varepsilon} = 10^{-3}$ s^{-1} and hold periods of 300 s in the:
a) compressive phase;   b) tensile phase

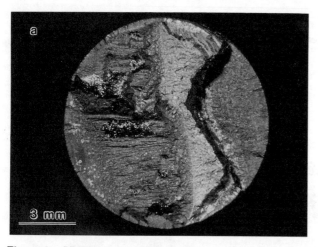 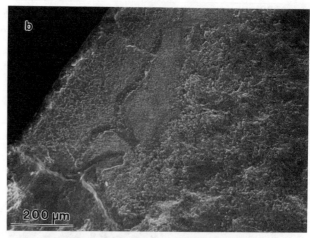

Figure 9: SRR99
Fracture surface after a LCF-test at 980°C with $\Delta\varepsilon_t = 0.7\%$, $R_\varepsilon = -1$ and $\dot\varepsilon = 10^{-3}\,s^{-1}$
a) overview;   b) surface crack (SEM)

 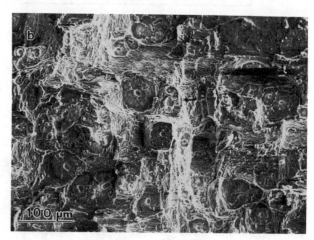

Figure 10: SRR99
Fracture surface after a LCF-test at 980°C with $\Delta\varepsilon_t = 0.7\%$, $R_\varepsilon = -1$ and $\dot\varepsilon = 10^{-3}\,s^{-1}$ and hold periods of 300 s in the compressive phase:
a) overview;   b) fatigue lines in the inclined surface (SEM)

Figure 11: SRR99
Fracture surface after a LCF-test at 980°C with $\Delta\varepsilon_t = 0.7\%$, $R_\varepsilon = -1$ and $\dot\varepsilon = 10^{-3}\,s^{-1}$ and hold periods of 300 s in the tensile phase:
a) overview;   b) detail of the fracture surface, the square regions are typical for creep fracture (SEM)

Fracture mechanisms

The light micrographs in Figure 5 show the evolution of the damage mechanism under LCF loading without hold times. The formation of a complex oxide layer (mainly containing nickel and aluminium) after few cycles depletes a surface near region of Al, resulting in a γ'-free layer. Both layers grow steadily with increasing value of N, but at some preferential sites distributed homogeneously over the specimen surface, the formation of nickel oxide speeds up. At these sites we observe protusions on the surface, under which small cracks perpendicular to the specimen axis are hidden. In the metallographical specimens these cracks are clearly identifiable as "spikes" running ahead of the normal corrosion front. These surface cracks grow laterally and may link to form the main crack, which grows more rapidly into the specimen. This point corresponds to the failure criterion defined above for N_A. The resulting fracture surface is nearly plane and perpendicular to the specimen axis, but the ligament usually shows a shear component, Figure 9a. Near to the specimen border, the fracture surface deflects and sets free some small surface cracks, which show a "thumb nail" shape, Figure 9b.

The initial stages of damage under LCF loading with hold periods are identical. However, the behaviour of the surface cracks after being formed changes radically in these cases. With hold periods in the compressive phase, the surface cracks stop growing laterally at some point. Sharp inclined cracks are generated at their extremities, some of them grow more rapidly and determine the rupture. The fracture surface consists of two or more inclined surfaces which are nearly ((111)) oriented, Figure 10a. The steps observed on these surfaces suggest clearly the cyclic nature of the growth these cracks, Figure 10b.

With hold periods in the tensile phase, surface cracks do form but become inactive. Instead of them, cracks originating from bulk pores grow and become predominant. The surface fracture is nearly plane and perpendicular to the specimen axis, Figure 11a, but it shows typical features of creep fracture in these alloys, Figure 11b. In the two other loading types we did not observe any damage at the bulk pores.

Discussion

The evolution of the γ/γ'-microstructure in single crystal superalloys is classically considered to be a diffusional process which depends strongly on the misfit between the γ and the γ' phases as well as on the applied stress (9). Our results concerning the evolution of the γ/γ'-microstructure in SRR99 and CMSX-6 under LCF loading with hold periods are in accordance with this approach. Both alloys have a negative misfit and show platelets perpendicular to the specimen axis when hold periods are introduced in the tensile phase, since in this case the specimen was submitted the far most long time to a tensile stress. *Mutatis mutandis* they present platelets parallel to the specimen axis when the predominant stress is compressive. These structures show significant long range internal stresses, which could be associated to the asymmetry observed in the σ-ε-loops.

However, this classical approach does not consider the strain component. Since the dislocation structure may influence diffusional processes, strain might play a significant rôle. In fact, LCF tests of long duration generate platelet structures which can not be explained by a stress argument alone. The observed softening during those tests seems to be connected to the formation of matrix channels which allow the reversible glide of dislocations. In this case, the long range internal stresses should not be very high. This would be consistent with the observed symmetrical shape of the σ-ε-loops.

When considering the LCF behaviour of superalloys, hold periods in the compressive phase are more damaging than those in the tensile phase, which is usually associated with the high mean stress levels resulting from assymetric loading (2). The single crystal superalloys SRR99 and CMSX-6 show at 980°C the same behaviour as reported in the literature. However, the shift in the mean stress seems to be connected with the changes observed in the γ/γ'-microstructure and the consequent formation of internal stresses. The radically different fracture mechanisms observed under these loading types are currently being investigated in more detail. We expect to derive from these investigations more concrete clues to the behaviour of superalloys under cyclic loading.

Acknowledgements

This work was partially supported by the *Bundesministerium für Bildung und Forschung* (03 M3038 D), by the *Deutsche Forschungsgemeinschaft* (Sfb 339, B7 and C1) and by the European Community (BriteEuram BRE2-CT92-0176). We are grateful for the experimental support given by Bärbel Bogel, Uwe Chrzanowski, Klaus Naseband, Wolfgang Wedell and Jörg Wuttke.

References

1. Josef Ziebs, Jürgen Meersmann, Hans-Joachim Kühn, Norbert Hülsmann and Jürgen Olschewski, "Multiaxial thermomechanical behaviour of IN 738 LC,", in: Proceedings of the 4th International Conference on Biaxial/Multiaxial Fatigue, vol. 2, (European Structural Integrity Society, 1994), 247-259.

2. W.J. Plumbridge and E.G. Ellison, "Low-cycle-fatigue behaviour of superalloy blade materials at elevated temperature," Materials Science and Technology 3(9)(1987), 706-715.

3. Alfred Donner, "Gerichtete Erstarrung - ein bauteilgerechtes Herstellverfahren für Turbinenschaufeln," Metall, 42(2)(1988), 128-132.

4. Pedro D. Portella, Konrad Breitkreutz and Veronica Bierwagen, "Quantitative Analyse von Gefügeänderungen in einkristallinen Superlegierungen," in: *Sonderbände der Praktischen Metallographie*, vol. 23, ed. W.-U. Kopp et al. (Munich, BY: Carl Hanser Verlag, 1992), 93-106.

5. Uwe Chrzanowski, Hellmuth Klingelhöffer, Pedro D. Portella and Konrad Breitkreutz, "Gefügeveränderungen der einkristallinen Nickel-Basis-Superlegierung SC 16 bei Kriechbeanspruchung und deren quantitative Auswertung," in: *Sonderbände der Praktischen Metallographie*, vol. 27, ed. M. Kurz and M. Pohl (Oberursel, HS: DGM-Informationsges. mbH, 1995), 255-258.

6. M.J. Goulette, P.D. Spilling and R.P. Arthey, "Cost effective single crystals," in: Proceedings of the 5th International Symposium Superalloys, ed. M. Gell *et al.* (Warrendale, PA: The Minerals, Metals & Materials Society, 1984), 167-176.

7. Klaus Naseband, Siegfried Ledworuski, Hans-Joachim Kühn and Holger Frenz, "High-temperature low-cycle-fatigue testing," in: Proceedings of the 6th International Fatigue Congress 1996 (London: Elsevier Applied Science, in press).

8. Klaus Naseband and Rolf Helms, "Low-cycle fatigue testing programme," (Report BCR contract 2119/1/4/269/85/3, BAM, 1989).

9. André Pineau, "Influence of uniaxial stress on the morphology of coherent precipitates during coarsening - elastic energy considerations," Acta metall. 24 (1976), 559-564.

MICROSTRUCTURAL BEHAVIOUR OF A SUPERALLOY UNDER REPEATED OR ALTERNATE L.C.F. AT HIGH TEMPERATURE

Valérie Brien*,# Brigitte Décamps ** and Allan J. Morton***

* Laboratoire d'Etudes des Microstructures, UMR CNRS-ONERA 104, BP 72, 92322 Châtillon Cedex, France
\# Now at Department of Materials, Imperial College, Prince Consort Road, London SW 7 2BP, UK
** Laboratoire de Métallurgie Structurale, URA CNRS 1107, Université Paris-Sud, Bâtiment 413, 91405 Orsay Cedex, France
*** CSIRO, Division of Material Science and Technology, Private bag 33, Rosebank MDC, Clayton, Victoria, Australia 3169

Abstract

Total strain controlled Low Cycle Fatigue tests have been performed at 950°C (1223K) on the AM1 superalloy and its microstructural behaviour has been studied by Transmission Electron Microscopy. The mechanisms of deformation have been studied mainly on repeated fatigue $R_\varepsilon=0$ but also in alternate one ($R_\varepsilon=-1$) along the cubic [001] axis. In simple tension localisation of dislocations networks at the matrix/precipitate interface perpendicular to the <001> loading axis occurs. These dislocations are shown to always relax the misfit. In fatigue tests, the larger the total strain imposed is or the greater the number of cycles is, the more the deformation spreads out to the vertical channels. In small total strain, type N coarsening is observed as early as 200 cycles. For $R_\varepsilon=-1$, compression is shown to deform the vertical channels without formation of networks. Alternate fatigue leads to type N coarsening for very few cycles.

Introduction

Industrial gas turbines blades in use undergo very complicated mixed effects of creep, fatigue and environment. The work presented here concerns mechanical fatigue which occurs more especially in the base of these blades when transitions between either normal runnings or idling of the engine occur. In fact, the fatigue experienced is of thermomechanical type ; but to get basic information, a test temperature of 950°C (1223K) was chosen. As the initiation and growth of cracks in fatigue has been the object of a certain amount of studies in the past, and as the knowledge of the behaviour of nickel based superalloys under fatigue before any cracks appears is rather limited ; the aim of this work was concentrated on the mechanisms of deformation during the damaging stage which is also extremely important since localisation of plastic deformation can condition the further stages. The type of fatigue tests and experimental conditions were chosen as close as possible to real conditions of the material in use : plastic low cycle fatigue, total strain imposed tests, temperature T=1223K (950°C), frequency f=0.25Hz, loading along the turbine blades main axis : [001]. Indeed, most studies indexed in literature (cf. References 1 to 12) concern $R_\varepsilon=-1$ fatigue (R_ε is the ratio of the minimum strain to the maximum strain for each cycle). So, although most of our tests have been done at $R_\varepsilon=0$, some tests with $R_\varepsilon=-1$ have been performed in order to make comparisons and see the influence of a reversed loading.

Another part of the present work (already published : Décamps, Brien and Morton 1992, Reference 13) compared microstructures after strain-controlled fatigue tests and stress-controlled fatigue tests (same level of initial stress).

Experimental and Method

Mechanical tests have been performed using a tension hydraulic INSTRON machines on samples cut in AM1 superalloy provided by SNECMA (Table I gives the nominal composition of the alloy). Figure 1 presents the metallographic structure of the AM1 alloy. Is is pointed out that the two phases composing the alloy are coherent despite a misfit of $\delta = -10^{-3}$ at the temperature of our fatigue tests (Royer 1995, Reference 14). Tests were carried out at 950°C (1223K) at 0.25Hz in air along the crystallographic axis [001]. For $R_\varepsilon=0$ tests, the range of $\Delta\varepsilon^t$ was 0.6-2.2%, the number of cycles varied from 1 to 2.10^4. Some of the tests were continued to rupture.

Table I : Percentage composition in weight of the AM1 superalloy

Ni	Co	Cr	Al	W	Ta	Mo	Ti
Base	6.5	7.5	5.3	5.5	8	2	1.2

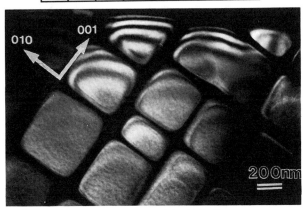

Figure 1 : Metallographic structure of the AM1 alloy

The γ' precipitates are cuboids aligned along the cubic crystallographic directions and represent 65% of the total volume at 950°C.

The present work has been mainly concentrated on the $R_\varepsilon=0$ fatigue and on exploration of the influence of the amplitude $\Delta\varepsilon^t$ and of the number of cycles N on the structural mechanisms of deformation. Also, a compression test and two $R_\varepsilon=-1$ fatigue tests have been carried out. The tables II and III (cf. end of article) index the mechanical details of the different tests carried out in this study. The compression test was performed at the same level of stress as the tension one done at $\Delta\varepsilon^t=1.6\%$ in tension.

All observations of microstructures were made by Transmission Electron Microscopy either on a JEOL 2000 EX, a JEOL 200 CX or a PHILIPS CM20. Thin foils were thinned down first mechanically then by the classic two jets electrochemical method using a 90% acetic acid, 10% perchloric acid solution. Foils were fully orientated in order to follow the projection of the loading axis [001], noted p on the micrographs.

Results

$R_\varepsilon=0$ fatigue

For N=1, $\Delta\varepsilon^t=1.25\%$ the microstructure of deformation (that corresponds to a simple tension test) is very anisotropic. The anisotropy is characterised by a preferred localisation of the dislocations in the channels perpendicular to the [001] axis (cf. figure 2) : γ/γ' {001} interfaces are covered by polygonal networks of dislocations.

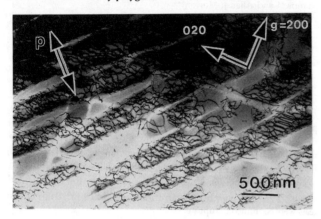

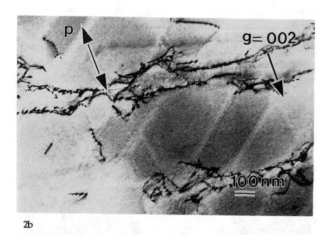

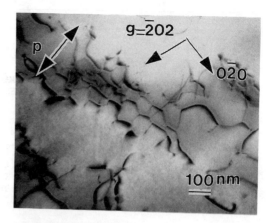

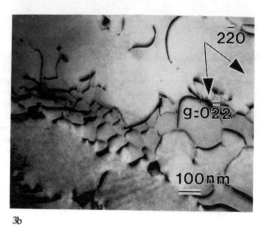

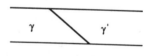

Figure 2 : Zone A microstructure. N=1, $\Delta\varepsilon^t=1.25\%$. One can notice the preferential activation of the channels perpendicular to [001] in comparison to the two other channels. Foil cut perpendicular to [111]. a/ Zone axis = [001]. b/ Zone axis between [110] and [010].

A careful study has been achieved in order to get the localisation of the extra plane associated to the either edge or 60° mixed dislocation localised at the γ/γ' interfaces.

This study requires the full orientation of the dislocations, the sign of the Burgers vector and the relative positions of the two phases. To get the first information, Bright field images with g.b=2 for each dislocation have been taken in order to get the asymmetrical contrast (Hirsh, Howie, Nicholson, Pashley and Whelan 1965 - Reference 15). FS/RH convention has been used (Bilby, Bullough and Smith 1955 - Reference 16) and Burgers vectors are determined by the classic extinction criterion (Howie and Whelan 1962 - Reference 17). The second information is obtained using stereomicroscopy. This study shows that all dislocations localised at the interfaces present an extra plane in the precipitates. So, they are on the right side of the interface to relax the existing misfit between γ and γ'. Dislocations at the interface form polygonal networks generally with 6 sides : they are classic honeycombs found in fcc materials ; interfacial networks of this type are encountered after creep of the same type of superalloys (Fredholm 1986 - Reference 18). Figure 3 present an exemple of such a relaxing network.

Figure 3 : Network relaxing the interfacial misfit. $\Delta\varepsilon^t=1.5\%$, N=25. Arrows indicate the double asymmetrical contrast. Foil parallel to [111], a/ Zone axis =[101], b/ Zone axis =[111], c/ Schema showing the relative position of the two phases : cut seen from the left of photos.

When the number of cycles increases, an expansion of the deformation to the vertical channels is noted. However, the resulting microstructure still presents a strong partition of the plastic deformation regarding the horizontal and vertical channels with an obvious preference for horizontal ones (cf. figure 4). This homogeneity is noted as well if is very big, the bigger is, the sooner the homogeneisation is observed (smaller N).

On the contrary, if the number of cycles N increases again, anisotropy is no longer observed : the microstructure of deformation is then completely homogeneous (cf. figure 5).

All the dislocations, either from one interface or the other have been shown to relax the misfit, except for large number of cycles, where some of the dislocations presented the other sign.

So, the deformation microstructures in many foils from each fatigue test from this map have been observed and two major domains were found (cf. map in figure 6). The first called H is a domain where the deformation microstructure is homogeneous , the second is a domain, called A, gives rise to anisotropic deformation microstructure. This anisotropy is either micromechanical, or comes from the morphology of the precipitates (coarsening). Indeed, for smaller $\Delta\varepsilon^t$ (like 0.6%) the dislocations never extend to the secondary channels and the microstructure always stays anisotropic. Moreover, for large number of cycles, N>200, this micromechanical anisotropy results in oriented coarsening of type N (classical notation to describe coarsening perpendicular to the stress axis), i.e. exactly like in tension creep (cf. figure 7).

So, the anisotropy of the deformation microstructure is enhanced by small values of $\Delta\varepsilon^t$, whereas large values of N increase the tendency for homogeneous microstructures (except for small $\Delta\varepsilon^t$). These two rules and the localisation of the two domains are summarised in the previous map. The type N coarsening domain is noted as well, and the greater intensity of shadowing of the domain reflects the stronger tendency for coarsening.

Very rarely, shearing (except for big $\Delta\varepsilon^t$ like 2.2% i.e. for big stresses) has been noted in microstructures, and all interfacial dislocations generally relax the γ/γ' misfit.

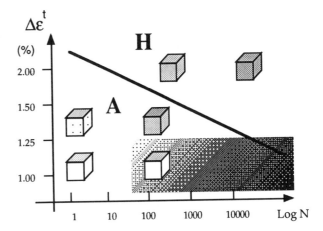

Cubic precipitate : the intensity of the coloration of the faces embodies the level of deformation localised at this interface

Type N Coarsening domain (the darker is the region the more intense is the coarsening)

Figure 6 : Map of microstructures of deformation in $R_\varepsilon=0$ fatigue versus $\Delta\varepsilon^t$ and N. This map is found to be divided in two main zones : zone A where the microstructures of deformation are anisotropic and zone H where the microstructures of deformation are homogeneous.

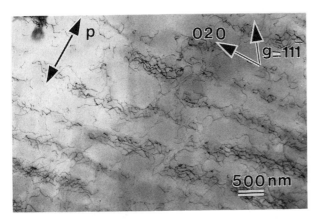

Figure 4 : Zone A microstructure. $\Delta\varepsilon^t=1.5\%$, N=25. Foil parallel to [111], Zone axis=[101], deformation is localised essentially in channels perpendicular to the loading axis, forming networks at interfaces.

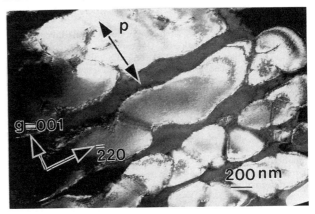

Figure 7 : Zone A microstructure. N=200, $\Delta\varepsilon^t=0.63\%$. Type N oriented coarsening of the AM1 alloy under $R_\varepsilon=0$ LCF for a great number of cycles at small $\Delta\varepsilon^t$. Foil cut perpendicular to [111]. Zone axis = [112].

<u>$R_\varepsilon=0$ fatigue discussion</u>

To explain the present observation we need to consider the presence, evolution and distribution of internal stresses within the fatigued specimen.

The first one comes from the misfit between the γ phase and the γ' phase. This misfit $\delta \approx (a-a')/2a$ is equal to -10^{-3} at 950°C, and gives rise to compression (tension) stresses in the γ (γ') phase. a (a') is the γ (γ') phase parameter. These compression stresses are at the origin of the partition of the plastic deformation since the resulting local stresses in horizontal channels and in vertical ones are different.
Tritely, the resulting stresses in the vertical channels are weaker than in horizontal ones since the applied and the misfit stresses are of opposite signs. This results in the localisation of the deformation in the horizontal channels.

This preferential localisation of deformation noted also by Benyoucef (1994 - Reference 19) after tension can be explained by the existence of the misfit. This argument had already been put forward by this writer.

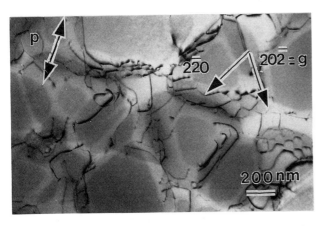

Figure 5 : Zone H microstructure : the 3 γ channels exhibit an equivalent density of dislocations. N=1300, $\Delta\varepsilon^t=1.28\%$. Foil cut perpendicular to [111]. Zone axis = [111].

After a while, since deformation occurs only in these channels, and since they have such a Burgers vector that they reduce the misfit, misfit stresses will be very low in horizontal channels. That phenomena destroying the source of the partition, added to the natural hardening due to the presence of dislocations themselves induces the extension of deformation to the vertical channels. This expansion is assisted at high stresses by shearing of γ' precipitates.

A calculation of the number of pure edge dislocations needed to fully reduce the misfit at one interface can be done via the Brooks formula.

$$\frac{\sqrt{2} d_\gamma \cdot \delta}{|b|} = \frac{\sqrt{2}(400 \text{ to } 500) \cdot 10^{-3}}{\frac{\sqrt{2}}{2} \cdot 3{,}58 \cdot 10^{-1}} = 2 \text{ to } 3 \text{ dislocations}$$

where

d = length of the γ' precipitates = 400 to 500 nm

δ = absolute value of the misfit between the two phases = 10^{-3}

b = Burgers vector of the considered dislocation : a/2 [110]

a = cubic lattice parameter of γ phase

So it means that in $R_\varepsilon=0$ fatigue after a small number of cycles, the misfit stresses are relaxed, only in the channels perpendicular to [001], since only {001} interfaces are concerned by the location of dislocations.

One has to take into account a second type of stresses : a stress coming from the strongly composite structure of such a superalloy. We make use of the plasticity composite model of Mughrabi (1979 - Reference 20) used for inhomogeneous distribution of plastic deformation adopted by monophase metals after fatigue loadings. This idea has also been exploited for nickel base superalloys by different authors (Kuhn 1991-Mughrabi 1992-Véron 1995 - References 21 to 23).

The model has been considered here in the vertical direction (the horizontal effect has been neglected). One can show (to be published by Brien et al.) that these stresses called compatibility stresses find their existence in the presence of plastic deformation in vertical channels. Our calculation gives estimate of these stresses : they are found to equal 80% of ($\sigma a - \sigma o$) where σo is the elasticity limit of γ phase inside the superalloy (including Orowan stresses, stress coming from misfit and hardening stress coming from the solid solution) and σa is the applied stress. So, this type of stress doesn't exist in the beginning of the fatigue, but as soon as deformation occurs in vertical channels, compatibility stresses arise in vertical channels and delay them to deform. These second stresses assist the anisotropic partition of the plastic deformation.

Explanations of the type of coarsening observed can be explained by a recent coarsening model developed by Véron (Reference 23) for superalloys deformed under creep conditions. This model is based on the existence of gradients of elastic energy between the interfaces provoked by the anisotropic relaxation of the misfit stresses by dislocations.

These elastic energy gradients are of the same type as those observed by Véron after tension creep : {001} interfaces are relaxed as {010} and {100} ones are not. Moreover the fact that at levels of stresses where the anisotropic micromechanical feature disappears rather early, no oriented coarsening is proves that the micromechanical anisotropy is really the needed condition for the type N oriented coarsening in repeated fatigue.

It is also worth noting, that some of $R_\varepsilon=0$ tests exhibit bands of deformation, but that phenomena will be discussed separately.

R$_\varepsilon$=-1 fatigue

The superalloy presents a peculiar microstructure after the R_ε=-1 fatigue.

After one cycle (one tension followed by one compression), as well as deformation localized in interfaces perpendicular to [001] due to one simple tension, the deformation localizes in channels parallel to [001] (cf. figure 8). The dislocations observed in these vertical channels belong to only one slip system : here 1/2<-101>{1-11}, as a consequence can not form networks, and they are also founf to be of the correct sign to relax the misfit between γ and γ'. A compression test has been order to confirm that the new observation comes directly from the compression (and not from mixed effects of a compression following a tension). This compression test presents a microstructure where only the channels parallel to [001] are affected by deformation : the dislocation found in each channel belong to only one slip system.

After cycling, here 113 cycles, the precipitates are coarsened in plates perpendicular to the loading axis. So, R_ε=-1 fatigue produces very fast type N coarsening (cf. figure 9).

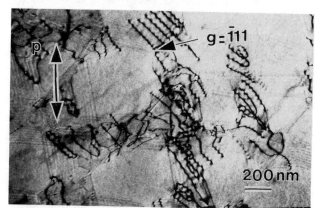

Figure 8 : R_ε=-1 fatigue microstructure after 1 cycle. N=1, $\Delta\varepsilon^t$=2x1.25%. Localisation of deformation in the channels containing the [001] axis as well as in the perpendicular one is observed. Foil cut perpendicular to [111]. Zone axis = [101].

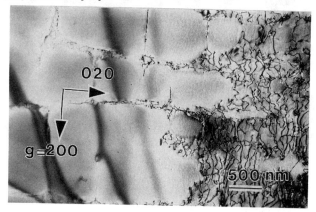

Figure 9 : R_ε=-1 fatigue microstructure after cycling 113 cycles. N=113, $\Delta\varepsilon^t$=2x1.25%. All foils exhibit type N coarsening. Foil cut perpendicular to [001]. Zone axis = [001].

R$_\varepsilon$=-1 fatigue discussion

The localisation of the deformation in vertical channels can be explained on the base of exactly the same arguments put forward for R_ε=0 fatigue behaviour. Indeed the internal misfit stresses in vertical channels added to the applied stress result in a much higher local stress in vertical channels than in horizontal ones.

A major difference between R_ε=0 and R_ε=-1 fatigue is the much higher efficiency of diffusion effects in alternate fatigue. Coarsening is observed much earlier for R_ε=-1 fatigue : only 113 cycles ; the type of coarsening is of N type : exactly the same type as in repeated fatigue or in tension creep. So, alternate fatigue has a very strong effect on diffusion. Explanations of the type of fatigue observed can be explained by the coarsening model already mentionned. Two hypothesis coming from experiments observations are made : 1/ Simple tension (compression) localises the deformation only in horizontal (vertical) channels ; 2/ For each cycle, the previous phenomena applies and that the plastic deformation has the same absolute value for tension and compression. So, it means that twice as much volume of γ phase is affected by deformation in compression than in tension, and after N cycles in alternate fatigue one horizontal channel contains globally twice as much deformation as a vertical channel. Moreover, as all the dislocations studied are found to relax the misfit : it means that the misfit in the horizontal channels will be reduced twice as quickly as that in the vertical channels : the elastic energy gradients then observed are of exactly the same type than in $R_\epsilon=0$ fatigue.

So, on the base of rather rough but reasonable hypothesis and reasoning, it is possible to explain the type N coarsening observed in alternate fatigue.

Conclusion

• A map of microstructures of deformation has been established in $R_\epsilon=0$ fatigue versus the number of cycles N and the total strain for the fatigue ; different zones are pointed out on this map : an homogeneous H domain, an anisotropic A domain as well as a coarsening domain. These domains are accounted for on the base of the influence of internal stresses like misfit stresses and compatibility stresses coming from the strong composite structure of the superalloy. The interpretation of this map leads us to structural mechanisms of deformation which rule $R_\epsilon=0$ fatigue deformation.

• The behaviour of the AM1 superalloy is very different in $R_\epsilon=0$ fatigue and in $R_\epsilon=-1$ fatigue. We point out theses differences and try to explain them.

• It is shown that alternate strain-controlled fatigue ($R_\epsilon=-1$) at 950°C (1223K) leads very quickly to a rafted structure perpendicular to the loading axis [001] (less than an half an hour) and that the type of coarsening is exactly the same as for creep tension tests. The alternation of applied stress is shown to have a very strong effect on the speed of diffusion.

• This work shows also that the observed type of coarsening in either $R_\epsilon=0$ or $R_\epsilon=-1$ fatigue can be explained using a recent coarsening model developed for superalloys deformed by creep.

Acknowledgements :

This work has been supported by the "CPR Stabilités Structurales des superalliages monocristallins" SNECMA and the CNRS. The as cast material were provided by SNECMA which is acknowledged for its interest in this work and for having performed all rupture tests in $R_\epsilon=0$ fatigue. The authors wish to thank Luc Rémy (Ecole des Mines-Evry-France) for discussions and for providing us fatigue tests facilities.

References :

Ref.1: Gabb T.P., Miner R.V. & Gayda J., Scripta Met. Vol. 20, pp. 513-518, 1986
Gabb T.P. & Welsh G.E. Scripta Met. Vol.20 pp.1049-1054, 1986
Gabb T.P., Gayda J. & Miner R.V. Met. Trans. A. Vol.17A, Part II, p497, Mars 1986
Ref.2: Milligan W.W., Jayaraman N. & Bill R.C. Mat. Sci. & Eng., 82 pp.127-139, 1986
Ref.3: Gabb T.P., Welsh G. & Miner R.V., Scripta Met. Vol.21, pp.987-992, 1987
Ref.4: Gabb T.P. & Welsh G., Acta Met., pp.2507-2516, 1989
Ref.5: Gabb T.P., Welsh G., Miner R.V. & Gayda J., Mat. Sci. & Eng., A108, pp189-202, 1989
Ref.6: Fritzmeier L.G. & Tien J.K., Acta Met., Vol. 36, N°2, pp. 283-290, 1988
Ref. 7: Chierragati R. & Rémy L. Mat. Sci. & Eng., A141, Part I pp1-9, Part II pp11-22, 1991
Ref.8: Fleury E., Thèse de Doctorat de l'Ecole Nationale Supérieure des Mines de Paris "Endommagement du superalliage monocristallin AM1 en fatigue isotherme et anisotherme" Octobre 1991
Ref. 9: Glatzel U. & Feller-Kniepmeier M., Scripta Met. et Mat., Vol.25, pp 1845-1850, 1991
Ref. 10: Hanriot F., Cailletaud G. et Rémy L. A.S.T.M. High temperature constitutive modelling-Theory and application, Book N° H00667, 1991.
Ref.11: Buffière J.Y., Thèse de Doctorat de l'Institut National Polytechnique de Grenoble "Contribution à l'étude du comportement anisotrope d'un superalliage par essais de flexion", 1993
Ref.12: Fleury E. & Rémy L. Mat. Sci. & Eng., A167, pp 23-30, 1993
Ref.13: Décamps B., Brien V. and Morton A.J., Scripta Met, 1992
Ref.14: Royer PhD thesis of Grenoble 1995
Ref.15: Hirsh P.B., Howie A., Nicholson R.B., Pashley D.W, Whelan M.J., "Electron Microscopy of thin crystals" (New York Krieger), 1965
Ref.16: Bilby, Bullough and Smith 1955
Ref.17: Howie A. and Whelan M. J., Proc. R. Soc. A., 267,206, 1962
Ref.18: Fredholm A., Ayrault D., Strudel J.L, Colloque National "Superalliages Monocristallins" Villard de Lans, 26-28 Feb. 1986
Ref.19: Benyoucef M., PhD thesis of Toulouse, 2 Feb 1994.
Ref.20: Mughrabi H., Microscopics mechanisms of Metals Fatigue ICSMA5 Vol.3, p1636, 1979
Ref.21: Kuhn H.-A., Biermann H., Ungar T., Superalloys 1992, the Minerals, Metals, Materials Society, 1992.
Ref.22: Mughrabi H., Biermann H. and UngarT., JMEPEG, Vol.2(4),pp557-564, 1992
Ref.23: Véron M., PhD thesis of INPG (Grenoble), 17 Jan.1995

Table II : Mechanical details of the repeated LCF tests

$\Delta\epsilon$ tot. %	Cycle / Total number of cycles	Maximal Stress in σ_{tens}^{max} MPa	Maximal Stress in compression σ_{comp}^{max} MPa	Plastic deformation per cycle %	Plastic cumulated deformation % (+/-)
0.63	1	458	9	0.01	2 (1)
	stabilized cycle 200	415	43		
0.70	1	608	-	0.04	200
	1/2 lifetime 5630	428	178		
0.79	1	663	-	0.06	1027
	1/2 lifetime 17117	438	212		
0.79	1	642	-	0.06	1279
	1/2 lifetime 21314	395	277		
1.09	1	815	-	0.12	401
	1/2 lifetime 3345	447	453		
1.27	1 1	739	-	0.153	0.153 (0.002)
1.33	1 25	661 637	143 157	0.247	0.47 (0.06)
1.25	1 25	732 632	57 151	0.16	2.10 (0.46)
1.26	1 1300	759 497	47 238	-	≤ 125 (15)
1.28	1 2 1300	785 764 451	50 67 357	-	125 (15)
1.6	1 5 25 200	704 663 611 520	39 76 123 205	0.80	160 (estimation)
1.6	1 5 25	707 670 620	56 81 126	0.80	20 (estimation)
2.2	1 2 3 4 5 200	975 950 930 920 910 610	85 110 126 140 150 470	1.10	100 (estimation) (33)

Table III : Mechanical details of the alternate LCF tests

$\Delta\varepsilon^{tot}./2$ %	Cycle Total number of cycles	σ_{tens}^{max} Mpa	σ_{comp}^{max} Mpa	Plastic deformation per cycle %	Total cumulated plastic deformation per cycle %
1.26	1	932	952	(+)0.253 (-)0.449 (+) 0.111	0.81
1.26	1 Stabilized 113	960 874	887 887	0.77	80 to 90

INTERACTION BETWEEN CREEP AND THERMO-MECHANICAL FATIGUE OF CM247LC-DS

C. C. ENGLER-PINTO Jr. [1], C. NOSEDA [2], M. Y. NAZMY [2] and F. RÉZAÏ-ARIA [1]

1) Swiss Federal Institute of Technology, MX-D Ecublens, CH-1015 Lausanne, Switzerland
2) ABB Power Generation Ltd., Materials Technology, CH-5401 Baden, Switzerland

Abstract

The interaction between creep and thermo-mechanical fatigue (TMF) is investigated on the CM247LC-DS alloy. Out-of-phase TMF tests are performed between 600°C and 900°C on tubular specimens (1 mm wall). Creep rupture experiments are conducted at 900°C on the *virgin* (as received) alloy under 280 MPa on both solid and tubular specimens. Sequential creep-TMF and TMF-creep experiments are performed. Creep cavities and internal cracks are initiated on carbides or residual eutectics. In addition, oxidation assisted crack initiation is observed at the surface of the specimens during creep. No noticeable change on the γ-γ' structure is observed during TMF cycling. Under creep, however, a γ'-raft structure develops perpendicularly to the stress axis. This γ'-rafting softens the pre-crept alloy and influences its non-isothermal cyclic stress-strain by increasing the inelastic strain. Short cracks initiated during pre-creeping drastically reduce the TMF crack initiation period. On the other hand, short cracks (100 to 300 μm surface crack length) initiated during pre-TMF cycling do not affect the creep lifetime. A linear creep-fatigue damage cumulative law can predict the residual life of TMF specimens which have been pre-crept.

Introduction

Directionally solidified (DS) nickel-based superalloys are candidate materials for blading of land based gas turbines. The introduction of DS and single-crystal (SC) superalloys has enhanced creep strength, oxidation resistance, temperature capability and turbine blade durability relative to the earlier conventionally cast equiaxed alloys (1, 2). In both SC and DS blades the [001] crystallographic orientation is set parallel to the blade principal axis. As this direction presents the lowest modulus of elasticity, lower thermal stresses are generated during operation.

The life-limiting factors for turbine blades are thermo-mechanical fatigue, creep and oxidation in the airfoil and high temperature low-cycle fatigue in the root section (3). During the transient regimes of start-up and shut-down operations the blades are submitted to low cycle thermo-mechanical fatigue due to the temperature gradients generated by temperature variations or internal air cooling. Rotor blades are also submitted to creep due to centrifugal forces during the stationary regime of operation. Creep damage accumulation at the maximum operation temperatures can alter the TMF resistance, the damage mechanisms and the mechanical behaviour of the alloy. Consequently, the interaction between creep and thermo-mechanical fatigue is of primary importance for the design of advanced DS blades. This paper aims to assess the interaction between thermo-mechanical fatigue and creep on the CM247LC-DS superalloy.

In general, the effect of creep is studied by adding a dwell period to the isothermal (4-6) or non-isothermal fatigue cycle (7, 8). In this investigation a sequential approach consisting of TMF tests on pre-crept specimens and vice-versa is adopted.

Alloy and experimental procedures

Alloy

The specimens are taken from CM247LC-DS cast slabs purchased from AETC (UK). The nominal chemical composition of the alloy is given in Table I. The alloy is multi-step heat treated: solutioning (1260°C, 2h) and ageing (1080°C/4h + 870°C/20h). Full solutioning increases the volume fraction of the fine γ' precipitates which enhance positively the mechanical properties of the alloy at high temperatures.

Table I Chemical composition of CM247LC-DS (weight percent).

C	Al	Co	Cr	Fe	Ta	Ti	W	Hf
0.073	5.63	9.3	8.1	0.022	3.19	0.7	9.5	1.4

TMF experiments

TMF tests are performed in air on hollow specimens (7, 9). The dimensions of the specimens are given in Figure 1. The specimens were machined from the DS slabs by electro-erosion with the elongated grains aligned with the specimen's axis. The internal surface of the specimen is honed and the external surface is mechanically polished with different diamond pastes (down to 1 μm) in the longitudinal direction. The tests are performed in a conventional closed-loop servo-hydraulic testing machine adapted for non-isothermal fatigue experiments. The specimen is heated by a high frequency generator (6 kW, 200 kHz). A bi-color infra-red pyrometer directed to the gauge length of the specimen is used to monitor the temperature.

The TMF test is conducted under total strain control. The minimum (T_{min}) and maximum (T_{max}) temperatures of the out-of-phase cycle are respectively 600°C and 900°C. Linear heating and cooling (10°C/s) cycles are used. Tests start at the mean temperature (750°C) with nil mechanical strain. Some tests were regularly interrupted in order to take replicas of the specimen surface. Details of the testing procedures are given elsewhere (7).

Creep experiments

In order to determine the creep rupture time at 900°C and 280 MPa, three solid cylindrical specimens (5 mm diameter and 70 mm length) were machined from the DS slabs by electro-erosion along the longitudinal grains. One creep test was also performed on a TMF specimen. A reference creep rupture time (t_r) for the virgin alloy is defined as the average of the lifetimes of all tests.

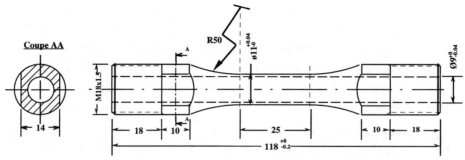

Figure 1: TMF specimen (all dimension in mm).

Sequential creep and TMF experiments

Tubular specimens are TMF cycled to a certain fraction of life at different mechanical strain ranges and are then crept to total fracture at 900°C and 280 MPa, Table II. The surface of the specimens is not re-polished before final creeping.

Some TMF specimens were pre-crept in tension for 241 h at 900°C and 280 MPa, introducing an inelastic strain of about 1 % to 1.4 %, Table II. After pre-creeping and prior to TMF testing, the internal and external surfaces of the specimens are re-polished in order to suppress the oxide-scale. One pre-crept specimen is used for the determination of the non-isothermal cyclic stress-strain behaviour. Replicas of the external surface were taken after pre-creeping and after pre-TMF cycling.

Results

TMF stress-strain curves

Non-isothermal cyclic stress-strain curves are performed on one virgin and one pre-crept specimen by step increasing of the mechanical strain range. The mechanical strain is increased each time a stabilised hysteresis loop is achieved. The non-isothermal stress-strain cyclic curves are reported in Figure 2. The stress and strain ranges measured during the TMF tests at half life are plotted on this graph.

Table II Creep and pre-creeping test results (longitudinal grain).

Temperature, °C	Stress, MPa	Time to rupture, h	Pre-creep time, h	Elongation, %	Specimen
900	280	456	—	26.59	Solid
		472	—	23.10	
		519	—	24.44	
		561	—	21.99	Tubular
		—	241	1.37	
		—	241	1.40	Tubular
		—	241	0.92	

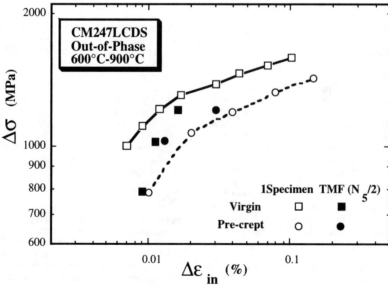

Figure 2: Non-isothermal stress-strain cyclic curves.

TMF-life curves for the virgin and pre-crept conditions

Table III summarises the TMF results obtained for the virgin and pre-crept specimens. The dissipated energy is calculated by numerical integration of the stress-inelastic strain loop at the half life. The TMF life, N_5, is defined as the number of cycles when a decrease of $0.05\sigma_{max}$ is observed in the stress range ($\Delta\sigma$). The crack initiation life N_i is defined as the number of cycles to form a crack of 0.3 mm at the surface (about 0.1 mm depth). N_i is obtained by crack measurements on the surface replicas under scanning electron microscopy (SEM).

The TMF life N_5 is plotted as a function of $\Delta\varepsilon_m$, $\Delta\sigma$, $\Delta\varepsilon_{in}$ and the dissipated energy respectively in Figures 3 to 8.

Table III Results of the TMF out-of-phase tests on CM247LC-DS.

T_{max}, °C	Alloy	$\Delta\sigma$, MPa	σ_{max}, MPa	$\Delta\varepsilon_m$, %	$\Delta\varepsilon_{in}$, %	Energy, MJ/m^3	N_i	N_5	N/N$_5$ (of virgin)
900	Virgin	794	527	0.8	0.009	0.045	—	12382	1
		1024	838	1.0	0.011	0.061	923	2590	1
		1215	759	1.2	0.016	0.130	510	1125	1
	Pre-crept	1025	711	1.0	0.013	0.088	295	974	0.38
		1208	728	1.2	0.030	0.265	97	631	0.56
	Pre-TMF	1030	664	1.0	0.008	0.076	—	—	0.30
		1025	769	1.0	0.009	0.058	—	—	0.49
		1224	900	1.2	0.010	0.088	—	—	0.51

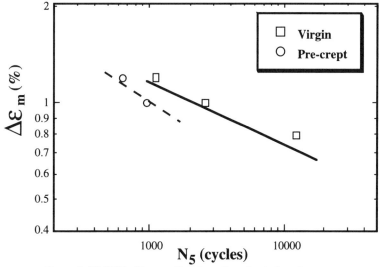

Figure 3: TMF life, N_5, as a function of mechanical strain range.

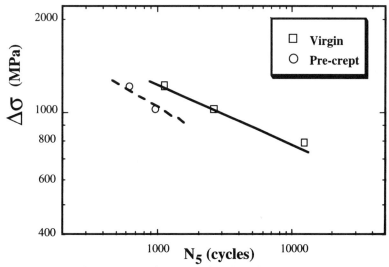

Figure 4: TMF life, N_5, as a function of stress-range.

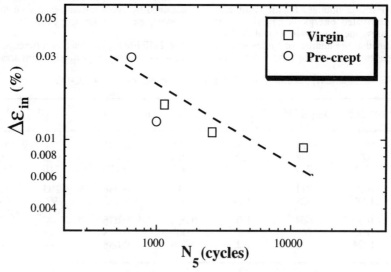

Figure 5: TMF life, N_5, as a function of inelastic strain range.

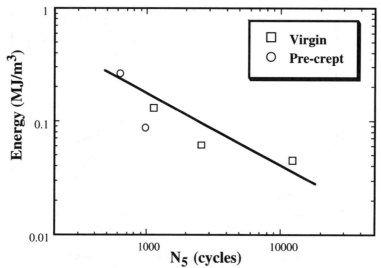

Figure 6: TMF life, N_5, as a function of dissipated energy.

Creep

The creep curve of one of the virgin solid specimens is shown in Figure 7. The results of all creep rupture tests are given in Table II. The tubular specimen presents a longer lifetime in comparison to the solid specimens. The mean rupture lifetime (t_r) is 502 h.

Damage mechanisms

SEM observations on the longitudinal sections of solid and tubular specimens after creep reveal that both the bulk and the surface are damaged. In the bulk, cavities nucleate at the interface of carbides or residual eutectics and on fractured carbides. Internal cracks initiate from these cavities. Surface cracks initiate by creep-oxidation interaction process at the grain boundaries. In addition, the γ'-precipitates in virgin alloy (Figure 8a) are rafted perpendicularly to the stress axis (Figure 8b).

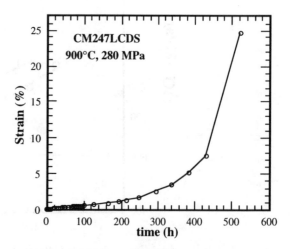

Figure 7: Creep curve of CM247LC-DS at 900°C.

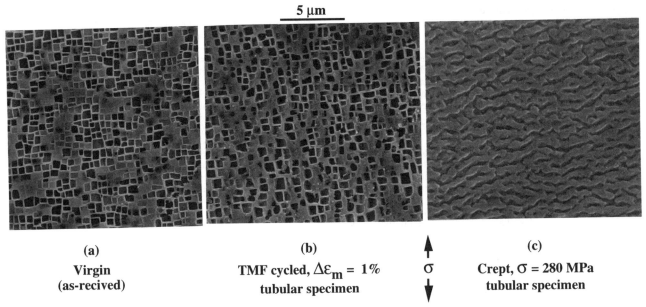

Figure 8: Microstructure of: (a) virgin alloy, (b) after TMF cycling and (c) γ'-rafting in crept alloy at 900°C.

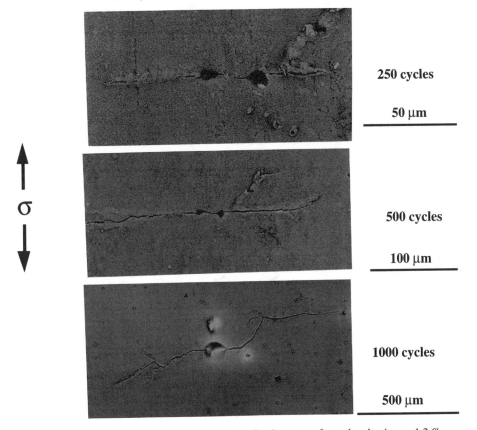

Figure 9: Main crack evolution observed on replicas for the test performed under $\Delta\varepsilon_m$ = 1.2 %.

Observations of the replicas taken from the virgin TMF cycled specimens show that cracks initiate on interdentritic cast porosities. The cracks then propagate in a transgranular manner. Figure 9 shows the evolution of the main crack for the test performed at $\Delta\varepsilon_m$ = 1.2 %. No change is observed on the γ-γ' structure after TMF cycling (Figure 8c).

Discussion

As stated before, a turbine blade is submitted to creep and thermo-mechanical fatigue during each operation cycle. The sequential creep-TMF and TMF-creep procedure adopted in the present investigation is very useful to dissociate the influence of the damage accumulation introduced in either creep or TMF separately.

Among the various types of damage developed during creep, the surface damage is of primary importance for the residual TMF resistance evaluation. The surface re-polishing has suppressed the oxide scale but has not eliminated the overall damage introduced at the surface during creep (oxide penetration, surface cracks and γ' depleted zone). The surface cracks initiated in creep are generally intergranular. These cracks reduce significantly the TMF life of the alloy by reducing the crack initiation period.

It is well known that the high resistance of nickel based superalloys at very high temperatures is due to the γ-γ' microstructure. During tensile creep the γ' precipitates on the virgin alloy are rafted perpendicularly to the loading axis. Rafting is observed on alloys which have a negative relative lattice mismatch: $\delta = 2(a_{\gamma'} - a_{\gamma})/(a_{\gamma'} + a_{\gamma})$, where $a_{\gamma'}$ and a_{γ} are the lattice parameters of the γ' and γ phases (10-12). The internal back stresses developed in γ phase while pre-creeping (12) seems to enhance the inelastic strain under out-of-phase TMF loading (compressive straining while heating) as showed in Figure 2.

It is generally assumed that the fatigue life under isothermal or non-isothermal cycling is governed by the inelastic strain range, Figure 5. The modification of the γ-γ' structure (hence the mechanical properties of the alloy) should be considered as an additional parameter which shortens the TMF life. The crack initiation life, N_i, plotted as a function of the inelastic strain range on Figure 10 clearly demonstrates the effect of pre-creeping on the reduction of the crack initiation period and on the enhancement of the inelastic strain.

Differently to the previous results, pre-TMF cycling has no effect on the creep rupture lifetime. Post-fracture SEM investigations have shown that the initial γ-γ' structure of the virgin alloy is not altered by TMF cycling. There are experimental evidences that rafting decreases the creep lifetime of the single crystal superalloys (12). Therefore, one should expect that the creep rupture lifetime of the CM247LC-DS is not reduced since its microstructure and hence its mechanical properties are not changed by pre-TMF cycling. It should be emphasised, that the short transgranular cracks which initiate from the surface during pre-TMF cycling reduce the creep load bearing section of the specimens but do not influence the creep damage mechanisms.

Figure 11 shows the variation of the residual TMF life, N/N_5, as a function of the residual creep rupture lifetime, t/t_r. This figure shows that creep-TMF interactions can be described by a linear damage summation law only when pre-creeping precedes TMF cycling, and not in the reverse case (creep tests on pre-TMF cycled specimens). Therefore, care should be taken when creep-fatigue cumulative damage models are employed for life prediction. It should be emphasised that the synergistic effects of creep and TMF acting simultaneously have yet to be investigated.

Conclusions

Creep-TMF interaction in CM247LC-DS is investigated. Sequential pre-creeping/TMF cycling and pre-TMF cycling/creeping tests are conducted. The γ'-raft structure developed during creep softens the alloy and enhances the inelastic strain during subsequent TMF cycling. Pre-creeping decreases drastically the TMF life. The main reason for this reduction in life is the initiation of cracks by creep-oxidation interaction, which decrease the TMF crack initiation period. Pre-TMF cycling has no apparent effect on the creep rupture lifetime.

Acknowledgements

Authors would like acknowledge the Swiss "Commission Pour l'Encouragement de la Recherche Scientifique" for supporting this investigation (project No. 2681.1). Prof. B. Ilschner, the head of the Mechanical Metallurgy Laboratory is very acknowledged for scientific discussions and support.

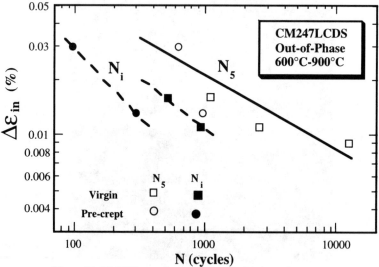

Figure 10: TMF life as a function of the inelastic strain range

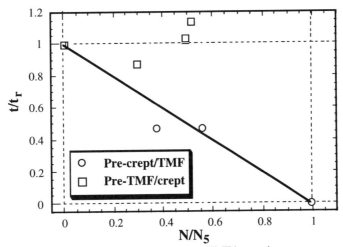

Figure 11: Residual life in creep-TMF interactions.

References

1. M. Gell, D. N. Duhl, and A. F. Giamei, "The development of single crystal superalloy turbine blades," (Paper presented at the Fourth International Symposium on Superalloys, Champion, Pennsylvania, USA, 21-25 September 1980), 205-214.

2. K. Harris, G. L. Erickson, and R. E. Schwer, "Development of the CMSX series of single-crystal alloys for advanced technology turbine components," (Paper presented at the TMS-AIME Fall Meeting, St. Louis, Missouri, 1982).

3. C. G. Date et al., "TMF design considerations in turbine airfoils of advanced turbine engines," (Paper presented at the Creep-fatigue interactions at high temperatures, Winter Annual Meeting of the American Society of Mechanical Engineers, ASME, Atlanta, Georgia, 1991), 59–64.

4. C. Levaillant et al., "Creep and creep-fatigue intergranular damage in austenitic stainless steels: discussion of the creep-dominated regime," (Paper presented at the Low Cycle Fatigue, Bolton Landing (on Lake George), New York, 1985), 414-437.

5. L. Rémy et al., "Evaluation of life prediction methods in high temperature fatigue," (Paper presented at the Low Cycle Fatigue, Bolton Landing (on Lake George), New York, 1985), 657-671.

6. P. Rodriguez, K. Bhanu, and R. Sankara, "Nucleation and growth of cracks and cavities under creep-fatigue interaction," Progress in Materials Science, Vol. 37 (1993), 403-480.

7. C. C. Engler-Pinto Jr et al., "Thermo-mechanical fatigue behaviour of IN738LC," (Paper presented at the Materials for Advanced Power Engineering 1994, Liège, Belgium, October 3-6 1994), 853-862.

8. G. T. Embley and E. S. Russell, "Thermal mechanical fatigue of gas turbine bucket alloys," (Paper presented at the First Parsons International Turbine Conference, Dublin, Ireland, June 1984), 157–164.

9. J. L. Malpertu and L. Rémy, "Influence of test parameters on the thermal-mechanical fatigue behaviour of a superalloy," Metallurgical Transactions A, Vol. 21A (Feb. 1990), 389-399.

10. A. Fredholm and J. L. Strudel, "On the creep resistance of some nickel base single crystals," (Paper presented at the Superalloys 1984, Champion, Pennsylvania, USA, 1994), 211-220.

11. H. Biermann et al., "Internal stresses, coherency strains and local lattice parameter changes in a creep-deformed monocrystalline nickel-base superalloy," High Temperature Materials and Processes, Vol. 12 (Nos. 1-2) (1993), 21-29.

12. H. Mughrabi, H. Biermann, and T. Ungar, "Creep-induced local lattice parameter changes in a monocrystalline nickel-base superalloy," Journal of Materials Engineering and Performance, 2(4) (August) (1993), 557-564.

CREEP AND FATIGUE PROPERTIES OF A DIRECTIONALLY SOLIDIFIED NICKEL BASE SUPERALLOY AT ELEVATED TEMPERATURE

M. Maldini [*], M. Marchionni [*], M. Nazmy [o], M. Staubli [o] and G. Osinkolu [*]

[*] CNR-ITM, Cinisello B. (Milano), Italy
[o] ABB Power Generation, Baden, Switzerland

Abstract

The creep and fatigue properties of the directionally solidified nickel base superalloy CM247LC DS have been investigated.
Constant creep tests have been carried out on specimens with different orientations in the temperature range of 700-1000°C at different loads to obtain times to rupture up to 35000 h. The comparison of the creep properties with the IN738LC conventionally cast alloy and with the oxide dispersion strengthened MA6000 alloy has shown the better creep performance of the alloy CM247LC DS in the temperature range of interest for application in land based gas turbines.
A series of cyclic load creep tests has allowed to study the effect of load variations on the creep. The effect of cycling stress is to increase the strain rate, compared with the constant load creep tests and then to reduce the rupture life of the alloy.
The LCF tests, performed at the temperatures of 850°C and 950°C in longitudinal strain controlled conditions, have evidenced a fairly stable cyclic response. Basquin and Coffin-Manson relationships can adequately predict the fatigue life of the alloy.
The CM247LC DS alloy exhibits a better fatigue life than IN738LC and MA6000 alloys.

Introduction

The constant demand of increasing the operating efficiency and durability of land based gas turbines requires the development of alloys with always better mechanical properties at high temperature. Recently, the directional solidification technique has allowed to produce new alloys with excellent high temperature resistance [1,2].
In this investigation the new directionally solidified (DS) nickel-base superalloy CM247LC DS has been studied to evaluate the creep and fatigue properties of the material, to compare its properties with commercial ODS MA6000 and cast IN738LC alloys and to analyse the creep and fatigue damage mechanisms.

Material characterisation and experimental procedures

The CM247LC DS alloy is a directionally solidified Ni-base superalloy containing a high volume fraction of the reinforcing γ' phase. Table I shows the chemical composition of the alloy in wt%.

The alloy was supplied by Howmet in form of directionally solidified slabs with the following dimensions: L = 160 mm, W = 100 mm, T = 20 mm.
The alloy microstructure consists of columnar grains with boundaries parallel to the <001> growth direction, with an average width of 0.5 mm and a grain aspect ratio of about 10. Fig. 1 shows the microstructure a) in longitudinal (L) and b) in long transverse (LT) direction.
These slabs have received the following solution and γ' precipitation heat treatment:
1221°C/2h + 1232°C/2h + 1246°C/2h + 1260°C rapid fan quench in argon,
1079°C/4h rapid fan quench in argon + 871°C/20h rapid fan quench in argon.
The specimens for creep tests under monotonic and cyclic loading had cylindrical symmetry of 5.6 mm gauge diameter and 28 mm gauge length. Three thermocouples, tied in the gauge length, allowed to control the temperature gradients (<0.5°C) during creep. Capacitive transducers, connected to extensometers clamped to the shoulders of the specimen, continuously monitored the creep strain.
In the cyclic load creep tests, performed on a servo electro-mechanical machine at 950°C, the load was cyclically changed between two values. Fig. 2 schematically represents the stress wave form as a function of time.
The low cycle fatigue (LCF) tests were performed in air on a closed loop electro-hydraulic system using cylindrical specimen having 8 mm gauge diameter and 12 mm gauge length. An induction coil heated the specimens to the test temperatures of 850°C and 950°C. At 950°C both the L and LT orientations of the alloy were studied. LCF tests were conducted under longitudinal strain controlled conditions at a constant strain rate of $3 \times 10^{-3} s^{-1}$. A fully reversed triangular wave form was used to achieve the various strain amplitudes. During the fatigue tests the stress response and the stress-strain hysteresis loops were recorded at regular intervals.
All the specimen fracture surfaces were examined by scanning electron microscopy (SEM).

Table I - Chemical composition of CM247LC DS alloy, in wt%.

C	Al	Co	Cr	Mo	Ta	Ti	W	Nb	Hf	Ni
0.07	5.55	9.5	8.1	0.5	3.2	0.69	9.5	0.05	1.4	Bal.

Figure 1 - Microstructure aspect of CM247LC DS alloy in a) L and b) LT direction.

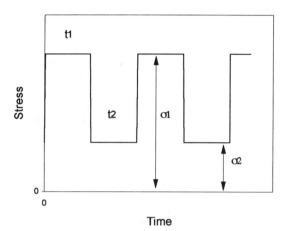

Figure 2 - Stress wave form vs. time.

Experimental results and discussion

Constant load creep results

Nickel base superalloys often creep with very short primary stage, while the majority of the creep life is tertiary creep. The studied alloy has confirmed this behaviour as shown in Fig. 3, where typical creep curves obtained at different stresses and testing temperatures for the longitudinal orientation are plotted. The primary stage contribution to the creep strain is only important for the tests run at the highest stresses and the lowest temperatures.

As pointed out by different papers [3,4], the dominant tertiary creep in nickel base superalloys, is not associated to fracture mechanisms, but to an increase of the flux of mobile dislocations with the strain due to an increase either of mobile dislocation density or the recovery rate.

A strain rate vs. strain plot (Fig. 4) shows that a large portion of the tertiary stage always manifests a linear dependence of the strain rate vs. strain, in agreement with the results obtained on other directionally solidified nickel base superalloys [5].

Fig. 5 shows the stress rupture resistance of CM247LC DS for the L and LT studied orientations in comparison with the ODS MA6000 with equivalent orientations and with the IN738LC conventionally cast [6-8]. While it is evident the better creep performance of the studied alloy compared with the IN738LC conventionally cast for the stress/temperature explored, the MA6000 alloy shows a better creep resistance at the lowest stresses/highest temperatures for the longitudinal orientation, but a worse behaviour for the transverse orientation.

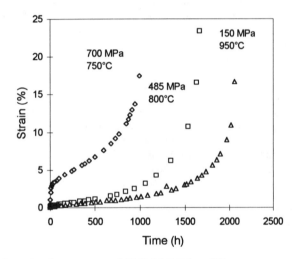

Figure 3 - Creep curves of CM247LC DS at different stresses and temperatures.

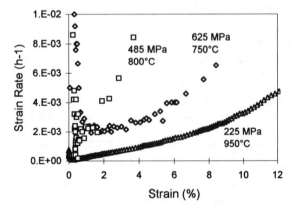

Figure 4 - Strain rate versus strain at different stresses and temperatures in creep tests.

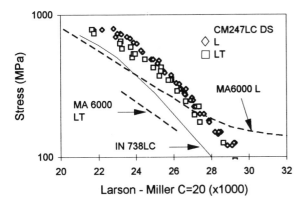

Figure 5 - Comparison of Larson - Miller plots of CM247LC DS, IN738LC and ODS MA6000 alloys.

TABLE II - Rupture data of cyclic creep tests: σ_1 = 225 Mpa, σ_2 = 22 Mpa, T=950°C.

Test Number	Stress waveform $t_1 \div t_2$	ε_f (%)	R.A. (%)	Rupture time, t_r (h)	Total time on-load (h)
1	*	26.7	54.4	172	172
2	*	43.4	61.5	168	168
3	24-0 hr	29.0	55.6	141	141
4	1-0 hr	23.7	56.0	143	140
5	1-1 hr	31.0	62.2	203	102
6	30-30 min	20.0	52.0	232	116
7	5-5 min	19.7	57.0	232	116
8	24-24 hr	28.8	62.3	253	136
9	30-5 min	24.0	60.0	176	152

* Constant load creep tests (σ = 225 MPa, T=950 °C).

Cyclic load creep results

All the cyclic creep tests have been performed at 950°C on specimens with L orientation. Most of these tests have been run at a maximum and minimum stress of 225 MPa and 22 MPa respectively, with different hold times at the highest and lowest stress. The minimum load of 22 MPa was imposed to keep the specimen in alignment and no creep was expected to occur during this load level. The loading and unloading time was about one minute. Table II shows the details of the experimental results.

Fig. 6 shows the variation of creep strain with elapsed time for the tests of Table II.

We can observe that the cyclic stressing decreases the rupture life of the alloy and the rupture properties, even in the tests without hold time at lowest stress. In addition both the unloading hold time and the number of unloading and reloading are important to determine the time to rupture.

The ductility of the alloy, in terms of reduction of area and fracture strain, remains unaffected by the different loading conditions.

The effect of the off-load period is to produce a creep rate acceleration in the following on-load period as evident in Fig. 7 that shows the strain rate behaviour during on-load period in the cyclic creep tests 3 and 8 of Table II compared with two constant

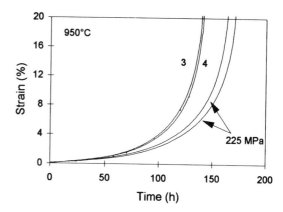

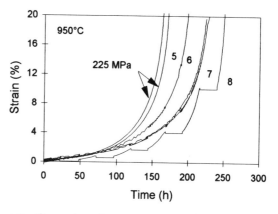

Figure 6 - Comparison between constant load creep tests at 225 MPa and the cyclic creep tests of Table II.

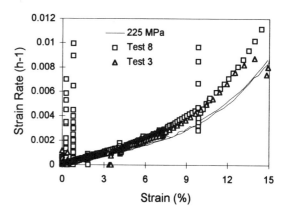

Figure 7 - Comparison of the strain rate vs strain for the cyclic creep tests number 3 and 8 of Table II and two constant load creep tests at 225 MPa.

creep tests at 225 MPa. Both cyclic creep tests show a higher creep strain rate compared with the constant load tests. For test 8, on each reloading, the strain rate quickly decreases from a high value to a minimum which is observed to be less than that obtained under constant load creep and then it increases gradually up to overtake the strain rate measured in the constant load creep tests.

Additional work is needed to rationalise the observed experimental results.

TABLE III - Rupture data for stress cycling between two stress levels

Stress wave form	Strain to rupture, ε_f (%)	Reduction of Area, R.A. (%)	Rupture time, t_r (h)	Linear Fraction Rule
300 MPa - 4 h 175 MPa - 68 h	29.0	56.0	158.0	0.57
300 MPa - 4 h 225 MPa - 20 h	30.3	53.7	88.5	0.88
300 MPa - 4 h 250 MPa - 10 h	24.4	61.8	44.0	0.71

Few specimens were tested in cyclic creep between a higher stress, $\sigma_1 = 300$ MPa, with a hold time, t_1 and a lower stress σ_2 ranging between 175 and 250 MPa with a hold time t_2. The holding times t_1 and t_2 are approximately 10% of the rupture life under constant load creep tests for the respective stress levels. The results are summarised in Table III for different values of minimum loads. The value of the constant of the Robinson's Life Fraction Rule is also shown: at the rupture, the sum of the creep life fractions consumed at each load condition always results in a value less than unity.

A typical creep curve for $\sigma_1 = 300$ MPa ($t_1 = 4$ hrs) and $\sigma_2 = 225$ MPa ($t_2 = 20$ hrs) is presented in Fig. 8 in form of the strain rate against strain. Similar data, under constant creep loading at 300 MPa and 225 MPa, are also included for comparison.

On reducing the stress from 300 MPa to 225 MPa, the transient creep is characterised by a strain rate decreasing down to a minimum value, after that it starts an increase staying always below the curve of the constant load creep test. A similar type of transient has been reported previously both in pure metals and solid solutions [9] and recently in Ni-base superalloys [10].

After reloading the specimen, the strain rate quickly decreases from a high value to a minimum followed by a rapid increase in the strain rate. At the end of the transient the strain rate stays above the constant load creep curve.

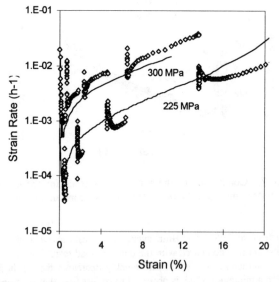

Figure 8 - Strain rate versus strain for cyclic stressing between 300 MPa and 225 MPa (Table III) in comparison with constant load creep tests at 300 MPa and 225 MPa (lines).

Low Cycle Fatigue

In the strain controlled LCF tests the cyclic stress response curves represent the variation of the stress amplitude versus the progressive number of cycles. The stress range levels were almost constant through the test with only approximate 3% softening until close to failure.

The variation of the stress amplitude with the plastic strain is established to determine the cyclic stress-strain curves and the data are represented by Hollomon type equation of the form:

$$\frac{\Delta\sigma_a}{2} = K * \left(\frac{\Delta\varepsilon_p}{2}\right)^m \qquad (1)$$

where $\Delta\sigma_a$ is the stress amplitude, $\Delta\varepsilon_p$ the plastic strain component, K the cyclic strength coefficient and m the cyclic strain exponent. The parameters are determined by least square method and listed in Table IV. The table also contains the fracture strain and the ultimate tensile strength (UTS) under monotonic loading. The values of K scale with those of the UTS. It is noteworthy that the value of the cyclic strain exponent is about 0.2, independent of the test temperature and the direction of the loading axis with respect to the columnar grain orientation.

The number of cycles to failure, N, has been defined in the point where the rapid variation of the stress level in the $\Delta\sigma$-n curve occurs. This variation corresponds to a crack length from 0.5 to 1mm depending on the total strain imposed. Fig. 9 shows the influence of test temperature and specimen grain orientations on the fatigue life of CM247LC DS alloy. A fatigue life decrease is observed when temperature increases and when the material is tested in LT direction. For the same imposed strain range, the fatigue life in L direction is from 6 to 7 times longer than in LT direction. The results can be partially explained by the higher Young modulus, E, for the material in the LT direction (Table IV). Fig. 10 shows the elastic and plastic strain components at half life versus the number of cycles to failure (N) at 850°C and 950°C.

On the basis of the elastic strain-life data, the fatigue resistance at 850°C is superior to that at 950°C, whereas the reverse seems to be the case when comparing the plastic strain-life curves.

Table IV - Cyclic stress-strain parameters and tensile properties.

Temp. °C	Grain orient.	Tensile fracture strain, mm/mm	UTS MPa	K MPa	m	E GPa
850	L	0.207	965	1127.2	0.217	90.6
950	L	0.356	740	692	0.195	84.3
950	LT	0.118	680	680.4	0.207	106.5

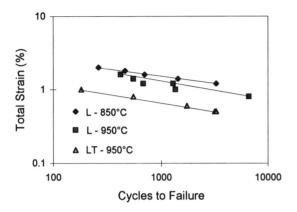

Figure 9 - Total strain range versus the number of cycle to failure.

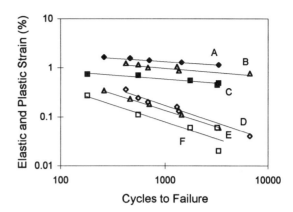

Figure 10 - Plastic and elastic strain components versus the number of cycles to failure: A= $\Delta\varepsilon_e$, L- 850°C; B= $\Delta\varepsilon_e$, L- 950°C; C= $\Delta\varepsilon_e$, LT- 950°C; D= $\Delta\varepsilon_p$, L- 950°C; E= $\Delta\varepsilon_p$, L- 850°C; F= $\Delta\varepsilon_p$, LT- 950°C.

However, the plastic strain-life curve presents some scatters at the lower imposed strain ranges.
The fatigue life of the alloy can be described by the empirical relationships of Basquin [11] and Coffin-Manson [12] respectively as follows:

$$\frac{\Delta\varepsilon_e}{2} = \frac{\sigma_f}{E}(N)^{-b} \qquad (2)$$

$$\frac{\Delta\varepsilon_p}{2} = \varepsilon_f(N)^{-c} \qquad (3)$$

where σ_f and b are the fatigue strength coefficient and fatigue strength exponent respectively, while ε_f and c are the failure ductility coefficient and fatigue ductility exponent respectively. The values of these parameters are listed in Table V.
It should also be noted that a better correlation is obtained for the elastic strain-life data. The relationships between the elastic, plastic strain components and the number of cycles to failure are also obtained by using equations 2 and 3.
The fatigue resistance of the directionally solidified CM247LC DS is compared at 850°C and 950°C (Fig. 11), on the total strain-life basis, with ODS MA6000 and IN738 LC conventionally cast alloys which are designed for similar applications. CM247LC DS exhibits superior fatigue resistance than that of other alloys in the whole testing temperature range. A similar behaviour was found by Vasser and Remy [13].

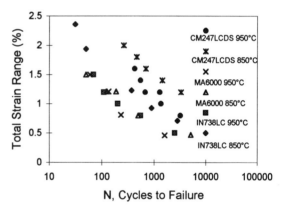

Figure 11 - Comparison of faigue life of CM247LC DS with MA6000 and IN738LC alloys.

Metallography

A typical creep fracture surface obtained on longitudinal stressed specimen is shown in Fig. 12. The presence of crystallographic facets associated with casting pores, suggests that the fracture originates from pores and then propagates on (001) planes. Examples of such pores are shown in Fig. 13 as observed from the longitudinal section of the specimen. Cracks propagating from the surface of the pores are also visible.
Some of these cracks were also observed to be initiated due to decohesion at carbide and matrix interface or due to carbide fracture, as illustrated in Fig. 14.
The observed fracture process, as it should be expected, agrees with the results on single crystals of Ni-base superalloys [14].
The fatigue failure mechanisms in Ni-base superalloys involve crack initiation at near surface flaws. Attempts were therefore made to detect some of the features that are often associated with fatigue failure. The SEM micrograph in Fig. 15 shows the fracture surface of a specimen tested at 850°C with a total imposed strain range of 1.2%. The micrograph shows a micropore at the

Table V - Fatigue parameters of Basquin and Coffin-Manson relationships.

Temperature °C	Grain orientation	σ_f GPa	b	ε_f mm/mm	c
850	L	176	0.15	0.075	0.68
950	L	140.7	0.17	0.13	0.75
950	LT	99	0.17	0.048	0.69

specimen surface that could act as a zone of stress concentration and eventually as the site of fatigue crack initiation. The pores could also act as origin of fatigue crack initiation, as it seems to be indicated by the SEM micrograph in Fig. 16 for the L specimen fatigued at 850°C under an imposed total strain range of 1.6%. No slip off-sets were discernible on the specimen surface.

Fatigue striations are occasionally observed and secondary cracks are frequently found in L direction specimens as shown by SEM micrograph in Fig. 17. Also in LT specimens transgranular crack propagation and secondary crack are observed.

The fracture aspect and the occurrence of secondary cracks depend on the amount of the imposed strain range. At lower imposed strain range, i.e. $\Delta\varepsilon_t = 0.5\%$, no secondary cracks are observed and the fracture surface is macroscopically flat. At higher strain ranges, fatigue striations are also observed with numerous secondary cracks.

The fatigue crack initiation phase consists of cracking along the crystallographic direction inclined to the applied load, i.e. stage I type and a typical example of initiation zone is shown in Fig. 18. The subsequent crack growth direction is perpendicular to the applied load (stage II type).

The fatigue failure mechanism can thus be envisaged as follows: crack initiates along the slip bands that intersect the defects e.g. pores on the specimen surface; although the precise role of oxygen can not be specified, the diffusion of this element along the deformed zone could also facilitate shear decohesion while the pore induces a local stress concentration. Once the crack is formed, its subsequent growth direction is perpendicular to the applied load.

Figure 12 - Typical fracture surface showing the presence of holes as sites for microcrack initiation.

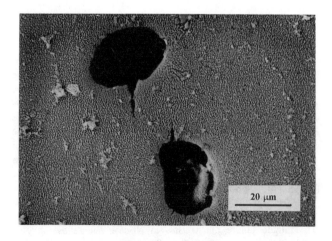

Figure 13 - Pores as observed on longitudinal section by SEM.

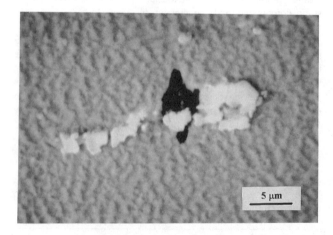

Figure 14 - Microcrack formation at carbide sites by SEM.

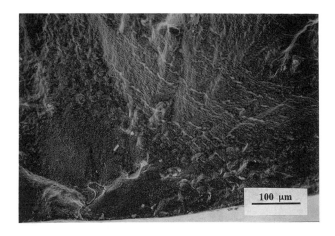

Figure 15 - Aspect of fatigue fracture: crack initiation at surface pore; L- 850°C, $\Delta\varepsilon_t = 1.2\%$, N= 3300.

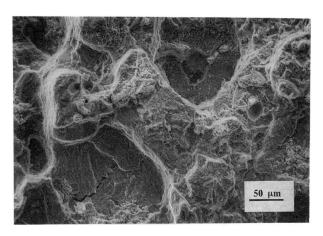

Figure 16.- Presence of some pores in the interior of the specimen: L- 850°C, $\Delta\varepsilon_t = 1.6\%$, N= 700

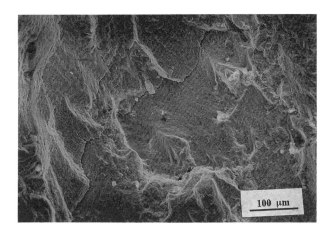

Figure 17 - Presence of fatigue striations and secondary cracks: L- 950°C, $\Delta\varepsilon_t = 1.6\%$, N= 420.

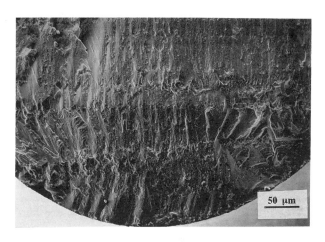

Figure 18 - Exemple of crack initiation zone: LT- 950°C, $\Delta\varepsilon_t = 0.5\%$, N= 3300.

Conclusions

The alloy CM247LC DS possesses a clear cut advantage, in creep properties, over that of IN738LC conventionally cast and MA6000 alloys.

After a step like load increase or decrease, the strain rate stabilises respectively to a higher or lower value, if compared with the constant load creep tests.

Irrespective of the cyclic stress wave form imposed on the alloy, a strain rate acceleration is always observed during on-load periods if constant load creep test is taken as a reference. Such strain rate acceleration could be an important factor to be taken into account when modelling the behaviour of the alloy under variable loading.

Creep fractures show that microcracks initiate at the shrinkage pores and at carbides and matrix interface.

In strain control LCF tests, the alloy exhibits fairly stable cyclic stress response, with only a slight stress softening. The fatigue life of the alloy can be satisfactorily predicted by the empirical relationships of Coffin-Manson and Basquin with the latter giving a better correlation.

The fatigue ductility parameters obtained from these relationships and the tensile properties of the alloy give reasonable correlation.

The fatigue crack initiation is mainly due to the shrinkage pores located at the surface or sub-surface of the specimen.

The alloy exhibits superior fatigue resistance than that of ODS MA6000 and IN738LC Ni-base superalloys.

Acknoledgements   The authors would like to acknowledge the skilful technical assistance of Mr. Bianchessi, Mr. E. Picco, Mr. D. Ranucci and Mr. E. Signorelli. G. A. Osinkolu undertook this work with the support of "ICTP Programme for Training and Research in Italian Laboratories", Trieste, Italy. The work was partially performed in the European programme COST 501 Second Round.

REFERENCES
1. M. McLean, Directionally Solidified Materials for High \Temperature Service (London: The Metals Society, 1983).
2. K. Schneider, High Temperatue Materials for Power Engineering, (Dordrecht, Netherland: Kluver Academic Publishers, 1994), 1155-1164.
3. B.F. Dyson and M. McLean, "Particle-Coarsening, σ_o and Tertiary Creep", Acta Metall., 31 (1983) 17-27.
4. A. Barbosa et al., Superalloys 1988 (Warrendale, PA:The Metallurgical Society, 1988), 683-692.
5. M. Maldini and V. Lupinc, "A Representation of Tertiary Creep Behaviour in a Single Crystal Nickel-Based Superalloy", Scripta Metall. , 22 (1988), 1737-1741.
6. R.C. Benn and S.K. Kang, Superalloys 1984 (Warrendale, PA:The Metallurgical Society, 1984), 319-326.
7. E. Arzt and R. Timmins, Creep Properties of ODS Superalloys, A Summury of Results Obtained by the ODS Group in Cost 501, Petten, Netherland: Commission of The European Communities Institute of Advanced Materials Joint Research Centre Petten, 1991).
8. V. Lupinc et al., "Nuove superleghe a struttura direzionale per palette di turbine agas avanzate", La Metallurgia Italiana, 81 (1989), 825-834.
9. M. Biberger and J.C. Gibeling, " Analysis of Creep Transients in Pure Metals Following Stres Changes", Acta Metall. Mater., 43 (1995) 3247-3260.
10. M. Maldini, CREEP: Characterization, Damage and Life Assessment (Materials Park, Ohio:ASM International, 1992), 111-116.
11. O.H. Basquin, "The experimental law of endurance tests", ASTM 10, (1910) 610-625.
12. L.F. Jr. Coffin, "Fatigue at high temperature", Fatigue at Elevated Temperatures ASTM STP 520, (1972) 520,5-34.
13. E. Vasseur, and L. Rémy, "High temperature low cycle fatigue and thermal-mechanical fatigue behaviour of on oxide-dispersion - strengthened nickel-base superalloy", Mater. Sci. and Eng., A184 (1994) 1 - 15.
14. S. H. Ai, V. Lupinc, and M. Maldini, "Creep Fracture Mechanisms in Single Crystal Superalloys", Scripta Metall. Mater., 26 (1992) 579-584.

SPECIFIC ASPECTS OF ISOTHERMAL AND ANISOTHERMAL FATIGUE OF THE MONOCRYSTALLINE NICKEL–BASE SUPERALLOY CMSX–6

H. Mughrabi, S. Kraft and M. Ott

Institut für Werkstoffwissenschaften, Lehrstuhl I,
Friedrich-Alexander-Universität Erlangen-Nürnberg
Martensstrasse 5, D–91058 Erlangen, Federal Republic of Germany

Abstract

An extensive study of the high–temperature isothermal and thermomechanical fatigue (TMF) behaviour of the "light" nickel–base superalloy CMSX–6 has been performed. Special emphasis was placed on a detailed microstructural interpretation of some specific aspects of the cyclic deformation behaviour. In the case of the isothermal fatigue tests, the main points of interest were the dependence on cyclic strain rate, effects of cyclic softening and directional coarsening, the cyclic stress–strain behaviour, mean stress and cyclic stress asymmetry effects and the dependence of the fatigue behaviour on different initial γ/γ' morphologies. In the TMF tests, the studies focussed on the dependence of fatigue life on the strain–temperature cycle shapes, on directional coarsening effects for different cycle shapes, on the microstructural processes during a single cycle and on the effects of the strongly varying plastic strain rates within a cycle in a total–strain controlled test. A critical comparison between the isothermal and the TMF behaviour permits several conclusions to be drawn. In particular, it follows that while isothermal tests can provide a valuable guideline for the understanding of TMF, they are inadequate for a more detailed interpretation.

1. Introduction

Monocrystalline γ'–hardened nickel-base superalloys, exhibiting superior high–temperature strength properties, are nowadays commonly used for advanced turbine blading of aircraft jet engines [1]. High–temperature creep strength and thermomechanical fatigue (TMF) resistance are considered to be the mechanical properties of major concern. While numerous detailed studies on the high–temperature creep behaviour of nickel–base superalloys have been performed in the past, cf. for example [2–7], much less work has been done on the isothermal [8–10] and, in particular, on the anisothermal TMF fatigue properties of monocrystalline superalloys, cf. [11–13]. Knowledge of the isothermal fatigue behaviour can provide a valuable guideline for the understanding of TMF but is, by itself, inadequate in order to deal satisfactorily with the much more complex deformation and damage processes occurring during TMF.

The goal of the present work is to gain a deeper understanding of the microstructural processes that govern high temperature fatigue and, in particular, thermomechanical fatigue, of the "light" monocrystalline nickel-base superalloy CMSX–6. For that purpose, the studies focussed on specific aspects of isothermal fatigue believed to be relevant also to TMF and on the material behaviour under selected well–defined TMF test conditions. The results of these investigations are contrasted against each other, and some general conclusions are drawn.

2. Experimental Procedure

In the present work monocrystalline rods of the γ'–hardened nickel–base superalloy CMSX–6 (composition in wt.%: 9.76 Cr, 3.01 Mo, 1.96 Ta, 4.70 Ti, 5.23 Co, 4.81 Al, bal. Ni) with orientations that lay within 10° near [001] were used. These rods had been supplied by Thyssen Guß AG, Feingußwerk Bochum, in the cast and heat-treated state. After machining and electropolishing, the fatigue test specimens had gauge lengths of 12 mm and a diameter of 9 mm. The microstructure consisted of fairly regularly arranged cuboidal γ' particles with 470 nm edge length, occupying a volume fraction of ≥ 0.55. The constrained misfit parameter was determined by high-resolution X-ray diffraction as $\delta \approx -10^{-3}$.

The fatigue tests were performed on uncoated specimens on a servohydraulic test machine (MTS 880) that had been equipped with high frequency induction coil heating and programming facilities for thermomechanical fatigue tests [14]. A high–vacuum chamber permitted tests in either air, arbitrary gaseous environments or in high vacuum. The isothermal and anisothermal fatigue tests in this study were all performed in closed–loop total strain control at prescribed total strain range $\Delta\varepsilon_t$. The corresponding plastic strains ε_{pl} could be calculated in all cases by taking into account the (temperature–dependent) elastic strains, determined via the temperature–dependent Young's moduli. The latter had been measured previously in the testing machine.

The TMF–tests were performed with in–phase (IP), out–of–phase (OP) and clockwise and counterclockwise diamond (CD, CCD) temperature–total strain cycles. These cycle shapes are shown in plots of temperature T versus total strain ε_t in fig. 1. In general, a total strain range of $\Delta\varepsilon_t = 10^{-2}$, an upper temperature of 1100 °C (or in some cases 900 °C) and a lower temperature of 600 °C were used. The cycle time was $t_c = 300$ s, resulting in a total strain rate $\dot\varepsilon_t = 6.67 \cdot 10^{-5}$ s^{-1}. The tests were always started at zero strain at the lowest possible temperature. For further details, see [15,16].

3. Experimental Results and Discussion

3.1 Isothermal Fatigue

In the following we report and discuss some noteworthy features of the isothermal fatigue behaviour.

3.1.1 Cyclic Deformation Behaviour
Cyclic deformation curves obtained at different temperatures between 950 °C and 1100 °C and at $\Delta\varepsilon_t = 10^{-2}$, $\dot\varepsilon_t = 5 \cdot 10^{-3}$ s^{-1}, are displayed in fig. 2 in the form of peak tensile and compressive stresses versus the num-

ber of cycles N (fig. 2a) and mean stress amplitude $\Delta\sigma/2$ and mean stress σ_m versus N (fig. 2b), respectively. As expected, the stress levels decrease with increasing temperature. At the same time, the fatigue lives decrease, mainly as a consequence of the increasing oxidation. An interesting effect is found with regard to the mean stresses σ_m which develop during cyclic deformation. Up to 1050 °C, these mean stresses are compressive; at the highest temperature investigated (1100 °C), a tensile mean stress is measured. These results indicate a cyclic stress asymmetry which changes sign around 1075 °C. We return to stress

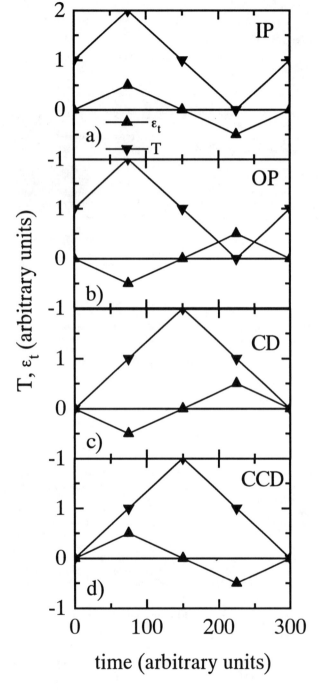

Figure 1: Temperature T and total strain ε_t versus time t for different TMF cycle shapes. a) In-phase TMF (IP); b) Out-of-Phase TMF (OP); c) Clockwise diamond TMF (CD); d) Counterclockwise diamond TMF (CCD).

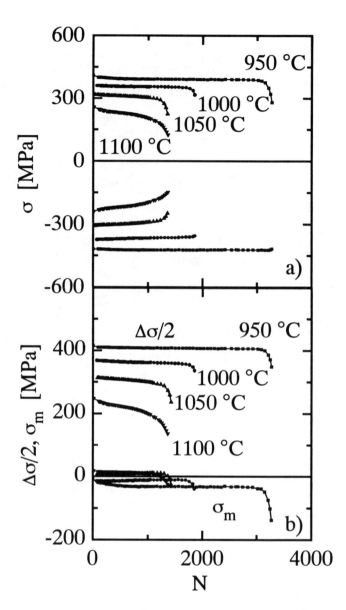

Figure 2: Isothermal cyclic deformation curves for different temperatures. a) Tensile and compressive peak stress σ versus number of cycles, N; b) Mean stress amplitude $\Delta\sigma/2$ and mean stress σ_m versus number of cycles, N.

asymmetry effects again in section 3.1.5.

3.1.2 Dependence of the Cyclic Deformation Behaviour on Cyclic Strain Rate The effect of cyclic strain rate on the cyclic deformation behaviour was studied at a temperature of 1100 °C at $\Delta\varepsilon_t = 10^{-2}$ for the total strain rates $\dot{\varepsilon}_t$ of $6.67 \cdot 10^{-5}$ s^{-1} and $5 \cdot 10^{-3}$ s^{-1}. Figure 3 shows some corresponding hysteresis loops. Due to the enhanced strain-rate sensitivity at high temperatures, the peak stress at the higher strain rate was initially about 25 % higher than at the lower strain rate. In addition, very severe cyclic softening related to a "directional coarsening" of the γ/γ' microstructure was observed at the lower but not at the higher strain rate, compare fig. 4a. As a consequence of these effects, the elastic strain amplitude was much larger at the higher strain rate and hence the plastic strain amplitude significantly smaller than at the lower strain rate. Accordingly, in the sense of a Manson–Coffin-type behaviour, fatigue life was

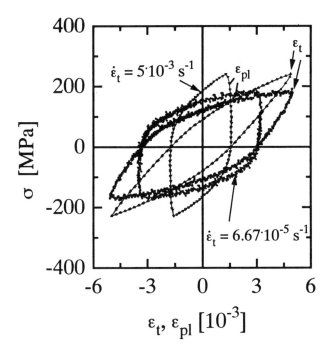

Figure 3: Hysteresis loops at $T = 1100$ °C and total strain range $\Delta\varepsilon_t = 10^{-2}$ for two total strain rates $\dot\varepsilon_t$ in the form of stress σ versus total strain ε_t and plastic strain ε_{pl}. The loops correspond to $N = 10$ (low strain rate) and $N = 200$ (high strain rate).

significantly larger at the higher than at the lower strain rate. We shall return to the coarsening effects in the next section.

3.1.3 Effects of "Directional" Coarsening at Low Cyclic Strain Rate

It is well known that coarsening of the γ/γ' microstructure can occur during deformation at sufficiently high temperature. In the case of the superalloy SRR 99, it has been shown [8,9] that, whereas during fatigue at 950 °C ($\Delta\varepsilon_t = 1.2\cdot 10^{-2}$) at a relatively high total strain rate ($\dot\varepsilon_t = 10^{-3}$ s^{-1}), coarsening effects are negligible, severe coarsening leading to an inclined raft–like γ/γ' structure, accompanied by pronounced cyclic softening and reduced fatigue life, occurs at a lower strain rate ($\dot\varepsilon_t = 10^{-5}$ s^{-1}). In our own studies at a rather high temperature (1100 °C), compare fig. 4, qualitatively similar results were obtained. Some coarsening of the initially cuboidal γ' precipitates (fig. 5a) was observed even after fatigue at the higher strain rate, with a tendency of raft formation perpendicular to the stress axis (fig. 5b), presumably in response to the positive mean stress. The changes in the γ/γ' microstructure after fatigue at the lower strain rate were more pronounced, both with regard to the amount of coarsening and with respect to the change of the morphology (fig. 5c). The edges of the coarsened γ' precipitates were found to lie more or less parallel to the traces of the {111} glide planes. Thus, "soft" γ–channels had developed along the glide planes. In the related work of Portella et al. [8,9], these soft γ–channels were even more extended.

It is interesting to discuss to what extent this kind of directional coarsening is induced by annealing at the high temperature and/or the deformation process. In order to be able to perform the comparison for larger times, an additional test was conducted in vacuum at the higher strain rate, and the data were also used. The results shown in fig. 4a, plotted in the form of the cyclic stress amplitude against time (rather than against the number of cycles) for the two cyclic strain rates, compare fig. 4b, sheds some light on this question. As expected, the initial stress amplitude at the lower strain rate is lower than that at the higher strain rate. What seems surprising, is that the subsequent softening (decrease of stress amplitude as a function

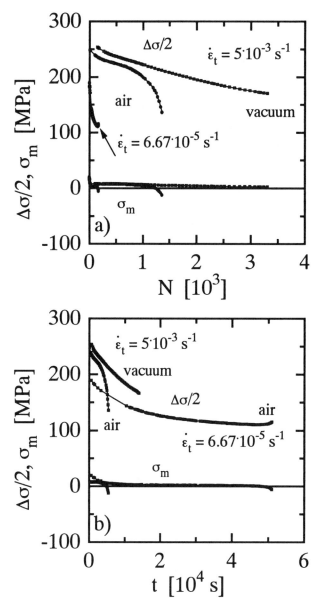

Figure 4: Cyclic hardening curves at $T = 1100$ °C and total strain range $\Delta\varepsilon_t = 10^{-2}$ for two total strain rates $\dot\varepsilon_t$. The test at the higher strain rate was performed in air and also in vacuum. Mean stress amplitude $\Delta\sigma/2$ and mean stress σ_m versus a) number of cycles N, and b) the time t.

of time) is comparable for the two strain rates. In fact, the rate of softening is even somewhat larger for the higher strain rate. This suggests that, in both cases, the time of exposure to the high temperature is important in promoting the coarsening process, but that, in addition, the amount of strain accumulated during that time also supports the coarsening process, leading to a slightly larger softening rate at the higher strain rate.

3.1.4 Cyclic Stress–Strain Behaviour

The cyclic stress–strain behaviour, was studied in a multiple step test ($5\cdot 10^{-3} < \Delta\varepsilon_t < 2.4\cdot 10^{-2}$) up to plastic strain amplitudes of about $\Delta\varepsilon_{pl}/2 = 4\cdot 10^{-3}$ at $\dot\varepsilon_t = 10^{-2}$ s^{-1} at a temperature of 950 °C. A set of typical "saturated" hysteresis loops is shown in fig. 6. The cyclic stress–strain behaviour exhibits a significant stress asymmetry, the compressive saturation stresses being significantly

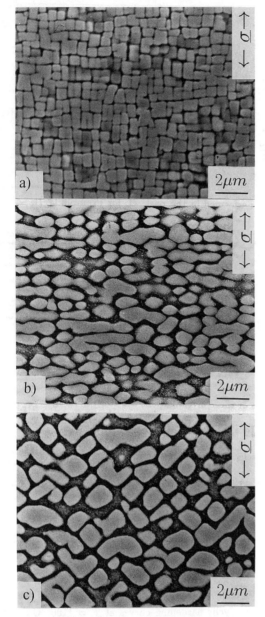

Figure 5: Coarsening of the γ' precipitates during isothermal fatigue at $T = 1100$ °C, $\Delta\varepsilon_t = 10^{-2}$. a) Initial γ' precipitates; b) After fatigue in air at $\dot{\varepsilon}_t = 5 \cdot 10^{-3}$ s^{-1}; c) After fatigue in air at $\dot{\varepsilon}_t = 6.67 \cdot 10^{-5}$ s^{-1}, γ etching, the γ phase appears dark.

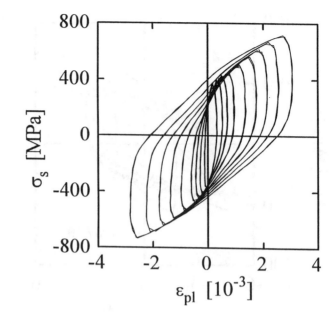

Figure 6: Saturated hysteresis loops measured in a multiple step test at $T = 950$ °C, $\dot{\varepsilon}_t = 10^{-3}$ s^{-1}. Plot of stress σ versus plastic strain ε_{pl}.

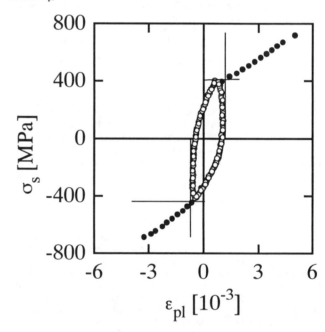

Figure 7: Cyclic stress–strain curves in tension and compression, obtained from hysteresis loops of multiple step test (fig. 6) at $T = 950$ °C. Plot of saturation stress σ_s versus plastic strain amplitude $\Delta\varepsilon_{pl}/2$. The mean value of σ_s is also plotted.

higher than the tensile saturation stresses, in particular in the range $2 \cdot 10^{-3} < \Delta\varepsilon_{pl}/2 < 3 \cdot 10^{-3}$ as shown in figs. 7 and 8. This stress asymmetry is in accord with that presented earlier in fig. 2b for a comparable temperature.

It should be noted that the hysteresis loops are slightly displaced to positive plastic strains, as shown in fig. 8 for the innermost loop. This is related to the fact that, in the first few cycles, more plastic strain is accumulated in tension than in compression, cf. fig. 9 and section 3.1.5. Since this plastic strain bias toward tension could distort the picture shown in fig. 8, the data were also evaluated, using the centre of the inner hysteresis loop as the origin. This procedure reduced the magnitude of the cyclic stress asymmetry somewhat but did not change the essence of the conclusions drawn earlier.

The shapes of the saturated hysteresis loops shown in fig. 6 differ in the tensile and compressive branches; the tensile branch exhibits a point of inflection in most cases. The same is true for the tensile and compressive cyclic stress–strain curves (fig. 7). In this context, it appears appropriate to point out that, in monotonic tensile testing in the same temperature range, the stress–strain curve consists of several clearly distinguishable stages and also exhibits a point of inflection. Of course, this analogy must be confined to a small range of strain, comparable to that in cyclic deformation. In the case of the monotonic deformation,

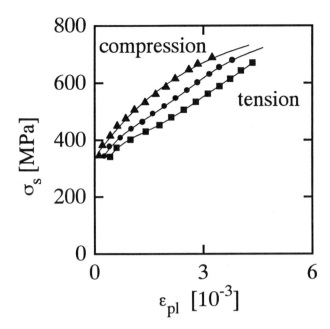

Figure 8: Comparison of cyclic stress–strain curves in tension and compression. Note cyclic stress asymmetry and point of inflection on tensile cyclic stress–strain curve.

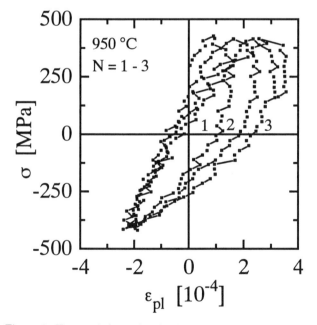

Figure 9: Hysteresis loops for the first 3 cycles at $T = 950$ °C, $\Delta\varepsilon_t = 10^{-2}$, $\dot{\varepsilon}_t = 5 \cdot 10^{-3}$ s^{-1}. Note accumulation of tensile plastic microstrain.

the different stages of deformation were related to a sequence of deformation processes in the γ-channels perpendicular and parallel to the stress axis, respectively, and to cutting of the γ' precipitates [17].

3.1.5 Cyclic Stress Asymmetry In this section we wish to discuss in a little more detail the effects of cyclic stress asymmetry and mean stress presented earlier (figs. 2, 4, 6, 7 and 8). Three findings are noteworthy with regard to total strain-controlled tests:

a) Below a temperature of about 1050 °C the cyclic stresses are larger in compression than in tension, leading to a negative mean stress.

b) Above that temperature, namely at 1100 °C in our work, the situation reverses: the tensile stresses now exceed the compressive stresses and a positive mean stress is found.

c) The tensile cyclic stress–strain curve below 1050 °C exhibits a point of inflection.

For the cyclic stress asymmetry below 1050 °C ($\sigma_m < 0$), the cyclic deformation experiments at constant total strain range $\Delta\varepsilon_t$ show that more plastic deformation occurs in the tensile half cycles of the first few cycles than in the compressive half cycles. Figure 9 shows an example for $\Delta\varepsilon_t = 10^{-2}$, $\dot{\varepsilon}_t = 5 \cdot 10^{-3}$ s^{-1}. Here the first three cycles are plotted in the form of stress σ against plastic strain ϵ_{pl}. While the plastic strain reached in compression is virtually constant, the tensile plastic strain increases with continued cycling. As a consequence the tensile elastic strain and hence the tensile stress decrease, thus leading to an increased compressive mean stress, in agreement with the results shown in fig. 2. At present, the explanation of the physical origin of the different cyclic stress asymmetries observed below and above about 1050 °C, is not straightforward. TEM studies indicate that dislocation activity in cyclic deformation is confined to the soft γ-channels. Hence, an explanation must be sought in the mechanisms of dislocation motion in the γ-channels whose deformation is, however, constrained by the two-phase γ/γ' morphology. For γ'-cuboids, it is expected that, under the combined action of the external stress and the coherency stresses due to a negative lattice mismatch δ, deformation in the tensile and compressive half-cycles occurs preferentially in the γ-channels that lie perpendicular and parallel to the stress axis, respectively, compare [18]. Such considerations could possibly explain the fact that a point of inflection is found in the tensile but not in the compressive cyclic stress-strain curve. For other γ/γ' morphologies, e.g. raft-like structures, geometric details such as the width of the γ-channels measured parallel to the glide planes would have to be taken into account. While corresponding detailed calculations are not available at present, it is felt that in order to explain the change of the cyclic stress asymmetry, some additional mechanism must be considered.

3.1.6 Effects of Prerafting on the Fatigue Behaviour Another point of interest concerns the effect of the γ/γ' rafts, which are introduced perpendicular to the stress axis during high-temperature creep, on the fatigue behaviour. Such rafts are also found in turbine blades that have been subjected to service conditions [19]. Lupinc et al. [20] have investigated the effect of γ/γ' rafts that were introduced by prior high-temperature creep on the high-temperature fatigue crack propagation. These authors showed that, for specimens with rafts lying perpendicular to the stress axis, the crack propagation was facilitated, compared to specimens containing cuboidal γ' precipitates. In order to test the effect of such γ/γ' rafts introduced by high-temperature creep, Ott and Mughrabi [21] carried out isothermal fatigue tests ($\Delta\varepsilon_t = 0.9 \cdot 10^{-2}, T = 950$ °C) in air on specimens having initially the cuboidal γ' particle morphology and on other specimens in which γ/γ'–raft structures which were either perpendicular or parallel to the stress axis had been introduced by a small (< 0.5 %) prior tensile or compressive creep strain, respectively. These γ/γ' microstructures are shown in fig. 10. Figure 11 shows some results of the fatigue tests. It was found that the fatigue life was enhanced, when the raft structure was parallel to the stress axis. In this case, a deflection of the crack into a direction along the raft structure was observed. On the other hand, for the raft structure perpendicular to the stress axis, fatigue life was reduced, and very smooth crack growth was facilitated perpendicular to the stress axis, as found also by Lupinc et al. [20]. In high vacuum, the fatigue lives were generally larger by a factor of about two. Otherwise, similar results were obtained.

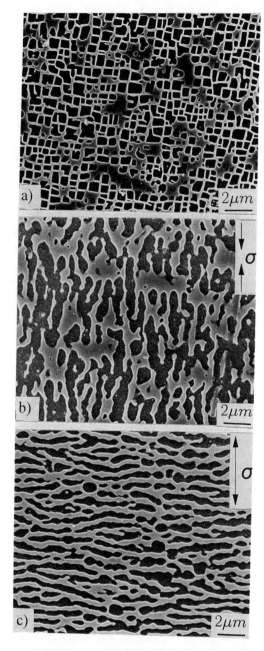

Figure 10: Different γ/γ' microstructures of fatigue samples. a) Initial cuboidal γ' precipitates; b) γ/γ' raft structure parallel to stress axis after small compressive creep strain at $T = 1050$ °C; c) γ/γ' raft structure after small tensile creep strain at $T = 1050$ °C, γ' etching, the γ' phase appears dark.

Figure 11 shows also that in the predeformed specimens mean stresses σ_m develop in the direction of predeformation. While these mean stresses can also affect fatigue life in the sense observed, it is felt on the basis of the studies of the modes of crack propagation that the microstructural effects described above are dominant.

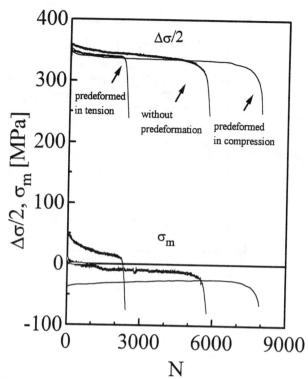

Figure 11: Cyclic deformation curves at $T = 950$ °C, $\Delta\varepsilon_t = 0.9 \cdot 10^{-2}$, $\dot{\varepsilon}_t = 0.9 \cdot 10^{-2}$ s^{-1} of specimens with the γ/γ' microstructures shown in fig. 10. Plots of mean stress amplitude $\Delta\sigma/2$ and mean stress σ_m versus the number of cycles N.

3.2 Anisothermal Thermomechanical Fatigue

3.2.1 Cyclic Deformation, Fatigue Lives Figure 12 summarizes the TMF deformation behaviour for an upper temperature of 1100 °C in plots of the tensile and compressive peak stresses (fig. 12a) and the mean stress amplitude $\Delta\sigma/2$ versus the number of cycles (fig. 12b).

Out–of–phase (OP) TMF leads to the shortest fatigue lives, in-phase (IP) TMF to the longest; diamond cycles (CD, CCD) yield intermediate fatigue lives. The explanation is based on the fact that fatigue failure was found to occur predominantly by mechanical shear, favoured by large tensile stresses. In OP (IP) tests, a significant tensile (compressive) mean stress develops, and the tensile stresses are hence largest (smallest), the tensile peak stress being reached at the lower (upper) temperature. In the following, emphasis will be laid on the results of the OP TMF cycle shape which is considered to be the most damaging cycle shape in the present study.

3.2.2 Directional Coarsening, Effects on Fatigue Life In OP TMF tests with an upper temperature of 900 °C (lower temperature 600 °C) only a very small plastic strain amplitude developed. Failure occurred predominantly by local mechanical shear. SEM showed that the cuboidal γ' particle shape had been completely preserved. In all other TMF tests with upper temperatures of 1100 °C, directional coarsening was observed, leading to more or less well developed γ/γ' raft structures perpendicular (IP, CCD) or parallel (OP, CD) to the stress axis [22]. Examples of the γ/γ' raft structure found for the IP and the OP TMF tests are shown in fig. 13. The orientation of the rafts perpendicular (parallel) to the stress axis in the IP (OP) tests is as expected in the sense that, at the high temperature at which coarsening occurs, the specimen is under a tensile (compressive) stress.

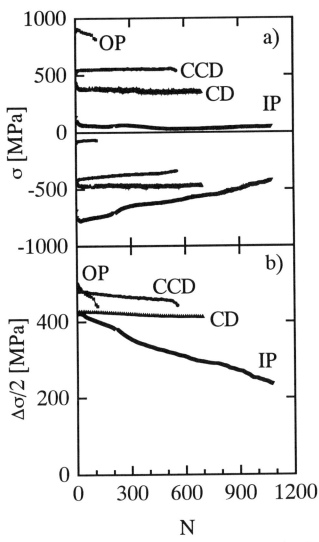

Figure 12: Cyclic deformation curves of TMF tests with different cycle shapes (IP, OP, CD, CCD) for upper and lower temperatures of 1100 °C and 600 °C, respectively, for $\Delta\varepsilon_t = 10^{-2}$, $\dot{\varepsilon}_t = 6{,}67 \cdot 10^{-5}$ s^{-1}.

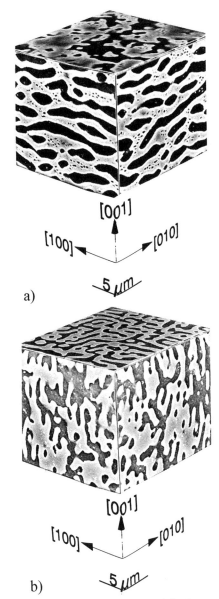

Figure 13: Directionally coarsened γ/γ' raft structures after IP and OP TMF, cf. fig. 12. a) IP; b) OP.

It would, of course, be interesting to know whether these raft structures affect fatigue life and, if so, in what manner. For that purpose, specifically designed TMF tests would have to be performed on specimens with different well defined initial γ/γ'-microstructures. The observation of a particular γ/γ' raft structure in a specimen that has exhibited an extended or reduced TMF life does not permit the conclusion that TMF life was enhanced or reduced by the formation of this paticular γ/γ' raft structure. Thus, Engler-Pinto Jr. et al. [23] found that, in SRR 99, specimens subjected to a so called thermal-fatigue-based (TFB) cycle, a well-developed raft structure was formed parallel to the stress axis and that the TFB fatigue life was larger than that observed after OP TMF or thermal fatigue. They concluded that a raft structure parallel to the stress axis seems to have a beneficial effect on the TMF life. On the other hand, the results of Kraft et al. obtained on the alloy CMSX-6 [21], as shown in figs. 12 and 13, show clearly that, in their work, fatigue life was shortest for the OP TMF test during which a raft structure developed parallel to the stress axis and longest for the IP TMF test which led to a raft structure perpendicular to the stress axis. In other words, these results simply show that the raft structures observed after anisothermal fatigue are a consequence of the type of anisothermal fatigue test performed but provide no unequivocal evidence on whether and how these raft structures affect anisothermal fatigue life. Thus, in the OP TMF tests discussed previously, any beneficial effect that the raft structure parallel to te stress axis may have had (compare, for example, the beneficial effect discussed earlier for isothermal fatigue), was obviously more than compensated by the damaging effect of the high tensile stress during the cold phase of the OP TMF cycle.

3.2.3 Microstructural Processes During a Single Cycle   Valuable insight into the processes during a single TMF cycle is gained by following the course of stress σ, plastic strain ε_{pl} and temperature T versus time t during one cycle, see fig. 14. In TMF OP tests with an upper temperature of 900 °C, the shape of the stress-strain hysteresis loop is complex, and the plastic strain is barely measureable, as shown in fig. 14a, for cycles number 1, 100 and 500. Nontheless, there is a cumulative accumulation of

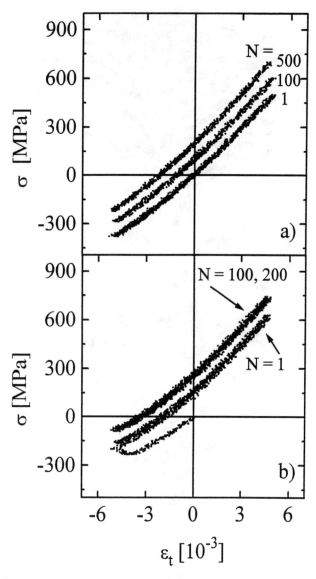

Figure 14: Single OP TMF cycles, plots of σ versus ε_t. a) Upper temperature $T = 900$ °C, cycle number 1, 100 and 500; b) Upper temperature $T = 1100$ °C, cycle number 1, 100 and 200.

very small cyclic microplastic strains in the sense that, within 100 cycles, a permanent negative strain of about $-1 \cdot 10^{-3}$ develops, accompanied by the build-up of a tensile mean stress of 150 MPa. No further changes are measurable up to 500 cycles. In TMF OP tests with an upper temperature of 1100 °C (fig. 14b), significant microplastic yielding in compression and the development of an appreciable tensile mean stress seem to occur in the hot phase during the first quarter cycle. After 100 cycles, the specimen has suffered a permanent negative plastic strain of about $-3 \cdot 10^{-3}$ and has acquired a tensile mean stress of about 300 MPa. No significant further changes occur within the next 100 cycles.

3.2.4 Variation of Cyclic Plastic Strain Rate During a Cycle

We now discuss the changes of the plastic strain rate $\dot{\varepsilon}_{pl}$ within an OP TMF cycle. The corresponding hysteresis loop is shown in fig. 15a. For a triangular linear $\varepsilon_t(t)$ course, the $\varepsilon_{pl}(t)$ course is strongly non-linear, compare fig. 15b. As a consequence, the instantaneous plastic strain rate $\dot{\varepsilon}_{pl}$ (slope of ε_{pl}) changes continuously, being close to zero most of the time and approaching, in

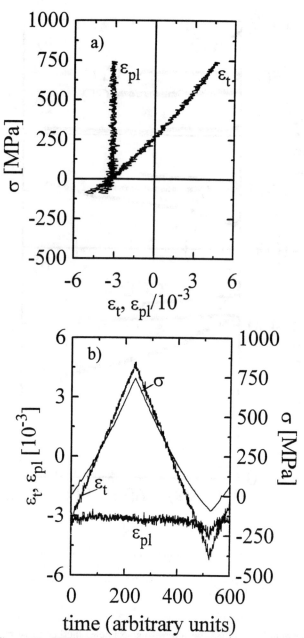

Figure 15: OP TMF test, as in fig. 12. a) Hysteresis loop, plot of stress σ versus total strain ε_t and versus plastic strain ε_{pl}; b) Plot of stress σ, total strain ε_t and plastic strain ε_{pl} versus time t within one cycle.

the case of an OP TMF test, a significant value only during microyielding in the approach to the maximum temperature in the compression phase [14]. This variation of instantaneous plastic strain rate within a cycle is more severe than during isothermal fatigue. A detailed discussion of these effects has been given recently [14]. Some consequences are as follows. In a typical TMF test on a bulk specimen, the cycle time is in the order of several 100 seconds. Thus, the mean cyclic strain rate is very low, beeing significantly less then 10^{-4} s^{-1} for the mean total strain rate $\dot{\varepsilon}_t$ and another one or two magnitudes lower for the mean plastic strain rate $\dot{\varepsilon}_{pl}$. As shown for the isothermal tests, compare figs. 4 and 5, the cyclic deformation behaviour for very low strain rates differs significantly from that at higher strain rates and leads to drastic coarsening and softening effects. Thus, a comparison of the cyclic deformation of TMF tests with data obtained in typical thermal fatigue tests on wedge-shaped specimens with much shorter cycle times is problematic. Further complications arise from the fact that the damage mechanisms and the process of crack propagation are quite different in the two cases, compare [24]. One step to improve the comparability would be to perform the TMF tests on hollow specimens in the hope to increase the strain rate by about one order of magnitude, compare [25]. Such work is in progress.

4. Conclusions

On the basis of the conducted experiments and, in particular, by a comparison of the results observed for isothermal and for anisothermal thermomechanical fatigue, several conclusions can be drawn. Here we note:

a) There are fundamental differences between isothermal and anisothermal fatigue behaviour. Hence, any similarities must be viewed critically.

b) Directional coarsening of the γ/γ' microstructure is qualitatively different in isothermal and anisothermal fatigue, and even for similar γ/γ' raft structures accompanying changes of fatigue lives can be quite different. The microstructural effects seem more important in isothermal than in anisothermal fatigue.

c) At high temperatures, drastic cyclic softening, caused by marked coarsening of the γ/γ' microstructure, occurs in particular at low strain rates.

d) In total-strain controlled tests, the instantaneous plastic strain rate varies in a complex manner during a single cycle. Because of the high strain-rate sensitivity at high temperatures, the consequences of such behaviour deserve more attention.

e) Mean stresses must be considered in both isothermal and anisothermal fatigue, being more important in the latter case.

f) In TMF tests, the temperature changes simultaneously with the mechanical strain. As a consequence, the effects of strain rate are more pronounced than in isothermal tests.

g) In TMF tests on bulk specimens, the cycle times are in the order of minutes, and the (plastic) strain rate is accordingly much lower than in thermal fatigue tests with typical cycle times in the order of seconds. In addition, the nature of fatigue damage is different in both cases. Thus, a direct comparison is problematic.

Acknowledgements

This work is supported by the Bundesminister für Bildung, Forschung und Technologie (BMBF) in a collaboration between several institutes and industrial companies under contract number BMFT-MatFo 03M3038F and by the Deutsche Forschungsgemeinschaft under contract number Mu502/8-3. Sincere thanks are expressed to all partners involved in this project, and in particular to Dr. Wortmann and Dipl.-Ing. Buchmann of MTU München, and Dr. Goldschmidt, formerly with MTU, München, now with Siemens AG/KWU, Mühlheim.

References

1. M. Gell, D.N. Guhl, D.K. Gupta, and K.D. Sheffler, "Advanced Superalloy Foils", Journal of Metals, 39 (1987) 11-15.

2. P. Caron and T. Khan, "Improvement of Creep Strength in a Nickel-Base Single Crystal Superalloy by Heat Treatment", Mater. Sci. Eng., 61 (1983) 173-184.

3. M.V. Nathal, R.A. MacKay, and R.V. Miner, "Influence of Precipitate Morphology on Intermediate Temperature Creep Properties of a Nickel-Base Superalloy Single Crystal", Metall. Trans., 20A (1989) 133-141.

4. T.M. Pollock and A.S. Argon, "Directional Coarsening in Nickel-Base Crystals with High Volume Fractions of Coherent Precipitates", Acta metall. mater., 42 (1994) 1859-1874.

5. M. Feller-Kniepmeier and T. Link, "Correlation of Microstructure and Creep Stages in the ⟨001⟩ Oriented Superalloy SRR 99 at 1253 K", Metall. Trans., 20A (1989) 1233-1238.

6. J. Hammer and H. Mughrabi, "High Temperature Creep and Microstructure of the Monocrystalline Nickel-Base Superalloy SRR 99", in Proc. of EUROMAT 1989, eds. H.E. Exner, and V. Schumacher, Vol. 1: Advanced Processing and High Temperature Materials, co-eds. D. Driver, and H. Mughrabi (Oberursel: DGM Informationsgesellschaft, 1990) 445-450.

7. W. Schneider, J. Hammer, and H. Mughrabi, "Creep Deformation and Rupture Behaviour of the Monocrystalline Superalloy CMSX-4 — A Comparison with the Alloy SRR 99", in "Superalloys 1992", eds. S.D. Antolovich et al. (Warrendale, PA: The Minerals, Metals and Materials Society, 1992) 589-598.

8. P.D. Portella, C. Kirimtay, and K. Naseband, "Kriech- und LCF-Verhalten der einkristallinen Superlegierung SRR 99" in Proc. of 13. Vortragsveranstaltung der AG Hochtemperaturwerkstoffe (Düsseldorf: VDEh, 1990) 145-152.

9. P.D. Portella, A. Bertram, E. Fahlbusch, H. Frenz, and J. Kinder, "Cyclic Deformation of the Single Crystal Superalloy SRR 99 at 980 °C", in Proc. of FATIGUE '96, May 6-10, 1996, Berlin (Oxford: Elsevier Science Ltd., 1996). In press.

10. T.P. Gabb and G. Welsch, "The High Temperature Deformation in Cyclic Loading of a Single Crystal Nickel-Base Superalloy", Acta metall., 37 (1989) 2507-1516.

11. E. Fleury and L. Rémy, "Thermal-Mechanical Fatigue Behaviour of AM1 Superalloy Single Crystals" in "High Temperature Materials for Power Engineering", eds. E. Bachelet et al. (Dordrecht, Netherlands: Kluwer Academic Press, 1990) 1007-1016.

12. D.A. Boisimer and H. Sehitoglu, "Thermo-Mechanical Fatigue of MAR-M247", Parts I and II, Trans. ASME, 112 (1990) 68-79, 80-89.

13. J.Y. Guédou and Y. Honnorat, "Thermomechanical Fatigue of Turbo-Engine Blade Superalloys", in "Thermomechanical Fatigue Behavior of Materials", ed. H. Sehitoglu (Philadelphia, PA: American Society for Testing and Materials, ASTM STP 1186, 1993) 157-171.

14. H.J. Christ, H. Mughrabi, S. Kraft, F. Petry, R. Zauter, and K. Eckert, "The Use of Plastic Strain Control in Thermomechanical Fatigue Testing", in Proc. of Int. Symp. on "Fatigue under Thermal and Mechanical Loading — Mechanisms, Mechanics and Modelling", May 22-24, 1995, Petten, Netherlands, (Dordrecht, Netherlands: Kluwer Academic Press, 1996). In press.

15. S. Kraft, R. Zauter, and H. Mughrabi, "Aspects of High–Temperature Low–Cycle Thermomechanical Fatigue of a Single Crystal Nickel–Base Superalloy", Fatigue Fract. Engng. Mater. Struct., 16 (1993) 237–253.

16. S.A. Kraft and H. Mughrabi, "Thermomechanical Fatigue of the Monocrystalline Nickel–Base Superalloy CMSX–6", in "Thermo–Mechanical Fatigue Behavior of Materials", 2nd Volume, eds. M.J. Verrilli, and M.G. Castelli, (Philadelphia, PA: American Society for Testing and Materials, ASTM STP 1263, 1996). In press.

17. S. Kraft, "Verformungverhalten und Mikrostruktur der einkristallinen Nickelbasis–Superlegierung CMSX–6 nach isothermer und thermomechanischer Beanspruchung", Doctorate Thesis, Universität Erlangen–Nürnberg, 1996.

18. H. Mughrabi, H. Feng, H. Biermann, "On the Micromechanics of the Deformation of Monocrystalline Nickel–Base Superalloys", in Proc. of the IUTAM Symposium on "Micromechanics of Plasticity and Damage of Multiphase Materials", Aug. 29–Sept. 1 1995, Sévres, eds. A. Pineau, and A. Zaoui (Dordrecht, Netherlands: Kluwer Academic Press, 1996). In press.

19. S. Draper, D. Hull, and R. Dreshfield, "Observations of Directional Gamma Prime Coarsening During Engine Operation", Metall. Trans., 20A (1989) 683–688.

20. V. Lupinc, G. Onofrio, and G. Vimercati, "The Effect of Creep, Oxidation and Crystal Orientation in High Temperature Fatigue Crack Propagation in Standard and Raft–Like Gamma Prime CMSX–2", in "Superalloys 1992", eds. S.D. Antolovich et al. (Warrendale, PA: The Minerals, Metals and Materials Society, 1992) 717–726.

21. M. Ott and H. Mughrabi, "Dependence of the Isothermal Fatigue Behaviour of a Monocrystalline Nickel–Base Superalloy on the γ/γ' Morphology", in Proc. of FATIGUE '96, May 6–10, Berlin, eds. G. Lütjering, and H. Nowack, (Oxford: Elsevier Science Ltd., 1996). In press.

22. S. Kraft, I. Altenberger, and H. Mughrabi, "Directional γ/γ' Coarsening in a Monocrystalline Nickel–Base Superalloy During Low–Cycle Thermomechanical Fatigue", Scripta metall. et mater., 32 (1995) 411–416.

23. C.C. Engler–Pinto Jr., F. Meyer–Olbersleben, and F. Rézaï-Aria, "Thermomechanical Fatigue Behaviour of SRR 99", in Proc. of Int. Symp. "Fatigue under Thermal and Mechanical Loading — Mechanisms, Mechanics and Modelling", May 22–24, 1995, Petten, Netherlands (Dordrecht, Netherlands: Kluwer Academic Press, 1996). In press.

24. F. Meyer–Olbersleben, C.C. Engler–Pinto Jr., and F. Rézaï-Aria, "On Thermal Fatigue of Nickel–Based Superalloys", in "Thermo–Mechanical Fatigue Behavior of Materials ", 2nd Volume, eds. M.J. Verrilli, and M.G. Castelli, (Philadelphia, PA: American Society for Testing and Materials, ASTM STP 1263, 1996). In press.

25. R. Zauter, H.J. Christ, and H. Mughrabi, "Some Aspects of Thermomechanical Fatigue of AISI 304L Stainless Steel", Parts I and II, Metall. Trans., 25 (1994) 401–406, 407–413.

Prediction of Oxidation Assisted Crack Growth Behavior within Hot Section Gas Turbine Components

Graham Webb, Tom Strangman, Norm Frani, Chet Daté, Lloyd Wilson and Rajiv Rana
AlliedSignal Engines
111 S. 34th Street
Phoenix, Arizona 85072-2181 USA

Dennis Fox
NASA-Lewis Research Center
21000 Brookpark Road
Cleveland, Ohio 44135 USA

ABSTRACT

This paper presents a physically-based deterministic methodology for prediction of the crack growth resistance of superalloys within gas turbine engine environments. The model combines experimentally determined temperature and pressure dependent superalloy oxidation rates with crack growth rates obtained from thermal-mechanical fatigue (TMF) specimen testing to develop a crack growth rate law for oxidation assisted crack growth behavior. This information is then used in conjunction with standard linear elastic fracture mechanics (LEFM) methods to predict propagation lives. The propagation analysis procedure is combined with previously presented procedures for the prediction of coating cracking during TMF loading (Strangman [1]). The combination results in an initiation plus propagation lifing strategy for superalloy gas turbine components subjected to thermal-mechanical fatigue. This analytical procedure has been found to predict the measured crack sizes obtained from superalloy components after engine testing with acceptable accuracy.

INTRODUCTION

Thermal Mechanical Fatigue (TMF) cracking of hot section gas turbine components remains a significant barrier to the establishment of long term gas turbine engine component durability. Cracking is most frequently observed on high cost superalloy precision cast airfoil components (i.e. blades and vanes), and usually results from initial cracking of the environmentally resistant airfoil coatings [2-5]. Once initiated, these cracks can propagate into the underlying superalloy substrate as a result of environmentally assisted fatigue crack propagation. The current work presents experimental and analytical methods for prediction of such behavior under the influence of an open loop out-of-phase TMF load cycle wherein a compressive strain hold is imposed (e.g. Figure 1), such cycles being representative of the time dependent response of turbine airfoil surfaces. Such waveforms are observed to result in cracking and subsequent retirement of gas turbine airfoil components. In previous work, a procedure was developed and presented for the prediction of coating cracking during out-of-phase TMF cycling [1]. In this investigation, the ductility, creep resistance, and thermal expansion mismatch between coating and substrate were found to determine the coating's resistance to cracking during out-of-phase TMF cycling. These variables were then used to estimate coating stresses resulting from the imposed strain cycle. Due to its extremely low creep resistance, the environmentally protective intermetallic coating experiences significant stress relaxation in the TMF critical airfoil locations which experience compressive thermal strains while operating. This phenomena shifts the cyclic mean stress, resulting in the development of tensile stresses within the coating (Figure 2). These stresses, if of sufficient magnitude, can subsequently overload (or fatigue) the coating material resulting in the formation of coating cracks. Once present, coating cracks are observed to propagate into the underlying substrate during further engine cycling as a result of environmentally assisted crack growth (Figure 3).

Metallurgical examination of such cracks (Figure 4) reveals severe oxidation of the crack faces due to the exposure of bare substrate material to the high temperature/pressure turbine environment. More importantly, oxidation is also observed to affect the material directly ahead of the growing crack tip (hereon termed "process zone", Figure 5). This phenomena will obviously modify the superalloy's intrinsic resistance to crack growth, and thus TMF crack propagation is observed to involve environmental, as well as mechanical crack growth mechanisms. As will be demonstrated, this environmental damage mode can dominate the crack growth process at the high temperatures and pressures typical of the gas turbine engine environment. Physical descriptions which ignore crack tip oxidation are shown to result

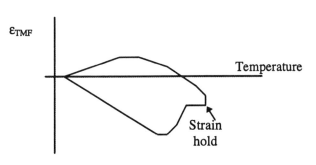

Figure 1. Schematic representation of an open loop out-of-phase thermal-mechanical fatigue strain cycle with a compressive strain hold.

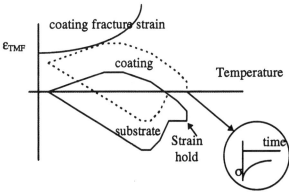

Figure 2. Schematic representation for the development of tensile coating cracks during TMF cycling of the underlying substrate material.

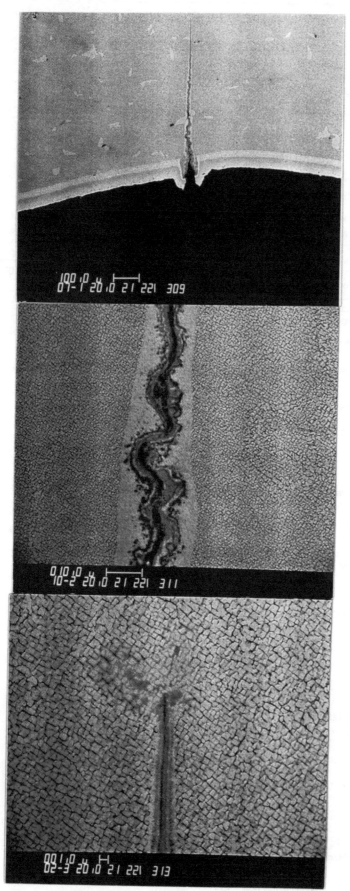

Figure 3. Coating initiated cracking of a CMSX-3 gas turbine airfoil component after experiencing open loop out-of-phase TMF cycling. Coating cracks are observed to propagate into the superalloy substrate as a result of environmentally assisted fatigue crack growth.

Figure 4. Oxidation of CMSX-3 crack faces during TMF cycling. Note depletion of gamma prime phase adjacent to crack surfaces.

Figure 5. Oxidation affected crack tip process zone observed within CMSX-3 gas turbine airfoil component after experiencing turbine thermal cycling.

Table I. Compositions ranges (in wt%) of Ni-base superalloys tested

Alloy	Ni	Al	Ti	Cr	Co	Ta	W	Mo	Hf	Cb	Re	B	C	Zr
IN738LC	bal.	3.2-3.7	3.2-3.7	15.7-16.3	8.0-9.0	1.5-2.0	2.4-2.8	1.5-2.0	-	0.6-1.1	-	0.007-0.012	0.09-0.13	0.03-0.08
MarM 247	bal.	5.3-5.7	0.9-1.2	8.0-8.3	9.0-11.0	2.8-3.3	9.5-10.5	0.5-0.8	1.20-1.60	-	-	0.01-0.02	0.13-0.17	0.03-0.08
CMSX3	bal.	5.5-5.8	0.8-1.2	7.2-8.2	4.3-4.9	5.8-6.2	7.6-8.4	0.3-0.7	0.07-0.15	-	-	0.003 max	0.02 max	0.008 max
SC180	bal.	5.0-5.4	0.9-1.1	5.0-6.0	9.7-10.3	8.0-9.0	4.8-5.2	1.4-2.0	0.08-0.12	-	2.8-3.2	0.002 max	0.02 max	0.008 max

in anti-conservative estimates of superalloy crack growth resistance, particularly for short cracks (high aspect ratio shallow cracks).

Although widely recognized [6], very little experimental data has been generated within the open literature to describe this phenomena, or its effect on TMF crack growth during cycling of gas turbine engine components. Much of this may be attributed to the experimental difficulty of conducting material crack growth measurements under high temperature and high pressure environments, in addition to the waveform specific nature of this phenomena. Recognizing this current deficiency, the current experimental work was conducted with the goal of providing a methodology to capture the primary factors influencing superalloy degradation during gas turbine airfoil TMF cycles using basic laboratory measurements. Towards this end experiments have been conducted to estimate both environmental and mechanical TMF crack growth components. Environmental crack growth is predicted upon the basis of experiments which evaluate superalloy oxidation rates as a function of temperature and oxygen activity (pressure). This information is then used to create a model for defining the environmental crack growth law.

Once defined, the environmental crack growth is analytically combined with measured values of the substrate materials crack growth resistance during out-of-phase TMF crack growth rate experiments at various temperatures and stress R-ratios. Such experiments, conducted using simple triangle out-of-phase waveforms in which minimum and maximum loads are cycled out-of-phase of temperature, estimate the superalloy's resistance to mechanical crack growth. Upon definition of the operating environment (temperatures, pressures, and time sequencing) it is possible to estimate a waveform dependent crack growth rate law specifically applicable to the case under analysis. This crack growth resistance is subsequently combined with a numerically simulated or assumed crack scenario to calculate the component TMF crack propagation life for an envisioned (or known) failure scenario. When combined with the predicted coating crack initiation life, component TMF life predictions are created.

The following description shall define the experimental and analytical procedures used to create the above component TMF lifing methodology applicable for superalloys used to manufacture airfoils within the hot section of gas turbine engines.

EXPERIMENTAL PROCEDURES AND RESULTS

Determination of Environmental TMF Crack Growth Rates

As described in the previous section, model environmental crack growth laws were created using measured rates of superalloy oxidation under both transient (i.e. linear) and steady state conditions as a function of temperature and oxygen pressure. When initially exposed to a high temperature oxygen containing environment, all of the metallic elements in the alloy can be converted to their oxides. Initially NiO and low density spinels are the fastest forming oxides and scale growth is linear with respect to time. As the transient oxide scale thickens, the oxygen activity at the metal interface decreases, permitting chromia and other more thermodynamically stable oxides (e.g. Al_2O_3 and HfO_2) to accumulate. As these dense adherent oxides grow under and through the initial oxide layer, the oxidation kinetics become limited by the rate of diffusion of oxygen and aluminum and/or chromium and parabolic growth kinetics become established in which the oxide scale grows proportional to the square root of time.

Thus, when un-cracked, the initial transient oxide scale results in the formation of a more protective oxide scale and the rate of oxidation decreases. However, the oxide scale does not remain un-cracked in the process zone ahead of a growing TMF crack. The low toughness transient oxide scale is assumed to rupture during every engine cycle exposing fresh metal which re-establishes the transient (i.e. linear) rate of oxidation at the crack tip. For this reason it is reasonable to assume that the rate of environmental crack growth is proportional to the rate of oxidation of the superalloy material for the turbine operating environment (i.e. temperatures/pressures). Due to this, experiments were conducted to measure the rate of oxide scale growth for a variety of superalloy materials (Table I) during both transient and parabolic oxidation. Oxidation kinetics were estimated from weight change experiments for specimens of known surface area. Although only oxidation data for CMSX-3 will be utilized further within the current paper, the information from all alloys tested is also provided to permit dissemination of this information within the public domain.

All samples were ground to 600 grit using SiC abrasive paper. Specimens were cleaned with detergent, distilled water, acetone, and ethyl alcohol. Oxidation experiments were conducted in a vertical tube furnace at 800°, 900°, 1000°, 1100°, and 1200°C. Quartz furnace tubes with an internal diameter of 2.2 cm were used. Samples were suspended from a sapphire hook. The oxidation kinetics were measured using thermogravimetric analysis (TGA). Sample weight change was continuously measured throughout the experiment using a recording microbalance. Exposure time for each sample was six hours. The furnace environment was either ambient air or high purity oxygen maintained at a gas flow rate of 100 cm³/min, corresponding to a velocity of 0.44 cm/sec. Samples were suspended in the tube furnace in the air, or flowing gas environment. Temperature was instantaneously applied to the specimens by raising the heated furnace around the sample. The experimental data obtained from these experiments reveals that the rate of superalloy oxidation is linear with respect to time for several minutes, and then slows to the more protective parabolic oxidation rate. Although limited, this data suggests that the transition time from linear to parabolic rate kinetics exhibits a baseline value of 0.17 hours in stagnant air. Transient (i.e. linear) oxidation rates are further found to be strongly dependent upon the activity of gaseous oxygen present within the tube furnace (Figure 6). In contrast, parabolic oxidation rates are observed to be independent of oxygen pressure. The temperature dependence of the linear and parabolic oxidation rates for the four superalloys are presented within Figure 7. Fitting of an Arrhenius-type kinetic equation to the experimental data results in the following equations for the linear and parabolic oxidation rate constants:

$$\Re_{linear} = \sqrt{P}\, \beta_{linear} \exp(-Q_{linear}/kT) \quad (1)$$

and

$$\Re_{parabolic} = \beta_{parabolic} \exp(-Q_{parabolic}/kT) \quad (2)$$

where: $\Re_{linear}, \Re_{parabolic}$ are the rate constants in mg/cm²/hr and mg/cm²/hr$^{1/2}$

β's are substrate dependent linear (transient) and parabolic (steady state) intercepts
Q's are substrate dependent linear and parabolic activation energies
k is Boltzmann's constant
T is the crack tip metal temperature (°K)
P is the turbine operating pressure (atm)

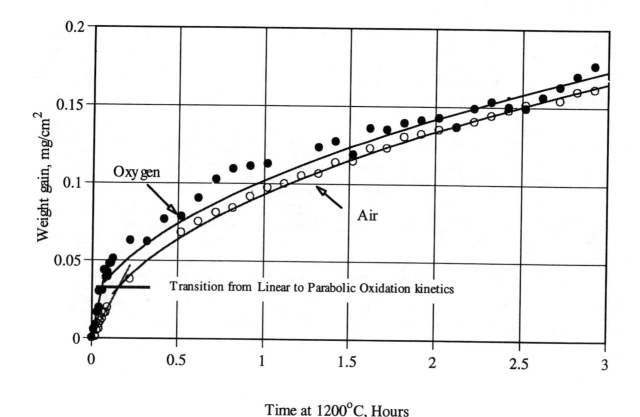

Figure 6. Effect of oxygen activity on transient and steady state oxidation of CMSX-3 at 1200°.

These equations describe the kinetics for weight gain in mg/cm² of exposed surface area, which may be analytically transformed into an equivalent environmentally degraded crack tip process zone. This is accomplished by determining the volume change associated with a weight gain of 1 mg/cm². For the CMSX-3 alloy, the conversion of the individual elements to their oxides results in a calculated scale thickness of 3.56 μm. Using this calculated value, the environmental crack propagation is obtained by assuming that the oxide layer forming at the crack tip process zone is completely cracked upon completion of the TMF cycle. These methods form the basis for the environmental crack growth rate law used within the analytical formulation.

Determination of TMF Mechanical Crack Growth Rates

Crack growth rate measurements of the single crystal superalloy CMSX-3 were conducted for an experimental 538°-982° C triangle out-of-phase waveform cycle. CMSX-3 test material was cast, HIPped, and heat treated in the form of 12.2 mm diameter bars nominally 152.4 mm in length in which the primary [001] crystallographic axis is within 10° of the bar longitudinal axis as verified by Laue back reflection method. Secondary crystallographic orientations were also determined. After casting the bars were HIPped at 1300°C at 100 MPa, cooled, then solutioned at 1200°C for 4 hours followed by forced cooling (>55° C/min to below 982°C). Following solution heat treatment, the bars were given a "psuedo-coat" heat treatment of 1050°C for 4 hrs followed by a low temperature age of 800°C for 16 hours. Single edge notch crack growth rate specimens were machined from the heat treated bars. Both pin loaded (rotation permitted) and threaded specimens were created. Specimen notches were oriented to produce cracking along <010> crystallographic direction. Tests were conducted in load control mode using a servohydraulic test frame with the temperature controller of the induction furnace built into the feedback loop. Crack length measurements were obtained using potential drop methods at the maximum stress-minimum temperature waveform endpoint for each cycle. R-ratios ($\Delta K_{min} / \Delta K_{max}$) of 0.05 and -1 were applied. Crack growth rate measurements were obtained using increasing, decreasing, and constant K testing using computer algorithms for controlling the servohydraulic feedback loop as the crack length changes.

Fatigue cracks are observed to propagate normal to the direction of applied stress, however overload fractures are highly crystallographic. Figure 8 presents the experimentally derived models for the average crack growth rate of CMSX-3 for triangle out-of-phase wave form for the two R-ratios experimentally evaluated. The R-ratio (or mean stress) is observed to have a strong influence on the slope of the crack growth rate curve for this alloy. Such effects are related to the differences in the effective ΔK actually applied to the crack tip as a result of anticipated crack closure mechanisms [7] resulting from the differences in mean stress (plastic wake effect) and the influence of crack surface oxidation (figure 4).

MODELING AND PREDICTIONS

Using the experimentally collected data a waveform specific crack growth law for the superalloy can now be constructed. This is accomplished using performance information for the turbine operating conditions anticipated to occur during the mission profile (gas temperatures and pressures). Component thermal-mechanical fatigue driving forces are estimated using 3-D transient heat transfer and stress analysis of the component (Figure 9). Initial estimates are provided assuming thermo-elastic constitutive behavior which can be further refined using time dependent viscoplastic constitutive equations for localized regions of interest [8]. Using such analytical tools it is possible

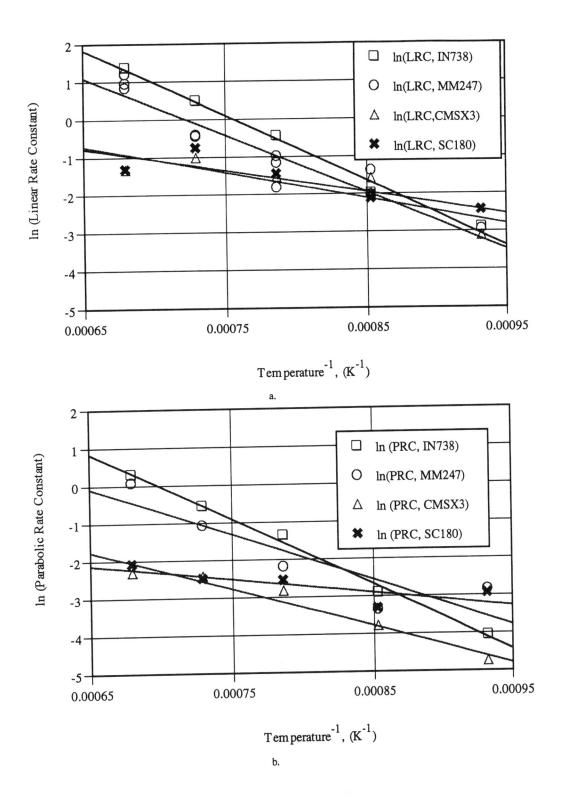

Figure 7. Graphical Arrenhius-type description of the experimental data used for determination of a). Linear (Transient) and b). Parabolic (Steady State) rate constants for the superalloys listed within Table I when tested in stagnant air.

to define the maximum and minimum time/path dependent strain excursions experienced on the surfaces of the component during the anticipated gas turbine cycle. Using this information, the crack propagation life of the component is estimated by growing the crack analytically into the through-thickness component stress field along the anticipated crack path. For the specific case examined (gas generator turbine vane), the temperatures where the maximum and minimum stresses occur correspond closely to the experimentally measured TMF waveform (i.e. 538°–982°C). For this reason it was assumed that the mechanical crack growth resistance of the CMSX-3 superalloy

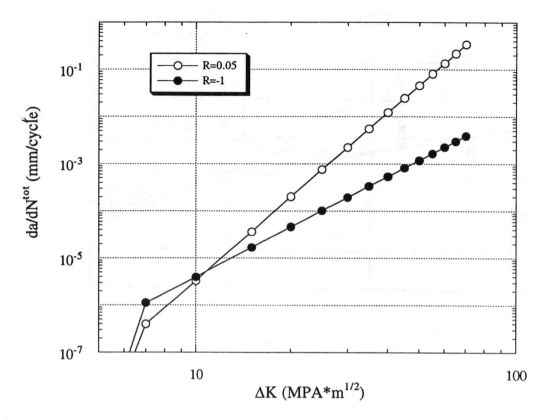

Figure 8. Experimentally derived models for the average TMF crack growth rate of CMSX-3 superalloy for triangle out-of-phase waveform cycling between temperatures of 538°C-982° C.

component during the vane TMF cycle can be represented by the experimentally measured TMF crack growth behavior (Figure 8). The effect of R-ratio changes during crack growth into the component are evaluated by interpolation between the two different crack growth curves at R=0.05 and R=-1. Environmental crack growth rates for the mission profile are determined based upon knowledge of the gas temperatures and pressures for the various mission cycle points. From this information it is possible to calculate the environmental crack growth increment

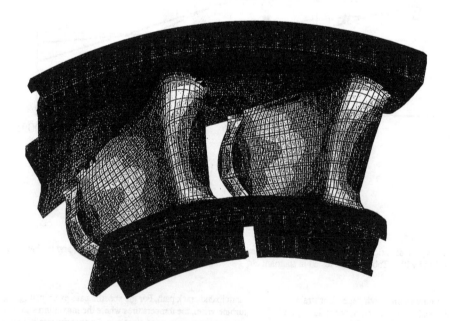

Figure 9. Three dimensional model of vane airfoil component employed to evaluate thermal and mechanical driving forces for TMF crack growth resistance of superalloy for a specific mission profile.

Table II. Comparison of mechanical and environmental crack growth rate of CMSX-3 for the laboratory and gas turbine engine environment for a short crack (0.07 mm) and a long crack (7.62 mm). All crack growth rates in mm/cycle.

Environment and waveform	Mechanical		Environmental
	0.07 mm crack	7.62 mm crack	both crack sizes
Laboratory Stagnant Air, 2.5 min dwell at 1 atm 538C-982C OP TMF	1.8796E-07	4.1910E-04	3.2766E-10
Engine, 2.5 min dwell at 10 atm 538C-982C OP TMF	9.40E-08	1.2903E-04	1.5037E-06

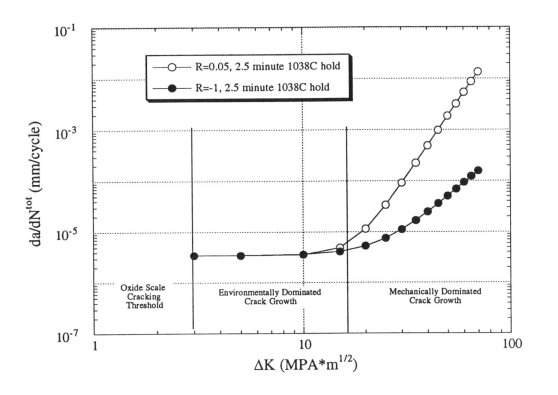

Figure 10. Predicted crack growth resistance for CMSX-3 superalloy vane for 538°C-982°C out-of-phase TMF engine cycle with a 2.5 minute dwell at maximum power.

per engine cycle. This calculated crack growth increment is linearly added to the mechanical crack growth rate to predict the superalloy crack growth resistance for the specific component operating conditions (Figure 10):

$$da/dN_{Total} = da/dN_{env} + da/dN_{mech} \quad (3)$$

Use of the above described methodology to predict the damage observed within vane component after various engine tests is found to successfully predict the measured crack depths within a factor of two. As such the model is deemed to be acceptable for engineering purposes.

An example of the predictive capabilities of this TMF component lifing methodology will demonstrate the importance of considering the environmental contribution to crack growth. Table II indicates the predicted mechanical and environmental crack growth rates for two different crack depths in both ambient air and high pressure turbine environments. From this comparison it can be observed that environmental crack growth only becomes significant only for small crack depths in the turbine environment. As the crack grows further (for a uniform stress field), the contribution of the environment to the total crack growth becomes insignificant. Further examination of Table II reveals that the environmental crack growth rate predicted to occur during ambient pressure high temperature crack growth rate testing is insignificant as compared to that predicted to occur within the high temperature/high pressure gas turbine engine environment. For these reasons, use of superalloy crack growth rate data obtained in ambient air without correcting for environmental influences can lead to significant over-estimates of component propagation life (Figure 11). These comparisons indicate that the use of superalloy TMF crack growth resistance data obtained entirely from laboratory experiments conducted in ambient pressure may grossly underestimate the true crack growth behavior within the gas turbine environment.

SUMMARY AND CONCLUSIONS

This paper has presented experimental and analytical methods for the estimation of superalloy crack growth resistance for out-of-phase TMF cycling experienced by gas turbine airfoils during engine cycling. The model uses experimentally derived estimates of crack tip transient and steady state oxidation to predict the amount of superalloy transferred into oxide ahead of a growing TMF fatigue crack. When combined with experimentally measured crack growth rates obtained from laboratory cycling, an analytical procedure was established for estimation of the crack growth resistance for the specific operating conditions under examination. This procedure is combined with previously presented methods for the prediction of coating induced crack initiation to estimate component TMF life for a specific mission profile. Use of these procedures for the prediction of TMF induced damage within superalloy gas turbine components of acceptable accuracy (factor of 2).

REFERENCES

[1] T.E. Strangman, "Thermal-Mechanical Fatigue Life Model for Coated Superalloy Turbine Component", Superalloys 1992, ed by S.D. Antolovich, R.W. Stusrud, R.A. MacKay, D.L. Anton, T. Khan, R.D. Kissinger, and D.L. Klarstrom, (The Minerals, Metals & Materials Society, Warrendale, PA, 1992), 795-804.

[2] H.L. Bernstein and J.M. Allen, "Analysis of Cracked Turbine Blades", Journal of Engineering for Gas Turbines and Power, 114 (1992), 293-301.

[3] M.I. Wood and G.F. Harrison, "Modelling the Deformation of Coated Superalloys under Thermal Shock", (Paper presented at the ASM Conference on Life Assessment and Repair Technology for Combustion Hot Section Components, Phoenix AZ, 1990).

[4] J. E. Heine, J.R. Warren, B.A. Cowles, "Thermal Mechanical Fatigue of Coated Blade Material" Final Report WRDC-TR-89-4027 under Contract F33615-89-C-5027, 1989.

[5] I. Linask and J. Dierberger, "A Fracture Mechanics Approach to Turbine Airfoil Design", ASME Paper No. 75-GT-79, 1975.

[6] S. Floreen and R. Raj, " Environmental Effects in Nickel-Base Alloys" Flow and Fracture at Elevated Temperatures, ed. by R. Raj, (The American Society for Metals, Metals Park Ohio, 1985), 383-405.

[7] H.L. Ewalds and R.J.H. Wanhill, Fracture Mechanics, (Edward Arnold Ltd, London, U.K. 1984), pp. 174-177.

[8] R.D. Krieg, J.C. Swearengen, W.B. Jones, "A Physically Based Internal Variable Model for Rate Dependent Plasticity", Unified Constituitive Equations for Creep and Plasticity ed by A. K. Miller, (Elsevier Applied Science, 1987).

ACKNOWLEDGEMENTS

Partial funding of this development was provided by the U.S. Army T800 engine program. The efforts of Mr. Chris Desidier, Mr. Harry Eckstrom and Mr. Dennis Chamblee of the AE materials test laboratory are gratefully acknowledged. All casting used thoughout this study were procured from the Howmet Whitehall Casting Division, Whitehall, MI.

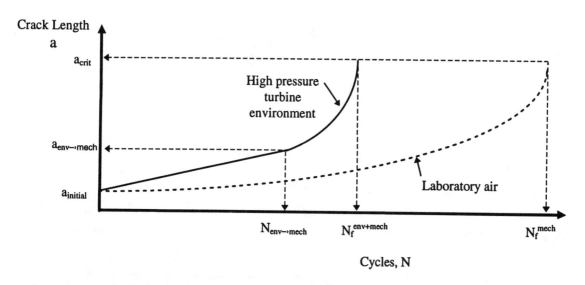

Figure 11. Comparison of crack growth rate behavior in laboratory air and the gas turbine environment.

CREEP STRENGTH AND FRACTURE RESISTANCE OF DIRECTIONALLY SOLIDIFIED GTD111

David A. Woodford

Materials Performance analysis, Inc.
1737 Union Street
Suite 543
Schenectady, NY 12309

Abstract

This work describes the application of a new approach (Design for Performance) to high temperature alloy development, design analysis, and remaining life assessment, based on short time high precision testing. The material tested was a directionally solidified nickel based alloy, GTD111, tested in the longitudinal, transverse and diagonal orientations relative to the growth direction. It was found that the creep strength comparison at 900C was dependent on stress and loading procedure, and was not necessarily enhanced by the preferred alignment of grain boundaries and crystal orientation. By contrast, the fracture resistance at 800C was improved in the longitudinal direction compared with transverse and diagonal orientations in terms of susceptibility to gas phase embrittlement (GPE) by oxygen. The new conceptual framework allows account to be taken of GPE, and other embrittling phenomena, which may develop in service, leading to rational life management decisions for gas turbine users. Additionally, straightforward design analysis procedures may be developed from the test data, which allow separate measurements of creep strength and fracture resistance to be used for performance evaluation.

Introduction

High temperature design of components in energy conversion systems is dependent to a large extent on data generated from creep to rupture testing. The same test is used as a basis for creep strength evaluation (i.e. the resistance to time-dependent deformation) and component life assessment (i.e the resistance to fracture). This leads to a number of conceptual problems. For example, the minimum creep rate and the time to rupture usually show a reciprocal dependence suggesting that they are separate measures of the same property. However, the latter is used as a measure of specimen, and hence component, life. Thus, fracture and life prediction are measured in terms of lapsed time from some arbitrary origin. This leads to fundamental ambiguity when assessing the remaining life after some prior creep exposure (1). And, in the limit, a specimen taken from a failed component invariably will have a finite creep rupture life.

It has been argued that long time creep to rupture tests are necessary to simulate the evolution of microstructure and damage during service. However, to simulate such evolution it must be recognized that there is a hierarchy of increasing test complexity and expense. For most applications there are effects of periodic unloading and temperature changes, multiaxial stresses, superimposed cyclic stresses, environmental attack, and synergism among them all. Clearly, the duration of a creep test is itself of little consequence relative to long term performance in most applications. Taking just one example, a typical jet engine operates with all these factors in complex thermal mechanical cycles of between one and ten hours; a 10,000 hour creep test at a fixed stress and temperature may not be a useful investment in time and money for this application.

Thus, there are flaws in the testing details, in the interpretation of results, and in the conceptual justification of long time testing. What is needed is a testing methodology that not only provides a procedural acceleration, is cheaper and more efficient, but also one that circumvents these flaws. Such a methodology has been termed ***Design for Performance*** (2). This approach decouples the creep strength and fracture resistance. The former is evaluated based on a high precision short time stress relaxation test (SRT), and the latter is determined from a constant displacement rate tensile test (CDR) run at a temperature where the material in service is most vulnerable to fracture.

Rather than attempt to incorporate microstructural evolution and damage development in the test methodology, the new approach measures the consequences of such changes in short time tests. The material in a component is treated as if it has a definable creep strength and fracture resistance at any stage in its service. The creep strength is defined in terms of a stress vs. strain rate relationship, although the analysis may be presented in terms of a predicted stress vs. time curve, as shown in the present paper, for comparison with traditional representations. The fracture resistance is defined in terms of tensile displacement at failure in a smooth or notched specimen at a temperature where the material is most sensitive to embrittlement due to environmental attack or microstructural changes. In practice, this temperature usually corresponds to a ductility minimum, and often to a temperature where maximum strains are developed in components during thermal fatigue(3).

The separation of creep strength and fracture resistance is particularly important in directionally solidified alloys because creep strength (expressed in terms of creep rates or time to specific strains) may be much less sensitive to orientation than fracture resistance (expressed as tensile ductility or crack propagation resistance) (4). The present study reports results on the GE alloy DSGTD111 (a modification of the aircraft engine blade alloy, Rene'80) for various specimen orientations

Experimental

Material and Specimen

Miniature specimens used in this study were designed to be machined from thin sections of gas turbine blades, and were 41mm long with a reduced section of 32mm by 1.9mm. Six specimens were cut from a new blade shank at each of three orientations: parallel, perpendicular, and at 45° to the growth direction. For the constant displacement rate tests (CDR) the same small specimens were used.

Stress Relaxation Testing (SRT)

The tests reported in the present paper, at a temperature of 900C, were loaded at a strain rate of .01%/sec. to a set strain which was then held constant. With the capacitive extensometer this was maintained to $\pm 5 \times 10^{-6}$ for the test duration which was usually one day.

The stress vs. time response was converted to a stress vs. creep strain-rate response by differentiating and dividing by the modulus measured on loading(2). The accumulated inelastic strain during relaxation was usually less than 0.1%, so that several relaxation runs at different temperatures and from different stresses could be made on a single specimen with minimal change in the mechanical state. Thus, an enormous amount of creep data was generated in a short time. By appropriate cross-plotting, pseudo stress-strain plots were constructed which may then be used to generate strain vs. time or stress vs. time curves.

Constant Displacement Rate Testing (CDR)

This test involved tensile testing to fracture at 800C under closed loop strain control on the specimen. The temperature was chosen to be in the range of greatest susceptibility to gas phase embrittlement (GPE) (4), and the displacement rate was 1%/hour. For the three orientations the susceptibility to GPE by oxygen was evaluated in terms of high temperature prior exposures in air.

Results

Figure 1 shows relaxation results on a transverse specimen of GTD111 at 900C from four total strain levels. The corresponding stress vs. inelastic strain (creep) rate curves are shown in figure 2. This plot may then be used to construct pseudo stress vs. strain curves from vertical intercepts at various strain rates (see figure 3). In practice, the construction is made by fitting polynomial expressions to the stress vs. strain rate curves,

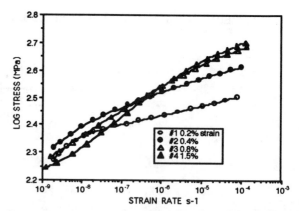

Figure 2 Stress vs. creep rate curves at 900C for transverse specimen

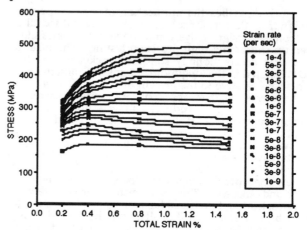

Figure 3 Pseudo stress vs. strain curves as a function of creep strain rate for transverse specimen

and solving at specific strain rates. These curves are similar to isochronous creep curves crossplotted from conventional creep curves and often used in high temperature design. The isostrain rate curves may be used in a slightly different way where the time is estimated by dividing the inelastic strain (total strain less elastic strain) by the inelastic strain rate. This then gives a stress, strain and time representation.

The same testing procedure and analysis for the longitudinal orientation are shown in figures 4-6. In this case the measured modulus was 87,200 MPa compared with 137,000 for the transverse orientation

One approach to analysis for comparison with conventional approaches is to use the actual stress vs. creep rate curve which is closely representative of 0.5% inelastic strain. The test run from 0.8% total strain was chosen as an appropriate standard in terms of the total inelastic strain during loading and relaxation. Also, it should be noted that the pseudo stress vs. strain curves are quite flat at this strain so the actual prestrain level is not especially critical. Once the appropriate curve is chosen, the time is calculated directly from this inelastic strain (0.5%) divided by the creep rate as a function of stress. Of course, other strains could be taken from the pseudo stress vs. strain curves and the stress and strain rate picked from those.

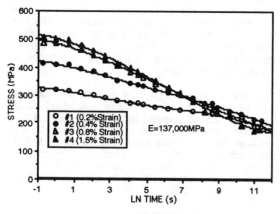

Figure 1 Stress relaxation at 900C from four strain levels for transverse specimen

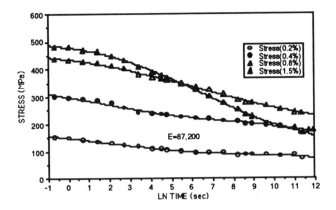

Figure 4 Stress relaxation at 900C from four strain levels for longitudinal specimen

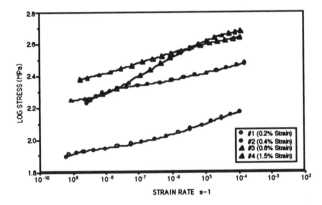

Figure 5 Stress vs. creep rate curves at 900C for longitudinal specimen

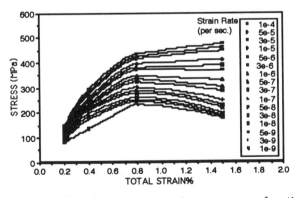

Figure 6 Pseudo stress vs. strain curves as a function of creep strain rate for longitudinal specimen

The data may also be presented in the form of strain vs. time curves using horizontal cuts at fixed stresses. However, using an actual test run allows a small computer program to generate the stress vs. predicted time curve for 0.5% creep from a polynomial curve fit to the actual stress vs. creep rate curve. Of special interest is the ability to extend the curve to several thousand hours from a single run of 20 hours (see figure 7). This stems from the calculation of the longest time for 0.5% creep being based on the lowest creep rate in the relaxation test. For example, a creep rate of 10^{-9}/sec. gives a predicted time of 1,390 hours. This might be thought of as a pseudo time rather than an extrapolated time, since actual creep rate data are used.

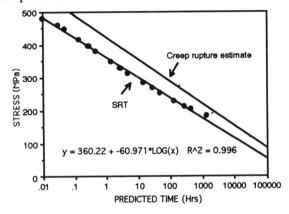

Figure 7 Comparison between creep rupture estimate at 900C and stress vs. predicted time to 0.5% creep from SRT for transverse specimen

Unfortunately there are no published long time creep data on this alloy. However, an estimate is made of the rupture behavior derived from a published Larson-Miller parameter plot of the conventionally cast alloy (6). While recognizing that different specimen sizes were involved, the comparison is good. It shows that the SRT data may be used as a basis for setting design stresses, either directly or with appropriate scaling.

Whether there is good quantitative agreement between the two approaches is dependent on the sensitivity to thermal mechanical history. The high strength cast superalloys are quite insensitive to prior thermal-mechanical history in terms of creep strength(2) and hence agreement is expected to be good.

What is important to recognize is that if these predictions are close to long term data, then they may be used in current design protocol either directly or with an appropriate scaling factor. If that is the case, then it is appropriate to recognize that figure 7 was derived directly from the stress vs. strain rate curves and that the latter should be amenable to direct use. The equivalent creep rate for 0.5% creep strain in 100,000 hours is 1.4×10^{-11}/sec. Thus, a design stress could be taken either from figure 7 in terms of time or directly from figures 2 and 5 in terms of creep rate; 10^6 hours is directly equivalent to 1.4×10^{-11}/sec. As an example, figure 8 compares creep rates for the transverse and longitudinal orientation. The particular representation shown in figure 8 is semi-logarithmic or exponential. It may be that a power function is more appropriate for some alloys. The fit is about as good for the data shown as the exponential fit and yields stress exponents of 17.62 and 11.88 for the longitudinal and transverse orientation specimens, respectively. These are appropriate exponents to be used, for example, in crack tip creep deformation analysis, rather than exponents calculated from minimum creep rate data.

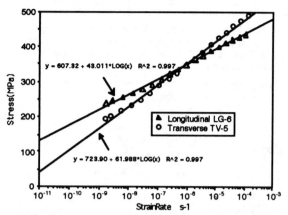

Figure 8 Stress vs. creep rate curves for 0.5% creep strain at 900C for two orientations

Figure 9 shows CDR results at 800C on the three GTD111 orientations compared with data on a widely used cast superalloy, IN738, taken from a test slab casting using the same miniature specimens. All show reasonable ductility at this intermediate test temperature. Although the yielding stresses are higher for GTD111 for all orientations, the pronounced fall in flow stress at higher strains for the transverse and diagonal orientation, below corresponding values for IN738, is interesting but, as yet, unexplained.

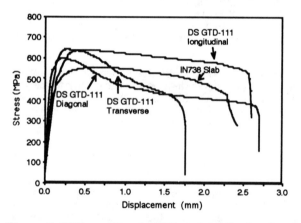

Figure 9 CDR results at 800C and 1%/h. for the three orientations

To compare the effect of orientation on embrittlement by oxygen (GPE), specimens were exposed at 1000C for 24 hours in air. In the longitudinal orientation, with very few grain boundaries intersecting the surface, the ductility was reduced by about 30% (see figure 10). However, the other specimens failed with little or no plasticity. Figure 11 shows that, for the diagonal orientation, even 5 hours exposure at this temperature leads to an appreciable loss in fracture resistance. This figure also shows that a specimen which had previously been exposed at 850C for about a week of SRT testing was unembrittled.

Discussion

It has become increasingly clear that the traditional approach to materials development and design, involving long time creep testing, is not able to explain instances of part failure, especially for nickel based superalloys operating under non-steady conditions in aggressive environments.

The new approach proposed here offers several innovations in concept: separation of creep strength and fracture criteria, short time high precision evaluation of the consequences of thermomechanical exposures, setting limiting critical property values to establish unambiguously appropriate criteria for end of part life.

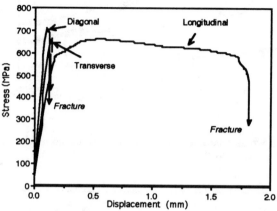

Figure 10 Effect of exposure in air at 1000C for 24h. on CDR results for the three orientations at 900C

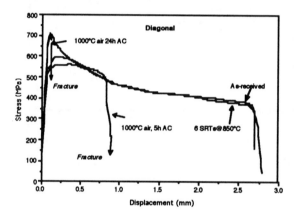

Figure 11 Effect of exposures at 1000C on CDR results for diagonal specimen

The apparent advantage in creep strength in the longitudinal direction for low strain rates (see figures 8) is, in part, a consequence of the lower modulus, e.g. for a fixed strain on loading the starting stress is less and there is less inelastic strain accumulated during relaxation for the longitudinal orientation. This advantage was not seen in unpublished work for tests started at a fixed stress. Thus, when comparing the creep strength for different orientations in anisotropic material the effect of loading procedure must be recognized. No direct comparison has so far been made with conventionally cast GTD111, although it is expected to be closely similar to the transverse and diagonal orientations.

The stress vs. strain-rate curves readily provide a basis for alloy development and optimization of creep strength. They may also be used directly to set design stresses. If a more traditional representation is desired, one approach is to cross plot the data in terms of pseudo stress vs. inelastic strain curves at fixed inelastic strain rates. Sets of such curves are shown in figures 3 and 6. Normally, the designer would see similar curves crossplotted from creep curves at constant time intercepts to give isochronous stress vs. strain curves. From curves, such as those shown in figures 3 and 6, creep curves can be constructed, if desired, by taking horizontal sections at fixed stresses. At the intersection points, for the indicated strain, the time is calculated by dividing that strain by the inelastic strain rate. A more convenient approach may be to take vertical cuts so that a plot of stress vs. time for a fixed creep strain may be plotted and extrapolated as desired. For example, figure 4 shows such a plot for the transverse orientation which is of the same form as that often used to establish design stresses based on extrapolation of long time creep tests. It should be noted that these are ways to manipulate the relaxation data for presentation: there is no intrinsic advantage in using any of these representations.

For cast superalloys little effect of prior exposure, including 1000C treatments, has been observed for creep strength(2). Even a factor of five on creep rate may not have sufficient effect on stress to cause concern because of the strong stress dependence of creep rate. However, a strong effect on fracture resistance based on the CDR test, especially from GPE, should create major concern. A suggested approach to quantifying this concern is to define a minimum acceptable CDR displacement at failure. For example, for the miniature specimens tested at 800C and 1%/hour this might be set at 0.5mm. Such a criterion would be valid for material in new components and also for material taken from components at any stage of their operating life. Additional refinements of such a failure criterion for service exposures might allow certain regions of the blade material to drop below this level, provided the bulk of the blade retained a good fracture resistance. For example, trailing edge sections of a small IN738 blade were found to be the only region embrittled after 65,000 hours service(7).

Conclusions

1. A new "Design for Performance" methodology may be cheaper, faster and fundamentally superior to traditional development and design procedures for high temperature applications which are currently based on long time testing.

2. A high precision short time stress relaxation test (SRT) is used to evaluate the creep strength in terms of stress vs. creep rate covering up to five orders of magnitude in creep rate.

3. In tests at at 900C, the longitudinal orientation has higher creep strength at low stresses and lower creep strength at high stresses.

4. A constant displacement rate test (CDR) is used to evaluate fracture resistance at the intermediate temperature of 800C.

6. Based on this new approach, there appears to be no clear advantage in directional solidification for this alloy in terms of creep strength.

7. By contrast, resistance to fracture, especially after high temperature exposure in air, is improved substantially for longitudinally oriented specimens of GTD111 compared with diagonal and transverse oriented specimens.

8. Proposed design criteria for both creep strength and fracture resistance based on the new methodology may be used directly for life management of operating components.

Acknowledgements

This work was made possible by the financial and technical support of ARCO Alaska and, in particular, David Stiles. I am especially indebted to Donald Van Steele for the detailed experimental work, and to Kyle Amberge for writing the analysis programs and helping with data reduction and graphing.

References

1. D. A. Woodford, "The Remaining Life Paradox," Int. Conf. Fossil Power Plant Rehabilitation, ASM International, 1989, p. 149.

2. D. A. Woodford, "Test Methods for Development, Design, and Life Assessment of High Temperature Materials," Materials and Design, vol. 14, 1993, p.231-242.

3. D. A. Woodford and D. F. Mowbray, "Effect of Material Characteristics and Test Variables on Thermal Fatigue of Cast Superalloys," Mat. Sci. and Eng., vol. 16, 1974, p. 5-43.

4. D. A. Woodford and J. J. Frawley, "The Effect of Grain Boundary Orientation on Creep and Rupture of IN738 and Nichrome," Met. Trans., vol. 5, 1974, p.2005-2014.

5. D. A. Woodford and R. H. Bricknell, Environmental Embrittlement of High Temperature Alloys by Oxygen," Treatise on Materials Science and Technology, Academic Press, vol. 25, 1983, p. 157-196.

6. P. W. Schilke, A. D. Foster, J. J. Pepe and A. M. Beltran, "Advanced Materials Propel Progress in Land-Based Gas Turbines," "Advanced Materials and Processes, vol. 4, 1992, p. 22-30.

7. D. A. Woodford, D. R. VanSteele, K. J. Amberge and D. Stiles,"Decision Support for Continued Service of Gas Turbine Blades," Materials Performance, Maintenance and Plant Life Assessment, Met. Soc. Can. Inst. Mining, Metallurgy and Petroleum, 1994, p.85-100.

The Influence of Inclusions on Low Cycle Fatigue Life in a P/M Nickel-Base Disk Superalloy

Eric S. Huron
Paul G. Roth

General Electric Aircraft Engines (GEAE)
Cincinnati, OH 45215

Abstract

The high alloy content of advanced nickel-base disk superalloys calls for powder metallurgy (P/M) processing to minimize segregation produced by conventional cast & wrought processing. Although the technology has developed to allow reliable application of P/M materials, of concern are ceramic inclusions which are intrinsic to the process due to the use of ceramic crucibles in producing the raw meltstock and in the powder atomization process itself.

For robust disk design, the impact of ceramics on Low Cycle Fatigue (LCF) must be assessed. Actual production material can be evaluated but the low frequency of larger inclusions means that impractically large volumes of material must be tested. To address this problem, tests were run on powder seeded with controlled distributions of ceramics. The alloy was the current GEAE disk material Rene' 88 DT (R'88DT). Two types of ceramic seeds were added at two size/density combinations. Small seeds (-270/+325 mesh) were added to -270 mesh powder at rates (numbers per unit volume) predicted via probabilistic calculations to cause surface initiations. Large seeds (-80/+100 mesh) were added to the powder at rates predicted to cause internal initiations. The powder was consolidated and processed using production parameters. LCF tests were made on samples taken from fully heat treated forgings at 204° C and 649° C at two stress levels. Roughly half of the bars were shotpeened to study the ability of this processing to suppress surface initiations.

The impact of the seeds was significant and was a function of seed type and size, temperature, and bar surface condition. At 204° C life decreased with increasing seed size up to a maximum life reduction of 33-50%. Shotpeening only slightly improved the lives of small seeded bars. At 649° C, the seeds had dramatic impacts. Small seeds reduced life by 1-2 orders of magnitude and large seeds by a further 1-2 orders of magnitude. Shotpeening did suppress surface initiations and significantly improved life. The relative impacts between 204° C and 649° C and the impact of shotpeening depended on failure mechanisms -- at 204° C most of the failures initiated at facets, explaining a relatively minor seeding impact at this temperature while at 649° C almost all failures initiated at the seeds (or at intrinsic inclusions in the unseeded material).

The results of this testing have been used to develop an incubation model incorporated in GEAE's Probabilistic Fracture Mechanics (PFM) methodology used for design life calculations.

Introduction

Advanced nickel-base superalloys for disk applications are highly alloyed to meet design strength and temperature requirements. Given the high alloy content of these materials, powder metallurgy (P/M) processing is preferred over conventional cast & wrought processing to prevent excessive segregation.

Ceramic inclusions are inherent in the P/M process. Inclusions are introduced from the refractory crucible material used to produce the starting Vacuum Induction Melt (VIM) ingot, the secondary VIM crucible used in the powder atomization, the nozzle and guide tube arrangement for the atomization stream, and by reactions between the molten stream and the gases in the atomization chamber.

The impact of ceramics on fatigue life must be quantified. Historically this has been attempted through LCF tests on actual production forgings, but inclusion content is seldom adequately characterized and varies from one forging lot to the next, and to capture larger inclusions, large volumes of material must be tested; also, many tests fail at parent metal initiation sites with no inclusion present. For all these reasons, prohibitively large numbers of LCF tests are required, and interpretation of the resulting data is usually difficult and controversial.

Testing of intentionally seeded material offers significant advantages. The impact of a much larger range of inclusion sizes may be assessed. Seeding distributions may be statistically described, and competition with grain initiated failures may be managed by adjustment of seeding rates.

This paper presents a seeded LCF study of P/M R'88DT, GEAE's current advanced disk alloy (References 1, 2). Loose powder was doped with controlled distributions of two types of ceramic seeds. The seeded powder was consolidated via extrusion, isothermally forged into pancakes, and heat treated. LCF bars were taken from the pancakes and tested.

Table I. Composition of R'88DT (Weight Percent)

Co	Cr	Mo	W	Al	Ti	Nb	B	C	Zr	Ni
13	16	4	4	2.1	3.7	0.7	0.015	0.03	0.03	bal

The study focused on three points in question:

- The effect of small inclusions inherent in powder screened to -270 mesh.

- The effect of larger inclusions representative of an unexpected contamination event.

- The influence of shotpeening on surface crack initiation at inclusions.

Portions of the program were performed with the assistance of SNECMA at the Gennevilliers, France facility.

Procedure

R'88DT powder obtained through GEAE's normal production sources was screened to -270 mesh. The chemistry is given in Table I.

The -270 mesh size was chosen to limit the size of naturally occurring inclusions in the material to better assess the influence of the added ceramic seeds. Two types of seeds were added, termed Type 1 and Type 2, consistent with past studies of inclusion behavior (Reference 3). A hard, blocky seed (Type 1) was added to represent crystalline or glassy, non-deforming ceramics, such as might occur due to breakage of an alumina crucible. The Type 1 was simulated using an Al_2O_3/MgO casting firebrick mold material supplied by SNECMA. A soft, deformable, agglomerate seed (Type 2) was added to represent reactive inclusions originating from patching putties or furnace assembly cements, or by reaction of the molten stream with trace gasses in the atomization chamber. The Type 2 was simulated by a moist Al_2O_3-based paste with a phosphorous/silicon-containing binder, sold under the trade name RAM 90 by Combustion Engineering Refractories. The RAM 90 was given the recommended drying cycle of 54° C/5.5hrs + 110° C/16.5hrs, then fired at 1149° C/4hrs/furnace cool. Both seed types were crushed in a mortar and pestle and sieved to appropriate distributions as discussed below.

Blending and extrusion experiments were made to verify the suitability of the Type 2 seed material. Carefully weighed amounts of sieved RAM 90 were added to powder samples of slightly finer mesh size; the samples were aggressively blended and then sieved to recapture the RAM 90. High recoveries showed the blending did not cause excessive disintegration of the ceramic. In addition, a laboratory extrusion using a 8.6 cm dia. x 15.24 cm long can was filled with -200 mesh R'95 powder with a few grams of -80/+100 and -170/+325 RAM 90 material. After extrusion at 1066° C, metallography (Figure 1), showed that the RAM 90 inclusions had the desirable characteristics of a Type 2: a multi-particle makeup with a tendency to elongate in the direction of deformation.

Control of seeding rates presented a challenge. Given that exact counts of hundreds of thousands of seeds are not practical, seeds were added by weight. Several means were used by SNECMA to obtain average seed weights, including: measurements of samples of 1000 large seeds and 10000 small seeds; use of a Coulter counter to measure size of individual particles; and computations based on assumed ellipsoidal shapes and mean diameter measurements. Multiple trials with the various methods suggested that the count-plus-weigh method was most accurate.

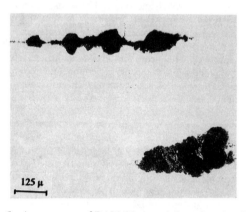

Figure 1: Appearance of RAM-90 material used to simulate Type 2 seeds after extrusion. The elongation of the seeds parallel with the deformation direction resembles that of intrinsic Type 2 inclusions found in production powder.

Two seed sizes were used. Small mesh seeds, at 680/kg goal density, were intended to represent the upper end of the distribution that would normally be present in -270 mesh powder, but at an artificially high frequency to promote a high percentage of surface intersections. Large -80/+100 mesh seeds, at 36/kg goal density, were intended to represent inclusions arising from contamination.

The general seeding procedure was to add a weighed quantity of ceramic to 5 kg of powder, mix for four hours in a Turbula blender, then pour the batch into an extrusion can. This was repeated 8 times to end with 40 kg in the can. After filling, the can was evacuated over 24 hours using a diffusion pump system attached to the fill tube, then sealed by crimping the tube.

Eight extrusions were made: one baseline (unseeded), two seeded with small Type 1 seeds, two seeded with small Type 2 seeds, one seeded with large Type 1 seeds, and one seeded with large Type 2 seeds. SNECMA was responsible for compaction and extrusion of the powder, sectioning into mults, and performing the isothermal forging. Stainless extrusion cans of 19 cm diameter x 28.5 cm long were compacted and extruded at a 5.5:1 reduction ratio at 1052° C.

Each extrusion yielded a total of five mults. The mults were isothermally forged using a laboratory isothermal press at 1050° C. Forgings were heat treated through a 1149° C supersolvus solution followed by air cooling. Pancakes were equipped with mid-thickness thermocouples and the interior cooling rates ranged from 79° C/min for the mid-radius to 88° C/min near the rim.

The material was compared to production R'88DT and found to be representative in microstructure (ASTM 7/8 ALA 4 grain size), tensile strength (1062 MPa 0.2% yield strength and 1524 MPa ultimate at 204° C, 1007 MPa yield strength and 1503 MPa ultimate at 649° C), and creep strength (155 hours to 0.2% at 704° C/690 MPa).

Eight LCF specimens were machined from each 18 cm diameter by 2.5 cm thick pancake (Figure 2). The LCF specimen gage section was relatively low in aspect ratio, 1.016 cm diameter by 1.905 cm tall (Figure 3), to help promote internal failures, with large radii from shoulder to gage to minimize stress concentration. Welded IN718 shanks and button heads were used. Roughly half of the test bars were shotpeened by GE using CCW 31 shot at 7A intensity and 125% coverage. Strain-controlled 0-max-0 LCF tests were performed at two strain levels at 204° C (0.52% and 0.70%) and two strain levels at 649° C (0.60% and 0.80%). The first 100 cycles of the test were done using a specialized procedure to minimize serrated yielding (Reference 4). This procedure utilized a triangle waveform in strain control at 1.5 cpm coupled with a 0.02% increase in strain range. After the first 100 cycles, the next 24 hours were run at the target strain range in strain control at 30 cpm, and any cycles over the first 24 hours (about 42,000 cycles) were conducted in load control at 300 cpm. After testing fractography of the bars was performed using the SEM. Inclusions and grain facet initiation sites were characterized for size, type, and location.

The final matrix covered two temperatures, two target stress levels (achieved via strain control) at each temperature, two seed types, two seed sizes, and peened and unpeened surface finish as variables (refer to Table II).

Results

Post-test fractography showed that the chosen ceramics were reasonably effective in simulating Type 1 and 2 inclusions found in production material. The median initiation site sizes (based on boxed areas) were 4,770 square microns for the small seeds and 40,000 square microns for the large seeds. These values fell below the median sizes of the loose seeds due to inclusion orientation relative to the stress plane of the specimens. Baseline material failed at grains or at intrinsic inclusions with an average size of about 2,130 square microns.

At low temperature, some seeded bars failed at grain facets indicating a competition between ceramics and grain facets for crack initiation. The influence of seeding rates on observed behavior are further discussed below. Figure 4 presents a comparison between typical initiation sites at grain facets, intrinsic inclusions, Type 1 seeds, and Type 2 seeds.

Post-test metallographic study was conducted on ten small seeded specimens looking for evidence of secondary cracking. Significant

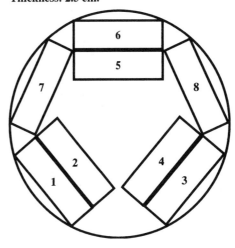

Figure 2: Specimen sectioning plan for subscale pancakes. The tangential layout of the specimens was used to provide a consistent deformation history at the gage section of the bars. The test matrix was blocked against the inner row position vs. the outer row position to ascertain whether location had any statistical impact.

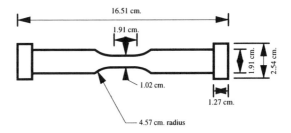

Figure 3: Low Cycle Fatigue specimen design.

numbers of secondary grain-initiated cracks were found in specimens tested at 204° C and very little secondary cracking in specimens tested at 649° C (Figure 5). Also, it was found that inclusions were most often tangentially oriented in the pancake forgings if oriented at all. This countered conventional wisdom which held that they would be oriented radially.

Given a distribution of inclusion areas, a distribution of aspect ratios, a seeding rate, the diameter and length of a cylinder, and the assumption of tangential orientation, it is possible to estimate by Monte Carlo simulation the expected distribution of the largest inclusion area normal to the stress field in a test bar volume. The estimated distribution should approximate the observed distribution of initiation areas if (as is expected when all failures are subsurface) the specimens tend to fail from the largest contained inclusions.

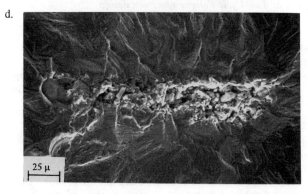

Figure 4: Micrographs of typical initiation sites. a) Grain facet initiation. b) Intrinsic inclusion in unseeded baseline material. c) Typical Type 1 seed. d) Typical Type 2 seed.

Application to the small and large seeding distributions with seeding rates of 27 inclusions per cubic centimeter for the small seeded material and 5 inclusions per cubic centimeter for the large yielded the comparisons shown in Figure 6.

Note that size is plotted against the probability of exceeding that size. Exceedance probability is estimated for experimental data by ranking the data from largest to smallest, then using the expression:

$$(n-i+0.5)/n$$

where n is the number of points and i is the ranking of the i'th data point.

The impact of the seeds was significantly different from the 204° C tests to the 649° C tests. Figure 7 summarizes the Type 1 data compared to baseline unseeded data at 204° C (Table II). Note that none of the large Type 1 seeded bars were peened. There was some spread between large seed and small seed data, but overall the lives were remarkably similar. The lowest lives were for unpeened Large Type 1 seeded bars. A near surface inclusion of 477 micron length by 191 micron width (91,107 square microns boxed area) located gave the lowest life at high stress. At low stress the lowest life was due to a much smaller seed (124 x 102 microns) unfavorably oriented on the surface.

Figure 8 summarizes the Type 2 data at 204° C (Table II). Note that all of the large Type 2 seeded bars were peened. The low lives of these bars were slightly above the lowest lives for the unpeened large Type 1 seeded bars. At high stress unpeened small seeds located on the surface (208 x 43 microns and 107 x 41 microns boxed areas) gave lives as low as peened large seeds located on the surface (460 x 140 microns and 201 x 104 microns boxed areas). At low stress only one peened large seeded bar failed at the surface, and resulted in the lowest life. A very large seed (1234 x 584 microns boxed area) gave the longest life at low stress, but was located internally. For the small seeded unpeened bars, the Type 2 appeared to have a slightly more negative effect on life than the Type 1. Overall, the peened lives were grouped tightly together and were slightly improved over unpeened lives at high stress due to suppression of lower life surface or near surface seed failures. Most failures were at grains, and although peening changed the location of the grain failures from surface to internal, the life was not dramatically enhanced by peening. An overall comparison of surface vs. internal initiations did show that internal initiation is associated with longer life.

Figures 9 and 10 provide the results at 649° C (Table II). Note the wide degree of scatter compared to 204° C. The relative life impact was much larger. It did not take a particularly large seed to cause a low life at 649° C. In Figure 9, an unpeened small seeded bar with a planimeter area of only 74 x 51 microns fell well below the unseeded baseline. At this same range of stress on the plot, the lives for the large seeded bars tended to fall in order of seed size - 3776 cycles for 376 x 152 microns, 5326 cycles for 239 x 122 microns, and 6126 cycles for 127 x 86 microns, while an internal seed of 584 x 147 microns gave a life of 17,038 cycles. At the low stress, however, surface failures due to 234 x 86 micron and 279 x 121 micron seeds gave very low lives. For the Type 2 seeds (Figure 10), which were all peened, most of the large seeded bars failed internally, but the lives were still low. For most of the data, peened lives were similar to unpeened lives, but there were a few

204° C

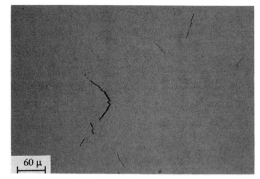

649° C

Figure 5: Comparison of secondary cracking in 204° C and 649° C specimens. The 204° C specimens had numerous grain-initiated cracks competing to failure while the 649° C specimens displayed very little secondary cracking.

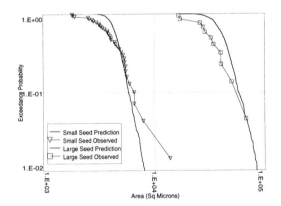

Figure 6: Input size distributions for small and large seeds. Type 1 and Type 2 distributions were nearly equivalent for each seed size.

low lives shown by unpeened seeded bars. Peening was apparently able to prevent these occasional surface failures, but if a bar was inclined to fail internally, obviously peening provided little benefit.

For the Type 1 seeds, very low lives were observed for surface failures from the unpeened large seeded bars. If the large seeds happened to be located internally, the LCF lives were not as strongly affected. Two low lives were also observed from surface failures at high stress for unpeened small seeded bars. Note that

for the Type 2 seeds, most of the failures were internal, but even internally located seeds caused significant life reductions. When the unpeened large Type 1 and peened large Type 2 data were considered together, peening appeared to have a slightly beneficial effect at high stress because it prevented very low lives due to surface failures.

Some large seeded bars failed at grains for R'88DT (even at 649° C), which indicated either that grains were more damaging than seeds, or perhaps that given the low seeding density, no seeds were present in the gage section of those particular bars.

At 204° C, the large Type 1 vs. large Type 2 seeds did show some differences in grain vs. inclusion failures. The two lowest lives at both high and low stress for the large Type 2 seeds were surface failures, even though the bars were peened. This suggested that the nominal 40,000 square micron seed size was above a ceiling of peening protection. In fact, at the low stress the lowest life for the large peened Type 2 was due to a seed of only 406 x 51 micron. A survey of the data found that for large Type 1 bars, over 70% of the bars failed at the surface, while for the unpeened large Type 2 bars, only about 15% failed at the surface. For the unpeened large Type 1 bars, over 60% failed at grains, while for the large Type 2 bars, only about 25% failed at grains.

In general the Type 1 seeds were very representative of Type 1 seeds intrinsically present in P/M superalloy powder, but the Type 2 seeds gave mixed results. Some showed classic Type 2 behavior, but many looked like Type 1 or mixed Type 1/Type 2. Jablonski (Reference 5) used Al_2O_3 and SiO_2 seeds in a study on the effect of inclusions in Low Carbon (LC) Astroloy. The seeds were not described as Type 1 and Type 2, but fractography showed that the Al_2O_3 seeds were non-reactive (Type 1) and the SiO_2 seeds reacted with the matrix and formed a "granular" inclusion. Therefore SiO_2 may provide a better simulation of Type 2 inclusions than the RAM 90. It has been suggested that many Type 2's in production involve interactions between the atomization nozzle and the metal stream (Reference 3). If this is true, more realistic simulation of Type 2 seeds will be difficult.

Discussion

GEAE's traditional lifing approach, based on LCF curves, is evolving for fracture critical powder metal components by incorporating Probabilistic Fracture Mechanics (PFM) analysis. Supporting this move is a growing validation database which convincingly demonstrates that probabilistics work given the right inputs.

Figure 11 shows comparisons between PFM-predicted and observed failure distributions for five specimen sets: Cylindrical specimens from this study and hourglass specimens from a USAF-funded program (Reference 6); unseeded baseline, large seeded and small seeded; all peened and all tested at the same conditions of strain and temperature. The comparisons demonstrate the capability of the PFM model over the range of probabilities 0.01 to 0.99 when provided the right inputs of inclusion distribution and behavior. It is easily accepted that were the rate of flaw occurrence orders of magnitude lower than the selected seeding rates (i.e., representative of actual production powder cleanliness), the failure probabilities would be proportionately lower in the absence of other failure mechanisms.

Table II. Test Matrix and Summaries of Observed Lives and Failure Origins

Test Temp	Seeding Parameters	Peened	Stress	Number of Bars	Minimum Life	Maximum Life	Median Life	Grain Origins	Inclusion Origins	Other Origins	Percent Grains	Surface Origins	Near Surface Origins	Internal Origins	Percent Surface
204C	Baseline	Yes	High	15	67480	83808	83808	15	0	0	100%	0	1	14	0%
			Low	15	267084	360842	317617	14	0	1	93%	1	0	14	7%
		No	High	5	48180	64017	52838	5	0	0	100%	5	0	0	100%
			Low	4	247822	302506	263565	4	0	0	100%	4	0	0	100%
	Type 1 Small (680/kg)	Yes	High	15	69385	85067	75201	14	1	0	93%	1	1	13	7%
			Low	15	267530	358258	304016	15	0	0	100%	0	0	15	0%
		No	High	4	43086	59964	51080	4	0	0	100%	3	1	0	75%
			Low	5	240091	317258	276005	4	1	0	80%	5	0	0	100%
	Type 2 Small (680/kg)	Yes	High	14	72526	94648	79473	14	0	0	100%	0	0	14	0%
			Low	15	258918	357909	312821	15	0	0	100%	0	0	15	0%
		No	High	5	29284	54050	52453	1	4	0	20%	5	0	0	100%
			Low	5	216456	305988	237117	5	0	0	100%	5	0	0	100%
	Type 1 Large (36/kg)	No	High	9	26703	59074	45677	6	3	0	67%	7	1	1	78%
			Low	8	87513	305186	189908	5	3	0	63%	7	1	0	88%
	Type 2 Large (36/kg)	Yes	High	10	34451	91644	77589	2	8	0	20%	2	1	7	20%
			Low	10	114809	350287	276678	3	7	0	30%	1	1	8	10%
649C	Baseline	Yes	High	15	161661	590762	291091	0	15	0	0%	0	0	15	0%
			Low	15	520998	9529000	3926400	0	15	0	0%	0	0	6	0%
		No	High	6	244058	447111	357255	0	6	0	0%	0	0	6	0%
			Low	3	430022	2734736	2513086	0	3	0	0%	0	0	3	0%
	Type 1 Small (680/kg)	Yes	High	15	52723	115550	81277	0	15	0	0%	0	1	14	0%
			Low	14	158024	628103	292015	0	14	0	0%	0	0	14	0%
		No	High	5	8006	124028	58768	0	4	1	0%	2	0	3	40%
			Low	5	130027	605028	276926	0	5	0	0%	0	1	4	0%
	Type 2 Small (680/kg)	Yes	High	15	68520	232180	101601	0	15	0	0%	0	0	15	0%
			Low	14	275640	1057073	472588	0	14	0	0%	0	1	13	0%
		No	High	5	10174	152172	76818	0	4	1	0%	2	0	3	40%
			Low	5	422392	2015326	764733	0	5	0	0%	0	1	4	0%
	Type 1 Large (36/kg)	No	High	14	2572	95583	6338	2	12	0	14%	8	0	6	57%
			Low	8	8411	174889	124725	0	8	0	0%	2	1	5	25%
	Type 2 Large (36/kg)	Yes	High	10	5887	64339	29963	0	10	0	0%	0	1	9	0%
			Low	10	25372	227432	116941	0	10	0	0%	0	1	9	0%

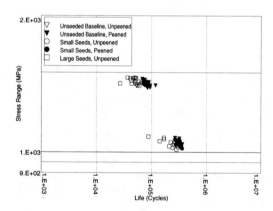

Figure 7: LCF results for Type 1 seeded material (vs. baseline) at 204° C.

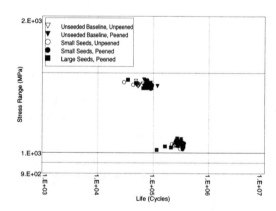

Figure 8: LCF results for Type 2 seeded material (vs. baseline) at 204° C.

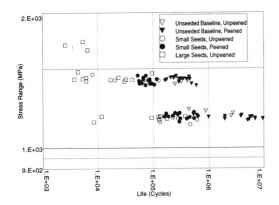

Figure 9: LCF results for Type 1 seeded material (vs. baseline) at 649° C.

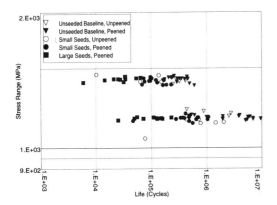

Figure 10: LCF results for Type 2 seeded material (vs. baseline) at 649° C.

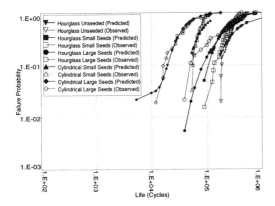

Figure 11: Peened specimens tested at one temperature/strain condition. Comparison of observed vs. probalistically-predicted failure distributions.

Significant efforts are being made to ensure the right inputs. For example, Heavy Liquid Separation (HLS) analysis has been developed to quantify and control inclusion content (Reference 7). Also, the seeded fatigue program discussed in this paper provides a model for crack initiation at inclusions and is continuing.

It is well known that internal initiated failures tend to have longer lives than surface initiated failures, at least at high temperatures (Reference 8). There is a growing body of evidence that buried ceramic inclusions in both P/M Rene' 95 and P/M Rene' 88 DT initiate sharp cracks only very late in life. Figures 12 and 13 compare observed lives with fracture mechanics calculations for both both alloys. Both data sets demonstrate that sharp crack fracture mechanics seriously underestimates total life. To ignore the reality of the incubation phenomenon usually results in very conservative PFM predictions.

It is hypothesized that incubation represents crack initiation in grains near stress concentrating inclusions, and therefore that standard LCF algorithms (e.g. rainflow counting, mean stress corrections and Miner's rule) can be used to calculate incubation given complex cycle missions.

Four points are offered supporting this interpretation:

- While fracture mechanics provides lower bound predictions for most inclusion initiated failures, it is in many cases very conservative (Figures 12 and 13). This is most evidently true for buried inclusions, but it is also true of surface inclusions.

- Given that incubation is observed for inclusions of all sizes, the phenomenon probably cannot be attributed to crack growth threshold.

- It is also observed that incubation life is stress range dependent for the strain controlled tests (again, Figures 12 and 13). This is not the case for the fracture mechanics predictions which are approximately max-stress controlled for negative stress R-ratios.

- There is direct evidence that incubation is a real phenomenon in that most inclusions in a failed fatigue specimen have no readily discernible cracks emanating from them. This is so for both surface intersecting and buried inclusions.

The relative impact of inclusions on LCF life was shown to change dramatically for R'88DT between 204° C and 649° C. Several factors can be offered to explain the transition: Changes in deformation behavior of the matrix R'88DT, changes in behavior of the inclusions, or changes in the role of the environment. These factors could work independently or be interrelated.

While the deformation mechanisms of blade superalloys has been well documented, the mechanisms of disk superalloys have not. Waspaloy, which has some similarities to R'88DT, displays planar slip at low temperatures and wavy slip at high temperatures (Reference 9). Planar slip occurs when dislocations tend to be tightly confined to the same plane, where they can combine to form stress concentrations. Wavy slip occurs when thermal activation allows cross-slip onto many planes and results in a more homogeneous dispersal of dislocation damage in the material. For a material operating in the planar slip mode, the effect is magnified for larger grain sizes because more dislocations (and higher stress concentration) can be established without the interference of grain boundaries. It is suggested that this is one reason why ASTM 7-8 R'88DT shows such a high percentage of grain failures at low temperatures.

The effect of inclusions at varying temperatures was studied by Jablonski (Reference 5). In this study -500 mesh LC Astroloy was seeded with Al_2O_3 (Type 1) and SiO_2 (Type 2). The material was tested as HIP'ed (plus heat treated) with a grain size of 30 microns.

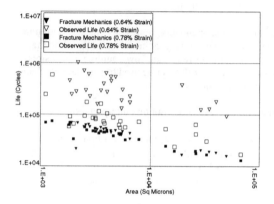

Figure 12: P/M R'88DT -- Fracture mechanics calculations compared to data.

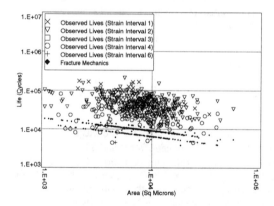

Figure 13: P/M Rene' 95 -- Fracture mechanics calculations compared to data.

The seeds were 75-150 microns in diameter and were added at 4 and 40 ppm for the Al_2O_3 and 2 and 20 ppm for the SiO_2, based on PFM calculations for the number of seeds required to cause failure. At room temperature, the fractures initiated at the ceramic-metal interface (debonding), and LCF life was not strongly affected by the seeds, but was controlled by Stage I cracking (crystallographic crack propagation along grain facets). At 500° C, the fractures initiated at cracked seeds, and LCF life was strongly impacted by the seeds. Lives were found to be inversely proportional to stress intensity factors based on seed sizes and distances from the surface.

Jablonski concluded that at low temperatures the inclusions debonded, while at high temperatures they cracked very early in life. It was hypothesized that the cracked inclusions presented much sharper notches and higher stress concentrations than the debonded inclusions. Debonding was not significant in the current study, at either temperature.

Work by Hyzak and coworkers (References 10, 11, 12) on two P/M alloys AF-115 and AF2-1DA also described a transition in initiation behavior between low and elevated temperatures. These alloys were not seeded, but the AF-115 alloy contained pores, inherent nonmetallic inclusions, and HfO_2 particles, and the AF2-1DA alloy contained inherent nonmetallic inclusions.

At elevated temperatures, most initiations occurred at inclusions or pores, while at low temperatures, most initiations were Type 1 grain crystallographic cracking at or near the surface. The data also showed an increased tendency for internal initiation at elevated temperatures, with surface initiations becoming prevalent only at high stresses.

Hyzak explained his results in terms of relative differences in crack initiation vs. propagation as follows:

- At low temperature, planar and heterogeneous slip, given its relative ease in forming at a surface, was always sufficiently pronounced to initiate failure, even if a larger, more angular inclusion may have been present internally.

- At elevated temperature and high stress, cracks were initiated at many stress concentrating pores and smaller ceramics. For cracks at or near the surface, rapid propagation occurred, due to environmental effects and the fact that near-surface flaws have higher stress intensity factors than internal flaws of the same sizes.

- At elevated temperature and low stress, numerous but small surface and near-surface pores and inclusions did not concentrate stress sufficiently to cause initiation, and larger, more angular buried flaws could win the competition for failure.

Obviously the environment must play some role at higher temperatures. However, its exact role is a point of controversy. Hyzak (Reference 12) claimed that the environment promoted rapid crack propagation from surface initiations at high temperature and high stress, but the environment has also been claimed to retard surface crack growth by crack tip blunting by oxide formation (Reference 13). It would be beneficial to study environmental effects further using seeded material and vacuum testing at different frequencies.

The transverse polishing work described above is consistent with Hyzak's discussion. The presence of numerous secondary cracks at 204° C probably reflects planar heterogenous slip. With many cracks vying for failure, lower lives and small scatter are not surprising. At 649° C, the relative absence of secondary cracking (and none in the matrix) probably reflects wavy homogeneous slip. Damage is less likely to accumulate in planes and lead to microcrack formation. Stress concentrating inclusions become more significant.

Incubation life is found to decrease with increasing inclusion size (Figures 12 and 13). The magnitude of the stress concentration of an inclusion does not directly depend on the size of the inclusion, but the spatial extent of the concentration does. Larger inclusions concentrate more damage than smaller inclusions, leading more quickly to microcrack formation.

While shotpeening was found to be very effective in suppressing surface initiations at both temperatures of the current study, and while it may be presumed that the suppression is due to the residual stresses imparted by the peening (beneficial both for incubation and propagation), the mechanism has not been quantified.

Conclusions

1. Seeded LCF testing is a powerful technique for studying the impact of ceramic inclusions on P/M alloy life.

2. A 204° C, for all test conditions, over 80% of the bars failed at grain facets. Overall, the seeds had very little impact on life, although the large seeded bars did show a smaller percentage of grain facet initiations: For the large Type 1 seeds, about 40% of the bars failed at seeds, and for the large Type 2 seeds, about 75% failed at seeds. Seed initiated failures did yield slightly lower lives.

3. At 204° C, peening had a negligible effect on LCF life, but it strongly influenced the location of the failures. For unpeened bars, 91.1% of the bars failed on the surface, while for peened bars, only 4.6% of the bars failed on the surface. In most cases this resulted in changing the failure site from a surface grain to an internal grain, but the lives were similar regardless of the location.

4. At 649° C, for all test conditions, the majority of the bars (98.7%) failed at inclusions, with life strongly correlated to inclusion size. Unpeened small seeded bars yielded lives 1/10th to 1/4 those of unpeened unseeded baseline. Peened small seeded bars yielded lives between 1/4 to 1/2 those of unpeened baseline. The lives of the large Type 1 seeded bars, which were all unpeened, were about 6% those of the unpeened baseline. The fraction was the same for the large Type 2 seeded bars, which were all peened.

5. At 649° C, peening did appear to suppress surface initiations. For the small seeded material, unpeened bars occasionally failed at surface seeds at high stress, while no peened bars failed on the surface. For the large seeds, all the Type 1 seeded bars were left unpeened, and many failed on the surface. All the Type 2 seeded bars were peened, and all failed either internally or near-surface, but the lives were not improved. The data suggested that Type 2 seeds may be somewhat more deleterious than Type 1.

6. Internal failures initiated at inclusions at 649° C were underpredicted by sharp crack fracture mechanics. It is hypothesized that the incubation prior to sharp crack formation represents fatigue crack initiation in the matrix about the inclusions. This interpretation is the basis of an empirical model which can be used for life prediction. The observations in this paper are consistent with discussions of slip band formation by other authors.

Acknowledgments

The success of this program was shared as a team effort. The SNECMA team, particularly Pierre-Etienne Mosser, Jean-Yves Guedou, Jean-Charles Lautridou, Gilles Klein, and Jean-Michel Franchez, provided careful material processing. The GEAE team included Tom Knost, Bob Kissinger, Stan Friesen, George Zuefle, Nick Denton, Nancy Sullivan, Paul Domas, Herb Popp, Mike Sauby and Jean Murray Stewart. Don Weaver of Wyman-Gordon Powder Systems (formerly Cameron Powder) produced the R'88DT powder and Denny Dreyer of Special Metals Princeton Operation supplied the material used for the Type 2 seed. Metcut Research Associates, particularly Dave Wingerberg and Bill Stross, provided careful machining and testing support. Careful microscopy by Tom Daniels, Ron Tolbert, Judy Mescher, Luann Piazza, and Ivan Miller was also appreciated. Finally, the authors wish to thank GEAE for permission to publish this work and the superalloy powder suppliers for their continuous improvements in overall powder cleanliness levels.

References

1. U. S. Patent No. 4,957,567, D. D. Krueger, R. D. Kissinger, R. G. Menzies, and C. S. Wukusick, General Electric Company.

2. D. D. Krueger, R. D. Kissinger, and R. G. Menzies, "Development and Introduction of a Damage Tolerant High Temperature Nickel-Base Disk Alloy, Rene'88DT," in Superalloys 1992, ed.by S. D. Antolovich et. al, TMS-AIME, Warrendale, PA, 1992, 277-286.

3. D. R. Chang, D. D. Krueger, and R. A. Sprague, Superalloy Powder Processing, Properties, and Turbine Disk Applications, in Superalloys 1984, ed. by M. Gell et. al., TMS-AIME, Warrendale, PA, 1984, .251.

4. E. S. Huron, "Serrated Yielding in a Nickel-Base Superalloy," in Superalloys 1992, ed. by S. D. Antolovich et. al., TMS-AIME, Warrendale, PA, 1992, 675-684.

5. D. A. Jablonski, "The Effect of Ceramic Inclusions on the Low Cycle Fatigue Life of Low Carbon Astroloy Subjected to Hot Isostatic Pressing," Mat. Sci. Eng., 48 (1981), 189-198.

6. P. G. Roth, "Probabilistic Rotor Design System (PRDS) Phase I Interim Report," USAF, Aero Propulsion and Power Directorate, Wright Laboratory, Air Force Systems Command, Wright-Patterson Air Force Base, Ohio, Report WL-TR-92-2011, (1992).

7. Paul G. Roth, J. C. Murray, J. E. Morra, and J. M. Hyzak, " Heavy Liquid Separation: A Reliable Method to Characterize Inclusions in Metal Powder," Characterization, Testing and Quality Control, Advances in Powder Metallurgy and Particulate Materials, 2, Metal Powder Industries Federation, Princeton, NJ (1994).

8. S. Bashir, P. Taupin, S. D. Antolovich, "Low Cycle Fatigue of As-HIP and HIP+Forged Rene 95," Met. Trans. A, 10A (1979), 1481-1490.

9. B. A. Lerch, N. Jayaraman, and S. D. Antolovich, "A Study of Fatigue Damage Mechanisms in Waspaloy from 25 to 800C," Mat. Sci. Eng., 66 (1984), 151-166.

10. J. M. Hyzak and I. M. Bernstein, "The Effect of Defects on the Fatigue Crack Initiation Process in Two P/M Superalloys: Part 1. Fatigue Origins," Met. Trans. A, 13A (1982), 33-43.

11. J. M. Hyzak and I. M. Bernstein, "The Effect of Defects on the Fatigue Crack Initiation Process in Two P/M Superalloys: Part 2. Surface-Subsurface Transition," Met. Trans. A, 13A (1982), 45-52.

12. J. M. Hyzak, "The Effect of Defects on the Fatigue Initiation Process in Two P/M Superalloys, AFWAL-TR-80-4063, WPAFB, Ohio, July (1980).

13. L. F. Coffin, Jr., "The Effect of Frequency on the Cyclic Strain and Fatigue Behavior of Cast Rene 80 at 1600F," Met. Trans., 5 (1974), 1053-1060.

CRACK INITIATION STUDIES OF MA 760 DURING HIGH TEMPERATURE LOW CYCLE FATIGUE

A. Hynnä*, V.-T. Kuokkala** and P. Kettunen*
Tampere University of Technology
*Institute of Materials Science
**Center for Electron Microscopy
P.O.Box 589, FIN-33101 Tampere, FINLAND

Abstract

The objective of this work was to investigate the high temperature low cycle fatigue properties of a mechanically alloyed and oxide dispersion strengthened Ni-base superalloy MA 760. The push-pull fatigue tests at different strain rates using specimens of two different grain orientations (longitudinal, L and long transverse, LT) were carried out at temperatures of 650 °C, 950 °C and 1050 °C. The control parameter in all tests was the total strain amplitude. After the tests, detailed fractographical and microstructural studies using optical microscopy, scanning electron microscopy and analytical electron microscopy were carried out. In this paper, the main emphasis is placed on the fatigue crack initiation mechanisms and their dependence of fatigue test parameters and grain orientations. Also, a brief overview of the mechanical behavior of MA 760 during high temperature cyclic straining will be presented.

Introduction

Oxide dispersion strengthened (ODS) superalloys form one of the new materials groups developed in search for materials with better properties in extreme operating conditions. They maintain their high strength and good corrosion resistance at higher temperatures better than conventional superalloys, and are therefore of great interest, for example, to designers of gas turbine engines (Ref. 1). ODS superalloys are produced by mechanical alloying process (Ref. 2) followed by conventional processing and heat treating processes, including extrusion, hot rolling, recrystallization- and γ'-precipitation heat treatments. The excellent properties of ODS-alloys are due to the very fine dispersion of stable, incoherent oxide particles, formed during the mechanical alloying process. These particles act as barriers to the movement of dislocations (Ref. 3). ODS Ni- base superalloys can also be strengthened by coherent γ'-precipitates. The γ'-strengthening is acting effectively at lower temperatures than dispersion strengthening, which becomes dominant at temperatures above 1000 °C (Ref. 4). After recrystallization heat treatment, the grain structure of ODS- alloys is very coarse and the grain aspect ratio is large. Together with serrated grain boundaries, the microstructures that develop result in properties that are necessary for high temperature service, where the amount of transverse grain boundaries should be minimized (Ref. 5).

The design criteria for components made of ODS- alloys are not up to date and sufficient due to the lack of experimental data. In addition, the high temperature micromechanisms and microstructures produced by the deformation processes are not well enough established. This is especially true for the MA 760, a modified more corrosion resistant version of the well known alloy MA 6000. In this paper, we report the low cycle fatigue properties of MA 760 at elevated temperatures putting the main emphasis on the crack initiation and crack propagation mechanisms.

Material and experimental techniques

The test material, a mechanically alloyed Ni-base ODS superalloy MA 760, was delivered by INCO Alloys Ltd., Hereford, U.K. The material was produced in the form of bars of rectangular cross-sections. The bars were of two different cross-sections, SCS (200mm×600mm) and LCS (320mm×950mm). The SCS-bars were completely produced at Hereford, whereas the final recrystallization heat-treatment of the LCS-material was performed at Asea Brown Boveri, Baden, Switzerland. Compositions of the delivered bars are presented in Table I. The final heat treatment of the SCS bar was 0.5 h at 1100 °C, furnace cooling with 60 °C/h to 600 °C, followed by uncontrolled furnace cooling. The heat treatment procedure for the LCS bar was the same with the exception that the solutioning temperature was 1120 °C.

Table I Compositions of the delivered bars

Element	SCS, [wt-%]	LCS, [wt-%]
C	0.043	0.042
Si	0.04	0.05
Fe	1.02	1.04
Cr	19.66	19.79
Al	5.97	5.93
Mo	1.92	1.96
S	0.003	0.003
Zr	0.14	0.14
N	0.284	0.286
O	0.54	0.6
W	3.5	3.5
B	110 ppm	110 ppm
P	<0.005	<0.005
Ni	Bal.	Bal.
Y_2O_3	1.03	1.03

The SCS material was aimed for mechanical testing in the L-direction and the LCS material in the LT-direction. The orientation of the fatigue specimens relative to the grain orientation and the extrusion direction are presented in Figure 1.

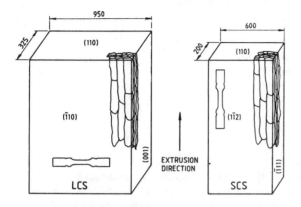

Figure 1. The orientation of the fatigue specimens in the SCS- and LCS-bars relative to the grain orientation and extrusion direction. Fatigue specimens are not in scale.

The push-pull fatigue tests were carried out with a MTS TestStar™ microcomputer based servohydraulic materials testing system. In all tests, $\Delta\varepsilon_t$ was used as a control parameter. A symmetrical triangular wave shape (R= -1) was applied using the strain rates of 0.5 s^{-1} (fast, F) or 0.05 s^{-1} (slow, S). The tests were carried out at three different temperatures specified in Table II. A more detailed description of the testing techniques is presented in Ref. 6.

Results and discussion

Mechanical behavior

A summary of the fatigue test parameters and test results is shown in Table II. The stress amplitudes σ_{max}, σ_{min} and $\Delta\sigma/2$ were measured from the cycle recorded at half life $N_f/2$. As a failure criterion, a 20 % stress drop was used.

During cyclic straining, there is no observable cyclic hardening or cyclic softening in MA 760, except during the crack growth period at the very end of each test. Also the loop shapes stay practically unchanged during cyclic straining. These observations indicate that the alloy is in a cyclically stable state from the initial cycles onwards, i.e., a stable dislocation configuration is generated at the beginning of the test. Also other microstructural changes during fatigue appear to be negligible.

The ductility of MA 760 changes markedly over the temperature region covered in this study. Therefore, in the Coffin-Manson plot, the relative positions of the data points at 650 °C change dramatically when total strain amplitude is replaced by plastic strain amplitude. If the comparison of fatigue lives between different temperatures is made purely on the basis of Coffin-Manson plots, one may enter into erroneous conclusions. Therefore, the comparison should be based on $\Delta\varepsilon_t$ or $\Delta\sigma/2$ vs. N_f plots. On the basis of $\Delta\varepsilon_t$ vs. N_f plots, the increasing creep component shortens the fatigue life. At comparable test parameter values the fatigue lives of the LT-samples are in most cases shorter than the fatigue lives of the L-samples. More information of the mechanical behavior of the alloy MA760 can be found in Ref. 7.

Table II Numerical values recorded in the fatigue tests.

Test No.	Specimen orientation	T [°C]	$\dot{\varepsilon}$ [%s^{-1}]	$\Delta\varepsilon_t$ [%]	$\Delta\varepsilon_{pl}$ [%]	$\Delta\varepsilon_e$ [%]	σ_{max} [MPa]	σ_{min} [MPa]	$\Delta\sigma/2$ [MPa]	N_f [cycl.]
1	L	650	0.5	1.0	0.1	0.9	850	-950	900	792
2	L	650	0.5	0.8	0.04	0.76	705	-780	742.5	3925
3	L	650	0.5	0.6	0.01	0.59	560	-575	567.5	16320
4	L	950	0.5	1.0	0.51	0.49	310	-340	325	570
5	L	950	0.5	0.8	0.35	0.45	275	-305	290	960
6	L	950	0.5	0.6	0.2	0.4	250	-270	260	4575
7	L	950	0.05	1.0	0.49	0.51	290	-300	295	415
8	L	950	0.05	0.8	0.34	0.46	270	-285	277.5	880
9	L	950	0.05	0.6	0.19	0.41	240	-265	252.5	3945
10	L	1050	0.5	0.8	0.55	0.25	160	-180	170	395
11	L	1050	0.5	0.8	0.55	0.25	160	-180	170	595
12	L	1050	0.5	0.6	0.35	0.25	170	-195	182.5	1025
13	L	1050	0.5	0.35	0.11	0.24	150	-160	155	6925
14	L	1050	0.5	0.25	0.01	0.24	160	-180	170	67300
15	L	1050	0.05	0.8	0.55	0.25	170	-190	180	305
16	L	1050	0.05	0.6	0.36	0.24	160	-180	170	620
17	L	1050	0.05	0.35	0.1	0.25	160	-170	165	3350
18	LT	950	0.5	1.0	0.66	0.34	285	-300	292.5	600
19	LT	950	0.5	0.6	0.3	0.3	250	-275	262.5	2105
20	LT	950	0.5	0.45	0.15	0.3	240	-260	250	6770
21	LT	1050	0.5	1.0	0.8	0.2	140	-160	150	105
22	LT	1050	0.5	0.6	0.42	0.18	145	-155	150	585
23	LT	1050	0.5	0.45	0.25	0.2	145	-150	147.5	1125
24	LT	1050	0.05	1.0	0.79	0.21	140	-145	142.5	110
25	LT	1050	0.05	0.6	0.4	0.2	135	-155	145	580
26	LT	1050	0.05	0.45	0.26	0.19	145	-150	147.5	1280

Crack initiation

During low cycle fatigue, several different damage accumulation processes are active simultaneously, and the crack initiation takes place at the weakest link in the microstructure. In the L-samples, the initiation sites fall into several categories. The most frequent initiation site seems to be a longitudinal grain boundary crossing the specimen surface; in this case the initiation site can be considered to be at the specimen surface. In Figure 2, a SEM image of a typical example falling into this category is presented. The magnification of this image is rather low and the boundary indicated by an arrow is not easily visible, but the opened boundary is more clearly visible in Figure 3 taken from the surface of sample 7. Convincing evidence of crack initiation at the opened boundary is also given in Figure 3, which clearly illustrates secondary cracking at the boundary crossing the specimen surface. Crack initiation at these boundaries is enhanced by oxidation effects. The high energy level, carbides and γ'-film along the boundaries (Ref. 8) promote the oxidation of the boundaries.

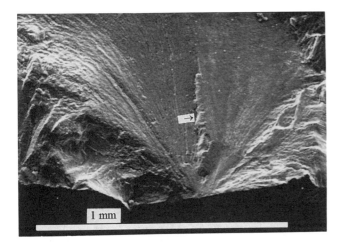

Figure 2. Fracture initiation site in sample 7.

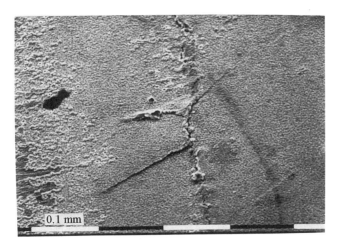

Figure 3. Longitudinal opened boundary on the surface of specimen 7. Crack initiation takes place at the opened boundary.

Occasionally individual small grains or a group of small grains were also observed to act as initiation sites in L-samples. These grains lie near the specimen surface, in some cases extending to the surface itself. However, in all observed cases the distance to the free surface was no more than 0.1 mm. The reason for the easy crack initiation at these sites can be understood by noting that the boundaries around small grains are very susceptible to the opening phenomenon (Ref. 9). It is believed (Ref. 10) that one reason for the susceptibility of the fine grains to fatigue damage is related to the differences in the crystallographic orientation between the large and small grains, resulting in high local stress concentrations at fine grain boundaries during loading. Also, if the high angle boundaries associated with the small grains extend to the specimen surface, they will be favorable sites for oxidation damage. In the L-samples crack initiation at subsurface inclusions was detected in two cases. In sample 6 an Al-rich inclusion containing 5% copper was found at the initiation site, and in sample 16 a Cr-rich inclusion had initiated the crack.

The crack initiation sites in the LT-samples were only, except for test 25, at transverse boundaries crossing the specimen surface as illustrated in Figure 4. The initiation sites were similar to those found in the L-samples, the only difference being the orientation of the boundaries. Only in one case initiation at an inclusion clearly inside the specimen was observed. In this case, the observed Al-rich defect was an exceptionally large one, approximately 0.5 mm in all three dimensions, and situated in a narrow grain with a width of approximately the same as the diameter of the inclusion.

Figure 4. Crack initiation at transverse grain boundary crossing the surface in sample 19.

On the basis of the crack initiation studies it may be concluded that the boundaries are the weakest link in the alloy MA 760. When the initiation of a crack was associated with an inclusion or small grains, no drastic effect on the fatigue life, expressed e.g. as Coffin-Manson-type plots, can be found. The fatigue lives of the LT-samples are shorter to those of the L-samples, a phenomenon clearly related to the damage accumulation during the crack initiation period. This leads to the conclusion that the damage accumulation at the sites of the intersecting boundaries and free surface in LT-samples is faster than in L-samples. One possible explanation for this behavior is, if oxidation effects are assumed to be identical, that there exist higher stress concentrations at crack initiation sites in the case of LT-samples. Indeed, this may be the case, because in each of the LT-samples there existed at least one packet of narrow grains crossing the free surface, as illustrated also in Figure 4. The boundaries of these grains were of the high angle type, an observation based on the large contrast differences between individual grains. In fact, the crack initiation site in every LT-sample (except sample 25) was associated with these packets of fine grains. It can be assumed that the local stress concentrations are higher at these sites compared to the situation in L-samples, where the initiation site was associated with a single boundary of low angle type. This explanation could account for the smaller strain rate dependence (a smaller role of time dependent damage accumulation) in LT-samples compared to the L-samples. Also, it is possible that the damage accumulation due to oxidation at these high angle boundaries is faster than at (in most of the cases) low angle boundaries in L-samples.

Surface studies

The aim of the surface studies was to investigate the effects on the specimen surface that could be involved in the crack initiation processes. A common phenomenon observed was the opening of the boundaries at the specimen surface, acting often as a site for crack initiation. Other surface effects include oxidation of the boundaries of small grains and oxidation of inclusions. However, obviously the damaging effects of these changes are small compared to the opened boundaries, because they were only occasionally involved in the crack initiation, although the opening susceptibility of the boundaries around small grains seems to be higher than that of the boundaries between large grains. The oxidation of these sites, as well as the overall oxidation of the surface, was negligible at 650 °C but became more pronounced at higher temperatures.

A general feature in all samples fatigued at 950 °C and at 1050 °C was the presence of unevenly spaced slip lines (max. length 2 mm measured along the specimen surface). Also the presence of short, ca. 0.1 -0.5 mm, slip line piles was observed. The amount of the lines clearly increased with decreasing strain rate. Also the increasing temperature had the same, although not so pronounced, effect. This behavior has been previously reported by Elzey and Arzt (Ref. 11) for MA 6000, but only at high plastic strain amplitudes. Generally the dispersoid particles are regarded to promote slip dispersal compared to non-ODS alloys. However, it should be pointed out that the density of the slip lines in the gauge length was not generally uniform. As a rule, most of the specimen surface was completely free of slip lines and the lines concentrated into a relatively narrow band near the fracture surface. In the tests conducted at 950 °C, only relatively few single lines in the bands of width less than one millimeter were detected, and generally they were associated with the boundaries of fine grains, as illustrated in Figure 5. The band width increased with increasing temperature and decreasing strain rate. The existing lines form a site for microcrack initiation, but normally these microcracks do not determine the fatigue life of the sample. Rather it seems that the primary crack is initiated at a point where an individual slip line crosses an opened boundary, as illustrated in the Figure 6. From the observations mentioned above, the deformation at 950 °C seems to be much more concentrated into fewer slip lines than at 1050 °C. According to these results the formation tendency of the slip lines becomes marked at 1050 °C (above γ' solidus). Also the lowering of the strain rate seems to have an effect of the same trend.

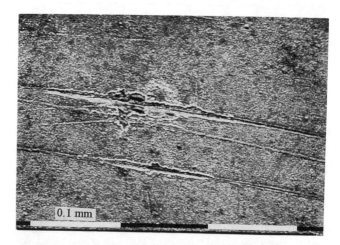

Figure 5. Slip line formation associated with the fine grains in sample 5.

Figure 6. The intersection of slip lines and an opened boundary in sample 5.

The slip lines can extend themselves over considerably long distances along the surface and secondary stage I cracks initiate at these sites. The cracks grow transgranularly and the longest observed cracks extended 50 μm into the grains. In the tests conducted at 950 °C, this short crack growth behavior revealed one of the deformation mechanisms to be γ' cutting. This is illustrated in Figure 7, in which the tip of a 15 μm deep crack is presented. Clearly at the very tip of the crack, cut γ' particles are present, indicated by an arrow. Occasionally slip lines were observed in the sectioned surface of the samples and only in the etched condition. In these cases the lengths of these lines varied from 10 to 30 μm. Also in these areas of intense deformation, the process of γ' cutting was clearly detectable. The slip lines were not observed in series L650F tests, but in all other tests they were detected occasionally in the longitudinal section of the samples. It is worth mentioning that this could also be an etching artifact caused by the γ' dropping-off from the etched surface. However, milder etching conditions did not reveal a marked increase of the slip lines and therefore it is assumed that the dropping-off does not play a significant role. A general view of the area of an abnormally high density of slip lines is presented in Figure 8.

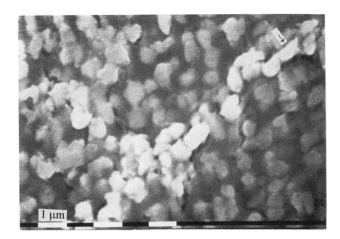

Figure 7. Tip of a microcrack in sample 6.

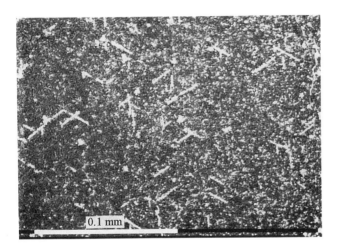

Figure 8. Slip lines in sample 6.

The strong [110] texture permits a more detailed analysis of the active slip planes to be performed. The nearly longitudinal slip lines in Figure 8 are perpendicular to the plane (110) and therefore they are traces of either conjugate $(1\bar{1}1)$ or cross $(1\bar{1}\bar{1})$ slip planes. If it is assumed that the slip plane is $(1\bar{1}1)$, then the other possible slip planes are critical (111), primary $(11\bar{1})$ and cross $(1\bar{1}\bar{1})$. If the second alternative holds, then the other possible planes are critical (111), conjugate $(1\bar{1}1)$ and primary $(11\bar{1})$. Normally the indexing of the slip line patterns is based on determining the loading axis and side surface orientations by, e.g. Laue X-ray technique. This tedious procedure is commonly applied to single crystals, but it can be applied to large grained materials also. However, the Laue method was not used in this study, but instead an approximate method was applied. The procedure in this method was as follows: 1) measurement of the angles between traces of three {111}-type slip planes on a longitudinal section of unknown indeces, 2) calculation of the angles between the trace vectors of {111}-type planes on different hkl-planes, and 3) comparison of the calculated angles with the measured ones using a computer program developed for this purpose (Ref. 12). The angles of the traces of the other two planes with the almost vertical one were both measured and found to be 58 degrees. In the calculations, error limits of ±1 degree for the measured angles were used and the value of 10 for the highest index of the unknown plane was chosen. It was found that the indices of the unknown plane were of type {10 10 9}. The calculations indicated that the two inclined slip planes must be the primary and the critical slip plane with angles of 57.3 degrees with the vertical {111}-type plane.

The Schmid factor for the primary slip system $(11\bar{1})[101]$, as well as for the critical slip system $(111)[10\bar{1}]$ is $1/\sqrt{6} \approx 0.408$. On the other hand, for the conjugate slip system $(1\bar{1}1)[110]$ and for the cross slip system $(1\bar{1}\bar{1})[101]$ the factor is zero, so that only the primary and the critical slip systems can be active in the SCS material. The vertical slip lines observed are thus probably an indication of the dislocation reactions between the primary and the critical slip planes.

Conclusions

1. Most of the cyclic life of MA 760 is spent in the fatigue initiation stage. The crack initiation is enhanced by the oxidation of the boundaries crossing the free surface of the sample, which in most of the cases act as crack initiation sites.

2. The inferior fatigue life, as well as the minor strain rate effects on the fatigue life of the LT-samples compared to the L-samples are due to higher stress concentrations at the fatigue initiation sites. This, in turn, results from the differences in the crack initiation sites. In the L-samples the initiation site in most cases is the low angle boundary crossing the specimen surface. In the LT-samples, the initiation site most commonly is a packet of narrow grains separated by high angle boundaries and crossing the specimen surface.

3. The tendency for heavier slip line formation in the specimen surface increased with increasing temperature and decreasing strain rate. At lower temperatures the deformation was concentrated into fewer slip lines indicating that the homogeneity of the deformation increases with increasing temperature. The crack initiation site was commonly the intersection point of a slip line and a boundary weakened by oxidation.

4. Slip line observations on the cross-sections of the L-samples fatigued at 950 °C show that the operative slip systems during cyclic straining are primary and critical slip systems. Also slip lines belonging to the cross or conjugate systems were found, but the Schmid factor in these systems is close to zero. Therefore it is concluded that these lines are a reaction product between the primary slip plane and the critical slip plane.

References

1. G. W. Meetham, "Superalloys in Gas Turbine Engines", Met. and Mat. Tech., 14(1982), 387-392.

2. J. S. Benjamin, "Mechanical Alloying", Sci. Am., 234(5)(1978), 40-48.

3. Y. Kaieda, "Trends in Development of Oxide-Dispersion-Strengthened Superalloys", Trans. Nat. Res. Inst. Metals, 28(3)(1986), 18-24.

4. M. J. Fleetwood, Mechanical Alloying - The Development od Strong Alloys", Mat. Sci. Tech., 2(1986), 1176-1182.

5. H. Zeizinger and E. Arzt, "The Role of Grain Boundaries in High Temperature Creep Fracture of an Oxide Dispersion Strengthened Superalloy", Z. Metallkde., 79(1988), 774-781.

6. A. Hynnä, High Temperature Low Cycle Fatigue Behavior of ODS Superalloy MA 760 - Mechanical Behavior, Fracture- and Deformation Characteristics, Microstructural Changes and Dislocation Configurations, (Doctoral Thesis, Tampere University of Technology), Tampere University of Technology Publications 140 (Tampere, Finland, 1994), 98.

7. A. Hynnä, V. T. Kuokkala and P. Kettunen, "Mechanical Behavior of Superalloy MA 760 during High Temperature Cyclic Straining", (Paper to be presented at The Sixth International Fatigue Congress "Fatigue '96" Berlin, FRG, 6 - 10 May 1996).

8. A. Tekin, J. W. Martin, "Fatigue Crack Growth Behaviour of MA 6000", Mat. Sci. Eng., 96(1987), 41-49.

9. E. Arzt, "High Temperature Properties of Dispersion Strengthened Materials Produced By Mechanical Alloying: Current Theoretical Understanding and Some Practical Implications", New Materials by Mechanical Alloying Techniques, ed. E. Arzt and L. Schultz (Oberursel, FRG: Deutsche Gesllschaft für Metallkunde e.V., 1989), 185-200.

10. D. M. Elzey, E. Arzt, "Crack Initiation and Propagation during High-Temperature Fatigue of Oxide Dispersion-Strengthened Superalloys", Metall. Trans. A, 22A,(1991), 837-851.

11. D. M. Elzey, E. Arzt, "Oxide Dispersion Strengthened Superalloys: The Role of Grain Structure and Dispersion During High Temperature Low Cycle Fatigue", Superalloys 1988, ed. S. Reichman et al. (Warrendale PA: The Metallurgical Society, 1988), 595- 604.

12. V.-T. Kuokkala, "Indexing of Slip Line Patterns on Single Crystalline FCC Specimens" (Report Tampere University of Technology, Institute of Materials Science 62/1993), 7.

CYCLIC DEFORMATION OF HAYNES 188 SUPERALLOY
UNDER ISOTHERMAL AND THERMOMECHANICAL LOADINGS

Michael G. Castelli[1] and K. Bhanu Sankara Rao[2]

[1]NYMA, Inc., [2]National Research Council
National Aeronautics and Space Administration
Lewis Research Center
Cleveland, Ohio 44135

Abstract

A detailed assessment of the macroscopic and microstructural cyclic deformation behavior of Haynes 188™ has been conducted under various isothermal and thermomechanical conditions over the temperature range 25 to 1000°C. A fully reversed mechanical strain range of 0.8% was examined with constant mechanical strain rates of $10^{-3} s^{-1}$ and $10^{-4} s^{-1}$. Particular attention was given to the effects of dynamic strain aging (DSA) on the stress-strain response. Detailed transmission electron microscopy was conducted to examine the deformation substructures and establish correlations with the cyclic macroscopic behaviors. Although DSA was found to occur over a wide temperature range between approximately 300 and 700°C, the microstructural characteristics and deformation mechanisms responsible for DSA varied considerably and were dependent upon temperature. In general, the operation of DSA processes led to a maximum of the cyclic stress amplitude at ~650°C, and was accompanied by pronounced planar slip, the generation of stacking faults and high dislocation density. DSA was evidenced through a combination of phenomena, including serrated yielding, an inverse dependence of maximum cyclic hardening with strain rate ($\dot{\varepsilon}$), and an instantaneous inverse $\dot{\varepsilon}$ sensitivity. The TMF cyclic hardening behavior exhibited unique behaviors in comparison to the isothermal response, predominantly at the minimum TMF temperature extremes.

Introduction

The cobalt base superalloy, Haynes 188™, is currently used in many aeronautics and aerospace applications, including extensive use in military and commercial aircraft turbine engines as a combustor liner material. The choice for use of this material is primarily based upon a good combination of its monotonic yield and tensile strength properties, excellent fabricability, weldability and good resistance to high temperature oxidation for prolonged exposures [1]. In the vast majority of aero applications, the various components are subject to repeated thermal and mechanical loadings, thus, making high temperature cyclic deformation and failure a primary concern for long term structural integrity and component life. Initial examinations of the high temperature cyclic behavior of Haynes 188™ revealed relatively complex behaviors. Various combinations of temperature, strain, and strain rate exert a strong influence on the cyclic stress response through the interaction of various time- and temperature-dependent phenomena, such as inelastic flow, dynamic strain ageing (DSA), creep and environmental damage, and dynamic precipitation.

Thus, with the ultimate goal of accurately modeling and predicting such behaviors, it is critical to have a comprehensive understanding of both the macroscopic cyclic deformation behavior under general thermomechanical loading conditions, and the micromechanisms which influence such behaviors, as revealed in the deformation substructure. Further, this understanding must include not only material behaviors at maximum operating temperatures, but also the behaviors at the lower temperatures encountered during transients.

Several past efforts aimed at characterizing the effects of temperature and mechanical loads on the fatigue behavior of Haynes 188™ have generally been focused upon relatively select conditions. These include elevated temperature stress-free exposures [2], the influence of temperature and creep-fatigue interaction on strain-life relationships [3], and the examination of thermomechanical and bithermal fatigue behavior and damage modes for developing suitable methods for life prediction [4-6]. Additionally, loading frequency effects on crack growth behavior [7] and the effects of axial/torsional loading on cyclic deformation [8] have been examined. With these efforts, however, the cyclic stress response and fatigue lives over a wide temperature range have not been examined in depth; and further, there is little to no information available on the dislocation arrangements and deformation induced microstructures of Haynes 188™ during elevated temperature fatigue loadings.

Toward the end of characterizing both the macro and substructural deformation behaviors during high temperature fatigue loadings, a series of isothermal fatigue, and in-phase and out-of-phase thermomechanical fatigue (TMF) tests were conducted in the range 25 to 1000°C at multiple strain rates on Haynes 188™. As the thermal stresses within a component are often highly localized and deformation constrained by surrounding material (i.e., deformation controlled), strain controlled loadings were examined using a fully reversed mechanical strain range of 0.8%. Detailed examinations of the deformation substructures were performed leading to a correlation between the micro and macro phenomenological cyclic deformation behaviors. Particular attention was given to the effects of dynamic strain aging (DSA) on the stress-strain response. The temperature dependence of the TMF deformation behavior is examined over several temperature intervals and compared with those observed under isothermal conditions. Macroscopic hardening behaviors unique to the TMF conditions are discussed and interpreted in accord with the isothermal deformation substructures.

Material and Experimental Details

The Haynes Alloy 188™ was originally supplied as hot rolled and solution annealed bar of 19 mm dia. Solution annealing was carried out at 1175°C for 1 hr followed by water quenching, which produced an average grain size of ASTM 6 in the longitudinal direction. The material had the chemical composition (wt%) of 22.43Ni, 21.84Cr, 13.95W, 1.24Fe, 0.01C, 0.75Mn, 0.40Si, 0.012P, 0.002S, 0.034La, 0.002B with the balance being Co.

Fully-reversed, total axial strain controlled isothermal LCF tests were performed on 6 mm dia. and 12.5 mm parallel gage length specimens at different temperatures in the range 25 to 1000°C, employing a triangular waveform and an axial strain range ($\Delta\varepsilon$) of 0.80%. The strain rates ($\dot{\varepsilon}$) applied in the isothermal tests included $10^{-4}s^{-1}$ and $10^{-3}s^{-1}$. In-phase (IP) and out-of-phase (OP) TMF tests with a mechanical $\dot{\varepsilon}$ of $10^{-4}s^{-1}$ were conducted on parallel sided tubular specimens featuring a 8.4 and 11.4 mm inner and outer dia., respectively, and a longitudinal gage length of 12.7 mm. Here, IP loadings were defined as the condition where maximum tensile strain coincided with the maximum temperature, and the maximum compressive strain with the minimum temperature; whereas, under OP conditions the maximum tensile strain coincided with the minimum temperature, and the maximum compressive strain coincided with the maximum temperature. In the TMF tests, the control of total axial strain was accomplished by continuously providing for compensation of the thermal strain, such that the mechanical strain range ($\Delta\varepsilon^m$) was 0.80 %.

Strains were measured with a high temperature water-cooled extensometer. All test specimens were heated with direct induction heating and cooling was accomplished through the use of water cooled grips; no forced air was used. Temperature gradients over the gage section were maintained within ±1% of the nominal test temperature for the isothermal tests and ±1.5% of the nominal dynamic test temperature for the TMF tests. The temperature was monitored and controlled with K-type thermocouples spot welded on the specimen surface. Cracks were not found to preferentially initiate at the spot weld locations during testing.

In general, a minimum of four specimens were tested at each of the isothermal temperatures. For a given set of conditions, duplicate tests showed very similar hysteresis behavior and the extent of variation in cyclic stress hardening/softening was relatively small. For the sake of brevity, the results reported here are limited to data from single specimens tested at each temperature. The TMF tests consisted of only single tests for each condition.

Microstructural examinations were conducted on failed specimens to document the microstructural changes that occurred during testing. Energy dispersive analysis of X-rays (EDAX) and the determination of lattice parameters by X-rays, were used to identify various precipitates. Deformation induced substructures were studied by transmission electron microscopy (TEM). Samples for TEM examination were obtained from thin slices cut at a distance of 1 mm away from the fracture surface. The slices were mechanically thinned down to 250 µm, and then electropolished in a solution containing 10% perchloric acid and 90% methanol, at 22 V and 5°C, in a twin jet apparatus. Thin foils were examined in an electron microscope operating at an acceleration voltage of 120 keV.

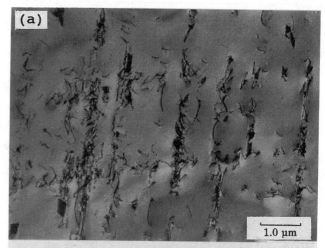

Figure 1: Representative TEM micrographs depicting (a) dislocation dipoles and bundles in two slip planes at 25°C and (b) planar slip bands exhibiting loops, dipoles and stacking faults at 300°C.

Results and Discussion

The primary focus of the research was to examine and characterize both the substructural and macroscopic deformation behaviors under high temperature fatigue loadings, and with this, to establish a correlation between the two. Therefore, the discussion will begin by examining the microstructure in the as-received condition and those developed as a result of the isothermal fatigue loadings. Subsequently, the macroscopic isothermal and TMF deformation behaviors will be described and discussed in view of the features detailed in the deformation substructures.

Isothermally Developed Deformation Substructures

In the as-received condition (solution annealed state), the Haynes 188™ alloy contained randomly distributed undissolved blocky particles identified as tungsten rich M_6C through energy dispersive X-ray and X-ray diffraction analyses [9]. It is apparent that these precipitates nucleated in the melt and were not dissolved during the solution treatment. These bulk precipitates will be referred to hereafter as primary M_6C

Microstructures developed during isothermal fatigue at 25 and 300°C with $\dot{\varepsilon}=10^{-3}s^{-1}$ are shown in Fig. 1a and 1b, respectively.

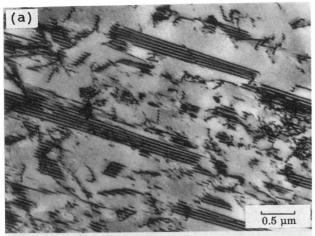

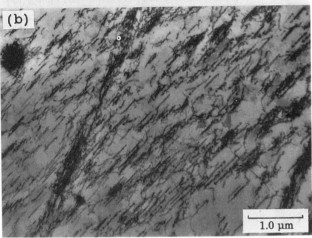

Figure 2: Representative TEM micrographs exhibiting the predominant staking faults and planar slip generated under (a) 400°C, and (b) 550°C isothermal fatigue conditions.

The features revealed in this figure are representative of those which developed over the general temperature range from 25 to 300°C. Based upon the macroscopic deformation data to be discussed, this temperature range represents a regime below the DSA domain, that is, the low temperature regime in which DSA effects appeared not to be active. In this temperature regime, deformation substructures were generally made up of distinctly spaced slip bands incorporating dislocation bundles and dipoles, and tangles at the intersection of multiple slip bands such as illustrated in Fig. 1a. As illustrated in Figs. 1a and b, planar slip occurred with the simultaneous operation of two slip systems. With increasing temperature in this range, a relative increase in dislocation density within the slip bands was evident. Thus, the substructure resulting from fatigue cycling at 300°C exhibited intense slip bands in which dislocation debris in the form of loops, dipoles and stacking faults accumulated, as illustrated in Fig. 1b. Here, it was often difficult to distinguish individual dislocations in the slip bands because of extensive tangling. Although dislocations were generated in two slip systems, the density of one system in a given grain was usually higher than that of the other, probably due to the difference between shear stresses of different slip systems. In this temperature regime, no significant differences were evident between the microstructures generated under $\dot{\varepsilon}=10^{-3}s^{-1}$ and $\dot{\varepsilon}=10^{-4}s^{-1}$ isothermal fatigue conditions.

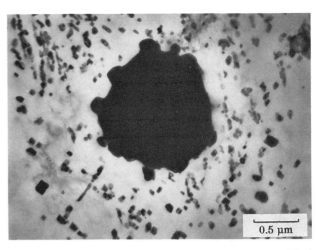

Figure 3: TEM micrograph depicting very fine $M_{23}C_6$ precipitated about the larger primary M_6C at 750°C.

The second temperature domain discussed is the general regime where DSA effects were evidenced in the macroscopic deformation response, that is approximately 350 to 700°C. DSA has been proposed to occur due to the aging of mobile dislocations by solute atmospheres either during a quasiviscous type of motion of dislocations or when the dislocations are temporarily held up at local obstacles in the glide plane. When DSA operates, in order to maintain the imposed strain amplitude, additional dislocations are generated, leading to observed increases in dislocation densities. In the low to mid temperatures of the DSA domain (i.e., ~350 to 550°C), dislocations frequently formed planar arrays, extensive pile-ups at grain boundaries, and developed numerous stacking faults. These features are represented by the deformation substructures shown in Figs. 2a and 2b taken from isothermal fatigue specimens tested at 400 and 550°C, respectively, with $\dot{\varepsilon}=10^{-3}s^{-1}$. In general, the planar slip and stacking faults were developed over the entire range from 25 to ~700°C; however, the propensity for their occurrence appeared to reach a peak in the range of 400 to 550°C. The origin of the planar slip in the DSA regime is believed to result from the combined effects of the elastic interaction between the solute atoms and the dislocations, and the Suzuki segregation on stacking faults [10]. Towards the end of the DSA regime at 600 and 650°C, dislocations assumed a configuration of well knit networks with densities exhibiting a maximum, in comparison to those developed at all other isothermal test temperatures. At 650°C (in a majority of the grains) the deformation was more homogeneous and the predominant coplanar slip was not apparent. As similar to the low temperature regime, no significant differences were evident between the microstructures generated under $\dot{\varepsilon}=10^{-3}s^{-1}$ and $\dot{\varepsilon}=10^{-4}s^{-1}$ conditions.

The final temperature regime, designated as the high temperature domain, consisted of those temperatures ranging from approximately 750 to 1000°C. In the isothermal fatigue tests at 750°C, precipitation of a chromium rich $M_{23}C_6$ carbide occurred. Substantial numbers of these very fine $M_{23}C_6$ could be seen intragranularly precipitated around the larger primary M_6C as shown in Fig. 3. The $M_{23}C_6$ served to pin deformation induced dislocations and generally led to an inhomogeneous distribution of dislocations (predominantly at $M_{23}C_6$ sites), as illustrated in Fig. 4. This pinning effect was considerably reduced with increasing temperature to 850°C where the

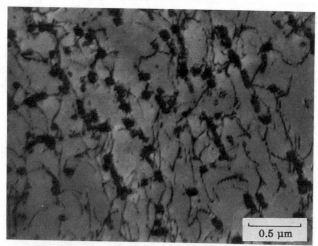

Figure 4: TEM micrograph revealing $M_{23}C_6$ dislocation pinning effect at 750°C.

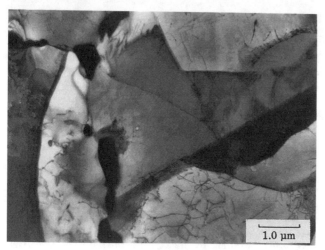

Figure 5: TEM micrograph revealing secondary precipitation of M_6C on grain boundaries at 1000°C.

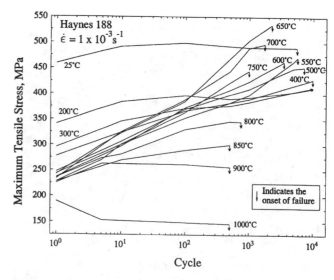

Figure 6: Cyclic maximum tensile stress during isothermal fatigue with $\Delta\varepsilon=0.8\%$ and $\dot{\varepsilon}=10^{-3}s^{-1}$.

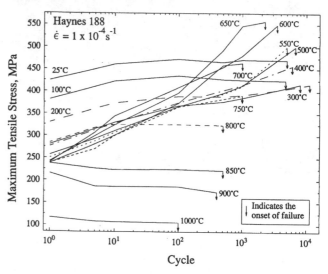

Figure 7: Cyclic maximum tensile stress during isothermal fatigue with $\Delta\varepsilon=0.8\%$ and $\dot{\varepsilon}=10^{-4}s^{-1}$.

dislocations appeared much more uniformly distributed. At temperatures higher than 850°C, dynamic recovery by thermally activated dislocation climb gained importance leading to the formation of cells and subgrains with sharp walls. Above 900°C, M_6C was found to be the predominant phase both in the intra- and intergranular regions, as secondary M_6C was found to precipitate as discrete particles on grain boundaries, shown in Fig. 5 from a specimen tested at 1000°C with $\dot{\varepsilon}=10^{-3}s^{-1}$. At the extreme temperatures of this regime, the subgrain interiors became virtually free of dislocations with the subgrain walls generally displaying orderly arrangements of dislocations.

In this high temperature domain from approximately 750 to 1000°C, modest differences could be noted between the microstructures resulting from cycling at the two different strain rates. Generally speaking, at the lower $\dot{\varepsilon}$ (i.e., $\dot{\varepsilon}=10^{-4}s^{-1}$), the substructural features associated with thermal recovery effects, such as growth of the $M_{23}C_6$, a decrease in intragranular dislocation density, the formation of cells and subgrains and the secondary precipitation of M_6C became apparent at slightly lower temperatures in comparison to the high $\dot{\varepsilon}$ tests. This, most likely due to the significantly increased time at test temperature (a factor of 10).

Macroscopic Deformation and Stress Response

Isothermal Fatigue.
The cyclic stress responses exhibited during isothermal fatigue with $\dot{\varepsilon}=10^{-3}s^{-1}$ and $\dot{\varepsilon}=10^{-4}s^{-1}$ are shown in Figs. 6 and 7, respectively. The curves illustrate the temperature dependence of the cyclic peak tensile stress (σ_{max}) as a function of progressive cycles for both strain rates. The downward arrows terminating the stress response curves represent the onset of failure of the specimen. That is, the point at which crack nucleation and growth impaired the load-carrying capacity of the specimen. As in the discussion of the deformation substructures, the temperature dependent stress response will be considered in three general domains, these are low (25 to ~300°C) mid (350 to ~700°C) and high (T ≥ 750°), where the mid temperature regime represents the DSA domain. In the low temperature regime for both strain rates, the alloy exhibited a brief period of initial moderate hardening, followed by an extended period of essentially stable stress response which persisted until the onset of failure. In general, at both strain rates, σ_{max} decreased with increasing temperature to

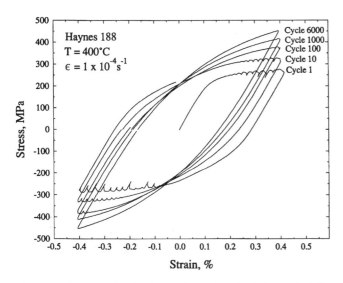

Figure 8: Progressive hysteresis loops illustrated serrated flow experienced in Haynes 188™.

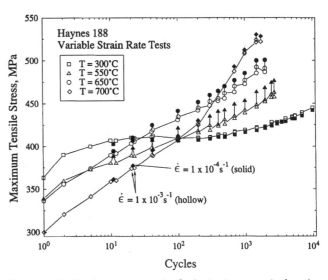

Figure 9: Cyclic stress response in the instantaneous strain rate sensitivity tests (10^{-3} to $10^{-4} s^{-1}$).

300°C, consistent with the trend exhibited by the first-cycle stress. Thus, the cyclic hardening exhibited at approximately 300°C represents a local minimum which occurs just prior to the regime where DSA becomes active.

In the mid temperature domain (350 to ~700°C), which varied slightly depending on $\dot{\epsilon}$, continuous cyclic hardening generally occurred until the onset of failure. Here σ_{max} displayed an increase with increasing temperature to approximately 650°C where the most pronounced hardening behavior was observed. Also note that the number of cycles required to attain σ_{max} decreased with increasing temperature. Above 650°C, the σ_{max} exhibited a fall (relatively large in the $\dot{\epsilon} = 10^{-4} s^{-1}$ tests) with increasing temperature. Calculations of the inelastic strain range developed in the cycle at half-life showed a rapid decrease with a raise in temperature in this range to 650°C at both strain rates. In tests performed at $\dot{\epsilon} = 10^{-3} s^{-1}$, the deformation behavior displayed serrations in the plastic portions of stress-strain hysteresis loops between 300 and 850°C, while at the lower $\dot{\epsilon}$ serrated flow was observed over the much narrower range of 400 to 650°C. This behavior is exemplified by the progressive hysteresis loops shown in Fig. 8 from a test conducted at 400°C with $\dot{\epsilon} = 10^{-4} s^{-1}$. Here, serrated flow is experienced predominantly in the early cycles.

The occurrence of DSA was suggested by several observations, including i) the maximum cyclic hardening occurred at an intermediate temperature, i.e., 650°C, ii) the deformation behavior revealed serrated flow, and iii) both the σ_{max} and the inelastic strain at half-life showed an inverse temperature dependence with strain rate in the mid temperature domain. A comparison of σ_{max} values attained at identical temperatures in Figs. 6 & 7 revealed an increase in σ_{max} with decreasing $\dot{\epsilon}$ between 400 and 700°C, establishing the negative dependence of σ_{max} on $\dot{\epsilon}$. Note that above and below this temperature range, σ_{max} decreased with decreasing $\dot{\epsilon}$. The occurrence of a negative dependence of maximum stress on the $\dot{\epsilon}$ is a typical manifestation of fatigue deformation accompanying DSA [11-13].

In an effort to obtain further insight into the macroscopic aspects of DSA during isothermal fatigue, the instantaneous strain rate sensitivity (ISRS) of cyclic stress was also examined. The ISRS is to be distinguished from the strain rate sensitivity of σ_{max} discussed above, as the ISRS corresponds to an immediate material response brought about by a sudden change in loading rate, with the condition that the substructure has not undergone any significant changes from one loading to the next. Whereas, the strain rate sensitivity of σ_{max} is related to the overall stress hardening as influenced by $\dot{\epsilon}$ under cyclic conditions. The ISRS was investigated by conducting instantaneous strain-rate-change tests at several temperatures; these tests were accomplished by conducting cyclic tests with a nominal $\dot{\epsilon}$ of $10^{-3} s^{-1}$ and periodically changing the $\dot{\epsilon}$ to $10^{-4} s^{-1}$ for one cycle and then reverting to the faster rate. This "change-up" was implemented at periodic intervals throughout the test. Representative results from these tests are given in Fig. 9. The data indicated that the ISRS of the stress amplitude was positive at 300°C, strongly negative at 550 and 650°C and only slightly negative at 700°C. At 750°C, the strain rate sensitivity was negative in the first 200 cycles, unaltered between 200 and 400 cycles and became positive above 400 cycles. The ISRS was positive above 750°C. At 550 and 650°C, when the strain rate was decreased to $10^{-4} s^{-1}$, the stress amplitude increased significantly and then returned to its regular progression upon increasing the strain rate to $10^{-3} s^{-1}$, indicating the change in stress amplitude is truly a reversible strain rate effect but not associated with a change in microstructure. The ISRS became progressively negative with increasing number of cycles in the range 400-700°C. In conjunction with the observation of several other manifestations, the strain rate change test results suggest the operation of DSA between 400 to 750°C. It is important to note, however, that the domain of serrated flow was much broader than that of the DSA, if the negative ISRS is considered as the compelling indicator of DSA.

The main features of dislocation substructure in the DSA range (~350 to 700°C) were the pronounced occurrence of stacking faults and co-planar distribution of dislocations as shown in Fig. 2, and very high dislocation densities and very fine precipitation of carbides towards the maximum end of the temperature range. The microstructural origin of DSA effects has been studied in various stainless steels [12,14,15]. In these materials, the dislocation substructure changed from a so-called wavy slip mode to a predominantly planar slip mode in the temperature range of maximum DSA and became wavy again at higher temperatures. In the present study, the observation of multiple slip below 300°C, the occurrence of co-planar slip bands with high dislocation density in the DSA range, and the cells and subgrains above the DSA range are all in qualitative agreement with the results reported on austenitic stainless steels. The abnormally high rates of cyclic hardening observed in the DSA regime could be associated with either the development of greater dislocation densities or precipitation of the $M_{23}C_6$ phase. The increase in the work hardening rate in the DSA regime has been attributed to the increased rate of accumulation of dislocations [13,15-17]. During DSA, due to aging of mobile dislocations by solute atmospheres, the material will continuously generate new dislocations in order to maintain the desired density of mobile dislocations at the imposed strain rate. Alternatively, $M_{23}C_6$ precipitation will hinder the dislocation movement of existing dislocations and enhance the production of additional dislocations to accommodate the applied strain. In fact, the interrupted cyclic deformation tests conducted on Hastelloy X by Miner and Castelli [16] showed a progressive and considerable increase in dislocation density over a greater number of cycles in the DSA range compared to that at room temperature. In Hastelloy X, the dislocation densities at failure also steadily increased on raising the temperature from 200 to 600°C in the DSA range.

In the present study, in the tests below 700°C, carbide precipitation did not occur; therefore, the marked hardening experienced below this temperature, such as at 500, 550, 600, and 650°C can be attributed to the effects of DSA of dislocations by solute atoms [12,13,15,18]. At these temperatures, deformation enhanced diffusion of chromium has been considered to impede the motion of mobile dislocations and DSA leads to changes in dislocation density without precipitate formation. At 700 and 750°C, the cyclic hardening appears to have been influenced by the combined effects of DSA and precipitation hardening. At dislocations, the chromium combines immediately with the rapidly diffusing carbon to form the carbides which provide the strong locking of dislocations. The carbides are very effective in pinning the dislocations when their size is extremely small. Rapid coarsening of the precipitates reduces the potential for the occurrence of DSA above 750°C.

In the high temperature domain (T ≥ 750°C), a dramatic reduction in cyclic hardening was observed in comparison to that which occurred in the DSA domain. Tests conducted at $10^{-3}s^{-1}$ experienced moderate hardening (decreasing with increasing temperature) to ~900°C, above this point, cyclic softening became evident. At $10^{-4}s^{-1}$, this general trend was also exhibited; however, softening effects commenced at 850°C. In the regime between 750 and 900°C, microstructural observations indicated that in the high $\dot{\epsilon}$ tests a greater number of dislocations remained anchored by the $M_{23}C_6$ precipitates when compared to those in the low $\dot{\epsilon}$ tests. Dislocation pinning by carbides appeared to impede thermal recovery in the higher $\dot{\epsilon}$ tests and led to modest cyclic hardening, albeit at a declining rate with increasing temperature. In general, the decreased dislocation densities and subgrain structures observed at the very high temperatures, such as those illustrated in Fig. 5, clearly suggest that the thermal recovery processes became dominant, and as a result, cyclic softening and a strong positive strain rate dependence was observed.

<u>Thermomechanical Fatigue.</u> The cyclic stress responses experienced under IP TMF conditions are shown in Fig. 10, where Fig. 10a contains data from the lower temperature conditions (ΔT: 350-550 & 400-650°C), and Fig. 10b contains data from the higher temperature conditions (ΔT: 500-750 & 600-

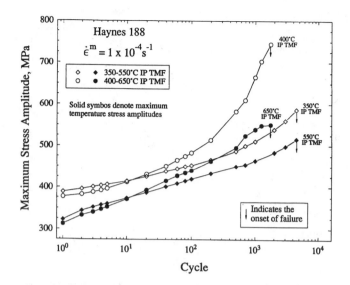

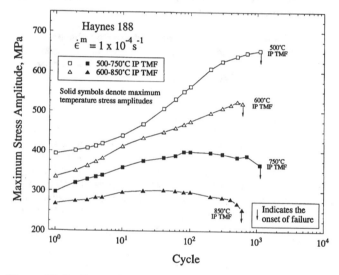

Figure 10: Cyclic maximum stress amplitudes for (a) ΔT: 350-550 & 400-650°C, and (b) ΔT: 500-750 & 600-850°C, IP TMF conditions with $\Delta\epsilon^m=0.8\%$ and $\dot{\epsilon}^m=10^{-4}s^{-1}$.

850°C) examined. OP TMF conditions were also conducted with similar temperature intervals (ΔT), but are not shown in Fig. 10 for the sake of clarity. Note that each TMF test must be represented by two curves, one showing the stress amplitude at the maximum temperature (for IP TMF this is tensile) and the other showing the stress amplitude at the minimum temperature (for IP TMF this is compressive). Further, for viewing simplicity, all stress amplitudes are shown as being positive (tensile). In the TMF test with ΔT=350-550°C, the maximum and minimum temperatures were within and below the DSA range, respectively (for $\dot{\epsilon}=10^{-4}s^{-1}$), while in the 500-750 and 600-850°C tests the maximum and minimum temperatures were above and within the DSA range, respectively. The test with ΔT=400-650°C is entirely within the DSA range.

In the isothermal fatigue tests, tensile and compressive stress amplitudes in any given cycle were essentially equal, whereas in the TMF tests, mean stresses developed as a consequence of the dynamic temperature conditions; the cyclic progression of these mean stress values are given in Fig. 11 where additional OP TMF tests are also included. IP TMF tests (such as those

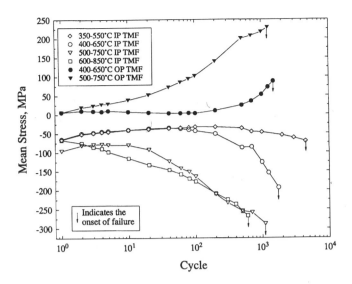

Figure 11: Cyclic mean stresses developed during TMF tests.

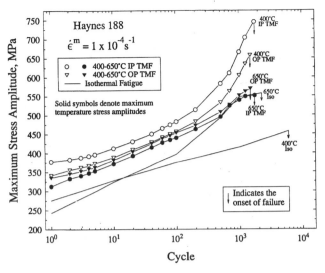

Figure 12: Cyclic maximum stress amplitudes for IP and OP TMF with ΔT=400-650°C and corresponding isothermal fatigue tests at the temperature extremes.

shown in Fig. 10), developed a compressive mean stress, that is, the compressive stress amplitudes were higher than the tensile stress amplitudes, whereas, OP TMF conditions led to tensile mean stresses. The mean stress was mild in the 350-550°C test; however, the other TMF tests experienced significant mean stresses which generally tended to increase with cycling. The mean stresses developed were particularly large when the maximum temperature of the TMF cycle was above and the minimum temperature was within the DSA range. This result was due to the fact that DSA enhanced hardening was prevalent at one extreme of the cycle and softening due to thermal recovery effects was prevalent at the other extreme. Also note in Fig. 11 that the IP TMF mean stress values were representative of those experienced under OP TMF conditions (for identical ΔT) when accounting for the tension/compression difference.

In all the TMF tests, the maximum stresses achieved just prior to the onset of failure at the maximum temperatures of the cycles were almost identical to those attained in corresponding isothermal tests; however, the number of cycles needed to attain the maximum in each of the TMF tests was less than those in the isothermal tests. In contrast, the maximum stresses achieved at the minimum temperature extremes in the TMF tests were not well represented by the isothermal tests performed at corresponding temperatures. This may suggest that the TMF macroscopic behavior is most influenced by the substructural features associated with the peak temperature, as was found with the nickel base superalloy, Hastelloy X [11,16].

In general, the low-temperature peak stresses experienced greater increases in the TMF tests showing significant deviations from the corresponding isothermal tests. This effect was most pronounced in the tests conducted with a temperature interval of 400-650°C. Results for this ΔT condition for both IP and OP TMF are presented in Fig. 12 along with data from isothermal fatigue tests conducted at 400 and 650°C with $\dot{\varepsilon}=10^{-4}s^{-1}$. As shown here, the maximum hardening rates and magnitudes experienced by the 400°C TMF peaks far exceeded those displayed isothermally at 400°C. This indicates that the TMF stress hardening at 400°C could not have been anticipated from the isothermal data. Further, the stress values achieved far exceeded (by 25 to 35%) those experienced isothermally at 650°C. This is a significant result considering that the 650°C isothermal values represent the maximum for the entire isothermal data base. This isothermally unbounded behavior is clear evidence of thermomechanical path dependence, as the material behavior observed at 400°C is profoundly influenced by the deformation substructure developed over the full ΔT. Although the minimum temperature peak stresses in the 500-750°C cycles showed the excessive increase due to DSA effects when compared to the isothermal test at 500°C, the amount of hardening was far less than that obtained in the 400-650°C TMF tests, because part of the cycle lied in the range where thermal recovery effects became operative. In the TMF tests with ΔT=600-850°C, where the main fraction of the cycle involves loading above the DSA range, it is apparent that the thermal recovery effects have become substantial, causing the hardening mechanisms associated with DSA to be less effective at the lower temperature extreme of the cycle.

Summary and Concluding Remarks

Isothermal and thermomechanical fatigue (TMF) experiments under fully reversed 0.8% mechanical axial-strain control with strain rate ($\dot{\varepsilon}$) values of $10^{-3}s^{-1}$ and $10^{-4}s^{-1}$ were conducted on Haynes 188™ over the temperature range 25-1000°C. Detailed characterizations of both the macroscopic and microstructural deformation behaviors were conducted.

The fatigue deformation and stress response indicated the operation of dynamic strain aging (DSA) processes, nominally between the temperatures of 300 and 700°C. DSA was evidenced through a combination of phenomena, including serrated yielding, an inverse dependence of maximum cyclic hardening with $\dot{\varepsilon}$, and an inverse instantaneous $\dot{\varepsilon}$ sensitivity. A maximum isothermal cyclic stress amplitude was experienced at 650°C for both strain rates examined. At the microstructural level, the occurrence of DSA was found to be associated with pronounced planar slip, generation of stacking faults and relatively high dislocation densities. Stress hardening behaviors exhibited under TMF conditions at the maximum temperature cycle extremes agree closely with those exhibited under isothermal conditions. However, TMF minimum temperature extremes are not accurately predicted from the isothermal data. The TMF deformation results indicate strong thermomechanical path dependence.

References

1. Haynes Alloy 188™, (Report-Haynes International, Inc., Kokomo, Indiana, 1991).

2. D.L. Klarstorm and G.Y. Lai, Superalloys 1988, (The Metallurgical Society, 1988) 585-594.

3. G.R. Halford, J.F. Saltsman, and S. Kalluri, "High Temperature Fatigue Behavior of Haynes 188," (Presented at Advanced Earth-to-Orbit Propulsion Technology Conference, MSFC, Huntsville, AL, May 10-13, 1988; Report NASA CP-3012) 1(1988), 497-507.

4. G.R. Halford, et al., "Thermomechanical and Bithermal Fatigue Behavior of Cast B1900+Hf and Wrought Haynes 188," ASTM STP 1122, (1992), 120-142.

5. S. Kalluri and G.R. Halford, "Damage Mechanisms in Bithermal and Thermomechanical Fatigue of Haynes 188," ASTM STP 1186, (1993), 126-143.

6. V.M. Radhakrishnan, S. Kalluri, and G.R. Halford, "An Analysis of Isothermal, Bithermal, and Thermomechanical Fatigue Data of Haynes 188 and B1900+Hf by Energy Considerations," (Report NASA TM-106359, Sept., 1993).

7. D.A. Jabolonski, J.V. Barisella, and R.M. Pelloux, Metall. Trans. A, 8A(1977), 1893-1900.

8. P.J. Bonacuse and S. Kalluri, "Cyclic Axial-Torsional Deformation Behavior of a Cobalt-Base Superalloy," (Report NASA TM-106372, 1992).

9. K. Bhanu Sankara Rao, et al., "A Critical Assessment of the Mechanistic Aspects in Haynes 188 During Low Cycle Fatigue in the Range 25 to 1000°C, Metall. Trans. A, accepted for publication.

10. K. Bhanu Sankara Rao, M.G. Castelli and J.R. Ellis, "On the Low Cycle Fatigue Deformation of Haynes 188 Superalloy in the Dynamic Strain Aging Regime," Script Met. 33(1)(1995).

11. M.G. Castelli, R.V. Miner and D.N. Robinson, "Thermomechanical Deformation Behavior of a Dynamic Strain Aging Alloy," ASTM STP 1186, (1993), 106-125.

12. R. Zauter, F. Petry, H.-J. Christ and H. Mughrabi, "Thermomechanical Fatigue of the Austenitic Stainless Steel AISI 304L," ASTM STP 1186, (1993), 70-90.

13. K. Bhanu Sankara Rao, et al., High Temp. Mater. and Processes, 7(1986), 171-177.

14. V. S. Sreenivasan, et al., Int. J. Fat. 13(1991), 471-478.

15. K. Bhanu Sankara Rao, "Influence of Metallurgical Variables on Low Cycle Fatigue Behavior of Type 304 Stainless Steel," (Ph.D. Thesis, IGCAR-Kalpakkam and University of Madras, India, Jan. 1989).

16. R.V. Miner and M.G. Castelli, "Hardening Mechanisms in a Dynamic Strain Aging Alloy, Hastelloy X, During Isothermal and Thermomechanical Cyclic Deformation," Metall. Trans. A, 23A(1992), 551-561.

17. R.E. Reed Hill, Rev. High. Temp. Mater., 2(1974), 217-242.

18. K. Bhanu Sankara Rao, et al. Metals, Materials and Processes, 2(1990), 17-36.

LONG TIME CREEP RUPTURE OF HAYNES™ ALLOY 188

Robert L. Dreshfield
National Aeronautics and Space Administration
Lewis Research Center
Cleveland, Ohio 44135

Abstract

The creep of Haynes Alloy 188 sheet in air was studied at temperatures of 790°, 845° and 900 °C for times in excess of 30,000 h as part of a program to assure that Haynes Alloy 188 could be used in critical components of a solar dynamic power conversion system for Space Station Freedom. The rupture life and time to 0.5 and 1.0 percent creep strain of creep rupture specimens which had lives from about 6,000 to nearly 59,000 h are reported. Both welded and as-received specimens were tested. The welded specimen had essentially the same lives as the as-received specimens. Comparison of this data with previously published results suggests that this material was similar in behavior to that previously studied by Haynes except at 790 °C where the current sheet is somewhat stronger in stress rupture. Therefore the previously published data for Haynes Alloy 188 may be used in conjunction with this data to estimate the lives of components for a long life solar dynamic power system.

Three creep rupture tests were discontinued after 16,500 to 23,200 h and tensile tested at room temperature or 480 °C. The elongation of all three specimens was substantially reduced compared to the as-received condition or as aged (without applied stress) for 22,500 h at 820 °C. The reduction in elongation is thought to be caused by the presence oxidized pores found across the thickness of the interrupted creep specimens. The implications of the severe loss of ductility observed in tensile tests after prolonged creep test needs further study.

Introduction

A solar dynamic power conversion system was planned for Space Station Freedom to provide additional electrical power to the space station as it grew.[1] Solar dynamic power systems are designed to convert solar energy to electric energy using heat engines. A schematic diagram of a space-based solar dynamic power system in shown in Fig. 1.

One component of the power conversion system (the receiver) both receives the solar energy for use while it is exposed to solar radiation and stores the energy as latent heat of fusion in molten salt to provide energy while the space station is in the shadow of Earth. The material chosen for some of the critical parts of the receiver, including the salt containment canisters and hot gas manifold, was Haynes alloy 188.[2]

Because Space Station Freedom was to be designed for a 30 year life, with only one replacement of the power module, long term creep rupture data was required to validate the designs of those critical parts.

Review of available data failed to identify data in the appropriate temperature range exceeding 10,000 h. Since the design life would be for 260,000 h, it was necessary to acquire data for times for at least 26,000 h to assure the design, initially based on the existing data, would meet durability goals.

The work reported here was initiated to provide such data. A limited study was conducted to evaluate the creep rupture behavior of 1.3 mm thick Haynes Alloy 188 at 790° and 845 °C for times varying from 10,000 to 30,000 h. The 10,000 h data was intended to provide a link with existing data[3] while the high time data would be about 10 percent of the target life. Both as-received and welded material were evaluated. During the course of the work a few tests at 900 °C were substituted for some of the lower temperature tests. Time to 0.5 and 1.0 percent creep and rupture properties were determined.

Materials And Test Methods

The material tested was from a single heat of Haynes Alloy 188 in the bright annealed, solution treated condition which had previously been procured to study the effects of long time molten salt and elevated temperature vacuum exposures on the alloy.[2,4,5] The sheet had a nominal thickness of 1.4 mm. The composition as provided by the vendor was: 0.11 C, 0.72 Mn, 0.38 Si, 21.69 Cr, 23.03 Ni, 1.95 Fe, 14.02 W, 0.048 La, 0.002 S, 0.13 P, and the balance Co. It was reported by the vendor to have a grain size of ASTM No. 6.5.

The test specimens which were machined from the sheets had a gage section 9.5 mm wide and 55 mm long. They were pin loaded using 6.4 mm diameter pins 12.8 cm on centers. The overall length of the specimen was 16.5 cm. V grooves about 0.05 mm deep were machined into one side of the specimen in a 15.9 mm wide shoulder to minimize slippage of knife edge extensometers. The welded samples were tungsten inert gas welded (TIG), using sheet material as starter stock, such that the weld was in the center of the gage section. After welding, the starter tabs were removed and the weld was blended to be smooth with the base metal surfaces. The thickness of the welds were typically about 5 percent greater than the base metal. No post-weld heat treatment was performed.

Creep rupture tests were performed in air on constant load frames. Loads were based on extrapolation of existing data[3] for failure times from 10,000 to 30,000 h at temperatures of 790° and 845 °C. Because of concern for higher temperature excursions in the receiver, a few shorter duration tests were initiated at 900 °C. For the lightest loads, direct loading was used, while for other tests lever arm machines

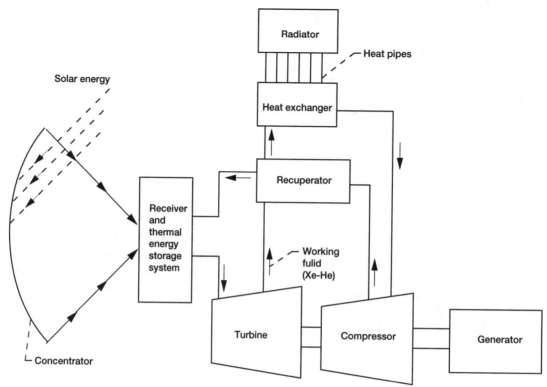

Figure 1.—Schematic representation of space based solar dynamic power system. (ref. 3).

having lever ratios from 4:1 to 20:1 were used. Deformation was measured using an extensometer with a LVDT transducer. All of the extensometer motion was assumed to occur in the gage section. The data was recorded using a PC based data acquisition system. The resolution of the creep measurement was approximately 0.01 percent. Recording time was typically set at 24 to 36 h increments for most of the test. Shorter times were used during the first few days of the tests.

Three tensile tests on creep exposed specimens were performed using a screw loaded testing machine. Initial strain rate was controlled by using a constant crosshead speed equivalent to 0.0016 percent strain per second. Extension was measured using an extensometer with a 12.7 mm gage length.

Results And Discussion

Creep-Rupture Tests

The test matrix and results are shown in Table I. The creep-rupture results are summarized in Fig.2(a). The data with arrows were discontinued to make the test frames available for another program and were subsequently tensile tested. "H" represents our estimate of 10,000 h life based on the Tackett[3] data. The lines are 1st order regression lines of the experimental data. Figure 2(a) shows that the time to rupture at 845° and 900° are in good agreement with the expectation derived from Tackett's work. At the lowest temperature studied, 790°, the material in this study appears to be somewhat stronger than that studied by Tackett.

It should be noted that the material studied by Tackett was in the black annealed, pickled and stretcher leveled condition and the current material was provided in the bright annealed condition. A private communication with Mr. M. Rothman of Haynes Alloys International revealed that the strain introduced by stretcher leveling has the potential to improve the lower temperature creep rupture strength. This was not observed in this study.

The stress rupture data at 790° and 850 °C are presented in Fig.2(b) with the as-received material identified as the open symbols and the welded alloy identified as the filled symbols. The presence of the welds did not appear to have a significant effect on the rupture life, and this is borne out by visual examination of the failures (Fig. 3). None of the welded samples failed in the weld. However as the welds were typically 5 percent thicker, with some up to 10 percent, it is possible that the weld could be up to 10 percent weaker than the base metal without reducing the lives of the specimens.

The same heat of material used in this investigation was also used by Whittenberger.[2,4,5] He performed stress rupture tests in vacuum at 775° at stresses to produce failure in 10 to 1000 h. The 790° data from this study are compared to his data in Fig. 2(c). While the linear extrapolations of each data set show some deviation, the data are in reasonable agreement. One might expect some curvature in isothermal rupture curves with the lower stress tests failing at somewhat shorter times than expected from extrapolation of the higher stress data.

Both elongation and reduction in area (RA) are presented for the failed Haynes Alloy 188 stress rupture specimens in Table I. At the two lower temperatures the elongations varied from 15 to 30 percent without any apparent dependence on the initial applied stress. At 900 °C the elongations varied from 67 to 96 percent which are significantly greater than those measured at the lower temperatures. Comparison of the elongations for the welded to the as-received indicates a tendency for somewhat less elongation for the welded specimens. For example at 845 °C-42 MPa the two as-received samples had elongations of 19.5 and 21.5 percent while the two welded samples had elongations of 21.5 and 12 percent. Similar behavior can be seen at 51 MPa. It is

Table I Creep Test Matrix

Stress, MPa	Temperature, °C	Weld	0.5% creep, h	1% creep, h	Elong., %	Red. in Area, %	Rupture Life, h
74	790	N	760	5620	23.5	11.3	58664
74	790	N	1380	6170	NA.	NA.	d23230
77	790	N	360	1980	14.5	12.9	48476
77	790	N	103	401	NA.	NA.	d16518
86	790	N	190	1190	20.5	12.8	31288
86	790	N	90	5400	19.5	12.9	26386
103	790	N	50	216	22.5	18.3	13441
42	845	N	1400	7480	19.5	6	27968
42	845	N	1480	8340	21.5	9.8	31213
42	845	Y	2400	10300	22	6.8	33370
42	845	Y	4940	12230	12.5	4.5	28540
44	845	N	1660	11290	30.5	10.3	39950
44	845	N	400	3300	23	11.3	26250
44	845	Y	NA.	NA.	21.5	13.1	21473
44	845	Y	1800	6307	NA.	NA.	d22115
51	845	N	1035	4860	23	10	18814
51	845	Y	720	3830	22.5	12.4	24111
51	845	N	870	2600	30.5	10.3	13940
51	845	Y	816	3200	15	10.9	15860
28	900	N	1570	3466	96	17.4	17807
28	900	N	2470	3650	86	21.4	12346
28	900	Y	1056	2455	76.5	16.6	11764
31	900	N	506	1611	70.5	18.8	8898
35	900	N	350	975	67.5	20.6	6186

d discontinued prior to fracture at the time indicated

thought that the reduced elongation for the welded specimens may, in part, be a manifestation of the increases thickness of the weld. The lower stress at the weld would be expected to result in less local creep and final extension at failure.

For all test conditions the reduction in area (RA) is generally much less than the elongation. This discrepancy between elongation and RA appears to increase with test temperature and decreasing stress. For example the ratio of elongation to RA is about 4 at 900 °C and 1.9 at 790 °C, similarly the ratio is about 4.5 at 28 MPa and 1.6 at 86 MPa. Such differences suggest that (1) Haynes Alloy 188 sheet is not necking, as confirmed in Fig. 3 and (2) the volume of the gage section is not constant during the creep-rupture test. It is suggested that the gage volume is likely increasing as a result of internal oxidation and/or grain boundary cavitation and oxide intrusion from the external surfaces. Grain boundary opening would be expected to be greater at the lower temperatures, while oxidation phenomena might be more prevalent at the higher temperatures.

The time to 0.5 and 1.0 creep at 790°, 845° and 900 °C as a function of stress are shown in Fig. 4. Estimates based on Tackett's[3] data of the stress required for 0.5 and 1.0 percent creep in 1000 h are shown as "H". The current 0.5 percent creep results are in excellent agreement with Tackett. The 1 percent creep data from this study, Fig. 4(b) agrees with the expectation from Tackett at 845° and 900 °C. At the lowest temperature, the 1 percent creep data appear to indicate that this material is slightly stronger than Tackett's.

Typical creep curves are shown in Fig. 5. Comparing the creep curves at each temperature, one can see that there is little or no period of "steady-state" creep. As the temperature increased, it appears that the shape of the curves above about 0.5 percent tend to change from being concave-down to concave-up. This change might reflect the change

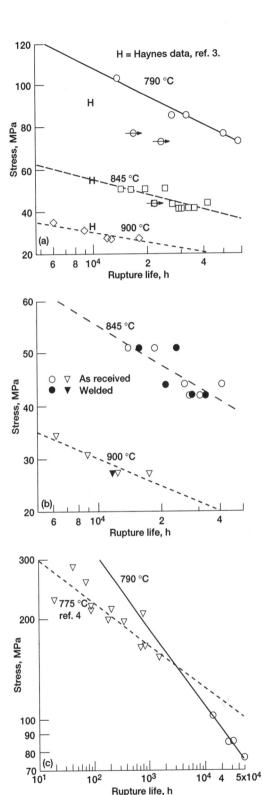

Figure 2.—(a) Stress rupture of Haynes alloy 188. (b) welds had little effect on rupture life. (c) Low temperature rupture data compares well with Whittenberger's.

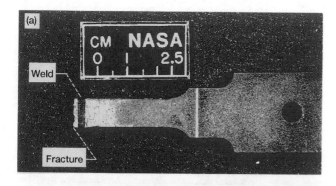
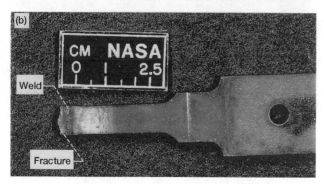
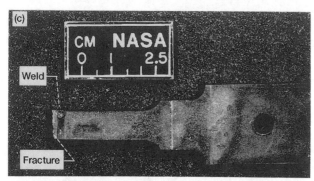
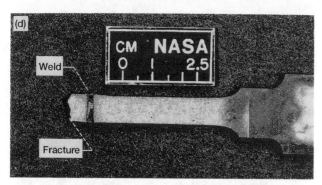

Figure 3.—Welded stress rupture samples failed in the base metal. (a) 845 °C, 42 MPa, 33370 h; (b) 845 °C, 51 MPa, 24111 h; (c) 845 °C, 42 MPa, 28540 h; (d) 900 °C, 28 MPa, 11764 h.

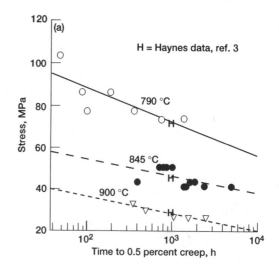
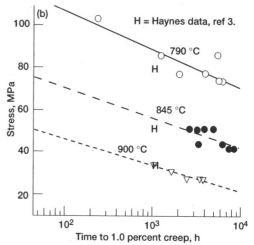

Figure 4.—Creep of Haynes alloy 188. (a) 0.5 percent creep. (b) 1 percent creep.

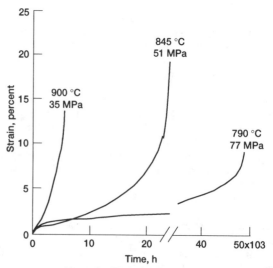

Figure 5.—Typical creep curves.

Table II Summary of Tensile Tests

Condition	Creep temp, °C	Creep time, h	Creep strain, %	Tensile temp. °C	0.2% yield, MPa	UTS, MPa	Elong., %
as rec'd	845	21150	7	room	466	692	2.5
weld	790	23230	3	room	389	525	0.5
as rec'd	790	16518	0.5	480	390	735	5.5

from continuous work-hardening at the lowest temperature to dynamic recovery overwhelming the effects of work hardening at the highest temperature observed. The very high ductilities observed at 900 °C support the idea that significant recovery is occurring at that temperature.

Tensile Tests

Because other programs at the Lewis Research Center required the use of creep test frames three creep rupture tests were discontinued prior to fracture. These specimens were then tensile tested, Table II. Two were tested at room temperature and one at 480 °C.

The tensile tests results are compared with as-received material and tests performed on the same lot of material after 22,500 h exposure to air at 820 °C[5] in Fig. 6. In Fig. 6(a), it can be seen that the room temperature ultimate tensile strength of the creep tested specimens was reduced 30 to 40 percent compared to either as-received material or that which had been aged without applied stress. The test performed at 480 °C compared well with both as-received and aged material from the earlier work. Figure 6(b), however, shows that the elongation, measured at either test temperature, was severely degraded by the prior creep exposure. The room temperature elongation was reduced to 0.5 and 2.5 percent from an as-received value in excess of 45 percent and an 820 °C aged value of 16 percent. At 480 °C the ductility was only 5.5 percent compared to an as-received value of 29 percent and an aged value of 19 percent.

Metallographic Evaluation

The microstructure of the as-received material shown in Fig. 7(a) consists of a twinned cobalt-base matrix and dispersed particles which are assumed to be M_6C and lanthanides. The microstructures observed in this study are consistent with those reported by Herchenroeder.[7]

The structure after 13441 h at 790 °C and 103 MPa is shown in Figs. 7(b) and (c). The grain boundaries and prior twin boundaries are decorated with precipitates and the intergrannular precipitates are coarsened. Oxide spikes originating at the sheet surface have penetrated to a depth of about 60 µm.

After 22,115 h at 845 °C and 44 MPa the oxide spike penetration is about 100 µm, Fig. 7(d). In addition, there are oxidized pores across the full thickness near the fracture. The presence of oxidized pores is even more pronounced in Fig. 7(e) which shows the structure of a specimen which failed after 6,186 h at 900 °C. It is probable that internal damage is responsible for the loss of tensile ductility observed in the interrupted creep specimens. The general structure after 22,115 h at 845 °C is

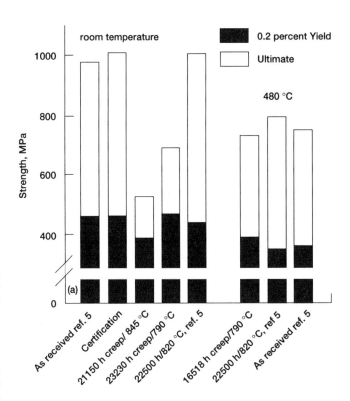

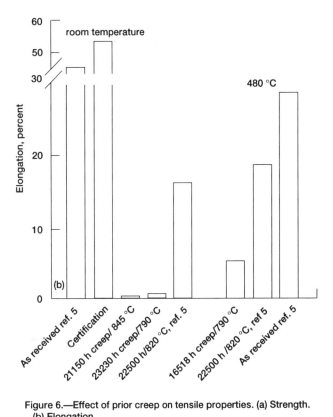

Figure 6.—Effect of prior creep on tensile properties. (a) Strength. (b) Elongation

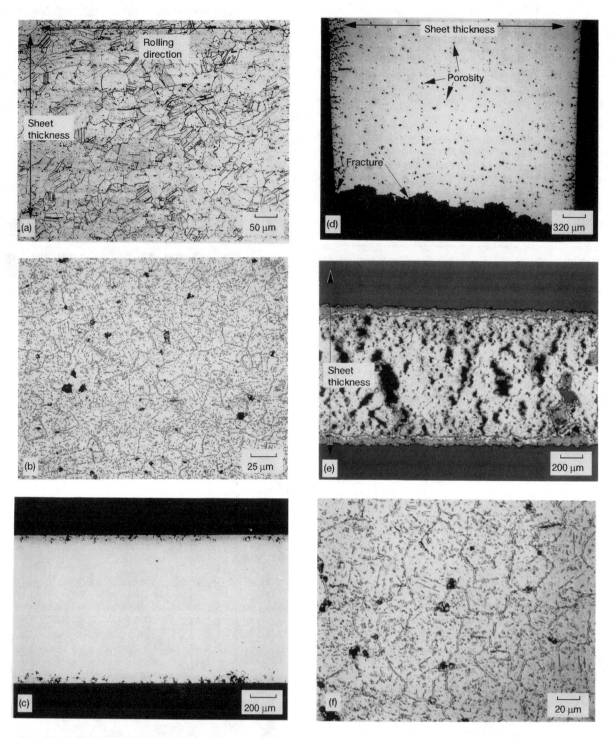

Figure 7.—Selected photomicrographs. (a) As received. (b) Transverse section 13441 h at 790 °C etched. (c) Transverse section 13441 h at 790 °C unetched. (d) Longitudinal section 22115 h, 845 °C unetched. (e) Longitudinal section 6186 h, 900 °C unetched. (f) Longitudinal section 22115 h, 845 °C etched.

shown in Fig. 7(f). It resembles Fig. 7(b), but the precipitates are coarser as a result of the longer exposure at higher temperature.

X-ray diffraction analyses were performed on residues from extractions using electrolytic HCl solution. The major phase identified was a Laves phase. An M_6C carbide could also be identified in specimens tested at both 790° and 845 °C. A very small amount of $M_{23}C_6$ was present in a specimen which failed after 21,150 h at 845 °C.

Selected cross sections were also examined in a scanning electron microscope. Figure 8 is a back-scatter electron image of a specimen which failed after 33,370 h at 845 °C and 421 MPa. The blocky light phase was apparent in specimens tested at all three temperatures and had the following approximate stoichiometry: $(Co_{.27}Ni_{.11}Fe_{.01}Cr_{.28}W_{.33})_{6.9}C$. A similar appearing phase was found in a specimen tested at 790 °C, but it did not appear to contain C. It had the following composition: $Co_{.37}Ni_{.1}Fe_{.01}Cr_{.23}W_{.29}$. It is thought that this phase is probably a Laves phase as one can interpret it as being of the form M_2B, with W being on the B site while all other elements are on A sites. It is noted, however that if one ignores the C content in the M_6C, the composition of that carbide and the Laves phase are similar. Internal Cr rich oxides and carbo-nitrides were identified in specimens tested at 900 °C and in a specimen tested at 845 °C for 33,370 h.

Summary And Conclusions

A study was performed to evaluate the creep-rupture behavior of Haynes Alloy 188 sheet at temperatures from 790° to 900 °C and times in excess of 30,000 h. Comparison of the rupture data to previously published results suggests that the heat of material studied here is equivalent to the Haynes data at 845° and 900 °C and somewhat superior at 790 °C. The time to 0.5 and 1 percent creep strain was comparable to that previously published by Haynes. While the sheet finishing technique had been changed, the slight difference between the lower temperature stress rupture behavior can not be ascribed to that change because of the limited data in this study.

The TIG welded specimens in the present work had virtually the same creep rupture lives as non-welded material. Thus it appears that the engineering creep-rupture behavior of welded Haynes Alloy 188 is the same as non-welded material.

Three creep specimens were removed from test prior to fracture and subsequently tensile tested. Tests at both room temperature and at 480 °C showed extreme loss in ductility. The room temperature tests showing only about 1 percent elongation and the 480 °C test having only about 5 percent. This loss is believed to be associated with internal creep damage across the thickness of the sheet. While it is not likely to be of significance for the proposed Space Station application where the alloy will be exposed either to salt, vacuum or an inert gas, it would be prudent to study this phenomenon in greater detail if a long-life solar dynamic system is to be deployed.

In closing it is appropriate to comment that a 2.0 kW solar dynamic conversion system using many components of the system which was originally developed for Space Station Freedom has been successfully tested in a vacuum tank using a solar simulator at the Lewis Research Center.[8]

References

1. T.L. Labus, R.R. Secunde and R.G. Lovely, "Solar Dynamic Power for Space Station Freedom," (NASA TM–102016, NASA, 1989).

2. J.D. Whittenberger, "Tensile Properties of HA 230 and HA 188 after 400 and 2500 Hour Exposures to LiF-22CaF$_2$ and Vacuum at 1093 K," J. Mater. Eng., 1990, No. 12:211–226.

3. J.W. Tackett, "The Creep Rupture Properties Of HAYNES alloy No. 188," (Report 8020 Stellite Division, Cabot Corp., 1971).

4. J.D. Whitenberger, "Mechanical Properties of Haynes Alloy 188 after Exposure to LiF-22CaF$_2$, Air, and Vacuum at 1093K for Periods up to 10,000 Hours," J. Mater. Engr. And Perf., 1(4))(1992) pp. 469–482.

5. J.D. Whittenberger, "77 to 1200 K Tensile Properties of Several Wrought Superalloys after Long-Term 1093 K Heat Treatment in Air and Vacuum," J. Mater. Engr. and Perf., 3(1)(1994) pp. 91–103.

6. Anon, "Haynes Alloy No. 188," (Cabot Corp., 1983).

7. R.B. Herchenroeder, "HAYNES ALLOY No. 188 Aging Characteristics," Structural Stability in Superalloys, anon, 1968, 460ff.

8. R.K. Shaltens and R.V. Boyle, "Initial Results from the Solar Dynamic (SD) Ground Test Demonstration (GTD) Project at NASA Lewis,"(IECEC '95-421, 1995).

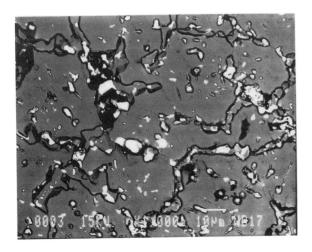

Figure 8.—Backscatter electron image 845 °C 33370 h.

EFFECTS OF GRAIN BOUNDARY MORPHOLOGY AND DISLOCATION SUBSTRUCTURE ON THE CREEP BEHAVIOR OF UDIMET 710

Y. Zhang[+] and F.D.S. Marquis[*]

[+]Department of Metallurgy, Chongqing University, P.R. China
[*]College of Physical, Chemical, and Materials Science and Engineering,
South Dakota School of Mines & Technology, Rapid City, SD 57701, USA

Abstract

An extensive investigation of the effects of grain boundary morphology (GBM) and dislocation substructure (DS) on the creep behavior of a Udimet 710 alloy was carried out. The interactions between: (1) dislocations and transgranular particles, and (2) dislocations and grain boundary carbides were also investigated. A significant understanding of the strengthening mechanisms and the mechanisms of fracture in this alloy was achieved. Two types of specimens, one with straight grain boundaries (SGB) and the other with zigzag grain boundaries (ZGB), were obtained by specially designed thermomechanical processes. Specimens of each of these types were creep tested at 1123K under an initial stress of 35kg/mm^2.

Creep in a grain was primarily caused by dislocation climbing over γ' particles. The migration of grain boundaries was controlled by their gliding for the specimens with straight grain boundaries. However, for specimens with zigzag grain boundaries, creep in a grain was caused by both dislocations climbing over and dislocations cutting through γ' particles. The migration of grain boundaries was controlled by both gliding of grain boundaries and the "straightening" motion of grain boundaries in zigzag steps. Zigzag grain boundaries increased the creep rate in a grain and decreased the gliding rate of grain boundaries. This effectively retarded the coalescence and growth of cracks and greatly enhanced the fracture resistance in the tertiary stage of creep observed at constant load. The creep life and plasticity to failure were increased in specimens with zigzag boundary morphology.

1. Introduction

It has been pointed out that zigzag grain boundaries obtained by special heat treatment techniques (involving slow cooling through the γ' solvus) could enhance the elevated and high temperature strength properties for most superalloys (1-6). Among those properties were the retardation of creep fracture processes and increments in creep failure plasticity of alloys (7).

No systematic studies of the effect of zigzag grain boundaries on dislocation substructures in creep, on creep resistance, and on fracture processes have been reported. Thus, in order to determine the fracture model and the strengthening mechanism of this alloy, it is very important to study the interactions between dislocations and γ' particles within a grain or dislocations and carbides (or other particles) at grain boundaries.

2. Materials and Experimental Procedures

The composition of the Udimet 710 alloy investigated is shown on table 1.

Table 1. Composition of Udimet 710 Alloy (WT%)

Ni	=	57.77	V	=	.26
Co	=	14.77	Fe	=	.19
Cr	=	10.80	C	=	.053
W	=	5.34	P	=	.006
Mo	=	5.20	S	=	.003
Al	=	3.79	Pb	<	.001
Ti	=	1.82			

Two types of specimens were produced in this alloy, those with zigzag grain boundaries and those with straight boundaries. Zigzag grain boundaries were obtained by the following isothermal treatments: 1473 K/ 3h/ A.C. + 1343 K/ 3h/ A.C. + 1123 K/ 8h/ A.C. Straight grain boundaries were obtained by conventional heat treatments: 1473 K/ 2h/ A.C. + 1323 K/ 4h/ A.C. + 1123 K/ 8h/ A.C., where A.C. means air cooling. A flow chart of these thermal processes is shown on figure 1(a) and (b).

Specimens of each kind of these two different types of grain boundary morphologies were creeped at 1123 K, at initial stress of 35 kgf/mm^2.

Specimens for transmission electron microscopy (TEM) were prepared from 0.3 mm thick slices cut with an electric spark cutting machine from the creeped specimens. The specimens were subsequently ground to 0.03 mm and electrolytically thinned at 263 K using an electrolyte consisting of 10% perchloric acid and 90% ethanol mixture. The thin foils were examined using a JEM-1000 electron microscope operated at 1000 KV.

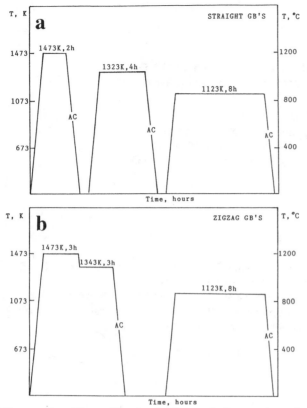

Figure 1: Flow Chart of Thermal Processing: (a) conventional (SGB) and (b) isothermal (ZGB).

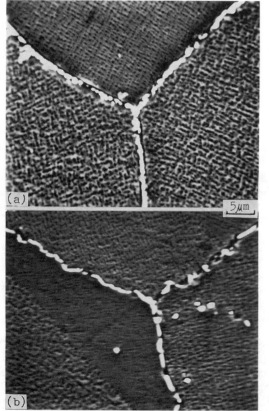

Figure 2: Typical microstructures after thermal processing: (a) straight GBs, (b) zigzag GBs.

3. Results

3.1 Effect of Zigzag Grain Boundaries on Creep Life

Typical microstructures obtained before creep are shown on figure 2. Figure 3 shows creep curves of specimens

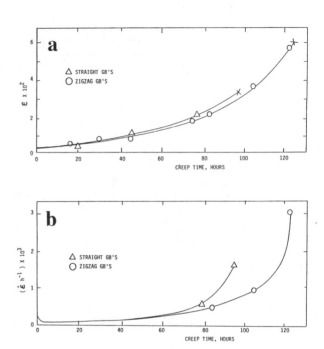

Figure 3: Creep behavior at 1123K under an initial stress of 35 kg/mm^2: (a) creep strain, (b) creep strain rate.

with different grain boundaries after being pulled to fracture. Typical microstructures of these specimens are shown on figure 4.

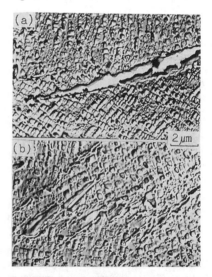

Figure 4: Typical microstructures after creep rupture: (a) straight GBs, (b) zigzag GBs.

The results show that zigzag grain boundaries lengthen the time of the tertiary stage of creep. The general creep

392

life to failure increases by 26% and the creep plasticity by 73% in specimens with zigzag grain boundaries. The average life span and percentage of elongation increases about 39.3% and 49.5%, respectively, at 1173 K under 22 kgf/mm².

3.2 Transcrystalline Dislocation Substructures

The dislocation substrucrures within grains are illustrated in figure 5 (a), (b), (c), (d), and (e) for specimens with straight grain boundaries.

It was observed that during the creep process dislocations moved mainly within the γ' matrix. Many three-dimensional dislocation networks and dislocation configurations after climbing over γ particles existed at γ/γ' interfaces. The dislocation density increased with increasing creep time in all specimens with straight grain boundaries. No traces were observed which showed that dislocations cut through γ' particles. However at γ/γ' interfaces regular two-dimensional dislocation networks were clearly seen as shown in figure 5 (d).

Figure 6 shows the dislocation substructures of specimens with zigzag grain boundaries obtained during creep stages.

Observation revealed that the dislocations in the interfaces between cubic-shaped γ' particles and between small spherical γ' particles could be simulataneously observed when suitable orientation of grains and electron diffraction conditions were chosen (figure 6 [a]).

Many three-dimensional networks of dislocations were observed at γ/γ' interfaces and the dislocation density was higher at the interfaces between small spherical γ' particles. Trace analysis showed that dislocations cut through some cubic-shaped γ' particles. Other dislocations were observed to climb over γ' particles by a mechanism controlled by diffusion. At some reflection conditions it could be observed that dislocation couples exist in γ' particles, as illustrated by arrows in figure 6 (a) and (e). However, due to the high antiphase boundary energy, the space between two dislocations of a dislocation couple was too short to distinguish them on some crystallographic planes. As a result, only a slightly wider dislocation line could be observed, as illustrated by arrows in figure 6(b) and (d). In addition, the increase of dislocation density was much larger than that in specimens with straight grain boundaries. These specimens exhibit lengthening of creep time, especially in the tertiary stage of creep. This can be well understood by comparing figures 2 and 3.

For specimens with zigzag grain boundaries, changes in configurations of dislocation substructures show that the creep in a grain is controlled by both dislocation climbing over and cutting through γ' particles, at the beginning of the tertiary stage of creep. The transcrystalline deformation rate within grains was larger than that for specimens with straight grain boundaries. Cracks in the grain boundaries had initiated at the beginning of the tertiary creep. Even at later tertiary stage of creep, the deformation in grains still dominated the creep of the alloy.

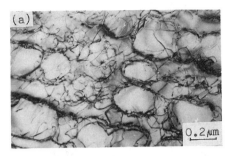
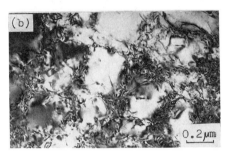
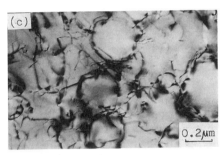

Figure 5: Dislocation substructures in grains with straight GBs: (a) 19.3h, (b) 44.8h, (c) and (d) 79.7h, (e) 94h.

3.3 Dislocation Substructures at and Near Grain Boundaries.

For specimens with straight grain boundaries, the carbide morphology was mainly transcrystalline, although some M_6C needle carbides were distributed at boundaries at the early stage of creep, as illustrated in figure 7 (a) (7-8).

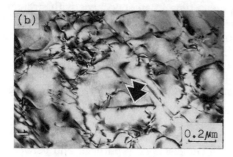

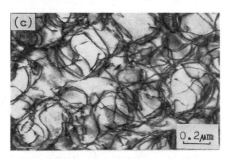

Figure 7: Dislocation substructures at and near GBs for specimens with straight GBs after creep for 19.3h.

Figure 7(b) shows that in specimens with straight grain boundaries, after creep for 19.3 h (still under conditions of steady state creep), a very large number of dislocation pileups were observed in regions near grain boundaries. Thus a few isolated cracks could be observed by means of optical microscopy at a few of these interfaces (figure 8).

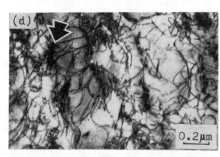

Figure 8: Crack at GB for specimen with straight GBs creeped for 19.3h.

It was observed that with the lengthening of creep time, the carbides distributed along longitudinal grain boundaries (parallel to the external force) move forward to transverse grain boundaries (perpendicular to the external force). This is consistent with other observations (9). After moving, the carbides distributed at transverse grain boundaries became strip-shaped. The amount of

Figure 6: Dislocation substructures in grains with zigzag GBs: (a) 17h, (b) 43.8h, (c) 73.3h, (d) 117.3h, (e) 120h.

carbides increased and the space between carbide particles decreased as shown in figure 9.

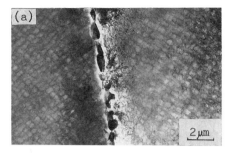

Figure 9: Dislocation substructures at and near GBs for specimens with straight GBs creeped for 79.7h.

The dislocation density near grain boundaries decreases in the tertiary stage of creep as shown by the comparison of figure 9(b) with figure 7(b). This process of dislocation relaxation is due to mechanisms occurring at dislocation pileups near grain boundaries. The head dislocations in the pileups climb to grain boundaries, which results in sliding of these grain boundaries. In these processes the unmatching displacements of sliding boundaries cannot be accommodated by transcrystalline microplasticity (i.e., by fractional quantities of dislocation motion within grains). Meanwhile, the carbides distributed at transverse grain boundaries have almost joined into thick plates covering all the interfaces, which provides a low energy path for crack growth. As a result, cracks grew so rapidly that specimens fractured. Figure 10 clearly shows the shape of the interfaces and the M_6C carbide morphology, i.e., their size, interparticle spacing, and tri-dimensional distribution.

Figure 10: Morphology of zigzag GBs.

Thus, we can evaluate the changes occurring at grain boundaries. Specimens with straight boundaries and specimens with zigzag grain boundaries have different interfacial conditions. Therefore, the migrating and the precipitating of carbides at grain boundaries and the alternation of dislocation substructures near interfaces must be different during creep process.

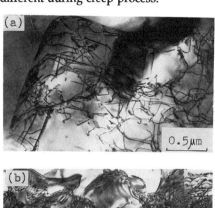

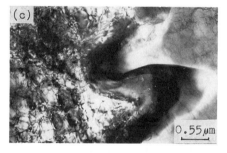

Figure 11: Dislocation substructures at and near interfaces in specimens with zigzag GBs: (a) 17h, (b) 43.8h, (c) 73.3h, (d) 117.3h.

Figure 11 shows the structure of zigzag grain boundaries and the configurations of dislocations substructures near interfaces. It was observed that they have the following characteristics, as compared with straight interfaces:

1. With the lengthening of creep time, the wavelength (λ) of zigzag grain boundaries increases and amplitude (h) decreases. This is in agreement with previous work (7). Meanwhile, the dislocation density near interfaces becomes heavier. On the other hand, the dislocation density near straight interfaces decreases with lengthening of creep time.

2. For stable conditions of creep and specimens creeped simultaneously, the dislocation density in regions near zigzag grain boundaries is lower than that near straight ones (figure 11[a] and figure 7[b]). However, during the fast accelerating period of the tertiary stage of creep, not only is the dislocation density in regions near zigzag grain boundaries much heavier than that near straight ones, but dislocation networks become finer as well (figure 11[c], [d], and figure 9[b]).

3. With the lengthening of creep time, the curvature of zigzag grain boundaries (h/λ) decreases, and the migration of carbides from vertical grain boundaries to transverse ones occurs mainly in zigzag steps of an interface. This makes coarse, irregular, bar-shaped carbides space themselves and become chain-shaped. This is very noticeable in late creep.

4. Discussion

4.1 Interactions Between Dislocations and γ' Precipitates Within a Grain.

After climbing over γ' precipitates, dislocations in a grain continue to move in matrix at 1123 K, under a stress of 35 kg/mm^2 for specimens with straight grain boundaries. Dislocations at γ/γ' interfaces were of the type a/2 <110> and during creep some of these dislocations moving on different slip systems react with each other. This gives origin to the formation of three-dimensional networks of dislocations that could be clearly seen at γ/γ' interfaces. While dislocations climb successively over γ' precipitates to continue motion, the dislocation density in grains continues to increases. However, at the stable creep stage, more networks of two-dimensional dislocations appear at γ/γ' interfaces (figure 12).

Figure 12: Regular networks of transgranular dislocations in ST creep for specimens with straight GBs (19.3h).

It is very possible that two groups of dislocations of type a/2 <110> react easily with each other at γ/γ' interfaces so that the dislocations are arranged evenly. This must result in dislocations climbing over γ' precipitates. In this process, the sliding distance on new slip planes of those dislocations after climbing becomes limited. This is not favorable to the deformation within grains.

It was observed that the coarsening of spherical γ' particles within grains and the decreasing of their volume fraction are much faster in specimens with straight grain boundaries than in specimens with zigzag grain boundaries. Consequently, the friction stress of dislocation motion within grains decreases rapidly. This is another reason why the climbing rate of dislocations increases. During the late tertiary stage of creep, the fine spherical γ' particles within grains were observed to coarsen rapidly and the climbing rate of dislocations was observed to increase. However, the free space of dislocation gliding was still limited by the regular two-dimensional networks of dislocations which limited the creep deformation within grains.

For specimens with zigzag grain boundaries, the dislocations within grains not only could climb a great distance over γ' particles but could also cut through cubic-shaped γ' particles to move continuously. This is clearly related to the deformation of cubic-shaped γ' particles.

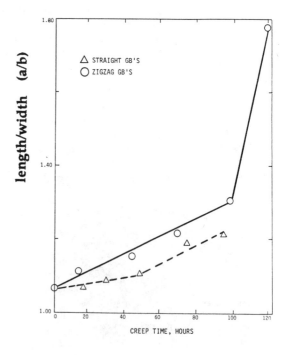

Figure 13: Effect of creep time on the length/width ratio (a/b) of γ' particles.

As shown in figure 13 the length/width ratio (a/b) of

cubic-shaped γ' precipitates in specimens with zigzag grain boundaries is larger than in specimens with straight ones. Specifically, during the late tertiary stage of creep it is very clear that the former length/width ratio rises rapidly. However, the general volume fractions of cubic-shaped γ' precipitates is slightly lower than in specimens with straight grain boundaries. The interactions between dislocations and lengthened cubic-shaped γ' precipitates makes it possible for the dislocations to climb over many more γ' particles in order to continue motion. For specimens with zigzag grain boundaries, the motion of dislocations climbing in matrix and dislocations cutting through the γ' particles alternates. This results in (1) significantly higher dislocation density at γ/γ' interfaces than that in specimens with straight grain boundaries for the same creep time, and (2) the increase in the gliding distance on a new slip plane of those dislocations after climbing. In addition, there exists no inhibition to motion from regular two-dimensional networks of dislocations. This means that the free space of dislocation gliding is longer. Thus, the deformation in grains is much larger in specimens with straight grain boundaries. Even in the fast accelerating period of tertiary stage of creep, the deformation still primarily occurred within grains.

The present results show that the change in the grain boundary microstructure results in the change of creep mechanism in a grain. Dislocation climbing dominates for specimens with straight grain boundaries. Both dislocation climbing and cutting are important for specimens with zigzag grain boundaries.

4.2 Characteristics of Zigzag Grain Boundary Motion. Effect on Dislocation Substructure.

Because of the difference of boundary conditions, the motion characteristics of zigzag grain boundaries are different from those of straight boundaries. The present results show that the amplitude (h) of zigzag grain boundaries decreases and the wavelength (λ) increases with the lengthening of creep time. This means that the migration of zigzag grain boundaries is characterized by boundary "straightening," which results in a "reversed" migration from zigzag grain boundaries to straight grain boundaries. The distortion field gives rise to additional resistance to dislocation motion. For the same creep period the rate of dislocation climbing forward to a grain boundary and the rate of dislocation sliding in specimens with zigzag grain boundaries is lower than that in specimens with straight grain boundaries. This is why a zigzag grain boundary retards cavity initiation. Thus, the motion of a zigzag grain boundary is characterized not only by the gliding of the grain boundary, but also by boundary "straightening." However, the motion of a straight boundary is primarily characterized by the gliding of the boundary.

During the tertiary stage of creep, the sliding rate (U) for specimens with zigzag grain boundaries was observed to follow the Raj and Ashby equation (13):

$$U = \alpha \lambda / h^2 \quad (1)$$

Since λ increases with decreasing h, this must result in the increase of sliding rate (U) of a zigzag boundary in the late tertiary stage of creep. As a result, the initiation of cavitation becomes much easier. It is very interesting to notice that during the tertiary stage of creep, while the sliding rate of a zigzag grain is accelerating, the dislocation density near a boundary increases and the dislocation networks were finer than that near a straight one. These might be related to the deformation of grains and the migration characteristics of grain boundaries. Maclean (14) pointed out that the ability of alloys to bear external stresses is mainly determined by their dislocation density and dislocation distribution. This is consistent with the formation of higher density dislocation networks near zigzag grain boundaries. While the initiation of cavities or isolated cracks makes the area bearing external stresses decrease, the specimen with zigzag grain boundaries can bear greater stresses than those with straight grain boundaries. Thus, the stress factor of creep is increased. On one hand this shows that the sliding of zigzag grain boundaries is accommodated by the motion of dislocations which are produced step by step from grains. On the other hand, the motion of zigzag grain boundaries changes the configurations of dislocation substructures as a result of the creep resistance to damage being effectively enhanced.

It is known from the above that both the grain boundary sliding accommodated by creep mechanisms of dislocation climbing in a grain, as well as dislocation cutting and the "reversed" motion of a zigzag grain boundary, make the stress concentration caused by grain boundary sliding relax so that the coalescence and the growth of cracks becomes difficult. This is the essential reason why creep life increases.

4.3 Interactions Between Dislocations and Grain Boundary Carbides.

Carbides, which are mostly coarse, irregular-shaped particles, became distributed at concave or raised positions and stretched out along a zigzag grain boundary. The formation of zigzag grain boundaries is believed to decrease considerably the free energy of the system, thus making the zigzag grain boundary a low energy interface. The interactions between dislocations and coarse, irregular carbides make the activation of dislocation sources in a boundary become more difficult. Meanwhile, they might raise the creep activation energy of the atoms in the boundaries and decrease the diffusion coefficient of the atoms in the boundaries so that the creep stress factor is raised. This makes the migration rate of carbides decrease from vertical boundaries to transverse ones, and the motion of the carbides becomes limited at each short zigzag step in a zigzag grain boundary. As a result the coarse, irregular carbides develop into chain shapes (figure 14).

The small carbide sizes and the small spacing between neighbor carbides is believed to weaken the grain boundary. In spite of this, while the carbides in zigzag grain boundaries migrate, the diffusional path of atoms is limited in a short close step in the grain boundary. In addition, the chain-shaped carbides existing at transverse grain boundaries can retard boundary sliding. This effectively retards crack growth. During tertiary creep, the curvature (h/λ) of zigzag grain boundaries decreases very rapidly, the atomic diffusional path lengthens, and the gliding rate of a boundary increases very rapidly. These cracks reach unstable conditions and grow rapidly, which results in final fracture.

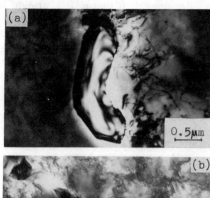

Figure 14: Carbide morphologies at zigzag GBs: (a) 17h, (b) 73.3h.

5. Summary of the Behavior of Zigzag and Straight GBs.

The present results, together with an ongoing investigation (15), show that the most significant differences in the behavior of specimens with zigzag grain boundaries versus those with straight grain boundaries are the following: (1) With increasing creep time, the wavelength λ of zigzag grain boundaries increases, the amplitude h decreases, and the adjacent dislocation density increases. The dislocation density adjacent to straight grain boundaries decreases with increasing creep time; (2) For steady state creep and specimens creeped simultaneously, the dislocation density in regions near zigzag grain boundaries is larger than that near straight grain boundaries. During tertiary creep, in regions near zigzag grain boundaries, the dislocation densities become much higher and the dislocation networks become finer than those near straight grain boundaries; (3) With increasing creep time, the curvature of zigzag grain boundaries h/λ decreases and the migration of carbides from longitudinal grain boundaries to transverse grain boundaries occurs mainly in zigzag steps of the interface. This makes coarse irregular bar-shaped carbides space themselves and become chain-shaped. This is very noticeable in late creep.

6. Conclusions

1. Different morphologies of grain boundaries give rise to different creep mechanisms in a grain. Dislocation climbing over γ' precipitates dominates for specimens with straight grain boundaries. Both dislocation climbing over and cutting through γ' particles are important for specimens with zigzag grain boundaries. The interaction between dislocations and γ' precipitates force the bulk-grains to have a higher contribution for the deformation of specimens with zigzag grain boundaries.

2. Different morphologies of grain boundaries lead to different characteristics of grain boundary migration and different configurations of dislocation substructures. The migration of a straight grain boundary occurs mainly by boundary sliding. The migration of a zigzag grain boundary occurs mainly by boundary "straightening," i.e., a "reversed" migration from a zigzag grain boundary to a straight grain boundary.

3. Only a very small amount of grain boundary sliding was observed in specimens with zigzag grain boundaries. The dislocation density in regions near zigzag grain boundaries is higher and dislocation networks are finer than that near the straight ones, for the same time of tertiary creep stage. The sliding rate of a zigzag grain boundary is much lower than that of straight grain boundary. This becomes the key to enhance effectively the creep resistance and to lengthen the creep life of this superalloy.

4. The interactions between dislocations and coarse, irregular-shaped carbides at zigzag grain boundaries make the motion of carbides limited in each step of these boundaries. This presents the coalescence of carbides at traverse grain boundaries at early stages of creep and is very effective in retarding the crack growth and lengthening the time of the tertiary stage of creep. The creep life and plasticity to failure were increased in specimens with zigzag grain boundary morphology.

References

1. M. Yamazaki, Journal of the Japan Institute of Metals, 30 (1966), 1032.

2. M. Kobayashi et al., Journal of the Iron and Steel Institute of Japan, 58 (1972), 859.

3. J. M. Larson and S. Floreen, Metall. Trans., 8A (1977), 51.

4. Y. Ruizeng et al., Acta Metallurgica Sinica, 20 (1984), A34.

5. M. Zhiping et al., Research in Metallic Materials, 8 (1982), 5.

6. G. Guixiong, <u>Acta Metallurgica Sinica</u>, 19 (1983), A467.

7. Z. Yaping and L. Hong, <u>Journal of Chongqing University</u>, 7 (1984), 141.

8. <u>The Metallographic Diagram Collection of Superalloys</u>, 1979, 134.

9. S. Kihara, J. B. Newkirk, A. Ohtomo, and Y. Saiga, <u>Metall. Trans.</u>, 11A (1980), 1019.

10. W. Zhi, (Thesis, Dept. of Metallurgy, Chongqing University, 1984).

11. R. A. Sterens et al., <u>Materials Science and Engineering</u>, 50 (1981), 271.

12. L. Guoxon, <u>The Principle of Physical Metallurgy</u>, 306.

13. R. Raj and M. F. Ashby, <u>Metall. Trans.</u>, 2 (1971), 1113.

14. D. Maclean, <u>Rep. Prog. Phys.</u>, 29 (1966), 1.

15. F.D.S. Marquis, "Effects of Zigzag Grain Boundaries on Creep Behavior of Udimet 710," 121st TMS Annual Meeting, San Diego, CA, February, 1992.

DYNAMIC STRAIN AGING BEHAVIOR OF INCONEL 600 ALLOY

Soon H. Hong, Hee Y. Kim, Jin S. Jang* and Il H. Kuk*
Dept. of Materials Science and Engineering,
Korea Advanced Institute of Science and Technology
373-1 Kusung-dong, Yusung-gu, Taejon, 305-701, Korea
* Korea Atomic Energy Research Institute,
P.O. Box 105, Yusung, Taejon, 305-600, Korea

Abstract

The dynamic strain aging behavior during tensile tests of Inconel 600 alloy has been investigated in the temperature range of 25-800°C over the strain rate range of 10^{-5}-$10^{-3}s^{-1}$. The serrations in flow curves were observed in the temperature range of 150-600°C. Four different types of serrations, identified as A1, A2, B and C serrations, were observed depending on the temperature, strain rate and strain. A1 type serration, a periodic rise and drop of stress with small amplitude, was observed in the temperature range of 150-245°C. A2 type serration, a rise of stress followed by a drop of stress, was observed at a higher temperature range of 245-400°C. B type serration, a successive oscillation of stress, was observed in the temperature range of 245-500°C. C type serration, characterized as abrupt irregular stress drops, was observed in the temperature range of 600-700°C. The activation energies for the serrated flow were calculated as 105, 81 and 150kJ/mol for A1, A2 and B type serration, respectively, from the analysis of the critical strains for onset of serrations. The rate controlling mechanism for dynamic strain aging is suggested as the migration of substitutional atoms for A1 type serration, carbon diffusion through dislocation core for A2 type serration and carbon diffusion through lattice for B type serration.

Introduction

Inconel 600 alloy is a Ni-base austenitic solid solution alloy which has been used as steam generator tubes in pressurized water reactors of nuclear power plant due to its good mechanical strength, thermal conductivity and corrosion resistance. There have been extensive studies on the stress corrosion cracking behavior of Inconel 600 alloy because the main failure made of steam generator tube is due to stress corrosion. The susceptibility of Inconel 600 alloy to stress corrosion cracking depends on composition, microstructure, environment and stress state. Recently, it is reported that stress corrosion cracking is related to the microdeformation[1, 2] and creep[3]. Therefore, it is important to understand the deformation behavior of Inconel 600 alloy.

Several researchers[4-6] have observed the serrated plastic flow and high work hardening rates with negative strain rate sensitivity in Inconel 600 alloy, however the controlling mechanism for the serration is not clearly understood yet. It is generally accepted that the serration in stress-strain curve is attributed to the dynamic strain aging of solute atoms. Hayes and Hayes[5] reported that the activation energy for the onset of serration is 54kJ/mol in Inconel 600 and suggested that dynamic strain aging phenomena are primarily caused by the diffusion of carbon rather than the substitutional atoms (primarily Fe and Cr) on the basis of the similar activation energy for onset of serration in Ni-C alloys. However, the temperature regime showing the serration in Inconel 600 is much higher than that showing the serration in Ni-C alloys[7]. Kocks[6] reported that the serration behavior of low carbon content below 0.05at.% in Inconel 600 was identical to that of commercial Inconel 600 alloy. These results are contradictory to the results of Hayes and Hayes[5].

In this paper, the controlling mechanisms of serrated flow in Inconel 600 alloy were analysed. The serrated flow in Inconel 600 alloy during tensile tests has been investigated at temperatures of 25-800°C with strain rates ranging 10^{-5}-$10^{-3}s^{-1}$. The stress-strain curves and types of serration were analysed. The activation energies of each type of serration were evaluated by the critical strain method and the controlling mechanisms for each type of serration will be discussed.

Experimental Procedures

Inconel 600 alloy was melted in a vacuum induction melting furnace and cast into ingots of 1180kg by Sammi Special Steel Co., LTD. The chemical composition of melted alloy is given in Table I. The cast ingots were forged at 1250°C and were hot extruded into tubes at 1200°C. The extruded tubes were cold worked into 1.06mm thick tubes by the pilgering with 96% reduction. The tubes were annealed at 1030°C for 20 minutes. The thermal treatment of annealed tubes was conducted at 700°C for 10 hours in Ar atmosphere to precipitate the carbides at grain boundaries. The average grain size was measured as 38μm after thermal treatment. The tensile specimens were machined from thermal treated Inconel 600 tubes and were tested in the temperature range of 25-800°C over the strain rate range of 10^{-5}-$10^{-3}s^{-1}$. The serration behaviors depending on temperature, strain rate and strain were analysed from the stress-strain curves. The critical strains for onset of serration were measured from the stress-strain curves and the activation energies for onset of serration were evaluated by the critical strain method.

Table I Chemical composition of Inconel 600

Element	Ni	Cr	Fe	Ti	Si	Mn	C	N
wt. %	76.88	15.55	6.98	0.28	0.16	0.23	0.02	0.0004

Results and Discussion

The observed stress-strain curves of Inconel 600 alloy with varying temperature and strain rate are shown in Figure 1. The serrated stress-strain curves were observed at temperatures between 300°C and 600°C under initial strain rate of $10^{-4}s^{-1}$ as shown in Figure 1(a). It is obvious that the work hardening rate is high within the temperature regime showing serrated stress-strain curves. The inverse strain rate dependence of flow stress is exhibited at temperature showing the serration as shown in Figure 1(b). Typical segments of the stress-strain curves during tensile tests of Inconel 600 are shown in Figure 2. Four different types of serration were reported and are classified as type A, type B, type C and type E, depending on the strain rate, temperature and strain[8]. Type A serration is characterized as periodic serration which is an abrupt rise of stress followed by a drop of stress in the stress-strain curve. Type B serration is characterized as successive oscillations of stress in the stress-strain curve, while type C serration is characterized as abrupt irregular stress drop. Type E serration is characterized as a random irregular drop of stress. Smooth flow curves without any serration were observed at temperatures below 150°C. Type A serration characterized as a periodic rise and drop of stress with small amplitude was observed at a lower temperature range of 150-245°C. Different type A serration characterized as a rise of stress followed by a drop of stress was observed at a higher temperature range of 245-400°C. Two different type A serrations were observed in the temperature range of 150-245°C and 245-400°C, and were designated as A1 type serration and A2 type serration, respectively. Type A serration changed into type E after certain amount of strain. Type B serration was observed at temperatures above 245°C and the amplitude of stress drop increased with increasing temperature up to 600°C. It is observed that type B serration was superimposed on type A serration at a temperature range of 245-400°C. Type C serration was observed at higher temperatures above 600°C. The temperature regime showing the different types of serrations varied systematically with strain rate. The regimes of various types of serration depend on the strain rate and temperature as shown in Figure 3.

It is well established that a critical amount of plastic strain is needed to initiate the serration on the stress-strain curves. The previous results[9] on the serrated flow suggest that the dependence of critical strain for onset of serration (ε_c) on strain rate and temperature is generally expressed as following Eq. (1),

$$\varepsilon_c^{m+\beta} = K\dot{\varepsilon}\exp(Q/RT) \qquad (1)$$

where Q is the activation energy for onset of serration, m and β are the strain exponents related with the vacancy concentration C_v and mobile dislocation density ρ_m, i.e. $C_v \propto \varepsilon^m$, $\rho_m \propto \varepsilon^\beta$, and K is constant. The activation energies, Q, for onset of each type of serration were measured by three different methods. First, Q was evaluated from the slope of a plot $log\varepsilon_c$ vs 1/T at a constant strain rate in Eq. (1). It can be obtained the exponent m + β from the slope of a plot of $log\dot{\varepsilon}$ vs $log\varepsilon_c$ at a constant temperature. Second, Q is evaluated from the plot of $log\dot{\varepsilon}$ vs $log\varepsilon_c$. The strain rates corresponding to fixed value of critical strains were obtained from Figure 4. The replot of the obtained strain rates with varying temperature yield straight lines and Q is obtained from the slope of the lines. Third, Q is evaluated using McCormick's strain aging model[10] expressed as Eq. (2),

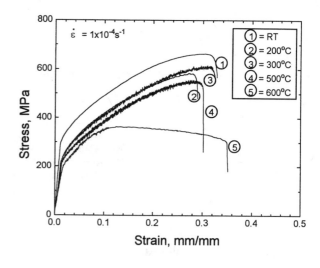

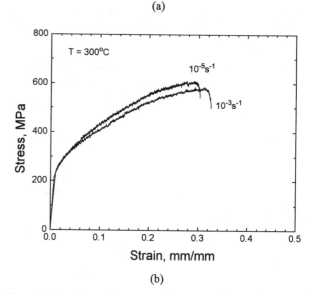

Figure 1: The observed stress-strain curves of Inconel 600 alloy. (a) The stress-strain curves with varying the temperature at the initial strain rate of $10^{-4}s^{-1}$. (b) The stress-strain curves with varying the initial strain rate at constant temperature of 300°C.

$$\frac{\varepsilon_c^{m+\beta}}{T} = \left(\frac{C_1}{\phi C_0}\right)^{3/2} \frac{\dot{\varepsilon}kb\exp(Q/kT)}{LNU_m D_0} \qquad (2)$$

where C_0 is initial concentration of solute in the alloy, C_1 is local concentration of the solute at dislocations, L is obstacle spacing, U_m is maximum solute-dislocation interaction energy, D_0 is frequency factor, b is Burgers vector and N and φ are constants. Q is obtained from the slope of a plot of $log(\varepsilon_c^{m+\beta}/T)$ vs 1/T plot[11].

The critical strain for onset of serration, ε_c, was measured from the stress-strain curves. The variation of ε_c with varying strain rate and temperature for A1, A2 and C type serrations are shown in Figure 4. The values of m+β were calculated as 2.15 for A1 type, 1.02 for A2 type and 1.04 for B type serration. The activation energies for onset of each type serration were calculated from the slopes a plot of $log\varepsilon_c$ vs 1/T as shown in Figure 5. The activation

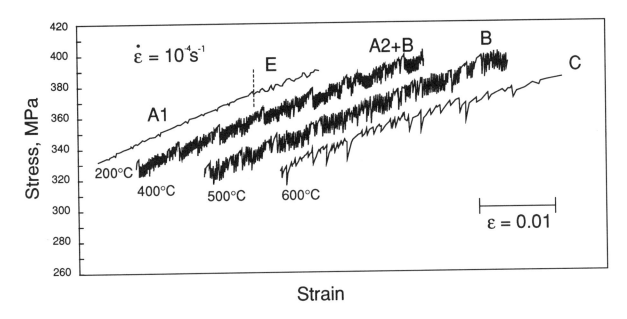

Figure 2: Various types of serrated flow observed in stress-strain curves of Inconel 600 alloy during tensile tests with strain rate of $10^{-4} s^{-1}$ at various temperature.

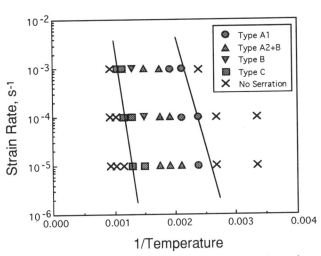

Figure 3: The observed range of temperature showing the various types of serrations in stress-strain curves of Inconel 600 alloy at different strain rates.

Table II The activation energies calculated by different methods for Inconel 600

	Activation Energy for Onset of Serration, kJ/mol		
	A1 type serration	A2 type serration	B type serration
From $log\varepsilon_c$ vs 1/T plot	95	84	144
From $log\dot{\varepsilon}$ vs 1/T plot	127	72	163
From $log(\varepsilon_c^{m+\beta}/T)$ vs 1/T plot	93	87	144
Average Q	105	81	150
Average m+β	2.15	1.02	1.04

energies were also calculated by other methods described above. It is calculated from a plot of $log\dot{\varepsilon}$ vs 1/T using intercept method and from a plot of $log(\varepsilon_c^{m+\beta}/T)$ vs 1/T using Cottrel-Bilby equation as shown in Figure 6 and Figure 7. The calculated values of m+β and Q for each type serration are listed in Table II.

It is generally known that the values of m+β ranged 0.5-1 for dynamic strain aging is due to the interstitial solute, whereas the values of m+β ranged 2-3 for dynamic strain aging is due to the substitutional solute in solid solution alloys[9]. The m+β values were measured as 0.63 in Ni-C[7] and 0.72 in Fe-C[12], while 2.4 in Au-Cu[13], 2-3 in Cu-Zn[14] and 3.33 in Al-Mg-Zn[15]. Comparing the calculated m+β values in Table II with those reported in solid solution alloys, it is suggested that the onset of A1 type serration was controlled by substitutional element in Inconel 600 alloy, while the onset of A2 type serration and B type serration were controlled by interstitial element in Inconel 600 alloy. The activation energy measured as 105kJ/mol for the onset of A1 type serration is close to the activation energy for vacancy migration of 106kJ/mol in Ni. The activation energy for vacancy migration in Ni is calculated from the difference of the activation energy for self diffusion of 280kJ/mol[16] and the activation for vacancy formation of 174kJ/mol[17]. Therefore, it is suggested that A1 type serration is controlled by the migration of the substitutional solute atoms in Inconel 600 alloy. The major substitutional elements in Inconel 600 are Cr, Fe, Ti and Mn as shown in Table I. Ti is considered as the most possible element for dynamic strain aging related with A1 type serration in temperature regime of 150-245°C, because the difference in atomic size of solute and solvent atom of Ti is the largest.

The activation energy for onset of A2 type serration measured as

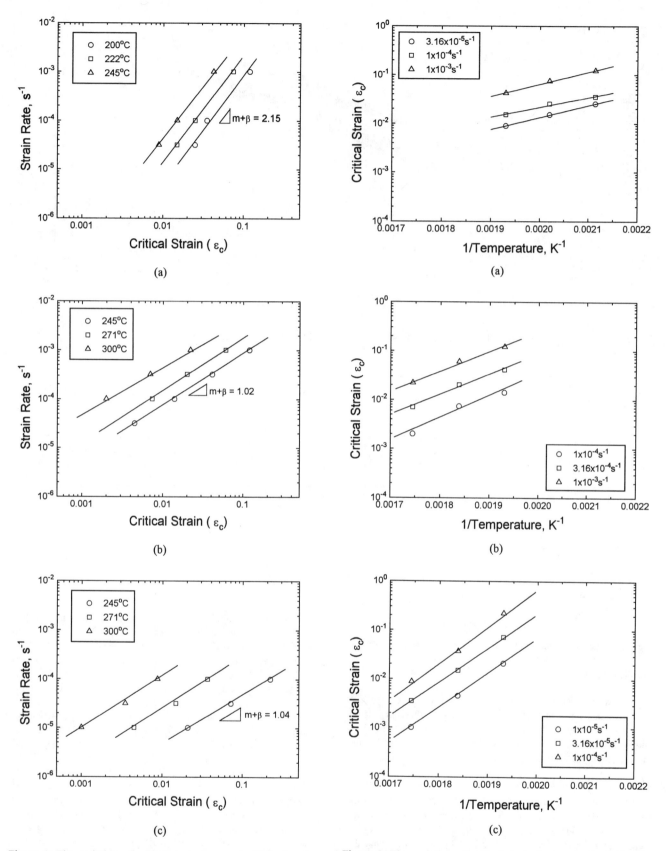

Figure 4: The variation of critical strain for onset of serration in stress-strain curve of Inconel 600 alloy with varying strain rate. (a) A1 type serration, (b) A2 type serration, (c) B type serration.

Figure 5: The variation of critical strain for onset of serration in stress-strain curve of Inconel 600 alloy with varying temperature. (a) A1 type serration, (b) A2 type serration, (c) B type serration.

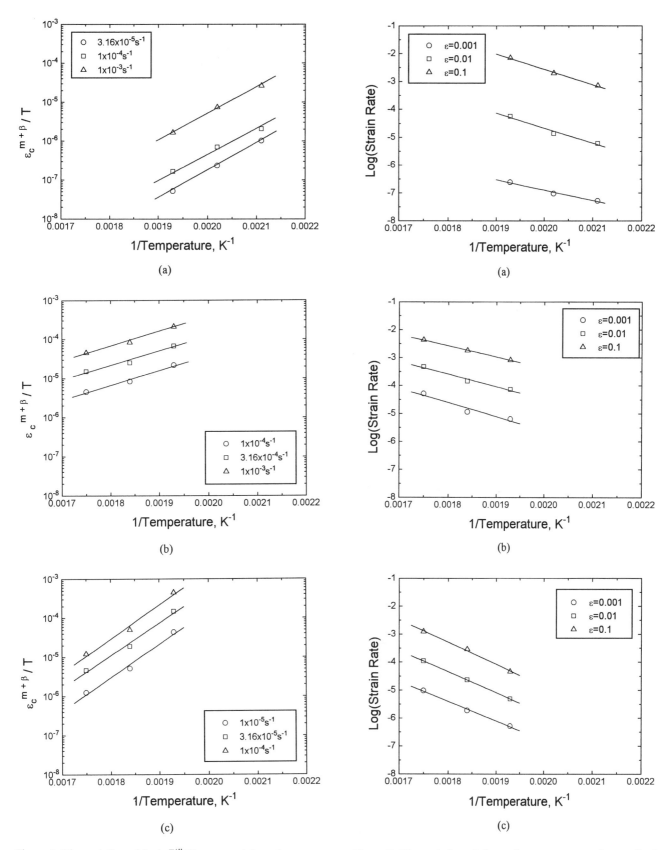

Figure 6: The variation of $log(\varepsilon_c^{m+\beta}/T)$ measured from the stress-strain curve of Inconel 600 alloy with varying temperature. (a) A1 type serration, (b) A2 type serration, (c) B type serration.

Figure 7: The variation of the strain rate corresponding to fixed value of critical strains in Figure 4 with varying temperature. (a) A1 type serration, (b) A2 type serration, (c) B type serration.

81kJ/mol. The A2 type serration observed in this study is comparable to the serration in Inconel 600 observed by Hayes and Hayes[5] and in Ni-C observed by Nakada and Keh[6]. Hayes and Hayes[5] reported the m+β is 0.74 and Q is 54kJ/mol in Inconel 600 alloy in the temperature range of 316-538°C, while Nakada and Keh[7] reported that that m+β is 0.63 and Q is 63kJ/mol in pure Ni-C alloys in the temperature range of -50-300°C. It is suggested that carbon is the rate controlling element because it is the major interstitial element in Inconel 600. It is reported that the activation energy for lattice diffusion of C in Ni is 139kJ/mol[18]. Since the ratio of activation energy for dislocation pipe diffusion to that for lattice diffusion, $Q_{disl}/Q_{lattice}$, is generally known as 0.65 in fcc metals[19], The Q_{disl} is calculated to be 90kJ/mol and is quite similar to the measured activation energy for onset of A2 type serration. Also, the measured activation energy for onset of A2 type serration is similar to the activation energy for onset of serration in Ni-C reported by Nakada and Keh[7]. Therefore, the rate controlling mechanism for onset of A2 type serration is suggested as the carbon diffusion through dislocation core.

The activation energy for onset of B type serration measured as 150kJ/mol, which is similar to the activation energy of 139kJ/mol for carbon diffusion in Ni[18]. The controlling mechanism of B type serration is suggested as the carbon diffusion in Inconel 600. Kocks et al.[6] reported that the serration behavior of Inconel 600 with low carbon less than 0.05at.% was identical to that of commercial Inconel 600. They suggested that the dynamic strain aging in Inconel 600 is more likely controlled by solute atoms rather than carbon. However, the low values of m+β indicate that the substitutional atom - carbon compounds could retard the diffusion rate of carbon and are responsible element for A2 and B type serrations, as suggested in austenitic stainless steels[20, 21]. The possible substitutional elements are Cr and Ti due to their strong tendency to form carbides. The amount of carbon in Inconel 600 alloy needed to pin the dislocation for onset of serration could be very low when the substitutional atom - carbon compounds interact with dislocations.

Conclusions

The serrations in stress strain curves due to dynamic strain aging were observed in the temperature range of 150-600°C with a strain rate of 10^{-5}-$10^{-3}s^{-1}$ in Inconel 600 alloy. Four different types of serration, A1, A2, B and C type serration, were observed depending on the temperature, strain rate and strain. A1 type serration, a periodic rise and drop of stress with small amplitude, was observed in the temperature range of 150-245°C. A2 type serration, a rise of stress followed by a drop of stress, was observed at a higher temperature range of 245-400°C. B type serration, successive oscillations of stress, observed in the temperature range of 245-500°C. C type serration, characterized as abrupt irregular stress drops, was observed at a temperature of 600°C. The higher work hardening rate and negative strain rate sensitivity were observed within the temperature regime showing serrated stress-strain curves. Based on the analysis of m+β and Q, the rate controlling mechanisms are suggested as the migration of substitutional atoms for A1 type serration, carbon diffusion through dislocation core for A2 type serration and carbon diffusion through lattice for B type serration. The diffusion of the substitutional atom - carbon compounds could be the rate controlling mechanism for A2 and B type serration.

References

1. S. M. Bruemmer and C. H. Henager, Jr., "High Voltage Electron Microscopy Observations of Microdeformation in Alloy 600 Tubing," Scr. Metall., 20 (1986), 909-914.

2. S. M. Bruemmer, "Microstructure and Microdeformation Effects on IGSCC of Alloy 600 Steam Generator Tubing," Corrosion, 44 (1988), 782-788.

3. G. A. Was, J. K. Sung and T. M. Angeliu, "Effect of Grain Boundary Chemistry on the Intergranular Cracking Behavior of Ni-16Cr-9Fe in High Temperature Water," Metall. Trans., 23A (1992), 3343.

4. R. A. Mulford and V. F. Kocks, "New Observation on the Mechanisms of Dynamic Strain Aging and of Jerky Flow," Acta Metall., 27 (1979), 1125-1134.

5. R. W. Hayes and W. C. Hayes, "On the Mechanism of Delayed Discontinuous Plastic Flow in an Age-Hardened Nickel Alloy," Acta Metall., 30 (1982), 1295-1301.

6. U. F. Kocks, R. E. Cook and R. A. Mulford, "Strain Aging and Strain Hardening in Ni-C Alloys," Acta Metall., 33 (1985), 623-638.

7. Y. Nakada and A. S. Keh, "Serrated Flow in Ni-C Alloys," Acta Metall., 18 (1970), 437-443.

8. B. J. Brindley and P. J. Worthington, "Yield Point Phenomena in Substitutional Alloys," Metall. Rev., 145 (1970), 101-114.

9. A. Van den Beukel, "On the Mechanism of Serrated Yielding and Dynamic Strain Aging," Acta Metall., 28 (1980), 965-969.

10. P. G. McCormick, "A Model for the Portevin-Le Chatelier Effect in Substitutional Alloys," Acta Metall., 20 (1972), 351-354.

11. I. S. Kim and M. C. Chaturvedi, "Serrated Flow in Inconel 625," Trans. JIM, 28 (1987), 205.

12. P. G. McCormick, "The Portevin-Le Chatelier Effect in a Pressurized Low Carbon Steel," Acta Metall., 21 (1973), 873-878.

13. A. Wijler, M. M. A. Vrijhoef and A. Van den Beukel, "The Onset of Serrated Yielding in Au(Cu) Alloys," Acta Metall., 22 (1974), 13-19.

14. D. Munz and E. Macherauch, "Dynamische Reckalterung von α-Messing," Z. Metllk., 57 (1966), 552-559.

15. K. Mukherjee, C. D'Antonio, R. Maciag and G. Fisher, "Impurity-Dislocation and Reported Yielding in a Commercial Al Alloy," J. Appl. Phys., 39 (1968), 5434.

16. A. M. Brown and M. F. Ashby, "Correlations for Diffusion Constants," Acta Metall., 28 (1980), 1085.

17. J. Friedel, Dislocations, (Oxford: Pergamon, 1964), 102.

18. C. J. Smithells, Metals Reference Book, Vol. 2, 4th ed. (Plenum Press, 1967), 649.

19. D. D. Pruthi, M. S. Anand and R. P. Agarwala, "Diffusion of Chromium Inconel-600," J. Nucl. Mater., 64 (1977), 206.

20. L. H. De Almeida, I. Le May and S. N. Monteiro, Strength of Metals and Alloys, (Pergamon Press, 1986), 337.

21. S. Venkadesan, C. Phaniraj, P. V. Sivaprasad and P. Rodriguez, "Activation Energy for Serrated Flow in a 15Cr-15Ni Ti-Modified Austenitic Stainless Steel," Acta Metall. Mater., 40 (1992), 569-580.

THE APPLICATION OF NEURAL COMPUTING METHODS

TO THE MODELLING OF FATIGUE IN NI-BASE SUPERALLOYS

J.M.Schooling[†] and P.A.S.Reed[#]

[†]Dept. of Materials Science and Metallurgy, University of Cambridge, U.K.
[#]Dept. of Engineering Materials, University of Southampton, U.K.

Abstract

The current financial climate is driving a move towards increased use of computer modelling techniques in alloy design and development in order to reduce cost. In this paper the potential for use of neural computing methods in the prediction of fatigue resistance in Ni-base superalloys is assessed. Initial work has been conducted on the Stage II (Paris regime) behaviour, as the literature indicates that this is the simplest region of the fatigue crack growth curve to predict, with an approximately linear relationship existing between $\log(da/dN)$ and $\log(\Delta K)$, and the crack growth rates being principally affected by temperature, Young's modulus and yield strength. These three parameters were chosen for initial data collection and modelling. The predictions made are of fatigue life, calculated from the slope and intercept values of the linear portion of the log-log fatigue crack growth curve. A test dataset has been successfully predicted along with the trends in the data. The effect of adding ultimate tensile strength and electron valencies as inputs to the model is assessed. It is shown that validation of models produced against metallurgical experience, and careful construction of the database are important conditions for effective use of neural network models for fatigue life predictions.

Introduction

In recent years there has been an increased interest in the ability to model the mechanical properties of alloys from compositional and processing data in order to reduce the cost and time required for alloy development. Increased expectations of engine performance, in terms of speed and range, have led to the need for higher performance alloys, while the financial climate has forced a reduction in the cost of new products.

Good fatigue crack growth resistance is an essential property for superalloys operating in the high temperature stage of gas turbine engines; hence over the last thirty years many studies on the fatigue crack growth behaviour of Ni-base superalloys have been conducted. The general trends in fatigue crack growth behaviour with variations in test conditions (load ratio, temperature and environment), microstructure and processing route are well established. In general such alloys exhibit two stages of long crack fatigue behaviour: Stage I facetted crack growth, occurring along crystallographic {111} planes corresponding to persistent slip bands; and more homogeneous Stage II crack growth resulting in essentially flat fracture surfaces [1]. Inspection intervals for turbine components are often determined on the basis of Stage II fatigue crack growth rates, assuming an initial crack length equal to the smallest detectable flaw size. Such flaws might be surface scratches, coating cracks or casting defects such as pores. In this paper Stage II crack growth is concentrated on as the regime of most interest with respect to lifing procedures.

The fatigue process is sufficiently complicated that, in spite of all the work which has been undertaken in this area, a comprehensive fundamental behavioural model is lacking. Hence a reliable empirical method for predicting fatigue life, using existing data, remains a desirable goal.

The complexity of the fatigue process and the noise associated with fatigue test results has meant that even traditional empirical methods, such as regression analysis, have failed to produce a sufficiently comprehensive, robust model which accurately predicts trends in the data and yet does not model the noise.

Artificial neural networks are powerful computing devices designed to mimic the structure and learning capabilities of the brain, consisting of a large number of simple computational elements (or nodes) which are extensively interlinked via weighted connections. Such networks have the ability to learn rather than being programmed, can pick out complex patterns or trends in data, and can deal with noisy or irrelevant data points in an input dataset. In this paper the possibilities for using such a neural network as a tool to model fatigue using existing fatigue crack growth data are evaluated.

Neural Networks

Artificial neural networks are computational tools based on the structure and function of the brain [2]. They are composed of simple computational elements (called neurons or nodes) which imitate the most basic function of a biological neuron. These artificial neurons are then connected to others by a series of connections broadly analogous to, although much simpler than, those in the brain. The nodes are arranged in layers, with each node being connected to every node in the adjacent layers (figure 1). The simplest node sums N weighted inputs, performs a non-linear function on the sum, and then passes the result to the nodes in the next layer [3]. Data flow forwards only through the network.

There are two phases in the development of a neural network model [4]. Initially the network is 'taught' using a number of example datasets - this process is referred to as training. The training datasets consist of a series of inputs paired with the corresponding output (the 'target' output). An input dataset is applied, the network calculates an output and compares it with the target. The error between output and target is calculated, and the weights in the network are then adjusted using an algorithm (in the case of the network used in this work, a 'back propagation' algorithm) in order to improve the output. The data are fed through repeatedly until the network output is deemed sufficiently accurate (the solution has 'converged'). After this training the network is tested on a set of previously unseen data.

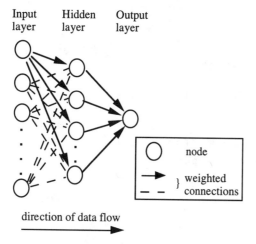

Figure 1 - Schematic diagram of a neural network

It is possible to vary the number of nodes in the hidden layer of the network (the 'architecture'), and care must be taken when deciding the network architecture, as it is possible to overmodel or undermodel the data. If too complex a network is chosen, then the training data will appear to be excellently modelled, however the network will be modelling the noise in the data as well as the trends, so that when the model is tested it will not be robust enough to cope with the new data presented to it. On the other hand, if too simple a network is chosen it will fail to model the trends in enough detail, and predictions will again be very inaccurate. Examples of this type of behaviour are shown in figure 2. In this work, therefore, the network is trained over a number of different architectures and the results are compared.

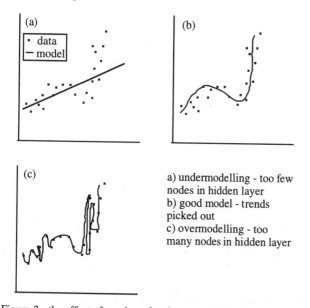

Figure 2 - the effect of number of nodes on modelling accuracy

A point to note is that a neural network learns from experience, and hence while it may interpolate between data with some confidence, it cannot accurately extrapolate into regions for which it has no information, and any attempts at such predictions should be treated with extreme caution.

The main advantage of neural networks over conventional regression analysis techniques is that the network finds an optimum solution without the need to specify the relationships or the form of relationships between variables. The ability of networks to generalise and find patterns in large quantities of often noisy data is also a major advantage. However the number of sets of training data required to establish a robust network is dependent on the number of input variables (the 'dimensionality' of the data). Previous studies have suggested that the size of the training data should be between three and ten times the input dimensionality[5].

Neural networks in materials science

It has been demonstrated in the literature that neural networks may be used with some success to model material properties and material behaviour [6, 7, 8, 9]. While it is acknowledged that material behaviour is best understood by carrying out experimental programmes, it is not always possible to describe the behaviour in terms of a simple mathematical expression, and hence quantitative modelling of behaviour is difficult, and will become more so as modern materials are further developed with increasingly complex behaviour.

Feedforward neural networks can be very useful in picking out patterns of behaviour and property relationships from a quantity of experimentally produced data, and their ability to generalise can make them useful in predicting the behaviour of a potential new material before it is made. Work on the strength properties of Ni-base superalloys [6] has shown that trends in behaviour with varying composition, temperature and material condition can be modelled, and within error limits absolute values of strength may be estimated.

This paper describes work carried out on the modelling of the Stage II fatigue properties of such alloys using a supervised feedforward neural network as described above.

The datasets

A number of input variables have been considered for presentation to the network. These are: temperature, yield stress, Young's modulus, ultimate tensile strength and Nv number.

Sixty-four sets of input data were gathered from a Rolls-Royce database [27], for which fatigue tests are conducted at a frequency of 0.25 Hz using a 1-1-1-1 trapezoidal wave form. The network trained using 34 sets and tested itself on the remaining thirty sets.

Before presentation to the network the data for each input variable was normalised between -0.5 and +0.5, according to the equation:

$$n_i = \frac{(x_i - x_{min})}{(x_{max} - x_{min})} - 0.5 \qquad (1)$$

where n_i is the normalised value of datapoint i and x_i, x_{max} and x_{min} are the actual values of datapoint i, and the maximum and minimum valued datapoints. This is to prevent a variable from swamping the network simply by virtue of having a large absolute value rather than as a result of the effect of its variation on the result.

Each set of input data is presented to the network along with a corresponding expected output. In this case the chosen output was a 'life' calculated from crack-C and crack-n values where crack-n is the slope of the log da/dN vs. log ΔK fatigue curve in the Stage II region and crack-C is the intercept of this line on the da/dN axis (see figure 3). All fatigue data were produced at a load ratio R=0.1, and lives were calculated using a prediction program for a semi-elliptical crack [10]. These 'life' values were also normalised.

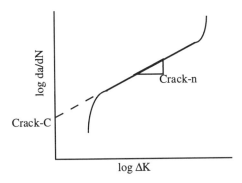

Figure 3 - Schematic to show crack-C and crack-n

When choosing a dataset to train the network on it is important to try and ensure that the data are spread evenly over the areas of input space which are of interest. Otherwise the network may be attempting to extrapolate from well defined regions of input space into regions about which nothing is known, and predictions may be unreliable. The range of values of input and output data is shown in figure 4.

Training and testing the network

The network used is based on a Bayesian statistical framework, which allows the probability of a model being a true representation of the data to be assessed[11]. Initial weights are set using a random number generator which is started using a 'seed'. This seed can have any positive value, and using different seeds may produce slightly different models. In order to ensure that all potential model types are generated the network is trained using a number of seed values for each architecture.

The network finds the optimum solution by minimising a penalised likelihood, in effect trying out a number of solutions to find the best relationship. As well as calculating the weights for the connections the network calculates a number of other parameters:

σ_{nu} - a measure of the noise allowed in the network's prediction of the data. Initially this is set at a fairly large value, and the program modifies the inferred noise level as it develops the model to more accurately fit the data. The value of this parameter varies with the number of nodes, generally decreasing to a limit as the number of nodes increases and a more complex model is formed such that the relationship between inputs and outputs is better modelled. The value of σ_{nu} is a fraction of the range of the output data, e.g. $\sigma_{nu}=0.2$ means that the inferred noise in the data is 20% of the range of the target output dataset.

training energy - a measure of the error in predicting the training dataset targets (this usually follows a similar pattern to σ_{nu}).

test energy - a measure of the error in predicting the previously unseen test dataset. This initially decreases with increasing number of nodes to a point, but if overmodelling starts to occur, and the network has effectively 'learnt' the training dataset, then it will be unable to cope with noise in the test set and test energy will start to increase.

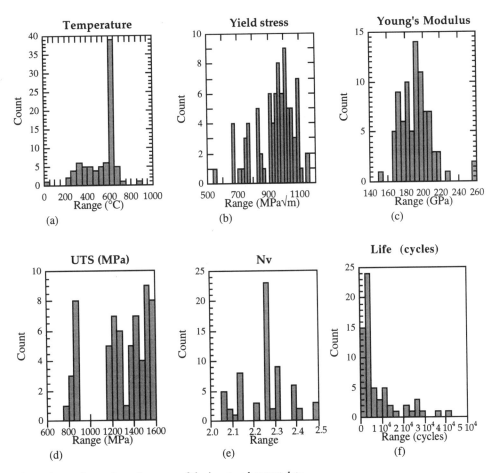

Figure 4 - graphs to show the range of the input and output data

relevance - this is a value which is produced for each input to the network, and is a measure of how important that input is in the model produced. A highly important input would have a relevance of the order of 1, whereas an input which is perceived to be irrelevant would be assigned a relevance of the order of 10^{-3} or less.

Energy values are calculated from the following equation:

$$energy = \frac{1}{2}\Sigma(o-t)^2 \qquad (2)$$

where *o* is the output value and *t* is the target value.

In order to be useful as a cost-saving design tool, the number of inputs to the network should ideally be kept to the minimum possible. For this reason it was decided to 'start small' and present the network with few variables, gradually adding other possibly important variables to see whether they had a positive or detrimental effect on the predictive ability of the network. Another advantage of this approach is that it keeps the input dimensionality low, thus requiring less data to produce a robust model.

The basic database

A review of literature on Stage II Paris regime fatigue crack growth and crack tip opening displacement (CTOD) models for crack growth indicates that the most important properties affecting fatigue crack growth are likely to be yield strength and Young's modulus [12, 13, 14, 15, 16, 17, 18, 19]. CTOD theory suggests the following form of equation for fatigue crack growth rate:

$$\frac{da}{dN} \propto \frac{\Delta K^2}{\sigma_{ys} E} \qquad (3)$$

where *da/dN* is the crack growth rate, ΔK is the stress intensity factor range, σ_{ys} is the yield stress and E is the Young's modulus [17, 18]. The literature also suggests that test temperature is an important factor affecting fatigue crack growth rates [20, 21]. Hence initial modelling used an input dataset consisting of temperature, σ_{ys} and E.

The network was trained for 2-8 hidden nodes. For each number of nodes seeds of 0, 20, 40, 60 and 99 were used.

A plot of σ_{nu} vs. number of nodes is shown in figure 5a. Two different types of model are generated by the network, depending on the initial random weights (determined by the 'seed'). The first type tend to have relatively high σ_{nu} values (0.18, or 18% of the range of the output data), this value being independent of the number of nodes. This indicates that an over-simple model is being produced which is just as accurately described with two nodes as with five or seven. Training and test energies for these models followed a similar pattern and were also high, with values of about 0.5 in both cases. This type of model will be referred to as the simple model.

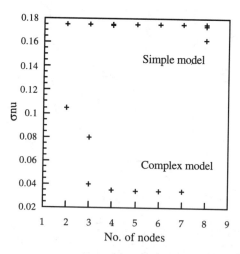

Figure 5a) - Graph of noise vs. no. of nodes for the simple dataset

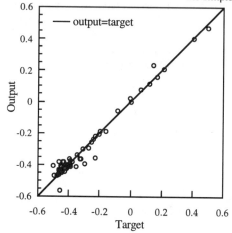

Figure 5b) - Graph of network output vs. target output for a type-B model

The second set of models on the other hand display a decreasing noise level, reaching a minimum of about 0.034 at four hidden nodes, indicating the formation of a more complex model. Training and test energies again followed a similar pattern, with values of training energy around 0.015 and test energy around 0.03. When the relevance values of the input variables for the two different types of model were compared it was found that the simple models all had very low relevance for Young's modulus (e.g. $1*10^{-4}$, c.f. 0.6 for yield stress) whereas the more complex models attributed a much higher relevance to Young's modulus (e.g.1.4, c.f. 2.2 for yield stress), more in accordance with CTOD theory and experimental evidence. A graph of predicted output vs. target output for a complex model is shown in figure 5b. It can be seen that predictions are good, with most of the predictions lying very close to the line for output = target.

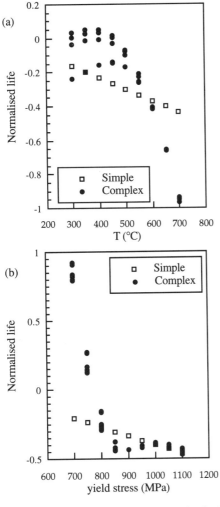

Figure 6 - Predictions of life generated using the simple database for (a) varying temperature and (b) varying yield stress

It is important to test the physical validity of a neural network model, to ensure that the network really has converged, and has not found a local minimum in the data. Thought experiments were conducted using a model dataset in which the input variables took on the median value of the original dataset (T=600 °C, σ_{ys}=950 MPa, E=195 GPa) and then each variable in turn was varied in order to see what trends the neural network had found. The best six complex models, ranked on test error, were chosen and the best simple model was used for comparison. The prediction of life with variation in temperature is shown in figure 6a. The prediction follows the expected trend of decreasing life with increasing temperature [21], with the effect becoming more marked as temperature increases. This can be understood both in terms of the decrease in yield stress (typically about 10% for a given alloy over this temperature range) and also the likely change in deformation mechanism. At higher temperatures a transition from planar to wavy slip is expected, with increased cross slip occurring and slip becoming inherently less reversible, resulting in more damage accumulation per cycle. The simple model produces a much simpler linear trend, indicative of undermodelling. The prediction of life variation with Young's modulus for the complex models shows a roughly linear increase in life with increasing Young's modulus, in accordance with what might be expected from CTOD theory and from experiment [22], while the simple models show life to be independent of E, as would be expected from the low relevance assigned to it.

When predictions of varying yield stress are examined, however, an unexpected result is obtained, as life is predicted to decrease rapidly as yield stress increases from 700-800 MPa, followed by a slight increase to 1000 MPa and then a further gentle decrease in life (figure 6b). CTOD theory on the other hand would indicate an increase in life with increasing yield stress in a similar manner to that seen in the Young's modulus predictions. There are two possible reasons for this apparent anomaly. It is possible that it may be explained by the fact that these predictions were made for a temperature of 600 °C, and the data contain only two alloys with a yield stress less than 800 MPa at this temperature, both of which have a Young's modulus of about 170 GPa, rather than the 195 GPa used for prediction. Hence attempting to predict life in the data range used may effectively be an extrapolation into an unknown area. While such extrapolations may not necessarily be totally misleading, this example illustrates that care is needed if attempting them. It also indicates the importance of constructing, where possible, a database which would cover the possible range of data of interest if neural network modelling is to be used as a design tool. That is not to say however that a neural network could not be used to reduce experimentation - if an unexpected result is found, then one or two experiments in the area of interest may suffice to test the proposed model, without an entire test matrix being required. The results could then be added to the training dataset.

Another possible explanation is revealed by further examination of the input data. If the data for life are plotted versus yield stress for temperatures between 550 °C and 650 °C (figure 7) it can be seen that there is in fact a reduction in life with increasing yield stress at low yield stress ranges. The points which cause this are all points for one particular alloy in the database which is a casting alloy, as opposed to a wrought alloy, and hence has a very different microstructure compared to the other alloys in the database. Examination of the actual test data revealed that the tests had been conducted on specimens containing only two grains, and the cracks had deflected at high angles from the normal to the principal stress, giving low apparent crack growth rates. The existence of only two grains in the specimens and the high deflection angle of the cracks indicate that the cracks may have propagated in Stage I, where CTOD models do not apply, as slip is inhomogeneous, being concentrated in intense bands. The low crack growth rates may have occurred as a result of crack closure [21] and/or shielding due to crack deflection [26], both of which reduce the effective stress intensity at the crack tip. This result indicates the importance of careful construction of the database to cover the problem of interest, without including spurious data.

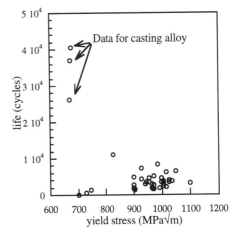

Figure 7 - Life versus yield stress for all data in the temperature range 550-650 °C.

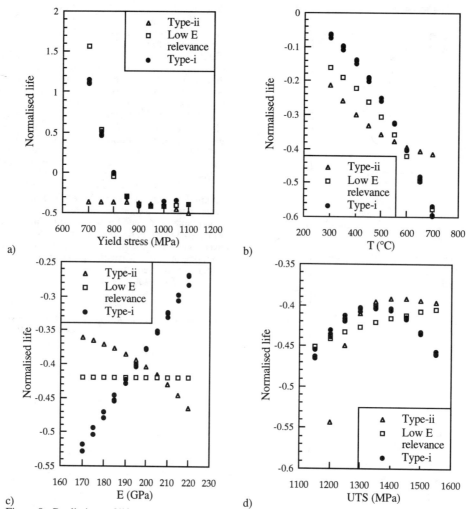

Figure 8 - Predictions of life generated using a database containing UTS for (a) varying yield stress (b) varying temperature (c) varying Young's modulus (d) varying UTS

Refining the data base

The model produced so far has been somewhat simplistic, with only three input variables used, and while some success has been obtained it was thought useful to add further parameters to see whether network performance improved. However, collecting data can be problematic as in order to use a given dataset it is necessary to have values for all the inputs. It would have been informative to look at grain size and other microstructural parameters, as Stage II fatigue is expected to be microstructurally sensitive to some extent [20]. Unfortunately, microstructural information is difficult to obtain from reports, papers and standards to a sufficient degree of accuracy to be useful in the network. Hence material property data and such data as are calculable from thermodynamic phase calculation programs such as MTDATA are used as inputs, as these are more readily available/determinable.

Accounting for cyclic hardening

In equation 3, the yield stress in the denominator is that at the crack tip. This may or may not be similar to the bulk yield stress, as repeated yielding of material in tension and compression may cause cyclic hardening or softening. Hence the cyclic yield stress may differ from the bulk yield stress. Data for cyclic yield stress are not readily available. However Manson and Hirschberg [23] proposed an empirical relationship between 'hardenability' and the ratio of ultimate tensile strength (UTS) to yield stress. Materials with a ratio greater than 1.4 are proposed to cyclically harden, those with a ratio less than 1.2 to cyclically soften, and those with a ratio between 1.2 and 1.4 to be cyclically stable. Therefore UTS was added to the database as an input to the network to provide a measure of deformation characteristics.

The addition of UTS as an input produced three types of low noise models, which had comparable noise levels and training and test energies to those produced previously. In this instance a number of the models (about half) assigned a low relevance to E. Of those that did not, again the best six models were chosen, ranked by test error, and were then compared with a model with low relevance for E. When predictions of variation in life with yield stress are made (using a mean UTS value of 1360 MPa), it may be seen that the models fall broadly into two groups, those that predict a large drop in life with yield stress increasing from 700-800 MPa (type-i) and a group which indicates a more linear dependence of life on yield stress (type-ii) (figure 8a). Above 800 MPa the two sets of models agree reasonably well. While the type-ii models at first seem promising, examination of predictions for temperature and Young's modulus calculation cast doubt on their validity (figure 8b &c). The prediction of temperature dependence shows a different form from previous models, and indicates a lessening of the effect of temperature as temperature increases, whereas observations on increased dislocation mobility with temperature would indicate an increasing effect of temperature as demonstrated by type-i. Dependence of life on Young's modulus for type-ii models is the reverse of that predicted by CTOD theory and experimental observation.

The dependence of life on UTS is shown in figure 8d. The type-i models predict an evident but modest dependence of life on UTS, indicating some importance of UTS, although it seems less important than temperature and E. The type-ii models however indicate a strong dependence of life on UTS below about 1250 MPa, above which value the two types of model agree reasonably well. When the values of predicted life are examined it is noted that below 1250 MPa the values fall well outside the range of the dataset (-0.75 as opposed to a limit on the data of -0.5). Again it is possible that the model is trying to extrapolate too far from the known database, and so such predictions should be regarded with care, as in the case of the yield stress predictions described above. Examination of the database reveals that at 600 °C the alloys with low UTS also have very low Young's modulus (about 170 GPa as opposed to the value of 195 GPa used for predictions).

Of the three models, type-i best reflects the actual trends in the data. The existence of two complex models with low noise and test energy indicates that within the experimental noise in the data there is more than one way to mathematically model the data. However, only one of these models makes physical sense. This difference in prediction by two sets of models which are ostensibly similar shows the importance of thoroughly examining a neural network model in the light of metallurgical knowledge.

The effect of alloy instability

It has been observed that some alloys which are unstable to the formation of the detrimental sigma phase have an increased resistance to fatigue crack propagation [24]. It is possible to express this instability in terms of an electron valency number (Nv) of the matrix, which can be readily calculated from alloy composition [25]. This parameter, Nv, was used as an input to the network in order to investigate this dependence. Nv has the advantage, along with yield stress, of being a means of entering compositional information without needing to enter the composition itself (which would vastly increase the input dimensionality and hence the number of datasets needed to produce a robust model).

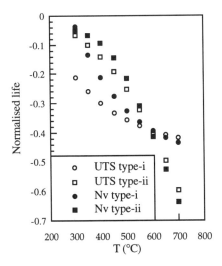

Figure 9 - comparison of predictions using data-base containing UTS and database with Nv added

Addition of Nv to the database again produces two types of complex model, similar in form to those described above (e.g. figure 9, Nv=2.27). The relevance attributed to Nv is quite high, and life is predicted to increase roughly linearly with Nv by both models (figure 10). The difference in life is not predicted to be very large as the range of Nv values in the database is quite small. This indicates that, although there is an effect of Nv on fatigue resistance, the benefits obtained by trying to alter Nv within the Ni-based alloy system would not be great as this class of alloys tends to exhibit a restricted range of Nv values.

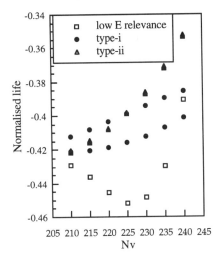

Figure 10 - Prediction of variation of life with Nv

Summary and Conclusions

A possible limitation on use of neural networks as predictive tools for design purposes is the amount of data required - large numbers of inputs require large, complete datasets which are difficult to compile. However success has been achieved in modelling Stage II fatigue crack growth behaviour in Ni-based superalloys. A basic dataset containing temperature, yield stress and Young's modulus models trends adequately, while refining the database to include ultimate tensile strength improves the performance.

The importance of careful construction of a database has been stressed, in order as far as possible to cover areas of input space that may be interesting, as extrapolative predictions must be treated with caution. Experimental results that may be of use in a database should be carefully logged to ensure that relevant inputs, such as microstructural information, are recorded accurately. It has been seen that it is important to validate any neural network model produced to ensure that its predictions are in line with experience.

It has been shown that it is possible to use a neural network to investigate and model trends in fatigue crack growth behaviour with variation in material properties, based either on proposed mechanisms or on observed empirical trends. A trend for increased Stage II fatigue life with increased instability to sigma-phase formation (indicated by electron valency number, Nv) has been shown to exist, although the observed effect is small due to the small range of Nv in Ni-based superalloys.

While further work on validation of such networks is required before they can be used in design, if such work is done neural networks may prove to be of great benefit in future alloy design programmes.

Acknowledgements

Professor C.J.Humphries is thanked for provision of research facilities at the University of Cambridge. Rolls-Royce plc are thanked for financial support for this work and for the provision of data. Thanks are also due to Joy Jones of the University of Cambridge for helpful discussion and advice.

References

1. S.R.Holdsworth & W.Hoffelner, "Fracture mechanics and crack growth in fatigue", High Temperature Alloys for Gas Turbines, R.Brunetaud et al eds., D.Riedel (1982), p345-368

2. P.D.Wasserman, Neural Computing Theory and Practice, Chapter 1, Van Nostrand Reinhold, 1989

3. R.P.Lippman, "An introduction to computing with neural nets", IEEE Acoustics, Speech and Signal Processing Magazine, 4 (2) (1987), pp4-22

4. J.A.Powell, "Preface: Neural networks, neuro-fuzzy and other learning systems for engineering application and research", Neural Networks, Neuro-Fuzzy and Other Learning Systems for Engineering Applications and Research Conference, Institute of Civil Engineers, London (1994), pp.i-iv

5. T.J.Stonham, "Neural networks, how they work, their strengths and weaknesses", Neural Networks, Neuro-Fuzzy and Other Learning Systems for Engineering Applications and Research Conference, Institute of Civil Engineers, London (1994), pp1-7

6. J.Jones, "Neural network modelling of the tensile properties of Ni-base superalloys" (CPGS dissertation, University of Cambridge, 1994)

7. H.K.D.H.Bhadeshia, D.J.C.Mackay & L.-E.Svensson, "Impact toughness of C-Mn steel arc welds - Bayesian neural network analysis", Mat. Sci. and Tech, 11 (1995), pp1046-1051

8. A.Mukherjee, S.Schmauder, M.Rühle, "Artificial neural networks for the prediction of mechanical behaviour of metal matrix composites", Acta Metall. Mater., 43 (1995), 11, pp4083-4091

9. U.Bork, R.E.Challis, "Nondestructive evaluation of the adhesive fillet size in a T-peel joint using ultrasonic lamb waves and a linear network for data discrimination", Meas. Sci. & Tech., 6 (1995), pp72-84

10. T.M.Edmunds, "Fatigue crack propagation program SA16" (Rolls-Royce in-house package)

11. D.J.C.MacKay, Bigback User Manual, 1994

12. L.A.James & W.J.Mills, "Effect of heat-treatment and heat-to-heat variations in the fatigue-crack growth response of Alloy 718", Eng. Fract. Mech., 22 (1983), pp797-817

13. C.Laird & G.C.Smith, "Crack propagation in high stress fatigue", Phil. Mag., 7 (1962), pp847-857

14. K.Tanaka, T.Hoshide & N.Sakai, "Mechanics of fatigue crack propagation by crack-tip plastic blunting", Eng. Fract. Mech., 19 (1984), p805

15. J.N.Hall, J.W.Jones & A.K.Sachdev, "Particle size, volume fraction and matrix strength effects on fatigue behaviour and particle fracture in 2124 aluminium-SiCp composites", Mat. Sci. & Eng., A183 (1994), pp69-80

16. A.J.McEvily, "Current aspects of fatigue", Metal Sci., 1977, p274

17. R.J.Donahue, H.McI.Clark, P.Atanmo, R.Kumble & A.J.McEvily, "Crack opening displacement and the rate of fatigue crack growth", I. J. of Frac. Mech., 8 (1972), pp209-219

18. D.Broek, Elementary Engineering Fracture Mechanics, Chapter 9, Klewer Academic, 1989

19. J.F.Knott, "Models of fatigue crack growth", Fatigue Crack Growth - 30 Years of Progress, R.A.Smith ed., Pergamon 1984, pp31-52

20. S.D.Antolovich & N.Jayaraman, "The effect of microstructure on the fatigue behaviour of Ni-base superalloys", Fatigue - Environment and temperature effects, J.J.Burke & V.Weiss eds., Plenum 1983, pp119-144

21. J.E.King, "Fatigue crack propagation in nickel-base superalloys - effects of microstructure, load ratio and temperature", Mat. Sci. & Tech., 3 (1987), pp750-764

22. M.O.Spiedel in High temperature materials in gas turbines, M.O.Spiedel & P.R.Sahm eds., Elsevier 1974, pp207-251

23. S.S.Manson & M.H.Hirschberg, "Fatigue behaviour in strain cycling in the low- and intermediate-cycle range", Fatigue - an interdisciplinary approach, J.J.Burke, N.L.Reed & V.Weiss eds., Syracuse University Press 1964, pp133-178

24. A.James, Personal communication, Rolls-Royce, 1995

25. J.P.Michalisin & D.L.Pasquire, "Phase transformations in Ni-base superalloys", Proc. Conf. on Structural Stability in Superalloys, Seven Springs, PA, Sept 1968, pp134-170

26. S.Suresh, Fatigue of Materials, Camb. Uni. Press (1991), pp255-258

27. Rolls-Royce design database

NEURAL NETWORK MODELLING OF THE MECHANICAL PROPERTIES OF NICKEL BASE SUPERALLOYS.

J. Jones[*] and D.J.C. MacKay[#]

[*] Department of Materials Science and Metallurgy, University of Cambridge, Cambridge, UK.
[#] Department of Physics, University of Cambridge, Cambridge, UK.

Abstract

Modelling techniques are being developed with the aim of reducing the cost and time associated with the development of new alloys for critical aerospace components. In this paper the yield and tensile strengths of commercial polycrystalline wrought Ni-base superalloys are modelled using an artificial neural network technique. Neural networks of this type are capable of realising a great variety of non-linear relationships of considerable complexity. They are 'trained' using existing experimental data which is presented to the network in the form of input-output pairs, thus the optimum relationship is found between the tensile properties and those parameters which are considered to be of importance. Through a series of tests it was found that with appropriate training a neural network can reliably reproduce metallurgical experience and knowledge. These results demonstrate that neural network models could be successfully used in the development of new alloys, reducing the amount of experimental work required and thus the time taken for a new alloy to be introduced into service.

Introduction

The yield and tensile strengths of nickel-base superalloys are fundamental design parameters in the manufacture of aerospace components. Today's highly alloyed, complex superalloys derive their outstanding tensile properties from a combination of several principal strengthening mechanisms (1): solid solution (2-5), precipitation (6-8) and grain boundary (9,10) strengthening. Consequently, there are a large number of variables that can influence these tensile properties. This complexity in the nickel superalloy system makes them very useful in the aerospace industry, but at the same time very difficult to design. Quantitative models can be very useful in this respect, assisting the development of new alloys by reducing the need for experimental work. Although many mechanisms are known to contribute to the strength of Ni-base superalloys it has not yet been possible to incorporate these mechanisms into a single model capable of predicting the tensile properties of complex industrial alloys. Already there is a vast experimental database for these properties, both in the published literature and in industrial data banks. This paper presents an investigation into the development of quantitative models of the tensile properties of Ni-base superalloys using an artificial neural network.

An artificial neural network is a computer intensive blind modelling technique which has the ability to recognise patterns in complex and often noisy experimental data (11,12). They first emerged as self learning tools in the early 1940's inspired by the architecture and learning capacity of the brain. Considerable development has occurred since then resulting in a wide range of neural network applications such as image and speech recognition, on-line control engineering, medicine, business and finance and many others (13).

Neural Network Approach

Neural networks are parameterized non-linear models used for empirical regression and classification modelling. Their flexibility enables them to discover more complex relationships in data than traditional statistical models which often assume a linear dependence of the predicted 'output' variable on the given 'input' variables. Neural networks are able to implement more general (and more complex) non-linear relationships thereby finding the optimum relationship automatically, irrespective of the complexity with which the variables may be associated. Only the input and output data must be presented to the network, which is a major advantage compared to traditional regression analysis where a prior choice of relationship must be made.

A neural network is composed of an interconnected array of simple processing units, referred to as nodes, arranged in a pattern analogous to the network of neurons and synapses in the brain.

The strength of the neural network lies in the large number of connections, each having an associated weight or strength which is adapted during training to improve performance. While the behaviour of a single node is relatively simple, the systematic behaviour of the nodes can be very sophisticated.

A network of the type shown in fig.1 has been used in this analysis. It consists of n input variables which are considered to control the output and two layers of nodes; a hidden layer of variable size and the output node.

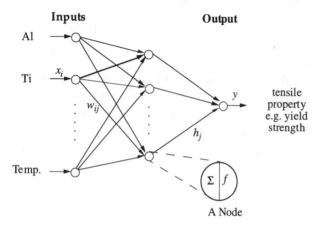

Figure 1: Schematic diagram of a neural network.

Each hidden node receives a contribution from every input node, x_i modified only by the connection weights, w_{ij}. The contributions are summed and passed through a non-linear 'sigmoidal' transfer function giving a node output h_j:

$$h_j = \tanh\left(\sum_i w_{ij} x_i + \theta_j\right) \quad (1)$$

where θ_j is a constant bias term. The non-linear nature of this function is necessary for the algorithms used during the training process (12). This sum, h_j is then passed on to the output node. At the output layer a linear transfer is performed giving an output y:

$$y = \sum_j w_{jl} h_j + \theta_l \quad (2)$$

As with regression analysis the neural network solution can be summarised by a series of coefficients or weights and a specification of the type of function relating the inputs to the output.

The connection weights are determined through a supervised 'training' process where real experimental data sets are presented to the network as model input - target output pairs. In response, the weights are adapted through a minimisation of a regularised sum of squared errors, $M(w)$ where:

$$M(w) = \beta E_t + \alpha E_w \quad (3)$$

where

$$E_t = 0.5 \sum_n (t - y)^2 \quad \text{and} \quad E_w = \sum_k \frac{w_k^2}{2}$$

where t is the target output, y is the network output, n is the number of training data set, w_k is a network weight, k is the number of weights of connections coming from all the input nodes. α and β are regularising constants. The objective function $M(w)$ consists of two terms E_t and E_w. E_t is a measure of the error in the network, that is the difference between the target output, t and the network output, y summed over all the data points in the training set. The second term E_w is designed to reduce the detrimental effect of any input variables present which do not affect the output. Further details of this procedure are described in more detail elsewhere (12,14,15). Thus the optimum non-linear interpolant that fits the data well is found.

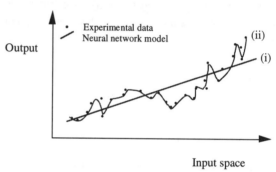

(a)   (i) Insufficient number of hidden nodes,
      (ii) Excessive number of hidden nodes.

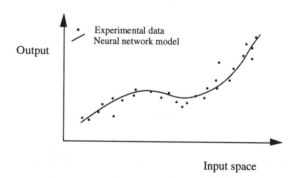

(b) Optimum network size.

Figure 2: Effect of hidden layer size on generalisation.

The integrity of a trained network lies in its ability to generalise, that is its performance on the actual problem once training is complete. Blind modelling of this type can be susceptible to over representation in which even interpolation becomes dangerous. In effect, the network learns the data used in the training process but not the general principles controlling the output. Obviously the complexity of the model is dependent on the number of hidden layer nodes and the regularisation constants α and β. The effect of hidden layer size on generalisation is shown schematically in fig.2. If too few nodes are used, fig.2(a), then the network may not train or will form a very simplistic model which does not represent the data well. Conversely, if too many nodes are used, fig.2(a), specific examples in the training set will be learnt and the resulting network will not generalise. Therefore an optimal number of nodes must be selected, fig.2(b).

Traditionally the danger of over representation is reduced by dividing a given data base into two representative sets. The network is then trained on one set and subsequently tested on the other to assess the networks generalisation. In addition, MacKay (14,15) has developed a neural network method which incorporates a Bayesian statistical framework allowing objective choices of the network parameters such as hidden layer size and regularisation constants. Therefore, the model complexity can be controlled thus enabling a network to be trained and used more efficiently and successfully. Using this method it is also possible to automatically identify which of the many possible input parameters are in fact important in the regression.

Once the network has been fully trained, estimation of the output for any given set of inputs is very rapid. These predictions are accompanied by error bars which depend on the specific position in input space, quantifying the model's certainty about its predictions. This is a further advantage over traditional regression analysis where a single global error bar is calculated.

Neural Network Modelling of Mechanical Properties

A large computer data base was assembled for the yield strength and ultimate tensile strength (UTS) for a range of commercial, wrought, polycrystalline alloys tested over a range of temperatures (16). The input parameters were the alloy chemistry (e.g. wt.% Al, Ti, Cr etc.) and temperature. Table I shows the range of the 16 input variables, the yield strength and UTS.

Two networks were employed, with 16 input variables each corresponding to one of those listed in table I, one giving an output of yield strength and the other UTS. For comparison purposes, each of the 16 inputs and the output was scaled linearly between +0.5 and -0.5, this prevents domination by a few large input variables. This normalisation process can be expressed quantitatively as:

$$x_n = \frac{x - x_{min}}{x_{max} - x_{min}} - 0.5 \quad (4)$$

where x_n is the normalised value of x, x_{min} and x_{max} are the minimum and maximum values of x in the entire data set as listed in table I.

The database consisted of ~200 points of which 100 were used to train the network and the remaining data points reserved to test the trained network. The training and test data were randomly taken from the master database. The size of the hidden layer was varied between 2 and 8 hidden nodes. Every network size was trained several times, each with a different set of initial random weights.

The performance of each network was quantified by calculating the root mean squared (RMS) error in the predicted output for both the training and test set. The RMS error is given by:

$$E = \sqrt{\frac{\sum[y-t]^2}{n}} \quad (5)$$

where n is the number of data points, t is the target experimental output and y is the output of the trained network.

Table I: Range of input and output data included in the data base.
All compositions are given in wt.%.

Input	Minimum	Maximum	Mean	Input	Minimum	Maximum	Mean
Ni	38	76	57	Mn	0	0.5	0.16
Cr	12	30	18.7	Si	0	0.5	0.13
Co	0	20	8.4	C	0.03	0.35	0.08
Mo	0	10	4.2	B	0	0.16	0.01
W	0	6.0	0.5	Zr	0	0.2	0.02
Ta	0	1.5	0.05	Temp. (°C)	21	1093	614
Nb	0	6.5	0.43				
Al	0	4.9	1.7	**Output** (MPa)			
Ti	0	5.0	2.1	Yield strength	28	1310	-
Fe	0	40	7.1	UTS	35	1620	-

Results and Discussion

As expected the performance on the training set for both networks is improved with hidden layer size, figs.3&4. The network outputs approach the target experimental results as the complexity of the network increases. However, more importantly the RMS error in the test set also decreases with increasing network size reaching an asymptotic limit at about 6 hidden nodes in both models, beyond which the model does not improve with increasing complexity. Therefore the optimum solution, with best generalisation, was selected as models with 6 hidden nodes. The yield and tensile strength data of these alloys have been modelled within an error of 60 and 70 MPa respectively. The level of agreement achieved in both the training and test set is shown in figs.5&6, each predicted value is accompanied by a ± 1σ error bar. The magnitude of the error bar is dependent on the position in input space and is a measure of how well the model is defined for that given set of input variables. A proportion of the noise in the models is believed to be real noise, that is, deviations in the tensile data which are not fully explained by the input parameters. It is believed that this could be reduced by inclusion of additional microstructural input parameters such as the γ′ volume fraction, size and grain size (17).

The trained models were then subjected to a variety of metallurgical tests to check they had learnt the correct science rather than nonsensical trends. Fig.7 illustrates the model perceived "significance" of each input variable in both datasets. This can be considered as a partial correlation coefficient, that is it indicates the fraction of variation explained by each input variable.

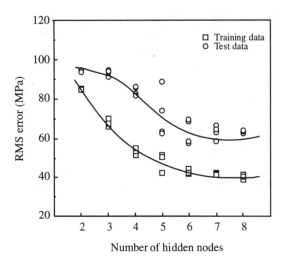

Figure 3: Performance on the yield strength training and test set as a function of the number of hidden nodes.

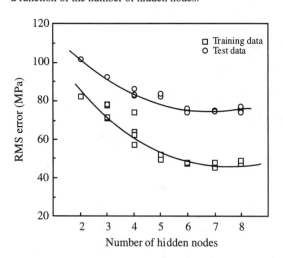

Figure 4: Performance on the UTS training and test set as a function of the number of hidden nodes.

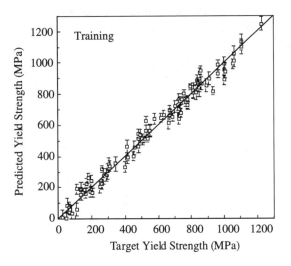

(a): Training set.

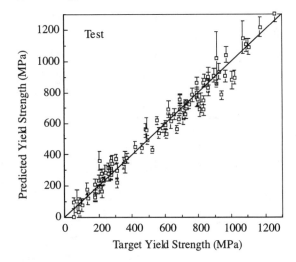

(b): Test set.

Figure 5: Performance of the optimum yield strength model on the training and test set with ±1σ error bars.

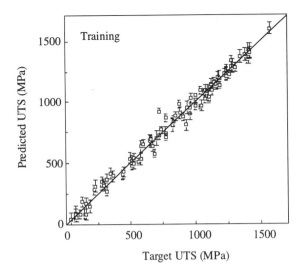

(a): Training set.

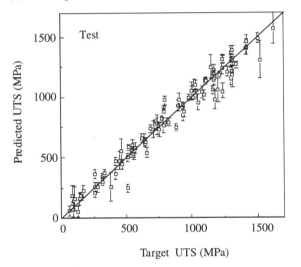

(b): Test set.

Figure 6: Performance of the optimum UTS model on the training and test set with ±1σ error bars.

Not surprisingly temperature is found to be the most significant input variable; the variation in temperature from 25-1100°C in the dataset is expected to lead to significant changes in microstructure as at elevated temperatures (>600°C) superalloys are chemically dynamic systems, the phases constantly reacting and interacting (18). The significant alloying elements identified by the network successfully reflect the active strengthening mechanisms in the superalloy system. Precipitation strengthening of the γ matrix by γ' is the dominant strengthening mechanism in the majority of superalloys and hence the predominant γ' formers Ti, Al and Nb are found to be the most significant alloying elements. Significant contributions are also made by Mo and W the solid solution strengthening elements, at elevated temperatures (>0.6Tm, where Tm is the absolute melting temperature) γ' strengthening is also diffusion dependent therefore Mo and W are also beneficial due to their low diffusivity in the nickel rich matrix. Minor effects of C, B and Fe are also detected.

The physical significance of the models was further tested by performing a series of predictions to investigate the sensitivity of yield strength and UTS to compositional and temperature variations. A selection of 'experiments' are shown in figs.8(a-d). All predictions are accompanied with ±1σ error bars. These predictions are similar to the matrix of experimental alloys prepared and tested during the design of a new alloy.

The temperature dependence of the tensile properties of superalloys is correctly predicted as shown in fig.8(a). The softening of the γ matrix is offset by the remarkable reversible increase in strength of γ' with increasing temperature (19,20).

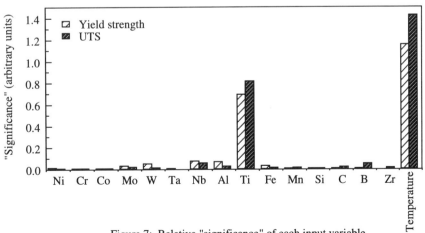

Figure 7: Relative "significance" of each input variable.

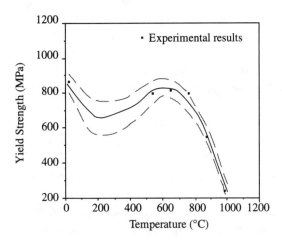

Figure 8(a): Predicted temperature dependence of the yield strength of a γ/γ' superalloy.

It was perhaps surprising that Ti received such a high significance value compared to Al the other major γ' forming element, fig.7. The relative strengthening effect of Ti and Al was investigated by systematically altering the ratio of Al to Ti atoms in a superalloy. In the test series the total Al + Ti content was kept constant, thus the total fraction of γ' is approximately unaltered. An increase in Ti/Al ratio is predicted to increase both the yield and tensile strength as shown in fig.8(b). This predicted trend is in agreement with experimental results of Miller and Ansell (21) on a Ni-15Cr-Mo alloy. Substitution of Ti atoms for Al atoms in the γ' lattice increases both the γ' lattice parameter, thereby increasing the lattice misfit (22) and the anti-phase boundary energy (23) of γ', both of these effects result in an increased γ' strengthening contribution as described by coherency and order hardening theories.

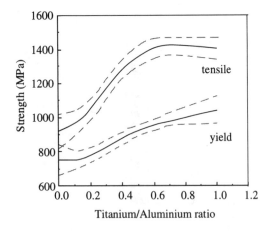

Figure 8(b): Predicted sensitivity of room temperature yield and tensile strength to variations in Ti/Al ratio.

Many researchers have investigated experimentally the role of Co in superalloys for both economic and strategic reasons. The effect of reducing Co levels in Waspaloy to zero was predicted as shown in fig.8(c), comparable experimental results of Maurer *et al.* (24) are also given. It should be noted that the data of Maurer *et al.* was not included in the training or test data bases, therefore the predictions made are a good test of the model's generalisation ability. It can be seen that Co levels have very little effect on the tensile properties of Waspaloy in good agreement with the experimental results. Any small changes in yield strength have been attributed to changes in fine γ' volume fraction and stacking fault energies (24,25).

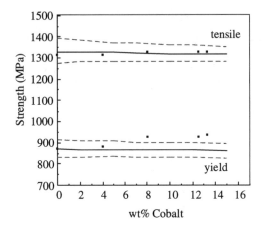

Figure 8(c): Predicted sensitivity of room temperature yield and tensile strength to variations in Co levels.

The predicted effect of Mo on the room temperature tensile properties of a typical Ni-base superalloy, namely Astroloy is shown in fig.8(d). It is evident that additions of Mo increase both the yield and tensile strength of the alloy, a 5wt.% increase in the Mo content leads to a predicted increase of ~200MPa in yield and tensile strength. There are several well documented theories to account for the strengthening effect of Mo relating to the atomic diameter oversize (3,4) and elastic modulus change (5) when an alloying element is introduced into the matrix and it's position in the periodic table (26), all of which have the net effect of impeding dislocation motion thereby increasing the strength. Decker (2) estimated the potency of solid solution elements present in the γ matrix and concluded that Mo contributes strongly, as predicted.

Fig.8(d) also illustrates the useful nature of the quantitative error bars, they serve as a measure of how well the model is defined for a particular set of inputs. In this example it is clear that the predictions are less useful at the higher Mo contents, due to the addition of Mo well outside the range of the training data (table I).

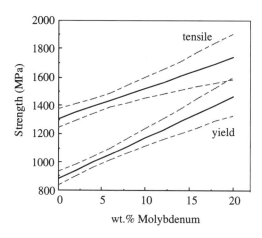

Figure 8(d): Predicted sensitivity of room temperature yield and tensile strength to additions of Mo.

Conclusions

An artificial neural network modelling technique has been successfully applied to Ni-base superalloy tensile properties. A quantitative model has been produced which can predict the yield and tensile strengths within error bounds. Despite the empirical nature of the modelling process the resultant models are found to be consistent with current metallurgical theory and experience. These results reveal the potential of these models as predictive alloy design tools, reducing the cost and time associated with experimental work.

Acknowledgements

The authors wish to acknowledge the financial support of Rolls-Royce plc. and the Engineering and Physical Sciences Research Council. They would also like to thank Professor C.J. Humphreys for the provision of research facilities. Thanks are also due to Dr. R.C. Thomson, Dr. H.K.D.H. Bhadeshia and Dr. D.M. Knowles for useful discussion and advice.

References

1. E.W. Ross & C.T. Sims, "Nickel Base Alloys", Superalloys II, eds. C.T. Sims et al., (New York, John Wiley & Sons, 1987), 97-133.

2. R.F. Decker, "Strengthening Mechanisms in Nickel Base Superalloys", Climax Molybdenum Company Symposium, Zurich, May 5-6 1969, 147-170.

3. R.M.N. Pelloux & N.J. Grant, "Solid Solution and Second Phase Strengthening of Nickel Alloys at High and Low Temperatures", Trans. AIME, 218, (1960), 232-237.

4. E.R. Parker & T.H. Hazlett, "Principles of Solution Hardening", Relation of Properties to Microstructure, American Society of Metals, 30, (1954), 30-70.

5. R.L. Fleischer, "Substitutional Solution Hardening" Acta Metall., 11, (1963), 203-209.

6. A.J. Ardell, "Precipitation Hardening" Met. Trans., 16A, (1985), 2131-2165.

7. E. Nembach & G. Neite, "Precipitation Hardening of Superalloys by Ordered γ'-Particles", Prog. in Mat. Sci., 29, (1985), 177-319.

8. B. Reppich, "Some New Aspects Concerning Particle Hardening Mechanisms in γ' Precipitating Ni-Base Alloys I Theoretical Concept", Acta. Metall., 30, (1982), 87-94.

9. A. Lasalmonie & J.L. Strudel, "Influence of Grain Size on the Mechanical Behaviour of Some High Strength Materials", J. of Mat. Sci., 21, (1986), 1837-1852.

10. W. Mangen & E. Nembach, "The Effect of Grain Size on the Yield Strength of the γ' Hardened Superalloy Nimonic PE16", Acta. Metall., 37, (1989), 1451-1463.

11. R.P. Lippmann, "An Introduction to Computing with Neural Nets", IEEE Acoustics, Speech and Signal Processing Magazine, 4, (2), (1987), 4-22.

12. D.R. Hush & B.G. Horne, "Progress in Supervised Neural Networks", IEEE Signal Processing Magazine, 1993 (January), 8-39.

13. N. Ryman-Tubb, "Implementation, the Only Sensible Route to Wealth Creating Success: a Range of Applications", Neural Networks, Neuro-Fuzzy and Other Learning Systems for Engineering Applications and Research, ed. J.A. Powell, (London, DRAL, 1994), 37-44.

14. D.J.C. MacKay, "Bayesian Interpolation", Neural Computation, 4, (1992) 415-447.

15. D.J.C. MacKay, "A Practical Bayesian Framework for Backpropagation Networks", Neural Computation, 4, (1992), 448-472.

16. "High Temperature High-Strength Nickel Base Alloys", Distributed by Nickel Development Institute, Birmingham, England.

17. J. Jones, "Neural Network Modelling of the Tensile Properties of Ni-Base Superalloys", (CPGS dissertation, University of Cambridge, 1994).

18. P.W. Keefe, S.O. Mancuso & G.E. Maurer, "Effects of Heat Treatment and Chemistry on the Long Term Phase Stability of a High Strength Nickel-Based Superalloy", Superalloys 1992, ed. S.D. Antolovich et al., (The Metallurgical Society, 1992), 487-496.

19. P. Beardmore, R.G. Davies & T.L. Johnston, "On the Temperature Dependence of the Flow Stress of Nickel-Base Alloys", Trans. AIME, 245, (1969), 1537-1545.

20. R.G. Davies & N.S. Stoloff, "On the Yield Stress of Aged Ni-Al Alloys", Trans. AIME, 233, (1965), 714-719.

21. R.F. Miller & G.S. Ansell, "Low Temperature Mechanical Behaviour of Ni-15Cr-Al-Ti-Mo Alloys", Met. Trans. A, 8, (1977), 1979-1991.

22. A.K. Jena & M.C. Chaturvedi, "The Role of Alloying Elements in the Design of Nickel-Base Superalloys", <u>J. of Mat. Sci.</u>, 19, (1984), 3121-3139.

23. D. Raynor & J.M. Silcock, "Strengthening Mechanisms in γ Precipitating Alloys", <u>Met. Sci. Journal</u>, 4, (1970), 121-129.

24. G.E. Maurer, L.A. Jackman & J.A. Domingue, "Role of Cobalt in Waspaloy", <u>Superalloys 1980</u>, ed. J.K. Tien, (American Society for Metals, 1980), 43-52.

25. R.N. Jarrett & J.K. Tien, "Effects of Cobalt on Structure, Microchemistry and Properties of a Wrought Nickel-Base Superalloy", <u>Met. Trans. A</u>, 13A, (1982), 1021-1032.

26. B.E.P. Beeston & L.K. France, "The Stacking Fault Energy of Some Binary Nickel Alloys Fundamental to the Nimonic Series", <u>J. Inst. of Metals</u>, 96, (1968), 105-107.

SOLIDIFICATION & CASTING TECHNOLOGY

EXTENDING THE SIZE LIMITS OF CAST/WROUGHT SUPERALLOY INGOTS

Ann D. Helms, Charles B. Adasczik, and Laurence A. Jackman
Teledyne Allvac
2020 Ashcraft Avenue
Monroe, North Carolina 28110

Abstract

Larger, more aggressive jet engine and industrial gas turbine designs have increased the demand for large diameter premium superalloy billets for rotating component applications. Forging suppliers are requesting larger diameter billet with structures and properties equivalent to smaller diameter billet. This requires larger diameter ingots since grain size in forging billet is strongly dependent on the amount of work imposed to the starting ingot structure. However, ingot diameter has been limited by segregation tendencies. This paper summarizes the development of larger diameter superalloy ingots for these applications. Much of the work focuses on alloy 718, but advances gained in this alloy system have been applied to alloys 706, 720, and Waspaloy. Extensive process development has been necessary to establish robust practices with defined process windows for each alloy. All steps of the total melt process must be evaluated but special emphasis has been placed on the final melt process where several melt parameters need to be evaluated. Sometimes thermal treatments of electrodes are necessary to prevent melt rate cycles during remelting. Development programs now underway are also presented.

Introduction

Large diameter solid solutioned strengthened and low hardener superalloy ingots have been routinely melted for many years. However, meeting the demands of jet engine and land based turbine industries for large diameter ingots of high hardener superalloy systems is challenging because of the strong tendency these alloys have towards macrosegregation and microsegregation. Also, their lower ductility can lead to internal cracks from thermal gradients created during heating and cooling operations. Development programs to meet this challenge must take advantage of all available technology. This presentation reviews recent programs to develop larger diameter ingots in gamma prime and in gamma double prime nickel-base superalloy systems. For each alloy system, two alloys that cover a wide range of hardener levels are discussed. The four alloys and their chemical compositions are given in Table I.

Table I. Nominal Alloy Compositions in Weight Percent

Element	718	706	Waspaloy	720
C	.025	.015	.035	.015
Ni	53	41	58	57
Fe	18	38	-	-
Cr	18	16	20	16
Mo	3	-	4.25	3
W	-	-	-	1.25
Co	-	-	13	15
Nb	5.4	3.0	-	-
Ti	1.0	1.75	3.0	5.0
Al	0.5	0.25	1.3	2.5
B	.004	.004	.006	.015

Alloy 718

Alloy 718 is the most frequently used alloy for rotor quality aerospace applications. Due to the need for inclusion-free material and the tendency for segregation during melting and solidification, the primary production routes for rotor-quality material have evolved to vacuum induction melting (VIM) followed by either vacuum arc remelting (VAR), electroslag remelting (ESR), or both (ESR+VAR). Current production of double melted (VIM+VAR) ingot is limited to 508mm (20 inch) diameter for high niobium levels. Larger double melted (VIM+VAR) ingots (610mm - 24 inch) are produced, but their use is limited to non-rotor applications with reduced niobium. Typically, the niobium is at least 0.30% lower for these applications to prevent freckle formation. Freckles are regions of positive macrosegregation that result from the flow of solute-rich interdendritic liquid in the mushy zone during solidification. ESR ingots are limited to less than 432mm (17 inch) diameter due to an increased tendency for freckle formation during ESR. Triple melted (VIM+ESR+VAR) 718 ingots offer the benefits of reduced inclusions from the ESR process along with a more solid electrode for VAR processing[1]. It has been a standard route to produce up to 508mm (20 inch) diameter ingots for a number of years.

Larger and more aggressive engine designs such as the PW4084, Trent 800, and GE90 have led to a demand for larger diameter rotor-quality billet. Requirements are such that this diameter billet must have microstructure and sonic inspection capabilities comparable to smaller billet. To meet this demand, a program was undertaken to develop a 610mm (24 inch) diameter ingot. Since the double melted (VIM+VAR) 610mm diameter ingot is limited to lower niobium contents, it is not an option for these applications. High niobium is necessary to meet the mechanical property requirements. Double melting is restricted by the lack of integrity of the VIM electrode. Cracks and solidification shrinkage cause instabilities during subsequent remelting. A more viable route seemed to be scaling up by triple melting where a solid ingot is generated in ESR that serves as the electrode in VAR. The enhanced process control obtained with the high integrity electrode results in a more stable VAR process. This reduces the tendency to form positive segregation such as freckles. Various melt parameters were investigated by evaluating billet structures to optimize the process. Numerous 610mm (24 inch) ingots have been melted with no occurrences of freckles in production material. The niobium content is the same high level used for smaller rotor quality ingots.

Macrostructural evaluation of billets produced by the triple melted 610mm (24 inch) ingot route has revealed white spot frequencies typical of 508mm (20 inch) triple melt product. White spots are light etching areas that are depleted in hardening elements[2]. Moderate tree ring patterns are sometimes observed as shown in Figure 1 for a top end slice. Figure 2 shows a typical macrostructure. These billet samples have been etched with Canada etchant to reveal melt related segregation.

More refined microstructures are possible for a given diameter billet when processed from a larger diameter ingot due to the greater amount of reduction. As a result, ultrasonic inspection capability is the same or better for a given diameter billet. Frequencies of occurrence for ultrasonic defects in triple melt material are low relative to double melted (VIM+VAR) products[3]. They are comparable for the 508 (20 inch) and 610mm (24 inch) triple melted ingot routes (Figure 3). In this figure, the defect frequencies are for fine grain billet 254mm (10 inches) in diameter and greater; they have been normalized to the frequencies for double melted billet from 508mm ingots. Rejectable indications are typically cracks initiated from clusters of oxides, carbides and/or carbonitrides; often they are associated with white spots.

Carbide sizes can become larger and carbide distributions less desirable with increasing ingot diameter. Small 127mm (5 inch) diameter ingots of alloy 718 made in Teledyne Allvac's pilot plant have smaller carbides than larger production ingots. For example, the 127mm ingots have no carbides exceeding 20 microns, while a 160mm^2 (0.25 inch^2) metallography sample from the center of billet from a 508mm (20 inch) diameter ingot can have over 25 carbide particles exceeding 20 microns. However, carbide ratings on billet from 610mm (24 inch) diameter ingots and 508mm (20 inch) diameter ingots have been found to be equivalent. Typical carbides are shown in Figures 4 and 5 for billet produced from ingots 508mm (20 inches) and 610mm (24 inches) in diameter. Reduction ratios of the billet shown are equivalent.

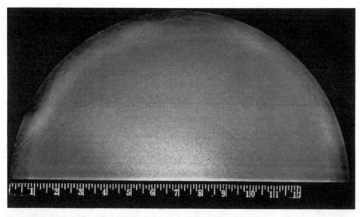

Figure 1: Moderate tree ring pattern in a billet slice from the top end of a 610mm diameter ingot of triple melted alloy 718. Canada etch. Scale is inches.

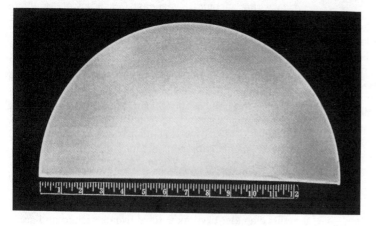

Figure 2: Typical macrostructure for a billet slice from a 610mm diameter ingot of triple melted alloy 718. Canada etch. Scale is inches.

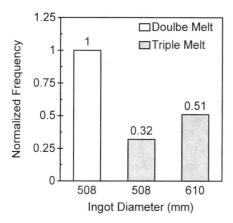

Figure 3: Relative frequencies of ultrasonic indications in alloy 718 billets 254mm and larger from 610mm triple melted ingots and 508mm double and triple melted ingots. Data have been normalized to the frequencies for double melted 508mm ingots.

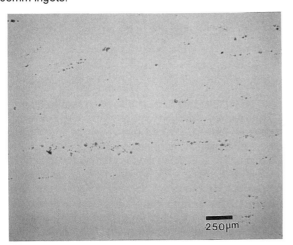

Figure 4: Typical carbide distribution for the bottom center location of 305mm diameter fine grain billet from 508mm diameter ingot.

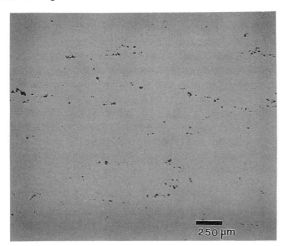

Figure 5: Typical carbide distribution for the bottom center location of 356mm diameter fine grain billet from 610mm diameter ingot.

Another concern with larger diameter alloy 718 ingots is increased microstructural segregation as manifested by banding with varying delta phase and niobium content. One method to evaluate this tendency is by determining delta solvus start temperatures by precipitating delta phase in a forged-down sample and establishing the temperature at which delta phase begins to dissolve. Evaluation of the 610mm (24 inch) diameter ingot product has shown no increased tendency towards microsegregation relative to the 508mm (20 inch) diameter triple melted product. Delta solvus start temperatures are equivalent.

Recently, the need has arisen for billet diameters and weights that can not be achieved from 610mm (24 inch) diameter ingots. Therefore, a program directed towards extending triple melted ingots to 686mm (27 inches) in diameter is underway. After several trials evaluating significant melt parameters, a process has been developed for successfully producing these large ingots. No freckles have been observed on any of the several production ingots made by this process.

Alloy 706

Following the development of alloy 718, a patent was issued by the International Nickel Company in 1972 for alloy 706. One of the objectives of the alloy 706 development program was to establish an alloy that could be melted to larger cross sections than alloy 718 without segregation problems[4]. This was achieved primarily by reducing the niobium (Nb) well below that for alloy 718 as shown in Table I. In the late 1980's and into the 1990's, this feature led to the use of large diameter alloy 706 ingots for heavy duty industrial gas turbine wheels. These forgings attain weights over 9,980 kgs. (22,000 lbs.) with diameters up to 2,210mm (87 inches) and thicknesses exceeding 406mm (16 inches)[5]. Teledyne Allvac provides ingots approaching 18,140 kgs. (40,000 lbs.) that are up to 914mm (36 inches) in diameter for these forging applications. Diameters as large as 1,016mm (40 inches) in diameter have been successfully melted.

Initial melting of these large diameter alloy 706 ingots involved double melting. Vacuum induction melting was followed by either VAR or ESR. However, it was found that segregation in the form of freckles sometimes occurred with double melting. In both 718 and 706, freckles are enriched in Nb and usually contain Laves Phase and numerous carbides and/or carbonitrides. An extreme example of freckles in a double melted (VIM+ESR) 706 ingot 864mm (34 inches) in diameter is shown in Figures 6 and 7[3]. Because of this problem, a triple melt process consisting of VIM+ESR+VAR was adopted.

It is necessary to control carefully the chemistry of large diameter 706 ingots to meet properties in final parts. Some of these chemistry controls, such as low silicon, carbon, and sulfur, also help to reduce the tendency of the electrodes to crack during heating and cooling and during melting.

Macrostructures for large diameter alloy 706 ingots are free of positive segregation such as freckles. An example is presented in Figure 8, which shows a typical transverse cross section for a 914mm (36 inch) ingot that has been double upset and drawn to a billet about 813mm (32 inches) in diameter.

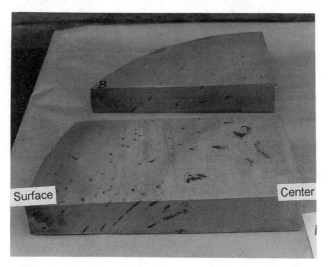

Figure 6: Large freckles outlining the pool profile near the top end of an 864mm diameter ingot of VIM+ESR alloy 706. The topmost transverse face of the plate is on the ground.

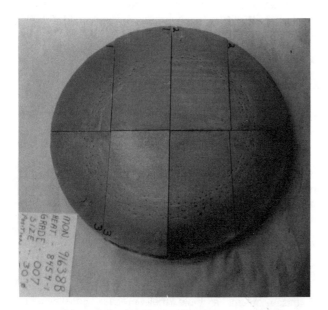

Figure 7: Pattern of freckles near the bottom end of the same 864mm diameter ingot of alloy 706 pictured in Figure 6.

Figure 8: Typical macrostructure of a billet slice from a 914mm diameter ingot of alloy 706. Scale is inches.

Waspaloy

Another superalloy that has continued to be developed for cast plus wrought applications is Waspaloy. Millions of pounds have been produced each year by the VIM+VAR process as 508mm (20 inch) and 610mm (24 inch) diameter ingots and by the VIM+ESR process for ingots up to 508mm (20 inches) in diameter. The design of larger aircraft engines has necessitated process development for production of even larger ingot sizes so that parts can be manufactured from larger diameter billets. In some cases, parts previously manufactured by investment casting can be replaced by wrought parts, which also has driven the development of larger ingots.

The ESR process was found to be limited to production of Waspaloy ingots no larger than 508mm (20 inch) diameter due to macrosegregation. ESR parametric studies were performed in attempts to produce 610mm (24 inch) diameter ingots. Although results were sometimes encouraging, it was concluded that the process window was too narrow and not viable as a production process. Ingots were sometimes found to contain freckle type macrosegregation without any indication from variations in critical ESR parameters. The ESR charts had normal traces for melt rate and voltage swing for both good and bad (freckled) sections of 610mm (24 inch) ingot. State of art ESR processing could not meet the demand for larger ingot sizes.

Extending the VIM+VAR process to 762mm (30 inch) diameter ingots for Waspaloy was readily accomplished. A 686mm (27 inch) diameter, 10,900 kg. (24,000 lb.) electrode was cast by the VIM process and then VAR processed under controlled conditions to explore the process window. This led to melting procedures that reproducibly provide satisfactory billet structures. Billet 457mm (18 inch) in diameter is routinely produced from 762mm (30 inch) VIM+VAR ingots with macrostructures comparable to billet produced from 610mm (24 inch) VIM+VAR ingots.

Alloy 720

Alloy 720 is a gamma-prime strengthened nickel-base superalloy like Waspaloy but the titanium (Ti) plus aluminum (Al) content is significantly higher as shown in Table I. Therefore, alloy 720 is more prone to positive segregation, such as freckles, than Waspaloy. Also, the decreased ductility of alloy 720 increases the risks of thermal cracking and melt rate cycles. In the 1980's, the largest standard ingots were double melted (VIM+VAR) to 432mm (17 inches) in diameter. As the requirement for larger disks and improved microstructures in the early 1990's intensified, the capability to produce triple melted (VIM+ESR+VAR) ingots 508mm (20 inches) in diameter was developed. These are now routinely provided with no incidence of freckling or other positive segregation. Now there is a need for even larger ingots to provide billet for larger disks and for further grain structure refinement. Consequently, an effort is underway to determine the feasibility of producing a triple melted ingot 610mm (24 inches) in diameter. It is recognized that a number of ingots must be successfully melted and fully evaluated to verify this process.

Melt Processing

ESR vs VAR

The VAR process results in a more controlled solidification of the ingot compared to the ESR process and is thus able to produce larger ingots of segregation sensitive alloys. ESR is complicated by the use of a CaF_2 based slag to melt the electrode by resistance heating. The slag freezes on the water cooled copper crucible and forms a skin between the ingot and crucible. Thus, the slag cap insulates the top of the ingot and the slag skin insulates the OD of the ingot. This results in a lower heat transfer rate during the ESR process and causes deeper molten pools and longer solidification times compared to VAR[6]. In VAR, the ingot is cast directly into the water cooled copper crucible and heat conduction is essentially unimpeded. Heat is radiated from the top of the molten pool between the electrode and crucible and transferred directly from the ingot to the crucible. Heat transfer between the ingot, which shrinks away from the crucible during solidification, is further enhanced by pressurizing the gap with helium. Heat is more efficiently conducted by the helium gas than if it were radiated from the ingot to the crucible. These factors contribute to shallower pools and reduced segregation in VAR relative to ESR.

VAR Controls

The VAR process is also preferred for larger diameter ingots of segregation prone alloys due to more reliable process controls. The process is controlled by regulating the electrode melting and ingot casting conditions. Electrode melting is automatically computer controlled using two algorithms. The first adjusts the electrode position in order to maintain a constant gap between the electrode and the molten metal pool, based on the arc voltage signal. F. J. Zanner[7] established a model relating the electrode gap to the frequency of drop shorts, which are caused by molten metal being transferred from the electrode to the ingot pool. A molten metal droplet can temporarily bridge the electrode gap and appear as an instantaneous drop in voltage or drop short. As the gap increases, fewer of the molten droplets bridge the electrode gap and the drop short frequency decreases. The computer continually adjusts the electrode speed to maintain a constant drop short frequency and therefore, a constant arc gap. The second algorithm is used to adjust the arc current continually in order to maintain a constant electrode melt rate. The melt rate is calculated by the computer using electrode weights from the VAR furnace load cell weighing system.

The ingot casting conditions are controlled by establishing adequate crucible cooling conditions. These include water temperature, water flow rate, and helium pressure in the gap between the ingot and crucible. The combined effect of computer process control of the electrode melting and establishment of appropriate ingot casting conditions enables production of alloys in ingot sizes previously unobtainable.

Melt Rate Cycles

During ESR and VAR, anomalies associated with melt rate can occur. One of these is commonly referred to as a melt rate excursion or melt rate cycle. It is distinct from other anomalies by its unique signature as shown in Figure 9, where the deviation in melt rate from the nominal melt rate is plotted versus time. A melt rate cycle starts with a gradual increase from the nominal steady state melt rate, followed by a rapid increase to a peak, a sudden decrease to a minimum, then gradual recovery to the nominal melt rate. When remelting ESR ingots as electrodes in VAR, the onset of the melt rate cycle is typically indicated by several pressure spikes up to 20-30 microns. Other melt rate anomalies usually are associated with a transient decrease in melt rate from steady state, such as that caused by a glow discharge or constricted arc in VAR.

Melt rate cycles in ESR and VAR are caused by imperfections in the electrode. The lack of appropriate thermal treatments of highly alloyed electrodes can result in internal transverse cracks. The crack interrupts the heat conduction along the length of the electrode from the end that is melting. This concentrates the heat below the crack, which causes the melt rate to increase as the process becomes more efficient. When the crack interface is reached, the end of the electrode is relatively cold, making the process less efficient. Therefore, the melt rate suddenly decreases. Finally, the melt rate gradually increases until the steady state temperature gradient is re-established in the electrode and the nominal melt rate is reached. A longitudinally sectioned and macroetched VIM electrode with a thermal crack is shown in Figure 10. The electrode is from a VAR melt that was aborted because of melt rate cycles. The melt rate cycle shown previously in Figure 9 is from the electrode in Figure 10. The longitudinal cracking in Figure 10 is pipe cavity from solidification shrinkage. The transverse crack

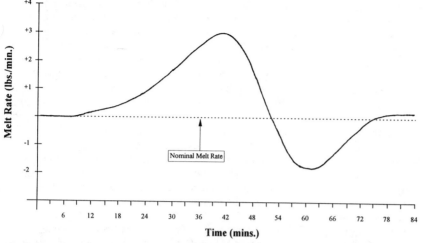

Figure 9: Melt rate cycle during VAR of the VIM electrode shown in Figure 10.

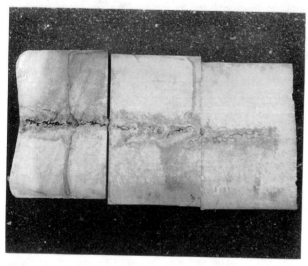

Figure 10: Longitudinal sections of VIM electrode aborted during VAR because of melt rate cycles, one of which is presented in Figure 9. Note transverse thermal crack in left section. Canada etch.

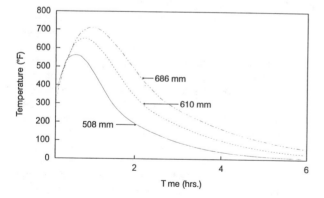

Figure 11: Calculated temperature difference from surface to center versus time when alloy 718 ingots at room temperature are put into a furnace at 1093°C.

near the left end of the electrode is the thermal crack. The left end of the electrode represents the surface being melted. The portion of the electrode shown was cut into three pieces to facilitate etching.

Evaluation of billets by macroetch and ultrasonic inspection has found acceptable structures for some melt rate cycles. Based on these evaluations, limits have been established for the severity of melt rate cycles for different alloy types and ingot sizes. In addition, the VAR and ESR control settings can be selected to react to the melt anomaly and reduce the variations in certain process parameters or ignore the melt rate cycle altogether, depending on the specific alloy and ingot size.

Thermal Treatments

Melt rate cycles can be avoided or their effects minimized by appropriate thermal treatments of electrodes. Thermal treatments improve the ductility of the electrode so that it is less susceptible to cracking from internal stresses. These internal stresses are generated by thermal gradients during melting and from heating and/or cooling of the electrodes. In addition to improving ductility, appropriate thermal treatments decrease residual stresses from heating and cooling. As diameters become larger, residual stresses become more of a concern because of increased thermal gradients. This is illustrated in Figure 11, which shows the calculated temperature difference between the center and surface locations as a function of time when three different diameter alloy 718 ingots at room temperature are placed in a 1093°C (2000°F) furnace. The larger temperature gradient for the 686mm (27 inch) ingot will generate greater internal stresses and, therefore, have more tendency for cracking. Concern about thermal gradients extends beyond melting to heating and cooling for subsequent homogenization and conversion.

Process Development

Extensive process development is necessary to establish a robust process for producing larger diameter ingots for a specific alloy. Each step of the total melt process must be evaluated but special emphasis is placed on the final melt process where several melt parameters need to be evaluated. Both aim levels and limits (upper and lower) must be determined for these parameters. It is not sufficient to concentrate only on the steady state portions of the final melt process; start-up and hot topping procedures must also be examined. One of the final melt parameters for which the process window must be identified is melt rate. Positive segregation such as freckles occur when the upper melt rate limit is exceeded. When the melt rate extends below the bottom limit, excessive solidification white spots occur. An example is presented in Figure 12, which shows a transverse macro slice for a 305mm (12 inch) billet forged from a 686mm (27 inch) alloy 718 ingot with a low melt rate. The surface region contains numerous solidification white spots.

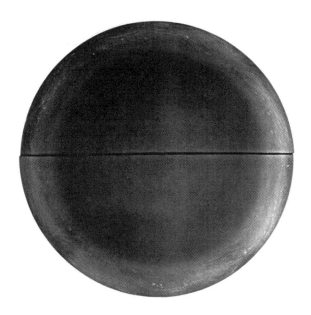

Figure 12: Solidification white spots in a 356mm (14 inch) billet slice from a 686mm (27 inch) ingot of 718 that was melted at a very low melt rate.

Modeling has been helpful for analyzing the feasibility of remelting large diameter ingots and for optimizing melt parameters. Performing simulations under different melt conditions can reduce the number of actual ingots that must be melted to establish a process. It is hoped as the ESR and VAR models are further developed and verified, they will become more useful in process development programs.

Conclusions

The demand of larger ingots for various superalloy systems is being met by continuous advances in melting controls, equipment, procedures, and understanding. For some alloy systems, triple melting is necessary to meet macro-segregation requirements. Maximum ingot diameters that can be melted without segregation problems have been extended in the 1990's to the sizes shown in Table II.

Table II. Present Ingot Diameter Capabilities that have Recently Been Extended

Alloy	Melt Process	Maximum Diameter (mm)		
		Standard Product	Special Product	Under Evaluation
718	VIM+ESR+VAR	610	686	
706	VIM+ESR+VAR	914		
Waspaloy	VIM+VAR	610	762	
720	VIM+ESR+VAR	508		610

Acknowledgments: The authors gratefully acknowledge Ramesh Minisandram for providing the modeling results and Scott Vallandingham for providing information on alloy 720.

References

1. A. Mitchell, "The Present Status of Melting Technology for Alloy 718", Superalloy 718 - Metallurgy and Applications, ed. E. A. Loria, The Minerals, Metals & Materials Society, 1989, 1-15.

2. L. A. Jackman, G. E. Maurer, and S. Widge, "New Knowledge About 'White Spots' in Superalloys," Advanced Materials & Processes, 5 (1993), 18-25.

3. J. M. Moyer et al., "Advances in Triple Melting Superalloys 718, 706, and 720", Superalloys 718, 625, 706 and Various Derivatives, ed. E. A. Loria, The Minerals, Metals & Materials Society, 1994, 39-48.

4. H. L. Eiselstein, "Properties of Inconel Alloy 706", Materials Engineering Congress, Cleveland, 1970.

5. P. W. Schilke, J. J. Pepe, R. C. Schwant, "Alloy 706 Metallurgy and Turbine Wheel Application", Superalloys 718, 625, 706, and Various Derivatives, ed. E. A. Loria, The Minerals, Metals & Materials Society, 1994, 1-12.

6. K. O. Yu, J. A. Domingue, "Control of Solidification Structure in VAR and ESR Processed Alloy 718 Ingots", Superalloy 718 - Metallurgy and Applications, ed. E. A. Loria, The Minerals, Metals & Materials Society, 1989, 33-48.

7. F. J. Zanner, "Vacuum Consumable Arc Remelting Electrode Gap Control Strategies Based on Drop Short Properties", Metallurgical Transactions B, American Society for Metals and the Metallurgical Society of AIME, Volume 12B, December 1981, 721-728.

COUPLED MACRO–MICRO MODELLING
OF THE
SECONDARY MELTING OF TURBINE DISC SUPERALLOYS

P.D. Lee, R. Lothian, L.J. Hobbs, and M. McLean
Department of Materials, Imperial College
London, UK

Abstract

A model of the macroscopic fluid flow, heat transfer and electromagnetic effects during secondary melting was developed and coupled to microstructural feature prediction. The importance of the different driving forces on the macroscopic model was explored using the test cases of vacuum arc remelting (VAR) and electron beam button melting (EBBM) of IN718. The predictions of the surface flows and final microstructure in the EBBM process were compared to experimental results.

Introduction

The demands of the aerospace industry have led to the development of a range of alloys with specific properties to satisfy the design requirements for different parts of the gas turbine engine. For turbine blades, the need is for a balance of very high temperature (up to ~ 1100°C) creep strength and corrosion resistance, whilst turbine discs have lower operating temperatures (~ 800°C). However, the demands for increasing power, through higher turbine speeds with associated increased centrifugal forces, and reduced engine weight have required significant increases in stresses carried by the discs. The development of alloys for turbine disc applications has concentrated on increasing the yield strength of the materials bringing a concomitant decrease in toughness, which has made the fracture characteristics of these materials very sensitive to the presence of defects. Consequently, it is now standard practice to improve the quality of turbine disc alloys, initially produced by vacuum induction melting (VIM), by subjecting them to one or more secondary melting processes. These processes include vacuum arc re-melting (VAR), electro-slag re-melting (ESR) and electron beam cold hearth refining (EBCHR). The quality improvements obtained are a finer microstructure, less macrosegregation and a reduced inclusion content.

In the past, the design and optimisation of secondary melting processes for new alloys has involved a great deal of expensive trial and error. Numerical modelling can, in principle, provide a less expensive route to reliable and efficient alloy processing. This paper evaluates the viability of combining macroscopic process models with empirical and theoretical treatments of microstructure formation for the accurate prediction of the ingot microstructure and defect susceptibility.

Theory

Macro-Model

There are many factors involved in the VAR process. Correspondingly, a detailed description of the real installation being modelled and accurate thermophysical data for the alloy involved must be supplied to any model which aspires to give accurate predictions. Factors to be dealt with include: heat transfer, fluid flow, magnetohydrodynamics (including both the Lorentz force and Joule heating), and solidification.

The modelling of the heat transfer, fluid flow, and solidification was performed using the commercial finite volume code Fluent with subroutines added to handle the spatially and temporally varying heat flux and surface tension gradient boundary conditions. The calculation of the Lorentz force and Joule heating was also coded in, but proved an insignificant effect in EBBM, although it is crucial in VAR. The equations being solved are listed below.

Energy Equation

The energy equation to be solved is:

$$\frac{\partial}{\partial t}(\rho h) + \nabla \cdot (\rho \mathbf{u} h - k \nabla T - \tau \cdot \mathbf{u}) = \frac{\partial p}{\partial t} + \mathbf{u} \cdot \nabla p + S_h, \qquad (1)$$

where h is the static enthalpy, p is the static fluid pressure, ρ is the density, k is the thermal conductivity, t is time, $\mathbf{u}$ is the fluid velocity, T is temperature and τ is the viscous stress tensor. S_h represents heat sources, including Joule heating (VAR), electron beam heating (EBBM), energy losses by radiation and conduction and latent heat evolution. The last of course, plays a crucial role in solidification modelling.

Momentum and Continuity Equations

The velocity of liquid, $\mathbf{u}$, is given by the momentum equation[1]:

$$\frac{\partial}{\partial t}(\rho \mathbf{u}) + \nabla \cdot (\rho\ \mathbf{u} \otimes \mathbf{u} + p\mathbf{I} - \tau) = \rho \mathbf{g} + \mathbf{F}, \qquad (2)$$

where $\mathbf{I}$ is the 3x3 identity tensor, $\mathbf{g}$ is the gravitational acceleration, and $\mathbf{F}$ is the sum of all other body forces (e.g. the Marangoni force or the Lorentz force).

The stress tensor τ has components:

$$\tau_{ij} = \left[\mu\left(\frac{\partial u_i}{\partial x_j} + \frac{\partial u_j}{\partial x_i}\right)\right] - \frac{2}{3}\mu\frac{\partial u_l}{\partial x_l}\delta_{ij}, \qquad (3)$$

where μ is the viscosity.

To solve for the momentum and pressure, the mass conservation equation must also be satisfied:

$$\frac{\partial \rho}{\partial t} + \nabla \cdot (\rho \mathbf{u}) = S_m. \quad (4)$$

S_m is the mass added to the continuous phase from any dispersed phase or as a source (e.g. due to droplets from the consumable electrode). In the present work, S_m has been taken to be zero, since filling is not modelled.

Electromagnetic Effects

The magnetic induction **B** in a magnetised fluid is calculated from the induction equation:

$$\frac{\partial \mathbf{B}}{\partial t} = \nabla \times (\mathbf{u} \times \mathbf{B}) + \eta \nabla^2 \mathbf{B}. \quad (5)$$

Here $\eta = (\mu_e \sigma)^{-1}$ is the magnetic diffusivity, where μ_e is the magnetic permeability (assumed in this work to take its vacuum value μ_0) and σ is the conductivity of the fluid. In the steady state, the LHS of (5) is zero. Furthermore, we may estimate the relative magnitudes of the terms on the RHS. The ratio of the first term, representing advection of field by the flow, to the second, representing field diffusion through Joule dissipation is given by the magnetic Reynolds number:

$$R_m = \frac{Lv}{\eta}. \quad (6)$$

The length scale L is typically 1 m in industrial scale remelting processes and $\eta \approx 1 \text{ m}^2/\text{s}$, so the diffusion term dominates if the typical flow velocity v is much less than 1m/s. R_m is correspondingly smaller on the smaller scales of pilot VAR and EBBM melts. It has been normal to neglect the advection term in such studies and this approach is justified *a posteriori* if computed velocities are indeed small. With these assumptions, **B** decouples from the fluid flow and may be calculated from the equation:

$$\nabla^2 \mathbf{B} = 0. \quad (7)$$

This equation was solved subject to the imposition of normal current components on the boundaries of the ingot.

The current density, **J**, Lorentz Force, $\mathbf{F}_e$, and Joule heating, Q_{joule}, were calculated from the equations:

$$\mu_0 \mathbf{J} = \nabla \times \mathbf{B}, \quad (8)$$

$$\mathbf{F}_e = \mathbf{J} \times \mathbf{B} \quad (9)$$

and

$$Q_{joule} = \frac{\mathbf{J} \cdot \mathbf{J}}{\sigma}. \quad (10)$$

In practice, a 2-D axisymmetric model, which considerably reduces computing time, was adopted.

Micro-Models

As summarised by McLean[2] the primary dendrite arm spacing, λ_1, in superalloys has been shown by many authors to be a function of the thermal gradient and the growth velocity both experimentally and theoretically. Recently Lu and Hunt[3] developed a numerical model of cellular and dendritic growth to predict the cell and dendrite spacing as well as the undercooling at the tip. Hunt and Lu[4] provided an analytic expression fitted to their numerical model results that predicts the minimum stable half-spacing of dendrites.

Cast in terms of non-dimensional parameters[*], Hunt and Lu found for the case of $G' > 1 \times 10^{-10}$ and $0.068 < k < 0.69$ that the dimensionless dendrite half spacing, λ', is given by:

$$\lambda' = 0.7798 \times 10^{-1} V'^{(a-0.75)} (V' - G')^{0.75} G'^{-0.6028}, \quad (11)$$

where:

$$a = -1.131 - 0.1555 \log_{10}(G') - 0.7589 \times 10^{-2} [\log_{10}(G')]^2. \quad (12)$$

To determine λ_1 the thermal gradient and front velocity were calculated using the macro-model. As a cell passes through the liquidus temperature the nearest neighbours' temperatures were used to calculate the thermal gradient normal to the isotherm. Hunt and Lu specify G as the solid thermal gradient, referring to the dendrite tip as the solid. In the macro model the mushy zone is treated as a continuum, making the thermal gradient close to the liquidus temperature an appropriate approximation for G.

The value of V used to determine λ_1 was the velocity of the liquidus isotherm, V_{liq}. This velocity was determined using a central difference approximation of the following derivative:

$$V_{liq} = \left.\frac{\partial n}{\partial t}\right|_{T=T_{liq}} = \frac{1}{\left.\frac{\partial t}{\partial n}\right|_{T=T_{liq}}} = \frac{1}{\frac{\partial t_l}{\partial n}}, \quad (13)$$

where t_l is the time at which the local temperature attains the liquidus value and n is the direction normal to the liquidus isotherm.

The ripening or local solidification time t_s, is defined as the time for which the dendrite is in the mushy zone. The secondary dendrite arm spacing, λ_2, is related to t_s by:

$$\lambda_2 = -7.0 + 12.5 \, t_s^{0.33} \, [\mu m]. \quad (14)$$

The carbide size, λ_{carb}, has also been shown experimentally to be a function of the local solidification time by Jardy et al.[5], who suggests the following formula for IN718:

$$\lambda_{carb} = \sqrt[3]{80.9 + 9.16 t_s} \, [\mu m]. \quad (15)$$

Comparison of Different Driving Forces

In order to estimate the relative importance of Lorentz, Marangoni and buoyancy forces, the relevant dimensionless numbers are constructed using the typical magnitudes listed in Table I.

To compare the relative importance of Marangoni and buoyancy forces, we evaluate the dimensionless quantity:

$$M \equiv \frac{\frac{\partial \gamma}{\partial T}}{\frac{\partial \rho}{\partial T} L^2 g}. \quad (16)$$

---

[*] The non-dimensional parameters are $G' = \frac{G \Gamma k}{\Delta T_o^2}$, $V' = \frac{V \Gamma k}{D \Delta T_o}$ and $\lambda' = \frac{\lambda \Delta T_o}{\Gamma k}$, where ΔT_o is the undercooling for a planar front given by $\Delta T_o = \frac{m C_o (k-1)}{k}$. G, V, Γ, D, m, C_o, and k are respectively the solid temperature gradient, velocity, Gibbs-Thomson coefficient, liquid diffusion coefficient, liquidus slope, bulk composition, and distribution coefficient[4].

Table I. Nominal properties used to calculate the relative importance of the driving forces during EBBM and VAR of nickel based superalloys.

Property	Symbol	EBBM	VAR	Units
Density	ρ	8000	8000	kg/m^3
Characteristic length	L	0.01	0.1	m
Surface tension gradient	$\partial\gamma/\partial T$	-1×10^{-4}	-1×10^{-4}	N/m/K
Density gradient	$\partial\rho/\partial T$	0.1	0.1	kg/m^3/K
Beam/arc current	I	0.35	5000	A
Beam/arc radius	a	0.005	0.1	m
Pool superheat	ΔT	200	200	K

In EBBM, M≈1, indicating that these forces are of comparable magnitude. In VAR, since M≈0.01, buoyancy forces are much greater than Marangoni forces.

We now estimate the typical value of the Lorentz force in the two processes. Assuming the current is spread evenly over a disc of radius a, we find that the average force is given by:

$$F_e = \frac{\mu_0 I^2}{3\pi^2 a^3} \quad (17)$$

We compare this with the typical buoyancy force:

$$F_b = \frac{\partial \rho}{\partial T} \Delta T \, \mathbf{g} \quad (18)$$

and find that $\frac{F_e}{F_b} \approx 2\times10^{-4}$ in EBBM, demonstrating that only buoyancy and Marangoni forces are important in determining the fluid flow. Conversely, in VAR, $\frac{F_e}{F_b} \approx 5$ and electromagnetic forces are comparable to buoyancy throughout the melt pool. Note that magnetic forces are dependent on the square of the current in the melt, so loss of half of the current would reduce these forces by a factor of four. Hence, it is important to know what proportion of the current reaches the ingot in a real installation. Evaluation of the Joule heating indicates that it is not significant in either process.

VAR Model

Validation

To validate the macromodel it was compared to previously published work simulating the VAR process (shown schematically in Figure 1). The model was run with the values given by Jardy and Ablitzer[6] in their study of fluid flow during the pilot scale VAR remelting of Zy4. They measured the pool shape experimentally (drawn as the dashed line in Figure 2) and used it as a boundary condition to determine the flow by solving the Navier-Stokes, Maxwell's, and Fourier equations. They assumed steady state and treated the problem as axisymmetric. With these conditions they calculated a maximum velocity of 0.013 m/s when half of the melting current reaches the pool, or $\alpha = 0.5$. In this study the location of the solidification front was not pre-specified, as it was by Jardy and Ablitzer, therefore values of the heat transfer coefficient, h_c, and the latent heat had to be approximated and are listed in Table II. A grid of 30×100 control volumes was used to simulate the 0.1×0.5 m domain.

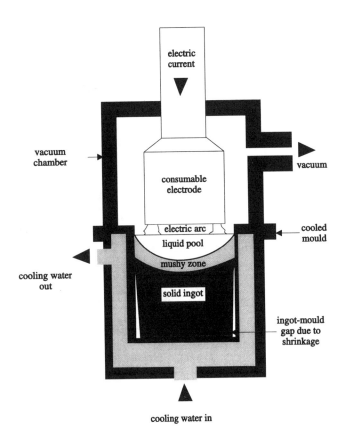

Figure 1. Schematic diagram of a VAR installation.

Table II. Approximated values used for VAR comparison simulation to Jardy and Ablitzer[6] and for calculating typical microstructural features.

Property		Value	Units
Density	ρ	6000–0.1T	kg/m^3
Ingot radius	L	0.10	m
Electrode (arc) radius	r	0.08	m
Arc current	I	5000 × α	A
Liquidus temperature	T_l	2073	K
Solidus temperature	T_s	2023	K
Latent heat	L	200,000	J/kg
h_c for top 0.05 m of ingot/crucible contact	h_c	3000, (T_{amb}=273 K)	W/m^2
h_c for rest of ingot/crucible contact	h_c	100, (T_{amb}=273 K)	W/m^2
Emissivity	ε	0.2, (T_{amb}=273 K)	
Gibbs-Thomson coefficient	Γ	2×10^{-7}	mK
Liquid diffusion coefficient	D	5×10^{-9}	m^2/s
liquidus slope	m	–4	K/wt.%
bulk composition	C_o	20	wt.%
distribution coefficient	k	0.2	

The flow patterns and molten pool shapes predicted in the present study are shown in Figures 2(a) to 2(c) for α values of 0, 0.5, and 1 respectively. The maximum velocity predicted with α = 0.5 is 0.015 m/s, in excellent agreement with Jardy and Ablitzer's value of 0.013 m/s, with the difference possibly due to the pool shape being calculated rather than imposed.

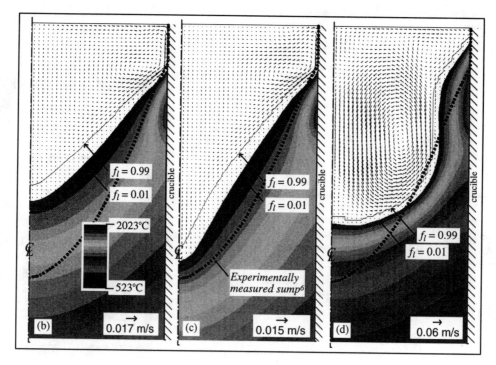

Figure 2. Calculated pool shape and velocity fields in a VAR ingot for: (a) $\alpha=0.0$; (b) $\alpha=0.5$; and (c) $\alpha=1.0$. The pool shape measured by Jardy and Ablitzer[6] is shown as the heavy dashed line.

Comparing the pool shapes calculated in the present study to that experimentally measured by Jardy and Ablitzer, the experimentally measured pool shape appears to fall between the $\alpha = 0.5$ and $\alpha = 1.0$ predictions (see Figures 2(b) and 2(c)). The calculated pool shape is strongly dependent on the heat transfer coefficient which was not known, hence the difference in pool depth can be attributed to any error in the approximate values used for h_c.

Microstructural Predictions

The importance of the melt pool shape on the final properties is illustrated by comparing the microstructural predictions for the two extremes of VAR with no EMF calculation (Figure 2(a)) and with the full melting current transferred (Figure 2(c)). Using the macro-model to determine G and V and the approximate values for the material properties as listed in Table II, the values for λ_1 were calculated using equation (11) and are shown in Figure 3.

When $\alpha=0$, the flow at the bottom of the sump is slow and the isotherms become stratified, with small thermal gradients, as is seen by the distance between the f_l=0.01 and f_l=0.99 contours. This leads to large λ_1 predictions. The strong counter-clockwise flow when $\alpha=1$ produces much higher gradients, and hence significantly smaller λ_1 predictions.

Using equation (14) the predictions for λ_2 are shown in Figure 4. Because the simulation was steady state, the value for t_s was calculated by taking the distance between T_s and T_l parallel to the pull direction and dividing by the pull velocity. This reaches an extreme condition when the solidification front is perpendicular to the pull direction, as is the case at a radius of 78 mm for the α=1.0 simulation. The primary dendrites are growing normal to the pull velocity, giving a long ripening time for the secondary arms. For most of the ingot, the EMF increases the thermal gradients,

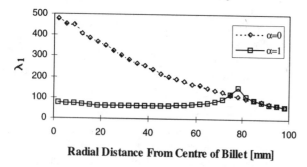

Figure 3. Predicted λ_1 values in a VAR ingot for $\alpha=0.0$ and $\alpha=1.0$.

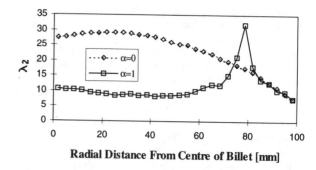

Figure 4. Predicted λ_2 values in a VAR ingot for $\alpha=0.0$ and $\alpha=1.0$.

reducing the size of the mushy zone, and hence producing a finer microstructure or smaller λ_2 values. However, the interaction of the EMF on the pool shape produces a region of growth normal to the pull direction, giving an inversion of the general trend. Comparing of the predicted λ_2 values for the two cases in Figure 4 illustrates the importance of including the EMF and of calculating the pool shape rather than specifying it *a priori* since both have strong influences on the final ingot microstructure.

Electron Beam Button Melting Model

Experimental Methods

The forming of a button by electron beam melting into a copper cooled hemispherical crucible of 168 mm diameter can be summarised by the following steps:

  i. crucible and electrode preheating;
  ii. electrode tip shaping
  iii. melting of the electrode with the molten metal dripping into the crucible and a fraction of the power being applied to the partially formed button.
  iv. controlled solidification of the fully formed button.

The model results were compared to two different types of experimental runs performed at the National Physical Laboratory, Teddington, UK. The first type was designed to allow characterisation of the surface flow velocities. The second type were the standard runs designed to give a controlled solidification pattern that concentrates inclusions into a central raft for simplified cleanliness characterisation. The two types of runs differ significantly only in the final stage, iv.

As described by Quested *et al.*[7], the first type of run approximates a stationary beam at the centre of the button. However, since the electron beam, even though partially defocused, distributes power over a circle with a radius less than 20 mm, the beam was scanned circumferentially at a high frequency on the surface of the button. The centre of the beam circumscribes the centre of the button at a radius varying linearly from 0 to 20 mm three times over the period of 60 s. Al_2O_3 particles were added to the surface and the motion of these particles was tracked using a video camera to provide rough quantitative surface flow velocities.

The second type of run was a normal practice solidification cycle used to concentrate inclusions at the centre of the button, often forming a visible raft. Quested *et al.*[7] suggested that the raft is caused by radial flow inward during stage iv, where the beam scans from the outer radius into the centre of the button. This inward radial motion is at a constant speed over a period ranging from 40 to 120 s. Simultaneously the current is decreased linearly from full power to 0 mA.

The buttons were examined metallographically after casting, to measure the microstructural features. The secondary dendrite arm spacing measurements, were obtained from longitudinal sections of the button which were polished and then etched in Marble's reagent (HCL 50 ml, saturated cupric sulphate solution 25 ml and distilled water 25 ml). Using an SEM, the length of five consecutive secondary arms was measured and averaged to determine the spacing. The primary dendrite arm spacing, measurement technique and results are given by Ellis[8].

Problem Formulation

The heat transfer and fluid flow was modelled in EBBM assuming that the flow was axisymmetric as was the distribution of heat from the electron beam. The model was solved on a grid of 40x40 control volumes for transient flow using an implicit solution with 0.5 s time steps. A steady state solution was obtained to use as an initial condition assuming a highly defocussed beam centred half way out the radius. The geometry and boundary conditions used are shown in Figure 5. The material properties used are listed in Table III or given in Figure 6 as a function of temperature.

Table III. Values used to simulate EBBM processing of IN718. (Note that f(t) indicates the value is a function of time whilst f(T) indicates a function of temperature.)

Property	Symbol	Value	Units
Button radius		37	mm
Button depth		25	mm
Beam current	I	f(t)	mA
Beam voltage	V	25	kV
Beam focal radius	r_σ	10	mm
Beam location radius	r_b	f(t)	mm
Density	ρ	f(T)	kg/m^3
Specific heat capacity	Cp	f(T)	J/kg/K
Viscosity	υ	5x10^{-3}	kg/m s
Surface tension gradient:	$\partial\gamma/\partial T$	f(T)	N/m/K
Liquidus temperature	T_l	1609	K
Solidus temperature	T_s	1533	K
Latent heat	L	270,000	J/kg
Ingot/crucible h_c	h_c	f(T_{ingot}), (T_{amb}=500 K)	W/m^2
Ingot/crucible emissivity	ε	0.3, (T_{amb}=500 K)	
Button top emissivity	ε_{top}	0.25, (T_{amb}=273 K)	

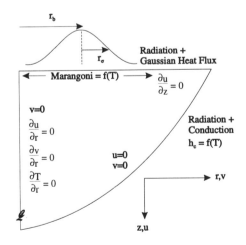

Figure 5. Schematic diagram showing the geometry and boundary conditions used to model the EBBM process.

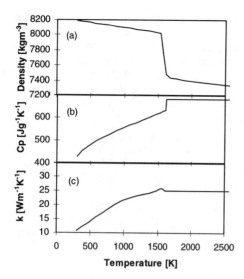

Figure 6. The temperature dependent material properties used in the EBBM simulations. (a) Density; (b) specific heat capacity; and (c) the thermal conductivity.

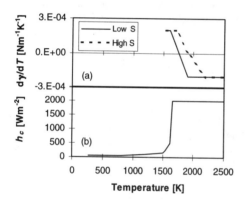

Figure 7. The temperature dependent values for (a) $\partial\gamma/\partial T$ and (b) the heat transfer coefficient used in the EBBM simulations.

The values for $\partial\gamma/\partial T$ were estimated using values provided by Mills[9] and assuming the behaviour is similar to that of sulphur in Fe-Ni-Cr alloys as determined by McNallan and Debroy[10]. The values were included as a piece-wise linear fit. The values used are shown in Figure 7(a) for the two cases of low surfactant concentration (6 ppm S, <10 ppm O), and high surfactant concentration (20 ppm S, 8 ppm O), the CPQ and CPZ compositions of IN718 as given by Quested et al.[7] respectively.

The values for the heat transfer coefficient between the ingot and mould wall, h_c, were calculated from measurements of the heat flux into a copper crucible made during the plasma remelting of IN718 into a 125 mm diameter cylindrical ingot. The heat flux was divided into radiative, convective, and contact components, with the h_c value representing the contact portion.

Starting with a steady state solution, the process was modelled with the transient boundary condition of the electron beam moving across the surface providing a Gaussian distribution of heat flux, Q, characterised by:

$$Q(R) = Q_o e^{-\frac{R^2}{r_\sigma^2}}, \qquad (19)$$

where R is the distance from the beam centre, r_σ is the beam focal radius, and Q_o is the total flux. Given that the beam circumscribes the centre of the button at a radius of r_b, the averaged heat flux, $Q_\theta(r)$, can be obtained by integrating Q by $d\theta$, giving:

$$Q_\theta(r) = \frac{1}{\pi}\int_0^\pi Q(R)\,d\theta = Q_o\, e^{-\frac{r^2+r_b^2}{r_\sigma^2}} I_o\!\left(\frac{2rr_b}{r_\sigma^2}\right), \qquad (20)$$

where I_o is the modified Bessel function of the first kind and order 0.

<u>Surface Flow Predictions</u>

Using the model as outlined in the previous section, two experimental runs were simulated, both with the same boundary conditions but with the low and high sulphur content being represented by the variation in $\partial\gamma/\partial T$ as a function of T. The electron beam motion was the same for both cases:

i. 30 seconds of r_b varying from 25 to 5 mm over 2 second cycles at a power of 6 kW;
ii. 30 seconds with no heat flux (during which time the Al$_2$O$_3$ particles were added);
iii. and finally 60 seconds of r_b varying from 20 to 0 mm over 20 second cycles at a power of 6 kW.

For the case of low S the surface flow was outwards from the centre of the beam with the particles reaching a maximum velocity of approximately 0.06 m/s. The particles travelled near to the edge of the pool, but a small inward surface flow was present in the outermost region of the pool. The predicted flow pattern 78 seconds into the simulation (seconds after the pool surface had become fully molten) is shown in Figure 8(a). The predicted surface flow is outwards with a maximum value of 0.16 m/s. The predicted value is higher than the maximum velocities observed experimentally, however, the Al$_2$O$_3$ particles were not tracked in the high velocity region. As observed experimentally, a small recirculating inward flow near the edge is predicted, with a time dependent size and peak velocity. This flow is caused by the positive value of $\partial\gamma/\partial T$ at temperatures less than 1750 K, and the size of inward flow is a function of the location of this isotherm, which is in turn dependent on the stage of remelting and the value of r_b.

For the high S experiments, the particles were observed to move inwards from the edge of the pool peak velocities of 0.19 m/s, moving at highest velocity shortly after leaving the edge of the pool, slowing to velocities of approximately 0.1 m/s half way towards the centre. The particles stop before reaching the centre but on the inside of r_b. Figure 8(b) shows the predicted flow pattern 78 seconds into the simulation. This period is when the velocities were recorded experimentally. The predicted surface flow pattern is identical to that observed experimentally, with the flow going from the outside into the centre, but turning down just before reaching the centre. The maximum velocity predicted is 0.14 m/s, lower than that observed experimentally (0.19 m/s). This suggests that the value for $\partial\gamma/\partial T$ may be greater than that used in the simulation, or the inversion point to a negative value could be at a higher temperature.

Comparing the two cases, a reduction in the rate of change of $\partial\gamma/\partial T$ from a positive to negative value and a 150°C increase in the inversion point, dramatically changed the flow patterns and location of the liquidus front. The size of the mushy zone is also altered, and hence the microstructural features will be different. A comparison of Figure 8(a) to Figure 8(b) illustrates the dominance of Marangoni flow in the EBBM process.

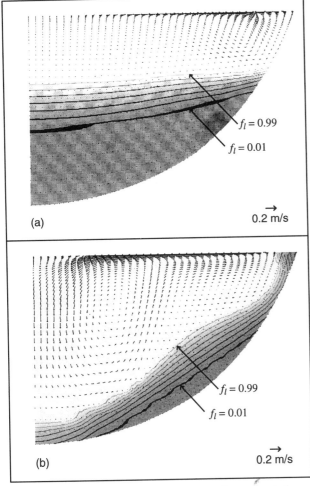

Figure 8. Predicted flow patterns using a 'central beam' with (a) a low sulphur content and (b) a high sulphur content.

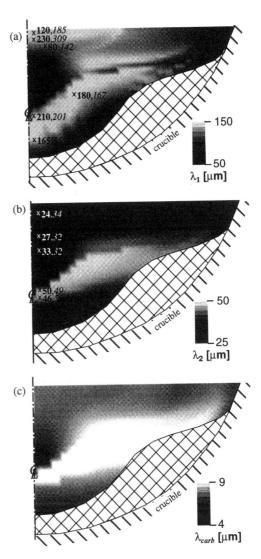

Figure 9. Predicted microstructural features for a button with a 120 s controlled solidification stage. (a) λ_1, (b) λ_2, and (c) λ_{carb}. Experimentally measured values are in bold, predicted values are in italics.

Microstructural Predictions

The microstructural features of electron beam melted buttons were measured for buttons undergoing a *controlled solidification* stage iv processing. During this stage the electron beam moves inwards whilst the power is reduced. This process was modelled with r_b going from 35 mm to zero over 120 seconds whilst the power decays linearly from 2.6 kW to zero. The resulting predictions for λ_1 are shown in Figure 9(a) calculated using equation (11) with the material properties for Γ, D, m, C_o, and k given in Table II. No predictions could be made in the bottom region of the button (cross-hatched area in Figure 9) because this area was already mushy in the steady state solution used as the initial condition for the model.

The values measured experimentally by Ellis[8] are listed on the figure (in bold) beside the predicted values (in italics). Near the top of the button at the centre line the model prediction fails. This is the region to solidify last and the only area where the gradients are sufficiently low that a liquidus isotherm is predicted to enter from the top of the button due to radiative heat loss competing with conduction through the button. Experimentally this region is occasionally found to be equiaxed, indicating that the columnar dendrites can not grow in from the sides sufficiently quickly. The λ_1 model assumes that the growth is near steady state, and this assumption does not appear to hold in this region. When the columnar dendrites were found to extend to the top, the spacing was smaller than predicted, suggesting that the dendrites could not adjust their spacing in this relatively small distance. Pratt and Grugel[11] have shown experimentally that the λ_1 adjust to order of magnitude changes in pull velocity during directional solidification experiments, suggesting either a slow response time to changes in thermal conditions or that the stable growth regime is large, adding a hysteresis effect.

The predictions for λ_2 are shown in Figure 9(b) calculated using equation (14). The correlation of predicted values to experimental is good. Predicted values for λ_{carb} are shown in Figure 9(c) calculated using equation (15). The carbide spacing was not measured experimentally.

Conclusions

Either the Lorentz or Marangoni force can be the governing factor for flow during secondary remelting processes, depending on the length scale and thermal gradients. The relative importance of these forces can be estimated by dimensionless analysis. A macro-model of the fluid flow and heat transfer was coupled to microstructural models to show that the changes in flow, caused by the different driving forces, have a large impact on the final microstructural features of the superalloy.

Many microstructural predictions can be made with straight forward correlation, for example λ_1, λ_2, and λ_{carb}. However, in the case of λ_1 a more complex model may be required to account for the response time and hysteresis of λ_1 to changes in thermal conditions.

The comparison of predicted to observed surface flows in electron beam melted buttons confirms that the Marangoni force is the main driving force for fluid flow in the EBBM process. Minor variations in the dependency of the surface tension on temperature can cause dramatically different flows, as shown by increasing the concentration of the surfactant sulphur from 6 to 20 ppm.

Acknowledgements

The authors would like to thank: Peter Quested, David Hayes and Ken Mills at the National Physical Laboratory for their assistance with both the experiments and the provision of material property data; the IRC in Materials for High Performance Applications, Birmingham, for providing experimental measurements of superalloy ingot to crucible heat fluxes; and Inco Alloys Ltd. for supplying both material and information. PDL would like to acknowledge the financial support of the EPSRC. RML would like to acknowledge the financial support of the DRA, Farnborough.

References

1. C. Hirsch, <u>Numerical Computation of Internal and External Flows</u>, Vol. 1 (John Wiley & Sons 1988).

2. M. McLean, <u>Directionally Solidified Materials for High Temperature Service</u>, (The Materials Society 1983), 28-33.

3. S.-Z. Lu, J.D. Hunt, "A Numerical Analysis of Dendritic and Cellular Array Growth: the Spacing Adjustment Mechanisms", J. Crystal Growth 123 (1992), 17-34.

4. J.D. Hunt and S.-Z. Lu, "Numerical Modelling of Cellular/Dendritic Array Growth: Spacing and Undercooling Predictions", <u>Modelling of Casting, Welding and Advanced Solidification Processes VII</u>, Ed. M. Cross and J. Campbell, (TMS 1995), 525-532.

5. A. Jardy, D. Ablitzer, J.F. Wadier, "Magnetohydrodynamic and thermal behavior of electroslag remelting slags." Met. Trans. 22B (1991), 111-120.

6. A. Jardy and D. Ablitzer, "On Convective and Turbulent Heat Transfer in VAR Ingot Pools", <u>Modelling of Casting, Welding and Advanced Solidification Processes V</u>, Ed. M. Rappaz et al., (TMS 1990) 699-706.

7. P.N. Quested, D.M. Hayes, K.C. Mills, "Factors affecting raft formation in electron beam buttons", Mat. Sci. and Eng., A173 (1993), 371-377.

8. J. D. Ellis, "Quality Assurance by Electron Beam Button Melting", (Ph.D. thesis, Imperial College of Sci., Tech. and Med., 1992).

9. K.C. Mills, private communication with author, National Physical Laboratory, Teddington, 16 October, 1995.

10. M.J. McNallan and T. Debroy, "Effect of Temperature and Composition on Surface Tension in Fe-Ni-Cr Alloys Containing Sulfur", Met. Trans. 22B (1991), 557-560.

11. R.A. Pratt, R.N. Grugel, , "Microstructural Response to Controlled Accelerations During the Directional Solidification of Al-6 wt.% Si Alloys", Materials Charact. 31 (1993), 225-231.

LIQUID DENSITY INVERSIONS DURING THE SOLIDIFICATION OF SUPERALLOYS AND THEIR RELATIONSHIP TO FRECKLE FORMATION IN CASTINGS.

P. Auburtin, S. L. Cockcroft, and A. Mitchell

Department of Metals and Materials Engineering
University of British Columbia, Vancouver B.C. V6T 1Z4 Canada
Tel : (604) 822 3677   Fax : (604) 822 3619

Abstract

The interdendritic segregation along the mushy zone of five directionally solidified and quenched alloys (MAR-M002, MAR-M247, IN718, T1 and C-276) has been measured by SEM/EDAX techniques and the corresponding concentration profiles are presented. These profiles have also been translated into liquid density profiles by a numerical model. Chemical compositions and estimated liquid densities of freckles found in MAR-M002, IN718, T1 and C-276 are also reported. It was deducted from the observations in this study that freckles tend to initiate relatively close to the tip of the dendrites, about 15 to 20°C below the liquidus temperature of the alloy, where the fraction liquid is typically of the order of 0.4 to 0.6. The Rayleigh number as a criterion for freckling is described and it is shown that freckles could result from density inversion gradients of the order of 0.03(g/cm^3)/°C, whereas 0.005(g/cm^3)/°C is probably too low. The importance of incorporating minor alloying elements such as C, or Si when studying liquid buoyancy related to freckling has also been outlined. Finally, emphasis is put on the crucial role played by carbon in some freckle-prone alloys as a powerful trigger for freckling, both through segregation and through precipitation of heavy elements into carbides.

Introduction

Freckles, also known as channel segregates, are thin trails of equiaxed grains and/or eutectic enriched metal which induce highly undesirable inhomogeneities in the properties of the final casting. Although much research about freckles has been published in the past 25 years (an extensive literature review is available in (1)), freckles are still a recurring casting problem for many alloy systems. It is now widely accepted that freckles in the final casting are the result of specific fluid flow patterns, known as thermosolutal convection, in the liquid during casting. This flow was shown to be driven by a density inversion phenomenon (2-5). This freckling mechanism has been described in numerous studies on analog systems, especially Pb-Sn and NH$_4$Cl-H$_2$O (5-9), and in computer simulations (2,10). However, very little research has been published on freckling in actual industrial alloys. Thus, to date, knowledge on freckling in superalloys remains largely qualitative.

In an attempt to reduce the probability of freckle occurrence, a few criteria involving temperature gradient G and growth rate R have been developed for use in computer casting models. However, these criteria remain empirical and none of them is entirely satisfactory (11). So far, casters usually relied on high thermal gradients to avoid freckles. While this method is relatively efficient for small castings, high thermal gradients are not achievable in bigger castings such as large directionally solidified (DS) or single crystals (SX) blades for industrial gas turbines. It seems therefore necessary to develop a criterion based on the actual physical mechanisms involved in freckle formation. Among various published parameters used to characterize the probability of freckle occurrence (7,12-14), the Rayleigh number Ra, as described by Sarazin & Hellawell (14), appears to be the most closely related to the physical conditions in the casting. It can be expressed in the following dimensionless form:

$$Ra = g \cdot \frac{d\rho}{dz} \bigg/ \frac{\eta D}{h^4} \qquad (1)$$

where g = gravitational constant
       ρ = density
       z = vertical coordinate
       η = dynamic viscosity
       D = thermal diffusivity
       h = characteristic linear dimension.

The numerator corresponds to the driving force in the liquid for fluid flow (to produce freckles) due to density inversion whereas the denominator represents the restriction to fluid flow (and freckles) due to viscosity, diffusivity and permeability in the liquid. The parameter h^4 has been linked to the dendritic array in the mushy zone with the following expressions: $h^4 = \lambda_1^4$ or $h^4 = K.\lambda_1^2$ (where λ_1 is the primary dendrite arm spacing and K is the permeability). Ra greater than a critical value, Ra^*, can then be considered a physical criterion for freckle formation. However, the application of this criterion in numerical models requires the knowledge of its various parameters. η and D can usually be approximated with reasonably good precision. Numerical equations for the permeability of K are still a topic for research (especially in high or low fraction liquid) but various expressions are now available in the literature (15,16).

The main unknowns in the Rayleigh number criterion remain the density inversion factor (dp/dz) as well as the critical Ra^* for various industrial alloys. The aim of the present research is to evaluate the order of magnitude of the density inversion factor for various freckle-prone alloys.

Experimental procedure

Choice of alloys

In this study, four alloys exhibiting freckles were chosen : DS superalloy MAR-M002, IN718, tool steel T1 and corrosion resistance alloy C-276. A fifth alloy, DS superalloy MAR-M247, was also chosen for the fact that it is not prone to freckling despite a composition very similar to that of MAR-M002. The nominal composition and melting range of these alloys is presented in Table I.

Sample casting

The alloys under study were directionally solidified and quenched (DSQ) in a vacuum induction furnace. The samples (6mm diameter × 50-60mm length) were contained in an alumina tube which was withdrawn at a rate of 2.5×10^{-5}m/s through a thermal gradient of 10°C/mm at the solidification front, leading to a dendrite network with a primary spacing of about 250-350μm. Before complete solidification, the samples were quenched from the steady-state growth regime in order to reveal the solidifying structure, without complications from diffusion during cooling.

Sample analysis

For each DSQ sample, several cross-sections located between the top and the bottom of the mushy zone were finely polished and analyzed using scanning electron microscope and energy dispersion analysis spectrometry (SEM/EDAX) techniques. A temperature, prior to quenching, was calculated for each cross-section (this temperature is easily calculated knowing the liquidus temperature of the alloy T_{Liq}, the thermal gradient G and the distance from the cross-section to the tip of the dendrites).

When the dendrite outlines could not be located directly under the SEM, the sample cross-sections were etched (with Kallings II or Marble's etch) and the dendrites were identified with micro-hardness diamond marks before repolishing the sample.

On each cross-section, at least four measurements of the quenched interdendritic liquid composition were made by EDAX. Other measurements, such as the composition at the dendrite centers and of carbides, were also carried out. SEM/EDAX measurements were always performed on unetched finely polished surfaces for maximum precision and minimum contamination. For each alloy composition, all the major alloying elements (down to 0.5wt%, except for carbon in T1, due to EDAX limitations) were included in the analyses. All the chemical compositions presented in this article are averages over 4 or more direct EDAX measurements. It is to be noted that the standard deviation observed on these multiple measurements always lay well within the expected precision of EDAX, namely ±5% for major alloying elements (content greater than 5wt%) and up to ±15% for minor alloying elements (content lower than 5wt%).

Moreover, micrographs of etched cross-sections at various depths in the mushy zone of the samples were also recorded. A typical example of these micrographs is shown in Figure 1. The quenched interdendritic liquid (tertiary arms and dark eutectic precipitate) (areas of EDAX analysis) can be clearly distinguished from the dendrites (primary and secondary arms).

The dendritic array was manually outlined on several micrographs. Computer image analysis was then used to measure the area of the outlined dendrites relative to the total area of the micrograph. Thus the fraction liquid was estimated at various temperatures in the mushy zone for several alloys.

Density evaluation

In order to estimate the density variations in the interdendritic liquid, chemical compositions and temperature for a given cross-section were mathematically translated into density by "METALS", a model developed by National Physical Laboratories (NPL, UK). This model is based on a weighted average of the molar volumes of each pure element forming the alloy. This approximation is now a widely accepted approach (19-21). The basic equations for this model can be written as follows.

At the temperature T, the molar volume in the liquid phase MV^i_L of each pure element i (of melting point T^i_{mp}) is given by :

$$MV^i_L(T) = MV^i_L(T_{Liq}) \times (1 + \alpha^i_L \times (T - T_{Liq})) \quad (2)$$

with $MV^i_L(T_{Liq}) = MV^i_L(T^i_{mp}) \times (1 + \alpha^i_L \times (T_{Liq} - T^i_{mp})) \quad (3)$

where α^i_L = expansion coefficient of pure element i in the liquid state.
T_{Liq} = liquidus temperature of the alloy.

Table I : Standard compositions (in wt%) and melting range of chosen alloys (17).
(T1 melting range estimated from reference (18))

Alloy	Nominal Composition (wt%)	T_{Sol}-T_{Liq} (°C)
MAR-M002	5.5Al, 0.15C, 10Co, 9.0Cr, 1.3Hf, 2.5Ta, 1.5Ti, 10W, Bal.Ni	1249-1365
MAR-M247	5.5Al, 0.15C, 10Co, 8.4Cr, 1.4Hf, 0.6Mo, 3Ta, 1Ti, 10W, Bal.Ni	1280-1360
IN718	0.5Al, 0.03C, 0.4Co, 19Cr, 3Mo, 5.5Nb, 52.5Ni, 1Ti, Bal.Fe	1260-1336
T1	0.75C, 4Cr, 0.3Mn, 1.1V, 18W, Bal.Fe	1320-1440
C-276	0.01C, 15.5Cr, 6Fe, 0.4Mn, 16Mo, 4W, Bal.Ni	1325-1370

For a given total weight W of an alloy of known composition, the number of mole a^i of each element is also known. Thus, the density of the liquid alloy, at any given temperature T, can be calculated as follows:

$$\rho_L(T) = W / (\Sigma (a^i \times MV^i_L(T))) \qquad (4)$$

This model is held to be accurate to about 5%. It was tested in the present study and showed very good agreement with various liquid densities reported in the literature (19,21,22). All the liquid densities reported in this article were evaluated by this model.

Results

DSQ samples

DSQ samples were mono- or bicrystals with dendrites oriented at 10° or less to the longitudinal axis. The growth front was always flat across the samples. Average compositions measured by EDAX along the samples showed no noticeable differences from the nominal compositions before melting, indicating no "zone refining effect".

No freckle was observed in any of the DSQ samples.

The segregation profiles measured by EDAX in the interdendritic liquid along the mushy zone for MAR-M002, MAR-M247, IN718, T1 and C-276 are presented in Figures 2 to 6. Only the alloying elements are reported (the balance is Ni (or Fe for T1)). These profiles are plotted against temperature in the melting range of each alloy.

The measured fraction liquid in the top part of the mushy zone is also reported on Figures 2 to 4 for alloys MAR-M002, MAR-M247 and IN718 (dashed lines). Fraction liquid measurements were not carried out for alloys T1 and C-276 due to difficulties in consistently outlining the dendrites throughout entire micrographs.

The density profiles in the interdendritic liquid for each alloy along their melting range were computed from the composition profiles shown in Figures 2 to 6. The results for all five alloys considered is presented in Figure 7.

The solid curves in Figures 2 to 7 are third degree polynomial regressions relative to the plotted data. The error bars in Figure 7 represent the worst possible effect of EDAX imprecision on the density calculations, when light and heavy alloying elements are shifted in opposite directions by their full expected imprecision.

Industrial castings exhibiting freckles

In order to develop a criterion for freckling, industrial castings with freckles were studied. Freckles could be easily observed by etching. They also showed specific characteristics : microporosity in MAR-M002, high concentration of niobium carbides in IN718, and larger concentration of carbides in T1.

Freckle and matrix compositions and estimated liquid densities for industrial castings of MAR-M002, IN718, T1 and C-276 are presented in Table II.

Density inversion provisions

Based on the Rayleigh number, it is possible to write :

$$\frac{d\rho}{dT} = \frac{d\rho}{dz} \times \frac{dz}{dT} = Ra \times \frac{\eta D}{g\lambda_1^4} \times \frac{1}{G} \qquad (5)$$

Assuming a critical Rayleigh number $Ra^* = 1$ as reported in (14), and substituting numerical values (in SI units) provided by (14) and "METALS", and a thermal gradient $G = 10°C/mm$ (i.e. $10^4 K/m$), yields :

In the case of Pb-10wt%Sn :

$$\frac{d\rho}{dT} = 1 \times \frac{2.5 \times 10^{-3} \cdot 1.1 \times 10^{-5}}{9.8 \cdot (3.0 \times 10^{-4})^4} \times \frac{1}{10^4} = 0.035 \, (g/cm^3)/°C$$

In the case of Pb-2wt%Sb :

$$\frac{d\rho}{dT} = 1 \times \frac{3.0 \times 10^{-3} \cdot 1.0 \times 10^{-5}}{9.8 \cdot (3.0 \times 10^{-4})^4} \times \frac{1}{10^4} = 0.038 \, (g/cm^3)/°C$$

In the case of Ni-based alloys (numerical data for pure liquid nickel at 1500°C) :

$$\frac{d\rho}{dT} = 1 \times \frac{4.4 \times 10^{-3} \cdot 1.0 \times 10^{-5}}{9.8 \cdot (3.5 \times 10^{-4})^4} \times \frac{1}{10^4} = 0.030 \, (g/cm^3)/°C$$

Thus, based on an estimation of the Rayleigh number criterion, freckles could result from density inversion gradients of the order of $0.03 (g/cm^3)/°C$.

Discussion

Measurement validation

The measured freckle composition for IN718 is in excellent agreement with that reported elsewhere (23). The proper calibration of the SEM/EDAX apparatus used in this study was confirmed by analyzing large sections of the samples. The global composition of these averaging sections corresponded very well to the nominal alloy composition. The nominal alloy composition (Table I) can also be found again in the interdendritic liquid at the top of the mushy zone (see right end of the curves in Figures 2 to 6).

Freckle initiation position

For any given alloying element in any given freckle-prone alloy considered in this study, it is possible to plot the freckle composition value (given in Table II) on the corresponding interdendritic segregation curve presented in one of Figures 2 to 6. When this is done for every element in a given alloy, all the plots exhibit very similar abscissa coordinates. This common abscissa value represents therefore the temperature (i.e. the depth in the mushy zone) at which interdendritic liquid is drawn to feed freckle plumes.

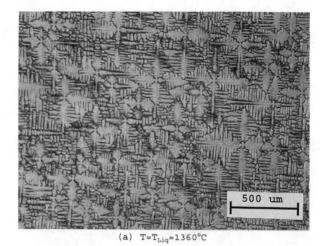

(a) $T=T_{Liq}=1360°C$

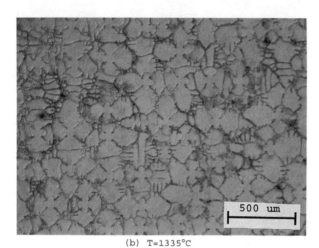

(b) $T=1335°C$

(c) $T=1321°C$

Figure 1 : Etched cross-sections at various depths along the mushy zone in DSQ MAR-M247, showing the dendritic array and the quenched interdendritic liquid.

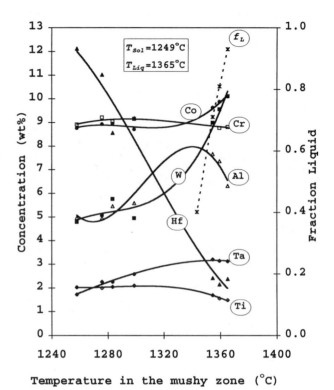

Figure 2 : Interdendritic liquid segregation and fraction liquid profiles along the mushy zone in DSQ MAR-M002.

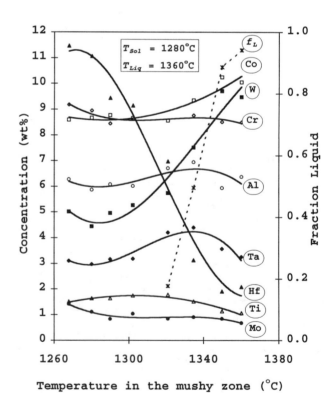

Figure 3 : Interdendritic liquid segregation and fraction liquid profiles along the mushy zone in DSQ MAR-M247.

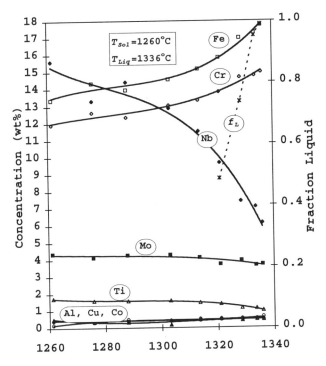

Figure 4 : Interdendritic liquid segregation and fraction liquid profiles along the mushy zone in DSQ IN718.

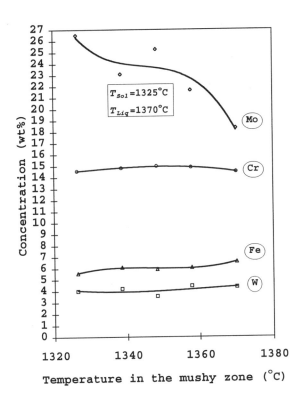

Figure 6 : Interdendritic liquid segregation profiles along the mushy zone in DSQ C-276.

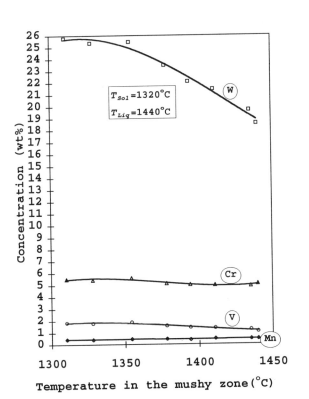

Figure 5 : Interdendritic liquid segregation profiles along the mushy zone in DSQ T1.

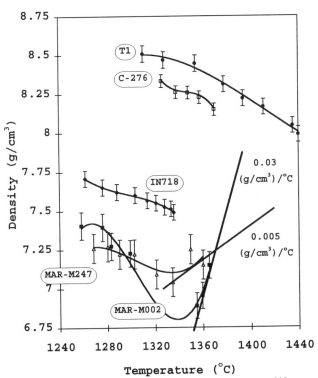

Figure 7 : Interdendritic liquid density profiles (estimated by numerical model) along the mushy zone in five DSQ alloys

Table II : Freckles and matrix compositions (measured by EDAX), and liquid state density range (estimated by numerical model) for various industrially cast alloys.

Alloy	Freckle/Matrix Compositions (in wt%)	Liquid Density (in g/cm^3) at T_{Liq}	at T_{Sol}
MAR-M002	Freckle: 7.5Al, 8.9Co, 8.0Cr, 4.6Hf, 60.7Ni, 2.7Ta, 1.8Ti, 5.8W	6.81	6.92
	Matrix : 6.8Al, 10.2Co, 9.2Cr, 1.8Hf, 58.0Ni, 2.6Ta, 1.3Ti, 10.1W	7.04	7.15
IN718	Freckle: 0.2Al, 16.9Cr, 0.5Cu, 16.8Fe, 4.1Mo, 10.1Nb, 50.2Ni, 1.2Ti	7.57	7.64
	Matrix : 0.3Al, 18.5Cr, 0.4Cu, 19.4Fe, 3.4Mo, 5.6Nb, 51.6Ni, 0.8Ti	7.49	7.57
T1	Freckle: 5.7Cr, 68.7Fe, 1.6V, 24.0W	8.29	8.41
	Matrix : 4.6Cr, 73.9Fe, 1.2V, 20.3W	8.09	8.20
C-276	Freckle: 16.7Cr, 6.5Fe, 19.8Mo, 53.3Ni, 3.7W	8.12	8.16
	Matrix : 16.1Cr, 6.6Fe, 17.3Mo, 55.5Ni, 4.5W	8.12	8.16

These positions correspond to temperatures about 1340°C, 1320°C, 1370°C and 1363°C for MAR-M002, IN718, T1 and C-276 respectively. It is interesting to note that in the case of MAR-M002, IN718 and C-276, freckles would thus initiate relatively close to the top of the mushy zone (about 15-20°C below T_{Liq}). Thus, freckles are only partially shifted toward the eutectic composition. This observation is further confirmed when considering data on binary analog systems. In Pb-2wt%Sb (T_{Liq}=315°C), freckle composition averaged 4wt%Sb (corresponding to T=300°C according to the phase diagram, i.e. 15°C below the liquidus temperature of the alloy), whereas eutectic composition is 11.2wt%Sb (8). Similarly, in Pb-17wt%Sn (T_{Liq}=290°C), freckle composition averaged 29wt%Sn (corresponding to T=270°C according to the phase diagram, i.e. 20°C below the liquidus temperature of the alloy), whereas eutectic composition is 61.9wt%Sn (8).

Moreover, freckle initiation position corresponds to a liquid fraction f_L of the order of 0.4-0.6 (as seen on Figures 2 to 4). This is consistent with the fact that fluid flow leading to freckles will develop more easily in regions of higher permeability.

Density inversion in MAR-M002 and MAR-M247

As shown on Figure 7, the curves representing the density of the interdendritic liquid along the mushy zone of MAR-M002 and MAR-M247 exhibit a noticeable minimum, slightly below T_{Liq}. In the case of MAR-M002, this minimum coincides exactly with the position of freckle initiation at T=1340°C as described above.

Moreover, as illustrated by the two tangents on Figure 7, it is possible to evaluate the density inversion gradient $d\rho/dT$ for these two alloys : the density inversion gradient between the tip of the dendrites and the point of lowest density is of the order of 0.03 (g/cm^3)/°C for MAR-M002 and 0.005 (g/cm^3)/°C for MAR-M247. According to the provisions previously calculated with the Rayleigh number, 0.03 (g/cm^3)/°C could be enough to produce freckles whereas 0.005 (g/cm^3)/°C should be too low. This is consistent with the fact that MAR-M002 is usually prone to freckles and MAR-M247 is not.

It should be mentioned that the density inversion observed for MAR-M002 and MAR-M247 is directly linked to carbide precipitation during solidification. In these alloys, the carbides' average metal composition was measured to be 55Ta+16W+16Hf+13Ti (wt%). Thus, carbide precipitation, by removing a significant amount of heavy elements from the interdendritic liquid (especially Ta), enhances its buoyancy since it is assumed that solid carbides do not participate in local liquid buoyancy. As Ta segregates preferentially toward the interdendritic liquid, thus making it heavier, it was suggested elsewhere (24) as a worthwhile addition to some alloy chemistries in order to eliminate freckles. This is probably valid for non-carbide forming alloys. However, in view of the present research in the case of carbide forming alloys, Ta may on the contrary favor and possibly trigger freckle formation. On the other hand, carbide precipitation in some alloys may also help prevent freckles by obstructing interdendritic channels, thus reducing the propensity to fluid flow and plume formation.

Density profile in IN718

As seen in Figure 7, the calculated density of the interdendritic liquid is increasing down the mushy zone of IN718. Although in apparent contradiction with the freckling theory, this observation is consistent with the density calculations estimating freckles to be heavier than the surrounding matrix (see Table II). Assuming that the theory is true (i.e. freckles do arise from buoyant plumes), this observation yields four possible explanations:

(a) First, it is possible that the density model is not accurate enough. However, the weighted average theory is the only acceptable to date and further work would be required to estimate possible deviations from the ideal mixing behavior. In any case, although deviation correction may influence the average calculated density of liquid alloys, it should only have a small effect on the relative variations of liquid density along the mushy zone.

(b) Secondly, since freckles in IN718 exhibit a large concentration of niobium carbides, it is possible that solute carbon should be taken into account in the density calculations. Other minute alloying elements, such as Si, may also play a significant role. Although EDAX measurements are difficult or impossible for such elements, segregation concentrations can be estimated and densities can be recalculated by the model. Such an estimation is presented below for IN718 with a nominal content of 0.03wt%C and 0.3wt%Si with overall compositions as reported in Figure 4 at the liquidus temperature (1336°C) and at the freckle initiation temperature (1321°C) :

Liquidus Compo. + 0.03C + 0.3Si : $\rho_L(1336°C) = 7.44 g/cm^3$
1321°C Compo. + 0.03C + 0.3Si : $\rho_L(1321°C) = 7.50 g/cm^3$
1321°C Compo. + 0.25C + 0.3Si : $\rho_L(1321°C) = 7.46 g/cm^3$
1321°C Compo. + 0.25C + 1.0Si : $\rho_L(1321°C) = 7.36 g/cm^3$

It can be seen that segregation of carbon alone is insufficient to create density inversion. However, additional segregation of Si could result in a density inversion sufficient to form freckles. It is to be noted that NbC precipitation in IN718 would not favor density inversion like in MAR-M002, because removal of niobium from the interdendritic liquid has very little influence on alloy buoyancy due to very similar densities.

(c) A third possible explanation is that the macrosegregates observed in "freckled" IN718 are indeed heavier than the surrounding matrix. In this case, they would actually be center-segregates rather than freckles. After forging of an ingot, the pockets of center-segregates appear thin and elongated in the direction of the ingot, much like freckles.

(d) A fourth explanation is a combination of the preceding two. Historically, IN718 used to contain relatively high amounts of carbon and silicon. Freckles could then develop as shown by calculation. However, more recently, the silicon content was substantially reduced and it is possible that IN718 ingots are now subject to center-segregation rather than freckling.

In any event, more detailed analysis of the segregation of minor alloying elements such as C or Si is required for further investigation. Moreover, in order to eliminate it, it is necessary to unambiguously distinguish which macrosegregation mechanism is at work.

Density profile of tool steel T1

Similarly to IN718, the density profile of T1 in Figure 7 does not exhibit any density inversion. Also similarly to IN718, this is consistent with the estimated density of freckles in T1 being greater than that of the surrounding matrix (see Table II). Assuming that freckles in T1 are not actually center-segregation, the phenomenon at work in T1 is probably a combination of the mechanisms in IN718 and MAR-M002. Freckles in T1 most likely result from a relatively high carbon segregation and from the precipitation of tungsten carbides, strongly depleting the interdendritic liquid of heavy tungsten. It was estimated with the density model that segregation of carbon from a nominal 0.8wt% up to 3wt% or more is indeed sufficient to create a density inversion susceptible to form freckles. This higher carbon content as a mechanism is further confirmed by the fact that freckles in T1 exhibited noticeably bigger and more numerous carbides (about 3 times as much) than the surrounding matrix.

Other light elements such as Mn, P or Si may also play a significant role on freckle formation in tool steels. Moreover, unlike in MAR-M002 or MAR-M247 where carbides eventually became embedded in the dendrite cores and were therefore not interfering with measurements of the interdendritic liquid composition, carbides in T1 remained in the liquid. Due to the lack of data on the carbide precipitation temperatures in T1, it was not possible to evaluate pro- and post-quench carbides. It is nevertheless interesting to note that the matrix surrounding interdendritic carbides contained only about 9wt%W, making it much lighter than any interdendritic liquid with an average tungsten content of 18wt% or higher. This confirms the definite effect of heavy element carbide precipitation on the potential for freckle formation. The estimated freckle initiation position is much further below T_{Liq} in T1 than in the other alloys in this study (70°C instead of 20°C). This could be due to a relatively low carbide precipitation temperature and identify carbon precipitation as a trigger for freckling.

Density profile in C-276

Figure 7 shows a slight increase in density of the interdendritic liquid down the mushy zone of C-276. Table II shows no density difference between freckle and matrix. Assuming freckling rather than center-segregation, and since C-276 does not normally form carbides, this confirms again the importance of measuring and taking into account the segregation patterns of seemingly minor light elements such as Al, B, C, Si or Zr.

Absence of freckles in the DSQ samples

Freckles did not appear in any of the DSQ samples. This is believed to be due to the relatively small cross-sectional area of the samples (of the order of 25mm²).

Indeed, it has been reported in the literature (5) that adjacent freckles are usually evenly spaced (about 5-10mm apart, yielding about 25-100mm² cross-sectional area per individual freckle). It is believed that this represents a minimum required area, in order to support the fluid flow patterns associated with freckle formation (upward freckle plume and slow downward feed flow).

In the present case, the DSQ samples were evidently too thin to support such fluid flow cells, and freckles could not develop.

Summary and Conclusion

Freckle compositions in selected industrial alloys, as well as interdendritic segregation and fraction liquid profiles along the mushy zone of DSQ samples were measured by SEM/EDAX techniques. The corresponding liquid densities were estimated by a numerical model.

(a) It was found by comparing actual freckle compositions to the segregation, that freckles in the studied alloys would tend to initiate about 20°C below the liquidus temperature where the fraction liquid in the mushy zone typically is 0.4 to 0.6.

(b) Density inversions were observed for MAR-M002 and MAR-M247. These inversions were linked to the precipitation of heavy elements (Ta, W, Hf) into carbides, lightening the interdendritic liquid and acting as a trigger for freckle formation.

(c) A range for the minimum required density inversion to form freckles was also determined : MAR-M002 is known to exhibit freckles and shows a density inversion of 0.03(g/cm³)/°C, whereas MAR-M247 does not usually form freckles and shows a density inversion of 0.005(g/cm³)/°C. Therefore, freckling can be linked to density inversions of the order of 0.01(g/cm³)/°C or greater.

(d) In the case of IN718, C-276 and tool steel T1, all reportedly freckle-prone, no density inversion was observed. A first possibility is that the reported freckles in these alloys are actually center-segregates, having the same appearance as freckles after ingot forging. A second possibility is that some alloying elements which were not included in the EDAX analysis (too small an atomic weight and/or too low a concentration) are indeed crucial to the determination of the liquid density.

(e) Estimates of the concentration of some of these minor elements in the mushy zone showed that segregation of carbon and silicon in IN718 (with high Si) could produce sufficient density inversion for freckling. In T1, carbon segregation and tungsten carbides precipitation could also yield enough density inversion to form freckles.

Acknowledgment

The authors would like to acknowledge the support from the Nickel Development Institute. They would also like to thank National Physical Laboratories for providing the much needed "METALS" model. Finally, the help from the following companies was also most appreciated : Asea Brown Boveri, Aubert & Duval, Inco Alloys International, Rolls-Royce, Sandia, Special Melted Products, Special Metals Corporation, Teledyne/Allvac Vasco.

References

1. P. Auburtin, "Determination of the influence of interdendritic segregation during the solidification of freckle-prone alloys", (M.A.Sc. thesis, Univ. Of British Columbia, 1995), 3-41.

2. J.C. Heinrich, S. Felicelli, and D.R. Poirier, "Vertical solidification of dendritic binary alloys", Computer Methods in Applied Mech. and Eng., 89(1991), 435-461.

3. D.G. Neilson and F.P. Incropera, "Effect of rotation on fluid motion and channel formation during unidirectional solidification of a binary alloy", Int. J. Heat Mass Transfer, 36(2)(1993), 489-505.

4. G.B. McFadden et al., "Thermosolutal convection during directional solidification", Met. Trans. A, 15A(12)(1984), 2125-2137.

5. J.R. Sarazin, A. Hellawell, and R.S. Steube, "Channel convection in partly solidified systems", Phil. Trans. R. Soc. Lond. A, 345(1993), 507-544.

6. R.J. McDonald and J.D. Hunt, "Fluid motion through the partially solid regions of a casting and its importance in understanding A type segregation", Trans. Metall. Soc. AIME, 245(09)(1969), 1993-1997.

7. S.M. Copley et al., "The origin of freckles in unidirectionally solidified castings", Met. Trans., 1(08)(1970), 2193-2204.

8. L. Wang, V. Laxmanan, and J.F. Wallace, "Gravitational macrosegregation in unidirectionally solidified lead-tin alloy", Met. Trans. A, 19A(11)(1988), 2687-2694.

9. M.H. McCay and T.D. McCay, "Experimental measurement of solutal layers in unidirectional solidification", J.Thermophysics, 2(3)(1988), 197-202.

10. W.D. Bennon and F.P. Incropera, "The evolution of macrosegregation in statically cast binary ingots", Met. Trans. B, 18B(09)(1987), 611-616.

11. A.L. Purvis, C.R. Hanslits and R.S. Diehm, "Modeling characteristics for solidification in single-crystal, investment cast superalloys", JOM, 01(1994), 38-41.

12. T.M. Pollock and W.H. Murphy, "The breakdown of single crystal solidification in high refractory nickel-base alloys", To be published.

13. M.C. Flemings and G.E. Nereo, "Macrosegregation : Part I", Trans. Metall. Soc. AIME, 239(09)(1967), 1449-1460.

14. J.R. Sarazin and A. Hellawell, "Channel flow in partly solidified alloy systems", Advances in Phase Transition, 10(1987), 101-115.

15. D.R. Poirier, "Permeability for flow of interdendritic liquid in columnar-dendritic alloys", Met. Trans. B, 18B(03)(1987), 245-255.

16. M.S. Bhat, D.R. Poirier, and J.C. Heinrich, "Permeability for cross-flow through columnar dendritic alloys", Met. Trans. B, 26B(10)(1995), 1049-1056.

17. "Properties and selection : irons, steels and high performance alloys", ASM International Metals Handbook 10th edition, ed. J.R. Davis et al., ASM International, 1(1990), 950-1006.

18. A. Grellier, private communication, Aubert & Duval, 28 July 95.

19. A.F. Crawley, "Densities of liquid metals and alloys", Int. Metallurgical Reviews, 19(1974), 32-47.

20. K.C. Mills and P.N. Quested, "Measurements of the physical properties of liquid metals", Liquid Metal Processing and Casting Conf. Proc., Am. Vacuum Soc., ed. A. Mitchell et al., Sept. 1994, 226-240.

21. T. Iida and R.I.L. Guthrie, The physical properties of liquid metals, (Clarendon Press, 1988)

22. A. Sharan, T. Nagasaka and A.W. Cramb, "Densities in liquid Fe-Ni and Fe-Cr alloys", Met. Trans. B, 25B(12)(1994), 939-942.

23. K.O. Yu et al., "Macrosegregation in ESR and VAR processes", Journal of Metals, 01(1986), 46-50.

24. T.M. Pollock et al., "Grain defect formation during directional solidification of nickel base single crystals", Superalloys 1992, TMS, ed. S.D. Antolovitch et al., (1992), 125-134.

THE EFFECT OF PHOSPHORUS, SULPHUR AND SILICON ON SEGREGATION, SOLIDIFICATION AND MECHANICAL PROPERTIES OF CAST ALLOY 718

J. T. Guo and L. Z. Zhou

Institute of Metal Research, Chinese Academy of Sciences,
Shenyang 110015, P. R. China

Abstract

Effects of trace elements P, S and Si on the microstructures and mechanical properties in cast alloy 718 have been studied. The results show that P, S and Si promote the segregation of Nb and formation of Laves phase. It was found that P and Si are significant Laves formers, with P enrichment in the Laves phase of about 1.0wt%, and Si enrichment of 2.06wt%. DTA results demonstrate that, as P, S contents in the alloys increased, the solidification temperature range increased obviously and the complete solidification was further delayed. On the contrary, Si raised the γ/Laves eutectic reaction temperature. Tensile testing indicated that P, S and Si had harmful effects on the tensile strength and ductility at RT and 650°C. When P, S and Si content exceeded 0.013wt%, 0.014wt%, 0.34wt% in the alloy, respectively, smooth stress-rupture life and elongations were reduced markedly.

Introduction

One of main problems in as-cast superalloys is the segregation of elements. How to reduce segregation is an important subject. In alloy 718, Nb is a primary element associated with Laves phase and carbides. It strongly segregates to interdendritic regions and promotes the formation of Laves phase, significantly affecting microstructure stability and mechanical properties[1]. Given the same cooling rate, the higher the Nb content of the alloy, the greater the volume fraction of Laves phase formed in the solidification microstructure. P, S and Si are impurity elements in superalloys, and are generally regarded as harmful, although little has been reported on the effect of phosphorus on the properties of superalloys. Recent studies showed P determined the Laves forming behavior of Nb [2]. R. G. Thompson et al reported that P segregated uniformly to grain boundary surfaces when it was present as an intentional additive [3]. In the solidification of alloy IN738, it has been shown that P increased the segregation of Al and Ti to the interdendritic region and the formation of γ/γ' eutectic which reduced stability of alloys[4]. Much attention has focused on sulphur in high-titanium-content alloys. The formation of Ti_2SC phase could reduce the effect of sulphur to some extent, but in alloy 718, titanium content is lower (about 1.0wt-%), and more work is necessary on the problem. Silicon may be used as a refining addition during melting, but its presence in the final alloys is considered detrimental and therefore upper limits are usually fixed at a low level in most alloy specifications. This research was aimed at the behavior of trace elements P, S and Si and their effects on the segregation of Nb during solidification, and the mechanical properties under different P, S and Si contents.

Materials and Experimental Proceeds

First, in order to eliminate the disturbance of other main elements by variation, a low impurity master heat IN718 was melted in vacuum induction furnace with the composition shown in Table 1.

Table 1   Chemical Composition of Master Heat IN718 (wt-%)

C	Ni	Cr	Mo	Nb	Ti	Al
0.031	53.19	19.07	3.04	5.20	1.05	0.53

B	Si	S	P	Mn	Fe
0.0055	<0.05	0.0038	0.0008	<0.03	bal.

This heat was divided, remelted and doped with different P, S and Si contents(in the form of Ni-P, FeS and elemental Si, respectively), and subsequently cast by lost wax process to mechanical property specimen ingots. Casting temperature were 1420℃, with a shell-model preheat temperature of 900℃. P, S and Si contents in the alloys are shown in Table 2.

Table 2 P, S and Si Contents in the Cast Alloys (wt-%)

Alloy	P	S	Si
1	0.0008	0.0038	<0.05
2	0.0055	0.0038	<0.05
3	0.008	0.0038	<0.05
4	0.013	0.0038	<0.05
5	0.032	0.0038	<0.05
7	0.0008	0.014	<0.05
8	0.0008	0.051	<0.05
9	0.0008	0.0038	0.34
10	0.0008	0.0038	0.95

The above cast specimen ingots after heat treatment were machined into mechanical property specimens with gauge diameter of 5mm and gauge length of 25mm. Tensile properties at room temperature, 650℃ and 650℃/620MPa rupture life were measured. The heat treatment process was: 1090℃/1h/AC + 950℃/1h/AC + 720℃/8h/FC 50℃/h → 620℃/8h/AC. The microstructures were studied by metallographic methods and an electron microprobe was mainly used to observe and measure element segregation. Fracture surfaces were studied by SEM.

The DTA experiments were performed on a Setaram HTC1800 instrument with a $\alpha-Al_2O_3$ crucible. The DTA cell was calibrated using pure Ni(>99.99%). The precision of this system, using Pt-Rh thermocouples, was determined to be better than 5℃. Specimens weighing 0.3g were heated at a fast rate of 100℃/min to 1450℃ under a flowing helium atmosphere and were held for five minutes before them were cooled at a rate of 15℃/min to 1000℃. The solidified samples were subsequently cooled at a fast rate(150℃/min) to ambient temperature. The DTA curves were recorded for the temperature range from 1450℃ to 1000℃. Reaction temperatures were determined by finding the temperature at which the DTA curves deviated from the local baseline.

Results

Microstructure and segregation

The microstructures of specimens in this study varied with composition, with the P, S and Si contents increasing in the alloy. The amount of Laves phase precipitating from liquid increased, especially for P-doped alloys and Si-doped alloys, giving the alloys a tendency to form well defined dendritic structures(Fig.1).

P and Si were heavily enriched in the final solidified γ/Laves eutectic phase or Laves phase during solidification process(Figures 2, 3). In the high-P-content alloy(0.032wt%P), the P content in the Laves phase reached 0.982wt%, about 30 times over average of that doped in the alloy. Nb content in the Laves phase also increased with the increasing P content(Table 3). In the high-Si-content alloy(0.95%Si), the Si content in the Laves phase reached 2.06%. Table 4 shows that S promoted the segregation of Nb, with the more S doped in the alloy, the more heavily Nb segregated in the interdendritic regions and Laves phase.

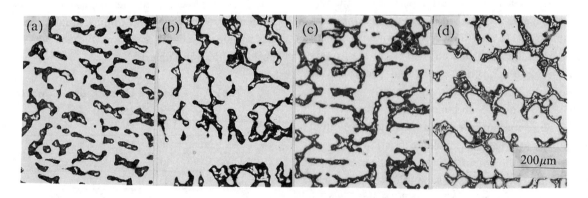

Figure 1: Effect of P, S and Si contents on microstructures of cast alloy 718
(a) No.1, base-alloy (b) No.5, 0.032wt%P (c) No.8, 0.051wt%S (d) No.10, 0.95wt%Si

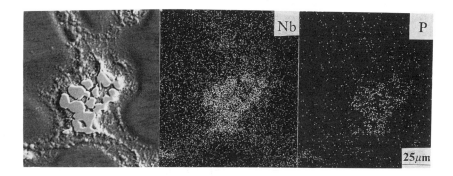

Figure 2: Elemental scanning images in alloy No.5 (0.032wt%P)

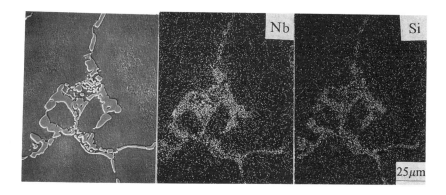

Figure 3: Elemental scanning images in alloy No.10 (0.95wt%Si)

Table 3 Average Composition of Laves Phase in the Alloys Doped with P (wt-%)

Alloy	P	Nb	Fe	Ti	Cr	Mo	Ni
1	0.036	26.619	9.957	1.608	10.649	3.558	bal.
3	0.243	29.904	9.455	1.927	10.059	3.792	bal.
5	0.982	30.886	11.391	0.950	12.818	4.738	bal.

Table 4 The Analysis of Nb Segregation in Alloys Doped with S

Alloy No.	S Wt-%	Dendritic Core	Interdendritic Regions	Laves
1	0.0038	3.16	5.00	15.98
7	0.014	1.71	5.70	20.88
8	0.051	1.55	6.36	21.68

DTA

The DTA cooling curves of cast alloy 718 with different P, S and Si contents have been determined. The data obtained from the DTA cooling curves were tabulated in table 5. It was found that P or S had little effect on the incipient solidification temperature($\gamma_{liquidus}$) and solidification temperature of MC, while Si obviously reduced $\gamma_{liquidus}$. The γ / Laves eutectic reaction temperatures varied with P, S and Si contents in the alloys. With increasing of P and S contents, the final solidification(γ / Laves eutectic reaction) temperature was reduced and the solidification temperature range increased. In contrast, Si raised the γ / Laves eutectic reaction temperature and reduced the solidification temperature range. The S-doped alloy solidified at following sequence: L → L +γ → L +γ+ Sulphide → L +γ +Sulphide + MC → γ +Sulphide + MC + γ / Laves(Eutectic).

Table 5 Characteristic Temperatures Obtained from the DTA Cooling Curves

Alloy	$\gamma_{liquidus}$	MC	Sulphide	Eutectic	$\Delta T^{1)}$
1	1329	1229	–	1157	172
3	1321	1225	–	1139	182
5	1321	1225	–	1129	192
8	1329	1225	1200	1143	186
10	1307	–	–	1186	121

1) ΔT Solidification temperature range (℃)

Mechanical properties

Tensile testing at room temperature and 650℃ with different P, S and Si contents was conducted. The results showed that P, S and Si were harmful to tensile strength and reduced ductility in the alloys. The smooth stress–rupture life and elongation with different P, S and Si contents have also been determined at the conditions of 650℃ / 620MPa. It showed that smooth stress–rupture life and elongation were reduced with increasing of P, S and Si contents. When phosphorus, sulphur, silicon contents in the alloys are in excess of 0.013wt%, 0.014%, 0.34%, respectively, life and elongation dropped rapidly.

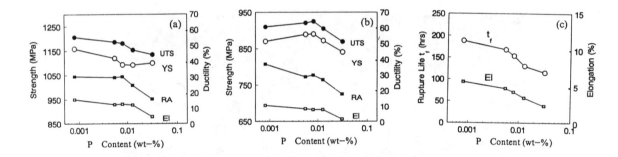

Figure 4: Effect of P content on mechanical properties of cast alloy 718
(a) at room temperature (b) at 650℃ (c) at 650℃ / 620MPa

Table 6 Effect of S Content on Tensile and Rupture Properties of Cast Alloy 718

Alloy No.	S Wt-%	YS, MPa RT	YS, MPa 650℃	UTS, MPa RT	UTS, MPa 650℃	El, % RT	El, % 650℃	RA, % RT	RA, % 650℃	SR$^{1)}$ Life,h	SR$^{1)}$ El,%
1	0.0038	1159	870	1207	910	16.0	10.8	30.6	37.0	188.5	5.7
7	0.014	1102	823	1164	864	8.7	8.7	24.0	36.0	78.0	3.6
8	0.051	1040	808	1105	841	7.3	8.2	21.0	21.5	19.0	2.0

1) SR Stress–rupture properties

Table 7 Effect of Si Content on Tensile and Rupture Properties of Cast Alloy 718

Alloy No.	Si Wt-%	YS, MPa RT	YS, MPa 650℃	UTS, MPa RT	UTS, MPa 650℃	El, % RT	El, % 650℃	RA, % RT	RA, % 650℃	SR Life,h	SR El,%
1	<0.05	1159	870	1207	910	16.0	10.8	30.6	37.0	188.5	5.7
9	0.34	1100	878	1179	889	8.7	6.7	24.0	32.0	136.5	4.2
10	0.95	982	795	988	858	2.0	1.8	2.6	7.8	57.0	2.4

Discussion

The alloy IN718 solidifies in the following sequence: L → L+γ → L+γ+MC → γ+MC+γ/Laves. This is verified by DTA curves of the alloys. The solubility of P, S, Si in nickel is very low with a maximum of about 0.3wt%[5], 0.005wt%[6], <0.5wt%, respectively, so that P, S and Si should mostly enrich in the final solidification regions and result in heavily segregation. The higher the P, S and Si contents doped in the alloys, the more they would be enriched in interdendritic areas.

The segregation of P, S and Si elements promotes the segregation of Nb, this could be seen from above. Nb is a primary element in the Laves phase and the Laves phase is generally accepted to be the form (Ni, Fe, Cr)$_2$(Nb, Mo, Ti) [1]. Therefore, the more Nb segregated in interdendritic regions, the higher volume fraction of Laves formed.

The effects of P, S and Si on the mechanical properties mainly lie in the following: P, S and Si segregate in the final regions, and promote the segregation of Nb to form Laves phase. Laves phase has been generally accepted as being deleterious to the mechanical properties of the alloy. It was associated with reduced tensile strength and ductility, because it is a brittle phase, even at elevated temperature, it is likely to act as a preferred crack initiation and propagation site [1]. The greater volume fraction of Laves phase in the alloys presents a higher probability for the crack to grow through the brittle phase or γ/Laves interfaces. In addition, Laves consumes large amounts of Nb depleting the matrix of Nb which is the principal hardening element. Stress-rupture fracture surfaces were studied by SEM. It was found that the fracture surfaces included grain boundary, eutectic interface, γ/MC interface. With the increasing of P, S and Si contents in the alloys, they showed greater tendency to brittle fracture. That would explain why the high P, S and Si alloys with gross Laves showed reduced rupture strength and ductility.

Conclusions

1. P, S and Si are enriched mostly in the final solidification regions, and P, Si become formers of Laves phase.
2. P, S and Si promote the segregation of Nb and the formation of Laves phase.
3. The solidification temperature range is widened with increasing of P and S contents in the alloy. On the contrary, Si raised γ/Laves eutectic reaction temperature and reduces solidification temperature range.
4. Tensile strength and ductility at room temperature and 650℃ are reduced with increasing of P, S and Si contents in the alloy.
5. Smooth stress-rupture life and elongations are reduced obviously markedly with increasing of P, S and Si contents in the alloy.

Acknowledgments

The authors wish to acknowledge Prof.Tresa Pollock and Radavich for helpful comments concerning this manuscript. The portions of the research were funded by the National Science Foundation.

Reference

1. J. J. Schirra, R. h. Caless, R. W. Hatana, "The Effect of Laves phase on the Mechanical Properties of Wrought and Cast + HIP Inconel 718"(Paper presented at the International Symposium on the Metallurgy and Application of Superalloys 718, 625 and Various Derivatives, Pittsburgh, Pennsylvania, 23-26 June 1991),375
2. C. Chen, R. G. Thompson and D. W. Davis, "A Study of Effects Phosphorus, Sulfur, Boron and Carbon on Carbide Formation in Alloy 718"(Paper presented at the International Symposium on the Metallurgy and Application of Superalloys 718, 625 and Various Derivatives, Pittsburgh, Pennsylvania, 23-26 June 1991), 81.
3. R. G. Thompson, M. C. Koopman, and B. H. King, "Grain Boundary Chemistry of Alloy 718-Type Alloys"(Paper presented at the International Symposium on the Metallurgy and Application of Superalloys 718, 625 and Various Derivatives, Pittsburgh, Pennsylvania, 23-26 June 1991), 53-65.
4. Y. X. Zhu, "The Effect of P on the Cast M38(IN738) Alloy" (Report No.85147, Institute of Metal Research, Chinese Academy of Sciences , 1985).
5. R. T. Holt, W. Wallace, "Impurities and Trace Elements in Nickel-base Superalloys," International Metals Reviews, 21(1976),1-24.
6. Taylor Lyman, ed.,Metals Handbook(Prepared under the direction of the Metals Handbook Committee. 1948), ASM Published, 1232.

Inclusion/Melt Compatibility in Pure Nickel and UDIMET 720

M. Halali, D.R.F. West and M. McLean

Department of Materials, Imperial College of Science, Technology and Medicine,
Prince Consort Road, London SW7 2BP, UK

ABSTRACT

The inclusion content and distribution in the alloy UDIMET 720 has been studied using levitation melting. Samples of the as-cast alloy have been melted and the inclusions that concentrate on the surface have been characterised. The specific behaviour of some typical inclusions such as Al_2O_3, TiN, and the spinel Al_2MgO_4 has been studied by deliberately adding particles onto the surface of the molten levitated sample. Light microscopy and SEM together with EDS analysis have been employed to study the surface and sections of the solidified samples. Most of the inclusions display "non-wetting" behaviour; the few "wetting" inclusions were complex oxides of aluminium, titanium, silicon, and calcium. It is found that most of the wetting complex oxide inclusions are associated with sulphur. Other wetting inclusions identified (TiN, chlorides and fluorides of alkali metals and calcium) are generally secondary inclusions that form on solidification. Although sulphur is associated with most of the wetting complex oxide inclusions, the addition of 15 ppm sulphur does not significantly change the wetting characteristics of the non-wetting inclusions.

INTRODUCTION

As turbine disc alloys have developed with increasing yield strengths to satisfy the requirement for increasing stresses associated with reduced engine weight and increased rotor speeds, there has been a concomitant increase in the sensitivity of these materials to fracture at defects such as inclusions. This has led to a demand for improved cleanness (or decreased inclusion content) using a range of primary and secondary melting procedures such as VIM, VAR, ESR, and EBCHR. Detection of these sparse inclusions is beyond the scope of current NDE and quality assurance and is likely to be achieved through control of the alloy processing.

In studying the intrinsic interactions between inclusions and the melt, it is important that the system is not distorted by the presence of ceramic crucibles. A few cold crucible and containerless procedures, including electron beam button melting and magnetic suspension (levitation) melting[1-12], have been used to avoid this problem and for fundamental studies. Most of these modern techniques for inclusion investigation and removal rely on the fact that many inclusions, particularly those with larger sizes (e.g. 100 μm) tend to be retained on the surface of the molten alloys. There is debate as to whether this phenomenon is due primarily to flotation or to the effect of surface tension. Levitation melting [3] was used in the present study to characterise the wetting behaviour of inclusions in the molten nickel-base alloy UDIMET 720. The inclusions studied include those found in the as-supplied alloy, both primary inclusions present in the melt and secondary inclusions which form on solidification due to falling solubility limits. A good example of the first is alumina which can be present in the source metals or can be debris from the crucible linings during alloy production. TiN is an example of secondary type inclusions which are generally very small. The size of the primary inclusions can vary over a wide range; alumina particles larger than 100 μm can be found. Typically TiN particles are ≈1 μm diameter although these may congregate in clusters. A better understanding of the physico-chemical relationships between the superalloy and the inclusions is required to develop reliable clean processing procedures.

EXPERIMENTAL DETAILS

Levitation experiments were carried out on both UDIMET 720 and pure nickel using the apparatus shown schematically in Figure 1. Power was supplied by a 15kW, 450kHz Radyne generator and an inert atmosphere was maintained inside silica tubes 14 mm OD. In most experiments helium was used to give additional control over the temperature of the molten sample by virtue of its high thermal conductivity; otherwise argon was used. Nickel samples were cut from specpure nickel rods and alloy samples from a cast UDIMET 720 ingot supplied by Rolls Royce.

The levitation coil was made from copper tubing 3/16" OD and 0.030" wall thickness. Of 30 coil geometries considered, that shown in Figure 1 gave stable drops with the optimum combination of melt size and temperature. Temperature could be measured by a disappearing filament pyrometer from the bottom prism, top window and mirror, or directly from the sample. The first two methods were most convenient but had an uncertainty of ≈30°C, so the third was used when possible because of improved accuracy.

Prior to every experiment surfaces were (i) ground with silicon carbide paper, (ii) ultrasonically cleaned in a dilute solution of "Decon 90" at 60°C for 10-15 minutes, (iii) washed sequentially in water, distilled water and acetone, (iv) dried and (v) stored in a desiccator. Each silica tube was washed in a hot solution of dilute

Table 1   Summary of results of levitation melting experiments involving the addition of ceramic powders and surface active salts

sample type	temperature (°C)	duration (minutes)	type of addition	inclusion distribution and remarks
Ni	1550	2	Al_2O_3	Powder concentrated at surface; no alumina detected in cross section. by optical microscopy.
Ni	1750	2	TiN	No powder on surface; no TiN detected in cross-section
Ni	1700	5	Al_2O_3 KCl	Powder concentrated at surface; no alumina detected in cross section by SEM.
UDIMET 720	1740	3		Inclusions detected: Surface: Al_2O_3, titanium aluminate, NaCl and KCl, complex oxides of Al, Si, Ca, Mg, Ti. (Most frequent alumina with a maximum size of 70μm.) Section: TiN, calcium and aluminium silicates, CaS, sodium and potassium halides, $CaCl_2$, titanium aluminate, calcium aluminium silicate, complex oxides. (Most frequent: aluminium silicates, 3-20μm.) Most inclusions in section contained sulphur.
UDIMET 720	1650	5	TiN	No surface TiN apparent. .Surface inclusions mainly alumina: Section inclusions included: TiN, titanium aluminate, NaCl, titanium silicate, CaO, aluminium silicate. Most inclusions in section contained sulphur.
UDIMET 720	1650	1	SiO_2	Almost all the inclusions remained on the surface.
UDIMET 720	1650	2	$MgAl_2O_4$	Almost all the inclusions remained on the surface. Surface inclusions: mainly alumina, spinel, and MgO. Section inclusions included alumina, aluminium silicate, zirconium sodium silicate, SiC, CaO, CaS, TiN, complex oxides of Na, Mg, Si, Ca, Al. Alumina was the most frequent inclusion. Most of the surface inclusions in section contained sulphur.
UDIMET 720	1625	3	CaO	Powder concentrated at surface. Surface inclusions included alumina and calcium aluminate. Section inclusions included CaS, alumina, aluminium calcium silicate, magnesium silicate, (Fe_2O_3), calcium silicate, complex oxides of Al, Si, Ca, Mg, Zr, Ti, and halides of Na, and K. Most inclusions in section contained sulphur.
UDIMET 720	1650	6	KCl	Solid alumina -rich crust formed after 1 minute. Inclusions detected by the SEM: CaS, titanium aluminate, alumina, calcium aluminium silicate, and complex oxides of Al, Ca, Si, Ti.
UDIMET 720 (EBBM))	>1700	1		Surface was very dirty, inclusions were mainly alumina and Fe_2O_3. Large agglomerated inclusions (400μm) detected. Most frequent inclusion was Fe_2O_3. The smaller Fe_2O_3 particles concentrated in a narrow band near the surface.
Ni-3%Al	1650	10	KCl	Alumina-rich crust formed but different in character from those on UDIMET 720.

nitric acid, rinsed, and dried before each run. In some runs, where the sample was quenched in a copper cup, the copper cup was cleaned in dilute nitric acid, rinsed with distilled water and dried.

The sample and the copper cup (when used) were weighed separately. These and the powder were then placed on boron nitride trays in separate compartments of the turntable which was raised to the level of the coil using an alumina tube. The chamber was flushed with inert gas for 5 minutes, the gas flow rate was set and the generator turned on. Time was measured from complete melting of the sample which usually took 20 to 50 seconds depending on power, gas flow rate, and the weight of the sample. Various fine ceramic powders could be introduced to the molten levitated drop by bringing the sample holder close to the drop (Figure 1, inset) and allowing the gas stream to blow the powder onto the surface of the drop.

Some samples were quenched in the copper cups; the rest were solidified *in situ* which was achieved by suddenly increasing the power setting of the induction generator which disturbed the levitation conditions removing the drop from the hot zone and simultaneously increasing the gas flow rate (to ≈30 litres/min). *In situ* solidification took 5 to 10 seconds. After cooling, the samples and copper cups were weighed. The surfaces were examined by SEM. prior to sectioning, using thin silicon carbide slitting wheels (0.010" thick). They were then ground and polished to 1μm using diamond compounds. Each solidified drop was systematically investigated by optical microscopy, SEM with EDS analysis and X-ray diffraction to determine the nature of inclusions and their distribution across the section.

RESULTS AND DISCUSSION

A summary of the observations made on specimens subjected to various levitation melting conditions are summarised in Table 1. Figures 2 and 3 show photographs of selected samples which have been solidified *in situ*. Where surface inclusions are evident they are concentrated in a circumferential band consistent with the intersection of convective flow patterns with the surface as proposed by Marechal *et al.* [3]. Several series of levitation experiments on pure nickel and UDIMET 720 with various deliberate additions of particles have been undertaken and the inclusion distribution characterised. The criterion for non-wetting of inclusions was the retention of the powder on the surface. However, it is possible to have particles that both remain on the surface and are entrained in the metal. This is discussed below.

1. Levitation melting of pure nickel.

Since nickel is the chief constituent of UDIMET 720, reference experiments were made using pure nickel. When nickel was levitated on its own inclusions neither concentrated on the surface nor were apparent in the cross-section consistent with the high purity of the starting material. This indicates that the experimental procedure does not introduce extraneous inclusions.

Three different types of powder - alumina, TiN, and the Al_2MgO_4 spinel - were added to pure nickel in separate levitation runs. The particle sizes used were as follows: two types of alumina were used in separate experiments with particle sizes in the approximate ranges of 50-200μm and <10μm; the spinel was 30-100μm and the TiN was predominantly <1μm with a some particles up to 5μm:

- Alumina was added to levitated samples at several different temperatures between 1500° and 1800°C: the durations of the runs varied from 10 seconds to 12 minutes. Most of the alumina, which was in the form of particles of ≈30 (±10) μm diameter, remained on the surface with no discernible change in size or morphology. No alumina particles could be identified by SEM of the sectioned samples in agreement with Ellis.

- When the spinel, of similar particle size to the alumina, was added the particles also remained on the surface. This behaviour differs from that reported by Ellis. There are important differences in the two studies; the source of spinel particles and the methods of introducing powders to the sample. Ellis introduced the powder into a hole bored in the solid sample before levitation; this allows very precise gravimetric studies but there is the inherent risk of particle agglomeration and sudden powder loss. Also, Ellis used a maximum melt duration of 1 minute measured from the point of levitation; in the present study, the sample was levitated for 2 minutes measured from complete melting.

- When titanium nitride was added, this time in the form of a very fine powder with a mean particle size of ≈2μm, the particles disappeared from the surface. Also, few titanium nitride particles were detected in the sectioned sample. It is likely that most of the powder dissolved in the metal rather than being retained as TiN

2. Levitation melting of UDIMET 720

The as-supplied alloy was studied by SEM before and after levitation to identify as many types of inclusions as possible. The inclusions detected by SEM on the surface were assumed to be non-wetting; those found in the sections of samples were assumed to be wetting. The temperature of the levitated melts varied between 1585 and 1750°C and the duration of the runs between 5 seconds to 7 minutes. The temperature and duration of runs did not have a significant effect on the inclusion distribution.

a) *Surface inclusions:* Alumina was the most frequent type of surface inclusion observed occurring in a range of sizes from a few to 150μm. Other inclusions detected were titanium aluminate and complex oxides of aluminium, silicon, calcium, magnesium, and titanium. Halides of reactive metals were occasionally identified.

The alumina particles detected could have been intrinsic to the alloy originating from oxidation reactions and agglomeration to form large inclusions. They may also be extrinsic debris from the ceramics and refractories used in melting. The complex oxides are deduced to be agglomerates of simpler compounds such as oxides and spinels that are most likely to be picked up from refractory linings during melting. The origin of the halide particles is not clear. Most manufacturers use a proportion of scrap in alloy production; if part was from ESR material, halide contamination from the slag may have persisted. It unlikely that the halides would be stable in the melt; more likely they precipitate as secondary inclusions.

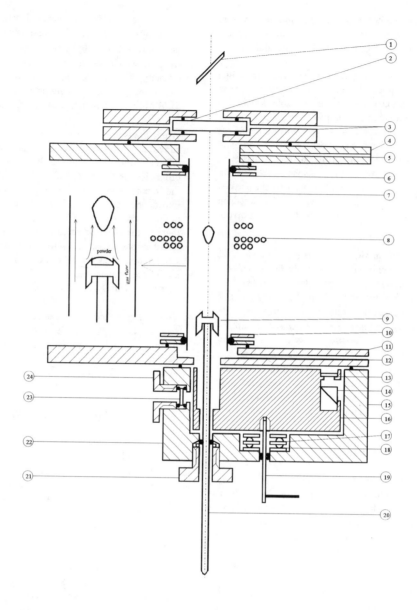

1) mirror
2) top optical window sealing plates
3) top optical window
4) gas outlet
5) top flange
6) top sealing arrangement
7) silica tube
8) levitation coil
9) sample holder
10) bottom sealing arrangement
11) bottom flange
12) gas inlet
13) prism optical window
14) prism
15) turn table case
16) turn table
17) ball bearing
18) ball bearing washers
19) turn table handle
20) alumina push rod
21) alumina push rod sealing gland
22) alumina push rod washer
23) bottom optical window
24) bottom optical window sealing gland

Figure 1: Schematic diagram of the levitation chamber. (not to scale)

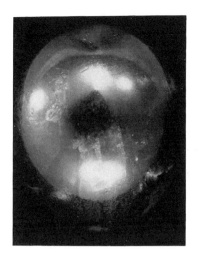  

        a   b   c

Figure 2: Samples of pure nickel after levitation melting:
(a) pure nickel, (b) pure nickel with Al_2O_3 and (c) pure nickel with Al_2O_3 and KCl additions.

Scale: ▬▬▬▬▬ 5 mm

b) *Section inclusions:* Although a large proportion of inclusions came to the surface, some particles remained inside the samples. The inclusions detected in the section were generally smaller than those on the surface and there was a wider variety of inclusion types in the section. The most frequently occurring inclusions were different compositions of aluminium silicate with sizes between 3 and 20μm. Other inclusions detected were alumina, TiN, calcium silicate, aluminium silicate, CaS, halides of reactive metals, and complex oxides of aluminium, silicon, calcium, titanium and magnesium. Most inclusions within the sections contained sulphur.

One experiment was carried out on material cut from the inclusion-rich cap produced during electron beam button melting. After levitation melting the surface of the drop had a much higher coverage by particles than was obtained from the ingot alloy indicating a high cleaning efficiency during levitation melting.

3. Addition of inclusions to UDIMET 720

Since it was not possible to isolate specific chemistries of the complex oxides found in the commercial alloy, deliberate additions were made of a range of relatively simple compounds, including alumina, spinel, silica, lime, magnesia and titanium nitride.

- When the oxide particles were added to a levitated sample, all the particles seemed to remain on the surface of the sample; different temperatures and durations of levitation did not change the behaviour significantly. Metallographic examination of the surfaces and sections of solidified samples confirmed the visual observation during levitation melting.

- On addition of TiN, all of the powder apparently disappeared. A very few TiN particles were detected on the sectioned drops. Titanium has a high solid solubility in nickel (7 at. %). Nitrogen, on the other hand, is quite insoluble in nickel. It is possible that the particles have dissolved in the alloy, titanium being retained in solution.

4 Addition of surface active compounds to UDIMET 720

During levitation melting induced electrical currents at the surface generate vigorous eddy currents which carry inclusions to the surface and then return a proportion of them to the bulk. If an addition can be made which would retain the inclusions on the surface, the bulk of the sample will have a reduced inclusion content. The ideal addition would both modify the surface characteristics of the melt and inclusion to enhance their non-wettability and would fuse the inclusions retained on the surface. Halides of reactive metals appear to have both of these properties.

In several experiments salts such as NaCl, KCl, KF, CaF_2, and $CaCl_2$ were added to levitated samples. After a period of between few seconds to just over a minute, a solid crust formed on the molten samples. The time taken for the crust to form depended on the chemistry and concentration of the salt. Salts of the more reactive elements such as KF took only a few seconds to form the crust; maintaining the sample in the state of levitation resulted in wrinkling of the crust and elongation of the sample. The formation of the crust reduced the convection currents and cooling rates making *in situ* solidification impracticable. Solidification was achieved by quenching in a cold copper crucible. Metallographic

Figure 3   **Samples of UDIMET 720 after levitation melting:**
(a) as-supplied, (b) with alumina addition, (c) with Al_2O_3 and KCl additions and (d) Ni-3% Al addition.

Scale ━━━━━ 5mm

examination of the sections showed them to be considerably cleaner than those without surface active additions. Figure 4 shows parts of sections of the UDIMET 720 alloy samples in different conditions: as-supplied, levitated on its own, and levitated with the addition of KCl. It is quite evident that the sample with the halide addition has many fewer internal inclusions than the other two samples.

X-ray diffraction showed the crust to consist mainly of alumina. The formation of the crust could result from various effects. One possible mechanism, by oxidation of aluminium in the alloy by the residual oxygen in the cooling gas which is at a level of a few ppm, can be ruled out since the crust does not form in the absence of the reactive additions.

No crust was observed when halides were added to levitated samples of pure nickel. Levitated pure nickel samples were sprayed by a mixture of alumina and halides and still no scale was formed. Thus the formation of the crust must be related to the presence of the inclusion particles and/or the aluminium content of the alloy.

When samples of an alloy of Ni-3 wt% Al were levitation melted and sprayed with the halide, a solid crust was formed. However, it differed both in appearance and in thermal properties from the crusts described above. It could be solidified *in situ in* a helium flow rate of 20 litres per minute while the crusts on UDIMET 720 were not solidified at helium flows of 30 litres per minute due to poor thermal conductivity of the scale.

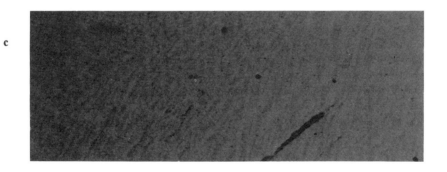

Figure 4 Sections of UDIMET 720:
(a) as-supplied, (b) as supplied and levitated and (c) as supplied and levitated with KCl additions.

Scale: ━━━━━ 100 μm

It appears that the addition of the halides, by changing the surface characteristics of the molten sample, result both in the entrapment of inclusions impinging on the surface and oxidation of some of the more reactive constituents of the alloy such as aluminium. Due to its high boiling point the compound CaF_2 is considered as a particularly appropriate addition to enhance inclusion removal from the molten alloy.

Samples of the alloy were doped with 15 to 20 ppm sulphur as a surface active element. Visual observation and a study by SEM showed no significant change in the wetting characteristics of any of the inclusions. However, this does not prove that sulphur has no effect on the surface properties of the alloy. Quantitative evaluation would be required to determine the effect of sulphur on the surface characteristics of the inclusions. As an indication of the effect of sulphur on the inclusions, as mentioned above most of the wetting inclusions contained some sulphur.

CONCLUSIONS

1. During levitation melting of both pure nickel and UDIMET 720, powder additions of alumina and spinel concentrated on the surface; TiN dissolved in the melt.

2. In melting as-supplied UDIMET 720 inclusions of alumina, titanium aluminate and complex oxides of aluminium, silicon, calcium, magnesium and titanium were retained at the drop surface; small inclusions, predominantly variants of aluminium silicate with associated sulphur, were observed within the drops.

3. The addition of alkali metal halides enhances the efficiency of collection of inclusions at the surface of the levitated drop and dramatically reduces the inclusion content within the drop. A thin solid crust forms around the molten drop and it is proposed that the crust is a composite of inclusions retained at the surface and bonded by the oxidation products of the reactive metals in the alloy.

ACKNOWLEDGEMENTS

The authors thank EPSRC for support of this work (Grant Number), Rolls Royce plc for provision of the UDIMET 720 alloy, INCO Alloys Ltd for technical support and Professor K.C.Mills for advice during the execution of the work.

REFERENCES

1- J. ADAMS et al, Proceedings of 1st International Conference on Processing Materials for Properties, Honolulu - USA, 7-10th Nov 1993.

2- R. HASEGAVA, Met. Tech. (JPN), **61**, (11), (1991), 12-15.

3- L. MARECHAL et al, Proceedings of Conference *Light Metals 1993*, Denver - USA, 21-25th Feb 1993.

4- M.J. KOCZAK et al, Proceedings of Conference *Advanced Materials for Future Industries*, Chibo Japan, 11-14th Dec 1991.

5- H.R. LOWRY et al, Proceedings of Conference *Metallurgical Processes for the Year 2000 and Beyond*, Las Vegas - USA, 27th Feb - 3rd Mar 1989.

6- R.I. ASFAHANI et al, J. Met., 37, (4), (1985), 22-26.

7- R. MORALES et al, Proceedings of *2nd Japan-Sweden Joint Symposium on Ferrous Metallurgy*, Tokyo, Japan, 12th Dec 1978.

8- A. MITCHELL, *Progress in understanding clean metal production*, Dept. of Metals and Materials Engineering, University of British Columbia, Vancouver - Canada.

9- S. CHAKRAVORTY et al, *Characterisation of inclusions in a nickel based superalloy by EBBM*, Division of Materials Applications, National Physical Laboratory, Teddington, Middlesex, UK.

10- J.D. ELLIS, *Quality Assurance by Electron Beam Button Melting, Ph.D. Thesis*, University of London 1992 (Dept. of Materials, Imperial College, London - UK.)

11- L. BARNARD et al, Iron & Steel Making, **20**, (5), (1993), p. 334.

PRIMARY CARBIDE SOLUTION DURING THE MELTING OF SUPERALLOYS

F. Beneduce,[*] A. Mitchell, S. L. Cockcroft, and A. J. Schmalz

Department of Metals and Materials Engineering
University of British Columbia
Vancouver, B.C., Canada, V6T 1Z4

[*] Instituto de Pesquisas Tecnologicas
Sao Paulo, Brasil

Abstract

The MC carbides which are precipitated during the freezing of most superalloys are very stable chemical compounds, as are the equivalent nitrides. The latter have been shown to persist through melting and casting processes when the nitrogen levels are above the nitride saturation solubility, but it is not clear whether or not the MC carbides can behave in the same way when exposed to liquids containing carbon at less than the carbide solubility limit. This study was directed towards defining this situation.

A series of alloys containing various MC carbides was studied in respect of the carbide solution reaction through a melting and casting sequence. The "original" MC carbides could be distinguished from the "new" carbides in the final casting by means of their size distribution and also their composition. We found that in all cases except one, the carbides in the final cast product were produced during the final solidification and were not relics of the initial carbide structure before melting. The exception was found in the case of HfC, but could have been due to the presence of a carbo-nitride of Hf, and not to the reactions of a pure carbide.

We conclude that a conventional melting and casting procedure, as practiced in the investment casting or ingot making processes will serve to remove all memory of the initial carbide distribution from the product. Rapid melting (>10^3 °C/s) with negligible superheat followed immediately by a rapid solidification process (>10^3 °C/s) would be required to retain any original carbides. The problem of carbide retention is hence restricted to some RST and powder production processes.

Introduction

The question of whether or not the superalloy MC carbides can be retained through a melting and casting process has not been studied previously although there is one literature report (1) which suggests that in the case of IN 100 this is a possibility. In addition, there exists anecdotal evidence in the casting industry that the master alloy carbide structure has an influence on that of the cast product, although the specific influence has not been precisely reported.

The MC carbides which are formed in the liquid/solid region of a superalloy casting depend on the alloy composition. In most cases they are a mixed Nb,Ti carbide in which the ratio of Ti to Nb depends on the alloy and ranges from approximately 1:2 in high Nb alloys such as IN 718, to pure Ti in the Nb-free alloys such as IN 100. In alloys which contain Hf, we find carbides which are essentially of Hf only, and also in Ta-rich alloys it is possible to identify pure Ta carbides. Since the MC carbides and TiN are isomorphous, they readily form solid solutions between them and amongst themselves. The question of the extent to which the "carbides" are in fact carbides and not carbo-nitrides has been addressed elsewhere (2), but since in most MC-containing alloys the nitrogen content is much smaller than the carbon content and in addition is principally present as TiN, we will initially assume that the MC carbides may be treated as pure carbide compounds for the purposes of thermochemical calculations.

Using assumed values for the diffusion coefficients of the appropriate elements in the liquid alloy, it is shown in Appendix I that the anticipated dissolution time for typical carbide particle sizes is extremely short and that on theoretical grounds we would not expect to see any carbide retention through an equilibrium melting process. However, since a practical system can possibly include substantial unsteady-state temperature fluctuations it is necessary to examine the problem from an experimental as well as a theoretical viewpoint.

Experimental

A series of alloys was chosen so as to exhibit the limiting range of MC carbides found in industrial superalloys. For convenience, one of these was an experimental composition, but the remaining three were common industrial alloys. The compositions of the alloys are given in Table I, together with the composition of the principal MC carbide.

The alloys were solidified under vacuum in a directional solidification furnace which has been described elsewhere (3). The resulting cylindrical ingots (2.5cm Dia x 20 cm long) were trimmed to eliminate the ill-defined head structure and expose the coarse directional dendritic structure of the bulk ingot. The MC carbide size distribution and compositions were then characterised by a combination of quantitative metallography and SEM/WDS techniques. Typical DS structures were obtained with primary dendrite spacings of the order of 200 μm, as shown in Figure 1 for the sample of alloy MAR M247.

The alloy bars were then remelted by electron beam in the apparatus shown schematically in Figure 2. The experimental objectives were to capture the bars' melting surfaces by vacuum quenching; to collect the liquid drops from the melting surface by a rapid chilling technique; and to collect the molten liquid in an ingot pool, but with a much different local solidification time than that experienced in the original fabrication of the DS bars.

Examination of these three materials would then provide a determination of the carbide path through the melting process.

Table I Chemical Compositions in Wt. % or ppm of the Nickel Base Superalloy Melting Stocks

	Alloy (%)			
	IN100	IN100Hf	MM247	IN718
C	0.168	0.180	0.152	0.060
Cr	9.9	9.5	8.3	19.5
Co	15.0	15.0	10.0	<0.100
W	-	-	9.90	<0.100
Mo	3.26	3.00	0.64	3.05
Nb	-	-	-	5.00
Ti	4.45	-	9.80	1.02
Al	5.46	5.50	5.61	0.55
B	0.015	0.015	0.015	0.004
Zr	0.050	0.060	0.030	<0.001
Fe	-	0.50 max.	-	bal.
Ni	bal.	bal.	bal.	52.8
V	0.97	0.95	-	<0.100
Ta	-	-	3.14	<.05
Hf	-	4.50	1.40	-
O(ppm)	5	-	1	3
N(ppm)	9	-	3	16
Carbide	TiC	HfC	TaC	NbC

Figure 1: Optical micrograph of columnar zone of DS MM247.

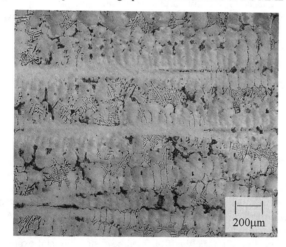

Figure 2: Schematic EB melting procedure of DS samples.

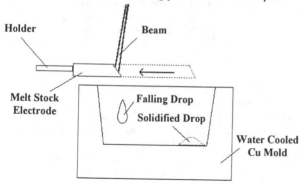

Figure 3a: Backscattered electron image in the columnar zone of DS MAR M247.

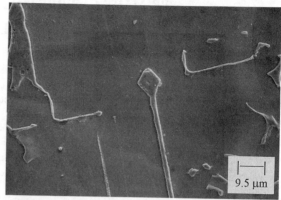

Figure 3b: Backscattered electron image in the columnar zone of EB MAR M247.

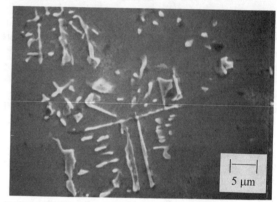

Figure 3c: Backscattered electron image in the columnar zone of DS IN718.

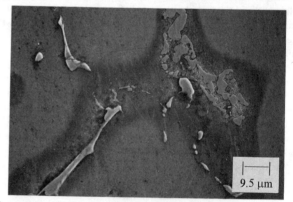

Figure 3d: Backscattered electron image in the chill region of an EB IN718 drop.

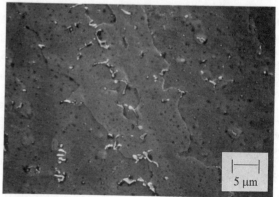

Figure 4: Backscattered electron image of a region near the chill plate in EB IN100.

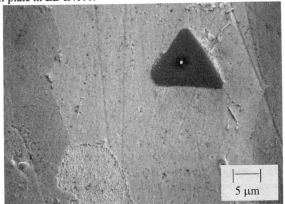

Figure 5a: Optical micrograph in the top region of the drop of EB IN100Hf.

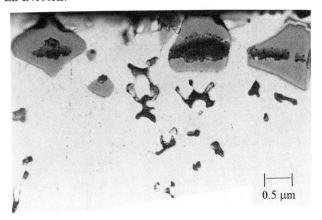

Figure 5b: Optical micrograph in the region near the chill plate of EB IN100Hf.

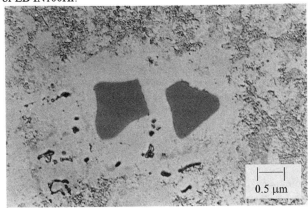

Figure 5c: Optical micrograph in the columnar region of DS IN100Hf.

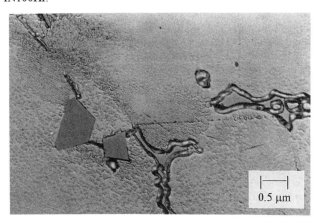

Results

The results for the three commercial alloys are considered first. It was clear from an examination of the carbide size distributions that there was a great difference between the initial carbide size and that in the remelted material. Typical fields for two of the alloys are shown in Figure 3. Large carbide particles were specifically sought in all of the samples which had been EB processed, including in the search the free surface of the remelted samples, but none were found in alloys IN718, IN100 or MAR M247.

Alloy IN100 showed a clear retention of nitride particles, as shown in Figure 4.

The alloy IN100Hf showed large particles in and near the free surfaces of the remelted sample, as shown in Figures 5(a) and (b), which are in the same size range as those in the initial DS bar (Fig 5(c)). They are considerably larger than the primary dendrite arm spacing in the remelted samples and are clearly particles which have been retained through the melting process.

The carbide particles were analysed by EPMA in all cases. The results are summarised in Table II.

The analyses are complicated by the small size of the particles in the EB samples, and in some cases it was necessary to reduce the EPMA accelerating voltage to 10kV in order to minimise matrix interference. In the case of the alloy IN100Hf, the apparent nickel and cobalt content of the carbides is probably due to the matrix interference in the analysis, but the content of chromium and vanadium is thought to be valid.

Table II Analyses of the Carbide Particles in Both DS and EB Structures

	Alloy (%)							
	IN 100		IN 100Hf		MM 247		IN 718	
	DS	EB	DS	EB	DS	EB	DS	EB
Cr	2	16	-	3.39	4	8	18.2	20.7
Co				2.54				
W					14	14		
Mo	19	17			6	4	4.9	3.2
Nb							10	3
Ti	71	60			31	22	1.6	0.7
Al							1	0.5
Fe							15	19.1
Ni			2.91	20.81			49.3	53
V	4	5	-	1.06				
Si	4	2						
Ta					39	39		
Hf			97.09	73.25	6	25		

Discussion

As emphasized in Appendix 1, the solution rate of the small carbide particles found in the initial DS samples should be high. The melting period on the tip of the electron-beam melted sample was difficult to estimate but lies between 5 and 10 seconds, which is well above the millisecond period estimated in Appendix 1 for the liquidus temperature. Examination of the melting interface indicates no carbides of the original size (in alloys IN718, IN100 and MAR M247) in the region which was processed above the MC precipitation start temperature i.e. in the region where carbide solution should have taken place. In the case of TiC, the carbide particles are less dense than the parent alloy and might have shown some flotation behaviour, but this was not observed in the vertically-grown DS samples. They are also not found to be present in any raft on the EB sample free surface. The remaining MC carbides in the alloys tested are all much more dense than the parent alloy.

The nitride behaviour is of some interest also. In the alloy IN718, the nitrogen content was sufficiently low (15ppm) that the nitride solution temperature was at the first eutectic melting point. It would be anticipated therefore, that nitride particles would not be seen in the structure, and also that the "carbide" particles would be carbides and not carbonitrides. No TiN particles were observed in any of the IN718 samples in other than the eutectic pools. In the alloy IN100, however, the higher Ti content renders TiN much less soluble and as expected, TiN particles are observable in all samples. The EB processed alloy shows TiN particles in a size range which indicates that they passed directly through the process without melting, and also had a propensity to form a "raft" on the EB alloy's free surface. These massive particles are also seen to be an effective nucleant for the MC carbide (Figure 4) in spite of the rapid chill. TiN (density 5.22 Kg/m^3) particles have been observed previously (4) to float in liquid IN100 which was not superheated to a temperature above their solution temperature in an alloy with approximately 10ppm nitrogen (the liquidus saturation solubility of TiN in IN100 is 4.5ppm N) and so their part in a raft formation in the EB button at low superheat is not unexpected.

As indicated above, the carbide behaviour in the three commercial alloys follows the expected solution process. The freezing process in the EB regime was sufficiently fast as to cause the composition of the carbide to deviate considerably from the equilibrium value. It can be seen from Table II that in all cases, the chromium content is considerably elevated in the EB material from the DS value. This effect is probably due to the fast chill process which restricts the diffusional gradient required to supply the minor elements to the precipitating particle and instead substitutes the next most stable carbide former from the matrix, i.e. Cr. The effect is less well-defined in the alloy IN100Hf because the high atomic weight of Hf restricts the accuracy of analysis, particularly in small particles, but is nonetheless visible in the results shown in Table II.

The apparent retention of carbide particles through the melting process in the case of alloy IN100Hf is interesting. These particles are seen (Figure 5(a)) to collect at the upper free surface of the EB button, which must be due to either a particle-pushing mechanism in the fast-freezing regime of the EB procedure or alternatively to a surface tension effect which keeps the particles on the free surface once they have been exposed on the surface of the melting DS feed bar. It can not be due to any bouyancy force, since HfC has a density of 12.2 Kg/m^3, well in excess of that of the parent alloy.

Examination of the freezing interface in the manufacture of the initial DS sample showed that the carbides are initially precipitated at a temperature 30°C below the equilibrium liquidus temperature and should, therefore dissolve during the EB melting/freezing sequence in a manner closely following that of the other alloys studied. They are, however, seen in optical section to have colour gradients internally which suggest that they are not in fact pure carbides, but some combination of HfC and HfN. The larger particles were examined by light-element WDS and found to have a small nitrogen content;

C (wt%) 6.5	N(wt%) 4.5

leading to an average particle composition which is quite close to Hf(NC). The extreme thermochemical stability of HfN would presumably lead to a very small solubility of this compound in the alloy (following the precedent of TiN in the similar matrix), and the large atomic size of Hf possibly stabilises the solid solution between the carbide and the nitride to an even greater extent than is found for the equivalent system based on Ti. The "MC" precipitates in the experimental alloy IN100Hf would therefore appear to be very stable carbonitrides and have solution characteristics which are quite different from the pure carbides. In order to test this hypothesis it would be necessary to re-make the IN100Hf alloy with an extremely-low nitrogen content (below 1ppm) so as to precipitate the pure carbide. In all practical alloys based on Hf additions, however, it seems probable that the "MC" precipitates are in fact carbonitrides and that they will accordingly have a finite probability of being carried through a melting process.

Implications for Melting Practice

It appears from the above work that the commercial superalloys would not exhibit retention of pure MC carbides through the conventional melting and casting processes even if the melt has little superheat and the melting step is rapid. In high energy drip-melting processes such as welding, surfacing or rotating electrode powder processes where the melting/freezing cycle is completed in a millisecond time scale, there is a possibility of carbide carry-over which could be identified by means of the Cr content in the product carbide compositions. Care must be taken, however, in applying these conclusions to systems in which the nitrogen content is high enough that the structure contains primary carbo-nitrides as well as carbides. The solubility of the carbonitrides is evidently significantly smaller than that of the pure carbide (particularly so in the case of Hf-containing alloys) as is that of the pure nitride, TiN, and it is very possible that these compounds will be carried through the melting/casting process without solution. In that case, not only will the particle size distribution of the initial material be important in determining the properties of the final product, but the particles will also influence the product carbide distribution through their effect on carbide nucleation during freezing.

Acknowledgments

The authors are grateful for financial support of this work through the Instituto de Pesquisas and the Nickel Development Institute. They would also like to acknowledge the generous support of the Special Metals Corporation, the Cannon-Muskegon Corporation and the Pratt-Whitney Corporation in supplying materials and analytical assistance.

References

1. R. Mehrabian et al., "Interdendritic Fluid Flow and Macrosegregation; Influence of Gravity," Met. Trans., 1, (1970), 1209-1220.
2. S. L. Cockcroft, A. Mitchell and A. J. Schmalz, "Primary Carbide and Nitride Precipitation in Superalloys Containing Niobium," J. High Temperature Materials and Processes, 15 (1), (1996), 245 - 255.
3. Y. Haruna, A. Mitchell and A. J. Schmalz, "Some Observations on the Recycling of Superalloys by the EBCHM Process," J. High Temperature Materials and Processes, 14 (3), (1995), 173 - 191.
4. Y. Haruna, (M.A.Sc. Thesis, University of British Columbia, 1994).

Appendix 1

THE SOLUTION RATE OF NbC IN LIQUID ALLOY IN718

The development of an equation describing the solution rate of a particle in a liquid is given by Geiger and Poirier (1A);

$$t = 2\rho_s D_p / 3\rho_l \lambda k_m (C_0 - C_\infty) \qquad (A1)$$

where; t is the time for complete solution (seconds).
D_p is the diameter of the particle (m).
C_0 is the weight fraction of Nb at the particle surface.
C_∞ is the weight fraction of Nb in the liquid distant from the particle surface.
k_m is the mass transfer coefficient for the process (assumed to be 0.02 cm/sec).
λ is a particle shape factor (assumed to 1.5).
ρ_s and ρ_l are particle and liquid densities respectively (Kg/m^3).

Substituting values into this relationship appropriate for NbC in alloy IN718 at 1630 K, for a particle of 10 microns dimension, the complete solution time is estimated to be 150 msec. Since the parameter values are very similar for the other carbides considered in this study, it is likely that their solution times will also be comparable. The samples remelted by EB in the present study were estimated to have a residence time in the temperature zone above the MC precipitation temperature of approximately 2 seconds.

References

1A. G. H. Geiger and D. R. Poirier; Transport Phenomena in Metallurgy, (New York, NY: Addison-Wesley, 1973), 541.

UNDERCOOLING RELATED CASTING DEFECTS IN SINGLE CRYSTAL TURBINE BLADES

M. Meyer ter Vehn, D. Dedecke, U. Paul* and P.R. Sahm

Gießerei-Institut, RWTH Aachen, Intzestr. 5, 52056 Aachen, Germany

* now at: Siemens KWU, Wiesenstr. 35, 45473 Mülheim/Ruhr, Germany

Abstract

During the single crystal solidification of turbine blades casting defects such as stray grains are often formed in the inner and outer shroud due to a curvature of the liquidus isotherm and a resulting thermal undercooling of the melt. An intensive study of the thermal conditions in typical shroud regions for two different blade geometries was carried out using macroscopic solidification modeling and temperature measurements during casting experiments. It was tried to reduce undercooling related defects by temporarily changing the withdrawal velocity of the mold and by positively influencing the heat flux with ceramic insulation. In addition to that the application of grain continuators was investigated in order to eliminate stray grains while casting large blades for use in stationary turbines.

Introduction

Avoiding structural inhomogeneities during single crystal solidification has always been a major point of interest to the investment casting engineer involved with the production of high temperature components, such as single crystal turbine blades. In order to use the full potential of these components it is necessary to optimize the Bridgman process for a resulting defect-free as-cast structure and for better results during the subsequent heat treatment.

Casting defects can be caused by a macroscopic curvature of the liquidus isotherm while it is passing through extreme enlargements in the cross-section of the component (e.g. transition from the blade to the outer shroud) [1]. A concave shaped liquidus isotherm can result in isolated, thermally undercooled regions of melt. Depending on the number of nuclei in the melt and the extent of thermal undercooling this may lead to heterogeneous nucleation close to or at the ceramic mold and the formation of spurious grains, Figure 1a. If a high angle boundary is formed the casting will not be accepted [2].

If nucleation is suppressed secondary dendrites will grow rapidly lateral towards the far corners of the shroud as soon as the primary dendrites reach the transition from blade to shroud resulting in a fine dendritic microstructure in this region, Figure 1b. This rapidly solidified structure is believed to have a slightly increased tendency to recrystallize during heat treatment because of its distorted lattice (mosaic structure) [1]. A convex shaped liquidus isotherm may hinder sufficient feeding and therefore cause open interdendritic microporosity on the upper side of the shroud.

It has often been recognized that thermal undercooling in the shroud of a turbine blade causes changes in the conditions for dendritic growth [3]. This paper concentrates on macroscopic undercooling phenomena during single crystal solidification of turbine blades, in particular on the effect of blade geometry and of changes in the process parameters on the resulting microstructure.

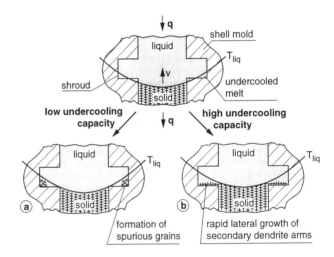

Figure 1: Depending on the undercooling capacity of the given alloy a macroscopic concave curvature of the liquidus isotherm can cause the formation of stray grains (a) or result in fast lateral growth of secondary dendrites (b).

Experimental procedure

Macroscopic solidification modeling was used to detect critical component areas that are crucial with regard to maintaining the single crystal dendrite structure during solidification and heat treatment. Temperature measurements during solidification with thermocouples in these regions enabled a reconstruction of the liquidus isotherm. Metallographic inspection of macro- and microstructure was used to analyze type and extent of casting defects. Correlation of the metallographic results with the simulated and measured course of the liquidus isotherm gave information about the origin of casting defects.

Two different blade geometries were investigated. The first, which was approximately 18 cm high, resembled a typical airfoil for use in aerospace engines, Figure 2. It had a simple ceramic core, and was cast in a 4 blade cluster with the Ni-base superalloy CMSX-6. The second geometry, which was approximately 30 cm high, was designed to give generally valid information about the solidification of large turbine blades for land-based turbines, Figure 3. It has an inner shroud only and was cast solid and tip down as a single blade with the alloy SC 16. Alloy compositions of CMSX-6 and SC 16 as well as other alloy data are given in Table I.

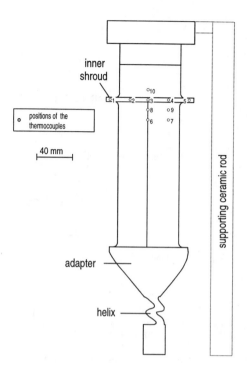

Figure 3: Casting set-up of the large dummy turbine blade.

Macroscopic solidification modeling of the small blade was carried out with the finite element code CASTS (Computer Aided Solidification TechnologieS) [4], which has been developed at the Gießerei-Institut and ACCESS (Aachen Center for Solidification in Space). The digital description of furnace and blade including shell mold as well as the subsequent FE mesh generation were performed with commercial software (IDEAS supertab). Due to the cluster's cyclic symmetry the mesh consists of ¼ of the cluster including starter, helix, central runner and gating system. It has 47,704 volume elements (tetraeders, pentaeders and hexaeders) and 3,441 surface elements, Figure 4.

CASTS calculates the 3D temperature distribution taking into account the release of latent heat and the heat transfer at interfaces between different materials [5]. Because of the nature of the bridgman process natural convection was neglected. Since thermal radiation is the dominant heat transmission mechanism in the vacuum casting process the net radiation method [6] was used to calculate heat exchange between all radiating surfaces. The transient temperature distribution and the macroscopic shape and progress of the liquidus isotherm were later visualized with the CASTS-tool COLOR3D.

The SC casting experiments were performed in a vacuum bridgman furnace (Leybold IS8/III). A round chill plate of 15 cm diameter and a 17.6 cm diameter baffle were used for the clusters of the smaller blades. The chill plate for the larger blades had the shape of a parallelogram, just like the corresponding baffle. Al_2O_3/SiO_2-ceramic molds were manufactured by investment process. Up to 20 thermocouples per mold were positioned in and close to the shrouds of the blades. For both geometries a withdrawal velocity (v_{mold}) of 3 mm/min and a heater temperature of 1550°C were standard process parameters of the Bridgman furnace.

In order to reduce or even eliminate casting defects by smoothing the curvature of the liquidus isotherm the with-

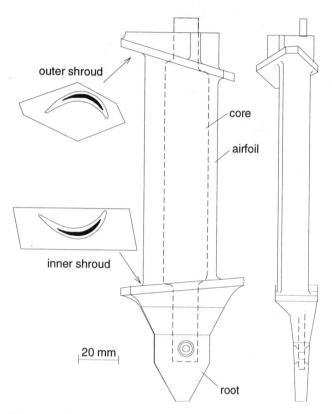

Figure 2: The small dummy blade has an inner and an outer shroud. The blade around the core has a thickness of about 1.5 mm. Four blades were assembled around a central runner rod in a cluster and thermocouples were placed in the outer shrouds.

Table I: Nominal compositions (wt.-%), densities and approximate liquidus temperatures for CMSX-6 and SC 16.

	CMSX-6	SC 16
Cr	9.8	16.0
Co	5.0	-
Mo	3.0	3.0
Ta	2.0	3.5
Al	4.8	3.5
Ti	4.7	3.5
Hf	0.01	-
C	0.02	-
N	10 ppm	-
O	9 ppm	-
Ni	bal.	bal.
density ρ (g/cm3)	7.98	8.21
liquidus temp. (°C)	1337	1338

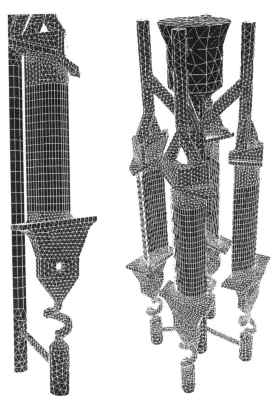

Figure 4: FEM-mesh of the small dummy blade.

drawal rates were varied (1 to 4 mm/min) while the solidification front was passing the critical shroud areas. In addition to that it was tried to minimize the thermal undercooling in the shroud of the larger blades by applying insulation to the ceramic mold just below the shroud. Another means to avoid casting defects is the use of grain continuators [7]. A grain continuator is a simple rod (diameter: 6 mm) with which the single crystal orientation selected by the helix is transferred from just above the helix to critical regions in the shroud.

Results and Discussion

Small dummy blade

FEM solidification modeling. For a constant withdrawal velocity (v_{mold} = 3 mm/min) the simulation shows a concave curved liquidus isotherm in the outer shroud of the small airfoil during solidification, Figure 5a. Part of the shroud on the leading edge is well below the liquidus-temperature, but since it is isolated from the solidification front in the airfoil and since the alloy has a certain capacity to undercool, it will remain liquid until secondary dendrites are able to grow into the platform or until a stray grain will nucleate due to excessive undercooling.

On the other side of the outer shroud, facing the central runner rod of the cluster, another large thermally undercooled zone can be identified, Figure 5b. Looking at the bottom side of the shroud it becomes distinct that this zone is connected with the undercooled volume in the overhanging shroud area at the leading edge, Figure 5c. This area is difficult to access with thermocouples because of the position of the airfoil within the cluster. Therefore, only the simulation can give detailed information about the solidification sequence in this part of the component. A view onto the trailing edge of the blade shows another small isolated undercooled area in the outer shroud, Figure 5d.

Temperature measurements. Temperature measurements in the outer shroud during the solidification experiments with standard parameters revealed an undercooling of about 10 K below an approximate liquidus-temperature of 1337°C in the overhanging shroud area at the leading edge (thermocouple No. 10), Figure 6a. In another set of casting experiments undercoolings of about 12 K were measured at the opposite side of the shroud (trailing edge, thermocouple No. 19), Figure 6b. The cooling curves of the thermocouples in the far corner at the leading edge show a small recalescence upon the sudden release of latent heat, whereas no recalescence was observed on the other side at the trailing edge. The extent of this recalescence depends on the amount of heat being released and the conditions of heat flux.

The curvature of the reconstructed liquidus isotherms, some of which are plotted in Figure 7 for both sides of the shroud (leading and trailing edge), is in good agreement with the simulated results. Figure 7 also illustrates the positions of the thermocouples. The vertical temperature gradients necessary for the reconstruction of the isotherms were measured with thermocouples No. 6,7,8 and 9 on the leading edge and thermocouples No. 11 and 20 on the trailing edge.

Metallography. Metallographically prepared cross-sections of the outer shroud show an area of primary dendrites around the core, which have grown from the airfoil into the shroud, Figure 8a. A deviating fine dendritic structure formed by fast lateral growth of secondary dendrites into the areas that were undercooled is visible in the regions far away from the core. In the far corner of the leading edge, where a considerable undercooling was measured, the structure is very fine.

Some castings also showed a stray grain in this area as well as in other undercooled regions, Figure 8b. Comparing all measured cooling curves with the corresponding micrographs it seems that lateral growth of secondary dendrites without nucleation of stray grains is accompanied with the occurrence of small recalescences, whereas no recalescences were observed, when stray grains were formed.

Variation of withdrawal velocity. A change of the withdrawal velocity at approximately 70 mm below the outer shroud down to 2 mm/min was carried out in order to minimize structural inhomogeneities by reducing the concave curvature of the liquidus isotherms before the shroud started to solidify. Furthermore, it was tried to deliberately provoke stray grains by raising the withdrawal velocity up to 4 mm/min in order to define critical parameters for the given geometry and the given alloy.

By reducing the withdrawal velocity it was possible to lower thermal undercooling in the overhanging shroud area at the leading edge to 2 K, Figure 9. Still, the above described fine dendritic structure was observed. By raising the velocity cooling rates and extent of undercooling in the critical regions increased, Figure 10. Nevertheless, this had no clear effect on the formation of spurious grains. This may be explained by the fact that even though it was possible to influence the extent of undercooling the overall curvature of the liquidus isotherm did not change much. Simulated results illustrate the situation: By raising the velocity from 2 mm/min to 4 mm/min the solidification front is shifted from the heater zone into the baffle, but no significant effect on the curvature of the isotherm is noticeable, Figure 11. Still, a view onto the

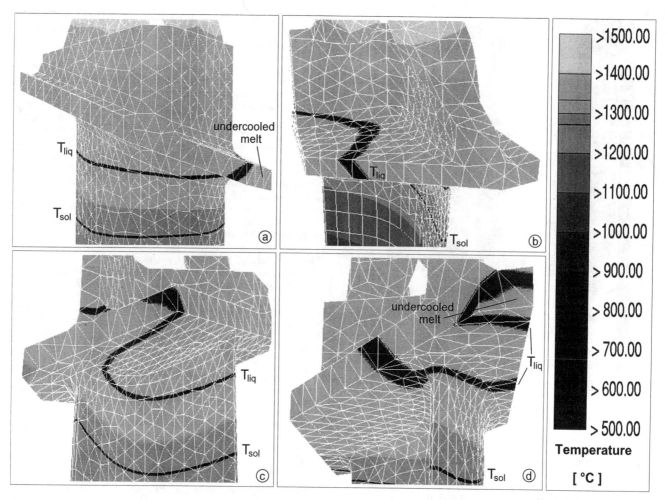

Figure 5: Temperature distribution in the outer shroud of the small blade. Several undercooled regions of melt are visible at the leading edge (a), towards the central runner rod (b and c) and at the trailing edge (d).

trailing edge of the outer shroud shows that the undercooled zone is becoming smaller as the velocity is changed from 3 mm/min to 2 mm/min, Figure 12. By lowering the undercooling in the critical areas the chance of spurious grain formation is reduced.

The occurrence of high angle boundaries due to undesired nucleation seemed to be rather the exception at the higher withdrawal velocity (v_{mold} = 4 mm/min) even though higher undercoolings were measured, Figure 13. It remains to be investigated, however, if the fast grown structure will cause recrystallisation during heat treatment. The alloy CMSX-6 proved to be relatively well "undercoolable" which is confirmed by other recent experiments by the authors investigating the maximum undercooling capacity of superalloys in ceramic molds.

Large dummy blade

Simply because of their larger size, undercooling phenomena caused by the curvature of the liquidus isotherm create a bigger problem during the single crystal solidification of turbine blades for stationary gas turbines. Hence the intention of temperature measurements during the solidification of large dummy turbine blades was to transfer present results onto a more extreme geometry and to find additional appropriate means to reduce casting defects.

Variation of process parameters. Using a constant withdrawal rate of 3 mm/min measured temperature curves showed high undercoolings of up to 40 K below an approximate liquidus-temperature of 1338°C in the outer regions of the shroud (alloy: SC 16), Figure 14, which generated a fine-grained structure, Figure 15. The reconstructed liquidus isotherm was of a concave shape and had a relatively stable position about 10 to 20 mm above the baffle in the heater zone of the Bridgman furnace. With the intention of pushing up the liquidus isotherm further into the heater zone, the withdrawal velocity was decreased to 1 mm/min 70 mm before the shroud passed the baffle (as it was done with the small dummy blades). The experiments resulted in a slightly convex curvature of the liquidus isotherm and in a decrease of recalescence. No significant effect on the undercooling was detected, but the fine-grained zone was much smaller and only a few large stray grains were observed in the far corners of the shroud.

By applying a SiO_2-Al_2O_3-insulating-blanket 40 mm below the shroud of the big dummy blade it was intended to reduce the heat flux between heating and cooling zone. It was possible to minimize the formation of small grains because the

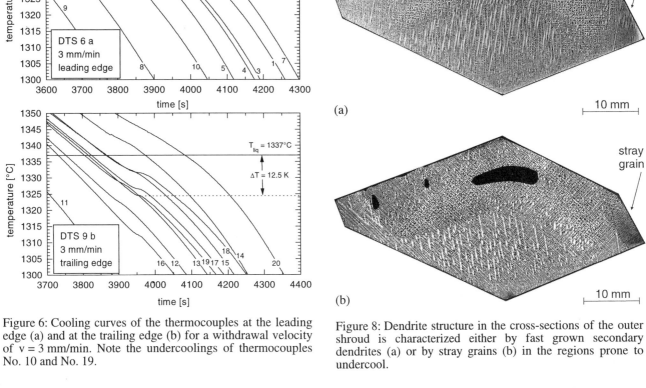

Figure 6: Cooling curves of the thermocouples at the leading edge (a) and at the trailing edge (b) for a withdrawal velocity of v = 3 mm/min. Note the undercoolings of thermocouples No. 10 and No. 19.

Figure 8: Dendrite structure in the cross-sections of the outer shroud is characterized either by fast grown secondary dendrites (a) or by stray grains (b) in the regions prone to undercool.

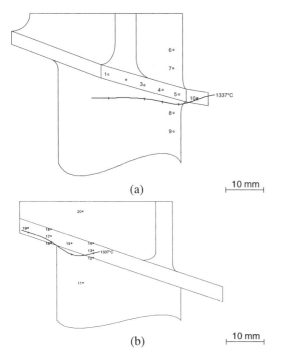

Figure 7: Reconstructed liquidus isotherms (1337°C) for the leading edge (a) and the trailing edge (b) are in good agreement with the simulated results.

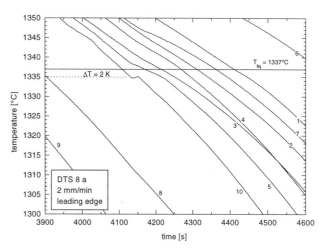

Figure 9: Changing the withdrawal velocity down to 2 mm/min reduces the undercooling to 2 K in the overhanging area at the leading edge.

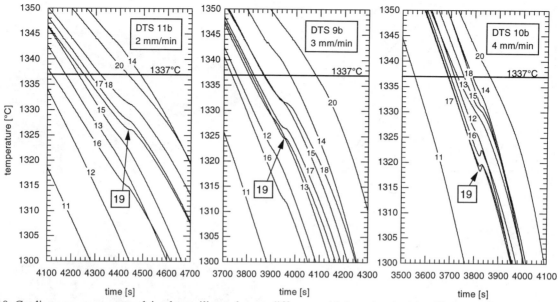

Figure 10: Cooling curves measured in the trailing edge at different withdrawal velocities. Raising the velocity causes an increase of the cooling rate and of the extent of undercooling.

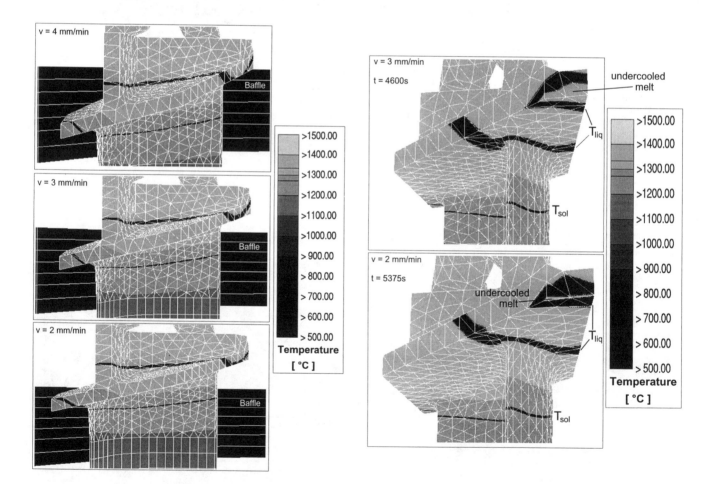

Figure 11: The higher the withdrawal velocity the more the liquidus isotherm will be shifted into the baffle. Still, the effect on the curvature of the isotherm is relatively small.

Figure 12: The volume of undercooled melt at the trailing edge increases as the withdrawal velocity goes up.

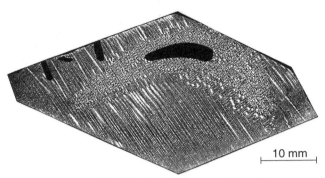

Figure 13: Even though the highest undercoolings were measured at a withdrawal velocity of 4 mm/min no stray grains were formed, instead the structure is characterized by excessive lateral growth of secondary dendrites.

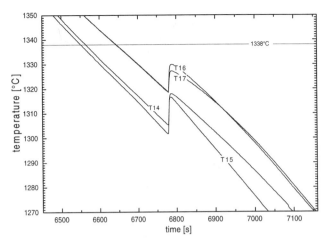

Figure 14: Cooling curves obtained with thermocouples in the outer regions of the shroud of the large dummy blade. The positions of the thermocouples 14, 15, 16 and 17 in the shroud are illustrated in Figure 15.

temperature loss out of the edges of the outer platform and therefore the undercooling effect decreased. The combination of reducing the withdrawal rate and application of insulation produced the best structure with almost no spurious grains. The recalescences were very small and a fine dendritic structure of fast grown secondary dendrites was dominant in the edges of the shroud comparable to the structure observed in the outer shroud of the small dummy blades, Figure 16.

Grain continuators. Two types of continuators were tested: a single version and a twin version, which was supposed to transfer the single crystal into two corners of the shroud. In order to make successful use of grain continuators it is important not to produce any misorientated grains while the single crystal orientation is transferred to the critical regions of the shroud. Investigations concentrated on possible nucleation sites at the lead-off angle just above the helix. Using different angles ($\alpha = 0°$, 45° and 60°, Figure 17a) it was always possible to transfer the single crystal from the adapter to the main body of the continuator, which was about 200 mm long. No defects were detected over the whole length of both the single and the twin version of the grain continuators.

Next, the solidification front in the grain continuator has to progress faster than in the blade, so it will reach the shroud before it is undercooled. Temperature measurements with thermocouples placed in the same vertical positions below the

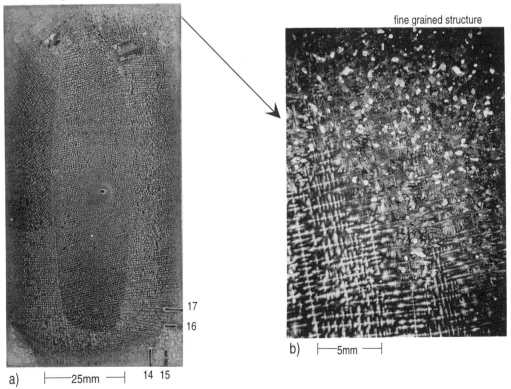

Figure 15: The dendrite structure in the platform shows large **misoriented grains** (a). The large undercoolings in the corners of the shroud caused a fine grained structure (b).

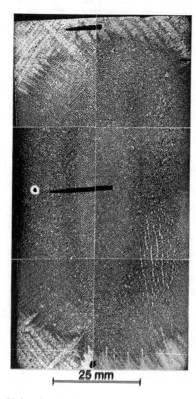

Figure 16: Using insulation in order to minimize the heat flux and additionally optimizing the withdrawal rate resulted in a fine structure of secondary dendrite arms in the corner areas.

shroud in the blade and in the continuator showed that the liquidus isotherm in the contiuator is clearly ahead. The bypassed single crystal orientation was transferred successfully into the critical regions of the shroud. Cross-sections of the platform show the positive influence of the continuator, Figure 17b. Primary dendrite axes are visible in the center as well as on the side of the shroud where the continuator was attached. Between the two areas the structure is characterized by lateral grown secondary dendrites. The formation of spurious grains was completely suppressed on this side. On the opposite side spurious grains could not be avoided. In a realistic casting set-up of several large blades in a cluster the fact that one blade influences the thermal conditions of another because of its shadow may be used to get good results with just a few precisely positioned grain continuators.

Further investigations will have to deal with the possible chance of the formation of low angle boundaries in the shroud in case that the orientation of the dendrite arms will slightly twist as the single crystal is growing inside of the grain continuator.

Conclusion

The combined use of numerical solidification modeling, temperature measurements and structural analysis has proven to be a practical tool while working on explaining and avoiding casting defects.

The variation of process parameters showed good results for the defect-free solidification of "small" airfoils. It was possible to positively influence the strong effect of extreme enlargements in the blade on the microstructure formation in

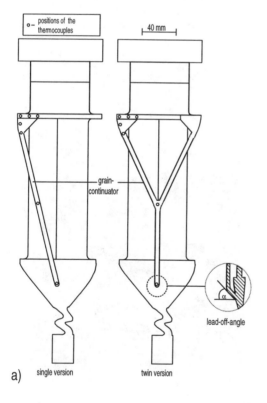

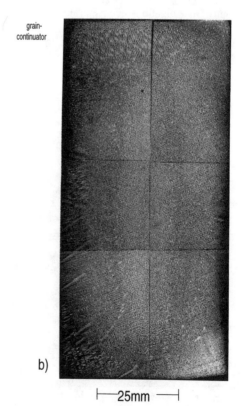

Figure 17: Single and twin version of the grain continuator casting set-ups (a). On the micrograph of single version casting (b) note the primary dendrite axes in the upper left hand corner where the continuator had been attached.

the critical shroud regions by optimizing the profile of the withdrawal velocity. The application of grain continuators is not required for the chosen small geometry and the alloy CMSX-6.

For larger geometries and alloys that are not easily undercooled, in contrast, grain continuators seem to be unavoidable. They proved to be capable of minimizing undercooling related defects and are sometimes the only practicable means to avoid spurious grains in outer shroud regions. They have the disadvantage that they imply a relatively high technical expenditure and that they can cause low angle boundaries in the shroud due to crystal growth distortion during passage in the continuator. In this field further research is necessary in order to fully control the grain continuators' mode of operation.

Acknowledgments

This study was carried out as part of a project sponsored by the German Federal Ministry of Education and Research (BMBF 03M 3038 E7). The authors also like to thank Thyssen Guß AG Investment Casting (Bochum, Germany), MTU Munich (Germany) and ACCESS e.V. (Aachen, Germany) for support and discussion.

References

1. D. Goldschmidt, U. Paul and P.R. Sahm, "Porosity Clusters and Recrystallisation in Single-Crystal Components", in Proc. of the 7th International Symposium on Superalloys, eds. S.D. Antolovich et al., Seven Springs, PA, Sept. 20-24, 1992, p. 155-164

2. B. Paine, "Product Property of Superalloys by DS and SX Castings", AETC Symposium 1986, AE-Group

3. S.L. Cockcroft, M. Rappaz, A. Mitchell, J. Fernihough, A. Schmalz, "An Examination of Some of the Manufacturing Problems of Large Single-Crystal Turbine Blades for Use in Land-Based Gas Turbines", in Proc. of Materials for Advanced Power Engineering, eds. D. Coutsouradis et al., Liège, 3-6 October 1994, p. 1145-1154

4. P.R. Sahm, W. Richter, F. Hediger, "Das rechnerische Simulieren und Modellieren von Erstarrungsvorgängen bei Formguß", Giesserei-Forschung, 35, 1983, Heft 2, p. 35-42

5. F. Hediger and N. Hofmann, "Process Simulation For Directionally Solidified Turbine Blades of Complex Shape", in Proc. of the 5th Intern. Conference on Modeling of Casting, Welding and Advanced Solidification Processes, Davos, Switzerland, Sept.16-21, 1990, eds. M. Rappaz et al., Warendale: TMS 1991, p. 611-619

6. R. Siegel and J.R. Howell, Thermal Radiation Heat Transfer (McGraw-Hill Book Company, New York, 2nd Edition, 1981)

7. A. Yoshinari, K. Iijima, I. Takahashi and H. Kodama, "Solidification Simulation for Large Single Crystal Buckets in Heavy Duty Gas Turbines", in Proc. of Modeling of Casting and Solidification Processes, eds. C.P. Hong et al., 1991, p. 201-212

THE COLUMNAR-TO-EQUIAXED TRANSITION IN NICKEL-BASED SUPERALLOYS

J. W. Fernihough, S. L. Cockcroft, A. Mitchell and A. J. Schmalz

The Department of Metals and Materials Engineering
The University of British Columbia
Vancouver, Canada
V6T 1Z4

Abstract

The thermal conditions leading to the columnar-to-equiaxed transition have been measured in nickel based superalloys AM1 and MAR-M200+Hf. The experimental methodology involved melting and directionally solidifying bars in alumina cylinders induction heated with a Mo susceptor and cooled with a chill located at the bottom. The temperature data obtained from embedded thermocouples clearly revealed a thermal recalesence in the equiaxed zone which was not found to be present in the columnar zone. Thus, the onset of a thermal recalesence was found to coincide well with the CET. The thermocouple data has been processed to yield estimates of the solidification rate (R) and temperature gradient (G) at the solidification front. Two approaches have been used: one based on a straightforward regression analysis of the data; the other, based on tracking the dendrite tip temperature. The latter has been done using the KGT model with the pseudo binary approximation to determine the dendrite tip undercooling ΔT_C as a function of growth rate. A comparison of the results obtained for the two techniques is presented. Finally, a modified Hunt CET model has been developed and employed to estimate the nucleation undercooling consistent with the occurrence of the CET at the G and R conditions determined from the isotherm tracking analysis of the thermocouple data.

Introduction

The nucleation of spurious grains during the directional solidification of single crystal castings is held to be related to the columnar-to-equiaxed transition (CET) in so-far-as both phenomena are related to the kinetics of heterogeneous nucleation and dendrite growth. In conventional casting processes, this transition occurs in association with decreasing temperature gradients and increasing growth rates and can occur in a variety of locations depending on the casting process, but generally arises in the last regions of a casting to solidify in association with the dissipation of superheat. The exact conditions under which the transition takes place and the parameters that influence it have not been extensively studied in high temperature systems, although much work has been carried out on low temperature analogue systems.

The objective of this study was to measure accurately the thermal conditions present at the time of the transition in alloys typical of those employed in the casting of DS and SX aeroengine blades. The thermal conditions identified in the present study provide only one set of limiting conditions for the avoidance of spurious grains during SX casting. This study is part of a larger programme aimed at developing the data necessary for input to microstructural models capable of predicting the evolution of microstructure, and ultimately the avoidance of spurious grains, in directional solidification processes given a knowledge of the macro and micro heat flow conditions.

Experimental

The alloys examined in the present study were nickel-based superalloys AM1 and MAR-M200+Hf. The compositions of the two alloys studied are presented in Table I.

Table I
Alloy Composition

	AM1	MAR200+Hf
Ni	bal	bal
Cr	7.52 wt%	8.40 wt%
Co	6.64 wt%	9.52 wt%
Mo	2.06 wt%	0.03 wt%
Al	5.34 wt%	4.98 wt%
Ti	1.25 wt%	1.92 wt%
W	5.59 wt%	11.9 wt%
Ta	7.80 wt%	0.10 wt%
Nb	0.01 wt%	0.86 wt%
Hf	0.004 wt%	1.18 wt%
C	80 ppm	0.14 wt%
B	3 ppm	150 ppm
Zr	10 ppm	400 ppm

The experiments, conducted at UBC, involved melting and directionally solidifying bars in alumina cylinders induction heated with a Mo susceptor and cooled with a chill located at the bottom. In the experiments, great care was taken to measure accurately the evolution of temperature in the region of the sample encompassing the CET. This was done by conducting a series of solidification experiments in which a thermocouple probe was frozen at different depths in the sample adjacent to the CET, both in the columnar region and in the equiaxed region. The probe was comprised of two Type-D W-3%Re /W-25%Re thermocouples vertically off-set from

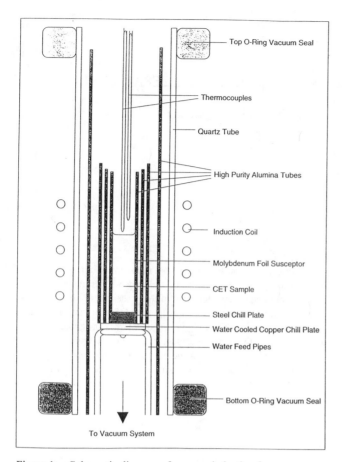

Figure 1: Schematic diagram of vacuum induction furnace.

one another by 12mm. A typical experimental set-up showing the probe and the sample to be solidified is shown in Figure 1. Effort was made to maintain consistent initial and cooling conditions run-to-run. At the completion of a series of runs, the sample was sectioned and etched for metallographic examination. The CET was generally found to occur approximately 2/3 of the way up the bar. Alloy composition was measured before and after completion of the runs to ascertain the extent to which alloy loss had occurred.

Prior to conducting the CET experiments the thermocouples were calibrated using 99.9% pure nickel in order to assess the run-to-run and absolute accuracy of the measurements. The run-to-run standard deviation was found to be approximately 2°C and the absolute accuracy, 16°C, which falls within the 1% error inherent in the thermocouple as specified by the manufacturer.

Results and Discussion

Microstructure

A polished and etched cross section of a typical CET sample is shown in Figure 2. As can be seen, the transition from columnar to equiaxed is abrupt, but occurs within a range of axial locations across the sample covering approximately 2mm. No zone of mixed equiaxed and columnar structure was observed in any of the experiments. This result is in agreement with the findings of Wienberg et. al [1,2] and Hunt and Flood[3] in their examination of the CET in low temperature alloys, but differs from the findings of Pollock et. al. [4] in their examination of grain defect formation

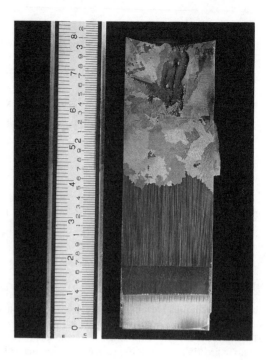

Figure 2: Sectioned and etched sample showing typical CET observed in AM1.

in directionally cast nickel base single crystals. Further, examination of the dendrite morphologies reveals that the growth is dominated by primary arms right up to the CET, whereas Pollock and co-workers reported a transition preceding the CET from primary to secondary and tertiary growth[4]. Thus, based on this comparison it would appear that the growth conditions, and perhaps the nucleation conditions, prevalent in this study differ substantially from the those in the study by Pollock et. al.[4]. The differences in thermal conditions between the two studies are examined in more detail later.

Thermal

Typical data showing the variation in temperature with time obtained in a sample of MAR-M200+Hf cooled with the probe placed at various locations is shown plotted in Figure 3. As can be seen, there is a distinct change in the evolution of temperature that occurs between 41 and 46mm from the chill, from continuous cooling, indicative of columnar growth, to interrupted cooling (thermal recalesence), typical of equiaxed growth. Fundamentally, the latter behaviour may be attributed to the failure of the local thermal gradient to remove the latent heat associated with equiaxed growth at a rate sufficient to maintain continuous cooling. Whether this phenomenon plays a role in the CET or is a consequence of it remains to be established.

An expanded view showing temperature histories obtained with one of the probe thermocouples located in the columnar zone and the other located in the equiaxed zone is shown in Figure 4. In this figure, there is clearly a thermal recalesence observed in the equiaxed zone which is not present in the columnar zone. A review of all of the thermal data from the various experiments revealed that the onset of a thermal recalesence was found to coincide well with the transition from columnar to equiaxed growth.

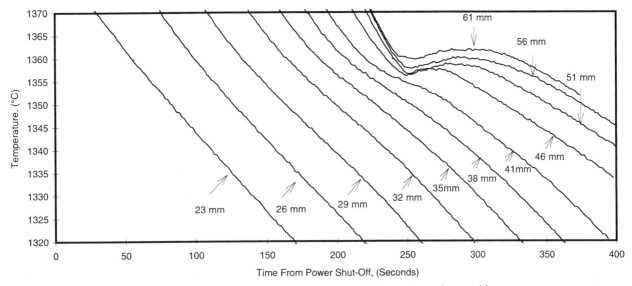

Figure 3: Variation in temperature with time recorded with the thermocouple probe located at various positions.

Since there was a 12mm vertical spacing between the probe thermocouples it was not possible to use the data from a single solidification sequence to pinpoint exactly the thermal conditions leading to the CET. Hence, the need to examine multiple solidification sequences with the probe located at various positions. While this approach yielded an improved picture of the variation in temperature in the vicinity of the CET it had a number of drawbacks. Principally, it was found that there were differences in the initial conditions from run-to-run. A sense of the magnitude of the run-to-run variation can be seen in Figure 5, which is a plot of the thermal profiles at 25s intervals derived on the basis of multiple solidification sequences. As can be seen in the example shown, there is roughly a 20°C variation in the initial temperature at between 35 and 45mm from the chill. The CET lies within this range. However, it should also be noted that there is some reduction in this variation with increasing time particularly as the liquidus temperature is approached, suggesting that latent heat effects dominate over the variability observed in the superheat.

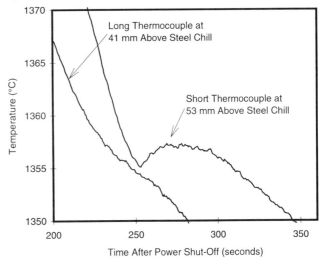

Figure 4: Variation in temperature with time recorded with the thermocouple probe located such that the thermocouples straddled the CET.

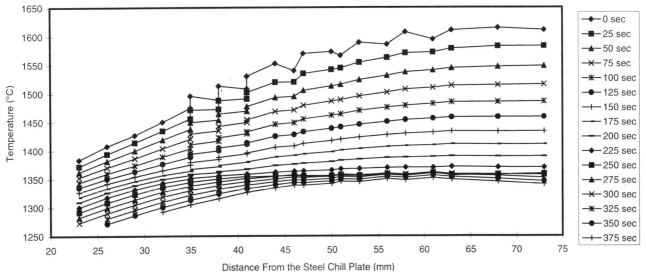

Figure 5: Thermal profiles at 25 second intervals as derived from multiple solidification sequences.

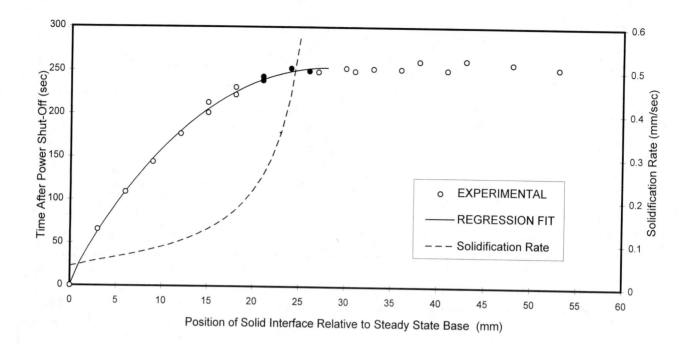

Figure 6: Progress of solid interface position with time and solidification rate.

Analysis

Given the circumstances described above - i.e. variability in axial position of CET across sample and in initial temperature distribution from run-to-run - extraction of the solidification rate, R, and thermal gradient, G, just prior to the CET presents a significant challenge. Consequently, two procedures were employed to analyze the thermal data. Both methods are essentially isotherm tracking techniques that use as input the thermocouple data. In the first, the isotherm, or temperature, tracked is taken to be the liquidus temperature less the undercooling observed at the onset of equiaxed solidification (as recorded by a thermocouple in the equiaxed regime adjacent to the CET). The columnar front position as a function of time was subsequently regression fit with a polynomial, the derivative of which, could then be used to estimate the growth rate, R. The other important solidification parameter, the thermal gradient, G, was estimated by regression fitting the thermal data at the time of the CET (as determined from the first appearance of a thermal recalescence).

In the second method, a more accurate estimate of the columnar tip velocity is determined by allowing the isotherm tracked to vary with velocity as dictated by solutal undercooling. The KGT model[5] modified for multicomponent alloys[6] was adopted for this task. The procedure used to accomplish this was to iteratively obtain improved estimates of the dendrite tip temperature, $T_{tip,i+1}$, by calculating a solutal undercooling, $\Delta T_{C,i}$, based on the isotherm velocity of the original estimate of the dendrite tip temperature, $T_{tip,i}$. The details of the approach are presented elsewhere [7]. This approach has the advantage that there is a unique isotherm traveling at a rate which corresponds to its undercooling below the equilibrium liquidus temperature. Thus, there is an additional constraint and the error is reduced[7].

Typical results obtained using the first method (referred to as regression analysis) are shown plotted in Figure 6. The discrete points represent the time at which the columnar front passes the location of the probe thermocouples and the solid line represents the regression fit to the discrete values. The solid data points represent the locus of CET positions as estimated based on thermal behaviour (transition from continuous cooling to thermal recalescence).

The method does not lend itself particularly well to an estimate of the solidification rate, R, just prior to the CET as the growth rate is changing rapidly within the limits of the estimated CET position. Utilization of the physical position of the CET instead does not significantly improve the accuracy owing to the variability in the axial position of the CET across the sample and from run-to-run. The same problem arises in attempting to estimate the thermal gradient, G.

The resulting R's and G's obtained for the CET using the two techniques for alloys AM1 and MAR-M200+Hf are presented in Table II. As can be seen from the data presented in Table II the two techniques yield similar results with estimated errors of approximately 50% for the regression method and around 30% for the method based on the solutal undercooling.

Arguably the results of this study are not readily transferable to the casting of single crystal turbine blades owing to the fact that breakdown of directional growth has been investigated under conditions of relatively high growth velocity and low thermal gradient. Nonetheless, some or all of the phenomena responsible for the morphological transition under the conditions examined in this study will be at play in the formation of spurious grains in single crystal turbine blades including dendrite growth kinetics and heterogeneous nucleation conditions.

Table II

Growth Rate and Thermal Gradients from Regression and Isotherm Tracking Analysis

Sample	Regression		Dendrite Tip Undercooling	
	R (mm/s)	G (°C/cm)	R (mm/s)	G (°C/cm)
AM1-1	0.33-0.44	2.6-4.0	.39 ± .05	4.6 ± 2.2
AM1-2	0.19-0.33	6.4-10.8	.31 ± .03	6.9 ± 1.7
AM1-3	0.22-0.40	3.8-9.1	.23 ± .05	4.5 ± 1.3
AM1-4	0.45-0.15	0.0-2.7	.39 ± .06	5.0 ± 2.1
AM1-5	0.22-0.46	4.1-9.0	.33 ± .05	5.6 ± 1.2
MAR-1	0.33-0.60	4.7-7.2	.37 ± .04	4.3 ± 1.1
MAR-2	0.33-0.50	4.7-7.2	.39 ± .04	7.5 ± 2.3
MAR-3	0.32-0.49	4.5-7.9	.39 ± .06	6.0 ± 1.5
MAR-4	0.34-0.50	0.07-2.8	.35 ± .05	4.1 ± 2.2
MAR-5	0.28-0.45	3.3-8.1	.36 ± .04	6.7 ± 2.6

Note: MAR is an abbreviation for MAR-M200+Hf

To investigate nucleation phenomena in these two alloys, a modified Hunt[8] CET model has been developed and employed[7,9] to estimate the nucleation undercooling necessary to give rise to a CET under the growth conditions determined from the isotherm tracking analysis of the thermocouple data. As the Hunt[8] model is based on an assessment of the extended volume fraction of equiaxed grains ahead of the columnar dendritic front an estimate of the gradient in the liquid has been used instead of the previously estimated gradient at the tip. These results of the analysis are shown in Table III for nucleation populations of 4 and 24 per cm^3. These values represent the range in number of equiaxed grains per cm^3 observed in the samples.

Table III
Heterogeneous Nucleation Undercooling

Sample	Average R (mm/s)	G_{Liquid} (°C/cm)	ΔT_N (°C)
AM1, N_0=4	0.32 ± 0.05	1.60 ± 0.59	3.5 ± 0.6
AM1, N_0=24	0.32 ± 0.05	1.60 ± 0.59	3.7 ± 0.4
MAR-M200+Hf, N_0=4	0.38 ± 0.04	1.47 ± 0.39	3.5 ± 0.4
MAR-M200+Hf, N_0=24	0.38 ± 0.04	1.47 ± 0.39	3.8 ± 0.3

As is evident from the data in Table III the undercooling predicted for the two alloys is similar with the difference falling within the uncertainty. The heterogeneous nucleation undercoolings of 3.5-3.8°C predicted using this technique are only slightly higher than the thermal recalesence measured in the experiments (1.0-3.2°C). Furthermore, they are consistent with those found in the platform of directionally solidified turbine blades[10] which would explain the tendency to form spurious grains in the platform section of blades.

Summary and Conclusions

The thermal conditions leading to the CET in single crystal alloys AM1 and MAR-M200+Hf have been examined in detail using data obtained from directionally solidified samples. A number of techniques have been utilized in an attempt to estimate the growth conditions (R and G) leading to the CET from the thermocouple data. In comparison to other investigations [4] [10], the solidification conditions examined in this study in association with the CET are substantially different and have focused on higher solidification rates and lower gradients. In respect of the work of Pollock *et. al.*[4], this may well explain the difference in growth morphology and transition morphology found in the two studies. Consequently, the two studies may well be examining two different mechanisms which can give rise to a breakdown in directional growth - one operating at high growth velocities and low gradients the other a low velocities and high gradients.

Analysis of the results with a Hunt-based CET model predict heterogeneous nucleation undercoolings of the order of those found in the platform section of directionally solidified blades. Thus, it is possible that a Hunt-based CET model can be used to help predict the onset of spurious grain formation in the platform section of SX turbine blades providing data of the type measured in this investigation is available for the alloy systems under investigation.

Finally, measurements of the type made in this study are extremely difficult to make even in low temperature systems. While the results are arguably approximate from the standpoint of their uncertainty, and, their inability to be directly applied to single crystal castings, they are an important contribution as they lay the foundation for the determination of critical data needed for the development of process analysis tools for the design of casting systems.

References

[1] Mahapatra, R.B., Weinberg, F. "Columnar to Equiaxed Transition in Tin-Lead Alloys." Metallurgical Transactions B, Vol. 18B, June 1987, pp.425-432.

[2] Ziv, I., Weinberg, F. "The Columnar to Equiaxed Transition in Al 3pct Cu", Metallurgical Transactions B, Vol. 20B, Oct. 1989, pp.731-734.

[3] Hunt, J.D. and Flood, S.C. "Columnar to Equiaxed Transition", Metals Handbook, 9th Edition, Vol 15. ASM International 1988, pp.130-136.

[4] Pollock, T.M., Murphy, W.H., Goldman, E.H., Uram, D.L. and Tu, J.S., "Grain Defect Formation During Directional Solidification of Nickel Base Single Crystals", Proceedings of a Conference: Superalloys 1992. pp. 125-134, Pub. The Minerals, Metals and Materials Society, 1992.

[5] Kurz, W., Giovanola, B. and Trivedi, T.R.: Acta Metall., vol. 34 (1986), No. 5, pp. 823-30.

[6] Rappaz, M., David, S.A., Vitek, J.M. and Boatner, L.A: Metall. Trans A, vol. 21A(1990), pp.1767-82.

[7] Fernihough, J.W., Ph.D. Thesis, "The Columnar to Equiaxed Transition in Nickel Based Superalloys AM1 and MAR-M200+Hf", The University of British Columbia, 1995.

[8] Hunt, J.D., "Steady State Columnar and Equiaxed Growth of Dendrites and Eutectic", Materials Science and Engineering, vol. 65 (1984) 75-83.

[9] Cockcroft, S.L., Rappaz M., Mitchell A., Fernihough, J.W. and Schmalz, A., "An Examination of Some of the Manufacturing Problems of Large Turbine Blades for use in Land Based Gas Turbines", Materials for Advanced Power Engineering 1994, ed. D. Coutsouradis, publ. Kluwer Press, Belgium.

[10] Goldschmidt, D., Paul, U., Sahm, P.R., "Porosity Clusters and Recrystallization in Single Crystal Components', Proceedings of a Conference: Superalloys 1992. pp. 125-134, Pub. The Minerals, Metals and Materials Society, 1992.

CLOSED LOOP CONTROL TECHNIQUES FOR THE GROWTH OF SINGLE CRYSTAL TURBINE COMPONENTS

M. Eric Schlienger
Sandia National Laboratories
Albuquerque, NM

Abstract

Analysis of processes used for the production of single crystal turbine components reveals significant shortcomings. Inadequate consideration has been made of the fact the system is cooling dominated and that the amount of cooling tends to increase as the emissive cooling area expands during the process. Experimental evidence suggests that during processing, this increased cooling causes the solidification interface to move away from the baffle and become curved. The motion of the interface results in a decrease in the solidification gradient. The combination of these actions can result in variations in PDAS, grain misalignment and the production of defects. It is shown that despite this tendency, microstructural stabilization may be achieved through the use of the heat of fusion as an internal process heat source.[1]

Introduction

Directionally solidified and single crystal turbine components are seeing increased utilization in both aviation and land based power turbines[2]. Due to the complexity of the process, production of these parts is an expensive and time consuming operation. The process for any given part often entails a withdrawal cycle of several hours, during which the single crystal parts are grown. When the mold expense is combined with the processing costs associated with the casting and growth of these components, such parts represent a substantial investment as they emerge from the furnace. Therefore, casting or solidification defects which impact product yield represent a considerable expense, lowering margins and raising cost. Further, concern has been voiced that with the continued expansion of the use of DS & SX parts, the application of these parts may soon become foundry capacity limited. The combination of the capacity and expense issues provides a driving force for the development of techniques that improve process yield and throughput. It is towards this end that this work explores processing techniques and the single crystal growth process as it occurs in industrial furnaces.

Solidification Apparatus

For this work, single crystal test bars of CMSX-4 were produced in ten different process cycles under varying conditions. The test bars were produced in a new single crystal furnace manufactured by Retech Inc. and being built for Rolls Royce. The furnace is of a modified Bridgeman design[3], the solidification portion is shown schematically in Figure 1.

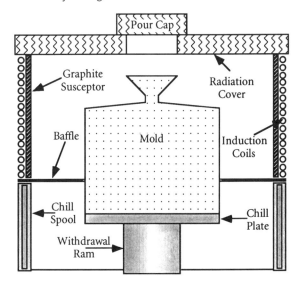

Figure 1: Typical configuration of single crystal turbine blade solidification apparatus.

During processing, metal is melted in an induction crucible. Once the metal has been heated to the appropriate temperature, the pour cap is removed and the metal is poured into the pre-heated mold. The mold was heated, prior to and during the solidification process, by an inductively heated graphite susceptor. The pour cap is replaced immediately after pouring. The mold, which is placed on a water cooled chill plate, is held in place after the pour for a span of time sufficient to allow dendritic solidification to become established as a result of the heat flux through the molten metal and into the chill plate. During this period, a crystal selector[4] inhibits slower growing dendrites such that only a single dendrite orientation is allowed to propagate through the part. After the hold time, the withdrawal process is initiated and the mold is lowered in a controlled fashion. As the mold is moved past the baffle, it is subjected to a temperature gradient intended to maintain dendritic growth and the single crystal morphology which has been established.

Current Control Techniques

Numerous articles have been written on the proper technique for the growth of single crystal turbine blades[5,6]. Perhaps the most prevalent control mechanism that has been proposed is expressed by the equation[7,8]:

$$k = \frac{G}{R} \quad (1)$$

Where k is a constant dependent on the alloy, G is the gradient at the solidification front and R is the growth rate. The ratio G/R is regularly used to describe the solidification environment. It implies that optimum crystal growth will occur when G and R are varied in such a fashion as to keep the above relationship equal to a constant. In practice this requires measuring the gradient and solidification rate on a continuous basis. Past implementations have made the simplifying assumption that the solidification rate is equivalent to the withdrawal rate. Unfortunately this is only true in a macroscopic sense and does not take into account the effects of the rapid geometrical changes that are typical of turbine blades.

A more typical process involves a mold heater which is held at a constant temperature, and the parts are withdrawn from the hot zone of the furnace to the cold, at a constant velocity. A variation of this technique involves modification of the withdrawal velocity to accommodate geometrical features of the part.

Analysis of G/R Techniques

To evaluate the basis of (1), the solidification phenomena that occur at the liquid solid interface must be examined. After the work of Kurz and Fisher[9], Fick's second law may be re-written for a steady state, moving reference frame as:

$$\frac{\partial^2 C}{\partial z^2} = \frac{-V}{D} \frac{\partial C}{\partial z} \quad (2)$$

Where C is the solute concentration in the liquid, z is the distance in front of the solidification interface, V is the solidification velocity and D is the solute diffusivity in the liquid. The solution to this differential equation is of the form:

$$C = A + Be^{-\left(\frac{Vz}{D}\right)} \quad (3)$$

In order to evaluate the boundary conditions which will yield a complete solution, a further simplification is made. For the purposes of this analysis, it is assumed that the diffusion of solute in the solid material is very slow when compared to the solidification rate and the liquid diffusion rate. In fact, the difference in the solute diffusion rates of the liquid to the solid is typically on the order of 3 to 4 orders of magnitude and therefore for the purpose of this analysis it may be assumed that there is no compositional variation in the solid past the initial transient. With this assumption in mind, the solution is therefore only valid for positive values of z, that is to say, within the liquid.

In order to obtain the complete solution, A and B in (3) must be evaluated. This is achieved by examination of the boundary conditions. Assuming eutectic solidification, and liquid of nominal composition C_0, steady state solidification must ultimately produce a solid of the same composition. As such, a solid of composition C_0 must be forming from a liquid of composition C_0/k. There is then a boundary layer in advance of the interface into which solute is being rejected such that ultimately, at the solidification interface itself the composition of the liquid is C_0/k. Evaluation of the boundary conditions yields the complete expression for the composition of the liquid as a function of the distance in front of the solidification interface:

$$C = C_0 + \left(\frac{C_0}{k} - C_0\right)e^{-\left(\frac{Vz}{D}\right)} \quad (4)$$

Equation 4 implies that there is an exponentially decreasing solute concentration gradient in advance of the solidification interface. If the liquidus temperature is assumed to decrease with composition, then the liquidus temperature of the alloy within the diffusion boundary layer increases in a logarithmic fashion at positions away from the solidification interface.

For solidification to occur, heat must be extracted from the system. In the case of the growth of single crystals, as described, the heat flux is in the -z direction. Since the alloy has a finite thermal conductivity, this flux creates a gradient in the material through which it flows. It is possible that for the liquid in advance of the solidification front, the liquidus temperature, due to the compositional variation, may increase more rapidly than the liquid metal temperature increase that occurs as a result of the temperature gradient. This situation results in a region of liquid metal which is undercooled with respect to the local liquidus temperature in front of the solidification interface. This phenomena is termed constitutional undercooling. Convex perturbations of the solid / liquid interface into the molten metal can extend rapidly under these conditions. This extension occurs because the perturbation is growing into a region where the interface is increasingly undercooled with respect to the local liquidus temperature. Therefore if a planar interface is assumed, then in order to ensure that a planar interface remains stable, the temperature gradient in the region of the solidification front must be sufficiently high to preclude any constitutional undercooling.

In order to evaluate the temperature gradient which must be maintained to preclude interface instabilities, it is first necessary to obtain an expression for the liquidus temperature in advance of the interface. It may be shown that the expression for the liquidus temperature in advance of the interface is:

$$T_L = T_{L(C_0)} - \left(T_{L(C_0)} - T_{S(C_0)}\right)e^{-\left(\frac{Vz}{D}\right)} \quad (5)$$

In addition, the expression for the temperature of the molten metal in advance of the interface, T_q, is a function of the gradient and written as:

$$T_q = T_{s(C_0)} + Gz \qquad (6)$$

Equation (7), which is the difference between (5) and (6), therefore describes the undercooling of the liquid metal as a function of the distance in advance of the interface.

$$T_U = \left(T_{L(C_0)} - T_{s(C_0)}\right)\left(1 - e^{-\left(\frac{Vz}{D}\right)}\right) - Gz \qquad (7)$$

It is reasonable to argue that if there is no position in advance of the interface where the liquid metal is undercooled, then a planar interface will be stable. If (7) is differentiated the result represents the slope of the undercooling function; if at all times this slope is less than zero, then none of the liquid will be undercooled.

$$\frac{dT_U}{dz} = \frac{\left(T_{L(C_0)} - T_{s(C_0)}\right)V}{D} e^{-\left(\frac{Vz}{D}\right)} - G \qquad (8)$$

The exponential in (8) has its maximum value at z = 0. It may therefore be written that the condition for the stability of a planar interface is:

$$G \geq \frac{\left(T_{L(C_0)} - T_{s(C_0)}\right)V}{D} \qquad (9)$$

This necessarily neglects the fact that the suppression of the melting point that occurs due to the curvature of an interface can have a local stabilizing effect that will allow some perturbations to exist[10]. however for practical processing conditions, such perturbations are not significant. This equation may be re-written by combining the alloy dependent values of T_L and T_S into a single constant K. The result, equation (10), is in fact the same as (1) with the only difference being that the interface velocity V has been used instead of the solidification rate R.

$$K = \frac{G}{V} \qquad (10)$$

As an example, the gradient required to insure a planar solidification in an Aluminum alloy with 2.5 atomic percent copper is calculated. This system is used because it is well characterized. Given a diffusivity of .003 mm²/s, a difference between liquidus and solidus of 100°K, and a solidification rate of .085 mm/s, then the gradient necessary to achieve a planar interface is approximately 2800°K/mm. Gradients of this magnitude are not typically achievable in most equipment that is available for the commercial production of single crystal turbine blades.

If, on the other hand, the solidification environment were to be controlled such that the dimensions of the dendrites are maintained at a length roughly equivalent to the diffusion boundary layer, then maximum utilization of the zone of constitutional undercooling could be achieved. Given that there is a composition gradient in advance of the solidification interface, it may be said that to a gross approximation the shortest dendrite that could be obtained can be no shorter than the diffusion boundary layer thickness. If an equivalent diffusion boundary layer thickness, δ_C, is chosen such it has a constant concentration gradient and the same total solute content as the infinite layer expressed by the exponential. Then the diffusion boundary layer thickness, δ_c, is given by:

$$\delta_c = \frac{2D}{V} \qquad (11)$$

If the assumption is made that the dendrite tips are at the liquidus temperature and that the roots are at the solidus temperature of the alloy, then the gradient may be expressed as:

$$G = \frac{\left(T_{L(C_0)} - T_{s(C_0)}\right)}{\delta_c} \qquad (12)$$

Substituting in (11) yields:

$$G = \frac{V\left(T_{L(C_0)} - T_{s(C_0)}\right)}{2D} \qquad (13)$$

which is but a factor of 2 different than the planar interface solution of (9). Evaluation of (13) with the same physical parameters as before requires a gradient in the neighborhood of 1400°K/mm, which again represents a somewhat challenging proposition. If a more reasonable gradient typical of production equipment where used to evaluate (13) the results are not acceptable. For example, if a gradient of 10°K/mm were assumed, a typical part would require 36 hours to produce. Similar parts are regularly produced in today's production equipment in about 1.5 hours.

In production, G/R techniques are limited by many of the same considerations as constant velocity techniques. In particular, material limitations preclude any significant variation in temperature and as such G is difficult to control. Further, while R may be effected as a result of changes in the withdrawal velocity, it will be shown that holding G/R constant results in a variation in withdrawal rate which is opposite to that required to produce a consistent microstructure.

Given that G/R control schemes are based on the assumption that the solidification can take place in an environment where the dimension of the diffusion boundary layer is significant when compared to the dendrite length, it is unlikely that any such scheme may be implemented in a production environment. A better approach would be an attempt at maintaining a constant microstructure, thus insuring a uniform heat treat response. As such, attempts at controlling the primary dendrite arm spacing would provide a better control scheme. It has been shown[11], and verified experimentally[12] that the PDAS follows:

$$\lambda_p = K_p G^{-.5} V^{-.25} \qquad (14)$$

Where λ_p is the primary dendrite arm spacing, K_p is a correlation constant which is based upon the solidification characteristics of the alloy, G is the gradient, and V is the solidification velocity. This equation implies that a more effective control scheme, applicable to a production environment, could be achieved through a control algorithm of the form:

$$K = G^2 V \qquad (15)$$

It should be noted that using this criteria, a decrease in G would be met with an increase in V, a result which is exactly opposite to the G/R case. No reported attempts at such a control scheme have been found.

Constant Velocity Techniques

Within some foundries, the production of single crystal turbine components may include several erroneous assumptions. The first of these is the assumption that the solidification interface is located at the baffle between the hot and cold zone of the furnace. The fallacy of such an assumption is easily understood.

Within the vacuum environment of a single crystal furnace, the predominant heat transfer mode is emissive. Given a ΔT between the heat source and sink, the flux will be proportional to ΔT^4. The total flux is dependent on the surface areas of the radiator and sink. In production furnaces the mold heater temperature and area are constant. In a similar fashion, the cold zone temperature and area are constant. However, the surface area of the part residing within the hot and cold zone respectively, changes throughout the withdrawal process[13]. Given that a flux balance must exist, and considering the very much greater ΔT which exists below the baffle, then as a greater surface area is pulled below the baffle a commensurate increase in flux into the part must occur above the baffle. Such an increase may only be accommodated by a decrease in the average temperature of the part of the mold that is above the baffle. This has several ramifications and has been born out with finite element modeling utilizing an emissive heat transfer code called "Coyote" developed by Dave Gartling of Sandia National Laboratories, Division 1511.

First, consider that there is a heat flux through the part. This flux results in a temperature gradient across the part. The top of the part is held at a temperature which is roughly that of the mold heater. Therefore, the only manner in which the average temperature of that portion of the mold which is above the baffle may be lowered manifests itself as a decrease in the mold temperature above the baffle. Associated with this change in temperature must be a variation of the location of the solidification interface relative to the baffle. Therefore, under conditions of constant withdrawal rate, the location of the solidification interface moves up in the furnace relative to the baffle. In addition, due to the ΔT^4 nature of emissive heat transfer, this action results in an ever increasing flux into the portion of the mold below the solidification interface. Since the maximum flux through the part occurs at the baffle, the motion of the interface away from the baffle results in a situation where an ever increasing percentage of that flux is entering the part below the solidification interface, and therefore, the gradient associated with the solidification environment continuously decreases, and as a result so does the PDAS within the part. This has been verified both experimentally and with the model.

Unfortunately material limitations associated with the mold and the vaporization of volatile alloying constituents, precludes an increase in the temperature of the furnace hot zone. As a result increasing the heat flux through an increase in the ΔT of the hot portion of the mold is not feasible.

Process Considerations

As described above, the solidification interface moves as a result of the changing emissive surface areas above and below the baffle. The motion of this interface away from the baffle not only serves to decrease the solidification gradient, but also results in a curved morphology at the solidification interface. This curvature is the result of non vertical components of the heat flux and is best understood by considering the case of a cylindrical bar.

Interface Morphology

A cylindrical bar, solidified within a standard furnace environment will have a solidification interface whose position changes as the withdrawal progress progresses. Initially the interface may be driven slightly below the baffle, however ultimately, the interface will reside above the baffle. Symmetry suggest that no matter what the location of the solidification interface to the baffle, the heat flux at the center of the bar must be strictly vertical. However, if the surface temperature of the bar differs from that of the integrated emissive view, then a horizontal component of heat flux must exist at the part surface. Therefore, at all locations except the baffle, flux lines must exhibit an increasing curvature as a function of the distance from the center of the part. As a result, an isothermal plot would show curved isotherms. This then implies that as the solidification interface moves away from the baffle, it develops a curved morphology[14]. If the interface is below the baffle, the center of the part is hotter than the edge and the interface is concave. Similarly, if the interface is above the baffle, then the center of the part is cooler and the interface is convex. It should be noted that such curvature has been predicted by the model and experimentally verified with BEKP (to be described later) on the solidified parts.

Since the crystallographic growth occurs against flux lines and perpendicular to the isotherms, it should be apparent that the crystallographic misorientation, which is commonly seen to increase as the process progresses, is an indication that the interface has moved and become curved. Further, visualization of these interface curvatures leads one to suspect that when the interface is driven below the baffle, sliver grains[15] are more likely to nucleate, whereas when the interface is above the baffle and the gradients are dropping, dendrite tips which are broken off as a result of solute convection jets[16] are more likely to impact the mold wall and stabilize, resulting in freckle chains[17].

Process Improvement

The previous discussions are intended to illustrate that there is room for improvement in many current practices for the production of single crystal turbine components.

The process occurs under conditions where G/R arguments are questionable and the process is driven by the cooling. The changing geometrical view factors can be shown to result in an increase in PDAS, growth misalignment, and the potential for an increased incidence of defects. The most obvious solution to these concerns is to find a mechanism whereby the solidification interface may be maintained at the baffle. This objective could be accomplished by reducing the effectiveness of the cold portion of the furnace. Such a reduction could be accomplished by a movable reflective insulator, used to shield the chill spool as the area for emissive heat input above the baffle decreases. Alternatively the addition of an additional heat flux to the system could serve to stabilize the location of the solidification interface; however, as previously mentioned, significant increases in the mold heater temperature are not an option. Fortunately, there is contained within the process, a source of heat which may be utilized.

Utilization of Heat of Fusion

Consider a test bar, 1.25 cm in diameter. If this test bar is solidifying under a gradient of 10°K/cm and has properties similar to those of Nickel with a thermal conductivity of 1.58 Watts / cm °K then the heat through the part is roughly 19.4 watts. If the bar is solidifying at a rate of .006 cm/s and has a heat of fusion of 71.5 cal / gm and a liquid density of 7.9 gm / cc then the heat released as a result of solidification is approximately 4.2 cal/sec or 17.5 watts. Clearly the heat of solidification (17.5 watts), when compared to the heat necessary to produce the gradient (19.4 watts), is significant. The implication is that heat of fusion may be used as an internal heat source for the stabilization of the solidification environment.

Experimental Procedure

The test bar molds utilized for these experiments were provided by Rolls Royce. Furnace parameters were recorded throughout the process. Each of the ten different molds used for these trials was comprised of 8 individual test bars. Among the parameters varied were, mold pre-heat temperature, pour temperature and withdrawal mode (closed loop or constant velocity). For this paper, only the effects of withdrawal mode, and their implications on process control will be discussed. On each mold, one test bar included thermocouple wells which allowed junction placement at the center of the test bar. This feature also insured accurate knowledge of the location of the junction. In this fashion, the test bars were instrumented with as many as 12 thermocouples. Preliminary work indicated that all of the test bars on a given mold were subjected to the same solidification conditions. Subsequent to the casting and growth, representative test bars from each mold were sectioned into a total of 287 transverse samples. Each sample was polished, then etched and a digital optical micrograph was obtained. The location along the test bar from which the polished samples were taken was carefully recorded. Image analysis software was developed to automatically count dendrites and determine the average primary dendrite arm spacing. Due to the direct impact on heat treatment response and the desire to evaluate solidification conditions during the process, primary dendrite arm spacing measurements were chosen as a metric of process response and overall quality.

Image Analysis

In order to accurately determine the dendrite statistics, a computer program was written to process the digital images and determine the location of the dendrite cores. Once these locations were established, it became a simple matter to ascertain the dendrite statistics. The digital images were stored as a 640 x 480 array of 8 bit gray scale pixels. These images were read into memory and stored as 16 bit integer numbers for ease of processing. The value of the individual pixels represented the brightness of the image. As there was a sharp demarcation in brightness between the dendritic and interdendritic regions it was not difficult to differentiate between the two types of material.

CMSX-4, the alloy used for these experiments, is cubic in nature. The cubic structure is apparent in the cross-sectional micrographs of the test bars. In many of the test bars, the primary dendrites are aligned into fairly regular arrays. If rectangular sampling areas are used, this regular alignment of primary dendrites, when coupled with the dendrite spacing calculation, brings up the possibility of aliasing. Aliasing occurs when the boundaries of the sampling area are not exactly equivalent to an integral number of dendrite spacings. The result of this situation is realized by envisioning a rectangular area sliding across a regular array of dendrites. A some point, a new column of dendrites will slide into the sampling area. However, if a column does not simultaneously slide out the other side, then the number of dendrites within the sampling area has increased. This occurs despite the fact that the test sample maintains a constant dendrite spacing. The same possibility occurs in the case of vertical displacement of the sampling area. As such, aliasing can contribute to significant errors in the calculation of PDAS and the question of how to determine the spacing of primary dendrite arms needed to be addressed. However, aliasing concerns may be eliminated by uitilizing a circular sampling area.

The method used to calculate the primary dendrite arm spacing allocates each dendrite counted an equivalent area of image. As the crystal is cubic, the area allocated to each dendrite is assumed to be in the shape of a square. Since each dendrite is allocated a square area, then if the dendrites were in a regular array the distance between them would be the square root of the allocated area. The circular sampling area chosen is 464 pixels in diameter (D) and is in the center of the image. The diameter of the circle is then converted into microns using the image scale factor (S). The number dendrites within this circle (N) is determined and the area of the circle (A) is calculated. The primary dendrite arms spacing μ, is then calculated using equation (16).

$$\mu = \sqrt{\frac{\pi (DS)^2}{4N}} \qquad (16)$$

The counting software was written under the assumption that the vertical and horizontal axes on the computer screen would correspond to the <001> directions of the secondary dendrite arms. This was not a requirement, but when combined with some care in the photography, allowed significant simplification of the dendrite recognition routine.

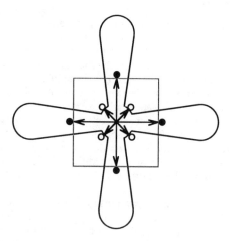

Figure 2: Dendrite Location Algorithm

Figure 2 is an illustration of a primary dendrite which will be used to describe the dendrite counting software. The program scans the rows of pixels sequentially from left to right, top to bottom. The software checks to see if the pixel currently addressed is within the sample circle, if it is not, the next pixel is checked. If the current pixel is within the sample circle then the image is interrogated at four points, each located a fixed displacement along the directions of the secondary arms. These points are shown in figure 2 as filled circles and are located at the points where the horizontal and vertical arrows impinge on the dashed square. If any of these four points possesses a brightness that is less than the brightness threshold for a dendrite, the program steps to the next pixel in the row. If all four points are brighter than the dendrite brightness level, then the software scans from the left point to the right and from the top point to the bottom. If at any time during these scans a pixel is detected with a brightness level that is lower than the dendrite brightness threshold, the program steps to the next pixel in the row. After the above procedure has occurred, the software then proceeds to verify if the location is a dendrite core. To accomplish this the program begins at the current pixel and tests outward at four points at angles 45° between the secondary arm directions. These point are indicated in figure 2 as open circles. The distance from the current pixel to these points is an adjustable parameter and may be entered by the user or automatically determined by the software. If all of these four points are of a brightness that is lower than the interdentritic brightness threshold, then the current pixel is considered to be near the core of a dendrite.

Once a potential dendrite location has been identified, four diagonal scans are made and the distance from the center dendritic material to the interdendritic boundary is determined for each scan. From these four distances an average distance is calculated. The average distance is then subtracted from the actual distances and each of these four results are squared and then summed together. If the current pixel is considered to be the center of the dendrite core, then the resultant number is a measure of the asymmetry of the dendrite.

The location of the current pixel is then compared to any dendrites which may have been found previously. If the dendrite core as specified by the current pixel is within twice the primary search distance (twice the length of a side of the dashed rectangle of figure 2), of a dendrite which was found previously, then the two are considered to be the same dendrite. In such a situation, the asymmetry of the dendrites at the two core locations are compared and the location and asymmetry value of the core location with the lowest asymmetry is retained within the list of found dendrites.

After the computer has found all locations which match the selection criteria of a dendrite core, the user is then able to add or subtract additional locations. This feature proved useful, but in general, the errors made by the computer were on the order of one to two percent of the total count.

Once all dendrite locations are identified, the statistics are calculated. The primary dendrite arm spacing is determined as previously described. In addition the standard deviation in the primary dendrite arms spacing is also calculated.

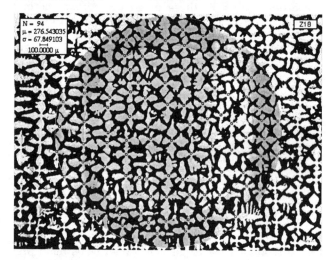

Figure 3: Example

Figure 3 is an example of a processed micrograph. The dendrite cores, as located by the software, are indicated by small circles. The identifier in the upper right hand corner identifies the test bar, in this case "Z" and the slice number, "18". The box in the upper left hand corner displays the number of dendrites "N", the primary dendrite arm spacing "μ", the standard deviation in the spacing σ, and a micron marker. This date was additionally written to a file for subsequent processing.

Data Reduction

Given the temperature and location of each thermocouple along the part, least squares techniques can be used to generate equations for both temperature and location, each as a function of the other. In this fashion, thermocouple data was processed continuously throughout each withdraw cycle yielding equations in the form of (17) and (18) below.

$$T = f^3(x) \quad (17)$$

$$x = f^3(T) \quad (18)$$

Thermocouples with values nearest the solidification temperature were weighted more heavily than those with temperatures farther away from the solidification band. This procedure increased the confidence that the resultant curve was representative throughout the solidification region. Once equation (17) had been obtained, it was differentiated with respect to x; the result being equation (19), an expression for the temperature gradient as a function of location along the part.

$$G = \frac{\partial T}{\partial x} = \frac{\partial (f^3(x))}{\partial x} \quad (19)$$

Due to the dynamic nature of the solidification process, these equations were reformulated several times a minute. Given (18) and (19), a real time measure of the solidification environment is possible.

In order to characterize the solidification conditions, a temperature, chosen to be half way between the liquidus and solidus, is entered into equation (18). The result is the position along the test bar of the center of the solidification band. This location, when combined with the distance that the part has been withdrawn, allows the determination of the location of the solidification interface with respect to the baffle. The location obtained from (18) is also used in conjunction with (19) to obtain the gradient at the solidification interface.

These data were combined with the dendrite measurements to provide information regarding the PDAS and the associated gradient and solidification rate for each slice.

Under conditions of constant withdrawal velocity, data analysis revealed that the motion and characteristics of the solidification interface vary significantly throughout the process. In addition to the interface position, the gradient within the part, around the solidification interface, was also seen to change during the growth process. Analysis of the thermocouple data revealed that the solidification gradient is a function of the position of the solidification interface relative to the baffle. Further, the motion of the solidification interface within the furnace environment results in solidification rates which vary as well. Hence, within this process, the physical environment results in solidification conditions which become a location dependent function. When solidification conditions for each slice (as expressed by $1/G^{.5}R^{.25}$) are plotted vs. the measured primary dendrite arm spacing (PDAS), a best fit equation (20) results.

$$PDAS = 231 + \frac{330}{G^{.5}R^{.25}} \quad (20)$$

Utilizing (20) and the gradient and rate information obtained from the thermocouple data as well as the withdrawal position, it is possible to predict the PDAS as a function of position along the test bar. Post procesing of the thermocouple and withdrawal position data resulted in 5th order polynomials for both the solidification gradiant and solidification rate as a function of position along the bar. These equations were substituted into (20) for G & R an the results are plotted in figure 4. (It should be noted that for these experiments a nominal pour temperatures of 1500°C and a susceptor temperatures of 1475°C were utilized.)

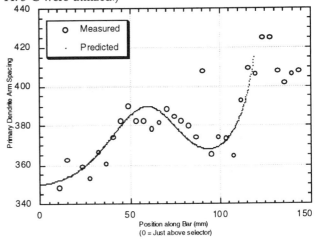

Figure 4: Measured and Predicted PDAS as a Function of Test Bar Position.

Figure 4 exhibits a good correlation between the calculated and experimental PDAS data and indicates that the relationship for PDAS as expressed in (14) is in fact valid. It may therefore be inferred that there exists a set of measurable furnace parameters capable of predicting at a minimum, the variation in PDAS that may be occurring during the process. This data also implies that thermocouple data may be effectively used as a means of determining the location of the solidification interface and solidification rate in real time.

It should be noted that the peak in PDAS occurs at the point where the solidification interface moved to its "setpoint position" of 7mm above the baffle. The subsequent reduction in PDAS is due to the controller action. The final increase in PDAS occurs as a result of the solidification process moving beyond the usefulness of the thermocouple array. Optimization and improved process tuning should mitigate this effect.

Interface Morphology Measurements

As previously stated, variations in solidification conditions are due to the changing emissive environment that occurs as the ratio of mold surface areas exposed to heating and cooling conditions changes. This results in varying heat fluxes across the solidification interface and as the process progresses, promotes the motion of the solidification interface up into the hot zone of the furnace.

As the solidification interface moves away from the baffle between the hot and cold zone of the furnace, it develops a curved morphology due to the increasing horizontal components of the heat flux. This curved interface morphology results in a variation of the crystallographic orientation of the part and, in severe cases, is cause for rejection. Such curvature was verified using Backscattered Electron Kikuchi Patterns (BEKP) in the SEM. These patterns arise when the electron beam within an SEM is focused on a single spot. Electrons, which are reflected back out of the surface, form diffraction patterns characteristic of the crystalline structure and its orientation. BEKPs were used to chart the orientation shift which occurred across the solidified test bar sections. Since the preferred growth is in the <0 0 1> direction, and dendrite growth is perpendicular to the solidification isotherms, then by measuring the angle that the <0 0 1> crystallographic direction makes with the sample surface, the shape of the solidification interface is revealed (assuming steady state conditions). Such measurements revealed both a curved interface and an asymmetry that was probably due to the uneven emissive environment that results from the presence of the mold tree. The degree of curvature was also seen to increase as the process progressed. As mentioned, the combination of this curvature and extended dendrites which may occur as a result of the lower gradients could be responsible for the formation of freckle chains and serve as an additional driving force for the development of real time process control.

Process Stabilization

In order to gain better control of the process, it was proposed that an increase in the withdrawal rate, must be (to a point) balanced by an increase in the solidification rate. This action results in an associated increase in the heat released as a result of solidification. This heat is sufficient to change the fluxes through the part and thereby lower the position of the solidification interface while increasing the gradient. These coupled phenomena may be effectively used as a method of microstructural stabilization in a closed loop control scheme while simultaneously improving crystallographic orientation.

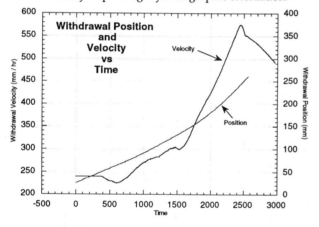

Figure 5: Closed Loop Control Response

For the closed loop process trial illustrated in Figure 5, a technique was developed which maintained the position of the solidification interface at the baffle. Utilizing the derived least squares polynomial equation for position as a function of temperature (18), and coupling it with the withdrawal distance, the location of the solidification interface relative to the baffle may be determined. During the growth process, the velocity was continuously adjusted in an attempt to maintain the center of solidification band at the baffle. In this fashion, an increase in the solidification rate resulted in enough excess heat to effectively drive the equilibrium position of the solidification interface down into regions of higher gradient and more vertical heat flux. Such a scheme ensures that the solidification zone is maintained at the highest achievable gradient within the available thermal environment. As may be seen from the figure, the technique results in an ever increasing withdrawal velocity. Such a result is expected, as the heat released on solidification must be of sufficient magnitude to overcome the heat losses associated with an ever increasing emissive radiator below the baffle. It is of interest to note that the final velocity utilized is well beyond that which is typically employed for this process, and yet good crystals were achieved.

Although this method does not explicitly control primary dendrite arm spacing, it does provide for a flatter interface which will result in fewer orientation defects. Further, utilization of this technique does result in solidification conditions which begin to approximate the previously mentioned $K = G^2V$ condition. In addition, the positioning of the solidification zone at the location of highest gradient should minimize those defects which may be associated with a low gradient environment. It may therefore be inferred that by utilizing the heat of fusion as an internal heat source, the effect of increasing cooling is offset and a more regular microstructure should result. Experimental results reveal that these expectations are in fact valid.

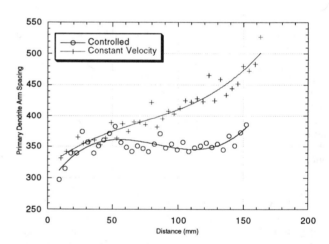

Figure 6: PDAS Measurements for Constant Velocity & Closed Loop Processing vs. Test Bar Location

Figure 6 is a comparison of the PDAS results for two process runs. The associated test bars utilized alloy from the same heat and in the same furnace, with identical pour rates, metal pour temperatures, and mold heater temperatures. Both molds were initially processed at a withdrawal rate of 240 mm/hr. However, during the closed loop trial a computer varied the withdrawal rate as described above and illustrated in figure 5. As Figure 6 demonstrates, a significant stabilization of the PDAS may be achieved as a result of real time closed loop control techniques.

Conclusion

It has been shown that the changing emissive environment within today's production single crystal furnaces results in a gradient of decreasing magnitude that moves away from the baffle as the withdrawal process progresses. It has been additionally demonstrated that the heat of fusion may be used as an internal heat source to counteract this effect. As such, and contrary to published G/R schemes, the decrease in gradient which occurs as the process progresses should be counteracted with an increase in withdrawal rate. Such a rate increase is consistent with the expression for primary dendrite arm spacing:

$$\lambda_p = K_p G^{-.5} V^{-.25}$$

and experimental results have demonstrated that this results in a stabilization of the microstructure. Commensurate with the microstructural stabilization is a flattening of the solidification interface morphology which occurs as a result of the more vertical heat flux in the vicinity of the baffle. This condition results in a reduction of orientation misalignment between the <001> and the stress axis of the part. Finally, these conditions should reduce defects such as freckles by maintaining the optimum solidification environment achievable for the configuration.

In practice, a fully instrumented mold would be subjected to a closed loop scheme which maintains the solidification interface at the baffle. The resultant velocity profile would be recorded and played back for non-instrumented parts. It is expected that final velocities may be well above what is normally considered prudent.

As a side bar, it should be noted that an increase in the amount of metal above the useful portion of the part can to some degree mitigate the effect of increased cooling. The flip side of this statement is that cost reduction efforts associated with reduced pour volumes may have a negative impact on process yields unless process modifications, such as those described herein, are implemented.

Acknowledgments

This work is supported by the U.S. Department of Energy under contract number DE-AC04-94AL85000.
The author would like to acknowledge Retech Inc. and Rolls Royce for their support in this effort. In particular Alan Moulden, Alan Patrick and Ray Snider of Rolls Royce are recognized for their support and insight, Thanks are also extended to Joe Michael and Randy Schunk of Sandia National Labs for their support and assistance with the BEKP and modeling efforts respectively.

---

[1] M.E. Schlienger, "Understanding and Development of Advanced Techniques for the Processing of Single Crystal Turbine Components", (Ph.D. Thesis, Oregon Graduate Institute of Science and Technology, February 1995).

[2] VerSnyder, F.L. and Guard, R.W., "Directional Grain Structures fort High Temperature Strength", ASM Transactions, 52, 1959.

[3] Erickson, J.S., Oxczarski, W.A. and Curran, M.C., "Process Speeds Up Directional Solidification", Metal Progress, March, (1971)

[4]. Higginbothom, G.J.S, "From Research to Cost Effective Directional Solidification and Single Crystal Production - An Integrated Approach", Materials and Design, 8 (1987)

[5] Cole, G.S. and Cremesio R.S., "Solidification and Structure Control in Superalloys", The Superalloys, (New York, NY: John Wiley and Sons, 1972) 479-508.

[6] Staub, F. and Walser B., "An Alternative Process for the Manufacture of Single Crystal Gas Turbine Blades", Sulzer Technical Review, March, (1988), 11-16.

[7] Yoshinari A., Morimoto s. and Kodama H., "Single Crystal Growth Technology of Nickel Base Alloys", Proceedings 1st Intl. SAMPE Symposium, Nov 28, - Dec 1, 1989

[8] Nakagawa Y.G. "Development of Solidification Technology for Superalloys in Japan" Superalloys (1985) Japan-US Seminar

[9] Kurz W. and Fisher D.: Fundamentals of Solidification, (Aedermannsdorf, Switzerland, Trans Tech Publications, 1986), 157

[10] Kurz W. and Fisher D.: Fundamentals of Solidification, (Aedermannsdorf, Switzerland, Trans Tech Publications, 1986), 192

[11] Kurz W. and Fisher D.: Fundamentals of Solidification, (Aedermannsdorf, Switzerland, Trans Tech Publications, 1986), 213

[12] Wills V.A. and McCartney D.G. "A Comparative Study of Solidification Features in Nickel-Base Superalloys: Microstructural Evolution and Microsegregation", Mat. Sci. and Eng., A145(1991), 223-232.

[13] Hediger F. and Sahm, P.R., "Simulation of DS and SC Processes and Their Experimental Verification", Modeling of Casting and Welding Processes IV, (Warrendale, PA, TMS Publications, 1988), 645-657.

[14] Morimoto, S., Yoshinari, A., Niyama, E., "Effects of Thermal Variables on the Growth of Single Crystals of Ni-Base Superalloys", Superalloys 1984, (Warrendale, PA, TMS Publications, 1984), 177-184

[15] Pollock T.M., et al. "Grain Defect Formation During Directional Solidification of Superalloys", Superalloys 1992, (Warrendale, PA, TMS Publications, 1992), 125 - 134

[16] Copley, S.M., et al., "The Origin of Freckles in Unidirectionally Solidified Castings", Metallurgical Transactions, 1(8), (1970), 2193-2204.

[17] Giamei, A.F. and Kear, B.H., "On the Nature of Freckles in Nickel-Base Superalloys", Metallurgical Transactions, 1(8), (1970), 2185-2192.

AUTONOMOUS DIRECTIONAL SOLIDIFICATION (ADS), A NOVEL CASTING TECHNIQUE FOR SINGLE CRYSTAL COMPONENTS

I.A. Wagner and P.R. Sahm

Gießerei-Institut, RWTH Aachen

Intzestraße 5, 52072 Aachen, Germany

Abstract

Autonomous directional solidification (ADS) for producing single crystal turbine blades utilizes the peculiarity of certain Ni-base superalloy-ceramic shell combinations to highly undercool. The single crystal directional solidification process is initiated by nucleation at typically 20 K undercooling temperatures. The results obtained from work with cylindrical bars delivered the basis for the successful single crystal solidification of solid and hollow aero-engine blades up to 18 cm in length. First trials of casting large turbine blades for application in stationary gas turbines have been done, accompanied by a numerical process simulation.

Compared to the conventional techniques, the essential advantages of an optimized ADS process are a shorter processing time and a finer microstructure, enabling a shorter duration of heat treatment.

Introduction

The casting of single crystal (SC) turbine blades for aero-engine application is widely used today [1]. The conventional technique is well developed but not optimized into the casting of large turbine blades for application in stationary gas turbines [2,3]. In addition, new alloys have been developed [4]. Due to the limitations of the conventional technique and the enormous production costs, alternative processes attract special interest.

At the Foundry-Institute of the Technical University of Aachen, the Autonomous Directional Solidification (ADS) process has been developed. The process is based on the peculiarity to undercool the superalloy in the ceramic shell mold to a certain level, leading to a rapid solidification after nucleation at the bottom of the mold. It was first presented in 1981 [5]. The authors then obtained several directionally solidified turbine blades.

Subsequent work in this area resulted in the development of a shell mold system which enabled efficient thermal undercooling of several Ni-base superalloys [6]. The quality of the single crystals was investigated using various techniques. A γ-ray diffraction analysis of the specimen showed an excellent quality of the single crystal microstructure [7]. The distortions of the main crystallographic orientation measured less than 5°.

These results are the basis for the current and ongoing research on the process engineering, characterization and control of the dynamic solidification technique ADS.

Experimental

Shell mold

Two different ceramic system approaches have been followed. The basic material was always alumina with the binders either AlOOH or silicasol. The shell molds were produced by standard investment casting procedures. The thickness of the shell molds was about 6 mm. The assemblies were dewaxed in a steam autoclave and fired at temperatures between 1200°C and 1450°C. After preheating the ceramic shell mold within the heating device, the separately melted superalloy was poured into the shell mold. The heater was then switched off, see Figure 1 for the basic set-up.

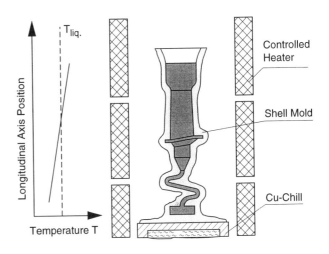

Figure 1: Schematic illustration of the ADS process with a typical temperature distribution. Thermal undercooling of the melt is used to accelerate the solidification.

Numerous experiments have been carried out in an attempt to investigate the microstructure as a function of the process parameters, experimental set-up, geometry and size of the specimen.

Cylindrical specimens

Single bars of 15 mm in diameter and 20 cm in length were cast for fundamental investigations. Temperatures were measured using precisely positioned Pt-Rh30/Pt-Rh6 thermocouples. The wires which were 0.2 mm in diameter were protected by high purity alumina tubes. The temperature distribution during solidification was measured along the vertical axis of the specimens. Two different furnaces suitable for directional solidification control were used.

A Bridgman-type furnace of industrial scale was used for most of the experimental runs to investigate the general process parameters and the resulting solidification behavior in detail. Up to 20 thermocouples could be accommodated per run. Additionally, a laboratory type liquid metal cooling (LMC) furnace, based on an old idea of P.R. Sahm and M. Lorenz [8], was used for preparing the cylindrical specimens, Figure 2. Two thermocouples were positioned in the lower part of the specimens through the bottom closed shell mold. Here, after recalescence, a liquid Ga-In bath was lifted up in order to accelerate and control the directional solidification of the interdendritic melt. The experiments were carried out in an argon atmosphere using the alloys CMSX-6 and recently CMSX-4, see Table I.

Secondary dendrite spacing and microporosity were determined to characterize the microstructure.

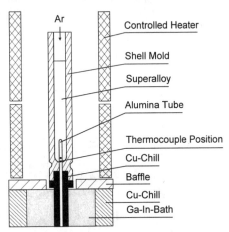

Figure 2: Schematic of the LMC furnace. The bottom closed shell mold is mounted on a copper chill plate. The Ga-In bath can be uplifted, together with the furnace.

Turbine Blades

In order to investigate their solidification behavior and microstructure, a number of turbine blades, similar to the blades for aero-engine applications, were cast using the industrial type furnace. Two blades were cast in each mold; assembled around a central sprue. The shell molds were additionally insulated with alumina material especially at critical corners of the turbine blades in order to influence the solidification rate. Most of the blades were solid, without cores. But also hollow blades, using silica cores, were produced. In most cases the CMSX-6 alloy was used. In order to also assess the effect of alloy composition same investigations were carried out using the CMSX-4 alloy, Table I.

Table I Nominal composition of the alloys used in the present work

Alloy	Cr	Co	Mo	W	Ta	Ti	Al	Hf	Re	Ni
CMSX-6	9.8	5.0	3.0	-	2.0	4.7	4.8	0.1	-	Bal.
CMSX-4	6.4	9.6	0.6	6.4	6.5	1.0	5.6	0.1	2.9	Bal.
SC 16	16	-	3.0	-	3.5	3.5	3.5	-	-	Bal.

Dummy turbine blades for stationary gas turbines of more than 20 cm in length with extreme cross sectional transitions at the inner and the outer shroud were cast, using SC 16, see Table I.

The dummy blade experiments were accompanied by a numerical process simulation. The institute-own software CASTS [9] was used. CASTS has been developed, meanwhile a well established tool, for the simulation of the Bridgman process. It describes the complete heat transfer balance in the vacuum furnace, including view factor radiation calculation [10-12], and including an algorithm that calculates the solidification of undercooled melts on a macroscopic scale [13]. This module is necessary for a realistic calculation of the dynamics of the process. The experimental set-up of the cylindrical specimen and of the dummy blade was generated as an FEM-mesh. The helix of the dummy blade set-up was replaced by a rectangular rod with an equal cross section. Thus the symmetric mesh in vertical direction was reduced to a quarter. The furnace was divided into four heating zones, enabling a better predetermination of vertical temperature gradients.

First experimental cylindrical specimen results were used to match the process parameters of ADS to the numerical simulation with CASTS. The computer was fed with the relevant thermophysical data, including the emissivity coefficients of the radiating materials, the heat transfer coefficients between the interfaces of the materials, the initial temperature distribution and the cooling rate of the heaters. The optimized values especially for the emissivity coefficients were used as an input for the process simulation of dummy blades.

Results

Radiation was observed to be the only form of heat transfer mechanism in the vacuum furnace and, consequently, the melt filled shell mold cooled down at a rate dependent on the temperature of the graphite heaters.

Cylindrical specimens

For both of the shell mold-binder systems utilized, a maximum undercooling of about 80 K was achieved for the superalloy CMSX-6. For the production of single crystal microstructure both systems were suitable. Undercooling of more than about 30 K led to a substantial grain refinement. The grains were of a size equal to the secondary dendrite spacing of dendritic specimens, as shown in Figure 3. The experimental results were described by a theory which predicts the fragmentation of secondary dendrites depending on the level of undercooling and the cooling rate dT/dt [14].

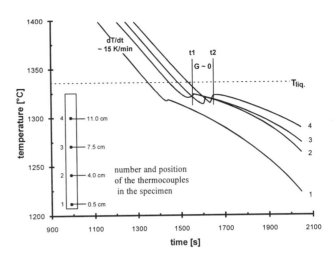

Figure 4: Typical temperature distribution of a cylindrical specimen, cast with the initial process parameters. The thermal gradient G (dT/dx) was approximately zero at between t1 and t2.

Figure 3: Grain refined microstructure, obtained at an undercooling above 30 K.

The process limiting critical undercooling condition requires a controlled initiation of the nucleation before the critical value is reached. The first series of experiments showed that just switching off the heating device led to a cooling rate below 20 K/min. During solidification the longitudinal thermal gradient of about 1 K/mm broke down completely, Figure 4. There was no directional heat flow as required for a reliable directional solidification. The secondary dendrite spacing was about 90 to 100 µm, and the microporosity was about 1%. In spite of these unfavorable conditions, single crystal microstructure was produced, even without using grain selectors, Figure 5a. Subsequently the cooling rate was increased to about 100 K/min and the thermal gradient up to about 6 K/mm during solidification, Figure 5b. The secondary dendrite spacing was thereby reduced to below 50 µm, and the measured microporosity was between 0.3 and 0.4%.

Figure 6 shows the measured growth velocity as a function of the vertical positioning. The unexpected process dynamics provided several options for producing single crystals of desired the <100> orientation. If the lower part of the specimen is separated from the main component using a helix and where the heat flow within the helix is very small, several well known methods like the usage of seed crystals etc. are possible. The nucleation can be initiated with or without an undercooling at the bottom of the shell mold. The shell mold can be closed or opened at the bottom, since it has no remarkable influence on both the undercooling and the solidification of the main specimen. Although the melt became thermally unstable when the temperature in the specimen became lower than that within the helix, deleterious effects of convective flows, for example, were not observed.

Defects like freckles, spurious grains and misalignment of the dendrites as well as porosity clusters were observed, however, when no remarkable vertical temperature gradient was maintained during the solidification of the residual melt. This is comparable to numerous observations of conventional single crystal solidification experiments, e.g. [15-17]. In order to attain a finer microstructure with a better alignment of the dendrites, the solidification of the residual melt has to be accelerated and controlled. A simple means of realizing this in the Bridgman furnace was to withdraw the mold from the heating zone at a given speed and within a specific period of time.

The most efficient way of accelerating the process after recalescence is to use the liquid metal cooling (LMC) technique, P.R. Sahm and M. Lorenz 1972 [8]. The shell mold was not wetted by the cooling metal. This enabled an easy processing in the laboratory type furnace (see Figure 2). Figure 7 shows the cooling curves measured in an LMC furnace experiment for the alloy CMSX-4, which were equal to those, obtained for CMSX-6. The low recalescence peak is due to the relatively high cooling rate and the thickness of the alumina tube which protected the thermocouples. The level of undercooling was comparable to that obtainable using the industrial type furnace. The cooling rates were varied between 100 K/min and 2.100 K/min depending on the time and the velocity of lifting the Ga-In bath. The resulting secondary dendrite spacing was measured to be about 50 µm for the lower cooling rate and between 14 µm and 17 µm for the highest, quench like cooling rates. The microstructure is shown in Figure 8. The primary dendrite spacing decreased from 200 µm to 65 µm. The porosity decreased below to 0.1%.

The LMC experiments show the potential of improving microstructure and the possibility of controlling the solidification.

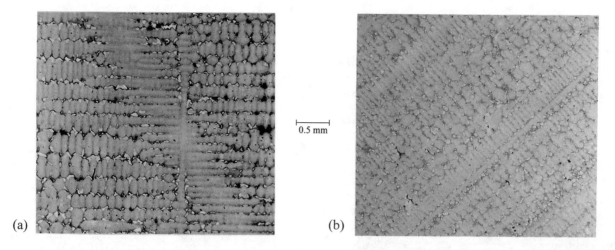

Figure 5: Single crystal microstructure produced with initial (a) and improved process parameters (b). Crystallographic orientations far from <100> produced extensive porosity clusters.

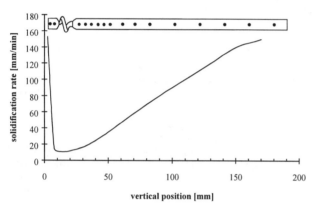

Figure 6: The growth velocity as a function of the vertical positioning demonstrates the dynamics of the process. The growth velocity is calculated at the times of maximum undercooling of 16 thermocouples. The positions are marked on the cylindrical bar.

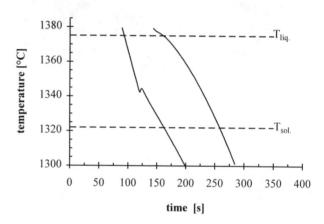

Figure 7: Cooling curves measured at the bottom of a cylindrical specimen in a laboratory type LMC furnace for the alloy CMSX-4. After recalescence a Ga-In bath was uplifted to accelerate the solidification of the residual interdendritic melt.

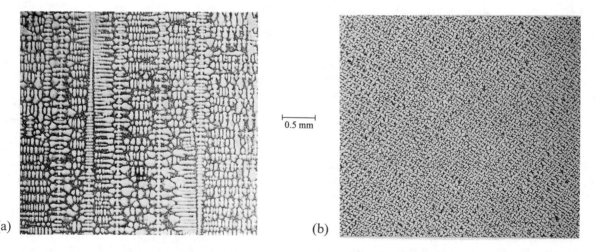

Figure 8: Longitudinal (a) and cross sectional (b) microstructure of characteristic specimen, processed in the LMC furnace.

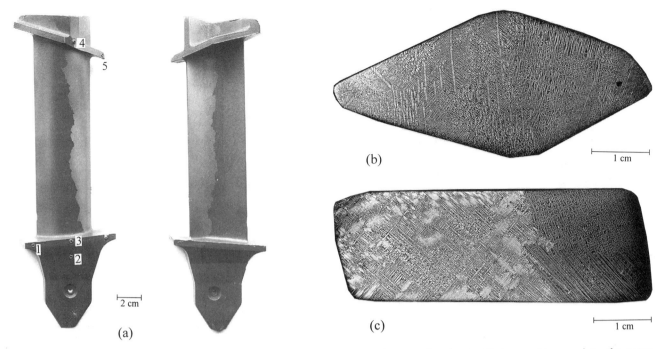

Figure 9: Macro-etched aero-engine blade consisting of two grains growing together in the blade, reaching up into the outer shroud (a); at the back view, the positions of the thermocouples are marked at the cross sectional transitions of the shrouds (b) and (c).

Aero-engine blades

A major aim of the present work is the application of the investigated process parameters to the casting of turbine blades. In order to cast complex shaped single crystal blades in mold assemblies several aspects of the process have to be considered. ADS uses free dendritic growth with extended mushy zone regions. Temperature gradient and solidification rate are not constant. Conventional feeding techniques on the other hand, prevent macro- and micro-porosity. Radiation is the main cooling mechanism, and the surface to volume ratio of the mold influences local cooling rates and temperature gradients. It is very important to optimize the geometrical arrangement of the blades in the assembly. Rotating the blades in the vertical axis is a factor which influences the undercooling at critical corners due to the asymmetric heat flow in vertical and radial directions. A series of turbine blades solidified as single crystals with only a few faults, like the grain nucleation at an outer corner, see, for example Figure 9. Figure 10 shows the cooling curves at the positions marked in Figure 9a. The solidification time of the whole blades was estimated from the cooling curves to about 15 minutes. Conventionally cast, the solidification of similar blades lasts more than 60 minutes.

Porosity in the cross sectional transitions was minimized by means of a predetermined positioning of each blade in the mold assembly and by improving on the thermal gradient during solidification. After recalescence had taken place at the lower thermocouple position, the mold was, for example, withdrawn 50 mm from the heating zone. The movement was stopped smoothly to prevent undesired nucleation and/or dendrite fragmentation.

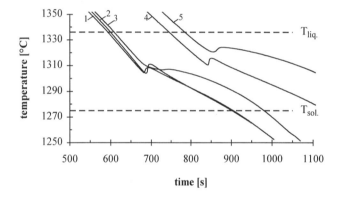

Figure 10: Cooling curves measured during solidification of an aero-engine blade, at the marked positions of Figure 9a. From the cooling curves, the solidification of the whole blade was estimated to about 15 minutes.

Extended freckling in combination with porosity clusters was found in turbine blades with crystallographic orientations far away from <100>. The <100> direction was found to be less prone to freckling.

Turbine blades of the same geometry were cast with ceramic cores. The platinum pins and the silica cores had not induced grain nucleation. Consequently the applicability of ADS for such thin walled castings was proved. Two of these tries are shown in Figure 11. Essentially it was observed that the castings were at least of the same quality as that of the blades without cores. The cooling curves were equal to the ones shown in Figure 10.

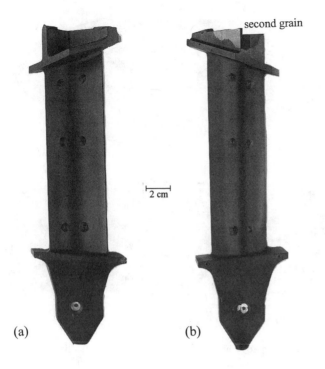

Figure 11: Aero-engine blades including ceramic cores. The platinum pins are visible on the surface. Single crystal specimen (a) and (b) single crystal blade with a second grain above the upper shroud. Platinum pins and ceramic cores had not induced grain nucleation.

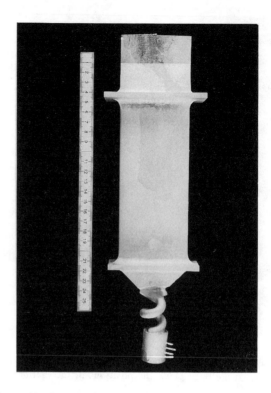

Figure 12: Dummy blade consisting of a few large grains. The single crystal was lost at the lower shroud. Down falling dendrite fragments from the top of the casting can additionally cause nucleation in the foil region.

During the development of process parameters for casting aero-engine blades no grain defect was found which could not be eliminated by local insulation of the shell mold at critical corners, variation of the experimental set-up and controlling the down cooling of the graphite heaters.

The recent aero-engine blades were cast in CMSX-4 in the same mold assembly. The main difference to using CMSX-6 was a reduced proneness to freckle defects. The undercoolability and the resulting microstructure were comparable. The mechanical stresses between blade and mold were increased because of the higher level of volume shrinkage during solidification. In an extreme case, the blade broke into two pieces during solidification. The shell mold did not crack as observed in successful castings. The effect of too high mechanical stresses during solidification can be prevented by varying the geometrical arrangement of the blades in the assembly and a smooth variation of the shell mold composition.

Dummy blades for stationary turbines

The casting of single crystal turbine blades for stationary gas turbines is one of the goals the ongoing research. The first experiments highlighted the general difficulties of casting large turbine blades. Figure 12 shows an example of a macro-etched dummy blade. The single crystal microstructure was lost in the lower shroud. High metallostatic pressure in the corners of the thin shroud with high cooling rates led to uncontrolled nucleation. Then only a few grains grew through the airfoil. The upper shroud with a surface of 10 cm x 5 cm and a thickness of 5 mm facing the cold top of the furnace cooled down at high rates resulting in high undercooling at the corners. More than 80 K were measured at 4 mm from the outer corner, Figure 13.

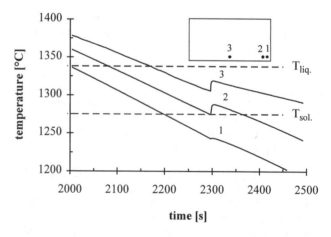

Figure 13: Cooling curves measured within the shroud of the dummy blade at the marked positions. The high level of undercooling resulted in grain refined microstructure.

Figure 14: Grain refined (a) and dendritic (b) microstructure in the upper shroud solidified at different levels of undercooling.

The grain refined microstructure as shown in Figure 14a is, as mentioned before, a reliable indication for the degree of undercooling. The recent castings have been done using local mold insulation at the shroud transitions and variation of the heater control. The amount of undercooling was reduced below the amount of grain refinement, as shown in Figure 14b. The results of these experiments provide, among other results, a basis for an optimized furnace design for industrial manufacturing of ADS single crystals.

Numerical process simulation

To reduce both the developmental time span and the cost for the expensive large blade castings with ADS, numerical simulation was also employed.

The FEM mesh of the dummy blade in the vacuum furnace is shown in Figure 15. The presented calculation was done for conditions similar to that of the recent experimental runs. The cooling rate of the lowest heater was increased above a rate, that was possible in the experiments. The effect was similar to withdrawing the mold for a certain distance from the heating into the cooling zone, as described above. The high surface to volume ratio of the starter enabled additional improvement of the cooling rate in the lower part of the blades. The critical undercooling for grain nucleation was set to 20 K.

The calculated curvature of the liquidus temperature within the shrouds shows the undercooled regions before solidification, Figure 16. Grain nucleation and the growth of the grains during solidification are shown in Figure 17. The level of undercooling in the corners was below 20 K when the solidification front reached the lower shroud. The initial grain grew rapidly from the center into the outer regions of the thinwalled shroud. Before the solidification front reached the upper shroud, a second grain was nucleated in the corner of the shroud. A third grain nucleated at the top of the feeder, resulting in a volume deficit at the transition between blade and feeder, which was observed in the experiments (see Figure 12). Cooling curves, calculated for three positions within the upper shroud, are shown in Figure 18. The nucleated grain grew within one time step through the shroud.

For a critical undercooling of 30 K to induce nucleation, the whole blade solidified as single crystal.

Systematic changing of variables is necessary to investigate their influence on single crystal solidification and the resulting microstructure. The correlation of the experimental process parameters to the resulting microstructure can be used as input for an algorithm which calculates an optimization of the process [11,13].

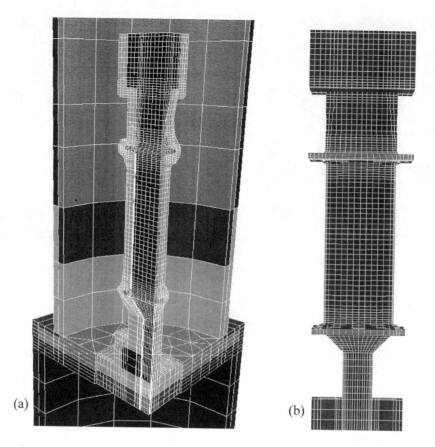

Figure 15: FEM mesh of the experimental set-up for casting large turbine blades (a) and enlarged view of the dummy blade including starter and feeder (b).

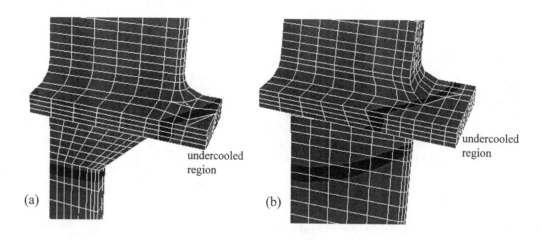

Figure 16: The curvature of the calculated liquidus isotherm at the lower (a) and the upper shroud (b) of the blade shows the undercooled regions.

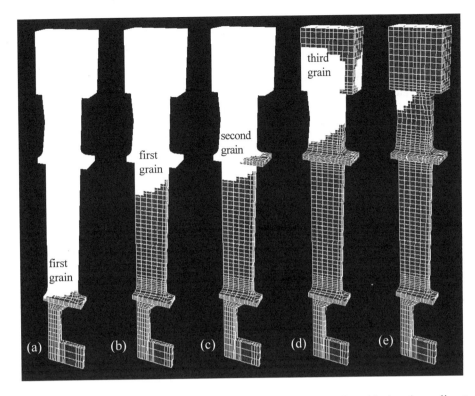

Figure 17: Grain nucleation and propagation of the solidification front are shown. The critical undercooling to induce nucleation was 20 K. The first grain nucleated at the bottom of the starter and grew through the lower shroud (a) and through the main part of the blade (b). Before the solidification front reached the upper shroud, a second grain was nucleated in the corner (c). The feeder solidified as a third grain (d). The resulting porosity at the top of the blade is visible as white area in (e) and matches well with the experimental observations.

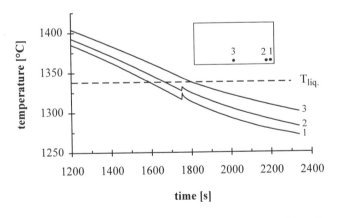

Figure 18: Cooling curves calculated for the marked positions in the upper shroud according to the measured ones.

Conclusions

1. Autonomous directional solidification (ADS), a novel casting technique was established. Single crystals were produced by solidification of undercooled superalloy melts.

2. Single crystal turbine blades up to 18 cm in length have been solidified successfully. Although not defect-free as required for industrial application, the results are promising and show that single crystal microstructure can be produced by ADS.

3. Further development in process engineering and a more suitable furnace are necessary to optimize the process for the reliable production of larger turbine blades with an improved single crystal microstructure.

3.1 An important future activity is to design a furnace which is optimized with regard to the requirements of the ADS process. It is necessary to use foil heaters, which enable higher cooling rates in more than three controllable regions.

3.2 Additionally, the LMC technique should be taken into consideration. A controlled withdrawal unit then would not be necessary.

3.3 Numerical process simulation will be used to work out some details and the dimensions of the required equipment.

4. Future activities require the testing of the characteristic mechanical properties like creep strength, low and high cycle fatigue after heat treatment. This work has been initiated meanwhile.

Acknowledgment

The authors like to thank the Deutsche Forschungsgemeinschaft (DFG) for the financial support of the present work.

References

1. P.R. Sahm, „Gerichtete Erstarrung - eine entwicklungsfähige Werkstofftechnologie", Radex Rundschau, 1/2 (1980), 20-29.

2. W. Eßer, „Directional Solidification of Blades for Industrial Gas Turbines" (Paper presented at „Materials for Advanced Power Engineering 1994", Liège, Belgium, 3-6 Oct. 1994), 641-659.

3. R. Singer, „Advanced Materials And Processes For Land-Based Gas Turbines" (Paper presented at „Materials for Advanced Power Engineering 1994", Liège, Belgium, 3-6 Oct. 1994), 1707-1729.

4. G.L. Erickson, „A New, Third-Generation, Single-Crystal, Casting Superalloy", JOM, 47 (April 1995), 36-39.

5. B. Lux, G. Haour and F. Mollard, „Dynamic undercooling of Superalloys", Metall, 35 (1981), 1235-1239.

6. J. Stanescu and P.R. Sahm, „Single Crystal Solidification of a Nickel-Based Superalloy" Ingenieur-Werkstoffe, 2 (1990), 64.

7. J. Stanescu, P.R. Sahm, J. Schädlich-Stubenrauch and A. Ludwig, „Autonomous Directional Solidification for Single Crystal Turbine Blades" (Paper presented at the 40th Ann. Techn. Meet.: Investment Casting Institute, Las Vegas, NV 1992).

8. P.R. Sahm and M. Lorenz, „Strongly coupled growth in faceted-nonfaceted eutectics of the monovariant type", Journal of Materials Science, 7 (1972), 793-806.

9. P.R. Sahm, W. Richter and F. Hediger, „Das rechnerische Simulieren und Modellieren von Erstarrungsvorgängen bei Formguß", Gießerei-Forschung, 35 (2) (1983), 35-42.

10. U. Reske, A. Bader, N. Hofmann and P.R. Sahm, „Numerische Simulation der gerichteten Erstarrung nach dem Bridgman-Verfahren" Gießerei-Forschung, 43 (3) (1991), 101-106.

11. N. Hofmann, U. Reske, H. Vor and P.R. Sahm, „Numerische Simulation der gerichteten Erstarrung nach dem Bridgman-Verfahren II" Gießerei-Forschung, 44 (3) (1992), 113-120.

12. N. Hofmann and F. Hediger, „Process Simulation For Directionally Solidified Turbine Blades Of Complex Shape" (Paper presented at „Modeling Of Casting, Welding And Advanced Solidification Processes-V", Davos, Switzerland, 16-21 September 1990, 611-619.

13. A. Ludwig, I. Steinbach, N. Hofmann, M. Balliel, M. van Woerkom and P.R. Sahm, „Modeling Of Undercooling Effects During The Directional Solidification Of Turbine Blades" (Paper presented at „Modeling Of Casting, Welding And Advanced Solidification Processes-VI" Palm Coast, Florida, 21-26 March 1993), 87-94.

14. M. Schwarz, A. Karma, K. Eckler and D.M. Herlach, „Physical Mechanism of Grain Refinement in Undercooled Melts", Physical Review Letters, A73 (1994), 1380.

15. T.M. Pollock, W.H. Murphy, E.H. Goldman, D.L. Uram and J.S. Tu, „Grain Defect Formation During Directional Solidification Of Nickel Base Single Crystals" (Paper presented at the 7th International Symposium on Superalloys, Seven Springs, 20-24 September 1992), 125-134.

16. D. Goldschmidt, U. Paul and P.R. Sahm, „Porosity Clusters And Recrystallization In Single-Crystal Components" (Paper presented at the 7th International Symposium on Superalloys, Seven Springs, 20-24 September 1992), 155-164.

17. U. Paul and P.R. Sahm, „Untersuchung der einkristallindendritischen Erstarrung von Superlegierungen auf Nickel-Basis für die Turbinenschaufelherstellung", Gießerei-Forschung, 45 (3) (1993), 19-27.

THE ENGINEERING APPLICATIONS OF A HF-FREE DIRECTIONALLY SOLIDIFIED SUPERALLOY IN THE AVIATION INDUSTRY OF CHINA

Sun Chuanqi, Li Qijiuan, Wu Changxin, and Tian Shifan
Beijing Institute of Aeronutical Materials, Beijing 100095, PRC
John F Radavich
Purdue University, West Lafayette, 47907 IN, USA

Abstract

Most directionally solidified (DS) superalloy blades and vanes operating at or above 950°C in jet engines usually contain Hafnium (Hf) to improve mechanical properties and performances. The addition of Hf increases the cost of these alloys and may cause a number of problems which limit their usage. At BIAM, a Hf-free DS superalloy, DZ4, was developed in the late 1970's. The improved mechanical properties, and good castability of this alloy, combined with low cost, make the DZ4 alloy suitable for production applications. In this paper, compositional and microstructural characterizations will be discussed. Mechanical properties, and engineering applications for the alloy will be presented.

Introduction

In the early 1960's when the first American DS superalloy, PWA 664, was developed using the Mar M200 composition, alloy properties were greatly improved through the use of directional solidification. Unfortunately, during the production of DS hollow blades, hot tearing problems appeared. In the early 70's, an alloy, PWA 1422 (Mar M200 +2% Hf) was invented by Duhl et al [1], which solved the hot tearing problem. The effect of the Hf addition is to change the morphology of the large script-like primary MC carbide into fine dispersed Hf-rich MC2 carbides [2]. The addition of Hf also resulted in a great increase in the amount of γ/γ' eutectic phase. For example, in alloys containing 1.5- 2.4 wt % Hf, the amount of γ/γ' eutectic may increase from ~ 6% to 20% on average. As the γ/γ' eutectic is mainly located at the longitudinal grain boundaries and interdendritic regions, it toughens the grain boundaries. Production experience has proven that the Hf-addition is effective in eliminating cracking in directionally solidified blades, especially DS hollow blades. Currently, most alloys used for DS blades, contain Hf (see Table I) for improving castability. However, there are shortcomings brought about by the addition of Hf, namely; 1) cost of Hf containing alloy is much higher than Hf-free alloys, and the recycling of Hf containing alloy revert remains difficult; 2) the incipient melting temperature of Hf containing alloys is significantly lowered due to the presence of low melting phases such as Ni_5Hf, which in turn effect high temperature properties, especially properties above 1000°C. In order to overcome the shortcomings of Hf-containing DS superalloys, research and development work was started in the late 70's at BIAM. The target goals were as follows: 1) Alloy properties must be essentially maintained at levels equivalent to that of alloy PWA 1422; 2) The alloy should not contain Hf, but retain good castability; 3) Consideration must be given to low specific weight and low cost. In the early 80's, a Hf-free DS superalloy DZ4 was developed at BIAM in China which met the target property and cost goals. The alloy composition, phase data and physical parameters are presented in Table II. The alloy has entered production, and since the late 80's the engineering applications for the DZ4 alloy have been increasing.

Table I Composition of Typical DS Superalloys Widely Used in Jet Engines

Alloy	Composition wt. %														Density g/cm^3	Relative Cost
	C	Cr	Co	W	Mo	Nb	Ta	Re	Al	Ti	Hf	B	Zr	Ni		
DZ4	0.13	9.5	6	5.5	3.8				6.2	1.9		0.015		Bal.	8.15	1X
PWA 1422	0.14	9.0	10	12		1			5	2.0	1.5	0.015	< 0.1	Bal.	8.5	2X
DS Rene' 150	0.06	5	5	1	1		6	3	5.5		1.5	0.015		Bal.	8.5	12X
DS MM002	0.14	9	10	10	< 0.5		2.5		5.5	1.5	1.5	0.015	0.05	Bal.	8.5	2.5X
DS Rene' 80	0.12	14	9.5	4	4				3	4.0	1.5	0.015	< 0.1	Bal.	8.2	2X

Table II Chemical Composition, Phase Data and Physical Parameters of Alloy DZ4

Composition wt %									
C	Cr	Co	W	Mo	Al	Ti	Al+Ti	B	Ni
0.1/0.16	9.0/10.0	5.5/6.0	5.1/5.8	3.5/4.2	5.6/6.4	1.6/2.2	>7.6	0.012/0.025	Bal.

γ' total %		a_γ	$a_{\gamma'}$	$(a_\gamma - a_{\gamma'})/a_\gamma$	N_v	Density	T_m °C
wt.	at.	(Å)	(Å)	%		gm/cm^3	incipient
60.78	62.82	3.588	3.578	0.27	2.16	8.19	1290

* calculated according to a nominal composition

Table III Chemical Composition of Alloy DZ3 and the Experimental Alloy DZ4

Alloy	Composition wt %									
	C	Cr	Co	W	Mo	Al	Ti	Zr	Ce	B	Ni
DZ3	.14	10.4	5	5.3	4.1	5.6	2.6	0.1	0.01	0.01	Bal.
DZ4	.14	9.5	6	5.4	3.8	6.2	1.8	-	-	0.015	Bal.

Alloy Design

The alloy composition is designed to eliminate the γ/γ' eutectic, maintain a high γ' volume fraction, and achieve good castability. The solution temperature must be as high as possible so that the alloy can meet the requirements for high mechanical properties, and structural stability for long term service. At the same time the alloy should have a low density and not contain rare expensive elements. By optimizing the γ' forming element content (Ti, Al and Zr), an alloy (DZ3) with 3-5 wt. % γ/γ' eutectic (see Table III) was developed [3]. DTA and isothermal quench experiments have been done for alloys DZ3 and DZ4[4]. The results of isothermal solidification tests show that at 1370°C, alloy DZ4 is essentially liquid except for a few primary crystals which appear at the periphery adjacent to the mold (shell). At 1360°C, about 6.8 % of the melt has solidified as γ primary crystals and precipitated, while at 1350°C ~ 70.8 % of the total volume has solidified. After isothermal solidificationat at 1330°C for 15 min. approximately 12.3% of the melt is liquid and MC carbides begin to precipitate out. At 1280°C only 0.65% of the liquid phase remains, which contains mainly M$_3$B$_2$ and carbides [5]. Thermal solidification tests show the results of the absence of the γ/γ' eutectic phase at 1370°C. The DZ4 alloy remains essentially solid below 1270°C. There is no secondary γ' precipitation in the alloy until isothermal solidification at 1180°C. After 15 minutes at 1150°C, secondary γ' appears at the interdendritic regions, but at the dendritic stems there is no sign of secondary γ' phase. The liquidus (T_L) and solidus (T_S) for alloys DZ3 and DZ4 are presented in Table IV.

Table IV Liquidus and Solidus of DZ3 and DZ4

Alloy	Heating			Cooling		
	T_L	T_S	$T_L - T_S$	T_L	T_L	$T_L - T_S$
DZ4	1365	1280	85	1356	1296	60
DZ3	1345	1180	165	1347	1206	141

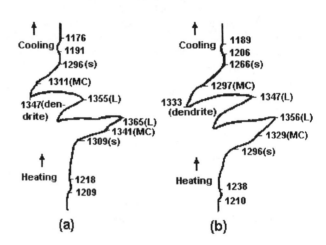

Figure 1. DTA Curves for Alloy DZ4 (a) and Alloy DZ3 (b)

Castability

Comparing DZ3 with DZ4, (Table IV) shows [5] that Zr has been removed from alloy DZ4 and the Ti found in DZ3 has been partially replaced by the equivalent amount of Al. There are also minor adjustments in Mo and W. It is believed that lowering the amount of elements, such as Ti and Zr, which concentrate in the γ/γ' eutectic, will reduce the amount of this phase while maintaining a high volume fraction of γ'. Experimental results indicate that no γ/γ' eutectic phase appears in the DZ4 alloy in the as cast condition, no matter how fast or slow it is cooled, or even when the Al + Ti content is at the upper limits of the specification. The elimination of the γ/γ' eutectic further narrows the temperature differential between the Liquidus (T_L) and Solidus (T_S) in the alloy. This is regarded as the main reason castability is improved for the Hf-free superalloy. It is clear from Table IV, that the temperature differential between T_L and T_S for DZ4 is approximately half that of DZ3 alloy. In practice the differences in ΔT between two alloys is reflected in their relative castability. The DZ3 alloy with a large ΔT, is prone to hot tearing, and a relatively high incidence of cracking has been observed during the casting of

DS hollow blades. Hollow blades cast from DZ4 alloy, with a small ΔT, do not experience hot tearing problems.

Microstructure

The phase constituents in the as cast condition are as follows: γ matrix, intermetallic γ', MC carbides and a minor amount of M_3B_2. In the heat treated or service condition, γ' transforms into fine cubes (0.3 - 0.5 μm) dispersed uniformly distributed throughout the matrix, and M_6C and $M_{23}C_6$ carbides may also precipitate. The absence of γ / γ' eutectic in DZ4 is one of the unique microstructural features of this alloy which contains a high volume fraction of γ' (see Figure 1).

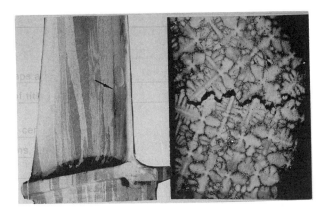

Figure 2. Cracking in DS Hollow Blades Cast From DZ3 Alloy.

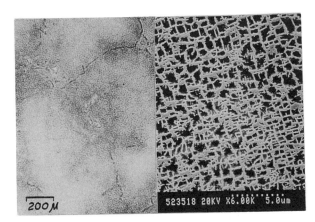

Figure 3. Microstructure of DZ4 Alloy As Cast (Left) After Heat Treatment (Right)

Mechanical Properties

As shown in Figures 4 and 5, alloy tensile strength and stress rupture life increase at high and intermediate temperatures as the solution temperature increases. Solutioning DZ4 at 1270°C does not lower alloy properties. In some alloys with Hf, solutioning at 1250°C, significantly reduces alloy properties, because incipient melted regions are produced in the microstructure. Therefore, the elimination of the γ / γ' eutectic enables solution heat treatment to be performed at high temperatures without producing incipient melting, and optimizing mechanical properties.

The DZ4 alloy property data accumulated during engineering development, are presented in Tables V and VI. Data listed in Table V and Figure 6 show that the rupture properties of DZ4 are superior to other DS superalloys, such as PWA 1422, Rene' 125 and DS MM002, for the entire spectrum of service temperatures. The superiority of alloy DZ4 becames more apparent at temperatures above 1000°C. Work done by Wakusick [6] showed that castability is related to the γ' volume fraction of the alloy, and as the γ' volume fraction increases, the castability is degraded. Often achieving high temperature properties and good castability are conflicting goals. In past alloy development programs, in order to improve alloy castability without Hf, high temperature properties had to be sacrificed. In alloy DZ4, however, γ' volume fraction is as high (62 vol. %), or higher then that of other typical DS superalloys, while excellent castability is maintained, due to the elimination of the γ / γ' eutectic through compositional design.

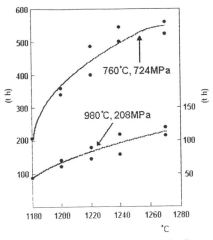

Figure 4. Effect of Solution Temperature on the Stress-Rupture Properties of DZ4 at 760 and 980°C

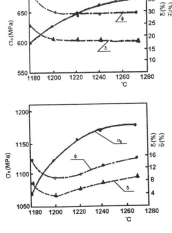

Figure 5. Effect of Solution Temperature on the Tensile Properties at 760°C (a), and 980°C (b)

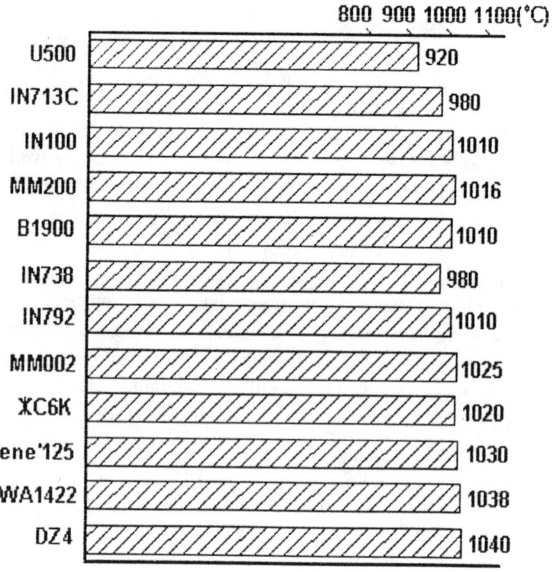

Engineering Performance and Applications

Stress-Rupture Properties

A complete mechanical property characterization of alloy DZ4 was conducted during alloy development. Alloy properties were determined directly from cast parts. Thin wall property tests are shown in Table VI [7], where transverse and longitudinal stress-rupture properties at 760 and 950°C are listed for DZ4 and IN-100 using specimens cut from hollow DS first stage blades. Both DZ4 and IN-I00 have the same rupture ductility but DZ4 exhibits rupture life 10 times higher than that of IN-100. Table VI shows the comparison of thin wall rupture properties for DZ4 and IN-100 using sheet samples of 1.0 mm thickness cut from the airfoils of second stage turbine blades. Both the rupture life and ductility of alloy DZ4 are superior to that of IN-I00.

Figure 6. Temperature Capability of Alloy DZ4 and Some Typical Superalloys Widely Used in Turbine Engines (Normalized for 100 Hour Stress Rupture Life at 137MPa)

Alloy	Temperature (°C)
U500	920
IN713C	980
IN100	1010
MM200	1016
B1900	1010
IN738	980
IN792	1010
MM002	1025
XC6K	1020
Rene'125	1030
PWA1422	1038
DZ4	1040

Table V Specific Stress-Rupture Strength (Rupture/Alloy Density) for DZ4 and Other Superalloys

T °C Alloy	Specific 100 Hr. Rupture Strength MPa.cm^3/g							
	760	800	850	900	950	980	1000	1040
DZ4	98.7	82.9	63.7	43.2	29.9	25.2	22.1	17.3
DZ22	93.7		61.7			24.0		15.9
PWA 1422	88.2					23.3		14.8
DZ3	95.5	77.4	60.4	42.3	29	25.3	21.5	16.8
K403		64.1	49.5	36.2	26.5		18.1	
IN100		70.2	54	40.1	26.3		18.8	
DZ5	92.3	77.2	62.7	41	30.1	25.3	21.7	
DZ17G	91	75.8		41.6	27.7	22.7	19.6	
DS B-1900	85	72.6	56	41.4	27.3	22.6	19.6	
DS IN-100	85.9				22.1			
DS MM002	87.1	74.8	59.9	42.3	29.4		21.3	16.4

Table VI Transverse and Longitudinal Stress-Rupture Properties for Hollow DZ4 DS and IN-100 Equiaxed Blades

Specimen Machining Location	Alloy	Specimen Orientation	Specimen Size mm	Stress Rupture Test Data				
				Temp C	Stress MPa	Life hrs	Elong %	RA %
 (upper)	DZ4	Transverse DS	diam. = 3	760	647	39:25 41:55	3.33 3.36	9.59 10.18
				950	235	252:00		
	IN100	Equiaxed	diam. = 3	760	647	4:25 4:00	6.67 2.00	17.90 5.10
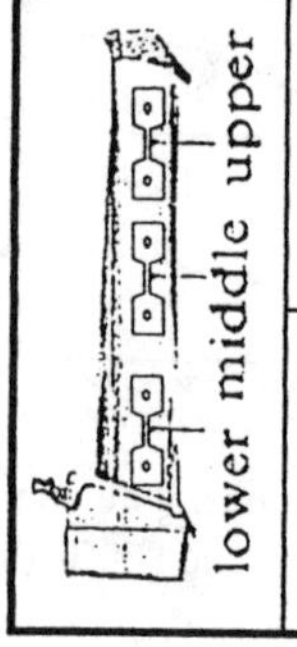 lower middle upper	DZ4	Axial DS Top Middle Lower	thick. = 1	760	647	595:40 781:50 615:45	17.80 17.80 20.00	
	IN 100	Equiaxed Top Middle Lower	thick. = 1	760	647	99:35 48:00 28:30	9.99 7.77 7.77	

Vibratory Fatigue Performance of Blades

Figure 7 presents a comparison of S-N curves [7] obtained in vibration fatigue tests using blades made from DZ4 alloy and IN-100, which DZ4 has replaced in a jet engine. At room temperature the fatigue limits for 1×10^7 cycles for DZ4 and IN-100 are 225-254 MPa and 147-176 MPa, respectively.

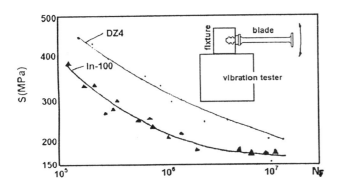

Figure 7. Vibration Fatigue Curves for Blades Made From DZ4 and IN-100 Alloys

Thermal Mechanical Fatigue

Thermal mechanical fatigue test results are shown in Table 8 and Figure 8 [7]. As presented in Figure 8, service life can be estimated from the stresses imposed on the component. For example, when the service stress on a DZ4 alloy turbine blade is 196 MPa, its fatigue life can be estimated at 10,000 cycles or more by using the thermal mechanical fatigue curve in the figure.

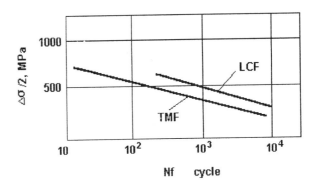

Figure 8. Thermal Mechanical Fatigue and Low Cycle Fatigue Curves for DZ4 at 980°C

Specification Properties

In order to compare overall properties versus an existing production DS alloy, PWA 1422 was selected for benchmarking. This alloy is employed in first stage high pressure turbine blades in F100 jet engines.[8] DZ4 was tested at specification conditions called out in the PWA 1422K specification and presented in Table IX. Also creep curves at 980°C / 193 MPa for both PWA 1422 and DZ4 are plotted and compared in Figure 9. It is evident that the two alloys have basically the same level of mechanical properties.

Table IX Comparison of Mechanical Properties for the Alloy DZ4 and PWA 1422

PWA 1422 Mechanical Property Test Conditions	PWA 1422 Specification Requirements	DZ4 Alloy Test Results	PWA 1422 Referenced Data
Tensile Properties at Room Temp	UTS > 1030 MPa Y S > 980 MPa Elong (4d) > 13%	1040 - 1197 MPa 932 - 961 MPa 5 - 7%	1106 MPa 942 MPa 6.1%
Stress-Rupture at 760C/690MPa	Life > 48 hrs Creep at 48 hrs < 4%	800 - 1000 hrs < 1.9%	700 hrs
Stress-Rupture at 980C/220MPa	Life > 32 hrs Creep at 32 hrs < 2%, Creep at Rupt. > 10%	56 - 70 hrs 1.5% 16 - 20%	129 hrs
Stress-Rupture at 1093C/82.8MPa (for vanes or buckets)	Life > 40 hrs Creep at Rupt. > 10%	50 - 61 hrs 22 - 32%	
Stress-Rupture at 1270C/68MPa (transverse-not required)		64.7 MPa 196 hrs	170 hrs

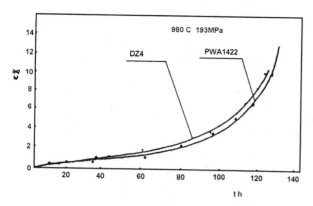

Figure 9. Creep Curves at 980°C /193 MPa for Alloy DZ4 and PWA 1422

Engineering Applications

In the present paper the engineering applications of the Hf-free DZ4 alloy, now in full scale production in China, were reported.

Annual consumption of Hf-free DZ4 alloy has reached 40 tons per year since 1992, and is growing rapidly with increased DZ4 usage for blades and guide vanes in modified and newly designed aeroengines and land based turbines. It now has entered service in marine, locomotive and petroleum industry applications.

The DS blades and vanes made from Hf-free DZ4 alloy include both hollow and solid castings and their sizes vary from tens of mm up to 200 mm. In Figure 9, two buckets are presented, one of them typically is a fabricated component made by vacuum brazing of DS airfoils and the other one (on the right) is integrally DS cast from Hf-free DZ4 alloy. In making complex components like this, the requirements for castability are even more stringent. Parts cast from alloy DZ4 have never experienced casting cracking. Accumulated service life for turbine blades made from DZ4 alloy has exceeded 8000 hours on more than 100 jet turbine engines flying over 3 years. Routine examinations indicate no blade distress has occurred to date. Measurements of blade elongation has revealed that the airfoils are just about to enter the steady state creep stage. As shown in Figure 11, the microstructure of a blade which has accumulated over 8000 flight hours, exhibits no change in γ' morphology and size. This in turn proves that the blade still has considerable life remaining. Over 30,000 blades, 188mm in length, have been produced. Production casting yields with DZ4 are higher than that of equiaxed blades owing to the absence of metallurgical defects such as porosity and inclusions. In addition, the reproducibility is also higher. Replacement of an older alloy in a hollow first stage turbine blade by DZ4 (see Figure 12), has resulted in service life being extended by 10 times that of the original. Hollow blades and vanes with sophisticated cooling passages have been successfully employed in advanced aeroengines. It is worth noting, that no blades have been rejected for hot tearing among tens of thousands of blades cast with the production process. No hot tearing appears even when the ceramic core deforms during the casting of a 160mm long blade with a thin and narrow airfoil, leading to a very thin wall (0.1mm wall thickness).

Figure 10. Blade and Guide Vanes Made from Hf-Free Alloy DZ4

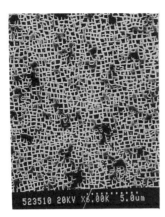
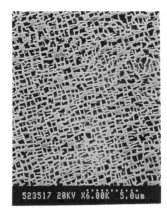

Figure 11. Microstructure of DZ4 Blade After 8000 Hours of Service
As Heat Treated (Left) and After 8000 Hours in Service (Right)

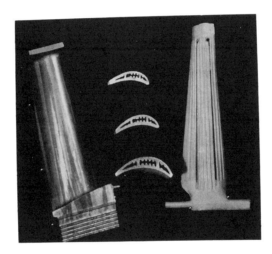

Figure 12. A Hollow Blade Made from DZ4

Conclusions

1. The DZ4 Hf-free directionally solidified superalloy is characterized by low cost, low density and high specific strength.

2. The elimination of γ/γ' eutectic phase in the DZ4 alloy greatly increases the incipient melting temperature and reduces the temperature range between the liquidus and solidus making it possible to increase high temperature performance and at the same time improve castability.

3. The Hf-free DZ4 alloy possesses mechanical properties essentially equivalent to that of the advanced Hf-containing DS superalloy PWA 1422

4. The many applications of the Hf-free DZ4 superalloy indicate that it has successfully met its program goals of high castability and improved high temperature properties.

References

[1] D.N.Duhl and C.P.Sullivan, "Some Effects of Hafnium Additions on the Mechanical Properties of a Columnar Grained Nickel Base Superalloy", J. Metals, 23 (7) (1971), 38-40
[2] Zheng Yunrong, "The Effects of Hafnium Additions on the Solidification of Cast Superalloys", Journal of Metals of China. Vol. 22, No. 2. April 1986, 119-124
[3] Wilfried Kurz, "Microsegregation in Directionally Solidified Superalloys", Proc of the 13th International Congress on Combustion Engines (CIMACT) Vienna 7th-11th May 1979
[4] Zheng Yunrong and Chen Hong, "Solidification Characteristics of Two Kinds of Directionally Solidified Superalloys, DZ3 and DZ4", Aeronautical Materials, 1 (1985), 10-16
[5] Sun Chuanqi et,al. "A Study of the Castability of DZ4 DS Superalloy", Aeronautical Materials, 3 (1984), 1-6
[6] C.S. Wakusick et al, US Patent 4169742, 1979
[7] Sun Chanqi et al, "Report on Superalloy DZ4", Beijing Institute of Aeronautical Materials, Unpublished Internal Report
[8] Pratt & Whitney, TM Specification PWA 1422F Nov. 1972

On the Castability of Corrosion Resistant DS-Superalloys

J. Rösler, M. Konter* and C. Tönnes
ABB Power Generation Ltd., Baden, Switzerland
*ABB Corporate Research, Baden, Switzerland

Introduction

In the past it has been well established that DS versions of conventional, corrosion resistant superalloys such as IN792 often suffer from poor castability. Main issues are insufficient resistance against grain boundary (GB) cracking, freckle formation, equiaxed grain formation and a negative heat treatment window which can impair the strength advantage over the conventionally cast counterpart. This highlights the need for compositions that are specifically tailored for DS-applications.

One compositional modification to avert grain boundary cracking, which is a consequence of shrinkage stresses during casting, is the addition of Hafnium /1, 4/. However, Hf-addition has some drawback by itself especially when large castings for stationary gas turbine applications are considered. It not only reduces the heat treatment window (temperature difference between incipient melting point and γ'-solvus), but also increases the reactivity of the alloy with the shell mould which can lead to wrinkle formation and inclusions. Compositional changes that improve GB-cracking resistance without addition of Hf are therefore particularly attractive.

For the above reason, it is the goal of this paper to investigate measures for DS-castability improvement in non-Hf containing superalloys. Furthermore, the focus is on materials with Cr-content in the 12% range which yields acceptable resistance in corrosive environments along with an attractive strength potential. After investigating fundamental aspects of DS-castability on model alloys in the first part, experience with alloy ABB 2 DS is reported in the second part. This alloy has been developed along the lines proposed here. For comparison, reference is also made to well established "low-Cr" DS-alloys such as CM247LC DS.

Fundamental aspects of DS-castability

Grain boundary cracking

Grain boundary cracking is frequently encountered during casting of hollow DS-blades. It is a consequence of shrinkage stresses between superalloy and ceramic core and is believed to occur at temperatures close to the melting point /4/ ("hot tearing"). To assess the mechanism of GB cracking, cylindrical DS-specimens were subjected to compressive loading along the longitudinal axis in a Gleeble 1500 testing machine /5/. Tests were performed in vacuum at temperatures between 1100°C and 1250°C. The specimens were heated by electric resistance heating. All specimens were tested in the as-cast condition unless otherwise stated. Measuring the specimen diameter D, the circumferential tensile strain acting perpendicular to the grain boundaries is given by

$$\varepsilon = \frac{D}{D_0} - 1 \qquad (1)$$

(D_0: initial specimen diameter). In fig. 1 ε_{crit}, which is defined as the circumferential strain at first crack occurrence, is plotted as a function of temperature. The composition of the investigated model compositions is given in tab. I. A remarkable finding is that the fracture strain strongly decreases towards higher temperatures above approx. 1150°C. This is consistent with the notion of "hot tearing" as the relevant fracture mechanism. Furthermore, ε_{crit} is strongly composition dependent. For Var. a, a circumferential strain of approximately 2% is sufficient for GB cracking to occur at 1250°C. In contrast, Var. b and CM247LC DS, which is known for its GB cracking resistance, endure approximately 10% strain before fracture. The major difference between Var. a and Var. b is the substitution of γ/γ'-eutectic forming elements by tungsten.

Table I Nominal composition of investigated model alloys

Alloy	Ni	Co	Cr	W	Al	Ti	Ta	B	C
Var. a	bal.	9.0	12.0	6.0	4.0	3.6	4,8	0.015	0.07
Var. b	bal.	9.0	12.0	9.0	3.5	2.3	5.5	0.015	0.07

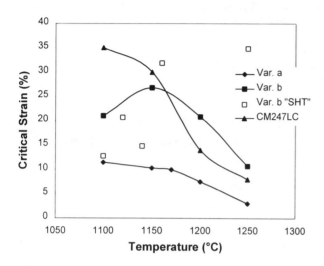

Figure 1: Plot of the critical circumferential strain to fracture, ε_{crit}, as a function of test temperature. All materials are tested in the as-cast condition unless otherwise stated ("SHT": solution heat treatment).

To further clarify the microstructural factors responsible for the resistance against GB cracking, Var. b was 100% solution heat treated and retested (fig. 1). Note, that ductility is now increasing with temperature. At 1250°C, ε_{crit} reaches 3 times the level of the as-cast counterpart.

Both results, the ranking of Var. a versus Var. b and the beneficial effect of solutioning, suggest that interdendritic segregation is an important factor[1]. Reducing the degree of interdendritic segregation, either by chemical modification or heat treatment, apparently leads to improved hot-tearing resistance. In fig. 2a, the as-cast microstructure of Var. b is depicted showing in fact a remarkably homogeneous appearance with an eutectic content of less than 2%. In contrast, the eutectic content of Var. a and IN792, which is known for its cracking propensity, is about 5%-6% (tab. II).

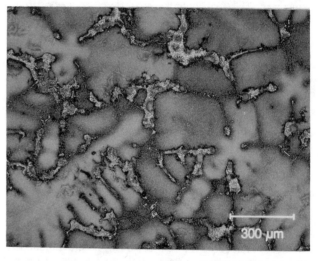

b

Figure 2a,b: As-cast microstructure of Var. b (a) in comparison to the as-cast microstructure of CM247LC DS (b), representing Hf-containing alloys.

---

[1] For all tested materials, the level of interstitial segregants has been carefully monitored and kept at comparable level to exclude tramp element effects on the propensity to grain boundary cracking.

The role of segregation may be best understood by noting that acceptable ductility, at temperatures close to the melting point, requires:

- Homogeneous deformation over a large material volume fraction.
- High intrinsic ductility of the volume element undergoing deformation.

Alloys with high amount of interdendritic segregants promote strong melting point gradients from dendrite core to the last liquid to solidify. This in turn causes a strong gradient in homologous temperature $T/T_{m,l}$ ($T_{m,l}$: local melting point) and, hence, localization of creep deformation. It also increases the likelihood of remaining liquid pools which can be separated by shrinkage strains and act as crack initiation sites. Thirdly, high amounts of interdendritic segregants increase the volume fraction of eutectic islands at grain boundaries which also appear to be preferential fracture sites (fig. 3) because they lead to additional strain localization and/or reduced ductility. Therefore, it is concluded that for Hf-free DS-alloys the amount of eutectic forming elements (i.e. interdendritic segregants) should be as low as possible to meet the above requirements.

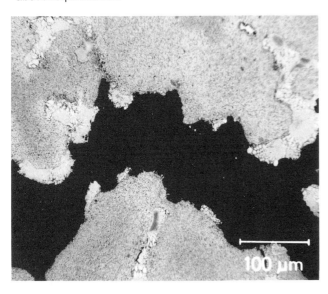

Figure 3: Microsection through a grain boundary crack of Var. a.

Although discussion of the Hf-effect is beyond the scope of this paper, it is noted that Hf-containing materials often form relatively broad eutectic regions rather than islands at grain boundaries (fig. 2b). Because the total shrinkage strain can then be distributed over a relatively large volume fraction of almost constant melting point, the beneficial effect of Hf when added in sufficient quantity is plausible. Prerequisite is, however, that the eutectic region is sufficiently ductile. To achieve this latter requirement in non-Hf containing materials is probably difficult due to the potential embrittling effect of Ti and Ta at levels necessary to promote sufficiently high eutectic volume fractions.

To cross-check the Gleeble results, blades of different size and complexity have been cast in Var. a, b, CM247LC DS and IN792DS. The results are summarized in tab. II. They show that the casting results are consistent with the ranking suggested by fig. 1. Remarkably good results are achieved with the "lean" Var. b and the Hf-containing material CM247LC. Examples of sound and cracked blades are given in fig. 4.

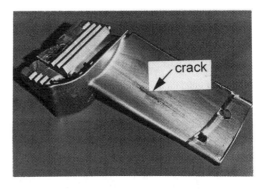

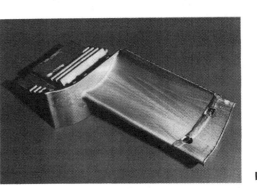

Figure 4a,b: 270mm test blade cast in Var. a (fig. a) and Var. b (fig. b). Note the absence of grain boundary cracking for Var. b.

Freckle formation

In addition to resistance against GB cracking, DS superalloys must have a high resistance against freckle formation. Freckles are defects that appear as elongated "chains" of equiaxed grains with high eutectic phase content caused by strong enrichment in Ti, Ta, and Hf (fig. 5). Their formation is attributed to the flow of interdendritic liquid /2, 3, 6/ which is driven by an instable density distribution (interdendritic liquid lighter than liquid with nominal composition). It is a phenomenon that is dependent on casting condition and alloy chemistry /3/. Long solidification times do promote freckle formation. Hence, it is of particular importance for the production of large, industrial gas turbine parts to select freckle-resistant DS-alloys.

Table II Assessment of alloy sensitivity to grain boundary cracking for two model compositions and two commercially available alloys. Results from Gleeble experiments are given as ranking (1: best; 3: worst). Comparison is made to the average γ/γ'-eutectic content in the as-cast condition.

Alloy	Gleeble ranking	DS-blade 220 mm	DS-blade 270 mm	DS-blade 350 mm	γ/γ'-eutectic content (%)
Var. a	3	sound	cracks	cracks	5
Var. b	1	sound	sound	sound	1.5
IN792	-	cracks	cracks	cracks	6
CM247LC	2	sound	sound	sound	8

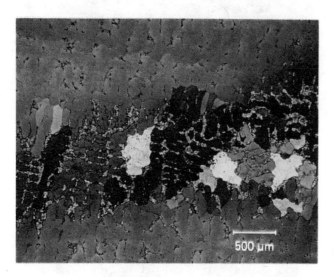

Figure 5: Freckle defect in a DS casting (longitudinal section). Equiaxed grains and a strong enrichment in the eutectic phase are visible.

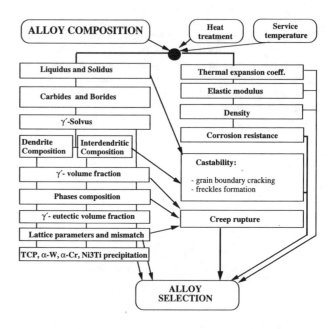

Fig. 6: Computer software flow diagram showing the calculated properties.

Absence of freckle formation regardless of the casting condition is expected for alloys where the interdendritic liquid exhibits a higher density, ρ_{id}, than the liquid with nominal composition, $\bar{\rho}$. This requires careful balance of elements that strongly partition to the dendritic or interdendritic regions (W, Ti, Ta). While $\rho_{id} > \bar{\rho}$ may not be met for other reasons, the tendency to freckle formation must be considered as an important parameter within the overall alloy development strategy.

To obtain quantitative information on freckle resistance, a software program (fig. 6) as described in detail elsewhere /7/ was used to calculate the partitioning of element i between dendritic and interdendritic volume:

$$k_i = x_{i,d} / x_{i,id} \qquad (2)$$

(k_i: partitioning coefficient; $x_{i,d}$, $x_{i,id}$: weight fraction of element i in dendrite core and interdendritic space respectively). Results obtained for Var. a and Var. b are shown in table III.

Tab. III Calculated partitioning coefficients, k_i, for Var. a and Var. b. Also included is the liquid metal density, ρ_i, taken for further calculations.

	Cr	Co	W	Al	Ti	Ta
k_i, Var. a	1.07	1.1	1.47	0.93	0.77	0.64
k_i, Var. b	1.09	1.1	1.44	0.95	0.75	0.64
ρ_i (g/cm³)	6.46	7.7	17.6	2.4	4.13	15.0

Given the nominal composition, $\bar{x}_i$, and the dendritic volume fraction, f_d, $x_{i,id}$ can be expressed as

$$x_{i,id} = \bar{x}_i / (f_d(k_i - 1) + 1) \quad (3)$$

Assuming furthermore that the liquid behaves as an ideal solution, the solidification density coefficient, SDC, defined as the quotient between the density of the interdendritic liquid and the density of the liquid with nominal composition /8/, can be derived:

$$\text{SDC} = \sum_i \rho_i x_{i,id} / \sum_i \rho_i \bar{x}_i \quad (4)$$

(ρ_i: density of pure element i in the liquid state). It characterizes the tendency to freckle formation. As discussed above, absence of freckle formation is expected for SDC > 1. Taking f_d = 0.45 as typical value, "no-freckle" lines defined by SDC = 1 are plotted in fig. 7 as a function of the Ti, Ta and W content. The area to the right of each "iso-tungsten" line corresponds to chemical compositions that are predicted to be strictly freckle-safe. It is seen that Var. a and Var. b do not quite meet this criterion which means that further alloy optimization should take place, for instance substituting tungsten by tantalum. It is also noted that a Ta/Ti ratio > 1 is required if tungsten levels in excess of 4wt.% are added for strength reasons.

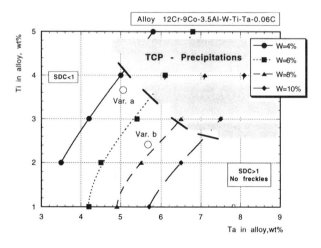

Figure 7: Plot of "no-freckle" lines (SDC = 1) in dependence of the Ti and Ta content for various tungsten levels. The boundary for TCP formation is also indicated.

Heat treatment window and mechanical properties

The ability to completely solution the γ'-phase is an important factor to avoid weak links in the microstructure /9/ and to exploit the full strength potential of DS-superalloys. For industrial applications, the heat treatment window should be at least 20K. This requirement is often not fulfilled using DS-versions of materials that were originally designed for conventional casting. Fig. 8a shows the differential thermal analysis (DTA) curve of IN792DS as example. It is noted that incipient melting starts at a temperature where γ' dissolution is still ongoing. This problem is further aggravated in large DS-castings for industrial gas turbine applications since cooling rates are relatively low.

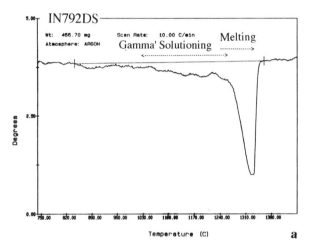

a

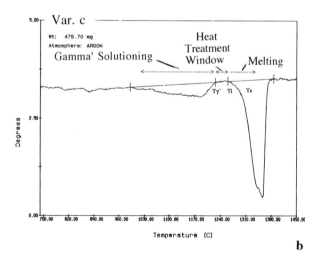

b

Figure 8a,b: Differential thermal analysis (heating rate: 10°C/min) of IN792DS (a) and Var. c (b). For Var. c, a heat treatment window of approximately 30°C between γ'-solutioning and incipient melting is apparent.

Table IV: Compilation of the composition and heat treatment window (HTW) for selected model alloys.

Alloy	Ni	Co	Cr	W	Al	Ti	Ta	B	C	HTW (°C)
Var. c	bal.	9.0	12.0	9.0	3.5	3.0	5.0	0.015	0.07	32
Var. d	bal.	9.0	12.0	6.4	4.1	3.3	5.0	0.015	0.07	< 0
Var. e	bal.	9.0	12.0	6.4	4.2	2.0	6.8	0.015	0.07	28

In tab. IV, a number of model compositions are given for which the heat treatment window was measured by DTA (fig. 8b). Comparing Variants d and e, it is apparent that replacement of Ti by Ta helps to enlarge the heat treatment window. For obvious reasons, it is also beneficial to replace γ'-forming elements (Al, Ti, Ta) by refractory elements such as tungsten (compare Var. c with Var. d).

Note, that the requirements for GB cracking resistance and acceptable heat treatment window point in the same direction, namely reduction of the alloy segregation level by formulation of "lean" compositions. It turns out that a Ta/Ti-ratio > 1.5 along with a Ti-content < 3.5% is generally preferable. Although corrosion resistance tends to decrease with decreasing Ti-content, acceptable levels can still be maintained following these guidelines /8/, especially since homogeneity adds to the environmental resistance. This strategy is quite different from previous alloy designs for stationary gas turbine applications where high Ti-contents at moderate Ta-levels were selected in the quest to maximize corrosion resistance.

Experience with ABB 2 DS

ABB 2 DS is an alloy that was developed along the lines discussed in the previous sections. The composition is derived from Var. b with further modifications to ensure SDC > 1. The typical composition range is given in /10/. Important features are a relatively high Ta/Ti- ratio >1.5 which was found to be necessary to achieve requirements on GB cracking resistance, freckle resistance and heat treatment window. Compared to conventional alloys such as IN738LC or IN792, the tungsten level of approx. 7.5%-9.5% is also relatively high. This was found to be beneficial from a strength point of view. In fig. 9, creep and fatigue properties are shown in comparison to IN738LC and CM247LC DS. Despite a Cr-content of approx. 12%, ABB 2 DS shows remarkably good properties. They are comparable to superalloys with significantly lower Cr-content. The creep strength advantage over IN738LC is approx. 50°C. This is partly attributed to the excellent heat treatment response of the alloy. For industrial applications, a two stage heat treatment turns out to be sufficient for complete solutioning (fig. 10).

Using a complex blade of approx. 300mm length, the castability of ABB 2 DS was evaluated in comparison to state-of-the-art alloys (CM247LC , CM186LC). Results are summarized in tab V. It is noted that ABB 2 DS shows excellent castability. Despite the complexity of the blade, which is reflected in the GB cracking phenomenon observed for CM186LC DS, 100% casting yield based on standard defect acceptance criteria has been achieved on a statistically significant number of parts. Very good castability is also observed for CM247LC DS. Freckles are generally shallow and can be machined off. On the one hand, this result confirms the excellent castability of CM247LC DS in stationary gas turbine applications. On the other hand, it demonstrates that DS-castable compositions can successfully be formulated in the 12%-Cr range.

Table V: Summary of the casting results for three DS-alloys (n.e.: not evaluated).

	ABB 2 DS	CM247LC DS	CM186LC DS
GB cracking	No	No	Yes
Equiaxed grains	No	No	n.e.
Freckles	No	Yes	n.e.

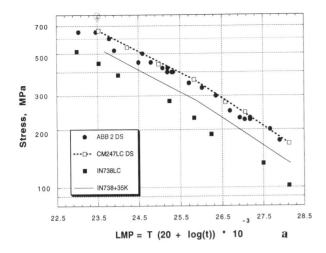

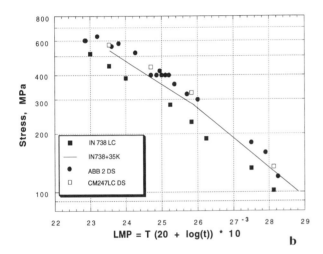

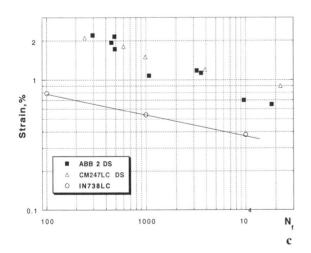

Figure 9a-c: Mechanical properties of ABB 2 DS in comparison to IN738LC and CM247LC DS.
a, b: Longitudinal (a) and transverse (b) creep strength in Larson-Miller representation.
c: LCF data in longitudinal direction at 850°C.

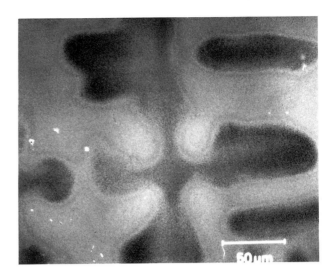

Figure 10: Microstructure of ABB 2 DS after solutioning at 1220°C/2h, 1240°C/2h.

Conclusions

Fundamental aspects of DS-castability were investigated with emphasis on Hf-free compositions. It was concluded that the amount of interdendritic segregants has to be reduced to improve resistance against grain boundary cracking and to enlarge the heat treatment window. Furthermore, a new parameter, the solidification density coefficient SDC, has been proposed to describe the freckle resistance of DS-alloys quantitatively. The excellent performance of alloy ABB 2 DS in terms of castability and high temperature strength is taken as evidence for the validity of the concepts discussed here.

Acknowledgment

Part of this work was sponsored under the Swiss Priority Program for Materials. The authors would also like to thank Prof. J.-J. Chene and Dr. H. Guo for stimulating discussions and the performance of the Gleeble experiments.

References

/1/ D.N. Duhl and C.P. Sullivan, "Some Effects of Hafnium Additions on the Mechanical Properties of a Columnar-grained Nickel-base Superalloy," Journal of Metals, (July 1977), 38.

/2/ M.C. Flemings and G.E: Nereo, Trans. Met. Soc. AIME, 212 (1967), 1449.

/3/ T.M. Pollock et al., "Grain Defect Formation during Directional Solidification of Nickel Base Single Crystals," Superalloys 1992, ed. S.D. Antolovich et al. (Warrendale, PA, The Metallurgical Society, 1992), 125.

/4/ M.R. Winstone, J.E. Northwood, "Structure and Properties of Directionally Solidified Superalloys in Thin Sections," Solidification Technology in the Foundry and Cast House (Coventry, England: The Metals Society, 1980), 298.

/5/ J.-J. Chene and H. Guo, "Characterization of Hot Ductility Behavior of Directionally Solidified Superalloys," Materials Research for Engineering Systems (Bern, Switzerland: Technische Rundschau, 1994), 20.

/6/ C.T. Sims, N.S. Stoloff and W.C. Hagel, eds., Superalloys II (New York, USA: John Wiley & Sons, 1987), 193.

/7/ M. Konter, "Computer Aided and Experimental Development of Superalloys for Gas Turbines," Ph.D. Thesis, CNIITMASH, Moscow, 1992.

/8/ J. Rösler and M. Konter, "Development of a High Strength Corrosion Resistant DS Superalloy," Materials for Advanced Power Engineering (Netherlands: Kluwer Academic Publishers, 1994), 1213.

/9/ P. J.-L. Meriguet, eds., Advanced Blading for Gas Turbines, COST501, Round II, Work Package 1 (Brussels, Belgium: European Commission, 1995), 61.

/10/ J. Rösler, DE Patent File 4,323,486 A1, BRD.

ADVANCED AIRFOIL FABRICATION

James R. Dobbs and Jeffrey A. Graves
General Electric Company
Corporate Research and Development
Schenectady, NY 12301

Sergey Meshkov
Rybinsk Motor-Building Design Bureau
Rybinsk, Yaroslavl Region, RUSSIA

Abstract

Improved turbine engine performance can be obtained by incorporating advanced materials which can tolerate higher turbine inlet temperatures. Higher inlet temperature in itself does not guarantee improved performance if a concomitant increase in cooling air is required. Such parasitic losses would reduce efficiency and can actually result in a net decrease in overall performance. Therefore the greatest likelihood for achieving a significant performance boost or durability increment will be obtained through a coordinated program that strives to improve both airfoil material temperature capability while maintaining or improving cooling efficiency.

This paper describes a manufacturing process under development by General Electric which addresses these issues. In addition, Rybinsk Motor-Building Bureau, Russia, has been independently pursuing a similar technology and has been instrumental in demonstrating proof of concept of several key technological issues.

Under development is a fabrication scheme for producing advanced turbine airfoils which will allow the use of novel materials and the flexibility of tailoring the microstructures while maintaining or improving cooling efficiency. In addition, these goals are targeted to be met at a cost which is comparable to current high performance airfoils. The process involves using electron beam physical vapor deposition for applying an outer skin on an inner spar made of conventionally cast single crystal Ni-base superalloy or an intermetallic.

Introduction

Background

During the past 30 years turbine airfoil temperature capability has increased on average by about 4°F per year. Two major factors which have made this increase possible are 1) advanced processing techniques which either improved alloy cleanliness (which leads to improved properties) or enabled the production of tailored microstructures such as directionally solidified or single crystal material, and 2) alloy development resulting in higher use-temperature materials primarily through the addition of substantial additions of refractory elements such Re, W, Ta, and Mo.

Although the incipient melting temperature of nickel alloys has been steadily increased (thus contributing to a higher use-temperature), eventually, as the incipient melting temperature approaches the melting point of the nickel alloys, the increased performance benefits versus the cost of developing and processing the alloys undergoes diminishing returns[1]. Superalloy advocates rightfully point out that predictions of Ni-base superalloys' imminent demise due to use-temperature requirements approaching the alloys' theoretical temperature capability have been voiced for many years. During this time Ni alloys remained the dominate material in the hot section of modern turbine engines and continued to post increases in temperature capability. In spite of this robustness, it must be acknowledged that inevitably the limit of nickel alloys will indeed be reached, at which point avenues other than alloying and process refinement must be sufficiently mature for transition into a production setting.

Current state-of-the-art superalloy turbine blade surface temperatures are near 1150°C (2100°F) while the most severe combinations of stress and temperature corresponds to an average bulk metal temperature approaching 1000°C (1830°F)[2]. Recent in-house studies at GE have documented several "double wall" concepts that provide 15-25% reductions in cooling airfoil requirements without a rise in surface temperature. Unfortunately, such airfoil concepts are extremely difficult to produce in directional castings because of the small dimensions and fragility of the ceramic core. Blade cost can be extremely high when the casting procedure results in a poor yield of successful cores. However, it may be possible to produce double-walled airfoils economically by an unconventional process using electron beam physical vapor deposition (EB-PVD)[3] for applying an outer skin on a single crystal Ni-base superalloy or intermetallic spar as the interior wall.

Spar-Skin Concept

The conceptual airfoil shown in Figure 1 has a central spar of a single crystal superalloy with a relatively simple inner cooling plenum geometry. This simplicity allows a substantial increase, as much as 25-33%, in the directional solidification casting yield as compared to current complex single crystal castings.

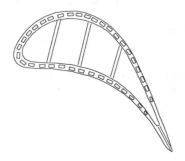

Figure 1 Conceptual Hollow Airfoil Designed for Affordable Fabrication

A general step-by-step procedure for the formation of the spar/skin blade is shown in Figure 2. In brief, the process, which is being termed Optimal Manufacturing for Efficiency Gains in Airfoils (OMEGA), involves casting the inner spar, machining cooling passages on the spar surface, filling the grooves with a sacrificial filler, depositing an outer skin, and leaching out the filler. The savings from decreased casting costs can offset the added cost for producing the airfoil outer skin configuration. Part of the added cost is for forming channels approximately 0.1 cm (0.040") deep by 0.255 cm (0.1") wide on the outer surface of the spar casting that eventually serve as the cooling channels between the double walls. As envisioned, the channels will be cast directly in the spar wall, although conventional machining, waterjet machining, ECM, or photoengraving techniques for forming the channels are alternative low-cost processes being evaluated. Cooling channels produced in these ways are very accurately positioned with respect to the wall, so metal currently added to castings to account for core shifts can be eliminated, producing a lighter structure. Dimensions of the cooling channel can be made to the designer's needs, rather than being dominated by the fragility of ceramic core materials thus dramatically improving design flexibility for the cooling channels.

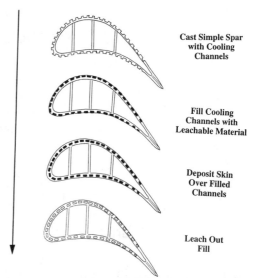

Figure 2 Overview of processing steps for fabrication of spar-skin turbine airfoil.

The spar and outer skin can be either a monolithic superalloy or an intermetallic/superalloy laminate. The strengthening phase may be a silicide intermetallic (e.g. Cr_3Si), a topologically close-packed phase such as σ, or a compound (e.g. TiC)[2]. Intermetallic laminate composites as the skin can operate at temperatures higher than conventional Ni-base single crystal superalloys, and the projected specific strengths of the superalloy-toughened intermetallic matrix composites (IMCs) far exceed those of Ni-base superalloys at the maximum operating temperatures of most turbine airfoils. The temperature capability gain and possible weight reduction due to specific strength improvement can be augmented by the more efficient cooling schemes possible from the skin deposition process. One attractive feature of the spar/skin concept is that its implementation can range from a blade with a simple cooling geometry using a conventional superalloy spar and skin to an advanced blade with complex cooling and an intermetallic spar and skin which is not attainable by current casting techniques. This flexibility allows for an evolutionary approach to integrating advanced materials and advanced cooling concepts.

Development Strategy and Status

The fundamental strategy adopted for developing this processing scheme addresses the following key points: 1) demonstration of lab-scale feasibility of individual processing procedures including spar casting, channel filling, and skin deposition, 2) estimation of cost differential compared to exiting fabrication schemes, 3) analytical predictions of performance gains or durability improvements using the OMEGA process and incorporating existing as well as new blade designs, 4) spar/skin concept validation, and 5) rig or engine demonstration.

The major programmatic accomplishment to-date has been the design and fabrication of the EB-PVD unit. While cost analysis and initial performance modelling results have also been completed, the bulk of the material development and characterization work still lies ahead. What follows is a brief description of progress made in each of the development areas.

1) Lab-Scale Feasibility

One of the attractive features of the OMEGA process is that most of the processing steps consist of fairly conventional techniques. In particular, the casting requirements for the single crystal spar are less demanding than for blades which are currently in commercial production since there is no need for small, fragile, and complex cores to create the cooling channels. Obviously as the complexity of the OMEGA blade increases so do the design challenges. In one scenario both the spar and skin will be IMC laminates, at which point the techniques for casting the IMC spar will need to be developed. However in the near-term version of the OMEGA blade, casting of the Ni-base single crystal spar with the cooling channels should not require additional development. If casting of the surface grooves proves to be impractical, machining of these surface grooves on the cast spar can be accomplished with standard techniques.

Basic feasibility was addressed by EDM'ing ½"-O.D., 3/8" I.D. tubes from N4 single crystals with eight 0.060" diameter semi-circular grooves machined in a symmetric pattern in the surface. 60-mil diameter Kovar tubes were placed in the channels to act as a filler and the assembly coated with Rene 142 by plasma spraying. After skin deposition the Kovar was etched out in an acid bath. The samples were then subjected to 100 cycles in a combustion burner rig. The cycles consisted of a 1-hour hold in the flame at 1800°F followed by a 5-minute hold in blowing room air. The cooling air was passed up the I.D. and out impingement holes at the top of the specimen then down the cooling channels in a single circuit. After completion of the test, metallography indicated no cracking or distress in either the spar or skin had occurred. While neither the use of Kovar tubes as the filler or the use of plasma spraying as the technique of applying the skin are optimal, this first test did demonstrate basic feasibility.

Channel Filling: The technological issue to be resolved in connection with the cooling channels is to identify a filler material that can completely fill the machined channels while retaining its dimensional stability after curing and during the subsequent skin deposition.

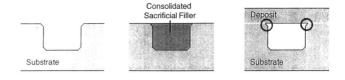

Figure 3 Sketch of channel filling and skin deposition. Care must be taken when designing the surface groove geometry to avoid creating stress concentration sites.

Depending on the method chosen for skin deposition, the filler material must be able to withstand temperatures of up to 1000°C (1830°F) during processing. Such temperatures are proposed to ensure large grains are formed in the deposited skin, thus providing adequate creep resistance. In addition, once the skin has been deposited the filler must be capable of being completely dissolved from the small passages. Potential materials under consideration include ceramics, metals, and ceramic-metal mixtures. Some filler consolidation processes being examined are cold-isostatic pressing (CIP) plus sintering, injection plus sintering, hot isostatic pressing (HIP), and CIP plus electroconsolidation.

Thus far no ideal material has been identified. Ceramics tend to have poor adhesion and a low coefficient of thermal expansion (CTE). The low CTE leads to separation from the channel wall during skin deposition. Metal filler adhesion depends on the consolidation process. In particular, pressureless sinters lead to poor adhesion while the pressure-assisted densification routes give good adhesion. In addition to identifying a filler material, the remaining obstacles include finding a reproducible method of exactly filling the channels so that machining is not required to obtain a flat surface. Also, design of the deposit-skin interface is an issue to insure that no stress intensification occurs which may initiate cracks (Figure 3).

Skin Deposition: The issues concerning skin deposition center around the desired architecture of the skin. The greatest performance gains will be achieved by using high-temperature intermetallics as the spar and skin material. As an evolutionary first step toward this goal, current work also includes efforts on single crystal superalloy spars (produced by conventional casting techniques) and polycrystalline superalloy skins applied by either EB-PVD or plasma. Although not possessing higher temperature capabilities than current materials, a superalloy spar/skin blade will benefit from reduced cooling requirements due to the advanced cooling passage geometries and thus greater engine efficiencies, while at the same time having similar manufacturing costs (as detailed in the next section) as conventional blades. While EB-PVD will allow the more complex architectures to be deposited, reproducible compositional control of materials composed of a wide range of vapor pressure constituents has not been achieved. For that reason plasma sprayed skins are being examined as a near-term alternative while the mechanics of the EB-PVD system are refined. Once the spar/skin concept has been validated using superalloys, the more advanced spar and skin laminated materials will be pursued. It is at this stage that the use of EB-PVD would be required since successive layers of alternate materials can be deposited in a multi-pool unit by controlling which melt pool is being vaporized. Such material layering would not be feasible in a plasma system.

In the following sections, the progress and issues arising in each of these deposition techniques are outlined.

EB-PVD: The EB-PVD process heats and evaporates a molten pool with the condensate collected on a substrate near the source. It is a high-rate process since local temperatures in the liquid pool near the electron beam source may be as high as 4400°C (7950°F). The molten pool is supplied with fresh material either from an adjacent liquid or solid source or from a bottom-fed rod stock.

The rate of metal deposition by EB-PVD may be as high as 25 µm (1 mil) per minute. This is two orders of magnitude faster than the fastest rates of sputter deposition processes, and it is this feature that brings the cost of EB-PVD processing of high-temperature components within the range of current aircraft engine blade commercial manufacturing costs. On the other hand, while pure elements and multi-component alloys with similar vapor pressures are fairly easy to evaporate, alloys which contain elements with vastly different vapor pressures are more difficult because vapor concentrations over the pool and the resultant condensate are much less predictable. In addition, small fluctuations in pool depth, solid feed rate, or pool temperature may cause major fluctuations in the vapor concentration over time.

A major milestone was achieved in mid-1994 when an advanced manufacturing-scale evaporator, dedicated to metallic development processes became operational at GE-CRD. A sketch of the unit is shown in Figure 4. The 300 Kw unit is capable of computer-controlled rotation and oscillation of substrates over two separate evaporation zones each of which can have three pools. The controlled oscillation and rotation will permit control of thickness on complex airfoil shapes, as well as the application of microlaminated structures. The advantage of using multiple pools is that elements with low vapor pressures (such as Mo, Ta, Nb, and Ti) can be isolated and melted independent of the higher vapor pressure elemental constituents. With this arrangement, processing parameters such as ingot feed rate and electron beam power can be tailored to the individual pools and thereby optimize the deposition rate while closely controlling chemistry. In addition, there are two 8 KW, electron-beam over-source substrate heaters for accurate substrate temperature control. Access for thermocouple or IR substrate temperature measurement is also provided. A Mikron Corporation Model 9004 Imaging Radiometer is used to measure the pool temperatures and uniformity of heating. The data will be used in an on-going modelling program for understanding both the molten metal pool dynamics and for design of advanced crucibles. Evaporation rates of several kilograms per hour have been achieved with collection rates exceeding 25µm/min. Oxygen content in deposits was found to be less than 50 ppm due to a unique vacuum load chamber and seal system.

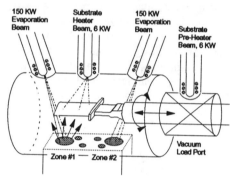

Figure 4 Sketch of EB-PVD unit showing multiple ingot feed pools, dual evaporation beams, and substrate heaters.

As a further aid to compositional control, melting is being performed using the "through-tungsten" approach. In this technique, tungsten is incorporated in the melt pool not as an constituent in the deposit but as a molten pool stabilizer. During melting, the tungsten, because of its extremely low vapor pressure, permeates the melt and acts to regulate the volatilization of all the elements in the pool to some predictable, uniform rate. This results in both greater deposition rates and a more stable composition.

While correcting the ubiquitous mechanical problems which inevitably arise in custom-built equipment, work is progressing on perfecting the multiple pool and through-W approach so as to allow for the deposition of superalloy compositions within tight compositional tolerances. The first trials have used a single pool to deposit a relatively simple target alloy chemistry of 71-Ni, 10.6-Cr, 13.2-Al, 5.3-Co (at.%). 40-mil thick deposits have been made and are currently being analyzed for chemical uniformity both in the as-deposited and homogenized condition. Once this composition can be deposited, more complex alloys will be attempted. As the alloy compositions approach those of commercial superalloys, development of multi-pool depositing will be required to handle the wide range of vapor pressures present.

Plasma Spraying: Plasma spray processes utilize the energy contained in a thermally ionized gas to melt partially and propel fine powder particles on to a surface such that they adhere and agglomerate to produce a coating. The principal advantage of plasma its is ability to deposit complex compositions reliably. In theory, any material that melts without dissociating or subliming can be sprayed using a plasma arc.

Plasma deposits were performed using the RF-plasma spray facility at GECRD. Because of the superior heating characteristics of RF compared to DC plasma, the RF process allows the use of larger size powder for spraying. The resulting larger grain size and lower oxygen content, together with the cleaner plasma environment, yields mechanical properties which are superior to those possible with DC plasma. This is based on GECRD's experience[4] with Rene 80 where yield stress was found to be comparable to cast material and no severe ductility loss was noted. As for creep resistance, the RF plasma material had ruptures lives at 1500°F of 120 hrs on average compared to 1000 hrs for the cast material. It is obvious that even with the larger grain sizes possible in RF plasma, improved creep resistance remains the critical concern for plasma-deposited material.

It is acknowledged that plasma spraying is not capable of producing intermetallic laminate skin structures, and in the as-sprayed condition the long-term creep and stress rupture properties may be insufficient for the demands placed on turbine blades. Inclusion of plasma material in the current program is justified in that it does provide a convenient means of applying superalloy skins on spars and thus allows for an initial evaluation of the spar/skin concept while the EB-PVD technique matures. It is expected that many of the most important issues which could eventually arise from the plasma work will be present in the EB-PVD material as well. Some of these issues which are likely to be common to skins applied by both plasma and PVD are skin adhesion, the effect of a skin material on the fatigue properties of the single crystal spar, thermal expansion mismatch effects, and electrochemical differences. Some issues which are more pertinent to plasma skins are porosity and grain size (since plasma is restricted to monolithics and can't rely on laminates for creep resistance).

The Ni-base alloys chosen for the developmental work are single crystal Rene N5 for the spar material and Rene 142 as the skin. Rene 142 is the directionally solidified version of N5 with the addition of grain boundary strengtheners. A complete characterization of the plasma-deposited Rene 142 material has been a first priority. Material attributes of interest include basic tensile properties, heat treatment, porosity, chemistry, etc. These properties must be determined because published data on Rene 142 is available primarily for material in the directionally solidified condition. Since the plasma sprayed skin has no crystallographic texture and may be somewhat different compositionally, it is crucial that the material properties in the sprayed condition be documented.

The first attempts at depositing Rene 142 were unexpected in that unacceptably high porosity levels (~7%) were noted. Numerous additional runs were performed with different processing conditions but to no avail; porosity levels remained at about the 7% level. It has become necessary to perform a full design of experiments (DOE) matrix in order to find the optimum conditions to achieve as-sprayed low-porosity material. As of this writing the DOE results were not available.

Another study is underway to find a heat treatment schedule that can be applied to a Rene N5/Rene 142 composite spar/skin component. Although compositionally similar, the heat treatment time-temperature profile for the Rene N5 and 142 are significantly different (due primarily to the presence of the grain boundary strengtheners in Rene 142). This means that a modified heat treatment must be developed which can be applied to the composite structure without degrading the properties of either constituent.

Finally, Rene 142 coatings have been applied to N5 single crystals to characterize the interfacial stability and strength as well as the fatigue, creep, and stress rupture properties of the "composite".

When the PVD process matures to the point where complex compositions can be deposited reliably, the knowledge-base derived from the plasma studies can be transferred to the EB-PVD material. In addition to the experience gained in the plasma studies it is possible that the plasma sprayed spar/skin material may be a viable structure in its own right if some of the issues such as small grain size can be controlled or accounted for in the design of the skin.

2) Cost

One of the first program priorities was to generate a detailed flowchart of the envisioned manufacturing procedure. From this, potential suppliers were consulted in order to develop a realistic cost assessment to ensure that the process was economically feasible. Costs were evaluated at the line item level. The basis of the cost comparison was an aero airfoil design approximately equivalent to a typical fighter engine first stage blade. The external airfoil shape was maintained and a cooling cavity design possible via the Omega process was substituted. Heat transfer and cooling effectiveness of the new cooling cavities were not determined, but were conservatively estimated. Actual cavity design would probably be less rigorous and subsequently less costly. Based on these parameters, initial Cost estimates for the Omega process showed that highly cooled multi-walled airfoils could be produced for approximately 70% of the cost of a similar investment casting.

In addition to initial costs of producing airfoils, there may be additional savings in terms of life-cycle costs of IMC blades. Used single crystal blades cannot presently be refurbished to original properties; tip cracks above the tip cap can be repaired only to equiaxed properties, but cracks below the cap are not repairable. For IMC-skinned airfoils, machining or etching the skin back to the plane of the single crystal spar may allow spar recovery through re-deposition. This would allow a significant increase in repairability and thus spar life.

4) Spar/Skin Concept Demonstration

The Rybinsk Motors Building Design Bureau (RKBM) of Rybinsk, Russia has provided a successful demonstration of the shell-spar cooling design concept for the production of gas turbine airfoils (Figure 5). Such components have

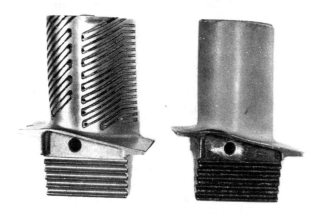

Figure 5 D277 engine turbine blade. On the left is the spar casting and on the right the finished blade after skin deposition.

been designed and used in both aerospace and industrial gas turbine engines, including the D277 turbofan engine and RD-38 lift engine. Current engine experience has been extremely successful, exceeding 1000 hours in Rybinsk aerospace gas turbine applications. RKBM has achieved these accomplishments as a result of over ten years of development experience using investment cast spars with Mcraly skins applied using a modified Movchan EB-PVD coater[5]. In addition to its significant experience with the production of highly cooled turbine airfoil components, RKBM also has extended this concept to the production of internally heated compressor blades for the TVD-1500 engine, to impart anti-icing capabilities.

Using the shell-spar technology RKBM is able to increase the cooling effectiveness of the turbine blades above that provided by combined convective and film cooling schemes used by present production turbine blade designs (Figure 6). Such cooling effectiveness is believed to be

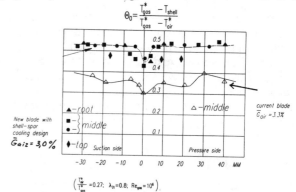

Figure 6 Experimental results comparing cooling effectiveness of current blades (open symbols) to shell-spar blades (filled symbols) on RD-38 engine.

similar to that provided using transpiration cooled airfoils[6,7,8]. In addition, the technology enables designers to reduce temperature gradients present within the cooled airfoils by optimization of the cooling hydraulics. Using this attribute it is possible to maximize the components durability to thermal-mechanical fatigue distress. A third advantage of the concept is the inherent flexibility it provides for rapid optimization of the airfoil cooling design. By machining in the external cooling passages of the hollow spar, various cooling designs may be rapidly and inexpensive produced for rig testing prior to committing to a final optimized configuration for investment casting. It is anticipated that this characteristic of the technology can reduce the design cost of cooled airfoil components. In addition, the increased cooling effectiveness enables the use of low cost superalloys and casting processes to offset the inherently high manufacturing cost of the EB-PVD airfoil skin.

To-date, engine tests of shell-spar components have been conducted using Mcraly airfoil skins. The Rybinsk modifications to the Movchan EB-PVD coater permit the rapid deposition of a multi-layer airfoil skin in which the chemical composition of the skin is gradually varied to produce controlled thermal compatibility with the underlying superalloy spar (Figure 7).

Mechanical verification of the shell-spar structure is obtained using internally cooled hollow shell-spar test specimens to which cyclic mechanical loads are applied both longitudinal and transverse to the channel orientation. When combined with the component structural analysis, the fatigue strength of the relatively weak Mcraly skin was verified to resist the components anticipated thermal and mechanical fatigue loading. Such procedures are also effective for development and evaluation of new spar-skin material systems.

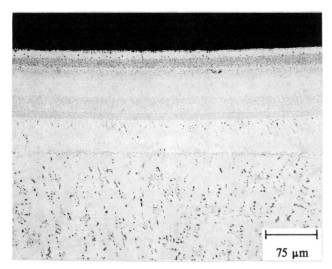

Figure 7 Multi-layered MCrAlY skin used on D277 turbine blade. Skin composition is gradually modified to match thermal compatibility between the superalloy spar and the environmentally protective external MCrAlY coating.

With the recent discovery of their mutual interest in shell-spar technology, RKBM has recently joined the ongoing efforts of GE shell-spar program and plans are being made to accelerate the introduction of this new manufacturing technology in the turbine airfoil supplier base. The organizations pursuing this airfoil concept have a common goal of development, optimization, and widespread availability of the shell-spar airfoil technology within the international gas turbine industry.

5) Engine Demonstrator

The first introduction of the OMEGA process in a General Electric engine is scheduled to occur in 1997. Utilizing the OMEGA process, leading-edge, rear-frame struts for the XTC-76 IHPTET demonstrator will be manufactured by EB-PVD with a NiCoCrAl-base alloy with refractory-element additions as the skin on a conventionally cast N5 spar. For this first application, the fill material is envisioned to be a simple Monel, which previous experiments have shown can be fully etched from the small cooling passages. It is anticipated that this demonstrator will provide valuable engine-environment information for guiding future development of this technology.

Summary

A proposed process for fabricating advanced cooling geometry turbine airfoils with tailored microstructures by electron beam physical vapor deposition has been described. The major accomplishments of the program thus far are 1) completion of the EB-PVD unit, 2) deposition of simple NiCoCrAl alloys and Rene 142 on N4 and N5 single crystals, 3) successful completion of burner-rig testing of R142/N4 without distress to spar or skin, and 4) predicted economic feasibility and performance benefits from the new procedure. Additionally, with the recent teaming of Rybinsk Motor-Building Design Bureau, development of this technology could be greatly accelerated because of their experience in both design and implementation of the spar-skin concept amassed over the past decade.

References

1. F.O. Soechting, "A Design Perspective on Thermal Barrier Coatings", in Thermal Barrier Coating Workshop, NASA Conference Proceedings, CP 3312, Cleveland, OH, March 27-29, 1995, pp. 1-15.

2. M.R. Jackson, B.P. Bewley, R.G. Rowe, D.W. Skelly, and H.A. Lipsitt: "High-Temperature Refractory Metal-Intermetallic Composites", JOM, Vol. 48, No. 1, pp. 39-44.

3. D.H. Boone, "Physical Vapour Deposition Process", Mat. Sci. Tech., Vol. 2, 1986, pp. 220-204.

4. P.A. Siemers, M.R. Jackson, J.R. Rairden, and S.D. Savkar: "Rapid Solidification Plasma Deposition (RSPD) for Fabrication of Advanced Aircraft Gas Turbine Components - Volume II", AFWAL-TR-86-4071, 1986.

5. S. Ashley, Mech. Eng., 1995, pp. 66-69.

6. H. Mueller-Largent and D.J. Frasier, Final Report NAPC-PE-225-C, 1991.

7. P.S. Burkholder, M.C. Thomas, D.J. Frasier, J.R. Whetstone, K. Harris, G.L. Erickson, S.L. Sikkenga, and J.M. Eridon, Institute of Materials 3rd International Charles Parsons Turbine Conference, Materials Engineering in Turbines and Compressors, Newcastle Upon Tyne, U.K., (April 25th-27th 1995).

8. Fullagar K.P.L., Broomfield R.W., Hulands M., Harris K., Erikson G.L., Sikkenga S.L., 39th ASME/IGTI International Gas Turbine & Aero Engine Congress & Expo, The Hague Netherlands, June 13th-16th, 1994).

THERMAL ANALYSES FROM THERMALLY-CONTROLLED SOLIDIFICATION (TCS) TRIALS ON LARGE INVESTMENT CASTINGS

Patrick D. Ferro
Sanjay B. Shendye
Precision Castparts Corporation
Portland, Oregon USA

Abstract

Thermally controlled solidification (TCS) has been developed as a patented process for casting complex components with minimal shrink. The principle of TCS is based on the controlled advancement of the solidification front in an investment casting mold. Benefits of the TCS process include reduced as-cast shrink levels, and enhanced abilities to cast thin walls and feed complex configurations.

The development of the TCS process began at Precision Castparts Corporation in 1988, using a modified directional-solidification furnace. In 1992, PCC procured a laboratory-scale TCS furnace and tested the capabilities of the TCS process with pour weights up to 45 kg. In 1995, PCC completed the construction of a TCS furnace with the capability for pour weights up to 370 kg.

One of the primary differences between castings made by the TCS process and castings made by directional solidification or single crystal processes is in the resultant as-cast grain size. Directional soldification and single-crystal processes produce as-cast microstructures with a large average grain size, and the TCS process results in an as-cast microstructure with grain sizes typical of that observed on conventionally-cast parts. Additionally, the TCS process results in a consistent and uniform grain size in all areas of a cast part.

To correlate observed as-cast results with known processing parameters, several techniques using thermocouple data have been investigated. For example, the ratio of the thermal gradient at the solidification front (G) to the solidification front advancement rate (R) is an indicator of the as-cast microstructure. Relatively large G/R ratios tend to produce microstructures with large, unidirectional grains and with relatively little shrink. By comparison, relatively low G/R ratios tend to result in microstructures with a generally higher degree of shrink, and a finer structure depending upon the solidification front advancement rate R. Under optimal conditions, the TCS process uses G/R ratios that result in microstructures with relatively small, equiaxed grains and with minimal shrink.

Besides use of the G/R ratio to characterize solidification conditions during TCS trials, additional thermal analyses have been investigated. One analysis uses the equivalent time calculation to quantify the thermal conditions during solidification at several physical locations on the mold.

Background

Directional solidification and single crystal casting techniques have been part of the superalloys industry since at least 1960 [1]. The objective of directional-solidification and single-crystal casting techniques is to produce components with minimal grain boundaries, and to improve high-temperature properties of cast components.

The TCS process evolved from the directional-solidification and single-crystal casting techniques. The TCS casting process is different from directional-solidification and single-crystal casting techniques in that TCS produces components with equiaxed grain structure and with minimal shrink.

The scope of this paper includes examples of thermal analyses that have been used to characterize the TCS process during early furnace trials. The purpose of thermal analyses of TCS thermocouple data is to provide quantitative methods to correlate observed metallurgical results with input process parameters. The two thermal analyses presented here are the G/R analysis and an analysis that calculates the equivalent time at a constant temperature during solidification. An example from one TCS casting trial is provided.

Procedure

The TCS process utilizes a cylindrical resistance-heater, or retort, which heats up a shell to establish a vertical gradient in the shell prior to pouring. The retort has a working zone of 110 cm diameter and 100 cm height, and has the capability to heat shells up to 1480°C. The diameter of the chill plate is 105 cm. Figure 1, from a PCC-held patent [2], shows a schematic diagram of the TCS furnace.

Prior to TCS casting, the invested and de-waxed shells are heated prior to transferring to the TCS furnace. When a shell is transferred from the burnout furnace to the TCS furnace, the shell is lowered onto the copper chill plate, ensuring that the shell is centered on the chill plate. Centering the shell allows uniform heat distribution on the shell during preheat and withdrawal.

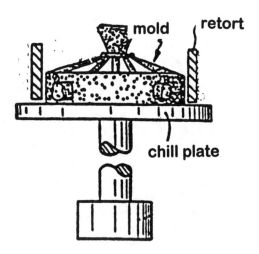

Fig. 1: Schematic diagram of the TCS furnace, indicating how molds are withdrawn from a cylindrical hot zone [2].

With the shell in the vacuum TCS chamber, the TCS heater is lowered to surround the shell. The shell is heated up to a thermal profile prior to pouring metal. Thermocouples are used on the shell to control the final shell thermal profile before pouring. A temperature differential on the order of 110°C or more, from the top of the shell to the bottom of the shell (near the chill plate), may be observed.

While the shell is being preheated, the alloy is vacuum-induction melted and stabilized at the final pouring temperature. The alloy is poured into the shell over the lip of the crucible. After the alloy has been poured into the shell, the heater is lifted up and away from the shell at a programmed rate, called the withdrawal rate.

Results

Thermocouples on the shell indicate the solidification conditions in the shell during the TCS withdrawal. One of the main objectives of monitoring the shell and alloy temperatures during solidification is to quantify the thermal conditions that correspond with minimal shrink conditions in the casting. Two analyses used to quantify the thermal conditions during solidification are 1) the G/R analysis and 2) the equivalent time analysis.

G/R Analysis

The G/R analysis refers to the ratio of the temperature gradient (G) across the solidification front to the solidification front advancement rate (R). G/R has been referenced in the solidification literature [3] as an indicator of as-cast microstructure. Relatively high G/R ratios tend to correlate with single-crystal and columnar types of microstructures, and generally with minimal shrink. Conversely, relatively low G/R ratios tend to correlate with equiaxed microstructures with generally high shrink levels.

In the present work, Type S thermocouples were placed on the outside of the shell to provide an indicator of solidification conditions inside the shell. Previous work has shown that shell thermocouples are at least representative of thermal conditions in the metal, at the solidification front. Shell thermocouples give an output which is generally lower than thermocouples in the mold cavity, and also generally appear to lag thermocouples in the mold cavity. The advantages of using shell thermocouples include (1) a high degree of re-usability and (2) easier to use in a production environment. Thermocouples are generally placed in a vertical orientation on a shell to allow for G/R calculations at different locations and at different times during solidification.

In practice, G/R is calculated by the following approximation:

$$\frac{G}{R} = \frac{G^2}{GR} \quad (1)$$

$$\frac{G^2}{GR} = \frac{\left(\frac{T_{1,t} - T_{2,t}}{z1 - z2}\right)^2}{\left(\frac{T_{1,t-\Delta t} - T_{1,t}}{\Delta t}\right)} \quad (2)$$

$$[=] \quad °C \, s \, cm^{-2}$$

The approximation shown in equation (2) allows for the estimation of G/R using data from two thermocouples, at locations 1 and 2, at time t. Thermocouples 1 and 2 are located at distances z1 and z2, measured in centimeters from the chill plate.

Figures 2 to 5 show examples of graphs giving G/R as a function of deviation from the liquidus (referred to as delta liquidus) for one mold. The chemical composition of the alloy used in this experiment (in wt %) was 22 Cr, 17.9 Fe, 8.5 Mo, 1.3 Co, 0.55 W, 0.67 Si, 0.02 Mn, 0.01 C, 0.004 P, 0.002 S, bal. Ni. The liquidus and solidus for the alloy was 1357 C and 1329 C. The thermocouple data were obtained from a test mold which comprised of four 24 x 84 cm panels, each less than 1 cm nominal thickness. Type S thermocouples were located at five locations along the length of the shell, at 75, 60, 46, 28 and 11 cm from the chill plate for thermocouples 1 through 5 respectively. The graphs provide calculated G/R values from thermocouple data received from thermocouples 1 and 2 (fig. 2), thermocouples 2 and 3 (fig. 3), thermocouples 3 and 4 (fig. 4) and thermocouples 4 and 5 (fig. 5).

G/R as a function of temperature, such as those shown in figs. 2 to 5, are useful for determining baseline ranges of parameters which result in minimal shrink in a casting. One method for quantifying and comparing data from TCS casting trials involves determining the G/R value when the upper thermocouple, for a given location, reads 1260°C.

Comparing the graphs shown in figs. 2 through 5 indicates that G/R values were generally highest at lower positions in the casting. For example, the G/R data shown in figs. 4 and 5 are generally higher than comparable data shown in figs. 2 and 3, especially at relatively higher temperatures i.e. at lower delta liquidus, where delta liquidus is the difference in the liquidus temperature of the alloy and the temperature recorded by the thermocouple on the top end of the part. Since figs. 2 and 3 correspond with higher relative positions on the casting (farthest distances from the chill plate), the data indicates that G/R levels were higher at lower positions in the casting, for the experimental data shown.

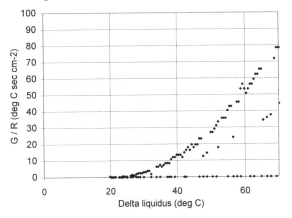

Fig. 2.: G/R as a function of shell temperature (delta liquidus) during solidification. Thermocouples 1 and 2 were 75 and 60 cm from the chill plate respectively.

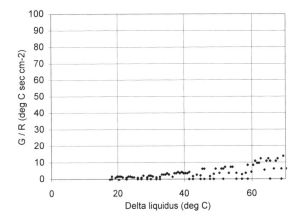

Fig. 3: G/R as a function of shell temperature (delta liquidus) during solidification. Thermocouples 2 and 3 were 60 and 46 cm from the chill plate respectively.

Table I. Example of Thermal Data from a TCS Trial

Thermo-couple number	Distance from chill plate	G/R at liquidus - 100°C (°C s cm^{-2})	t_{equiv} at 1200°C (minutes)	Temp. before pour, dev. from liquidus	Peak temp. after pour, dev. from liquidus
1	75 cm	160	43.8	25°C	21°C
2	60 cm	26	43.5	27°C	19°C
3	46 cm	1438	42.3	33°C	22°C
4	28 cm	341	35.3	86°C	28°C
5	11 cm	--	25.7	136°C	34°C

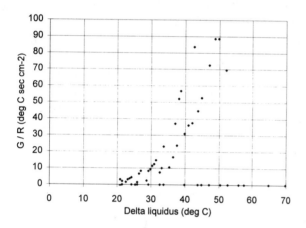

Fig. 4: G/R as a function of shell temperature (delta liquidus) during solidification. Thermocouples 3 and 4 were 46 and 28 cm from the chill plate respectively.

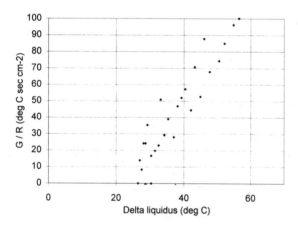

Fig. 5: G/R as a function of shell temperature (delta liquidus) during solidification. Thermocouples 4 and 5 were 28 and 11 cm from the chill plate respectively.

Equivalent time analysis

The objective of the equivalent time analysis is the same as that for the G/R analysis: to determine a thermal 'signature' that corresponds with minimal shrink and optimal grain size in TCS castings. The basis for the equivalent time calculation is an exponential function of temperature given by:

$$t_{equiv,1200C} = \sum_{t=0}^{20\min} \left[\left(\frac{\exp^{-\frac{\Delta H}{RT}}}{\exp^{-\frac{\Delta H}{R(1200C)}}} \right) \cdot \Delta t \right]$$

$$= 29382 \sum_{t=0}^{20\min} \left[\left(\exp^{\frac{-15154}{(°K)}} \right) \cdot \Delta t \right] \quad (3)$$

[=] minutes at 1200°C

In equation (3), t_{equiv} is the equivalent time at 1200°C as measured by a thermocouple on the shell. The activation energy for solidification (ΔH) that was used in eq'n (3) was 126 kJ mole^{-1}. Data from thermocouples for the first twenty minutes after pour is used in the calculation. Depending on where the thermocouple is located on a given shell, t_{equiv} as calculated with eq'n (3) may range between approximately 10 minutes and 60 minutes.

Other data used to provide a means for comparing thermal parameters for TCS casting trials include shell thermal profile before pour and peak temperatures reached by individual shell thermocouples after pour. Table I provides an example of thermal data summarized from a TCS casting trial.

Discussion

TCS is a relatively new casting process which has undergone steady progress since its inception nearly ten years ago. For pour weights up to 45 kg TCS is now a reliable process that gives consistent results for production parts.

The present paper has discussed TCS development for larger parts, with pour weights up to 370 kg. One of the challenges in developing TCS for large parts is in the determination of which aspects of the TCS process for small parts applies to the TCS process for large parts. The heavier and taller parts of the large parts TCS process requires different heat management methods before, during and after pour, compared with the small parts TCS process. For example, the mold thermal profile before for small parts may be defined by specifying required temperatures at two thermocouples, one each at the top and the bottom of a small parts mold. For some large TCS parts, the required mold thermal profile before pour may be defined by a gradient between the lowest adjacent thermocouples on a given mold, and with the upper thermocouples all at the same temperature which is higher than the lowest two thermocouples. As a result, further development is proceeding on many large parts that are candidates for the TCS process.

In the examples presented, the analyzed thermal data (G/R and t_{equiv}) were not correlated with microstructural information such as degree of shrink and grain size. One reason why this data was not correlated with microstructural information was due to the relative stage of development of the large parts TCS process. With continuing parts pouring trials, the analyzed thermal data will be correlated with metallurgical results to determine relationships between input process parameters and resultant microstructure.

Summary

Thermally controlled solidification (TCS) process trials are ongoing for large investment casting parts. The TCS process depends on the controlled advancement of the solidification front, to provide castings with minimal shrink. Other benefits of the TCS process include enhanced abilities to cast thin walls and to feed complex configurations.

Two analysis methods for correlating observed as-cast results with known processing parameters are provided. Each analysis method depends on a function of mold temperature after pour and provides a quantitative method of comparing casting trials. TCS trials on large parts, up to 370 kg pour weight, are ongoing.

Acknowledgment

The author acknowledges Mark Eimon, Matt Kernal and Dean Salvadore for their skill and knowledge in thermal data collection and retrieval, and Roger Goodman for many discussions about the operation of the TCS furnace.

References

1. F.L. Versnyder, M.E. Shank, "The Development of Columnar Grain and Single Crystal High Temperature Materials Through Directional Solidification", Mater. Sci. Eng., v. 6 (1970), pp. 213-247.

2. Ronald R. Brookes, United States Patent no. 4,724,891; Feb. 16, 1988.

3. W. Kurz, D. Fisher, *Fundamentals of Solidification*, Trans Tech Publishing, Aedermannsdorf, Switzerland (1984).

ALTERNATE MATERIALS

Gas Turbine Engine Implementation of Gamma Titanium Aluminide

C. M. Austin and T. J. Kelly
GE Aircraft Engines
Cincinnati OH 45215

Abstract

Gamma titanium aluminide will be introduced into commercial service during 1997, replacing superalloys, following a relatively brief application development and engine testing program. This paper will describe gamma and discuss its transition from laboratory curiosity to engineering material.

Introduction

Gamma titanium aluminide, TiAl, is the base for an emerging class of low-cost, low-density alloys with unique properties. Cast processing [1] is being used for production of initial components. The progress of this effort has been reported previously [2, 3]; a general review of gamma has also been published by Kim [4].

It is interesting to note in the context of these proceedings that the implementation of gamma has been led by superalloy metallurgists after initial development by titanium metallurgists. This may be a result of better familiarity with the applications for which gamma is most suited as well as the greater acceptance of cast processing by the superalloy community.

Physical Metallurgy

"Gamma" alloys are actually mixtures of the neighboring aluminide phases Ti_3Al (alpha-two, hexagonal) and TiAl (gamma, tetragonal), Figure 1. The morphology of these constituents is complex, and depends in complex ways on aluminum level, processing and heat treatment. The so-called duplex structure is generally preferred, Figure 2, which results from heat treatment in the $\alpha + \gamma$ phase field. On cooling, the structure transforms to a mixture of gamma grains and colonies of lamellar gamma and alpha-two. This is superficially similar to typical alpha-beta titanium alloys.

The lamellar constituent actually begins to form during cooling within the $\alpha + \gamma$ phase field. There is little interfacial energy involved in the formation of gamma laths within the alpha phase, and this is the most expedient way in which to maintain the equilibrium proportions of the two phases. Upon cooling below the eutectoid, additional gamma phase forms, either through thickening of existing laths or creation of new laths. The alpha phase orders to become alpha-two; at this point, the alpha-two laths make up a small fraction of the lamellar constituent.

The presence of gamma grains, as opposed to laths, must arise from other processes. Extended periods low within the $\alpha + \gamma$ phase field can lead to coarsening of gamma laths. One form of coarsening gamma can be seen along the edges of lamellar grains in Figure 2. Alternatively, gamma grains can form in the interdendritic, aluminum-rich areas of the microstructure during cooling from the casting operation and persist through subsequent heat treatment provided temperatures are kept low and/or aluminum is high.

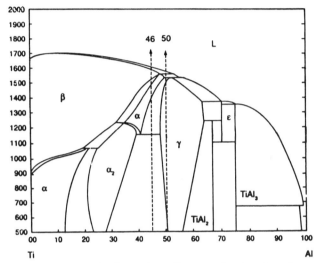

Figure 1: Titanium-aluminum phase diagram, showing two-phase regime of engineering gamma alloys (temperatures in °C, composition in atomic percent).

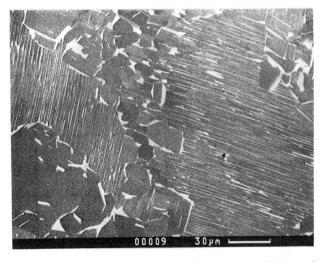

Figure 2: Microstructure of a gamma alloy with the duplex grain structure; the lighter phase is alpha-two.

Figure 3: Microstructure of as-cast gamma showing typical columnar, lamellar grains.

The management of gamma grain creation and persistence is particularly important in the thermal processing (HIP and heat treatment) of cast material. As-cast structures consist almost entirely of columnar grains of the lamellar constituent, with a small number of gamma grains in the interdendritic regions, Figure 3. The first thermal exposure in the $\alpha + \gamma$ field should be at low temperature, in order to at least preserve the gamma grains, if not grow them.

A number of minor microstructural features are often found. The most prominent is the appearance of plates of alpha-two in gamma grains after a thermal treatment constituting an upquench; this results from the need for gamma grains to reject titanium to satisfy the lever law. Just as gamma laths will appear in alpha grains on cooling, alpha plates will form in gamma on heating. The difference is that there are four habit planes in the latter case.

The duplex structure provides a useful balance of properties, involving several factors in addition to the usual benefit of controlled grain size. Neither fully gamma or fully lamellar structures are as ductile. The lamellar constituent is associated with toughness and fatigue crack growth resistance, a ramification of crack branching and deflection. The ductility of the gamma phase is thought to depend on having minimum aluminum; barring excessive segregation, this is assured by the presence of alpha-two.

Alloys

The objectives in gamma alloy development include increasing ductility, oxidation resistance, tensile strength, creep resistance and processability. The alloying elements of greatest interest are described below:

Ductility:	Cr, Mn, V
Oxidation resistance:	Cr, Nb, Ta, Zr
Tensile strength:	Cr, Ta, W, B
Creep resistance:	Cr, W, Ta, C, Si
Processability:	Mo, B (wrought)

As for superalloy development during the early (and perhaps late) years, the actual effects of various elements is a subject of some debate. An extensive cast alloy development program funded at GE by the U. S. Navy has sought to quantify the effects of many elements on castability and properties, but has only partially succeeded due to the complexity of the system and non-linear effects.

The most significant compositional factor is aluminum level. Properties and structure vary greatly over the range of interest, about 45 to 49% (atomic), Figure 4. Properties can vary dramatically even over a reasonable specification range. Some properties are clearly related to structural effects of aluminum, notably toughness. Tensile properties do not appear to depend on any single, easily observed structural feature in cast material.

GE is implementing an alloy that contains chromium and niobium [5]. Both contribute to oxidation resistance and creep strength, while chromium increases ductility. The alloy is known as "48-2-2" but is nominally Ti-47Al-2Cr-2Nb in atomic percent. The specification range is currently 46.5 to 48.2%, which is narrow with respect to process capability but wide with respect to property variations.

The aforementioned cast alloy development program has identified a Cr-Nb-Ta alloy with significantly greater creep strength than 48-2-2 and other current alloys, Figure 5.

Processing

For a variety of reasons, current implementation efforts by GE and others are pursuing cast processing. First, the limited ductility of gamma alloys requires that deformation processing be done isothermally or in a thick can to prevent surface chilling; this sort of processing is used for only the most critical parts in aircraft engines where the expense can be justified. Second, the sorts of application for which gamma is appropriate are parts that are currently cast, and are generally not amenable to wrought processing. The specific components where wrought processing might be desirable are those that are most critical, which near-term efforts are correctly avoiding until more experience is gained.

The casting of gamma presents no special challenges to the conventional titanium VAR casting process, but a great deal of knowledge must be accumulated before the production of gamma components becomes as routine as conventional alloys. There are certain processing steps that pose cost and quality issues.

The production and qualification of ingots is more difficult than for titanium or nickel alloys: 1) triple-melt VAR fails to adequately mix aluminum and titanium uniformly, 2) analysis of aluminum to the required accuracy is difficult, 3) preparation of casting VAR electrodes by the usual titanium process (forging of a large VAR ingot) is not possible, and 4) the supplier base in this area is small. Oremet and Howmet are actively developing modified processes for gamma. At some point, revert material must be qualified.

The casting operation itself can make use of standard titanium foundry practices, with modified mold configurations to prevent cracking. Precision Castparts, Howmet, IMI-Tiline and others have all had significant casting experience with gamma. A large variety of shapes and sizes have been produced and at this point no significant limitations are seen. As for any alloy, significant development is required for each new component to find casting parameters that produce a sound part.

Thermal processing (HIP and heat treatment) has been modified recently to reduce cost and the sensitivity of gamma to aluminum variations. The use of standard HIP cycles, such as those used for conventional titanium castings, reduces cost and cycle time. Low temperature heat treatments and correct sequencing can "lock in" gamma grains and maintain more uniform structure and properties over the aluminum specification range.

Structural castings usually require in-process repair, generally by welding. Common defects include HIP dimples, tears and small unfilled areas. For gamma to be viable, equivalent repair processes are required. Perhaps surprisingly in view of its ductility, gamma castings have been shown to be readily weld repairable using matching filler metal. Fabrication welding (EB, GTA) has also been demonstrated using special post-weld procedures.

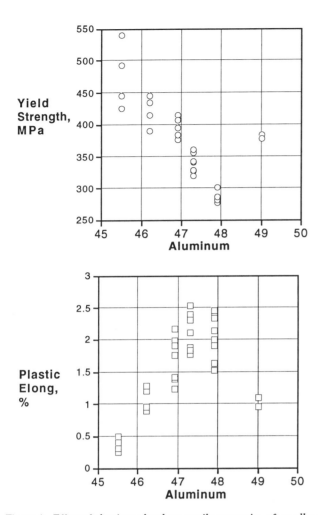

Figure 4: Effect of aluminum level on tensile properties of an alloy with 2Cr and 2Nb, investment cast, various heat treatments.

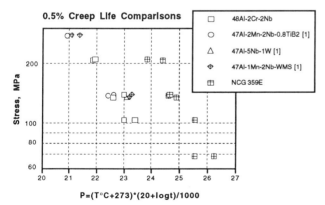

Figure 5: Creep strength of various gamma alloys, including the recently developed cast alloy NCG 359.

Properties

Density: The density of gamma is half that of nickel alloys and about 10% lower than titanium alloys.

Ductility: Gamma alloys can have a fair degree of ductility, about 2% at room temperature and 3 to 4% at typical operating temperatures. The wide aluminum range over which the gamma phase is stable means that titanium and aluminum atoms have some tolerance for being on each other's sites, a helpful attribute for slip, twinning, and the accommodation of the debris of their interactions. The alpha-two phase field is also wide. As discussed above, neither phase is very ductile in isolation, however. The presence of two phases does several things: 1) grain growth is restricted, 2) the aluminum contents of the gamma and alpha-two phases are pinned to the low and high ends of their respective ranges, 3) lamellar grains are formed in which interfaces may promote or accommodate extra deformation modes, and 4) oxygen or other impurities are gettered within one phase or at interfaces.

At this point, it can be concluded from analysis [6] and engine testing [3] that gamma has sufficient ductility to survive normal manufacturing, assembly and engine operations. Concerns remain for special situations, including field service operations and impact damage. The effects of long-term exposures to engine environments are, of course, largely unknown and will remain so for some time.

Strength: Gamma has relatively low yield strength, about 300 to 500 MPa (45 to 70 ksi). In one sense this is good: gamma might otherwise have less ductility to the extent that fracture is cleavage, i.e., controlled by normal stress. Allowable stress will be limited by defect behavior in most situations, so the low tensile strength may not be a significant limitation.

Fatigue strength is very good relative to yield strength, with ratios of endurance limit to yield strength approaching 100%. The slope of S-N curves is relatively flat. This behavior is a simple consequence of the limited ductility of intermetallics; metal fatigue is, after all, a consequence of plasticity.

Creep strength is good up to about 800°C. The specific creep strength is better than cast Alloy 718 beyond about 600°C, but never quite matches gamma-prime strengthened alloys.

Modulus: Perhaps the most unique property of gamma is its high modulus, 175 GPa (25 msi). The specific modulus is 50% higher than any commonly used alloy.

Thermal expansion: The coefficient of expansion, alpha, of gamma is somewhat higher than titanium, and substantially lower than nickel alloys. Depending on the temperature range of interest, gamma is a candidate for specialty low expansion nickel alloys such as the 90X series.

Defect tolerance: The fracture toughness of gamma alloys range from 15 to 25 MPa$\sqrt{m}$. For reasonable assumptions for defect sizes in castings, this will restrict allowable stresses to somewhat below yield strength. The fatigue crack growth curves for gamma have high Paris-region slopes, dictating that a threshold criteria be applied [7]. Gamma alloys exhibit true threshold behavior at about 5 to 8 MPa$\sqrt{m}$, which further limits allowable stress range in components subject to vibratory loads. The threshold is low relative to nickel alloys, but is higher than titanium alloys, which do not show true thresholds.

Oxidation and corrosion resistance: On the basis of laboratory testing, alloys with a Cr+Nb+Ta of 4% or more have excellent oxidation resistance up to about 800°C [8]. Oxidation/corrosion burner rig tests have shown equal or better resistance than nickel alloys such as René 80. Correlation between laboratory tests and service behavior is not established, however. As is common for titanium-base materials, laboratory tests indicate great sensitivity to hot salt stress corrosion cracking, but it remains to be seen if this is of significance in service.

Coatings have been identified for protection from the environment, but the current view is that a coating will not be required.

Ignition: Various types of tests have shown that gamma is much less likely to ignite than titanium alloys. The ability to contain a titanium fire, as is often required of compressor cases, is only somewhat better than titanium alloys.

Applications

There are several application areas that provide substantial payoff given the properties above. Gamma has 50% higher density-adjusted stiffness than all other commonly used engineering alloys. Applications are envisioned in a wide range of structural parts on this basis, including low temperature applications. COE is lower than nickel alloys, which is also of benefit in structures that determine clearances at seals and airfoil tips.

Useful creep strength extends to above 750°C, exceeding that of 718 and René 220. This temperature regime corresponds to compressor discharge air in advanced engines, to which many components are exposed. While specific creep strength is lower than gamma-prime strengthened cast alloys such as IN 100 or René 77, many components are not strength-limited. The best example is a low pressure turbine blade, which has a thickness set by manufacturing and aeromechanical requirements.

The ignition resistance of gamma is a required feature for some of the applications noted above, and may be the key feature of others. Some components are made of steel or nickel only because titanium would present a fire hazard.

Finally, there are some parts for which substitution for gamma may represent a cost reduction, which is an unusual feature for an advanced material. This stems from the low theoretical raw material cost and similar processing requirements.

Evaluation

GE initiated efforts with several titanium casting suppliers to produce cast gamma components in 1989. Being able to show design customers actual components proved to be a strong factor in gaining acceptance. Still, the risk and expense involved in implementing gamma was clearly high. In 1991, an engine project with a challenging weight goal found that gamma could provide more than half of the needed weight reduction. This led to a program to test gamma low pressure turbine blades in a factory CF6-80C2 engine.

Certain modifications to the blade and disk design were required for this test. LPT blades are leaned to counterbalance aerodynamic and centrifugal forces. This reduces bending stresses at the dovetail. The lower density of gamma required greater lean. The different thermal expansivity of gamma and the 718 disk material would prevent even loading across the tangs of a multiple-tang dovetail, so a single tang dovetail was designed. The different modulus to density ratio shifted a bladed disk system vibration mode to near-redline rpm; damper pins were used to avoid that problem (normally, the aeromechanical design would be changed to avoid such modes).

Blades were produced by a non-production process consisting of over-sized castings that were ECM'ed and otherwise machined to shape. Many processes were given their first tests during this program, and a great deal of attention was required from ingot-making through finish machining and assembly. It was during the last step that the authors became aware of the greatest benefits of ductility, though no blades were damaged.

The engine was first run to full power in June, 1993. Due to the nature of gamma, most observers in attendance believed if the blades survived that event, they would survive the rest of the test. This proved to be the case, though not without a few tense moments. The engine was run through many cycles for various other purposes, then put through 1000 simulated flight "C" cycles. The post-run condition of the rotor, Figure 6, was excellent, showing no distress related to the use of gamma. Wear occurred at the damper pins and at the shroud interlocks, but this was considered normal.

The rotor was disassembled, the blades were refurbished, and the rotor reassembled. An additional 500 C cycles were imposed during 1994 without incident. Demonstrating the ability to handle blades exposed to service conditions was considered a major milestone.

Transition

During the period of the CF6-80C2 test, GE adopted a risk-phased approach to gamma implementation. A small non-critical component was selected, the GE90 transition duct beam, Figure 7. Normally made from René 77, this part prevents duct panels from buckling during engine surges. Gamma beams completed rig and engine testing during 1995 and have been certified. Production orders will soon be placed and gamma beams will be installed in GE90's starting in early 1997.

GE has also successfully engine tested several air/oil seal supports in an advanced F404 derivative engine.

Low pressure turbine blades provide the greatest weight-savings of any gamma application, up to 45 Kg (100 lb) per stage. A NASA Aerospace Industry Technology Program was awarded during 1995 that will design and produce GE90 Stage 6 blades for certification testing. This is a heavily cost-shared program being conducted by GE, Chromalloy, Howmet, Oremet and PCC; the purpose of this type of program is to encourage early adoption of new technology by addressing the specific barriers to implementation.

For gamma, the specific barriers being addressed by this program are 1) ingot making, particularly with respect to aluminum control and cost, 2) net-shape casting and subsequent processing, 3) design, including the accommodation of casting limitations and the unique properties of gamma, 4) blade tip rubs, in terms of survivability and proper shroud cutting behavior, and 5) resistance to foreign object damage. FOD is a particular concern, since gamma will have substantially lower capability than superalloys. An impact test program has been devised that will utilize special cast-to-shape specimens, Figure 8, having gage sections closely representing the profiles of candidate leading edge designs.

In addition, numerous metallurgical issues and unknowns remain for gamma. The evolution of the microstructure through solidification and heat treatment is complex, and the structure itself is complex. Compositional and processing variations generally change many aspects of the structure simultaneously, which has confounded attempts to isolate effects on properties.

Significance

The introduction of new materials occurs at infrequent intervals. Among the many "revolutionary" materials that the U.S. Integrated High Performance Turbine Engine Technology (IHPTET) program has fostered, gamma appears to be the one that will have the greatest impact. This is an outcome of several factors. First, the material has significant, multiple payoff areas. Second, an engineering, component-oriented approach was taken that demonstrated parts-making capability and focused on specific design requirements. Third, and most important, gamma components can be produced at a small cost premium over conventional materials.

Acknowledgments

The success of gamma is the result of efforts by a large number of people, from those that first explored the titanium aluminide system to those that designed, produced and tested engine components. The support of the U.S. Air Force, Navy and NASA is gratefully acknowledged.

Figure 6: CF6-80C2 stage 5 rotor made up from gamma blades, after 1000 simulated flight cycles in a factory engine test.

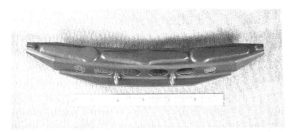

Figure 7: GE90 transition duct beam made from gamma and slated for introduction in 1997.

Figure 8: Simulated airfoil test specimen designed to evaluate impact damage at leading edges.

References

1. B. London and T. J. Kelly, *Microstructure/Property Relationships in Titanium Aluminides and Alloys*, ed. Y. W. Kim and R. R. Boyer, TMS, Warrendale, PA, 1992, 285-296.

2. C. M. Austin and T. J. Kelly, *Structural Intermetallics*, ed. R. Darolia et al, TMS, Warrendale, PA, 1993, 143-150.

3. C. M. Austin and T. J. Kelly, *Gamma Titanium Aluminides*, Ed. Kim, Wagner and Yamaguchi, TMS, Warrendale, PA, 1995, 21-32.

4. Y. W. Kim, *Journal of Metals*, 46, No. 7, 1994, 30-40.

5. S. C. Huang, U. S. Patent 4,879,092 "Titanium Aluminum Alloys Modified by Chromium and Niobium and Method of Preparation", 1989.

6. P. K. Wright, *Structural Intermetallics*, ed. R. Darolia et al, TMS, Warrendale, PA, 1993, 885.

7. J. M. Larsen et al, *Gamma Titanium Aluminides*, Ed. Kim, Wagner and Yamaguchi, TMS, Warrendale, PA, 1995, 821-834.

8. D. W. McKee and S. C. Huang, "Oxidation Behavior of Gamma Titanium Aluminide Alloys under Thermal Cycling Conditions," *Corrosion Science*, 33 (1992), 1899-1914.

Designing with Gamma Titanium
CAESAR Program
Titanium Aluminde Component Applications

Dwight E. Davidson

Pratt and Whitney Aircraft Engines
Government Engines and Space Propulsion Division
Advanced Engines Programs
West Palm Beach, Florida 33410

Abstract

The United States Government embarked on the Integrated High Performance Turbine Engine Technology (IHPTET) initiative to double the thrust to weight (Fn/Wt) performance from baseline turbine engines produced in the mid 1980's. In order to achieve these improvements, increases in engine systems temperature and pressure and decreases in weight must be simultaneously accomplished. Under IHPTET, the United States Air Force (USAF) funded the Component and Engine Structural Assessment Research (CAESAR) Program, whose goal is to demonstrate the rules and tools of the design intent for transitioning new materials. Of these materials, this article will critically review gamma titanium aluminide (TiAl), and comment on design issues, material property comparisons and manufacturing challenges required for successful static and rotating component demonstration.

Introduction

The CAESAR Program provides for the structural assessment of several key technologies in a multi-company endeavor involving Pratt and Whitney Aircraft (PWA), General Electric Aircraft Engines (GEAE), Rolls Royce Inc US/UK (RR) and Allison Advanced Development Company (AADC). These technologies are to be tested in an Advanced Turbine Engine Gas Generator (ATEGG) core. IHPTET has 3 phases. The CAESAR core will subject the TiAl blades to IHPTET Phase I environmental conditions. Following this test, the blades will be subjected to a planned structural test for up to 2000 Total Accumulated Cycles (TAC) of durability in an engine environment. The complete test plan will yield durability data at conditions expected for production use, thus providing a base from which to assess various TiAl material systems and correlate
new design criteria. The CAESAR Program TiAl technologies are:

- Attached Compressor Blades
- Combustor Fuel Nozzle Swirlers
- Compressor Variable Vane Shrouds
- Engine Nozzle Tiles

This article will critically review TiAl component development pertaining to the High Pressure Compressor (HPC). Areas to be covered are:

- Environmental Requirements
- Material Definition
- Design Considerations
- Lifing Methods
- Manufacturing Challenges

Clearly, the HPC blade, which is a rotating application, required a more rigorous structural analysis than the shroud, a static component. Thus, the HPC blades paved the way in establishing design criteria, manufacturing know-how and pretest validations.

Environmental Requirements

HPC Attached Blade

The HPC Attached Blades are designed to meet the F119 full mission requirements (8600 TAC and 5000 hours of service life). The blades are capable of achieving metal temperatures of 1000°F for up to 1000 hours and 1300°F for up to 10 hours. The CAESAR program engine test, which will subject these blades for up to 2000 TAC, could impose up to 1/4 of the primary life (HCF and Creep) requirement in a short amount of time. After post engine testing, environmental effects (oxidation, material stability, crack growth resistance etc) on the blades will be measured and correlated into existing lifing and design criteria.

HPC Variable Vane Shroud

The Variable Vane Shrouds will also be required to meet the F119 full mission requirement. The CAESAR/F119 engine test could impose a metal temperature of 700°F for up to 1000 hours and 1000°F for up to 10 hours. For this application the primary property of interest is wear. To date, an initial wear coating containing cobalt, chromium and tungsten has passed a 1200°F, 500 cycle thermal shock test. In all variable vane stages the vane stem material is titanium; therefore, for environmental considerations, protecting against titanium to titanium wear is a requirement.

Material Definition

PWA TiAl blades for the CAESAR program were cast by HOWMET Corporation using Ti-47Al-2Mn-2Nb-0.8%TiB2[1] (47XD) with a duplex microstructure.

To minimize cost and schedule delays typically encountered for a development program, all of the CAESAR component castings were produced from tooling made from 3 dimensional stereo lithographic models. Components were investment cast using the "lost wax" technique and single Vacuum Arc Remelting (VAR) process. In all cases the castings were overstocked between .020-.080 inches and machined into final configuration. All of the castings were HIP'ed at 2300°F and 25 ksi for 4 hours followed by an 1850°F and 50 hours heat treat.

HOWMET performed tensile and creep qualification test from each heat of material provided for each component. In all cases, tensile and creep data from every component heat met or exceeded design specifications. Table I presents comparative tensile properties of 47XD and wrought IN100.

Table I. Comparative Material Property of 47XD and IN100

47XD	70F	Temp	1200F
Tensile Strength (ksi)	71		70
Yield Strength (ksi)	55		47
Elong. (% in 4D)	1.2		2.6
Modulus (lb/in^{2}E6)	24		21
Density (lb/in^3)		0.143	
IN100			
Tensile Strength (ksi)	220		190
Yield Strength (ksi)	155		155
Elong. (% in 4D)	14		14
Modulus (lb/in^{2}E6)	31		26
Density (lb/in^3)		0.284	

Design Considerations

All of the TiAl components tested had a specific goal of either; offering better performance, reducing system weight or lowering cost. In the case of the HPC Attached Blade, better performance in the form of temperature capability was of primary benefit. Weight was a secondary benefit, while cost was hard to evaluate, especially under a development program. Since the Variable Vane Shrouds has a near term transition path to the F119 program, cost and weight were the driving factors.

HPC Attached Blade

The F119 development engines use attached nickel blades, which offer a means to swap and upgrade airfoils during testing. Two features on the basic nickel design were modified on the TiAl blades, the airfoil shape and the attachment bearing surface. In both cases the bearing surface area and the airfoil thickness were increased 20% and 40% respectfully. In addition, the thicker airfoils reduced the overall flow area between each airfoil. To compensate for this

---

[1]Chemisty is given in atomic percent

reduced flow, slight airfoil chordwise rechambering and spanwise bowing were introduced into the geometry. In essence this was done to minimize performance losses associated with thicker airfoils.

The total HPC blade height, including airfoil, is 1.2 inches. The airfoil height alone is 0.9 inches. while the chord length is 0.7 inches. The airfoil root thickness starts at 0.11 inches and decreases to 0.03 inches at the tip. The airfoils are bowed tangentially against the direction of rotation. The bowed airfoil root intersect the platform of the blade with a -5 degrees incidence angle and ends up with a +18 degree incidence angel at the blade tip. One hundred ten TiAl blades are designed to fit into a nickel based superalloy (IN100) rotor. See Figure 1.

Figure 1. HPC Attached Blade

TiAl blades, with a density half that of nickel based superalloy blades, realized a total of 0.5 pounds of static weight savings. Secondary weight savings, due to rotational effects, could have been realized in the nickel disk, but were reintroduced back into the rotating system to allow for greater creep resistance at higher operating temperatures. In all, the TiAl attached blades weight savings to the CAESAR program were minimal; however, the temperature capabilities of the airfoils were increased several hundred degrees. (See section on Lifing Methods)

HPC Variable Vane Shroud

The F119 Engineering, Manufacturing, and Development (EMD) engines use a ceramic matrix composite (CMC) as its Variable Vane Shroud bill of material. Because of the CMC high production cost, associated with consolidating plys of glass and epoxy fibers, and its low overall industry production requirement, a less costly replacement was required.

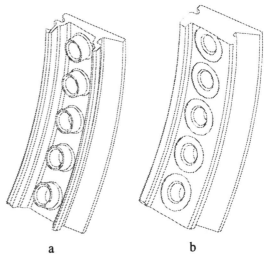

a   b

Figure 2. Design Comparison of HPC Variable Vane Shroud Fabricated from a) TiAl b) CMC

The HPC Variable Vane Shrouds are cast as segments with a 36 degree included arc. Ten pieces are required per assembly. Total radial thickness is 0.75 inches while the total axial length is 2.0 inches. Allowed minimum wall thickness after machining is 0.10 inches. At assembly the variable vane stems are fitted though each shroud hole and bolted together from the under side. No threaded features are machined into the TiAl hardware.

Table II Comparative Geometric Properties of 47XD versus CMC Shroud

TiAl Shroud		CMC Shroud
0.143	Density (lb/in^3)	.087
2.24	Segment Volume (in^3)	4.07
0.32	Segment Weight (lb)	0.35
3.32	Stage Weight (lb)	3.64

Since the CMC shroud application and its location, would not push the material to its temperature envelope, would not allow for the weight increases by going to a nickel alloy, and were too hot for the use of conventional titanium, TiAl castings were the logical candidate. TiAl cast shrouds offered multiple material sources, multiple material types (alloy and microstructure), more ductility, higher strength and higher strain to failure than CMC for this application. With its higher density than CMC a TiAl based shroud had to do a better job achieving its weight goals. This was done by hollowing out pockets of material in noncritical areas of the component. See Table 2 and Figure 2a and Figure 2b.

The hollowed out design of the HPC Variable Vane Shroud required no further rigorous analysis. The other critical design feature to consider was its wear characteristics. TiAl shrouds with a coating, would be competing with CMC shrouds that do not require a wear coating at all.

Lifing Methods

For most component design, a rigorous structural analysis is required to ensure that they would survive testing and that they would meet full life. The classic types of design information that are required to perform such an analysis are creep data, low cycle fatigue data, high cycle fatigue data and fatigue crack growth data. The basis and ultimately the criteria for these types of components can be derived from specimen testing.

HPC Attached Blade

Figure 3 was derived from fatigue crack growth specimen test data. Standard compact tension specimen were used to generate a da/dN versus ΔK data set. From the da/dN versus ΔK data, Stress versus Cycles (S-N) curves were produced for various combinations of crack size, crack shape, R-ratio and temperature.

Figure 3 compares the elastic stress versus cycles for a surface flaw size of .015 inch, R-ratio = 0.1 at a temperature of 1200°F of IN100 to that of a 47XD alloy. IN100 strengths range from 160 ksi @ 1000 cycle (typically associated with K_{IC} on a da/dN versus ΔK curve) capability, to 80 ksi @ 9000 cycle (typically associated with ΔK threshold, ΔK_{th}, on a da/dN versus ΔK curve) capability. TiAl strengths range from 70 ksi @ 1000 cycles capability to 45 ksi @ 100,000 cycle capability. Regardless of temperature, R ratio, crack size or crack type, all other combinations of fatigue crack growth S-N curves yielded the same type of results.

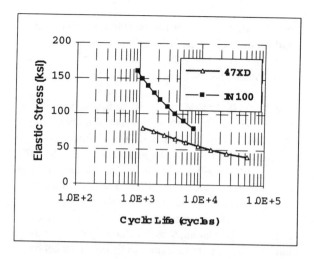

Figure 3 Elastic Stress versus Cycles Surface Flaw Model with 0.015 Initial Crack Size at 1200°F, R Ratio = 0.1

At first glance, IN100 appears superior to 47XD. If however, the results in Figure 3 are analyzed more closely by normalizing for the density difference, the results would allow a design engineer to compare more accurately material systems to each other for rotating or static applications.

Figure 4 compares the density normalized elastic stress versus cycles for a surface flaw size of .015 inch, R-ratio = 0.1 and a temperature of 1200°F for 47XD and a IN100. Across all stress levels and cycles, 47XD is as good as if not better than the IN100 in a rotating application (density normalized).

The data presented in Figure 4 were tested from the HOWMET 47XD material system which has a duplex microstructure. It has been reported in recent literature [ref 1,3,4] that a lamellar based microstructure has a higher ΔK_{th} than a duplex based material system. Therefore one could speculate that improvements in the existing data presented in Figure 4 are possible.

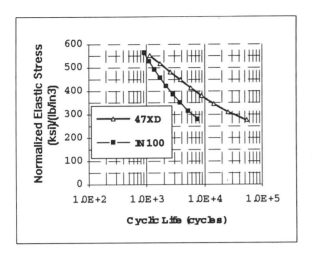

Figure 4 Density Normalized
Elastic Stress versus Cycles
Surface Flaw Model with 0.015 Initial Crack Size
at 1200°F, R Ratio = 0.1

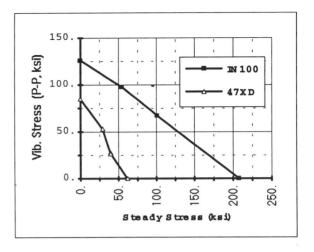

Figure 5. Modified Goodman Diagram Comparing
IN100 to 47XD at $10E^7$ cycles and 1200°F

Another criteria that is used in designing hardware that operates in high vibratory environments, is a Modified Goodman Diagram. Figure 5 was derived from smooth cylindrical High Cycle Fatigue (HCF) test specimen data. These specimens were tested at R-ratios of -1.0, 0.1, and 0.5 and over a range of temperatures between 70°F to 1400°F.

Figure 5 compares the allowable elastic vibratory stress of a IN100 to that of 47XD. IN100 strengths for an R-ratio = -1.0 (endurance limit stress @ $10E^7$ cycles:i,e, fully reversed bending) at temperature of 1200°F is 125 ksi. The endurance strength limit of 47XD is 85 ksi. In addition, the IN100 steady stress allowable and the TiAl steady stress allowables are 210 ksi and 60 ksi respectfully. A line drawn between these points determines a high cycle fatigue endurance limit for a range of steady stress and vibratory stress. Again, IN100 looks better than 47XD in HCF capability.

If the results in Figure 5 are analyzed more rigorously by normalizing the elastic allowable steady stress of the material. This would yield a normalized specific steady stress allowable. These results would allow a design engineer to compare any material systems in HCF for rotating application.

Figure 6 is a plot of allowable elastic vibratory stress versus normalized specific steady stress for 47XD and IN100. In this case, cast 47XD material properties are not quite as good as the IN100 material. The graph presented was taken from a sample of 1200°F test data. As temperature continues to increase to 1400°F the IN100 fatigue capability decays rapidly towards zero. As temperature continues to increase towards 1400°F the 47XD fatigue properties remain constant and in fact increase slightly. For cast properties 47XD is not quite as good as IN100 up through 1200°F. Above 1250-1300F°F it then become desirable to switch to TiAl because of the intrinsic density benefits.

A secondary benefit for using TiAl can be derived from other literature [ref 9,10]. Typical alloys exhibit HCF and LCF endurance behavior that track closely with the yield strength of the material. Wrought TiAl when processed via extrusion and/or forgings has reported yield strengths of 100 ksi or higher at room temperature. If these data points are incorporated into the existing Modified Goodman Diagrams one could estimate that wrought TiAl is as good as if not better than IN100 across all temperature ranges and stress levels for rotating applications.

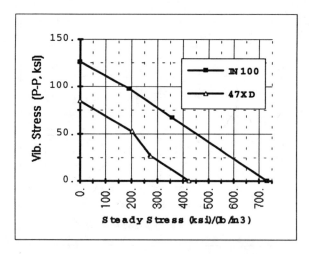

Figure 6. Modified Goodman Diagram Vibratory Stress versus Density Normalized Steady Stress at 1200°F and $10E^7$ cycles

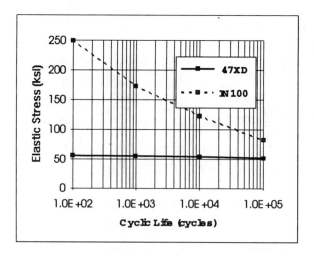

Figure 7. Low Cycle Fatigue Elastic Allowable Stress versus Cycles at 1200°F

Another criteria that is used to design hardware that operates in severe cyclic environments, is Low Cycle Fatigue (LCF). Figure 7 was derived from smooth cylindrical LCF test specimen data. These specimens were tested at R-ratios of 0.1, and 0.5 and over a range of temperatures from 70°F to 1400°F.

Figure 7 compares the allowable elastic stress versus cycles of IN100 to that of 47XD. IN100 cyclic capability to crack initiation for an R-ratio of 0.1 is 175 ksi @ 1000 cycles and 1200°F. 47XD cyclic capability to crack initiation for an R-ratio of 0.1 is 55 ksi @ 1000 cycles and 1200F. In addition IN100 based LCF endurance limit @ 10E5 cycles is 80 ksi. Gamma titanium endurance limit @10E5 cycles is 50 ksi. A line drawn between these points determines the remaining cycles and stress levels required for crack initiation. As in prior figures IN100 looks superior to 47XD.

It is truly in LCF resistance where the commercial utilization of nickel based superalloys has won out in the last 30 year. This curve is indicative of the ability of nickel based superalloys to be able to tolerate hammer blows, dings, abusive handling, etc without the worry of premature failure.

If we density normalize the LCF data as was done with the crack growth and HCF normalized data, little improvement in TiAl LCF properties are realized. See Figure 8.

Figure 8, however, does offer some additional insight [ref 9,10]. Typically, alloys exhibit HCF and LCF endurance behavior that track closely with the yield strength of the material. Wrought TiAl, when processed via extrusion and/or forgings has reported yield strengths of 100 ksi or higher at room temperature. If these data are incorporated into the LCF design data base, one could estimate that TiAl LCF capability would be as good as if not better than nickel based superalloys across all temperature ranges where endurance ($> 1E^4$ cycles) becomes a driving product requirement. Such applications would be industrial gas turbines, long range unmanned reconnaissance aircraft, aircraft auxiliary power unit, etc.

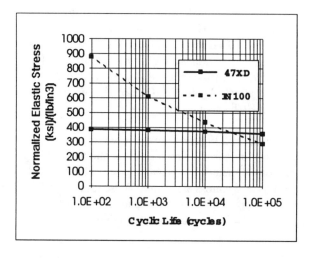

Figure 8. Low Cycle Fatigue Density Normalized Elastic Allowable Stress versus Cycle at 1200°F

This leads us to the last important design criteria, which helps designers make application decisions, creep based life behavior. Figure 9, a Larson-Miller plot, was derived from smooth round and smooth flat tensile creep specimen. These specimens were tested over a range of temperatures from 70°F to 1400 °F.

Figure 9 compares the allowable creep stress of IN100 to that of 47XD. IN100 0.2% /10hr/1200°F creep capability is 125 ksi. 47XD 0.2%/10hr/1200°F creep capability is 45 ksi. In addition IN100 0.2%/1000hr/1200°F creep capability is 100 ksi. 47XD 0.2%/1000hr/1200°F creep capability is 30 ksi. A curve drawn between these points determines an overall stress allowable versus time for material usage at temperature. As in the prior graphs IN100 appears better than 47XD.

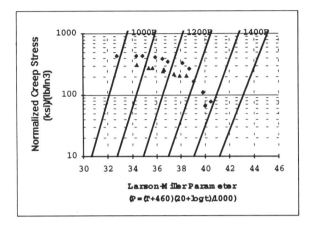

Figure 10. Density Normalized Larson Miller Creep Parameter
Δ = TiAl   ◊ = IN100

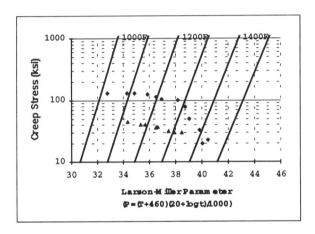

Figure 9 0.2% Larson Miller Creep Parameter
Δ = TiAl   ◊ = IN100

Density normalization reveals a much clearer picture. Figure 10 shows that TiAl has comparable normalized stress levels to that of IN100 at 1300°F and above and near comparable stress levels as IN100 at lower temperatures.

A secondary benefit of TiAl can be derived from other literature [ref 5-8] on its creep response. Most alloys exhibit improvements in creep capability when the material is hot worked during consolidation. Wrought TiAl, when processed via extrusion and/or forging have reported creep capabilities several hundred degrees beyond the 47XD creep capabilities shown in Figure 10. Thus, it can be anticipated that substantial increases in creep strength will be achieved in TiAl alloys through chemical modifications and microstructural control.

Manufacturing

HPC Attached Blade

The four participants in the CAESAR Program, GEAE, PW, RR , AADC, each used its own TiAl material system and manufacturing methodologies. GEAE and AADC provided their blades in wrought form, while PW and RR provided their blades in cast form.

Specifically, AADC electro-chemically etched (ECM'ed) its airfoils and ground its dovetails. RR EDM'ed its entire blade, while both PW and GEAE milled their airfoils and grounds their dovetails. An unusual item was that each of the four contractors used widely different final stress relief schedules.

Elaborating on PW manufacturing details, PW used conventional carbide milling tools with conventional speeds and feeds. The dovetails where ground using diamond tools with very fine grit sizes. For comparison, United Technology Research Center (UTRC), determined a machinability rating for 47XD that lies somewhere between IN100 and conventional titanium alloys.

All of the CAESAR participants have agreed to perform cleaning tests for TiAl. Concerns that have been raised are:

a) can TiAl tolerate multiple acid/alkali cleaning, without forming cracks, over the service life of the parts
b) does TiAl pick up surface hydrogen, which can help form cracks during machining, thus making the parts unusable.

HPC Variable Vane Shroud

As highlighted in "Design Considerations", for the shrouds, cost was the primary driver; specifically, the raw material cost of the CMC shroud. Switching to cast TiAl shrouds produced an overall cost benefit.

The secondary cost benefit for TiAl shrouds came from manufacturing. As illustrated in Figure 11a and Figure 11b, the CMC based shroud had to have every surface feature machined. By contrast the TiAl cast shroud only has the critical features machined (contact areas, through holes and outside diameters). The rest of the part remains in the as-cast state. This reduces the number of setups, the number of tools and time on the machining centers.

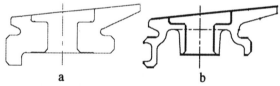

Figure 11. Manufacturing Comparison of HPC Variable Vane Shroud Fabricated from a) CMC b) TiAl

The one cost detriment for cast TiAl shrouds, was the cost of a wear resistant coating for the holes. The overall cost summary however still yielded a favorable cost analysis for switching to TiAl shrouds.

Finally, the commercialization of TiAl (blade or shrouds, static or rotating) will be driven by Life Cycle Cost benefits gained through either weight reductions or value analysis. Near term cost of cast TiAl ingots, casting tooling and machining will keep the cost higher than conventional titanium and conventional nickel hardware, until scale up of wrought TiAl to full sizes can be achieved repetitively.

Conclusions and Observations

Extensive progress and improvements in titanium aluminides have been made in the last decade. Industry has progressed from first generation cast alloys with moderate strength, poor ductility and process control problems, to second generation alloys in both cast and wrought form that have better overall properties and better process control. These second generation alloys have yielded multiple hardware demonstrations most notably at General Electric but also at Pratt and Whitney, Rolls Royce and many of the other gas turbine engine manufacturers in the United States and abroad.

Based on the comparative material properties, TiAl has superior fatigue crack growth and creep resistance over IN100. As shown, improvements in both weight and performance can be achieved in rotating applications. TiAl, with its limited ductility and LCF capabilities compared to IN100 will see little weight advantage in most static applications, but should see modest end use in hot static applications where impact and foreign object damage are not of great concern. Regardless of application, TiAl's broader use in the commercial world (manned use) will still be limited until issues concerning validation of design system and durability, low toughness, limited ductility and low threshold stress levels are fully understood.

As far as unmanned applications, TiAl offers an improved risk picture. With material properties comparable to nickel based alloys, TiAl should see quick and continued expansion in unmanned design applications, especially for rotating and static hardware where long endurance is envisioned. Industrial gas turbine products, long endurance Unmanned Aerial Vehicles (UAV), IHPTET and Advanced Concept propulsion systems that have repetitive long mission requirements are prime examples. Only after gaining experience and correlating design systems will TiAl be introduced for broad "manned" applications.

All CAESAR TiAl hardware applications were processed from existing foundry and machining centers. No new capital equipment was required to produce these parts. It is therefore envisioned that the expanded use of this alloy will rely heavily on industrial scale up. Industry will have to learn to deal with producing full size cast ingots and learn innovative foundry techniques to increase yields for cast and wrought hardware.

Acknowledgments

The author would like to acknowledge the research and development activities of his colleagues D. R. Clemens, C. I. Lobo at Pratt and Whitney, Dr. D.L. Anton from United Technology Research Center, D.E Thomson and M.F. Huffman from USAF Wright Labs, Structures Branch, from which these discussions have been based.

References

Published Papers

1. Venkataswara Rao, K.T., Y. W. Kim , R.O. Ritchie, Fatigue Crack Growth and Fracture Resistance of Two Phase TiAl Alloy in Duplex and Lamellar Microstructure, Scripta Metallugica et Material, 1995.

2. T. G. Nieh, J.N. Wang, Stress Change and its Implication for the Reduction of Primary Creep in Gamma TiAl, Scripta Metallugica et Material, 1995.

3. J.M. Larsen, B.D. Worth, S.J. Balsone, A.H. Rosenberger, J.W. Jone, Mechanism and Mechanics of Fatigue Crack Initiation and Growth in TiAl Intermetallic Alloys, Fatigue 1996.

4. J.M. Larsen, B.D. Worth, S.J. Balsone, J.W. Jone, An Overview of the Structural Capability of Available Gamma Titanium Alloys, TMS World Conference 1995.

Unpublished Papers

5. T.G.Nieh, et al Developement of Micro-Toughened Gamma TiAl and its Composites LLNL 1996.

6. C.T Lui, et al Development of Micro-Toughened Gamma TiAl for High Temperature Service ORNL 1996.

7. V.K Sikka, et al Processing of TiAl Intermetallics Alloys and Micro-Toughened Composite ORNL 1996.

8. P.J. Maziasz, et al Microstructual Analysis of Gamma TiAl Alloys Base on Ti-47Al-2Cr-2Nb, ORNL 1996.

9. D.R. Clemens, R.A. Anderson, Creep and LCF/HCF Behavior of TiAl Material for NASP, PW 1990.

10. D.R. Clemens, C.I.Lobo, Creep and HCF/LCF Behavior of TiAl Material for CAESAR, PW 1993.

11. P.L. Martin, Creep and Process Control of TiAl Material for MSF, RRSC 1995-1996.

Microstructural Effects on Fatigue Crack Growth of Cast and Forged TiAl Alloys

D.R. Clemens*, C.I. Lobo*
& D.L. Anton◆
* Pratt & Whitney, W. Palm Beach, FL
◆ United Technologies Research Ctr., E. Hartford, CT

Abstract

Fatigue crack growth and tensile experiments were conducted on three TiAl alloys in the cast duplex 47-XD, cast nearly fully lamelar 45-XD and wrought fully lamellar 47-5 at 25 and 650°C. The microstructures were fully characterized defining the extent of lamelar colony size, lath spacing presence of other phases. The tensile experiments showed the two lamellar structures to be stronger while the duplex alloy maintained a significantly higher fracture toughness at elevated temperatures. Fatigue crack growth threshold tests showed negligible difference between the three alloys and microstructures, having a K_{th}=6 to 8 ksi√in. The need is exposed for an alternate failure design criteria to take advantage of γ-TiAl, where low densities and high threshold stress intensities hold advantage in light weight and durability not attainable in conventional nickel base alloys.

Introduction

Titanium aluminides based on the γ phase TiAl are being seriously considered for various applications in aero gas turbine engines. One serious consideration in implementation of these alloys is their low ductility and damage tolerance. Before consideration can be given to their implementation in highly stressed flight critical applications such as compressor or turbine blades, vanes or disks; a thorough understanding of flaw growth characteristics will need to be attained.

In spite of this urgent requirement, relatively little attention has been given to the microstructural factors affecting fatigue crack growth in these complex alloys. A number of reviews have focused on the microstructural influences on tensile properties and fracture toughness [1-3]. These studies of fatigue crack growth characteristics in γ-TiAl have concentrated on either the duplex or fully lamellar structures without directly comparing the two [4-10]. Recent results have shown that the fully lamellar structures are more resistant to fatigue crack growth, especially at lower stress intensities, than the duplex microstructures [5-7]. The orientation of the lamellar colony with respect to the crack plane was found to play a strong role in crack growth rate. Trans-lamellar cracks were observed to propagate rapidly between lamellae only to stop and blunt at lamellar interfaces [5]. Those cracks which ran along lamellar interfaces propagated at a much more rapid rate. Fatigue crack closure was found to be minimal under ambient conditions and more significant at elevated temperatures due to oxidation induced closure [4,6].

A study is described here, which begins to fulfill the requirement of investigation and characterization of fatigue crack growth and fracture in two cast and one forged γ alloys being considered for rotating aircraft engine components. Through observation of fracture and fatigue crack growth characteristics with γ-TiAl microstructural variations, a clearer understanding of the influence of these factors are obtained.

Design Criteria

Fracture

For current materials, fracture critical components are designed and life managed under the Engine Structural Integrity Program (ENSIP) program which requires the damage tolerance of materials with respect to intrinsic or service generated defects must be demonstrated. This typically involves using an inspectable flaw size (determined by NDE) and calculated stress levels to determine remaining fracture (crack growth) life. This life is then used to set an inspection interval. However, for brittle materials, there is a narrow interval between the threshold stress intensity, ΔK_{th}, and the fracture toughness, K_{IC}, the crack growth rate curves are extremely steep, especially at room temperature [11]. Thus, the fraction of total design life resulting from crack propagation is small, and the inspection intervals, if any, are too frequent to be practical.

As a result, a new method of addressing ENSIP for brittle materials might be required, such as the use of ΔK_{th} as a life prediction criteria. In such a case, the service loads would have to be kept below Kth for a given initial flaw size to prevent the rapid propagation to failure of cracks initiated at flaws [7]. Designing below threshold would theoretically result in infinite life ($>10^5$ cycles), thus eliminating the need for design intervals.

Fatigue

For current materials, Low Cycle Fatigue (LCF) is also managed under the auspices of the ENSIP program. Low cycle fatigue is generally caused by engine start/stop or throttle cycles which occur at low frequency. Crack initiation is defined as a 1/32" (1 mm) long surface crack. P&W and the USAF define the acceptable probability of occurrence of an LCF crack as 1 crack occurring in a sample size of 1000 (1/1000 or B.1) having a 1/32" (1 mm) long crack at the predicted minimum life. Stress vs. life data can be regressed with a variety of equations; one of the most popular is the strain range/mean stress design system approach. This current approach covers a wide range of stress concentrations, temperature effects, stress ratios and the mean stress shift due to yielding. These methods, however, are geared to data which exhibits the following: low cycles to failure at high stresses and high cycles to failure at low stresses. Usually, an endurance limit approach, such as is popular with HCF, is avoided as too conservative. Data may exhibit some degree of scatter at individual stress levels, but not excessively. With the brittle gamma though, the data is relatively flat (high stresses result in few to many life cycles) and exhibits material scatter than scans 4 orders of magnitude (Figure 1).

As a result, new methods of addressing ENSIP for brittle materials in LCF might be required. In particular, similar to the crack growth rate case of designing below the threshold stress intensity, an LCF design system might involve designing to below an endurance stress limit. In this method, one would still design to a 3 sigma minimum endurance limit, and this would hopefully be considered a 'safe' limit. However, in reality, since the gamma LCF curves are so flat, and especially since there

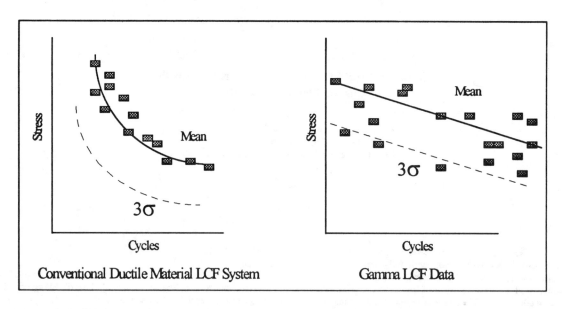

Fig. 1 Gamma LCF date vs. conventional design methods.

is a wide amount of scatter in life with very little difference in stress, the endurance limit approach might not be conservative enough if there's only limited testing, as is currently the case. Also in the endurance limit approach, there is a risk of being too conservative in the high stress, low life area, which might reduce the weight advantages of using gamma.

Experimental Procedures
Three alloys were selected for study[1]: 47-XD (Ti, 47Al, 2Mn, 2Nb, 0.8B), 45-XD (Ti, 45Al, 2Mn, 2Nb, 0.8B) and 47-5 (Ti, 47Al, 2Cr, 2Nb, 0.8B). The XD alloys are investment cast alloys which were HIP'ed to close casting porosity in the α/γ two phase region and stabilized in the α_2/γ two phase region. The 47-5 alloy was two step forged and solution heat treated in the α region and stabilized in the α_2/γ two phase region.

The microstructures were characterized under both optical and SEM microscopy to determine grain size, and α_2/γ lathe spacing and grain boundary configuration. Preliminary tensile tests were conducted in air at ambient and 650C to guide FCG loading. Fatigue specimens were machined in a compact tension configuration. Fatigue crack growth tests were conducted in a servo-hydraulic testing machine using an R-ratio of 0.1 at 20 Hz at ambient and 650C. Fatigue crack growth measurements were made optically. The fracture surfaces were characterized utilizing SEM and crack propagation features correlated with stress intensity factor.

Results and Discussion
Microstructural Characterization
Results of the metallographic examination are summarized in Table I. The 47-XD alloy contained a duplex microstructure, DP, consisting of both single phase γ grains as well as grains composed of α_2/γ lathes. The lathe grains composed approximately 40% of the structure and were 100 μm in diameter while the single phase γ grains were 30-50 μm in diameter and dispersed among the lathe colonies. Along with these two majority phases, a fine dispersion of borides was observed uniformly distributed through out the structure. These borides took on three morphologies, discrete (0.1-0.3 μm in diameter), filamentary (1 x 0.05 μm) and blocky (ranging as large as 1 x 5 μm). Fig, 2 gives a typical view of this microstructure. The 45-XD specimens were nearly fully lamellar microstructure, NFL, having a grain size of 50 μm. A few single phase γ grains were observed, but they composed less than 5% of the materials volume and were on the order of 10 μm in diameter. The borides were all either discrete or filamentary in these specimens, on the same order in size as those in the 47-XD alloy, with no blocky borides found. Fig. 3 is an optical micrographs of this alloy detailing boride distribution. The 47-5 alloy is shown in Fig. 4. Here the grains range from 200 μm. The microstructure is in the fully lamelar, FL, form. The borides in this alloy were descrete to acicular and having dimensions ranging from 1x1 to 1x6 μm. Additionally, a number of boride stringers, remnant from forging and clearly visible in Fig. 4, were found throughout the structure.

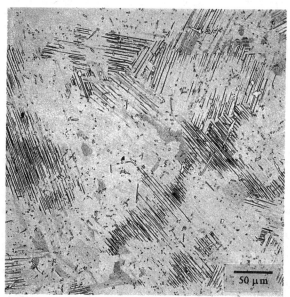

Fig. 2 Micrograph of 47-XD alloy showing the duplex cast microstructure of single phase γ grains with α_2/γ lathe colonies, along with filamentary and discrete borides.

Tensile and Fatigue Crack Growth Testing
The results of tensile and fatigue crack growth threshold results are given in Table II. Under tensile conditions the 45-XD alloy is considerably stronger than either the cast 47-XD or the wrought 47-5 alloys.

---
[1] All Compositions are given in atomic per centages.

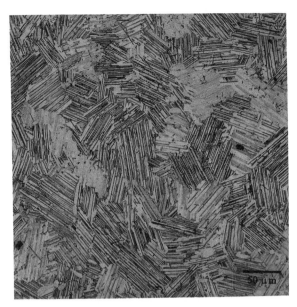

Fig. 3 Micrograph of 45-XD alloy showing the nearly fully lamellar cast microstructure of with α_2/γ lathe colonies with only small remnants of single phase γ grains, along with filamentary and discrete borides.

Fig. 4 Micrograph of 47-5 alloy showing the nearly fully lamellar wrought microstructure of with α_2/γ lathe colonies with only small remnants of single phase γ grains, along with discrete and blocky borides.

Table I Summary of Alloy Microstructures

Alloy	Micro.	Vol. % Lathe	Lathe Spacing (μm)	Lathe GS (μm)	γ GS (μm)	Discr. (μm)	Boride Filament (μm)	Blocky (μm)
47-XD	DP	40%	2.0	100	30-50	0.1-0.3	1x0.5	1-5
45-XD	NFL	>95%	3.0	50	10	0.1-0.3	1x0.5	
47-5	FL	100%	0.5	200			6x1	

Surprisingly, both cast alloys displayed higher elongation than the wrought alloy. Threshold stress intensity for all of these alloys was comparable at 25 and 650°C, with the wrought alloy displaying a slightly enhanced resistance to slow crack growth at ambient conditions. Fatigue crack growth curves represented as da/dn vs. ΔK are given for these three alloys at 25 and 650°C in Figs. 5-7.

Alloy 45-XD, a NFL structure, maintained significantly greater yield and ultimate strengths than either of the other alloys in the conditions tested. This was followed by the FL 47-5 alloy in strength at all but the 650°C ultimate strength where the NFL 47-5 and DP 47-XD alloys were comparable. These results are consistent with data reported in the literature [12-14] which have shown that, fully lamellar structures can maintain strengths approaching and surpassing duplex material, if the grain size is maintained below 500-1000 μm. If, however, the NFL and FL lamellar colony size is minimized, tensile strength and ductility are enhanced. The addition of boron to these cast alloys is known to minimized grain growth to a large extent and resulted in a relatively fine grain, 50 to 200 μm, fully lamellar structure.

As expected, the fully lamellar 47-5 alloy displayed a lower strain to failure at both ambient and elevated temperatures than the two cast alloys, due to its some what larger grain size.

Cyclic fracture toughness for the three alloys and microstructures studied here were comparable under ambient conditions; while the 47-XD alloy in the duplex microstructure displayed a near doubling in toughness at 650°C to nearly 20 ksi$\sqrt{}$in. The FL and NFL alloys (alloys 47-5 and 45-XD respectively) maintained a moderate fracture toughness of approximately 10-13 ksi$\sqrt{}$in., consistent with the literature values of other alloys in the NFL and FL structure[1,2].

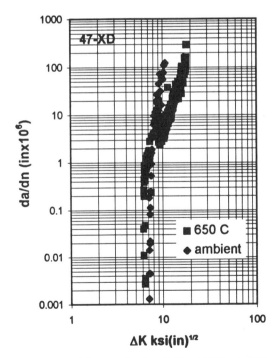

Fig. 5 da/dn curve for alloy 47-XD at 25°C and 650°C.

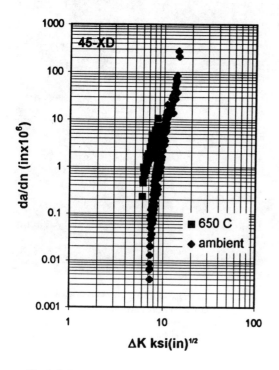

Fig. 6 da/dn curve for alloy 45-XD at 25°C and 650°C.

Fig. 7 da/dn curve for alloy 47-5 at 25°C and 650°C.

Table II Tensile and Fatigue Crack Growth Threshold Results

Alloy	Temp. (°C)	Yield Stress (ksi)	Ultimate Stress (ksi)	Failure Strain (%)	Fracture Toughness (ksi$\sqrt{in.}$)	Threshold Stress Intensity (ksi$\sqrt{in.}$)
47-XD	25	57	73	1.3	10	7.5
47-XD	650	49	75	3.9	19	6.5
45-XD	25	85	100	1.5	13	7.3
45-XD	650	70	96	2.8	9	6.2
47-5	25	65	83	0.8	10	7.9
47-5	650	62	74	0.8	9	6.2

The threshold stress intensity was very consistent between the these three alloys. At ambient temperatures, where little closure is anticipated, ΔK_{th} ranged between 7.3 and 7.9 ksi$\sqrt{in}$. At elevated temperature, where closure due to oxidation of the crack tip has been observed [4], ΔK_{th} ranged between 6.2 and 6.5 ksi$\sqrt{in}$. The literature has indicated the fully lamellar structure to be more resistant to crack growth than the duplex structure [8]. This was not observed here, and may be attributed to the larger grain sizes, approximately 400 μm, tested in [8] for the fully lamellar structure. This is somewhat surprising given the wide differences in processing and microstructure. Other researchers have shown substantial differences in K_{th} dependent on microstructure [7].

Fractography
Four of the failed compact tension (CT) fatigue crack growth specimens were selected for fractographic examination. The specimens were examined at crack lengths yielding approximately equal stress intensities. The specimens are described in the Table III. The fractographic comparisons of 47-XD alloy provide information about the effects of temperature. The three room temperature fracture examinations provide information about the effect of alloy/microstructure/thermal mechanical processing on microscopic fracture mode.

The 47-XD alloy microstructure was clearly mirrored in the fracture topology. Some of the fine grains exhibited a lamellar structure, and others showed no apparent structure. The later grains are assumed to be equiaxed γ grains. At 26°C, these fine grained areas occasionally exhibited intergranular fracture (see Fig. 8a). The large lamellar colonies invariably showed transgranular fracture, either translamellar, interlamellar or both in combination [7]. Fig. 8b highlights the details of translamellar crack growth. Intralamellar crack branching is evident (at the center of Fig. 8b and running into the plane of the photo). Secondary cracking produced a serrated appearance suggesting crystallographic slip/fracture.

Table III Specimens Selected for Fractographic Analysis

Alloy	Temp. (°C)	da/dn (in./cycle)	ΔK (ksi $\sqrt{in.}$)
47-XD	26	2×10^{-5}	10
47-XD	650	4×10^{-6}	10
45-XD	26	4×10^{-6}	10
47-5	26	4×10^{-6}	10

(a) (b)

Fig. 8 Fracture surface of 47-XD at room temperature showing (a) translamellar fracture (upper left), intergranular fracture adjacent to single phase γ or β grain (center) and intralemelar fracture (lower right) and (b) details of translamellar fracture showing intralamelar crack branching into the photo plane.

At 650°C in the 47-XD alloy, the fracture path was primarily translamellar with regions of transgranular fracture evident where a number of fine grains occurred. Such a region is depicted in Fig. 9. Additionally, a number of TiB_2 needles are apparent in these grain boundaries.

Fractographic details of cast 45-XD appeared very similar to those observed in 47-XD at both ambient and elevated temperatures with predominantly translamelar fracture occurring, as shown in Fig 10. Also shown is the presence of TiB_2 needles lying parallel to the crack propagation path within a $α_2/γ$ lamellar colony. Intergranular fracture was again noted in fine grains adjacent to prior α grain boundaries.

Fig. 9 Fracture surface of 47-XD at 650°C showing slightly oxidized intergranular fracture in a region of fine grains.

Compared to the cast 47-XD specimens, the wrought 47-5 fully lamellar microstructure exhibited a much coarser fracture appearance. The difference was due to the larger grain size in the wrought alloy. An apparent forging induced texture, parallel to the direction of crack propagation (perpendicular to the forging axis) was also noted. Intergranular fracture in areas of finer grains was much less common in the wrought fully lamellar microstructure. When observed, these finer grains were extremely fine and showed no evidence of lamellar structure. These grains may be grain boundary β or γ grains. Such grain boundary phases were observed previously in optical metallography (see Fig. 4).

Transgranular crack growth of the lamellar colonies predominated with interlamellar fracture resulting in a prominent "faceted" appearance (owing to the large $γ/α_2$ lathe colonies). Intralamellar fracture could be observed on these facets producing a distinct texture with 120° symmetry (see Fig. 11). The 120° symmetry was composed of intralamellar fracture as detailed previously in Fig. 8.

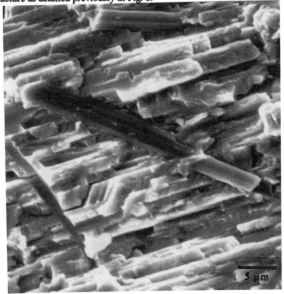

Fig. 10 Fracture surface of 45-XD at 25°C showing extent of translamelar fracture with TiB_2 needles.

Fig. 11 Fracture surface of 47-5 at 25°C showing effect of texture in mixed translamelar and intralamelar fracture.

Conclusions

It has been concluded that alloy microstructure, boride morphology and chemistry for the alloys studied here have only a second order effect in determining fatigue crack growth threshold stress intensity. Even in the case where dynamic fracture toughness is elevated as in the 47-XD alloy at 650°C, threshold stress intensity was not significantly impacted. Through control of the α_2/γ colony lath size (through the introduction of TiB_2 precipitates in this case), similar properties in both cast Duplex, cast Nearly Fully Lamellar and wrought Fully Lamellar alloys can be achieved. Translamellar cracking is dominant with evidence of intergranular fracture occurring adjacent to duplex α grains or refined grains of α or β. Much further study is needed as different researchers are reporting differing effects of microstructure on crack growth. Some of these discrepancies may be attributable to the fact that a variety of alloys have been studied that were processed under a variety of conditions.

The relatively small range between K_{th} and K_{IC} leaves very little allowance for in service crack growth, therefore, design considerations will need to be made for operation well below K_{th}. In order to design and fabricate structurally reliable components, new failure criteria and design guides will need to be developed. This is particularly critical in the implementation of γ-TiAl, where low densities and high threshold stress intensities hold advantage in light weight and durability not attainable in conventional nickel base alloys.

References

1. Y.-W. Kim, Acta. Metall. Mater., 40, 1121-1134 (1992).
2. Y.-W. Kim and D.M. Dimiduk, Proc. JIMIS-7 on High Temperature Deformation and Fracture, Japan Institute for Metals, Sendai, pp. 373-382 (1993).
3. Y.-W. Kim, Mat. Sci. and Eng., A193, pp. 519-533 (1995).
4. W.O. Soboyejo, J.E. Deffeyes and P.B. Aswath, Mat. Sci. and Eng., A138, pp. 95-101 (1991).
5. D.L. Davidson and J.B. Campbell, Metall. Trans., 24A, pp.1555-74 (1993).
6. S.J. Balsone, B.D. Worth, J.M. Larsen and J.W. Jones, Scripta Met. et Mat., 32, pp. 1653-58 (1995).
7. J.M. Larsen, B.D. Worth, S.J. Balsone and J.W. Jones, in *Gamma Titanium Aluminides*, Y-W. Kim et.al. eds., TMS/ASM Int., USA (in press).
8. A.W. James and P. Bowen, Mat. Sci. and Eng., A153, pp. 486-492 (1992).
9. K.S. Chan and D.L. Davidson, in *Structural Intermetallics*, R. Darolia et. al. eds., TMS, Warrendale, PA, USA, pp. 223-230 (1993).
10. A.W. James and P. Bowen, in *Titanium '92*, F. Froes et. al. eds., TMS, Warrendale, PA, USA, pp. 1139-1146 (1993).
11. S.J. Balsone, J.W. Jones and D.C. Maxwell, in *Fatigue and Fracture of Ordered Intermetallics*, W. Soboyejo et.al. eds., TMS, Warrendale, PA (1994) pp. 308-317.
12. Y. Nagakawa et.al., Japanese Inst. Metals, (1993) p. 334.
13. D. Larsen et. al., in *Intermetallic Matrix Composites*, D. Anton et. al. eds., MRS, Pittsburgh, PA, 1990, pp. 285-92.
14. Y-W. Kim, JOM, July (1994) pp.30-39.

NIAL ALLOYS FOR TURBINE AIRFOILS

R. Darolia, W. S. Walston and M.V. Nathal*

GE Aircraft Engines, Cincinnati, OH 45215

* NASA Lewis Research Center, Cleveland, OH 44135

Abstract

NiAl alloys offer significant payoffs as structural materials in gas turbine applications due to high melting temperature, low density and high thermal conductivity. Significant improvements in the material properties, processing and design methodology have been achieved. High strength NiAl alloys which compete with Ni-base superalloys have been developed. NiAl alloys have been successfully manufactured into a variety of turbine components. A high pressure turbine vane has been successfully engine tested. However, limited ductility and toughness as well as poor impact resistance continue to be critical issues which will impede near term production implementation.

Introduction

One of the greatest challenges currently facing the materials community is the need to develop a new generation of materials to replace nickel-base superalloys in the hot sections of gas turbine engines for aircraft propulsion systems. The present alloys, which have a Ni-base solid solution matrix surrounding Ni_3Al-base precipitates, are currently used at temperatures exceeding 2000°F (1100°C), which is over 80% of the absolute melting temperature. Since Ni_3Al melts at 2543°F (1395°C) and Ni at 2651°F (1455°C), it is clear that significantly higher operating temperatures, with the attendant improvements in efficiency and thrust-to-weight ratio, can only be attained by the development of an entirely new higher melting temperature material system. This problem is a primary reason for the high level of interest in high temperature intermetallic compounds.

NiAl offers many advantages, 1) density of 0.21 lb/in^3 (5.9 g/cm^3), approximately 2/3 of nickel-base superalloys, 2) thermal conductivity which is 4 to 8 times those of nickel-base superalloys, 3) high melting temperature (1638°C) which is approximately 450°F (250°C) higher than nickel-base superalloys, 4) excellent oxidation resistance, 5) simple ordered body centered cubic derivative (CsCl) crystal structure and small slip vectors for potentially easier plastic deformation compared to many other intermetallic compounds, 6) lower ductile-to-brittle transition temperature relative to other intermetallics and 7) relatively easy processability by conventional melting, powder, and machining techniques.

Many recent reviews provide additional information on the physical and mechanical properties of NiAl.[1-6] The purpose of this paper is to build on these reviews while emphasizing more recent results that pertain to actual application of NiAl in jet engines. We will emphasize single crystal NiAl since this technology has come the farthest towards this goal. Some alternatives to single crystals will also be addressed.

General Characteristics of NiAl

NiAl melts congruently at ≈ 2980°F (1638°C) and has a wide single phase field which extends from 45 to 60 at.% Ni. This feature is different from the majority of other intermetallic compounds which are either line compounds or have a very narrow phase field. The ordered bcc B2 (CsCl prototype) crystal structure of NiAl consists of two interpenetrating primitive cubic cells, where Al atoms occupy the cube corners of one sublattice and Ni atoms occupy the cube corners of the second sublattice. The ordering energy is believed to be very high which makes dislocation mobility rather difficult. The crystal structure of the strengthening Heusler phase, Ni_2AlTi, represents a further ordering of the B2 structure.

Single Crystal NiAl

A relatively large effort has been ongoing at GE Aircraft Engines since the late 1980's on the development of single crystal NiAl alloys. Initial efforts successfully improved the room temperature tensile ductility of relatively weak NiAl alloys, while more recent efforts have been focused on alloys with improved high temperature strength. The following sections will review the physical and mechanical properties of single crystal NiAl alloys and recent successes in utilizing these alloys in turbine airfoil applications.

Physical Properties

One of the primary advantages of NiAl alloys is lower density compared to superalloys. The density of stoichiometric NiAl is 0.21 lb/in^3 (5.90 g/cm^3) compared to about 0.31 lb/in^3 (8.6 g/cm^3) for many first generation single crystal superalloys. Because the alloying additions made to NiAl tend to be less than 5 at.%, the effect on density is minor as shown in Figure 1. Current single crystal NiAl alloys of interest have a density of approximately 0.22 lb/in^3 (6.0 g/cm^3), which offers a 30% reduction compared to single crystal superalloys.

The coefficient of thermal expansion (CTE) is important for structural applications since thermal stresses depend directly on the magnitude of the CTE. The CTE of NiAl is comparable to that typical for Ni-base superalloys, as shown in Figure 2. The elastic modulus of single crystal NiAl is highly anisotropic; the modulus values are around 13.8, 26.8 and 39.2 x 10^6 Msi. for the <100>, <110> and <111> orientations, respectively.[7,8] Stoichiometry and minor alloying additions have little influence on both the elastic modulus and the CTE.[7]

Thermal conductivity is another important property in turbine airfoil design because a high thermal conductivity can create a more uniform temperature distribution in a turbine airfoil and reduce the "life limiting" hot spots. The thermal conductivity of binary NiAl is approximately 5X that of single crystal superalloys at 1000°C. Figure 3 shows the effect of alloying additions on the thermal conductivity of NiAl.[7] While alloying additions decrease the thermal conductivity of binary NiAl, a 3X benefit is still maintained over single crystal superalloys at 1000°C.

Mechanical Properties

Tensile Ductility

For binary NiAl, <001> oriented "hard" single crystals fail at room temperature after only elastic deformation, while other ("soft") orientations, such as <110> and <111>, usually possess up to 2% plastic deformation.[9] Impurities and the test specimen surface condition can have a dramatic effect on measured ductility.[10,11] Potential factors limiting ductility in single crystals include inadequate dislocation sources, low dislocation mobility, inhomogeneous slip, and low fracture stress. Each of these have been postulated to explain the brittle failure of NiAl single crystals.

Results of a large alloy development effort at GE Aircraft Engines have shown that microalloying with Fe, Ga, and Mo can significantly improve the room temperature (RT) tensile ductility of NiAl single crystals tested along a <110> direction (soft orientation), increasing the failure strain from a typical value of 1% for stoichiometric NiAl to as high as 6% for NiAl+0.25% Fe.[1,12] A subsequent study was conducted to determine the source of scatter in the room temperature tensile ductility of a NiAl + 0.1% Fe alloy.[10] RT tensile plastic elongations varied from about 0.5% for as-ground specimens to about 8% for electropolished specimens when the grinding marks were completely removed, as shown in Figure 4. The surface effect is related to high notch sensitivity of NiAl due to low ductility and low fracture toughness.

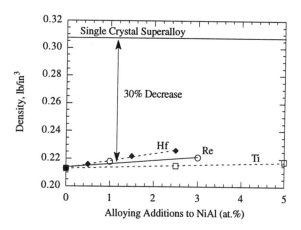

Figure 1. Measured density of NiAl as function of alloying content.

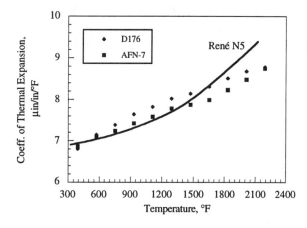

Figure 2. The coefficient of thermal expansion of two NiAl alloys compared to single crystal superalloy René N5.

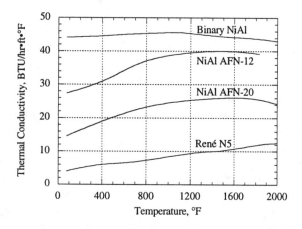

Figure 3. The thermal conductivity of three NiAl alloys as a function of temperature. Data for a typical single crystal Ni-base superalloy is provided for comparison.

The ductile-to-brittle transition temperature (DBTT) for "soft" binary NiAl crystals is 400°F (200°C), while "hard" crystals become ductile just below 750°F (400°C).[9] Binary NiAl single crystals become very ductile above the DBTT, and both "hard" and "soft" crystals exhibit more than 20% elongation at 750°F (400°C).[9,13] However, NiAl single crystal alloys which possess reasonably good high temperature strengths are much less ductile than binary NiAl, have high DBTT's which increase as the strength is increased, and possess low fracture toughness values. Figure 5 shows ductility as a function of temperature typical of strengthened NiAl alloys.[14] In "strong" NiAl alloys, even with minor alloying additions, the DBTT can increase beyond 1470°F (800°C) for the <110> orientation, as shown in Figure 5. Of significant scientific interest is the shift in orientation dependence of the DBTT. In binary NiAl, the <110> has a DBTT about 350°F (200°C) lower than the <001> orientation, however, in strong NiAl alloys, this behavior is reversed.[14]

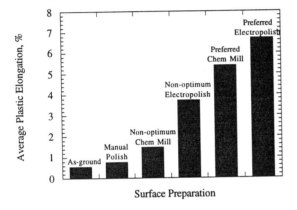

Figure 4. The effect of surface finish on the RT ductility of a Fe-containing NiAl alloy.

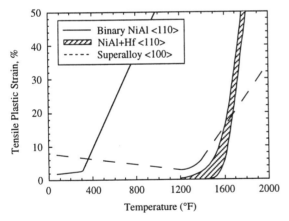

Figure 5. Tensile ductility of strengthened NiAl single crystal alloys as a function of temperature compared to the ductility of binary NiAl and a typical Ni-base superalloy.

Fracture Toughness

Like most intermetallic compounds, the fracture toughness of binary NiAl is low. The fracture toughness is dependent on the heat treatment, crystallographic direction with respect to loading direction, and the orientation and geometry of the notch. Typically a value of 8 $ksi\sqrt{in}$ is obtained from a specimen oriented in the <100> direction in a four point bend test with a single edge through notch, whereas a value of 4 $ksi\sqrt{in}$ is obtained in the <110> orientation.[15,16] Recently,[17] the fracture toughness in the <110> oriented single crystal NiAl was shown to be improved by minimizing strain-age embrittlement by fast cooling through the temperature range 400°C - 20°C. Fracture toughness in the range 13 - 17 $ksi\sqrt{in}$ was obtained in <110> oriented double cantilever beam toughness specimens. Additions of Fe and Ga, which increased the RT ductility of <110> single crystals, improved the RT fracture toughness by up to 20%.[13] However, in high strength alloys the fracture toughness is typically as low as 3-5 $ksi\sqrt{in}$.[2] The high DBTT and low fracture toughness of the strengthened NiAl single crystal alloys remain an issue which must be resolved through further alloy development as well as innovative designs.

Ductility and Fracture Toughness Requirements

An exact level of ductility or toughness requirement has not been established in the design of a turbine blade or vane. Very limited experience exists on components made out of materials with low ductility and fracture toughness. However, some amount of ductility or toughness is desirable for processibility, handling and assembly, component reliability and attachment to a superalloy part. For example, plastic deformation will be helpful in relieving high contact stresses between the airfoil and the turbine disk in the attachment region, especially at radii in the dovetail. A fraction of one percent room temperature plastic elongation is probably sufficient for relieving point loading. Also, since the NiAl part will be attached to a Ni-base superalloy part which could have different thermal expansion characteristics, up to about 2% plastic elongation (based on typical thermal expansion mismatch between NiAl and Ni-base superalloy and temperatures of use) may be required to avoid premature fracture under thermal transient conditions. Additionally, innovative designs and attachment concepts are required. Ni-base superalloys typically have a room temperature toughness of 40-50 $ksi\sqrt{in}$. A new material must have a minimum toughness in the range of 15-20 $ksi\sqrt{in}$, based on typical defect sizes and to obtain acceptance in the design community.

Implications of a Low Toughness Material for Design

Due to its low fracture toughness, NiAl has a low defect tolerance. Internal defects, such as inclusions and porosity, may originate from processing, while machining may introduce defects such as scratches, grinding marks, and cracks. Based on typical stresses encountered and the fracture toughness of the NiAl alloys, the defect size needs to be controlled to no higher than 25 μm. In addition to reducing the defect size, design methodologies which can account for the size, type, and location of defects likely to be encountered in the part need to be established. The design should consider allowable stresses based on the fracture toughness and the typical defect size likely to be introduced in the part. Also, a design methodology based on a probabilistic, and not a deterministic, approach needs to be developed. This type of approach is being emphasized in a current Air Force program being conducted at GE Aircraft Engines and NASA Lewis Research Center.[18] The design data base should include not only properties determined on laboratory specimens and sub components, but also properties from component testing in a variety of simulated conditions on parts with actual configurations. These parts should be made utilizing the same processes which will be used in production to reliably reproduce the defect distributions of the manufactured hardware. The design of the components should utilize the minimum properties possible, and not the average properties. Design safety margins will depend on the criticality of the part in the total system, and allowed safety factors are likely to be greater than for ductile metals until an experience base for alloys with limited ductility and damage tolerance is established. The long range goal should be to improve fracture toughness as well as produce cleaner material. These design considerations should be applicable to all intermetallic components.

High Temperature Strength

The high temperature strength of NiAl has been improved by solid-solution strengthening, precipitation strengthening, dispersion strengthening and by elimination of grain boundaries by the single crystal processing route.

Elements such as Co, Fe, and Ti have a high solubility in NiAl, and can provide significant solid solution strengthening.[19-21] Solid solution strengthening can be significant even at low level additions of the ternary element, especially with addition of group IVB and VB elements such as Ti, Hf, Zr, V, and Ta. For example, an addition of 0.2% Hf to NiAl increases the room-temperature tensile strength of a <110>-oriented specimen from a typical value of 210 MPa for stoichiometric NiAl to 600 MPa.[12] Elements such as Cr, Mo and Re have a low solubility (<1%) in NiAl. When added beyond their solubility limit, these elements precipitate a disordered *bcc* phase, which strengthens NiAl.

Additions of group IVB or VB elements beyond their solubility limits in NiAl produce several ternary intermetallic compounds, including the $L2_1$ Heusler (ß', Ni_2AlX) phase and the primitive hexagonal Laves phase (NiAlX), where X can be Hf, Ti, Ta, Zr and Nb. These phases contribute significantly to the strengthening of the NiAl alloys.[19,22,23] The close relationship between the B2 and $L2_1$ crystal structures results in coherent or semi-coherent ß' precipitates. Si pick up (600-1000 wppm) from the mold materials during single crystal processing also results in several possible precipitate microstructures that are not implied from the ternary phase Ni-Al-X phase diagrams. Both $Ni_{16}X_6Si_7$ (G-phase) and NiXSi can form in addition to the expected Heusler and Laves phases. The presence or absence of these various precipitates can change after heat treatment or creep exposure, depending on the level of Si in the alloy. All of these phases tend to be very fine in size and resistant to coarsening, and appear to contribute to the strength of these alloys.[24-26]

By combining several of the strengthening mechanisms mentioned above, significant improvements in high temperature strength of single crystal NiAl alloys have been made such that their strengths equal single crystal superalloy René N4 strength levels (Figure 6). In this figure, tensile rupture stress for single crystal NiAl alloys is plotted against the Larson-Miller parameter. While there is a strong orientation dependence on strength in the weaker NiAl alloys, the stronger NiAl alloys show little orientation effect. The <110> oriented specimens were slightly stronger than the <001> oriented specimens, although this difference was not statistically significant. Accounting for the lower density of NiAl, these NiAl alloys are viable replacements for the state-of-the art single crystal Ni-base superalloys, as shown in Figure 7. Figure 8 shows a comparison of the rupture life of several NiAl alloys at 982°C. A SEM micrograph showing the ß+ß' microstructure of one of the high strength alloys is presented in Figure 9.

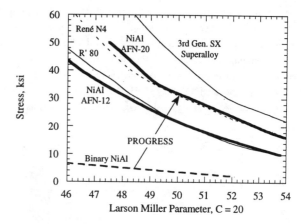

Figure 6. The rupture strength of NiAl alloys is equivalent to the first generation single crystal superalloy René N4.

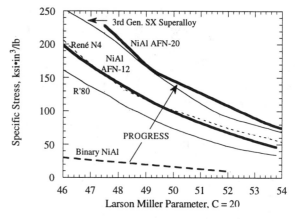

Figure 7. The density compensated rupture strength of NiAl alloys is equivalent to the latest single crystal superalloy René N6.

Creep Behavior

The creep resistance of the NiAl alloys has followed the improvements in their high temperature tensile and stress rupture strengths. Significant improvement in creep resistance is obtained with the addition of ß' forming elements to NiAl. Figure 10 shows the progress made in creep strength over the last several years. While the binary NiAl (D5) is very weak,[27] a small addition of Hf (D117) can dramatically improve the strength. Further alloying modifications (AFN-20) result in creep strength nearly equivalent to first generation single crystal superalloys. The creep exponents for the NiAl alloys appear to be similar to the superalloys, which implies that the precipitates are contributing to strengthening in much the same manner as the γ' phase in the γ/γ' microstructure of the superalloy.

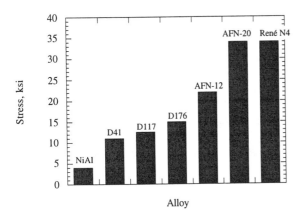

Figure 8. Stress level for 100 hr rupture life at 1800°F (982°C) for various NiAl alloys.

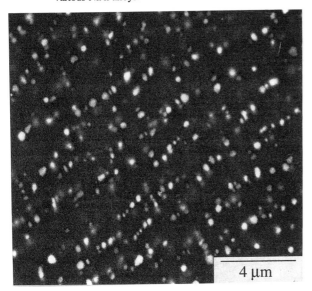

Figure 9. ß+ß' microstructure typical of the strengthened NiAl alloys.

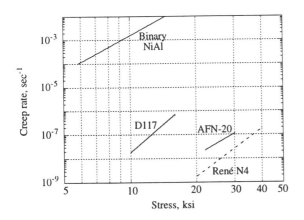

Figure 10. 1800°F creep strength of NiAl alloys D5, D117 and AFN-20 compared to first generation single crystal superalloy René N4.

Fatigue Behavior

Low cycle and high cycle fatigue tests have been conducted on several strengthened single crystal NiAl alloys. LCF tests were run in strain control at 1200°F (649°C) and 2000°F (1093°C) at an A ratio of infinity, and HCF tests were run in load control at 2000°F (1093°C) with an A ratio of 0.25 and 1.0. Compared to Ni-base superalloys, the LCF behavior of the strengthened NiAl alloys was found to be excellent. For example, at 1200°F, the fatigue lives were greater than even the third generation single crystal superalloys. The fatigue life was about 125,000 cycles at 586 MPa. 1200°F is about 200°F below the DBTT of the alloy in this orientation, and was chosen to determine if the alloy would have poor fatigue behavior below its DBTT due to lack of ductility and poor defect tolerance. The LCF behavior of NiAl alloy AFN-12 at 1093°C is shown in Figure 11 and compared to René 80 and René N4. While the rupture strength of AFN-12 is only equivalent to René 80, the fatigue strength is comparable to René N4. The HCF behavior of NiAl alloy AFN-12 also compares very favorably to single crystal superalloys. Figure 12 shows the 2000°F (1093°C), A = 1, HCF strength of AFN-12 compared to René N4. The NiAl specimens were brazed to superalloy threads, and Figure 12 shows that one of the specimens failed at the braze joint.

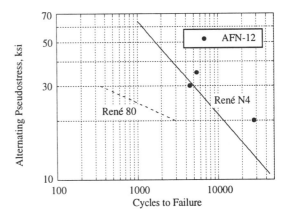

Figure 11. LCF behavior of NiAl alloy AFN-12 at 2000°F (1093°C), A = ∞ compared to René 80 and René N4.

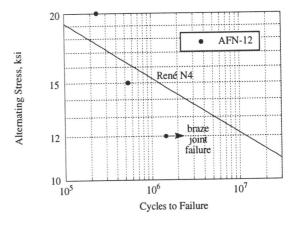

Figure 12. HCF behavior of NiAl alloy AFN-12 at 2000°F (1093°C), A = 1 compared to René N4.

Impact Resistance

A series of ballistic impact tests on NiAl panels was conducted to evaluate high strain sensitivity of the strengthened NiAl alloys. These panels were tested as a function of thickness, temperature, angle of impact, impact velocity, impacting particle size and particle strength. Panels were tested in the uncoated condition, and with a 5 mil thick thermal barrier coating (TBC). To better simulate engine conditions, panels were also impacted at 1700-1800°F with Al balls of varying room temperature yield strengths to simulate superalloys at elevated temperatures.

Testing conditions were varied to simulate impact speed conditions for blades, vanes and combustor shingles. The test data are schematically summarized in Figure 13. Based on dozens of tests, it was concluded that the NiAl panels would not survive the impact velocities (>1000 ft/sec) typically encountered by high pressure turbine blades. However, at lower impact speeds simulative of vane conditions (450 ft/sec), NiAl alloys performed well with no damage under most conditions. Additionally, tests with much thicker (20 to 40 mil) plasma sprayed TBC coatings simulative of combustor applications, showed that NiAl is a viable candidate for combustor applications with no damage under combustor impact conditions. These tests have confirmed the vulnerability of NiAl alloys to high strain rate impact events, especially for blade applications. Further material and design innovations will be required to make single crystal NiAl a viable blade material.

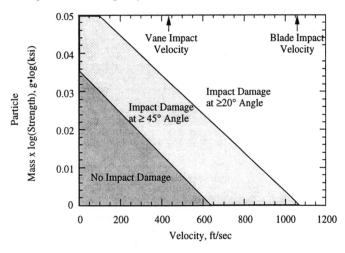

Figure 13. Summary of ballistic impact behavior of a strengthened NiAl alloy. Tests were conducted on thin wall panels at 1800°F.

Processing and Manufacturing

Due to the high ductility and low flow stress of binary NiAl above 750°F (400°C), conventional thermomechanical processing, such as hot isostatic pressing (HIP) and hot pressing of powder compacts, extrusion, and swaging have been successfully applied to NiAl. However, directional solidification and single crystal growth are currently the preferred processing routes for turbine blades and vanes, since the elimination of grain boundaries is necessary to obtain adequate high temperature creep strength. Single crystal bars of NiAl alloys up to 1.5 in. x 1.5 in. in cross section have been produced by a modified Bridgman technique.[1,2,28] Containerless float zone processes, Czochralski crystal growth, and modified edge-defined, film-fed growth have also successfully produced NiAl single crystals.[28]

The major challenge in producing single crystal blades of NiAl alloys is the high processing temperature, which is roughly 600°F (300°C) higher than the most advanced Ni-base superalloys. The higher processing temperatures approach the limits of existing ceramic mold and core materials in terms of structural capability and reactivity with molten NiAl. While large NiAl single crystal bars and airfoil shapes can be made using the modified Bridgman process, casting of hollow, thin-wall structures has been demonstrated only with tubes. The limited low-temperature plasticity of most NiAl alloys, combined with the strains generated when the metal shrinks around the ceramic core during cooling from the casting temperature, results in severe cracking of the casting.

The current approach to make a turbine blade consists of casting a solid near net-shape airfoil which can be split into two halves to machine the internal cooling cavities, channels and slots. The two halves are then bonded together by an activated diffusion bonding process. Finally, the 'fabricated' airfoil is finish machined which includes drilling of cooling holes. Blade castings sufficiently large enough to fabricate airfoils for advanced engines have been produced. Examples of a large single crystal NiAl casting are shown in Figure 14.

Figure 14. Single crystal NiAl castings of a large HPT blade shape.

A need to join NiAl to itself or to a superalloy exists in order to overcome processing difficulties or to add a superalloy's toughness to a NiAl component. For example, a NiAl airfoil bonded to a superalloy dovetail could alleviate much of the ductility requirements associated with the root attachment, although at a weight penalty. To this end, several techniques based on diffusion bonding, transient liquid phase processes and brazing have been evaluated.[29,30] Self generated filler metal produced by high temperature vacuum annealing of NiAl has also been utilized to bond NiAl to itself and to superalloys with positive initial results.[30,31] Tensile strengths equivalent to the base NiAl alloy and failure away from the weld zone have been achieved. Burner rig thermal shock test results, however, have been mixed. Further development in this area is required.

NiAl alloys have been successfully manufactured into a variety of turbine components. Many conventional and non-conventional low-stress material removal techniques (grinding, electro-discharge machining (EDM), electrochemical machining, chemical milling, electrostream drilling, ultrasonic machining, and abrasive waterjet) have been successfully used on NiAl alloys. High pressure turbine (HPT) vane airfoils have been machined from single crystal NiAl slabs using EDM. The recrystallized layer generated by the EDM process was removed by subsequent chemical milling. Several processes were demonstrated on this component including 14 mil cooling holes, trailing edge slots, brazing and thermal barrier coating (TBC) application. Figure 15 shows several of these airfoils prior to assembly as a HPT nozzle. Figure 16 shows one of the HPT nozzles that was recently successfully engine tested.

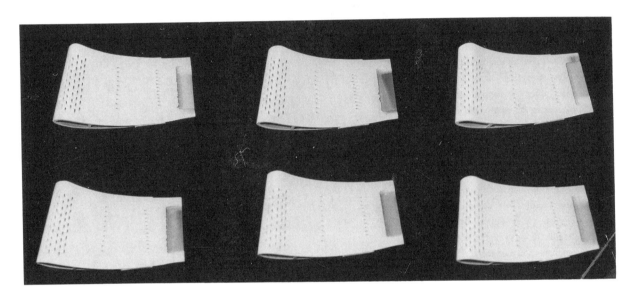

Figure 15. Single crystal NiAl alloy HPT vanes prior to nozzle assembly.

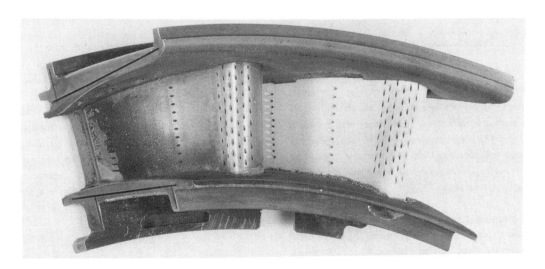

Figure 16. Fully manufactured HPT nozzle. Left vane is a single crystal NiAl alloy and right vane is a single crystal superalloy.

Component and Engine Tests

During the manufacturing of the NiAl HPT vanes, several component tests were performed prior to the engine test. These tests addressed the concern of thermal shock resistance because of the low ductility of the alloy. In addition, these tests served as 'proof tests' deemed necessary to weed out parts which might fail prematurely due to a manufacturing defect or handling damage. Component tests were also used to validate several new features utilized to design this vane. These component tests simulated engine thermal transient conditions as much as possible. Prior to an engine test, a full scale combustor rig test was successfully carried out. A fully assembled nozzle assembly is shown in Figure 17. In this assembly, the NiAl vanes were tested alongside single crystal superalloy vanes. After the engine test, no distress was observed in the NiAl vanes while the superalloy vanes showed some spallation of the thermal barrier coating.

The successful engine test of the HPT vane has demonstrated that it is possible to design, manufacture and engine test complicated turbine engine parts made out of limited ductility intermetallic alloys in highly stressed and harsh environments. In essence, technical feasibility of such materials has been demonstrated. However, production implementation will still be a difficult challenge primarily due to limited ductility and damage tolerance.

Figure 17. Fully assembled HPT nozzle containing NiAl alloy vanes.

NiAl Matrix Composites as Alternatives to Single Crystal NiAl Alloys

The single crystal alloys described in previous sections have exhibited very high creep strengths. However, these alloys are not very tough, and extensive design changes will be required to accommodate these low values. It is preferable to have materials that are both tough and creep resistant. Figure 18 displays both creep strength and toughness of various NiAl systems and compares them to approximate goals. This figure indicates that two composite strategies have potential for reaching a balance of strength and toughness.

NiAl based directionally solidified eutectics (DSE's) have shown some potential to reach an appropriate balance of properties. NiAl with α-Cr, α-Mo or Cr-Mo are two phase eutectics that have been extensively studied,[6] and Oliver and co-workers[32,33] have developed a three-phase eutectic with both α-Cr and the Ta-rich Laves phase co-existing. They have found that the α-Mo and the Cr/Mo eutectic alloys are reasonably tough, and the Ta-rich Laves phase eutectics are very good in creep. The next step of progressing to the three-phase Laves plus α-Cr showed a substantial gain in toughness at only a small sacrifice in creep strength. This indicates that further alloy development along this line has considerable potential for reaching the balanced creep/toughness goals. Transverse properties, thermal stability and expensive processing methods have limited the application of the past Ni-base superalloy eutectic alloys and NiAl based DSE's may have similar constraints. However, cost/benefit trade-offs can change over time, and NiAl DSE's may have cost advantages over competing ceramic composites,. A modest level of effort is appropriate to more fully explore the feasibility of achieving an appropriate balance of properties in these alloys.

Another promising alternative is provided by AlN strengthened NiAl, which is produced by high energy milling in liquid N_2, a process termed cryomilling.[34] Nitrogen is absorbed into the NiAl powders during cryomilling and forms high volume fractions of AlN in the form of nanometer sized particles. The most desirable microstructure obtained thus far has been obtained by hot extrusion, where the very fine nitride particles are clustered and strung out parallel to the extrusion axis. It is possible to vary the milling conditions to obtain nitride contents ranging from 0 to 30 vol.%. It has been found[35] that improvements in creep resistance scaled with the volume fraction of AlN, yet room temperature fracture toughness was not affected by AlN content. This combination of properties is also displayed in Figure 18, where the cryomilled material showed equivalent creep strength and slightly better toughness compared to the single crystal NiAl alloys. Another advantage of the cryomilled material is its relatively low DBTT. Measurements of fracture toughness as a function of temperature, Figure 19, indicate that unalloyed NiAl has the lowest DBTT, achieving 20 $ksi\sqrt{in}$ at about 750°F (400°C). The cryomilled NiAl + 30 vol.% AlN exhibited a DBTT shift to about 1000°F (550°C), which is considerably lower than that seen in single crystal NiAl alloys. A DBTT of 1000°F is low enough that the entire turbine blade could be designed to be operated in the ductile regime, thus providing adequate toughness.

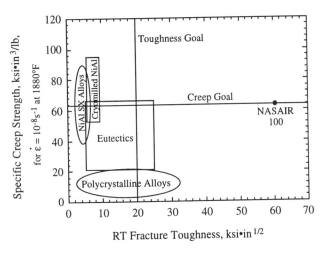

Figure 18. Creep/toughness plot showing typical ranges for different NiAl systems.

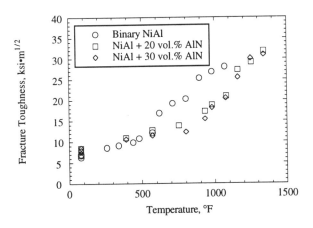

Figure 19. Fracture toughness of cryomilled NiAl containing AlN particles as a function of temperature.

Summary

Significant progress in material properties, processing, and design methodology has been made for the NiAl technology. High strength NiAl alloys which compete with Ni-base superalloys have been developed. Single crystal turbine airfoils with complex design features have been fabricated by a combination of casting and various machining processes. In addition, component and engine tests have been carried out successfully. However, low fracture toughness, poor impact resistance and high DBTT continue to be a concern when they are considered as replacements for ductile, high toughness Ni-base superalloys.

Three generic problems exist for most high temperature intermetallics, and the alloy developer typically encounters at least two of them for a given intermetallic system:

1. Low ductility and toughness, especially at low temperatures.
2. A balance of sufficient creep strength, ductility and toughness.
3. Environmental stability.

For example, the first two items remain important issues for NiAl and for $MoSi_2$, and all three still present various degrees of difficulties for Nb_5Si_3-based eutectics.[36,37] As more experience is gained with these alloy systems, three conclusions emerge:

1. Ni-based superalloys are remarkably good materials with properties that are extremely difficult to surpass.
2. Any new material which is developed to replace superalloys will probably carry with it vestiges of one or more of the above-mentioned three problems, and therefore:
3. Use of new intermetallic-based alloy systems will be very selective and gradual, and it is unlikely that there will be a rapid and widespread displacement of Ni-based superalloys.

Acknowledgments

The authors gratefully acknowledge the support of their employers in this research over the past many years and also thank several US agencies for continuing support: the Air Force, the Air Force Office of Scientific Research, Navy and NASA.

References

1. R. Darolia, "NiAl Alloys for High Temperature Structural Applications", JOM, 43 (3) (1991), 44-49.

2. R. Darolia, D. F. Lahrman, R. D. Field, J. R. Dobbs, K. M. Chang, E. H. Goldman and D. G. Konitzer, "Overview of NiAl Alloys for High Temperature Structural Applications", Ordered Intermetallics - Physical Metallurgy and Mechanical Behavior, C. T. Liu, et al, eds., Vol. 213, (Kluwer Academic Publishers, Netherlands, 1992), 679-698.

3. R. Darolia, "NiAl for Turbine Airfoil Applications", Structural Intermetallics, R. Darolia, et al, eds., TMS, 1993, 495-504.

4. D. B. Miracle, "The Physical and Mechanical Properties of NiAl", Acta metall. mater., 41 (3) (1993), 649-684.
5. D. B. Miracle and R. Darolia, "NiAl and Its Alloys", Intermetallic Compounds: Principals and Practice, J. H. Westbrook, et al, eds., (John Wiley and Sons, New York, 1993), 53-72.
6. R. D. Noebe, R. R. Bowman and M. V. Nathal, "Review of the Physical and Mechanical Properties of the B2 Compound NiAl", Inter. Mater. Rev., 38 (4) (1993), 193-232.
7. W. S. Walston and R. Darolia, "Effect of Alloying on Physical Properties of NiAl", High-Temperature Ordered Intermetallics V, I. Baker, et al, eds., MRS, 1993, 237-242.
8. R. J. Wasilewski, "Elastic Constants and Young's Modulus of NiAl", Trans. AIME, 236 (1966), 455-457.
9. D. F. Lahrman, R. D. Field and R. Darolia, "The Effect of Strain Rate on the Mechanical Properties of Single Crystal NiAl", High-Temperature Ordered Intermetallics IV, L. Johnson, et al, eds., MRS, 1991, 603-607.
10. R. Darolia and W. S. Walston, "Effect of Specimen Surface Preparation on Room Temperature Tensile Ductility of a Fe-Containing NiAl Single Crystal Alloy", accepted for publication in Intermetallics, 1996.
11. D. R. Johnson, S. M. Joslin, B. F. Oliver, R. D. Noebe and J. D. Whittenberger, 1st Int. Conf. on Processing Materials for Properties, H. Henien, et al, eds., TMS, 1993, 865.
12. R. Darolia, D. Lahrman and R. Field, "The Effect of Iron, Gallium and Molybdenum on the Room Temperature Tensile Ductility of NiAl", Scr. Met., 26 (1992), 1007-1012.
13. R. J. Wasilewski, S. R. Butler and J. E. Hanlon, "Plastic Deformation of Single Crystal NiAl", Met. Soc. AIME, 239 (1967), 1357-1364.
14. R. Darolia, W. S. Walston and D. F. Lahrman, Unpublished Research, GE Aircraft Engines, 1992.
15. K. M. Chang, R. Darolia and H. A. Lipsitt, "Cleavage Fracture in B2 Aluminides", Acta Metall., 40 (1992), 2727-2737.
16. R. Darolia, K. M. Chang and J. E. Hack, "Observation of High Index {511} Type Fracture Planes and Their Influence on Toughness in NiAl Single Crystals", Intermetallics, 1 (1993), 65-78.
17. J. E. Hack, J. M. Brzeski, R. Darolia and R. D. Field, "Evidence of Inherent Ductility in Single Crystal NiAl", High-Temperature Ordered Intermetallics V, I. Baker, et al, eds., MRS, 1993, 1197-1202.
18. R. Darolia, W. S. Walston, R. D. Noebe and M. Nathal, "Design Methodology for Intermetallics", F33615-94-C-2414, 1994-present.
19. W. S. Walston, R. D. Field, J. R. Dobbs, D. F. Lahrman and R. Darolia, "Microstructure and High Temperature Strength of NiAl Alloys", Structural Intermetallics, R. Darolia, et al, eds., TMS, 1993, 523-532.
20. M. Rudy and G. Sauthoff, "Dislocation Creep in the Ordered Intermetallic (Fe,Ni)Al Phase", Mat. Sci. and Eng., 81 (1986), 525-530.
21. M. Rudy and G. Sautoff, "Creep Behavior of the Ordered Intermetallic (Fe,Ni) Al Phase", High-Temperature Ordered Intermetallics, C. C. Koch, et al, eds., MRS, 1989, 327-333.
22. G. Sauthoff, "Intermetallic Phases - Materials Developments and Prospects", Z. Metallkde., 80 (1989), 337-343.
23. R. S. Polvani, W. Tzeng and P. R. Strutt, "High Temperature Creep in a Semi-Coherent NiAl-Ni$_2$AlTi Alloy", Metall. Trans., 7A (1976), 33-40.
24. I. E. Locci, R. D. Noebe, R. R. Bowman, R. V. Miner, M. V. Nathal and R. Darolia, "Microstructure and Mechanical Properties of a Single Crystal NiAl Alloy with Zr or Hf-rich G-phase Precipitates", High-Temperature Ordered Intermetallics IV, L. A. Johnson, et al, eds., MRS, 1991, 1013-1018.
25. I. E. Locci, R. Dickerson, R. R. Bowman, J. D. Whittenberger, M. V. Nathal and R. Darolia, "Microstructure and Mechanical Properties of Cast, Homogenized and Aged NiAl Single Crystal Containing Hf", High-Temperature Ordered Intermetallics V, I. Baker, et al, eds., MRS, 1993, 685-690.
26. A. Garg, R. D. Noebe and R. Darolia, "Crystallography of the NiHfSi Phase in a NiAl (0.5 Hf) Single Crystal Alloy", accepted for publication in Acta Metall., 1995.
27. W. D. Nix and R. H. Dauskardt, "High Temperature Deformation and Fracture Resistance of Single Crystals of NiAl and NiAl-Based Intermetallic Alloys", (Stanford University, Annual Technical Report, AFOSR AF-F49620-95-1-0163, 1995).
28. E. H. Goldman, "Single Crystal Processing of Intermetallics for Structural Applications", High-Temperature Ordered Intermetallics V, I. Baker, et al, eds., MRS, 1993, 83-94.
29. E. H. Goldman, "Advanced NiAl Turbine Blade", (GE Aircraft Engines, F33615-90-C-5938, Interim Report, 1992).
30. T. J. Moore and J. M. Kalinowski, U.S. Patent 5,284,290.
31. R. D. Noebe, A. Garg, D. Hull, J. Kalinowski, R. Darolia and W. S. Walston, "Joining of NiAl to Ni-Base Superalloys", HITEMP Review, NASA CR-10178, 1995, 29-1 to 29-20.
32. D. R. Johnson, X. F. Chen, B. F. Oliver, R. D. Noebe and J. D. Whittenberger, "Processing and Mechanical Properties of In-situ Composites from the NiAl-Cr and the NiAl-(Cr,Mo) Eutectic Systems", Intermetallics, 3 (1995), 99-113.
33. D. R. Johnson, X. F. Chen, B. F. Oliver, R. D. Noebe and J. D. Whittenberger, "Directional Solidification and Mechanical Properties of NiAl-NiAlTa Alloys", Intermetallics, 3 (1995), 141-152.
34. J. D. Whittenberger, "Characteristics of an Elevated Temperature AlN Particulate Reinforced NiAl", Structural Intermetallics, R. Darolia, et al, eds., TMS, 1993, 819-828.
35. M. G. Hebsur, J. D. Whittenberger, C. E. Lowell and A. Garg, "NiAl-Base Composite Containing High Volume Fraction of AlN Particulate for Advanced Engines", High-Temperature Ordered Intermetallic Alloys VI, J. A. Horton, et al, eds., MRS, 1995, 579-584.
36. M. R. Jackson, B. P. Bewlay, R. G. Rowe, D. W. Skelly and H. A. Lipsett, "High-Temperature Refractory Metal-Intermetallic Composites", JOM, 48 (1) (1996), 39-44.
37. P. R. Subramanian, M. G. Mendiratta and D. M. Dimiduk, "The Development of Nb-Based Advanced Intermetallic Alloys for Structural Applications", JOM, 48 (1) (1996), 33-38.

CERAMIC GAS TURBINE TECHNOLOGY DEVELOPMENT

M.L. Easley and J.R. Smyth
AlliedSignal Engines
AlliedSignal Aerospace Company
Phoenix, Arizona

ABSTRACT

Under the U.S. Department Of Energy/National Aeronautics and Space Administration (DOE/NASA) funded Ceramic Turbine Engine Demonstration Program, AlliedSignal Engines is addressing the remaining critical concerns slowing the commercialization of structural ceramics in gas turbine engines. These issues include demonstration of ceramic component reliability, readiness of ceramic suppliers to support ceramic production needs, and enhancement of ceramic design methodologies.

The AlliedSignal/Garrett Model 331-200[CT] Auxiliary Power Unit (APU) is being used as a ceramics test bed engine. For this program, the APU first-stage turbine blades and nozzles were redesigned using ceramic materials, employing the design methods developed during the earlier DOE/NASA funded Advanced Gas Turbine (AGT) and Advanced Turbine Technologies Application Project (ATTAP) programs. The present program includes ceramic component design, fabrication, and testing, including component bench tests and extended engine endurance testing and field testing. These activities will demonstrate commercial viability of the ceramic turbine application. In addition, manufacturing process scaleup for ceramic components to the minimum level for commercial viability will be demonstrated.

Significant progress has been made during the past year. Engine testing evaluating performance with ceramic turbine nozzles has accumulated over 910 hours operation. Ceramic blade component tests were performed to evaluate the effectiveness of vibration dampers and high-temperature strain gages, and ceramic blade strength and impact resistance. Component design technologies produced impact-resistance design guidelines for inserted ceramic axial blades, and advanced the application of thin-film thermocouples and strain gages on ceramic components. Ceramic manufacturing scaleup activities were conducted by two ceramics vendors, Norton Advanced Ceramics (East Granby, CT) and AlliedSignal Ceramic Components (Torrance, CA). Following the decision of Norton Advanced Ceramics to leave the program, a subcontract was initiated with the Kyocera Industrial Ceramics Company Advanced Ceramics Technology Center (Vancouver, WA). The manufacturing scaleup program emphasizes improvement of process yields and increased production rates.

Work summarized in this paper was funded by the U.S. Dept. Of Energy (DOE) Office of Transportation Technologies, part of the Turbine Engine Technologies Program, and administered by the NASA Lewis Research Center, Cleveland, OH under Contract No. DEN3-335.

NOMENCLATURE

AE	AlliedSignal Engines (Phoenix, Arizona)
AGT	Advanced Gas Turbine
APU	Auxiliary Power Unit
ASME	American Society of Mechanical Engineers
AS-800	AlliedSignal Ceramic Components Silicon Nitride
ATTAP	Advanced Turbine Technology Applications Project
AZ	Arizona
BN	Boron Nitride
C	Celsius
CA	California
CC	AlliedSignal Ceramic Components (Torrance, California)
CLP	Closed-Loop Processing
CT	Connecticut
CTEDP	Ceramic Turbine Engine Demonstration Project
deg	Degrees
DOE	U. S. Department of Energy
DYNA3D	Three-Dimensional Finite Element Computer Code
EGT	Exhaust Gas Temperature
F	Fahrenheit
ft	Feet
GN-10	AlliedSignal Ceramic Components Silicon Nitride
HIP	Hot Isostatic Pressed
HS25	Haynes Alloy No. 25
Hz	Hertz
in	Inch
kg	Kilograms
KICC	Kyocera Industrial Ceramics Company (Vancouver, Washington)
ksi	Thousands of Pounds Per Square Inch
lb	Pounds
mm	Millimeters
MOD	Modification No.
N/A	Not ApplicableNAC Norton Advanced Ceramics (East Granby, Connecticut)
NASA	National Aeronautics and Space Administration
No.	Number
NT154	Norton Advanced Ceramics Pressure Slip Cast Silicon Nitride
NT164	Norton Advanced Ceramics Pressure Slip Cast Silicon Nitride
OH	Ohio
ORNL	Oak Ridge National Laboratory (Oak Ridge, Tennessee)
POF	Probability Of Failure
rpm	Revolutions Per Minute
sec	Seconds
Si_3N_4	Silicon Nitride
SN-251	Kyocera Industrial Ceramics Company Silicon Nitride
SN-252	Kyocera Industrial Ceramics Company Silicon Nitride
SPC	Statistical Process Control
TN	Tennessee
UDRI	University of Dayton Research Institute (Dayton, Ohio)
U.S.	United States
WA	Washington State
2-D	Two-dimensional
3-D	Three-dimensional

INTRODUCTION

This paper summarizes progress during 1995 in the U.S. Dept. of Energy/National Aeronautics and Space Administration (DOE/NASA) sponsored 331-200[CT] Ceramic Turbine Engine Demonstration Program conducted by AlliedSignal Engines, Phoenix, AZ, a unit of AlliedSignal Aerospace Company, in developing the needed technologies for ceramic gas turbine engines. The Ceramic Turbine Engine Demonstration Project (CTEDP) is sponsored by the U.S. Department of Energy (DOE) to develop the technology for an improved automobile propulsion system under Title III of U.S. Public Law 95-238, "Automotive Propulsion Research and Development Act of 1978." The CTEDP program is authorized under DOE/NASA Contract DEN3-335, with the National Aeronautics and Space Administration (NASA) Lewis Research Center (Cleveland, OH) providing program management and administration.

The thrust of the CTEDP/331-200[CT] program is to "bridge the gap" between ceramics in the laboratory and near-term commercial heat engine applications. The intent is to use this application as a stepping stone to transition the technology into the automotive marketplace where its benefits can have the greatest impact on reducing fuel consumption and gaseous emissions.

As part of this overall effort, the CTEDP program will provide essential and substantial early field experience demonstrating the reliability and durability of ceramic components in modified, available, real engine applications, including manufacturing scaleup for competitive production. These efforts will lead to accelerated commercialization of advanced, high-temperature engines in hybrid vehicles and other applications. Additional efforts supported by this project include the DOE-sponsored Propulsion Systems Materials program, which has the objectives of improving the manufacturing processes for ceramic turbine engine components and demonstrating application of these processes in the production environment.

The 331-200[CT] ceramic engine test bed is based on the production AlliedSignal/Garrett Model 331-200[ER] auxiliary power unit (APU), with the first-stage turbine modified to incorporate ceramic nozzles and blades (Fig. 1). This work will simultaneously ready ceramic technology for the aircraft APU application, while gathering extensive laboratory and field experience, and develop ceramic component design methods and fabrication techniques. In this way, the 331-300[CT] Ceramic Turbine Engine Demonstration Project will effectively support the expansion of ceramics technology into automotive designs.

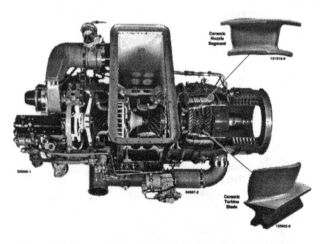

Fig. 1. Ceramic Turbine Engine Demonstration Test Bed Is Based on Proven AlliedSignal/Garrett Model 331-200 Auxiliary Power Unit, With Ceramic First-Stage Turbine Components.

This strategy will augment the maturing ceramics technology by developing the infrastructure and engineering disciplines within the technology to overcome those barriers that prevent its commer-cialization. Currently, the principal barriers to the commercial-ization of ceramics are seen as:
- Immature supporting technologies,
- Underdeveloped production capability, and
- Inadequate demonstration.

The overall CTEDP/331-200[CT] program plan provides the approach to resolve each of these issues. The following discussions describe the progress to date in the various project activities and outline the go-forward plans to meet the program objectives. This work was initiated in 1993 and the project progress through the end of 1994 has been reported by Easley, Smyth, and Rettler.[1,2]

TECHNOLOGY DEVELOPMENT

Ceramic technologies supported under the CTEDP/331-200[CT] Program in 1995 included:
- Impact design methods refinement
- Ceramic blade attachment technology
- Ceramic nozzle proof testing techniques
- Oxidation/corrosion characteristics of ceramic materials
- Ceramic proof testing methodology.

All of these technologies were identified as critical to the success of ceramics in commercial gas turbine applications. A description and discussion of the progress in each of these technologies follow.

Impact Design Methods Development

Development of design methods capable of accurately predicting structural impact damage for any ceramic component from carbon particles (combustor carbon) is the end goal of this activity. This work will result in design guidelines for impact-resistant, axial ceramic turbine blades. During the past year this activity resulted in the development of design guidelines for axial ceramic inserted blades, and began evaluation of carbon impact on an integral ceramic bladed rotor.

Development of the axial inserted ceramic turbine blade impact design guidelines included an analytical study of various blade designs similar to the 331-200[CT] turbine blade but differing in critical geometric parameters. The study results were calibrated with actual impact tests on 331-200[CT] ceramic turbine blades using 0.1 and 0.2 inch (2.5 and 5.1 mm) diameter spherical carbon projectiles. The impact testing, performed at the University of Dayton Research Institute (UDRI) in Dayton, OH, determined the blade strain response at two locations adjacent to the impact site for the baseline (MOD 0) 331-200[CT] ceramic turbine blade design. This data, plotted in Fig. 2, is compared against the predicted threshold values from the DYNA3D three-dimensional finite element analysis com-puter code using a carbon pulverization model developed earlier under this program. Good correlation of the test data and predicted values was noted.

Having validated the accuracy of the DYNA3D computer code predictions with the carbon pulverization model, a study was then performed to determine the sensitivity of various turbine blade geometric parameters to the peak stress in the blade generated during impact with a 0.2 inch (5.1 mm) diameter carbon sphere at a relative velocity of 560 ft/sec (171 m/sec). This impact velocity was selected as representative of the high range of critical impact velocity for a 0.2 inch (5.1 mm) diameter combustor carbon particle on the MOD 0 331-200[CT] ceramic blade, based on UDRI impact test data. The analytical work used a Taguchi L4 array in which blade thickness, taper ratio, and twist were evaluated. Of these parameters evaluated, the peak blade stresses were most affected by blade thickness.

A follow-up study was then performed, to characterize the ceramic blade stresses with respect to airfoil tip thickness. Relative particle impact velocities of 1300 and 1600 ft/sec (396 and 488 m/sec) were used, bounding the typical axial turbine design tip speeds. The results are shown in Fig. 3, compared with the 10-percent probability of failure (POF) strength value for AS-800 silicon nitride. Effects due to high strain rate fracture of the AS-800 material were not considered.

The analysis results showed that the current MOD 0 331-200[CT] ceramic blade tip thickness needs to be increased to withstand the impact from a 0.2 inch (5.1 mm) diameter carbon particle. Additional options for robust, impact-resistant turbine blade designs and applications also need evaluation; these include (but are not limited to) the following:
- Closer stator nozzle spacing, to reduce the maximum size of carbon particles that may pass through

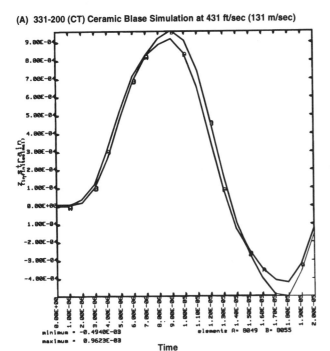

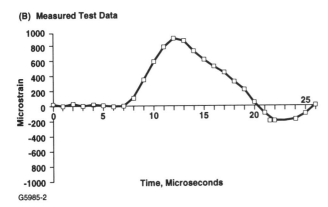

Fig. 2. Comparison of UDRI Impact Test Data With DYNA3D Computer Code Predictions Shows Good Correlation.

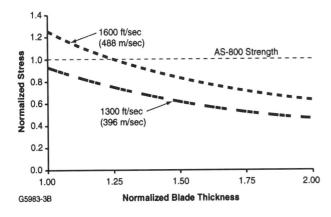

Fig. 3. Analysis Shows Current (MOD 0) Ceramic Turbine Blade Tip Thickness Needs to Be Increased to Withstand Impact From 0.20 inch (0.51 mm) Diameter Carbon Particle.

- Designs with lower tip speeds, and possibly more stages, to reduce impact damage severity and increase survivability
- Clean burning/carbon free combustors
- Utilization of non-carbon-producing fuels (i.e., methanol).

Blade Attachment Technology

The goal of this activity is to provide effective design solutions for robust, axial inserted ceramic blade attachments for production gas turbines. These design solutions necessarily require a thorough understanding of the blade attachment contact stresses and the environmental and geometric factors that control these stresses. A contact test rig has successfully been used for evaluation of ceramic contact interfaces of various configurations.

Contact rig tests were performed during 1995 in which the contact interface was simulated with a MIL-B type ceramic specimen, loaded with a radiused Astroloy indenter. During these room temperature tests, the indentor was rubbed across the face of the ceramic bar with a pinch load of 300 pounds (136 kg), simulating the bearing stress on a ceramic turbine blade. The stroke (rub) length was varied over the range from 0.020 (0.51 mm) down to 0.006 inch (0.15 mm), which was the shortest controllable stroke length for the contact test rig. For comparison, the translation expected in a typical turbine blade attachment is approximately 0.003 inch (0.076 mm). Friction in the interface was modified during testing with various coatings, including cobalt oxide (an oxide layer on Haynes Alloy HS25), boron nitride (BN), and Lubeloc 7401* (a proprietary high-temperature antifriction com-pound). Following exposure to the contact loading, each ceramic specimen was then tested for retained room-temperature fast fracture strength.

The tests conducted in 1995 confirmed that the room temperature fast fracture strength of NT154 silicon nitride ceramic was significantly degraded by rubbing contact in which the interface friction loads were relatively high. Rub length also had a strong influence on the retained strength of the ceramic specimens. This relationship is shown in the test data plotted in Fig. 4. Within the range of testing, it was found that the

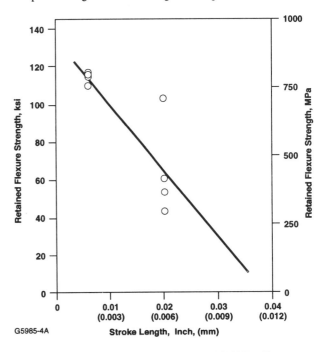

Fig. 4. Contact Rig Test Data Indicates NT154 Fast Fracture Strength Is Degraded by Rubbing Contact.

---

*Lubeloc 7401 is a proprietary high-temperature antifriction compound produced by EM Corporation, Connecticut.

number of cycles, the cumulative travel, and the type of lubricant (coating) did not significantly affect the retained strength.

Cyclic spin tests were also conducted, in which various anti-friction treatments were applied to several types of blade compliant layers, and assembled with ceramic test blades in a slotted metalllic (engine hardware) turbine rotor disk. In this test series, 100 spin test cycles were performed on each rotor assembly at speeds from 4,000 to 42,000 rpm (representing 10 percent to 100 percent design speeds), after which the rotor was disassembled and the blades and compliant layer specimens were examined. The compliant layer configurations tested included pre-oxidized Haynes Alloy HS25 (Baseline), pre-oxidized HS25 with boron nitride (BN) applied via aerosol to both sides, and pre-oxidized HS25 coated with Lubeloc 7401 on the disk side only. The cyclic spin test results are listed in Table 1. Based on these tests, neither the boron nitride (BN) or Lubeloc 7401 coatings performed better than the baseline pre-oxidized HS25.

Table 1. Compliant Layer Cyclic Spin Test Results.

Compliant Layer	Test Temperature	
	Room Temperature	1200F (649C)
Pre-oxidized HS25 (Baseline)	No wear through	No wear through
Pre-oxidized HS25 with BN	Wear through	(Not tested)
Pre-oxidized HS25 with Lubeloc 7401	No wear through	Severe galling

Ceramic Nozzle Proof Testing Techniques

The focus of this effort is to identify cost-effective proof testing techniques to ensure the quality of ceramic hardware prior to engine installation. The goals are to develop proof testing methods that are accurate, yet inexpensive, that may potentially be used by the ceramic component manufacturers and users. Work during 1995 resulted in the design and fabrication of three test rigs for proof testing the 331-200[CT] MOD 2 ceramic nozzles. One of the rigs, intended to test the nozzles with respect to attachment loads, was functionally evaluated with actual ceramic nozzles and used to proof test a set of parts received in preparation for engine testing. Evaluations of the other two rigs, the airfoil test rig and the fillet test rig, are planned for early 1996.

Ceramic Proof Test Methods

This activity, in collaboration with the ongoing DOE-sponsored Phase II Life Prediction Methodology For Ceramic Components of Advanced Heat Engines Program, is being performed by AlliedSignal Engines and managed by the DOE Oak Ridge National Laboratory (ORNL). This work will establish the reliability of proof testing of ceramic components with respect to volume flaws. (Additonal work to establish ceramic component reliability with respect to surface flaws is being performed by AlliedSignal Engines under the DOE/ORNL-sponsored Phase II Life Prediction Program, and is not reported here.) This complementary work under both programs will prove extremely useful in defining the types of proof tests and the test criteria for future ceramic components.

Ceramic tensile test specimens have been procured to determine the effect of room-temperature proof testing on elevated-temperature fast fracture and time-dependent failure modes for volume-distributed flaws. The goal is to determine if proof testing affects the component integrity. The test specimens have been divided into two groups: one group initially received proof testing, with criteria selected to fail 30 percent of the specimens; the other group has not been proof tested. All of the specimens will be further tested to failure in high-temperature tensile fast-fracture and tensile stress-rupture, to determine whether the failure populations were truncated or modified by the proof testing.

CERAMIC COMPONENTS QUANTITY FABRICATION DEMONSTRATION

The purpose of this activity is to develop the required capabilities of domestic ceramic engine component suppliers to support engine production. This activity will move the ceramic component fabrication processes out of the laboratory and into a production environment in which consistent, high-quality components can be made economically.

Initiated in late 1993 and scheduled to continue through 1997, this subcontract work is focusing on the suppliers of the 331-200[CT] ceramic components for the Ceramic Turbine Engine Demonstration Program. The suppliers are being challenged to scale-up and improve their demonstrated fabrication processes to achieve production readiness. AlliedSignal Ceramic Components (CC, in Torrance, CA), and Norton Advanced Ceramics (NAC, in East Granby, CT) participated in this activity during 1995, with Kyocera Industrial Ceramics Company (KICC, in Vancouver, WA) initiating subcontract work at the end of the year.

At the beginning of this activity, production processes for the volume manufacture of ceramic gas turbine components did not exist at the subcontractor facilities. The laboratory-based processes in use were typically capable of only producing less than 50 pieces per month, at overall yields of less than 25 percent acceptable parts. At the conclusion of the manufacturing scaleup and demonstration subcontract activity, each of the participating manufacturers will have achieved the goals listed in Table 2.

Table 2. Ceramic Component Manufacturing Scaleup Goals.

Item	Goal
Individual Process Capability	500 Parts/Month
Overall Yield	75 Percent
Overall Process Capacity	100 Parts/Month (Demonstrated)

By the end of 1994, CC demonstrated the feasibility of AS-800 silicon nitride as a potentially lower-cost turbine component material with respect to both material properties and shape capability. This demonstration enabled CC to replace GN-10 with AS-800 as the ceramic material for the planned 1995 scaleup demonstration. AS-800 is a gas-pressure-sintered silicon nitride that has the potential for lower component fabrication costs compared to GN-10, which is a glass-encapsulated, hot isostatic pressed (HIPped) material. In addition, AS-800 has superior as-processed surface properties, better elevated-temperature fast-fracture and stress-rupture properties, and exhibits very high Weibull characteristics in each of these strength categories compared to GN-10. During 1995, CC made several AS-800 process improvements, increasing capacity for slip preparation, pre-sintering, and drying. In addition, CC deployed short-run statistical process control (SPC) in the AS-800 slip preparation and pre-sintering processes.

NAC focused their 1995 activities on the elimination of iron inclusions and demonstration of closed-loop processing (CLP) of NT164 silicon nitride components. This work was completed by mid-year, at which time NAC announced their business decision to discontinue production of gas turbine engine ceramics and subsequently terminated participation in this program.

AlliedSignal Engines initiated ceramic manufacturing scaleup and demonstration activities with KICC during the third quarter of 1995. Before the end of the year, KICC had successfully formed one lot of SN-252 silicon nitride ceramic nozzles using their hybrid molding process, achieving a green-forming yield of 70 percent.

ENGINE DEMONSTRATIONS

Ceramic turbine engine demonstration activities began under the present program during 1992, with selection of the AlliedSignal/ Garrett Model 331-200[ER] APU, a fully-developed gas turbine in production with current applications in the Boeing 757 and 767 aircraft, for modification into the ceramic turbine engine demonstration test bed.

Within the 331-200[CT] test bed, the existing metallic first-stage axial turbine was redesigned to incorporate ceramic turbine nozzle segments and inserted axial ceramic turbine blades. This activity, completed in 1993, included detailed design of the ceramic components and modified metallic support structures, test hardware fabrication, component testing to verify the component performance characteristics, and engine demonstration testing. The first engine tests of the ceramic design were conducted during 1993.[1] These initial test results led to incremental redesign efforts, achieving improvements in both the ceramic nozzles and ceramic turbine blades which were evaluated during 1995.

During the past year, the MOD 0 (baseline) ceramic nozzles completed evaluation in endurance tests; MOD 1 ceramic nozzles were received and evaluated for performance in rig tests; and MOD 2 ceramic nozzles were designed and fabricated. Design modifi-cations to the ceramic blade were also completed, resulting in the MOD 1 blade design. These tasks are summarized in the paragraphs following.

<u>MOD 0 Ceramic Nozzle Engine Testing</u>

500-hour duration engine endurance test of the MOD 0 ceramic nozzles was completed in early 1995. Of the 23 ceramic nozzles in this test, 21 had been used in previous tests. The high-time MOD 0 nozzle segments have successfully accumulated over 895 hours engine operation in the accelerated mission engine test cycle. The 500-hour ceramic nozzle engine test was successfully completed without incident. Post-test inspections identified two nozzle segments with thermally-induced cracks similar to those seen earlier in proof testing.[2] Despite these cracks, the nozzles had remained intact and did not result in engine failure.

These thermally-induced cracks, found in several MOD 0 ceramic nozzles during proof tests and later engine testing, indicated that the MOD 0 nozzle design was too highly stressed. The MOD 0 ceramic nozzle stress analysis, performed in prior years,[1] had considered only nominal temperature conditions during the engine starting transient and ignored any potential for temperature maldistributions (hot streaks) from the combustor during the engine start. Additional studies performed since then have revealed that such hot streaks can significantly increase the transient stress levels in the ceramic nozzles. Since the MOD 0 nozzles were all proof tested to approximately 60 ksi (414 MPa) prior to assembly for engine testing, it was assumed that the cracked nozzles from the 500-hour engine test experienced stresses greater than during the proof testing, or 50 percent higher than specified for the original stress analysis. Due to this finding, the stress analyses performed during the MOD 1 and MOD 2 ceramic nozzle redesign efforts employed worst-case light-off temperatures, to ensure more accurate modeling of actual engine operation stresses.

<u>MOD 1 Ceramic Nozzle Engine Testing</u>

MOD 1 ceramic nozzles made of SN-252 silicon nitride were received from Kyocera. These parts were proof tested using the hot gas flow rig developed for the MOD 0 nozzle configuration.[2] All the parts completed the proof test intact. Following proof testing, the MOD 1 nozzle segments were assembled and evaluated in a cold flow rig. This rig is used to determine the corrected flow that the nozzle assembly will pass at the choked-pressure condition, to gage the effective area of the nozzles. Figure 5 compares the choked-flow values for the MOD 0 and MOD 1 ceramic nozzles. A reference test was also performed with the MOD 0 nozzle segments, artificially sealed to prevent gas leakage through the platform gaps and also sealed on the inner hub and outer shroud surfaces. The MOD 1 configuration had lower flow and therefore lower leakage than the MOD 0 design, but higher flow than the "sealed" reference test. This result suggested that the MOD 1 ceramic nozzle assembly has less leakage and should show higher performance in the engine than the MOD 0 assembly.

The MOD 1 ceramic nozzle assembly was then installed into the test bed and the 331-200[CT] engine was operated to gather performance data for comparison with previous tests conducted on the (Baseline) metallic nozzle equipped and MOD 0 ceramic nozzle equipped engine builds. These test results are summarized in Fig. 6. Note that the MOD 1 ceramic nozzle equipped engine exhibited a consistently lower turbine discharge exhaust gas temperature (EGT) than the MOD 0 ceramic nozzle equipped

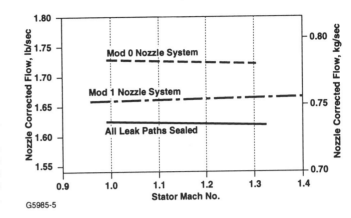

Fig. 5. Flow Test Results Indicate MOD 1 Ceramic Nozzle Design Has Decreased Leakage.

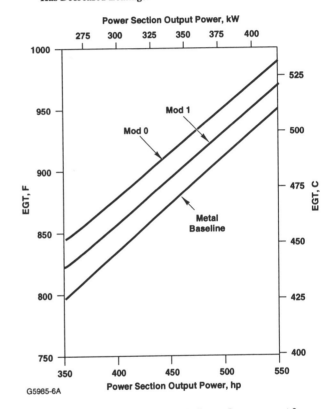

Fig. 6. Engine Performance Data Indicates Improvement for MOD 1 Ceramic Nozzle Design.

engine when producing the same power. This demonstrates that the MOD 1 ceramic nozzle assembly is an improvement over the MOD 0 ceramic nozzle with respect to engine performance, though it still falls short of the baseline metallic nozzle equipped engine.

<u>MOD 2 Ceramic Nozzle Design</u>

In 1995, the MOD 1 ceramic nozzles were again redesigned, to further improve the producibility of the ceramic nozzle system. In the MOD 2 design, the nozzle attachment features were simplified, and the performance and durability of the ceramic nozzle system was improved. Figure 7 shows the evolution of the 331-200[CT] ceramic nozzle through the three design iterations to date. The MOD 1 design has thicker airfoil and platform sections than the initial MOD 0 design for decreased stress, and the MOD 2 design has simplified attachment features for improved performance and manufacturability.

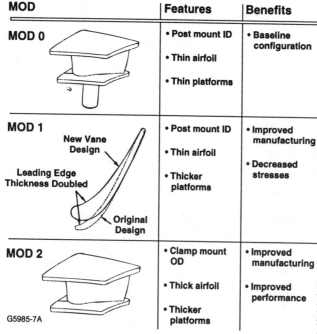

Fig. 7. 331-200[CT] Ceramic Turbine Nozzle Design Evolution.

The MOD 2 ceramic nozzle design resulted in a change in the way the ceramic nozzle segment is supported by the metallic structures. In the MOD 0 and MOD 1 designs, the ceramic nozzle mounting was cantilevered from a resilient attachment to the ceramic post on the nozzle hub platform. This design worked well in engine testing from a structural perspective; but designing a robust seal for the outer shroud platform and machining the post feature on the ceramic nozzle presented difficulties. The MOD 2 nozzle is mounted on the outer shroud platform, which significantly improves the design of the shroud and hub platform seals; and fabrication is eased by removal of the ceramic post feature from the hub platform. The MOD 2 design employs the MOD 1 airfoil geometry, for benefits in low airfoil stress and good formability.

An analysis was performed to predict the stresses induced in the MOD 2 ceramic nozzle airfoil at steady-state and transient operating thermal conditions. Table 3 shows a comparison of predicted stresses in the MOD 0, MOD 1, and MOD 2 ceramic nozzle designs. Note that the thermal stresses in the MOD 2 nozzle are even lower than for the MOD 1 nozzle. From these results, thermomechanical fracturing of the MOD 2 airfoil is not expected to present any problems during engine testing. However, contact stresses in the MOD 2 shroud platform are of concern. Attempts to analyze these stresses using a refined mesh for the finite element model were unsuccessful. A decision was made to evaluate the integrity of the MOD 2 nozzles with respect to contact stress in the attachment proof test rig.

Attachment proof rig tests were performed on the initial batch of NT154 MOD 2 ceramic nozzles. The nozzles were individually loaded to over 500 pounds (227 kg) across the attachment surfaces on the shroud platform. This load represents 125 percent of the maximum load expected in the engine. Initial testing resulted in no damage to the test nozzle, so the entire lot of 30 nozzles was then individually tested in the rig at the same conditions, resulting in only two indications of contact damage. A set of 23 undamaged nozzles was selected from the proof-tested articles and prepared for further evaluations, including flow testing and engine testing.

MOD 1 Ceramic Turbine Blade Design

Ceramic turbine blade development focused on enhancing the overall robustness of the 331-200[CT] first-stage inserted axial turbine blades. This included an appraisal of the shortcomings of the (baseline) MOD 0 ceramic blade design, as well as research into methods of controlling blade vibration and instrumentation to measure vibration at engine operating conditions.

- The blade first resonant mode fell within the range of engine operating speeds
- The blade attachment shank was thinner than good design practice suggested
- The specified 60 degree contact angle made the blade attachment feature bearing stresses sensitive to interface friction.

For the MOD 1 ceramic blade design, the airfoil was not changed from the MOD 0 baseline, and the MOD 0 design issues were addressed with revisions to the attachment geometry. The major features of the MOD 1 design include a thicker, shortened blade shank, which serves to increase the shank section modulus, raising the blade resonant frequencies above the engine operating speed range. A second major change was a reduction in the attachment contact angle from 60 to 45 degrees, which was intended to decrease the sensitivity of the attachment bearing stress to changes in interface friction.

Figure 8 shows vibration analysis results for the MOD 1 ceramic blade design. The more robust shank of the MOD 1 blade was principally responsible for an increase in the first resonant frequency to above 27,000 Hz. (It should be noted that this calculation was performed for NT154 silicon nitride.) Manufacture of the MOD 1 blade from denser silicon nitride materials such as AS-800 or SN-251/SN-252 would result in slightly lower resonant frequencies, decreasing the operating margin for the engine. A thorough evaluation is planned of the as-received MOD 1 ceramic blades with respect to resonant characteristics. Adjustments may be made to the scheduled engine operating speeds, to avoid any potential blade resonance condition.

Figure 9 shows steady-state operation stress analysis results for the MOD 1 ceramic blade. A peak stress of 42 ksi (210 MPa) is predicted to occur in the blade shank; this is approximately the same as the values for the baseline MOD 0 design. This analysis also assumed the blade material was NT154 silicon nitride, and the attachment interface was frictionless. Further analysis is planned to evaluate the effect of AS-800 blade material and to evaluate the effects of friction in the interface. The MOD 1 ceramic blade design has been completed and parts have been placed on order.

Ceramic Blade Vibration Damping Evaluation

A vibration damping method for the MOD 0 331-200[CT] ceramic blade was designed and evaluated during 1995. The selected method is similar to a concept successfully used with conventional metallic blades. A friction damper is added under the blade platforms, to control vibration excursions during transient operation through possible resonance conditions. The damper design was evaluated using MOD 0 ceramic

Table 3. Predicted 331-200[CT] Ceramic Nozzle Stresses.

Nozzle Location	Peak Transient Stress, ksi (MPa)			Peak Steady-State Stress, ksi (MPa)		
	MOD 0	MOD 1	MOD 2	MOD 0	MOD 1	MOD 2
	2-D Nominal	3-D Hot Streak	3-D Hot Streak	2-D	3-D	3-D
Airfoil Leading Edge	44 (303)	8 (55)	8 (55)	24 (166)	5 (35)	3 (21)
Airfoil Trailing Edge	41 (283)	16 (110)	15 (103)	20 (138)	8 (55)	6.5 (45)
Post Fillet	---	10 (69)	[N/A]	---	4 (28)	[N/A]
[N/A] = Not Applicable.						

**Vibration Test Data
Mode Shapes**

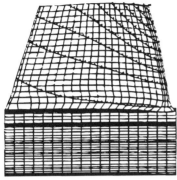

27,500 Hz Mode 1

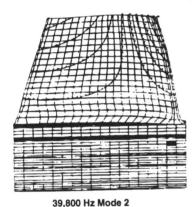

39,800 Hz Mode 2

G5985-9

Fig. 8. Vibration Mode Shapes For MOD 1 Ceramic Turbine Blade.

blades in a room-temperature airjet spin test setup.[1, 2] A total of six ceramic blades were used in the test setup: three blades had dampers installed under the platforms, and three were undamped. All of the blades were instrumented with strain gages, and a total of nine tests were performed. A rotor assembly with the six blades installed was spun up to the target speed in the airjet rig, while the blades were excited into resonance with gas jets (simulating the passing frequencies from the turbine nozzles as in engine operation). A significant variation in blade vibration amplitude from blade to blade was observed, making it difficult to determine whether the dampers effectively attenuated the response of the blades. However, the blades with dampers exhibited a 30 percent higher measured damping ratio compared to the undamped blades. It is expected that better results might be obtained with heavier dampers.

Future Work
Program activity is planned to continue during 1996 and forward into 1998 with preparation of ceramic component technology for commercialization. The schedule in Fig. 10 summarizes the overall program activities from its inception through 1998. The following future activities are planned:
- MOD 2 ceramic turbine nozzle development and qualification testing
- Initiation of field testing with the MOD 2 ceramic nozzles
- Initiation of MOD 1 ceramic turbine blade testing
- Initiation of separate engine endurance tests with ceramic nozzles and ceramic blades.

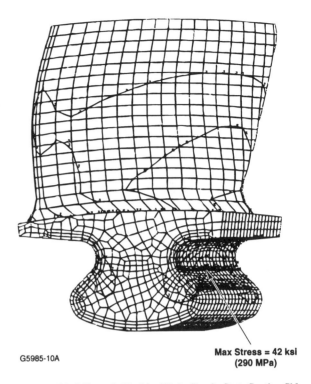

G5985-10A Max Stress = 42 ksi (290 MPa)

Fig. 9. MOD 1 Ceramic Turbine Blade Steady-State Suction Side Stress Plot.

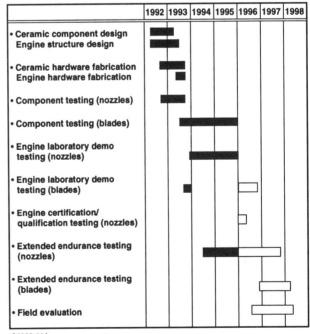

Fig. 10. 331-200[CT] Ceramic Turbine Engine Demonstration Program Overall Schedule.

SUMMARY AND CONCLUSIONS
The DOE-NASA sponsored 331-200[CT] Ceramic Turbine Engine Demonstration Project is planned to continue into 1998, with the mission of advancing ceramic gas turbine component technology towards

commercialization. This will be accomplished by enhancing critical ceramic design technologies, scaling up and demonstrating production-level ceramic component manufacturing capability of domestic ceramics manufacturers, and demonstrating ceramic engine component durability and reliability in extensive laboratory and field engine tests.

During the past year, the design technologies for ceramic turbine blades were further advanced with completion of ceramic axial turbine blade impact design methodology development, ceramic blade impact testing, and a parametric study that produced guidelines for designing impact-resistant axial ceramic turbine blades. Investigations into the contact stress environment of the ceramic axial turbine blade attachment confirmed the importance of coatings to reduce friction and minimize contact stresses and also revealed the sensitivity of ceramic damage to translation (rub) length.

In the ceramic component manufacturing scaleup and demonstration activities, AlliedSignal Ceramic Components and Norton Advanced Ceramics (NAC) completed work necessary to justify a change to new ceramic materials and production processes that were more amenable to quantity production and improved quality. Due to business reasons, NAC opted to leave the program. Kyocera Industrial Ceramics Company was invited to join the program and began forming components.

Successful engine testing with the (baseline design) MOD 0 ceramic nozzles continued. Coupled with additional testing of the redesigned MOD 1 ceramic nozzles, total cumulative ceramic nozzle engine test time increased to over 910 hours. Ceramic nozzle engine testing is summarized in Table 4. The MOD 1 nozzle testing confirmed that the new "robust" airfoil geometry performed at least as well as the original MOD 0 design in engine performance testing. Further improvements were incoporated into the MOD 2 ceramic nozzle design and parts fabrication was initiated. The MOD 2 nozzle design is intended to reduce fabrication costs and improve yields.

Table 4. 331-200[CT] Ceramic Nozzle Engine Testing Summary.

Nozzle Type	Test Time, Hours	Starts
MOD 0	897.6	2481
MOD 1	9.8	22
MOD 2	0.0	0
Totals	907.4	2503

Vibration dampers for the ceramic blades were designed and evaluated in airjet spin testing. These dampers, similar to friction dampers commonly used with metallic inserted blades, were found to add damping to the ceramic blade system. However, the measured damping was not sufficient to significantly reduce the ceramic blade airfoil strain values at resonance conditions. A new (MOD 1) ceramic turbine blade design was completed. The MOD 1 blade design features added improvements over the MOD 0 design, including: increased attachment stiffness; increased shank thickness, resulting in the blade resonant frequency raised out of the engine operating range; and new dovetail geometry, intended to reduce bearing stress sensitivity to interface friction.

Engine testing with ceramic nozzles is planned to continue, with the objective of validating the performance and integrity of the MOD 2 nozzle design, and qualifying the 331-200[CT] ceramic nozzle for initiation of field testing, planned to begin in 1996. Engine testing of the ceramic turbine blades is planned to begin in early 1996, with a series of tests including blades with high-temperature thin-film strain gages, to measure ceramic turbine blade stress values during engine operation.

The 331-200[CT] Ceramic Turbine Engine Demonstration Project has the vision of augmenting the development of ceramic technology in support of automotive gas turbine development. To achieve this goal, the program plan is to enhance the ceramic technologies required to support the design of gas turbine ceramic components, to refine and scale up the production capability of domestic ceramic component manufacturers, and to demonstrate the capabilities of the ceramic components, first in laboratory field tests and then in extensive field trials. The engine demonstration and field evaluations will provide the experience required to verify the improvements in ceramic design technology and component fabrication.

ACKNOWLEDGMENT

This paper was originally presented at the 41st ASME International Gas Turbine & Aeroengine Congress, Exposition and Users Symposium on June 13, 1996 in Birmingham, England.

REFERENCES

1. M.L. Easley and J.R. Smyth, "Ceramic Gas Turbine Technology Development," ASME Paper 94-GT-485, American Society of Mechanical Engineers, 1994.

2. M.W. Rettler, M.L. Easley, and J.R. Smyth, "Ceramic Gas Turbine Technology Development," ASME Paper 95-GT-207, American Society of Mechanical Engineers, 1995.

TITANIUM METAL MATRIX COMPOSITES FOR AEROSPACE APPLICATIONS

S. A. Singerman* and J. J. Jackson**

*Pratt & Whitney
West Palm Beach FL

**GE Aircraft Engines
Lynn MA

Abstract

Aerospace engine and airframe designers are constantly seeking lighter weight high strength materials to reduce weight and improve performance of powerplants and aircraft. Titanium metal matrix composites (Ti MMCs) have offered the promise of significant weight savings since their initial development in the early 1960s, but until recently, their inadequate quality and reproducibility combined with high processing and materials costs have prevented their introduction into production applications. This paper describes the state-of-the-art for Ti MMC aerospace fabrications, their potential payoffs and the recent advances in processing which are now leading to high quality, affordable Ti MMC components.

Introduction & Historical Perspective

Over the past 30 years, titanium metal matrix composites (Ti MMCs) have been under considerable development and evaluation for use in aircraft engine and airframe applications. For airframers, the high specific modulus of Ti MMCs has been the impetus,[1-3] while engine makers have sought to take advantage of their high specific strength, especially for compressor rotor applications.[4] With the development of titanium aluminide matrix alloys[5,6] which have temperature capabilities approaching 760°C (1400°F), Ti MMCs offer a potential 50% weight reduction in the hotter compressor sections now dominated by nickel based superalloys.

The introduction of Ti MMCs into high performance engine applications has been inhibited partly by the complexities of composite rotor fabrication. However, a more significant barrier is their high materials and implementation cost[7] which is mainly driven by low market volume. To overcome these barriers, Ti MMC components applications are now being emphasized by Pratt & Whitney (P&W) and GE Aircraft Engines (GEAE) under the Advanced Research Projects Agency (ARPA)/Air Force sponsored Titanium Matrix Composite Turbine Engine Component Consortium (TMCTECC) Program.[8] The high bypass commercial turbofan engines which will power long range aircraft into the next century can benefit greatly from the weight and operating cost reductions enabled by the selective use of Ti MMCs in their structures. These applications represent the size market needed to make Ti MMCs cost competitive (Ti MMCs at $1100 per kilogram, $500 per pound) for production introduction into engines or airframes.[8] The following describes the status of Ti MMCs in terms of their demonstrated capabilities, potential payoffs and progress towards achieving affordable manufacture of components for aerospace applications.

Processing & Properties

Ti MMCs which have demonstrated properties suitable for aerospace applications consist of conventional (Ti6Al4V, Ti6Al2Sn4Zr2Mo, etc.) and advanced (Ti$_3$Al, TiAl, etc.) titanium matrix alloys reinforced with 30-40 volume percent of continuous arrays of high strength (>3450 Mpa, >500 ksi), high modulus (380 Gpa, 55 msi) SiC fibers.§ These fibers are approximately 0.127 mm (5 mils) in diameter and produced by chemical vapor deposition (CVD) with a 4 μm (0.2 mil) carbon rich surface layer to enhance processability, fiber strength and achieve desired metal/fiber interface characteristics.[9,10]

Processing

For many years, Ti MMCs were primarily fabricated using foil/fabric processes consisting of alternating layers of woven fiber mats and 0.1-0.15 mm (4-5 mil) thick titanium alloy foils which were stacked up and vacuum hot press (VHP) or hot isostatic press (HIP) consolidated into multilayer composites. High foil costs associated with cross-roll processing of the preferred titanium alloys combined with high fiber costs and low volume demands caused Ti MMCs to only be considered for very high payoff applications. Additionally, a lack of reproducible quality for foil/fabric Ti MMC components precluded their introduction into any man-rated aerospace applications. More recently, innovative processes including tape casting,[11,12] induction plasma deposition (IPD),[13-15] electron beam physical vapor deposition (EBPVD) fiber coating[16] and fiber/wire co-winding[17] have been developed in order to increase alloy flexibility and improve quality of Ti MMCs and at the same time reduce their fabrication costs.

The availability of this assortment of approaches now allows composite manufactures to select the method most suited for a particular component configuration. For example, airfoils, ducts and certain unidirectionally reinforced parts (actuators, exhaust link and struts) are most easily assembled using tape cast or IPD processed Ti MMC monotapes (a single Ti MMC ply). Cylindrical shapes requiring cross-ply layups which include shafts and ducts/cases can be more easily assembled with IPD processed monotapes which can be produced in wide sheets and maintain fiber position during assembly more effectively than other methods. Rings for reinforcing rotor components can now be produced with tape cast strips, co-wound fiber/wire or coated fiber techniques more easily than with foil/fabric or IPD processes. Selectively reinforced structural applications are most cost effectively fabricated using coated fiber and pre-consolidated shapes made from tape cast or IPD monotapes. As a consequence of this increased flexibility for fabricating Ti MMC components, manufacturing costs are being dramatically reduced compared with previous foil/fabric components and quality significantly improved.

Having multiple fabrication options can aid the development of prototype components but it leads to a fractional market and resulting low volumes. One goal of the TMCTECC Program is to focus on a common material specification and mill product for fan applications in order to drive the cost of Ti MMCs to $1100 per kilogram ($500 per pound).

Properties

Some of the new processes cited above have now been developed to the point where a large enough materials property database exists to enable engine designers to make Ti MMCs serious candidates for weight reduction opportunities in advanced and growth versions of current engines. A comparison of properties for Ti MMCs and superalloys is shown in Table I.

Table I Comparative Properties of Ti MMCs and Superalloys

Property*	Conventional Ti MMC	Ti Aluminide Ti MMC	Superalloys
Density, g/cm^3	4.04	4.18	8.3
0° Stiffness, GPa	200	242	207
90° Stiffness, GPa	145	200	207
Max Use Temp- °C	538	760	1090
0°CTE,°C^{-1} x 10^{-6}	7.20	7.92	13.0
90°CTE,°C^{-1} x 10^{-6}	8.91	9.18	13.0

* 0° = Direction of Fiber, 90° = Transverse to Direction of Fiber

---

§ SCS-6 SiC fiber made by Textron Specialty Materials, Lowell MA.
Trimarc1 SiC fiber made by AMERCOM, Inc., Chatsworth CA.

Achieving desired properties in Ti MMC structures is strongly dependent on preventing fiber damage or degradation during component fabrication while maintaining uniform fiber spacing illustrated in Figure 1. When properly processed, conventional Ti MMCs can be reproducibly fabricated by tape casting or plasma deposition to achieve superalloy strengths up to 538°C (1000°F) at half their density as shown in Figure 2. Advanced titanium aluminide MMCs based on the Ti_3Al intermetallic are being developed which may enable Ti MMC use up to 760°C (1400°F).[5,6] The specific modulus of Ti MMCs, which is especially important for structural applications, is also nearly double that of superalloys as shown in Figure 3.

seeks to double the specific thrust of military engines and must rely heavily on the weight reduction potential of Ti MMCs and other advanced materials to do so.

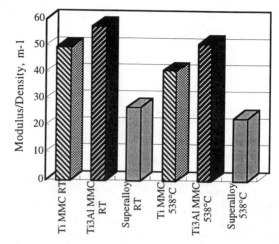

Figure 3: Comparison of specific modulus of Ti MMCs and typical superalloys.

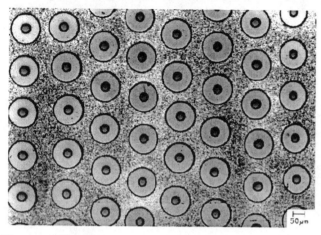

Figure 1: Typical microstructure of properly processed Ti MMC (Ti6Al2Sn4Zr2Mo/SCS-6) fabricated from plasma sprayed monotapes. *Note the uniformly spaced, non-touching fiber array which is critical to achieving predicted properties.*

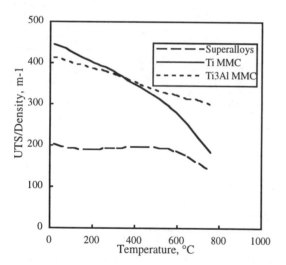

Figure 2: Comparison of specific strength of Ti MMCs and typical superalloys.

Of course, these excellent properties in the direction of fiber orientation can only be taken advantage of if the lower transverse properties of Ti MMCs do not fall below design requirements. Many applications have been identified that can cope with this anisotropy of Ti MMCs and in some instances take advantage of it.

Other mechanical properties critical to aerospace applications include low cycle fatigue (LCF) and fatigue crack growth (FCG). As shown in Figures 4 and 5, Ti MMCs exhibit LCF and FCG properties superior to superalloys when loaded in the fiber direction, even before considering their density benefit. However, transverse LCF and FCG for Ti MMCs are significantly lower than superalloys and therefore their use is restricted to those rotor applications which introduce low transverse cyclic stresses. Many rotor applications which satisfy this criteria have been identified by designers of the advanced military engines being developed under the Integrated High Performance Turbine Engine Technology (IHPTET) Program. This program

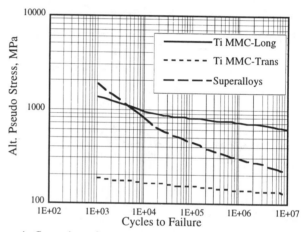

Figure 4: Comparison of room temperature low cycle fatigue capability of Ti MMCs with wrought IN718, a typical disk superalloy.

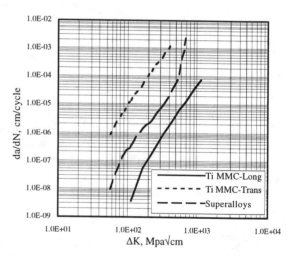

Figure 5: Comparison of room temperature fatigue crack growth behavior of Ti MMCs with PM Rene 88DT, a typical disk superalloy especially designed for improved crack growth resistance.

Potential Engine Applications & Payoffs vs Risks

Potential aerospace applications for Ti MMCs fall into several categories of payoff and risk. Figure 6 illustrates Ti MMC component opportunities and Table II lists the estimated payoffs, risks and relative costs associated with each. The applications are divided into categories of rotating and non-rotating parts which fall into various risk classifications. These risk classifications for Ti MMC components refer to the consequences of component failure or failure to perform per design intent. Weight savings are based on a comparison with the current part which may be titanium, nickel or steel alloys.

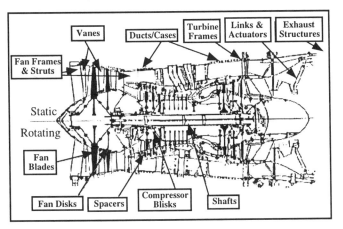

Figure 6: Potential engine applications for Ti MMCs.

Rotating Components

Rotating parts such as Ti MMC reinforced impellers, disks, integrally bladed rotors (IBRs) or blisks (bladed disks), blings (bladed rings) and blotors (bladed rotors) are high risk because they are inherently difficult to manufacture and their failures can destroy an engine. However, payoffs for Ti MMC rotors, in terms of engine performance and weight savings are the highest of any application. Rotor weight savings of from 30% to >50% can be achieved with Ti MMCs with the added advantage of a larger free-hoop radius than either monolithic titanium or nickel disks. Ti MMC shafts represent a moderate to high risk application because, while they are more easily fabricated than disks, their failure in service could also be catastrophic. Payoffs for Ti MMC shafts are only moderate (up to about a 30% weight savings) but they can lead to improved rotor dynamics. Non-load carrying spacers represent a low risk Ti MMC rotating application which may perform a critical engine function and thus justify their higher cost even though offering a small weight savings versus monolithic spacers.

Fan and compressor airfoils represent low to moderate risk rotating components since engines are designed to contain or otherwise cope with their failure and prevent engine destruction. Fabrication of Ti MMC containing airfoils is the least difficult of all rotating parts since their structures are usually two-dimensional layups with moderate curvature. In most cases, the payoff for Ti MMC airfoils on a direct substitution basis is relatively low (15-20% weight savings) since they would replace titanium blades and thus have difficulty justifying their higher cost. However, in advanced high bypass engine applications like the PW4084, the specific stiffness of Ti MMC reinforced hollow fan blades (RHFBs) combined with a weight savings make them attractive enough to pursue.[8] The payoff for Ti MMC airfoils in future "rubber engine" designs can be even more significant if the airfoil weight savings allows further reductions in disk and support structure weights.

Non-Rotating Components

Most non-rotating Ti MMC engine applications are of moderate to low risk. Pressure vessel and containment applications including ducts/cases represent typical examples of moderate risk parts. The modest payoff of 15-25% potential weight savings for these parts can be very significant due to the large size of these components. However, these weight savings must be traded off against the complexity and cost of their fabrication, which can be difficult due to the multitude of ports, attachments and other features prevalent. Stator vanes containing Ti MMC can not compete with current materials until titanium alloys for use above 760°C (1400°F) are fully developed and can be made into composites. To date, Ti MMCs based on the high temperature (>815°C, >1500°F) gamma TiAl intermetallic have not been successful due to the low ductility of these alloys combined with the significant SiC/TiAl coefficient of thermal expansion (CTE) mismatch.[18,19]

Structural components like struts and fan or turbine frames, whose primary function is to maintain engine shape and clearances under the wide range of mission loadings, are of moderate to low risk and rely heavily on material stiffness. Both weight savings and performance gains through reduced specific fuel consumption (SFC) can result from using Ti MMCs in these structures. On large bypass engines such as the GE90, fan frame weight savings of 10 to 15% are possible along with net cost savings of up to 35%. These cost savings result from the use of lower cost aluminum or polymeric vanes which can be substituted for complex fabricated monolithic parts as a consequence of the selected application of Ti MMC reinforced components.[20]

Table II Potential Engine Applications and Payoffs for Ti MMCs.

Category	Components	Engine Risk	Potential Payoff	Relative* Manufacturing Difficulty	Relative* Manufacturing Cost
Rotating Parts	Disks, IBRs (Blisks), Blings, Blotors, Impellers, etc.	High	• 30% to >50% Weight • Larger Free-Hoop Radius • Strength	5	5
	Shafts	Moderate to High	• 15% to 30% Weight • Enhanced Stiffness • Improved Rotor Dynamics	3	4
	Fan & Compressor Blades	Moderate to Low	• 15% to 20% Weight (substitution) • >30% Weight (new designs) • Improved Dynamics	2	3
	Spacers	Low	• 10% to 15% Weight • Dimensional Stability	4	4
Non-Rotating Parts	Ducts/Cases	Moderate	• 15% to 25% Weight • Stiffness	3	3
	Stator Vanes	Moderate	• 5% to 15% Weight • Stiffness	2	1
	Struts, Fan & Turbine Frames	Moderate	• 25% to 35% Weight • Stiffness • Improved SFC	2	1
	Links, Actuators	Low	• 15% to 45% Weight • Stiffness, Stability	2	2
	Exhaust Sidewalls & Structures	Low	• 25% to 40% Weight • Stiffness	1	2

* Relative Scale: 1 to 5 = Increasing Cost or Difficulty

Other non-rotating parts being considered for Ti MMCs are low risk exhaust components including links, actuators, sidewalls and structural members for advanced military and commercial applications like the High Speed Civil Transport (HSCT) engine. Ti MMC links and actuators offer only a small overall engine weight savings but sidewalls and structural components in exhausts can represent a major portion of engine weight. Studies at GE and P&W have shown that Ti MMCs could offer 20-40% weight savings for F120 exhaust nozzle structures compared to nickel components.

Fabrication Demonstrations

In the early 1970s, fabrication development of Ti MMCs for compressor fan blade applications was started.[21] Since that time, numerous Department of Defense (DOD) and engine company funded programs have been conducted to determine and demonstrate the feasibility of fabricating the wide array of Ti MMC components cited above. The following is a sampling of the results of those efforts.

Disks, Blisks, Impellers

Ti MMC reinforced disks offer lower density components and an increased hoop radius which can result in up to 50% weight reductions in compressors. This high payoff potential has resulted in many DOD sponsored programs aimed at demonstrating fabrication feasibility and performance capability.[4,21,22] Those programs relied primarily on foil/fabric approaches to produce Ti MMC rings and encountered significant manufacturing difficulties but were ultimately successful at producing rings which achieved predicted burst and other performance capabilities.[23]

The potentially lower cost and simpler Ti MMC ring making processes based on powder and fiber/wire co-winding have been successfully demonstrated by engine and composite makers. Using a powder process, P&W fabricated the 40.6 cm (16") diameter Ti MMC reinforced IBR shown in Figure 7. This rotor was proof spin tested and then tested in P&W's XTC-65 IHPTET demonstrator engine where it met all performance requirements.[24]

Figure 7: Ti MMC ring insert and corresponding integrally bladed rotor fabricated using powder process techniques and successfully tested in P&W's XTC-65 IHPTET demonstrator engine.

The simulated 17.8 cm (7") diameter IBR shown in Figure 8 was fabricated by Atlantic Research Corporation (ARC), Wilmington MA, using their fiber/wire co-winding process.[17] This component was subsequently spin tested to failure at 98% of its predicted burst capability. A third new approach to Ti MMC ring making has recently been demonstrated by 3M, St. Paul MN, which utilizes their EBPVD coated fiber technology.[16] Using Ti6Al4V coated SiC fiber consolidated into fully dense thin strips, a 16 ply, 10.2 cm (4") diameter Ti MMC ring was produced by 3M as shown in Figure 9. Future rings of larger diameter will be made by this approach to determine its capabilities. The minimization of debulking required for the 3M and ARC ring making approaches offers the distinct advantage of precise fiber location control, which can be critical to the effective use of Ti MMCs in rotors.

Figure 8: A simulated Ti MMC rotor fabricated by ARC using co-wound fiber/wire techniques and spin tested to failure at 98% of predicted capability.

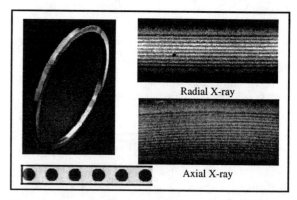

Figure 9: Demonstration Ti MMC ring fabricated by 3M using monotapes produced from coated SiC fibers. *Note the precise positioning of fibers in both the radial and axial directions which is critical to design.*

Shafts

Development of titanium composite shafts started in the early 1980s[25-27] and has progressed to the fabrication and testing of Ti MMC power turbine shafts for small engines[28] and low pressure turbine (LPT) shafts for advanced IHPTET engines.[29] Figures 10 and 11 show a GE27 power turbine shaft and an XTE-45 LPT fan shaft, respectively, fabricated by Textron Specialty Materials, Lowell MA, for GE Aircraft Engine. Ti MMC shafts of these types are typically fabricated with cross ply layups oriented from ±15° to ±45° to the shaft axis. These layups enhance shaft stiffness and torque capability while reducing weight compared to nickel or steel shafts.

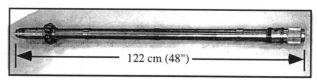

Figure 10: Early Ti MMC reinforced GE27 power turbine shaft fabricated by Textron using foil/fabric methods. *This shaft was out of balance due to difficulty controlling Ti MMC ply locations and wall thicknesses.*

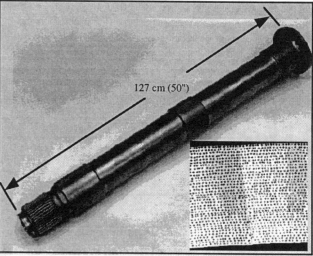

Figure 11: Low pressure turbine fan shaft for GE's XTE-45 demonstrator engine fabricated by Textron with IPD processed Ti MMC monotapes. *This shaft exhibited no balance problems and exceeded all predicted strength and fatigue capabilities in component tests.*

The 122 cm (48") long, 5.07 cm (2") diameter power turbine shaft shown in Figure 10 was fabricated with 13 plies of ±25° Ti MMC using foil/fabric methods. This shaft exhibited predicted bending stiffness and natural frequencies but was out of balance due to wall thickness variations which resulted from ply wrapping difficulties. The 127 cm (50") long, 12.1 cm (4.75") diameter LPT fan shaft shown in Figure 11 was fabricated more recently using 36 plies of ±15° oriented plasma sprayed Ti MMC monotapes. This shaft was subsequently LCF tested to a runout of 100,000 cycles which exceeded predicted capability. No significant balance problems were encountered with this shaft and natural frequencies were as predicted. A similar shaft is currently being fabricated for testing in the joint GE/Allison XTE-76 IHPTET demonstrator engine.

Blades

The ability to enhance airfoil stiffness and lower fan and compressor blade weights by elimination of shrouds required for vibration mode control has made Ti MMC reinforcement of these parts very attractive to engine designers. However, until recently, materials costs and fabrication issues (also cost drivers) have limited Ti MMC blade development. One program conducted by GE under Air Force sponsorship in the early 1980s demonstrated that large Ti MMC reinforced hollow fan blades like that shown in Figure 12 could be successfully fabricated using foil/fabric methods.[30,31] These F110 configured blades had preconsolidated 18 ply unidirectional (0°) oriented Ti MMC reinforced skins on the concave and convex airfoil surfaces and when component tested, met or exceeded all predicted strength, stiffness, LCF and HCF capabilities.

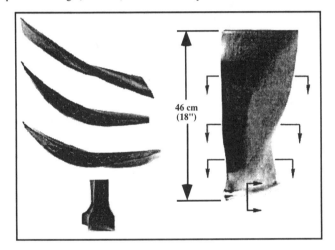

Figure 12: Hollow Ti MMC reinforced F110 fan blade fabricated by GE using preconsolidated foil/fabric skins produced by Textron. *This blade demonstrated manufacturing feasibility and exhibited predicted performance benefits in component tests.*

The encouraging results on the F110 fan blade and the potential availability of lower cost Ti MMCs has led to further development of hollow Ti MMC fan blades[32,33] for P&W military engines (similar in size to the GE blades) and for the high bypass PW4000 series commercial engine applications under the TMCTECC Program.[8] The PW4000 fan blade application uses 8 pounds of unidirectional Ti MMC tape to stiffen the airfoil wall. Four to eight plies of 50.8 cm by 101.6 cm (20" by 40") Ti MMC tape are placed on either side of a hollow core region and subsequently HIP consolidated. Figure 13 shows a PW4084 fan blade fabricated on the TMCTECC Program.

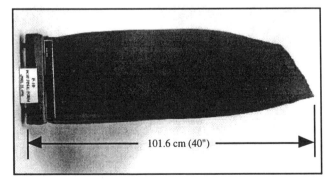

Figure 13: Hollow Ti MMC reinforced PW4084 fan blade fabricated on the TMCTECC Program.

Ducts/Cases, Spacers

Engine ducts or cases with Ti MMC reinforcements have been designed and fabricated by both GE and P&W for their respective IPHTET engines. GE design studies have shown that weight savings from 20-30% can be achieved with Ti MMC ducts where ducted gas temperatures are in the 427-538°C (800-1000°F) range and normally steel or nickel based ducts would be used. A prototype Ti MMC XTE-45 bypass duct which was fabricated for GE by Textron using IPD processed monotapes is shown in Figure 14. This duct utilized an 8 ply combination of unidirectional and cross-ply fiber layup to achieve design strength and stiffness requirements. A slightly different design case/duct which consisted of solid Ti MMC rods and struts was fabricated by Textron for P&W using foil/fabric and wire winding methods. This case was successfully proof tested at room temperature and exceeded all design predictions. In addition, a 16 ply conical shaped Ti MMC HPC spacer was fabricated by Textron.[34] This thin shell ring consisted of a ±15° tape cast plies and was successfully spin tested by P&W.

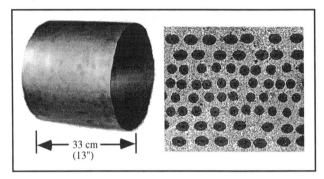

Figure 14: Prototype Ti MMC reinforced bypass duct fabricated for GE by Textron using plasma sprayed monotapes. *Note the combined off-axis and unidirectional plies incorporated to meet design requirements.*

Vanes, Frames, Struts

Ti MMCs for non-load bearing vanes in engine structures offer little or no payoff compared to monolithic or polymeric composite parts. However, where vanes help carry structural loads, the stiffness of Ti MMC can be of benefit. This is particularly true where a combined vane/frame structure is used as in the GE90 engine. Under the GE portion to the TMCTECC Program several fan frame outlet guide vane (OGV) designs with Ti MMC reinforcements are being fabricated for component testing and to demonstrate manufacturing feasibility and costs. One OGV design of interest which consists of airfoil skins selectively reinforced with preconsolidated tape cast 0° Ti MMC has been successfully fabricated as shown in Figure 15.

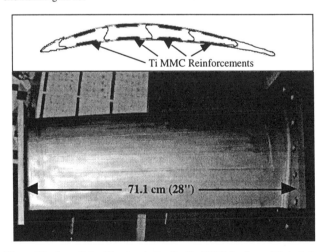

Figure 15: An outlet guide vane design for the GE90 fan frame fabricated on the TMCTECC program using preconsolidated tape cast Ti MMC inserts for the selectively reinforced airfoil skins. *This is one of several candidate designs being evaluated for potential weight and cost reductions.*

Another OGV design being evaluated for the GE90 is a king strut which utilizes a bicasting approach developed by Howmet, Whitehall MI. This process was successfully demonstrated by Howmet in fabricating the Ti MMC reinforced prototype CF6 fan frame strut shown in Figure 16. A third OGV design being evaluated consists of highly reinforced leading and trailing edge Ti MMC elements which can be readily fabricated to near net shape with required bow and curvature using 3M's coated fibers. A 66 cm (26") long leading edge element fabricated by 3M on the TMCTECC Program is shown in Figure 17. These various OGV designs along with others are being evaluated to identify the most cost effective and highest payoff design for potential production introduction into GE90 growth engine designs.

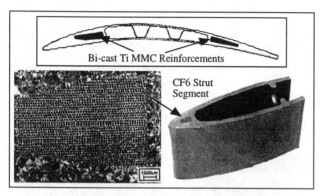

Figure 16: An outlet guide vane design for the GE90 fan frame fabricated on the TMCTECC program using preconsolidated tape cast Ti MMC inserts in a king strut design based on Howmet's bicasting process previously demonstrated in GE's CF6 fan frame struts. *This is one of several candidate designs being evaluated for potential weight and cost reductions.*

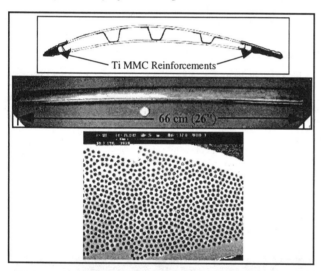

Figure 17: An outlet guide vane for the GE90 fan frame fabricated on the TMCTECC program using the 3M coated fiber process to produce near net shape leading and trailing edge inserts. *This is one of several candidate designs being evaluated for potential weight and cost reductions.*

Solid high pressure compressor (HPC) blades with an 8 ply, ±20° Ti MMC reinforcement have recently been produced by P&W using tape casting methods and successfully proof tested.

Links, Actuators, Nozzle Structures

Relatively low risk parts such as links and actuators for moving exhaust flaps have been used as the first flight demonstration applications for Ti MMCs. In 1992, three Ti MMC compression links as shown in Figure 18, were installed in a GE F110-100 engine exhaust and flight tested for 31 hours in an Air Force F16 aircraft with no visible distress. This flight testing was preceded by over 700 hours of factory engine tests which included over 3700 after burner lights. These Ti MMC links were fabricated by Textron using IPD processed monotapes and replaced IN718 links providing a 43% direct weight savings.[35] However, the high Ti MMC component cost prevented production implementation.

Over the past two years, P&W, ARC and Parker-Bertea have worked together to design and fabricate a 35.6 cm (14") long actuator piston rod for the F119 engine exhaust nozzle as shown in Figure 19.[36] Production quantities of Ti MMC reinforced piston cylinders with precisely located fibers were produced using ARC's fiber/wire co-winding process. These piston actuators, which offer a greater than 30% weight savings, exceeded all mechanical design requirements and have been qualified for use in production F119 engines. Ti MMC reinforced actuators are now also being considered for airframe applications.

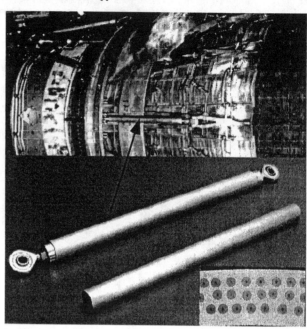

Figure 18: Ti MMC reinforced compression links for GE's F110-100 exhaust flaps fabricated by Textron using plasma sprayed monotapes. *These links were successfully fight tested for over 30 hours in an Air Force F16 with little evidence of distress.*

Figure 19: Ti MMC reinforced actuator piston rod fabricated by ARC/Parker-Bertea for P&W's F119 engine for the F-22 fighter aircraft.

Fabrication approaches for large flat structures, I-beam sections, box sections and other structural members reinforced with Ti MMCs were extensively evaluated on an Air Force sponsored program by GE for potential F120 engine exhaust structures.[37] While substantial potential weight savings were identified, the high cost of Ti MMCs curtailed fabrication efforts and led to the TMCTECC initiative now in progress.

Manufacturing Technology Status

Many fabrications processes exist that can meet the component fabrication needs for advanced aerospace applications. In order to focus the manufacturing infrastructure on a common approach for near term fan applications, the TMCTECC team has worked with the ARPA-sponsored High Performance Composites (HPC) Program[34] to develop baseline specifications for SiC/Ti6Al4V (TMC 2000 and 2001). These specifications are for green (unconsolidated) monotape and consolidated mill product. TMCTECC believes the key to establishing a high volume Ti MMC market is to agree to common material forms and common specifications.

During TMCTECC's Phase I activity, 3M, Textron, and ARC all produced material that met TMC 2000/2001 requirements. Textron and ARC are using a powder tape casting approach and 3M is EBPVD coating fibers. The current processes that produce the SiC/Ti6Al4V tape are generating very uniform microstructures and mechanical properties. In addition, the HPC Program is sponsoring the development and implementation of in-process monitoring sensors in the tape lines at ARC and Textron. Current capacity at the three suppliers totals more than 4535 kilograms (10,000 pounds) of tape per year. While the suppliers have not been able to achieve the TMCTECC Ti MMC material cost goal of less than $1100 per kilogram ($500 per pound) at these volumes, they have validated their cost models to show that the goal can be achieved at production volumes greater than 6803 kilograms (15,000 pounds) per year.

One way TMCTECC will be able to implement Ti MMC into its fan components is by using the strengths of the integrated product team philosophy. The designers, Ti MMC material suppliers, and component fabricators are working together to develop the optimum component based on performance, fabricability and cost. For GE and P&W to replace the current bill of material Ti products, the cost of the Ti MMC containing component must be less than the production model. To achieve this while using $1100 per kilogram ($500 per pound) Ti MMC material, the designer must understand how to maximize the composite benefits while minimizing its volume in the engine component. This leads the team to selecting simple shapes with little associated scrap during the fabrication process. With this approach, both the GE and P&W TMCTECC applications are projecting a 30% cost savings compared with the components being replaced.

Unless a company identifies an enabling use for Ti MMCs, they must buy their way into a production application. The current Ti MMC material cost of greater than $11000 per kilogram ($5,000 per pound) is not competitive with any anticipated production opportunities. TMCTECC believes that achieving the cost goal of $1100 per kilogram ($500 per pound) will lead to widespread Ti MMC use. The curve in Figure 20 shows the relationship between cost and volume projected by Ti MMC manufacturers. The projected DOD propulsion applications amount to less than 2268 kilograms (5,000 pounds) per year, so additional commercial applications are needed to achieve the volume that will lead to supplier capitalization and the resulting economies of scale.

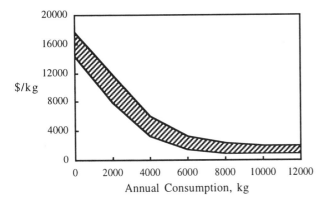

Figure 20: Projected cost of Ti MMCs as a function of market volume.

Summary

Over the past 20 years, the Ti MMC community has been able to demonstrate the benefits and fabrication feasibility of Ti MMC reinforced propulsion components. GE has gained flight experience with nozzle links and P&W is inserting actuator piston rods in the F119 engine for the F-22 fighter aircraft. The material has been able to deliver the projected benefits in the applications that GE and P&W have pursued. The current Ti MMC material fabrication processes are ready for production implementation and can routinely meet the TMC 2000/2001 specification requirements. However, the suppliers and end users have been unable to generate sufficient demand to yield an affordable Ti MMC material. TMCTECC was formed to take IHPTET developed Ti MMC material into production. The cost models and component demonstration articles are meeting the program milestones, but more work is required to achieve widespread implementation.

References

1. D.J. Dorr, "Two Sheet Available Fiber/Matrix Composite Design Development for Airframes," (Report AFWAL-TR-86-3065, Wright-Patterson AFB OH, 1986).

2. M. Shea, "Low Life Cycle Cost Landing Gear," (Report WL-TR-91-3044, Wright-Patterson AFB OH, 1991).

3. J.T. Niemann, et al, "Titanium Matrix Composites, Volume IV-Fabrication Development," (NASP Contractor Report 1145, NASP Materials and Structures Augmentation Program, NASP JPO, Wright-Patterson AFB OH, 1993).

4. R. Ravenhall, et al, "Composite Disk Validation Program," (Report WRDC-TR-90-2022, Wright-Patterson AFB OH, 1990).

5. T.E. O'Connell, "Production of Titanium Aluminide Products," (Report AFWAL-TR-83-4050, Wright-Patterson AFB OH, 1983).

6. J.J. Jackson, et al, "Titanium Aluminide Composites," (NASP Contractor Report 1112, NASP JPO, Wright-Patterson AFB OH, 1991).

7. J.A. Graves, A.H. Muir, Jr. and C.G. Rhodes, "Manufacturing Science for Titanium Aluminide Composite Engine Structures," (Report WL-TR-92-4100, Wright-Patterson AFB OH, 1992).

8. R. Anderson, et al, "Titanium Matrix Composite Turbine Engine Component Consortium," (Quarterly Status Reports for Contract F33615-94-2-4439, WL/MTPM, Wright-Patterson AFB OH, November 1994 through April 1996).

9. A. Kumnick, et al, "Filament Modification to Provide High Temperature Consolidation and Fabrication Capability and to Explore Alternative Consolidation Techniques," (Final Report for Contract N00019-82-C-0282, Naval Air Systems Command, Washington DC, 1983).

10. J.A. Cornie, R.J. Suplinskas and H. DeBolt, "Surface Enhancement of Silicon Carbide Filament for Metal Matrix Composites," (Summary Report for ONR Contract N00014-79-C-0691, Dept. of Navy, Arlington VA, 1981).

11. W.J. Grant and R. Lewis, "Continuous Powder/Fiber Tape Optimization Plan and Rationale for Powder/Fiber Tape Selection," (Report for Contract F33615-91-C-5728, Titanium Matrix Composite (TMC) Engine Components, Wright-Patterson AFB OH, 1993).

12. J.T. Niemann and J.F. Edd, "Fabrication of Titanium Aluminide Composites by Tape Casting," Titanium Aluminide Composites, P.R. Smith et al, eds., (Report WL-TR-91-4020, Wright-Patterson AFB OH, February 1991), 300.

13. P.A. Siemers, J.J. Jackson, "Ti$_3$Al/SCS-6 MMC Fabrication by Induction Plasma Deposition," Titanium Aluminide Composites, P.R. Smith et al, eds., (Report WL-TR-91-4020, Wright-Patterson AFB OH, February 1991), 233.

14. H. Gigerenzer, "Low Cost Metal Matrix Composites Preform Process," (Report WL-MT-63, Wright-Patterson AFB OH, 1992).

15. D.R. Pank and J.J. Jackson, "Metal Matrix Composite Processing Technologies for Aircraft Engine Applications", J. of Matls Engrg & Perf., 2(3)(1993), 341.

16. J. Sorenson, et al, "Continuous Fiber Metal Matrix Composites Model Factory," (Quarterly Status Report for ARPA Contract MDA972-90-C-0018, ARPA/CMO, Arlington VA, December 1995).

17. C.R. Rowe and S. Spear, private communications with authors, Atlantic Research Corporation, Wilmington MA, 9 March 1995.

18. G. DeBoer, et al, "Titanium Aluminide Composites," (Final Report for Contract F33657-86-C-2136, WL/POTO, Wright-Patterson AFB OH, 1993).

19. R. Holmes and D. Clemens, "High Temperature Metal Matrix Composite Compressor Rotor," (Final Report for Contract F33615-90-C-2003, Wright-Patterson AFB OH, 1995).

20. J. Madge, "TMCTECC Phase 1 Reinforced Fan Frame Trade Studies," (Report for Contract F33615-94-2-4439, WL/MTPM, Wright-Patterson AFB OH, 31 May 1995).

21. E.C. Stevens and D.K. Hanink, "Titanium Composite Fan Blades," (Report AFML-TR-70-180, Wright-Patterson AFB OH, 1970).

22. G. Richardson, "Life Prediction of Composite Disks," (Report TR-92-2107, WL/POTC, Wright-Patterson AFB OH, 1992).

23. A.M. Zoss, et al, "Advanced Turbine Rotor Design," (Reports TR-86-2102 and TR-92-2020, WL/POTC, Wright-Patterson AFB OH, 1986, 1992).

24. W. Doehnert "Manufacturing Technology for TMC Ring Inserts," (Reports for Contract F33615-91-C-5723, Wright-Patterson AFB OH, September 1991 to present).

25. S.R. Johnson and R. Ravenhall, "Metal Matrix Composite Shaft Research Program," (Report AFWAL-TR-87-2007, Wright-Patterson AFB OH, 1987).

26. D. Profant and G. Burt, "Manufacturing Methods for Metal Matrix Composite (MMC) Shafts," (Interim Reports for Contract F33615-80-C-5176, Wright-Patterson AFB OH, 1981).

27. D. Gray, et al, "Titanium Metal Matrix Composite Shafts," (Report AFWAL-TR-84-4124, Wright-Patterson AFB OH, 1984).

28. E.M. Sterling and J.E. Bell, "Advanced High Stiffness Power Turbine Shaft," (Report USA AVSCOM TR 90-D-5, U.S. Army, AATD, Fort Eustis VA, 1990).

29. R. Barnes, et al, "Exoskeletal Structures Program," (Report WL-TR-93-2034, Wright-Patterson AFB OH, 1993)

30. R. Ravenhall, "Advanced Reinforced Titanium Blade Development," (Report AFWAL-TR-82-4010, Phase I Final Report, Contract F33615-80-C-3236, Wright-Patterson AFB OH, 1982).

31. R. Ravenhall, "Advanced Reinforced Titanium Blade Development," (GE Report TM89-357, Phase II and III Final Report, Contract F33615-80-3236, Wright-Patterson AFB OH, August 1989).

32. D. Kasperski "Reinforced Hollow Fan Blade (RHFB) PRDA II," (Reports for Contract F33615-90-C-2040, Wright-Patterson AFB OH, August 1990 to present).

33. D. Romlein, "XTC-65/3 ATEGG Core Fabrication and Test," (Reports for Contract F33657-86-C-2013, Wright-Patterson AFB OH, 1986 to 1993).

34. S. Skemp, "ATEGG MMC Rotor Spacers," (Reports for Contract F33657-90-C-2212, Wright-Patterson AFB OH, 1992).

35. J.J. Jackson, et al, "Titanium Matrix Composite Engine Components," (Reports for Contract F33615-91-C-5728, WL/MLLM, Wright-Patterson AFB OH, October 1991 to September 1993).

36. R.D. Tucker, "Lightweight Engine Structures and Drum Rotor," (Report WL-TR-95-2070, Wright-Patterson AFB OH, 1995).

37. R.R. Oliver, et al, "Manufacturing Technology for Metal Matrix Composite Exhaust Nozzle Components," (Final Report for Contract F33615-91-C-5730, WL/MTPM, Wright-Patterson AFB OH, 30 September 1994).

38. P. Parish, "High Performance Composites," (Reports for ARPA Contract MDA972-93-2-0008, ARPA/CMO, Arlington VA, May 1994 to present).

POWDER METALLURGY
& WROUGHT MATERIALS

PHOSPHORUS - BORON INTERACTION IN
NICKEL-BASE SUPERALLOYS

W. D. Cao and R. L. Kennedy
Teledyne Allvac
2020 Ashcraft Avenue
Monroe, NC 28110
USA

Abstract

The effect of B on the properties of superalloys has been known for many years and is well documented in the literature. There is, however, very little published information on the effect of P content in Ni-base superalloys. This paper describes the results of an experimental program to study the effects of P content in Allvac 718 and Waspaloy, particularly as it interacts with B content.

Based on a series of sub-scale heats, a strong interaction between P & B on the rupture life of both alloys was discovered. This interaction has not been previously reported. Results showed that B alone had only a very small positive effect on stress rupture life of 718. However, P, which is generally considered to be a harmful impurity element, had a strong positive effect. Most interesting was the observation of a strong interaction between P & B. This effect was synergistic in that the combined effect was much greater than the sum of the individual elements acting independently. More than one order of magnitude increase in stress rupture life was observed over the entire range of B & P contents investigated. A 300-400% improvement in rupture life over typical commercial compositions was obtained with optimum P & B levels.

Dramatically different effects of P & B were observed in Waspaloy. When acting alone, P had only a small positive effect while B strongly improved rupture life. However, when both elements were present, increased P levels drastically deteriorated the beneficial effect of B. Conversely, when P was controlled to ultra-low levels, the beneficial effects of B extend to much higher levels than normal, and improvements in stress rupture life of 300-400% were possible.

The mechanisms of the observed P-B interaction in superalloys are discussed on the basis of the known behavior of P & B, alloy chemistry, and the observed fracture behavior.

Introduction

Minor elements have a profound effect on the performance of nickel-base superalloys and have been extensively studied in the last few decades. Several review papers [1-3] have summarized the major results up to 1984. Minor elements are generally divided into three categories: detrimental, beneficial and neutral elements. The detrimental elements such as S, Se, Te, As, Sb, Bi, Pb and Ag have been thoroughly studied and are tightly controlled in Ni-base superalloys. A large amount of research has also been conducted on the beneficial elements such as B, Mg, Hf, Zr and rare earth metals. Boron, in particular, has been regarded as a universal strengthening element and is added to almost all Ni-base superalloys to improve creep or stress rupture properties. Most investigators have attributed the beneficial effect of B to increased grain boundary cohesion due to grain boundary segregation of B [4-6], depletion or reduced agglomeration of precipitates in grain boundaries [7], reduced grain boundary concentration of detrimental elements, such as S, through site competition [6,8], and improved slip and dislocation transfer at grain boundaries to prevent the initiation of cracks [9,10].

One element, P, has not been studied in detail. There is also some controversy in the literature concerning the role of P in Ni-base superalloys. Most investigators regard P as a detrimental element which causes grain boundary embrittlement at high temperatures [11], reduces hot workability [12], and damages the resistance to high temperature corrosion [13]. However, early work by Bieber and Decker [14] suggested P, in small amounts, to be beneficial to malleability. Was and his co-workers [15,16] revealed that the addition of P in Ni-Cr-Fe alloy 600 improved the intergranular cracking resistance and creep resistance in Ar or high temperature water. Recent work [17] has demonstrated that increasing the P level of commercial 718 up to 0.022%, which is higher than the maximum level allowable in most specifications, significantly improved stress rupture properties. The increase in stress rupture life or reduction in average creep rate was more than one order of magnitude when P content was increased from a few parts per million to 0.022%. Previous investigators suggest the mechanism of the beneficial effect of P is increased grain boundary cohesion and reduced grain boundary concentration of detrimental elements by site competition [16,18].

It is well known that interaction may exist between minor elements, and this interaction may exert a tremendous effect on the properties of Ni-base superalloys. A typical example is the interaction between S and other beneficial elements such as Mg and rare earth metals. However, very little work has been performed on the possible interaction between the important minor element B and other elements. This work has shown that there is a strong interaction between P and B which can significantly influence the properties of Ni-base superalloys.

Experimental Procedure

Test Materials

A large number of experimental heats were produced covering a wide range of P and B levels. Phosphorus in 718 ranged from < 0.001 % to 0.035% (levels of all elements are in weight percentage) and B from < 0.001 % to 0.020%. Two C levels, < 0.01% and 0.03%, were used. The nominal aim composition of all other elements remained fixed at 18% Cr - 19.2% Fe - 2.9% Mo - 5.25% Nb - 0.95% Ti - 0.6% Al - Bal. Ni. Waspaloy P and B levels ranged from < 0.001 % to 0.022% and < 0.001% to 0.025%, respectively. The test alloys also had two C levels, < 0.01% and 0.035%, and a fixed nominal composition of 19.7% Cr - 13.4% Co - 4.25% Mo - 3.0% Ti - 1.3% Al - Bal. Ni. The 718 alloy with 0.006% P, 0.004% B, 0.03% C and Waspaloy with 0.005% P, 0.005% B and 0.035% C are considered to be typical "commercial" compositions and are the baseline for many of the comparisons made in this paper. Two to three heats were made for some critical chemistries to check the repeatability of test results, but no tabulation of all of the chemistries produced in this study is presented due to the large amount of data.

All test alloys were vacuum induction melted (VIM) with a weight of 23 Kg per heat and cast as 70 mm diameter electrodes. VIM electrodes were further refined by vacuum arc remelting (VAR) to 100 mm diameter ingots. VIM/VAR ingots were homogenized for 16 hours at 1190°C and rolled to 15 mm diameter bars within the temperature range of 1040°C to 920°C for 718 and 1060°C to 960°C for Waspaloy.

Heat Treatment

Specimen blanks for tensile and stress rupture tests were cut from 15 mm rolled bars and subjected to standard heat treatment as follows:

718 was solutioned for 1 hr at 954°C, air cooled and then aged for 8 hrs at 718°C, furnace cooled to 621°C, held at 621°C for 8 hrs, air cooled.

Waspaloy was solutioned for 4 hrs at 1010°C, water quenched and then aged for 4 hrs at 843°C, air cooled, aged for 16 hrs at 760°C, air cooled.

Mechanical Tests

Stress rupture tests of 718 were performed in air at 649°C at applied stress levels of either 669 MPa or 773 MPa. Waspaloy was tested at four different test temperatures / stress combinations: 649°C / 669 MPa, 732°C / 516 MPa, 760°C / 440 MPa and 816°C / 329 MPa. All tests were conducted according to ASTM E292. Two to three specimens were tested at each test condition.

Microstructure and Fractography

Microstructural observation. Microstructures were examined by optical and Scanning Electron Microscope (SEM) before and after stress rupture tests. Special attention was paid to grain structure and phase morphology. A meaningful comparison of microstructural changes between different alloys could not be obtained from broken stress rupture test specimens due to different exposure times and total deformations. A special experiment was therefore performed to generate specimens with equal temperature / stress / time history. Two compositions of both 718 and Waspaloy were selected; one was the nominal commercial composition, and the other alloy had the optimum P and B levels. Stress rupture test specimens from these two alloy pairs were tested at the same conditions: 718 at 649°C / 773 MPa for 36 hrs, and Waspaloy at 732°C / 516 MPa for 13 hrs. The specimens were unloaded after reaching specified test times, and micros were made from the gauge sections.

Fractography. The fracture surface of representative stress rupture test specimens was examined by SEM, and the fracture mode was identified to provide information on the effect of P and B on the fracture process of both alloys.

Experimental Results

Stress Rupture Properties

P and B Effects in Commercial Alloys. Figures 1 and 2 show the effect of changing P on stress rupture life of 718 and Waspaloy with nominally "commercial" levels of B (0.004% and 0.005%, respectively). It can be seen that increasing P had different effects in these two alloys. Rupture life of 718 significantly increased with increasing P level. An order of magnitude increase in life was achieved when P was raised from a few parts per million to 0.022%. However, in Waspaloy rupture life was unchanged or decreased with increasing P, depending upon test temperature. Figures 3 and 4 show the effects on stress rupture life of changing B at a constant, "commercial" level of P (0.006% and 0.005%, respectively). Different behavior was again revealed in these two alloys. Boron showed a significant, beneficial effect on stress rupture life of 718 up to the highest level tested (0.011%), but the beneficial effect of B saturated at about 0.005% in Waspaloy and further additions resulted in only moderate improvements

P and B Effects in High Purity Alloys. To prevent any possible interaction and to isolate the independent effects of P & B, a series of "high purity" heats were produced which varied P at the lowest possible levels of B, and C and B at the lowest levels of P and C. Results are summarized in Figures 5 through 8. Compared to Figures 1 through 4, there are differences in terms of P or B effect:

1. In high purity 718 (< 0.001% B), P also dramatically improved stress rupture life, but there appeared to be a critical P level (0.0075%) below which no significant benefit was observed (Fig. 5, 6). P produced an

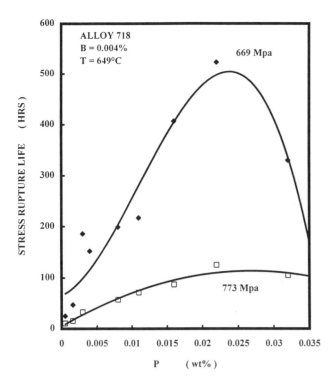

Figure 1. Stress rupture life of 718 as a function of P content.

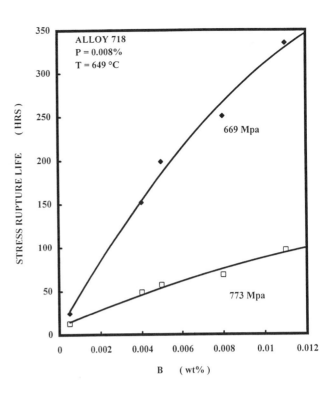

Figure 3. Stress rupture life of 718 as a function of B content.

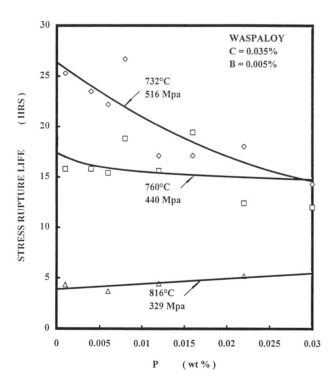

Figure 2. Stress rupture life of Waspaloy as a function of P content.

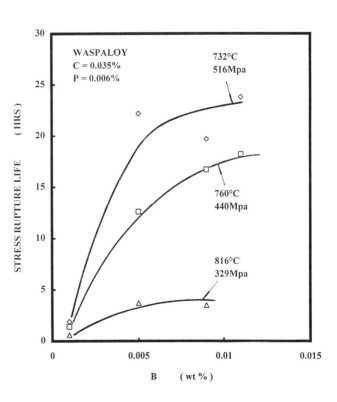

Figure 4. Stress rupture life of Waspaloy as a function of B content.

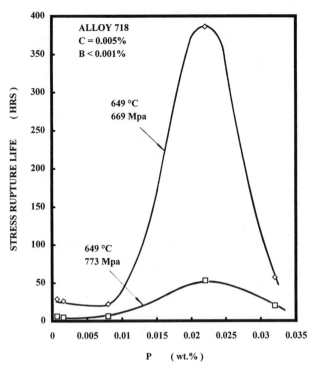

Figure 5. Effect of P level on stress rupture of Alloy 718 with very low levels of C and B.

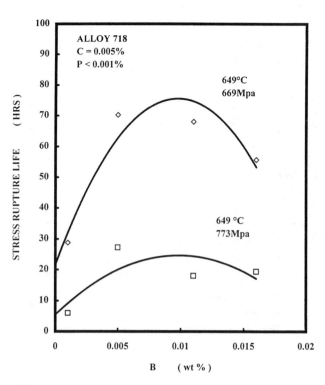

Figure 7. Effect of B level on stress rupture life of 718 with very low C and P levels.

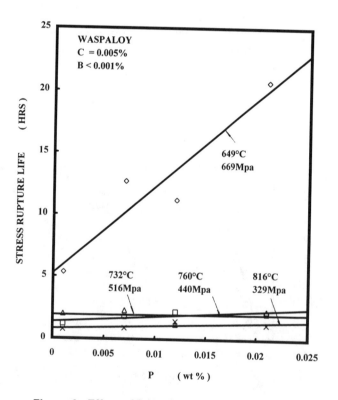

Figure 6. Effect of P level on stress rupture life of Waspaloy with very low C and B levels.

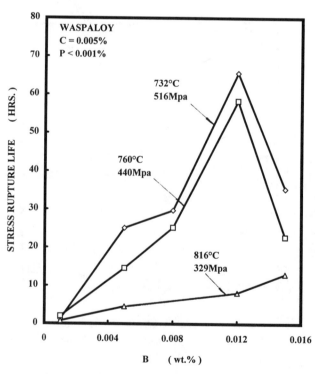

Figure 8. Effect of B level on stress rupture life of Waspaloy with very low P and C levels.

insignificant to weakly positive effect in high purity Waspaloy (< 0.001% B), depending upon temperature (Fig. 6). This contrasts to an insignificant to slightly detrimental effect in commercial Waspaloy (0.005% B, Fig. 2).

2. In the absence of P, B produced a relatively weak improvement in rupture life of 718 (Fig. 7). In Waspaloy, however, the absence of P dramatically accentuates the strengthening effect of B (Fig. 8).

These results show that P as well as B can be beneficial for stress rupture properties of high purity 718 and Waspaloy, but their effects in these two alloys are different. P is a strongly beneficial element in 718 and a weakly beneficial element in Waspaloy, while the reverse is true for B. Further, these results clearly suggest that there is an interaction between P and B and their effects on rupture life of both alloys.

Interaction. The combined or interactive effects of P and B on the rupture life of 718 and Waspaloy are shown in Figures 9 - 12. Figures 9 and 10 illustrate the strong synergistic effect of P and B in 718. If the P content was extremely low, rupture properties were rather poor, and B was almost ineffective as a strengthener. As the P content increased, B became a very potent strengthener. Optimum rupture life was achieved at P and B levels considerably above levels typical of today's commercial practice (0.022% P, 0.011% B vs 0.006% P, 0.004% B).

Strong synergistic effects of P and B were also displayed in Waspaloy (Fig. 11 and 12) although the effects were opposite with regard to P. At very low levels of P (< 0.001% P), significantly below levels of current commercial practice (nominally 0.005% P), B was an extremely effective strengthening element. With relatively small increases in P content, the strengthening effect of B rapidly diminished, and at typical commercial P levels, boron's effect appeared to saturate at about 0.005% B, i.e. typical commercial B levels. Optimum rupture properties for Waspaloy were achieved with < 0.001% P, 0.015% B vs typical commercial levels of 0.005% P, 0.005% B.

The variation in rupture properties for both 718 and Waspaloy over the entire P and B composition ranges investigated was greater than an order of magnitude while improvements over typical commercial compositions was more than 300%. Other properties were either unchanged (e.g. room and elevated temperature tensile) or slightly improved (e.g. stress rupture ductility) but are omitted here due to space limitations.

Fractography and Microstructure. For the most part, the stress rupture failure mode for 718 was transgranular, ductile dimple type. Waspaloy consistently failed in an intergranular mode over the entire range of P and B studied (Fig. 13). No significant differences in structure were apparent by either optical or SEM microscopy of alloys with different P and B levels. The size and quantity of delta phase in 718 and grain boundary carbides in Waspaloy did not vary with composition. Grain size of all samples was consistently fine: ASTM 12 for 718 and 10 for Waspaloy.

Interrupted Stress Rupture Tests. The significant difference in stress rupture properties for both 718 and Waspaloy with different P and B levels was also evident in the interrupted stress rupture tests in Table 1. Alloys with optimum levels of P and B were unbroken and had barely elongated under the same conditions which caused failure and high elongations in alloys of nominal commercial compositions. Metallography of these specimens showed extensive intergranular cracking over the entire gauge length in commercial Waspaloy (G947-1) as opposed to none at all in the optimum alloy (Fig. 14). No intergranular cracks were noted in either 718 sample.

Microhardness tests taken before and after the interrupted stress rupture tests are also shown in Table 1. Results from the undeformed shoulder regions show significant overaging at the test temperature for the normal commercial 718 (G457-1) but very little for the optimum composition (G727-2). Very little change and no difference between the different heats of Waspaloy was apparent.

Discussion

In superalloys, P has generally been treated as a harmful or, at best, a somewhat innocuous trace element, and most specifications limit it to relatively low maximum values (e.g. 0.015% max). This work has shown that this long standing belief is not necessarily true and that the effect of P is, in fact, strongly alloy dependent. It has further been discovered that there is a strong interaction between P and B and that the long recognized stress rupture strengthening effect of B can be very substantially enhanced by controlling P content.

In the absence of B, P has been shown to promote a moderate or weak rupture life improvement in 718 and Waspaloy. When combined with B, however, the results in these two alloys are distinctly different. Maximum rupture life in 718 was obtained at P levels well above those of typical commercial practice while in Waspaloy, optimum results were obtained with P content significantly lower than normal. Optimizing the P content of superalloys also extends the stress rupture strengthening effect of B. At typical commercial P levels, the effect of B appears to saturate at relatively low levels, and adding additional B produces little or no further gain in rupture life. At optimum P contents, however, peak rupture lives were obtained with B approximately three times normal commercial levels and more than double typical specification maximums.

The exact reasons for the above phenomena are uncertain at this time, but it is possible to postulate certain mechanisms which are consistent with what is already known about P and B and the data obtained in this work. Numerous studies have shown that both P (19-22) and B (17,19,22,23) strongly segregate to grain boundaries and compete for

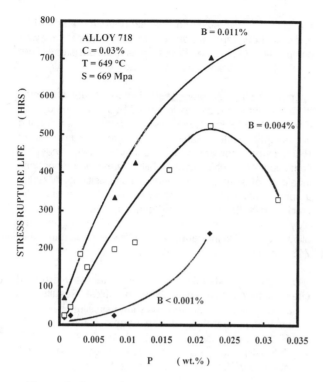

Figure 9. Interaction of P and B with stress rupture life of 718.

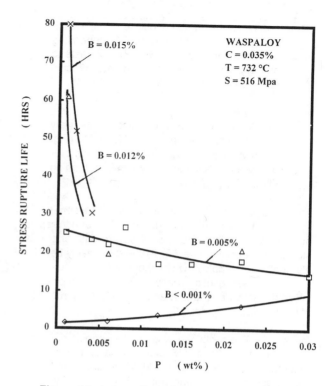

Figure 11. Interaction of P and B content with stress rupture life of Waspaloy.

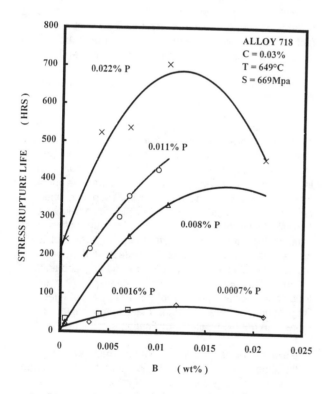

Figure 10. Interaction of P and B with stress rupture life of 718.

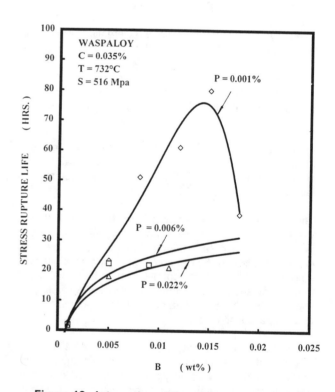

Figure 12. Interaction of P and B content with stress rupture life of Waspaloy.

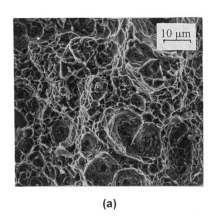

Fig. 13. Typical stress rupture failure mode in (a) Alloy 718, (b) Waspaloy

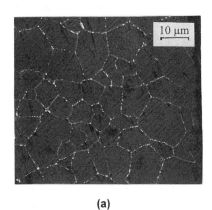

Figure 14. SEM photographs of gauge section of Waspaloy interrupted stress rupture samples: (a) Heat G947-1, < 0.001% P, 0.012% B, no intergranular fracture and (b) Heat G753-1, 0.006% P, 0.005% B, extensive intergranular fracture.

Table 1. Results of Interrupted Stress Rupture Tests

Alloy	Heat No.	Chemistry (wt%)			Stress Rupture Test				Shoulder Hardness (HV)	
		C	P	B	T °C	S MPa	t Hr	EL %	Before	After
718	G457-1	0.031	0.006	0.004	649	773	31.9*	24.0	478	455
	G727-2	0.033	0.022	0.011	649	773	36.0	1.2	476	472
WASPALOY	G753-1	0.035	0.006	0.005	732	516	13.3*	40.1	425	418
	G947-1	0.035	<0.001	0.012	732	516	13.3	2.2	422	419

* Specimens broke at indicated time.

grain boundary sites with P having the stronger tendency to occupy these sites (24,25). Both elements are believed to increase grain boundary cohesion and improve the resistance to high temperature intergranular fracture, but B appears to be much more effective in this regard. Both elements could also segregate to precipitate/matrix interfaces or vacancies to increase the stability of precipitate particles by reducing diffusion rates or interact with dislocations to retard dislocation movement.

In 718, the transgranular failure mechanism and the reduced rate of overaging, shown in Table 1, suggest transgranular dislocation creep as the controlling mechanism and would suggest increased stability of precipitate particles (probably γ") as the probable cause in the absence of further information as to the location of P and B in the alloy. For Waspaloy, very little change in hardness after interrupted stress rupture testing was shown (although the test time was very short), but a very obvious reduction in the degree of intergranular fracture was observed. This, coupled with the completely intergranular stress rupture fracture mode for Waspaloy, would suggest grain boundary strengthening as the controlling mechanism. It could be postulated that, due to site competition, when P is low, more B can segregate to the grain boundaries, thus improving rupture life. A CRADA program is currently underway with Oak Ridge National Laboratories to better define the mechanisms of stress rupture life improvements which have been observed in this study. With such an understanding, it should be possible to improve the rupture life of a number of existing superalloys and to develop entirely new alloys with superior properties.

Conclusions

The following conclusions have been drawn from this study:

1. This work has demonstrated that P content can play a key role in the stress rupture life of superalloys. Contrary to generally held beliefs, the effect may be positive in some alloys, suggesting a purposeful addition rather than the low maximum limit normally imposed.

2. A strong synergistic interaction between P and B was found. This interaction was alloy specific, but for both 718 and Waspaloy, rupture properties at optimum P and B contents exceeded the results of each element acting independently.

3. At optimum P contents, the usual effect of B on improving rupture life can be extended to B concentrations nearly three times normal.

4. In both alloys, more than an order of magnitude improvement in rupture life was observed over the entire compositional range investigated. Greater than 300% life improvements were measured for optimum P and B compositions compared to nominal commercial compositions.

5. The mechanisms for property improvements noted are still being studied, but data suggest they are alloy specific and may involve competitive grain boundary segregation, interaction of P and B atoms with precipitates, vacancies, or dislocations.

Acknowledgements

The authors would like to thank Teledyne Allvac for permission to publish this paper and also greatly appreciate its financial support to this project.

References

1. R.T. Holt and W. Walace, International Metals Reviews, 21 (1976), p. 1.

2. M. McLean and A. Strang, Metal Technology, 11 (1984), p. 454.

3. G. W. Mectham, ibid, 11 (1984), p. 414.

4. M. J. Mills, Scripta Met., 23 (1989), p. 2061.

5. J. Takasugi and O. Izumi, Acta Met., 33 (1985), p. 1247.

6. B. Landa and H. K. Birnbaum, ibid, 36 (1988), p. 745.

7. R. F. Decker and J. W. Freeman, Trans AIME, 218 (1960), p. 277.

8. J. Kameda and J. Bevolo, Acta Met., 41 (1993), p. 527.

9. S. S. Brenner and Hua Min-Jian, Scripta Met., 24 (1990), p. 671.

10. E. M. Schulson, T. P. Weihs, I. Baker, H. J. Frost and J. A. Horton, Acta Met., 34 (1986), p. 1395.

11. W. Yeniscavich and C. W. Fox, in "Effect of Minor Elements on the Weldability of High Nickel Alloys", Welding Research Council, (1969), p. 24.

12. M. Tamura, in "Superalloys, Supercomposites and Superceramics", Akademic Press, Inc., 1989, p. 215.

13. D.A. Vermilyea, C. S. Tedmon, Jr., and D. E. Broecker, Corrosion, 31 (1975), p. 222.

14. C. G. Bieber and R. F. Decker, Trans AIME, 221 (1961), p. 629.

15. J. K. Sung and G. S. Was, Corrosion, 47 (1991), p. 824.

16. G. S. Was, J. K. Sung and T. M. Angeliu, Met. Trans., 23A (1992), p. 3343.

17. W. D. Cao and R. L. Kennedy, in "Superalloys 718, 625, 706 and Various Derivatives," Ed. By E. A. Loria, TMS, (1994), p. 463.

18. A. W. Funkenbusch, L. A. Heldt and D. F. Stein, Met. Trans., 13A (1982), p. 611.

19. J. M. Walsh and B. H. Kear, ibid, 6A (1975), p. 226.

20. R. G. Thompson, M. C. Koopman and B. H. King, in "Superalloys 718, 625, 706 and Various Derivatives," Ed. By E. A. Loria, TMS, (1991), p. 53.

21. L. Letellier, A. Bostel and D. Blavette, Scripta Met. Et Mat., 30 (1994), p. 1503.

22. P. Caceras, B. Ralph, G. C. Allen and R. K. Wild, Surface and Interface Analysis, 12 (1988), p. 191.

23. D. J. Nettleship and R. K. Wild, ibid, 16 (1990), p. 552.

24. E. Hall and C. L. Briant, Met. Trans., 16A (1985), p. 1225.

25. R. M. Kruger and G. S. Was, Met. Trans., 19A (1988), p. 2555.

THE ROLE OF PHOSPHORUS AND SULFUR IN INCONEL 718 *

Xishan Xie, Xingbo Liu, Yaohe Hu, Bin Tang, Zhichao Xu, Jianxin Dong and Kequan Ni
University of Science & Technology Beijing,, Beijing 100083, China

Yaoxiao Zhu
Institute of Metal Research, Academy of Sciences, Shenyang 110015, China

Shusen Tien
Fushun Steel Plant, Fushun 113001, China

Laiping Zhang
No.3 Steel Works of Changcheng Special Steel Co., Sichuan 621704, China

Wei Xie
Shanghai No.5 Steel Works, Shanghai 200940, China

Abstract

Phosphorus and sulfur are generally regarded as the most common impurities and detrimental elements in nickel-base superalloys. For further understanding the role of P and S in Inconel 718 nine experimental heats were melted on the base of conventional Inconel 718 chemical composition with variation of P (from 10 to 130 ppm) and S (from 15 to 175 ppm) respectively. Phosphorus and sulfur both have almost no influence on strengths and ductilities at room temperature tensile test. Phosphorus also has no effect on the yield and ultimate strengths and elongation at 650°C tensile test. However, sulfur has an obviously decreasing effect on 650°C tensile elongation but no effect on yield and ultimate strengths.

Sulfur has a remarkable detrimental effect on stress rupture life and especially on ductility loss at 650°C. However, phosphorus is in total difference to sulfur effect, generally can increase stress rupture life and ductility both.

Microstructure observation on grain structure, δ-phase and strengthening phase γ" and γ' can not reveal the effect of P and S on the morphology and amount of precipitates. Fractography analyses show different patterns of stress rupture specimens with the variation of P and S contents. Experimental results lead us to consider the segregation behaviors of P and S at grain boundaries, however the effect of P and S should be different.

---

* This project is supported by the Chinese National Natural Science Foundation and the Ministry of Metallurgical Industry.

Background and Industrial Tests

The clean superalloy production is regarded to meet the strict demand of aero-engine and gas-turbine industries and cleanliness has been basically considered in two categories, i.e. inclusions and detrimental elements control[1]. Phosphorus and sulfur are generally regarded as the most common inpurities and detrimental elements in Ni-base superalloys. From the points of view on melting process improvement at industrial alloy production background and on basic understanding the role of P and S in Ni-base superalloys the present investigation was conducted in industrial and laboratory scales.

Niobium segregation and Laves phase formation during solidification from molten metal has been regarded as the most serious problem in Inconel 718 production. Results of segregation study during solidification by means of metalography and electron-probe analyses show that P and S most seriously aggravate dentritic segregation by decreasing solidus temperature and lowering the maxium content of Nb in γ solid solution and also moving the eutectic point (γ + Laves)to the higher content of Nb in the pseudo-binary phase diagram[2]. In result of these effects P and S (especially P) will favour the formation of isolated large-size blocky Laves phase instead of eutectic Laves[2,3]. Based on above mentioned experimental results the low segregation Inconel 718 with very low content of P (P<10ppm) VIM+VAR industrial ingots (with diameters 406 or 423mm) were melted in Chinese steel works[2,4].

The comparison of solidified structure sliced directly from the industrial production ingots (low segregation and conventional) shows the advantage of lower segegation and less amount of blocky Laves phase especially at the mid-radius and the center of low segregation Inconel 718

Table I. Example of the Mechanical Properties Comparison between Low Segregation and Conventional Inconel 718 Die-Forging Disks.

Alloy	20°C				650°C				650°C, 686MPa	
	YS (MPa)	UTS (MPa)	EL (%)	RA (%)	YS (MPa)	UTS (MPa)	EL (%)	RA (%)	T (hrs)	EL (%)
L.S.	1068	1268	25.0	37.5	1060	1197	14.8	27.0	55	14.4
Con.	1038	1289	20.8	34.5	1005	1172	21.2	33.3	47	22.8

* L.S.-- low segregation In 718; Con.-- conventional In 718.

ingots with very low contents of phosphrus (<10ppm). High temperature homogenization heat treatment tests show the easier tendency to reach homogenized structure, i.e. to eliminate Nb segregation to the lowest degree. However, Nb segregation of both ingots (low segregation and conventional) can be eliminated to meet industrial demand after 2 step high temperature long time homogenization treatment (first step for Laves phase solution and the second step for homogenization of Nb).

Mechanical property test results including room temperature tensile, 650°C tensile and stress rupture, stress controlled LCF, stress controlled fatigue and creep interaction properties determination at 650°C for die-forging disks(~500mm diameter) have not revealed advantages of low segregation Inconel 718 in comparison with conventional Inconel 718 as shown in Table I and Figure. 1.

For further understanding the role of P in comparison with the role of S in Inconel 718, nine experimental heats were melted on the base of conventional Inconel 718 composition with the variation of P (from 10 to 130ppm)and S (from 15 to 175ppm) respectively.

Meterials and Experimental Procedure

Nine experimental heats of Inconel 718 with variation of P and S were melted in 25kg VIM furnace and poured in 15kg ingots. Chemical composition and alloy designation are shown in Table II.

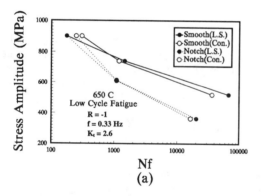

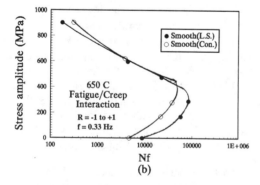

Figure 1 Stress controlled LCF(a) and fatigue/creep interaction(b) properties of low segregation and conventional IN718 disks.

Nine ingots were conducted with 2 step homogenization treatment i.e., 1160°C/24hrs →1180°C/24hrs/A.C. All the ingots were forged down to 40mm square bars and finally hot rolled to 18mm round bars for tensile tests at room temperature and 650°C, stress rupture and creep tests were conducted at 650°C, 686MPa and 725MPa respectively. Structural characterization and fractography analyses were carried out by means of optical, SEM and TEM microscopy.

Results and Discussion

The content of P and S in the range less than 130 ppm and 175 ppm respectively, both elements have almost no effect on tensile strengths (UTS and YS) and ductility (EL) at room temperature (see Fig.2 and 3). Phosphorus also has no effect on the strengths and ductilily at 650°C tensile test (see Fig.2b). However, sulfur has an obviously decreasing effect on 650°C tensile elongation as indicated in Fig.3b.

Table II. Chemical Composition of Test Alloys (wt%)

Alloy	C	Mn	Si	S	P	Ni	Cr	Mo	Al	Ti	Nb	B	Fe
P1	0.02	0.02	0.05	.001	.0010	52.52	18.69	3.01	0.52	1.01	5.20	.005	bal.
P2	0.03	0.02	0.05	.002	.0025	52.38	18.62	2.98	0.47	1.01	5.15	.005	bal.
P3	0.03	0.02	0.06	.003	.0033	52.79	18.56	3.01	0.52	1.00	5.24	.006	bal.
P4	0.02	0.02	0.05	.003	.0083	52.76	18.45	3.05	0.52	1.01	5.17	.006	bal.
P5	0.02	0.02	0.05	.003	.0130	52.90	18.76	3.01	0.50	1.01	5.17	.005	bal.
S1	0.04	0.02	0.05	.0015	.004	52.52	18.52	2.98	0.54	1.02	5.18	.005	bal.
S2	0.04	0.02	0.11	.0050	.004	52.70	18.56	2.95	0.49	1.01	5.22	.005	bal.
S3	0.02	0.02	0.06	.0145	.001	52.44	18.52	2.95	0.54	1.05	5.15	.005	bal.
S4	0.02	0.02	0.06	.0175	.001	52.64	18.45	2.97	0.55	1.01	5.11	.005	bal.

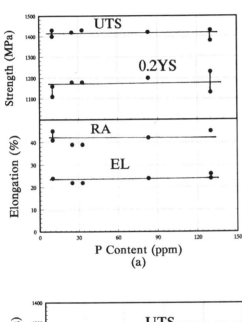

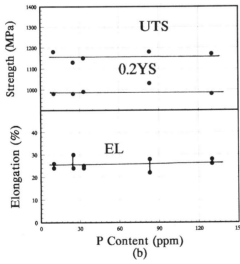

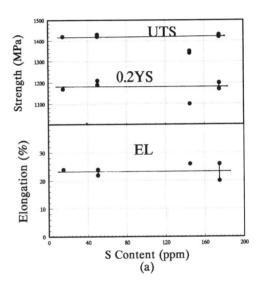

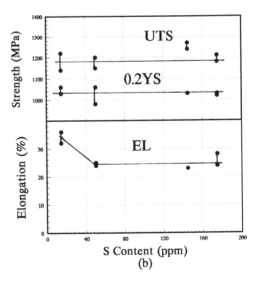

Figure 2 Effect of phosphorus on tensile properties at room temperature (a) and 650°C (b) of IN718.

Figure 3 Effect of sulfur on tensile properties at room temperature (a) and 650°C (b) of IN718.

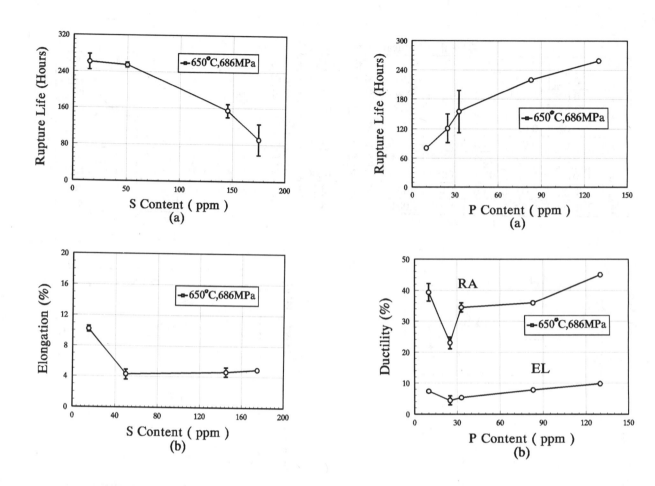

Figure 4 Effect of sulfur on 650°C, 686MPa stress rupture life (a) and ductility (b) of IN718.

Figure 5 Effect of phosphorus on 650°C, 686MPa stress rupture life (a) and ductility (b) of IN718.

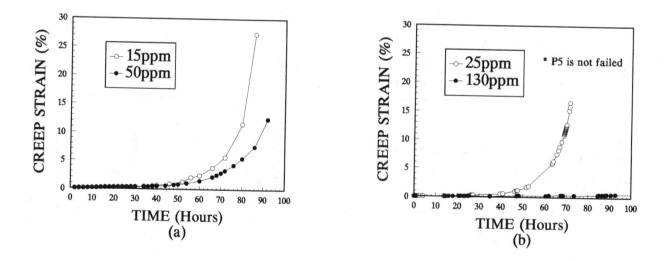

Figure 6 Effect of sulfur (a) and phosphorus (b) on 650°C, 725MPa creep curves of IN718.

Sulfur has a remarkable detrimental effect on stress rupture life and especially ductility loss even at a small amount of sulfur content in Inconel 718 as shown in Fig.4. However, phosphorus is in total difference to sulfur effect, P can increase stress rupture life and ductilities (see Fig.5). It implies that P has certain strengthening and ductility improvement effects. Special attention should be paid, that Alloy 718 with low content of P (~20ppm) is susceptible to the lowest ductilities at 650°C stress rupture test (Fig.5b shows the "lowest valley" of stress rupture ductility curve), which shows consistence with the experimental results of Cao and Kennedy[5].

Creep test results at 650°C show that S and P have no influence on secondary creep rates, However, phosphorus can prolong secondary creep stage tremendously (Fig.6b). In results of that Alloy 718 with higher content of phosphorus characterizes with longer creep failure life.

Microstructure observation on grain structure, δ-phase at grain boundaries, γ" and γ' precipitates in the grains can not reveal the effect of P and S on the grain size and morphology and amount of precipitates (see Fig.7 and 8).

Metalographic observation on the longitudinal sections of stress rupture failure specimens shows typical grain boundary cracks in all specimens with different contents of S and P respectively. These results show all the failures at 650°C stress rupture tests characterize with intergranular fracture. Alloy 718 with low content of S (Alloy S1 with15ppm sulfur) clearly shows deformed and prolonged grain structure (see Fig.9a) because of the longer stress rupture life and higher ductility of Alloy S1 in comparison with Alloy S4 (175ppm sulfur) as shown in Fig.9b. However, Alloy P5 with highest content of P(130ppm)

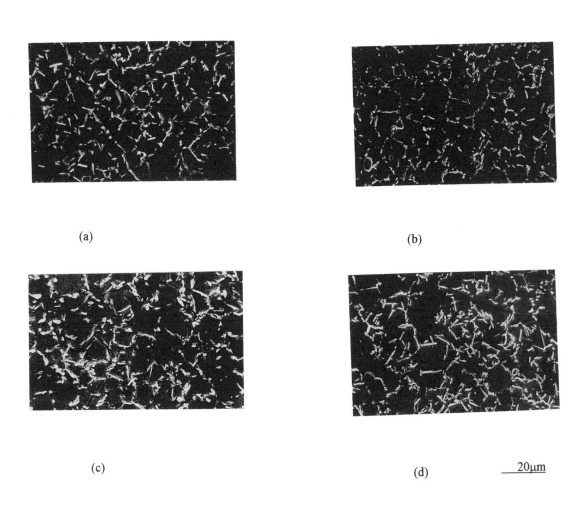

(a)　　　　　　　　　　　　　(b)

(c)　　　　　　　　　　　　　(d)　　20μm

Figure 7　Effect of sulfur and phosphorus on grain structure and δ-phase in IN718.
　　a - S1(15ppm S);　　b - S4(175ppm S);　　c - P2(25ppm P);　　d - P5(130ppm P)

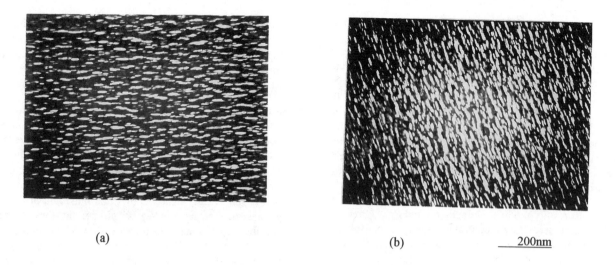

Figure 8 Effect of sulfur and phosphorus on γ" and γ' precipitates in IN718.
a -P5(130ppm P, 30ppm S); b - S4(10 ppm P, 175 ppm S).

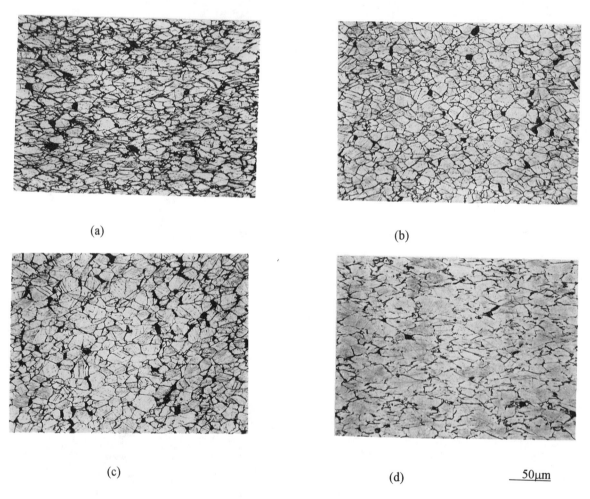

Figure 9 Effect of sulfur and phosphorus on fracture manner at longitudinal direction of 650°C stress rupture failed specimens.
a - S1(15ppm S);   b - S4(175ppm S);   c - P2(25ppm P);   d - P5(130ppm P)

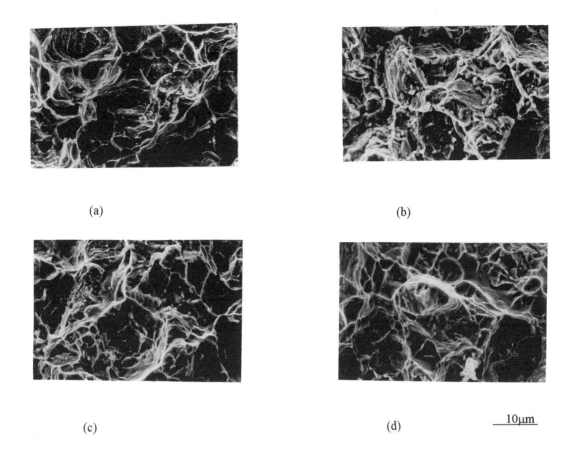

Figure 10 Fractography observation on the effect of sulfur and phosphorus in IN718.
a - S1(15ppm S); b - S4(175ppm S); c - P2(25ppm P); d - P5(130ppm P)

among our experimental heats characterizes typical deformed and prolonged grain structure, which is in total reverse in comparison with the S effect (see Fig.9d). SEM fractograpy observation direct on the fracture surfaces after 650°C stress rupture tests shows that Alloy S1 with the lowest content of S(15ppm) characterizes more ductile intergranular fracture mode (see Fig.10a and b). However, phosphorus promotes the fracture to be more ductile as shown in Fig.10c and 10d.

From above mentioned comprehensive experimental results it can be concluded that the effects of P and S on mechanical properties in IN 718 are as follows:

1. Phosphorus and sulfur both have almost no influence on strengths and ductilities at room temperature tensile test.

2. Phosphorus also has no effects on the yield and ultimate strengths and elongation at 650°C tensile test. However, sulfur has an obviously decreasing effect on 650°C tensile elongation but no effect on yield and ultimate strengths.

3. Sulfur has a remarkable detrimental effect on stress rupture (creep) life and especially on ductility loss (tertiary creep) at 650°C. However, phosphorus is in total difference to sulfur effect. Phosphorus can increase stress rupture life and ductility, as well as prolong secondary creep and develop tertiary creep both.

Microstructure and fractography observation reveals that phosphorus and sulfur have only the effect on these mechanical properties characterized with intergranular fracture, especially for high temperature stress rupture or creep failures, however the effects of P and S are in totally different manner.

Special attention should be paid, that P has certain high temperature strengthing and ductility improvement effects. Only a few experimental results in the recent years show that phosphorus can be benifical for high temperature stress rupture, creep and creep-fatigue properties in IN718[5], Ni-Cr-Fe[6] and 304L[7] austenitic alloys as we have in this investigation. To the best of today's knowledge, the reason of beneficial effect of P in these alloys is still not clear yet.

Phosphorus and sulfur seriously aggravate dendritic segregation during solidification and both are concentrated at grain boundaries after heat treatment of wrought IN718. These effects have been already experimentally determined[2-5,8]. Theoretical explanation is tried to make that based on the first principle interatomic potential study suggested by Wang et al.[9] to calculate the doping effects on the NiΣ11[110] tilt grain boundary by using Discrete Variational X_α Method (DVM). However, their results[10] indicate that both P and S elevate the binding energy of atom clusters at grain boundaries and the equilibrium lattice constant (ELC) at grain boundaries in nickel. The ELC of pure nickel grain boundary is 3.6000Å while the P-doping's and the S-doping's grain boundaries are 3.8555Å and 3.8823Å respectively. Therefore, the segregation of P and S at grain boundaries leads to the grain boundary embrittlement in nickel. However, Inconel 718 is a very complicated multi-component alloy system. An hypothesis is trying to take in consideration, that phosphorus may interact with the other elements and they may co-exist at the grain-boundaries to decrease grain boundary binding energy and to increase grain boundary cohesive force. In results of these, phosphorus may effectively strengthen grain boundary and to retard grain boundary crack til to failure at high temperature tests. A further research program is going to enlighten the P effect in IN718.

Conclusions

1. Phosphorus and sulfur both have almost no influence on strenghts and ductilities at room temperature tensile test. Phosphorus also has no effect on the yield and ultimate strengths and elongation at 650°C tensile test. However, sulfur has an obviously decreasing effect on 650°C tensile elongation but no effect on yield and ultimate strengths.

2. Sulfur has a remarkable detrimental effect on stress rupture life and especially on ductility loss at 650°C. However, phosphorus is in total difference to sulfur effect, generally can increase stress rupture (creep) life and ductility.

3. Microstructure observation on grain structure, δ-phase and strengthening phase γ'' and γ' can not reveal the effect of P and S on the morphology and amount of precipitates Fractography analyses show different patterns of failed stress rupture specimens with the variation of P and S contents. Experimental results lead us to consider the segregation behaviors of P and S at grain boundaries, however the effect of P and S should be different.

4. Sulfur is regarded as a detrimental impurity in Inconel 718 and it should be controlled at the lowest level in alloy production.

5. From the view point of serious segregation problem in Inconel 718 the production of low segregation Inconel 718, suggested to control P content at very low level (<10ppm), was successfully conducted. Industrial results have not shown any benefits for mechanical properties. However, laboratory experimental results show that very low level phosphorus leads stress rupture (creep) life loss.

6. Phosphorus seems to be benifical for stress rupture life and ductility improvement, however its mechanism is still not clear yet.

References

(1) A. Mitchell, "Progress in Understanding Clean Metal Production for IN 718", Superalloys 718, 625, 706 and Various Derivatives Ed. E.A. Loria, TMS(1994) P.109

(2) Y. Zhu et al., "Effect of P, S, B and Si on the Solidification on INCONEL 718 Alloy", Superalloys 718, 625, 706 and Various Derivatives Ed. E.A. Loria, TMS(1994) P.89

(3) C. Chen, R.G. Thompson and D.W. Davis, "A Study of Effects of Phosphorus, Sulfur, Boron and Carbon on Laves and Carbide Formation in Alloy 718", Superalloys 718, 625, 706 and Various Derivatives Ed. E.A. Loria, TMS(1991) P.81

(4) Y. Zhu et al., "A New Way to Improve the Superalloys", Superalloys 1992 Eds. S.D. Antolovich et al., TMS(1992) P.145

(5) W.D. Cao and R.L. Kennedy, "The effect of Phosphorus on Mechanical Properties of Alloy 718", Superalloys 718, 625, 706 and Various Derivatives Ed. E.A. Loria, TMS(1994)P.463

(6) G.S. Was, J.K. Sung and T.M. Angeliu, "Effects of Grain Boundary Chemistry on the Intergranular Cracking Behaviors of Ni-16Cr-9Fe in High-Temperature Water", Met. Trans. 23A(1992) P.3343

(7) Y.C. Yoon et al, "Effect of Phosphorus on the Creep-Fatigue Interaction in AISI 304L Stainless Steel, J. of the Korean Inst. of Met. & Mater. 30(1992) P. 1401

(8) R.G. Thompson, Personal Communication, Univ. of Alabama at Birmingham (1995)

(9) Wang Chongyu et al., "A First Principle Interatomic Potential and Application to Grain Boundary in Nickel", Phys. Letter A197 (1995) P.447

(10) Wang Chongyu, Wang Ligen and Wang Fuhe, "Binding Energy and Doping Effect of Grain Boundary on the First Principle Calculations for Transition Metal", Inter. Conf. of IUMRS in Asia (1993) P.109

THE EFFECTS OF INGOT COMPOSITION AND CONVERSION ON THE MECHANICAL PROPERTIES AND MICROSTRUCTURAL RESPONSE OF GTD-222

T. Banik, T. C. Deragon, and F. A. Schweizer

Special Metals Corporation
4317 Middle Settlement Road
New Hartford, NY 13413

Abstract

The current trend in the aerospace industry toward larger and fewer engines per aircraft has resulted in new demands in gas turbine materials. GTD-222, a new alloy selected for combustion casings due to its excellent elevated temperature properties, was evaluated to determine the potential effects of chemistry and conventional processing techniques (VIM+VAR) on the alloy's microstructural response and mechanical properties. It was concluded that homogenization processing was not a major contributor to improvements in elevated temperature properties. An increase in tensile properties was achieved with a solution heat treatment below the $M_{23}C_6$ solvus temperature. Stress rupture life and ductility were improved with increasing carbon content. In addition, increasing carbon content reduced the solidus and gamma-prime solvus temperatures and yielded a finer primary microstructure. It was concluded that reforging stock could be produced by VIM+VAR processing the alloy.

Introduction

The current trend in the aerospace industry toward larger and fewer engines per aircraft has resulted in new demands in gas turbine materials. The combustor case is one example of this trend. The new larger engines require materials which operate at higher temperatures and can be cost effectively produced into cases greater than one meter in diameter. Traditional alloys such as 718 and Waspaloy do not meet the temperature capability requirements. A cast alloy, GTD-222[1], was selected as a candidate for combustor case applications based on its elevated temperature properties. The challenge with GTD-222 focused on establishing cast/wrought production processes for the large ingot diameters necessary to produce billet sizes compatible with the requirements for large diameter seamless ring production.

GTD-222 was originally developed for cast gas turbine engine nozzle applications in 1989. The alloy composition is similar to Waspaloy with some of the titanium replaced with niobium and tantalum. Initial cast wrought evaluations of GTD-222 were conducted on material produced by vacuum induction melting (VIM) plus electroslag remelting (ESR) to 300 mm diameter ingots. The original properties generated from the VIM+ESR ingot demonstrated that GTD-222 alloy would meet the elevated properties required by GEAE design specifications. However, the ability to produce large diameter ingots of GTD-222, free of unacceptable macro and micro segregation, had to be demonstrated. Therefore, the next step was to determine the processing necessary to produce large diameter ingot products (≥ 500 mm) using a vacuum induction melt followed by vacuum arc remelting (VAR).

The purpose of the present investigation was to evaluate the potential effects of chemistry and conventional processing techniques in assessing the alloy's mechanical properties and microstructural response, specifically the effect of carbon, homogenization, and solution heat treatment temperatures. Modifications to the composition were focused on the carbon level. The GTD-222 alloy exhibits a primary carbide phase, MC, and a chromium rich, $M_{23}C_6$, grain boundary carbide. The MC carbides precipitate during and shortly after solidification. The $M_{23}C_6$ carbides precipitate during aging as a semi-continuous film and negatively impact the stress rupture strength. The baseline composition of GTD-222 and the three compositional modifications evaluated are presented in Table I. Two different homogenization cycles and a no homogenization cycle were evaluated. The two solution temperatures evaluated were 1093°C and 1149°C.

Table I   The compositional modifications (in weight percent) were focused on carbon content.

	Control Heat	Heat 8377	Heat 8376	Heat 8422
C	0.08	0.04	0.02	0.13
Cr	22.5	22.6	22.5	22.4
Co	19.1	19.0	19.1	19.1
W	2.0	2.0	2.0	2.0
Nb	0.8	0.8	0.8	0.8
Ti	2.3	2.3	2.3	2.3
Al	1.2	1.2	1.3	1.2
B	0.004	0.003	0.004	0.001
Zr	0.02	0.02	0.02	0.03
Ta	0.94	0.94	0.93	0.95
Ni	Bal	Bal	Bal	Bal

---

[1] GTD-222 is a patented nickel-based superalloy by General Electric Co.

Experimentation

Subscale heats were produced using a laboratory scale vacuum induction melting furnace. Four electrodes were produced and vacuum arc remelted to 152 mm diameter.

Each ingot product was sectioned to permit assessment of multiple homogenization procedures. Samples were exposed to homogenization cycles of 1178°C for 48 Hours, 1191°C for 27 Hours or no homogenization operation. A TA Instruments DSC 2910 Differential Thermal Analyzer was utilized to establish phase changes for each composition and to establish homogenization parameters. After homogenization, samples were rolled using a 75% reduction sequence. Solution anneal was performed at two temperatures, 1093°C or 1149°C. All samples were followed by an age cycle of 802°C for 8 Hours prior to testing. Mechanical property evaluations were performed at 760°C.

The laboratory sized ingots restricted mechanical property evaluations to one sample per condition. Metallographic evaluations were performed to assess carbide distribution and grain size.

Stress rupture testing was performed on SATEC M3 or SATEC C type stress rupture testing machines. Tensile testing was performed on an INSTRON 4208 tensile testing machine. Stress rupture and tensile testing furnaces incorporated standard laboratory cylindrical furnaces and proportional analog controllers for temperature control.

Results and Discussion

Differential Thermal Analyses

As indicated in Figure 1, the liquidus temperature remained approximately 1375°C for the different carbon compositions, whereas the solidus temperatures decreased with increasing carbon content. The stability of the liquidus temperature was expected due to the overall chemical composition similarity. At low carbon contents (0.02 wt %), the solidus temperature was determined to be 1318°C. At 0.13 wt % carbon, the solidus temperature decreased to 1300°C. The lower solidus temperature at increasing carbon levels is expected since carbon is a melting point depressant.

The GTD-222 alloy is a gamma-prime strengthened alloy. The gamma-prime solvus start and finish temperatures decreased with increasing carbon content, as indicated in Figure 2. The decrease in gamma-prime solvus temperature is due to the increase in primary carbides depleting the matrix of titanium, niobium and tantalum.

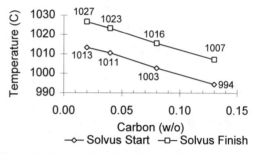

Figure 2: Increased carbon weight percent decreased gamma-prime start and finish solvus temperatures.

Mechanical Properties

Homogenization. The test data revealed no differences in mechanical properties among the various homogenization cycles. The yield and tensile strengths were generally 690 MPa and 900 MPa, respectively. Tensile ductility ranged from 8% to 20% for elongation in a 4D sample and 12% to 22% for reduction of area. A stress rupture life of approximately 45 hours and 2% to 20% elongation in a 4D testing section at 760°C and 462 MPa load were typical for all carbon compositions. The similarity of the mechanical properties for the homogenization trials may be the result of the laboratory scale ingot size. The faster cooling rate of the smaller, laboratory-size ingots (152 mm diameter) permit solidification with minimal segregation, whereas larger, production sized ingots (> 300 mm diameter) experience slower solidification of alloy lean areas resulting in increased segregation of alloy rich areas.

Solution Temperature. A graphical summary of the yield and tensile strengths for the 1093°C and 1149°C solution temperatures is presented in Figure 3a. Similarly, the tensile ductility results are presented in Figure 3b for the two solution temperatures. The anneal temperature (1093°C) generally exhibited slightly higher yield and tensile strengths for the 0.02, 0.04 and 0.08 wt % carbon composition samples when compared to the results for the 1149°C solution treated material. The tensile properties for the 0.13 wt % carbon composition material was unaffected by the solution temperature. Tensile ductility was generally lower for all compositions with the 1093°C solution temperature compared to the 1149°C solutioned samples.

The stress rupture life and elongation results for the 1093°C and 1149°C solution temperatures are presented in Figure 3c. The decrease in the stress rupture life between the 1093°C and

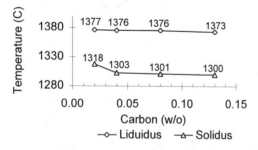

Figure 1: Increasing carbon weight percent did not affect the liquidus temperature but resulted in a decreasing solidus temperature.

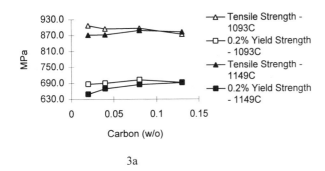

3a

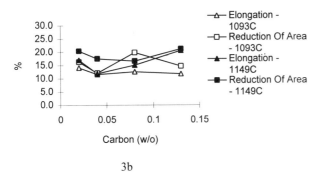

3b

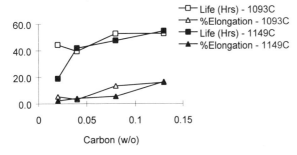

3c

Figure 3: a) The samples exposed to the 1093°C solution temperature generally exhibited increased yield and tensile strengths. Carbon content did not appear to significantly affect yield and tensile strengths.

b) The samples exposed to the 1093°C solution temperature generally exhibited lower tensile ductility. Carbon content did not appear to significantly affect tensile ductility.

c) The samples exposed to the 1093°C solution temperature did not appear to significantly affect stress rupture life and ductility. Carbon content appears to affect stress rupture life and ductility positively.

1149°C solution temperatures for the 0.02 wt % carbon composition may be due to an increase in $M_{23}C_6$ continuity with increasing grain size at the higher solution temperature. Stress rupture life and ductility of higher carbon levels (>0.02 wt %) appeared unaffected by solution temperature.

Carbon Level. When comparing mechanical property results from the varying carbon levels, the tensile properties appeared unaffected by carbon composition, as indicated in Figures 3a and 3b. An increase in rupture life and ductility appears evident with increasing carbon content (Figures 3c), probably due to the finer grain size created by the increased primary carbides. While the increased carbon content provides improved control of the grain size during heat treatment, with higher carbon levels, the propensity of carbides and the magnitude of stringer formation also increases. A trade-off between grain size control and stringer formation subsequently results.

Microstructure

The carbide distribution, as presented in Figure 4, significantly increased with increasing carbon content. The primary carbides consisted principally of niobium, tantalum and titanium. In production scale VIM+VAR processed material, increased carbon content has resulted in a coarse stringer-like carbide distribution.

The grain size of the samples correlated directly with the carbon content at both solution heat treatments. As indicated in Figure 5, an increase in carbon content resulted in a finer primary grain structure by inhibiting grain growth. The increase in grain size was most dramatic at the lowest carbon content of 0.02 wt %. Grain coarsening was the greatest for the 0.02 wt % carbon composition with increasing solution treatment temperatures from 1093°C to 1177°C. The 0.02 wt % sample increased in grain size from ASTM 4 to ASTM 00, while the grain size for the 0.13 wt % sample remained relatively constant at an ASTM 6.5 to 7.

Conclusion

The subscale investigations into GTD-222 processing revealed:

- The carbon content did not significantly affect the liquidus temperature, but gradually reduced the solidus and gamma-prime solvus temperatures with increasing carbon content.

- On subscale ingots, homogenization was not a major contributor to improvements in elevated temperature tensile and stress rupture strength.

- The lower solution temperature, 1093°C, exhibited slightly increased yield and tensile strengths and slightly decreased ductility.

- For the higher (>0.02 wt %) carbon compositions, solution temperature did not appear to have a significant effect on stress rupture properties.

- At 0.02 wt % carbon the 1093°C solution temperature produced significantly increased life compared to the 1149°C solution.

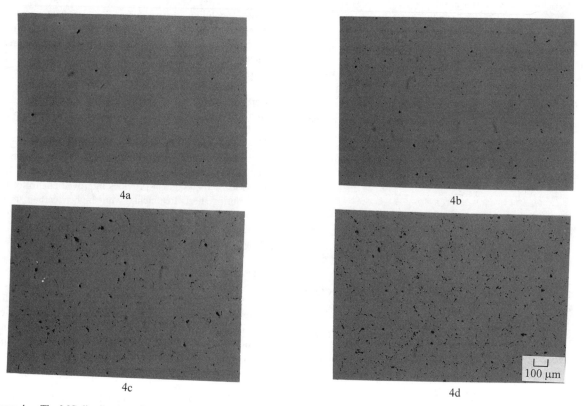

Figure 4: The MC distribution of VAR ingot material increased considerably with increasing carbon content. a) 0.02 wt % Carbon. b) 0.04 wt % Carbon. c) 0.08 wt % Carbon. d) 0.13 wt % Carbon.

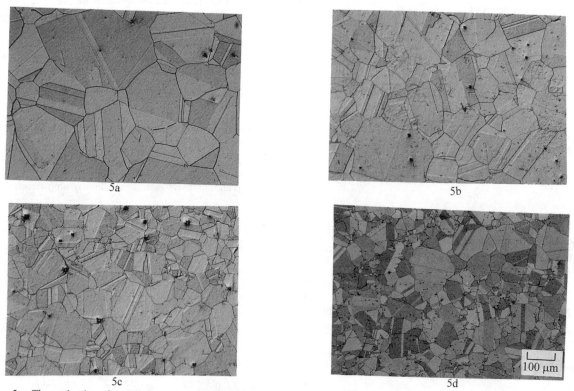

Figure 5: The grain size after roll-down processing and 1093°C anneal and age heat treatment was finer with increasing carbon content due to the increased MC frequency. a) 0.02 wt % Carbon. b) 0.04 wt % Carbon. c) 0.08 wt % Carbon. d) 0.13 wt % Carbon.

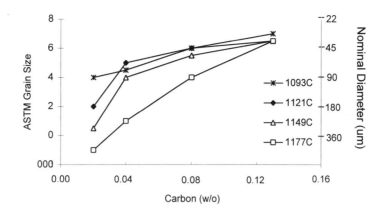

Figure 6: The grain size decreased with increasing carbon content and solution temperature.

- Increasing carbon levels did not have a significant influence on the tensile properties.

- Increasing the carbon content from 0.02 wt % to 0.13 wt % resulted in a finer primary microstructure. Heat treatments through the 1149°C solution temperature resulted in grain coarsening of 0 to 4 ASTM values. Grain coarsening occurred at a higher rate with lower carbon compositions.

Subscale investigations on GTD-222 indicated that VIM+VAR processing can be utilized to meet the property requirements necessary for the advanced engines. Subsequent full-scale material has been produced and confirmed the initial results.

Acknowledgments

The authors wish to express their thanks to Lance Myers for his metallographic support.

MICROSTRUCTURE MODELING OF FORGED WASPALOY DISCS

G. Shen, J. Rollins, and D. Furrer

Ladish Co., Inc., P.O. Box 8902, Cudahy, WI 53110-8902

Abstract

An effort has been undertaken to develop, validate and refine a modeling tool for the prediction of forged Waspaloy microstructures. Previous work on steel materials [1] has shown that microstructural modeling is possible and can be readily accomplished through use of metallurgically-based mathematical equations. Experimentation is required to develop and tailor the equations for the kinetics of the material of interest.

This paper outlines the steps required to determine the kinetics for microstructural evolution for Waspaloy, and their application to the metallurgically-based equations. Examples of the predictions and actual measured results show that this methodology is effective to develop practical tools to aid forging engineers develop and optimize processes for specific final component requirements.

Nomenclature

A	constant for grain growth in preheating
d	final grain size, μm
d_0	as-preheated grain size, μm
d_1	starting grain size for post-deformation grain growth, μm
d_i	initial grain size (before preheating), μm
d_{dyn}	dynamically recrystallized grain, μm
d_{m-dyn}	meta-dynamically recrystallized grain, μm
m	exponent for grain growth in preheating
n	exponent for Avrami equation
Q	activation energy, J/mol
Q_{gl}	activation energy for grain growth in preheating, J/mol
R	gas constant, 8.314J/(mol*K)
T	temperature, K
t	time, s
$t_{0.5}$	time for 50% meta-dynamic recrystallization
X_{dyn}	fraction of dynamic recrystallization
X_{m-dyn}	fraction of meta-dynamic recrystallization
Z	Zener-Hollomon parameter, 1/s
e	effective strain
$e_{0.5}$	effective strain for 50% dynamic recrystallization
e_c	critical strain for dynamic recrystallization
e_p	peak strain for dynamic recrystallization
$\dot{e}$	effective strain rate

Introduction

Forging methods and approaches for complex, high-technology components for aerospace and general industrial applications has evolved to greater levels of sophistication in recent years under economic and market pressures. Metal flow simulations which were once impossible, except for plastic and clay physical models, are now common-place using the latest, high-speed computers and latest generation of computer processing codes [2, 3]. The understanding and prediction of microstructures which develop during deformation processing has long been the "art" of forging metallurgists. Many tools, such as wedge tests were used in the past to understand the impact of temperature, strain and strain rate and to guide metallurgists in the processing direction of success [4]. These methods were good, but quantitative extraction of vital processing data and microstructural results, and the deconvolution of the separate effects of the various processing parameters on the measured results were not effectively accomplished.

Modern forging design and development relies heavily on computer process modeling. Process modeling provides a scientific tool for rapid evaluation of process changes for process optimization. Bulk metal flow modeling has been highly successful in elimination of shop tryouts for nearly all forging methods. Microstructure modeling is now the major emphasis in advanced forging design and development. The development and utilization of physical metallurgy based microstructural models has allowed for grain size and property prediction in steel [1], and in titanium [5].

An alloy of considerable importance to modern turbine engine applications is Waspaloy. It is used for numerous rotating and non-rotating turbine engine components, and is processed by numerous thermomechanical processing routes [6]. Many of the methods used for the processing of Waspaloy were developed by iterative shop trials. This method to define specific manufacturing processes for new components is not acceptable due to excessive development time and cost for trial component manufacture.

A model has been developed and implemented for the prediction of microstructures of forged Waspaloy discs [7, 8, 9]. This model considers grain growth in preheating prior to forging, dynamic recrystallization during forging, and meta-dynamic recrystallization and grain growth during post-forging cool down. The model has been integrated into finite element software to predict microstructures developed for different forging processes. The actual implementation of this model has been highly successful in the prediction of microstructures of numerous actual production components processed under widely varied processing conditions.

Experimental Procedure

Three sets of experiments were used in the model development.

(1) <u>Preheating Tests</u>: Heat treatment studies were conducted with different temperatures and hold times to model the as-preheated grain size for a forging operation. From these tests, the grain growth model for preheating of a particular billet pedigree was developed. This model was then used to establish the grain size just prior to subsequent forging operations.

(2) <u>Compression Tests</u>: Laboratory upset tests were conducted with different temperatures, strains, strain rates, starting grain sizes, and post-deformation hold times to characterize dynamic recrystallization during forging, and meta-dynamic recrystallization and static grain growth during post-forging cool down. An MTS compression stand was used for the simulation of isothermal forging conditions and a Gleeble test unit for non-isothermal forging conditions. From these compression tests with rapid post-deformation cooling, information related to dynamic recrystallization kinetics were obtained. The kinetics information included: peak strain for dynamic recrystallization, which is related to the critical strain for dynamic recrystallization; strain which corresponds to 50% dynamic recrystallization, fraction of dynamic recrystallization and the size of dynamically recrystallized grains. Information regarding meta-dynamic recrystallization and post forging grain growth were also obtained from compression tests with controlled post-forge hold times and cooling rates. This information included: time for 50% meta-dynamic recrystallization, fraction of meta-dynamic recrystallization, meta-dynamically recrystallized grain size, and grain growth at a given temperature and time after the completion of meta-dynamic recrystallization.

(3) <u>Pancake and Generic Forgings</u>: In addition to laboratory tests, large pancakes and generic component configurations were produced on production equipment under various forging conditions and methods to verify and refine the microstructure model.

Finite element analysis for each experiment provided detailed information for each test. This quantified specific data was used to develop the models for the microstructure evolution of Waspaloy during the forging process.

Results

PREHEATING MODEL

It is the as-preheated grains which go through deformation processes and have changes take place to their structure. Therefore, a model of the as-preheated grain size is the necessary first step in microstructure modeling. For Waspaloy, there is no grain size changes in sub-solvus preheating. However, grain growth occurs rapidly under super-solvus preheating conditions. An equation was developed for the prediction of the grain size in production preheating conditions of the following form:

$$d_0^m - d_i^m = A\, t\, \exp(-Q_{g1}/[RT]) \quad [1]$$

HOT WORKING MODEL

The processes that control grain structure evolution during hot working of Waspaloy were found to be dynamic recrystallization, meta-dynamic recrystallization, and static grain growth. The equations developed for the quantitative prediction of these phenomenon are summarized in Table 1.

TABLE 1. Mathematical Model for Microstructure Development in Waspaloy Forging.

Dynamic Recrystallization

$Z = \dot{e} * \exp(468000/RT)$

$e_c = 0.8 e_p$

<u>sub-solvus forging</u>

$e_p = 5.375 * 10^{-4}\, d_0^{0.54} Z^{0.106}$

$e_{0.5} = 0.1449 d_0^{0.32} Z^{0.03}$

$X_{dyn} = 1 - \exp\{-\ln 2 [e/e_{0.5}]^{3.0}\}$

$d_{dyn} = 8103 Z^{-0.16}$

<u>in-solvus forging</u>

$e_p = 5.375 * 10^{-4}\, d_0^{0.54} Z^{0.106}$

$e_{0.5} = 0.056 d_0^{0.32} Z^{0.03}$

$X_{dyn} = 1 - \exp\{-\ln 2 [e/e_{0.5}]^{2.0}\}$

$d_{dyn} = 8103 Z^{-0.16}$

<u>super-solvus forging</u>

$e_p = 1.685 * 10^{-4}\, d_0^{0.54} Z^{0.106}$

$e_{0.5} = 0.035 d_0^{0.29} Z^{0.04}$

$X_{dyn} = 1 - \exp\{-\ln 2 [e/e_{0.5}]^{1.8}\}$

$d_{dyn} = 108.85 Z^{-0.0456}$

Meta-dynamic Recrystallization

$t_{0.5} = 4.54 * 10^{-5} d_0^{0.51} e^{-1.28} \dot{e}^{-0.073} * \exp(9705/T)$

$X_{m-dyn} = 1 - \exp\{-\ln 2 [t/t_{0.5}]^{1.0}\}$

$d_{m-dyn} = 14.56 d_0^{0.33} e^{-0.44} Z^{-0.026}$

Grain Growth

$d^3 - d_1^3 = 2 * 10^{26}\, t\, \exp(-595000/[RT])$

Dynamic Recrystallization

Dynamic recrystallization happens instantaneously during high temperature deformation. The fraction of dynamic recrystallization can be obtained by examining micrographs obtained from samples quenched after their deformation. Under production conditions pure dynamic recrystallization is difficult to achieve. However, a study of dynamic recrystallization is helpful for understanding the microstructural development in the forging. The amount of dynamic recrystallization is related to the as-preheated grain size, strain, temperature and strain rate in a hot deformation process. There are four important issues related to dynamic recrystallization: the peak strain, the strain for 50% dynamic recrystallization, the fraction of dynamic recrystallization, and the size of dynamically recrystallized grains.

Peak Strain - The strain corresponding to the peak stress (e_p) in the flow stress curve is an important measure for the onset of dynamic recrystallization. The occurrence of dynamic recrystallization modifies the appearance of these curves. At the strain rates typical for forging of Waspaloy, single peak stress-strain curves are most common. As a result of dynamic recrystallization, the stress diminishes to a value intermediate between the yield stress and the peak stress once past the peak strain. The reason for this curve following a single peak is that under the condition of high Z, the dislocation density can be built up very fast. Before recrystallization is complete, the dislocation densities at the center of recrystallized grains have increased sufficiently that another cycle of nucleation occurs and new grain begin to grow again. Thus, average flow stress intermediate between the yield stress and the peak stress is maintained. The equations developed for the peak strain are:

$$e_p = 5.375 \times 10^{-4} d_0^{0.54} Z^{0.106} \quad \text{(sub- and in-solvus)} \quad [2]$$

$$e_p = 1.685 \times 10^{-4} d_0^{0.54} Z^{0.106} \quad \text{(super-solvus)} \quad [3]$$

Strain for 50% Dynamic Recrystallization - Micrographs taken from quenched compression samples show that dynamic recrystallization progresses in a sigmoidal manner with respect to strain. The Avrami equation can be used to describe a sigmoidal curve for the fraction of dynamic recrystallization versus strain:

$$X_{dyn} = 1 - \exp\{-\ln 2 [e/e_{0.5}]^n\} \quad [4]$$

When the constants $e_{0.5}$ and n are determined, the relation for the fraction of dynamic recrystallization is determined. The strain for 50% recrystallization, $e_{0.5}$, can be obtained from compression tests with different magnitudes of strain for a given condition of temperature, strain rate, and as-preheated grain size, as shown in Figure 1. The exponent can be obtained by taking the logarithm of Equation 4. $e_{0.5}$ is related to as-preheated grain size, d_0 and Z by:

$$e_{0.5} = 0.1449 d_0^{0.32} Z^{0.03} \quad \text{(sub-solvus)} \quad [5]$$

$$e_{0.5} = 0.056 d_0^{0.32} Z^{0.03} \quad \text{(in-solvus)} \quad [6]$$

$$e_{0.5} = 0.035 d_0^{0.29} Z^{0.04} \quad \text{(super-solvus)} \quad [7]$$

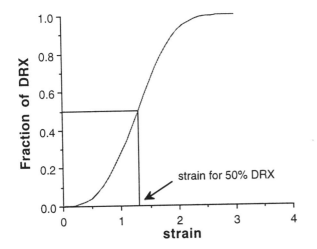

Figure 1. Schematic of strain corresponding to 50% dynamic recrystallization.

Fraction of Dynamic Recrystallization - After the strain for 50% dynamic recrystallization and the exponent for Equation 4 are determined, equations for the fraction of dynamically recrystallized grains can be formulated:

$$X_{dyn} = 1 - \exp\{-\ln 2 [e/e_{0.5}]^{3.0}\} \quad \text{(sub-solvus)} \quad [8]$$

$$X_{dyn} = 1 - \exp\{-\ln 2 [e/e_{0.5}]^{2.0}\} \quad \text{(in-solvus)} \quad [9]$$

$$X_{dyn} = 1 - \exp\{-\ln 2 [e/e_{0.5}]^{1.8}\} \quad \text{(super-solvus)} \quad [10]$$

Figures 2(a) and 2(b) summarize the experimental data and the fitted model for dynamic recrystallization at 1010°C and 1066°C respectively.

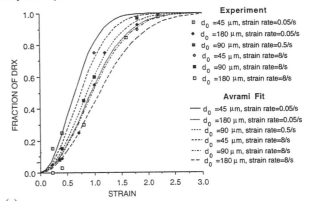

(a)

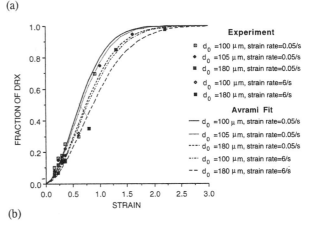

(b)

Figure 2. Measured (data points) dynamic recrystallization (DRX) kinetics for hot deformation of Waspaloy at (a) 1010°C and (b) 1066°C and fitted curves.

The critical strain for the start of dynamic recrystallization usually follows the relationship [1]:

$$e_c = 0.8 e_p \quad [11]$$

The Size of Dynamically Recrystallized Grain - The dynamically recrystallized grain size is the function of Zener-Hollomon parameter, Z, only. This is because Z defines the density of the subgrains and the nuclei. Though the dynamically recrystallized grain size is not related to strain, the strain has to reach the value of steady state strain to result in full dynamic recrystallization. The relationship between dynamically recrystallized grain, d_{dyn}, and Z, is shown in the following equations.

$$d_{dyn} = 8103 Z^{-0.16} \quad \text{(sub- \& in-solvus)} \quad [12]$$

$$d_{dyn} = 108.85 Z^{-0.0456} \quad \text{(super-solvus)} \quad [13]$$

Figure 3 shows how Equations 12 and 13 relate to experimental data. It is seen that there is a difference between subsolvus forging and supersolvus forging in terms of dynamically recrystallized grain size. The subsolvus forging results in finer grain sizes, while supersolvus forging results in coarse grain sizes. However, subsolvus forging needs large strains to finish dynamic recrystallization as shown in Equations 5 and 8.

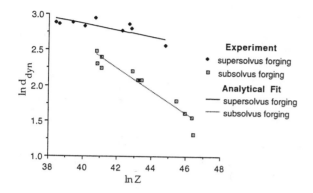

Figure 3. Logarithm of dynamically recrystallized grain size (in μm) versus ln Z obtained from compression tests.

Meta-Dynamic Recrystallization

Meta-dynamic recrystallization is important in the determination of the grain size obtained under practical forging conditions. Meta-dynamic recrystallization occurs when a deformation stops at a strain which passes the peak strain but does not reach the steady state strain for dynamic recrystallization [10], which is the case for most regions in a forged part. Under meta-dynamic recrystallization conditions, the partially recrystallized grain structure which is observed right after deformation [Figure 4(a)] changes to a fully recrystallized grain structure [Figure 4(b)] by continuous growth of the dynamically recrystallized nuclei at a high temperature. The meta-dynamically recrystallized grains are coarser than the fully dynamically recrystallized grains. However, understanding this phenomenon is the key issue in the control of the size and the uniformity of the grains produced under a production condition. The amount of meta-dynamic recrystallization is related to the as-preheated grain size, the strain, the temperature, the strain rate and the holding time in a hot deformation process. Important factors related to meta-dynamic recrystallization are: the time for 50% meta-dynamic recrystallization, the fraction of meta-dynamic recrystallization, and the size of meta-dynamically recrystallized grain.

Time for 50% Meta-dynamic Recrystallization - Meta-dynamic recrystallization is time dependent. For a given strain, strain rate, and as-preheated grain size, meta-dynamic recrystallization progresses in the following manner with respect to time:

$$X_{m-dyn} = 1 - \exp\{-\ln 2 [t/t_{0.5}]^n\} \qquad [14]$$

The $t_{0.5}$ can be obtained from compression tests with different holding times for a given temperature, strain, strain rate, and as heated grain size. The empirical $t_{0.5}$ for meta-dynamic recrystallization follows the following relationship:

$$t_{0.5} = 4.54 * 10^{-5} d_0^{0.51} e^{-1.28} \dot{e}^{-0.073} * \exp(9705/T) \qquad [15]$$

The experimentally obtained $t_{0.5}$ is shown in Figure 5. It is seen that when the strain is larger than 0.5, the $t_{0.5}$ is below 2 seconds for the processing conditions in this study.

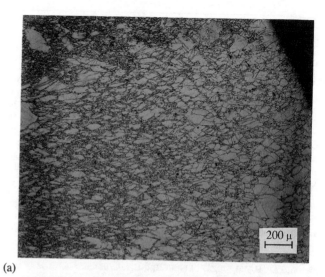

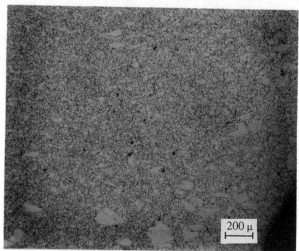

Figure 4. Micrographs obtained from Waspaloy samples with different cooling histories after forging: (a) rapidly cooled immediately after deformation, and (b) rapidly cooled after a 3 second hold at deformation temperature (1066°C).

The exponent, n, for meta-dynamic recrystallization is found to be 1 as shown in Figure 6. This number is typical for meta-dynamic recrystallization [11, 12].

Fraction of Metadynamic Recrystallization - The fraction of meta-dynamic recrystallization progresses according to the following equation:

$$X_{m-dyn} = 1 - \exp\{-\ln 2 [t/t_{0.5}]^{1.0}\} \qquad [16]$$

Figures 7(a) and (b) shows the fraction of meta-dynamic recrystallization versus time obtained from the experiments and the predictions at 1066°C with different as-preheated grain size, strain, and strain rate conditions. The meta-dynamic recrystallization finishes sooner for cases of larger strain, finer as-preheated grain size, higher strain rate and higher temperature.

Meta-dynamic Recrystallized Grain Size - The grain size obtained at the end of meta-dynamic recrystallization is found to have the following relationship with the strain, the as-preheated grain size, and Zener-Hollomon parameter:

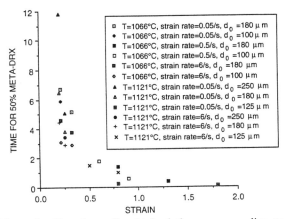

Figure 5. Experimentally obtained time corresponding to 50% meta-dynamic recrystallization.

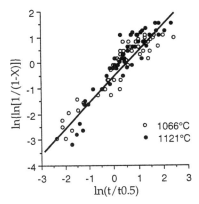

Figure 6. Determination of the Avrami exponent n for meta-dynamic recrystallization. "n" is the slope of the fitted line.

$$d_{m\text{-}dyn} = 14.56 d_0^{0.33} e^{-0.44} Z^{-0.026} \qquad [17]$$

A finer meta-dynamically recrystallized grain size is associated with cases with larger strain, finer as-preheated grain size, and higher values of the Zener-Hollomon parameter.

The meta-dynamic recrystallized grain size in ASTM number versus strain under conditions with different as-preheated grain sizes, temperatures and strain rates is shown in Figure 8.

Grain Growth

Under high temperature deformation conditions grain growth happens rapidly after the completion of meta-dynamic recrystallization. Grain boundary energy is the driving force causing grain boundary motion at high temperature. Grain boundary energy is comparable to the surface energy, it tends to minimize itself whenever possible by decreasing the grain boundary area. In general, grain growth will continue to occur at elevated temperatures until the balance between the grain boundary energy and the pinning effects of precipitates (precipitate size and spacing) is reached.

Grain growth is characterized by compression tests with different post-forging holding times. From the micrographs obtained from these tests, the microstructural evolution from partial dynamic recrystallization to full meta-dynamic recrystallization, and to grain growth were observed. The change in grain size versus time after the completion of meta-dynamic recrystallization is found to follow relationship:

$$d^3 - d_1^3 = 2*10^{26} \, t \, \exp(-595000/[RT]) \qquad [18]$$

The form of the equation (18) is well known for the characterization of grain growth. The reason for emphasizing that the d_1 is the grain size after complete meta-dynamic recrystallization is that after the completion of meta-dynamic recrystallization the dislocations have essentially disappeared and the driving force for grain size changes is the grain boundary energy only.

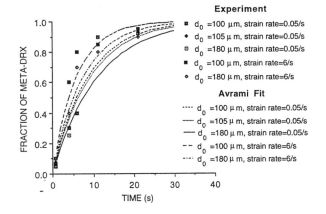

(a)

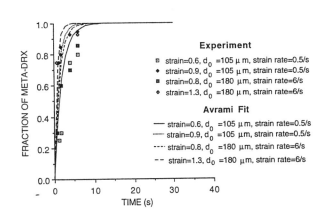

(b)

Figure 7. Fraction of meta-dynamic recrystallization versus time at 1066°C with a strain of (a) 0.22 and (b) 0.6 to 1.3.

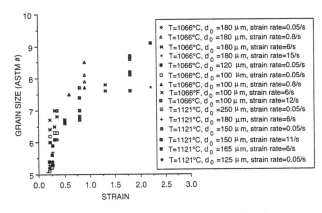

Figure 8. Meta-dynamically recrystallized grain size versus strain for various process conditions.

The experimentally obtained data and the model prediction for the short time grain growth are shown in Figure 9. It is seen from this figure that the grain growth at a temperature around 1121°C is very fast. There was not much difference in grain size after the completion of meta-dynamic recrystallization between the two samples obtained from compression tests at 1066°C and 1121°C (Figure 8). However, the grain growth results in a large difference in the final grain size for the two sets of tests conducted at 1066°C and 1121°C.

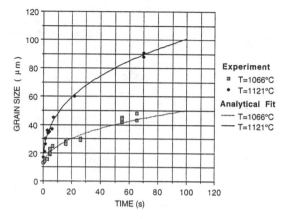

Figure 9. Grain growth versus time after the completion of meta-dynamic recrystallization in Waspaloy.

Summary of Equations Developed - It is seen from these equations that the major factors in the control of the grain size in the forging of Waspaloy are strain, temperature compensated strain rate and the as-preheated grain size.

Strains create localized high densities of dislocations. To reduce their energy, dislocations rearrange into subgrains. When the subgrains reach a certain size the nuclei of new grains form. The higher the strain, the greater the amount of dislocations and the greater the number of cycles of recrystallization. This results in a higher percentage of recrystallization.

The reason for deformation under high Z condition giving finer grain size is that an increase in Z results in the increase in the subgrain density which gives a higher density of nuclei. There are also more cycles of recrystallization present under high values of Z. Thus, the size of the recrystallized grains decrease [10].

At a given strain, the reason for as-preheated grain size playing an important role in the determination of the fraction of recrystallization and recrystallized grain size is that when polycrystalline metal is deformed, the grain boundaries interrupt the slip processes. Thus, the lattice adjacent to grain boundaries distort more than the center of the grain. The smaller the as-preheated grain, the larger the grain boundary area and the volume of distorted metal. As a consequence, the number of possible sites of nucleation increases. The rate of nucleation increases, and the size of the recrystallized grains decreases. Moreover, the uniformity of distortion increases with the decrease in as-preheated grain size. Therefore, having a fine as-preheated initial grain size is very important for obtaining a fine recrystallized final grain size.

INTEGRATION OF THE MICROSTRUCTURE MODEL WITH FEM CODES TO PREDICT FORGING MICROSTRUCTURAL RESULTS:

The equations developed were integrated with a non-isothermal visco-plastic finite element software code for microstructure prediction. The grain growth model was applied to billet preheating processes. The as-preheated grain size is predicted based on starting grain size and time at temperature. The as-preheated grain size is then used as the starting grain size for the deformation processes. The dynamic and meta-dynamic recrystallized grain size is calculated for a given forging and post-forging cooling condition. Grain growth in post forging cooling is modeled upon the completion of meta-dynamic recrystallization and the time at temperature.

These modeling tools were used to predict different Waspaloy forging processes. Examples of model implementation for a sub-solvus and a super-solvus generic forging and subsequent evaluation results are outlined below.

Sub-Solvus Forging: The model predictions and actual measured microstructure results from a generic part forged at sub-solvus temperatures were compared. The results of this effort are shown in Figures 10(a) -10(c). Figure 10(a) compares the average grain size observed from the actual forging and the model prediction. Figure 10(b) compares the ALA grain size observed from the actual forging and the model prediction. Figure 10(c) compares the fraction of recrystallization observed from the actual forging and the model prediction. Overall, the model predictions agree very well with the actual measured forging results. Necklace structures are predicted with these modeling tools and can be seen in the plots of the prediction of fraction recrystallization shown in Figure 10(c). The region where the fraction of recrystallization is between 0.2-0.8 has a necklaced structure. This prediction of necklaced structures was shown to be very accurate, and is being used in the design of the forging processes.

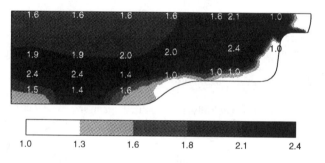

(a)

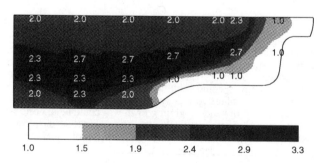

(b)

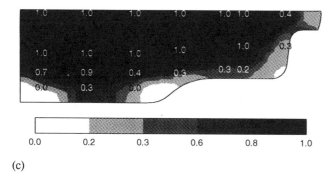

(c)

Figure 10. Partial component cross-sectional plots of predicted and observed microstructures for a generic sub-solvus forging: (a) average grain size, (b) ALA grain size, and (c) fraction of recrystallization. The microstructure values are normalized.

Super-Solvus Forging: The model predictions and actual measured microstructure results from a generic part forged in the super-solvus temperature range were also compared and evaluated. Results from this comparison effort are shown in Figure 11(a) - 11(c). Again the comparison is made in terms of average grain size, ALA grain size, and fraction of recrystallization. The predictions are again in very good agreement with the measured results.

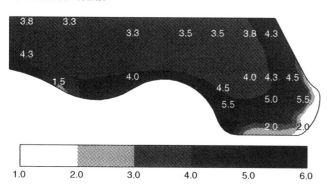

(a)

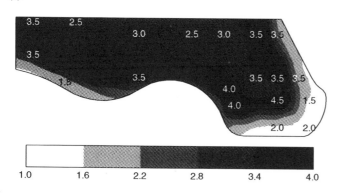

(b)

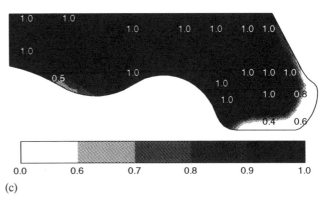

(c)

Figure 11. Partial component cross-sectional plots of predicted and observed microstructures for a generic super-solvus forging: (a) average grain size, (b) ALA grain size, and (c) fraction of recrystallization. The microstructure values are normalized.

Extensive efforts to further study the implementation and application of computer models to the prediction of microstructure results of forged Waspaloy under various processing temperature regimes, and numerous processing methods, has allowed considerable growth and refinement of this predictive model. The overall goal of this effort, for the real-world manufacturing environment, is to develop tools which are practical, but provide the best possible shop-floor correlation.

Efforts to refine the model have included developing an understanding and identifying the differences between material suppliers, and ingot-to-billet conversion methods (ie., mill-to-mill differences, and billet size differences). Significant differences between mills and billet sizes have been found, and these differences are defined accordingly in the model for an accurate prediction of the as-preheated microstructure and evolution during deformation processing.

Conclusions

Microstructure development in terms of dynamic recrystallization, meta-dynamic recrystallization and grain growth has been formulated. These equations were developed using preheating and compression tests. The integration of the microstructure model with finite element simulation codes made it possible for microstructure prediction of production parts with different shapes and different thermomechanical histories. The model was validated by generic experimental and actual production forging comparisons. Confidence in the refined model equations is at a level where the predictions are currently being used to guide all new part and existing part optimization efforts.

Acknowledgments

The authors would like to thank the Engineering Research Center at The Ohio State University for support of the initial experimentation and model development, and Ladish Co., Inc. for support of the model integration and optimization efforts, and for allowing this paper to be presented.

Additionally, the authors would like to thank R. Shivpuri at the IWSE Dept. of Ohio State University and S.L. Semiatin at Wright-Patterson Air Force Base for guidance regarding initial model development.

References

1. C.M. Sellars: in <u>Hot Working and Forming Processes</u>, C.M Sellars and G.J. Davies, eds., TMS, London, (1979), 3-15.

2. W.T. Wu, and S.I. Oh, "ALPIDT - A General Purpose FEM Code for Simualtion of Nonisothermal Forging Processes", <u>Proceedings of North American Manufacturing Research Conference (NAMRC) XIII</u>, Berkley, CA, (1985), 449-460.

3. Scientific Forming Technologies Corporation, <u>DEFORM 4.1 User Manual</u>, Columbus, OH, 1995.

4. A.J. Deridder and R. Koch, "Forging and Processing of High Temperature Alloys," <u>MiCon 78: Optimization of Processing, Properties, and Service Performance Through Microstructural Control</u>, ASTM STP 672, Halle Abrams et al., Eds., American Society for Testing and Materials, (1979), 547-563.

5. G. Shen, D. Furrer, and J. Rollins, " Microstructure Development in a Titanium Alloy," The Proceedings of the Symposium on <u>Advances in Science and Technology of Titanium</u>, TMS, Anaheim, CA, Feb. 4-8, 1996.

6. D. Stewart, "ISOCON Manufacturing of Waspaloy Turbine Discs," <u>Superalloys 1988</u>, S. Reichman et al. Eds, (1988), 545-551.

7. G. Shen, "Modeling Microstructural Development in the Forging of Waspaloy Turbine Engine Disks," (Ph.D. thesis, The Ohio State University, 1994).

8. G. Shen, S.L. Semiatin, and R.Shivpuri, "Modeling Microstructural Developemtn during the Forging of Waspaloy", <u>Met. Trans. A</u>, 26A, (1995), 1795-1802.

9. Balaji, T. Furman, J. Rollins, and R. Shankar, "Prediction of Microstructure in Forged Waspaloy Components Using Computer Process Modeling," <u>Proceedings of North American Forging Technology Conference</u>, ASM/FIA, Dallas, 1994.

10. H.J. McQueen and J.J. Jonas, "Recovery and Recrystallization During High Temperature Deformation," <u>Treatise on Materials Science and Technology</u>, Vol. 6, Academic Press, New York, (1975), 393-493.

11. J.J. Jonas, "Recovery, Recrystallization and Precipitation under Hot Working Conditions," <u>Proceedings of the Fourth International Conference on Strength of Metals and Alloys</u>, Nancy, France, Aug. 30 - Sept. 3, (1976), 976-1002.

12. C. Devadas, I.V. Samarasekera, and E.B. Hawbolt, "The Thermal and Metallurgical State of Strip during Hot Rolling: Part III, Microstructural Evolution," <u>Met. Trans. A</u>, 12A, (1991), 335-349.

EFFECT OF COOLING RATE FROM SOLUTION HEAT TREATMENT ON WASPALOY MICROSTRUCTURE AND PROPERTIES

J.R. Groh
Engineering Materials Technology Laboratories
GE Aircraft Engines
Evendale, OH

Abstract

Achievement of critical dimensions after wholesale removal of stock from large, ring rolled Waspaloy forgings and fabrications has been complicated by residual stresses induced via rapid quench from the solution heat treat cycle. Part movement during machining can be reduced significantly by substituting air cool for the oil quench commonly specified to achieve high mechanical strengths. However, excessive time in the sub-γ' solvus temperature range during slow cooling results in undesirable precipitate coarsening with an attendant degradation in strengths. Therefore, cooling rates must be achieved which strike a balance between manufacturing ease and property controlling microstructural features.

This report provides microstructural and mechanical property data for flash welded, shaped Waspaloy bar stock subjected to various cooling rates from a 1018°C/4 hour solution cycle followed by a stabilize and age heat treat. Average cooling rates evaluated in this investigation ranged from 5.5 to 145°C/minute through the γ' precipitation temperature range of 982 through 760°C. Oil quench data from production hardware are also provided for reference. A significant improvement in 538 and 760°C yield strengths, 760°C tensile strengths, and 704°C creep resistance were significantly improved by increasing cooling rates. Ductility simultaneously decreased at 760°C but did not exhibit dependence at 538°C. Minimum cooling rates of approximately 40°C/minute were necessary to approach the properties and microstructural features typical of oil-quenhed Waspaloy.

Introduction

Waspaloy has been used successfully in turbine case applications which require a balance of fatigue and creep resistance with tensile capability. Additional considerations for this application include grain structure for weldability and residual stresses from fabrication and heat treat which may result in part movement during machining.

Primary γ' has been used to control Ni-base microstructures during forge and heat treat cycles which significantly affect mechanical properties (ref. 1). Balanced properties are achieved via temperature selection relative to the γ' solvus which is a strong function of hardener content. Excessive exposure to sub-solvus temperatures reduces the amount of hardener available for precipitation hardening. Super-solvus operations late in the process coarsens grain size which degrades tensile and fatigue strengths.

Destructive mechanical tests of a large casing which had been solution heat treated in vacuum followed by argon cooling for manufacturing ease revealed unacceptable creep and tensile properties. Slow cooling from the sub-solvus solution temperature at 5.5°C/minute, nominal, through the sub-γ' solvus range had generated a coarsened γ' precipitate structure. Subsequent stabilization and aging cycles further overaged this structure. Resulting properties showed a 10-100X reduction in time to 0.2% creep and 50-170 MPa reduction in tensile strengths. Strain-control low cycle fatigue behavior was not affected (ref. 2).

Simply switching to rapid oil quench from the solution temperature to improve properties was not considered an acceptable option as unknown residual stress patterns result. This project was performed to determine the minimum cooling rate which is necessary to achieve a microstructure which yields an acceptable balance of mechanical properties. The resulting property trends, coupled with heat transfer models of the quench from solution, have provided data used to identify a cooling method for production hardware to obtain mechanical properties similar to oil quenched forgings with minimal residual stress.

Materials and Experimental Procedure

Material and Heat Treatment

Material for this study was shaped bar stock processed by Teledyne Alvac from VIM+VAR ingot (heat BR12, γ' solvus 1043°C) to the requirements of AMS5706. Heat chemistry follows in table I. Grain structure was equiax with an average grain size of ASTM 6-7.

Cooling rate blanks measured 53 mm in length with a chromel-alumel thermocouple centered in the roughly 25 mm square end and embedded 12 mm deep. Most solution heat treatments were performed in air at 1018°C for 4 hours. Various combinations of insulation batting and a steel retort were used to achieve cooling rates which varied from an average of 20 to 145°C/minute between 982 and 760°C. An additional sample was prepared at 5.5°C/minute in a vacuum furnace to enable programming of this relatively slow cool to simulate the hardware discussed above. All material blanks were then subjected simultaneously to the standard stabilize (843°C/4 hours) and age (760°C/16 hours) heat treat cycles specified in AMS5707.

TABLE I
Waspaloy Heat BR12
(Weight %)

C	Mn	Si	P	S	Cr	Al
.058	.04	.06	.005	.0004	19.45	1.37
Ni	**Mo**	**Co**	**Ti**	**Al+Ti**	**Bi**	**Pb**
Bal	3.89	12.38	2.91	4.28	<.0001	<.0001
Se	**B**	**Cu**	**Fe**	**Zr**	**Nb**	**Ta**
<.00005	.006	.02	1.02	.06	.06	.02

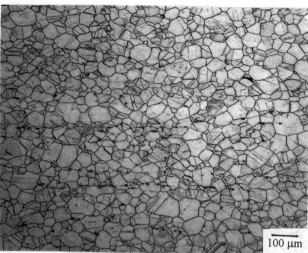

Figure 1 Equiaxed grain structure of shaped bar material used in this investigation. ASTM 6-7 average grain size.

Metallography
A 12mm slice was removed from one end (transverse to rolling) of each fully heat treated blank for standard mechanical polishing techniques through 0.3 μ diamond. Subsequent electropolishing and electroetching were performed in the same 10% perchloric ($HClO_4$)+ methanol solution with the former at 30 volts for 15 seconds and the latter at 6V for 5 seconds. Specimen cross-sections were nominally 12mm square. Electropreparation time was maintained constant regardless of microstructure to enable unbiased microstructural comparisons.

High resolution scanning electron microscopy (field emission electron gun) was used to document γ' and carbide morphologies.

Depending on precipitate size, the area percent γ' was measured at either 25K or 50K magnification. A manual point count was performed in accordance with ASTM E562 (ref. 4) to quantify observations. SEM evaluation was also performed on test material from an oil quenched AMS5707 component with an average ASTM 5.-5.5 grain size from a different heat, included as reference.

Mechanical Testing
All specimens were machined parallel to the rolling direction by Accurate Metallurgical Services. Gage sections measured 4 mm in diameter by 19 mm with an overall specimen length of 50 mm (GE drawing #4013195-001). The test gage of each specimen was immediately adjacent to the embedded thermocouple lead used to verify cooling rate.

Mechanical tests consisted of 538 and 760°C tensile per ASTM E21 and 704°C/427 MPa creep per ASTM E139 with the stress axis parallel to the rolling direction. Average tensile and time to creep data from production oil quenched AMS5707 components with an average ASTM 5.-5.5 grain size from various heats have been included for reference. Oil quench was not part of this study due to the limited amount of material.

Experimental Results

Metallography
Optical micrograph of grain structure is provided in Figure 1. High resolution SEM photomicrographs at 50,000X of the 5.5, 25, 40, 63, 145°C/minute and oil quenched samples are provided in Figure 2. Age γ' size decreased from 0.16μm to the 0.06-0.07 μm range by increasing from 5.5 to 40°C/minute. Faster quench did not achieve the 0.04 μm recorded for oil quenched material.

Area percent of age γ' was normalized relative to the point count data generated for the oil quench from solution+stabilize and aged specimen. The measured area % data continued to increase through 63°C/minute before leveling. Primary γ' size was reduced by increasing quench rate but also leveled at 40°C/minute. Primary γ' fraction and grain sizes were not significantly affected by cooling rate. The balance of microstructural features are provided in Table I.

TABLE II
AMS5707 Waspaloy
Microstructural Features

Cooling Rate	Age γ' size (μm)	Normalized % Age γ'	Primary γ' size (μm)
5.5°C/min	0.16	0.35	0.40
25°C/min	0.12	0.35	0.35
40°C/min	0.07	0.52	0.30
63°C/min	0.06	0.82	0.30
145°C/min	0.06	0.82	0.30
Oil (Reference)	0.04	1.	0.30

Mechanical Properties
The results of tensile and a summary of creep data are presented in tables III and IV, respectively. Figure 3 depicts the ultimate and 0.2% yield strength data as a function of cooling rate. Figure 4 shows the influence of cooling rate on tensile ductilities. Plots of creep strain time pairs for each cooling rate evaluated are provided in Figure 5. With the exception of 538°C UTS and ductilities, regression models for each property versus cooling rate are provided in Table V.

TABLE III
AMS5707 Waspaloy
Tensile Data

Cooling Rate (°C/min)	Test Temp (°C)	UTS (MPa)	0.2% YS (MPa)	Elongation (%)	Reduction of Area (%)
5.5	538	1156	714	24.0	32
25	538	1193	845	21.9	30.3
40	538	1207	794	21.9	30.5
63	538	1163	840	23.4	35.8
145	538	1175	834	23.4	31.7
Oil (ref.)	538	1189	807	20.0	28.4
5.5	760	884	585	36.0	61.1
25	760	984	728	26.6	50.9
40	760	982	651	21.9	57.4
63	760	1065	820	12.5	41.4
145	760	1172	920	15.6	28.7
Oil (ref.)	760	931	759	17	28.2

Figure 2   High resolution SEM micrographs of AMS5707 Waspaloy cooled from solution at indicated cooling rates followed by stabilize and age. Micrograph representing a solution+oil quench, stabilize and age heat treated component is also provided for reference.

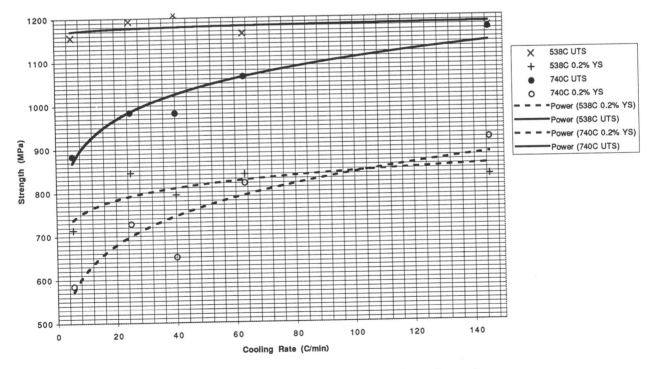

Figure 3   Plot of cooling rate vs. tensile strength data. Strengths continue to improve with increasing cooling rate at 760°C. At 538°C only the yield strength shows an initial improvement, ultimate was not affected at this temperature.

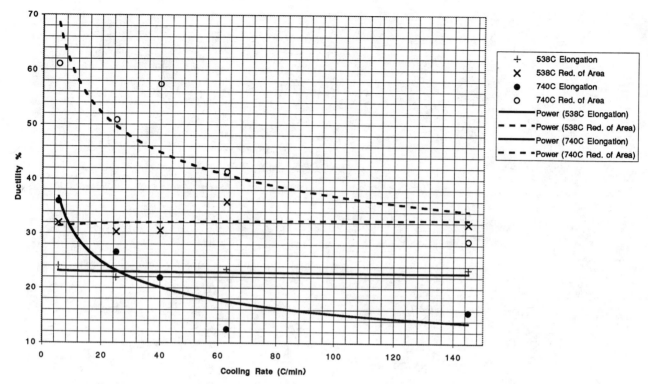

Figure 4 Plot of cooling rate vs. tensile ductility data. Ductilities continue to improve with increasing cooling rate at 760°C. Neither measure showed a relationship to cooling rate at 538°C.

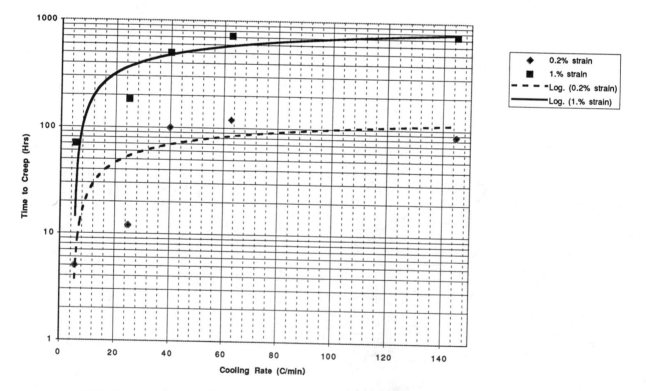

Figure 5 Plot of cooling rate vs. time to 0.2 and 1.0% creep for 704°C/427 MPa. Creep resistance improved with cooling rate through ~40°C/minute then tapered to a level approaching oil quenched Waspaloy of this grain size.

TABLE IV
AMS5707 Waspaloy
704°C/427 MPa Creep Data

Cooling Rate (°C/min)	Time to 0.1% ε (Hours)	Time to 0.2% ε (Hours)	Time to 0.5% ε (Hours)	Time to 1.% ε (Hours)
5.5	0.2	0.5	1.6	4.6
25		12	78	183
40	43	100	269	500
63	28	119	420	724
145	4	86	388	748
Oil (ref.)	54	450	not available	not available

Table V
Property vs. Cooling Rate Regressions

538°C 0.2% YS=680.09*(cooling rate)$^{0.0471}$	R^2=0.65
538°C UTS, %El, and % RA not f(cooling rate)	R^2<0.1
740°C 0.2% YS=454.9*(cooling rate)$^{0.1339}$	R^2=0.82
740°C UTS=753.3*(cooling rate)$^{0.0835}$	R^2=0.94
740°C % El=61.694*(cooling rate)$^{-0.3038}$	R^2=0.77
740°C %RA=98.534*(cooling rate)$^{-0.2131}$	R^2=0.72
704°C/427 MPa time (hrs)$_{0.2\%}$ = 32.95*ln(cooling rate)-52.458	R^2=0.59
time (hrs)$_{1.0\%}$=233.76*ln(cooling rate)-384.03	R^2=0.85

Discussion

Laboratory cooling rates did exhibit a pronounced effect on microstructure and resulting mechanical properties. In general, a greater percentage of age γ' was realized with increasing cooling rates which improved properties. Reduction in age γ' size coincides with a change in morphology from slightly cuboidal to spheroidal. A slight decrease in primary γ' size was also measured relative to the two slowest rates as the time of exposure to sub-solvus temperatures was reduced.

Age γ' percents were normalized to the oil quench specimen as the measured volume % data was higher than theoretical due to the specimen preparation techniques used in this study. Constant electro polish and etch of all specimens provided different depths of attack which is a function of structure. A more accurate metallographic determination could be obtained via replication techniques although the depth of etch variable may still be an issue. Determination of the actual volume fraction requires phase extraction.

Tensile strengths and ductilities at 760°C were influenced by cooling rate to a greater degree than at 538°C. The insensitivity of UTS, %El and %RA at 538°C was somewhat surprising as reference 5 documents cooling rate dependent behavior for Rene'88DT similar to that observed in this study at both 400 and 649°C. R'88DT is a relatively high hardener Ni-base powder alloy which differs in that solution heat treatment is accomplished above the γ'-solvus temperature.

A trend of improved creep resistance was observed for both 0.2% and 1.0% ε at 704°C/427 MPa over the 1000 hours tested. The trend in 0.1% creep lives versus cooling rate did not track well versus cooling rate (table IV). The primary creep rate of the 145°C/minute specimen was relatively fast which caused the time to 0.1% ε to be less than that recorded for the 40 and 63°C/min. samples. Higher creep strains showed a much better progression of time versus cooling rate. Cross-over occurred at roughly 0.25% strain with a relatively strong trend of increasing strain versus cooling rate for a given time above this level.

The recent investigation of R'88DT (reference 5) supports the results of this study. Similar improvements in tensile strengths were most significant between 7 and 40°C/minute. Further increases in cooling rate resulted in relatively modest tensile strength improvements. Marked further improvements in 704°C creep resistance were observed when oil quenched solution. Similar benefits would be expected for Waspaloy based on the aging γ' data provided in this report.

Waspaloy microstructural features and testing indicated that a significant improvement in properties relative to the argon cooling process could be realized at cooling rates as low as 40°C/min. Average rates of 63°C/min or faster are preferred if the resulting residual strains are tolerable during fabrication.

Since project completion, several casings were rapid fan air cooled from the solution cycle which achieved a satisfactory cooling rate of 44°C/minute. The benefit to mechanical properties was substantiated via mechanical testing prolong material at the conditions evaluated in this report (ref. 3). SEM metallography of very small sections removed from stock added areas of subsequent production hardware has been used as a non-destructive tests to confirm adequate cooling. Problems due to distortion during machining of this component were not encountered with this procedure.

Conclusions

1. Morphology of γ' precipitates was a strong function of cooling rate from the solution temperature. Structures approaching that achieved by oil quench were achieved at cooling rates as low as 40°C/minute.
2. Waspaloy 760°C tensile and 704°C creep properties were shown to be highly dependent on cooling rate from the solution temperature. At 538°C, tensile yield strength showed improvement with increasing cooling rate; however, the UTS and ductilities were not affected.
3. Models of tensile properties and creep resistance as a function of cooling rate were generated.
4. Minimum cooling rates of 40°C/minute are needed to approach the tensile and creep design curves generated from relatively small, oil quenched forgings.
5. The production rapid fan air cool cycle from the solution heat treatment temperature yielded γ' morphologies sufficient to provide mechanical behaviors comparable to those of the design curves generated from parts oil quenched from solution.
6. SEM evaluation of precipitate morphology is an effective method to confirm that the appropriate heat treatment and anticipated properties are achieved.

References

1. P.R. Bhowal, E.F. Wright, and E.L. Raymond, "Effects of Cooling Rate and γ' Morphology on Creep and Stress-rupture Properties of a Powder Metallurgy Superalloy", Metallurgical Transactions A, 21A (1990), 1709-1717
2. W.P. Rehrer, D.R. Muzyka, and G.B. Heydt, "Solution Treatment and Al+Ti Effects on the Structure and Tensile Properties of Waspaloy", Journal of Metals (1970), 32-39.
3. Jon Groh, internal GE report dated 4/5/95.
4. ASTM E562 "Standard Practice for Determining Volume Fraction by Systematic Manual Point Count" dated 1983.
5. E.S. Huron, internal GE report dated 2/1/95.
6. AMS5706, "Alloy Bars, Forgings, and Rings, Corrosion and Heat Resistant ; 1825 to 1900F solution Heat Treated"

EFFECT OF STABILIZING TREATMENT ON PRECIPITATION BEHAVIOR OF ALLOY 706

Takashi Shibata*, Yukoh Shudo*, Tatsuya Takahashi**, Yuichi Yoshino*, and Tohru Ishiguro**

*Technology Research Center, The Japan Steel Works, Ltd.,
1-3 Takanodai, Yotsukaido, Chiba 284, Japan
**Muroran Research Laboratory, The Japan Steel Works, Ltd.,
4 Chatsu-machi, Muroran, Hokkaido 051, Japan

Abstract

Ni-Fe-base superalloy 706 has recently been used for high temperature services. A stabilizing treatment between solution-annealing and age-hardening treatments has been proposed for this alloy to improve its creep rupture life. However, the relationship between stabilizing treatment and precipitation behavior has not been well understood. The precipitation behavior of Alloy 706 was investigated and related with creep rupture properties. Samples taken from a gas turbine disk forging were solution-treated at 980°C for 3h and stabilizing-treated in a range of 780 to 900°C for 1.5 h, followed by the double-aging at 720°C for 8h and at 620°C for 8h. Precipitation behavior of these samples was examined by TEM, and creep rupture tests were conducted at three conditions with varying temperature and applied stress.

Fine γ'-γ'' co-precipitates having the core of γ' being overlayed with γ'' on its top and/or bottom and large γ'-γ'' co-precipitate having the core of γ' being completely covered with γ'' were identified in the grain matrix together with large γ' and fine γ'' phases that have been found to precipitate in Alloy 706. η was identified at the grain boundary and found to be accompanied with a serrated grain boundary and denuded zone. Such precipitation behavior was significantly affected by stabilizing temperature, especially below 840°C, and so was creep rupture property accordingly. It is concluded that the best creep rupture properties are obtained when an optimum combination is established between intra-granular fine γ'/γ'' co-precipitates, intra-granular large γ'/γ'' co-precipitates, and the inter-granular η phase.

Introduction

Ni-Fe-base superalloys are age-hardened by the precipitation of coherent γ' and/or γ'' in the austenitic matrix γ (1). Alloy 706 is a relatively new material and was developed from Alloy 718 a representative wrought superalloy. Compared with Alloy 718, Alloy 706 has a chemical composition of no molybdenum, reduced niobium, aluminum, chromium, nickel and carbon, and increased titanium and iron. This excellent balance of chemical composition results in superior characteristics to Alloy 718 in segregation tendency, hot workability and machinability (2-4). Therefore, Alloy 706 is suitable for large forgings and has been used for high temperature services (5).

A stabilizing treatment between solution-annealing and age-hardening treatments has been proposed for Alloy 706 to improve its creep rupture life (6). The improvement in creep properties is attributed to the precipitation at grain boundary during the stabiling treatment (2-4, 7-9). We previously reported elsewhere that creep rupture properties of Alloy 706 were significantly affected by stabilizing treatment temperature (10). However, the relationship between stabilizing treatment and precipitation behavior has not been well understood. The present study is concerned with the stabilizing treatment in an effort to correlate creep rupture properties to the morphlogy of the precipitates that form during the heat treatment.

Procedure

Material

The experimental material was taken from a forged gas turbine disk manufactured frome a vacuum induction melted (VIM) and electro slag remelted (ESR) ingot that was diffusion treated and subsequently forged. The material was sectioned mechanically into samples of suitable sizes for the following experiments. The alloy composition used in this study are given in Talbe I.

Table I Chemical Composition of Alloy 706

mass %

Ni	Fe	Cr	Al	Ti	Nb	B	C	N	Si	Mn	P	S
42.0	37.1	15.65	0.26	1.54	2.96	0.0034	0.008	0.0046	0.05	0.02	<0.003	<0.0005

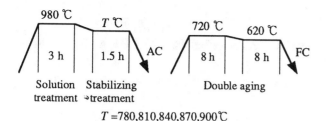

Figure 1 : Heat treatment program and conditions.

Heat Treatment

Heat treatment conditions are shown in Figure 1. The samples were solution-treated at 980℃ for 3h and then stabilizing-treated in a range of 750 to 900℃ for 1.5 h. Subsequently they were double-aged at 720℃ for 8h and at 620℃ for 8h. Alloy 706 is usually reheated to the stabilizing temperature after solution annealing (6). In this study, the stabilizing treatment was conducted in the cooling stage from the solution treatment without cooling the material to the room temperature from the point view of industrial advantage.

Microscopy

These samples were examined by scanning electron microscopy (SEM) and transmission electron microscopy (TEM) for their precipitation behavior. Thin film method was used for TEM sample preparation, and final thinning was achieved by electro-polishing. A 200kV TEM was used with micro-beam technique in both electron diffraction and energy disperssive X-ray spectroscopy (EDS), with the probe diameter being minimum 1nm.

Creep Rupture Test

Creep tests were conducted at three conditions : 600℃ / 686.5 MPa, 600 ℃ / 745.3 MPa and 650℃ / 686.5 MPa. The diameter of specimens was 6mm and the gauge length was 30mm. In order to ensure the uniformity of temperature, the specimens were held for 24h at the test temperatures before loading.

Results

SEM Observation

SEM micrographs of Alloy 706 stabilized at various temperatures and aged are shown in Figure 2. No precipitate was seen inside the grains or at the grain boundary for the samples stabilized at 900 and 870℃ as for the un-stabilized one. On the contrary, many precipitates were observed when the stabilizing temperature was below 840℃. Especially cellular precipitates that lay parallel to each other were observed clearly at the grain boundary.

Intra-granular Precipitate

In order to identify these precipitates that can not by SEM, TEM observation was conducted. It is reported that the intra-granular precipitate is either γ' (2,3,7,8) or γ'' (9,10) and that they form simultaneously (4,11,12). Therefore, TEM observation was directed to the grain interior.

TEM image of Alloy 706 stabilized at 810℃ and aged is shown in Figure 3, where all types of precipitates identified in this study are seen. The precipitates of various shapes and sizes exist inside the grain. The spots arising from long-range ordering were clearly observed in a selected area

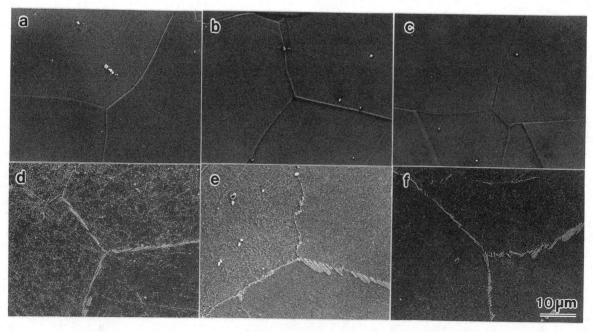

Figure 2 : SEM micrographs of Alloy 706 aged following by the stabilizing treatment at various temperatures : (a) unstabilized, (b) 900℃, (c) 870℃, (d) 840℃, (e) 810℃ and (f) 780℃.

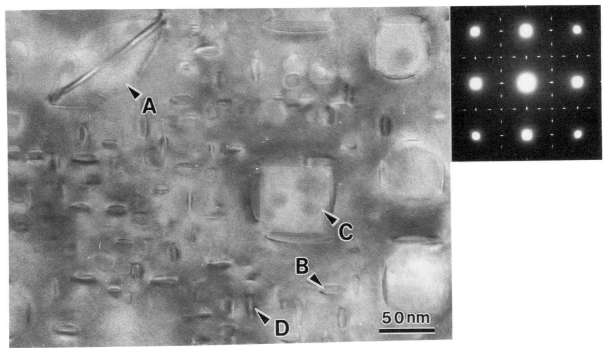

Figure 3 : TEM micrograph and selected area diffraction of Alloy 706 stabilized at 810℃ and aged.

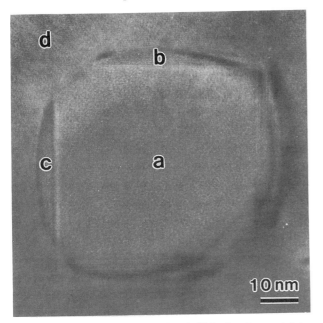

Figure 4 : High resolution image of the cuboidal γ'- γ"co-precipitate in Alloy 706 stabilized at 810℃ and aged.

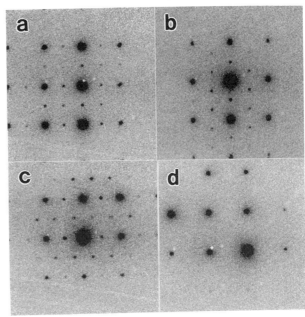

Figure 5 : Micro beam diffraction patterns at the location shown in Figure 4.

diffraction pattern given in Figure 3, indicating the presence of γ' and/or γ". Therefore, high resolution observation was made together with micro-beam techniques in order to identify each precipitate.

As a result, four types of precipitates were identified in the grain interior. The relatively large precipitates as indicated by arrow A were identified γ', there size being several hundred nanometers. The fine precipitates as indicated by arrow B were identified γ", there size being several nanometers. The precipitates as indicated by arrows C and D were identified two types of γ'- γ" co-precipitates which had not yet been reported with Alloy 706.

A high resolution image, micro-beam electron diffraction patterns and micro-beam EDS spectra are shown in Figure 4, 5 and 6, respectively, of the co-precipitates as indicated by arrow C. They indicate that this large cuboidal co-precipitate has the core of Ti rich γ' being completely covered with the Nb rich γ" thin skin, refferred to by R.Cozar and A.Pineau as co-precipitate of "compact morphlogy" in the modified 718 alloys (13-

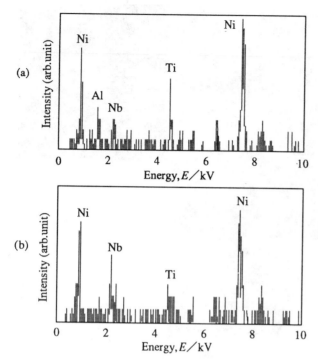

Figure 6 : Micro beam EDS spectra at the locations shown in Figure 4.

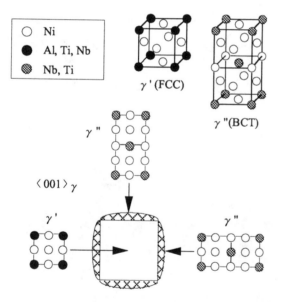

Figure 7 : Schematic diagram of the cuboidal γ'-γ'' co-precipitate in Alloy 706

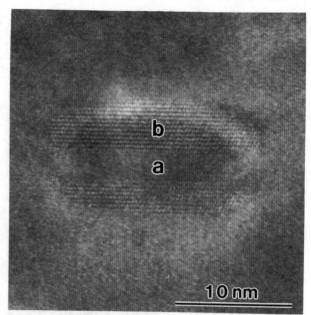

Figure 8 : High resolution image of the overlayed γ'-γ'' co-precipitate in Alloy 706 stabilized at 810℃ and aged.

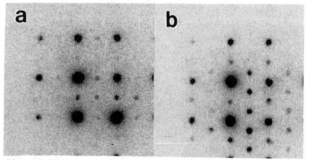

Figure 9 : Micro beam diffraction patterns at the locations shown in Figure 8.

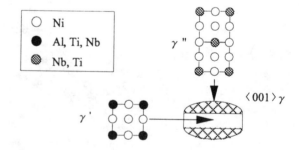

Figure 10 : Schematic diagram of the overlayed γ'-γ'' co-precipitate in Alloy 706.

20). The coherency between γ, γ' and γ'' is maintained as schematically illustrated in Figure 7. It is reported that this co-precipitate is obtained at a higher (Ti+Al) / Nb ratio than in regular Alloy 718. In fact, this ratio is greater of Alloy 706 than of Alloy 718. However, the chemical composition of Alloy 706 is out of the co-precipitate range of the S-R-diagram for Alloy 718 (13,14). The co-precipitate range may be shifted for Alloy 706. Further study is needed in this respect.

TEM information of the co-precipitates as indicated by arrow D are shown in Figure 8 and 9, respectively. Although it is too small to determine precisely its shape and inner structure, it appears to be either a disk or an ellipsoid. From both figures, this fine precipitate has the core of γ' being overlayed with γ'' on its top and/or bottom, refferred to by R.Cozar and A.Pineau as co-precipitate of "non-compact morphlogy", again (13-20). The coherency between γ, γ' and γ'' is maintained as schematically illustrated in Figure 10.

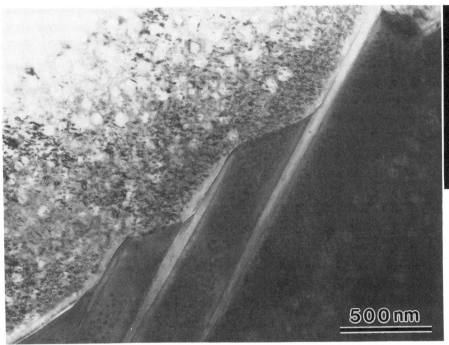

Figure 11 : TEM micrograph and selected area diffraction pattern of grain boundary in Alloy 706 stabilized at 810℃ and aged.

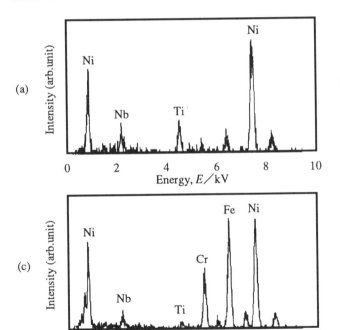
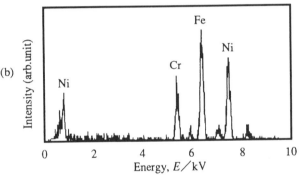

Figure 12 : EDS spectra at (a) η, (b) denuded zone and (c) the matrix.

Inter-granular Precipitate

Figure 11 shows TEM image of the cellular precipitates at the grain boundary of Alloy 706 stabilized at 810℃ and aged. These phases were identified η from the results of micro-beam electron diffraction and micro-beam EDS. These precipitates have a specific orientation relationship with the γ matrix ; $[011]_\gamma \text{ // } [2\overline{1}\overline{1}0]_\eta$ and $\{11\overline{1}\}_\gamma \text{ // } \{0001\}_\eta$. This relationship is consistent with the study reported by other (9), indicating a semi-coherency between η and γ. The cellular η appears parallel to each other in order to meet this orientation relationship.

As shown in Figure 11, the precipitation of η resulted in the formation of the serrated grain boundary and denuded-zone around it. Micro-beam EDS found a leanness of Niobium and Titanium in this zone as shown in Figure 12. The denuded zone was obscure after 840℃ stabilization, but it became wider and more distinct as the stabilizing temperature decreased. At 780℃, its width was greater than 100nm as shown in Figure 13.

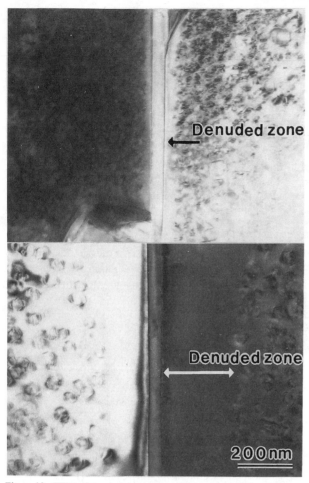

Figure 13 : TEM micrograph near the grain boundary of Alloy 706 aged after stabilizing at (a) 810°C and (b) 780°C.

fine precipitates formed at the aging treatment are grater as the stabilizing temperature higher. It is concluded that the cuboidal co-precipitate is dominant when stabilized at 780°C whereas the fine precipitates are when the stabilizing temperature is higher than 810°C.

The results obtained in this study has led to three groups of stabilizing temperatures in terms of the precipitation behavior ; 900 and 870°C (Group A), 840 and 810°C (Group B), and 780°C (Group C). The precipitation behavior of Group A is virtually the same as in the case where the stabilizing treatment is discarded. That is, the fine γ'' and overlayed γ'-γ'' co-precipitate arise inside the grain without any precipitation at the grain boundary. In the case of Group B, the fine precipitates are still dominant whereas the large cuboidal γ'-γ'' co-precipitate is also present inside the grain. At the same time, the η phase grows at the grain boundary, resulting in the grain boundary serration and the denuded zone that is still limited and narrow with this group. The large cuboidal co-precipitate is dominant in Group C, and the denuded zone becomes very wide. The precipitation behavior described here is summarized in Table II.

Creep Rupture Property and Stabilized Temperature

The creep strain vs. time curves for three test conditions are shown in Figure 15, 16 and 17 ; 600°C/686.5MPa, 600°C/745.3MPa and 650°C/686.5MPa, respectively. The creep rupture properties can be grouped into three classes ; here again Group A (900 and 870°C), Group B (840 and 810°C) and Group C (780°C). The creep rupture property does not appear to be affected by the stabilizing treatment for Group A. However, it was significantly affected when stabilizing temperature is below 840°C. The creep rupture property of Group B are markedly improved in creep rupture life and creep rupture elongation. On the contrary, it was much degraded in the Group C.

It is reported that there is not only η, but also δ at the grain boundary at Alloy 706 (7,9,12). However, no precipitate was identified δ for more than a hundred precipitates at grain boundary in this study. Only a few precipitates were identified γ' in the samples stabilized at 840 and 810°C. This is thought that η forms through the transformation of $\gamma \rightarrow \gamma' \rightarrow \eta$ as in A286 (1), leaving γ' as an intermediate phase. The grain boundary appears to become serrated as the η phase grows.

Precipitation Behavior and Stabilizing Temperature

In order to clarify at which stage of the whole heat treatment such precipitation as described above occurs, TEM studies were carried out for the samples before and after the aging treatment. The relatively large γ' and cuboidal γ'-γ'' co-precipitate were found to form in the stabilizing treatment at 840, 810 and 780°C. The precipitation of η and the accompanying phenomena were also found to occur in the same stabilizing treatment. On the contrary, the fine γ'' and overlayed γ'-γ'' co-precipitate formed during the double-aging treatment. But, when stabilized at 780°C, these fine precipitates were rarely observed even after aging. It is extremely difficult to determine the amount of these fine precipitates, especially the co-precipitate. From Figure 14, however, the amount of the

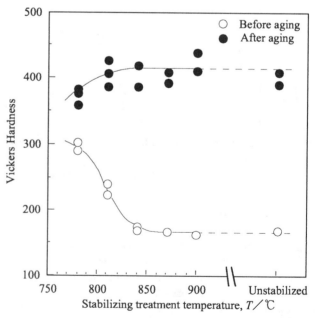

Figure 14 : Change in vickers hardness with the stabilizing treatment of Alloy 706.

Table II Summary Of Precipitation Behavior For Alloy 706 Stabilized At Various Temperatures (○ : pronounced △ : observable × : absent)

stabilizing temperature	large γ'	fine γ''	fine overlayed γ'-γ''	large cuboiral γ'-γ''	η	denuded zone
Unstabilized	×	○	○	×	×	×
900℃	×	○	○	×	×	×
870℃	×	○	○	×	×	×
840℃	○	○	○	△	○	△
810℃	△	○	○	△	○	△
780℃	△	△	△	○	○	○

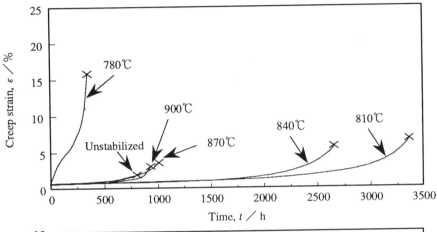

Figure 15 : Creep rupture curves of Alloy 706 tested at 600 ℃/686.5MPa.

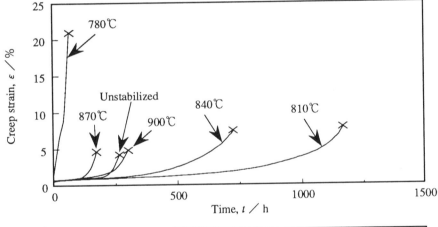

Figure 16 : Creep rupture curves of Alloy 706 tested at 600 ℃/745.3MPa.

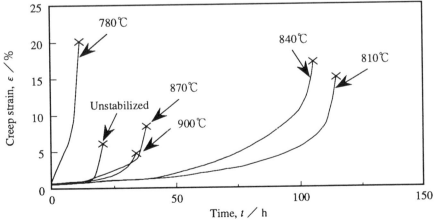

Figure 17 : Creep rupture curves of Alloy 706 tested at 650 ℃/745.3MPa.

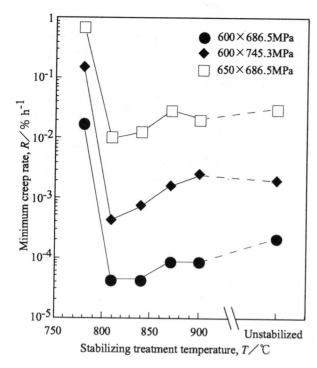

Figure 18 : Change in the minimum creep rate of Alloy 706 with stabilizing temperature.

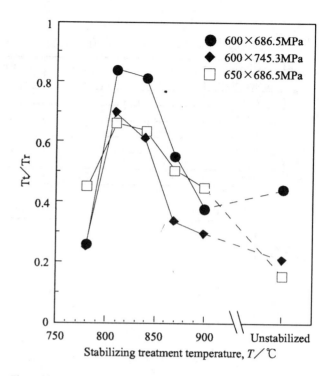

Figure 19 : The relative duration of tertially creep rupture life for Alloy 706 stabilized at different temperatures.

Discussion

As described, the classification of precipitation behavior agrees well with the classification of creep rupture properties, indicating a strong connection between them. Therefore, the results of the present study are discussed here in accordance with the classification of stabilizing temperature.

Group A ; stabilized at 900 and 870 ℃

The precipitation behavior is virtually the same as for the heat treatment program without stabilizing treatment. Therefore, creep rupture properties are not affected by the stabilizing treatment.

Group B ; stabilized at 840 and 810 ℃

Creep rupture properties are markedly improved. This improvement is contributed by the precipitations in the grain matrix and at the grain boundary.

The contribution of the precipitates in the matrix is depicted in Figure 18 that shows the minimum creep rate replotted from Figure 15, 16 and 17. The minimum creep rate is the smallest in the stabilizing temperature range for Group B. In this stabilizing temperature range, relatively large cuboidal γ' - γ'' co-precipitate is found to co-exist with the fine overlayed co-precipitates. It is reported that the cuboidal co-precipitates are more stable than single phase precipitates and are able to improve high temperature properties of Alloy 718 (13-20). The co-precipitates shold have the same effect in Alloy 706. The co-existed condition with large precipitates and fine ones may also contribute to the reinforcement of grain interior.

The contribution of the precipitation at the grain boundary is seen in the relative duration of tertiary creep to the whole creep life shown in Figure 19 that is replotted from Figure 15, 16 and 17. The relative tertiary creep duration is the greatest in the stabilizing temperature range of Group B. As seen in Table II, this improvement is mainly due to the reinforcement of grain boundary which is caused by the pinning of grain boundary by η, grain boundary serration, and indistinct denuded zone. Firstly, precipitates at grain boundary prevent effectively the grain boundary sliding. In Alloy 706, η phase has semi-coherency with the matrix and therefore effectively pins the grain boundary. Secondly, the grain boundary serration also prevents the grain boundary sliding effectively (21,22). That is, the geometrical change of grain boundary is considered to extend the tertiary creep duration. Thirdly, the indistinct denuded zone is considered to be retard effectively crack propagation as reported with Alloy 718 (23). In fact, this was supported by SEM micrographs of fracture surfaces shown in Figure 20. The fracture surface of Group B specimens consists of many micro-dimples in contrast to the smooth fracture surface of Group A specimens, although grain boundary fracture occurs in both groups.

Group C ; stabilized at 780 ℃

Creep rupture properties are extremely degraded in creep rupture life. However, creep rupture elongation is very large. The minimum creep

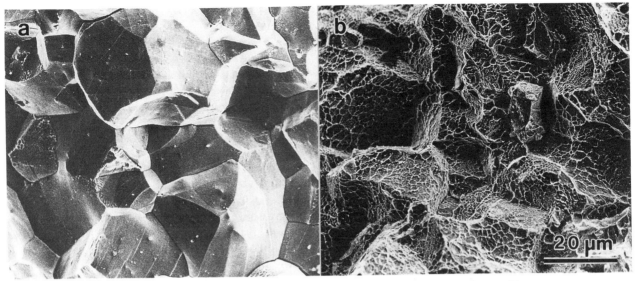

Figure 20 : SEM fractographs of ruptured Alloy 706 (a) unstabilized and (b) stabilized at 810℃.

rate is far larger than those of the other groups as seen in Figure 19. The relative tertiary creep duration is smaller than that of Group B as seen in Figure 20. These facts mean that there are two reasons for the degradation of creep properties.

As seen in Table II, the large cuboidal co-precipitate is dominant inside the grain in this stabilized temperature range. In general, the smaller the precipitates, the more effective they are in precipitation hardening and the more stable they are for heat. Therefore, the degradation of creep properties is attributed to the large precipitates dominant in the matrix. The wide denuded zone around η is another characteristic of the precipitation in this temperature range. It causes readily grain boundary sliding whereby overshadowing the beneficial effect of η precipitation.

Conclusions

Intensive TEM study and creep rupture tests were conducted in order to characterize the precipitates and to relate them to creep rupture properties for Ni-Fe-base superalloy Alloy 706 stabilized at various temperatures. Conclusions are summarized as follows.

(1) The precipitation behavior of Alloy 706 stabilized at 900, 870℃ is pracyically the same as in the case where the stabilizing treatment is discarded. That is, both fine γ'' and overlayed γ'-γ'' co-precipitate form in the grain matrix at the double-aging treatment, and no precipitation occurs at the grain boundary. Therefore, creep rupture properties is not affected by the stabilizing treatment.

(2) When stabilized at 840 and 810℃, the fine precipitates are still dominant inside a grain while large cuboidal γ'-γ'' co-precipitate is present together with them. At the same time, η phase precipitates at the grain boundary, causing serrated grain boundaries and narrow denuded zone around η. The best creep rupture properties are obtained with this complex stage of precipitation.

(3) In the case of Alloy 706 stabilized at 780℃, large cuboidal co-precipitate is dominant inside the grain, and the denuded zone iss very wide, leading to extremely degraded creep rupture properties.

It can be concluded that the desirable creep rupture properties are obtained when the best combination is achieved for the fine overlayed γ'/γ'' co-precipitates, large cuboidal γ'/γ'' co-precipitates, and η phase.

References

1. E.E.Brown and D.R.Muzyka, "Nickel-Iron Alloys", Superalloys II, ed., C.T.Sims, N.S.Stoloff, and W.C.Hagel (New York, John Willey & Sons, 1987), 165-188.

2. H.L.Eiselstein, "Properties of Inconel Alloy 706", ASM Technical Report, No.C 70-9.5 (1970), 1-21.

3. H.L.Eiselstein, "Properties of a Fabricable, High Strength Superalloy", Metals Engineering Quarterly, November(1971), 20-25.

4. E.L.Raymond and D.A.Wells, "Effects of Aluminum Content and Heat Treatment on Gamma Prime Structure and Yield Strength of Inconel Nickel-Chromium Alloy 706", Superalloys --Processing (Columbus, OH:Metals and Ceramics Information Center, 1972), N1-N21.

5. P.W.Schilke, J.J.Pepe, and R.C.Schwant, "Alloy 706 Metallurgy and Turbine Wheel Application", Superalloys 718,625,706 and Various Derivatives, ed., E.A.Loria (Pittsburgh, PA:TMS, 1994), 1-12.

6. Inconel 706 : Undated brochure obtained from The International Nickel Company, 1974.

7. J.H.Moll, G.N.Maniar, and D.R.Muzyka, "The Microstructure of 706, a New Fe-Ni-Base Superalloy", Metallurgical Transactions, 2(1971), 2143-2151.

8. J.H.Moll, G.N.Maniar, and D.R.Muzyka, "Heat Treatment of 706 Alloy for Optimum 1200°F Stress-Rupture Properties", Metallurgical Transactions, 2(1971), 2153-2160.

9. L.Remy, J.Laniesse, and H.Aubert, "Precipitation Behavior and Creep Rupture of 706 Type Alloys", Materials Science and Engineering, 38(1979), 227-239.

10. T.Takahashi et.al., "Effects of Grain Boundary Precipitation on Creep Rupture Properties of Alloy 706 and 718 Turbine Disk Forgings", Superalloys 718, 625, 706 and Various Derivatives, ed., E.A.Loria (Pittsburgh, PA:TMS, 1994), 557-565

11. K.A.Heck, "The Time-Temperature-Transformation Behavior of Alloy 706", ibid., 393-404.

12. G.W.Kuhlman et.al., "Microstructure - Mechanical Properties Relationships in Inconel 706 Superally", ibid., 441-449.

13. R.Cozar and A.Pineau, "Morphology of γ' and γ'' Precipitates and Thermal Stability of Inconel 718 Type Alloys", Metallurgical Transactions, 4(1973), 47-59.

14. E.Andrieu, R.Cozar, and A.Pineau, "Effect of Environment and Microstructure on the High Temperature Behavior of Alloy 718", Superalloy 718 - Metallurgy and Applications, ed., E.A.Loria (Pittsburgh, PA:TMS, 1989), 241-256.

15. J.P.Collier, A.O.Selius, and J.K.Tien, "On Developing a Microstructurally and Thermally Stable Iron - Nickel Base Superalloy", Superalloys 1988, ed., D.N.Duhl et.al. (Warrendale, PA: The Metallurgical Society, 1988), 43-52.

16. E.Gou, F.Xu, and E.A.Loria, "Effect of Heat Treatment and Compositional Modification on Strengthening and Thermal Stability of Alloy 718", Superalloys 718, 625 and Various Derivatives, ed., E.A.Loria (Pittsburgh, PA:TMS, 1991), 389-396.

17. E.Gou, F.Xu, and E.A.Loria, "Comparison of γ'/γ'' Precipitates and Mechanical Properties in Modified 718 Alloys", ibid., 397-408.

18. E.Andrieu et.al., "Influence of Compositional Modifications on Thermal Stability of Alloy 718", Superalloys 718, 625, 706 and Various Derivatives, ed., E.A.Loria (Pittsburgh, PA:TMS, 1994), 695-710.

19. X.Xie et.al., "Investigation on High Thermal Stability and Creep Resistant Modified Inconel 718 with Combined Precipitation of γ'' and γ''', ibid., 711-720.

20. E.Gou, F.Xu, and E.A.Loria, "Further Studies on Thermal Stability of Modified 718 Alloys", ibid., 721-734.

21. A.K.Koul and R.Thamburaj, "Serrated Grain Boundary Formation Potential of Ni-Based Superalloys and its Implications", Metallurgical Transactions, 16A(1985), 17-26.

22. H.L.Danflou, M.Marty, and A.Walder, "Formation of Serrated Grain Boundaries and Their Effect on the Mechanical Properties in a P/M Nickel base Superalloy", Superallys 1992, ed., S.D.Antolovich et.al. (Warrendale, PA: TMS, 1992), 63-72.

23. S.Li et.al., "The Effect of δ-Phase on Crack Propagation under Creep and Fatigue Conditions in Alloy 718", Superalloys 718, 625, 706 and Various Derivatives, ed., E.A.Loria (Pittsburgh, PA:TMS, 1994), 545-555.

DUAL ALLOY DISK DEVELOPMENT

D. P. Mourer*, E. Raymond** S. Ganesh*** and J. Hyzak****

*GE Aircraft Engines
Lynn, MA

**GE Aircraft Engines
Evendale, Oh.

***GE Power Systems
Schenectedy, NY

****Wyman Gordon
Worcester, Ma.

Abstract

A high integrity dual alloy disk (DAD) process was demonstrated under Navy contract funding (N00140-89-C-WC14). The forge enhanced bonding (FEB) joining process utilized for DAD manufacture demonstrated metallurgically clean bonds during scale up from test specimens through full scale disks. Use of advanced development alloys KM4 and SR3 for the dual alloy disk yielded mechanical properties meeting the program goals. Tensile strengths in excess of 1378Mpa(200 Ksi) were achieved at 649°C(1200°F) with creep capability demonstrated at up to 760°C(1400°F). Mechanical testing across the FEB joint resulted in failures in base metal(not the joint) with strengths/lives equivalent to base metal properties, confirming joint integrity.

Extensive use of finite element modeling of the FEB joining process and the subsequent heat treatment processes proved invaluable in shortening the development cycle as well as minimizing the overall technical risk. The modeling activity enabled definition of a preferred FEB joint configuration enhancing bondline material expulsion--critical for consistent joint integrity. The modeling also aided in defining an aggressive quench from solution heat treat to achieve mechanical property goals while avoiding quench related thermal cracking.

Technical Overview

A major challenge for any rotor system is to meet the overall goals of conflicting bore and rim mechanical property requirements. The bore requirements are driven strongly by tensile and lower temperature LCF requirements. The rim, however, requires creep and high temperature crack growth characteristics to meet the demands of this region of the disk. As rim temperatures increase, these demands become increasingly difficult to reconcile and challenge the metallurgist for a creative solution. A dual alloy approach is one that uniquely allows the bore and rim composition and microstructure to be tailored to meet the design requirements. GEAE's dual alloy disk process[1] was further developed and demonstrated on Navy contract funding (N00140-89-C-WC14)[2] as a means to meet the demands of advanced disk requirements.

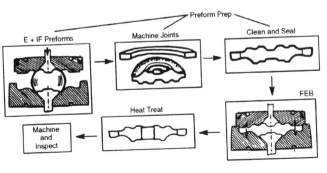

Figure 1. Process schematic of dual alloy disk manufacture via forge enhanced bonding. *A high integrity bond is assured by expelling joint material outside the component envelope.*

The potential payoff for a DAD has led to a number of prior studies over the years. A variety of approaches have been attempted and carried to varying levels of demonstration. Methods demonstrating at least partial success include HIP bonding [3,4,5,6] coextrusion and isothermal forging [7,8] and inertia welding [9]. Prior work was carefully reviewed during an exhaustive downselect to a final preferred method for DAD manufacture. A number of factors were utilized to rank candidate processes. Critical requirements included bond properties equal or greater to the weaker or the alloy pair, an inherently high integrity interface and a precisely controllable interface location.

The FEB process was selected as preferred by GEAE primarily for its inherent high joint integrity. This high integrity which is required in rotor quality material is insured by the high degree mechanical work at the joint and bondline material expulsion- both key factors in joint integrity. Additionally the FEB process is compatible with existing production facilities, has an inherently precise joint location, and is compatible with the planned supersolvus heat treatment approach.

A typical high pressure turbine disk (HPTD) shape was chosen as the full scale demonstration part configuration. The overall process selected involved assembling the bore and rim forging preforms together, followed by FEB of the preforms to physically join the preforms and yield a high integrity joint. This step was followed by annealing the dual alloy configuration above the gamma prime solvus (supersolvus) prior to aging to obtain optimum mechanical properties. An overall process schematic is shown in Figure 1.

The results of earlier work had selected KM4 and SR3, two advanced high strength superalloys[10] as the bore and rim alloys respectively. Compositions are shown in Table I.

Element (w/o)	KM4	SR3
Nickel	Balance	Balance
Cobalt	18	12
Chromium	12	13
Molybdenum	4	5.1
Aluminum	4	2.6
Titanium	4	4.9
Niobium	2	1.6
Carbon	.030	.030
Zirconium	.030	.030
Boron	.030	.015
Hafnium	0	0.2
% Gamma prime	54%	49%

Table I. **Nominal chemistries of alloys KM4 and SR3.** *These alloys were chosen as the baseline alloys for dual alloy disk manufacture.*

The extensive use of finite element modeling (FEM) to model the FEB and heat treat processes reduced the cost and time to develop the optimum process combination for subscale and full scale forgings as is discussed in the following sections.

FEB Process Modeling

The first area of modeling was the actual FEB process. It has been shown in the past that nickel base alloys similar to those in this study can exhibit abnormal or non homogeneous grain growth during supersolvus heat treatment[11] and this has been related to non superplastic deformation during isothermal forging. Since the intent was to supersolvus anneal the dual alloy disk, it was critical to determine the temperature/strain rate conditions that would render the alloys superplastic during both isothermal forging and the subsequent FEB operation.

As input to the FEM model of this process the specific heat, thermal conductivity, diffusivity, and coefficient of thermal expansion values were determined for each alloy as a function of temperature. In general these properties showed only a small variation between the alloys of this program. To define local strain rates required for superplasticity, flow stress specimens were used to establish the degree of superplasticity as a function of strain rate and temperature. The resultant flow stress data obtained in the range of 1093 to 1038°C (2000 to 1500°F) was used to establish target strain rates at or below about 0.01/second or 0.6/minute.

The alloy deformation behavior was used along with forging input data to develop local strain rate histories for the subscale and full scale FEB operations. The forging input data included forging ram strain rate, friction between the die and work piece, and heat transfer between the die and work piece. A typical FEM output is shown in Figure 2. Although the basic ram strain rate is 0.1/minute or 0.0017/second, the local strain rates in the FEB cavity can reach much higher strain rates, in this case up to 0.04/second. This information was used in a comparative manner by relating the resultant structures of the subscale plane strain forgings to the full scale HPTD and in absolute terms relative to the data from the flow stress specimens. In both cases FEB operations on subscale coupons performed in the right strain rate range yielded sound joints free of critical grain growth.

ALPID[12] was also used to examine the sixteen different cavity geometry's in a design of experiments study to establish correlation's and limitations between cavity parameters and FEB efficiency (as measured by the percent bond line removed per unit volume of metal expelled, and the forging loads for die fill). Forging loads and die stresses were part of the ALPID output, and draft angles to ensure part removal were based on prior experience. Prior FEM analysis indicated the symmetric die cavity design to be most efficient in terms of percent bond line removed per unit volume of metal expelled into the die cavity. As a result the study was based on a symmetric FEB cavity design.

A step-wise regression analysis was performed to establish the significance of correlation's between cavity parameters and percent bond line removed. Additionally, the effect of cavity geometry on forging load, die stresses, and part removal from the die was also considered in determining the most effective die cavity geometry. The conclusion was that for a given cavity volume, the most efficient bond line removal was a function of both the cavity width and entrance radius to the cavity and that die stresses and forging loads were compatible with the desired high degree of bondline material expulsion.

FEB process modeling using ALPID was also carried out on a full scale high pressure turbine disk geometry to study the effect of using bore and rim alloys with different flow stresses on the resultant FEB process. The objective was to evaluate the robustness of the FEB process and determine if FEB could be safely used to bond materials with flow stresses varying ≥2.5X, in this case, 6 and 16 ksi. The results of the model, whether the high flow stress material was the bore or rim, was that a successful joint could be made. The interface shapes within the sonic envelope of the forging were very similar, and in either case 82% of the material was expelled into the FEB ribs in one operation. Two FEB operations remove over 95% of the original bond line. The results of the model and a physical validation using materials with different flow stresses are shown in Figure 2. Although some of the fine detail of the joint interface was not precisely reproduced, the macro shape and other key factors such as amount of material expulsion- critical to joint integrity were in good agreement. This result coupled with the knowledge that the preferred alloys would have more similar flow stresses than that modeled in this extreme case further added confidence that analytical modeling would reasonably predict full scale FEB behavior. Additionally it indicated that relative flow stress variations would not limit the process.

Results of prior work had examined the effects of preform joint preparation, canning and sealing the assembled preforms and actual FEB cycles on the final bond integrity. These studies were used to select improved preform joining techniques for the program. The effectiveness of the FEB process in bond line cleanup was substantiated by bonding oxidized preforms (not cleaned or canned) in multiple FEB operations.

Test results using plane strain forging showed that even with preoxidized unbonded preforms, the FEB process can achieve excellent clean up and a sound joint. Tests were performed with a symmetric FEB angle and 2X and 3X FEB operations and the results are shown in Table II.

Initial State	Joint Angle	No. of FEB Steps	UTS Mpa	0.2% YS Mpa	Elong %	RA %	Failure Location
Oxidized	0°F	2X	1359	1104	9.5	8.7	Bond Line
Oxidized	0°F	3X	1580	1111	20.7	18.6	Base Metal
Oxidized	45°F°	2X	--	1152	2.5	1.0	Bond Line

Table II. Test results on pre-oxidized FEB coupons. *Excellent bonding is demonstrated even with this gross level of intentional bond line contamination.*

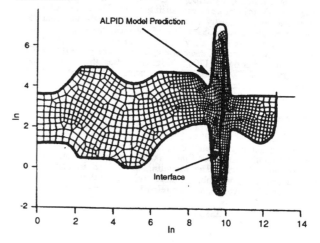

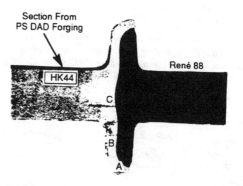

Figure 2. Mathematical model (upper) and physical verification (lower) of FEB material flow. *Excellent agreement is demonstrated validating the model fidelity even with a ≥2.5X differential in material flow stresses.*

The 0° (vertical) FEB cavity was determined to be an effective joint angle in expelling the material at the original interface. The process ends itself to rework or component salvage via use of a 3X FEB versus a 2X FEB or 1X FEB if the interface is severely oxidized. This should not be required in routine production but rework potential is a necessary consideration in any production environment. It was determined that bondline quality can be deduced from the quality of the material expelled into the FEB cavity; if the expelled bondline material is contamination free, the internal joint will be also.

Full scale FEB Modeling

The FEB cavity location and dimensions for the full scale forgings were finalized at 24.1 cm(9.5") radial location. This selection was based on the subscale modeling results, plane strain and mini disk forging results, and discussions with die design personnel.

To ease part removal from the die after FEB and minimize stress concentration at the cavity root, a decision was made to build segmented dies for the full scale forgings. This required the parting of the die block at the cavity center line and the fitting of a separate die block insert. The value of such design was demonstrated in the mini disk forging trials and in full scale die stress analysis. Modeling indicated that a fully segmented design could cause significant radial displacement and, plastic deformation of the top die punch where it contacts the bottom die along the die wall. Based on the analysis a decision was made to use a partially segmented top die and a fully segmented bottom die. A cross section of the top and bottom dies is shown in Figure 3. Finite element modeling of the FEB process is illustrated in Figure 4.

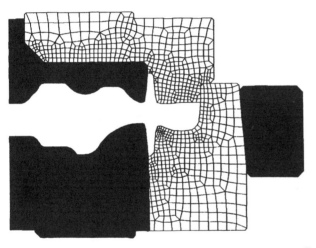

Figure 3. Cross section of the die stack finite element model. *This portion of the study was used to verify the acceptability of the preferred segmented die concept.*

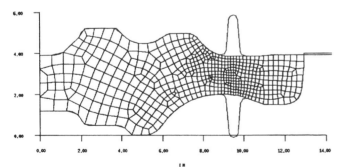

Figure 4. Finite element model results at the start of the FEB operation. *This model was used to validate the desired joint flow characteristics.*

The pre FEB shape is shown in Figure 4, and the final effective strains in the FEB area are shown in Figure 5. These models compared the predicted deformation of the full scale disk with the subscale mini disk and the plane strain forgings. The results indicated there would be sufficient preform interface removal and residual strain in the bond line to produce a clean bond. The analysis confirmed the subscale coupon study findings that 82% of the original bond line would be removed in one FEB operation and 95% would be removed by two FEB operations.

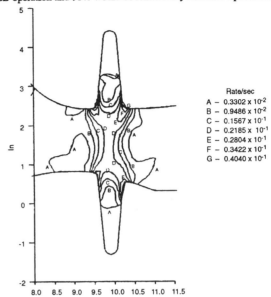

Figure 5. Finite element model results for the FEB operation. *The predicted strains were utilized to assure desired material flow in the joint region was achieved*

Heat Treat Modeling

The objective of this phase of the study was to evaluate the feasibility of a standard supersolvus heat treatment to meet bore, rim, and joint property goals without causing quench cracking, excessive residual stress, and/or abnormal grain growth. Heat transfer and thermal stress FEM modeling was used in conjunction with experimental trials to evaluate the alternate techniques[13,14,15,16]

A focus of FEM HT modeling was the solution annealing quench process. Before thermal modeling in the full scale high pressure turbine disk (HPTD), the NIKE/TOPAZ[17,18] computer programs from Lawrence Livermore Laboratories were used to model several simpler configurations in order to validate these programs for this application. The stress distributions and deformations in cylinders under the action of internal and external pressures, with and without interference, with and without thermal gradients were modeled. The results were in good agreement with the exact closed-form analytical solutions. With the programs thus validated, the full scale thermal modeling was initiated.

As a class these high strength, high gamma prime content alloys typified by SR3 and KM4 are prone to thermal stress cracking upon cooling from the supersolvus anneal[13]. This occurs because the thermal stress generated near the surface of a part during cooling due to the extreme thermal gradients exceeds the tensile strength of the alloy at a point when it also has a low ductility. On cooling tensile testing for the two alloys was performed to determine limits for quench rates. This data established the relative strength and ductility at various temperatures as the alloy cools from solution annealing and provided a comparative limit versus quench related stresses predicted by modeling. This approach has been previously described[13]

Preliminary studies had indicated that a minimum cooling rate of about 121°C(250°F/)minute was needed for the range of 1093 to 816°C(2000 to 1500°F) to meet the strength and creep goals. It was shown that as the cooling rate increases the hold time fatigue crack growth rate of the alloys tends to increase, but this varies with alloy. Tensile and creep properties have also been linked to quench rate variations[19]. It was also shown that residual stresses increase with increase in cooling rate, so the cooling rate distribution in the part must be kept in control.

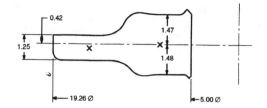

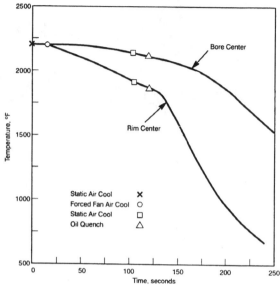

Figure 6. Heat treat trial results simulating the combination static air/fan air/oil quench procedure. *This data is representative of the final route chosen for dual alloy disk manufacture.*

Preliminary trials were conducted using a thermocoupled high pressure turbine disk of the approximate shape as the full scale dual alloy disk to determine what cooling rates could be expected with normal type quenchants. One such trial used a cycle of heating the disk to 1204°C (2200°F) in a furnace, transferring to a high velocity fan facility in 15 seconds, cooling in the fan for 90 seconds, transferring to an oil quench tank in 15 seconds, and oil quenching to room temperature. The cooling rates at the rim center and bore center were established in particular for the range of 1093 to 816°C (2000 to 1500°F) as shown in Figure 6. The cooling rates for the bore locations varied from 111-183°C(200-330°F/)minute and the rim values ranged from 111-444°C(200-800°F)/minute for this particular trial.

As noted the cooling rates increased to very high values towards the surface. Although the required nominal 121°C(250°F/)min quench rates could be met a particular concern was the high surface quench rates approaching 427°C(800°F)/min. Heat treat analysis and physical trials established that this quench rate would result in stresses that would exceed the material strength and cause thermal quench cracking. Thus cooling from the supersolvus temperature must be controlled to avoid thermal cracking problems. Prior work suggested that this surface quench rate/high stress condition could be mitigated by canning the part to reduce near surface quench rates. This was verified by both FEM and coupon trials to substantiate selection of the final detailed heat treat parameters. With quench cracking risk minimized the basic heat treat approach taken was similar to that outlined in Figure 6. Also the heat treat fixture designed specifically for the part has supports which were designed to minimize creep of the part at high temperature and eliminate any possible dead zones where oil might be trapped during the oil quench and provide an slower than desired quench.

The cooling tensile data along with FEM enabled the selection of the cooling rate necessary to satisfy mechanical property requirements while avoiding thermal stress. The effect of solution temperature and time on grain size, and incipient melting was studied taking into consideration that the normal production furnaces are presently capable of a tolerance of ±8°C(15°F). This data along with the on cooling temperature profiles and the resulting thermal stresses from FEM modeling allowed the design of an air, fan, and oil quench procedure in combination with canning the surface of the part with stainless steel to obtain desired mechanical properties without thermal cracking the part..

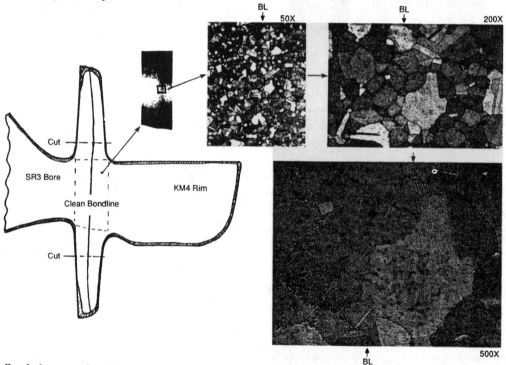

Figure 7. Metallurgical cross sections through the joint region of the dual alloy disk. *The structures observed confirmed the high degree of integrity or joint cleanliness as predicted by prior modeling.*

Demonstration Disk Manufacture

The general process used by Wyman Gordon in making the full scale disks is illustrated in Figure 1. Bore and rim preforms were isothermally forged. Fabrication of the near net shape can used during the FEB operation was performed by spinning over a mandrel. Using a template of the dimensions of the preform, the can was dimensionally inspected. After thorough cleaning the rim was lowered on to the bore and the assembly was lowered into the bottom can. The assembly was placed on to a welding turn table and the top can placed on the assembly. First the can was evacuated through the fill stem and then the desired vacuum in the interstice was maintained by crimping the fill tube. The assembly was then heated and moved into the isothermal press, the joint isothermally forged, and the finished forging ejected from the FEB dies with the knockout.

After each of the FEB operations both the top and bottom ribs (or expelled joint material) were sectioned and examined metallographically for joint integrity. Typically this examination was done at the top and base of the rib, and was shown to be a highly sensitive nondestructive measure of bond integrity. For assurance this prototype part was 2X FEB'ed to insure joint integrity. In addition to removing the ribs, preparation for heat treat included machining a four inch diameter center hole into the disk. The disk was recanned with a stainless steel near net shape can followed by a HIP cycle to make the assembly fully integral. The solution anneal was at 1188°C(2170°F) for 2 hours, with a transfer to a high velocity fan and fan cooling. This step was followed by a transfer to an oil quench bath and the forging was oil quenched to room temperature. Several dry runs were made with the equipment and procedures that culminated in successful heat treatment of the final demonstration disk per the intended practice.

After can removal and fluorescent die inspection for surface cracks the disk was then sectioned for full structural inspection and mechanical property determinations. After 2X FEB both the top and bottom of the top rib was clean as shown in Figure 7. The contour of the bond line in the 2X FEB ribs are also shown in Figure 7 along with the micro appearance of the bond line near the base of the top and bottom rib. The location from which the sample was taken as well as the structure at 50X, 200X, and 500X is shown. This established that the desired high integrity, defect free joint was obtained. In all cases the bond was clean and acceptable for finish operations and subsequent evaluation.

Figure 7 shows a picture of a macro slice taken from the DAD. The bond line is linear axially oriented, and in the exact radial location predicted. Figure 8 shows a montage of photographs at about 100X illustrating the grain sizes after heat treatment at various locations in the disk. The grain sizes were all in the desired ASTM 7-9 range. Higher magnification pictures of the structures in the same areas are shown in Figure 8. No large intragranular gamma prime precipitation is seen indicating that the solution temperature was supersolvus as planned. Figure 8 shows the structures of the grain boundaries in the various locations at higher magnification obtained in the high resolution SEM. Grain boundary precipitates of gamma prime, carbides, and borides can be seen in the boundaries, but not copious amounts. This along with the gamma prime size and shape also shown in Figure 8 indicates that the desired cooling rate was obtained during heat treatment.

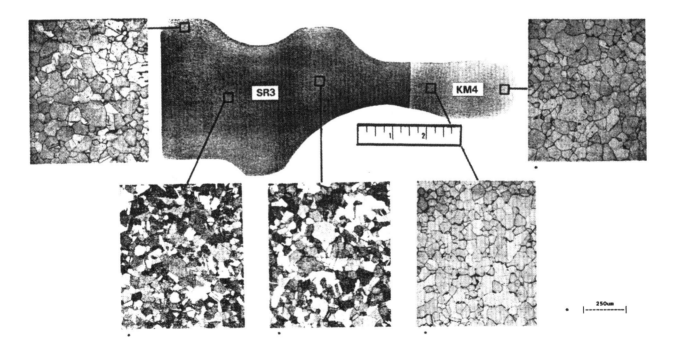

Figure 8. Micrographs illustrating grain size consistency throughout the dual alloy disk forging. *These grain sizes met the desired aims.*

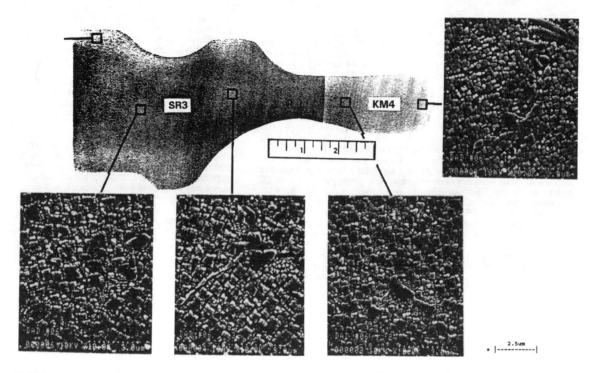

Figure 9. Micrographs illustrating gamma prime size consistency throughout the dual alloy disk forging. *These gamma prime sizes confirmed the aim cooling rates had been met.*

The disk was sectioned for mechanical property determinations. Tests were taken in the bore alloy, rim alloy, and across the joint. Figure 10 shows the tensile results relative to the high temperature goals of the program. Figure 10 shows that the rim alloy and the joint exceeded the tensile goals of the program, but the bore alloy was somewhat below goal.

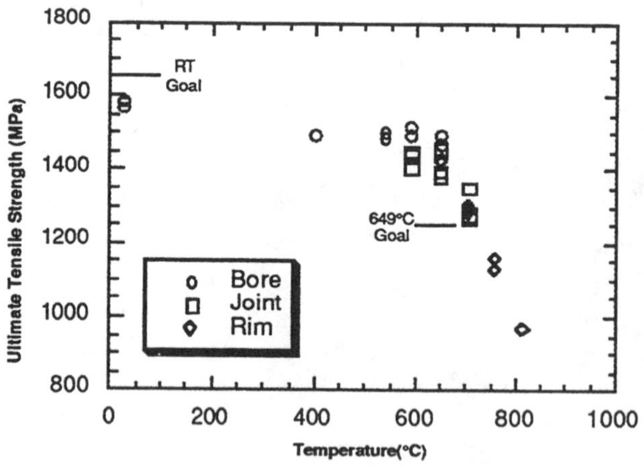

cases where joint material was tested, the failure occurred in the base metal with results within the scatter for the bore alloy and the rim alloy.

The analysis of both the structure and mechanical properties of demonstration disk attests to the capability of the FEB process to produce an excellent dual alloy disk.

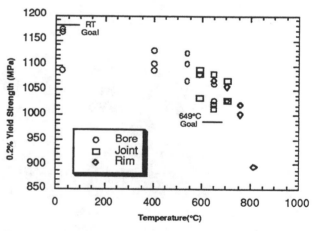

Figure 10. Ultimate tensile strength data for the dual alloy disk. *The strength met goals and was similar to single alloy forgings similarly processed. No loss in strength was observed in the joint region.*

Similar results were obtained with the 0.2% yield strength data. Additionally the notch tensile results indicate that all areas are in a notch strengthened condition.

The creep results relative to the goals are shown in Figure 12. Both the bore and rim creep results exceeded the goals of the program. In all

Figure 11. 0.2% yield strength data for the dual alloy disk. *The yield strength met goals and was similar to single alloy forgings similarly processed. No loss in strength was observed in the joint region.*

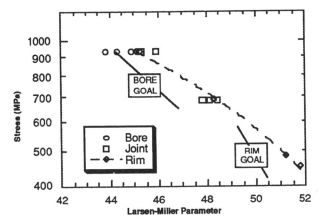

Figure 12. 0.2% creep data for dual alloy disk. *The creep strength met goals and was similar to single alloy forgings similarly processed. No loss in strength was observed in the joint region.*

Summary

A dual alloy disk of SR3 bore-KM4 rim was successfully forged using the FEB process to yield a full scale high integrity dual alloy shape. The established preform preparation process yielded a fully defect free bond line.

The dual alloy disk was successfully heat treated using the baseline treatment developed and incorporation of the canning protection to yield a disk free of thermal stress cracking, and meeting the structures and mechanical properties expected from SR3 and KM4 heat treated in a similar manner.

The combination of FEM modeling and subscale validation of key processing steps greatly accelerated the development effort and reduced the overall program risk.

Acknowledgments

This program's success was achieved by the support number of people at both Wyman Gordon and GE Aircraft unfortunately space permits mention of only a few. Ron Tolbert of GEAE is thanked for his tireless support of this effort. Key contributions by Mike Henry of GE Corporate Research were crucial to the success of the program. Modeling support critical to success was provided by Hugh Delgado and Ram Ramikrishnan of Wyman Gordon. Finally the contractual support of the US. Navy and personal support of the Navy DAD program manager, Andrew Culbertson are gratefully acknowledged.

References

1. J. M. Hyzak et al, U.S.Patent 5,161,950, "Dual Alloy Disk System" April 21, 1992

2. S. Ganesh et al, "Advanced Technology for Turbine Disk Materials (Dual Alloy Disk)," Final Report, September 30, 1994, GE Aircraft Engines, Cincinnati, Ohio. Navy Contract No. N00140-89-C-WC14

3. Weaver, D.M. and Reichman, U.S. Patent 4,063,939, "Composite Turbine Wheel and Process For Making Same"

4 Cross, K.R. U.S. Patent 4,096,615, "Turbine Rotor Fabrication"

5. Clark, J. et al, U.S. Patent 4,659,288, "Dual Alloy Radial Turbine Rotor with Hub Material Exposed in Saddle Regions of Blade Ring"

6. Ewing, Bruce, "A Solid -To-Solid HIP-Bond Processing Concept For The Manufacture of Dual -Property Turbine Wheels for Small Gas Turbines" ,in Superalloys 1980, ed. by John K. Tien et al., TMS–AIME, Warrendale, PA, 1980, pp.169-178

7. Hughes, Anderson and Athey R.L., "Fabrication and Heat Treatment of a Ni-base Superalloy Integrally Bladed Rotor for Small Gas Turbine Applications" Modern Developments in Powder Metallurgy, Volume 14, Special Materials, June 22, 1980.

8. Miller, J.A et al, U.S. Patent 4,479,293 "Process for Fabricating Integrally Bladed Bimetallic Rotors"

9. Stalker, K.W. and Janke, L.P. "inertia Welded Jet Engine Components" ASME paper 71-GT-33, 1971

10. D.D. Krueger, J.F. Wessels, and K.M. Chang, U.S.Patent 5,143,563, "Creep, Stress Rupture and Hold Time Fatigue Resistant Alloys" September 1, 1992

11. J.Y. Guide, J.C., and Y. Honoree, "N18, Powder Metallurgy for Disks, Development and Applications," J. Mat. Eng. and Performance, 2, 1993, pp. 551 and 556.

12. S.I. Oh, "Finite Element Analysis of Metal Forming Processes with Arbitrarily Shaped Dies" International Journal of Mechanical Science, Vol. 24, 1982, PP 479-493

13. R. A. Walls and P.R. Bhopal, "Property Optimization in Super alloys Through the Uses of Heat Treat Process Modeling" in Superalloys 1988, ed. by S. Reichman et al., TMS–AIME, Warrendale, PA, 1988.

14. T.E. Howsen and H.E. Delgado, "Utilization of Computer Modeling in Superalloy Forging Process Design," in Superalloys 1988, ed. by S. Reichman et al., TMS–AIME, Warrendale, PA, 1988.

15. J.M. Franchet et al., "Residual Stress Modeling During the Oil Quenching of An Astroloy Turbine Disk," in Superalloys 1992, ed. by S.D. Antolovich et al., TMS–AIME, Warrendale, PA, 1992, pp. 73–82.

16. R.I. Ramikrishnan, "Quench Analysis of Aerospace Components Using FEM," in Proceedings of the First Int. Conf. on Quenching and Distortion Control, Chicago, 22–25 September 1992, p. 235–242.

17. A. B. Shapiro, "TOPAZ 2D -A Two -Dimensional Finite Deformation, Finite Element Code for Heat Treat Analysis, Electrostatic, and Magenetostatic Problems" (Report No. UCID-20824, July 1986).

18. J.O. Hallquist, "NIKE 2D- A Vectorized, Implicit, Finite Deformation, Finite Element Code for Analyzing the Static and Dynamic Response of 2-D Solids" (Report No. UCID-19677, Rev 1, December 1986).

19. P.R. Bhowal, E.L. Wright, and E.F. Raymond, "Effect of Cooling Rate and γ' Morphology on Creep and Stress–Rupture Properties of a Powder Metallurgy Superalloy," Met Trans., 21A, pp. 1709–1717.

DEVELOPMENT OF HIP CONSOLIDATED P/M SUPERALLOYS FOR CONVENTIONAL FORGING TO GAS TURBINE ENGINE COMPONENTS

Gernant E. Maurer, Wayne Castledine,
Frederick A. Schweizer, and Sam Mancuso

Special Metals Corporation
New Hartford, NY 13413

Abstract

Recent investigations have identified thermal-mechanical processing conditions which significantly enhance the hot workability of P/M nickel-based superalloys. The process incorporates hot isostatic pressing (HIP) of powder at temperatures just below the solidus of the alloy. The hot ductility produced by this sub-solidus HIP (SS-HIP), has been found to be superior to that of powder metallurgy (P/M) superalloys that were either conventionally HIPed or extrusion-consolidated. The benefits of this new process have the potential of reducing processing costs and producing components with superior microstructures and properties.

The purpose of this study was to understand the mechanisms which are responsible for the excellent hot workability observed. A series of test samples was prepared from UDIMET* Alloy 720 material that was prepared using cast/wrought, P/M as-HIP, P/M as-extruded and PM SS-HIP processing. High-strain rate tensile tests were used in an attempt to quantify the differences in forgeability. Microstructures were examined to gain insight into the mechanisms envolved.

This study provides convincing evidence that SS-HIP P/M processing can produce billets of highly alloyed superalloys, such as UDIMET*Alloy 720 and Rene' 95, that can be subsequently thermomechanically processed by conventional open die forging or ring rolling. SS-HIP P/M offers the unique combination of cast wrought workability and PM homogeneity including freedom from cast/wrought defects such as freckles and white spots. Due to this unique combination of properties, SS-HIP P/M parts costs should compare favorably to cast/wrought while offering the advantage of ultrasonic inspectability similar to PM parts. An additional benefit is the potential for super-solvus heat treatment grain size control not possible in cast/wrought or conventional P/M products.

* UDIMET is a registered trademark of Special Metals Corporation

Introduction

Historically, P/M superalloy billets have not been amenable to high-strain rate deformation and, therefore, required P/M billets to be isothermally forged under superplastic strain rate conditions. The ability to use conventional forging methods to make P/M billets will result in lower conversion costs for production of the billet and the subsequent closed-die forged or ring-rolled components. In order to be able to conventionally forge a P/M superalloy, a process had to be identified that minimized the effects of prior particle boundaries (PPB's) that are known to have a strong influence on their mechanical properties.

The effect of PPB's in P/M superalloys has been extensively reviewed by Thamburaj and colleagues[1]. His conclusions are that PPB's are caused by carbon segregation and particle contamination. Both result in precipitation of PPB's, either as carbides, oxides, oxy-carbides, or possibly as oxy-carbonitrides. The volume of PPB's is dependent on many things including the atomization method, particle size, pre-consolidation handling and consolidation parameters. The cooling rate after consolidation and post consolidation heat treatments were also noted as factors affecting PPB's.

Much work has been done to understand and eliminate the effect of (PPB's) in PM products. The determental effects of PPB's on mechanical properties is well documented[1][2]. It is generally agreed that as-atomized powder surfaces contain carbides, borides, nitrides, and oxides. It is theorized that these precipitates may prevent optimum diffusion bonding during solid state consolidation and that the oxides are sites for carbide and boride nucleation and growth during heat treatments[1]. Whether the cause is a boundary heavily decorated with precipitates or remnants of oxide films, studies have shown that PPB's are a weak link in the microstructure[2].

Throughout the development of P/M superalloys many alloy and process development efforts found ways to reduce the effects of PPB's. Alloy designers have attempted to reduce the powder particle surface decoration by reducing carbon and nitrogen and using higher HIP temperatures. The benefits of putting PPB's into solution were offset by the grain growth that occurred. Hack et al[3] investigated HIP consolidation of a low carbon Astroloy alloy at temperatures above and below the boride solvus (1220^0C) of the alloy. The investigators identified the disadvantages of the higher HIP temperatures to be many. In addition to the difficulty in controlling the precise HIP temperature, they sited continuous grain boundary films upon subsequent precipitation of carbides and borides. Grain growth up to ASTM 4½ impeded ultrasonic inspection and reduced yield strength. Reduced forgeability was reported and attributed to the larger grain size and grain boundary films.

Aubin et al [4] looked closely at the particle surfaces from both argon atomized and rotating electrode atomized Astroloy. In the gas atomized material they found that oxycarbides were predominantly MC in nature and were generally very stable during heat treatment. The material that was not gas atomized contained less PPB's and more M_6C carbides that were less prone to form oxy-carbides. During the initial stages of HIP consolidation the formation of oxy-carbides can intensify before interdiffusion between powders occur. Aubin et al[4] also suggested, that during thermal treatment, sulfur would also have the opportunity to segregate to the free surfaces. The roles the sulfur would play and the phases that it would form are unclear. Since sulfur has been known to be a strong agent in reducing hot ductility, its role cannot be overlooked.

Jeandin et al [5] investigated the potential of liquid phase sintering of P/M Astroloy followed by HIP and forging. The forging results compared favorably with conventionally HIP processed material. Grain growth in material that has liquid phase present led to increased grain growth compared to conventionally HIPed material.

Grain growth in powder alloys is relatively restrained compared to cast/wrought material. Grain growth in fully annealed cast/wrought UDIMET Alloy 720 occurs as soon as the gamma prime solvus is reached and can readily reach grain sizes of ASTM 0-00[6]. P/M material, on the other hand, experiences minor grain growth after the gamma prime is in solution, limited apparently by the prior particle boundaries[7][8]. This has been suggested since the resultant grain size is approximately equivalent to the average powder mesh size. When P/M materials have been extensively worked grain sizes of up to ASTM 1-2 are possible[8].

Lu et al [9], studying the effect of HIP temperature and powder process method, showed that material with less PPB precipitates (rotating electrode processed (REP) powder) could recrystallize beyond the PPB. Their work, conducted significantly below the alloy solidus, showed no grain growth for argon atomized (AA) powder HIPed above the gamma prime solvus. His observation was that the AA powder had more decorated PPB's compared to REP powder. The carbon and oxygen levels in the AA powder were higher compared to the REP powder.

Extrusion consolidation is also used as a means to break up powder surface structures and to enhance subsequent diffusion[2]. Extrusion can produce a consolidated product with uniform fine grains but, subsequent working can only be achieved by isothermal forging due to limited high strain rate ductility and extensive cracking. The cause for this has been attributed to the PPB's.

The purpose of this study was to investigate the benefits of HIP processing a P/M superalloy at temperatures just below the solidus to enhance the workability of the HIP consolidation[11]. In general it is believed that by growing grain boundaries past PPB's and by rendering the remnant PPB as dispersed as possible, P/M materials will behave more like cast/wrought materials with the added benefit of not having the segregation problems associated with conventional materials.

Experimental Procedure

Powder for all of the P/M material investigated originated as vacuum induction melted (VIM) filtered master heats to ensure chemistry and then VIM and filtering prior to argon atomization. Powder was processed under argon throughout sieving (-150 mesh) and then vacuum can filled in 304 stainless steel. At the conclusion of the HIP cycles, material was cooled under pressure in the autoclave. Thermocouple readings from the HIP cycles were believed to be reading 16°C high and therefore temperatures reported here have been corrected accordingly.

HIP/cogged material was made from 460mm dia. compacts that were forged on a 4000 metric ton hydraulic press to 300, 230, and 150mm dia. billets. P/M HIP/extruded material was HIP processed at 1121°C to 460mm dia. and then pushed through a 254mm die producing approximately a 3:1 reduction ratio.

Cast/wrought material was made from VIM electrodes that were electroslag remelted and then vacuum arc remelted to 508mm dia. Ingots were forged on a 4000 metric ton press to 165mm dia. billet.

High strain rate tensile ductility testing was performed on standard threaded tensile bars with a 25.4mm gauge length and a diameter of 6.4mm. Test bars were heated to the test temperature, held for 15 min. and then pulled at 50 mm/mm/sec (2 in./in/sec).

An ultrasonic inspection technique which included a high sensitivity longitudinal scan and two opposing circumferential shear wave tests was used to inspect the various P/M billet forms. The technique was capable of detecting flaws as small as .08 mm in P/M material. Cast/wrought material was inspected utilizing a longitudinal scan to a No.2 (0.8mm) flat bottom hole.

Rolldowns were made from samples (50x50x100mm) that were rolled longitudinally at 1093°C in increments of 2.5mm per pass with 5 minute reheats between passes.

Results and discussions

HIP and grain growth

The effect of grain growth as a function of HIP temperature of P/M UDIMET Alloy 720 is presented in Figure 1. While only a limited number of temperatures were tested, the results suggest grain growth to occur in stages depending on the sequence at which barriers go into solution. As the HIP temperature increased, the first rapid change in grain size occurred at the gamma prime solvus (~1150 °C). As soon as the gamma prime is in solution, grain sizes increase from ASTM 8 to ASTM 5.5. The microstructure of material HIPed at just above the gamma prime solvus (Figure 2) suggest that the very fine powder particles have grown beyond their prior particle boundaries and that the larger powder particles are still restraining grain growth. Many of the large particle areas still have multiple grains within them. Grain growth up until 1254°C appears to be restrained by the average particle size in the microstructure (~270 mesh). Above 1254°C and to temperatures below the incipient melting point, rapid grain growth occurs beyond the prior particle boundary of the large particles to grain sizes of ASTM 3-4. Metallographic evidence of the material HIPed just below the solidus temperature (Figure 3 and 4) clearly shows that grain growth has occurred beyond the PPB's which are still clearly decorated in the intragranular microstrucure.

SS-HIP treatments applied to other widely used P/M superalloys appear to respond similarly in terms of grain growth (Figure 5). UDIMET Alloy 720, Rene´88 and Rene´95 all exhibited similar microstructures. Interestingly, the same remnants of PPB's can be seen in Rene´ 88 and Rene´ 95. The latter two alloys contain niobium (Table I) where UDIMET Alloy 720 does not. One might suspect that if the dissolution of a MC carbide is responsible for grain growth during the SS-HIP treatment then the generally more stable NbC precipitate would inhibit the reaction. Another suggestion is that

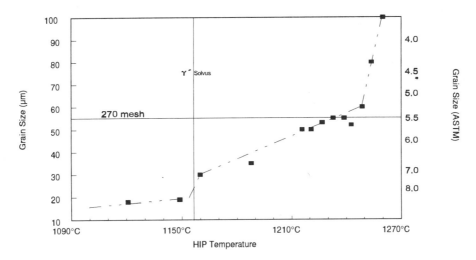

Figure 1: Grain size of hot isostatically pressed P/M UDIMET Alloy 720 at selected temperatures ranging from below the gamma prime solvus to the solidus.

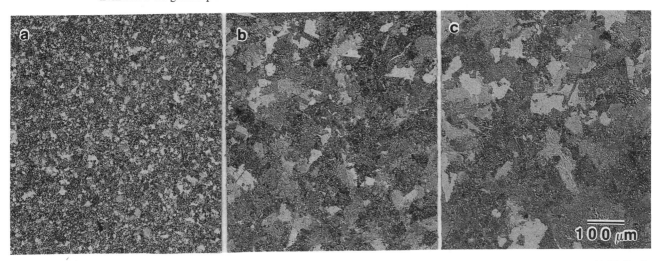

Figure 2: Typical microstructures of hot isostatically pressed UDIMET Alloy 720. Material is in the as-HIPed condition and etched with Kalling's reagent. a) 1149 °C b) 1188 °C c) 1254 °C

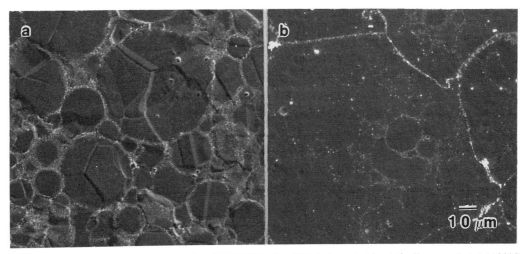

Figure 3: Scanning electron micrographs of UDIMET Alloy 720 that was hot isostatically pressed at a) 1188°C b) 1254°C. Samples were given a 1204°C solution to eliminate coarse cooling gamma prime. (Material was electrolytically etched in 5% HCL and methanol).

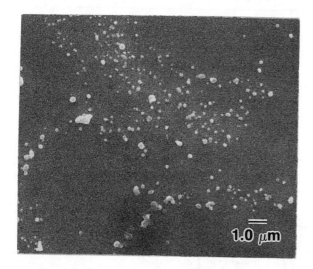

Figure 4: Scanning electron micrographs of UDIMET Alloy 720 that was hot isostatically pressed at 1254 °C Sample was given a 1204°C solution to eliminate coarse cooling gamma prime. (Material was electrolytically etched in 5% HCl and methanol.)

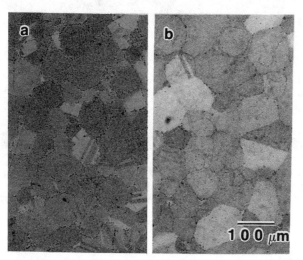

Figure 5: Micrographs comparing the SS-HIP microstructures of various P/M superalloys. a) Rene´ 88 (1254 °C) b) Rene´ 95 (1260 °C)

the nitrides (TiN) and carbo-nitrides in all three alloys would be similar and that all three alloys would experience similar dissolutions of the phase at the SS-HIP temperatures. Until some detailed analysis of the powder surfaces is documented (the types and amounts of precipitates), it will be difficult to precisely identify the mechanism.

The results suggest that PPB's of the fine particles are not sufficient to be a barrier to grain growth at temperatures just above the gamma prime solvus. However, the coarser PPB's of the larger particles are large enough and densely distributed enough to prohibit grain growth until about 1254°C when the desired grain growth occurs. While it is difficult to document the mechanism for the sudden grain growth, it may be that a portion of the PPB structure is dissolving sufficiently to allow grain growth to proceed. The components of phases that are found in PPB's include carbides, nitrides, borides and oxides (Al, Zr and Mg). It seems unlikely that the oxides go into solution but certainly MC carbides and possibly nitrides are candidates. Scanning electron microscopy of the PPB areas (Figure 4) reveal Ti rich phases in the PPB's at 1188°C which may have been precipitated during cooling of the samples and probably include carbides and carbo-nitrides. Extraction and x-ray diffraction analysis indicates that the remnant PPB areas of samples HIPed at 1254°C are dispersions of fine discrete particle thought to be titanium carbo-nitrides and zirconium oxide. The very fine oxides are not in the form of continuous films as are thought to be present on the as-atomized powder. Further work is needed to determine if the oxides become spheroidized during exposure at the SS-HIP temperature. If continuous films originally existed, then their breakup could be a mechanism that is contributing to the enhanced high strain rate ductility.

Another possible mechanism for the grain growth observed is that the SS-HIP temperature may be high enough to provide the energy needed to grow the grains beyond the pinning PPB's. Below this critical temperature less grain growth is seen. The growth of the smaller grains at lower temperatures would be explained by Kissinger et al[10] who suggested that smaller powder particles receive more deformation during HIP compared to the larger particles. The higher deformation imparted to these particles could explain why they recrystallize at a lower temperature. However, if the smaller particles were to have fewer oxides associated with them due to more rapid solidifcation and cooling they would have less resistance to grain growth also.

The microstructure in Figure 3 clearly shows how the SS-HIP treatment has caused grain growth well beyond a number of PPB regions. The remnant PPB areas contain phases that are extremely fine and relatively dispersed. It is unlikely that these areas are detrimental mechanically and could be areas that are locally dispersion strengthened. Since the high strain rate ductility of the SS-HIP material improves by at least 50% once the structure is worked and recrystallized into a fine grain structure the presence of remnant PPB areas does not appear to result in the material reverting back to a situation where the grain boundaries are coincident with detrimental PPB structures.

Table I Nominal Composition of Various P/M Superalloys (wt%)

Alloy	Ni	Co	Cr	Mo	W	Nb	Al	Ti	C	B	Zr	Other
UDIMET Alloy 720	Bal	15.2	16.0	3.0	1.25	-	2.5	5.0	.025	.018	.03	-
Rene´ 88*	Bal	13.0	16.0	4.0	4.0	0.7	2.2	3.75	.05	.015	.05	-
Rene´ 95	Bal	8.0	13.0	3.5	3.5	3.5	3.5	2.5	.06	.010	.05	-
IN-100**	Bal	16.0	12.5	2.8	-	-	5.0	4.3	.07	.02	.06	.9V
UDIMET Alloy 700	Bal	17.0	15.0	5.0	-	-	4.0	3.5	.03	.02	.02	-

* Rene´ is a registered trademark of the General Electric Company.
**Inconel is a registered trademark of the International Nickel Company

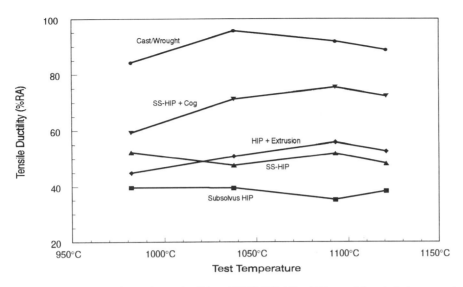

Figure 6: Comparison of high strain rate ductilities of UDIMET Alloy 720 materials tested at common hot-working temperatures. Samples were heated to test temperature, held for 15 min. and then pulled at 50 mm/mm/sec. (2 in./in./sec.). Samples were taken circumferentially from transverse slices.

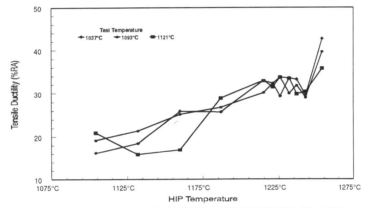

Figure 7: High strain rate tensile ductilities of P/M UDIMET Alloy 720 material versus HIP temperature. Samples were heated to test temperature, held for 15 min. and then pulled at 50 mm/mm/sec. (2 in./in./sec.).

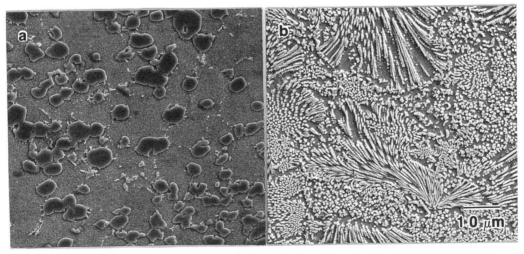

Figure 8: Scanning electron micrographs of UDIMET Alloy 720: a) cast/wrought billet b) SS-HIP (1254°C) furnace cooled. Material was electrolytically etched in chromate solution.

Hot workability

The high strain rate tensile results presented in Figure 6 show a relative comparison of the hot workability of UDIMET Alloy 720 processed in a variety of sequences. Test samples were taken from transverse slices and were generally oriented circumferentially. The corresponding grain size of each condition is presented in Table II. Despite having a very fine grain size of ASTM 9, the material HIPed in the conventional manner had the poorest workability throughout the test range. The material HIPed just below the solidus temperature (Sub-Solidus HIP, SS-HIP) had workability equivalent to P/M-extruded material even though the former had a grain size of ASTM 3-4 and the latter had a grain size of ASTM 9.5. Furthermore, when the SS-HIP material was either cogged or rolled until it had a grain size of ASTM 8-10 the ductility increased to approach that of conventional cast/wrought material.

Table II   Grain Size of UDIMET Alloy 720

Process	ASTM No.
Sub-Solvus HIP	8-9
Sub-Solidus HIP	3-4
Sub-Solidus HIP + Cog	8-9
HIP + Extrude	9-10
Cast/Wrought	9-10

Samples were not tested from longitudinal sections, however, it is expected that different results would be observed. The HIP/extruded material, for example, would be expected to have greater ductility along the longitudinal direction. Measuring hot workability in transverse sections is believed to be a better measure of the critical workability that is needed during forge shop operations since most operations begin with upset deformation.

The effect of HIP temperature on high strain rate ductility is further defined in Figure 7 where the reduction in area at 1066°C is compared for the same lot of P/M UDIMET Alloy 720. Improvement of ductility is reported for HIP temperatures from 1149°C to 1190°C. From 1190°C to 1254°C the tensile ductility appears to level out before an increase in ductility is experienced when the SS-HIP condition is reached. The ranges in which ductility is improved are somewhat similar to the grain growth behavior shown in Figure 1. It would be difficult to clearly sort out the mechanisms that improves ductility over the 1149°C to 1190°C range. Grain growth beyond the weak PPB's of the smaller powder particle could account for the improvement. However carbides and borides going into solution could also be beneficial.

The high strain rate ductility enhancement that is reached when the SS-HIP condition is met, clearly suggests that PPB's are deleterious to hot ductility. Despite the coarse grain condition, the ductility of SS-HIP material is superior to fine grain material where the grain boundaries are coincident with the PPB boundaries of the coarser powder. In the extruded P/M material the high strain rate ductility (Figure 6) was being limited by the PPB's because large portions of them are still coincident with the recrystallized grain boundaries. The ductility of the extruded material is superior to the ductility of the conventional HIPed material apparently due the presence of some new grain boundary areas and PPB areas that are somewhat more dispersed than the as-HIPed conventional material.

The superior high strain rate ductility of the cast/wrought material was consistent throughout this study. An examination of the microstructure (Figure 8) suggests that the very coarse gamma prime precipitate softens the gamma matrix and therefore accommodates plastic strain better. The grain boundaries of the cast/wrought material also appear more discrete and tend to be less continuous.

Throughout this investigation the role of isostatic pressing during the HIP process was questioned. Obviously the isostatic forces are necessary to bond the powder particles together but whether stress is necessary to accelerate the dissolution of phases and to allow the SS-HIP grain growth to occur is still not completely defined or understood. Samples of material HIPed at 1243°C were heat treated in air at the SS-HIP temperature condition (1260°C) and were tested mechanically and examined microstructurally. The heat treat cycle was successful in producing grain growth. However, Figure 9 clearly shows that not only was the ductility not improved, the material lost nearly half of its ductility. The cause for degradation in properties was immediately suspected to be a result of thermally induced porosity (TIP). Microstructural examination confirmed that porosity did occur and was most likely responsible for the loss in high strain rate ductility (Figure 10). TIP features in gas atomized powder is a common phenomena and is a result of trapped argon gas within particles. HIP is capable of essentially closing argon filled voids, but if the material is subjected to high temperatures, well above the gamma prime solvus, the intense internal gas pressure is capable of reopening the voids. Conventional heat treatments of P/M superalloys are at low enough temperatures that the reoccurrence of TIP is not a problem.

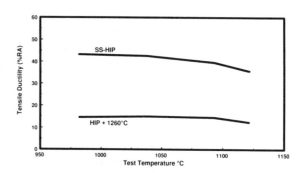

Figure 9: Comparison of the high strain rate tensile ductilities of P/M UDIMET Alloy 720 material that was SS-HIPed and material that was HIPed at 1243°C and heat treated at 1260°C.

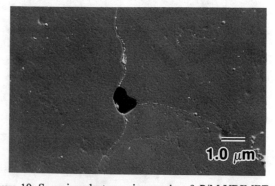

Figure 10: Scanning electron micrographs of P/M UDIMET Alloy 720 that was hot isostatically pressed at 1188°C and then heat treated at 1260°C. Void that is believed to be associated with TIP is shown. (Material was electrolytically etched in 5% HCl and methanol.)

While it appears that HIP is needed to achieve void free material at temperatures in the SS-HIP range, it is believed that the isostatic condition may be extending the useful process temperature range by increasing the incipient melting temperature. If there is any incipient melting during the HIP cycle, any melt-related void, that would normally occur during heat treatment at ambient temperature, would be readily closed. The effect of incipient melting on subsequent hot workability and mechanical properties has not been fully investigated.

The effect of cooling rates from the HIP temperature would be expected to influence both the gamma prime structures and also the size and morphology of the carbides and borides on the grain boundaries. This study did not try to sort out this effect but we would expect contributions to the high strain rate ductilities.

Thermal mechanical processing

The benefits of SS-HIP are most apparent after the as-HIPed coarse grain material is worked sufficiently to refine the grain size as is seen in HIP/cogged material (Figure 6). A series of rolldowns was prepared with different amounts of reductions to determine what amount of work is needed to maximize hot workability in SS-HIP material (Figure 11). It was found that with as little as 20% reduction the high strain rate ductility increased from 45% to nearly 90% at 1037°C. Microstructural examination of the rolldown did not reveal any readily apparent fine grain recrystalization. However, it would be expected that a fine grain substructure had developed but was not yet defined enough to be made apparent by etching.

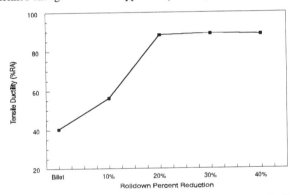

Figure 11: High strain rate tensile ductility of P/M UDIMET Alloy 720 SS-HIP material versus rolldown reductions.

This result is useful for developing conventional forging practices for SS-HIP material for producing cogged billets or to produce forge parts directly from SS-HIP material. The result demonstrates that if a forger were to initially get sufficient work into the SS-HIP structure, subsequent workability improves dramatically. Production trials have shown that SS-HIP material has hot workability that is superior to cast/wrought ingots. While the high strain rate tensile results in this paper may suggest that the ductility of cast/wrought billets is excellent, the presence of surface defects and imperfections on a cast ingot causes nuisances during forging operations. The SS-HIP material on the other hand is extremely uniform in bulk and comes with a well bonded stainless steel can that provides thermal insulation and lubrication during initial breakdown operations.

The versatility of SS-HIP material is fully demonstrated by the trial production of conventionally ring-rolled parts that were manufactured directly from SS-HIP material rather than from SS-HIP cogged billet. Ring rolling is a severe test since the strain rates are high and sensitivity to surface cooling is most pronounced. Ring rolling of the fine grain SS-HIP material was compared with the excellent results that are experienced with lower strength alloys like Waspaloy.

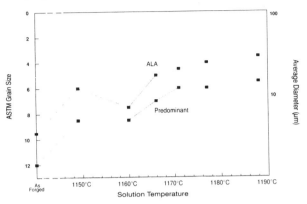

Figure 12: Grain size of ring rolled P/M SS-HIP UDIMET Alloy 720 as a function of solution heat treatment temperature. (Provided by Schlosser Forge Co.)

Ring-rolled parts have been evaluated to determine the grain growth response when supersolvus heat treated (Figure 12). The results define a wide range in temperatures that can be used to achieve controlled grain growth when desired. The grain size that can be achieved is larger than what can be achieved with conventional P/M material where uniform grain sizes greater than ASTM 6-8 are more difficult to achieve. The grain growth of SS-HIP material is much more uniform than cast/wrought material which is subject to banding due to solidification segregation. As with all superalloys, conditions were identified where certain strain conditions led to abnormal grain growth above the gamma prime solvus.

Figure 13: Micrograph of ring-rolled P/M SS-HIP UDIMET Alloy 720 that was solution heat treated to provide a uniform ASTM 4.5 grain size. (Kallings etch)

Figure 13 illustrates a uniform grain size of a conventional ring rolled part. The uniform grain size of ASTM 4.5 provides a microstructure that potentially has balanced creep, tensile, and fatigue properties. It is speculated that, by minimizing the effects of PPB's, enhancement in mechanical properties over conventionally processed P/M superalloys will be realized. Also, forgings made directly from SS-HIP billet offers a cost-effective approach to make superior parts with short lead times and improved process control.

Inspectability

A comparison of UDIMET Alloy 720 billet ultrasonic inspectability (Table III) highlights the advantages of P/M processing in reducing inspection noise levels and improving the detection of smaller defects. While one would expect better inspectability for finer grain sizes, cast/wrought appears to an exception to the rule. Despite having a grain size as small as P/M processed materials, ASTM 11, the detection limit of cast/wrought material has coarse macrostructural features, such as low angle grain macro grain boundaries, that are believed to cause gross noise levels.

Table III Inspectability of UDIMET Alloy 720

Form	Billet Dia. (mm)	ASTM G.S.	Noise Level FBH*
P/M Extruded	150-300	14	6% @ #1
P/M SS-HIP/Cog	300	10	25% @ #1
P/M HIP	165	10	15% @ #1
P/M SS-HIP	300	3	50% @ #1
Cast/Wrought	165	11	100% @ #2

* #1 = 1/64 = 0.40 mm

Conclusions

1) Grain growth restriction of P/M superalloys by PPB can be overcome by HIP at temperatures just below the alloy's solidus.

2) By eliminating grain boundaries that are coincident with PPB's, the high strain rate ductility of a P/M superalloy can be improved.

3) SS-HIP P/M UDIMET Alloy 720 exhibits sufficient high strain rate ductility that conventional forging and ring rolling of billet is practical.

4) Post forging super-solvus heat treatments exhibit uniform grain growth and resistance to abnormal grain growth.

5) SS-HIP material has superior ultrasonic inspectabily compared to cast/wrought material.

Acknowledgments

The authors wish to thank Prof. John Radavich for his insightful analysis of the microstructures and his helpful discussions regarding potential mechanisms. We would also like to thank Steven Sawochka and Schlosser Forge Company for providing us data and samples from ring rolling trials and Howmet Corporation in Whitehall, Michigan for providing HIP services.

References

1. R. Thamburaj, A.K.Kroul, W. Wallace, and M.C.deMalherbe, "Prior Particle Boundary Precipitation in P/M Superalloys," Modern Developments in Powder Metallurgy, Proceeding of the 1984 International Powder Metallurgy Conference, (Princeton, NJ: Metal Powder Industries and the American Powder Metallurgy Institute, 1984), 635-674.

2. R.Chang, D.D.Krueger, and R. A. Sprague, "Superalloy Powder Processing, Properties and Turbine Disk Applications," Superalloys 1984, Proceedings of the Fifth International Symposium on Superalloys, (Warrendale, PA: TMS-AIME, 1984), 245-273.

3. G.A.J.Hack, J.W.Eggar, and C.H.Symonds, "A Comparison of APK-1 Consolidated at HIP Temperatures Above and Below the Boride Solvus," Proceedings, Powder Metallurgy Superalloys Conference, Zurich, November 1980, 20.1-20.50.

4. C. Aubin, J.H. Davidson, and J.P.Trottier, "The Influence of Powder Particle Surface Composition on the Properties of a Nickel-based Superalloy Produced by Hot Isostatic Pressing," Superalloys 1980, Proceedings of the Fifth International Symposium on Superalloys, (Metals Park, OH: ASM International, 1980), 345-354.

5. M. Jeandin, B. Fieux and J.P.Trottier, "P/M Astroloy Obtained by Forging or Hiping Supersolidus Sintered Preforms," Modern Developments in P/M, Vol. 14, Special Materials, MPIF/APMI, 1981, 65-91.

6. K.R.Bain, M.L.Gamone, J.M.Hyzak, and M.C.Thomas, "Development of Damage Tolerant Microstructures in Udimet 720," Superalloys 1988, Proceeding of the International Symposium on Superalloys, (Warrendale, PA: TMS, 1988), 13-22.

7. C.H.Symonds, J.W. Eggar, G.J. Lewis, and R.J.Siddall, "The Properties and Structures of As-HIP Plus Forged Nimonic APK 1 (Low Carbon Astroloy)", Proceeding Powder Metallurgy Superalloys Conference, Zurich, 1980, 17.1-17.28.

8. J.M. Hyzak, R.P.Singh, J.E.Morra, and T.E.Howson, The Microstructural Response of As-HIP P/M U-720", Superalloy 1992, Proceeding of the International Symposium on Superalloys, (Warrendale, PA: TMS, 1992), 93-101.

9. T.C. Lu, T.T.Nguyen, Y, Bienvenu, J.H.Davidson, and O.Dugue, "The Influence of Powder Processing Variables on the Structure and Properties of HIPed Low Carbon Asroloy" (Publication unknown)

10. R.D.Kissinger, S.V.Nair, and J.K.Tien, "Influence of Powder Particle Size Distribution and Pressure on the Kinetics of Hot Isostatic Pressing (HIP) Consolidation of P/M Superalloy Rene 95," Superalloys 1984, Proceedings of the Fifth International Symposium on Superalloys, (Warrendale, PA: TMS-AIME, 1984), 285-294.

11. US Patent No. 5,451,244, "High Strain Rate Deformation of Nickel-Base Superalloy Compact", B. Wayne Castledine, Special Metals Corporation.

Hot-Die Forging of P/M Ni-Base Superalloys

C.P. Blankenship, Jr., M.F. Henry, J.M. Hyzak[†], R.B. Rohling and E.L. Hall

GE Corporate Research and Development
PO BOX 8, Schenectady, NY 12301

[†]Wyman Gordon Company
PO BOX 8001, N. Grafton, MA 01536

Abstract

Hot die forging was evaluated as a process for producing acceptable grain structures in P/M nickel base superalloy René 88DT. Laboratory and subscale experiments were carried out, and grain sizes in the range 10-20μm were produced after supersolvus heat treatment. There were some issues with die fill and cracking. Further process improvements would be required for successful implementation.

Introduction

Advanced gas turbine rotor components made from powder metallurgy (P/M) Ni-base superalloys are currently isothermally forged at slow strain rates and high temperatures (slightly below the γ' solvus) [1,2]. These processing parameters are chosen mainly to encourage superplastic deformation, resulting in low forging loads and low die stresses. This operation requires expensive tooling, an inert environment, and slow ram speeds for successful operation. Superplastic deformation in the workpiece also allows large geometric strains to be achieved during the forging operation without cracking. At the end of the forging operation, no substantial increase in dislocation density should be observed, as strain is accommodated by grain boundary sliding and diffusional processes. In the event that dislocations are generated, the high temperatures and slow stroke rates allow dynamic recovery to occur. The grain size after forging is typically 3-5μm. After forging and heat treatment, cooling and aging operations are performed. For applications that demand enhanced creep and time dependent fatigue crack propagation resistance, a coarse grain size (20-40μm) is required. This is achieved by heat treating above the γ' solvus.

Complex contoured forgings contain a range of strain and strain rates. If temperatures are too low, or local strain rates are too high, diffusional processes cannot keep up with the imposed strain rate. In this case, dislocations are generated. If a component is subsequently heat treated above the γ' solvus, this may lead to an occurrence referred to as critical grain growth (CGG), which leads to a bimodal grain size distribution [3]. Factors in addition to dislocation density, such as carbon/boron/nitrogen content and subsolvus annealing time also appear to influence the grain size distribution [3].

Numerous studies have illustrated the effect of retained strain[1] on final grain size [3-6]. Data from room temperature compression tests contained in Figure 1 summarize this trend for the powder metallurgy nickel base superalloy René 88DT.

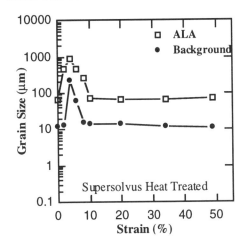

Figure 1. Grain size[2] after supersolvus heat treatment as a function of room temperature compression

---

[1] For convenience, the term "retained strain" is used to describe the dislocation density, or metallurgical strain present in the microstructure.
[2] ALA indicates the largest grain diameter in the section, while background grain size represents the average grain size.

Analogous behavior is observed in René 95, cast and wrought superalloys [7-9] and other alloy systems [10]. This behavior may be translated to elevated temperature deformation; however, strain rate and temperature replace strain as the primary variables that influence the amount of retained strain [11]. Prior studies [3,12,13] demonstrated that a window exists in strain-rate and temperature space in which critical grain growth can be avoided, thus producing a microstructure of uniform 20-40μm grains after supersolvus heat treatment.

Critical grain growth is thought to result from nucleation limited recrystallization followed by grain growth until the strain free grains impinge on one another. It occurs over a narrow range of retained strain. Slightly higher retained strain results in a higher nucleation density and a finer resultant grain size. Slightly lower retained strain is insufficient to trigger the recrystallization process. Thus the term critical grain growth was adopted to describe the observation that a critical amount of strain was required to lead to this undesirable microstructure.

Critical grain growth is not observed in high volume fraction γ′ alloys until heat treatment is performed above the γ′ solvus. It is therefore noted that, in this complicated alloy system, factors in addition to retained strain influence grain structure evolution. Particles that pin grain boundaries play an active role in controlling grain size, most notably, the coherent, high volume fraction γ′ phase. Carbides, borides and oxides are also reported to influence final grain size, especially if the alloy is heat treated above the γ′ solvus [3].

An alternative procedure to high temperature-low strain rate, isothermal forging is to forge the component fast and cold enough so that the retained strain everywhere in the part is above the amount that would lead to critical grain growth. If successful, this procedure would offer the added benefit of increased productivity and reduced cost for manufacturing turbine disks.

The objectives of this effort were to explore the fundamental metallurgical characteristics of high retained strain forging using laboratory experiments, and investigate the applicability of the process by demonstrating the procedure on a subscale hot die forging press.

Experimental Procedure

René 88DT extrusions were supplied by Special Metals and Wyman Gordon for this study. The compositions of each extrusion were typical of production René 88DT [1]. Right circular cylinders (RCC's) (10mm diameter, 15mm height) and double cone specimens (8.5mm end radii, 25.4mm middle radius, 21mm height) were machined from these extrusions. The extruded microstructure was characterized by recrystallized grains measuring 1-5 μm in diameter and 0.1-1 μm primary γ′ particles. Unrecrystallized powder particles measuring 30-50 μm in diameter were observed throughout the billet cross section. The apparent area fraction of these unrecrystallized regions varied throughout the cross section, but was on the order of 0.001 to 0.01.

Laboratory forging simulations were performed using a servohydraulic machine equipped with a clamshell furnace. Tests were run at constant true strain rates of 0.1 and 0.01 s^{-1}. After 50% nominal reduction in height, the samples were unloaded, removed from the furnace and air cooled.

Transmission electron microscopy (TEM) was performed on sections of cylinders in the "as-compressed" condition. Slices were made parallel to the forging direction, and mechanically ground to 100μm in thickness, followed by electropolishing in an 80% methanol 20% perchloric acid solution. The microstructure was characterized using a TEM operated at 300kV.

After a γ′ supersolvus heat treatment of 1150°C for 2 hours, metallographic sections were mechanically polished and etched with Walker's reagent. Average grain size was measured according to ASTM E112, except on samples where a bimodal distribution of grain sizes was encountered. In those cases, the abnormally large grains were avoided in measuring an average, or background grain size, and the large grains were measured individually leading to an "as large as" (ALA) grain size using ASTM E930.

Subscale forging trials were performed at Wyman Gordon using a 1500 ton, hot die press. An IN718 die set was configured to provide an aggressive shape to test the procedure. Die temperature was not an intentional variable, though it varied slightly from run to run. The nominal die temperature was held near 593°C. The mults were coated with lubricant, and wrapped. The press velocity was 13 mm/s for each test. Mult temperatures were chosen based on laboratory specimen results: 871°C, 927, 982, 1037°C. Initial mult geometries are shown in Table I.

Table I. Initial mult geometries for subscale forging experiments

Diameter (mm)	Initial Height (mm)	Nominal True Strain (final)	Nominal Strain Rate (s^{-1})
112	38	0.7	0.3-0.7
89	61	1.0	0.2-0.7

The forged disks were sectioned into quarters. The first quarter was placed in an 1150°C furnace and held for 2 hours. The second quarter was placed in a 1050°C furnace and stabilized for 15 minutes followed by a two hour ramp to 1150°C where it was held for two hours. The third quarter was given a subsolvus anneal of 1050°C for 8 hours followed by a two hour ramp to 1150°C and a 2 hour hold. Gamma prime solvus for René 88DT is approximately 1110°C. All sections were air cooled after heat treatment.

Results and Discussion

Laboratory Experiments

As described in the experimental procedure section, laboratory simulations of the forging operation were performed using cylindrical samples. The results of each campaign are summarized below. Table II contains the processing conditions and resulting grain sizes after supersolvus heat treatment.

Table II. Grain size after forging and supersolvus heat treatment (RCC's)

Strain Rate (s^{-1})	Temp (°C)	Grain Size (μm)	ALA Grain Size (μm)
0.1	871	11	85
0.1	927	11	55
0.1	982	13	70
0.01	871	11	65
0.01	927	13	55
0.01	982	12	75

Forging at low temperatures and high strain rates results in high forging loads and die stresses. Figure 2 compares the true stress-true strain curves for the 871°C / 0.1 s^{-1} compression condition to a curve from a compression test run at 1050°C / 0.003 s^{-1} (nominal isothermal forging conditions that result in superplastic deformation).

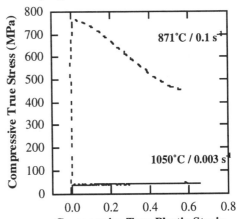

Figure 2. Flow stress data comparison: hot die forging results in 10X higher die stresses.

Figure 3 illustrates the grain structure that is produced after supersolvus heat treatment. Lightly decorated prior powder particle boundaries (MC and ZrO_2) can be seen, and no primary γ' is observed.

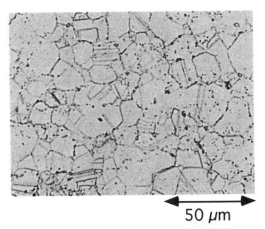

Figure 3. Grain structure of cylinder compressed at 871°C / 0.1 s^{-1} and heat treated at 1150°C for 2 hours.

TEM was performed on sections from samples compressed at 871°C / 0.1 s^{-1} and 927°C / 0.01 s^{-1} forging conditions. Figure 4 contains a bright field TEM image of the microstructure after compression at 871°C (no subsequent heat treatment).

Figure 4. TEM image of cylinder compressed at 871°C / 0.1 s^{-1}. Microstructure consists mainly of dense dislocation tangles.

Both microstructures contained significant amounts of retained metallurgical strain in the form of dislocation tangles, though the dislocation structures appeared more dense in the 871°C / 0.1 s^{-1} microstructure.

Production heat treatment cycles typically contain a stabilization at 1050°C on the way to 1150°C. Therefore, TEM samples were prepared from a specimen compressed at 871°C / 0.1 s^{-1} after the stabilization phase of the heat treatment (1050°C for 0.25 hours). Dense dislocation tangles were observed in some areas, while other regions were essentially strain free, as shown in Figure 5. This structure is representative of the recrystallization process. Recovery can be discounted, as it tends to occur continuously throughout the microstructure, rather than as discrete nucleation and growth events. The 1050°C heat treatment followed by a ramp to 1150°C appears to allow the nucleation and (limited) growth of recrystallized grains prior to passing through the γ' solvus. This sequence is preferred, as the grain structure can undergo its two major alterations one step at a time. Recrystallization and elimination of statistically stored dislocations can occur in the presence of the efficient pinning phase (γ'). The fine grain microstructure can then undergo a growth spurt after the dissolution of the major pinning phase without the added complication of another strong driving force (retained strain).

Subscale Forging Trials

Based on the results of the laboratory compression tests, four forging temperatures and two billet geometries were used to construct an eight run subscale forging matrix (shown in Table III). Conditions were chosen to be representative of hot die forging operations. Billet and die temperatures were significantly lower than those used in isothermal forging operations, and press velocities (strain rates) were significantly higher. These faster and colder process conditions were well outside the superplastic window (as illustrated by the flow stress curves and microstructures in the laboratory experiments). Two concerns in this new processing regime were die strength and cracking of the forged article.

Some of the forgings exhibited cracking in the rim region. In fact, some of the cracks ran a significant distance into the web. Cracking was more severe at the lower forging temperatures, and it was postulated that the low die temperatures contributed to the observed cracking.

Complete metal flow simulations were performed for 927 and 1037°C forging temperatures. Metal flow patterns were similar for the two billet geometries and forging temperatures, but local strain rates, strains and temperatures were quite different. Figure 6 contains calculated final strain contours for the 89mm and 112mm diameter billets forged at 927°C as an example of the modeling results.

Polished and etched cross sections were evaluated for uniformity of grain structure after heat treatment. As mentioned in the experimental procedure section, three heat treatment schedules were applied to sections of each forging:

1. 1150°C / 2 hours

2. 1050°C / 15 minutes + ramp to 1150°C in 2 hours + 1150°C / 2 hours

3. 1050°C / 8 hours + ramp to 1150°C in 2 hours + 1150°C / 2 hours

Figure 5. TEM image of recrystallization process after 15 minutes at 1050°C. The specimen was compressed at 871°C / 0.1 s^{-1}

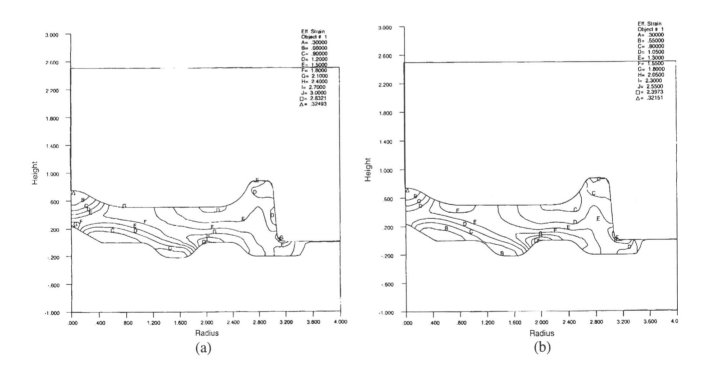

Figure 6. Effective strain contours (a) 89mm billet and (b) 112mm billet for forging conditions: 927°C and 13 mm/s press velocity.

The second procedure is a typical heat treat sequence for production forgings. The third is a procedure that involves an extended subsolvus anneal designed to reduce or eliminate retained strain before ramping to the supersolvus heat treatment temperature. The results of the grain structure evaluations are shown in Table III.

A high, medium, low, zero (H,M,L,0) relative rating scale was used to compare the amounts of cracking and CGG observed. For cracking, the number and depth of cracks determined the rating, and for critical grain growth the approximate area fraction of large grains determined the rating. Figure 7 contains an example of each critical grain growth level.

The grain structure was reasonably uniform in the forgings that did not contain CGG. The average grain size varied between 9 and 18 μm.

The most attractive, uniform grain structure was produced by the heat treatment schedule that included an extended subsolvus anneal. An example is shown in Figure 8.

The results tabulated in Table III were analyzed using statistical data analysis software to evaluate the trends in a quantitative manner. For the relative ratings, values of 0,3,6 and 9 were assigned for ratings of 0, L, M and H respectively. The following variables were evaluated for their effects on cracking, CGG and resultant grain size: forging temperature, billet diameter, and time at 1050°C during heat treatment. The results of the analysis are shown in Table IV. One result that warranted further investigation was that longer times at 1050°C before the supersolvus heat treatment were universally better for producing a uniform grain structure.

Table III. Summary of subscale forging results

S/N	Billet Temp (°C)	Billet Diameter (mm)	Degree of Cracking	Heat Treatment (hours at 1050°C)	Critical Grain Growth Rating	Grain Size (μm)
7	871	89	H	0	0	10
				0.25	0	12
				8	0	13
8	871	112	M	0	0	9
				0.25	0	11
				8	0	11
3	927	89	M	0	L	12
				0.25	0	14
				8	0	16
4	927	112	H	0	L	10
				0.25	0	14
				8	0	16
1	982	89	L	0	M	12
				0.25	0	14
				8	0	18
2	982	112	0	0	L	10
				0.25	L	12
				8	0	11
5	1037	89	L	0	H	13
				0.25	0	12
				8	0	14
6	1037	112	L	0	M	12
				0.25	L	10
				8	L	14

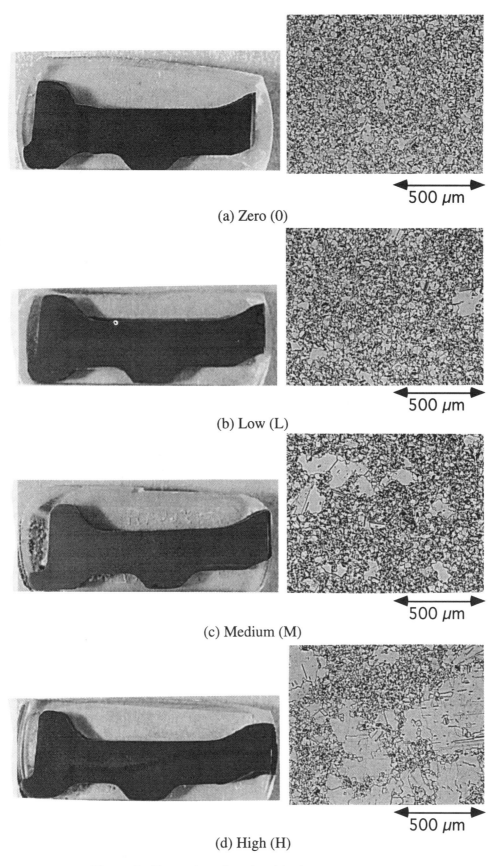

Figure 7. Examples of each critical grain growth level

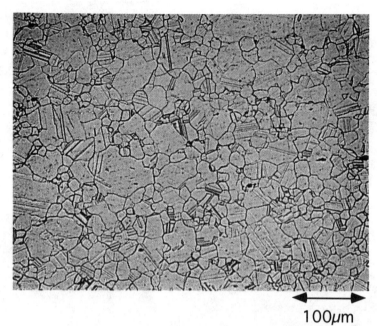

Figure 8. Uniform grain structure from near the axial and radial midpoint-- S/N 3 (927°C billet temperature, 89mm billet, extended subsolvus anneal, supersolvus heat treatment).

Table IV. Correlation of response variables with input conditions (95% confidence)

RESPONSE	INPUT
uniformity of grain structure improves with	reduction in forging temperature
uniformity of grain structure improves with	increase in time at 1050°C
amount of cracking decreases with	increase in forging temperature
average grain size (μm) decreases with	increase in starting billet diameter
average grain size (μm) decreases with	reduction in time at 1050°C

A TEM investigation was performed on subscale forgings S/N 7 (89mm diameter billet, 871°C forging temperature) and S/N 5 (89mm diameter billet, 1037°C forging temperature). For each of these forgings, samples were taken from identical locations (between web ring and rim) Foils were examined from the as-compressed and extended subsolvus annealed (1025°C / 8 hours) conditions.

Figure 9 illustrates subtle differences in the as-compressed microstructures for each forging. Significant recrystallization appears to have taken place during forging (dynamic), or during the cool down after forging (meta-dynamic). Some regions remain unrecrystallized, and these regions appear to constitute ~10% of the volume in each region that was analyzed. The recrystallized grain size of S/N 7 is ~0.5μm, and the recrystallized grain size of S/N 5 is ~1μm.

The location where the TEM foil was taken was consistent with the large grain band (CGG region) that formed in S/N 5 after direct 1150°C heat treatment. The recrystallized grain size was slightly larger than S/N 7, and the amount of retained strain in the unrecrystallized regions was slightly less than that of S/N 7 (from selected area diffraction patterns and TEM images). MC, boride, and oxide particles were observed in both microstructures, and their distributions were typical of most production René 88DT microstructures.

Figure 9 (a). Microstructure after forging at 871°C--S/N 7

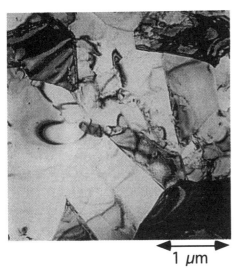

Figure 9 (b) Microstructure after forging at 1037°C--S/N 5

The subtle differences mentioned could be important, but it is difficult to formulate a consistent rationale that explains why the large grains appear in S/N 5 after direct supersolvus heat treatment, and not in S/N 7.

TEM was performed on these same forgings after extended subsolvus annealing. There was a significant reduction in the amount of strain retained in the microstructures. The grain sizes were measured as 3-5μm. The microstructures were essentially fully recrystallized, and low angle boundaries were observed in both samples.

The TEM results for microstructures given an extended subsolvus anneal indicate that recrystallization was nearly complete before the ramp to 1150°C was initiated. This heat treatment approach represents a modification to the original stated strategy. It relies on the forging operation to produce enough retained strain to allow complete recrystallization below the γ' solvus and therefore ensure that the microstructure is strain-free prior to heat treatment above the γ' solvus.

Summary

The results of the subscale hot die forging experiments summarized in Table III, coupled with the TEM results on laboratory specimens and contoured subscale forgings, indicate that two strategies are available for producing a uniform grain structure.

1. Ensure that there is no retained strain in the microstructure before crossing the γ' solvus.

2. Ensure that there is sufficient strain to promote a high nucleation density of recrystallization during the supersolvus heat treatment.

Moreover, there are at least two practical production methods for carrying out the first strategy. For example, current isothermal forging practices are aimed at using superplastic deformation to achieve the shape change without causing an increase in dislocation density. Therefore, subsequent supersolvus heat treatment may be given to a microstructure that is essentially free from retained strain. However, in practice, variability in the process may result in local areas being forged outside the superplastic window, which results in retained strain.

A second approach to the first strategy involves using lower forging temperatures and faster strain rates, typical of hot die forging practices. This practice introduces a high dislocation density into the microstructure of the forged article. The next step is to anneal the forging at a temperature below the γ' solvus, encouraging complete recrystallization prior to the supersolvus step of the heat treatment.

The second strategy also presents an opportunity to apply the hot die forging technique to produce a uniform grain structure. This process does not appear to be as robust as the extended subsolvus anneal approach. The data generated in this study indicate that the forging temperature must be below 927°C to avoid CGG for a press velocity around 13 mm/s. This temperature range coincides with the temperature which produced significant amounts of cracking. Further process development or canning would be required for successful application of this method.

The grain size typical of isothermal forging and supersolvus heat treatment of René 88DT is 20-40μm. As noted earlier, hot die forging, even with an extended subsolvus anneal, produced a grain size range of 10-20μm. The uniform, finer grain, supersolvus heat treated microstructure that is produced by colder, faster forging of René 88DT may be useful for a number of applications where strength and LCF performance are key design criteria.

Conclusions

1. Hot die forging in the range of 871-927°C produced a uniform grain structure with a standard supersolvus heat treatment. Die fill and cracking posed problems in this temperature range.

2. Hot die forging at higher temperatures produced uniform grain structures when combined with an extended subsolvus anneal (1050°C / 8 hrs) prior to the supersolvus heat treatment. Die fill and cracking tendencies were reduced under these conditions.

Acknowledgments

The authors would like to thank Mark Benz, Joe Corrado, Eric Huron, Richard Menzies and Mike Weimer for technical discussions; Bill Catlin, Jeff Thompson, John Hughes, Don Wemple and Ray Schnoor for experimental assistance; and Yvonne Mastracchio for assistance with the manuscript.

References

1. D.D. Krueger, R.D. Kissinger, R.G. Menzies, C.S. Wukusick, US patent number 4,957,567, General Electric Co.

2. D.D. Krueger, R.D. Kissinger, R.G. Menzies, *Superalloys 1992*, S. Antolovich et al., eds. TMS, Warrendale, PA (1992) 277-286.

3. A.E. Murut and C.P. Blankenship, Jr., unpublished research, General Electric Corporate Research and Development, Schenectady, NY, 1993.

4. S. Channon and H. Walker, *Trans. Am. Soc. Metals*, **45** (1953) 200.

5. J.S. Smart and A.A. Smith in *Physical Metallurgy Principles*, by R.E. Reed-Hill, PWS (Boston, 1973) 295.

6. C.P. Blankenship, Jr., W.T. Carter, Jr., A.E. Murut and M.F. Henry, *Scripta Metall.*, **31** (1994) 647.

7. M. Koryagina, I. Meshchaninov and S. Khayutin, *Fitz Metal. Metalloved.*, **49**, no. 4 (1980) 843.

8. C. White, *J. Inst. Metals*, **97** (1969) 215.

9. W.H. Couts, in *The Superalloys*, John Wiley and Sons (1972) 451.

10. P.A. Beck, *Phil. Mag.*, **3** (1954) 245.

11. M.F. Ashby and R.A. Verall, *Acta Metall.*, **21** (1973) 149.

12. R.D. Kissinger, unpublished research, General Electric Aircraft Engines, Cincinnati, OH, 1991.

13. E.S. Huron and S. Srivatsa, unpublished research, General Electric Aircraft Engines, Cincinnati, OH, 1993.

THE EFFECT OF HIGH TEMPERATURE DEFORMATION ON GRAIN GROWTH IN A PM NICKEL BASE SUPERALLOY

Michèle Soucail, Michel Marty, and Henri Octor
ONERA, Direction des Matériaux,
BP72, 92322 Châtillon, FRANCE

Abstract

The high temperature deformation conditions were investigated for the grain growth during subsequent supersolvus annealing in a PM nickel base superalloy N18. If the strain is sufficiently high, the strain rate imposed to the material appeared to be critical with regard to the grain growth. The development of very large grains (abnormal grain growth) during the supersolvus annealing is consecutive to a deformation at a strain rate in a narrow region of transition between stage II (superplastic behaviour) and stage III.

Introduction

The concern to improve the heat efficiency of aeroengines has led to a continuous increase in temperature in the compressor, the combustion chamber and the turbine. Thus it is desirable to develop materials which exhibit high tensile and creep resistance with excellent damage tolerance capability at high temperature. A significant improvement of high temperature creep and crack propagation properties of N18 (a powder metallurgy nickel base superalloy selected by SNECMA for compressor and turbine disks of the M88 engine (1)) was accomplished by increasing the grain size (2, 3). But these promising results require a perfect control of the grain size everywhere in the disk. In particular, abnormal grain growth, defined as the growth of a few grains to very large sizes and possibly the elimination of the surrounding smaller grains, must be avoided. The fabrication of a disk is a complex process. During the various operations, the microstructure of the alloy is modified. Recrystallisation, normal and abnormal grain growth in advanced alloys depend on numerous microstructural, mechanical, and thermal factors (4, 5, 6, 7, 8). Their respective roles are not clearly understood. We report in this paper our investigations concerning the influence of high temperature deformation conditions on the grain growth during subsequent annealing.

Experimental

The nominal composition of the investigated nickel base superalloy N18 is (in wt %) : Co (15.7), Cr (11.5), Mo (6.5), Al (4.35), Ti (4.35), Hf (0.45), C (0.015), B (0.015), and bal. Ni. N18 alloy was produced by powder metallurgy via argon atomisation and hot extrusion. It is a $\gamma - \gamma'$ superalloy with a maximum volume fraction of γ' of about 55 % and the γ' solvus temperature is at 1195°C.

The effect of a high temperature deformation on the final grain size was studied by means of tensile tests performed at constant ram speed and temperature (1100°C, 1120°C, or 1140°C) then followed by air quenching. Strain rate was varied between $10^{-4} s^{-1}$ and $5 \times 10^{-2} s^{-1}$ and strain between 0.05 and 0.9. The final treatment for grain growth was a supersolvus heat treatment : specimens were held for 4 hours in a furnace kept at 1205°C. The grain size was measured by quantitative image analysis of etched sections of tensile specimens. Because the microstructure was equiaxed, the equivalent diameter, $D = \sqrt{4S/\pi}$ where S is the grain area, was chosen to describe the grain size.

Results

Microstructural analysis.

After high temperature tensile deformation, N18 is found to exhibit a fine equiaxed $\gamma - \gamma'$ structure. The mean grain size is about 5 µm. The γ' phase is distributed between large primary precipitates (about 25% volume fraction) located at grain boundaries and finer secondary intragranular precipitates.
By annealing above the γ' solvus temperature, this initial fine grained structure was found to exhibit rapid growth because of the dissolution of the γ' phase.
Grain size after final supersolvus heat treatment was studied more specifically in the case of tensile deformation at 1120°C. Some typical microstructures, showing normal and abnormal grain growth, are presented in figures 1, 2, and 3.
The evolution of mean grain size versus strain rate in specimens deformed in tension to various strains at 1120°C and given a 4-hour supersolvus treatment at 1205°C is plotted in figure 4. In this case, when strain rate increased from $10^{-4} s^{-1}$ to $2 \times 10^{-2} s^{-1}$ mean grain size had a general tendency to decrease from about 45 µm to 30 µm, independently of the strain if this factor was greater than about 0.2. If the strain was less than 0.1, the final grain size was about 50 µm.
Nevertheless, after deformation in the neighbourhood of a critical strain rate, and a supersolvus heat treatment, very large grains of a few millimeters were observed. This exaggerated growth was detected independently of the applied strain when this factor was sufficiently high. For example, for strains of 0.3, 0.47 and 0.74 and a strain rate of $7 \times 10^{-3} s^{-1}$ at 1120°C, abnormal grain growth was observed ; the phenomenon was evidenced to occur only for a limited range of strain rate, between $6 \times 10^{-3} s^{-1}$ and $2 \times 10^{-2} s^{-1}$ at 1120°C (figure 4).

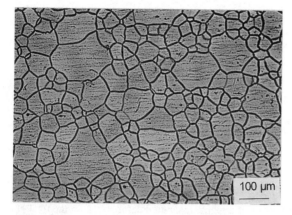

Figure 1. Microstructure of a tensile specimen tested at 1120°C, $\dot{\varepsilon} = 6 \times 10^{-4}$ s^{-1}, $\varepsilon = 0.3$, and supersolvus treated (1205°C-4h).

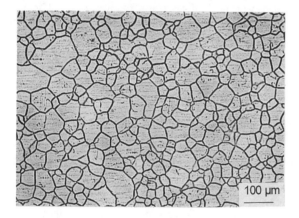

Figure 2. Microstructure of a tensile specimen tested at 1120°C, $\dot{\varepsilon} = 2 \times 10^{-2}$ s^{-1}, $\varepsilon = 0.3$, and supersolvus treated (1205°C-4h).

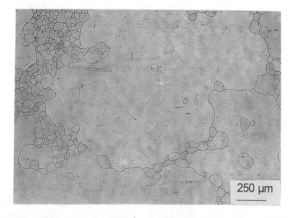

Figure 3. Microstructure of a tensile specimen tested at 1120°C, $\dot{\varepsilon} = 7 \times 10^{-3}$ s^{-1}, $\varepsilon = 0.74$, and supersolvus treated (1205°C-4h).

Similar observations were made after tensile deformations of about 0.5 at 1100°C or 1140°C. In these cases, abnormal grain growth occured in a limited range of strain rate too (around 3×10^{-3} s^{-1} and 1×10^{-2} s^{-1} respectively). Thus, the strain rate imposed to the material appeared critical as regards the development of very large grains during a subsequent supersolvus annealing.

When tensile strain was sufficiently large (more than about 0.2 at 1120°C), the strain rate rather than strain was identified as the most pertinent parameter, during high temperature deformation, influencing the grain growth during subsequent annealing.

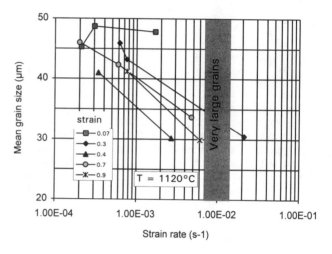

Figure 4. Variation of mean grain size versus strain rate after deformation at 1120°C and a supersolvus treatment (1205°C-4h).

Mechanical behaviour.

Stress - strain curves were calculated from the experimentally measured load and displacement data using the assumptions of constant volume and uniform deformation. The equivalent strain and the equivalent stress are defined as:

$\varepsilon = \ln(l/l_0)$ and $\sigma_1 = F l / S_0 l_0$ where l_0 and S_0 are the initial specimen length and cross section respectively, l is the specimen length at time t and F the applied load.

The tests were performed at constant ram speed v and consequently the strain rate during the test decreased according to $\dot{\varepsilon}_1 = v / l$. It must be noted that the temperature increase associated with the heat generated by plastic deformation was negligible (less than 4°C, deduced from the thermal equilibrium

Table I Effect of temperature on critical strain rate and strain rate sensitivity coefficients.

Temperature (°C)	$\dot{\varepsilon}_c$ (s^{-1})	m in stage II	m in stage III
1100	3.5×10^{-3}	0.66	0.37
1120	8×10^{-3}	0.73	0.34
1140	1×10^{-2}	0.61	0.35

The maximum stress and the final stress of each tensile test were plotted versus strain rate with logarithmic scales in figure 5. Two domains (stages II and III) with different strain rate sensivity coefficients m (the slopes of the straight lines) were identified regardless of the strain. The mechanical behaviour changes at a critical strain, rate $\dot{\varepsilon}_c$. Between 1100 and 1140°C this critical strain rate increases with temperature. The effect of temperature on values of m is not very significant (table I).
Then the coefficients m were used to correct the stress by taking the decrease of strain rate into account, according to :

$$\sigma = \sigma_1 \, (\dot{\varepsilon}/\dot{\varepsilon}_1)^m$$

Some final corrected stress-strain curves are shown in figure 6. At 1120°C stress, after a short period of work hardening, is constant or decreases slightly at the highest strain rates. The equivalent strain at the onset of steady state or at the peak stress is less than 0.1. The steady state stress increases with increasing strain rate.

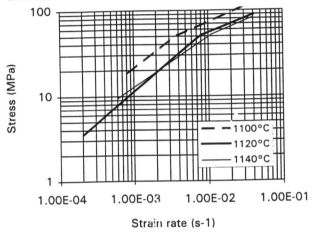

Figure 5. Variation of maximum and final stress of tensile test versus strain rate

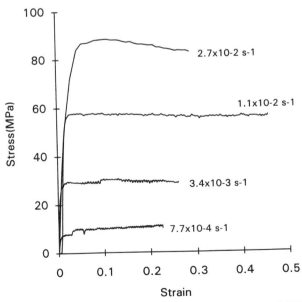

Figure 6. Typical corrected stress-strain tensile curves of N18 at 1120°C

Discussion

Nickel base superalloys produced by powder metallurgy, such as the alloy investigated here, show a resistance to grain coarsening beyond the limits set by the powder particles. Indeed the oxycarbides preferentially located at the prior particle boundaries are known to act as effective obstacles to grain boundary migration, because of small sizes and small spacings. That is the reason why the grain size in PM superalloys heat treated above the γ' solvus is generally not very large, similar to the powder particle size. After deformation at high temperature and a supersolvus annealing, the mean grain size in PM N18 was indeed generally smaller than 50 μm.

Dahlen and Winberg (6) were able to overcome, by critical strain annealing, this limit imposed by particle powders in Astroloy, another PM superalloy. In this case a small amount of deformation is applied at room temperature and the material is then annealed at high temperature. If the amount of deformation is not sufficient, no grain growth occurs and the grain size remains independent of deformation. A critical deformation is needed for the growth of a few grains that will consume all the others, the initial driving force being the decrease in dislocation density. As the amount of deformation increases beyond this critical value, the grain size will decrease. In PM Astroloy, Dahlen and Winberg (6) found that a strain of about 2% at room temperature produced coarse grain structures during subsequent annealing.

In our experiments at high temperature, if the strain is large enough, strain rate rather than strain (at a fixed temperature) is evidenced to control the final grain size after a subsequent supersolvus annealing. In particular, the growth of very large grains occurs after deformation (followed by a heat treatment) within a short interval of strain rates. This phenomenon was evidenced at different temperatures.

The rheological behaviour of N18 is in good agreement with these observations. Indeed, after a short period of work hardening, the stress remains constant and the steady state stress depends only on the strain rate. Besides, the growth of very large grains presents some similarities with critical straining (6, 7) : if we assume that the dislocation density is related to the stress, the abnormal growth depends, in the steady state, on the strain rate. Since deformation involves dynamic recovery at elevated temperature, large and various amounts of deformation can produce abnormal grain growth at a given strain rate.

Nevertheless, the decrease in grain size with the increase in strain rate from 10^{-4} s^{-1} to 2×10^{-2} s^{-1} at 1120°C, and the development of very large grains after sufficiently large deformations at a strain rate confined to a narrow range (figure 7), suggests more complex mechanisms. More precisely, the analysis of the strain rate sensitivity coefficient revealed a change in behaviour, at a strain rate which was a function of temperature. Two domains were identified. In the first one (stage II), at low strain rate, the coefficient m is high (about 0.7) and the material is superplastic. In the second one (stage III), at higher strain rate, the coefficient m is lower (about 0.3). Transition from stage II to stage III involves changes in the features of intragranular deformation, and more generally, in the mechanisms of plastic accomodation and related phenomena (dynamic recrystallization, for instance). The fact that the exagerated grain growth occurs in a limited range of strain rates corresponding to the change in the strain rate sensivity coefficient, suggests a change in the recrystallization mechanisms related to the initial deformation microstructure. The intervention of a new mode of intragranular dislocation-activity at the beginning of the stage III is supposed to be responsible for exagerated grain growth during subsequent annealing.

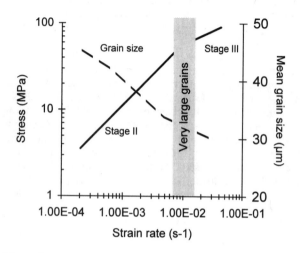

Figure 7. Relationship between mechanical behaviour and grain growth after deformation at 1120°C ($\varepsilon > 0.2$) and a supersolvus treatment (1205°C-4h).

Conclusion

If the strain is sufficiently high, the strain rate during deformation at elevated temperature of PM N18 has a major effect on the grain growth during a subsequent supersolvus heat treatment.
The final grain size has a tendency to decrease when the strain rate increases.
The development of very large grains can occur and it is related to the strain rate.
This abnormal grain growth during a subsequent supersolvus annealing is consecutive to a deformation at a strain rate in a narrow region of transition between stage II and stage III.
Finally an important practical implication of this work is that if constitutive equations can be established for stages II and III, it then becomes possible to predict the conditions under which abnormal grain growth will occur in various superalloys.

Acknowledgements

The authors gratefully acknowledge SNECMA for its financial support. They wish to thank Françoise Passilly for her assistance with metallography.

References

1. Y. Honnorat, "N18, damage tolerant nickel-base superalloy for aircraft turbine discs", Matériaux et Techniques, (1991), 19-29.

2. G. Raisson, J. H. Davidson, "N18, a new generation PM superalloy for critical turbine components", High Temperature Materials for Power Engineering Part II, eds. E. Bachelet and al., (Dordrecht, The Netherlands : Kluwer Academic Publishers, 1990), 1405-1416.

3. Y. Guedou, J. C. Lautridou, Y. Honnorat, "N18, P M Superalloy for Disks : Development and Applications", Superalloys1992, eds. S. D. Antolovitch and al. (Warrendale, PA : The Minerals Metals and Materials Society, 1992), 267-276.

4. M. Hillert, "On the theory of normal and abnormal grain growth", Acta Met., 13 (1965), 227-238.

5. M. N. Menon, F. J. Gurney, "Microstructural investigation of the growth of large grains in prealloyed powder extrusions of a nickel base superalloy", Met.Trans.A, 7A (1976), 731-743.

6. M. Dahlen, L. Winberg, "Grain coarsening of PM nickel-base superalloy by critical strain annealing", Met.Science, (1979), 163-169.

7. C. L. Briant, F. Zaverl, W. T. Carter, . "The effect of deformation on abnormal grain growth in tungsten ingots", Acta Met. et Mater., 42 (1994), 2811-2821.

8. M. Soucail, , M. Marty, H. Octor, "Development of coarse grain structures in a powder metallurgy nickel base superalloy N18", Scripta Met. et Mater., 34 (1996) 519-525.

THE INFLUENCE OF ALLOY CHEMISTRY AND POWDER PRODUCTION METHODS ON POROSITY IN A P/M NICKEL-BASE SUPERALLOY

Eric S. Huron (*)
Rebecca L. Casey (#)
Michael F. Henry (#)
David P. Mourer (@)

* - General Electric Aircraft Engines, Cincinnati, OH, 45215
\# - General Electric Corporate Research & Development, Schenectady, NY, 12345
@ - General Electric Aircraft Engines, Lynn, MA, 01910

ABSTRACT

Advanced nickel-base superalloys for use in gas turbine engines are produced using powder metallurgy (P/M) processing. The high alloy content of these alloys typically results in a high solvus temperature for the strengthening gamma-prime (γ') phase and so the heat treatment must be carried out at high temperature. This can allow entrapped gas from the atomization process to form Thermally Induced Porosity (TIP) with negative impact on low cycle fatigue.

A designed experiment was undertaken to investigate chemistry and atomization parameters on porosity. The alloy used was the advanced alloy KM4 (U. S. Patent 5,143,563), which contains about 54 volume fraction of γ' and has a γ' solvus of about 1170°C. The variables studied included carbon, boron, superheat, mesh size, and gas:metal (G:M) ratio.

In general, atomization parameters influenced loose powder porosity, with some secondary influence on TIP, while composition parameters influenced TIP and had little influence on loose powder porosity. Increased levels of boron strongly increased the amount of TIP but only slightly impacted loose powder porosity. Carbon had negligible impact on either type of porosity. Lower levels of loose powder porosity were associated with lower superheat, increased gas:metal ratio, finer powder size distributions, and reduced metal flow rate. Some of the responses were interactive. For example, higher metal flow rate promoted coarse mesh distributions.

The different levels of influence on loose powder porosity versus TIP appeared to be due to different mechanisms. For boron, which impacts TIP but not loose powder porosity, the apparent mechanism was that boron impacted grain boundary strength or promoted local incipient melting. For the atomization parameters which influence loose powder porosity, the apparent mechanism was a change in how much entrapped gas was retained during formation of the powder particles. The results are discussed in terms of significance for alloy development and process development.

INTRODUCTION

This experimental program studied factors controlling Thermally Induced Porosity (TIP) in powder metallurgy (P/M) alloys. It was actually the byproduct of a study intended to determine the influence of minor element chemistry on properties and processing of KM4, an advanced P/M disk alloy [1] developed for the Dual Alloy Disk (DAD) program[2]. Powder heats with various boron and carbon levels were produced at GE Corporate Research & Development Center (CRD) in Schenectady, New York, and were subjected to a variety of consolidation routes. An unusual amount of porosity was noted in many of the extrusions. The atomization parameters had been varied during the runs to support a concurrent fundamental atomization parameter vs. yield program, with the assumption that changes in those parameters would have little effect on the final powder. The initial data review indicated that this assumption had been wrong; in fact, the atomization parameters had a strong impact on porosity. Since the existing matrix of runs did not allow separation of chemistry impacts versus atomization impacts, an expanded matrix was developed, evolving into a core eight-run (L8) Designed Experiment with high and low levels of superheat, flow rate, and boron level. The L8 experiment, combined with the initial runs and several additional supporting runs, provided a range of data allowing a more complete assessment of what factors impacted porosity.

The results suggested that both alloy chemistry and atomization had major influences on both loose powder porosity and on TIP. The results are discussed in terms of possible mechanisms and practical impact on alloy design, powder production, and component fatigue life.

PROCEDURE

Multiple powder runs were produced from alloy KM4[1]. The nominal composition for the base alloy and the minor element variations studied are shown in Table 1. The product of each heat was screened to -140/+200, -200/+270, and -270 mesh cuts.

The powder was produced using a pilot scale atomization facility at the General Electric Corporate Research & Development

Table 1: Compositions for Initial KM4 Modification Heats (w/o)

base composition: Ni-18Co-12Cr-4Mo-4Al-4Ti -2Nb-0.030Zr-0.030B-0.030C		
minor element variations:		
Designation	Chemistry	Heat Number
base KM4	0.030B - 0.030C	T858
1-KM4	0.015B - 0.015C	T860
2-KM4	0.015B - 0.040C	T850
3-KM4	0.015B - 0.065C	T852
4-KM4	0.000B - 0.040C	T851
5-KM4	0.015B - 0.040C	T853 (a)
6-KM4	0.030B - 0.040C	T854
7-KM4	0.015B - 0.040C	T861 (b)
6-KM4	0.030B - 0.040C	T882 (c)
4-KM4	0.000B - 0.040C	T882 (d)

notes: a) duplicates 2-KM4
b) nitrogen atomization
c) remake of 6-KM4 at T851 conditions
d) remake of 4-KM4 at T882 conditions

Table 2: Atomization Runs for L8 Designed Experiment

Run#		Variables (3 at 2 levels = 2^3)			Resulting Nozzle Pressure
		Boron	Melt Flow Rate	Superheat	
			Aim / Actual	Aim	Aim / Actual
		(w/o)	(kg/min)	(C)	(bar)
1	T-884	0.030	>18 / 27.7	50C	37.4 / 33.5
2	T-854*	0.030	>18 / 30.8	200C*	37.4 / 30.6
3	T-885	0.030	<4.5 / 3.5	50C	23.8 / 23.8
4	T-886	0.030	<4.5 / 3.6	200C	23.8 / 23.8
5	T-888	0.015	>18 / 18.8	50C	37.4 / 30.6
6	T-889	0.015	>18 / 27.5	200C	37.4 / 31.3
7	T-890	0.015	<4.5 / 2.1	50C	23.8 / 17.0
8	T-891	0.015	<4.5 / 3.9	200C	23.8 / 23.8
9	T-882ə	0.030	11 / 11.5	200C	20.4 / 37.4
10	T-887	0.030	11 / 16.7	200C	20.4 / 22.1

Notes: * this run in DofE assumed to exist as T-854
ə this run in DofE assumed to exist as T-882

Center in Schenectady, New York. The powder heats were typically 10-15 kg in size. Charge material for the powder facility was prepared using Vacuum Induction Melted (VIM) ingots. The nominal 16 kg ingots were sandblasted and cropped to remove surface oxides and provide an ingot shape compatible with the atomizer.

The detailed layout and operation of this facility have been previously described elsewhere[3]. A 40 kW induction powder supply was used to heat the metal. Typical superheats were 200°C although some runs were performed at 50°C as part of the experimental design. Nominal melt flow rates were in the 2.2-4.5 kg/min range coupled with nominal gas flows of 18-27 kg/min. Powder was separated from the exiting gas flow via a cyclone. Once cooled, the powder was removed from the system (exposing the powder to air) and transferred to the sieving operation.

The initial experimental runs mixed variation in boron and carbon with variation in atomization technique indiscriminately, on the assumption that the atomization studies would not impact the quality of the powder, only the powder yield. The wide variations in porosity actually observed proved to be impossible to separate on the basis of atomization versus chemistry influence. Additional runs were added understand the effects. First, the two runs at the highest and lowest boron levels were repeated with the atomization conditions interchanged (Table 1), and eventually more runs were added to arrive at the experimental design shown in Table 2. The core Designed Experiment was a 2^3 design (three variables at two levels each, requiring eight runs) which included controlled boron, superheat, and flow rate levels to clearly distinguish effects. Another variable could have been added to make a $2^{(4-1)}$ partial factorial with only a slight increase in experimental noise, but the variables considered as fourth variables (particularly atomization chamber pressure) were likely to interact with the set-up parameters for the flow rate and superheat. One run was produced using nitrogen atomization gas to investigate a potential benefit in reduced TIP.

Atomization chamber pressure was considered as an additional variable but it could not be independently well controlled in high and low gas and melt flow rate experiments. The atomizer chamber was such that pressure built up from an initial value slightly over atmospheric pressure (about 1.1 bar) until the exhaust system released some pressure in the atomization tower during the runs, especially for the high melt flow rate runs. This was due to limited exhaust capability of the atomizer and the fact that high melt flows were coupled with high gas flows to maintain appropriate gas-to-metal ratios (G/M, or weight of gas per weight of metal, both per unit time). Hence all powder made at the high flow rates was atomized into an elevated pressure atmosphere (up to 1.5 bar greater than the low melt flow rate runs).

Powder was consolidated using either hot compaction plus hot extrusion or Hot Isostatic Pressing (HIP). The extrusion cans were fabricated as shown in Figure 1 from 300 series stainless steel. After filling a vacuum was slowly pulled on the cans, and held for a minimum time of 48 hours at room temperature. The vacuum level was stabilized at 0 microns with a maximum leakup rate of 5 microns in 15 minutes. The cans were then heated to 200°C and held under vacuum for 16 additional hours. The can stems were heated and pinched off at the end of the outgassing cycle and the pinched tubing was welded shut.

The extrusion cans were consolidated using a two-step process of hot compaction plus extrusion. Compaction was done at 1038°C after a four hour preheat into a blind die. The compactions were pushed into carbon steel sleeves placed in the liner (with copious lubricant) so that after compaction the extrusions could be processed in the same liner without remachining by simply removing the sleeves. Extrusion was done at 1049°C using the nominal 8.5 cm compacted can diameter can after compaction within a 9.05 cm liner, through a 3.68 cm die.

The HIP cans are shown in Figure 2. The HIP cans were produced using a standard procedure of cold evacuation for 36 hours minimum to a vacuum of 5 microns or better with a maximum leakup rate of 0.25 microns/minute or 15 microns/hour. This was followed by can heating to 550°C for at least 48 hours

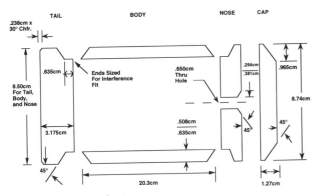

Figure 1: Extrusion can design

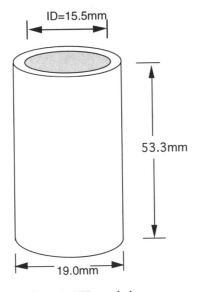

Figure 2: HIP can design

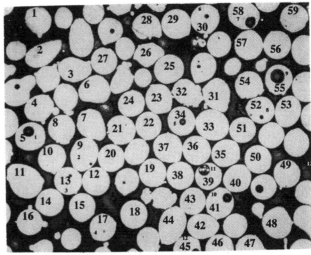

Figure 3: Schematic of pore counting measurement technique. Only particles with a diameter on the plane of polish within a tolerance range were included in the analysis. In this example, 12 particles with pores were counted in a total of 59 particles.

to a vacuum level of 5 microns or better with a maximum leakup rate of 0.25 microns/minute.

To provide material for LCF testing, additional extrusions were processed by Cameron Forge (now Wyman-Gordon). Extrusion cans of nominal 25 cm length x 9 cm dia. were produced from 304 stainless steel. The cans were evacuated with a cold static degas cycle. The cans were pumped down to a minimum aim vacuum level of approximately 10 microns with a leakup rate of about 3-7 microns/ minute. The cans were hot compacted and extruded using a 2500-ton laboratory press. Compaction was performed at 1038°C and extrusion was performed at 1043°C with a 5.5:1 reduction ratio and a 100 cm/minute ram speed.

A loose powder porosity evaluation procedure developed at CRD was key to successful data interpretation. The powder was sieved to appropriate size fractions and mounted by vacuum impregnation in epoxy. The mount was polished and three 200X micrographs were taken. The largest particle diameter from all three pictures was measured (this particle was assumed to be sectioned at mid-plane). Particle diameters that fell within a 10% tolerance of the diameter were counted. Next, those particles that fell within that counted population that contained pores or voids, of any size or shape, were counted (Figure 3). Typical values were on the order of 10-20 pores in 40-70 particles.

A density measurement procedure[4] for bulk consolidated material developed at CRD was another key to the success of this study. The procedure used the Archimedian principle of comparing the weight of the test sample in air vs. a liquid of known density, calibrated against parallel measurements using a solid of known density. Improved controls of the liquid temperature, surface finish of the sample (to control surface bubbles) and the use of an immersion fixture designed to minimize the impact of surface tension allowed high accuracy and precision. HIP'ed and extruded coupons were heat treated at a variety of temperatures for four hour exposures to examine TIP as a function of chemistry, atomization, and exposure. The material was evaluated metallographically and via the density procedure in the as-consolidated form and after TIP exposure. The baseline exposure condition was 1204°C/4hrs in an air furnace.

Limited low cycle fatigue (LCF) testing using the original CRD extrusions was performed. Alloys 3-KM4, 4-KM4, and 7-KM4 were included to see if either reduction of boron or N_2 atomization could eliminate the negative impacts of TIP. Sections of the extrusions were supersolvus solution heat treated at 1190°C and air cooled to give a cooling rate estimated at 111°C/min, then aged at 760°C for 8 hours. LCF bars were machined from the extrusion. An LCF specimen with a nominal gage section of 0.64 cm diameter by 1.91 cm long were used. Standard strain control procedures were used with a strain range of 0.80% at 649°C used as the test condition. Tests were run in strain control at 30 cpm for the first 24 hours of testing (about 43000 cycles), then in load control at 300 cpm to failure.

RESULTS

After sorting through some of the confounding influences, several key lessons concerning alloy chemistry and atomization influences were learned. The overall observation was that alloy chemistry (chiefly boron) impacted TIP porosity while several atomization parameters impacted loose powder porosity. The full data are given in Table 3, but the results can be more clearly understood

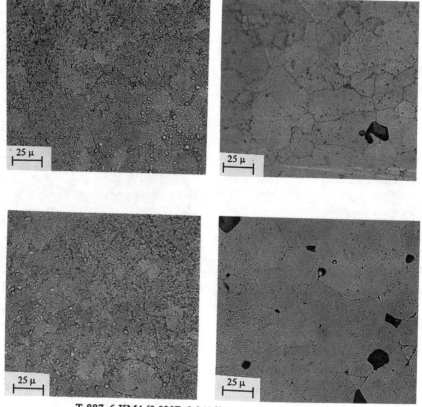

T-887, 6-KM4 (0.030B-0.040C): density change 0.60%.
left - as HIP'ed at 1175°C, right - TIP exposure 1204°C/4hr

Figure 4: Representative micrographs of material with varying degrees of TIP. These micrographs were from selected runs from the DoE shown in Table 2.

Table 3: Atomization Parameter and TIP Response Data

Atomizer Run#	T-860	T-850	T-852	T-851	T-853	T-854	T-861	T-858	T-882	T-883	T-884	T-885	T-886	T-887	T-888	T-889	T-890	T-891	L-982
Alloy#	1-KM4	2-KM4	3-KM4	4-KM4	5-KM4	6-KM4	7-KM4	KM4	6-KM4	4-KM4	6-KM4	6-KM4	6-KM4	6-KM4	2-KM4	2-KM4	2-KM4	2-KM4	KM4
Purpose	Exp. 1	Exp. 1	Exp. 1	Exp. 1	Exp. 1	Exp. 1	Exp. 1	"@851	"@854	DoE	DoE	DoE	DoE	DoE	DoE	DoE	DoE	DoE	Prod.
Atomizing Gas	Ar	Ar	Ar	Ar	Ar	Ar	N2	Ar	Ar	Ar	Ar	Ar	Ar	Ar	Ar	Ar	Ar	Ar	Ar
Metal Flow (kg/min)	3.1	16.3	13.5	7.1	14.3	30.8	13.2	2.6	11.5	11.5	27.8	3.5	3.6	16.7	18.8	27.5	2.1	3.9	n/a
Gas/Metal Ratio	6.8	2.1	2.0	8.8	2.9	1.2	2.8	9.0	4.8	3.4	1.7	6.6	6.4	1.5	2.5	1.6	8.0	6.1	n/a
Plenum Press. (bar)	23.8	21.8	18.7	37.4	30.6	30.6	34.0	23.1	37.4	30.6	33.5	23.8	23.8	22.1	30.6	31.3	17.0	23.8	n/a
C (weight percent)	0.015	0.040	0.065	0.040	0.040	0.040	0.040	0.030	0.040	0.040	0.040	0.040	0.040	0.040	0.040	0.040	0.040	0.040	0.030
B (weight percent)	0.015	0.015	0.015	0.000	0.015	0.030	0.015	0.030	0.030	0.000	0.030	0.030	0.030	0.030	0.015	0.015	0.015	0.015	0.030
POWDER SOURCE	CRD	CRD	CRD	CRD	CRD	CRD	CRD	CRD	CRD	CRD	CRD	CRD	CRD	CRD	CRD	CRD	CRD	CRD	WG
%-270	n/a	51.2	54.6	87.4	77.9	43.4	67.5	87.1	78.5	56.5	33.7	87.7	n/a	39.9	51.3	39.4	80.1	77.0	n/a
%-400	n/a	40.1	41.7	75.8	66.5	33.0	55.9	72.0	66.0	43.0	22.7	71.2	n/a	29.8	36.2	26.5	60.1	59.2	n/a
%-270/+400	n/a	11.1	12.9	11.6	11.4	10.4	11.6	15.1	12.0	14.0	11.0	16.5	n/a	10.1	15.1	12.9	20.0	17.9	n/a
% Porous (-270/+325)	19.3	35.9	45.3	27.3	16.2	44.7	11.1	9.7	12.1	14.7	24.3	17.4	10.8	30.8	30.0	19.5	10.7	13.1	5.2
% Porous (-325/+400)	7.5	29.6	26.6	25.1	13.1	25.2	18.7	2.9	20.1	14.3	21.1	24.7	11.3	28.0	24.0	16.5	7.4	9.5	5.1
EXTRUSIONS																			
As Extruded Density (g/cc)	8.0189	8.0160	8.0290	8.0304	8.0098	8.0063	8.0272	8.0203											
Extrusion TIP Level	-2.20%	-2.20%	-0.19%	-0.11%	-2.30%	-5.50%	0.04%	-0.19%											
HIP CANS																			
As HIP'ed Density (g/cc)	8.0336	8.0276	8.0296	8.0317	8.0321	8.0259		7.8898	8.0261	7.8803	8.0245	8.0310	8.0272	8.0245	8.0264	8.0271	8.0291	8.0265	8.0286
HIP Can TIP Level	-0.12%	-0.16%	-0.17%	-0.10%	-0.27%	-0.70%		-0.33%	-0.68%	-0.08%	-0.38%	-0.38%	-0.31%	-0.60%	-0.16%	-0.13%	-0.08%	-0.10%	-0.28%

by following the somewhat chronological outline discussed below.

The initial observation of high TIP levels was somewhat of a surprise. The first indication of high TIP levels in some of the extrusions was observed in a microstructure versus chemistry vs. thermal exposure study. Samples from each of the alloys were exposed to temperatures of 1163°C, 1171°C, 1179°C, 1188°C, 1196°C, and 1204°C for four hours. Representative micrographs are shown in Figure 4. These are compared to as-HIP'ed micrographs to show that consolidation was not the problem - good can integrity allowed full consolidation to be achieved in both HIP & extrusion. The porosity formed during the subsequent thermal exposure. After this initial study, 1204°C/4hr was chosen as the standard TIP evaluation heat treat cycle.

An early assessment using only the runs shown in Table 1 showed that 6KM4, with the worst TIP, had the highest melt flow rate, coarser size distribution, and low G/M, while 4KM4, with the best TIP, had low flow rate, finer size distribution, and a high G/M.

An additional comparison was made to an available production-scale KM4 extrusion, made with 0.030B-0.030C material, E462. This extrusion was produced from -140 mesh Cameron Powder Systems powder, extruded at Cameron Forge (both are now part of Wyman-Gordon) for the Dual Alloy Disk (DAD) program. The TIP values reported for E462 by Cameron were 0.23 to 0.26%. This compared favorably to a TIP level measured at GE CRD of 0.28% using a small HIP can of archived powder from the same heat (L-982). Overall for the DAD extrusions, the four heats, representing four chemistries, with boron levels above 0.030 w/o had an average TIP response of 0.24%, while the two heats with boron levels around 0.015 w/o had an average TIP response of 0.19%[2].

These initial trends began to suggest that boron impacted TIP. However, the apparent scatter in the initial data suggested that other factors were playing a role on overall porosity. The atomization runs performed at this point did not allow complete separation of chemistry impacts from atomization parameter impacts. Additional runs were made to produce the core Designed Experiment (DoE) shown in Table 2 to better identify these trends.

The first step of the DoE analysis, scatter plots, further established what controlled porosity. The overall results are summarized in Figure 5, with the complete data presented in Table 3. The most obvious trend was the influence of boron level on TIP. The other variables of superheat, flow rate, and Gas:Metal (G:M) ratio also showed trends. Each of the parameters will be discussed in detail below.

In general the porosity could be classified into two types. The first was porosity in the loose powder particles, referred to as "percent (%) porous". This was measured by the loose powder technique. The second was classical Thermally Induced Porosity (TIP), produced after consolidated samples were exposed to elevated temperature. This was measured by the density change technique. Overall, alloy chemistry trends tended to correlate with TIP porosity, while the atomization method trends tended to correlate with the percent porous powder porosity.

The most obvious trend was for boron. Boron strongly influenced TIP porosity (Figure 6). The relationship appeared to be non-linear. For the heat with 0.000 boron, the TIP porosity density change was less than 0.1%, while for heats with 0.015 boron, the TIP porosity density change data fell between about 0.28% and 0.18%, and for heats with 0.030 boron, the TIP porosity density change data fell between 0.7% and 0.3%. The 0.015B and 0.030B levels were markedly different. Metallography showed that the higher boron corresponded to increased amount of incipient melting (Figure 7). The boron apparently promoted TIP formation by weakening of grain boundary triple points. Boron did not appear to have an effect on loose powder porosity (Figure 8).

In contrast to boron, carbon had no apparent effect on either TIP porosity or loose powder percent porous.

The level of superheat showed a weak trend against both TIP porosity and loose powder percent porous, with higher superheat tending to increase porosity. This is shown in Figures 9 and 10.

The as-atomized size distribution of the powder showed a strong trend against percent porous, with finer powder resulting in reduction in percent porous powder (Figure 11). However, the size distribution showed a weaker trend against as-HIP density (Figure 12), with finer size distributions resulting in higher as-consolidated densities.

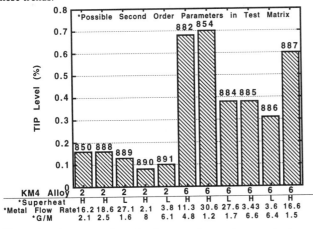

Figure 5: Summary of results for overall experiment. A clear separation between high boron heats (KM4 alloy 6) and low boron heats (KM4 alloy 2) versus TIP level was obvious.

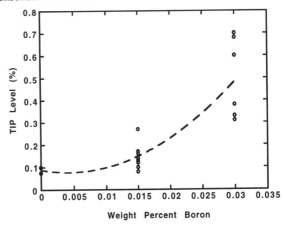

Figure 6: Influence of boron on TIP level. A sharp increase in TIP occurred as boron increased to 0.030 weight percent.

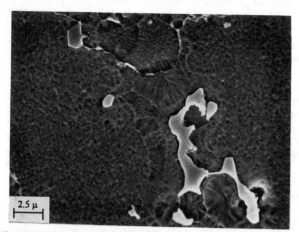

Figure 7: Micrograph showing incipient melting and pore formed at triple point. The boron apparently caused localized melting of the material.

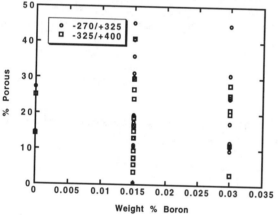

Figure 8: Boron had no statistical impact on loose powder porosity (% porous). This result was consistent for both fine (-325/+400 mesh) and somewhat coarser (-270/+325 mesh) size fractions.

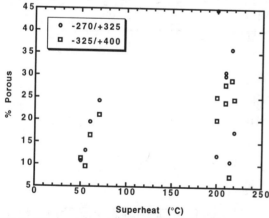

Figure 9: Increased superheat promoted a slight increase in loose powder porosity (% porous). Only the data from the L8 Designed Experiment (Table 2) are shown since these were the runs with controlled superheat variations.

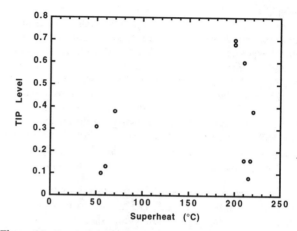

Figure 10: Increased superheat also promoted an increase in TIP level of consolidated material. As in Figure 9, only the data from the L8 Designed Experiment (Table 2) are shown.

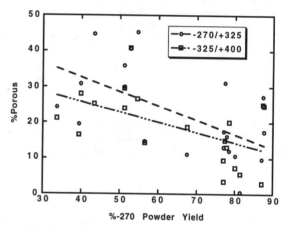

Figure 11: Relationship between relative fineness of powder heats and loose powder porosity. Heats with a finer size distribution (greater percentage of fine powder) showed a reduction in loose powder porosity.

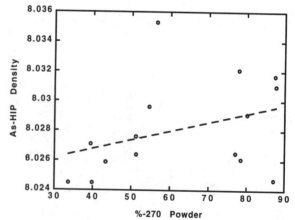

Figure 12: Finer powder distributions promoted a higher as consolidated density (and reduced TIP).

The metal flow rate appeared to have more influence than superheat on loose powder porosity. Metal flow rate showed a strong trend, with higher flow rate increasing porosity (Figure 13). The relationship between metal flow rate and TIP porosity was much less clear, although the data suggested that higher flow rate correlated weakly with higher TIP porosity (Figure 14). The influence of metal flow rate was related to mesh size, as the metal flow rate strongly impacted the mesh size distribution, with finer powder resulting from lower flow rate (Figure 15). Thus low flow rate promoted higher production yields of fine powder.

Metal flow rate is also related to the gas/metal ratio (G/M), so the trends observed for G/M were expected based on the flow rate trends. Increased G/M ratio was associated with lower porosity, with a strong relationship against porous powder (Figure 16), but no clear relationship with TIP porosity (Figure 17).

The nitrogen atomized heat, 7-KM4, was very low in porosity, and actually showed a very slight density increase after the TIP exposure. This was probably within test technique tolerance and

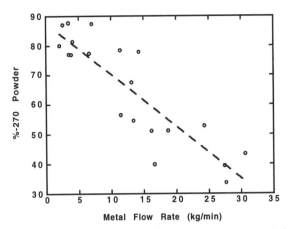

Figure 15: A very strong relationship between flow rate and yield of fine powder was observed. Low flow rates were preferred for finer powder distributions.

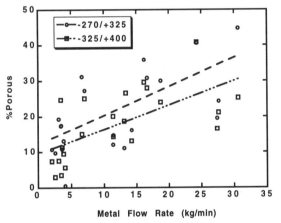

Figure 13: Relationship between metal flow rate and loose powder porosity. Higher flow rates led to significant increases in porosity.

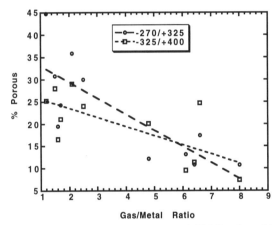

Figure 16: Relationship between gas:metal (G:M) ratio and loose powder porosity. Higher values of G:M reduced the amount of porosity. Note that metal flow rate is related to G:M so the trends between Figure 15 and Figure 16 are understandable. Only the runs from the L8 DoE (Table 2) are included.

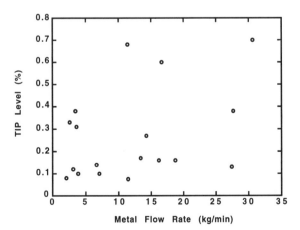

Figure 14: Relationship between metal flow rate and TIP level, showing a weak tendency for increased TIP with increased flow rate. Only the data from the L8 DoE are included.

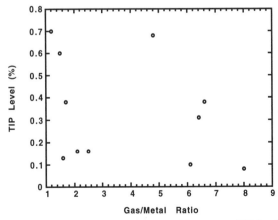

Figure 17: Gas:metal ratio had no real impact on TIP level, in contrast to its impact on loose powder porosity (Figure 16).

indicated essentially zero tip. No pores were observed metallographically in this alloy while they were readily observable in the argon-atomized alloys.

The next step of DoE analysis, tests for statistical significance, was performed using standard techniques[5]. The numerical analysis largely confirmed the trends observed from the effect scatter plots discussed above. Statistically significant relationships were found between boron and TIP (Figure 6), flow rate and percent fine powder (Figure 15), and flow rate and percent porous (Figure 13). Although many of the other trends were not rigorously shown to be statistically significant, this type of experiment has a high standard error and the trends identified from the scatter plots can still suggest potential approaches to achieve process improvements.

LCF tests were done at 649°C and 0.80% strain (Table 4). The alternating pseudostress values varied slightly due to experimental differences in establishing the modulus and initial strain range. The material chosen for testing represented the alloys from the initial extrusions likely to have the lowest porosity: alloys 3-KM4 (0.015B-0.065C), 7-KM4 (0.015B-0.040C-nitrogen atomized), and 4-KM4 (0.000B-0.040C). The lives were compared to results from the baseline 0.030B-0.030C composition 2 as shown in Figure 18. The 0.015B material was definitely improved relative to the 0.030B baseline, but the alloy with no boron showed a lower average life than the other two alloys (44717 average cycles for 4-KM4 compared to 67421 average cycles for 3-KM4 and 68410 for 7-KM4). The comparisons with the 0.030B baseline were somewhat compromised by the use of different atomization sources, so that the ceramic inclusion content would be expected to vary, but the 0.030B material displayed low lives due to pore initiation sites and the reduced boron definitely reduced this tendency. The differences between the 0.000B and 0.015B heats in the present study were small relative to the reduced lives (Figure 18) shown by the 0.030B data[2].

DISCUSSION

The impact of boron was an important result as it suggests a means to widen the heat treat window of advanced P/M disk alloys. Many of these alloys have solvus temperatures above

Table 4: LCF Test Results

Alloy	Test S/N	Alternating Pseudostress (MPa)	Nf	Average Nf
3-KM4 (0.015B-0.065C)	1-02	714.3	69457	
	2-02	717.8	42617	67421
	3-02	719.8	90191	
4-KM4 (0.000B-0.040C)	4-02	722.6	40507	
	5-02	714.3	61894	44717
	6-02	730.9	31750	
7-KM4 (0.015B-0.040C nitrogen)	7-02	722.6	78569	
	8-02	725.3	50234	68410
	9-02	722.6	76429	

Notes: 1) all tests conducted at 649C, 0.80% aim strain range
2) alternating pseudostress = (modulus)(strain range) / 2

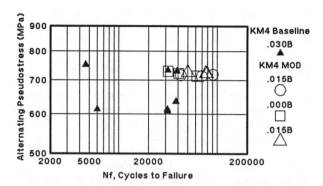

Figure 18: Low Cycle Fatigue (LCF) results for KM4. Data from 0.030B baseline composition[2] are compared to heats with reduced boron. The reduced boron led to increased life and reduced scatter.

1170°C, yet have added boron content to improve fatigue crack growth[6]. The alloys thus show TIP formation concurrent with boride melting as low as 1193°C. This heat treatment window may be too tight for reliable production. This study suggests that added boron to improve crack growth properties must be balanced against a low boron level desirable to improve producibility and improve LCF capability.

The results in terms of boron impact are also consistent with a study conducted on alloy APK1, essentially Astroloy[7]. In that study, boron was varied from 0.017 to 0.025 weight percent. Using Differential Thermal Analysis (DTA), equations for the solidus and M_3B_2 solvus temperatures were determined:

Solidus in degrees C = 1285 - 0.26B (B in ppm)

M_3B_2 solvus in degrees C = 1255 - 0.20B (B in ppm)

For 0.015 (150 ppm) boron these equations predict a solidus temperature of 1246°C and a boride solvus of 1125°C. For 0.030 (300 ppm) boron these equations predict a solidus temperature of 1206°C (a 40°C drop) and a boride solvus of 1195°C (a 30°C drop). If these equations directly applied to KM4, increasing the boron from 0.015 to 0.030 would be predicted to produce both incipient melting of the alloy itself and the dissolution of the boride phase, which probably also promotes local grain boundary incipient melting, for typical solution temperatures of 1190-1200°C and the TIP temperature of 1204°C.

Since excessive boron promotes incipient melting and a high level of measured TIP density change, reduction of boron is attractive. The minimum level to insure adequate properties must be established. Previous researchers[8] have claimed that amounts above the solubility limit are of little value. At a grain size of 10 microns (~ASTM 11), only 0.001 w/o boron was calculated as the minimum level. In the present study the very slightly lower LCF results of 4-KM4 (0.000B) suggested that even at relatively time-independent conditions of 649°C, 30 cpm, elastic LCF cycling, complete omission of boron had some effect. Further work focused on property testing is needed to establish an optimum boron content to balance TIP vs. high temperature performance.

The potential of alloy chemistry to influence porosity has been noted by others. An existing patent[9] claims that the addition of trace amounts of Mg or Ca functioned as activating agents which reduced TIP. The activating agents worked by rapidly diffusing to the surface of the metal powder particles and attracting oxygen.

The boron influence was stronger for TIP porosity than for the as-atomized loose powder porosity. The TIP porosity mechanism seems related to local weakening of the material allowing the residual entrapped gas to form a pore. Apparently the incipient melting mechanism is not kinetically likely to occur during droplet solidification and the powder solidification behavior is not influenced by boron content.

The fact that low melt flow rates increased yield and tended to correlate with reduced TIP is comforting as both results are desirable and the superalloy powder metal suppliers are working to improve yields. Reduced flow rate, low superheat, and high gas/metal ratio (to reduce powder size) all promoted lower porosity and tended to have greater impact on the as-solidified loose powder porosity relative to the impact of boron. The loose powder porosity mechanism appears related to the droplet formation. Apparently these atomization events influenced the stream behavior and how the droplets were formed. The lack of a strong impact of atomization chamber pressure is surprising. Other researchers[10] have held that reduced pressure during atomization reduced entrapped argon content. The details of the atomization condition in that study are unknown. The lack of apparent impact of pressure in the present study may reflect the fact that it was not deliberately varied in a controlled manner (or was not varied to a sufficient level to observe the effect). Also the short run time and subscale nature of the CRD atomizer mean that the pressure may not reach equilibrium during the runs.

As a final comment, this study supports the power of Designed Experiment approaches to solve engineering problems. In the initial runs atomization parameters were allowed to vary with the assumption the final properties would not be affected. A DoE was required to eventually sort through the various factors impacting porosity. The initial experiment could still have been made with lower risk if the variation had been systematic and the alloy chemistry variation was "blocked" appropriately, essentially a Designed Experiment. This was the approach taken for the third group of powder heats which when completed allowed very clear identification of trends. It is recommended that all experiments be blocked against both deliberately varied parameters and potential shifts in equipment-related or environmentally-influenced factors.

CONCLUSIONS AND RECOMMENDATIONS

1. Porosity could be classified into two types: porosity in the loose powder particles, referred to as "percent (%) porous"; and Thermally Induced Porosity (TIP), produced after consolidated samples were exposed to elevated temperature. TIP porosity was strongly influenced by alloy chemistry while loose powder porosity was most influenced by atomization parameters.

2. Boron strongly impacted the TIP porosity. The relationship appeared to be non-linear, with sharply increasing porosity for 0.030 boron. Boron did not have any apparent effect on the percent porous loose powder. The higher boron corresponded to increased amount of incipient melting, apparently promoting TIP formation by weakening of grain boundary triple points.

3. Several atomization parameters influenced loose powder porosity. Higher superheat, coarser powder size distribution, increased metal flow rate, and reduced G:M ratio all correlated with increased porosity.

4. Two important measurement techniques, measurement of porosity in loose powder (percent porous), and careful density measurements before and after TIP exposure, were key in obtaining useful conclusions.

5. Low Cycle Fatigue (LCF) testing results indicated that a reduction in boron from 0.030 to 0.015 or below weight percent had a beneficial impact on life, and was correlated with reduced severity of crack initiation due to TIP.

ACKNOWLEDGMENTS

The support of GE Aircraft Engines colleagues is appreciated, including technical guidance from Dan Krueger and the sample processing provided by Ron Tolbert and Bill Brausch. Swami Ganesh of GE Power Generation also provided key technical input to this program. GE Corporate Research and Development personnel also assisted with this project, notably Steve Miller, Larry Wojcik, Paul Dupree, and Paul Martiniano. John Hughes processed the extrusions at GE CRD while Don Weaver, Bill Konkel, and Noshir Bhathena were responsible for the Cameron extrusions (Cameron is now a part of Wyman-Gordon).

REFERENCES

1. D. D. Krueger, J. F. Wessels, and Keh-Minh Chang, U. S. Patent 5,143,563, General Electric Co., September 1, 1992.

2. S. Ganesh et. al., "Advanced Technology for 1500°F Turbine Disk Materials (Dual Alloy Disk) Program", Final Report, Contract N00140-89-C-WC14, Sept. 30, 1994.

3. S. A. Miller and R. S. Miller, "High Speed Video and Infrared Imaging of Close-Coupled Gas Atomization", in Powder Metallurgy in Aerospace, Defense, and Demanding Applications, ed. by F. H. Froes, Metal Powder Industries Federation, Princeton, New Jersey, 1993.

4. P. L. Dupree, T. F. Sawyer, and M. G. Benz, "Improved Method For Measuring Relative Density", GE Corporate Research & Development Report 90CRD240, May 1991.

5. R. H. Lochner and J. E. Matar, Designing For Quality, American Society for Quality Press, Milwaukee, Wisconsin, 1990.

6. J. Gayda. T. P. Gabb, and R. V. Miner, "Fatigue Crack Propagation of Nickel-Base Superalloys at 650°C", Low Cycle Fatigue, ASTM STP 942, ed. by H. D. Solomon et. al., American Society for Testing and Materials, Philadelphia, PA, 1988, 293-309.

7. G. A. J. Hack and J. W. Eggar, "Comparison of Nimonic Alloy APK1 Consolidated at HIP Temperatures Above and Below the Boride Solvus", in <u>Powder Metallurgy Superalloys: Aerospace Materials for the 1980's, Metal Powder Report Conference, Volume 2</u>, 18-20 November 1980, Zurich, pp. 20.1-20.50.

8. T. J. Garosshen, T. D. Tillman, and G. P. McCarthy, "Effects of B, C, and Zr on the Structure and Properties of a P/M Nickel Base Superalloy", Met. Trans. 18A, 1987, pp. 69-77.

9. J. M. Larson, I. S. R. Clark, and R. C. Gibson, "Porosity Reduction in Inert-Gas Atomized Powders", U. S. Patent 4,047,933, September 13, 1977.

10. R. J. Siddall, "Atomization into a Chamber Held at Reduced Pressure", U. S. Patent. 4,233,062, November 11, 1980.

Damage Tolerance of P/M Turbine Disc Materials

M. Chang[*], A.K.Koul[**], C.Cooper[***]

[*] Dept of Mechanical and Aerospace Engineering,
Carleton University, Ottawa, Canada K1S 5B6
[**] Structures, Materials and Propulsion Laboratory,
Institute for Aerospace Research,
National Research Council of Canada, Ottawa, Canada K1A 0R6
[***] Chief of Materials, P&WC, Longueuil, Quebec, Canada J4G 1A1

Abstract

The influence of heat treatments on the microstructure and mechanical properties of a powder metallurgy (P/M) turbine disc alloy MERL76, was studied. A damage tolerant microstructure (DTM) was designed for the alloy using a modified heat treatment (MHT) sequence. Comparisons of mechanical properties between the DTM and the conventional microstructure (CM) were made on the basis of tensile strength, low cycle fatigue (LCF) life, creep rupture life, and fatigue crack growth rates (FCGRs). The DTM significantly improved the creep rupture life (≈10 times) at 760°C but resulted in a marginal loss in tensile strength. The FCGR tests showed that FCGRs of the DTM at 700°C were much lower than the CM material by a factor of 10 at a frequency of 0.15 Hz whereas the FCGRs at a frequency of 1 Hz were comparable over a wide range of ΔK values. The LCF lives of the two microstructures at 700°C were also comparable.

Introduction

The United States Air Force (USAF) military standard MIL-STD-1783 [1], for new jet engine design and MIL-STD-1843 [2] for engine maintenance, demand that both crack initiation life as well as damage tolerance requirements for critical components such as turbine discs must be satisfied. Designing microstructures which satisfy damage tolerance requirements as well as crack initiation life at all fracture critical locations over a wide range of temperatures is a challenging task.

A damage tolerant microstructural design philosophy for turbine disc materials has been previously developed [3,4] with a view to improving their creep properties and elevated temperature fatigue crack growth resistance without sacrificing tensile strength and LCF crack initiation life relative to the conventional microstructure (CM). This philosophy was used to develop a damage tolerant microstructure (DTM) for Alloy 718 [3-5] and later other workers [6,7] extended this philosophy to P/M turbine disc materials such as René 88DT and N18. This microstructural design philosophy relies on super-solvus processing of the disc materials coupled with a strict control of microstructural features such as the grain size, grain boundary morphology, grain boundary carbides distribution, γ' precipitate distribution and dislocation sub-structures [8].

This paper examines the possibility of applying this philosophy to P/M MERL76 turbine disc alloy and developing a damage tolerant microstructure (DTM) for this material. Influence of the DTM on elevated temperature mechanical properties is studied and the results are compared with the CM properties.

Test Materials and Methods

A Gatorized[1] MERL76 turbine disc forging was used in this study and the forging conformed to the following chemical composition (wt.%):

Cr	Co	Mo	Al	Ti	Nb	Hf	C	B	Zr	Ni
12.2	18.3	3.2	5.0	4.3	1.4	0.4	0.02	0.02	0.06	bal.

The CM is produced through the standard heat treatment (SHT) sequence of solutioning at 1143°C/2hr/OQ followed by stabilization at 982°C/1hr/Fan Air Cool and aging at 732°C/8hr/AC. A modified heat treatment (MHT) was designed to produce the DTM through supersolvus solutioning above 1190°C followed by the same aging sequence. The stabilization treatment, which is used for stabilizing the primary γ' size in the SHT, was however dropped in the MHT because the specimens are furnace cooled after the solutioning treatment and this slow cooling also has a stabilizing effect on primary γ'. The DTM has a coarser grain size (ASTM 7) relative to the SHT material (ASTM 11) and it also contains a serrated grain boundary structure which was produced through the precipitation of γ' phase along the grain boundaries. The intragranular γ' distribution was also modified in the DTM.

A series of mechanical tests was performed on both CM and

---

[1] Gatorizing involves hot extrusion of powder metallurgy material followed by isothermal forging.

DTM, as described below.

Creep Rupture Test specimens, conforming to standard ASTM-E8 and E139 specifications, were tested in air at 760°C using a constant load which resulted in an initial stress of 350 MPa. Temperature was controlled to within ±1°C. Two creep specimens each were tested per heat treatment condition.

Tensile Tests, on specimens conforming to ASTM-E8 and E21 specifications, were conducted at RT and 700°C. The test strain rates were of the order of 3.7×10^{-4} s^{-1} and 3.7×10^{-5} s^{-1} in 700°C and room temperature tests respectively. Temperature was controlled to within ± 2°C.

Fatigue crack growth rate (FCGR) tests at 700°C were conducted in air on compact tension (CT) specimens, conforming to ASTM-E647 specifications, at frequencies of 1, 0.25, and 0.15 Hz using a trapezoidal waveform and a stress ratio $R = 0.1$. A maximum load of 6 - 7 kN was used, which produced a stress intensity range ΔK in the range of 10 to 100 MPa$\sqrt{m}$. A direct current potential drop (DCPD) technique, using a current of 20 ampere, was used to monitor crack propagation. Pre-cracking was carried out with load-shedding from 11 kN down to 7 kN in 4 steps at 700°C and 1 Hz. Three CT specimens were tested for each microstructural condition.

LCF Tests on smooth axial cylindrical low cycle fatigue specimens (⌀7.62 × 25.4 mm), conforming to ASTM-E606 specifications, were conducted in air at 700°C under strain control using a triangular waveform and fully reversed ($R = -1$) loading at a constant strain rate of 0.002 s^{-1}. The tests were conducted over a total strain range of 0.9 - 2.0%. Crack initiation life was defined as the number of fatigue cycles required to create a 5% drop in the steady state stress value. Three to five specimens at each strain range level were tested for each microstructural condition.

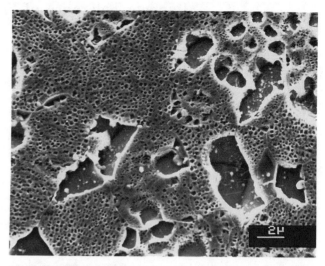

Fig-1. Conventional microstructure (CM) of P/M MERL76.

microstructures containing high γ′ volume fraction with IPBs [8]. It is believed that the two GCTs are associated with the γ′ and primary MC carbides solvus temperatures. The γ′ solvus temperature of MERL76 is about 1188 - 1191°C [10-12], depending on the Hf content, which matches the first GCT in Fig-2. The MC carbides solvus temperature of MERL76 is not well documented in the open literature. However, it was reported that, in a high-carbon (0.17) P/M IN100 (MERL76 is a low carbon derivative of P/M IN100), the MC solvus is about 1260°C [8,20]. Because of the low carbon content, the amount of MC carbides in MERL76 should be lower and the grain coarsening therefore starts at a lower temperature relative to the MC carbide solvus temperature of P/M IN100.

Results and Discussion

The conventional microstructure of MERL76 after the standard heat treatment is shown in Fig-1. Large blocky pits are γ′ particles or γ′ grains with interphase boundaries (IPBs) which were etched out as a result of etching in Kalling's reagent 2. These large γ′ precipitates with IPBs contribute little to the strength of the γ′ strengthened superalloys. It has been recognized that, during superalloys processing, super-solvus solution treatment for high γ′ volume fraction (>40%) alloys should be selected in a manner such that the temperature is high enough to dissolve all γ′ but at the same time excessive grain growth must be prevented in order to achieve a balance between tensile strength, creep, and fatigue properties [8,9].

Grain coarsening characteristics of MERL76 as a function of solution treatment temperatures and times are shown in Fig-2. Two grain coarsening temperatures (GCTs) were observed at 1185 ~ 1190°C and 1235 ~ 1250°C. This behaviour is known to occur in Ni-base P/M superalloys with micro-duplex γ - γ′

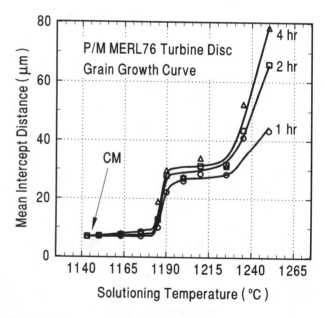

Fig-2. Grain coarsening characteristics of P/M MERL76.

At temperatures above 1185°C some porosity (<0.1 vol.%) was also observed. It is not clear whether these pores form as a result of γ' particles with IPBs being etched out or condensation of argon gas into bubbles during elevated temperature solutioning. Above 1210°C, evidence of intergranular incipient melting was observed by SEM. Evidence of incipient melting temperature (IMT) as low as 1196°C has been reported for this alloy [13]. The temperature window between the γ' solvus and the grain boundary IMT in this alloy is narrow, compared to other disc alloys in Table-1, because of the presence of high levels of Hf, B, and Zr which are well known melting point depressants. For example, at 0.4% Hf the IMT is 1207° while at 0.7% Hf it is only 1196°C [11].

In spite of the small temperature window, super-solvus processing for this alloy is still possible if the temperature distribution can be well controlled. In P/M alloys, the temperature window between γ' solvus and incipient melting should be large enough, to allow for super-solvus processing. After super-solvus solutioning, it is necessary to control the subsequent γ' precipitation along the grain boundaries to form serrated grain boundaries and intragranular matrix γ' precipitate size and distribution to optimize properties.

Table-1. The γ' solvus, GCT, IMT (°C) in turbine disc alloys

Alloy	f (%)	γ' solvus	GCT	IMT
MERL76 [10,11]	64	1190	1180	1196
IN100 [10,20]	61	1185	1200	1260
N18 [7,10]	55	1190	1140	>1200
AF115 [14]	55	1190	----	>1200
Rene95 [10,15]	48	1155	1175	>1190
Astroloy [10,16]	45	1145	1130	1220

f = γ' volume fraction;   GCT = grain coarsening temperature;   IMT = incipient melting temperature.

Solutioning time has no significant effect on the grain size of the alloy in the temperature ranges of 1150°C - 1180°C and 1190°C - 1225°C. But between 1180°C - 1190°C (the first GCT), however, longer solutioning time (4 hours) results in a larger average grain diameter (19 μm) than shorter solutioning time (2 hours, 13 μm). Solutioning at 1190°C for 2 or 4 hours produces similar grain sizes. This suggests that solutioning time should be about 2 hours if solution temperatures above 1190°C but below 1225°C are used. For comparison, one sample was treated at 1185°C/7hr/AC to check the effect of a very long solution time on grain growth. The result showed that the grain size was similar to that observed after 2 hours of solutioning. However, very large γ' grains were reduced in number with the longer solution time of 7 hours and the grains were more uniformly distributed. When solution temperatures are higher

Fig-3. Serrated grain boundaries in MERL76.

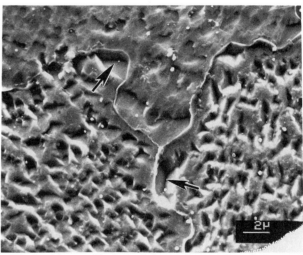

Fig-4. Almost elliptical γ' (arrows) precipitated along the grain boundaries form the serrations.

than 1225°C, the grain growth is rapid and solution time has a very strong effect on grain growth. Therefore, depending on the material heat and its incipient melting temperature, the super-solvus solution temperature must be strictly controlled over a range of 1180 to 1205°C for dissolving all γ' including γ' with IPBs.

Serrated grain boundaries in MERL76 can be successfully generated simply through furnace cooling from super-solvus solution temperature, Figs 3 and 4. All large angle grain boundaries are serrated. The relatively lower misoriented and twin boundaries, about 10% of the total grain boundaries, are not changed into a serrated morphology. These results are not totally unexpected because the grain boundary γ' - matrix misfit in this alloy is large enough to provide the necessary driving force for the formation of serrations along high angle boundaries [17,18].

The **γ' morphology and size distribution** of super-solvus solutioned and furnace cooled plus aged (732°C/8 hours) specimen, i.e. DTM specimen, are compared with the γ' morphology and size distribution in the conventional microstructure (CM) in Fig-5. The γ' size, morphology, and distribution are considerably different in the two microstructures. The CM contains micro-duplex γ - γ' grain structure where the γ' grains with IPBs are up to 6 μm in diameter. The CM also contains coarse intragranular primary and secondary γ' precipitates. In contrast, the DTM contains grain boundary γ' along each valley and peak of the serrated grain boundaries, Fig-4 and a uniform distribution of intragranular primary γ' with rosette morphology and a finer secondary γ' size, Fig-5(b).

Monotonic **tensile properties** of the two microstructures are compared in Fig-6. The DTM lowers the yield strength marginally by ~ 7% and the UTS by 16% relative to the CM at room temperature. Tensile ductility of the DTM material is about 15% relative to 22% observed in the case of the CM. The presence of a coarser grain size in the DTM could have contributed towards the decrease in yield strength at room temperature through the well known Hall-Petch effect. At 700°C, however, the tensile strength is about the same for both microstructures but the yield strength is reduced by 7% relative to the CM. The difference in tensile ductility may also be influenced by, other than the Hall-Petch effect, the presence of porosity in the DTM. The difference in room temperature hardness values between the two materials (46.5 HRC for CM and 44.5 HRC for DTM) is consistent with the trends in yield strength results. The elastic modulus of the DTM is marginally higher than that of the CM at both test temperatures.

Tensile fracture of the two microstructures at room temperature and 700°C is quite different. At room temperature, the CM specimen revealed a mixture of intergranular and transgranular fracture. The transgranular part is associated with the fracture of the large γ' grains with IPBs. At 700°C, the CM

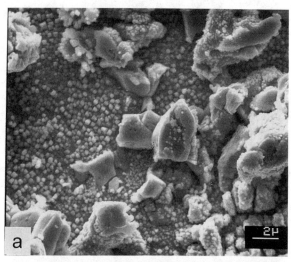

Fig-5. The γ' morphology in (a) CM and (b) DTM.

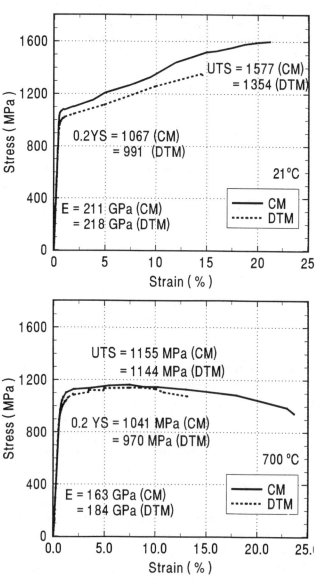

Fig-6. Comparison of the tensile properties of CM and DTM at room temperature and 700°C.

fractures intergranularly and many secondary intergranular cracks are also present. The DTM material, on the other hand, is characterized by a mixed mode fracture at both room and elevated temperatures, Fig-7.

Owing to the hardness difference between the matrix and the large γ' particles in the CM, deformation is not evenly distributed, stress builds up at IPBs and fractures the γ' grains. The large γ' particles are probably sheared due to their brittle character at lower temperatures. As temperature increases, the grain boundaries become the weak link in the overall structure and this is why the fracture mode in the CM at 700°C is predominantly intergranular. This transition in fracture mode also indicates that the grain boundaries of MERL76 need to be strengthened for elevated temperature applications. At lower as well as elevated temperatures, the IPBs between the matrix and large γ' particles provide the readily available crack initiation sites. In the case of the DTM, however, the serrated grain boundaries re-distribute and lower the stresses along the grain boundaries and promote the transgranular deformation.

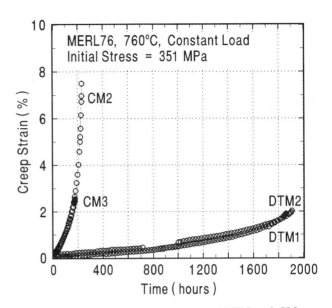

Fig-8. Creep rupture comparison of DTM and CM at 760°C and 350 MPa.

The DTM substantially improved the **creep rupture** life of MERL76, Fig-8, by a factor of 9 and reduced the minimum creep rate by a factor of 20 relative to the CM. The rupture strain in the DTM was lower. It is evident that the improvement in DTM creep rupture life occurs as a result of an extended secondary stage of creep and the loss in creep ductility occurs as a result of the suppression of the tertiary creep stage. This behaviour is particularly significant for industrial applications where designs are based on the creep life to a certain amount of creep strain such as 0.2% or 0.5% strain.

Creep fracture in both the CM and the DTM are predominantly intergranular. The differences in the CM and DTM lie in the cracking behaviour. The crack density is much higher in the CM than in the DTM material, Fig-9. In the CM, intergranular cracks develop very early during creep and the overall creep resistance of the material is poor. In the DTM, grain boundary sliding is hindered by a relatively coarser grain size and the presence of serrated grain boundaries. Cavity formation and coalescence are also delayed by the serrated grain boundaries due to their role in re-distributing the stresses. In addition, if vacancy diffusion distance to the grain boundaries is regarded as a rate controlling parameter for cavity nucleation and growth, the CM would be expected to have a higher cavity density than the DTM because of the finer grain size. Although the cavity link-up along the grain boundaries eventually leads to intergranular fracture in the DTM as well, the creep rupture life is substantially extended. The reduction in rupture strain in the DTM material is not totally unexpected because, in long term creep rupture tests, rupture strain is generally reduced due to longer exposure to an aggressive (oxidizing) test environment.

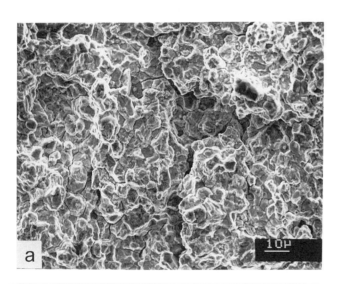

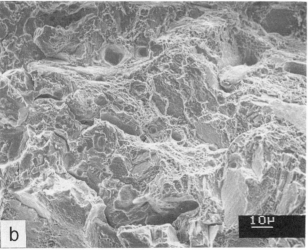

Fig-7. Tensile fracture of CM (a) and DTM (b) at 700°C.

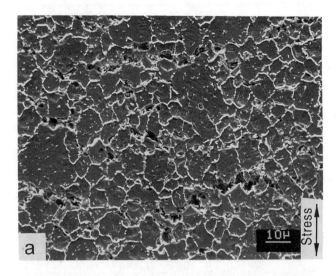

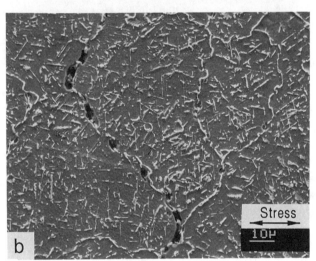

Fig-9. Creep cavities in both (a) CM and (b) DTM tested at 760°C and 350 MPa in laboratory air.

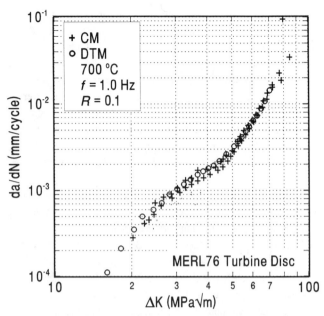

Fig-10. Comparison of FCGRs of CM and DTM at a frequency of 1 Hz. tests conducted at 700°C in laboratory air.

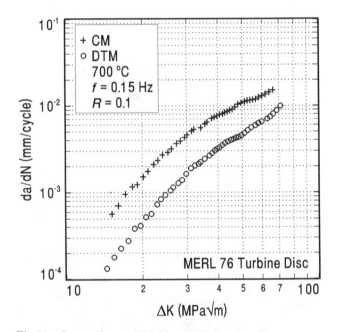

Fig-11. Comparison of FCGRs of CM and DTM at a frequency of 0.15 Hz. Tests conducted at 700°C in laboratory air.

The elevated temperature **Fatigue Crack Growth Rate** (FCGR) results of the two microstructures at two test frequencies of 1 Hz 0.15 Hz are compared in Figs 10 and 11 respectively. It is evident that the FCGRs of the CM and DTM are comparable at 1 Hz while the FCGR of the DTM is lower by a factor of 10 over a wide range of ΔK at a test frequency of 0.15 Hz. It seems that there are two distinct Paris regimes in the CM as well as DTM at 1 Hz, where the FCGR behaviour is represented by $da/dN = C\Delta K^n$, and the break point lies at ~45 MPa√m. In contrast, at 0.15 Hz the slope n is changing continuously in both microstructures and the da/dN versus ΔK curves are bending down. Strictly speaking, the application of the Paris equation is limited to short ΔK ranges in both cases.

The n values at comparable ΔK ranges decrease with increasing frequency in both microstructures, Table-2, indicating a change in the contribution of different deformation mechanisms and grain boundary sliding assisted environmental attack within the plastic zone towards crack extension. This is because at lower frequencies the creep contribution towards crack growth increases.

Table-2. Constants in Paris equation for MERL76 tests.

Frequency	CM		DTM		
	n	C	n	C	
1.00 Hz	2.7	0.9×10^{-7}	2.7	0.9×10^{-7}	$\Delta K = 20\text{-}40$
	4.9	1.4×10^{-11}	4.9	1.4×10^{-11}	$\Delta K = 40\text{-}80$
0.25 Hz	3.0	1.1×10^{-7}	2.9	6.7×10^{-8}	$\Delta K = 15\text{-}40$
	1.4	2.9×10^{-5}	1.8	3.3×10^{-6}	$\Delta K = 40\text{-}70$
0.15 Hz	2.9	2.5×10^{-7}	3.2	2.8×10^{-8}	$\Delta K = 15\text{-}30$
	1.4	4.3×10^{-5}	1.9	2.8×10^{-6}	$\Delta K = 30\text{-}70$

Note: da/dN in mm/cycle; ΔK in MPa$\sqrt{m}$.

The **fatigue fracture characteristics** of the two microstructures at 700°C are somewhat different, Fig-12. In the CM, intergranular crack propagation is observed over the entire range of ΔK studied. In contrast, a mixture of transgranular and intergranular crack propagation behaviour predominates in the DTM. This change in fracture behaviour is considered to be the main reason for improved FCGR resistance of the DTM relative to the CM. The presence of a larger area fraction of transgranular fracture in the DTM indicates suppression of grain boundary sliding and promotion of transgranular deformation within the plastic zone. This feature can be attributed to the presence of a coarser grain size and serrated grain boundaries in the DTM. In addition, the proportion of transgranular fracture also increases with increasing ΔK in the DTM.

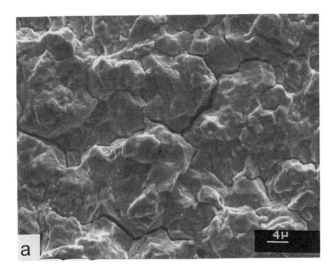
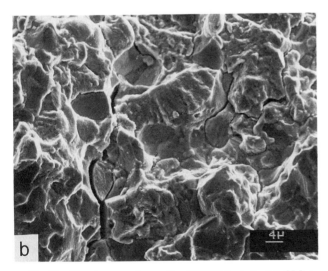

Fig-12. Fracture surface after FCGR tests in (a) CM and (b) DTM, at 700°C, 0.15 Hz.

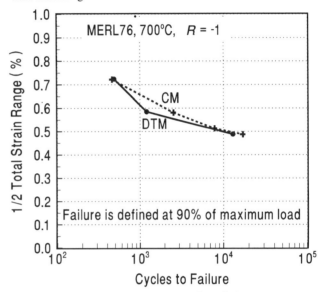

Fig-13. LCF life comparison between the two microstructures.

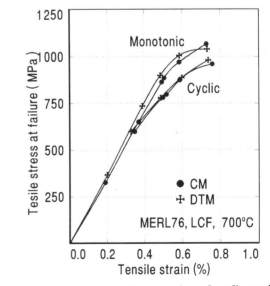

Fig-14. Comparison of monotonic and cyclic tensile stress in the two microstructures.

The **LCF** crack initiation lives, as a function of total strain range, for both CM and DTM are compared in Fig-13. With the exception of one data point in Fig-13, the LCF lives of both microstructures are comparable at all strain ranges. The monotonic and cyclic stress-strain curves for the CM and DTM are compared in Fig-14. Both materials revealed cyclic softening with little apparent difference in the cyclic stress strain behaviour of the two microstructures.

Similar to tensile fracture, the LCF cracks initiate transgranularly or intergranularly near the specimen surfaces in both cases. A number of secondary cracks initiate on the specimen surfaces in both cases and they lie perpendicular to the tensile axis. The fracture surface characteristics of CM and DTM are compared in Fig-15. In the CM, crack propagation and secondary cracks are intergranular. In the DTM, the crack propagation occurs in a mixed mode fashion. Also striations were observed in the DTM at a crack depth of about 0.5 mm from the crack initiation site. This feature was not observed in the CM.

Apart from concern about the loss in LCF life while designing the DTM, it is always important to ensure that the material is not rendered notch sensitive as a result of DTM. The LCF tests on notched Alloy 718 specimens have shown [19] that the DTM can produce longer crack initiation life than the CM.

4. Conclusions

A damage tolerant microstructure (DTM) has been designed for P/M MERL76 turbine disc material through super-solvus processing, furnace cooling from the solution treatment temperature followed by standard aging treatment. The DTM contains a uniform distribution of relatively coarser grain size than the conventional microstructure (CM) coupled with a serrated grain boundary structure. The primary as well as secondary γ' distributions in the DTM are more uniform than in the CM. Based on the mechanical properties assessment and fracture surface examination, the following conclusions can be drawn:

1. Test results show that the DTM is indeed flaw tolerant under different loading conditions including creep and fatigue loading conditions. In the DTM, a coarser grain size and the serrated grain boundaries are beneficial to both creep resistance and elevated temperature FCGRs. These microstructural features were observed to suppress intergranular deformation and fracture and promote transgranular deformation and fracture.

2. Creep rupture life of DTM is 9 times longer than the CM at a stress of 350 MPa.

3. At 700°C, the FCGRs of the DTM are reduced by a factor of 10 over a wide range of ΔK at a frequency of 0.15 Hz relative to the CM whereas the FCGRs at a higher frequency of 1 Hz are similar. The serrated grain boundaries are believed to significantly contribute towards fatigue crack growth resistance at lower frequencies.

4. The LCF lives of the DTM and CM are similar over a wide range of strain range values but the yield strength of the DTM is lowered by about 7%.

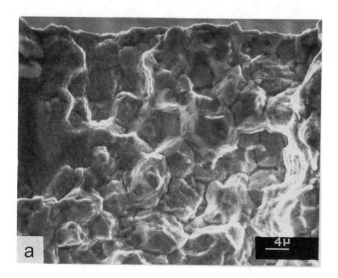

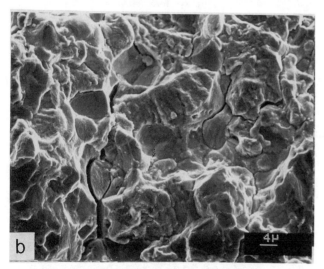

Fig-15. Crack initiation sites observed in LCF tests.
a: CM, $\Delta\varepsilon_t = 0.974\%$, $N_f = 17000$ cycles.
b: DTM, $\Delta\varepsilon_t = 0.975\%$, $N_f = 12900$ cycles.

Acknowledgement

This work was conducted under IAR-NRC project JHM05 with financial assistance provided by the Chief Research and Development, Department of National Defence, Canada, under financial arrangement 220794NRC05.

References

[1] MIL-STD-1783 (USAF): Military Standard, Engine Structural Integrity Program (ENSIP), by the Department of Defence, USA, Nov. 1984.

[2] MIL-STD-1843 (USAF): Military Standard, Reliability-Centred Maintenance for Aircraft, Engine and Equipment, by the Department of Defence, USA, Feb. 1985.

[3] A.K.Koul, P.Au, N.Bellinger, R.Thamburaj, W.Wallace, and J-P. Immarigeon, "Development of a Damage Tolerant Microstructure for Inconel 718 Turbine Disc Material", **SUPERALLOYS 1988**, Ed. D.N.Duhl et al., TMS-AIME, 1988, 3-12.

[4] M.Chang, P.Au, T.Terada, and A.K.Koul, "Damage Tolerance of Alloy 718 Turbine Disc Materials", **SUPERALLOYS 1992**, Ed. S.D. Antolovich et al., TMS-AIME, 1992, 447-456.

[5] J.Benson, R.Hunziker, and C.Williams, "Rejuvenation of Wrought IN-718 Diffuser Cases", **SUPERALLOYS 1992**, Ed. S.D.Antolovich, et al., TMS-AIME, 1992, 877-883.

[6] D.D.Krueger, R.D.Kissinger, and R.G.Menzies, "Development and Introduction of a Damage Tolerant High Temperature Nickel-Base Disk Alloy, Rene'88DT", **SUPERALLOYS 1992**, Ed. S.D.Antolovich et al., TMS-AIME, 1992, 277-286.

[7] JY.Guedou, JC.Lautridou, and Y.Honnorat, "N18, PM Superalloy for Disks: Development and Applications", **SUPERALLOYS 1992**, Ed. S.D.Antolovich et al., TMS-AIME, 1992, 267-276.

[8] A.K.Koul, J-P.Immarigeon, and W.Wallace, "Microstructural Control in Ni-base Superalloys", **Advances in High Temperature Structural Materials and Protective Coatings**, book, Ed. A.K.Koul, et al., NRC publication, CANADA, 1994, 91-125.

[9] W.Wallace, A.K.Koul, J-P.Immarigeon, P.Au, and J.C.Beddoes, "PM Matures as Route to Aerospace Alloys", **Metal Powder Reports**, June 1994, p.32.

[10] J.H.Davidson, G.Raisson, and O.Faral, "The Industrial Development of a New PM Superalloy for Critical High Temperature Aeronautical Gas Turbine Components", Proceedings of **Int. Conf. on PM Aerospace Materials**, Lausanne, Switzerland, 1991, 2-1 - 2-12.

[11] D.J.Evans and R.D.Eng, "Development of a High Strength Hot Isostatically Pressed (HIP) Disk Alloy, MERL76", **Modern Developments in Powder Metallurgy**, 14(1981), 51-63.

[12] R.H.Caless and D.F.Paulonis, "Development of Gatorized® MERL76 for Gas Turbine Disk Application", **SUPERALLOYS 1988**, Ed. by S.Reichman, et al., TMS-AIME, 1988, 101-110.

[13] P.D.Genereux and D.F.Paulonis: "Nickel Base Superalloy Articles and Method for Making", **European Patent, No. 0 248 757**, 1987.

[14] K.-M.Chang and H.C.Fiedler, "Spray-Formed High-Strength Superalloys", **SUPERALLOYS 1988**, Ed. by S.Reichman, et al., TMS-AIME, 1988, 485-493.

[15] J.M.Hyzak and S.H.Reichman, "Forming of Advanced Ni-base Superalloys", **Advances in High Temperature Structural Materials and Protective Coatings**, book, Ed. A.K.Koul, et al., NRC publication, CANADA, 1994, 126-146.

[16] T.C.Lu, O.Faral, Y.Bienvenu, and J.H.Davidson, "Study of the Evolution of the Microstructure of a Nickel Base Superalloy from the Atomised Powder to the Hip'ped Preform", Proceedings of **Int. Conf. on PM Aerospace Materials**, Lausanne, Switzerland, 1991, 5-1 - 5-11.

[17] A.K.Koul and G.H.Gessinger, "On the Mechanism of Serrated Grain Boundary Formation in Ni-Based Superalloys", **Acta Met.**, 31(1983), 1061-1069.

[18] A.K.Koul and R.Thamburaj, "Serrated Grain Boundary Formation Potential of Ni-Based Superalloys and its Implications", **Metall. Trans.**, 16A (1985), 17-26.

[19] M.Chang, P.Au, T.Terada, and A.K.Koul, "Damage Tolerance of Wrought Alloy 718 Ni-Fe-Base Superalloy", **J. of Mat. Eng. and Performance**, 3(3)(1994), 356-366.

[20] L.N.Moskowitz, R.M.Pelloux, and N.J.Grant, "Properties of IN-100 Processed by Powder Metallurgy", **SUPERALLOYS 1972**, AIME, 1972, z-1 - z-25.

Cooling Path Dependent Behavior of a Supersolvus Heat Treated Nickel Base Superalloy

R.D. Kissinger
GE Aircraft Engines, Cincinnati, OH 45215

Abstract

A key business initiative for jet engine manufactures is to reduce the engine development cycle time. Cycle time reductions lead to reduced introduction costs, and allow the manufacturer to get their product to market before their competitors. Two materials evaluation techniques are presented that allow for significant reductions in cost and time during the development of new turbine components. The on-cooling tensile test produces data that, when coupled with thermal/stress models, allows for the design of a forging shape and/or heat treatment parameters that preclude quench cracking and the accompanying manufacturing productivity losses. The two-step cooling path technique allows for the measurement of critical mechanical properties from small coupons, instead of heat treating and evaluating a full scale forging.

Introduction

The demand for improved efficiency of aircraft gas turbine engines by operating at higher compressor discharge temperatures has required an improvement in the creep capabilities of turbine disk alloys. Concurrently, disk alloys require improved resistance to fatigue crack growth to meet customers' residual life requirements, e.g. ENSIP (1). These demands have resulted in the development of powder metallurgy (PM) alloys which are solution heat treated above their gamma prime solvus temperatures to achieve a significantly coarser grain size as compared to PM alloys processed below their gamma prime solvus temperatures. While the supersolvus solution heat treatment has been shown to improve both creep strength and fatigue crack growth resistance (2), it presents processing challenges which must be overcome, the most significant being the balance between producibility and mechanical properties. It is well known that the mechanical properties of solutioned and aged nickel base superalloy disks are dependent on the cooling path (instantaneous cooling rate and the shape of the cooling curve) through the gamma prime solvus. In general, faster cooling rates produce higher tensile and creep strengths; however, if cooling rates are too fast during the quench from the solution temperature, quench cracking will occur. This is especially true for supersolvus heat treated alloys, which experience higher thermal gradients and thermally induced surface stresses during quenching.

Rotating PM turbine components vary significantly in size and shape, and have different mechanical property requirements as dictated by function, e.g. disk bore, disk rim, or shaft. It would not be practical from a cost and timing perspective to perform full scale trials for each component in order to determine a heat treatment that produces adequate mechanical properties without quench cracking. Heat treatment process modeling offers the ability to determine heat treatment parameters quickly and cheaply provided the input data and failure criterion are accurate. It is the purpose of this paper to present experimental techniques and data that has allowed successful heat treatment modeling of full scale disk hardware for a recently developed PM disk alloy, Rene'88DT (2). This paper will not discuss the details of heat treatment models or the application of these models to the heat treatment of forgings; additional details can be found in cited references. It is believed that the methods and conclusions described herein would be applicable to other gamma prime strengthened nickel base alloys.

On-cooling Tensile Tests

A critical processing step in the manufacture of a Rene'88DT turbine disk is the quench from the supersolvus solution temperature. A minimum cooling rate is required to achieve acceptable mechanical properties; however, an excessively rapid quench may generate thermally induced stresses sufficient to cause catastrophic failure (quench cracking). Ideally, a finite element model would be able to define a disk geometry and quench medium which would achieve the desired cooling rate and properties without cracking. If a model is to be successful in predicting quench cracking, then a failure criterion must be defined which identifies a stress above which the material will crack. The on-cooling tensile test is an attempt to define this criterion.

The test consists of taking a standard tensile specimen (0.25 inch diameter by 1.36 inch gage length) and heating it to the solution temperature (2100°F) which is above the gamma prime solvus of Rene'88DT (2030°F). The specimen is then cooled at a controlled rate to a specified test temperature, stabilized for approximately one minute, and then pulled in tension to failure at an initial strain rate of 0.05/minute. The 0.2% yield and ultimate tensile strengths are calculated from load vs. stroke curves.

The ultimate tensile strength (UTS) can be used as a failure criterion when attempting to predict quench cracking. For a given disk geometry, effective stresses can be calculated at specific locations as a function of decreasing temperature and plotted against the on-cooling data. If the local effective stress exceeds the measured UTS, there would be a high probability that the disk would crack at that location (3).

Turbine disks are usually heat treated horizontally on a support grate. Inherent to this procedure is a significantly different cooling rate on the top surface of the disk compared to the bottom surface. It would therefore be of interest to look at the effect of different cooling rates on on-cooling tensile properties. Two cooling rates have been investigated: 400°F/minute (bottom disk surface) and 2000°F/minute (top disk surface). These cooling rates were chosen based on data from near surface thermocouples (T/C's) embedded in a turbine disk (3).

Two-Step Cooling Path

A typical turbine disk solution heat treatment process can include a fan air cool (FAC) before the forging is oil quenched (3). The brief FAC reduces surface temperatures and thereby eliminates quench cracking because the surface material has gained sufficient strength to survive

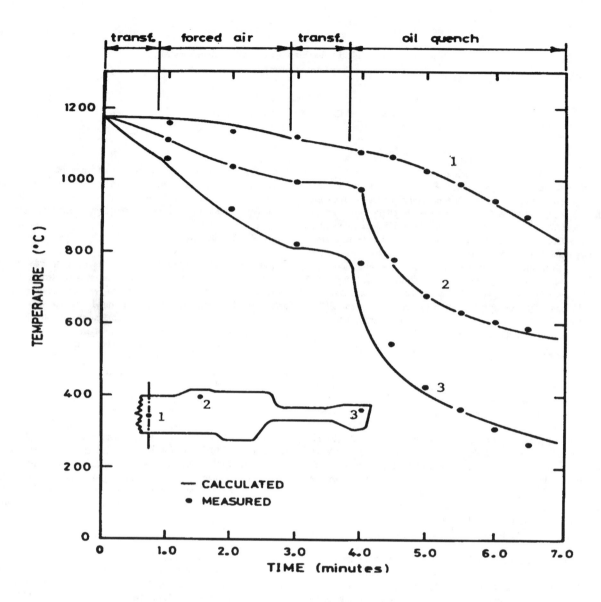

Figure 1. Typical cooling curves for different locations in a turbine disk (Ref. 3).

thermally induced stresses. The FAC segment produces a two-step cooling curve as shown in Figure 1. Depending on location, the FAC may produce a range of instantaneous cooling rates from 30°F/minute in the bore (location 1), or 200°F/minute in the rim (location 3), of an turbine disk given a two minute FAC. The oil quench produces a bore cooling rate of 170°F/minute and a initial rim cooling rate of 800°F/minute. The transition temperature, i.e. knee in the cooling curve, between FAC and oil quench cooling rates is strongly dependent on location, ranging from 1920°F at location 1 to 1450°F at location 3.

An experiment was designed to determine the effect of a two-step cooling curve on tensile, creep/rupture and fatigue crack growth (FCG) properties in the thicker sections of a large Rene'88DT forging. Cooling rate variables included slow (7°F/minute) and fast (30°F/minute) first step rates, a second step cooling rate of 360°F/minute, and 2000°F and 1900°F as locations of the transition temperature. The transition temperatures of 1900°F and 2000°F were chosen to be close to, but lower than, the gamma prime solvus of Rene'88DT.

Experimental Procedure

Material

The material for this study was taken from a production Rene'88DT turbine disk in the solutioned (2100°F for one hour) and aged (1400°F for 8 hours) condition. Rene'88DT is a powder metallurgy alloy that was argon atomized, extrusion consolidated, isothermally forged, supersolvus heat treated and aged to produce a nominal ASTM 7-8 grain size (2). The alloy chemistry (in weight percent) is 56.5% Ni, 16.0% Cr, 12.7% Co, 4.0% Mo, 4.0% W, 3.81% Ti, 2.23% Al, 0.7% Nb, 0.043% C, 0.016% B, and 0.05 Zr.

On-Cooling Tensile Test

The on-cooling tensile test was developed at Metcut Research Associates, Inc. in support of a GE purchase order and statement of work (5). The test consisted of three segments. First, the tensile specimen (0.25 inch diameter and 1.3 inch gage length) was given a controlled heat-up to a pre-test, subsolvus soak temperature of 1900°F, held for five minutes, and then heated at about 150°F/minute to 2100°F, and held for five minutes. Second, the specimen gage was subjected to a controlled cooling rate of either 400°F or 2000°F/minute from 2100°F to the selected test temperature, and stabilized for about 60 seconds. Test temperatures were in the range of 1300-2100°F. Finally, the specimen was pulled to failure on a 20,000 pound MTS closed loop servo controlled hydraulic system at a constant crosshead speed with an initial strain rate of 0.05/minute. An Ameritherm induction unit was the heating source, and the heating/cooling cycle was controlled by a Wavetek Model 75 waveform generator. The 400°F/minute cooling rate required a controlled power reduction of the induction coil, while the 2000°F/minute cooling rate required a complete shut-down of power to the coil plus forced air cooling of the top and bottom grips and specimen shoulders. The cooling air was supplied by copper tubes.

Temperature control during the 1900°F preheat, 2100°F solution, and cool down was achieved using an Ircon pyrometer sampling the center gage location. To assure pyrometer accuracy, two tensile specimens were instrumented with surface and internal T/C's to compare readings. (For internal T/C's, the specimens were machined with a hole centered along the specimen axes, and a sheathed T/C was inserted so as to read the temperature at the center of the gage length.) Testing procedures were established which assured accurate pyrometer temperature measurements of the specimen surface. Cooling rates were calculated from the average slope of the temperature vs. time plots as recorded from the pyrometer. After completion of the tensile testing, selected specimens were evaluated for grain size. Metallographic samples were taken from broken tensile specimens adjacent to the fracture surfaces or from the midsections of unfailed specimens. Optical and high resolution SEM micrographs were obtained from transverse surfaces.

Two-Step Cooling Path

Cooling Rates. Table I and Figure 2 summarize the various heat treatments evaluated. Cooling rates were calculated from T/C data obtained from a IN718 coupon with an embedded thermocouple (T/C). The 7°F/minute rate was obtained by air cooling the coupon in a ceramic jacket (hollow ceramic brick). The 30°F/minute rate was obtained by fan air cooling the coupon in a ceramic jacket. A standard shop floor fan was used. The 350°F/minute rate was obtained by fan air cooling the bare coupon after removing it from the ceramic jacket.

Heat Treatments. Jacketed Rene'88DT coupons were loaded into an air furnace. The coupons were stabilized at 1960°F before heating to 2100°F. Jacketed coupons were removed and air cooled or fan air cooled for a specified time, as determined by the IN718 T/C data, to reach the transition temperature. At the prescribed time, the coupon was removed from the ceramic brick and FAC to room temperature. Following the solution heat treatment, the coupons were aged at 1400°F for 8 hours and air cooled.

Testing. Tensile and creep/rupture tests were performed using standard ASTM testing procedures. FCG tests were performed at GE Aircraft Engines using standard potential drop testing techniques. Test specimens had a rectangular cross section of 0.400 inch by 0.168 inch, and a gage length of 1.25 inch. 20 cpm specimens (triangular wave form) at 750°F and 1200°F were EDM notched with a 5 mil x 10 mil surface flaw from which precracks were initiated. The notches were centered on one of the 0.400 inch faces, and specimens were precracked at 90 ksi/10 Hz at room temperature to a nominal 12 mil x 24 mil crack before testing at elevated temperatures. The net section maximum stress for the 750°F tests was 125 ksi; for 1200°F it was 100 ksi. 1200°F/90 second hold time tests (20cpm triangular wave form with a superimposed 90 second hold at the maximum stress of 100 ksi) used a 5 mil x 10 mil EDM notch and were precracked at 90 ksi/10 Hz at room temperature to a nominal 15 mil x 30 mil crack before testing at elevated temperature. Gage section temperatures for all FCG tests were controlled using tac welded T/C's located above one of the potential drop wires. All FCG tests used an R ratio of 0.05.

Experimental Results

On-Cooling Tensile Tests

Averages of duplicate on-cooling UTS and percent elongation data from 400°F/minute and 2000°F/minute tests are plotted graphically in Figures 3 and 4., respectively. 400°F/minute tests showed variations in cooling rate from 368°F to 440°F/min; 2000°F/minute tests varied from 1463°F to 2084°F/min. Only four tests exceeded the one minute stabilization time before starting the test, with the longest time being 90 seconds. Grain sizes measured from selected specimens ranged from ASTM 6.8-9.5, except for 400°F/minute tests at 2075 and 2100°F, where grain sizes were ASTM 5.5 and 3.0, respectively. High resolution SEM photos in Figure 5 compare gamma prime sizes as a function of cooling rate for tensile tests at 1500°F. For the 1500°F test, cooling rate significantly affected gamma prime size. The 400°F/minute cooling rate produced gamma prime precipitates that were approximately 0.1 μm in diameter; the 2000°F/minute cooling rate produced 0.03 μm diameter

Table I.
Two-Step Heat Treatment Parameters

Heat Treatment	First Step Cooling Rate (°F/minute)	Second Step Cooling Rate (°F/minute)	Transition Temperature (°F/minute)
A	30	350	2000
B	30	350	1900
C	7	350	2000
D	7	350	1900

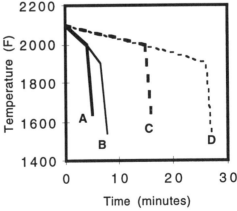

Figure. 2. Two-step cooling curves.

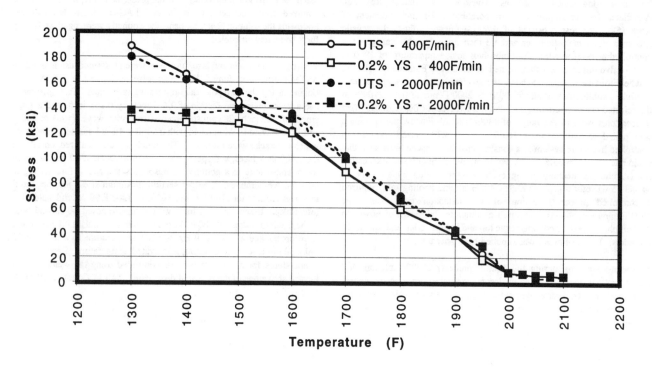

Figure 3. On-Cooling Tensile UTS and YS

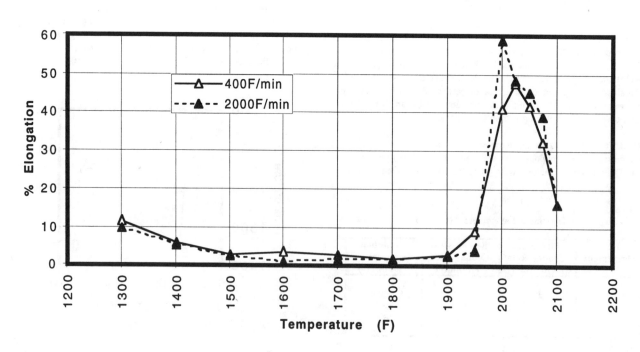

Figure 4. On-Cooling Tensile % Elongation

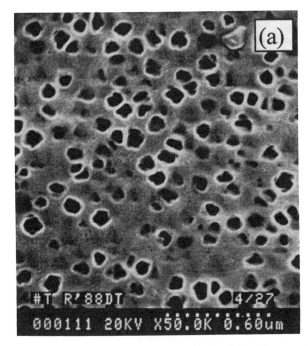 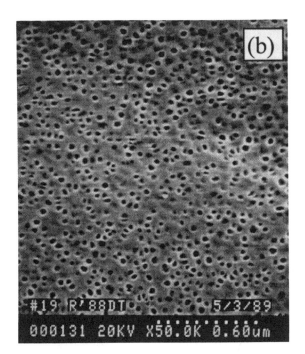

Figure 5. SEM micrographs of gamma prime precipitates from specimens tested at 1500°F (a) 400°F/min, (b) 2000°F/min.

gamma prime. For the 2000°F tests, the cooling rate did not affect gamma prime size because the stabilization time at a temperature only 30°F below the gamma prime solvus was sufficient to normalize any microstructural differences attributed to the differences in cooling rate from 2100°F to 2000°F.

Two-Step Cooling Path

Microstructure. Optical and TEM replica micrographs are shown for heat treatments A, B, C and D in Figures 6 and 7. Differences in cooling rates and transition temperatures significantly affected the cooling gamma prime size and distribution. In general, the size of the cooling gamma prime increased in order of heat treatments A, C, B, and D. The largest precipitates correlated with a slower initial cooling rate and lower transition temperature (D), and the smallest precipitates correlated with the faster initial cooling rate and higher transition temperature (A).

Mechanical Properties. Tensile, creep and FCG properties are given in Tables II-IV and Figures 8-10, respectively. Duplicate tensile and creep tests were performed for each heat treatment.

Table II.
1200°F Tensile Properties

Heat Treatment	0.2% Yield Strength (ksi)	Ultimate Tensile Strength (ksi)	Elongation (%)	Reduction of Area (%)
A	153.4	224.0	17	20
	153.6	224.0	19	20
B	136.9	216.9	19	20
	137.3	211.0	23	27
C	150.9	222.9	16	17
	153.4	223.2	18	20
D	134.0	210.0	23	25
	141.5	214.3	21	24

FCG data represent FCG rates measured from single tests at a ΔK of 30 ksi$\sqrt{}$inch for each heat treatment. SEM micrographs of fracture surfaces from 1200°F/20cpm and 1200°F/90 second hold specimens given heat treatment A are shown in Figure 11. Both micrographs were taken in locations where ΔK was approximately 25 ksi$\sqrt{}$inch.

Discussion

On-Cooling Tensile Data

Effect of Cooling Rate on On-Cooling Tensile Properties.
Comparisons of 400°F/minute and 2000°F/minute UTS and 0.2% yield stresses (YS) are shown in Figure 3. At 2000°F and above,

Table III.
1200°F / 123ksi Creep Properties

Heat Treatment	Time to 0.2% Strain (hours)	Minimum Creep Rate (sec-1 x E-10)
A	225 (1)	2.8
	1090	3.5
B	255	17.5
	255	6.9
C	795	2.1
	(2)	0.75
D	440	6.5
	300	7.1

(1) Specimen A2 exhibited an unusually large amount of primary creep strain (0.15%) compared to all other specimens in this study.

(2) Test discontinued at 1,316 hours before reaching 0.2% creep strain

Table IV.
Fatigue Crack Growth Rates at ΔK = 30 ksi√inch
(inch/cycle x E-6)

Heat Treatment	750°F 20 cpm	1200°F 20 cpm	1200°F 90 Second Hold
A	8.96	26.7	297
B	9.28	35.0	94.8
C	8.36	30.3	620
D	10.1	32.1	153

UTS and YS were not sensitive to cooling rate. This observation is in agreement with the previous description of similar gamma prime morphologies for the two different cooling rates when tested at 2000°F. Below 2000°F, the faster cooling rate generally produced a higher UTS and YS. Significant differences in UTS and YS for a given cooling rate were not seen until test temperatures were 1500°F and lower. A cross-over in UTS occurred between 1400°F and 1500°F, where the 400°F/minute UTS exceeded the 2000°F/minute UTS. It is not known why the cross-over in UTS occurred since all other tensile properties (YS, ductilities, and grain size) behaved in a predictable manner. Percent elongation is shown in Figure 4. Both cooling rates showed similar trends. Maximum elongation occurred at 2000°F-2025°F. A dramatic reduction in elongation occurred at temperatures above and below 2000-2025°F. The nil ductility temperature was estimated to be

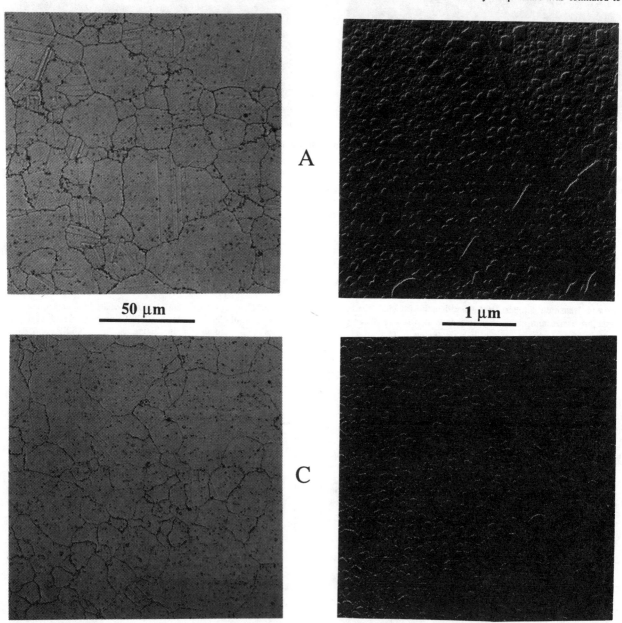

Figure 6. Optical (50X) and TEM replica (20,000X) micrographs of heat treatments A and C (2000°F transition temperature).

about 2120°F. There was a ductility trough in the temperature range of 1500-1900°F, where the average percent elongation was 2.5%. Ductilities did not begin to increase until 1400°F. The 2000°F/minute tests showed generally higher ductilities at temperature greater than 2000°F, but lower ductilities at temperatures less than 1950°F, when compared to the 400°F/minute data.

It was not surprising to see very large grains (ASTM 3.0-5.5) in the 400°F/minute specimens tested at 2075°F and 2100°F. These temperatures are above the gamma prime solvus and would allow dynamic grain growth to occur similarly to grain growth observed in Udimet 720 deformed at a slow strain rate at a supersolvus temperature (6). The grain growth was probably localized which would help to explain why large grains were seen in the 400°F/minute specimens but not seen in the 2000°F/minute specimens.

The on-cooling tensile data can be utilized by finite element models to predict thermally induced stresses and displacements. The approach allows the engineer to manage thermally induced stresses via a computer model, and to vary heat treatment parameters and/or forging geometries to keep local stresses from exceeding the failure criterion, e.g. UTS. Additional data not shown, but easily available from the test, include Young's Modulus and load vs. displacement curves at each test temperature. The on-cooling data, coupled with appropriate thermal/stress models, have successfully modeled the heat treatment of Rene'88DT forgings to preclude quench cracking, thereby reducing production problems and accompanying costs (4).

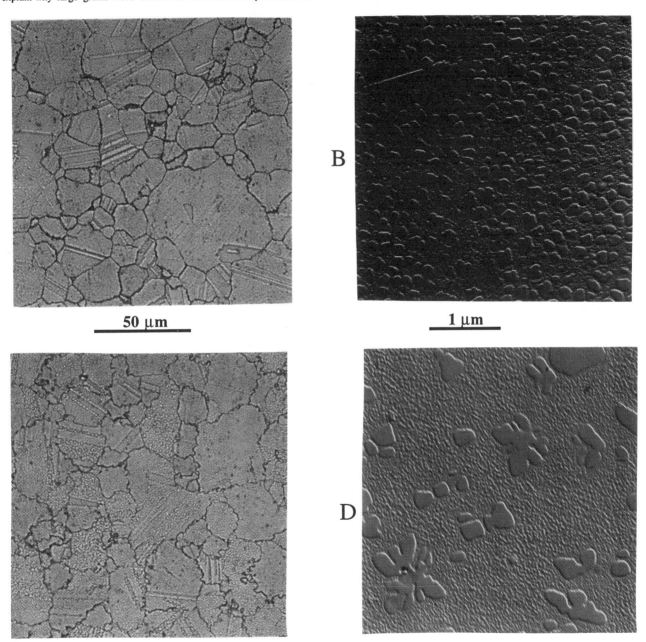

Figure 7. Optical (50X) and TEM replica (20,000X) micrographs of heat treatments B and D (1900°F transition temperature).

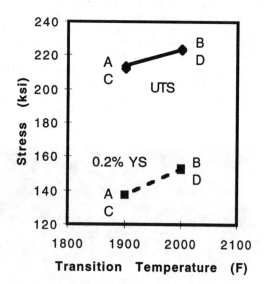

Fig. 8 - 1200F Tensile Data

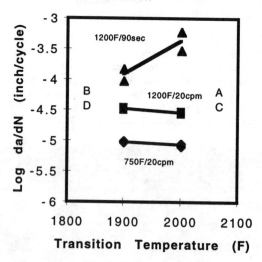

Fig. 10 - Fatigue Crack Growth

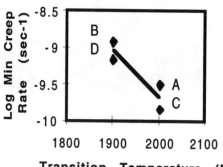

Fig. 9 - Minimum Creep Rate

Two-Step Cooling Path

Tensile, creep and FCG data are plotted as a function of transition temperature in Figures 8-10. Tensile strengths and minimum creep rates were not sensitive to the first step cooling rate, as noted by similar mechanical property values for heat treatment pairs A (30°F/minute) and C (7°F/minute), and B (30°F/minute) and D (7°F/minute). It was interesting to note that while heat treatments B and D had similar mechanical properties, their cooling gamma prime sizes and distributions were very different. The most important heat treatment variable was the transition temperature. Tensile strengths increased, and minimum creep rates decreased, as the transition temperature increased from 1900°F to 2000°F. The FCG data showed mixed behavior as a function of transition temperature. 20cpm data showed a slight decrease in crack growth rates as the transition temperature increased; however, the 1200°F/90 second hold time crack growth rates showed a significant increase (24X) compared to the 1200°F/20cpm data. The slower hold time crack growth rates for the 1900°F transition temperature were attributed to the higher creep rates. The higher creep rates allowed for significant stress relaxation at the crack tip, thereby lowering the effective stress concentration at the crack tip.

All 20 cpm fracture surfaces at 750°F and 1200°F showed predominately transgranular fracture paths. 1200°F/90 second hold surfaces were predominately intergranular, with a slight increase in transgranular behavior moving from heat treatment A to D. Hold time facet texture became coarser from A to D because of the larger gamma prime precipitates that formed at grain boundaries for the slower cooling rates. Examples of 1200°F/20cpm and 1200°F/90 second hold fracture surfaces for heat treatment A are shown in Figure 11.

The nucleation and growth of gamma prime during the quenching of a disk from a supersolvus solution temperature is a complex process which is difficult to model, especially in regions that cool slowly (7). Yet these regions can be critical to the function of the component, e.g. tensile strength in the thick bore section of a turbine disk. Rather than heat treat and evaluate an entire forging, the approach outlined for the two-step cooling path evaluation could be used to save considerable time and cost during the design and manufacture of a new disk forging. For example, one could determine via a thermal model the shape of local cooling curves at critical locations in the disk. Coupons could then be solutioned and cooled to duplicate the predicted cooling curves. Mechanical properties from the coupons would represent approximate values expected from full scale hardware. Trade studies could also be performed based on the engineering requirements for a given region in a disk. For example, if hold time crack growth was considered to be important, then the engineer would be able to quantify the reductions in tensile and creep strengths which resulted when a desired hold time crack growth rate was achieved.

Conclusions

On-Cooling Tensile Test

i. The on-cooling tensile test is an excellent method to determine strength and ductility data for input to finite element models, although care must be taken to assure accurate pyrometer temperature measurements during the test.

ii. Between 2000°F and 2100°F, Rene'88DT UTS and YS were relatively insensitive to temperature. Below 2000°F, tensile strength increased rapidly.

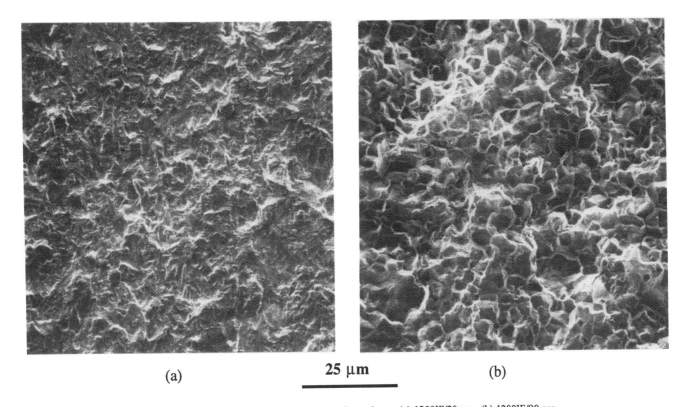

Figure 11. SEM micrographs of fracture surfaces from fatigue crack growth specimens (a) 1200°F/20cpm, (b) 1200°F/90 sec.

iii. Percent elongation was highest at 2000-2025°F. At higher and lower temperatures, ductilities dropped precipitously. The nil ductility temperature was estimated to be about 2120°F. Between 1900°F and 1500°F, a ductility trough existed, where percent elongation values averaged 2.5%. Below 1500°F, ductilities began to increase.

iv. On-cooling tensile data, in conjunction with thermal/stress models, have been used to successfully model the heat treatment process to preclude quench cracking of Rene'88DT forgings (see Ref. 4).

Two-Step Cooling Path

i. For the two-step cooling curves evaluated in this study, the transition temperature was found to be more important in determining mechanical properties than was the initial cooling rate.

ii. A higher transition temperature correlated with better tensile and creep/rupture properties and faster hold time fatigue crack growth rates.

iii. The approach outlined for the two-step cooling path technique can be applied to the evaluation of turbine components. Significant reductions in time and cost can be achieved by evaluating coupons heat treated to predicted cooling curves at critical locations in the full scale hardware.

Acknowledgments

The author acknowledges Mr. Ronald Tolbert and Mr. Thomas Daniels for their technical support of the work described in this publication.

References

(1) T.T. King, et al., "USAF Engine Damage Tolerance Requirements," AIAA-85-1209, 1985.

(2) D.D. Krueger, R.D. Kissinger and R.G. Menzies, "Development and Introduction of a Damage Tolerant High Temperature Nickel Base Disk Alloy, Rene'88DT," **Superalloys 1992**, S.D. Antolovich, et al., eds.,TMS-AIME, 1992, pp. 277-286.

(3) R.A. Wallis and P.R. Bhowal, "Property Optimization in Superalloys through the Use of Heat Treatment Modeling," **Superalloys 1988**, D.N. Duhl et al., eds., TMS-AIME, 1988, pp. 525-534.

(4) R.I. Ramakrishnan and T.E. Howson, "Modeling the Heat Treatment of Superalloys," **JOM** (June, 1992) pp. 29-32.

(5) Metcut Research Associates, Inc. Job #100-48042, W.J. Stross and M. J. Booker.

(6) J.M. Hyzak, R.P. Singh, J.E. Morra and T.E. Howson, "The Microstructural Response of As-HIP P/M U720 to Thermomechanical Processing," **Superalloys 1992**, S.D. Antolovich et al., eds., TMS-AIME, 1992, pp. 93-102.

(7) J.J. Schirra and S.H. Goetschius, "Development of an Analytical Model Predicting Microstructure and Properties Resulting from the Thermal Processing of a Wrought Powder Nickel-Base Superalloy Component," **Superalloys 1992**, S.D. Antolovich et al., eds., TMS-AIME, 1992, pp. 437-446.

DEVELOPMENT OF ISOTHERMALLY FORGED P/M UDIMET 720 FOR TURBINE DISK APPLICATIONS

Kenneth A. Green
Allison Engine Company

Joseph A. Lemsky
Ladish Co., Inc.

Robert M. Gasior
Dynamet Powder Products

Abstract

Udimet 720®* nickel-base superalloy is in current use as the material for the stage-two power turbine disk for the T800 engine. This alloy was selected because of its attractive level of tensile strength and resistance to low cycle fatigue (LCF). Advancements in novel isothermal forging technology have enabled the production of complex disk forgings called cluster forgings. A single cluster forging produces seven PT2 disks in a single forging operation. Coupled with recent advances in powder metallurgy (P/M) billet production featuring consolidation by HIP followed by extrusion, turbine disk forgings can be produced at a significant cost reduction as compared to single-piece hot die forged disks made from conventional cast/wrought billet. P/M Udimet 720 material displays excellent mechanical properties, competitive with other high strength disk alloys. Turbine disks produced from isothermally forged P/M Udimet 720 passed component test requirements established for the T800 engine. A significant cost reduction resulted from the use of this process.

Introduction

Udimet 720 is a high strength nickel-base superalloy currently in use as a turbine disk material. Udimet 720 is a relatively new disk alloy, having been developed originally as a wrought turbine blade alloy for industrial turbine applications (Ref. 1). More recently, Udimet 720 has been recognized as having outstanding strength and fatigue resistance when used in a fine grain form for turbine disk applications. From a mechanical property standpoint the alloy retains the high strength characteristics of alloys such as P/M alloy Rene 95, but has superior crack growth characteristics (Ref. 2). Overall, Udimet 720 has the best combination of mechanical properties in its class of high strength cast/wrought disk alloys and is used in a variety of applications in small- to medium-sized gas turbine engines.

Udimet 720 alloy in the cast/wrought form is in current use as the turbine disk alloy for all four stages of the T800 turboshaft engine currently being developed by the U.S. Army. The T800 engine is the powerplant for the Army's RAH-66 Comanche helicopter. When fielded, this aircraft system will represent the current state-of-the-art in its class, with the most advanced air vehicle, avionics, propulsion, and weapons systems available. Udimet 720 was selected for this application because of its high tensile strength characteristics as well as its resistance to LCF. Currently, the disks are produced as single-piece hot die forgings from 4.5 in. diameter cast/wrought billet.

In support of affordability goals, an effort was undertaken to implement lower cost disks in the stage-two power turbine (PT2) of the T800. Cost reduction was achieved by utilizing a recently developed isothermal forging process called cluster forging in which multiple disks are produced in a single large forging. In this application, a single cluster forging produces seven PT2 disks. Because multiple disks are produced in a single pressing operation, material utilization is improved and overall costs are reduced significantly. The high weight (nearly 300 lb) of the cluster forging requires a relatively large starting billet of 9.0 in. diameter. However, conventional cast/wrought Udimet 720 billet is difficult to produce economically in this size. To provide a cost-effective option to the bill-of-material, a P/M approach was utilized to produce uniform large diameter forging billet. Coupled with the cluster forging concept, the PT2 disk can be produced in a very cost-effective manner.

Material

Background

The composition of Udimet 720 has evolved over a period of several years from a wrought turbine blade alloy to high strength turbine disk applications. Udimet 720 is the most highly alloyed disk material currently being produced by the cast/wrought route. Current ingot manufacturing practice involves triple melting (VIM/ESR/VAR) followed by conversion to billet by a series of homogenization, upsetting, and cogging cycles. Conventional cast/wrought Udimet 720 billet is difficult to produce in the size required for cluster forgings because of yield losses that occur due to the alloy's propensity to crack during the billet manufacturing cycle. Microstructural variation caused by inherent solidification segregation in the cast ingot may persist in the eventual billet (Ref. 3). As a result, much effort has been given to the alloy's ability to be manufactured consistently and with high yield into large diameter billet. To control the propensity for segregation and formation of unwanted phases such as carbide- and boride-rich stringers, the composition has been adjusted to meet the strict quality and property requirements. In particular, levels of carbon, boron and chromium have been lowered from the original composition to address the various segregation, alloy stability and billet yield issues.

Composition Selection

In selecting the composition for P/M Udimet 720, Table I, careful consideration was given to factors that determine cost and performance. It was desirable to keep the alloying variables (carbon, boron and chromium) near the levels in current use cast/wrought Udimet

---

* Udimet is a registered trademark of Special Metals Corporation.

720 MC, which is the composition currently in use at Allison for turbine disk applications. The composition for P/M moderates the levels of carbon (0.025 wt.%) and boron (0.020 wt.%) with respect to Udimet 720 MC. Raising carbon and boron takes advantage of the high temperature strengthening contributions attributed to those elements. Chromium for the P/M version was maintained at the same level 16.0 wt.% as Udimet 720 MC. The P/M composition selected for this effort is also being evaluated by other engine manufacturers and offers the potential for universal acceptance which should ultimately have a favorable impact on cost due to commonality of specifications.

P/M Billet Processing

The billet production process, as shown in Figure 1, utilized in this effort by Dynamet Powder Products featured consolidation of powder by hot isostatic press (HIP) followed by extrusion into 9 inch diameter forging billet. The benefits of an extruded P/M billet are well documented with respect to minimizing the effects of nonmetallic inclusions and other defects (Ref. 4). The combination of HIP and extrude enhances microstructural uniformity and forgeability in a very cost-effective manner. By consolidation to full density during HIP, the need for high extrusion ratios commonly employed for hot compacted powder billet was avoided. This creates a more favorable situation with respect to the available supplier base for large diameter extrusions.

Powder Processing

Udimet 720 powder was produced by Dynamet via argon atomization of vacuum induction melted virgin material. The powder was screened to a -150 mesh. The powder particle distribution size exhibited a high percentage of fines, which were measured in excess of 50% -325 mesh. Water elutriation test results from the powder master blend totaled approximately 6.0 particles (1.6 hard, 3.6 friable and 0.8 soft) per pound of powder, indicating a very clean blend of powder.

HIP Consolidation

HIPed billet preforms were made by filling 20 in. diameter stainless steel cans with approximately 3200 lb of powder, followed by evacuation and sealing of the fill tubes. HIP parameters were selected based upon previous experience and verified by an experiment in which samples of Udimet 720 powder were exposed at HIP

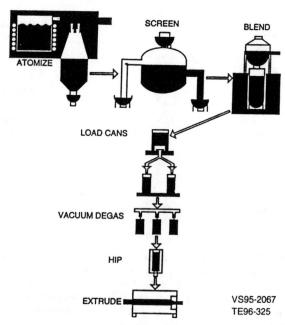

Figure 1: Process for producing HIP/extruded P/M billet.

temperatures of 2010°F, 2050°F and 2100°F. As expected, all samples exhibited full consolidation. The HIP cycle selected for the fullscale billets was based upon selection of a commercially common HIP cycle for economic considerations. A thermally induced porosity (TIP) test of the consolidated powder billet showed a density loss of less that 0.3% at 2200°F/4 hr.

Extrusion

The extrusion processing was performed in a 12,000 ton extrusion press at IXP (formerly Curtiss-Wright Corporation). The HIPed billet preforms were pushed through a 10.25 in. diameter die. The extruded billet was turned to 9.3 in. diameter, inspected and cut to individual forging mults. The as-extruded microstructure displayed a uniform fully hot-worked grain structure with grain size of ASTM 12. Ultrasonic inspectability of the billet was excellent, with no indications above the 30% of a No.1 flat bottom hole (FBH) limit. This shows the superiority of P/M material with regard to the ultrasonic inspectability of cast/wrought Udimet 720 billet which has a 100% of a No. 2 FBH minimum limit.

Table I Nominal Compositions by Wt % of High Strength Turbine Disk Alloys

Alloy	Form*	C	B	Cr	Co	Mo	W	Ti	Al	Zr	Other	Ni
U720	P/M	0.025	0.020	16.0	14.7	3.0	1.25	5.0	2.5	0.03	—	Bal
U720	W	0.035	0.033	18.0	14.7	3.0	1.25	5.0	2.5	0.03	—	Bal
U720 MC	W	0.015	0.017	16.0	14.7	3.0	1.25	5.0	2.5	0.03	—	Bal
Rene 95	P/M	0.07	0.010	13.0	8.0	3.5	3.5	2.5	3.5	0.05	3.5 Cb	Bal
Astroloy	P/M	0.042	0.02	14.7	17.0	5.1	—	3.5	4.1	—	—	Bal
N18	P/M	0.020	0.015	11.5	15.5	6.5	—	4.3	4.3	—	0.5 Hf	Bal

*P/M - powder metallurgy; W - wrought

PT2 Disk Forging Development

Forge Process

The forging method utilized for the manufacture of the PT2 disks was a multi-disk cluster design that was processed through a very controlled isothermal press forging operation. The cluster was designed to yield seven individual PT2 disks from each single large forging. This 1=7 approach allowed usage of cost-effective large diameter input billet stock and an extensive increase in production rates. In contrast, a conventional single-piece forge method would require more costly smaller diameter billet input stock and one at a time forge processing.

A signature of the isothermal forging press method is the ability to forge to a near-net shape, with tight envelopes over the ultrasonic shape. The effective manufacture of the highly contoured PT2 disk to a near-net, nonaxisymmetric cluster configuration was a significant challenge. Extensive computer modeling was performed by Ladish to simulate metal flow and enable an efficient design of the forge tooling. Flow stress data for the P/M Udimet 720 was used in the forging simulation model and identified a range of acceptable forging conditions. The flow stress data was found to be very comparable to conventional fine grain cast/wrought Udimet 720.

Subscale dies were first produced to validate the aggressive near-net shaped tool design, to confirm acceptable metal flow and complete die fill, and to validate the P/M Udimet 720 material which was produced by the novel processing route. Following successful validation with subscale forgings, full-scale tooling was manufactured and PT2 cluster forgings over 26 inches in diameter were successfully produced, as typified by the forging in Figure 2.

To prepare for an effective and uniform heat treatment, each large fullscale forging was excised to produce seven individual disks.

Heat Treat Selection

The current heat treat approach for fine grain Udimet 720 is a subsolvus solution followed by a two-step age. An empirical study was made to evaluate the effect of cooling rate on mechanical properties. Cooling rate was controlled by disk geometry (solid or bored) and quench medium (oil or air). Test disks were drilled and instrumented with thermocouples to determine cooling rate during quenching for each of the conditions. To correlate cooling rate with mechanical properties, tensile and creep tests were performed for each condition. As shown in Figure 3, the highest cooling rate, which was produced by oil quenching a bored disk, produced the highest tensile properties. Ductilities for all strength levels were excellent. Based on this study, a heat treat practice was selected and implemented for the PT2 disk forgings.

Evaluation of Forgings

Microstructure

Selected disks were sectioned at various locations and evaluated for microstructural uniformity. Because of the design of this particular cluster forging, the metal flow characteristics during forging are nonaxisymetric for six of the seven disks. Only the center disk of this cluster forging experiences true axisymetric metal flow during the forging operation. To determine whether this presents any microstructural variation, samples were taken

Figure 2: P/M Udimet 720 isothermal cluster forging.

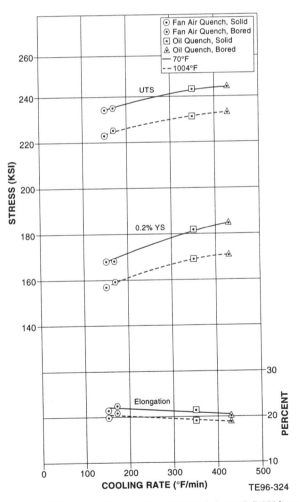

Figure 3: Effect of cooling rate from 2000°F/2 hr on P/M Udimet 720 tensile properties.

from various areas of cluster forgings. As shown in the photomicrographs in Figure 4, the microstructure displays an extremely uniform ASTM 12-14 grain size. The microstructure exhibits well distributed primary gamma prime with no evidence of a prior powder particle boundary structure. A higher magnification SEM photomicrograph, Figure 5, shows a high percentage of primary gamma prime and grain boundaries with discrete carbide and boride particles. No effect of location within the cluster forging can be detected in the microstructure, attesting to the extreme uniformity of the isothermal forging method.

Material Properties

An extensive material property characterization was conducted on specimens removed from representative disks processed to the final heat treatment schedule. Testing included tensile, creep/rupture, smooth- and notched-bar LCF, and fatigue crack growth rate testing. A sampling of the test data is presented in the following paragraphs.

Tensile samples were removed from the chordal direction from the disks and tested over a temperature range of 70°F to 1600°F. Ultimate tensile strength and 0.2% yield strength, shown in Figure 6, demonstrate excellent strength characteristics competitive with other high strength disk alloys. Ductility was excellent over the entire temperature range that was tested.

Smooth bar LCF tests were conducted at 800°F, 1000°F and 1200°F. Results of tests conducted at these temperatures at an R-ratio = 0, and frequency of 20 cpm in strain control are summarized in Figure 7 for 30,000 cycles mean initiation life. P/M Udimet 720 displayed higher mean lives than cast/wrought Udimet 720 at all three temperatures. The difference in mean lives between materials decreases with increasing temperature due to creep/fatigue interaction with the finer grain microstructure observed in the P/M Udimet 720 material. The data scatter was also better for the P/M material at 800°F and 1000°F, and only slightly higher at 1200°F than the cast/wrought material.

Fracture surfaces of LCF specimens were examined to assess the characteristics of the initiation site. Failures typically initiated from features shown in Figure 8, which are typical of most other argon atomized P/M superalloys. The predominant initiation features were microporosity formed by entrapped argon and grain boundary facets. Fewer than 30% of the specimens failed at nonmetallic inclusions. The inclusions were of two general types: discrete particles of zirconia and agglomerated particles of alumina, silica and titanium oxide. An analysis of the LCF data indicated that nonmetallic inclusions did not seem to adversely affect the overall behavior of the LCF specimens.

Fatigue crack growth rate testing was performed in air at 800°F and 1200°F using a frequency of 10 Hz and a triangular loading wave form. The results at 800°F, Figure 9, show equivalence with cast/wrought Udimet 720, a material which has demonstrated excellent damage tolerance characteristics (Ref. 2). The data shows a general trend of slightly lower crack growth rates at low stress intensity factors and higher crack growth rates at the higher stress intensity factors relative to the cast/wrought Udimet 720. The differences in crack growth rates are likely explained again by the finer grain size demonstrated by the P/M Udimet 720 microstructure. Dwell time tests were performed at 1200°F with a 5-minute hold at maximum stress. The effect of five minute dwell time on crack growth rate was measured to be

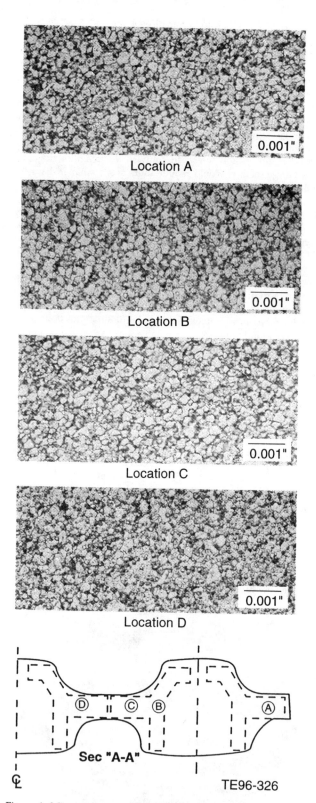

Figure 4: Microstructure of P/M Udimet 720 forging at various locations.

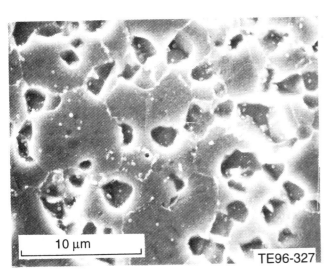

Figure 5: SEM photomicrograph of P/M Udimet 720 microstructure showing primary gamma prime (etched) and distribution of fine carbide and boride particles located primarily in grain boundaries.

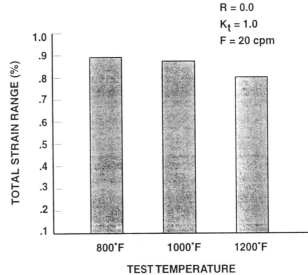

Figure 7: Total strain for 30,000 cycles mean life for P/M Udimet 720.

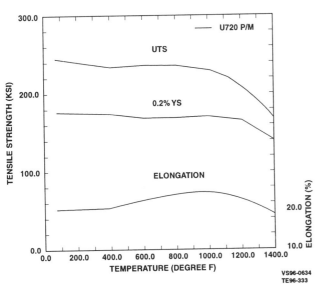

Figure 6: Tensile properties of P/M Udimet 720.

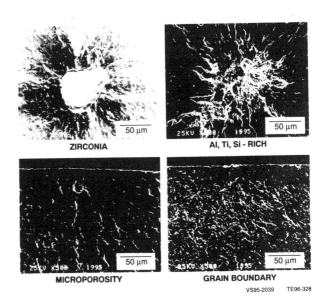

Figure 8: Typical low cycle fatigue initiation site features.

an order of magnitude acceleration in crack growth rate as compared with the cyclic-dependent test conditions (Ref. 5). This phenomenon is typical of high strength superalloys which demonstrate grain sizes in the size range of the P/M Udimet 720 represented in this work.

Component Testing

To qualify the P/M Udimet 720 forgings for development engine testing, component tests were required by the U.S. Army to certify the material and certain aspects of the design. The required tests included a cyclic spin test and an overspeed burst test. Several forgings were fully machined to the PT2 disk configuration. The PT2 disk, Figure 10, features an integral stub shaft on the aft face which connects to the power turbine output shaft. Turbine blades are attached with a single lob dovetail attachment.

Cyclic Spin Test

A fully-bladed wheel was utilized for the cyclic spin test which was conducted at 1010°F in an evacuated spin pit facility. The test cycle consisted of an excursion from idle speed to 23,375 rpm (105% Np) and back to idle. Each test cycle was approximately 20 seconds in duration. The test was run in two 15,000 cycle segments, with a teardown inspection at the end of each cycle. The P/M Udimet 720 disk displayed no cracking or other visible distress at the end of each test segment, successfully passing the cyclic life requirement.

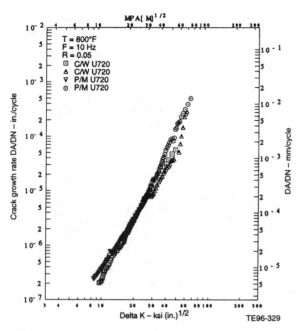

Figure 9: Fatigue crack growth rate comparison at 800°F.

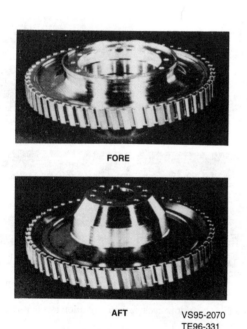

Figure 10: Fully machined T800 PT2 turbine disk made from P/M Udimet 720.

Burst Test

Overspeed capability is a significant feature of most power turbine designs. The T800 engine must demonstrate mechanical integrity at a minimum of 151% Np overspeed condition. The PT2 wheel is designed to intentionally fail in the blade rather than in the disk in an extreme overspeed situation. For this reason, the overspeed burst test was performed on a PT2 disk which was fitted with high strength "dummy" blades in place of the production cast turbine blades so as not to fail a blade before wheel burst. The disk used in the burst test was first spun to 34,848 rpm (151% Np) and held for five minutes to demonstrate minimum integrity speed. Following dimensional inspection, the disk was reinstalled in the spin pit and accelerated to burst. Disk burst occurred at 36,875 rpm (160% Np). The disk fragments were reassembled after the test and examined for fracture characteristics. Failure analysis indicated that the fracture initiated in the forward flange area near the bore and propagated through the web and into the rim. No metallurgical anomalies were found in the fracture initiation site. This test demonstrated adequate burst margin for the P/M Udimet 720 material and cleared the Army requirements for subsequent engine testing.

Cost Analysis

The primary goal of this effort was to achieve cost savings on the order of 20% over the current disk produced by hot-die forging of cast/wrought Udimet 720. To assess relative costs, estimates were obtained for the bill-of-material forgings and compared with cost projections for the cluster forged P/M Udimet 720 forgings. A comparison of costs are shown in Figure 11 for both T800 power turbine disk configurations. The cost savings for the P/M Udimet 720 disks exceeded the goals by significant amounts for lot sizes representing minimum production quantities. These savings could have a significant impact on expenditures over the entire life of the T800 production program.

Summary

P/M Udimet 720 has proven to be a viable, cost-effective material for use in highly stressed turbine disk applications. An approach for producing billet by the HIP/extrude process was identified which is shown to be very competitive with conventional cast/wrought Udimet 720 billet on a cost basis. The development of an aggressive geometry isothermal cluster forging was key in the approach to utilize cost-effective large diameter fine grain P/M billet for the production of small turbine disks. The individual disks excised from the cluster forging prove to be uniform in microstructure and mechanical properties, irrespective of orientation within the larger cluster forging. The mechanical properties demonstrated by P/M Udimet 720 in general meet or exceed the properties of the cast/wrought version of that alloy and are competitive with other contemporary P/M superalloys. The fatigue crack growth properties of P/M Udimet 720 demonstrate good crack growth resistance over a broad temperature range. However, fatigue crack growth could benefit from additional improvement in the area of high temperature (1200°F) dwell time. The P/M Udimet 720 properties meet the design criteria for the burst-limited PT2 turbine disk and representative disks have surpassed component test requirements established for continued T800 development engine testing. Cost projections obtained for isothermally cluster-forged P/M Udimet 720 PT2 disks indicate that a potential 25-30% cost savings can be achieved for production-size lots.

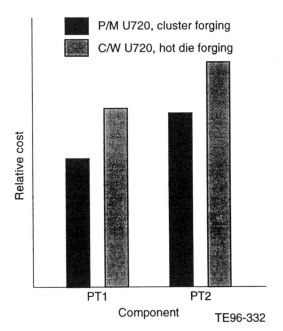

Figure 11: Comparison of cost estimates.

Acknowledgements

The authors wish to thank the U.S. Army Aviation and Troop Command (ATCOM), Messrs. N. Singh and S. Nahal, for permission to publish the results of this effort. The efforts of the following Allison Engine Company employees are greatly appreciated: Dr. C. Yin for test monitoring, Messrs. T. Hensler and D. Pridemore for metallographic evaluations, and Messrs. R. Stusrud, B. Ewing and Dr. M. Thomas for their support and guidance. The efforts of Messrs. J. Blair and J. Buzzanell at Dynamet are greatly appreciated. Also contributing at the Ladish Company were Messrs. D. Furrer, D. Russell and Dr. R. Shankar.

References

1. F. E. Sczerzenie and G. E. Maurer, "Development of Udimet 720 for High Strength Disk Applications," Superalloys 1984 (Warrendale, PA; The Metallurgical Society, 1984), 573-582.

2. K. R. Bain et al., "Development of Damage Tolerant Microstructures in Udimet 720", Superalloys 1988 (Warrendale, PA; The Metallurgical Society, 1988), 13-22.

3. G. Kappler et al., "Conventionally Processed High Performance Disc Material for Advanced Aeroengine Design," published in the proceedings of the 4th European Propulsion Forum, 1993, Bath, U.K., pages 9.1-9.9.

4. D. R. Chang, D. D. Krueger, and R. A. Sprague, "Superalloy Powder Processing, Properties and Turbine Disk Applications," Superalloys 1984 (Warrendale, PA; The Metallurgical Society, 1984), 245-273.

5. C. Yin, "Materials Design Data Generation of LCF and FCGR for P/M Udimet 720 Alloy at Elevated Temperatures," (Report TDR AH1800-005, Allison Engine Company, 1995).

EVALUATION OF P/M U720 FOR GAS TURBINE ENGINE DISK APPLICATION

Hiroshi HATTORI*, Mitsuhiro TAKEKAWA*, David FURRER**, and Robert J. NOEL**

*Ishikawajima-Harima Heavy Industries Co., Ltd
Aero Engine & Space Operations
3-5-1 Mukodai-Cho, Tanashi-Shi, Tokyo, 188 Japan

**Ladish Co., Inc.
P.O. Box 8902, Cudahy WI, 53110-8902 USA

Abstract

Advanced aircraft engine turbine disk material needs the balance of creep/LCF, damage tolerant capabilities and economic feasibility for commercial implementation.
To incorporate an improved disk material for IHI newly designed aircraft engine, U720 in powder metallurgy + HIP + Extrude + Isothermal Forge form have been processed and evaluated. This study shown that P/M U720 have capabilities of wide range of microstructure control to allow optimum mechanical properties for specific design requirement.

Introduction

Advanced aircraft gas turbine engines are pushing the capability of conventional materials and processes. For the increasing temperature of the high pressure compressor and turbine sections of new turbine engines, alloys such as Inconel718 and Waspaloy are losing their ability to meet the demanding weight and performance requirements. More highly alloyed nickel-base superalloys have been developed and demonstrated in both cast and wrought (C&W) and powder metallurgy (P/M) forms. Subsolvus Rene95 or AF115 are one of the representative P/M alloys for high performance engines HP/LP turbine disk. These disks are previously evaluated[1] and concluded that although these disk materials have high performance capabilities when we compare them with conventional alloy such as Inconel718 and Waspaloy, the potential will not meet the newly designed engines from the point of view of LCF/creep strength balance, damage tolerance capabilities and these cost.

Alloy U720 has been identified as an outstanding intermediate temperature material for potential future applications for both technical and economic reasons. U720 is a unique high temperature alloy in that both C&W and P/M forms are being commercially processed, although the P/M route offers greater microstructural and property flexibility, and control.

Developmental processing and evaluation efforts have been conducted to determine the mechanical property response of P/M U720 processed with various conventional and novel thermo-mechanical processing routes. The primary processing effects on grain size, primary and secondary gamma-prime size and morphology, and subsequent mechanical properties have been evaluated.

U720 has been shown to be capable of developing a wide range of properties through thermo-mechanical processing (TMP), grain size and microstructure control. The alloy content of U720 as compared to numerous advanced, very high temperature capable P/M nickel-base superalloys, also allows economic feasibility for commercial implementation.

Experimental Procedure

P/M U720 billet material was procured for TMP processing and metallurgical evaluations. The material investigated in this study was produced from -270 mesh powder which was subsequently HIP + extrusion consolidated into 165mm diameter billet form. The chemistry of the billet material used is shown in table I. Nominal chemistry for Rene95 and AF115 are also shown in the same table for reference.

Table I Chemical composition of alloy used in this study (wt%)

	Ni	Cr	Co	Mo	W	Ti	Al	Hf	Nb	Zr	C	B
P/M U720	Bal.	15.6	14.6	3.0	1.24	5.05	2.55			0.03	0.008	0.03
AF115 [1]	Bal.	10.5	15.0	2.8	5.9	3.9	3.8	0.80	1.80	0.05	0.05	0.02
Rene95 [1]	Bal.	13.0	8.0	3.5	3.6	2.5	3.5		3.5	0.05	0.06	0.01

Table II Solution and aging condition used in this study

Alloy	Process	Solution heat treatment	Aging
P/M U720	Subsolvus A	1377K/4 Hrs./Oil Quench	
	Subsolvus B	1377K/4 Hrs./Fan air cool	923K/ 24 Hrs. / Air cool
	Supersolvus A	1441K/4 Hrs./Fan air cool	+ 1033K/ 16 Hrs. / Fan air cool
	Supersolvus B	1441K/8 Hrs./Fan air cool	
AF115	Subsolvus	1448K/3 Hrs./Fan air cool	1033K/ 16 Hrs. / Air cool
Rene95	Subsolvus	1383K/1 Hrs./Oil Quench	1033K/ 8 Hrs. / Air cool

Forgings were produced by controlled isothermal forging methods. The use of isothermal forging combined with fine grain P/M material allows extremely high material utilization. The superplastic nature of P/M nickel-based superalloy billet materials allow near-net component processing. Isothermal forging techniques additionally offer very tight processing controls, which is required for many alloys, which have very narrow processing windows.

The forged materials were heat treated to produce variations in grain size, and secondary or cooling-rate gamma prime size. The heat treatment used to produce these variations incorporated combinations of subsolvus and supersolvus solution heat treatments, variations in solution hold times, and variations in cooling methods from the solution heat treatment cycle. The heat treatments used in this program are listed in Table II.

Mechanical properties evaluation were conducted and compared them with those of Rene95 and AF115.

Round bar tensile specimens were machined from heat treated forgings and tested at 673 and 923K. Round bar creep specimens were machined from the forgings and constant creep testing were conducted at the applied stress range of 600 ~ 1200MPa in anbient air at temperature up to 1023K. Axial smooth bar LCF specimens with 6.25mm dia. were machined and polished gage section. Axial strain control LCF tests were performed using 30CPM triangle wave and strain A ratio of 1.0 at the temperature of 673 and 923K, strain range of 0.8% ~1.2%.

To evaluate fatigue crack growth behavior, 1/2 CT specimens were machined from the forgings. Tests were performed using the sinusoidal wave at stress-ratio R value of 0.05 and frequency of 120CPM. In addition, to evaluate the time dependent fatigue crack growth behavior, tests at 923K, R value of 0.05, with 90sec.dwell at maximum load and 1.5sec. ramp up/down were performed.

Statistical database were generated for Rene95 and AF115[1] and supplied for comparison of mechanical capabilities with this study.

Metallographic analysis were conducted to allow correlation of the microstructures to the mechanical properties results.

Results/Discussion

Microstructure

Numerous alloy and property characterization efforts have been previously conducted on U720. These efforts range from initial alloy characterization in cast and wrought form[2], to efforts on alloy chemistry and processing condition effects on properties of cast and wrought U720[3,4,5]. A more limited number of investigations have been published regarding or comparing properties of P/M U720 materials[6,7]. Many of these previous investigations give good insight into the mechanical property potential of cast and wrought U720, while the current effort is focused on evaluating the property response of U720 in P/M HIP+Extrude+isothermal Forge form.

Microstructual feature of each materials produced in this study were shown in Figure 1. The grain size are listed in Table III, The subsolvus solution heat treatment produced fine grain structures due to primary gamma prime pinning effects, which resulted in limited gamma grain growth, The supersolvus solution heat treatments produced coarser grain size as a result of grain growth.

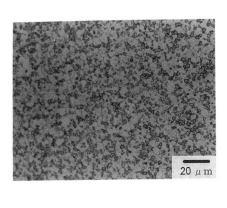 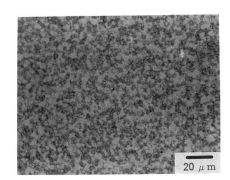

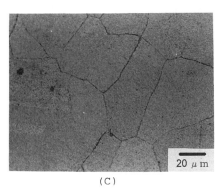 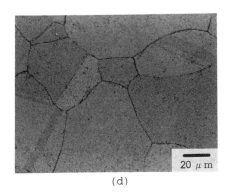

(c) (d)

Figure 1 Microstructure of each tested P/M U720; (a)Subsolvus A, (b)Subsolvus B, (c)Supersolvus A and (d)Supersolvus B

Table III Grain Size of tested P/M U720

	ASTM #
Subsolvus A	13
Subsolvus B	13
Supersolvus A	6
Supersolvus B	5

Tensile and creep properties

The average 0.2% yield strength and Larson-Miller parameter plot for creep resistance are presented in figure 2 and figure 3 respectedly , plotted with statistical data from Rene95 and AF115.

Subsolvus processed P/M U720 show equal or superior tensile capabilities to AF115 and approaching those of Rene95 at 673K, although those are lower at 923K and poor creep capabilities are shown, which are lower than Rene95. Subsolvus A, which were experienced faster cooling after solution, increase the yield strength than subsolvus B. There are no remarkable difference in creep capabilities between subsolvus A and B.

Supersolvus processed P/M U720 loose tensile capabilities but greatly enhanced creep strength, which are enough exceed to Rene95 and approaching to those of AF115, especially at higher temperature test condition.

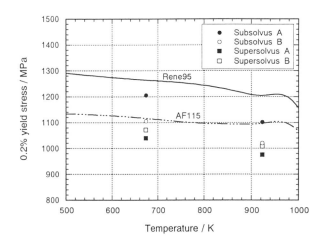

Figure 2 Tensile properties comparison between P/M U720 and AF115,Rene95

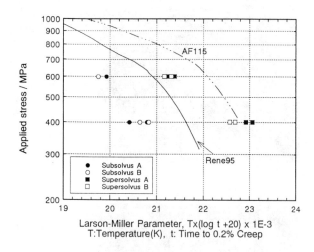

Figure 3 Larson-Miller comparison of 0.2% Creep time between P/M U720 and AF115, Rene95

LCF and fatigue crack growth properties

LCF and fatigue crack growth capabilities are summarized in figure 4, figure 5, and figure 6 respectively. Subsolvus processed P/M U720 show LCF capabilities equivalent to the AF115 and Rene95 at 673K. Although supersolvus U720 loose LCF potential, the crack growth rate were about two times or more slower than subsolvus U720 and AF115.

Figure 7 shows the time depend tendency of fatigue crack growth capabilities for both sub-/supersolvus U720. Higher time dependent tendency were seen in the supersolvus forgings.

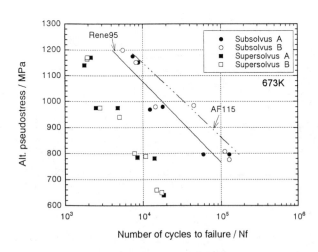

Figure 4 LCF properties comparison between P/M U720 and AF115, Rene95 at 673K.

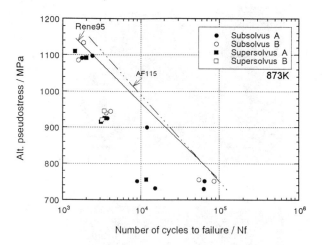

Figure 5 LCF properties comparison between P/M U720 and AF115, Rene95 at 873K.

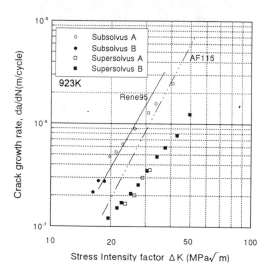

Figure 6 Fatigue crack growth properties comparison between P/M U720 and AF115, Rene95.

Discussions

The grain size influence on mechanical properties is well documented[8,9]. Efforts have been directed in other programs to utilize grain size to optimize properties such as damage tolerance[4,10]. The grain size effects are very apparent in the current effort when comparing all of the properties evaluated. A debit is seen in tensile strength and LCF for the intermediate grain size material as compared to the fine grain processed material. A corresponding increase in creep and crack growth rate is seen for the intermediate grain P/M U720.

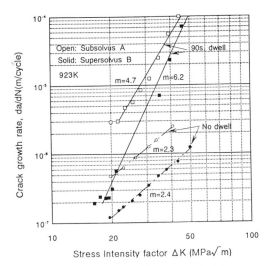

Figure 7 Dwell time effect on fatigue crack growth properties of P/M U720

SEM observation for the tested U720 in this study are shown in figure 8. The primary gamma prime morphology for the P/M U720 forgings processed in this effort were seen to be greatly modified by heat treatment, which is not surprising since sub- and supersolvus heat treatments were investigated. The supersolvus heat treatment cycles, which also included a second, subsolvus solution heat treatment produced uniform spherical gamma prime particles approximately 0.3-0.6 micron meters in diameter. The completely subsolvus processed forgings contained very large, angular primary gamma prime with an average particle size of 1-4 micron meters.

From previous work[11], it is known that cast and wrought U720 properties are greatly influenced by the cooling rate imposed after solution heat treatment. A near 15% increase in room temperature yield strength is seen from increasing the solution cooling rate from 0.27K/sec to 3.24K/sec for a variety of solution temperatures and aging conditions. The secondary, or cooling-rate gamma prime has also been previously shown to be greatly dependent on variations in cooling rate from the solution heat treat temperature. The secondary gamma prime size for U720 has been shown to be approximately 0.08 micron diameter spherical particle at 3.24K/sec cooling rates, and approximately 0.37 micron diameter irregular particles at 0.27K/sec.

A corresponding shift in properties and secondary gamma prime size is seen between the forgings produced in this program which were heat treated using oil quench and fan air quench cooling techniques; although the range in cooling rate variation for these materials is less than that noted above in the previous characterization effort and the size variation of secondary gamma prime is approximately 0.08 to 0.2 micron meters and around 9 % increase of 673K yield strength were seen in subsolvus A.

Fatigue crack growth rate of subsolvus P/M U720 is more than two times faster than those of supersolvus material at 923K. When we compare the time dependency of both grain size material, m-factor of each data in Paris-equation are also indicated in figure 7. Although there is little difference of m-factor for both material at 120CPM(no dwell time)test condition, higher effect of hold time on m-factor were seen in the corser grain material than fine grain material. The difference of tensile properties of this temperature range may affect this phenomena rather than creep-fatigue interaction. And these facts indicate the crack grew by intergranular fracture mode. More detail analysis would be necessary on the material microstructure/chemistry.

For advanced applications of P/M U720, processing must be optimized to achieve the best balance of properties, residual stress, overall disk performance, and cost. To allow for such process selection and optimization, efforts similar to the current program must be undertaken to fully understand alloy/process interactions.

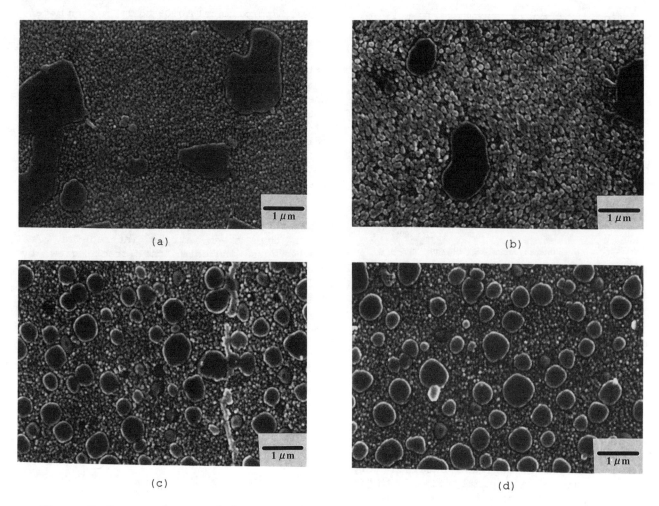

Figure 8 Gamma prime morphology and distribution for (a) Subsolvus A, (b) Subsolvus B (c) Subsolvus A, and (d) Subsolvus B heat treated P/M U720.

Conclusions

P/M U720 material has been shown to be readily processable by isothermal forging methods. Resultant microstructures and mechanical properties can be manipulated through variations in forging and heat treatment practices.

P/M U720 forgings processed to fine grain, high strength conditions show tensile properties above AF115 and approaching those of fine-grain Rene95; LCF properties equal or superior to fine-grain Rene95 and approaching those of AF115 at 673K, but are somewhat lower at 923K; and creep behavior which is below that of fine-grain Rene95 and AF115.

Intermediate grain size processed P/M U720 forgings resulted in reduced tensile and fatigue properties, but greatly enhanced creep response, which was shown to be superior to fine-grain Rene95 for all creep test conditions and superior to AF115 for high temperature creep test conditions.

Fine grain U720 showed equivalent fatigue crack growth capabilities to those of AF115. Intermediate grain U720 enhanced the crack growth resistance two times as fine grain P/M U720.

This evaluation program has shown that both grain size, and gamma-prime size and distribution can be altered during the processing of P/M U720 forging to allow optimization of mechanical properties and this material can be applied various design type of components.

Reference

1. H.Hattori at al. Unpublished research work, IHI

2. F.E.Sczerzenie and G.E.Maurer,Superalloys 1984(Warrendale,PA,The Metallurgical Society,1984),572-582

3. S.Bashir and M.C.thomas,Superalloy 1992(Warrendale,PA,The Minerals,Metals & Materials Society,1992),747-755

4. K.R.Bain,et al.,Superalloys 1988(warrendale,PA,The Metallurgical Society,1988),13-22.

5. C.J.Teague et al.,"Effect Of Quench From Solution Heat Treatment On The Properties And Structure Of Udimet 720",(Paper presented at the 1994 Fall TMS meeting, Rosemont, IL, Oct.26,1994).

6. S.J.Panel and Elliott,Superalloys 1992(Warrendale,PA,The,Minerals,Metals & Material Society,1992),13-22.

7. J.M.Hyzak et al.,Superalloys 1992(Warrendale,PA,The,Minerals,Metals & Materials Society,1992),93-102.

8. E.O.Hall,Proc.Phys.Soc.B,64,747(1951).

9. F.Wallow and E.Nembach,Scripta Materialia,vol.34,no.3,pp499-505,1996.

10. D.D.Krueger et al.,Superalloys 1992(Warrendale,PA,The Minerals,Metals & Materials Society, 1992),277-286.

11. J.Belonger,Unpublished Research,Ladish Co.,Inc.

THE MANUFACTURE AND EVALUATION OF A LARGE TURBINE DISC IN CAST AND WROUGHT ALLOY 720Li

D J Bryant, Dr G McIntosh*

Rolls-Royce plc, PO Box 31, Derby, DE24 8BJ, UK
* Wyman Gordon Ltd, Houstoun Road, Livingston, West Lothian, Scotland EH54 5BZ

Abstract

The new generation of large high thrust aero-engines, such as the Rolls-Royce Trent, lead to ever increasing demands on materials. In particular as the requirements on turbine discs become more stringent traditional alloys such as IN718 and Waspaloy become less viable choices. To meet the more demanding requirements highly alloyed, high strength materials will be required. One such alloy is alloy 720Li, which is available in the powder or conventionally processed cast and wrought form. Recent developments in the manufacture of good quality Cast and Wrought billet at relatively large diameters (230mm and 250mm) has led to it being a cost effective choice for meeting turbine disc requirements.

A successful development programme has already been completed approving alloy 720Li billet and forging routes for a relatively small diameter turbine disc (450mm diameter). This paper will discuss the potential ability to scale-up the technology to be applicable to large civil engines. A large (840mm diameter) Intermediate Pressure Turbine Disc (IPTD) was selected for the exercise. The increase in the size of the forging is illustrated by comparison of the forging mult weight increase from typically 160Kgs to 500Kgs for the IPTD.

The manufacture and test of IPTD forgings will be discussed in detail. A key to the success of the evaluation was the use of numerical process modelling by the forgemaster. A number of tools were available to aid the design of the forging route, which not only enabled microstructural and property targets to be met, but achieved at the first attempt with confidence.

Two forgings were successfully manufactured meeting all of the release requirements normally defined for the alloy. One of the forgings was subsequently selected for full cut-up evaluation of macro/microstructure and mechanical properties. The results from the macro/microstructural evaluation revealed the required structure had been fully generated, with a recrystallised grain size of ASTM 9-10. This was consistent with previous experience of the alloy. A full suite of mechanical testing was carried out from tensile testing through to the measurement of the crack propagation rate. In all cases the results were found to be comparable with previous experience for the alloy. The results will be compared to the forging manufactured in Waspaloy.

It was concluded that there was no debit in terms of mechanical properties, or degradation of macro/microstructure at the increased forging size, albeit with similar section thickness. The successful outcome of the programme has generated confidence that large disc forgings can be manufactured from existing cast and wrought alloy 720Li billet diameters.

Introduction

Alloy 720Li is a high strength nickel based alloy which has been identified as a turbine disc material to meet the ever increasing demands imposed by the new large, high thrust aero-engines. The alloy is available in both the powder form (1) and the conventional cast and wrought form. A successful development programme has previously been completed approving the alloy 720Li billet and forging routes for a relatively "small" diameter forging (450mm diameter) (2).

This paper will discuss the programme carried out to scale-up the technology to large turbine disc forgings. A large civil engine IPTD was selected as a suitable evaluation forging (840mm diameter) which is currently manufactured in Waspaloy.

The choice of this part led to an increase in billet input weight from previous experience (160 Kgs) to some 500 Kgs for the IPTD forging. This represented a significant challenge to the forgemaster, particularly as the billet diameter was restricted to 250mm diameter. The use of numerical process models was key in designing the forging route, ensuring 'right first time' manufacture was achieved.

In addition to testing the forgemasters capabilities, the large forging represents a significant metallurgical challenge. A particular concern is the effect of quench cooling rate on the disc properties. It is known that a reduced cooling rate will decrease material strength, as shown in Figure 1. Additional areas of concern regard the generation of the required microstructure and achieving the desired levels of ultrasonic inspection.

A full metallurgical evaluation of the disc forgings was completed to allow comparison to the existing experience of the alloy. Any degradation in terms of macro/microstructure and mechanical properties as a result of the scaling-up in size would be determined. Furthermore, a comparison was made to the disc as manufactured in Waspaloy. The billet material for the evaluation was produced by Teledyne Allvac and the forging carried out by Wyman Gordon Ltd.

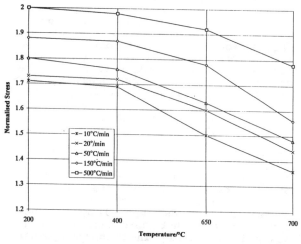

Figure 1: Effect of cooling rate from solution heat treatment temperature on ultimate tensile strength

Material

Alloy 720Li is a high strength nickel material, its composition is given in Table 1. The development of the alloy from a wrought blade alloy for land based turbine applications is well documented (3). Being a highly alloyed, high strength material it is inherently difficult to process through the cast and wrought route. Early attempts at processing the alloy centred on double melt (VIM-VAR) processing and was limited to small diameter billet, up to 160mm. The early material was prone to segregation showing a variable microstructure. Disc forgings produced from such material showed very variable structures which failed to meet microstructural targets (4).

Table I: Composition of alloy 720Li by weight %

	C	B	Cr	Co	Mo	W	Ti	Al	Zr	Ni
Nominal U720Li	0.15	0.15	16.0	15.0	3	1.25	5.0	2.5	0.03	Bal
Heat BT34-2	0.11	0.17	16.03	14.41	3.04	1.19	4.95	2.57	0.05	57.48

The development of conversion routes based on triple melt material (VIM-ESR-VAR) overcame many of the problems initially encountered. The addition of the ESR stage results in improved quality as well as a sound electrode to go into VAR. This in turn allows an improved VAR ingot in terms of structure and chemical segregation for final thermomechanical conversion to finished billet. The triple melt route also has the significant advantage of improving the cleanliness of the billet compared to double melt material (5).

The manufacture of good quality billet is key to the manufacture of acceptable forgings and is subject to continued development by the material suppliers. Figure 2 shows a macroslice from a 250mm diameter billet used to manufacture the IPTD. A featureless macrostructure is observed with no deleterious features evident.

The microstructure is typically ASTM 8 to 9, Figure 3 shows some grain size variation and banding of the γ' is also present due to persistent solidification segregation which has not been broken down through billet conversion. This standard of billet has previously been found to produce acceptable forgings.

Forging Route Development

Numeric Modelling Considerations:

The need to gain competitive advantage by reducing the 'development to market' cycle time in the forge industry has led to the introduction of a number of numerical tools for the forgemaster, which enable the simulation of both forge and heat treatment processes to be carried out with confidence, thus reducing or obviating the need for expensive and time-consuming prototyping trials. Each of these processes may be simulated to achieve various objectives, with increasing degrees of complexity:

a) Forge modelling objectives:

   i) Die fill / load prediction

   ii) State variable prediction (temperature, strain, strain rate ....)

Figure 2: 250mm diameter billet slice after macro-etching (immersion etched in acidified ferric chloride)

    iii) Microstructural prediction (qualitative or quantitative)

    iv) Property correlation

b) Thermal modelling objectives:
    i) Cooling rate optimisation

    ii) Quench stress reduction

    iii) Property correlation

    iv) Residual stress determination

A number of validated commercial packages are available to the forgemaster, and Wyman-Gordon Ltd use "Forge 2" for forge modelling (2-D axisymmetrical), and ''Topaz'' or ''Nike'' for thermal modelling.

However, accurate output is necessarily a function of the accurate input parameters, and Wyman-Gordon have developed in-house expertise at its Corporate Research and Development laboratories to generate the requisite parameters for both forge and thermal modelling. Validation of forge modelling predictions have been carried out on numerous component standardisation cut-up evaluations, over several alloy systems.

Rheological databases, which predict the flow stress behaviour over the hot-working process window, are developed for each alloy system under consideration via sub-scale isothermal compression testing. To optimise the accuracy of the flow stress data, which may otherwise be severely distorted by adiabatic heating effects, proprietary low friction lubrication processes have been developed which allow flow stress generation up to true strains of 1.0 - 1.2. Nevertheless, correction factors may still be required, particularly at higher strain rates.

An extract from the Alloy 720Li rheology database is given in Figure 4. Although each forge process is carried out at a nominal fixed temperature, non-isothermal temperature effects need to be examined to incorporate both surface chilling (resulting from the die - workpiece or die - atmosphere thermal differential) and localised heating (generated adiabatically during the forging process), which may result in grain growth or secondary recrystallisation. The rheological database is thus determined over a wide temperature range, as well as strain rate range, to encompass the anticipated forge process window. In the case of Alloy 720Li, flow stresses were calculated over the following ranges:

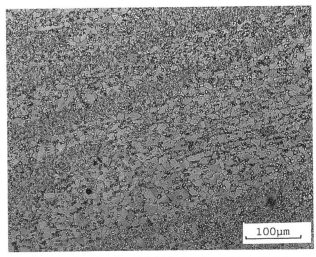

Figure 3: Typical billet microstructure (Etch: 10% bromine in methanol)

i) Temperature: 1038 - 1148°C (1900 - 2100°F)

ii) Strain rate: 0.001 - 10.0/ s

iii) Strain: 1.0

Interrupted testing is carried out at several strain intervals, with the partially deformed samples being rapidly quenched to "freeze-in" the dynamically evolved microstructure, which can then be examined by metallographic evaluation. A simulated heat treatment cycle may be subsequently applied, to follow microstructural changes from static processes. These data may then be incorporated into the model (by either qualitative or quantitative means) to allow a measure of microstructural prediction. To achieve a fully recrystallised grain structure with Alloy 720Li, a high degree of strain at hot working temperatures is required.

Both flow stress and structural effects of simulated die-chill can be evaluated in more detail by instituting a controlled rapid cooling cycle from the nominal forge temperature, and then performing in-situ tensile testing, and subsequently, metallographic analysis.

Individual quench facilities are characterised to determine local heat transfer coefficients for use with thermal modelling, which are validated via thermocoupled experiments. Residual stress modelling is validated via incremental depth measurements on as-quenched components.

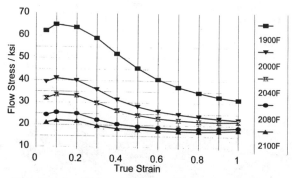

Figure 4: Alloy 720Li, flow stress data [S/R=1.0/s] (uncorrected for adiabatic heating)

Forge Route Development:

Cast and Wrought Alloy 720Li billet to Rolls Royce specification is available in sizes up to 250 mm diameter. Although nominally recrystallised to an ASTM 10 grain size, unrecrystallised grains ALA ASTM 0 may be present. The forge route must be designed to eliminate or break down such microstructural features by designing intermediate forge operations that locally increase the strain level, such that the component ultimately achieves an overall high degree of strain. Particular emphasis is placed on regions of the forging associated with the billet centre, where the highest proportion of large unrecrystallised billet grains are found.

The 250 mm billet diameter limitation resulted in an aspect ratio of 4.2:1, which necessitated a total of five forge operations to achieve the final shape.

To meet grain size, and associated mechanical property and ultrasonic inspection requirements, each forging operation is carried out at a sub-solvus temperature to prevent grain growth. Using the rheological database, the forge speed of each operation is chosen such that chill and adiabatic heating effects are balanced. To minimise flow stress increases and prevent chill cracking by forging at a "fast" speed, while preventing localised grain growth due to adiabatic heating (gamma prime dissolution & resultant grain growth) by forging at a "slow" forge speed. The Wyman-Gordon ltd 30,000T and 9,000T hydraulic presses allow the use of controlled forge speed profiles, and a constant ram speed was stipulated for each of the five forge operations (overall, within the range of 25 - 80 mm/s). Examples of strain and temperature prediction, dominant factors in the microstructural evolution of Alloy 720Li are given in Figure 5. Over the range of forge speeds (chosen to balance thermal effects), strain rate effects on microstructural evolution of fine grain Alloy 720Li are minimal.

Heat Treatment Route Development

Heat treatment was carried out as per Rolls Royce specification. Thermal modelling was used to design the pre-heat treatment shape such that quench stresses are reduced to prevent quench cracking, yet cooling rates are high enough to satisfy mechanical property requirements. On-cooling quench stresses and average cooling rates for the IPTD are given in Figure 6.

Residual stress modelling has been validated by Wyman-Gordon for Alloy 720Li on a High Pressure Turbine Disc (HPTD) (6) and agreed with Rolls-Royce. Predicted residual hoop stresses (following quenching) are given in Figure 7. Although this technique is not as yet used as a forging route design criterion, it has been carried out to extend Wyman-Gordon's expertise in this area.

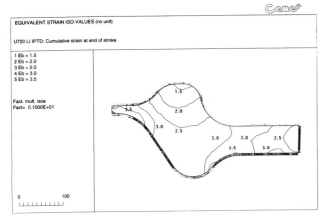

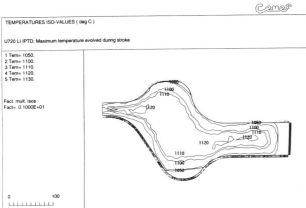

Figure 5: Predictions of strain and temperature generated on forging the IPTD

Manufacture:

A trial batch of two forgings was successfully manufactured to the following nominal parameters:

i) Starting Stock:

Nominal 500 kg @ 250 mm diameter
Cast and Wrought Alloy 720 LI
Supplied by Teledyne Allvac.

ii) Forge Process:

Five independent closed die forge operations (hot worked) on 9,000T and 30,000T hydraulic presses at Wyman-Gordon Ltd, Scotland
Process parameters assigned as per forge modelling simulations
Warm die philosophy
Sub-solvus forge temperature
Controlled ram velocity profiles

iii) Heat Treatment:

Pre-heat treatment shape defined via thermal modelling
Solution Treat: 4 hours @ 1105°C (2021°F); Oil Quench
Age: 24 hours @ 650°C (1202°F); Air Cool
Age: 16 hours @ 760°C (1400°C); Air Cool

Experimental Procedure

After forging and heat treatment, the forgings were machined to the final rectilinear shape to allow ultrasonic inspection. An integral test ring was also removed to allow release testing of the parts.

One of the forgings was then selected for cut-up evaluation. Forging macroslices were removed and surface ground to allow macro-etch inspection using acidified ferric chloride. Micro-specimens were prepared from locations throughout the forging. The samples were metallographically prepared and etched in bromine and methanol to reveal the general microstructure. The samples were then repolished and etched by means of electrolytic 10% phosphoric acid in water to reveal the γ' distribution.

Optical microscopy was carried out at each stage.

The bulk of the disc was then cut-up to provide specimens for the mechanical testing programme which is described in Table II.

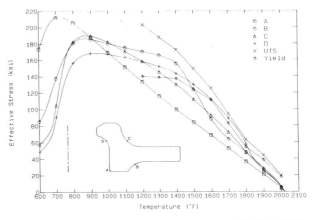

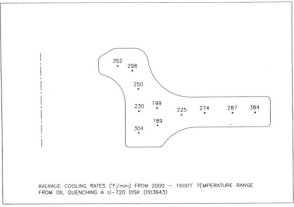

Figure 6: Predictions of on cooling quench stresses and average cooling rates generated in the IPTD in heat treatment

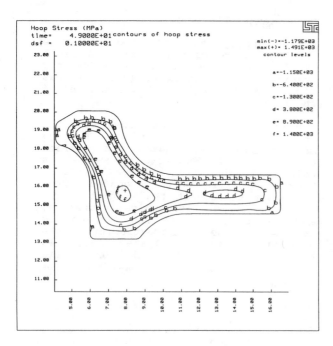

Figure 7: Predictions of the residual hoop stresses generated in the IPTD on heat treatment

Table II: Mechanical test programme

Test Type	Position	No.	Test Detail
Room Temperature Tensile	Test - Ring Bore - Axial Bore - Tangental Web - Radial Rim - Tangental Rim - Radial Bore - Radial	1 2 2 1 1 1 1	20°C " " " " " "
Elevated Temperature Tensile	Bore - Tangental Bore - Tangental Web - Radial Web - Radial Web - Radial Rim - Tangental Rim - Tangental Test Ring	1 1 1 1 1 1 1 1	200°C 400°C 400°C 500°C 600°C 500°C 600°C 650°C
Creep Strain	Test - Ring Web - Radial Rim - Tangental	1 1 1	625°C/730MPa " " " "
Plain Low Cycle Fatigue	Bore - Tangental Web - Radial Web - Radial Rim - Tangental Rim - Tangental	6 4 2 4 2	200°C 500°C 600°C 500°C 600°C
Crack Propogation	Bore Rim	2 2	200°C 500°C

Results

Release Testing

Normal release testing requirements were carried out on both forgings, the requirements being fully met in each case, results are given in Table III.

Good levels of ultrasonic inspectability were achieved in both forgings. Typical grass levels of -30dB or better were achieved against a No.1 FBH standard.

Both Macro etch and fluorescent penetrant inspection of the discs revealed no defects. These results were consistent with previous experience.

Table III: Results from release testing of forgings

Serial No.	Temp/°C	UTS/MPa	PS/MPa	%El	%RA
0001	20	1100	1581	16	17
0001	650	1058	1431	20	36
0002	20	1112	1594	16	17
0002	650	1041	1433	20	39

Macro/Microstructure

The etched forging macro slice is shown in Figure 8. An acceptable grain flow has been achieved. The macrostructure of the forging was consistently uniform across the full section, with no deleterious features evident.

The microstructures were found to be generally fully recrystallised uniform microstructure with an average grain size of ASTM 8-9, as shown in Figure 9. Some γ' banding with associated local grain size variation was visible. This microstructure resulting from the inhomogeneity associated with the cast and wrought route. This is an inherent feature of the cast and wrought processing, particularly of such a highly alloyed material.

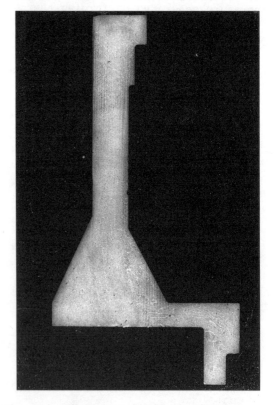

Figure 8: Etched forging macroslice (Etch: acidified ferric chloride)

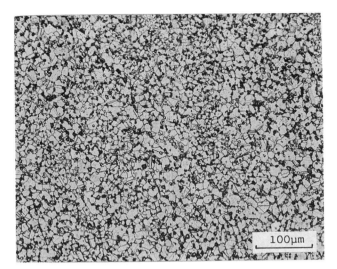

Figure 9: Representative microstructure from IPTD disc forging
(Etch: 10% bromine in methanol)

Some occasional (<1%) isolated unrecrystallised grains as large as ASTM1 were also identified, these grains were confined to the drive arm region of the disc forging. This corresponds with a relative low strain region of the forging as predicted in Figure 5. Modification of the forging route to locally increase the level of strain should eliminate such features. A typical example is shown in Figure 10. These features are typical of the process route and isolated examples are not thought to be detrimental. Figure 11 shows the typical γ' distribution found within the forging, the effects of the segregation discussed above are again evident in the γ' banding.

The macro/microstructure generated in the IPTD is consistent with previous experience for material forged via the hot die route. The IPTD disc forging shows a slightly coarser grain size compared to previous Rolls-Royce experience.

Tensile Properties

The tensile properties generated from the IPTD forging were compared against typical values for Waspaloy. As shown in Figure 12 the results from the IPTD forging compare favourably with previous experience for the alloy for both the ultimate tensile strength and yield strength. It is concluded that there has been no degradation in tensile properties in the scale-up to large disc forgings.

Considering the results from the IPTD alone, as would be expected there is some variation according to location within the disc. The rim area tending to have higher strengths than those from the bore. This largely reflects the different heat treatment response related to section thickness. The effect of grain flow during forging is also evident, the samples tested in the axial direction showing the lowest strengths.

The material showed good ductility across the temperature range investigated, all the values measured being greater than 15% for both reduction of area and elongation. These results are again consistent with previous experience for the alloy.

In comparison to the current IPTD disc material Waspaloy, the Alloy 720Li IPTD shows a 25-30% increase in strength. This would be expected in terms of alloy composition and grain structure (Alloy 720Li forging ASTM 9 or finer, compared with ASTM 5-7 for the Waspaloy forging).

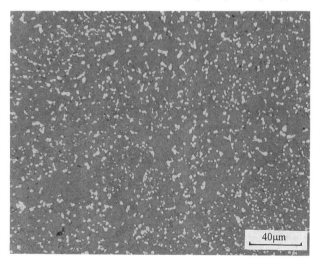

Figure 11: Typical γ' distribution
(Etch: electrolytic - 10% phosphoric in water)

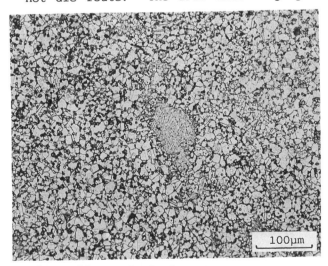

Figure 10: Typical microstructure showing isolated unrecrystallised grain
(Etch: 10% bromine in methanol)

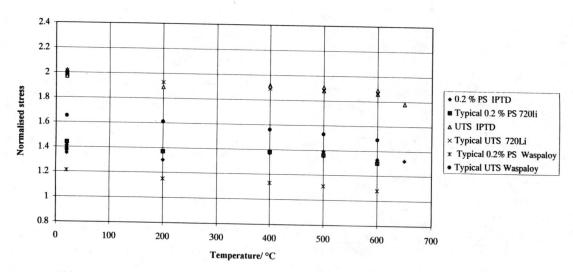

Figure 12: Tensile strength measured from IPTD forging compared with typical data for Alloy 720Li and Waspaloy

Creep Strain Testing

Creep strain testing was carried out on three locations within the IPTD forgings, the results are given in Table IV. In each case the results met the target requirements and were consistent with previous experience for the alloy.

Table IV: Results from creep strain testing

Location	Temp/°C	Stress/MPa	TPS(%)
2	625	730	0.036
4	625	730	0.025
6	625	736	0.010

Low Cycle Fatigue Testing

Plain load control LCF S-N curves were generated at 200°C and 500°C. All tests were carried out at 15 cycles per minute with a 0-maximum stress cycle (R=0). Figures 13 and 14 show the LCF curves at 200°C and 500°C respectively. In both cases the material shows good resistance to fatigue under these test conditions. The results are consistent with previous experience for the alloy, although only limited data is available at 200°C. At 600°C the curves appear very 'flat' and small increases in the applied stress can produce a much reduced life. The results can also be grouped into two 'families' dependent upon the location from where the samples were manufactured. The specimens manufactured from the rim location showing improved performance compared to those from the web of the disc. This may be indicative of the relative degrees of structural refinement observed in those areas.

Investigation of the fracture surface of the failed specimens typically showed that the initiation site was a titanium rich nitrides/carbonitrides. Figure 15 shows an SEM micrograph of a representative feature. These discrete titanium nitride/carbonitride were usually in the size range 10-40μm and are an inherent microstructural feature, and do not compromise the LCF life of the material.

Comparing these results to typical Waspaloy data, the Alloy 720Li shows a greater resistance to fatigue at both 200°C and 500°C. This behaviour is again as would be expected partly due to to the compositional and structural differences between the alloys.

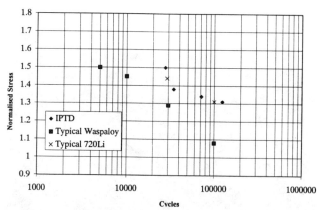

Figure 13: LCF behaviour at 200°C for IPTD forging compared with typical alloy 720Li and Waspaloy data

Crack Propagation Testing

Crack propagation testing was carried out at 200°C and 500°C and specimens prepared from the rim and bore of the disc. The results for the testing at 200°C are given in Figure 16, and it can

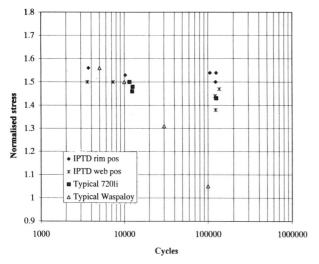

Figure 14: LCF behaviour at 500°C for IPTD forging compared with typical alloy 720Li and Waspaloy

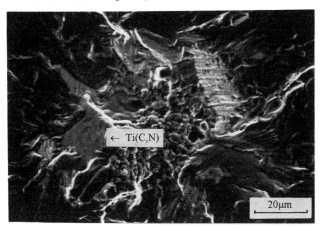

Figure 15: Sub-surface initiation site within failed LCF specimen.

be seen that the results are comparable to previous alloy 720Li experience. In comparison it can be seen that the Waspaloy shows a greater resistance to crack propagation than the Alloy 720Li. The coarse grain microstructure being the dominant feature, outweighing any differences in alloy composition.

The results from the crack propagation testing at 500°C, Figure 17, compared favourably with previous Alloy 720Li experience and was similar to Waspaloy behaviour.

Summary

Two large IPTD forgings have been successfully manufactured from 250mm cast and wrought billet. This represents a significant scaling-up in size from previous manufacturing experience. The forgings met all of the requirements specified in terms of macro/microstructure and mechanical properties. Excellent correlation was achieved with previous experience, an indication that there was no degradation in material performance in scaling up the forging size. The successful outcome of the evaluation demonstrates that alloy 720Li has the potential to meet future requirements for high strength large disc forgings.

The use of numerical forge and heat treatment models has been critical in designing the forging route, production standard forgings being produced at the first attempt. The design cycle has been significantly reduced from the traditional trial and error techniques. This successful design philosophy can be extended to other component geometries, and is not restricted to Alloy 720Li alloy.

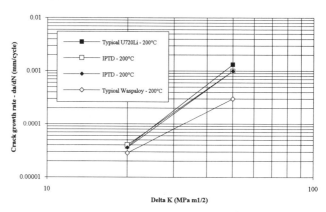

Figure 16: Results from crack propagation testing on IPTD forging at 200°C compared to typical alloy 720Li and Waspaloy data

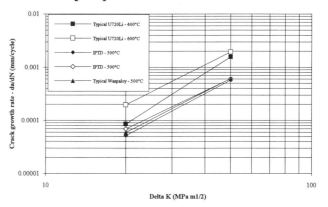

Figure 17: Results from crack propagation testing on IPTD forging at 500°C compared to typical alloy 720Li and Waspaloy data

Further Developments

Further developments of the manufacturing route have been identified by this study. The developments will lead to the definition of a cost effective production route for large forgings. In particular, the need to develop larger billet diameters whilst maintaining the current structural standards. This represents a stern challenge for the billet manufacturer, who have already come a long way in developing the current manufacturing routes to produce the current standard of Alloy 720Li billet.

Acknowledgements

The work was carried out as a collaborative effort between Rolls-Royce plc, Wyman Gordon Ltd and Teledyne Allvac. The authors would like to thank: P D Spilling, Dr A James, B Towill, S Blackham and D Crofts at Rolls-Royce plc for review of data and preparation of manuscript; Dr R Wallis at Wyman-Gordon (Houston) for thermal modelling/database generation; F Logan and Dr I Dempster at Wyman Gordon Ltd for technical input and component evaluation. Thanks also go to Dr L Jackman and S Vallandingham at Teledyne Allvac for assistance in the provision of billet material.

References

1) J M Hyzak, et al, The microstructural response of as-Hip P/M U720 to thermomechanical processing (Superalloys 1992, TMS Warrendale, PA) 93-102

2) A Plath, et al, Conventionally processed high performance disc material for advanced aero engine design. 4th European Propulsion Forum 1993, Bath, UK P9.1-9.9.

3) P W Keefe, S O Mancuso & G E Mawer, Effects of Heat Treatment and Chemistry on the Long-Term Stability of a High Strength Nickel Based Superalloy. (Superalloys 1992, TMS, Warrendale, PA) 487-496

4) C A Harwood, Rolls-Royce - Private Communication.

5) J M Moyer, et al, Advances in Triple Melting Superalloys 718, 706 and 720 (Superalloys 718, 625, 706 and Various Derivatives, TMS, Warrendale, PA) 39-48

6) R A Wallis and I W Craighead, Predicting Residual Stresses in Gas Turbine Components (TOM, October 1995) 69-71

ELECTROSLAG REFINING AS A CLEAN LIQUID METAL SOURCE FOR ATOMIZATION AND SPRAY FORMING OF SUPERALLOYS

Mark G. Benz*, William T. Carter, Jr.*, Felix G. Müller**,
Robin M. Forbes Jones†

*GE Corporate Research & Development, River Road, Schenectady, NY 12301
**ALD Vacuum Technologies, Rückinger Strasse 12, D-6455 Erlensee, Germany
†Teledyne Allvac, 2020 Ashcraft Avenue, Monroe, NC 28110

Abstract

A pilot plant has been constructed to demonstrate the concept of using a combination of electroslag refining (ESR) and an induction-heated, segmented, water-cooled copper guide tube (CIG) to melt, refine, and deliver a stream of liquid metal to a spray forming process. The basic ESR system consists of a consumable electrode of the alloy to be melted, a liquid slag, and a water-cooled copper crucible. The liquid slag is heated by passing an ac-electric current from the electrode through the slag to the crucible. The liquid slag is maintained at a temperature high enough to melt the end of the electrode. As the electrode melts, a refining action takes place—oxide inclusions are exposed to the slag and are dissolved. Droplets of molten metal fall through the slag and are collected in a liquid metal pool contained in the crucible below. By the addition of the induction-heated, segmented, water-cooled copper guide tube (CIG) to the bottom of the crucible, a liquid metal stream can be extracted from the liquid metal pool. This stream makes an ideal liquid metal source for atomization and spray forming. The pilot plant has been operated at a melt rate of 15 to 25 kg/min with the Ni-base superalloys Alloy 718, Rene' 95 and Rene' 88. Process optimization and cleanliness evaluation studies are in progress.

Introduction

Spray Forming

Spray forming is being considered as a low-cost alternative to *powder metallurgy* for the preparation of high-strength forged superalloy components. Spray forming is the term applied to a process in which a stream of liquid metal is gas-atomized and then immediately reconsolidated into a solid shape [1, 2]. In this process, the atomized droplets of liquid metal are cooled by the atomizing gas, so that they lose most (75%) of their heat of fusion while in flight from the atomizer. Upon striking the solid shape, the metal quickly loses the balance of its heat of fusion such that only a thin layer of liquid metal is maintained on the surface throughout the process. The spray forming process has been studied extensively for nickel-base superalloys [3, 4], and has been shown to be capable of producing a homogeneous fine-equiaxed-grain microstructure (30- to 40-μm grain diameter), with a relative density of greater than 98%.

Ceramic Inclusions - A Problem

Ceramic inclusions can reduce low-cycle-fatigue (LCF) life in both the *powder metallurgy* approach and the *spray forming* approach to the preparation of superalloys [5, 6]. Mechanistically, a large ceramic inclusion will crack early in life and act as a crack starter for the surrounding superalloy. Direct observation of this event is shown in Figure 1. The aluminum oxide particle cracked very early in life and the crack rapidly extended into the surrounding superalloy.

Figure 1. Polished surface view of an aluminum oxide particle acting as a crack starter in the superalloy Rene' 95 mechanically loaded under low-cycle-fatigue conditions. [7]

Applying the discipline of fracture mechanics to this case, it can be shown that the higher the mechanical loading during the fatigue exposure, the smaller the cracked oxide must be to achieve equivalent life. The results of such a computation are shown in Figure 2.

The challenge from a processing point of view, therefore, is to limit the size of the largest ceramic inclusion that reaches the

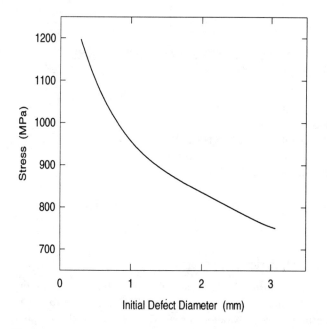

Figure 2. Stress required to propagate a crack from an initial defect size to the critical crack size in 10⁴ cycles. da/dn and K_{IC} taken as the average of values for Rene' 95, IN 100, Astroloy, Waspaloy, and INCO 901 at 650°C. [8]

final component. In the case of the powder metallurgy approach, this challenge is met by *sieving* the powder after *vacuum induction melting* and *gas atomization*; by careful handling during *canning* to place the powder in a container suitable for reconsolidation; by *vacuum degassing and sealing* to remove and prevent re-entry of the ambient atmosphere from the can; and by *extrusion* to reconsolidate the alloy to a fine grain billet [5].

ESR - A Solution

In the case of the spray forming approach, a novel melting system is being developed to make use of electroslag refining (ESR) to dissolve ceramic inclusions before the liquid metal is converted to droplets by gas atomization [9-15]. A 15-kg/min pilot scale system has been constructed and is being used to produce material for testing and evaluation [16-19].

Pilot Scale System

Concept

The pilot scale system being used to evaluate this approach is shown schematically in Figure 3.

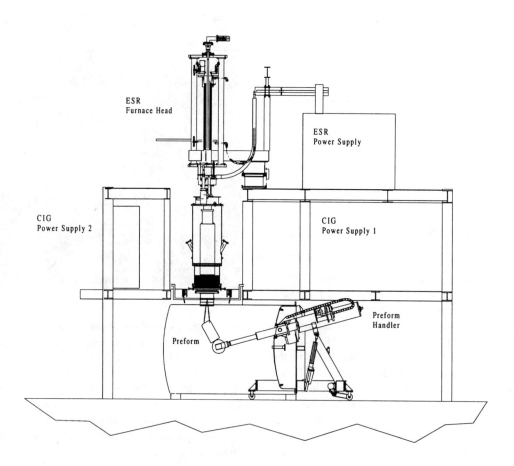

Figure 3. Schematic diagram showing the general arrangement of the major components of the ESR-CIG pilot plant configured to deliver liquid metal to the atomizer of a spray deposition system.

The basic concept makes use of electroslag refining (ESR) to melt and remove ceramic inclusions from the superalloy, combined with an induction-heated, segmented, water-cooled copper guide tube (CIG) to deliver a stream of liquid metal to a spray forming atomizer. ESR removes ceramic inclusions in the following manner—liquid slag (calcium fluoride, calcium oxide and aluminum oxide) is heated by passing an ac-electric current from a consumable superalloy electrode (prepared by vacuum induction melting), through the slag to a water-cooled crucible. The liquid slag is maintained at a temperature high enough to melt the end of the electrode. As the electrode melts, a refining action takes place—ceramic inclusions are exposed to the slag and are dissolved. Droplets of molten metal fall through the slag and are collected in a liquid metal pool contained in the crucible below. The CIG system then delivers a stream of liquid metal from the crucible to the spray forming atomizer without reintroducing inclusions.

ESR System

The pilot scale system is equipped with an ESR unit that is a shortened version of a conventional ESR furnace [20]. The consumable electrode is 355 mm in diameter × 1800 mm in length and weighs approximately 1140 kg. The major components of the system consist of an electrode drive mechanism, a 430-mm-diameter water-cooled copper melt crucible that is only 600 mm tall (rather than 3000 to 5000 mm for a conventional system), a coaxial current return path (to minimize electromagnetic disturbance of the liquid metal stream), a 20-kA 1.6-kVA-power supply, and a melt-rate controller. The melt-rate controller uses a video imaging system to sense the metal-plus-slag height in the crucible and to maintain this level constant. A constant height is required to maintain a constant liquid-metal flow rate.

CIG System

The transfer of liquid metal from the ESR crucible to the spray forming atomizer is accomplished by an induction-heated, segmented, water-cooled copper guide tube (CIG). This type of guide tube was first developed for the transfer of liquid Ti-base alloys from a plasma-arc melting furnace (PAM) to the atomizer of a powder production system [21, 22]. In the current case, the induction heating is provided by a two-coil system, as shown schematically in Figure 4. A "starter plate" of previously refined material is placed at the bottom of the crucible, above the CIG, to isolate slag from the CIG. A hole is melted through the starter plate during the ESR/CIG start procedure.

The upper coil is driven by a 240-kW medium-frequency power supply (approximately 10 kHz). Its main function is to prevent re-solidification of liquid metal that fills the guide tube as the ESR starter plate melts. The lower coil is driven by a 110-kW high-frequency power supply (100 to 200 kHz) and is used to initiate flow of liquid metal by melting out a plug in the lower portion of the CIG once the correct metal head height is established in the ESR crucible, and to maintain superheat in the

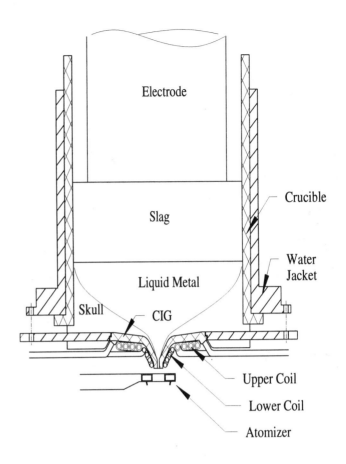

Figure 4. Schematic diagram showing the general arrangement of the major components of the CIG system used to transfer liquid metal from the ESR crucible to the atomizer of a spray deposition system.

stream during the balance of the run. The liquid metal flow can also be turned off and on by varying the power level in the lower coil. The water-cooled guide tube is segmented into eight individual segments to allow the induction heating to reach the liquid metal inside. The segments are insulated from each other over most of their length, but are joined electrically at the outside diameter. Each segment is independently water-cooled.

Spray Deposition System

The spray deposition system includes a scanning atomizer, a preform handler, and a water-cooled deposition chamber. The scanning atomizer is a gas atomizer (nitrogen or argon) capable of converting the liquid metal stream into a spray of liquid metal droplets and directing this spray over a range of angles. The preform handler is a four-axis robot capable of positioning and rotating the preform beneath the spray. The preform may be held at any orientation relative to the atomized spray, allowing a high degree of flexibility in the angle of attack between the spray and the preform.

Preliminary Results

Liquid Metal Stream

The pilot scale system is operational. Figure 5 shows a liquid metal stream pouring from a 5-mm-diameter orifice at a rate of approximately 15 kg/min.

Figure 5. Liquid metal stream pouring from the induction-heated, water-cooled copper guide tube (CIG). The guide tube orifice is 5 mm in diameter. The liquid metal pour rate is approximately 15 kg/min.

A wide range of liquid metal flow rates can be achieved with this system. From energy balance considerations the mass flow rate of liquid metal exiting the CIG system, $\dot{m}$, can be expressed as

$$\dot{m} = A \, \rho_m \, C_D \sqrt{\frac{2 \, \delta p}{\rho_m} + 2 \, g \left(h_m + \frac{\rho_s}{\rho_m} h_s \right)} \quad (1)$$

where A is the cross-sectional area of the outlet orifice of the CIG system, ρ_m and ρ_s are the densities of the liquid metal and the slag, δ_p is the difference in pressure between a point directly above the slag and a point directly below the outlet orifice, h_m and h_s are the corresponding heights of metal and slag, g is the acceleration caused by gravity, and C_D is an empirically determined drag coefficient. Measurements conducted with a ceramic-guide-tube system, under conditions where δ_p was controlled to maintain a constant equivalent metal head height [23], give a value of 0.84 for C_D. Liquid metal flow rates calculated using Equation 1, and the original data for the ceramic-guide-tube system are given in Figure 6.

Flow rates observed for the ESR-CIG system seem to follow this relationship quite closely.

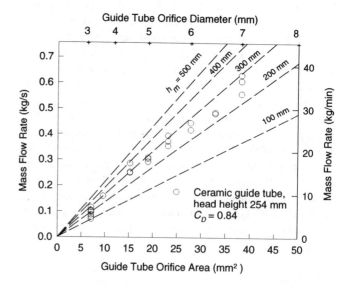

Figure 6. Liquid metal flow rate as a function of guide-tube orifice area and liquid metal head height calculated using Equation 1. Data from the ceramic-guide-tube system used to determine the drag coefficient are also plotted [23].

Solidified Melt

A cross section of the solidified melt remaining in the crucible and the CIG after a run is shown in Figure 7.

Clearly visible in this photograph is the starter plate, with its corresponding heat-affected zone; the refined superalloy material, which fills the crucible and flows through the starter plate to fill the CIG; and the large-grain dendritic growth of the solidified melt. Some remnants of the unmelted slag are visible at the periphery of the solidified melt directly above the starter plate.

Cleanliness Evaluation

The electron-beam button test method [24] has been used to evaluate the capability of the ESR-CIG process to remove oxide nonmetallic inclusions from the superalloy Alloy 718. Electrodes were prepared as feedstock for this evaluation by vacuum induction melting (VIM). The specific oxide area (SOA) determined for these electrodes varied from 0.158 to 0.497 mm²/kg. This high level of SOA was reduced to 0.007 mm²/kg by passing the metal through the ESR-CIG process and solidifying it in an ingot mold. This low level is excellent and compares very favorably to the nominal level of 0 to 0.0488 mm²/kg normally observed for Alloy 718 prepared commercially by the "triple melt" process (VIM + ESR + VAR). Although SOA gives a relatively good measure of the total oxide volume, it does not give a good measure of the largest inclusion size to be found (the value most important for determining fatigue life).

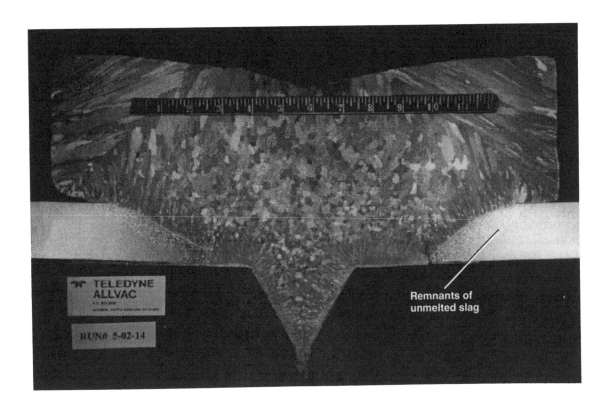

Figure 7. Solidified melt remaining in the crucible and the CIG after the run.

Spray Forming

Initial spray forming trials have produced preforms in the 250 to 325 mm diameter range. One such preform is shown in Figure 8.

Figure 8. Spray formed preform. The dimensions are 300 mm in diameter × approximately 500 mm long. The weight is 270 kg. The alloy is Rene' 95.

The maximum length produced to date is 550 mm. The nominal yield (metal atomized divided by preform weight) is running in the range of 60 to 80%, with an average of 70%. The microstructures are equiaxed with an average grain size of 40 to 50 µm (with some larger grains observed at the center location of some preforms). High levels of porosity are observed in the outer few millimeters of most preforms.

Process studies are in progress to optimize cleanliness, microstructure and yield. Electron-beam button tests are being used to measure the average level of cleanliness. Large-bar low-cycle fatigue tests are being used to determine the size of the largest inclusion that is able to survive the process.

Summary and Conclusions

A pilot plant has been constructed to demonstrate the concept of using a combination of electroslag refining (ESR) and an induction-heated, segmented, water-cooled copper guide tube (CIG) to melt, refine, and deliver a stream of liquid metal to a spray forming process. The basic ESR system consists of a consumable electrode of the alloy to be melted, a liquid slag, and a water-cooled copper crucible. The liquid slag is heated by passing an ac-electric current from the electrode through the slag to the crucible. The liquid slag is maintained at a temperature high enough to melt the end of the electrode. As the electrode melts, a refining action takes place—oxide inclusions are exposed to the slag and are dissolved. Droplets of molten metal fall through the slag and are collected in a liquid metal pool contained in the crucible below. By the addition of the induction-heated, segmented, water-cooled copper guide tube (CIG) to the bottom of the crucible, a liquid metal stream can be extracted from the liquid metal pool. This stream makes an ideal liquid metal source for atomization and spray forming. The pilot plant has been operated at a melt rate of 15 to 25 kg/min with the Ni-base superalloys Alloy 718, Rene' 95, and Rene' 88. Process optimization and cleanliness evaluation studies are in progress.

Acknowledgments

The authors would like to acknowledge the many helpful discussions with the following contributors during the early phases of this effort: T.B. Cox, P.L. Dupree, P.J. Frischmann, J.A. Graves, H.R. Hart, Jr., J. Koca, B.A. Knudsen, T.F. Sawyer, and R.J. Zabala of GE-CRD; E.S. Huron, R.G. Menzies, D.P. Mourer, and E. Raymond of GE Aircraft Engines; C.B. Adasczik, R.M. Davis, W.M. Leftwich, and R.L. Kennedy of Teledyne Allvac; M. Hohmann, F.W. Hugo, W. Renner, G. Schumann, and W.R. Zenker of ALD Vacuum Technologies; and A.G. Leatham, R.G. Brooks, J.S. Coombs, and J. Forrest of Osprey Metals, Ltd.

References

1. A.R.E. Singer and R.W. Evans: "Incremental Solidification and Forming," *Metals Technology,* 1983, vol. 10, pp. 61-68.

2. R.W. Evans, A.G. Leatham, R.G. Brooks: "The Osprey Preform Process," *Powder Metallurgy,* 1985, vol. 28, pp. 13-20.

3. A.R.E. Singer, D.J. Hodkin, P.W. Sutcliffe, P.G. Mardon: "Centrifugal Spray Forming of Large Diameter Tubes," *Metals Technology,* 1983, vol. 10, pp. 105-110.

4. M.G. Benz, R.J. Zabala, P.L. Dupree, B.A. Knudsen, W.T. Carter Jr., and T.F. Sawyer: "Spray-Formed Alloy 718," *Superalloys 718, 625, 706 and Various Derivatives, Proceedings of Third International Conference on Superalloys 718, 625, 706 and Various Derivatives, Pittsburgh, June 26 -29, 1994,* TMS, Warrendale, PA, 1994, pp. 99-105.

5. D.R. Chang, D.D. Kreuger, and R.A. Sprague: "Superalloy Powder Processing, Properties and Turbine Applications," *Superalloys 1984, Proceedings of Fifth International Symposium on Superalloys,* TMS-AIME, Warrendale, PA, 1984, pp. 245-273.

6. H.C. Fiedler, T.F. Sawyer, R.W. Kopp: "Spray Formed Rene′ 95," *GE Corporate Research and Development Center Report 87CRD034,* 1987.

7. M.F. Henry: Personal Communication, 1995.

8. G.W. Meetham: "Superalloy Processing and Its Contribution To The Development of The Gas Turbine Engine," *High Temperature Alloys for Gas Turbines,* D. Coutsouradis, et al., editors, Applied Science Publishers Ltd., London, 1978, pp. 837-858.

9. M.G. Benz and T.F. Sawyer: "Direct Processing of Electroslag Refined Metal," US Patent 5,160,532, November 3, 1992.

10. M.G. Benz and T.F. Sawyer: "Direct Processing of Electroslag Refined Metal," US Patent 5,325,906, July 5, 1994.

11. M.G. Benz and T.F. Sawyer: "Electroslag Refining of Titanium to Achieve Low Nitrogen," US Patent 5,332,197, July 26, 1994.

12. T.F. Sawyer, M.G. Benz, W.T. Carter Jr., and R.J. Zabala: "Method and Apparatus for Flow Control in Electroslag Refining Process," US Patent 5,348,566, September 20, 1994.

13. T.F. Sawyer, W.T. Carter Jr., and M.G. Benz: "Molten Metal Spray Forming Atomizer," US Patent 5,366,206, November 22, 1994.

14. M.G. Benz, T.F. Sawyer, W.T. Carter Jr., and P.L. Dupree: "Molten Metal Spray Forming Apparatus," US Patent 5,472,177, December 5, 1995.

15. W.T. Carter Jr., M.G. Benz, T.F. Sawyer, and M.E. Braaten: "Gas Atomizer with Reduced Backflow," US Patent 5,472,177, January 2, 1996.

16. W.T. Carter Jr., M.G. Benz, F.G. Müller, R.M. Forbes Jones, and A.G. Leatham: "Electroslag Remelting as a Liquid Metal Source for Spray Forming," *International Conference on Powder Metallurgy and Particulate Materials, May 14-15, 1995, Seattle, WA,* Metal Powder Industries Federation, Princeton, NJ 08540, 1995.

17. F.G. Müller, F.W. Hugo, M.G. Benz, T.F. Sawyer, and W.T. Carter Jr. "A New Process for the Production of Ceramic-Free Metals," *International Symposium on Electromagnetic Processing of Materials - 1994, Nagoya, Japan,* ISIJ, 1994, pp. 435-440.

18. W.T. Carter Jr., M.G. Benz, F.G. Müller, and R.M. Forbes Jones: "Electroslag Refining as a Liquid Metal Source for Spray Forming," *European Conference on Advanced PM Materials,* Birmingham, UK, 1995.

19. R.J. Zabala, M.G. Benz, W.T. Carter Jr., B.A. Knudsen, and P.L. Dupree: "A Ceramic-Free Spray Forming Facility," *European Conference on Advanced PM Materials,* Birmingham, UK, 1995.

20. G. Hoyle: *Electroslag Process, Principles and Practice,* Applied Science Publishers Ltd, London, 1983, pp. 1-215.

21. M. Hohmann, M. Ertl, A. Choudhury, and N. Ludwig: "Experience with Ceramic-Free Powder Production Methods," *1991- P/M in Aerospace and Defense Technologies,* Metal Powder Industries Federation, Princeton, NJ, 1991, pp. 261-272.

22. O.W. Stenzel, G. Sick, and M. Hohmann: "Process and Device for Forming a Poured Stream," German Patent 4,011,392 A1, October 10, 1991.

23. H.C. Fiedler, T.F. Sawyer, and M.G. Benz: Personal Communication, 1989.

24. C.E. Shamblen, D.R. Chang, and J.A. Corrado: "Superalloy Melting and Cleanlines Evaluation," *Superalloys 1984 - Proceedings of the Fifth International Symposium on Superalloys,* TMS-AIME, Warrendale, PA, 1984, pp. 509-519.

MICROSTRUCTURE AND PROPERTIES OF SPRAY ATOMIZED AND DEPOSITED SUPERALLOYS

Tian Shifan, Zaho Xianguo, Ren Liping, Liang Zhikai, Li Zhou and Mi Guofa
Institute of Aeronautical Materials Beijing, P.O. Box 81, 100095, P.R. China

John F. Radavich
Purdue University, West Lafayette, IN 47907 (USA)

Abstract

Spray atomization and deposition or Spray Forming (SF) has been a newly emerging science and technology in the field of materials development and production. SF makes it possible to combine the atomization of molten metals (Rapid Solidification) and the deposition of atomized droplets (Dynamic Droplets Compaction) into one metallurgical operation to produce directly from liquid metal near-net shaped preform materials that have uniform chemical composition, refined grain and microstructure and improved properties.

At BIAM, fundamental work on the research and development of SF superalloys has been going on since 1990. A spray atomization and deposition facility has been designed and built with a strong vacuum evacuation system and minor leakage chamber. Inside the chamber a special designed water cooled multifunctional substrate/collector has been installed. The problem of contamination by the ceramic inclusions has been satisfactorily solved. A series of Ni-base superalloys such as K403, K405, K417, GH95, and the intermetallic alloys IC-218, IC-6 etc. have been spray atomized and deposited. Various experiments and tests have been conducted on the samples taken from the disk-shaped performs (ψ220~250×50m/m) and column-shaped billets (ψ170×150m/m) including macrostructure inspection on vertically sectioned slices, density measurements, microstructure examination by optical SEM or TEM, X-ray diffraction, HIP, hot deformation behavior and mechanical testing.

On the basis of results from the investigation, the following conclusions can be made: Spray formed superalloy materials have low oxygen content, homogeneous chemistry, refined microstructure and improved forgebility. HIP eliminates microporosity to make the preforms fully dense, which reduces the scatter of mechanical properties. The properties of SF superalloys may be adjusted through hot deformation and subsequent heat treatment to meet the specific requirements of a particular applications.

Introduction

The Process Description

Spray Forming (SF) or spray atomization and deposition has been an emerging science and technology in the field of materials development and production in recent years [1]. It is a most effective and economical way of making large preforms weighing from a few kgs to several thousand kgs by means of rapid solidification. The SF process may be described principally as follows: (see also schematic representation in Figures 1 and 2). The molten stream of a superheated melt of metal (or alloy) is being broken and disintegrated at the exit end of the metal delivery tube by a stream of high speed gas and is atomized into a spray of highly dispersed fine droplets with a specific particle size distribution. Being accelerated by the momentum of the atomizing gas stream, these atomized droplets simultaneously undergo extensive heat exchange during flight. While droplets under certain critical size solidify into fine solid particles, large ones remain in liquid state with different undercooling. In between these two extremes, medium sized droplets become semi-solid with different contents of liquid phase (mushy state). These particles collide at high velocity (25~75 m/s) upon the substrate - or collector surface where they adhere and deform to form a thin semi-liquid layer on its top where it continuously solidifies and grows into a large piece of integral dense solid preform. Thus SF makes it possible to combine the atomization of molten metals (Rapid Solidification) and the deposition of atomized droplets (Dynamic Droplets Compaction) into a one-step metallurgical operation to produce directly near-net shaped preforms/materials that have uniform chemical composition, refined grain and microstructure and improved properties. Because of its potential technical and economical benefits, the SF process has received a great deal of attention and has reached the stage of industrial application [2~4].

R & D of SF Superalloys at BIAM

The technology of spray formed superalloys has entered the stage of practical applications. At BIAM, fundamental work on the research and development of spray formed superalloys has been going on since 1990

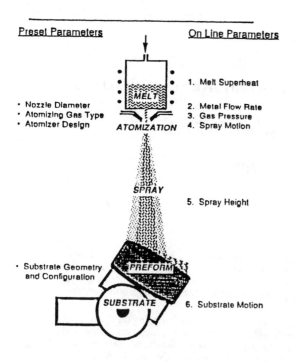

Figure 1. Schematic of the Osprey™ deposition process showing the formation of a disk/billet, and the process parameters which can be manipulated to optimize preform shape, structure and yield.

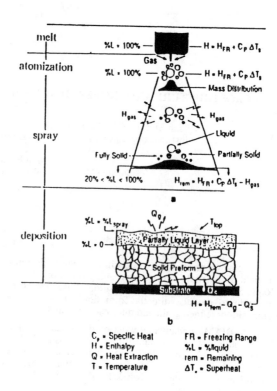

Figure 2. Schematic representation of spray casting showing the physical and thermal states of (a) the spray and (b) the deposit.

Table 1. Chemical Composition of Superalloys Processed by Spray Forming.

Alloy	C	Cr	Co	Mo	W	Al	Ti	B	Zr	Others	Density g/cm³	T_m °C	γ vol %	$T_{γ'}$ °C
Nimoniel 15	0.15	15.0	14.5	4.0	-	5.0	4.0	0.016	0.04		7.80	1315	50	1150
Mar M002	0.15	9.0	10.0	<0.5	10.0	5.5	1.5	0.015	0.05	Ta2.5,Hf1.5	8.5	1280	60	1250
Rene'-80	0.17	14.0	9.5	4.0	4.0	3.0	5.0	0.015	0.03		8.16	1320	-	-
IN-100	0.24	10.0	15.0	3.0	0.87	5.6	5.0	0.015	0.06		7.8	1340	62	
Merl-76	0.03	12.5	18.5	3.2	0.8	5.0	4.4	0.02	0.05	NB-1.4	7.83	-	64	1185
IN-718	0.04	18.0	-	2.90	-	0.57	0.95	0.006	0.0	Fe-20	8.24	1320		1020
Rene'-95	0.06	13.0	8.0	3.50	3.5	3.5	2.5	0.01	0.05	NB-3.5	8.3	1348	49	1160
IN 625	0.03	21.0	<1.0	9.0		0.3	0.2			Nb4,Fe<5	8.44	1350	-	1020
Waspaloy	0.07	19.5	13.5	4.2	-	1.4	3.0	0.01	0.06		8.22	1360	20	1020
Rene'-41	0.09	19.0	11.0	9.8		1.6	3.25	0.01			8.27	1371	-	1050
AF-115	0.05	10.7	15.0	2.8	5.9	3.8	3.9	0.02	0.05	Nb1.7,Ta1.5	8.33	-	55	
AF2-1DA	0.04	12.0	10.0	2.75	6.5	4.6	2.8	0.015	0.10		8.33	-	52	
Astroloy	0.02	15.0	17.0	5.0	-	4.0	3.5	0.03	0.045		8.0	-	45	1120
U-720	0.032	17.8	14.5	2.0	1.2	2.45	5.0	0.032	0.03		8.1	1245	-	1140
Rene'88DT	0.03	16	13	4	4	2.1	3.7	0.03	0.03	Nb-0.7	8.36		40	1105

[5]. A number of heavily alloyed high strength Ni base alloys have been spray atomized and evaluated. The chemical compositions are listed in Table 1. A spray atomization and deposition facility has been designed and built with a strong vacuum evacuation system and minor leakage chamber. Inside the chamber a special designed water cooled multifunctional substrate/collector [6] has been installed that can be tilted and eccentrically fixed. During the process deposition, it rotates around the axis of the preform growth and at the same time may move down along the axis to maintain the deposition distance to be optimum as the deposit builds up. Because the key process parameters can be adjusted, the desired shape of the preform can be made. The problem of contamination by the ceramic / inclusions has been satisfactorily solved using the clean melting technology from research and development carried out at BIAM. This creates the potential for the application of the spray formed superalloys for the critical rotating turbine components. A number of Ni-base superalloys such as K403, K405, K417, Rene 95 and the intermetallic alloys IC-218, IC-6, etc. have been spray formed, and the test results from this work were encouraging. Test results of two Ni-base superalloys with different alloying level and γ' vol. % have been selected for presentation in this paper.

Experimental Procedure

The materials used for the spray forming processing were vacuum melted master alloy as cast ingots and some revert. Disc shaped preforms were made in the experimental facility. The values of the primary experimental variables used during the experiment are listed in Table 2.

Table 2. Experimental Variables Used for SF

Variable Alloy Designation	A	B
Charge weight (kg)	18.7	11.3 ~ 14.5
Superheat (°C)	200	150 ~ 20
Atomizing gas	Pure N_2	Pure N_2
Atomization pressure (MPa)	2Ⅰ1.0/1.5	
Atomization time(s)	120	46 ~ 60
Deposition distance (mm)	340	390 ~ 400
Collector position	eccentric, tilted	eccentric, tilted
Collector movement	spin, synchronic descending	spin, synchronic descending
Number of preform for test	1	4

A vertically sectioned slice 10 ~ 15 mm thick was cut along the center line of the preform and prepared for macroetching to reveal any possible metallurgical defects. Then the macroetched slice was cut by electrosparking into small pieces for density measurements and metallographic examinations to detect the presence of any ceramic inclusions. Hot isostatic pressing was conducted on preform alloy A and one of the preforms of alloy B. Samples cut from alloy B were used for hot deformation tests at different strain rates and temperatures. The rest of the two preforms of alloy B were hot forged to a pancake on a 1250 ton forging press.

Chemical analysis of alloy composition and gas content were first run and a detailed microstructural study was carried out by optical metallography and SEM on alloy B. Mechanical property tests were made on samples having different thermal treatments.

Results and Discussion

Chemical Composition

The results of chemical analysis of alloys A and B are listed in Table 3.

The results show that SF has little effect on alloying elements. The gas content indicates almost no hydrogen in the preform. Since the time the alloy is in a liquid state is very short ($\sim 10^{-3}$ sec) and is exposed to an inert environment of pure N_2 (or Ar), the oxygen pick-up is very limited. The oxygen in the preform is very low and at the same level as in the master alloy. As the nitrogen was used as atomizing gas in these experiments, its content in both alloy A and B is relatively high especially for the alloy containing higher Cr%. The solubility of N_2 in Ni-base alloys is very low and generally exists in the form of a nitride. In the SF superalloys containing Ti and Nb, the nitride usually is in the form of (Ti,Nb) (CN). Owing to the high cooling rate during the atomization, many of the primary carbonitrides that precipitated out of the liquid were finely dispersed and had no detrimental effects on the mechanical properties.

Deformation Behavior

Hot deformation tests were conducted on the samples of alloy B at temperatures of 1120, 1100, 1080 and 1020°C and at a strain rate range $\varepsilon = 3.2 \times 10^{-2} \sim 2 \times 10^{-4}$. Results show that at all temperature and strain rate tests the alloy has good hot ductility. The hot ductility will be improved and flow stress reduced as the temperature increases and strain rate decreases. On the basis of the hot deformation tests, two

Table 3. Chemical Composition

| Alloy | C | Co | Cr | Mo | Al | Ti | other | γ' | H | O_2 | N_2 |
			wt.%					vol.%		ppm	
A	0.14	14.8	9.02	3.3	5.32	4.40	V0.73	60	1	14	200
B	0.06	9.95	14.7	5.11	2.87	2.66	Nb2.65	35	-	15	320

billets of the preform alloy B were forged with one upset >60%. No cracking in the forged billets was found.

Mechanical Properties

The mechanical properties for alloy A and B are summarized in Tables 4 and 5, respectively. Since alloy A may be used in cast condition, test data of the cast alloy A were listed for comparison. It is obvious that SF increases the yield and rupture strength remarkably. While HIP decreases the scatter of the properties and considerably improves the ductility, the yield strength is somewhat reduced. The SEM observation of the fracture surface of the test bars indicates that the porosity has been healed by HIPing. If the tensile properties of the as-HIPed powder metallurgy alloy of the same composition as alloy A: $\sigma_b = 1125$ MPa, $\delta_5 = 8.0\%$ which were reported in [7] are compared with those listed in Table 4, the properties of alloy A in both spray deposited and HIPed condition are superior.

Table 4. Tensile Properties of Alloy A

No. of test	condition	T test °C	σ_b MPa	$\sigma_{0.2}$ MPa	δ_5 %	ψ %
24	SF	20°C	1306	904	18.0	15.8
28	SF	20°C	1393	927	21.2	20.4
12	HIP+HT	20°C	1373	799	33.2	29.2
21	HIP+HT	20°C	1368	810	32.4	29.9
16	HIP+HT	700°C	1053	/	23.0	23.1
23	HIP+HT	700°C	1041	/	23.2	32.0
	as cast	20°C	990	765	11.5	19.0
	as cast	700°C	1000	774	13.0	2.0

*(Handbook of the Aeronautical Materials of China), vol. 2, p. 651.

Table 5. Mechanical Properties of Alloy B

No. Pref.	Test Condition	Tensile σ_b MPa	Tensile $\sigma_{0.2}$ MPa	δ %	ψ %	Stress Rupture 650°C/834MPa τ(h)	Impact A_k J
E01	SF	1356	898	26.3	36.6	164:50	47.0
E03	F+HT1	1389	941	26.2	32.9	-	55.8
		1396	953	25.3	32.8	-	52.6
E04	F+HT2	1470	1067	22.8	22.4	144:30	-
		1466	1060	21.6	23.4	205:10	62.7
Specification		1210	755	13	14	50	24

1) HT1-1140°C solution treated.
2) HT2-1080°C solution treated.

Data listed in Table 5 show the properties of alloy B in the as deposited condition were much higher than that required by the specification. Low temperature solution treatment improves the rupture and yield strength as well as the impact properties while maintaining the ductility and 650°C stress rupture property at a relatively high level. A high temperature solution treatment didn't improve the properties as much as the low solution treatment did.

Macrostructure and Density

Examination of macro etched sample shows the macrostructure to be dense with a little porosity, and no coarse dendritic grains were found as normally seen in cast alloys. Porosity is mainly concentrated near the bottom surface of the preform where the atomized droplets first impact against cooled substrate. Some porosity was found at the periphery and near the top surface of the preform at the very beginmning of the deposition process. The average density of the entire preform is about 98.8% of theoretical, and after HIP, the alloy density may reach 99.9% ~ 100%.

Microstructure

Optical studies revealed a fine uniform equiaxed grain structure in all alloys in the as-deposited condition. Alloys with high carbon contents such as alloy A will occasionally show PPB within 3-5 mm range of the bottom surface. Eutectic γ' islands normally found in high Al+Ti alloys are rarely seen in spray formed materials.

Higher resolution detailed structural studies of alloy B were carried out with the SEM, EDS, and X-ray diffraction techniques on the as-deposited, HIPed, HIPed+ forged, and heat treated samples. The as-deposited material contained some porosity which was closed by HIPing. The grains are uniform and range from 10-25 microns in size. Primary carbides are well distributed and have a range of sizes, some as small as carbides seen in powder alloys. In addition to these small particles, there are some larger TiN particles in the matrix.

The grain boundaries show discrete carbide or other phases formed during the solidification while small round γ' is present in the grains due to the cooling from above the γ' solvus temperature. Figure 3 shows the various grain structures and the γ' + primary carbides in the matrix.

HIPing at 1150°C(2100°F)/150 MPa/3 hours and forging at 1110°C(2030°F) does not greatly change the grain size even though the temperatures may be above the γ' solvus. Primary carbides and other stable phases in the as-deposited material tend to pin down the grain boundaries unless the material is subjected to temperatures higher than those used in this study.

Samples which were HIPed and heat treated at 860°C(1580°F)/8 hours + 788°C(1450°F)/16 hours show small discrette particles at the grain boundaries and the γ' is slightly larger than the as-cooled γ', Figure 4.

To appreciate the size difference of spray cast microstructures, an ingot sample of alloy B was evaluated for segregation and the nature of the microstructures formed during slow cooling. Figure 5 shows large as-cast grains, large primary carbides and dendritic segregation. Alloy B has appreciable Ti, and a Ti rich eta phase as well as accicular needles are found together in the dendritic segregation areas. These phases are not found in spray cast materials of the same composition. The primary carbides of alloy B ingot material form the normal Chinese script morphology while the γ' shape and size varies with the degree of Al+Ti segregation and the ingot cooling rate.

Phase Identification

Insitu EDS analyses of the small particles showed strong Nb+Ti and appreciable Mo contents, but the surrounding matrix excitation prevented phase identification by composition alone. Large TiN particles were easily identified by their high Ti content and their reddish clor when viewed optically.

Extraction of all the small particles in the spray deposited condition was carried out for X-ray diffraction studies. The floating residue had a reddish hue which confirmed the presence of TiN particles in the residue.

The X-ray diffraction study showed that only two phases were extracted, an MC phase with a lattice parameter of 4.388 Å and TiN with a parameter of 4.246 Å. Some of the TiN in the residue is the large TiN particles seen optically.

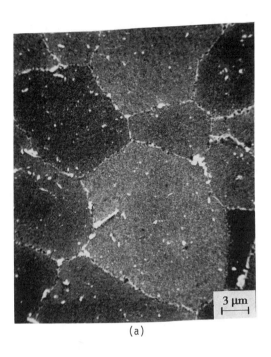

(a)

(b)

(c)

Figure 3. As Deposited
 (a) Grain Structure.
 (b) γ' Phase.
 (c) Extracted Carbides and Nitrides.

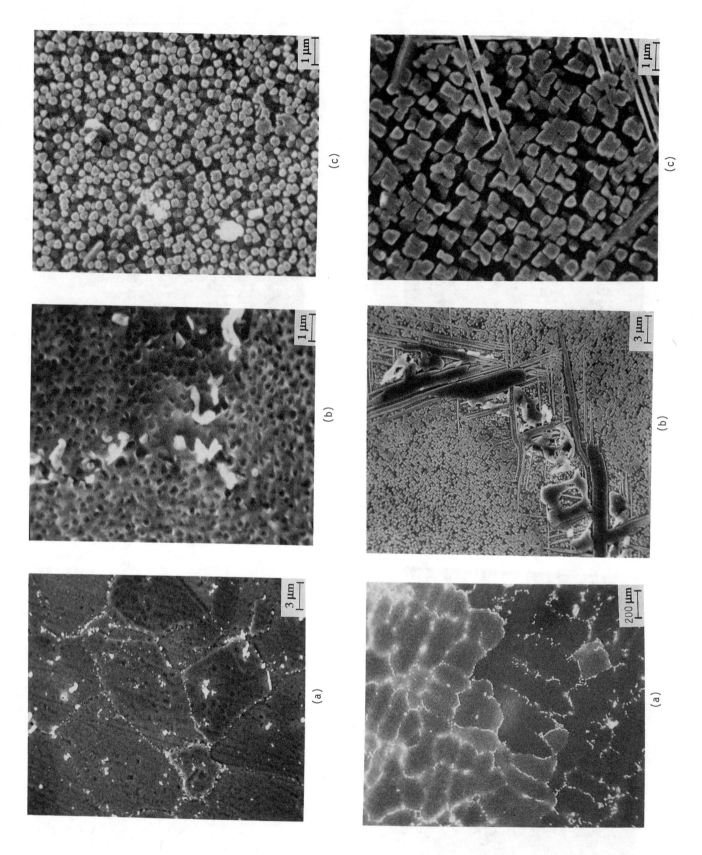

Figure 4. HIP+ Heat Treated.
(a) Grain Structure.
(b) Grain Boundaries.
(c) γ' Phase.

Figure 5. Ingot Structures.
(a) Grain Structure.
(b) Eta Phase.
(c) γ' Phase.

Long Time Stability

In order to verify the chemical homogeneity of the SF superalloys, alloy A was selected for a sigma phase stability test. This is because the alloy A composition with high Al+Ti is prone to the precipitation of TCP σ-phase at the interdendritic area where most alloying elements are normally segregated in the as-cast condition. A sample cut from the preform alloy A was aged at the peak temperature of σ-precipitation -850°C for 200 and 400 hrs. No σ precipitation was detected.

Discussion

The observed improvement in mechanical properties in this study is in line with similar reported mechanical property studies of spray cast materials by Chang and Fiedler (8) and Prichard and Dalal (9). In both of these studies the mechanical property behavior was emphasized and so less structural characterization was carried out. Spray casting involves both solid and liquid materials solidifying in a temperature range where, depending on composition, MC, M_6C, and M_3B_2 can form discrete or continuous structures.

Because of the varying rates of solidification during spray casting, various size particles can form, some of which approach the size of fine particles in powder materials. Some of the particles show distinct shapes which others have a splat-like nature. Because of the high nitrogen content shown in Table 3, small nitrogen rich particles should be present similar in size to the carbide particles. However, the very large TiN particles present in the spray cast material appear to be too large to form during the spray casting, and therefore, it appears that the TiN particles (which were not melted) are carried over from the master melt alloy. A study of the microstructural nature of the master alloy should be further studied to avoid large inclusions which may serve as crack initiation sites.

Conclusions

1. Spray formed superalloys have low oxygen content, homogeneous chemistry, refined microstructure and improved forgebility.
2. HIPing eliminates the microporosity and makes the preform fully dense, which reduces the scatter of mechanical properties.
3. The properties of SF superalloy may be adjusted through hot deformation and subsequent heat treatment to meet the specific requirements of a particular application.

Acknowledgements

The Aeronautical Science Foundation and the National Natural Science Foundation are greatly acknowledged for their financial support.

References

1. Spray Forming: Science, Technology Application, 1992, P/M World Congress, ed. Alan Lawley.
2. Internal Report of BIAM, 1994.
3. Proceedings of the 1st. ICSF, 1991.
4. Proceedings of the 2nd. ICSF, 1993.
5. Internal Report of BIAM, 1988.
6. Internal Report of BIAM, 1993.
7. N. J. Grant, Metall. Trans. Vol. 23A, No. 4 p1083 ~ 10.93.
8. K. M. Chang and H. C. Fiedler, "Spray Forming High Strength Superalloys," pp. 475-484, Superalloys 1988, TMS.
9. P. D. Prichard and R. P. Dalal, "Spray Cast-X Superalloy for Aerospace Applications," pp. 205-216, Superalloy 1992, TMS.

PRECAD®, A COMPUTER ASSISTED DESIGN AND MODELLING TOOL FOR SUPERALLOY POWDER PRECISION MOULDING

D. Lasalmonie[*], L. Le Ber[*], C. Dellis[*], R. Baccino[*], F. Moret[*], J.C. Garcia[**] and J.P. Buhle[**]

[*] CEA/CEREM - 17, rue des Martyrs, 38054 Grenoble Cedex 9, France
[**] TURBOMECA, 40220 Tarnos, France

Abstract

This paper presents a computer assisted design and modelling tool, PRECAD®, developed by CEA/CERUM and its appplication to superalloy powder precision moulding. This tool includes a computer assisted design (CAD) module that enables the design of part, container and cores used to produce net-shape product from superalloy powder. The meshes of the different components are automatically generated in this module, even for complex 3D-geometry. A coupled FEM module modelled the HIP process using meshes and limit conditions generated by the CAD module. Powder data base is now available for titanium base alloys, stainless steel and nickel base superalloy SY625 produced by TECPHY. Data base for steel container and cores is also available for a large range of temperature. Deformed mesh is visualized on the CAD module and compared with initial and desired final shapes. Container and cores geometry can be optimized in order to produced the net-shape geometries where necessary in the part.

Introduction

Hot Isostatic Pressing (HIP) is an established process for compacting powder materials far below their melting point. Processing of superalloy powders using HIP has been used for many years in different industrial applications, especially to produce high quality parts with complex geometry. In the HIP process, a steel container is filled with superalloy powder, evacuated and then submitted to simultaneous application of high pressure (about 100 MPa) and a temperature of about 1100 °C for several hours. During the process, volume is reduced by about 30% and porosity is completely eliminated. It is important to obtain net-shape HIP parts in order to reduce machining and material costs, particularly for superalloys, or even to produce non-machineable geometries. However, the final geometry of the product can not be simply deduced from the initial one, due to non isotropic deformation, because of container stiffness and temperature gradients during consolidation process. To produce net-shape parts, it is essential to be able to forecast the behaviour of the superalloy powder and steel container during HIP process in order to predict the final geometry of the product. Numerical simulation can provide much information about the final as well as intermediate stages of the HIP process.

PRECAD® a Computer Assisted Design and Modelling Tool for Powder Precision Moulding

A special tool has been developed by CEA/CEREM for the design and modelling of containers used to produce net-shape parts using Hot Isostatic Pressing (HIP). This tool is constituted of three modules : PRECAD®/D for the design of the parts and meshing, PRECAD®/M for the modelling of consolidation of the powder, and PRECAD®/B for the data base materials including powders and container. A general description of this tool is given figure 1.

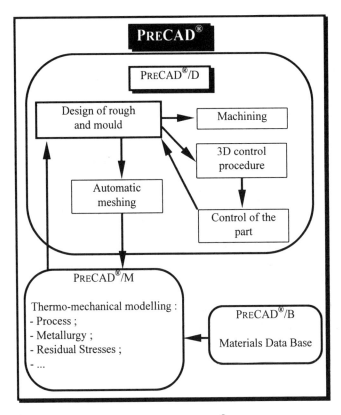

Figure 1 : General description of PRECAD® tool and its modules.

Computer assisted design module PreCAD®/D

PreCAD®/D is the Computer Assisted Design (CAD) module. It is used to design both the part to be realized and the container and cores necessary to produce this part "net-shape" from powder. The geometry of the design part can be provided directly by the end-user, in the form of a data file, using common CAD interfaces, such as IGES. An example of CAD image is shown figure 2.

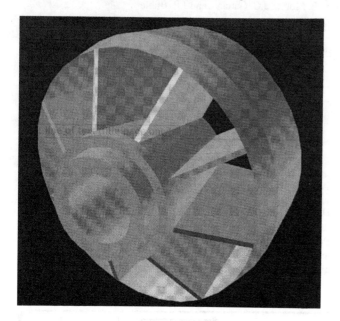

Figure 2 : CAD of the part to be realized.

To produce this part by net-shape HIPping, a container and some solid cores must be used. These different components constitute the mould that is shown in figure 3. Machining programmes can be generated in the CAD module, as well as 3D control programme.

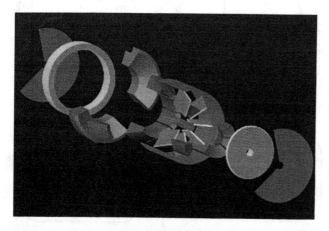

Figure 3 : Mould splinter.

Each part of the mould, powder and container are automatically meshed in this CAD module. A 3D mesh of the initial powder is shown in figure 4.

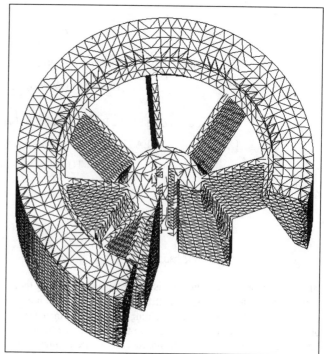

Figure 4 : 3D mesh of the initial powder.

Meshes, materials characteristics and limit conditions are then sent to the modelling module PreCAD®/M.

Modelling module PreCAD®/M

To simulate consolidation of powder, a macromechanical approach is used. Powder is considered as a continuous medium with relative density as an internal variable. The relative density is defined as the ratio of the apparent density to the density of the fully dense material. The behaviour of the powder during consolidation is modelled by constitutive equations based on continuous media mechanics for viscoplastic metals. The flow formulation for compressible viscoplastic materials proposed by Abouaf [1] is used. This model extends, for hot deformation, the approach of Green [2] for cold deformation.

Description of the model

Constitutive equations for porous materials are developed in the framework of continuous mechanic. It is an extension of the classical J_2 Mises theory allowing volume change during viscoplastic flow. The equivalent stress σ_{eq}, which includes the effect of pressure, is defined as :

$$\sigma_{eq} = \sqrt{fS_1^2 + \tfrac{3}{2} c \bar{S}_2^2} \qquad (1)$$

with

$S_1 = Tr(\tilde{\sigma})$ first invariant (2)

$\bar{S}_2^2 = Tr(\tilde{\bar{\sigma}}.\tilde{\bar{\sigma}})$ second invariant (3)

$\tilde{\bar{\sigma}} = \tilde{\sigma} - \tfrac{1}{3} S_1 \tilde{\delta}$ deviator of the stress tensor (4) where $\tilde{\sigma}$ is the Cauchy stress tensor and $\tilde{\delta}$ is the unity second order tensor. c and f are two functions representing the stress localisation induced by the porosity, and which a priori depend only on the relative density ρ. For full density (ρ=1), c=1 and f=0 so that the equivalent stress defined for the porous metal tends to the classical Mises equivalent stress. For small elastic strains, the strain rate D is partitioned into an elastic part and a viscoplastic part D_v. This last term is derived from a viscoplastic potential Ω, the normality rule being assumed.

$$\tilde{D}_v = \frac{\partial \Omega}{\partial \tilde{\sigma}} = \frac{\partial \Omega}{\partial \sigma_{eq}} \cdot \frac{\partial \sigma_{eq}}{\partial \tilde{\sigma}} \quad (5)$$

The equivalent strain rate D_v^{eq} is defined using the relation :

$$\rho D_v^{eq} \cdot \sigma_{eq} = \tilde{\sigma} : \tilde{D}_v \quad (6)$$

Using the expression (1) of the equivalent stress, and after some calculation one get :

$$D_v^{eq} = \frac{1}{\rho} \cdot \frac{\partial \Omega}{\partial \sigma_{eq}} \quad (7)$$

$$\tilde{D}_v = \rho \cdot \frac{D_v^{eq}}{\sigma_{eq}} \cdot \left(f \cdot S_1 \cdot \tilde{\delta} + \frac{3}{2} \cdot c \cdot \tilde{\sigma} \right) \quad (8)$$

D_v^{eq} and σ_{eq} are related using the creep law of the material. A simple Norton law is chosen in which the parameters A and n depend only on the temperature T.

$$D_v^{eq} = A(T) \sigma_{eq}^{n(T)} \quad (9)$$

Density variations are related to D_v through mass conservation :

$$\frac{\dot{\rho}}{\rho} = -Tr(\tilde{D}_v) \quad (10)$$

Description of the calculation

Finite Element Method (FEM) is used in PRECAD®/M module Simulation of HIP is available for plane strain, plane stress, generalized plane strain, axisymmetric and tridimensional analysis.

The modelling procedure enables the user to carry out an incremental thermo-mechanical non linear analysis. For each time step, a thermic procedure first calculates a thermal map of the mesh. This procedure enables linear and non linear computations with conduction, convection and radiation. The thermo dependant material parameters are then calculated and sent to the mechanical non linear procedure. There, the non linearity can result either from the material (visco-plasticity) or from large displacement, even from both [3].

The material non linearity are integrated using a Runge-Kutta 2.1 method. A flow chart of the elastic-viscoplastic finite element program is given figure 5 [4].

Thermal history of the material is modelled, giving the possibility to predict the mechanical properties as well as the microstructure of the consolidated powder at the end of HIP process.

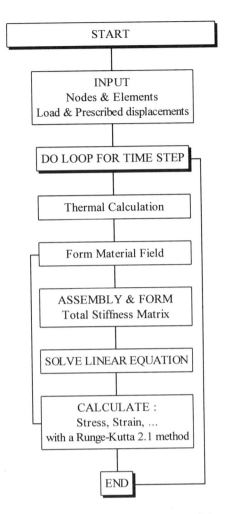

Figure 5 : Flow chart of the elastic-viscoplastic finite element program

Data base module PRECAD®/B

The use of simulation requires the constitution of material data files containing physical and thermo-mechanical properties of the materials (Young modulus, Poisson's ratio, thermal expansion coefficient, creep properties, thermal conductivity, ...) including coefficients of the porous constitutive equations for powder.

The material parameters for the model of Abouaf are introduced through four functions : A(T), n(T), c(ρ) and f(ρ).

Creep law parameters

The functions A(T) and n(T), describing the creep law being independent of the density of the material, are identified on the dense material using uniaxial compression tests. The temperature is introduced through an activation energy.

The experiments are performed at different temperatures. For each temperature, several axial deformation rates have been applied on the same specimen using the strain rate jump method. The resultant stresses have been recorded. The power creep law has been fitted on the experimental curves strain rate - stabilized stress with a good accuracy.

Function f

The identification of the function f(ρ) is obtained on interrupted HIP tests through the measure of the densification kinetic.

During the densification of cylindrical samples under the pressure P and the temperature T, the evolution of the relative density is related to the function f through the relation :

$$f(\rho) = \frac{1}{9}\left(\frac{\dot{\rho}/\rho}{A.P^n}\right)^{\frac{2}{n+1}} \quad (11)$$

Different HIP conditions are tested. For each HIP cycle, various step durations have been chosen and the relative density of the partially dense materials has been measured. A simple time derivation of the interpolated curves is then performed to obtain the master curves relating the densification rates with the relative densities. From these curves and knowing the creep parameters, it is possible to established the curve f(ρ).

Using this experimental procedure for 316LN stainless steel, it has been shown that the rheological function f is independent of both the temperature and the pressure even on a large range of T and P (temperature from 800°C to 1125°C and pressure from 15 to 100MPa) [5]. Thus the function f depends only on the relative density ρ. This experimental result is in agreement with the theoretical assumption assumed by Abouaf concerning the independence of this function with T and P.

Function c

The identification of the function c(ρ) is obtained from uniaxial compression tests on porous samples.

Uniaxial compression tests under imposed strain rate have been performed on the porous samples obtained with interrupted HIP cycles. We may define now a new function called s(ρ) as the ratio of the measured stress on the porous material characterized by the relative density ρ, over the stress of the dense material submitted to the same strain rate. Using the equations of the model, and after some calculation one obtain the following relations :

$$c(\rho) = s(\rho)^{\frac{-2n}{n+1}} - f(\rho) \quad (12)$$

with

$$s(\rho) = \frac{(\sigma_z)_\rho}{(\sigma_z)_{\rho=1}} \quad (13)$$

The experimental function s(ρ) is first determined, then the rheological function c(ρ) is calculated.

A general framework of the experimental determination is given figure 6.

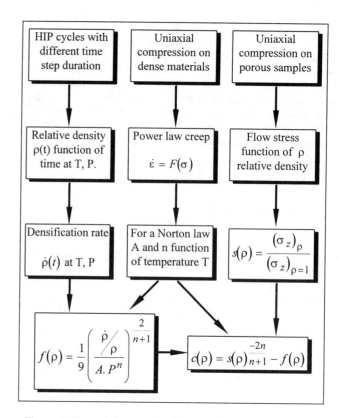

Figure 6 : General framework of the experimental determination.

Data Base for Superalloys

A large amount of experiments have been performed on superalloys powder, to identify the parameters for modelling. The most studied are Nickel base superalloys but some results are also available for Cobalt base superalloys. Such data are directly usable in PreCAD.

Nickel base superalloys

Several Nickel base superalloys have been already studied : René 95 [6], Astrolloy [7, 8] and MERL76 [9]. These materials are usually produced by gas atomisation, however Lafer [8] has also studied a rotating electrod atomized Astrolloy made by Tecphy. The chemical composition of these materials is given in table I.

Creep law of the material is usually of Norton type (equation 9), but a Sellars-Tegart relation is often appropriate to cover a large range of temperatures and strain rates.

$$D_v^{eq} = A(T)\left[sh(\alpha\sigma_{eq})\right]^n \quad (14)$$

Thermal dependency of the A parameter is usually take into account with an Arhenius law :

$$A(T) = A_0 \cdot \exp\left(\frac{-Q}{RT}\right) \quad (15)$$

The creep law parameters are A_0, n, α and Q that must be experimentally determined for each material.

Table I : Chemical composition of Nickel base superalloys studied in litterature.

René 95 [6], Gas atomized powder produced by TECPHY.

	Ni	Cr	Co	Mo	W	Al	Ti	Nb	Zr	B
(% weight)	Bal.	14.0	8.0	3.5	3.5	3.5	2.5	3.5	0.05	0.01

Astrolloy [7, 8], Gas (ATGP3 and AA) and Rotating electrod (RE) atomized powder produced by TECPHY.

	C	Co	Cr	Mo	Al	Ti	B	Zr	O_2	N_2
ATGP3 (% weight)	<0.035	17	15	5	4	3.5	0.03	0.05	-	-
AA (% weight)	0.02	16.5	15	5	4	3.5	0.025	-	<0.01	<0.005
RE (% weight)	0.013	16.88	14.85	5.05	3.96	3.37	0.031	-	<0.024	<0.025

Table II : Chemical analysis of the SY625 Ni-base superalloy produced by TECPHY (in % weight).

Ni	Cr	Mo	Nb	C	O	N	Si	Mn	S	P	Ti	Al	Zr	Fe	Cu	W	V	Co
Bal.	20.39	8.61	3.41	0.013	0.041	0.03	0.183	0.015	0.0012	0.007	0.01	<0.005	0.0065	0.204	<0.005	0.092	0.044	0.014

The experimental results are creep parameters of the materials at HIPping temperature (summarized table III) and rheological functions f and c of Abouaf's model (figure 7 and 8). These functions have been identified for relative density from 80% to 100%. The tapped density of the powder is nearly 64%, so functions have to be extrapolated for density ranging from 64% to 80%. Figure 7 and 8 show that there is a great discrepancy between results of Abouaf [7] and Lafer [8] on an argon atomized Astrolloy powder. Processing of powder (gas atomization or rotating electrod) is extremly important as shown by the exponents of power law creep (2.52 for AA and 4.41 for RE).

Table III : Creep parameters of some Nickel base Superalloys.

Ref.	A	n	α	Q/R
ATGP3 [7]	$3.42.10^{18}$	2.136	$0.864.10^{-2}$	67652
René95 [6] (950°C)	$1.27.10^{-12}$	3.34	-	-
AA [8] (1000°C)	$4.3.10^{-9}$	2.52	-	-
RE [8] (1000°C)	$3.26.10^{-14}$	4.41	-	-

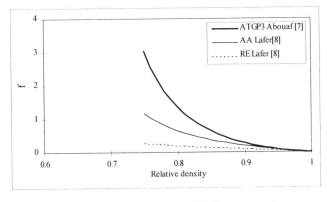

Figure 7 : "f" function for some Nickel base superalloys.

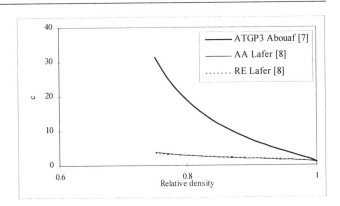

Figure 8 : "c" function for some Nickel base superalloys.

Experiments on Superalloy SY625 made by powder metallurgy

The alloy used in this study is the SY625, a solid solution-strenghtened Nickel base superalloy made by TECPHY, France. It combines strength, good corrosion resistance at elevated temperature (up to 540°C) and good weldability. It has thus found applications in marine environments, aerospace, oil and gas industry and nuclear environments.

Moreover the powder of alloy SY 625 take advantage of all the benefits of powder such as near net-shape forming and fine microstructure. All these properties of the SY625 PM are due to its composition (see table II) [10].

The Nickel is the major constituent. A chemical composition of 50 to 55 % (weight percent) allows the alloy to have a maximum yield strenght. The addition of chromium increases the corrosion resistance and with molybdenum and niobium, they enhance the solid solution strenghtening because of their high atomic volume.

The titanium content is kept as low as possible because it induces titanium carbides and nitrides precipitations at prior particle boundaries [11]. The nitrogen content is dependent upon atomization event and so it is more difficult to precisely control. The SY 625 PM is made without iron because it does not promote either the strenght or the corrosion resistance.

TECPHY made experiments on SY 625 PM at different temperatures of HIPping. Optimized temperature is determinated in order to have a maximum yield stress. The HIPping conditions are then 2 hours at 1120°C and 100MPa. The reference HIP cycle is shown figure 7.

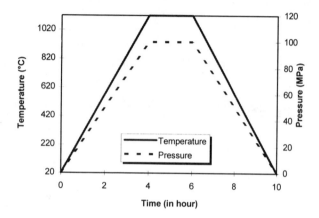

Figure 7 : Optimized HIP cycle for consolidation of SY 625 PM.

Uniaxial compression tests have been performed by TECPHY on SY625 dense material at different temperatures [12]. Creep parameters for a Norton law can be identified at 1100°C as shown in figure 8. The creep parameters are n=3.49 and log(A)=-21.70.

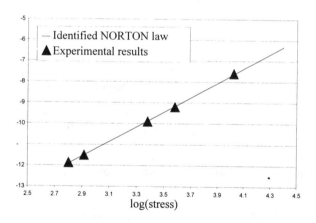

Figure 8 : Uniaxial compression tests at 1100°C on dense SY625.

Other experiments are on going at CEA/CEREM to identify the rheological functions f and c of Abouaf's model, using the experimental procedure presented earlier. Porous samples have been realized for relative density of 70% to 100%, to be abble to interpolate the f and c functions on all the range of density.

Validation Process of Modelling

The modelling module PRECAD®/M has been validated on real 3D-part for titanium alloys [13], by comparison between the modelled and the realized part (Figure 9). Dimensions have been measured on a 3D-control machine with an accuracy of 4µm. Initial geometry is measured, then modelled and the part is realized in parallel. Final geometry of the part is measured after decanning on internal and external geometries. Comparison of the two geometries, modelled and realized, is made with the CAD module. The accuracy of the modelling is ±50 µm for internal cavities (Figure 10 and table IV) [14]. So net-shape complex internal geometries can be realized without extra machining after HIP cycle. External geometries are finally machined using conventional machining. The same validation process is on going for nickel base superalloy SY625.

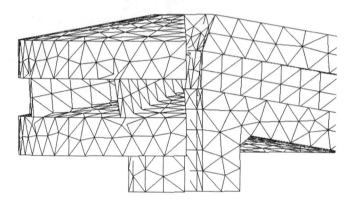

Figure 9 : 3D-mesh of powder for the validation part. External view (left) and representative section (right)

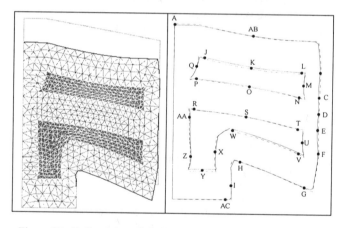

Figure 10 : Deformed mesh (left). Comparison between modelling and realized part on the representative section (right).

Table IV : Comparison of realized and modelled internal geometries (in mm)

	N	O	P	R	S	T
DX	0.00	0.00	0.00	0.00	0.00	0.00
DY	-0.09	0.14	-0.10	-0.19	0.08	-0.27

Conclusion

A computer assisted design and modelling tool for superalloy precision moulding, PreCAD®, has been presented. It allows design, meshing and modelling for the HIPping of 2D and 3D parts. Thermo-mechanical modelling is performed with the Green-Abouaf's viscoplastic model, using finite element method. Final geometry, temperature distribution and relative density map are available. Validation of the modelling by the comparison with an actually manufactured 3D parts indicates a ± 50 µm accuracy on internal geometries of a titanium alloy parts. Such areas of the part can thus be realized net-shape. Only the external geometry has to be machined after decanning. This tool is industrially used to optimize geometries of cores and containers to realize net-shape parts by HIPping of titanium and stainless steel powders. Data base is also available now for SY625 Nickel base superalloys, and validation process is on going.

Acknowledgement

This study has been supported by TURBOMECA. It has been done in the scope of IsoPREC® agreement between TURBOMECA, SEP, TECPHY and CEA/CEREM.

References

[1] Abouaf M., Chenot J.L., "Simulation numérique de la déformation à chaud de poudres métalliques", Journal de Mécanique Théorique Appliquée, Vol.5, p.121, 1986.

[2] Green R.J., "A Plastic Theory for Porous Solid", International Journal of Mechanical Science, Vol.14, pp.215-224, 1972.

[3] De Gayffier A., "Algorithme non-linéaire de CASTEM2000 (NONLIN et INCREM)", Rapport CEA, n° DMT/94-188, 1994

[4] Imatani I., "Studies on nelastic Constitutive Relationship for High Temperature Materials and its Application to Finite Element Analysis", Kyoto University, March 1990.

[5] Dellis C., Abondance D., Bouaziz O., Stutz P., "Rhéologie des Poudres Métalliques à Chaud", Acte du colloque SF2M sur les traitements des poudres et leurs conséquences, Paris, 18-20 mars 1996.

[6] Bouvard D., "Rhéologie des poudres métalliques au cours de la mise en forme à haute température", Thèse de Docteur ès-sciences, UJF - INPG, septembre 1989.

[7] Abouaf M., "Modélisation de kla compaction de poudres métalliques frittées", Thèse de Docteur ès-sciences, UJF-INPG, mai 1985.

[8] LAFER M., "Comportement de poudres métalliques et de composites particulaires lors de la compaction isostatique à chaud", Thèse de l'Université Joseph Fourier - Grenoble I, Spécialité Mécanique, avril 1992.

[9] Noharah A., Soh T., Nakagawa T, "Numerical simulation of hot isostatic pressing", Proceedings of the 3rd International Conference on Hot Isostatic Pressing, London, MPR Publishing Services, 1986.

[10] Davidson J.H., "The Influence of Processing Variables on the Microstructure and Properties of PM 625 Alloy", Proceedings of the International Symposium on Superalloys 718 and 625, Pittsburgh, 24-26 june 1991.

[11] Ferrer L., "Précipitation, recristallisation et déformation à chaud de l'alliage 625. Application aux traitements thermomécaniques", Thèse de l'Institut National Polytechnique de Toulouse, 1992.

[12] Pierronnet M., Raisson G., Private communication, TECPHY, september 1995.

[13] Abondance D., "Modélisation thermomécanique d'un procédé de mise en forme aux cotes de pièces complexes par compression isostatique à chaud de poudre de TA6V", Thèse de l'Université Joseph Fourier - Grenoble I, Spécialité mécanique, 06 février 1996.

[14] Abondance D., Dellis C., Baccino R., Bernier F., Moret F., De Monicault J.M., Guichard D., Stutz P. and Bouvard D., "Numerical Modelling of Near-Net-Shape Hipping of TA6V powder", Proceedings of Titanium 1995, Birmingham, October 1995.

COATINGS, JOINING & REPAIR

NONDESTRUCTIVE EVALUATION OF HIGH-TEMPERATURE COATINGS FOR INDUSTRIAL GAS TURBINES

G. L. Burkhardt, G. M. Light, H. L. Bernstein, J. S. Stolte
Southwest Research Institute
P.O. Drawer 28510
San Antonio, Texas 78228-0510

M. Cybulsky
ABB Combustion Engineering
1000 Prospect Hill Road
Windsor, Connecticut 06095

Abstract

Current techniques for determination of the thickness of coatings for gas turbine parts involve metallographic measurements on sections cut from the coated component. This technique is time consuming and expensive, and only a limited amount of data can be obtained for individual systems. In this paper, initial results are presented for a nondestructive technique using eddy current testing. Measurements have been carried out on flat and curved surfaces with and without thermal barrier coatings, and good agreement has been obtained with metallographic determinations in every case. Preliminary data suggest that a combination of eddy current and ultrasonic techniques may enable degradation behavior to be monitored.

Introduction

In order to achieve the high efficiencies demanded of modern combined-cycle generating plants, it is necessary to operate the gas turbine with high turbine entry temperatures. Currently, gas temperatures at the combustor exit are around 1300°C (2372°F), and the objective in turbine development projects such as those in the Advanced Turbine Systems (ATS) program of the U.S. Department of Energy is to devise methods to increase gas temperatures further and to achieve combined-cycle efficiencies of 60 percent. Since the materials used in the hottest parts of the engine are either nickel or cobalt base superalloys with an incipient melting temperature of around 1200°C (2200°F), the use of high gas temperatures requires extensive internal cooling of the blading. The use of ceramic coatings on the component surfaces to provide a thermal barrier offers even more thermal protection. The combined effect of thermal barrier coatings (TBCs) (1,2) and efficient cooling configurations can ensure that metal temperatures remain acceptable with even higher gas temperatures.

A key factor in the effective use of TBCs with advanced cooling configurations is the extent to which the reduction in metal temperature can be used at the design stage to increase the operating capability of the components. Clearly, such a process implies that the integrity of the coating system will be maintained throughout the operating life of the blade or vane. In order to reach this objective, it will be necessary to ensure reliability of the TBC coating system, which consists of an underlying bond coat (usually an overlay coating) and an outer ceramic coating (also called a top coat). This objective implies a consistent and reproducible process for the production of high-quality coatings.

An essential factor in this procedure is the ability to measure coating thickness. Currently, this can only be done by sectioning of coated components and making metallographic measurements of coating thickness.* Clearly, this is a time-consuming operation limited by the cost of providing large parts for destructive examination; and since few data are obtained, statistical evaluation is quite limited. Accordingly, work has been initiated to develop a nondestructive technique to enable thickness measurements to be made on coated parts which will serve as a quality control tool and will enable a large database of information to be obtained. Also, efforts have been made to develop a method for evaluating the degradation of the coating system to determine the condition of the coating in the engine and the remaining coating lifetime. A critical factor is the growth of the oxide between the bond coat and the top coat, which leads to failure of the TBC. Therefore, an ability to monitor Al_2O_3 growth rates may provide a basis for determining the onset of failure.

In this paper, techniques being developed are described, and initial results are presented from measurements made with TBCs produced by air plasma spray and with MCrAlY-type bond coats for both flat and curved surfaces.

Experimental Procedure

NDE Techniques

In the present program, two techniques are being investigated: the two-frequency eddy current testing (ET) method for determination of coating thickness and ultrasonic testing (UT) and ET techniques for evaluation of coating degradation.

Eddy Current. The ET method involves the generation of a localized electrical current in a conductive material placed in close proximity to a coil energized with an alternating current (3). The ET response is measured as the reactive and resistive components of the electrical impedance of the coil. The ET response is very sensitive to the spacing between the probe and the conductive material, as well as to the conductivity of the material. For specimens with conductive coatings having conductivities that differ from that of the substrate, the response will be sensitive to the conductivity of both coating and base material. The response will also be sensitive to the coating thicknesses and thus can be used for their measurement (3-6). In the case of TBCs, the thickness of the nonconductive ceramic can be readily determined since the response is simply a measure of the spacing between the probe and the layer of bond coat. However, since the thickness and/or conductivity of the bond coat will also affect the ET response, the bond coat thickness can also be measured if the conductivity is constant.

---

*Alternately, the weight gain of the coated component is used to determine the average coating thickness. However, this method is impractical to determine coating uniformity or in cases where different thickness is required at different locations.

An example of eddy current data for a coated specimen is shown in Figure 1. Here the data are plotted in the impedance plane with the resistive component of the signal on the horizontal axis and the reactive component on the vertical axis. These data were obtained from an MCrAlY coating similar to those described in this paper. The responses are shown for different thicknesses of MCrAlY coating and for different simulated ceramic coating thicknesses. The end points of the traces (A-D) are with zero ceramic thickness, but different MCrAlY layer thicknesses. As the MCrAlY thicknesses increase, the response shifts to the left horizontally. As the ceramic thickness increases, each trace moves from the end point toward the top of the figure. Values of simulated ceramic thicknesses are shown on one trace. From this figure, it is evident that ET is sensitive to thicknesses of both layers.

In order to independently determine the ceramic and bond coat thicknesses, a multiple-frequency method is needed, because additional frequencies (which result in different penetration depths of the eddy currents into the sample) will provide more independent information to help characterize the thickness of each coating layer. Two frequencies were used in the tests to provide four parameters for each measurement. To provide a calibration for measuring the coating thicknesses, a least-squares fit (7) to the ET data was made for data from specimens having different coating thicknesses. Separate fits were made for the bond coat and ceramic layers. The equations from these fits were then used to predict the coating thickness independently for each layer. The equations were of the form:

$$T = C_1 + C_2 \cdot R_1 + C_3 \cdot R_1^2 + C_4 \cdot X_1 + C_5 \cdot X_1^2 \\ + C_6 \cdot R_2 + C_7 \cdot R_2^2 + C_8 \cdot X_2 + C_9 \cdot X_2^2 \quad (1)$$

where T is the layer thickness, R_1 and R_2 are the resistive components and X_1 and X_2 the reactive components, respectively, of the two frequencies, and C_1, etc. are constants.

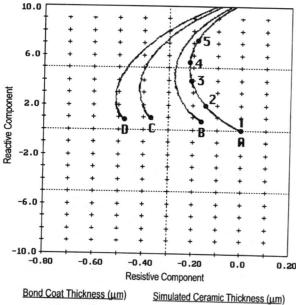

Bond Coat Thickness (μm)
A – Uncoated
B – 127
C – 254
D – 381

Simulated Ceramic Thickness (μm)
1 – 0
2 – 51
3 – 102
4 – 152
5 – 254

Figure 1: Example of an impedance plane plot of ET responses to different thicknesses of bond coat and to different simulated ceramic coating thicknesses.

The ET measurements were made using a SmartEddy 3.0 instrument in a portable (lunch-box style) computer with analysis of the data performed offline. The probe was a 0.06-inch-diameter absolute coil. Excitation frequencies were 1.8 and 4.0 MHz. For measurements on flat specimens, the probe was held by a collar, which aligned it with the specimen surface (Figure 2). For measurements on curved blade and vane specimens, the probe was placed in a fixture molded to position it properly against the specimen surface. Repeat measurements made after an interval of time confirmed that good repeatability could be obtained.

Ultrasonics. The ultrasonic method used to investigate degradation involved the application of a pulsed sound wave of high frequency perpendicular to the surface of the specimen (i.e, 0-degree longitudinal wave) and the measurement of the amplitude of the reflected ultrasound from the interface between the ceramic and bond coat.

The UT tests were performed with a Krautkramer USD-15 UT instrument and a 25-MHz, 6.4-mm (0.25-inch)-diameter focused transducer. The tests were conducted with the probe and specimen immersed in water to provide coupling of the ultrasound to the specimen. (Note that the immersion test was performed for simplicity in the lab; tests could be performed in the field using a gel-type couplant.) The amplitude of the signal reflected from the ceramic/bond coat interface was measured. Data were taken over the entire specimen surface by performing a raster scan using a laboratory scanner.

Materials

To establish the feasibility of using nondestructive methods to evaluate TBCs, two types of bond coat were evaluated. The first system was a NiCrAlYSiTa vacuum plasma deposited coating and an air plasma overcoat (coating 1). The second bond coat was a NiCoCrAlYSiTa deposited by vacuum plasma spraying (coating 2). A third coating (coating 3), similar in composition to coating 1 but with no air plasma overcoat, was used without a TBC to represent the overlay coatings used for blades. In all cases, the substrate was IN-738LC

To establish the quality of parts such as blades and vanes, it is important to be able to determine the coating thickness on curved airfoil surfaces. The thickness of a coating can vary on an airfoil from less than 100 μm (3.9 mils) to greater than 325 μm (12.8 mils) at the leading edge. In addition to coating thickness, the radius of curvature of the surface changes from relatively flat surfaces on the pressure and suction sides of the airfoil to areas with a high radius of curvature at, for example, the leading and trailing edges.

Figure 2: Eddy current probe and holder positioned on a specimen.

A special fixture was used (Figure 3) so that coatings ranging in thickness from 50.8 μm (2 mils) to over 508 μm (20 mils) could be applied on the surface of stage 1 blade and vane sections for a large gas turbine engine. Coated areas of various thicknesses were produced by removing sections of the fixture in turn and spraying the exposed surface while the rest of the surface was masked. After coating, the airfoil sections were heat treated and sectioned across the chord width into pieces approximately 9.5 mm (0.375 inch) wide. These sections were ground and polished, and the coating thicknesses were determined at various locations on the airfoil.

In order to simulate a degraded coating, static oxidation tests were performed on specimens with coating 2 at 1150°C (2100°F) for 15 hours and 75 hours. Figure 4 shows the typical layers which can form on an oxidized TBC coating.

Results

Flat Surface Thickness Tests

The ET results for coating 1 with a flat surface are shown in Figures 5 and 6 for the ceramic and bond coat layers, respectively. The fits to the ET data are plotted as a function of thickness measurements obtained from metallographic examination of coated sections. The results in both figures represent different fits to the same data, which were taken from specimens having varying thicknesses for both layers. The ceramic thickness measurements in Figure 5 were thus made in the presence of varying bond coat thickness and include the effects of this varying thickness on the accuracy of the ceramic measurement. Likewise, the results in Figure 6 include the effects of varying ceramic thickness on the bond coat thickness measurement.

The agreement between the ET and metallurgical results for the ceramic thickness (Figure 5) are very good, with an rms error (RMSE) of 13.3. The results for the bond coat have more scatter, with an RMSE of 17.6. A contributor to this variability is believed to be the effect of the coarse powder used for the air plasma sprayed top

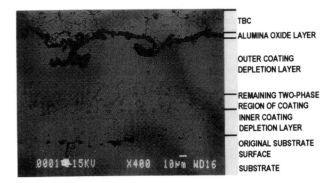

Figure 4: Cross section of coating 2 flat panels oxidized at 1150°C (2100°F)/50 hours. Note coating degradation, i.e., coating depletion layer and growth of oxide.

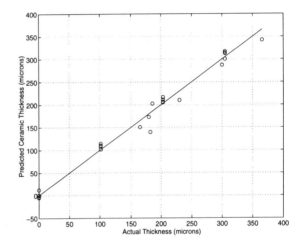

Figure 5: Eddy current thickness measurement of ceramic layer for flat samples with bond coat 1. Solid line represents ideal 1:1 relationship between actual and calculated values.

Figure 3: Fixturing for coating gas turbine blades above and vanes below.

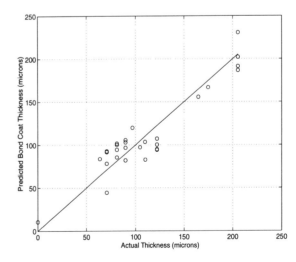

Figure 6: Eddy current thickness measurement of bond coat layer for flat samples with bond coat 1.

layer of this coating. It is believed that the result is localized variations in the conductivity of the coating which affect the ET response and reduce the accuracy of the thickness measurement.

Similar data are presented for coating 2 in Figures 7 and 8. The fit for the ceramic thickness (Figure 7) is very good (RMSE = 11.4) and is comparable to that for the ceramic layer with coating 1 (Figure 5). The results for the bond coat (Figure 8), however, are much better than for coating 1 (Figure 6), with an RMSE of 7.5. This result is attributed to the finer powder used in the final stages of the spray process for coating 2.

<u>Curved Surface Thickness Tests</u>

Next, the ability to measure coating thicknesses on actual components with curvature was investigated for a blade and a vane. Measurements were made at the locations shown in Figure 9. Data will be shown for location C-2 on the vane and on the blade. The results at these locations are typical of most other locations. An exception is the leading edge, for which the data analysis has not been finalized.

Results from the vane coated with bond coat 1 are shown in Figures 10 and 11 for the ceramic and bond coat layers respectively. The agreement between the ET results and the metallurgical results are good for the ceramic coat (RMSE = 16.0), but show more scatter for the bond coat (RMSE = 30.6). Some of the scatter may be caused by the coarse powder used in the air plasma overcoat, as was the case for the flat specimens with coating 1 (Figure 6). The fact that the scatter is larger for the vane is also attributed to the fact that the ET data were fit to a larger range of thicknesses, which could reduce the fit accuracy.

The results obtained from the blade coated with bond coat 3 are shown in Figure 12 for the bond coat layer. Agreement with the metallographic measurements is good (RMSE = 12.0). The error is larger than obtained with coating 2 (Figure 8), which may be a result of fitting the ET data to a larger range of thicknesses. Since the variability in the ET results is still reduced compared to the air plasma layer case (Figures 6 and 11), this provides further evidence that the additional variability observed in Figures 6 and 11 was caused by the coarse powder used for the top layer applied by air plasma spray.

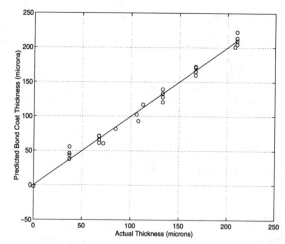

Figure 8: Eddy current bond coat thickness measurement for flat samples with bond coat 2.

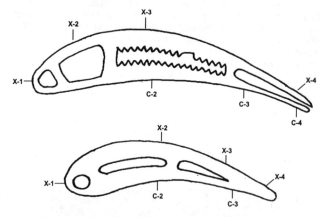

Figure 9: Coating thickness measurement locations for stage 1 vane above and stage 1 blade below.

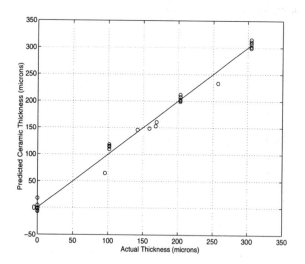

Figure 7: Eddy current thickness measurement of ceramic layer for flat samples with bond coat 2.

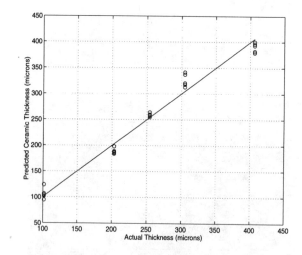

Figure 10: Eddy current ceramic coating thickness measurement at position C-2 for vane with bond coat 1.

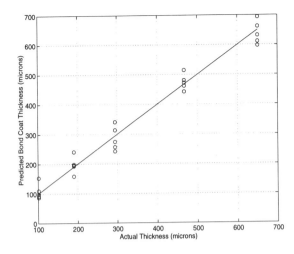

Figure 11: Eddy current bond coat thickness measurement at position C-2 for vane with bond coat 1.

Degradation Studies

The UT response after exposure of the samples to an oxidizing environment at high temperature showed that the amplitude of the signal reflected from the interface between the ceramic and the bond coat increased with time of exposure. Thus, the signal increased by 5 and 7.5 dB for exposure times [at 1100°C (2100°F)] of 15 and 75 hours, respectively (Figure 13). The signals shown in the figure are taken from a single location on each specimen, but are typical of the signals observed in scans over the entire specimen surface. The cause of the signals is believed to be an increase in the acoustic impedance mismatch at the ceramic/bond coat interface due to the growth of an alumina layer.

The ET signal also responds to the degradation associated with this exposure. Figure 14 shows the impedance plane display of ET signals from the degraded specimens. The endpoints of the signals represented by dots are with the probe on the specimen surface, and the vertical lines are generated as the probe is lifted off of the surface. The separation between the endpoints (dots) shows a definite

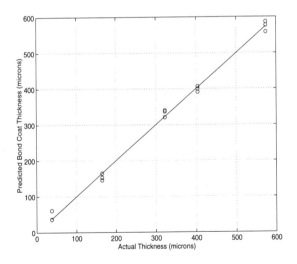

Figure 12: Eddy current bond coat thickness measurement at position C-2 for blade with bond coat 3.

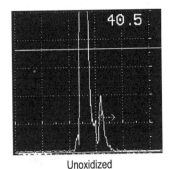
Unoxidized

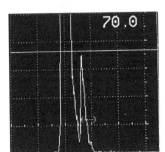
Oxidized for 15 Hours

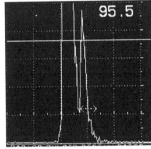
Oxidized for 75 Hours

Figure 13: Ultrasonic A-scan of a thermal barrier coating system oxidized at 1100°C (2100°F). Note that the second signal (two divisions amplitude in the unoxidized case) is from the thermally grown oxide on the bond coat. The signal height increases by 5 and 7 dB for 15- and 75-hour samples, respectively, over the signal from the unoxidized sample.

change with degradation. While the precise reason for the response is as yet unclear, factors such as changes in conductivity due to second-phase particles in the bond coat and formation of a depleted layer at the interface between bond coat and TBC may have an influence.

Discussion

Two-frequency ET results show that very good relationships are obtained between the ET data and actual coating thickness for both the bond coat and ceramic coatings. The bond coat thickness can be measured independently of varying ceramic thickness, and the ceramic thickness can be measured independently of varying bond coat thickness. This was demonstrated for two different bond coat systems, and for flat specimens, as well as for actual vanes and blades with curved surfaces. In most cases, the ET results agreed with actual metallurgical sectioning results within approximately ±15 μm (0.6 mil) for the ceramic layer and ±16 μm (0.6 mil) to ±40 μm (1.6 mils) for the bond coat layer. The higher variability in the bond coat layer was experienced where a coarse powder was used for the top layer of the bond coat.

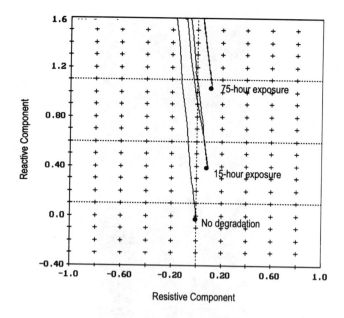

Figure 14: Impedance plane plot of ET responses to specimens with bond coat 2 and no degradation, 15-hour exposure, and 75-hour exposure. The vertical signal component is from probe liftoff.

The next step toward demonstrating a thickness-measurement technique for practical application is to investigate the variability in the coating process on the ET response. This would be done by using the relationships between ET data and thickness (described in this paper) as a calibration and performing blind tests on specimens which have been coated at different times and with different setups. Good results from these tests would indicate a robust measurement approach capable of field application.

Results from limited tests with degraded specimens subjected to high-temperature exposure show that UT appears to be sensitive to the growth in the alumina layer between the bond coat and ceramic. In addition, the ET response is affected by changes in electrical conductivity associated with degradation process and/or the formation of a depleted layer at the bond coat/TBC interface. Although these results are very preliminary and are based on limited data, they indicate promise for use of these techniques to sense degradation. Additional work needs to be performed to more precisely relate the responses of these techniques to the degradation process.

Conclusions

1. The application of the dual-frequency eddy current technique for the measurement of the thickness of as-sprayed overlay and ceramic coatings has been successfully demonstrated on flat surfaces and airfoils. The method provides a basis for a nondestructive tool for quality assessment of coatings on turbine parts.

2. The correlation between eddy current and metallographic measurements was better on TBC bond coats without an air plasma sprayed layer, presumably a result of electrical conductivity variations caused by the coarse structure of this layer.

3. Preliminary work shows that both ultrasonic and eddy current techniques may yield information on coating degradation.

Acknowledgment

Dr. T. B. Gibbons, ABB-CE, has made helpful comments on the text.

References

1. H. L. Bernstein and J. M. Allen, "A Review of High Temperature Coatings for Combustion Turbine Blades," Proceedings of the Steam and Combustion Turbine-Blading Conference and Workshop–1992, TR-102061, Research Project 1856-09, April 1993, 6-19 – 6-47.

2. C. H. Liebert and R. A. Miller, "Ceramic Thermal Barrier Coatings," Ind. Eng. Chem. Prod. Res. Dev. 23 (1984) 344–349.

3. H. L. Libby, Introduction to Electromagnetic Nondestructive Test Methods, (New York, NY: Wiley-Interscience, 1971) 46.

4. D. J. Hagemaier, "Applications of Eddy Current Testing to Airframes," Nondestructive Testing Handbook, Volume 4: Electromagnetic Testing (Columbus, OH: American Society for Nondestructive Testing, 1986) 369.

5. J. C. Moulder, E. Uzal, and J. H. Rose, "Thickness and Conductivity of Metallic Layers from Eddy Current Measurements," Rev. Sci. Instrum. 63 (1992), 3455-3465.

6. J. K. Hulbert and B. W. Maxfield, "Thin, Finite-Plate, Clad Thickness Determination Using Eddy Currents," Review of Progress in Quantitative Nondestructive Evaluation, Volume 7A, ed. D. O. Thompson and D. E. Chimenti (New York, NY: Plenum Press, 1988), 199-206.

7. W. E. Deeds, C. V. Dodd, and G. W. Scott, "Computer-Aided Design of Multifrequency Eddy-Current Tests for Layered Conductors with Multiple Property Variations," ORNL/TM-6858, Contract No. W-7405-eng-26, Oak Ridge National Laboratory, October 1979.

Effect of Postweld Heat Treatment on Ductility of Ni-Co-Cr Based Alloy Welds

Kwang-Ki Baek, Chae-Seon Lim and Joong-Geun Youn

Hyundai Industrial Research Institute
Hyundai Heavy Industries Co., Ltd., Ulsan, KOREA

Abstract

Properties of Gas Tungsten Arc weldments of a commercial, wrought alloy with nominal composition of 29Co-28Cr-2.75Si-Bal.Ni have been investigated. The weldments with a filler metal of matching chemistry are found to have a limited room temperature ductility in the as-welded condition. Metallographic evaluation revealed that the weld metal was characterized by a continuous eutectic phase, which was widely distributed along the dendrite microstructure. The eutectic phase is mainly consist of $(Si,Ti)_xNi_y$. Controlling welding parameters was not effective in improving the ductility of the weldments, whereas a proper Postweld Heat Treatment (PWHT) restored the ductility. The continuous eutectic phase in the as-welded deposit, which caused poor ductility of the weld metals, was considerably reduced or removed with a proper PWHT. The optimum PWHT "window" is proposed based on the experimental result.

Introduction

Increasing demands for industrial plants with higher efficiency proposed a new challenge in the materials engineering to come up with alloys which can meet more severe service condition, such as higher temperature, corrosive environment. This trend has led to a sharp increase in the use of high temperature alloys. For example, Ni-Co-Cr based superalloys have replaced the austenitic stainless steels for major components of various industrial boilers, incinerators.

Recently, a lot of industrial plants employed a new breed of burner, so called "low NO_x burner", which operated at higher temperature than the conventional one. This new design requires to replace the austenitic stainless steels with the Ni-Co-Cr based alloys for the metallic components to withstand high temperature and corrosive environments of the fuel rich zone. The alloy is useful due to its combination of good high-temperature strength, thermal stability, high-temperature corrosion resistance against environments containing sulfur, chlorides and vandalism. Moreover, its excellent room temperature ductility provides superior fabricability.

Welding of this type of alloy, however, presented the same kind of problems as those frequently encountered in the welding of the austenitic stainless steels, such as solidification cracking, microfissuring.[1-3] The controlling factors responsible for the solidification cracking are both the chemical composition of the filler metals and the welding parameters. The solidification cracking of weldments, in principle, can be successfully prevented by combination of the following two ways; selection of the filler metals with lower content of detrimental elements, such as P and S, and employment of low heat input and interpass temperature in the welding process.[1-6]

Another important issue in the welding of the alloy is degradation of the mechanical properties, such as drastic decrease of welds ductility in the as-welded condition.[1] The poor ductility resulted in a cracking during the guided bend test of the weldment, which has been attributed to the formation of a second phase with low melting temperature. The second phase, which was strongly affected by the segregation of alloying elements, are possibly minimized by a subsequent PWHT.[1]

In this study, therefore, the effects of PWHT on the properties of Gas Tungsten Arc (GTA) weldments of a commercial, wrought Ni-Co-Cr based alloy have been investigated focusing on the subsequent improvement in the mechanical properties of the weldments. The optimum condition of the PWHT, henceforth, is proposed for the application to the actual fabrication procedure, in which GTA welding of the alloy is required.

Experimental Procedure

Materials

The nominal chemical composition of the alloy and filler metal are shown in Table 1. Plate thickness was 6.4mm for bend test and tensile test, 12.7 mm for all-weld-metals tensile test. Matching filler metals are solid wires of 2.4mm diameter.

Welding & PWHT

Several batches of GTA weldments with different heat input were prepared. Microstructural examination of the weldments confirmed that neither fusion-zone solidification cracking nor microfissuring was encountered. GTA welding parameters are summarized in Table 2. For bend testing and tensile testing specimens, one side, flat position welding with a groove design of 60° single V with 4mm root gap was done, whereas a groove design of 20° single V with 12mm root gap with backing plate was used for all-weld-metal specimens.

Postweld heat treatment was carried out with a box furnace, monitoring temperature of the specimen surface with a K-type thermocouple attached on it. PWHT of the welds was conducted at the holding temperature in the range of 880℃ to 1095℃, followed by an air cooling.

Table 2. GTA Welding Parameters.

Current	115-140 A
Voltage	10-12 V
Heat Input	8.9-11.2 kJ/cm
Travel Speed	4.7-10.5 cm/min
Shielding Gas & Rate	Ar, 15 ℓ/min
Interpass Temperature	< 90 ℃

Table 1. Chemical Composition of Base Metal and Filler Metal (wt%).

	Ni	Co	Cr	Si	Mn	Ti	Fe	C	S	P
Base Metal	41.1	28.4	26.4	2.82	0.82	0.47	0.10	0.06	0.001	0.002
Filler Metal	39.4	28.7	27.5	2.81	0.71	0.49	0.10	0.06	0.001	0.002

Mechanical testing

Mechanical properties of the weldments were evaluated using an universal tensile tester with 100 ton capacity, and 2T guided bend tester (bending radius two times plate thickness). Before machining of the weldments to the final specimen configuration, specimens were heat treated for the specific PWHT conditions.

Microstructural evaluation

Representative samples of weldments having weld metal as well as the heat-affected zone were metallographically prepared to characterize microstructural features. The specimens were polished through 0.05 μm alumina and electroetched in a solution of (60mℓ HCl+35mℓ HNO$_3$+25mℓ Methanol) at 10~15V. Both an optical microscopy (for magnifications up to 500X) and a scanning electron microscope (SEM) equipped with an energy dispersive spectrometer (EDS) were used for the examination. The SEM was also used to examine the fracture surfaces of tensile and bend test specimens. An electron probe microanalyzer (EPMA) was used for chemical analysis of different phases found in the SEM observations.

Results and Discussion

Mechanical properties

Results of tensile and bend tests on the as-welded GTA weldments are summarized in Table 3. All tensile test specimens were fractured at weld metal, and the bend testing of weldments yielded overall cracks in the weld metal regardless of their welding parameters. Tensile properties of the weldments obtained from the all-weld-metal specimens are compared with those of the base

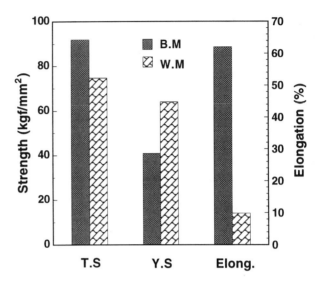

Fig. 1 Tensile properties of weld metal and base metal.

metal with solution-annealed conditions, as shown in Fig. 1. The tensile elongation is only 9% in the as-welded condition, whereas a fully solution-annealed alloy shows an elongation of 73%. This suggests a very brittle nature of the weld metal in the as-welded condition.

Table 3. Mechanical properties of the as-welded GTA weldments at RT.

Thickness (mm)	Tensile Strength[1] (kgf/mm^2)	Location of Rupture	2T Guided Bend Test[2]	
6.4	74.4	Weld Metal	Face	Cracked
			Root	No Crack
12.7	69.4	Weld Metal	Side	Cracked

1) ASME SFA-5.11 (25.4mm gage length, 6.4mm dia.), for transverse tensile test.
2) ASME QW-462 (38W×240L×6.4^{t}mm)

In an attempt to improve the weld metal ductility, various welding conditions, such as heat input were applied. It was reported that lower heat input, as in the present study, resulted in faster cooling of the weld pool, which would suppress segregation of the alloying elements due to narrower interdendritic spacing.[8,9] However, the mechanical test result of the welds with different welding condition, as shown in Table 4, indicates that the weld metal still lacked a proper ductility, thus, resulting in cracks at the weld metals during the bend test. This suggests that controlling welding parameters alone cannot improve the ductility of the weld metal in the as-welded condition.

Microstructural Analysis

On the fracture surface of the bend test specimen in the as-welded condition shown in Fig. 2, a continuous, film like second-phase at the dendrite interface mixed with dimples was observed.

Metallographic evaluation of the interdendritic second-phase found on the fracture surface was carried out on the overetched, cross-sectional weld metal. SEM photos of the overetched weld metal's microstructure, as shown in Fig. 3(a) and (b), show a very aggressive "wetting" action of the liquid phase. In Fig. 3(b), the magnified view of the weld metal reveals that the weld metal can be characterized by a series of continuous eutectic phases with considerable liquation of the phase along with interdendritic boundaries. EDS analysis of the matrix and the eutectic phase, as shown in Fig. 3(c) and (d), respectively, strongly suggests that the eutectic phase is mainly consists of $(Si,Ti)_x Ni_y$.

EPMA analysis of the eutectic phase, as shown in Fig. 4, confirms the EDS result that the eutectic phase was enriched with Si, Ti, Ni, C, as well as being depleted with Cr and Co. Therefore, rather brittle nature of the present, Ni-Co-Cr based weld metal in the as-welded condition could be originated by the formation of interdendritic eutectic phase of low melting temperature during solidification of the weld pool.

This type of eutectic phase with liquid film like nature can also promote the solidification cracking or microfissuring in the weld metal under highly stressed conditions.[10] Interdendritic microfissures of 30~80μm in length have been observed in the weldments deposited with high heat input in this study. In Fig. 5, the microfissuring is shown in the reheated zone of the weld, where the eutectic phase provided constitutional liquation reaction between the phase and matrix.[4,8,10]

Table 4. Bend test results with various welding heat Inputs.

Heat Input (kJ/cm)	Travel Speed (cm/min)	Interpass Temp. (°C)	Bend Test
8.9	10.5	< 90	Cracking
5	13.4	< 20	Cracking
10	7.3	< 185	Cracking

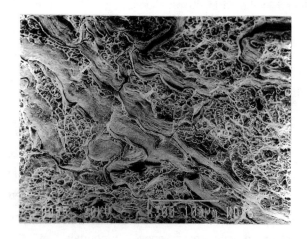

Fig. 2 Fracture surface of the bend test specimen.

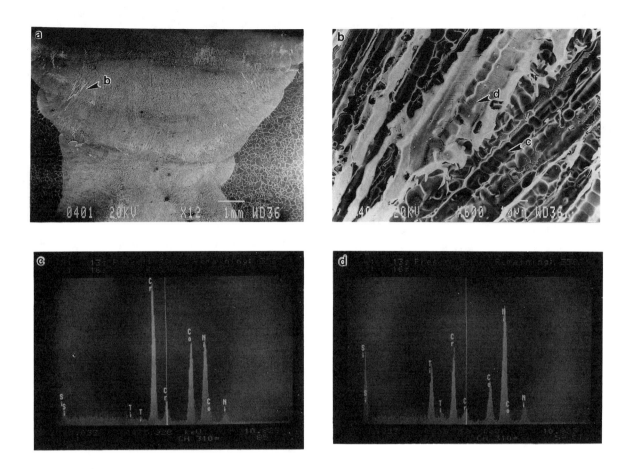

Fig. 3 (a), (b) SEM photos of the overetched weld metal's microstructure, (c), (d) EDS results of matrix and continuous eutectic phase, respectively.

PWHT

Based on the other reports that PWHT would enhance the ductility of the weld metal,[1] a proper condition for PWHT was investigated by varying holding temperature and time. To select proper range of PWHT temperature, previously reported transformation temperatures were reviewed from literatures.[11] The solution annealing temperature of the present Ni-Co-Cr based alloy is about 1100℃, whereas binary phase diagrams of Ni-Si and Ni-Ti suggested eutectic temperature of 1152℃ and 942℃, respectively.[11] Differential thermal analysis (DTA) on the weld metal, carried out previously by the authors, indicated the melting temperature of the eutectic phase was in the range of 940~1060℃.

These evaluations leaded to setting PWHT temperature of the weldments in the range of 880℃ to 1095℃. In Fig. 6, tensile properties of the all-weld-metals obtained before and after PWHT at 1095℃ for 1 hour are compared. PWHT has increased the room temperature elongation of the weld metal from 9% to about 44%. Fig. 6 also indicates that PWHT cause an increase in tensile strength as well as a decrease in yield strength of the weld metal.

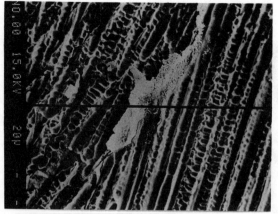
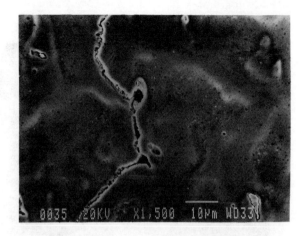

Fig. 5 Microfissuring in interpass region of weld metal.

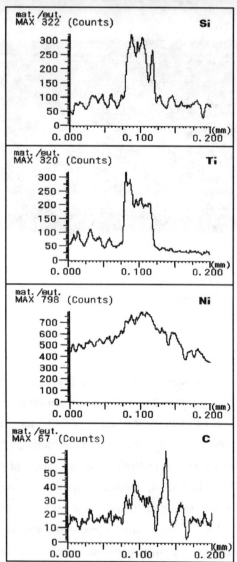

Fig. 4 SEM photos of the overetched weld metal's microstructure, and its EPMA results of line scanning of the eutectic phase.

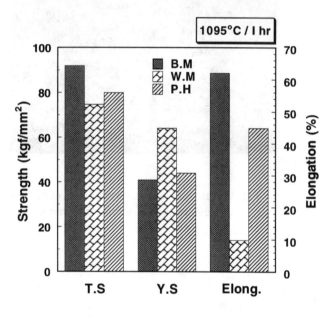

Fig. 6 Tensile properties of the as-welded and PWHT specimens.

A set of PWHT was also conducted for 1 hour at temperature in the range of 880℃ to 1095 ℃. In Fig. 7, the room temperature elongation of the weld metal is plotted as a function of PWHT condition. The elongation of the weld metal after PWHT is found to increase with PWHT temperature. Especially sharp increase in the elongation value after PWHT above 985℃ is quite noticeable.

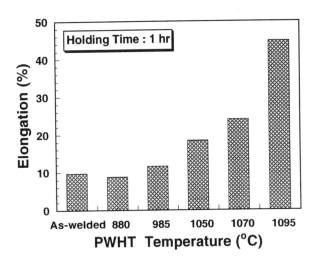

Fig. 7 Ductility change with various PWHT temperature (all-weld-metal tensile test).

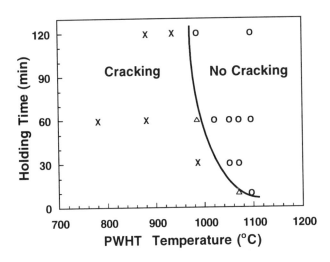

Fig. 8 2T guided bend test result with various PWHT conditions.

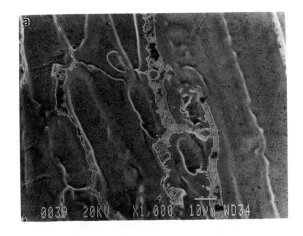

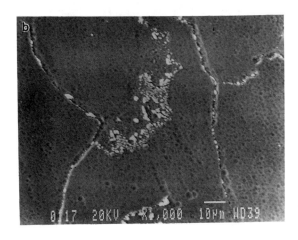

Fig. 9 SEM microstructure of weld metal in the as-welded (a) and PWHT (b) condition.

Bend test results obtained from the all-weld-metal specimens subjected to various PWHT conditions are plotted in Fig. 8, which indicates that higher temperature and longer time result in sounder weld metals without any crack during the bend test. Temperatures higher than 1200℃ should be avoided, however, due to the possibility of grain growth of the base metal and incipient melting. The melting range of the alloy is about 1225℃ to 1300℃ as determined by DTA. Lower temperature limit for the PWHT "window" is about 985℃, below which the longer holding time does not improve the weld metal ductility. Therefore, the optimum PWHT condition is 1050℃/1hr/AC, which enables the weld metal not to develope cracking during the bend test.

Microstructurally, the effect of PWHT is to break-up the continuous network of eutectic phase in the weld metal. SEM photos of the overetched weld metal observed before and after PWHT, as

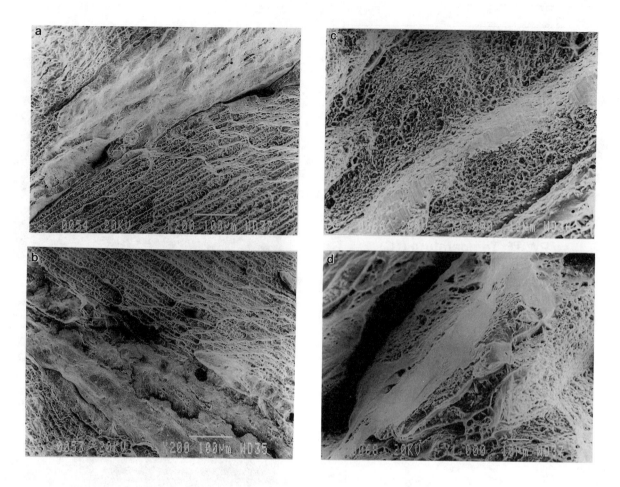

Fig. 10 Fracture surface of all-weld-metal tensile specimen in (a), (b) : as-welded and in (c), (d) : PWHT condition.

shown in Fig. 9(a) and (b), confirm this explanation. The continuous network of eutectic phase found in the weld metal in the as-welded condition is drastically reduced after PWHT at 1095℃ for 1 hour. On the other hand, a series of smaller carbides ($M_{23}C_6$) are formed along the interdendrite boundaries, which occurred during a rather slow, air cooling process.[1]

SEM fractographs of the tensile test specimens obtained before and after the PWHT, as shown in Fig. 10, also reveal the same kind of phenomenon as being observed in the microstructural evaluation. Fracture surface (200×) of the as-welded specimen revealed a very aggressive "wetting" action of the eutectic phase, of which very flat fracture mode with no dimple suggests that the interface between the phase and matrix would be a source of the poor ductility. On the other hand, only small amount of the phase is noticeable in the fracture surface (1000×) of the weld metal subjected to PWHT at 1095℃.

Conclusions

1) Evaluation on the Gas Tungsten Arc weldments of a wrought, nickel-base alloy with nominal composition of 28Cr-29Co-2.75Si-Bal.Ni showed that the weld metals exhibited a poor ductility in the as-welded condition.

2) Poor ductility of the weld metal in the as-welded condition is caused by a continuous eutectic phase, which is mainly consist of $(Si,Ti)_xNi_y$,

3) The eutectic phase was successfully reduced or removed with a proper PWHT, whereas controlling welding parameters was not effective in improving the ductility of the weldments.

4) PWHT of the alloy is essential if cold forming of the weldment is required after welding, or if adequate joint ductility is a design requirement. The recommended PWHT condition is 1050℃/1hr/AC.

References

1. T. S. Chester, S. S. Norman and C. H. William, Superalloy II, (New York, John Wiley & Sons Inc., 1987), 496.

2. C. D. Lundin, C. -P. Chou and C. J. Sullivan, "Hot Cracking Resistance of Austenitic Stainless Steel Weld Metals", Welding Jr., 59(8)(1980), 226(s)-232(s).

3. W. F. Savage and B. M. Krantz, "Microsegregation In Autogenous Hastelloy X Welds", ibid, 50(7)(1971), 292(s)-303(s).

4. G. E. Linnert, "Weldability of Austenitic Stainless Steel as Affected by Residual Elements", ASTM STP 418, (1967), 105-119.

5. T. G. Gooch and J. Honeycombe, "Welding Variables and Microfissuring in Austenitic Stainless Steel Weld Metal", Welding Jr., 59(8)(1980), 233(s)-241(s).

6. J. C. Lippold and W. F. Savage, "Solidification of Austenitic Stainless Weldments : Part III-The Effect of Solidification Behavior on Hot Cracking Susceptibility", ibid, 61(12)(1982), 388(s)-396(s).

7. R. P. George, "The Practical Welding Metallurgy of Nickel and High-Nickel Alloys", ibid, 36(7)(1957), 330(s)-334(s).

8. K. Eastering, Introduction to the Physical Metallurgy of Welding, (New York, Butterworth, 1983), 89.

9. J. C. Boland, "Generalized Theory of Super-Solidus Cracking in Weld (and Castings)", British Welding Jr., (Aug)(1960), 508-512.

10. S. C. Ernst, "Weldability Studies of Haynes 230 Alloys", Welding Jr., 73(4)(1994), 80(s)-86(s).

11. E. A. Brandes and G. B. Brook, Smithells Metals Reference Book, 7th Ed., (McGraw-Hill, New York, 1992), 405-409.

LOW CYCLE FATIGUE PROPERTIES OF LPM™ WIDE-GAP REPAIRS IN INCONEL 738

Keith A. Ellison, Joseph Liburdi and Jan T. Stover[†]

Liburdi Engineering Limited, Hamilton, ON, Canada, L9J 1E7
[†]Consulting Engineer, Greenville, SC, 29611

Abstract

Strain controlled, continuous cycling and 2 minute compressive dwell LCF tests were performed on simulated LPM™ wide-gap repair joints in the nickel-base superalloy Inconel 738 at 850°C over a total strain range ($\Delta\epsilon_T$) of 0.4 to 0.9 percent, 0.33 Hz, A = ∞, R= -1. The test specimens were uncoated hybrid bars fabricated with a gauge section consisting of a cast inner core and an equal cross-sectional area overlayer of Inconel 738 LPM™ repair material. For samples with near-surface porosity of 0.25 mm and under, the fatigue lives of the LPM™ samples were about equal to those of cast Inconel 738 when $\Delta\epsilon_T < 0.5\%$, and approximately one-half to three-fourths for strains above this level. A two minute compressive hold reduced the fatigue lives of both the hybrid samples and cast Inconel 738 by a factor of 5 to 10 as compared to the continuous cycling tests, due to the increased creep, oxidation and mean-stress effects. In all cases, the dominant fatigue cracks initiated at near-surface pores. At strains below $\Delta\epsilon_T = 0.5\%$ crack propagation was transgranular, but shifted to an intergranular mode at higher strains. For specimens with near-surface porosity of approximately 0.5 mm, the fatigue lives were much shorter and initiation of the dominant cracks took place at these features. The results of these and other mechanical property tests indicate that, using the appropriate non-destructive inspection procedures and defect size restrictions, the LPM™ repair limits can potentially be extended into the more fatigue-sensitive areas of components such as airfoil leading edges.

Introduction

One of the most common types of service damage experienced by turbine hot section components is cracking due to thermal-mechanical fatigue. During a typical cycle of operation consisting of start-up, steady-state operation and shut-down, components experience a complex series of thermal and mechanical stresses. Under the most severe conditions, these stresses can cause local plastic deformation which, after a certain number of engine operating cycles, leads to thermal fatigue cracking. When the damage progresses beyond allowable limits, the components must be replaced or repaired.

Repair of cracked components has traditionally been carried out by fusion welding or diffusion brazing. Each process has its limitations. Weld repairs are often difficult and limited to lower-stressed areas of blades, such as airfoil tips and seal edges, due to their lower strength and susceptibility to heat affected zone cracks [1]. The quality of diffusion braze repairs is largely dependant on the efficiency of specialized hydrogen or hydrogen fluoride pre-cleaning processes which are used to remove corrosion products from the cracks. In practice, the effectiveness of the thermochemical cleaning processes has been reported to be inconsistent and the repairs are often ineffective. Poor thermal fatigue properties may result from incomplete crack cleaning, lack of filler metal penetration, the presence of pores and voids, brittle phase formation and cracking of the braze alloy itself [2,3].

The LPM™ wide-gap repair method was developed in order to overcome the limitations of both welding and diffusion brazing. Since LPM™ is capable of bridging gaps of 1 cm or more, defects can be mechanically removed, avoiding the potential pitfalls associated with special chemical cleaning operations mentioned above. The LPM™ filler material is comprised of superalloy powders in tape, slurry or putty form, with an overall composition (in the case of Inconel 738) similar to that of the base alloy. As a result, the mechanical properties approach those of the substrate material [4-6]. Finally, there are no problems related to restraint cracking or distortion since the material is fused and bonded in a vacuum heat treatment operation. This paper describes the results of laboratory low cycle fatigue (LCF) testing and field experience with LPM™ repairs in the nickel-base alloy Inconel 738.

Experimental Procedure

Tests were performed on simulated LPM™[(1)] wide-gap repair joints in the nickel-base superalloy Inconel 738. The LCF samples were hybrid specimens fabricated with a gauge section consisting of a cast inner core of Inconel 738 and an equal cross-sectional area overlayer of the Inconel 738 LPM™ repair material, Figure 1.

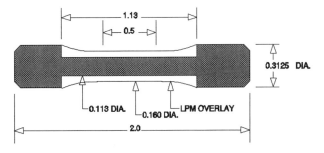

Figure 1. Sketch of the hybrid LPM™ overlay test specimen used to conduct the LCF tests.

[1] LPM™ (Liburdi Powder Metallurgy) is a patented and proprietary technology of Liburdi Engineering Limited.

The cast Inconel 738 cores were machined from the root sections of scrap turbine buckets. These buckets had been previously HIP'ed to eliminate casting porosity which could otherwise lead to greater scatter in the mechanical properties. For these experiments, the LPM™ wide-gap filler material was applied in slurry form using a proprietary water-based binder. Due to the unusual joint geometry, it was necessary to apply and heat treat the LPM™ slurry in two or three operations by varying the specimen orientation for each cycle. This resulted in conditions, especially changes in the base alloy microstructure (see below) which were regarded as being more severe than those which would be encountered in a typical repair.

The compositions of the cast Inconel 738 LC and LPM™ wide-gap filler material are given in Table I.

Table I   Composition of Inconel 738 Test Bars and LPM™ Wide-Gap Filler Alloy

Element	Nominal Composition (Weight Percent)	
	Inconel 738LC	Inconel 738 LPM™
Ni	Balance	Balance
Cr	16	15.3
Co	8.5	9.0
Al	3.5	3.5
Ti	3.5	2.2
W	2.6	1.7
Ta	1.75	2.0
Mo	1.75	1.14
Cb	0.85	0.55
Zr	0.12	0.08
C	0.10	0.08
B	0.015	0.95

After application of the LPM™ filler material, the LCF specimens were processed through a vacuum heat treatment. The complete details of the cycle are proprietary, but in the case of Inconel 738 LPM™ repairs, processing is completed at a maximum temperature of 1200°C without any subsequent diffusion cycles. At this temperature, full solutioning of the cast Inconel 738 base alloy occurs, and consolidation and bonding of the LPM™ material takes place by liquid phase sintering. Primary aging of the LCF specimens was completed at 1120°C for 2 hours, followed by secondary aging at 843°C for 24 hours (i.e. the standard heat treatment for Inconel 738).

The bars were then rough machined to allow inspection by FPI and radiography. Removal of the final 0.40 mm of the gauge diameter on the LCF bars was accomplished by a low stress grinding procedure as described in ASTM E606-92, Section X3.

LCF testing was conducted according to ASTM E606-92 at Mar Test, Cincinnati, OH. Strain controlled, continuous cycling and hold time LCF tests were completed at 850°C over a total strain range of 0.4 to 0.9 percent. The continuous cycling tests were conducted at a frequency of 0.33 Hz, A=∞, R=-1. For the hold time tests, a two minute compressive dwell was added to each cycle. Load and strain outputs were plotted to produce total strain/time records as well as load/time records and periodic load/strain loops were generated during each test.

Cycles to crack initiation (N_i) was taken to be (a) the cycle on the alternating load vs. cycles plot at which the load decayed to approximately 90% of its initial stable level for a test in which the load decreased smoothly, or (b) the cycle at which a non-uniform change in character (slope) of the plot occurred, whichever was smaller. Cycles to failure (N_f) was taken to be the cycle at which the load dropped to below 30% of its initial stable value.

The effects of pore size and location on fatigue lives were studied by comparing the results of pre-test radiographic and penetrant inspections with visual, scanning electron microscope (SEM) and metallographic analyses of the fatigue crack origins.

Results

Low Cycle Fatigue Testing

The results of the continuous cycling and hold time tests are shown in Figures 2-4. Also shown are literature data for the cast Inconel 738 alloy. The literature data came from several sources, including the European COST 50 project and GE M&P Laboratory [7-9]. Nine samples were tested in the continuous cycling mode and three were used in the two-minute hold-time tests. Of these, two samples (one from each type of test) had much shorter lives than the others and were treated as outliers. It was later discovered that these two samples had larger porosity which was not typical of the rest of the specimens, as described below. All of the samples failed either within the gauge section or within the uniform section of the bar.

A plot of total strain range vs. cycles to crack initiation is shown in Figure 2. When represented in this form, the fatigue lives of the Inconel 738 LPM™ samples were approximately one-half those of cast Inconel 738 for $\Delta \epsilon_T \geq 0.5\%$, and about equal for strains below this level. The two minute compressive hold reduced the fatigue lives of the LPM™ samples by a factor of 5 to 10 as compared to the continuous cycling tests. As shown by the referenced data, a similar reduction in life is experienced by cast Inconel 738 alloy under these conditions. The difference in life between these test types (continuous cycling vs. 2 min. compressive hold) increases at lower strains for both materials.

For the continuous cycling tests, constants in the fatigue life relationship

$$\Delta \epsilon_T = AN_i^a + BN_i^b \qquad (1)$$

where

$$\Delta \epsilon_e = elastic\ strain = AN_i^a \qquad (2)$$

and

$$\Delta \epsilon_p = inelastic\ strain = BN_i^b \qquad (3)$$

were determined by regression analysis, Figure 3. Over the range of the data, both $\Delta \epsilon_e$ and $\Delta \epsilon_p$ gave reasonable fits to the classical power law relationships. The four material constants were A = 1.024, a = -0.102 (correlation coefficient = -0.957); B = 3.841, b = -0.569 (correlation coefficient = -0.987).

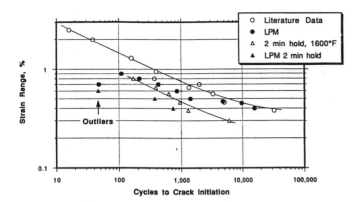

Figure 2. Plot of cycles to crack initiation vs. total strain range for LPM™ hybrid overlay samples compared against literature data for cast Inconel 738 alloy.

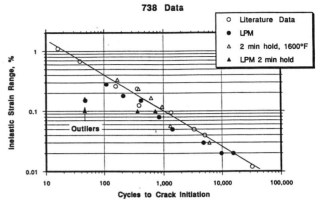

Figure 4. Plot of cycles to crack initiation vs. inelastic strain range for LPM™ hybrid overlay samples compared against literature data for cast Inconel 738 alloy.

A cyclic stress-strain plot for the triangular wave-form tests is shown in Figure 5. The stabilized stress values were determined at approximately 50% of fatigue life. A curve for the cyclic stress-strain data was fit using the Holloman relation:

$$\Delta\sigma/2 = k(\Delta\epsilon_p/2)^{n'} \quad (4)$$

A slight cyclic softening effect was observed which was quantified using the following expression:

$$Degree\ of\ hardening(\%) = \frac{\Delta\sigma(half\ life) - \Delta\sigma(start)}{\Delta\sigma(half\ life)} \times 100 \quad (5)$$

For the continuous cycling tests, the degree of softening was less than three percent. In contrast, the degree of softening in the trapezoidal, two minute compressive hold tests was much greater, at 3 to 15 percent. The increased effects of time-dependant damage (creep & oxidation) and the mean stresses generated in these tests were believed to be the major causes for the increased softening and reduced cycles to crack initiation [8].

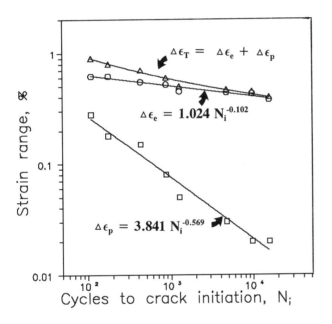

Figure 3. Fatigue life of the hybrid Inconel 738 overlay specimens on the basis of inelastic, elastic and total strain.

A plot of the inelastic strain range vs. cycles to crack initiation is shown in Figure 4. The inelastic strain range was determined near the half life of each sample by subtracting the elastic strain (calculated from the measured modulus of elasticity) from the total strain output by the machine control. For the two cases in which $N_i/N_f < 0.5$, the plastic strain was measured just prior to cycle N_i. For the remainder of the samples, the cyclic stress-strain response was approximately stable throughout 70 to 90 percent of the tests. When represented in this form, the fatigue lives of the Inconel 738 LPM™ samples were approximately three-fourths those of cast Inconel 738 for the higher strains, and again about equal for the lower strains.

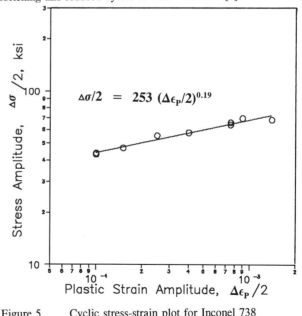

Figure 5 Cyclic stress-strain plot for Inconel 738 LPM™ hybrid overlay specimens at 850°C.

Figure 6. Schematic cross section of an LPM™ repair in Inconel 738 showing three microstructural zones: (1) an outer layer of LPM™ filler material, (2) a transition or diffusion zone and (3) the unaffected base alloy. The SEM images show typical microstructures within each zone.

Metallurgical Examination

Initial Structures. A schematic representation of a typical LPM™ repair is shown in Figure 6, which can be divided into three microstructural zones: (1) an outer layer of LPM™ filler material, (2) a transition or diffusion zone and (3) the unaffected base alloy. The Inconel 738 LPM™ filler microstructure consists of a uniform distribution of spherical gamma prime precipitates embedded in an austenitic gamma matrix. Scattered throughout this structure are eutectic gamma prime colonies, carbides, pores and irregular-shaped boride intermetallics. Previous tests have shown that the intermetallics, which form predominantly along the grain boundaries, are chromium-rich with minor amounts of tungsten and molybdenum and have a Cr_5B_3 structure.

The cast Inconel 738 base alloy contains a duplex gamma prime microstructure consisting of cuboidal primary and spherical secondary particles. The structure also contains large, randomly distributed MC carbides and $M_{23}C_6$ grain boundary carbides.

Within the diffusion zone, a gradual transition of structures is observed. Moving towards the base alloy, the number density of primary cuboidal gamma prime precipitates increases, while the volume fraction of eutectic gamma prime colonies and boride precipitates decreases. The depth of the diffusion zone typically lies between 0.25 to 0.50 mm. However, as noted earlier, the LPM™ filler was applied and heat treated in two or three build-ups. For this reason and also because of the fact that diffusion took place from all directions, the core of some of the hybrid LCF specimens was composed almost entirely of the transition structure. Consequently, the microstructure and properties determined in these tests are considered to represent a severe or worst case test of the LPM™ material.

Analysis of Fracture Surfaces Low power optical examination and SEM imaging revealed that the dominant fatigue cracks initiated at near-surface pores in all of the LCF specimens. This was as expected, since LCF failures are generally associated with surface-related defects or discontinuities. However, metallographic sections revealed that the majority of pores, even those at the surface, were less than 0.25 mm in diameter and had not initiated fatigue cracks. The dominant fatigue cracks initiated at isolated larger pores located within approximately one pore diameter of the surface. In addition to the dominant crack, numerous secondary initiation sites were observed along the gauge section of each specimen. In general, there were more crack initiation sites at the higher strain ranges.

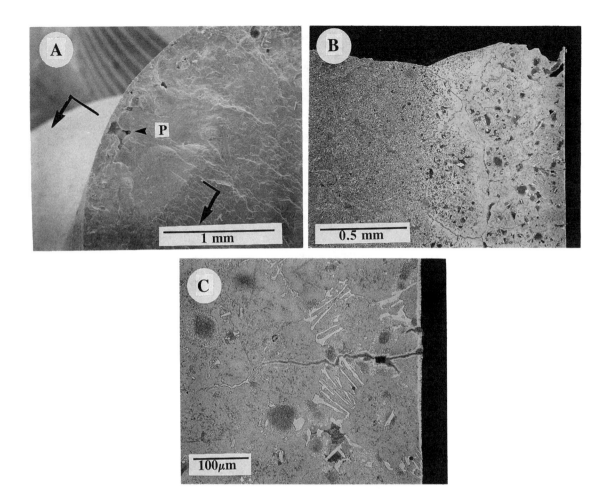

Figure 7. SEM fractograph (A) and optical micrographs (B,C) of the fatigue fracture surface on sample B13 ($\Delta\epsilon_T = 0.47\%$, N_i = 4741 cy., triangular wave form). The initiation site was a 0.18 mm x 0.18 mm pore (P) just beneath the surface of the uniform section. Arrows in (A) indicate the orientation of the metallographic section. This sample is representative of those tested with $\Delta\epsilon_T$ < 0.5% which had smooth fracture surfaces and transgranular crack propagation.

The fracture surfaces of the samples tested at strain ranges below $\Delta\epsilon_T = 0.5\%$ were macroscopically smooth, with Stage II crack propagation taking place normal to the loading axis, Figure 7. Metallographic examination of the dominant and secondary cracks showed that propagation was transgranular through the LPM™ material.

The remainder of the specimens tested at total strain ranges of 0.5% or higher had rougher and more convoluted fracture surfaces. In some cases the fatigue fracture surfaces contained steps, indicating that two dominant cracks had joined at some time during the growth process. As shown in Figure 8, the rough fracture surfaces at the higher strain ranges were associated with a change to an intergranular mode of Stage II crack propagation. This intergranular propagation was associated with fracturing of the Cr_5B_3 precipitates. Note that this change also occurred at a strain level corresponding to the divergence in cycles to crack initiation for cast Inconel 738 vs. the LPM™ hybrid overlay specimens in the continuous cycling tests (Figure 2).

Samples B3 and B4 (outliers) had larger isolated pores which were not typical of the rest of the specimens. Upon examination of the fracture surfaces, one of these samples was found to contain a 0.40 mm x 0.50 mm pore (Figure 8), and the other contained a 0.40 mm x 0.50 mm cluster of interconnected porosity, each of which acted as the primary crack initiation sites. Both features were surface connected. The porosity in the remainder of the samples, as determined by fractography, metallography, FPI and radiographic inspections was 0.25 mm or less in diameter and randomly distributed.

Discussion

Compared to cast Inconel 738, the LCF properties of the LPM™ hybrid overlay samples were considered to be very good, not only in the context of a powder metallurgy (P/M) repair material, but also in view of the difficulties associated with sample preparation and the effects which this had on the base alloy microstructure.

It is well established that the LCF capability of P/M superalloys is largely controlled by the presence of defects such as pores, inclusions and contaminated prior particle boundaries [10-12].

Figure 8. SEM fractograph (A) and optical micrographs of the fatigue fracture surface on sample B3 ($\Delta\epsilon_T = 0.6\%$, $N_i = 42$ cy., trapezoidal wave form). The arrows in (A) indicate the orientation of the polished cross section. This sample was one of the outliers with an initiation site corresponding to the 0.40 mm x 0.50 mm pore just beneath the surface of the uniform section. This sample is representative of those tested at $\Delta\epsilon_T \geq 0.5\%$ which had rough fracture surfaces and intergranular crack propagation.

LPM™ contains a random distribution of porosity and the dominant fatigue cracks in all tests initiated at the largest near-surface pores. As the total strain range increased, a larger number of pores were capable of initiating fatigue cracks. This behaviour is similar to the observations of Miner and Gayda [10] and Hyzak and Bernstein [11] on P/M René 95. The former authors also noted that the size of the defects initiating failure increased with decreasing total strain range.

Nevertheless, randomly distributed, spherical porosity of approximately 0.25 mm or less in diameter does not appear to have a limiting influence on the LCF lives of the LPM™ repair material. In the low strain range ($\Delta\epsilon_T < 0.5\%$), the LCF lives of the hybrid LPM™ specimens were approximately equal to those of cast Inconel 738, crack propagation was transgranular and was not associated with any specific microstructural feature in the LPM™ material. Based on similarity in fatigue lives, crack propagation rates in LPM™ vs. cast Inconel 738 may be comparable in this strain range. At the higher strain ranges ($\Delta\epsilon_T \geq 0.5\%$), the fatigue lives of the LPM™ hybrid overlay samples were about half to three-quarters those of cast Inconel 738. It was noted that this was accompanied by a switch to an intergranular mode of crack propagation in the hybrid overlay specimens. Thus, within this pore size range, the principle differences in LCF behaviour between the LPM™ and cast versions of Inconel 738 are believed to be associated with the presence of the grain boundary intermetallic phases in the repair material.

In contrast, the LCF lives of the hybrid overlay specimens with larger near-surface porosity were much shorter. In one case, the pore was roughly spherical, in the other case the defect was an area of interconnected porosity but each had a maximum dimension of approximately 0.5 mm. Qualitatively, the negative effects of larger surface pores on the LCF lives of the hybrid overlay LPM™ specimens are again similar to the behaviour of P/M René 95 alloy [10,12]. At $\Delta\epsilon_T = 0.66\%$, the cycles to failure in René 95 also decreased by almost an order of magnitude as defect areas increased from approximately 2×10^{-3} to 0.25 mm^2. Furthermore, for a given size, internal defects were found to be much less detrimental than those situated at the surface [12].

Experience has shown that the size and volume fraction of porosity within the LPM™ filler material is influenced to a large extent by the form (i.e. slurry, tape or putty) in which it is applied.

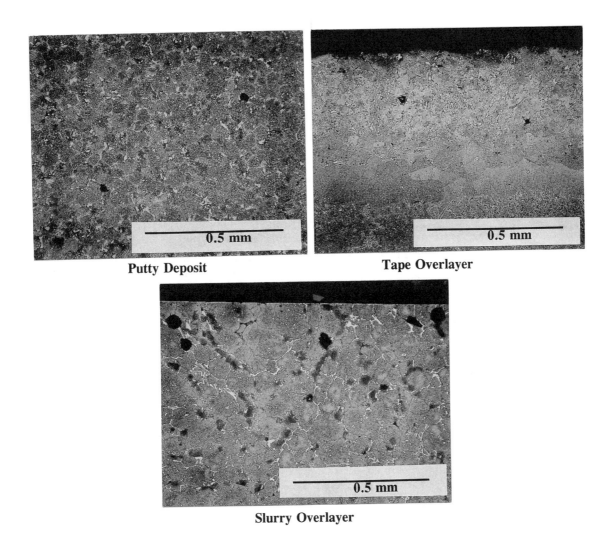

Figure 9. Examples of LPM™ slurry, tape and putty deposits. Note the lower porosity levels in the tape and putty repairs.

Compared to the slurry, the tape and (more recently developed) putty forms of the filler material tend to produce repairs with lower overall porosity. In a typical slurry application, the volume percent of porosity does not exceed four volume percent and is usually less than two percent. In contrast, tape and putty repairs rarely contain more than one percent porosity. In addition, the size distribution of defects within tape and putty repairs is shifted to lower values. Examples of each of these types of LPM™ build-up are shown in Figure 9.

Because of the sensitivity of LCF lives of P/M materials to the size and location of porosity, it will be necessary to specify the appropriate filler material and inspection limits in cases where LPM™ repairs are performed in higher stressed (critical) areas of turbine components. At fatigue sensitive locations, a maximum allowable pore size should be set somewhere between 0.25 to 0.5 mm. Standard penetrant and radiographic techniques are capable of detecting flaws down to about 0.38 mm and should therefore be useful in finding defects of about the critical size. In order to achieve this size limit, it is expected that the tape and putty forms of the LPM™ material will be used in preference to the slurry form.

The results of these laboratory tests are consistent with the results of previously documented field tests on LPM™-repaired Inconel 738 nozzle guide vanes (NGVs)[13]. The NGVs were from an engine used in the propulsion of a hydrofoil. This test environment was considered to be severe, in that the engine experienced approximately one major thermal fatigue cycle (start/stop) per hour of operation. The NGVs were originally removed from service due to extensive thermal fatigue cracking in the outer shrouds and airfoil trailing edge fillet radii. The most severe cracking occurred in the shroud "windows" where at least one 2.5 cm long, axial crack was found between airfoils in each of the components, Figure 9. Following LPM™ wide-gap repair, the NGVs were returned to service alongside new components for an additional 2409 hours. At the following inspection, damage was found to be similar on both the "new" (non-repaired) NGVs and those repaired by LPM™. None of the LPM™-repaired NGVs had re-cracked along the airfoil trailing edges and there was an equal probability of finding a major axial shroud crack in a repaired window or in an adjacent, non-repaired window. Together, the laboratory and field testing have shown that the LPM™ wide-gap method provides a viable solution to many of the problems encountered by existing repair processes.

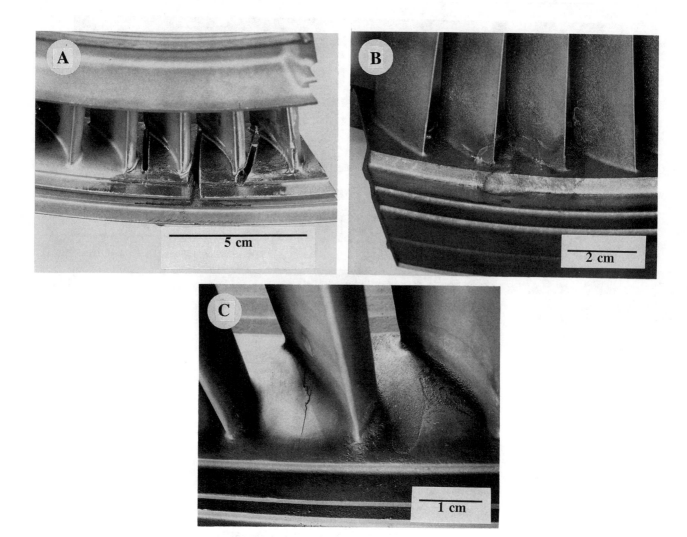

Figure 10 (A) Inconel 738 nozzle guide vane after bench grinding to rout out the crack defects in preparation for LPM™ application. (B) Vane segment after application and heat treatment of Inconel 738 LPM™ repair material. Note also the repaired trailing edge fillet radii. (C) Vane segment after 2409 hours of operation. Recracking was observed in an adjacent window to which LPM™ repair was applied, but not in the LPM™ filler.

Conclusions

1. The low cycle fatigue properties of Inconel 738 LPM™ hybrid overlay specimens were established at 850°C over a total strain range of 0.4 to 0.9 percent in continuous cycling and compressive dwell tests. When near-surface pore dimensions were less than 0.25 mm, the fatigue lives of the repair material were approximately equal to those of cast Inconel 738 at strains below 0.5 percent and one-half to three-fourths those of Inconel 738 at the higher strains.

2. Specimens containing near-surface pores with maximum dimensions of 0.5 mm had much shorter LCF lives. This indicates that, for low cycle fatigue conditions, a maximum allowable pore size should be set somewhere between 0.25 and 0.5 mm. The use of LPM™ in tape and putty form, which have lower overall porosity levels, should help to ensure that this porosity size limit is met.

Acknowledgements

The authors wish to thank the Electric Power Research Institute and co-funding utilities (Anchorage Municipal Light and Power, Duke Power, Florida Power Corp., Nevada Power, Salt River Project and Tri-States Generation and Transmission) for their financial support of this study and permission to publish this data. Dr. Henry Bernstein of the Southwest Research Institute assisted with the planning and analysis of the LCF tests. The many helpful suggestions and comments of Paul Lowden of Liburdi Engineering Limited and Scott Sheirer of Power Tech Associates Inc. are gratefully acknowledged.

References

1. "Gas Turbine Blade Life Assessment and Repair Guide," (Report GS-6544, Research Project 2775-6, EPRI, 1989), pp 7-1 to 7-3.

2. K. Schneider, B. Jahnke, R. Bürgel and J. Ellner, "Experience with Repair of Stationary Gas Turbine Blades - View of a Turbine Manufacturer," Materials Science and Technology, 1 (1985), 613 - 619.

3. P. Brauny, M. Hammerschmidt and M. Malik, "Repair of Air-Cooled Turbine Vanes of High Performance Aircraft Engines - Problems and Experience," ibid, 1 (1985), 719 - 727.

4. K. A. Ellison, P. Lowden and J. Liburdi, "Powder Metallurgy Repair of Turbine Components," Journal of Engineering for Gas Turbines and Power, 116 (1994), 237-242.

5. J. Liburdi, K. A. Ellison, J. Chitty and D. Nevin, "Novel Approaches to the Repair of Vane Segments," (Paper No. 93-GT-230 presented at the International Aeroengine Congress and Exposition, Cincinnati, Ohio, 24-27 May 1993).

6. K. A. Ellison, P. Lowden, J. Liburdi and D. Boone, "Repair Joints in Nickel-Based Superalloys with Improved Hot Corrosion Resistance," (Paper No. 93-GT-247 presented at the International Aeroengine Congress and Exposition, Cincinnati, Ohio, 24-27 May 1993).

7. R. Stickler, ed. Summary of the Physical and Mechanical Property Data of the Ni-Base Superalloy IN-738, (University of Vienna, Institute for Physical Chemistry, Materials Science, Report No. 81-UW-COST-B1, 1981).

8. V.S. Ostergren, "A Damage Function and Associated Failure Equations for Predicting Hold Time and Frequency Effects in Elevated Temperature Low Cycle Fatigue", Journal of Testing and Evaluation, 4 (5) (1976), 327-339.

9. F. Gabrielli, M. Marchionni and G. Onofrio, "Time Dependant Effects on High Temperature Low Cycle Fatigue and Fatigue Crack Propagation of Nickel-Base Superalloys", Advances in Fracture Research, (Proceedings of the 7th Conference on Fracture (ICF7), Vol. 2, Houston, Texas, March 1989), 1149-1163.

10. R. V. Miner and J. Gayda, "Effects of Processing and Microstructure on the Fatigue Behaviour of the Nickel-Base Superalloy René 95," International Journal of Fatigue, 6 (3) (1984), 189-193.

11. J. M. Hyzak and I. M. Bernstein, "The Effects of Defects on the Fatigue Crack Initiation Process in Two P/M Superalloys: Part I. Fatigue Origins", Metallurgical Transactions A, 13A (1982), 33-43.

12. D. R. Chang, D. D. Kruger and R. A. Sprague, "Superalloy Powder Processing, Properties and Turbine Disk Application," Superalloys 1984, ed. M. Gell et al. (Warrendale, PA: The Metallurgical Society, 1984), 245-273.

13. K. A. Ellison, "Field Testing of LPM™ Repairs on Allison 501KF Second Stage Vanes," (Report No. 92-6-50A, Liburdi Engineering Limited, 15 July, 1992).

SUBJECT INDEX

activation energy, 176, 401
advanced turbine engine gas generator (ATEGG), 545-553
aging, 173, 174, 176-178
alloy design, 36, 46
alloy development, 27-34
alloy vector, 62, 63
alumina scales, 72, 75
anisotropy, 315
APB energies, 108, 109
atom-probe field ion mocroscopy (APFIM), 249, 252, 253, 256, 257
atomization
    influence of parameters on porosity, 673
    influence on yield, 673
autonomous directional solidification (ADS), 497, 498, 505

boron, 590
backscattered Kikuchi patterns (BEKP), 493
blade, 576, 581, 583, 584
blade casting, 481
bond order(Bo), 61-66
borides, M3B2 and M5B3, 130-135
bridgeman process, 471, 472
bucket, 173, 178, 179
business drivers, 4

carbide eutectic reaction temperature, 260-262
carbides, 465
    MC and $M_{23}C_6$, 130, 133, 134, 137, 139-142, 377, 465
carbonitride, 465
castablity, 37, 46, 507-509, 512, 515
    freckles, 517-519
    grain boundary cracking, 515-517
cast and wrought, 713, 714, 716, 718
casting defects, 471, 478
casting, melting, 465
cellular precipitation, 9-18
cellular recrystallization, 9-18
ceramic, 571-578

ceramic matrix composite (CMC) shroud, 545-553
chemistry, 37, 46, 91-95, 97, 100, 607
    creep strength, 112
    effect of Co, 422
    effect of Mo, 422
    effect of Ti/Al ratio, 19, 422
    structure, 113
civil aerospace market, 3
closed die forging, 716-718
cluster forging, 697, 699
cluster variation method (CVM), 249, 252, 253, 257
co-precipitate, 629, 630, 634
coating, 9-18, 42
    aluminide, 42
    cracks, 345, 346
    CoNiCrAlY, 43
coherency strain, 211, 212, 215-217, 219
coherent phase equilibria, 219
cold-induction guide, 723
columnar, 466
columnar-to-equiaxed transition (CET), 481
component and engine assessment research (CAESAR), 545-553
composite processing, 579
compressive stress, 214-216
computer assisted design (CAD), 738, 742
computer simulation, 181, 183
controlled solidification, 531-535
conventionally cast turbine alloys, 20
convergent beam electron diffraction (CBED), 201, 206-209
cooling rates, 621-626
Cr-rich phases, 167
crack growth, 412, 415, 545-553, 555-557
cracking, 267-269
crack initiation, 319, 321, 324, 371, 372, 684, 763-768
crack propagation, 91-98, 100, 415
    paris regime, 409
creep, 191, 192, 197, 199, 201, 212, 204-209, 319, 320, 322-324, 328, 329, 511, 520, 521, 600, 602, 621, 622, 624-626, 631-634, 681
    activation energy, 114
    behavior, 25, 283-290, 391, 392, 399
    chemistry, 112
    cyclic, 328-330
    damage mechanisms, 331, 332
    Larson-Miller, 113
    life, 112, 391, 392, 397, 398
    modeling, 291-296
    predictions, 112
    rate, 353-357
    resistance, 297, 299, 301-303
    strain, 392
    strength, 58, 692, 695
    stress exponent, 114
    tertiary, 392, 398
creep anisotropy, 273, 283-290
    gamma prime single crystals, 275-279
    geometric factors, 280
    Hastelloy single crystal, 275-279
    high temperature, 276
    low temperature, 278
    model, 279
    role of cube slip, 280
    role of octahedral slip, 280
    threshold shear stress, 280
creep/fatigue interaction, 600
creep-rupture, 39, 40, 48, 49, 61, 62, 67-70, 384, 385
critcal grain growth, 653, 654, 658, 659
critical strain, 401
cyclic deformation, 305, 307-309-311, 319, 320
cyclic oxidation, 71-76, 78, 79
cyclic softening, 335-337, 343
cyclic strain rate, 335-337, 342

d-electron concept, 61, 62, 64
d-orbital energy level(Md), 61-66
damage tolerance, 137-139, 142, 677
damping, 234
deformation, 314

deformation mechanisms, 285
delta (d) Ni$_3$Nb orthorhombic phase, 164
delta (d), Ni$_3$Ta, 163
density, 37, 46
   liquid, inversion, 444, 447, 448
design, 353, 354, 357
differential thermal analysis, 259, 260, 262, 508
   hafnium effect, 265-267, 269
directional coarsening, 181, 183, 185, 189, 335, 336, 340
directional segregation, 163
directional solidification, 21-23, 61, 62, 65-67, 69, 70, 353, 357, 466, 467, 481, 488, 507, 509-512, 515-521, 531
discontinuous reaction, 119-127
discrete atom method (DAM), 211, 212, 214, 219
dislocation, 314, 393
   climbing, 391, 393, 397, 398
   cutting, 397
   edge, 218, 219
   networks, 393, 396-398
   pinning, 378
   reactions, 286
   structure, 285
   substrauctre, 297, 300-303
dispersion strengthening, 81
dual alloy disk, 637
ductility, 755, 757
ducts and cases, 581, 583
dynamic strain aging, 375, 401

eddy current, 747-752
elastic constants, 229, 231-233, 235, 262
elastic limit, 239-248
elastic moduli, 229-235
electrodes
   cracked electrodes, 427
   internal stresses, 427
electron beam, 467
engine applications, 581, 582
engine test, 542
engineering and manufacturing development (EMD), 545-553
environmental effects, 91, 97, 98, 100
environmental resistance, 19
eta (η), Ni$_3$Ti, 145, 150, 151, 163
evolutionary materials development, 5
experimental analyses

fractography analysis, 605
   microstrucutre observation, 603, 604
extrusion, 649, 697, 698

fatigue, 41, 58, 305-311, 313, 314, 319, 323, 324, 330, 331, 359-374, 409, 410, 412, 415, 511, 513, 520, 521, 545, 555, 556, 558-600, 684, 763, 764, 770, 771
   behavior, 565
   crack growth, 412, 415, 692-695, 682
   damage mechanisms, 332-333
   deformation, 375
   high cycle, 41
   high temperature, 520, 521, 335, 339
   hold periods, 305-307, 309-311
   isothermal, 335, 338, 339, 341-343, 378
   life, 335-337, 339-341
   life predictions, 409, 410, 413, 415
   propagation, 415
   resistance, 409
   stage II, 409, 410, 412, 415
   test technique, 376
finite element modeling (FEM), 203, 204, 207-209, 614, 618
first generation single crystal alloys, 24, 25, 37, 46
fluid flow, 435
fluorescent penetrant inspection (FPI), 163
forge bonding, 637
forgeability, 651
forging, 267, 653-658, 660-662
fourth generation single crystal alloys, 25
fracture, 353, 354, 357, 555, 556, 558, 559
fracture mechanisms, 305, 310, 311
fracture toughness, 562
freckles, 443, 447, 448
friction stresses, 245-248

gamma (γ) 38, 46, 239-248, 319, 322, 324
   channel, 297, 298, 301-303
gamma prime (γ'), 38, 46-49, 145-151, 163, 167, 173-177, 179, 319, 322, 323, 607-609
   aspect ratio, 299
   coarsening, 201-210, 288
   distribution, 55

   growth mechanism, 122-127
   intergranular, 119-127
   morphology, 288
   precipitate, 131, 132, 391, 393, 396-398, 621-623, 625, 626
   rafting, 25, 191, 193, 195, 197-199, 175, 176, 297-300, 319, 322, 323, 339-341, 343
   size, 55, 302, 303
   solvus, 262-268, 653, 654, 656, 661
   stability, 133
   volume fraction, 62, 64-67, 299
gamma/gamma prime (γ/γ')
   composition, 249, 252, 253, 256, 257
   eutectic 38, 48, 164, 264, 266, 269, 507-509, 756, 760
   interface, 133, 374, 393, 396, 397
   microstructure, 305, 306, 308, 309, 311, 335-337, 339-341, 343
   morphology, 201-210
gamma double prime (γ''), 164
gamma-ray diffraction, 221-224, 226
gamma titanium aluminide, 539, 555
gas phase embrittlement, 353, 354, 356, 357
gas tungsten acr welding (GTAW), 754, 756
gas turbine, 571-578, 747, 749
grain boundaries, 371, 372
   amplitude, 396-398
   carbides, 391
   continuator, 473, 477, 479
   effects, 132
   growth, 119-124, 646, 651, 663-666
   migration, 398
   morphology, 391, 398, 393, 394, 396-398
   refinement, 499, 503
   size, 653-655, 658, 660, 662-666
   straightening, 391, 397, 398
   strengthening, 19-21, 24
   wave length, 396-398
   zigzag, 392, 393, 395, 396

Hall-Petch, 148
heat of fusion, 491
heat transfer, 435
heat treatment, 29, 38, 47, 49, 91, 92, 94-98, 100, 145-151, 515, 519, 520, 627, 628, 687, 688, 695
   grain size, 663-666

supersolvus, 653-655, 658, 660-662
window, 38, 267, 268, 515, 519, 520
HfC, 465-467
high Mo containing superalloys, 275
high temperature deformation, 291-296
high temperature strength, 563
Hookean potential, 212
hot corrosion, 19, 20, 42, 49, 50, 61, 64, 67, 70
hot isostatic pressing (HIP), 268, 640, 645-652, 697, 698, 738, 740, 742
   sub-solidus, 645-652
   supersolvus, 645-652
hot shortness, 267
hot tearing, 268, 269, 508, 512
hot workablity, 649
hydrogen annealing, 75, 76, 79
hydrogen embrittlement, 53, 58
hydrostatic pressure, 214

image analysis, 173, 174, 491
image force, 216, 219
impact resistance, 566, 572
impact strength, 39
impurities
   detrimental element, 599
   phosphorus, 599
   sulfur, 599
in situ deformation tests, 242
incipient melting, 166, 263, 264, 267, 268, 509
inclusions, 359-368
   agglomeration, 463
   Al_2O_3, TiN, spinel, 475-464
   ceramic composition, 359-368
   effect on LCF life, 359-368
   seeding, 359-368
   wetting, 457-464
ingot, 427
inhomogeneity effect, 214, 215, 219
inspectability, 652
instability, 9-18
integrated high performance turbine engine technology (IHPTET), 545-553
interaction energy, 402
interface, 191-199
interface morphology, 490, 493
interfacial waves, 211, 214, 219
intermetallic compounds, 561
internal stress, 201-210

interstitial atom, 403
inverse strain rate sensitivity, 379
investment core/metal reactions, 163
investment shell/metal reactions, 163
isothermal forging, 640, 697, 699

KGT, 481

lattice mismatch, 201-210
lattice parameters, 221, 222, 224, 226, 227
Laves phase, 267
leviation melting, 457-464
life prediction, 359-368, 409, 410, 413, 415
   using seeded LCF specimens, 359-368
   probabilistic methods, 359-368
liquid metal cooling (LMC), 498, 499
liquidus, 260-261, 508, 608, 609
local order, 240, 243-248
long range order, 242-244
low angle boundary/strength, 19-21, 23, 24
low expansion alloy, 91, 97, 100
LPM™, 763-771

manufacturing methods, 579
materials/process development, 5
mean stress effects, 381
mechanical alloying, 369
mechanical properties, 241, 370, 451, 452, 454, 455, 580, 607-609, 706-708, 713, 718-721
   modeling, 477
   structural effects, 134, 135
mechanisms of deformation, 315
melting
   Electron Beam Button Melting (EBBM), 435
   electroslag remelting (ESR), 427, 431, 723
   liquid density inversion, 444, 447, 448
   triple melt, 427
   vacuum induction melting (VIM), 427, 428
   vacuum arc remelting (VAR), 427, 428, 431, 435, 545-553
   VIM+ESR, 427

   VIM+VAR, 427
microporosity, 269
microstructural design, 677
microstructural evolution, 305, 308, 309, 311
microstructural stability, 29-32
microstructure, 9-18, 129-136, 260, 262-264, 264, 266, 268, 283, 314, 435, 607, 609
   carbides, 428
   carbonitrides, 428
   delta phase, 427
   grain size, 427, 663-666
   Laves phase, 427
   modeling, 614
   prediction, 618
   treatment effects, 145-151
military aerospace market, 3
misfit, 191, 192, 197-199, 221-224, 226, 227, 313
misorientation, 21, 24
model alloys, 260, 261, 264, 265
modeling, 101-109, 409, 472, 474, 476, 738-740, 743
   computer, 409
   computer, 417
   neural network, 409, 410, 413, 415
   neural networks, 417
   numerical, 713
   temperature gradient, 431
   VAR, 431
monocrystalline nickel-base superalloy, 335
Monte Carlo simulation (MCS), 211, 212, 218, 249, 252, 253, 255-257
morphological evolution, 213-215
Mu (μ) phase, 133, 134, 163
multi-component alloys, 101-109

Navier-Stokes equation, 178
NbC, 465-467
neural network, 409, 410, 413, 415
neutron diffraction, 221, 226, 227
neutron diffuse scattering, 243
Ni_5Hf phase, 164
NiAl Eutectics, 568
nickel aluminide, 561
nickel base single crystal
   creep anisotropy, 276-277
   yield strength anisotropy, 281
nickel base superalloy, 71-79, 153,

211, 219, 409, 589, 627
NIKE/TOPAZ, 638
nitrides, 465
non-homogenous grain growth, 637
nondestructive evaluation, 747, 752
nondestructive inspection, 161
nozzle, 575
nozzle vane, 178, 179
numerical simulation, 498, 503

on-cooling tensile test, 687, 689-692, 694-695
Ostwald ripening, 213, 217
oxidation assisted crack growth, 345-347, 350-352
oxidation resistant alloy, 19, 20
oxidation, 42, 50, 51
oxide crust, 463
oxide dispersion strengthening, 369

phosphorus-boron interaction, 589
partitioning, 249, 252, 253, 255, 257
performance, 354, 357
phase digrams, 101-109
phosphoric acid grain etch, 170
phosphorous, 164, 451-455, 590
physical properties, 579
planar slip, 376
plastic deformation, 181-183
plasticity to failure, 391, 398
platinum aluminide coating, 42, 43
postweld heat treatment (PWHT), 753, 754, 757
powder metallurgy (P/M), 129-136, 265-267, 645-652, 653, 663, 677, 697-703, 737, 741, 763, 767
    alloys, 705
precipitation behavior, 628-631
prestrain, 191-199
primary dendrite arm spacing, 436
prior particle boundaries, 645-652
process control, 494
process modeling, 435
process/numerical modeling, 713-716, 721
proof test, 574

quenching experiments, 263

rapid solidification, 497
rapid solidification plasma

deposition, 81
recrystallization, 163, 663
    dynamic, 614, 615, 665
    meta-dynamic, 616, 617
remaining life, 354, 357
residual stress, 638
retained strain, 653, 656, 661
reverse partitioning, 25
revolutionary materials development, 6
rhenium alloying, 35, 37, 109, 283
ring rolling, 651
rotors, 581, 582
Rayleigh number, 443, 445, 447

second generation single crystal alloys, 24, 25, 37, 46
secondary dendrite arm spacing, 436
secondary reaction zone, 9-18
segregation, 163, 451, 452, 455
    interdendritic, 444-448
    macrosegregation, 427
    microsegregation, 427
    freckles, 427
    white spots, 427
serrated grain boundaries, 120, 121, 123, 125, 401, 679
serrated yielding, 379
shafts, 581-583
shape bifrcation, 212, 215, 217
sheet alloy, 81
short range order, 242-244
sigma (σ) phase, 106, 107, 163
silicon, 169, 451-455
single crystal (SC), 19-25, 27-35, 45, 61-63, 65-68, 229-235, 283, 297-303, 471, 477, 481, 487, 497, 501, 502
single crystal solidification, 163
single crystal superalloy, 9-18, 53-60, 191, 199, 267, 268
site occupancy behavior, 249, 250, 253, 256, 257
slip lines, 372, 373
solid solution alloy, 401
solidification, 107, 108, 259-262, 451, 452, 455, 471
solidification control, 488, 494
solidus, 508, 607-609
spray forming, 82, 723, 729-735
spurious grains, 473, 475, 481
stability, 111, 132-134
    Phacomp, 111

d-electrons, 111
    valency electrons, 111
    transition metals, 111
stabilization heat treatment, 138-142
stacking faults, 377
strain exponent, 402
stress hardening, 380
stress relaxation, 353-357
stress-rupture, 24, 25, 40, 48, 49, 355, 509, 510, 589, 590, 596, 600, 681
    transverse strength, 21-24
structures, 581, 584
subsolvus anneal, 655, 658, 660-662
sulfur, 71-79, 451-455
superplastic deformation, 637, 653, 655, 665
surface active
    salts, 463
    sulfur, 463
    alkali halides, 463

TaC, 465-467
    dispersoids, 81
Taguchi, 137, 140, 141
technology averse, 4
temperature effects, 213
temperature evolution, 221, 224-226
temperature measurement, 473, 474
tensile properties, 56, 57, 145-151, 600, 601, 621-626
    ductility, 562, 649
    modeling, 417
    notched strength, 57, 58
    strength, 39, 509, 692, 695
    stress, 215, 216, 218, 219
    stress-strain curve, 401
    test, 387
    ultimate tensile strength, 417
    yield strength, 417
tetragonal distortion, 221, 222, 226
textures, 229-233, 235, 236
thermal barrier coating, 747, 748, 751, 752
thermal conductivity, 55, 56
thermal mechanical fatigue (TMF), 317-324, 335, 336, 340-343, 345-347, 350-352, 380, 511
thermal gradient ratio, 532
thermally induced porosity (TIP), 650
    formation of, 667-676
    influence in LCF life, 674

thermograms, 260-263, 266, 267, 269
third generation single crystal alloys, 24, 25, 37, 46
TiAl attached blade, 545-553
TiAl variable vane shroud, 545-553
TiC, 465-467
time evolution, 221, 223, 224
TiN, 465-467
titanium metal matrix composites, 579-585
topologically-close-packed phase (TCP), 9-18, 36, 37, 39, 46
total accumulted cycles (TAC), 545-553
transmission electron microscopy, 242, 306, 309, 313

TTH diagram, 155, 156
TTP diagram, 159-161
turbine airfoil alloy, 19-25
turbine blade, 201-210, 471, 472, 497, 498
turbine components, 409
turbine disk, 697, 677, 699, 701, 705
turbine entry temperature (TET), 36, 45
two-step cooling path, 687, 689, 692, 695

ultrasonic, 747, 748, 751, 752
ultrasonic inspection, 652
undercooling, 471, 473, 474, 477, 497-500, 502, 503

variation coefficient, 177, 178
volume fraction, 224

weld, 383-386
weldability, 137, 138
wide-gap repair, 763
Widmanstatten phases, 149, 164
work hardening rate, 401

x-ray inspection, 171
x-ray line profiles, 203-209

Young's modulus, 56

Zener's anisotropy ratio, 211

ALLOY INDEX

29Co-28Cr-2.75Si-BalNi, 753-761
45XD, 555, 556, 558, 559
47XD, 545-553, 555, 556, 558, 559
5A, 9-18, 27-34
48-2-2, 540
454, 283

A-286, 163, 261, 265
ABB 2 DS, 515, 520, 521
AEREX™ 350, 145-151
AF 56 (SX 792), 37, 46
AF 115, 705, 729-735
AF2-1DA, 264, 265, 729-735
alloy 706, 153-162, 427, 429, 627-636
alloy 713LC, 261
alloy 718, 145, 150, 153, 155, 159, 427, 589-597, 592, 595, 627, 629, 634, 677, 684, 723
alloy 720, 427, 431
alloy 720Li, 713-717, 719-721
AM1, 37, 46, 54-58, 181-188, 221. 222, 224, 227, 313, 481
AM3, 37, 46, 54, 58, 239-248
AMS 5707, 621-623
AS-800, 574
Astroloy, 120, 121, 261, 268, 269, 665, 729-735, 741
ATGP 3, 741

B-1900, 262, 266

C 276, 444, 446, 448
C101, 163
C103, 163
CM186 LC, 520
CM186 LCR, 46, 49, 50
CM247 LC, 515-518, 520, 521
CM247 LCR, 46, 49
CM247 LC-DS, 319, 321, 322, 324, 327-334

CMSX-1, 265
CMSX-2, 25, 221, 222, 226, 227, 250, 252, 254, 265, 275, 277, 283
CMSX-2R, 37, 45, 46
CMSX-3, 346-349
CMSX-3R, 37, 45, 46
CMSX-4, 25, 28, 33, 71, 72, 75-78, 112, 208, 229, 230, 233-236, 252, 252, 254, 283-290, 297-303, 487, 498, 499, 502
CMSX-4R, 37, 39-40, 45, 46, 48, 49
CMSX-6, 71, 72, 76-78, 210-210, 229, 230, 234-236, 304-307, 311, 335, 472, 474, 498, 499, 502
CMSX-6R, 37, 46
CMSX-10, 28-33
CMSXR-10, 35-43, 46
CMSXR-11B, 37, 45-51
CMSXR-11C, 37, 45-51

DZ3, 508
DZ4, 507-513

GH 95, 729-735
GN-10, 574
GTD-111, 163, 353-357
GTD-222, 167, 607-609

Haynes alloy 188, 375-381, 383-389
Hastelloy, 275, 277
Hastelloy X, 84, 380
HS 25, 574
HS-188, 84

IC-6, 729-735
IC-218, 729-735
IN-100 112, 139, 261, 262, 265-269, 466-468, 510, 511, 729-735
IN-100Hf, 466-468
IN-625, 729-735
IN-6203, 163
IN-713 C, 49

IN-713LC, 112
IN-718, 107, 108, 356, 435, 444, 446, 448, 466-468, 580, 729-735, 748
IN-738, 163, 261, 262, 265-257, 347, 349
IN-738 LC, 37, 40, 45, 49-51, 61, 64, 66-69, 71, 72, 74, 76-78, 112, 173-179, 229, 230, 232, 234-236, 328-331, 520, 521
IN-792, 50, 515-517, 519, 520
IN-792 Hf, 48, 49, 163
IN-792 mod 5A, 112
IN-901, 163
IN-939, 45, 48, 50, 112, 163
Incoloy 901, 59, 265
Incoloy 903, 53, 91, 96
Incoloy 939, 137, 138, 142
Incoloy 907, 91
Incoloy 908, 91, 97, 100
Incoloy 909, 91, 93, 96
Inconel 713, 137
Inconel 718, 92, 95, 96, 98-100, 137, 143, 265, 267, 599
Inconel 738, 763-771
Inconel 783, 91, 92, 95-100
Inconel MA 760, 369-374
Inconel X-750, 100
Inconel 600, 401

K403, 729-735
K405, 729-735
K417, 729-735
KM4, 637, 667-674

MA6000, 328-331
MA6000E, 48
MAR-M001, 112
MAR-M002, 112, 444, 445, 447, 729-735
MAR-M200, 275, 277, 283, 507

MAR-M200+Hf, 481
MAR-M246, 265-267
MAR-M247, 137, 265, 267, 275, 277, 283, 347, 349, 444, 445, 447, 466-468
MC2, 37, 46, 239-248, 252, 253, 254
MERL76, 677, 729-735
MM-002, 37, 42, 265, 267-269, 509
MM-006, 262-265, 267
MM-007, 262, 265
MM-009, 265
MM-200H, 20
MXON, 283

N18, 134, 135, 663, 665, 666
NASAIR - 100, 265, 268
NCG 359, 540
Ni-13.3 at% Al-8.8 at% Mo, 191, 199
Ni_2CoCr, 240
$Ni_3(Al,Ta)$, 275, 276
Ni_3Al, 71, 76, 229, 234, 275, 276
NiAl, 561, 569, 568
NiCoCrAlYSiTa, 748
NiCr, 240-245
NiCrAlYSiTa, 748
Nim80, 112
Nim90, 112
nimoniel 15, 729-735
NiTaC-14B, 82
NT 154, 573, 576
NT 164, 574

PWA 1422, 265
PWA 664, 507
PWA 1422, 507, 509, 511, 513
PWA 1480, 25, 37, 45, 46, 49, 175, 177
PWA 1480LC, 163
PWA 1483, 37, 46, 163
PWA 1484, 25, 28, 33, 37, 46, 49, 275, 277
Pyromet, 265

R77, 112
R88DT, 626
RD-8A, 81, 82
René 41, 84, 729-735
René 80, 19, 25, 48, 84, 261, 262, 265, 527, 729-735
René 80 H, 48-50
René 88 DT, 129-126, 359, 580, 646, 653, 654, 660, 662, 678, 688, 695, 723, 729-735
René 95, 134, 135, 646, 654, 705, 723, 729-735, 741
René 108, 25
René 120, 265
René 125, 509
René 142, 21, 25, 33, 525
René N4, 19-25, 37, 45, 46, 49, 108, 525
René N5, 25, 27-34, 85, 527
René N6, 13, 17, 27-34, 37, 46, 265, 267, 268
RR 2000, 37, 46
RR 3000, 37

SC-16, 36, 37, 46, 48, 49, 61, 64, 66-68, 305, 306, 308, 472, 474, 498
SC 180, 28, 37, 46, 347, 349

SCS-6 SiC, 579, 580
SiC, 579, 582, 584
SN - 252, 574-576
SR 3, 637
SRR-99, 37, 46, 71, 72, 76-78, 112, 204, 208, 209, 229, 230, 234-236, 265, 268, 286, 288, 291-296, 304-311, 337, 341, 343, 344
Superwaspaloy, 53, 58-60
SX792, 163
SY 625, 737, 741-743

T1, 444, 446, 448
THYMONEL, 53-60
Ti-6Al-2Sn-4Zr-2Mo, 579, 580
Ti-6Al-4V, 579, 582, 584-585
TiAl, 6, 545-553, 579, 580, 582, 584
Ti_3Al, 579, 580
TMS-63, 250, 253, 254
TMS-71, 250, 252, 255
trimarc1 Sic, 580

U500, 112, 163
U700, 163
U720, 106, 107, 109, 705, 729-735
U720Li, 106, 107
UDIMET 710, 391, 399
UDIMET 720, 145, 150, 457, 645-652, 694, 697-703

Waspaloy, 137, 139, 143, 145, 150, 261, 265, 427, 429, 589-596, 613, 621-626, 713, 719-721, 729-735

AUTHORS

Adasczik, C. B., 427
Alden, D. A., 129
Anton, D. L., 555
Auburtin, P., 443
Austin, C. M., 539

Baccino, R., 737
Baek, K.-K., 753
Banik, T., 607
Bastie, P., 221
Beneduce, F., 465
Benz, M. G., 723
Bernstein, H. L., 747
Bertram, A., 229
Bhanu Sankara Rao, K., 375
Biermann, H., 201
Birks, N., 71, 145
Blachere, J. R., 71
Blankenship, C. P., Jr., 653
Bouse, G. K., 163
Bréchet, Y., 181
Brien, V., 313
Bryant, D. J., 713
Buffiere, J.-Y., 291
Buhle, J. P., 737
Burkhardt, G. L., 747

Cao, W. D., 589
Caron, P., 53, 239
Carter, W. T., Jr., 723
Casey, R. L., 667
Castelli, M. G., 375
Castledine, W., 645
Cetel, A., 273
Chang, M., 677
Changxin, W., 507
Chuanqi, S., 507
Clemens, D. R., 555
Clément, N., 239
Cockcroft, S. L., 443, 465, 481
Conjou, A., 239
Cooper, C., 677

Cornu, D., 53
Cybulsky, M., 747

Danflou, H., 119
Darolia, R., 81, 561
Daté, C., 345
Davidson, D. E., 545
de Monicault, J. M., 53
Décamps, B., 313
Dedecke, D., 471
Dellis, C., 737
Deragon, T. C., 607
Dobbs, J. R., 81, 523
Dong, J., 599
Dreshfield, R. L., 383
Durber, G., 111

Easley, M. L., 571
Ellison, K. A., 763
Engler-Pinto, C. C., Jr., 319
Erickson, G. L., 35, 45

Fährmann, E., 191
Fährmann, M., 191
Feller-Kniepmeier, M., 283
Feng, H., 201
Fernihough, J. W., 481
Ferro, P. D., 531
Forbes Jones, R. M., 723
Fox, D., 345
Frani, N., 345
Fratzl, P., 191
Frenz, H., 305
Fujiyama, K, 173
Furrer, D., 613, 705

Ganesh, S., 637
Garcia, J. C., 737
Gasior, R. M., 697
Glatzel, U., 283
Goulette, M. J., 3
Graves, J. A., 523

Green, K. A., 697
Groh, J. R., 621
Guo, J. T., 451
Guofa, M., 729

Halali, M., 457
Hall, E. L., 653
Han, J., 229
Harada, H., 249
Hashizume, R., 61
Hatala, R. W., 137
Hattori, H., 297, 705
Heck, K. A., 91
Helms, A. D., 427
Henderson, M. B., 291
Henry, M. F., 653, 667
Hermann, W., 229
Hobbs, L. J., 435
Hong, S. H., 401
Hu, Y., 599
Huron, E. S., 359, 667
Hynnä, A., 369
Hyzak, J. M., 637, 653

Ishiguro, T., 627

Jackman, L. A., 427
Jackson, J. J., 579
Jang, J. S., 401
Jones, J., 417
Jouiad, M., 239

Kahn, T., 119
Kashiwaya, H., 173
Kelly, M., 129
Kelly, T. J., 539
Kennedy, R. L., 589
Kettunen, P., 369
Khan, T., 53, 239
Kim, H. Y., 401
Kinder, J., 305
Kirchner, H. O. K., 239

Kissinger, R. D., 687
Kitazaki, N., 297
Klingelhöffer, H., 305
Kondo, Y., 297
Konter, M., 515
Koul, A. K., 677
Kraft, S., 335
Kuk, I. H., 401
Kuokkala, V.-T., 369

Lasalmonie, D., 737
Le Ber, L., 737
Lee, J. K., 211
Lee, P. D., 435
Lemsky, J. A., 697
Liburdi, J., 763
Light, G. M., 747
Lim, C.-S., 753
Liping, R., 729
Liu, X, 599
Lobo, C. I., 555
Lothian, R., 435
Louchet, F., 181

Macia, M., 119
MacKay, D. J. C., 417
Maldini, M., 327
Maloney, J. L., 145
Mancuso, S., 645
Marchionni, M., 327
Marquis, F. D. S., 391
Marty, M., 663
Maurer, G. E., 645
McIntosh, G., 713
McLean, M., 291, 435, 457
Meshkov, S., 523
Meter, G. H., 71
Meyer ter Vehn, M., 471
Mitchell, A., 443, 465, 481
Miyazaki, S., 61
Moret, F., 737
Morinaga, M., 61
Morton, A. J., 313
Mourer, D. P., 637, 667
Mughrabi, H., 201, 335
Müller, F. G., 723
Murakami, H., 249
Murata, Y., 61
Murphy, W. H., 9, 27

Namekata, J., 297
Nathal, M. V., 561
Nazmy, M. Y., 319, 327
Ni, K., 599
Noel, R. J., 705
Noseda, C., 319

O'Hara, K. S., 19, 27
Octor, H., 663
Ohi, N., 297
Okabe, N., 173
Okamura, T., 173
Osinkolu, G., 327
Ott, M., 335

Pan, L.-M., 291
Paris, O., 191
Paul, U., 471
Pettit, F. S., 71, 145
Pollock, T. M., 27, 191
Portella, P. D., 305

Qijiuan, L., 507

Radavich, J. F., 145, 507, 729
Rana, R., 345
Raymond, E., 637
Reed, P. A. S., 409
Rézaï-Aria, F., 319
Rohling, R. B., 653
Rollins, J., 613
Rösler, J., 515
Ross, E. W., 19, 27
Roth, P. G., 359
Royer, A., 221

Sahm, P. R., 471, 497
Saito, D., 173
Saito, Y., 249
Sanders, T. H., 119
Sarioglu, C., 71
Sass, V., 283
Saunders, N., 101
Schaeffer, J. C., 9
Schirra, J. J., 137
Schlienger, M. E., 487
Schmalz, A. J., 465, 481
Schneider, T., 201
Schooling, J. M., 409
Schweizer, F. A., 607, 645
Shah, D. M., 273

Shen, G., 613
Shendye, S. B., 531
Shibata, T., 153, 627
Shifan, T., 507, 729
Shollock, B. A., 291
Shudo, Y., 153, 627
Singerman, S. A., 579
Smialek, J. L., 71
Smith, J. S., 91
Smyth, J. R., 571
Sockel, H. G., 229
Soucail, M., 663
Sponseller, D. L., 259
Staubli, M., 327
Stinner, C., 71
Stolte, J. S., 747
Stover, J. T., 763
Strangman, T., 345

Takahashi, T., 627
Takekawa, M., 705
Tang, B., 599
Tien, S., 599
Tomasello, C. M., 145
Tönnes, C., 515

Véron, M., 181
von Grossmann, B., 201

Wagner, I. A., 497
Walston, W. S., 9, 27, 561
Webb, G., 345
West, D. R. F., 457
Wilson, L., 345
Wlodek, S. T., 129
Woodford, D. A., 353

Xianguo, Z., 729
Xie, W., 599
Xie, X., 599
Xu, Z., 599

Yoshino, Y., 153, 627
Yoshioka, Y., 173
Youn, J.-G., 753

Zhang, L., 599
Zhang, Y., 391
Zhikai, L., 729
Zhou, L. Z., 451, 729
Zhu, Y., 599